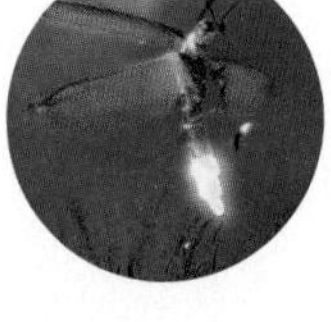

Chemistry

An Introduction to General, Organic, and Biological Chemistry

GLOBAL EDITION

Twelfth Edition

Karen C. Timberlake

PEARSON

Boston Columbus Indianapolis New York San Francisco Hoboken
Amsterdam Cape Town Dubai London Madrid Milan Munich Paris Montréal Toronto
Delhi Mexico City São Paulo Sydney Hong Kong Seoul Singapore Taipei Tokyo

dam Jaworski
or: Jeanne Zalesky
Marketing Manager: Jonathan Cottrell
oject Editor: Jessica Moro
Assistant Editor: Lisa R. Pierce
Associate Media Producer: Jackie Jakob
Supervising Project Manager, Instructor Media: Shannon Kong
Editorial Assistant: Lisa Tarabokjia
Marketing Assistant: Nicola Houston
Managing Editor, Chemistry: Gina M. Cheselka
Team Lead, Project Management: David Zielonka
Head of Learning Asset Acquisition, Global Editions: Laura Dent
Senior Acquisitions Editor, Global Editions: Priyanka Ahuja
Assistant Project Editor, Global Editions: Amrita Kar
Full Service / Compositor: Lumina Datamatics Inc.
Full Service Project Manager: Andrea Stefanowicz
Illustrations: Imagineering
Design Manager: Mark Ong
Interior Designer: Jerilyn Bockorick
Cover Designer: Lumina Datamatics Inc.
Photo Manager: Maya Melenchuk
Photo Researcher: Erica Gordon, Q2A / Bill Smith
Manager, Rights and Permissions: Timothy Nicholls
Associate Project Manager, Rights and Permissions: Michael V. Farmer
Text Permissions Research: Lumina Datamatics Inc.
Operations Specialist: Christy Hall
Cover Photo Credit: © Africa Studio / Shutterstock

Credits and acknowledgments borrowed from other sources and reproduced, with permission, in this textbook appear on p. C-1.

Pearson Education Limited
Edinburgh Gate
Harlow
Essex CM20 2JE
England

and Associated Companies throughout the world

Visit us on the World Wide Web at:
www.pearsonglobaleditions.com

Authorized adaptation from the United States edition, entitled Chemistry: An Introduction to General, Organic, and Biological Chemistry, 12th edition, *ISBN 978-0-321-90844-5, by Karen C. Timberlake, published by Pearson Education © 2015.*

ISBN 10: 1-292-06132-4
ISBN 13: 978-1-292-06132-0

British Library Cataloguing-in-Publication Data
A catalogue record for this book is available from the British Library

10 9 8 7 6 5 4 3 2 1

Typeset by Lumina Datamatics Inc.

Printed and bound by Courier/Kendallville

www.pearsonhighered.com

Brief Contents

Contents

1 Chemistry in Our Lives 33

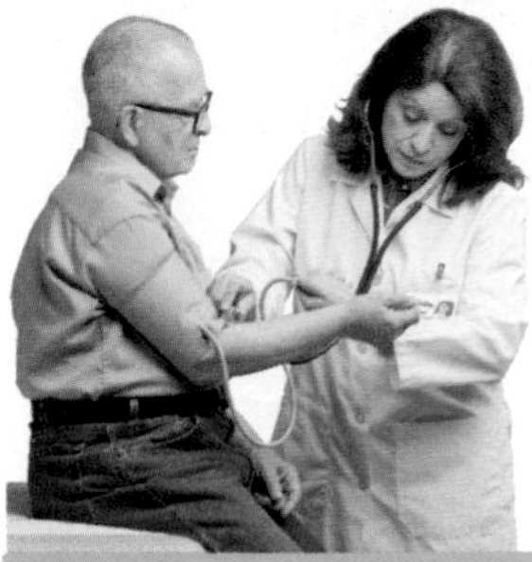

2 Chemistry and Measurements 55

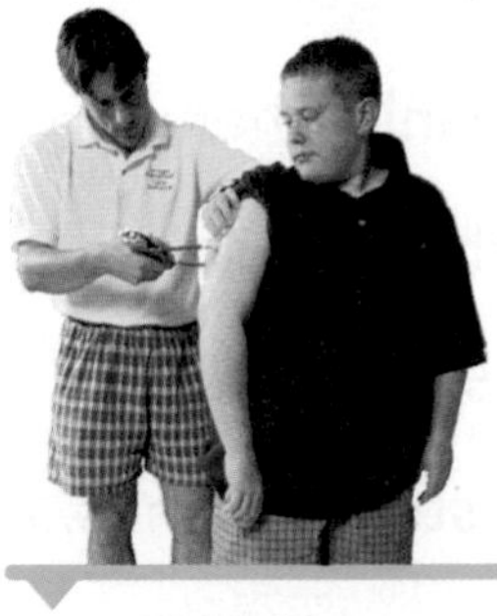

3 Matter and Energy 89

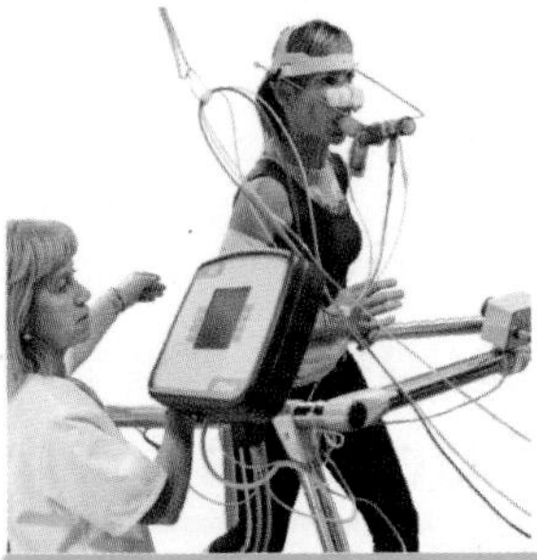

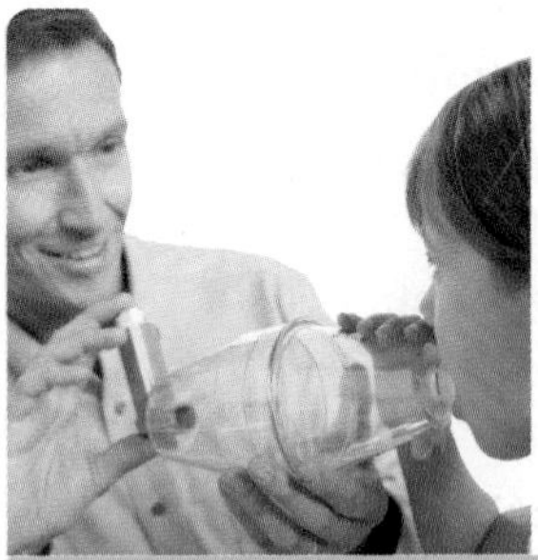

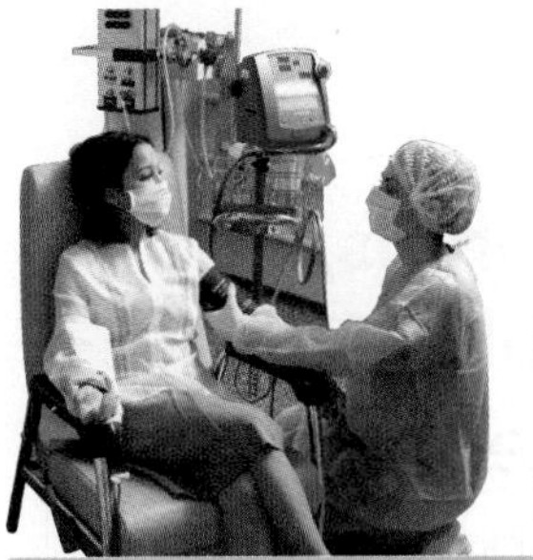

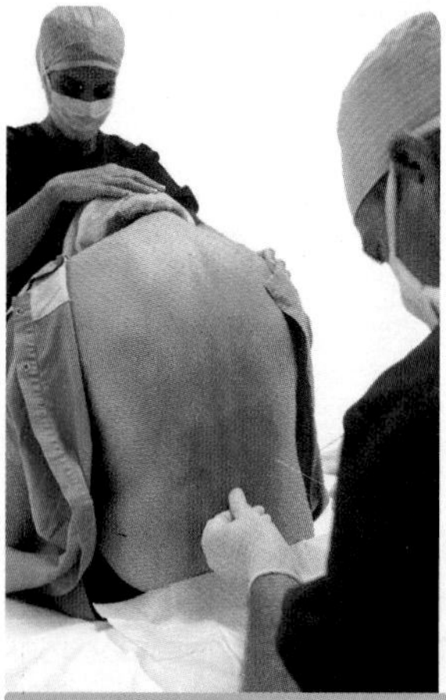

13 Carbohydrates 465

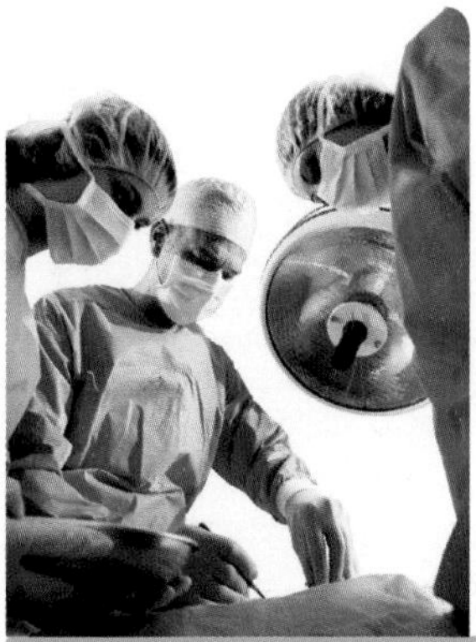

14 Carboxylic Acids, Esters, Amines, and Amides 504

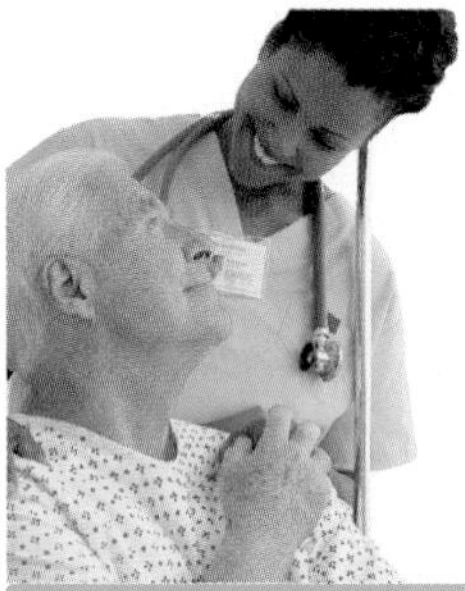

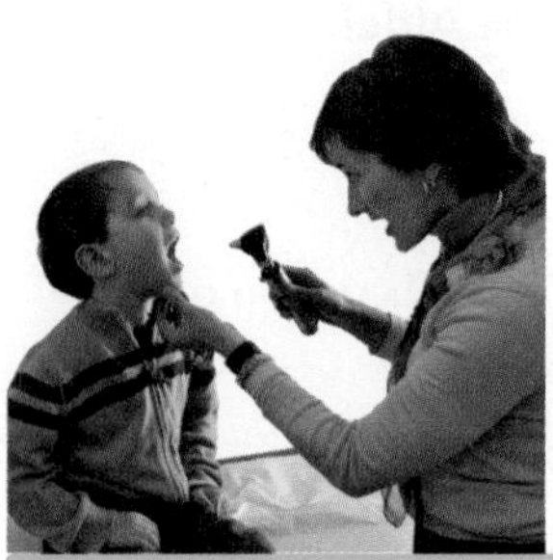

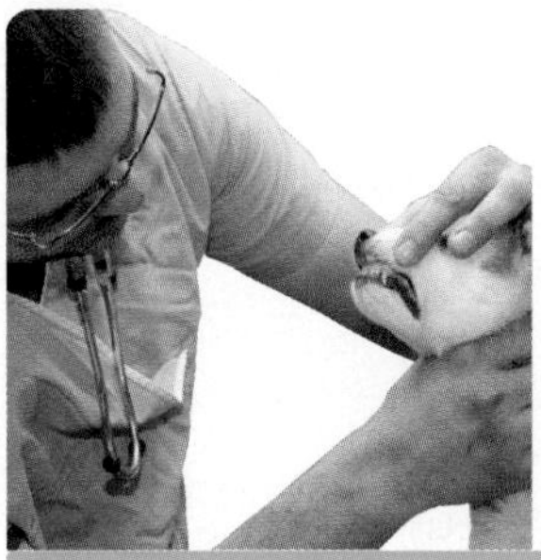

Applications and Activities

Explore Your World

Chemistry Link to **Health**

Chemistry Link to the **Environment**

Guide to Problem Solving

Key Math Skills

Core Chemistry Skills

About the Author

Karen at the Natural History Museum where she is a supporter of children's environmental programs.

KAREN TIMBERLAKE is Professor Emerita of chemistry at Los Angeles Valley College, where **she taught chemistry for allied health and preparatory chemistry for 36 years.** She received her bachelor's degree in chemistry from the University of Washington and her master's degree in biochemistry from the University of California at Los Angeles.

Professor Timberlake has been writing chemistry textbooks for 37 years. During that time, her name **has become associated with the strategic use of pedagogical tools that promote student success in chemistry and the application of chemistry to real-life situations.** More than one million students have learned chemistry using texts, laboratory manuals, and study guides written by Karen Timberlake. In addition to *Chemistry: An Introduction to General, Organic, and Biological Chemistry*, twelfth edition, she is also the author of *General, Organic, and Biological Chemistry, Structures of Life*, fourth edition, with the accompanying *Study Guide and Selected Solutions Manual*, and *Basic Chemistry*, fourth edition, with the accompanying *Study Guide and Selected Solutions Manual*, *Laboratory Manual*, and *Essentials Laboratory Manual*.

Professor Timberlake belongs to numerous scientific and educational organizations including the American Chemical Society (ACS) and the National Science Teachers Association (NSTA). In 1987, she was the Western Regional Winner of Excellence in College Chemistry Teaching Award given by the Chemical Manufacturers Association. In 2004, she received the McGuffey Award in Physical Sciences from the Textbook Authors Association for her textbook *Chemistry: An Introduction to General, Organic, and Biological Chemistry*, eighth edition, which has demonstrated excellence over time. In 2006, she received the "Texty" Textbook Excellence Award from the Textbook Authors Association for the first edition of *Basic Chemistry*. She has participated in education grants for science teaching including the Los Angeles Collaborative for Teaching Excellence (LACTE) and a Title III grant at her college. She speaks at conferences and educational meetings on the use of student-centered teaching methods in chemistry to promote the learning success of students.

When Professor Timberlake is not writing textbooks, she and her husband relax by playing tennis, taking ballroom dance lessons, traveling, trying new restaurants, cooking, and taking care of their grandchildren, Daniel and Emily.

DEDICATION

I dedicate this book to

- My husband for his patience, loving support, and preparation of late meals
- My son, John, daughter-in-law, Cindy, grandson, Daniel, and granddaughter, Emily, for the precious things in life
- The wonderful students over many years whose hard work and commitment always motivated me and put purpose in my writing

FAVORITE QUOTES

The whole art of teaching is only the art of awakening the natural curiosity of young minds.

—Anatole France

One must learn by doing the thing; though you think you know it, you have no certainty until you try.

—Sophocles

Discovery consists of seeing what everybody has seen and thinking what nobody has thought.

—Albert Szent-Györgyi

I never teach my pupils; I only attempt to provide the conditions in which they can learn.

—Albert Einstein

Preface

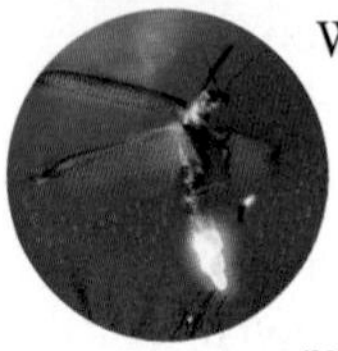

Welcome to the twelfth edition of *Chemistry: An Introduction to General, Organic, and Biological Chemistry*. This chemistry text was written and designed to help you prepare for a career in a health-related profession, such as nursing, dietetics, respiratory therapy, and environmental and agricultural science. This text assumes no prior knowledge of chemistry. My main objective in writing this text is to make the study of chemistry an engaging and a positive experience for you by relating the structure and behavior of matter to its role in health and the environment.

It is my goal to help you become a critical thinker by understanding scientific concepts that will form a basis for making important decisions about issues concerning health and the environment. Thus, I have utilized materials that

- help you to learn and enjoy chemistry
- relate chemistry to careers that interest you
- develop problem-solving skills that lead to your success in chemistry
- promote learning and success in chemistry

New for the Twelfth Edition

This new edition introduces more problem-solving strategies, including new Key Math Skills; new Core Chemistry Skills; new Analyze the Problem features; more Guides to Problem Solving; and more conceptual, challenge, and combined problems. New and updated features have been added throughout this twelfth edition, including the following:

- NEW AND UPDATED! **Chapter Openers** provide modern examples and engaging stories that illustrate how the chemistry you will be learning in each chapter relates to allied health professional experience.
- NEW! **Chapter Readiness** sections at the beginning of each chapter list the Key Math Skills and Core Chemistry Skills from the previous chapters, which provide the foundation for new chemistry principles in the current chapter.
- NEW! **Key Math Skills** review basic math relevant to the chemistry you are learning throughout the text. A **Key Math Skill Review** at the end of each chapter summarizes and gives additional examples.
- NEW! **Core Chemistry Skills** identify the key chemical principles in each chapter that are required for successfully learning chemistry. A **Core Chemistry Skill Review** at the end of each chapter helps reinforce the material and gives additional examples.
- **Analyze the Problem** features included in the solutions of the Sample Problems strengthen critical-thinking skills and illustrate the breakdown of a word problem into the components required to solve it.
- UPDATED! **Questions and Problems, Sample Problems, and art** are directly related to nursing and health applications to better demonstrate the connection between the chemistry being discussed and how these skills will be needed in professional experience.
- UPDATED! **Combining Ideas** features offer sets of integrated problems that test students' understanding by integrating topics from two or more previous chapters.
- UPDATED! **Chapter Reviews** now include bulleted lists and thumbnail art samples related to the content of each section.

Chapter Organization of the Twelfth Edition

In each textbook I write, I consider it essential to relate every chemical concept to real-life issues of health and environment. Because a chemistry course may be taught in different time frames, it may be difficult to cover all the chapters in this text. However, each chapter is a complete package, which allows some chapters to be skipped or the order of presentation to be changed.

Chapter 1, Chemistry in Our Lives, introduces the concepts of chemicals and chemistry, discusses the scientific method in everyday terms, guides students in developing a study plan for learning chemistry, and now has a new section of Key Math Skills, which reviews basic math needed for learning chemistry. The section on Writing Numbers in Scientific Notation was moved from Chapter 2 and is now part of the section of Key Math Skills in this chapter.

- A new chapter opener features the work and career of a forensic scientist.
- A new section, "Scientific Method: Thinking Like a Scientist," has been added, which discusses the scientific method in everyday terms.
- A new section, "Key Math Skills for Chemistry," reviews basic math required in chemistry, such as Identifying Place Values (1.4A), Using Positive and Negative Numbers in Calculations (1.4B) including a new feature Calculator Operations, Calculating a Percentage (1.4C), Solving Equations (1.4D), Interpreting a Graph (1.4E), and Writing Numbers in Scientific Notation (1.4 F).

- New sample problems with nursing applications are added. New Sample Problem 1.1 illustrates the use of scientific method in the nursing environment, and new Sample Problem 1.4 requires the interpretation of a graph to determine the decrease in a child's temperature when given Tylenol.
- New art includes a photo of a nurse making observations (scientific method) in the hospital, and a plastic strip thermometer placed on a baby's forehead to determine body temperature.

Chapter 2, Chemistry and Measurements, looks at measurement and emphasizes the need to understand numerical relationships of the metric system. Significant numbers are discussed in the determination of final answers. Prefixes from the metric system are used to write equalities and conversion factors for problem-solving strategies. Density is discussed and used as a conversion factor.

- A new chapter opener features the work and career of a registered nurse.
- New material is added that illustrates how to count significant figures in equalities and in conversion factors used in a problem setup.
- New abbreviation mcg for microgram is introduced as used in health and medicine.
- New Core Chemistry Skills are added: Counting Significant Figures (2.2), Rounding Off (2.3), Using Significant Figures in Calculations (2.3), Using Prefixes (2.4), Writing Conversion Factors from Equalities (2.5), Using Conversion Factors (2.6), and Using Density as a Conversion Factor (2.7).
- New photos, including pint of blood, Keflex capsules, and salmon for omega-3 fatty acids, are added to improve visual introduction to clinical applications of chemistry.
- Updated Guides to Problem Solving (GPS) use color blocks as visual guides through the solution pathway.
- Updated Sample Problems relate questions and problem solving to health-related topics such as the endoscopic camera, blood volume, omega-3 fatty acids, radiological imaging, and medication orders.

Chapter 3, Matter and Energy, classifies matter and states of matter, describes temperature measurement, and discusses energy, specific heat, and energy in nutrition. Physical and chemical changes and physical and chemical properties are now discussed in more depth.

- A new chapter opener features the work and career of a dietitian.
- Chapter 3 has a new order of topics: 3.1 Classification of Matter, 3.2 States and Properties of Matter, 3.3 Temperature, 3.4 Energy, 3.5 Energy and Nutrition, 3.6 Specific Heat, and 3.7 Changes of State.
- New Core Chemistry Skills are added: Classifying Matter (3.1), Identifying Physical and Chemical Changes (3.2), Converting between Temperature Scales (3.3), Using Energy Units (3.4), and Using the Heat Equation (3.6).
- New Questions and Problems and Sample Problems now have more applications to nursing and health, including Sample Problem 3.4, high temperatures used in cancer treatment; Sample Problem 3.5, the energy produced by a high-energy shock output of a defibrillator; and Sample Problem 3.7, body temperature lowering using a cooling cap.
- The interchapter problem set, Combining Ideas from Chapters 1 to 3, completes the chapter.

Chapter 4, Atoms and Elements, introduces elements and atoms and the periodic table. The names and symbols of element 114, Flerovium, Fl, and 116, Livermorium, Lv, have been added to update the periodic table. Atomic numbers and mass number are determined for isotopes. Atomic mass is calculated using the masses of the naturally occurring isotopes and their abundances. Trends in the properties of elements are discussed, including atomic size, electron-dot symbols, ionization energy, and metallic character.

- A new chapter opener features chemistry in agriculture and the career of a farmer.
- New Core Chemistry Skills are added: Counting Protons and Neutrons (4.4), Writing Atomic Symbols for Isotopes (4.5), Writing Electron Arrangements (4.6), Identifying Trends in Periodic Properties (4.7), and Drawing Electron-Dot Symbols (4.7).
- A new weighted average analogy uses 8-lb and 14-lb bowling balls and the percent abundance of each to calculate weighted average of a bowling ball.
- New nursing and medical examples to Sample Problems/Questions and Problems are added.
- Updated Chemistry Link to Health, "Biological Reactions to UV Light," adds information on using light for neonatal jaundice.
- New Sample Problems on size of atoms and metallic character are added.
- Updated photos and diagrams including a new diagram for the electromagnetic spectrum are added.

Chapter 5, Nuclear Chemistry, looks at the types of radiation emitted from the nuclei of radioactive atoms. Nuclear equations are written and balanced for both naturally occurring radioactivity and artificially produced radioactivity. The half-lives of radioisotopes are discussed, and the amount of time for a sample to decay is calculated. Radioisotopes important in the field of nuclear medicine are described.

- A new chapter opener about the work and career of a radiation technologist is added.
- New Core Chemistry Skills are added: Writing Nuclear Equations (5.2) and Using Half-Lives (5.4).
- New Sample Problems and Questions and Problems use nursing and medical examples, including Sample Problem 5.7 that uses phosphorus-32 for the treatment of leukemia and Sample Problem 5.9 that uses titanium seeds containing a radioactive isotope implanted in the body to treat cancer.

- New and updated Sample Problems and Challenge Problems are added.
- New Analyze the Problem features help students organize and clarify information in word problems.

Chapter 6, Ionic and Molecular Compounds, describes the formation of ionic and covalent bonds. Chemical formulas are written, and ionic compounds—including those with polyatomic ions—and molecular compounds are named. Section 6.1 is now titled "Ions: Transfer of Electrons," 6.2 is titled "Writing Formulas for Ionic Compounds," 6.3 is titled "Naming Ionic Compounds," and 6.5 is titled "Molecular Compounds: Sharing Electrons."

- A new chapter opener featuring the work and career of a pharmacy technician is added.
- "Ions: Transfer of Electrons" has been rewritten to emphasize the stability of the electron configuration of a noble gas.
- New Core Chemistry Skills are added: Writing Positive and Negative Ions (6.1), Writing Ionic Formulas (6.2), Naming Ionic Compounds (6.3), Drawing Electron-Dot Formulas (6.5), Writing the Names and Formulas for Molecular Compounds (6.5), Using Electronegativity (6.6), Predicting Shape (6.7), Identifying Polarity of Molecules (6.7), and Identifying Attractive Forces (6.8).
- A new art comparing the particles and bonding of ionic compounds and molecular compounds has been added.
- A new flowchart for naming chemical compounds in Section 6.5 shows naming patterns for ionic and molecular compounds.
- New Guide to Writing Formulas with Polyatomic Ions is added.
- A new Chemistry Link to Health, "Attractive Forces in Biological Compounds," is added.
- The interchapter problem set, Combining Ideas from Chapters 4 to 6, completes the chapter.

Chapter 7, Chemical Quantities and Reactions, introduces moles and molar masses of compounds, which are used in calculations to determine the mass or number of particles in a given quantity. Students learn to balance chemical equations and to recognize the types of chemical reactions: combination, decomposition, single replacement, double replacement, and combustion reactions. Section 7.5 discusses Oxidation–Reduction Reactions using real-life examples, including biological reactions. Section 7.6, Mole Relationships in Chemical Equations, and Section 7.7, Mass Calculations for Reactions, prepare students for the quantitative relationships of reactants and products in reactions. The chapter concludes with Section 7.8, Energy in Chemical Reactions, which discusses activation energy and energy changes in exothermic and endothermic reactions.

- A new chapter opener featuring the work and career of an exercise physiologist is added.
- New Sample Problems and Challenge Problems use nursing and medical examples.
- New Core Chemistry Skills are added: Converting Particles to Moles (7.1), Calculating Molar Mass (7.2), Using Molar Mass as a Conversion Factor (7.2), Balancing a Chemical Equation (7.3), Classifying Types of Chemical Reactions (7.4), Identifying Oxidized and Reduced Substances (7.5), Using Mole–Mole Factors (7.6), and Converting Grams to Grams (7.7).

Chapter 8, Gases, discusses the properties of gases and calculates changes in gases using the gas laws: Boyle's, Charles's, Gay-Lussac's, Avogadro's, and Dalton's. Problem-solving strategies enhance the discussion and calculations with gas laws.

- A new chapter opener featuring the work and career of a respiratory therapist is added.
- New Sample Problems and Challenge Problems use nursing and medical examples, including Sample Problem 8.3, calculating the volume of oxygen gas delivered through a face mask during oxygen therapy; and Sample Problem 8.9, preparing a heliox breathing mixture for a scuba diver.
- New Core Chemistry Skills are added: Using the Gas Laws (8.2, 8.3, 8.4, 8.5, 8.6) and Calculating Partial Pressure (8.7).

Chapter 9, Solutions, describes solutions, saturation and solubility, insoluble salts, concentrations, and osmosis. New problem-solving strategies clarify the use of concentrations to determine volume or mass of solute. The volumes and molarities of solutions are used in calculations of dilutions and titrations. Properties of solutions, osmosis in the body, and dialysis are discussed.

- A new chapter opener featuring the work and career of a dialysis nurse is added.
- New Core Chemistry Skills are added: Using Solubility Rules (9.3), Calculating Concentration (9.4), and Using Concentration as a Conversion Factor (9.4).
- A new Explore Your World "Preparing Rock Candy" is added to illustrate the formation of a saturated solution.
- Table 9.8 Solubility Rules for Ionic Solids in Water is updated.
- New photos include vanilla and lemon extracts for illustrating volume percent (v/v), and EMTs administering an isotonic NaCl solution.
- Molality and freezing point depression and boiling point elevation are removed.
- New Questions and Problems are written to provide matched problem pairs.
- The interchapter problem set, Combining Ideas from Chapters 7 to 9, completes the chapter.

Chapter 10, Acids and Bases and Equilibrium, discusses acids and bases and their strengths, conjugate acid–base pairs, the ionization of acids, weak bases, and water, pH, and buffers. The reactions of acids and bases with metals, carbonates, and bicarbonates are discussed. Acid–base titration uses the neutralization reactions between acids and bases to calculate quantities of acid in a sample.

- A new chapter opener describes the work and career of a clinical laboratory technician.
- New three-dimensional models of sulfuric acid, carbonic acid, bicarbonate, and carbonate are added.

- A new section "Acid–Base Equilibrium," which includes Le Châtelier's principle, has been added.
- A new diagram of water in tanks reaching equilibrium after water has been added to one tank is given while updated diagrams illustrate the decrease of reactants and increase of products to reach equilibrium.
- Key Math Skills are added: Calculating pH from $[H_3O^+]$ (10.5) and Calculating $[H_3O^+]$ from pH (10.5).
- New Core Chemistry Skills are added: Identifying Conjugate Acid–Base Pairs (10.1), Using Le Châtelier's Principle (10.3), Calculating $[H_3O^+]$ and $[OH^-]$ in Solutions (10.4), Writing Equations for Reactions of Acids and Bases (10.6), and Calculating Molarity or Volume of an Acid or Base in a Titration (10.6).
- A new Guide to Calculating $[H_3O^+]$ from pH has been added.
- New material on diprotic acids has been added.
- New visuals include the ionization of the weak acid hydrofluoric acid, a new photo of calcium hydroxide and information about its use in the food industry, dentistry, and preparation of corn kernels for hominy, as well as a new photo of sodium bicarbonate reacting with acetic acid.
- New material and art on gastric cells and the production of HCl has been added to Chemistry Link to Health, "Stomach Acid, HCl."

Chapter 11, Introduction to Organic Chemistry: Hydrocarbons, combines Chapters 10 and 11 of GOB, eleventh edition. This new chapter compares inorganic and organic compounds, and describes the structures and naming of alkanes, alkenes including cis–trans isomers, alkynes, and aromatic compounds.

- A new chapter opener describes the work and career of a firefighter.
- Chapter 11 has a new order of topics: 11.1 Organic Compounds, 11.2 Alkanes, 11.3 Alkanes with Substituents, 11.4 Properties of Alkanes, 11.5 Alkenes and Alkynes, 11.6 Cis–Trans Isomers, 11.7 Addition Reactions, and 11.8 Aromatic Compounds.
- A new guide to naming aromatic compounds is added.
- New Core Chemistry Skills are added: Naming and Drawing Alkanes (11.2) and Writing Equations for Hydrogenation and Hydration (11.7).
- New skeletal formulas have been added.
- The material on the addition of hydrogen halides and halogens and hydrogenation of alkynes is deleted.
- The section on polymers has been removed.

Chapter 12, Alcohols, Thiols, Ethers, Aldehydes, and Ketones, previously titled "Organic Compounds with Oxygen and Sulfur," describes the functional groups and names of alcohols, thiols, ethers, aldehydes, and ketones. The solubility of alcohols, phenols, aldehydes, and ketones in water is discussed. Section 12.4 is now titled "Reactions of Alcohols, Thiols, Aldehydes, and Ketones."

- A new chapter opener describes the work and career of a nurse anesthetist.
- New Core Chemistry Skills are added: Identifying Functional Groups (12.1), Naming Alcohols and Phenols (12.1), Naming Aldehydes and Ketones (12.3), and Writing Equations for the Dehydration and Oxidation of Alcohols (12.4).
- New Guides to Naming Alcohols, Naming Aldehydes, and Naming Ketones have been added.
- The classification of alcohols has been moved to Section 12.2 "Properties of Alcohols."
- The discussion of boiling points for alcohols and ethers and aldehydes and ketones has been deleted.
- A new table Solubility of Selected Aldehydes and Ketones has been added.
- A new Chemistry Link to Health, "Hand Sanitizers and Ethanol," has been added.
- New material on the use of phenol by Joseph Lister as the first surgical antiseptic is added.
- New tables for the solubility of alcohols and ethers and aldehydes and ketones have been added.
- New material on the oxidation of alcohols to aldehydes and carboxylic acids has been added.
- New material on the reduction of aldehydes and ketones has been added.
- The section "Chiral Molecules" is moved to Chapter 13, Carbohydrates.
- The interchapter problem set, Combining Ideas from Chapters 10 to 12, completes the chapter.

Chapter 13, Carbohydrates, describes the carbohydrate molecules monosaccharides, disaccharides, and polysaccharides and their formation by photosynthesis. Monosaccharides are classified as aldo or keto pentoses or hexoses. Chiral molecules, moved from Chapter 12 to Chapter 13, are discussed along with Fischer projections and D and L notations. An Explore Your World feature models chiral objects using gumdrops and toothpicks. Carbohydrates used as sweeteners are described and carbohydrates used in blood typing are discussed. The formation of glycosidic bonds in disaccharides and polysaccharides is described.

- A new chapter opener describes the work and career of a diabetes nurse.
- Chiral molecules are now discussed in Chapter 13 along with drawing Fischer projections.
- New examples of chiral molecules in nature are added to Chemistry Link to Health, "Enantiomers in Biological Systems."
- New Core Chemistry Skills are added: Identifying Chiral Molecules (13.2) and Identifying D- and L-Fischer Projections (13.3).
- Guide to Drawing Haworth Structures is updated.

Chapter 14, Carboxylic Acids, Esters, Amines, and Amides, discusses the functional groups and naming of carboxylic acids, esters, amines, and amides. Chemical reactions include esterification, amidation, and acid and base hydrolysis of esters and amides.

- A new chapter opener describes the work and career of a surgical technician.

- New Core Chemistry Skills are added: Naming Carboxylic Acids (14.1), Hydrolyzing Esters (14.4), and Forming Amides (14.6).
- Material on heterocyclic amines now describes only pyrrolidine and piperidine.
- New material on the use of aniline to make indigo is added.
- The section on "Alkaloids: Amines in Plants" in GOB, eleventh edition, is now a Chemistry Link to Health.

Chapter 15, Lipids, discusses the functional groups of alcohols and carboxylic acids found in fatty acids, and the formation of ester bonds in triacylglycerols and glycerophospholipids. Chemical properties of fatty acids and their melting points along with the hydrogenation of unsaturated triacylglycerols are discussed. Steroids, which are based on a group of connected multicyclic rings such as cholesterol and bile salts, are described. Chemistry Links to Health include "Omega-3 Fatty Acids in Fish Oils," "Trans Fatty Acids and Hydrogenation," "Infant Respiratory Distress Syndrome (IRDS)," and "Anabolic Steroids." The role of phospholipids in the lipid bilayer of cell membranes is discussed as well as the lipids that function as steroid hormones.

- A new chapter opener describes the work and career of a geriatric nurse.
- New Core Chemistry Skills are added: Identifying Fatty Acids (15.2), Drawing Structures for Triacylglycerols (15.3), Identifying the Products for the Hydrogenation, Hydrolysis, and Saponification of a Triacylglycerol (15.4), and Identifying the Steroid Nucleus (15.6).
- New notation for number of carbon atoms and double bonds in a fatty acid is added.
- New color-block diagrams for triacylglycerols, glycerophospholipids, and sphingolipids are added.
- New lipid panel for cholesterol, triglycerides, HDL, LDL, and cholesterol/HDL ratio is added.
- New photos include jojoba plant, use of triacylglycerols to thicken creams and lotions, and poisonous snake with venom that hydrolyzes phospholipids in red blood cells.
- The interchapter problem set, Combining Ideas from Chapters 13 to 15, completes the chapter.

Chapter 16, Amino Acids, Proteins, and Enzymes, discusses amino acids, formation of peptide bonds and proteins, structural levels of proteins, enzymes, and enzyme action. Amino acids are drawn as their ionized forms in physiological solutions. Section 16.3 describes the primary level of protein structure. Section 16.4 describes the secondary, tertiary, and quaternary levels of proteins. Enzymes are discussed as biological catalysts, along with the impact of inhibitors and denaturation on enzyme action.

- A new chapter opener discusses the career of a physician assistant.
- Abbreviations for amino acid names use three letters as well as one letter.
- New ball-and-stick models of several amino acids have been added.
- The updated Chemistry Link to Health, "Essential Amino Acids," is moved to Section 16.1.
- Amino acids are now drawn with the carboxyl or carboxylate groups showing single and double bonds to O atoms.
- New Core Chemical Skills are added: Drawing the Ionized Form for an Amino Acid (16.1), Identifying the Primary, Secondary, Tertiary, and Quaternary Structures of Proteins (16.3, 16.4), and Describing Enzyme Action (16.5).
- A new photo of a spiral staircase is added to illustrate the alpha helical secondary structure of proteins.
- New ribbon models of proteins for alpha helices, beta-pleated sheets, myoglobin, hemoglobin, denatured protein, prions in mad cow disease, and enzymes are added.
- New diagrams illustrating enzyme action and the effect of competitive and noncompetitive inhibitors on enzyme structure and action are added.
- Section 16.8 "Enzyme Cofactors" has been removed.

Chapter 17, Nucleic Acids and Protein Synthesis, describes the nucleic acids and their importance as biomolecules that store and direct information for the synthesis of cellular components. The role of complementary base pairing is discussed in both DNA replication and the formation of mRNA during protein synthesis. The role of RNA is discussed in the relationship of the genetic code to the sequence of amino acids in a protein. Mutations describe ways in which the nucleotide sequences are altered in genetic diseases. We also look at how DNA or RNA in viruses utilizes host cells to produce more viruses.

- A new chapter opener discusses the work and career of a histology technician.
- Chemistry Links to Health include "DNA Fingerprinting," "Many Antibiotics Inhibit Protein Synthesis," and "Cancer."
- Color-block diagrams clarify the relationship between mRNA codons and the amino acid sequence.
- New Core Chemical Skills are added: Writing the Complementary DNA Strand (17.3), Writing the mRNA Segment for a DNA Template (17.4), and Writing the Amino Acid for an mRNA Codon (17.4).

Chapter 18, Metabolic Pathways and Energy Production, describes the metabolic pathways of biomolecules from the digestion of foodstuffs to the synthesis of ATP. Students look at the stages of metabolism and the digestion of carbohydrates along with the coenzymes required in metabolic pathways. The breakdown of glucose to pyruvate is described using glycolysis, which is followed by the decarboxylation of pyruvate to acetyl-CoA. We look at the entry of acetyl-CoA into the citric acid cycle and the production of reduced coenzymes. We describe electron transport, oxidative phosphorylation, and the synthesis of ATP. The oxidation of lipids and the degradation of amino acids are also discussed.

- A new chapter opener discusses the career of a veterinary assistant.
- New Core Chemical Skills are added: Identifying the Compounds in Glycolysis (18.4), Describing the Reactions in the Citric Acid Cycle (18.5), Calculating the ATP Produced from Glucose (18.6), and Calculating the ATP from Fatty Acid Oxidation (18.7).
- Updated art for ATP structures and coenzymes NAD^+ and FAD is added.
- Updated art for diagrams of glycolysis, the citric acid cycle, and electron transport is added.
- Chemistry Links to Health include "ATP Energy and Ca^{2+} Needed to Contract Muscles," "Lactose Intolerance," "ATP Synthase and Heating the Body," "Stored Fat and Obesity," and "Ketone Bodies and Diabetes."
- New ribbon models for cytochrome *c* and leptin hormone are added.
- The interchapter problem set, Combining Ideas from Chapters 16 to 18, completes the chapter.

Instructional Package

Chemistry: An Introduction to General, Organic, and Biological Chemistry, twelfth edition, provides an integrated teaching and learning package of support material for both students and professors.

Name of Supplement	Available in Print	Available Online	Instructor or Student Supplement	Description
MasteringChemistry® (www.masteringchemistry.com)		✓	Supplement for Students and Instructors	The most advanced, most widely used online chemistry tutorial and homework system is available for the twelfth edition of *Chemistry: An Introduction to General, Organic, and Biological Chemistry*. MasteringChemistry® utilizes the Socratic method to coach students through problem-solving techniques, offering hints and simpler questions on request to help students *learn*, not just practice. A powerful grade book with diagnostics that gives instructors unprecedented insight into their students' learning is also available. MasteringChemistry reinforces key topics and skills while allowing students to see chemistry and its applications come to life.
Pearson eText enhanced with media within MasteringChemistry®		✓	Supplement for Students	For the first time, the twelfth edition of *Chemistry: An Introduction to General, Organic, and Biological Chemistry* features a Pearson eText enhanced with media within Mastering. In conjunction with Mastering assessment capabilities, new **Interactive Videos and 3D animations** will improve student engagement and knowledge retention. Each chapter will contain a balance of interactive animations, videos, sample calculations, and self-assessments/quizzes embedded directly in the eText. Additionally, the Pearson eText offers students the power to create notes, highlight text in different colors, create bookmarks, zoom, and view single or multiple pages. Icons in the margins throughout the text signify that there is a new **Interactive Video** or animation located within MasteringChemistry® for *Chemistry: An Introduction to General, Organic, and Biological Chemistry*, twelfth edition.
Instructor's Solutions Manual		✓	Supplement for Instructors	Prepared by Mark Quirie, the solutions manual highlights chapter topics, and includes answers and solutions for all questions and problems in the text.
Instructor Resource Materials–Download Only		✓	Supplement for Instructors	Includes all the art, photos, and tables from the book in JPEG format for use in classroom projection or when creating study materials and tests. In addition, the instructors can access modifiable PowerPoint™ lecture outlines. Also available are downloadable files of the Instructor's Solutions Manual and a set of "clicker questions" designed for use with classroom-response systems. Also visit the Pearson Education catalog page for Timberlake's *Chemistry: An Introduction to General, Organic, Biological Chemistry*, twelfth Edition, at www.pearsonglobaleditions.com/timberlake to download available instructor supplements.
TestGen Test Bank–Download Only		✓	Supplement for Instructors	Prepared by William Timberlake, this resource includes more than 1600 questions in multiple-choice, matching, true/false, and short-answer format.

Acknowledgments

The preparation of a new text is a continuous effort of many people. I am thankful for the support, encouragement, and dedication of many people who put in hours of tireless effort to produce a high-quality book that provides an outstanding learning package. The editorial team at Pearson has done an exceptional job. I want to thank Adam Jaworski, editor in chief, and Jeanne Zalesky, executive editor, who supported our vision of this twelfth edition and the development of new math remediation strategies with Chapter Readiness, which appear throughout the chapter along with their reviews at the end of each chapter and with the new Analyze the Problem feature that clarifies the components of a word problem for problem solving. I also appreciate the addition of new chapter openers related to health careers, more Guides to Problem Solving, new Chemistry Links to Health and the Environment, and new problems that relate to health and nursing.

I appreciate all the wonderful work of Lisa Pierce, project manager, who skillfully brought together reviews, art, web site materials, and all the things it takes to prepare a book for production. I appreciate the work of Wendy Perez, project manager, and Andrea Stefanowicz of Lumina Datamatics Inc., who brilliantly coordinated all phases of the manuscript to the final pages of a beautiful book. Thanks to Mark Quirie, manuscript and accuracy reviewer, and Lumina Datamatics Inc. copy editor and proofreaders, who precisely analyzed and edited the initial and final manuscripts and pages to make sure the words and problems were correct to help students learn chemistry. Their keen eyes and thoughtful comments were extremely helpful in the development of this text.

I am especially proud of the art program in this text, which lends beauty and understanding to chemistry. I would like to thank Wynne Au Yeung, art specialist; Mark Ong and Jerilyn Bockorick, interior and cover designers, whose creative ideas provided the outstanding design for the cover and pages of the book. Erica Gordon, photo researcher, was outstanding in researching and selecting vivid photos for the text so that students can see the beauty of chemistry. Thanks also to *Bio-Rad Laboratories* for their courtesy and use of *KnowItAll ChemWindows*, drawing software that helped us produce chemical structures for the manuscript. The macro-to-micro illustrations designed by Production Solutions and Precision Graphics give students visual impressions of the atomic and molecular organization of everyday things and are a fantastic learning tool. I also appreciate all the hard work in the field put in by the marketing team and Jonathan Cottrell, marketing manager.

I am extremely grateful to an incredible group of peers for their careful assessment of all the new ideas for the text; for their suggested additions, corrections, changes, and deletions; and for providing an incredible amount of feedback about improvements for the book. I admire and appreciate every one of you.

If you would like to share your experience with chemistry, or have questions and comments about this text, I would appreciate hearing from you.

Karen Timberlake
Email: khemist@aol.com

Pearson would like to acknowledge and thank Chun-Yu Ho (The Chinese University of Hong Kong) and Chitralekha Sidana (writer) for their valuable contribution to the Global Edition, and Debajyoti Mahanta (Gauhati University), Diganta Das (Gauhati University), Bhushan Popatkar (University of Mumbai) and S. Nehru (University of Madras) for reviewing the Global Edition.

Reviewers

Twelfth Edition Reviewers

Sheryl Drake
Parkland College

Lisa Sharpe Elles
Washburn University

Denise Gigante
Onondaga Community College

Sheldon Knoespel
Jackson Community College

Lauren McMills
Ohio University

Parris Powers
Volunteer State Community College

Mark Quirie
Algonquin College

Tanea Reed
Eastern Kentucky University

John Singer
Jackson Community College

Annie Lee Stanislaus
Rockhurst University

Mary Lamar
Eastern Kentucky University

Frederick Lempe
Mott Community College

Provi Mayo
Ivy Tech Community College

Barbara Mowery
York College

Deepa Perera
Muskingum University

Joe Selzler
Ventura College

R. George Shelton, Jr.
Austin Peay State University

Laura Slocum
Heathwood Hall Episcopal School

Autumn Sutherlin
Abilene Christian University

W. Doug Urban
North Central State College

Focus Group Participants

Diane Beecroft
Central Arizona College

Gina Fiorini
Bellevue College

Christine Hermann
Radford University

Melody Jewell
South Dakota State University

Clinton King
Snow College

Eleventh Edition Reviewers

Gary Burce
Ohlone College

Pamela Goodman
Moraine Valley Community College

Emily Halvorson
Pima Community College

Mark Quirie
Algonquin College

Richard Triplett
Des Moines Area Community College

Students learn chemistry using real-world examples

"Discovery consists of seeing what everybody has seen and thinking what nobody has thought."
—Albert Szent-Györgyi

Feature	Description	Benefit	Page
Chapter Opener 	Chapters begin with **stories** involving careers in fields such as nursing, agriculture, exercise physiology, and veterinary sciences.	Show you how health professionals use chemistry every day	354
Chemistry Link to Health	**Chemistry Links to Health** apply chemical concepts to relevant topics of health and medicine such as weight loss and weight gain, trans fats, anabolic steroids, alcohol abuse, blood buffers, kidney dialysis, and cancer.	Provide you with connections that illustrate the importance of understanding chemistry in real-life health and medical situations	36
Chemistry Link to the Environment	**Chemistry Links to the Environment** relate chemistry to environmental topics such as climate change, radon in our homes, and pheromones.	Help you extend your understanding of the impact of chemistry on the environment	102
Macro-to-Micro Art 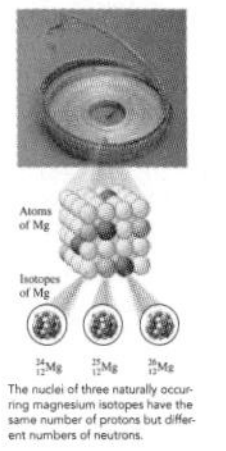	Macro-to-Micro Art utilizes photographs and drawings to illustrate the atomic structure of chemical phenomena.	Helps you connect the world of atoms and molecules to the macroscopic world	144

". . . never teach my pupils; I only attempt to provide the conditions in which they can learn."
—Albert Einstein

Feature	Description	Benefit	Page
Learning Goals LEARNING GOAL Write a balanced nuclear equation showing mass numbers and atomic numbers for radioactive decay.	Learning Goals at the beginning and end of each section identify the key concepts for that section and provide a roadmap for your study.	Help you focus your studying by emphasizing what is most important in each section	171
Writing Style **4.3 The Atom** All the elements listed on the periodic table are made up of atoms. An **atom** is the smallest particle of an element that retains the characteristics of that element. Imagine that you are tearing a piece of aluminum foil into smaller and smaller pieces. Now imagine that you have a microscopic piece so small that it cannot be divided any further. Then you would have a single atom of aluminum.	Timberlake's accessible writing style is based on careful development of chemical concepts suited to the skills and backgrounds of students in chemistry.	Helps you understand new terms and chemical concepts	137
Concept Maps 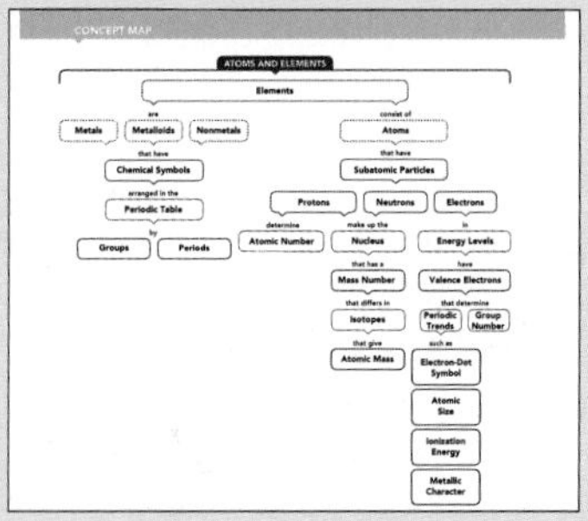	Concept Maps at the end of each chapter show how all the key concepts fit together.	Encourage learning by providing a visual guide to the interrelationship among all the concepts in each chapter	158
Key Math Skills KEY MATH SKILL Identifying Place Values	Key Math Skills review the basic math required needed for chemistry. Instructors can also assign these through MasteringChemistry.	Help you master the basic quantitative skills to succeed in chemistry	41
Core Chemistry Skills CORE CHEMISTRY SKILL Counting Protons and Neutrons	Core Chemistry Skills provide content crucial to problem-solving strategies related to chemistry. Instructors can also assign these through MasteringChemistry.	Help you master the basic problem-solving skills needed to succeed in chemistry	141
Art Program 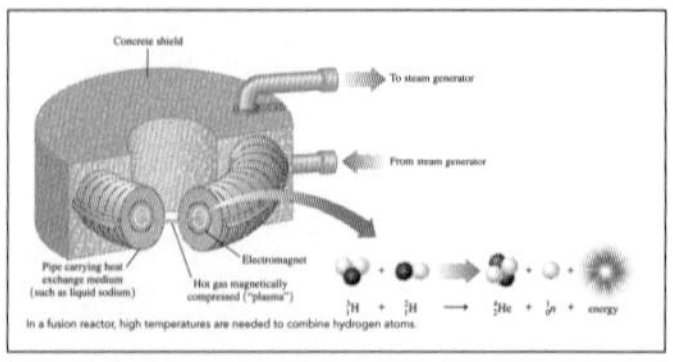	The art program is beautifully rendered, pedagogically effective, and includes questions with all the figures.	Helps you think critically using photos and illustrations	189
Chapter Reviews 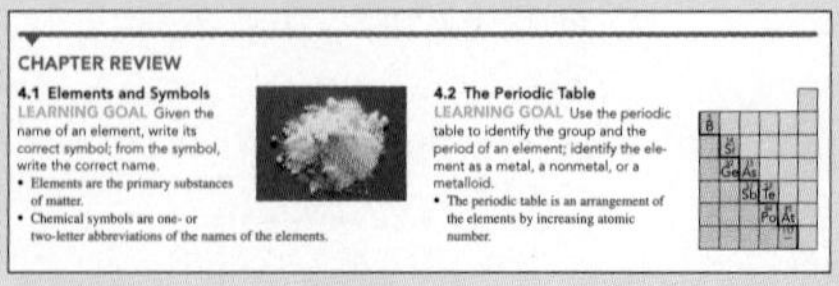	The Chapter Reviews include Learning Goals and visual thumbnails to summarize the key points in each section.	Help you determine your mastery of the chapter concepts and study for your tests	158

Many tools show students how to solve problems

"The whole art of teaching is only the art of awakening the natural curiosity of young minds."
—Anatole France

Feature	Description	Benefit	Page
Guides to Problem Solving	Guides to Problem Solving (GPS) illustrate the steps needed to solve problems.	Visually guide you step-by-step through each problem-solving strategy	74
Analyze the Problems	Analyze the Problem features now included in Sample Problem Solutions convert information in a word problem into components for problem solving.	Help you identify and utilize the components within a word problem to set up a solution strategy	104
End-of-Section Questions and Problems	Questions and Problems are placed at the end of each section. Problems are paired and the Answers to the odd-numbered problems are given at the end of each chapter.	Encourage you to become involved immediately in the process of problem solving	41
Sample Problems with Study Checks	Sample Problems illustrate worked-out solutions with step-by-step explanations and required calculations. Study Checks associated with each Sample Problem allow you to check your problem-solving strategies.	Provide the intermediate steps to guide you successfully through each type of problem	74
Understanding the Concepts	Understanding the Concepts are questions with visual representations placed at the end of each chapter.	Build an understanding of newly learned chemical concepts	193
Additional Questions and Problems	Additional Questions and Problems at the end of each chapter provide further study and application of the topics from the entire chapter.	Promote critical thinking	52
Challenge Questions	Challenge Questions at the end of each chapter provide complex questions.	Promote critical thinking, group work, and cooperative learning environments	53
Combining Ideas	Combining Ideas are sets of integrated problems that are placed after every 3 chapters.	Test your understanding of the concepts from previous chapters by integrating topics	351

teringChemistry® for Timberlake

Mastering system motivates students to learn outside of class and arrive prepared for lecture or lab. New Tutorials, Activities, Interactive Videos, and Dynamic Study Modules help each individual student develop proficiency in the most challenging aspects of the GOB course.

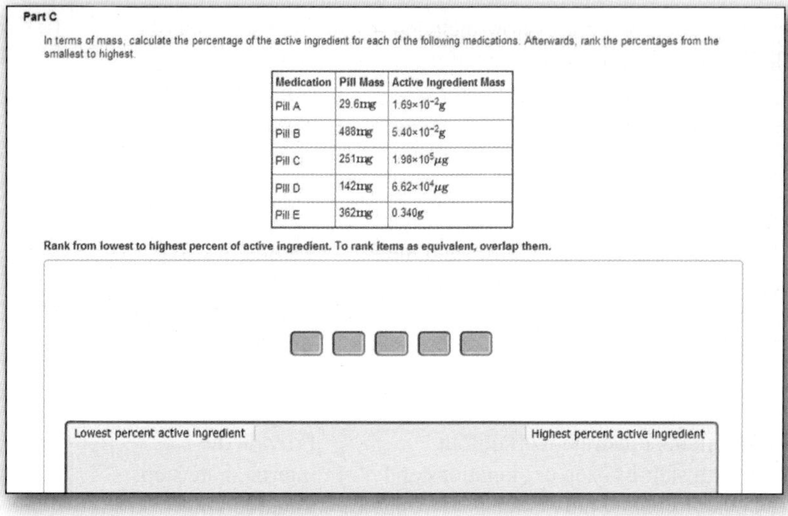
Part C

In terms of mass, calculate the percentage of the active ingredient for each of the following medications. Afterwards, rank the percentages from the smallest to highest.

Medication	Pill Mass	Active Ingredient Mass
Pill A	29.6mg	1.69×10^{-2}g
Pill B	488mg	5.40×10^{-2}g
Pill C	251mg	$1.98\times10^{5}\mu$g
Pill D	142mg	$6.62\times10^{4}\mu$g
Pill E	362mg	0.340g

Rank from lowest to highest percent of active ingredient. To rank items as equivalent, overlap them.

Lowest percent active ingredient | Highest percent active ingredient

New Tutorials in MasteringChemistry complement chemistry presented by Karen Timberlake in the text and help students develop problem-solving skills needed to succeed in this course and in their future allied health careers. New tutorials for the *Twelfth Edition* include Conversion Factors in Medicine, Develop a Problem Solving Strategy, and Think Like a Chemist.

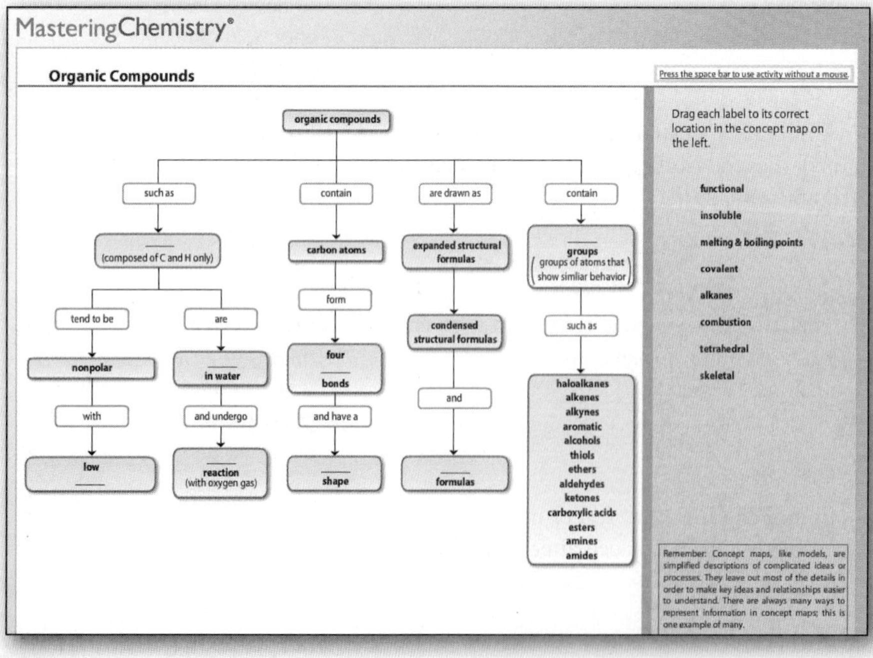

Concept Map Quizzes and New Construct-a-Concept Map Problems

Concept Map Quizzes based on the text help students make connections between important concepts while Construct-a-Concept Map problems prompt students to create their own connections between concepts, providing wrong answer feedback that helps clarify the links they propose between topics and ideas.

MasteringChemistry promotes interactivity and active learning in General, Organic, and Biological Chemistry. Research shows that Mastering's immediate feedback and tutorial assistance help students understand and master concepts and skills in chemistry—allowing them to retain more knowledge and perform better in this course and beyond.

Pause & Predict Videos

Video demonstrations prompt students to interact with experiments that illuminate the chemical concepts being discussed. During the video, a set of multiple choice questions challenges students to apply the concepts from the video and predict the outcome of the experiment on screen. After students make their prediction, the video continues and the students will see if the outcome they chose correctly corresponds to the actual demonstration. These videos are available in web and mobile-friendly formats through the eText, assignable in Tutorial problems in MasteringChemistry, and also available in the Study Area of MasteringChemistry.

Math Remediation

MasteringChemistry offers a variety of math remediation options for students to brush up on required quantitative skills. The **Mathematics Review chapter** contains practice tutorials. A new **Math Remediation Quiz** provides a comprehensive set of review questions with links to additional algorithmically generated practice problems in MathXL. Select tutorials also contain Math Remediation Links to additional MathXL® questions.

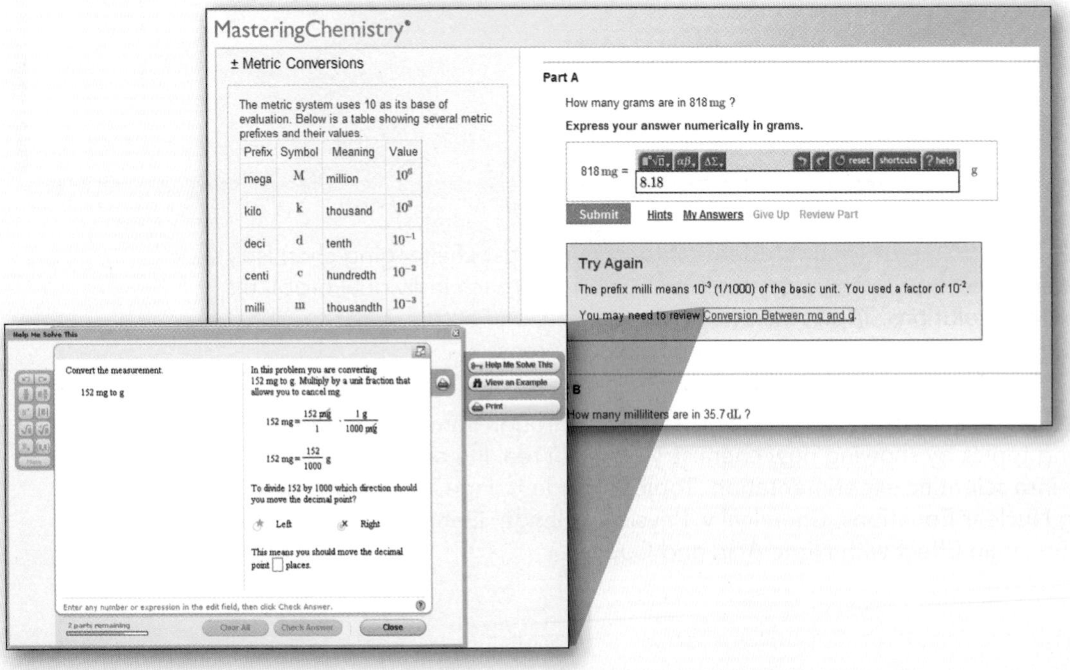

Engaging, Interactive, and Adaptive Media Specific to the GOB Course

For the first time, the twelfth edition of *Chemistry: An Introduction to General, Organic, and Biological Chemistry* features a Pearson eText enhanced with media. In conjunction with MasteringChemistry assessments, new interactive videos, sample calculations, video demonstrations, and animations will engage students by bringing chemistry to life and walking students through different approaches to solving problems.

NEW! **Interactive Videos**
Green play button icons appear in the margins throughout the text. In the eText, the icons link to new interactive videos or sample problems that the student can use to clarify and reinforce important concepts.

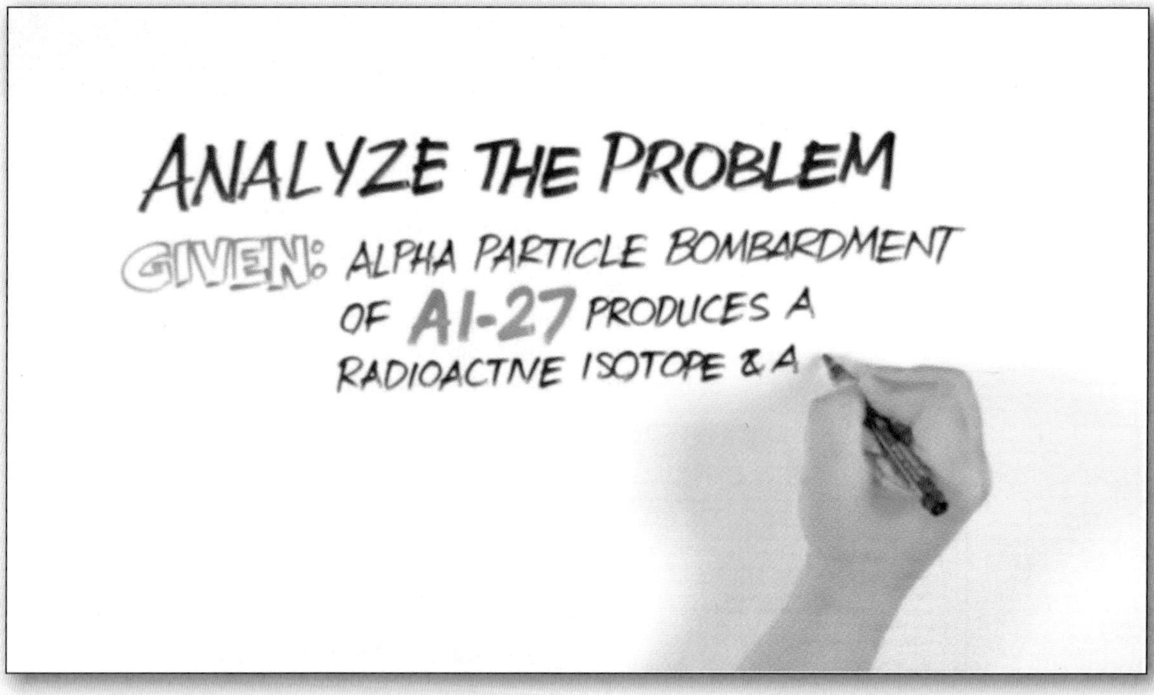

Sample Calculations walk students through the most challenging chemistry problems and provide a fresh perspective on how to approach individual problems and reach their solutions. Topics include Using Conversion Factors, Mass Calculations for Reactions, and Concentration of Solutions.

Live Video Demonstrations help students through some of the more challenging topics by showing how chemistry works in real life and introducing a bit of humor into scientific experimentation. Topics include Using Conversion Factors, Balancing Nuclear Equations, Chemical v. Physical Change, Dehydration of Sucrose, and The Fountain Effect with Nitric Acid and Copper.

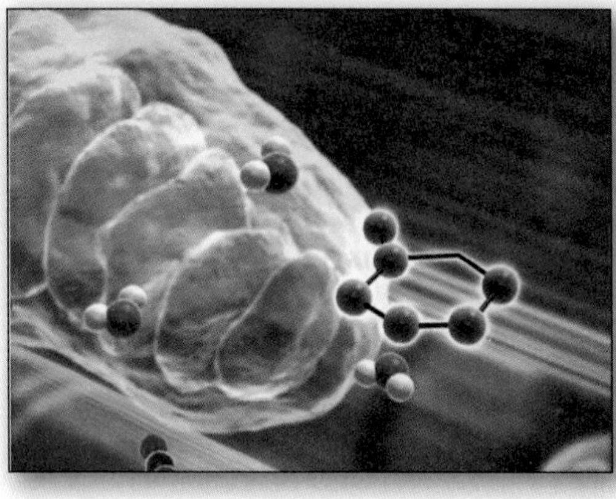

NEW! Animations

Interactive animations illustrate foundational topics and concepts and help students better visualize some of the more complex concepts in GOB Chemistry including Carbohydrate Digestion, Protein and Membrane Structure, Cellular Respiration, and Kinetic Molecular Theory of Gases.

NEW! Dynamic Study Module on Fundamental Math and Chemistry Skills

Skills, designed to enable students to study effectively on their own, help students quickly access and learn the basic math and science skills they need to be successful in the GOB course. The module can be accessed on smartphones, tablets, and computers. Results can be tracked in the MasteringChemistry® Gradebook. How it works:

1. Students receive an initial set of questions. A unique answer format asks students to indicate how confident they are about their answer.
2. After answering each set of questions, students review their answers. Each answer provides a comprehensive explanation of the concept students struggled with.
3. Once students review the explanations, they are presented with a new set of questions. Students cycle through this dynamic process of test–learn–retest until they achieve mastery of course math and science skills needed to do well in an introductory chemistry course.

New Interactive Videos are available in web and mobile-friendly formats through the eText, assignable in Tutorial problems in MasteringChemistry, and also available in the Study Area of MasteringChemistry. You can also scan the QR code featured on this page and on the back cover of the textbook for immediate access via your smartphone.

1

Chemistry in Our Lives

Sarah works as a forensic scientist where she applies scientific procedures to evidence from law enforcement agencies. Such evidence may include blood, hair, or fiber from clothing found at a crime scene. At work, she analyzes blood for the presence of drugs, poisons, and alcohol. She prepares tissues for typing factors and for DNA analysis. Her lab partner Mark is working on matching characteristics of a bullet to a firearm found at a crime scene. He is also using fingerprinting techniques to identify the victim of a crime.

A female victim is found dead in her home. The police suspect that she was murdered, so samples of her blood and stomach contents are sent to Sarah. Using a variety of qualitative and quantitative tests, Sarah finds traces of ethylene glycol. The qualitative tests show that ethylene glycol is present, while the quantitative tests indicate the amount of ethylene glycol the victim has in her system. Sarah determines that the victim was poisoned when she ingested ethylene glycol that had been added to an alcoholic beverage. Since the initial symptoms of ethylene glycol poisoning are similar to being intoxicated, the victim was unaware of the poisoning.

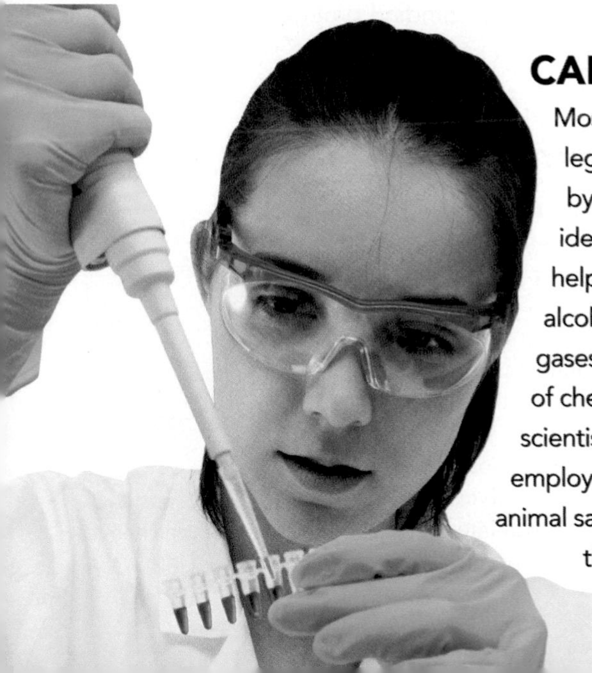

CAREER Forensic Scientist

Most forensic scientists work in crime laboratories that are part of city or county legal systems where they analyze bodily fluids and tissue samples collected by crime scene investigators. In analyzing these samples, a forensic scientist identifies the presence or absence of specific chemicals within the body to help solve the criminal case. Some of the chemicals they look for include alcohol, illegal or prescription drugs, poisons, arson debris, metals, and various gases such as carbon monoxide. In order to identify these substances, a variety of chemical instruments and highly specific methodologies are used. Forensic scientists also analyze samples from criminal suspects, athletes, and potential employees. They also work on cases involving environmental contamination and animal samples for wildlife crimes. A forensic scientist usually has a bachelor's degree that includes courses in math, chemistry, and biology.

LOOKING AHEAD

Now that you are in a chemistry class, you may be wondering what you will be learning. What questions in science have you been curious about? Perhaps you are interested in how smog is formed, what hemoglobin does in the body, or how aspirin relieves a headache. Just like you, chemists are curious about the world we live in.

- What does hemoglobin do in the body? Hemoglobin consists of four polypeptide chains each containing a heme group with an iron atom that binds to oxygen (O_2) in the lungs. From the lungs, hemoglobin transports oxygen to the tissues of the body where it is used to provide energy. Once the oxygen is released, hemoglobin binds to carbon dioxide (CO_2) for transport to the lungs where it is released.

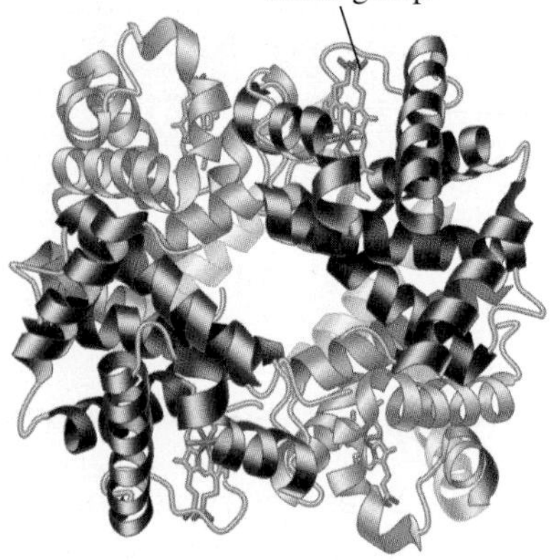

Hemoglobin transports oxygen to the tissues and carbon dioxide to the lungs.

- How does car exhaust produce the smog that hangs over our cities? One component of car exhaust is nitrogen oxide (NO), which forms in car engines where high temperatures convert nitrogen gas (N_2) and oxygen gas (O_2) to NO. In chemistry, these reactions are written in the form of equations such as $N_2(g) + O_2(g) \longrightarrow 2NO(g)$.

- Why does aspirin relieve a headache? When a part of the body is injured, substances called prostaglandins are produced, which cause inflammation and pain. Aspirin acts to block the production of prostaglandins, thereby reducing inflammation, pain, and fever.

Chemists working in research laboratories test new products and develop new pharmaceuticals.

Chemists in the medical field develop new treatments for diabetes, genetic defects, cancer, AIDS, and other diseases. Researchers in the environmental field study the ways in which human development impacts the environment and develop processes that help reduce environmental degradation. For the chemist in the laboratory, the nurse in the dialysis unit, the dietitian, or the agricultural scientist, chemistry plays a central role in understanding problems, assessing possible solutions, and making important decisions.

LEARNING GOAL

Define the term chemistry and identify substances as chemicals.

1.1 Chemistry and Chemicals

Chemistry is the study of the composition, structure, properties, and reactions of matter. *Matter* is another word for all the substances that make up our world. Perhaps you imagine that chemistry takes place only in a laboratory where a chemist is working in a white coat and goggles. Actually, chemistry happens all around you every day and has an impact on everything you use and do. You are doing chemistry when you cook food, add bleach to your laundry, or start your car. A chemical reaction has taken place when silver tarnishes or an antacid tablet fizzes when dropped into water. Plants grow because chemical reactions convert

Antacid tablets undergo a chemical reaction when dropped into water.

carbon dioxide, water, and energy to carbohydrates. Chemical reactions take place when you digest food and break it down into substances that you need for energy and health.

Chemicals

A **chemical** is a substance that always has the same composition and properties wherever it is found. All the things you see around you are composed of one or more chemicals. Chemical processes take place in chemistry laboratories, manufacturing plants, and pharmaceutical labs as well as every day in nature and in our bodies. Often the terms *chemical* and *substance* are used interchangeably to describe a specific type of matter.

Every day, you use products containing substances that were developed and prepared by chemists. Soaps and shampoos contain chemicals that remove oils on your skin and scalp. When you brush your teeth, the substances in toothpaste clean your teeth, prevent plaque formation, and stop tooth decay. Some of the chemicals used to make toothpaste are listed in Table 1.1.

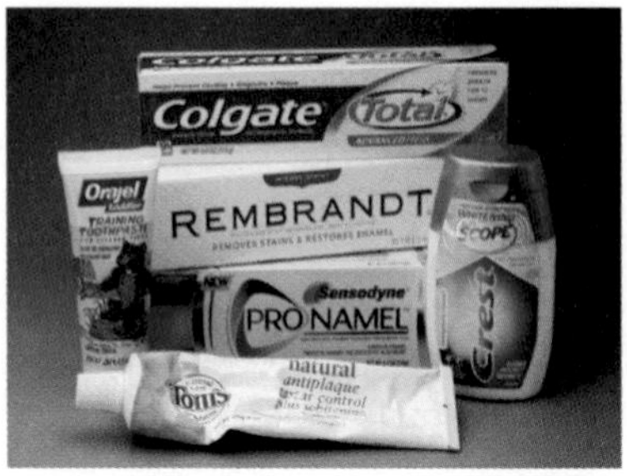

Toothpaste is a combination of many chemicals.

In cosmetics and lotions, chemicals are used to moisturize, prevent deterioration of the product, fight bacteria, and thicken the product. Your clothes may be made of natural materials such as cotton or synthetic substances such as nylon or polyester. Perhaps you wear a ring or watch made of gold, silver, or platinum. Your breakfast cereal is probably fortified with iron, calcium, and phosphorus, while the milk you drink is enriched with vitamins A and D. Antioxidants are chemicals added to food to prevent it from spoiling. Some of the chemicals you may encounter when you cook in the kitchen are shown in Figure 1.1.

TABLE 1.1 Chemicals Commonly Used in Toothpaste

Chemical	Function
Calcium carbonate	Used as an abrasive to remove plaque
Sorbitol	Prevents loss of water and hardening of toothpaste
Sodium lauryl sulfate	Used to loosen plaque
Titanium dioxide	Makes toothpaste white and opaque
Triclosan	Inhibits bacteria that cause plaque and gum disease
Sodium fluorophosphate	Prevents formation of cavities by strengthening tooth enamel with fluoride
Methyl salicylate	Gives toothpaste a pleasant wintergreen flavor

FIGURE 1.1 ▶ Many of the items found in a kitchen are chemicals or products of chemical reactions.

Q What are some other chemicals found in a kitchen?

QUESTIONS AND PROBLEMS

1.1 Chemistry and Chemicals

LEARNING GOAL Define the term chemistry and identify substances as chemicals.

In every chapter, odd-numbered exercises in the *Questions and Problems* are paired with even-numbered exercises. The answers for the magenta, odd-numbered *Questions and Problems* are given at the end of each chapter. The complete solutions to the odd-numbered *Questions and Problems* are in the *Student Solutions Manual*.

1.1 Write a one-sentence definition for each of the following:
a. chemistry
b. chemical

1.2 Ask two of your friends (not in this class) to define the terms in problem 1.1. Do their answers agree with the definitions you provided?

1.3 Obtain a bottle of multivitamins and read the list of ingredients. What are four chemicals from the list?

1.4 Obtain a box of breakfast cereal and read the list of ingredients. What are four chemicals from the list?

1.5 Read the labels on some items found in your medicine cabinet. What are the names of some chemicals contained in those items?

1.6 Name the key components of each of the following:
a. vinegar
b. alcohol disinfectant pads

LEARNING GOAL

Describe the activities that are part of the scientific method.

1.2 Scientific Method: Thinking Like a Scientist

When you were very young, you explored the things around you by touching and tasting. As you grew, you asked questions about the world in which you live. What is lightning? Where does a rainbow come from? Why is water blue? As an adult, you may have wondered how antibiotics work or why vitamins are important to your health. Every day, you ask questions and seek answers to organize and make sense of the world around you.

When the late Nobel Laureate Linus Pauling described his student life in Oregon, he recalled that he read many books on chemistry, mineralogy, and physics. "I mulled over the properties of materials: why are some substances colored and others not, why are some minerals or inorganic compounds hard and others soft?" He said, "I was building up this tremendous background of empirical knowledge and at the same time asking a great number of questions."[1] Linus Pauling won two Nobel Prizes: the first, in 1954, was in chemistry for his work on the nature of chemical bonds and the determination of the structures of complex substances; the second, in 1962, was the Peace Prize.

Linus Pauling won the Nobel Prize in Chemistry in 1954.

Scientific Method

Although the process of trying to understand nature is unique to each scientist, a set of general principles, called the **scientific method**, helps to describe how a scientist thinks.

1. **Observations.** The first step in the scientific method is to make *observations* about nature and ask questions about what you observe.
2. **Hypothesis.** Propose a *hypothesis*, which states a possible explanation of the observations. The hypothesis must be stated in such a way that it can be tested by experiments.
3. **Experiments.** Several *experiments* may be done to test the hypothesis.
4. **Conclusion.** When the results of the experiments are analyzed, a *conclusion* is made as to whether the hypothesis is *true* or *false*. When experiments give consistent results, the hypothesis may be confirmed. Even then, a hypothesis continues to be tested and, based on new experimental results, may need to be modified or replaced.

Chemistry Link to **Health**

Early Chemist: Paracelsus

Paracelsus (1493–1541) was a physician and an alchemist who thought that alchemy should be about preparing new medicines, not about producing gold. Using observation and experimentation, he proposed that a healthy body was regulated by a series of chemical processes that could be unbalanced by certain chemical compounds and rebalanced by using minerals and medicines. For example, he determined that inhaled dust, not underground spirits, caused lung disease in miners. He also thought that goiter was a problem caused by contaminated water, and he treated syphilis with compounds of mercury. His opinion of medicines was that the right dose makes the difference between a poison and a cure. Paracelsus changed alchemy in ways that helped to establish modern medicine and chemistry.

Swiss physician and alchemist Paracelsus (1493–1541) believed that chemicals and minerals could be used as medicines.

[1]Horgan, John, Profile: Linus Pauling, Scientific American, March 1993.

Using the Scientific Method in Everyday Life

You may be surprised to realize that you use the scientific method in your everyday life. Suppose you visit a friend in her home. Soon after you arrive, your eyes start to itch and you begin to sneeze. Then you observe that your friend has a new cat. Perhaps you ask yourself why you are sneezing and you form the hypothesis that you are allergic to cats. To test your hypothesis, you leave your friend's home. If the sneezing stops, perhaps your hypothesis is correct. You test your hypothesis further by visiting another friend who also has a cat. If you start to sneeze again, your experimental results support your hypothesis and you come to the conclusion that you are allergic to cats. However, if you continue sneezing after you leave your friend's home, your hypothesis is not supported. Now you need to form a new hypothesis, which could be that you have a cold.

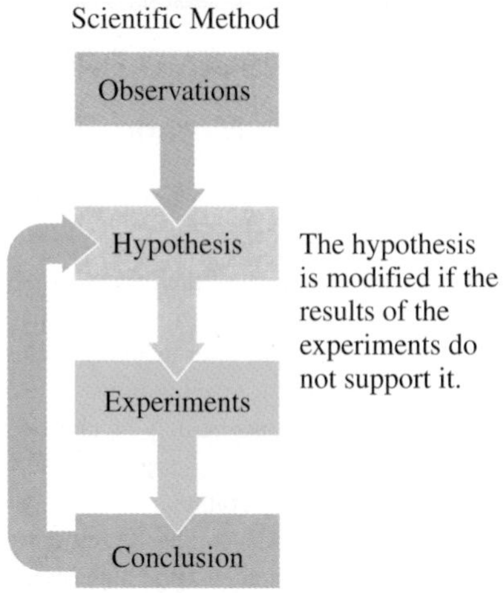

The scientific method develops conclusions using observations, hypotheses, and experiments.

Through observation you may determine that you are allergic to cat hair and dander.

SAMPLE PROBLEM 1.1 Scientific Method

Identify each of the following as an observation (O), a hypothesis (H), an experiment (E), or a conclusion (C):

a. During an assessment in the emergency room, a nurse writes that the patient has a resting pulse of 30 beats/min.
b. A nurse thinks that an incision from a recent surgery that is red and swollen is infected.
c. Repeated studies show that lowering sodium in the diet leads to a decrease in blood pressure.

SOLUTION

a. observation (O)
b. hypothesis (H)
c. conclusion (C)

STUDY CHECK 1.1

Identify each of the following as an observation (O), a hypothesis (H), an experiment (E), or a conclusion (C):

a. Drinking coffee at night keeps me awake.
b. If I stop drinking coffee in the afternoon, I will be able to sleep at night.
c. I will try drinking coffee only in the morning.

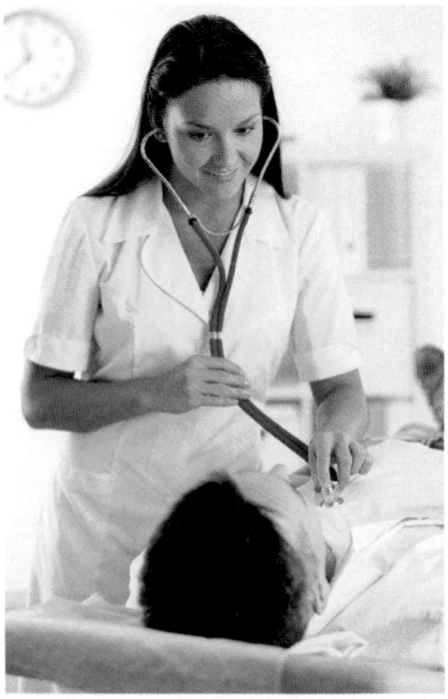

Nurses make observations in the hospital.

QUESTIONS AND PROBLEMS

1.2 Scientific Method: Thinking Like a Scientist

LEARNING GOAL Describe the activities that are part of the scientific method.

1.7 Identify each of the following as an observation (O), a hypothesis (H), an experiment (E), or a conclusion (C):
- **a.** One hour after drinking a glass of regular milk, Jim experienced stomach cramps.
- **b.** Jim thinks he may be lactose intolerant.
- **c.** Jim drinks a glass of lactose-free milk and does not have any stomach cramps.
- **d.** Jim drinks a glass of regular milk to which he has added lactase, an enzyme that breaks down lactose, and has no stomach cramps.

1.8 Identify each of the following as an observation (O), a hypothesis (H), an experiment (E), or a conclusion (C):
- **a.** Sally thinks she may be allergic to shrimp.
- **b.** Yesterday, one hour after Sally ate a shrimp salad, she broke out in hives.
- **c.** Today, Sally had some soup that contained shrimp, but she did not break out in hives.
- **d.** Sally realizes that she does not have an allergy to shrimp.

1.9 Identify each activity, **a–f**, as an observation (O), a hypothesis (H), an experiment (E), or a conclusion (C).
At a popular restaurant, where Chang is the head chef, the following occurred:
- **a.** Chang determined that sales of the house salad had dropped.
- **b.** Chang decided that the house salad needed a new dressing.
- **c.** In a taste test, Chang prepared four bowls of lettuce, each with a new dressing: sesame seed, olive oil and balsamic vinegar, creamy Italian, and blue cheese.
- **d.** The tasters rated the sesame seed salad dressing as the favorite.
- **e.** After two weeks, Chang noted that the orders for the house salad with the new sesame seed dressing had doubled.
- **f.** Chang decided that the sesame seed dressing improved the sales of the house salad because the sesame seed dressing enhanced the taste.

Customers rated the sesame seed dressing as the best.

1.10 Identify each activity, **a–f**, as an observation (O), a hypothesis (H), an experiment (E), or a conclusion (C).
Lucia wants to develop a process for dyeing shirts so that the color will not fade when the shirt is washed. She proceeds with the following activities:
- **a.** Lucia notices that the dye in a design fades when the shirt is washed.
- **b.** Lucia decides that the dye needs something to help it combine with the fabric.
- **c.** She places a spot of dye on each of four shirts and then places each one separately in water, salt water, vinegar, and baking soda and water.
- **d.** After one hour, all the shirts are removed and washed with a detergent.
- **e.** Lucia notices that the dye has faded on the shirts in water, salt water, and baking soda, while the dye did not fade on the shirt soaked in vinegar.
- **f.** Lucia thinks that the vinegar binds with the dye so it does not fade when the shirt is washed.

LEARNING GOAL

Develop a study plan for learning chemistry.

1.3 Learning Chemistry: A Study Plan

Here you are taking chemistry, perhaps for the first time. Whatever your reasons for choosing to study chemistry, you can look forward to learning many new and exciting ideas.

Features in This Text Help You Study Chemistry

KEY MATH SKILL

CORE CHEMISTRY SKILL

This text has been designed with study features to complement your individual learning style. On the inside of the front cover is a periodic table of the elements. On the inside of the back cover are tables that summarize useful information needed throughout your study of chemistry. Each chapter begins with *Looking Ahead*, which outlines the topics in the chapter. At the end of the text, there is a comprehensive *Glossary and Index*, which lists and defines key terms used in the text. *Key Math Skills* that will be helpful to your understanding of chemical calculations are reviewed. *Core Chemistry Skills* that are critical to learning chemistry are indicated by icons in the margin, and summarized at the end of each chapter. In the *Chapter Readiness* list at the beginning of every chapter, the *Key Math Skills* and *Core Chemistry Skills* from previous chapters related to the current chapter concepts are highlighted for your review.

Before you begin reading, obtain an overview of a chapter by reviewing the topics in *Looking Ahead*. As you prepare to read a section of the chapter, look at the section title and turn it into a question. For example, for section 1.1, "Chemistry and Chemicals," you could ask "What is chemistry?" or "What are chemicals?". When you come to a *Sample Problem*, take the time to work it through and compare your solution to the one provided. Then try the associated *Study Check*. Many *Sample Problems* are accompanied by a *Guide to Problem Solving*, which gives the steps needed to work the problem. In some *Sample Problems*, an *Analyze the Problem* feature shows how to organize the data in the word problem to obtain a solution. At the end of each chapter section, you will find a set of *Questions and Problems* that allows you to apply problem solving immediately to the new concepts.

ANALYZE THE PROBLEM	Given	Need
	165 lb	kilograms

Throughout each chapter, boxes titled "Chemistry Link to Health" and "Chemistry Link to the Environment" help you connect the chemical concepts you are learning to real-life situations. Many of the figures and diagrams use macro-to-micro illustrations to depict the atomic level of organization of ordinary objects. These visual models illustrate the concepts described in the text and allow you to "see" the world in a microscopic way.

At the end of each chapter, you will find several study aids that complete the chapter. *Chapter Reviews* provide a summary and *Concept Maps* show the connections between important topics. The *Key Terms*, which are in boldface type within the chapter, are listed with their definitions. *Understanding the Concepts*, a set of questions that use art and models, helps you visualize concepts. *Additional Questions and Problems* and *Challenge Problems* provide additional exercises to test your understanding of the topics in the chapter. The problems are paired, which means that each of the odd-numbered problems is matched to the following even-numbered problem. The answers to all the *Study Checks*, as well as the answers to all the odd-numbered problems are provided at the end of the chapter. If the answers provided match your answers, you most likely understand the topic; if not, you need to study the section again.

After some chapters, problem sets called *Combining Ideas* test your ability to solve problems containing material from more than one chapter.

Using Active Learning

A student who is an active learner continually interacts with the chemical ideas while reading the text, working problems, and attending lectures. Let's see how this is done.

As you read and practice problem solving, you remain actively involved in studying, which enhances the learning process. In this way, you learn small bits of information at a time and establish the necessary foundation for understanding the next section. You may also note questions you have about the reading to discuss with your professor or laboratory instructor. Table 1.2 summarizes these steps for active learning. The time you spend in a lecture is also a useful learning time. By keeping track of the class schedule and reading the assigned material before a lecture, you become aware of the new terms and concepts you need to learn. Some questions that occur during your reading may be answered during the lecture. If not, you can ask your professor for further clarification.

Many students find that studying with a group can be beneficial to learning. In a group, students motivate each other to study, fill in gaps, and correct misunderstandings by

Studying in a group can be beneficial to learning.

TABLE 1.2 Steps in Active Learning

1. Read each *Learning Goal* for an overview of the material.
2. Form a question from the title of the section you are going to read.
3. Read the section, looking for answers to your question.
4. Self-test by working *Sample Problems* and *Study Checks*.
5. Complete the *Questions and Problems* that follow that section, and check the answers for the magenta odd-numbered problems.
6. Proceed to the next section and repeat the steps.

teaching and learning together. Studying alone does not allow the process of peer correction. In a group, you can cover the ideas more thoroughly as you discuss the reading and problem solve with other students. You may find that it is easier to retain new material and new ideas if you study in short sessions throughout the week rather than all at once. Waiting to study until the night before an exam does not give you time to understand concepts and practice problem solving.

Making a Study Plan

As you embark on your journey into the world of chemistry, think about your approach to studying and learning chemistry. You might consider some of the ideas in the following list. Check those ideas that will help you successfully learn chemistry. Commit to them now. *Your* success depends on *you*.

My study plan for learning chemistry will include the following:

________ reading the chapter before lecture
________ going to lecture
________ reviewing the *Learning Goals*
________ keeping a problem notebook
________ reading the text as an active learner
________ working the *Questions and Problems* following each section and checking answers at the end of the chapter
________ being an active learner in lecture
________ organizing a study group
________ seeing the professor during office hours
________ reviewing *Key Math Skills* and *Core Chemistry Skills*
________ attending review sessions
________ organizing my own review sessions
________ studying as often as I can

Students discuss a chemistry problem with their professor during office hours.

SAMPLE PROBLEM 1.2 A Study Plan for Learning Chemistry

Which of the following activities would you include in your study plan for learning chemistry successfully?

a. skipping lecture
b. going to the professor's office hours
c. keeping a problem notebook
d. waiting to study until the night before the exam
e. becoming an active learner

SOLUTION

Your success in chemistry can be improved by

b. going to the professor's office hours
c. keeping a problem notebook
e. becoming an active learner

STUDY CHECK 1.2

Which of the following will help you learn chemistry?

a. skipping review sessions
b. working assigned problems
c. staying up all night before an exam
d. reading the assignment before a lecture

QUESTIONS AND PROBLEMS

1.3 Learning Chemistry: A Study Plan

LEARNING GOAL Develop a study plan for learning chemistry.

1.11 What are four things you can do to help yourself to succeed in chemistry?

1.12 What are the three safety precautions you can take while working in the laboratory?

1.13 A student in your class asks you for advice on learning chemistry. In which order will you advice the following steps:
- **a.** Work on the Q&A for each section and check the answers at the end.
- **b.** Read the chapter before the class lecture.
- **c.** Be an active learner during the class lecture.
- **d.** Review key core chemistry skills.

1.14 A student in your class asks you for advice on learning chemistry. Which of the following might you suggest?
- **a.** doing the assigned problems
- **b.** not reading the text; it's never on the test
- **c.** attending review sessions
- **d.** reading the assignment before a lecture
- **e.** keeping a problem notebook

1.4 Key Math Skills for Chemistry

LEARNING GOAL

Review math concepts used in chemistry: place values, positive and negative numbers, percentages, solving equations, interpreting graphs, and writing numbers in scientific notation.

During your study of chemistry, you will work many problems that involve numbers. You will need various math skills and operations. We will review some of the key math skills that are particularly important for chemistry. As we move through the chapters, we will also reference the key math skills as they apply.

A. Identifying Place Values

KEY MATH SKILL

Identifying Place Values

For any number, we can identify the *place value* for each of the digits in that number. These place values have names such as the ones place (first place to the left of the decimal point) or the tens place (second place to the left of the decimal point). A premature baby has a mass of 2518 g. We can indicate the place values for the number 2518 as follows:

Digit	Place Value
2	thousands
5	hundreds
1	tens
8	ones

We also identify place values such as the tenths place (first place to the right of the decimal point) and the hundredths place (second place to the right of the decimal place). A silver coin has a mass of 6.407 g. We can indicate the place values for the number 6.407 as follows:

Digit	Place Value
6	ones
4	ten**ths**
0	hundred**ths**
7	thousand**ths**

Note that place values ending with the suffix *ths* refer to the decimal places to the right of the decimal point.

B. Using Positive and Negative Numbers in Calculations

KEY MATH SKILL

Using Positive and Negative Numbers in Calculations

A *positive number* is any number that is greater than zero and has a positive sign (+). Often the positive sign is understood and not written in front of the number. For example, the number +8 can also be written as 8. A *negative number* is any number that is less than zero and is written with a negative sign (−). For example, a negative eight is written as −8.

Multiplication and Division of Positive and Negative Numbers

When two positive numbers or two negative numbers are multiplied, the answer is positive $(+)$.

$$2 \times 3 = +6$$
$$(-2) \times (-3) = +6$$

When a positive number and a negative number are multiplied, the answer is negative $(-)$.

$$2 \times (-3) = -6$$
$$(-2) \times 3 = -6$$

The rules for the division of positive and negative numbers are the same as the rules for multiplication. When two positive numbers or two negative numbers are divided, the answer is positive $(+)$.

$$\frac{6}{3} = 2 \qquad \frac{-6}{-3} = 2$$

When a positive number and a negative number are divided, the answer is negative $(-)$.

$$\frac{-6}{3} = -2 \qquad \frac{6}{-3} = -2$$

Addition of Positive and Negative Numbers

When positive numbers are added, the sign of the answer is positive.

$3 + 4 = 7$ The $+$ sign $(+7)$ is understood.

When negative numbers are added, the sign of the answer is negative.

$$(-3) + (-4) = -7$$

When a positive number and a negative number are added, the smaller number is subtracted from the larger number, and the result has the same sign as the larger number.

$$12 + (-15) = -3$$

Subtraction of Positive and Negative Numbers

When two numbers are subtracted, change the sign of the number to be subtracted and follow the rules for addition shown above.

$$12 - \mathbf{(+5)} = 12 \mathbf{- 5} = 7$$
$$12 - \mathbf{(-5)} = 12 \mathbf{+ 5} = 17$$
$$-12 - \mathbf{(-5)} = -12 \mathbf{+ 5} = -7$$
$$-12 - \mathbf{(+5)} = -12 \mathbf{- 5} = -17$$

Calculator Operations

On your calculator, there are four keys that are used for basic mathematical operations. The change sign [+/-] key is used to change the sign of a number.

To practice these basic calculations on the calculator, work through the problem going from the left to the right doing the operations in the order they occur. If your calculator has a change sign [+/-] key, a negative number is entered by pressing the number and then pressing the change sign [+/-] key. At the end, press the equals key [=] or ANS or ENTER.

Addition and Subtraction

Example 1: $15 - 8 + 2 =$

Solution: 15 [-] 8 [+] 2 [=] 9

Example 2: $4 + (-10) - 5 =$

Solution: 4 [+] 10 [+/-] [-] 5 [=] −11

Multiplication and Division

Example 3: $2 \times (-3) =$

Solution: 2 [×] 3 [+/-] [=] −6

Example 4: $\frac{8 \times 3}{4} =$

Solution: 8 [×] 3 [÷] 4 [=] 6

C. Calculating a Percentage

KEY MATH SKILL

Calculating a Percentage

To determine a percentage, divide the parts by the total (whole) and multiply by 100%. For example, if there are 8 chemistry books on a shelf that has a total of 32 books, what is the percentage of chemistry books?

$$\frac{8 \text{ chemistry books}}{32 \text{ total books}} \times 100\% = 25\% \text{ chemistry books}$$

When a value is described as a percentage (%), it represents the number of parts of an item in 100 of those items. If the percentage of red balls is 5, it means there are 5 red balls in every 100 balls. If the percentage of green balls is 50, there are 50 green balls in every 100 balls.

$$5\% \text{ red balls} = \frac{5 \text{ red balls}}{100 \text{ balls}} \qquad 50\% \text{ green balls} = \frac{50 \text{ green balls}}{100 \text{ balls}}$$

For example, if you eat three pieces (parts) of a pizza that contained six pieces (whole), we can calculate the percentage of the pizza that was eaten as

$$\frac{3 \text{ pieces of pizza}}{6 \text{ pieces of pizza}} \times 100\% = 50\%$$

Thus, if you eat three pieces (parts) of a pizza that contained six pieces (whole), you ate 50% of that pizza.

D. Solving Equations

KEY MATH SKILL

Solving Equations

In chemistry, we use equations that express the relationship between certain variables. Let's look at how we would solve for x in the following equation:

$$2x + 8 = 14$$

Our overall goal is to rearrange the items in the equation to obtain x on one side.

1. *Place all like terms on one side.* The numbers 8 and 14 are like terms. To remove the 8 from the left side of the equation, we subtract 8. To keep a balance, we need to subtract 8 from the 14 on the other side.

$$2x + \cancel{8} - \cancel{8} = 14 - 8$$
$$2x = 6$$

2. *Isolate the variable you need to solve for.* In this problem, we obtain x by dividing both sides of the equation by 2. The value of x is the result when 6 is divided by 2.

$$\frac{\cancel{2}x}{\cancel{2}} = \frac{6}{2}$$
$$x = 3$$

3. *Check your answer.* Check your answer by substituting your value for x back into the original equation.

$$2(3) + 8 = 6 + 8 = 14 \qquad \text{Your answer } x = 3 \text{ is correct.}$$

Summary: To solve an equation for a particular variable, be sure you perform the same mathematical operations on *both* sides of the equation.

If you eliminate a symbol or number by subtracting, you need to subtract that same symbol or number on the opposite side.

If you eliminate a symbol or number by adding, you need to add that same symbol or number on the opposite side.

If you cancel a symbol or number by dividing, you need to divide both sides by that same symbol or number.

If you cancel a symbol or number by multiplying, you need to multiply both sides by that same symbol or number.

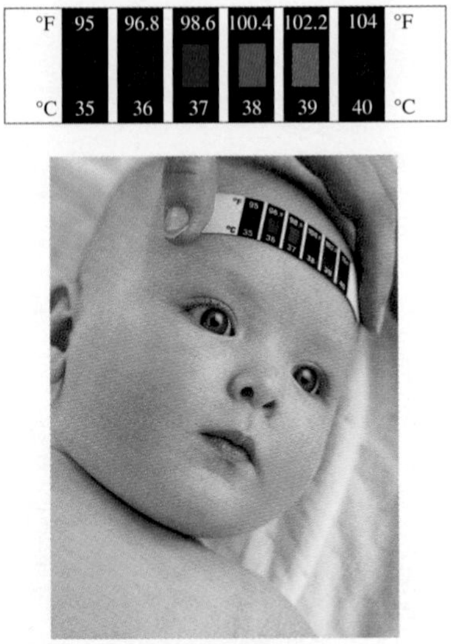

A plastic strip thermometer changes color to indicate body temperature.

When we work with temperature, we may need to convert between degrees Celsius and degrees Fahrenheit using the following equation:

$$T_F = 1.8(T_C) + 32$$

To obtain the equation for converting degrees Fahrenheit to degrees Celsius, we subtract 32 from both sides.

$$T_F = 1.8(T_C) + 32$$
$$T_F - 32 = 1.8(T_C) + \cancel{32} - \cancel{32}$$
$$T_F - 32 = 1.8(T_C)$$

To obtain T_C by itself, we divide both sides by 1.8.

$$\frac{T_F - 32}{1.8} = \frac{\cancel{1.8}(T_C)}{\cancel{1.8}} = T_C$$

Interactive Video

Solving Equations

SAMPLE PROBLEM 1.3 Solving Equations

Solve the following equation for V_2:

$$P_1V_1 = P_2V_2$$

SOLUTION

$$P_1V_1 = P_2V_2$$

To solve for V_2, divide both sides by the symbol P_2.

$$\frac{P_1V_1}{P_2} = \frac{\cancel{P_2}V_2}{\cancel{P_2}}$$

$$V_2 = \frac{P_1V_1}{P_2}$$

STUDY CHECK 1.3

Solve the following equation for m:

$$\text{heat} = m \times \Delta T \times SH$$

E. Interpreting a Graph

A graph represents the relationship between two variables. These quantities are plotted along two perpendicular axes, which are the x axis (horizontal) and y axis (vertical).

KEY MATH SKILL

Interpreting a Graph

Example

In the following graph, the volume of a gas in a balloon is plotted against its temperature.

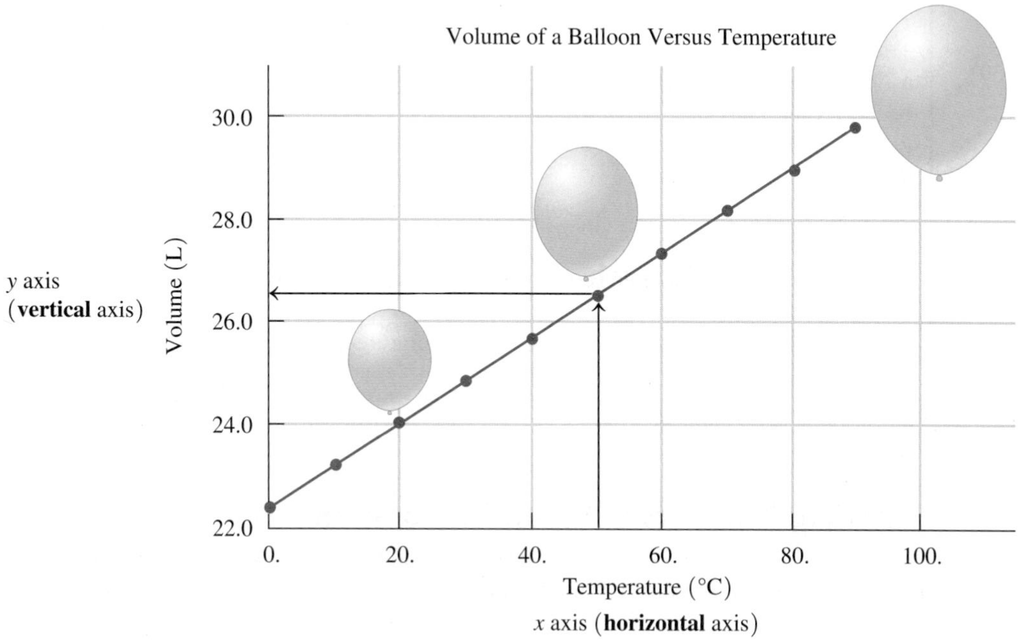

Title

Look at the title. What does it tell us about the graph? The title indicates that the volume of a balloon was measured at different temperatures.

Vertical Axis

Look at the label and the numbers on the vertical (y) axis. The label indicates that the volume of the balloon was measured in liters (L). The numbers, which are chosen to include the low and high measurements of the volume of the gas, are evenly spaced from 22.0 L to 30.0 L.

Horizontal Axis

The label on the horizontal (x) axis indicates that the temperature of the balloon was measured in degrees Celsius (°C). The numbers are measurements of the Celsius temperature, which are evenly spaced from 0. °C to 100. °C.

Points on the Graph

Each point on the graph represents a volume in liters that was measured at a specific temperature. When these points are connected, a straight line is obtained.

Interpreting the Graph

From the graph, we see that the volume of the gas increases as the temperature of the gas increases. This is called a *direct relationship*. Now we use the graph to determine the volume at various temperatures. For example, suppose we want to know the volume of the gas at 50. °C. We would start by finding 50. °C on the x axis and then drawing a line up to the plotted line. From there, we would draw a horizontal line that intersects the y axis and read the volume value where the line crosses the y axis.

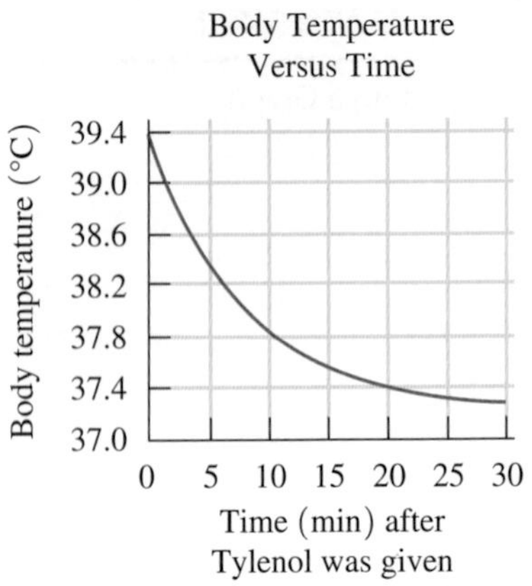

SAMPLE PROBLEM 1.4 Interpreting a Graph

A nurse administers Tylenol to lower a child's fever. The graph shows the body temperature of the child plotted against time.

a. What is measured on the vertical axis?
b. What is the range of values on the vertical axis?
c. What is measured on the horizontal axis?
d. What is the range of values on the horizontal axis?

SOLUTION

a. temperature in degrees Celsius
b. 37.0 °C to 39.4 °C
c. time, in minutes, after Tylenol was given
d. 0 min to 30 min

STUDY CHECK 1.4

a. Using the graph in Sample Problem 1.4, what was the child's temperature 15 min after Tylenol was given?
b. How many minutes elapsed before the temperature decreased to 38.0 °C?

KEY MATH SKILL

Writing Numbers in Scientific Notation

F. Writing Numbers in Scientific Notation

In chemistry, we use numbers that are very large and very small. We might measure something as tiny as the width of a human hair, which is about 0.000 008 m. Or perhaps we want to count the number of hairs on the average human scalp, which is about 100 000 hairs. In this text, we add spaces between sets of three digits when it helps make the places easier to count. However, we will see that it is more convenient to write large and small numbers in *scientific notation*.

Standard Number	Scientific Notation
0.000 008 m	8×10^{-6} m
100 000 hairs	1×10^{5} hairs

Humans have an average of 1×10^5 hairs on their scalps. Each hair is about 8×10^{-6} m wide.

A number written in **scientific notation** has two parts: a coefficient and a power of 10. For example, the number 2400 is written in scientific notation as 2.4×10^3. The coefficient, 2.4, is obtained by moving the decimal point to the left to give a number that is at least 1 but less than 10. Because we moved the decimal point three places to the left, the power of 10 is a positive 3, which is written as 10^3. When a number greater than 1 is converted to scientific notation, the power of 10 is positive.

$$2400.\ \text{m} = 2.4 \times 10^3\ \text{m}$$

← 3 places (2400.) | Coefficient (2.4) | Power of 10 (10^3) | Unit (m)

In another example, a number smaller than 1, 0.000 86, is written in scientific notation as 8.6×10^{-4}. The coefficient, 8.6, is obtained by moving the decimal point to the right.

Because the decimal point is moved four places to the right, the power of 10 is a negative 4, written as 10^{-4}. When a number less than 1 is written in scientific notation, the power of 10 is negative.

$$0.00086\ \text{g} = 8.6 \times 10^{-4}\ \text{g}$$

4 places → Coefficient Power of 10 Unit

Table 1.3 gives some examples of numbers written as positive and negative powers of 10. The powers of 10 are a way of keeping track of the decimal point in the number. Table 1.4 gives several examples of writing measurements in scientific notation.

TABLE 1.3 Some Powers of 10

Number	Multiples of 10	Scientific Notation	
10 000	$10 \times 10 \times 10 \times 10$	1×10^4	Some positive powers of 10
1 000	$10 \times 10 \times 10$	1×10^3	
100	10×10	1×10^2	
10	10	1×10^1	
1	0	1×10^0	
0.1	$\frac{1}{10}$	1×10^{-1}	Some negative powers of 10
0.01	$\frac{1}{10} \times \frac{1}{10} = \frac{1}{100}$	1×10^{-2}	
0.001	$\frac{1}{10} \times \frac{1}{10} \times \frac{1}{10} = \frac{1}{1\,000}$	1×10^{-3}	
0.0001	$\frac{1}{10} \times \frac{1}{10} \times \frac{1}{10} \times \frac{1}{10} = \frac{1}{10\,000}$	1×10^{-4}	

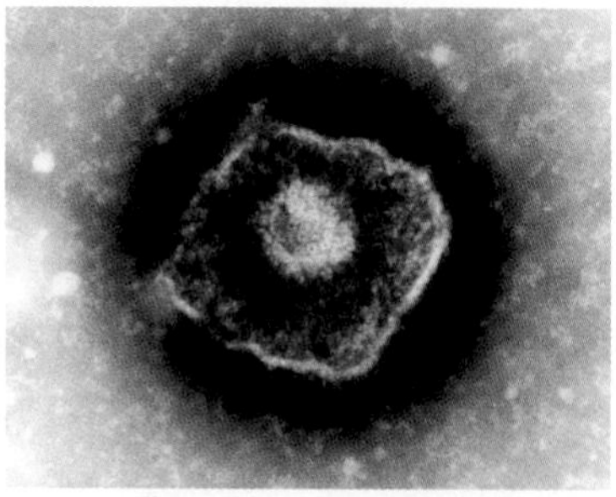

A chickenpox virus has a diameter of 3×10^{-7} m.

TABLE 1.4 Some Measurements Written in Scientific Notation

Measured Quantity	Standard Number	Scientific Notation
Volume of gasoline used in the United States each year	550 000 000 000 L	5.5×10^{11} L
Diameter of the Earth	12 800 000 m	1.28×10^7 m
Average volume of blood pumped in 1 day	8500 L	8.5×10^3 L
Time for light to travel from the Sun to the Earth	500 s	5×10^2 s
Mass of a typical human	68 kg	6.8×10^1 kg
Mass of stirrup bone in ear	0.003 g	3×10^{-3} g
Diameter of a chickenpox (*Varicella zoster*) virus	0.000 000 3 m	3×10^{-7} m
Mass of bacterium (mycoplasma)	0.000 000 000 000 000 000 1 kg	1×10^{-19} kg

Scientific Notation and Calculators

You can enter a number in scientific notation on many calculators using the [EE or EXP] key. After you enter the coefficient, press the [EE or EXP] key and enter only the power of 10, because the EXP function key already includes the $\times$ 10 value. To enter a negative power of 10, press the [+/-] key or the [-] key, depending on your calculator.

Number to Enter	Procedure	Calculator Display
4×10^6	4 [EE or EXP] 6	406 or 4^{06} or 4E06
2.5×10^{-4}	2.5 [EE or EXP] [+/−] 4	2.5−04 or 2.5^{-04} or 2.5E−04

When a calculator display appears in scientific notation, it is shown as a number between 1 and 10 followed by a space and the power of 10. To express this display in scientific notation, write the coefficient value, write $\times$ 10, and use the power of 10 as an exponent.

Calculator Display	Expressed in Scientific Notation
7.5204 or 7.52^{04} or 7.52E04	7.52×10^4
5.8−02 or 5.8^{-02} or 5.8E−02	5.8×10^{-2}

On many scientific calculators, a number is converted into scientific notation using the appropriate keys. For example, the number 0.000 52 is entered, followed by pressing the 2nd or 3rd function key and the SCI key. The scientific notation appears in the calculator display as a coefficient and the power of 10.

0.000 52 [2nd or 3rd function key] (Key) [SCI] (Key) = 5.2−04 or 5.2^{-04} or 5.2E−04 (Calculator display) = 5.2×10^{-4}

SAMPLE PROBLEM 1.5 Scientific Notation

Write each of the following in scientific notation:

a. 3500 **b.** 0.000 016

SOLUTION

a. 3500

STEP 1 **Move the decimal point to obtain a coefficient that is at least 1 but less than 10.** For a number greater than 1, the decimal point is moved to the left three places to give a coefficient of 3.5.

STEP 2 **Express the number of places moved as a power of 10.** Moving the decimal point three places to the left gives a power of 3, written as 10^3.

STEP 3 **Write the product of the coefficient multiplied by the power of 10.**

3.5×10^3

b. 0.000 016

STEP 1 **Move the decimal point to obtain a coefficient that is at least 1 but less than 10.** For a number less than one, the decimal point is moved to the right five places to give a coefficient of 1.6.

STEP 2 **Express the number of places moved as a power of 10.** Moving the decimal point five places to the right gives a power of negative 5, written as 10^{-5}.

STEP 3 **Write the product of the coefficient multiplied by the power of 10.**

1.6×10^{-5}

STUDY CHECK 1.5

Write each of the following in scientific notation:

a. 425 000 **b.** 0.000 000 86

Guide to Writing a Number in Scientific Notation

STEP 1 Move the decimal point to obtain a coefficient that is at least 1 but less than 10.

STEP 2 Express the number of places moved as a power of 10.

STEP 3 Write the product of the coefficient multiplied by the power of 10.

QUESTIONS AND PROBLEMS

1.4 Key Math Skills in Chemistry

LEARNING GOAL Review math concepts used in chemistry: place values, positive and negative numbers, percentages, solving equations, interpreting graphs, and writing numbers in scientific notation.

1.15 What is the place value for the bold digit?
a. 7.0**9**84
b. 2**6**.2860
c. **8**5.5258

1.16 What is the place value for the bold digit?
a. 97.**5**689
b. 375.88
c. 46.1**0**00

1.17 Evaluate each of the following
a. $20 - (-10) =$ _______
b. $-6 + (-38) =$ _______
c. $2 \times (-7) + 14 =$ _______

1.18 Evaluate each of the following
a. $-17 - (-12) =$ _______
b. $19 + (-86) =$ _______
c. $\frac{-78}{6} =$ _______

1.19 **a.** A cargo comprises of 80 kg of wheat and 20 kg of cotton. What is the percentage of cotton by weight in the cargo?
b. A ship contains 400 tons of wheat and 100 tons of cotton. What is the percentage of wheat by weight on the ship?
c. Which contains a higher percentage of wheat?

1.20 **a.** At a local hospital, 35 babies were born. If 22 were boys, what percentage of the newborns were boys? Express your answer to the ones place.
b. An alloy contains 67 g of pure gold and 35 g of pure zinc. What is the percentage of zinc in the alloy? Express your answer to the ones place.
c. A collection of coins contains 15 pennies, 14 dimes, and 6 quarters. What is the percentage of pennies in the collection? Express your answer to the ones place.

1.21 Solve each of the following for a:
a. $5a + 10 = 60$
b. $\frac{a}{8} = 6$

1.22 Solve each of the following for b:
a. $2b + 7 = b + 10$ **b.** $3b - 4 = 24 - b$

Use the following graph for questions 1.23 and 1.24:

Time for Cooling of Tea Versus Temperature

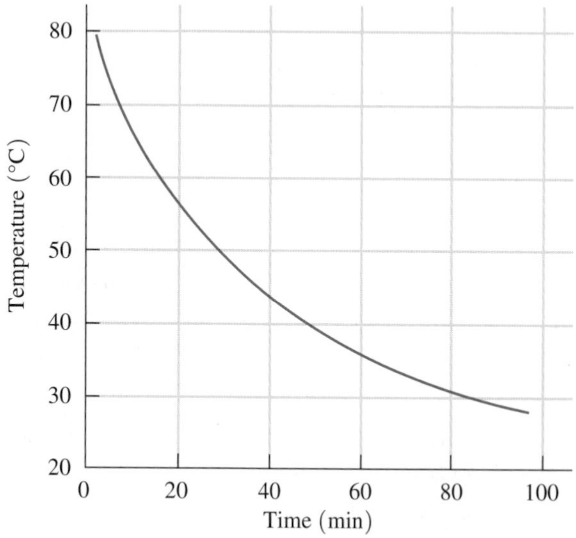

1.23 **a.** What does the title indicate about the graph?
b. What is measured on the vertical axis?
c. What is the range of values on the vertical axis?
d. Does the temperature increase or decrease with an increase in time?

1.24 **a.** What is measured on the horizontal axis?
b. What is the range of values on the horizontal axis?
c. What is the temperature of tea after 20 min?
d. How many minutes were needed to reach a temperature below 30 °C?

1.25 Write each of the following in scientific notation
a. 67000 **b.** 520
c. 0.0000081 **d.** 0.00027
e. 0.0094 **f.** 490000

1.26 Write each of the following in scientific notation:
a. 180 000 000 **b.** 0.000 06
c. 750 **d.** 0.15
e. 0.024 **f.** 1500

1.27 Which number in each of the following pairs is larger?
a. 6.1×10^4 or 4.2×10^2 **b.** 3.7×10^{-6} or 5.8×10^{-3}
c. 7×10^{-8} or 7×10^8 **d.** 0.000 69 or 8.3×10^{-1}

1.28 Which number in each of the following pairs is smaller?
a. 0.9×10^{-3} or 5.5×10^{-9} **b.** 1250 or 3.4×10^2
c. 0.5×10^3 or 5.0×10^2 **d.** 2.50×10^2 or 4×10^5

CONCEPT MAP

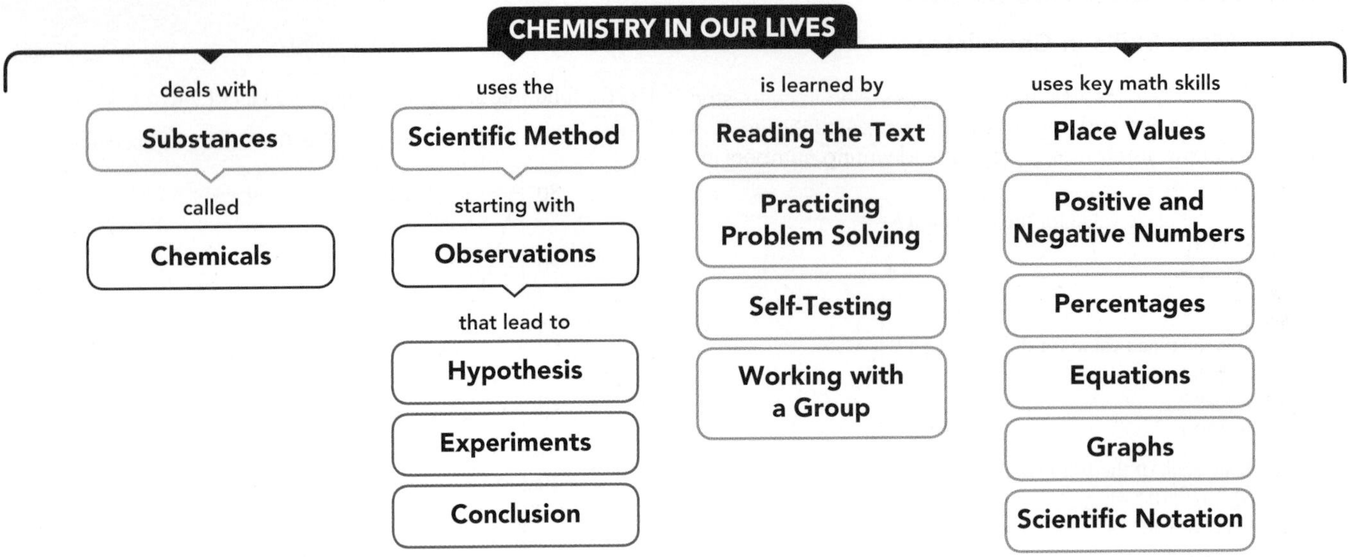

CHAPTER REVIEW

1.1 Chemistry and Chemicals

LEARNING GOAL Define the term chemistry and identify substances as chemicals.

- Chemistry is the study of the composition, structure, properties, and reactions of matter.
- A chemical is any substance that always has the same composition and properties wherever it is found.

1.2 Scientific Method: Thinking Like a Scientist

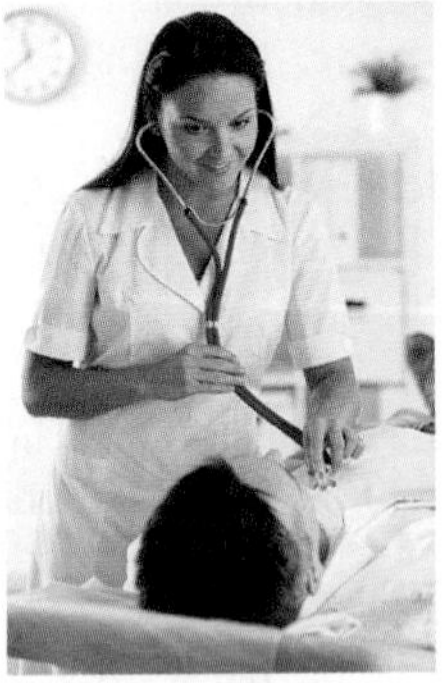

LEARNING GOAL Describe the activities that are part of the scientific method.

- The scientific method is a process of explaining natural phenomena beginning with making observations, forming a hypothesis, and performing experiments.
- After repeated experiments, a conclusion may be reached as to whether the hypothesis is true or not.

1.3 Learning Chemistry: A Study Plan

LEARNING GOAL Develop a study plan for learning chemistry.

- A study plan for learning chemistry utilizes the features in the text and develops an active learning approach to study.
- By using the *Learning Goals* in the chapter and working the *Sample Problems*, *Study Checks*, and the *Questions and Problems* at the end of each section, you can successfully learn the concepts of chemistry.

1.4 Key Math Skills for Chemistry

LEARNING GOAL Review math concepts used in chemistry: place values, positive and negative numbers, percentages, solving equations, interpreting graphs, and writing numbers in scientific notation.

- Solving chemistry problems involves a number of math skills: identifying place values, using positive and negative numbers, calculating percentages, solving algebraic equations, interpreting graphs, and writing numbers in scientific notation.

KEY TERMS

chemical A substance that has the same composition and properties wherever it is found.

chemistry The study of the composition, structure, properties, and reactions of matter.

conclusion An explanation of an observation that has been validated by repeated experiments that support a hypothesis.

experiment A procedure that tests the validity of a hypothesis.

hypothesis An unverified explanation of a natural phenomenon.

observation Information determined by noting and recording a natural phenomenon.

scientific method The process of making observations, proposing a hypothesis, testing the hypothesis, and making a conclusion as to the validity of the hypothesis.

scientific notation A form of writing large and small numbers using a coefficient that is at least 1 but less than 10, followed by a power of 10.

KEY MATH SKILLS

The chapter section containing each Key Math Skill is shown in parentheses at the end of each heading.

Identifying Place Values (1.4A)

- The place value identifies the numerical value of each digit in a number.

Example: Identify the place value for each of the digits in the number 456.78.

Answer:

Digit	Place Value
4	hundreds
5	tens
6	ones
7	tenths
8	hundredths

Using Positive and Negative Numbers in Calculations (1.4B)

- A *positive number* is any number that is greater than zero and has a positive sign (+). A *negative number* is any number that is less than zero and is written with a negative sign (−).
- When two positive numbers are added, multiplied, or divided, the answer is positive.
- When two negative numbers are multiplied or divided, the answer is positive.
- When a positive and a negative number are multiplied or divided, the answer is negative.
- When a positive and a negative number are added, the smaller number is subtracted from the larger number and the result has the same sign as the larger number.
- When two numbers are subtracted, change the sign of the number to be subtracted.

Example: Evaluate each of the following:

a. $-8 - 14 =$ _______

b. $6 \times (-3) =$ _______

Answer: **a.** $-8 - 14 = -22$

b. $6 \times (-3) = -18$

Calculating a Percentage (1.4C)

- A percentage is the part divided by the total (whole) multiplied by 100%.

Example: A drawer contains 6 white socks and 18 black socks. What is the percentage of white socks?

Answer: $\frac{6 \text{ white socks}}{24 \text{ total socks}} \times 100\% = 25\%$ white socks

Solving Equations (1.4D)

An equation in chemistry often contains an unknown. To rearrange an equation to obtain the unknown factor by itself, you keep it balanced by performing matching mathematical operations on both sides of the equation.

- If you eliminate a number or symbol by subtracting, subtract that same number or symbol on the opposite side.
- If you eliminate a number or symbol by adding, add that same number or symbol on the opposite side.
- If you cancel a number or symbol by dividing, divide both sides by that same number or symbol.
- If you cancel a number or symbol by multiplying, multiply both sides by that same number or symbol.

Example: Solve the equation for a: $3a - 8 = 28$

Answer: *Add 8 to both sides* $3a - \cancel{8} + \cancel{8} = 28 + 8$

$3a = 36$

Divide both sides by 3 $\frac{\cancel{3}a}{\cancel{3}} = \frac{36}{3}$

$a = 12$

Interpreting a Graph (1.4E)

A graph represents the relationship between two variables. These quantities are plotted along two perpendicular axes, which are the x axis (horizontal) and y axis (vertical). The title indicates the components of the x and y axes. Numbers on the x and y axes show the range of values of the variables. The graph shows the relationship between the component on the y axis and that on the x axis.

Example:

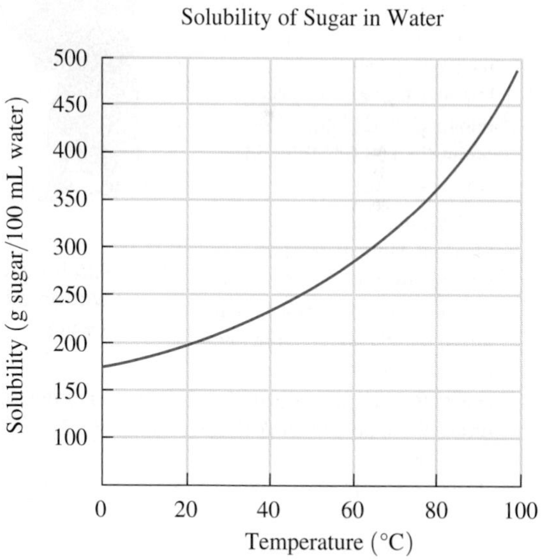

a. Does the amount of sugar that dissolves in 100 mL of water increase or decrease when the temperature increases?

b. How many grams of sugar dissolve in 100 mL of water at 70 °C?

c. At what temperature (°C) will 275 g of sugar dissolve in 100 mL of water?

Answer: **a.** increase

b. 320 g

c. 55 °C

Writing Numbers in Scientific Notation (1.4F)

A number is written in scientific notation by:

- Moving the decimal point to obtain a coefficient that is at least 1 but less than 10.
- Expressing the number of places moved as a power of 10. The power of 10 is positive if the decimal point is moved to the left, negative if the decimal point is moved to the right.
- A number written in scientific notation consists of a coefficient and a power of 10.

Example: Write the number 28 000 in scientific notation.

Answer: Moving the decimal point four places to the left gives a coefficient of 2.8 and a positive power of 10, 10^4. The number 28 000 written in scientific notation is 2.8×10^4.

Example: Write the number 0.000 056 in scientific notation.

Answer: Moving the decimal point five places to the right gives a coefficient of 5.6 and a negative power of 10, 10^{-5}. The number 0.000 056 written in scientific notation is 5.6×10^{-5}.

UNDERSTANDING THE CONCEPTS

The chapter sections to review are shown in parentheses at the end of each question.

1.29 A "chemical-free" shampoo includes the following ingredients: water, cocomide, glycerin, and citric acid. Is the shampoo truly "chemical-free"? (1.1)

1.30 A "chemical-free" sunscreen includes the following ingredients: titanium dioxide, vitamin E, and vitamin C. Is the sunscreen truly "chemical-free"? (1.1)

1.31 According to Sherlock Holmes, "One must follow the rules of scientific inquiry, gathering, observing, and testing data, then formulating, modifying, and rejecting hypotheses, until only one remains." Did Holmes use the scientific method? Why or why not? (1.2)

Sherlock Holmes is a fictional detective in novels written by Arthur Conan Doyle.

1.32 In *A Scandal in Bohemia*, Sherlock Holmes receives a mysterious note. He states, "I have no data yet. It is a capital mistake to theorize before one has data. Insensibly one begins to twist facts to suit theories, instead of theories to suit facts." What do you think Holmes meant? (1.2)

1.33 Classify each of the following statements as an observation (O) or a hypothesis (H)

a. At boiling point, water is converted into steam.
b. If temperature is related to metabolism activity in animals, decreasing the ambient temperature will result in decrease in metabolism.
c. Tomato plant grows well when exposed to better sunlight.

1.34 Classify each of the following statements as an observation (O) or a hypothesis (H): (1.2)

a. Analysis of 10 ceramic dishes showed that four dishes contained lead levels that exceeded federal safety standards.
b. Marble statues undergo corrosion in acid rain.
c. A child with a high fever and a rash may have chicken pox.

1.35 For each of the following, indicate if the answer has a positive or negative sign: (1.4)

a. Three negative numbers are added.
b. Two negative numbers are multiplied, and then divided by a positive number.

1.36 For each of the following, indicate if the answer has a positive or negative sign: (1.4)

a. A negative number is subtracted from a positive number.
b. Two negative numbers are divided.

ADDITIONAL QUESTIONS AND PROBLEMS

1.37 Why does the scientific method include a hypothesis? (1.2)

1.38 Why is experimentation an important part of the scientific method? (1.2)

1.39 Select the correct phrase(s) to complete the following statement: If experimental results do not support your hypothesis, you should (1.2)

a. try to verify the experimental parameters again and see if there is any error
b. discuss with your classmate and see if there is something wrong
c. consult with your professor
d. do more experiments

1.40 Select the correct phrase(s) to complete the following statement: A conclusion confirms a hypothesis when (1.2)

a. one experiment proves the hypothesis
b. many experiments validate the hypothesis
c. you think your hypothesis is correct

1.41 What should you do if your study group is unable to settle an argument in chemistry? (1.3)

a. simply insist your own opinion further
b. search for an answer on the web
c. go through the textbook again
d. do experiments on your own
e. consult with the course coordinator

1.42 Which of the following will help you develop a successful study plan? (1.3)
a. studying all night before the exam
b. forming a study group and discussing the problems together
c. working problems in a notebook for easy reference
d. copying the homework answers from a friend

1.43 Evaluate each of the following:
a. $6 \times (-9) =$ ______
b. $+7 + (-81) =$ ______
c. $\frac{-120}{-40} =$ ______

1.44 Evaluate each of the following:
a. $-67 - (-8) =$ ______
b. $\frac{135}{-15} =$ ______
c. $6 - 39 =$ ______

1.45 A bag of gumdrops contains 16 orange gumdrops, 8 yellow gumdrops, and 16 black gumdrops. (1.4)
a. What is the percentage of yellow gumdrops? Express your answer to the ones place.
b. What is the percentage of black gumdrops? Express your answer to the ones place.

1.46 On the first chemistry test, 12 students got As, 18 students got Bs, and 20 students got Cs. (1.4)
a. What is the percentage of students who received Bs? Express your answer to the ones place.
b. What is the percentage of students who received Cs? Express your answer to the ones place.

1.47 Use scientific notation to express each of the following numbers: (1.4)
a. 1200 000
b. 0.000 000 44
c. 0.066
d. 1000 000

1.48 Express each of the following numbers in scientific notation: (1.4)
a. 0.0056
b. 310
c. 890 000 000
d. 0.000 000 056

CHALLENGE QUESTIONS

The following groups of questions are related to the topics in this chapter. However, they do not all follow the chapter order, and they require you to combine concepts and skills from several sections. These questions will help you increase your critical thinking skills and prepare for your next exam.

1.49 Classify each of the following as an observation (O), a hypothesis (H), or an experiment (E): (1.2)
a. The bicycle tire is flat.
b. If I add air to the bicycle tire, it will expand to the proper size.
c. When I added air to the bicycle tire, it was still flat.
d. The bicycle tire must have a leak in it.

1.50 Classify each of the following as an observation (O), a hypothesis (H), or an experiment (E): (1.2)
a. A big log in the fire does not burn well.
b. If I chop the log into smaller wood pieces, it will burn better.
c. The small wood pieces burn brighter and make a hotter fire.
d. The small wood pieces are used up faster than burning the big log.

1.51 Solve each of the following for x: (1.4)
a. $2x + 5 = 41$
b. $\frac{5x}{3} = 40$

1.52 Solve each of the following for z: (1.4)
a. $3z - (-6) = 12$
b. $\frac{4z}{-12} = -8$

Use the following graph for problems 1.53 and 1.54:

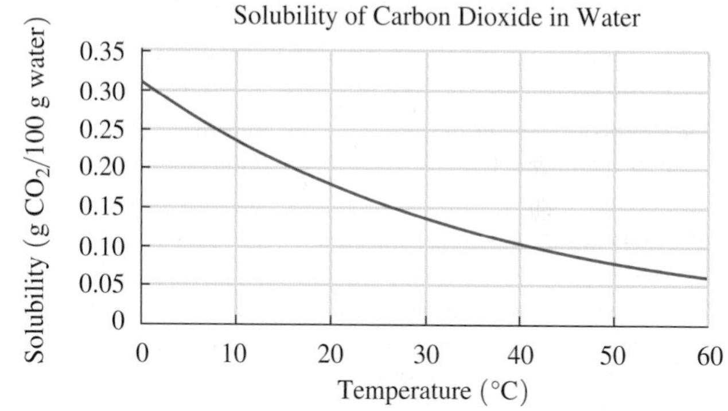

1.53 **a.** What does the title indicate about the graph? (1.4)
b. What is measured on the vertical axis?
c. What is the range of values on the vertical axis?
d. Does the solubility of carbon dioxide increase or decrease with an increase in temperature?

1.54 **a.** What is measured on the horizontal axis? (1.4)
b. What is the range of values on the horizontal axis?
c. What is the solubility of carbon dioxide in water at 25 °C?
d. At what temperature does carbon dioxide have a solubility of 0.2 g/100 g water?

ANSWERS

Answers to Study Checks

1.1 a. observation (O)
b. hypothesis (H)
c. experiment (E)

1.2 b and **d**

1.3 $m = \dfrac{\text{heat}}{\Delta T \times SH}$

1.4 a. 37.6 °C **b.** 8 min

1.5 a. 4.25×10^5 **b.** 8.6×10^{-7}

Answers to Selected Questions and Problems

1.1 a. Chemistry is the study of the composition, structure, properties, and reactions of matter.
b. A chemical is a substance that has the same composition and properties wherever it is found.

1.3 Many chemicals are listed on a vitamin bottle such as vitamin A, vitamin B_3, vitamin B_{12}, vitamin C, and folic acid.

1.5 Typical items found in a medicine cabinet and some of the chemicals they contain are as follows:

Antacid tablets: calcium carbonate, cellulose, starch, stearic acid, silicon dioxide
Mouthwash: water, alcohol, thymol, glycerol, sodium benzoate, benzoic acid
Cough suppressant: menthol, beta-carotene, sucrose, glucose

1.7 a. observation (O) **b.** hypothesis (H)
c. experiment (E) **d.** experiment (E)

1.9 a. observation (O) **b.** hypothesis (H)
c. experiment (E) **d.** observation (O)
e. observation (O) **f.** conclusion (C)

1.11 There are several things you can do that will help you successfully learn chemistry, including forming a study group, going to lecture, working *Sample Problems* and *Study Checks*, working *Questions and Problems* and checking answers, reading the assignment ahead of class, and keeping a problem notebook.

1.13 (b), (c), (a), (d)

1.15 a. 09 (9 hundredths) **b.** 6 (6 ones)
c. 80 (8 tens)

1.17 a. 30 **b.** −44 **c.** 0

1.19 a. 80% **b.** 20%
c. Both have the same percentage of wheat.

1.21 a. 10 **b.** 48

1.23 a. The graph shows the relationship between the temperature of a cup of tea and time.
b. temperature, in °C
c. 20 °C to 80 °C
d. decrease

1.25 a. 7×10^4 **b.** 5.2×10^2
c. 8.1×10^{-6} **d.** 2.7×10^{-4}
e. 9.4×10^{-3} **f.** 4.9×10^5

1.27 a. 6.1×10^4 **b.** 5.8×10^{-3}
c. 7×10^8 **d.** 8.3×10^{-1}

1.29 No. All of the ingredients are chemicals.

1.31 Yes. Sherlock's investigation includes making observations (gathering data), formulating a hypothesis, testing the hypothesis, and modifying it until one of the hypotheses is validated.

1.33 a. observation (O) **b.** hypothesis (H)
c. observation (O)

1.35 a. Negative **b.** positive

1.37 A hypothesis, which is a possible explanation for an observation, can be tested with experiments.

1.39 b and **c**

1.41 b, **c** and **e**

1.43 a. −54 **b.** −74 **c.** +3

1.45 a. 20% **b.** 40%

1.47 a. 1.2×10^6 **b.** 4.4×10^{-7}
c. 6.6×10^{-2} **d.** 1.0×10^6

1.49 a. observation (O) **b.** hypothesis (H)
c. experiment (E) **d.** hypothesis (H)

1.51 a. 18 **b.** 24

1.53 a. The graph shows the relationship between the solubility of carbon dioxide in water and temperature.
b. solubility of carbon dioxide (g CO_2/100 g water)
c. 0 to 0.35 g of CO_2/100 g of water
d. decrease

2 Chemistry and Measurements

During the past few months, Greg has been experiencing an increased number of headaches, and frequently feels dizzy and nauseous. He goes to his doctor's office where the registered nurse completes the initial part of the exam by recording several measurements: weight 88.5 kg, height 190.5 cm, temperature 37.2 °C, and blood pressure at 155/95. A normal blood pressure is 120/80 or below.

When Greg sees his doctor, he is diagnosed as having high blood pressure (hypertension). The doctor prescribes 80 mg of Inderal (propranolol), which is used to treat hypertension and is to be taken once a day.

Two weeks later, Greg visits his doctor again, who determines that his blood pressure is now 152/90. The doctor increases the dosage of Inderal to 160 mg, once daily. The registered nurse informs Greg that he needs to increase his daily dosage to two tablets.

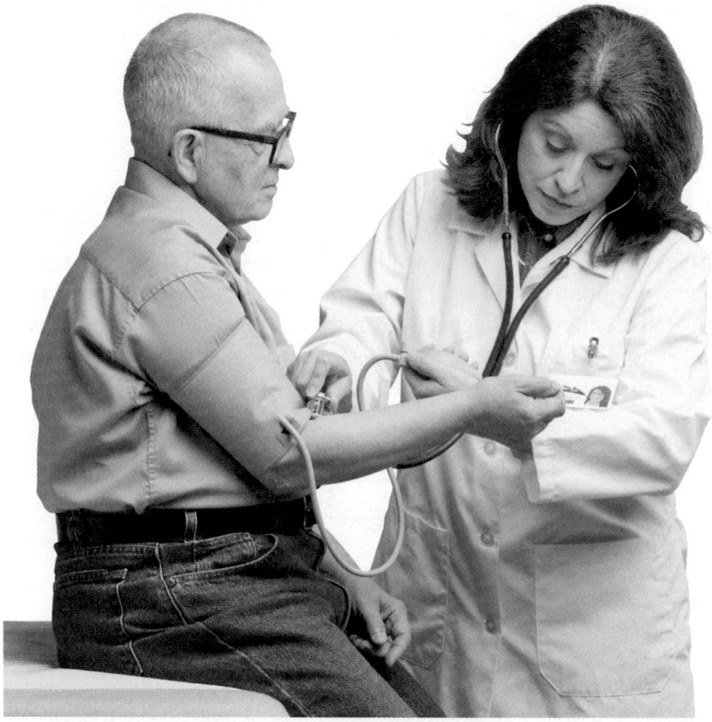

CAREER Registered Nurse

In addition to assisting physicians, registered nurses work to promote patient health, and prevent and treat disease. They provide patient care and help patients cope with illness. They take measurements such as a patient's weight, height, temperature, and blood pressure; make conversions; and calculate drug dosage rates. Registered nurses also maintain detailed medical records of patient symptoms and prescribed medications.

LOOKING AHEAD

Chemistry and measurements are an important part of our everyday lives. Levels of toxic materials in the air, soil, and water are discussed in our newspapers. We read about radon gas in our homes, holes in the ozone layer, trans fatty acids, and climate change. Understanding chemistry and measurement helps us make proper choices about our world.

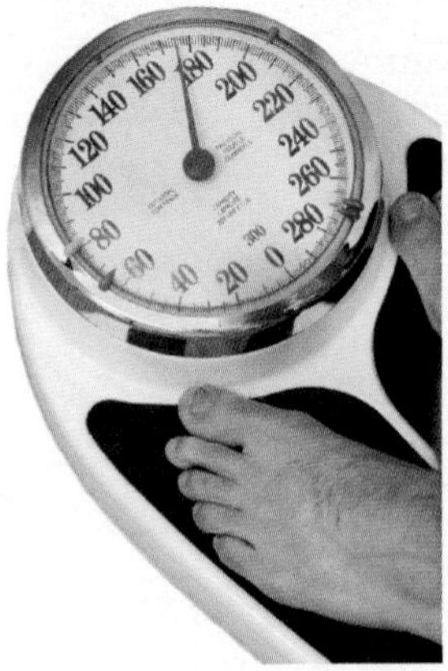
Your weight on a bathroom scale is a measurement.

Think about your day. You probably took some measurements. Perhaps you checked your weight by stepping on a bathroom scale. If you made some rice for dinner, you added two cups of water to one cup of rice. If you did not feel well, you may have taken your temperature. Whenever you take a measurement, you use a measuring device such as a balance, a measuring cup, or a thermometer. Over the years, you have learned to read the markings on each device to take a correct measurement.

Measurement is an essential part of health careers such as nursing, dental hygiene, respiratory therapy, nutrition, and veterinary technology. The temperature, height, and weight of a patient are measured and recorded. Samples of blood and urine are collected and sent to a laboratory where glucose, pH, urea, and protein are measured by the clinical technicians. By learning about measurement, you develop skills for solving problems and working with numbers.

CHAPTER READINESS*

KEY MATH SKILLS

- Identifying Place Values (1.4A)
- Using Positive and Negative Numbers in Calculations (1.4B)
- Calculating a Percentage (1.4C)
- Writing Numbers in Scientific Notation (1.4F)

*These Key Math Skills from the previous chapter are listed here for your review as you proceed to the new material in this chapter.

LEARNING GOAL

Write the names and abbreviations for the metric or SI units used in measurements of length, volume, mass, temperature, and time.

2.1 Units of Measurement

Scientists and health professionals throughout the world use the **metric system** of measurement. It is also the common measuring system in all but a few countries in the world. The **International System of Units (SI)**, or Système International, is the official system of measurement throughout the world except for the United States. In chemistry, we use metric units and SI units for length, volume, mass, temperature, and time, as listed in Table 2.1.

TABLE 2.1 Units of Measurement and Their Abbreviations

Measurement	Metric	SI
Length	meter (m)	meter (m)
Volume	liter (L)	cubic meter (m^3)
Mass	gram (g)	kilogram (kg)
Temperature	degree Celsius (°C)	kelvin (K)
Time	second (s)	second (s)

Suppose you walked 1.3 mi to campus today, carrying a backpack that weighs 26 lb. The temperature was 72 °F. Perhaps you weigh 128 lb and your height is 65 in. These measurements and units may seem familiar to you because they are stated in the U.S. system of measurement. However, in chemistry, we use the *metric system* in making our measurements. Using the metric system, you walked 2.1 km to campus, carrying a backpack that has a mass of 12 kg, when the temperature was 22 °C. You have a mass of 58.2 kg and a height of 1.65 m.

There are many measurements in everyday life.

Length

The metric and SI unit of length is the **meter (m)**. A meter is 39.4 inches (in.), which makes it slightly longer than a yard (yd). The **centimeter (cm)**, a smaller unit of length, is commonly used in chemistry and is about equal to the width of your little finger. For comparison, there are 2.54 cm in 1 in. (see Figure 2.1). Some relationships between units for length are

$$1\text{ m} = 100\text{ cm}$$
$$1\text{ m} = 39.4\text{ in.}$$
$$1\text{ m} = 1.09\text{ yd}$$
$$2.54\text{ cm} = 1\text{ in.}$$

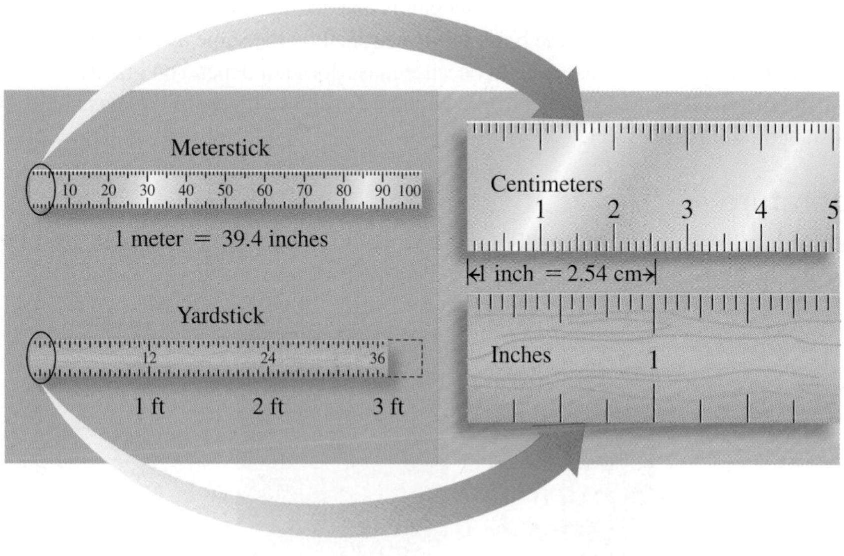

FIGURE 2.1 ▶ Length in the metric (SI) system is based on the meter, which is slightly longer than a yard.

Q How many centimeters are in a length of 1 inch?

Volume

Volume is the amount of space a substance occupies. A **liter (L)** is slightly larger than the quart (qt), (1 L = 1.06 qt). In a laboratory or a hospital, chemists work with metric units of volume that are smaller and more convenient, such as the **milliliter (mL)**. There are 1000 mL in 1 L. A comparison of metric and U.S. units for volume appears in Figure 2.2. Some relationships between units for volume are

$$1\text{ L} = 1000\text{ mL}$$
$$1\text{ L} = 1.06\text{ qt}$$
$$946\text{ mL} = 1\text{ qt}$$

The standard kilogram for the United States is stored at the National Institute of Standards and Technology (NIST).

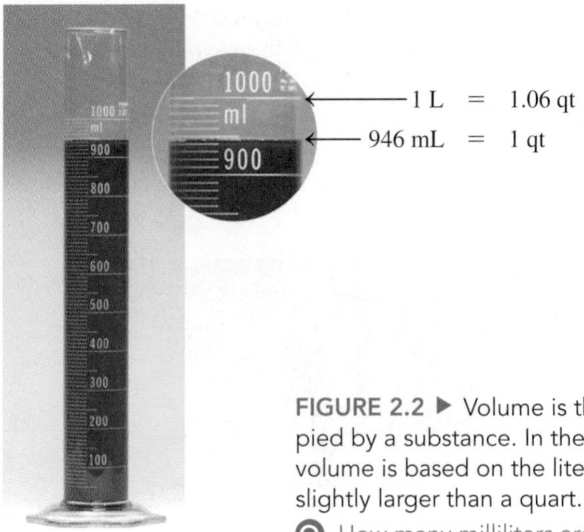

FIGURE 2.2 ▶ Volume is the space occupied by a substance. In the metric system, volume is based on the liter, which is slightly larger than a quart.

Q How many milliliters are in 1 quart?

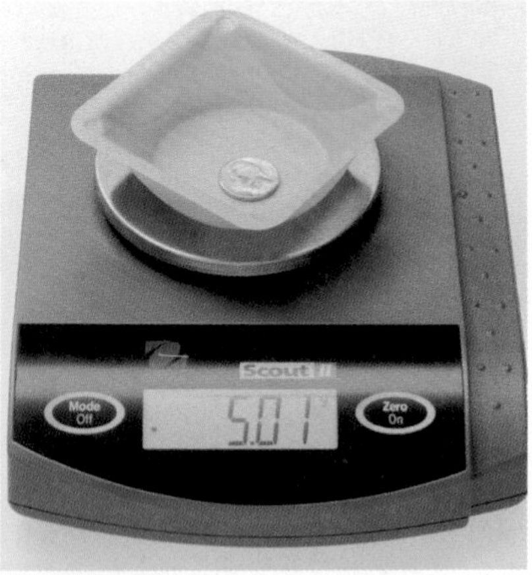

FIGURE 2.3 ▶ On an electronic balance, the digital readout gives the mass of a nickel, which is 5.01 g.

Q What is the mass of 10 nickels?

Mass

The **mass** of an object is a measure of the quantity of material it contains. The SI unit of mass, the **kilogram (kg)**, is used for larger masses such as body mass. In the metric system, the unit for mass is the **gram (g)**, which is used for smaller masses. There are 1000 g in 1 kg. It takes 2.20 lb to make 1 kg, and 454 g is equal to 1 lb. Some relationships between units for mass are

$$1\text{ kg} = 1000\text{ g}$$
$$1\text{ kg} = 2.20\text{ lb}$$
$$454\text{ g} = 1\text{ lb}$$

You may be more familiar with the term *weight* than with mass. *Weight* is a measure of the gravitational pull on an object. On the Earth, an astronaut with a mass of 75.0 kg has a weight of 165 lb. On the moon where the gravitational pull is one-sixth that of the Earth, the astronaut has a weight of 27.5 lb. However, the mass of the astronaut is the same as on Earth, 75.0 kg. Scientists measure mass rather than weight because mass does not depend on gravity.

In a chemistry laboratory, an electronic balance is used to measure the mass in grams of a substance (see Figure 2.3).

Temperature

Temperature tells us how hot something is, how cold it is outside, or helps us determine if we have a fever (see Figure 2.4). In the metric system, temperature is measured using

FIGURE 2.4 ▶ A thermometer is used to determine temperature.

Q What kinds of temperature readings have you made today?

Celsius temperature. On the **Celsius (°C) temperature scale**, water freezes at 0 °C and boils at 100 °C, while on the Fahrenheit (°F) scale, water freezes at 32 °F and boils at 212 °F. In the SI system, temperature is measured using the **Kelvin (K) temperature scale** on which the lowest possible temperature is 0 K. A unit on the Kelvin scale is called a kelvin (K) and is not written with a degree sign.

Time

We typically measure time in units such as years (yr), days, hours (h), minutes (min), or seconds (s). Of these, the SI and metric unit of time is the **second (s)**. The standard now used to determine a second is an atomic clock.

A stopwatch is used to measure the time of a race.

SAMPLE PROBLEM 2.1 Units of Measurement

State the type of measurement indicated in each of the following:

a. 45.6 g **b.** 1.85 L
c. 14 cm **d.** 6.3×10^2 mg

SOLUTION

a. A gram is a unit of mass.
b. A liter is a unit of volume.
c. A centimeter is a unit of length.
d. A milligram (mg) is a unit of mass.

STUDY CHECK 2.1

State the type of measurement indicated in each of the following:

a. 15 °C **b.** 25 s

Explore Your World

Units Listed on Labels

Read the labels on a variety of products such as sugar, salt, soft drinks, vitamins, and dental floss.

Questions

1. What metric or SI units of measurement are listed?
2. What type of measurement (mass, volume, etc.) does each indicate?
3. Write each quantity as a number and a metric or SI unit.

QUESTIONS AND PROBLEMS

2.1 Units of Measurement

LEARNING GOAL Write the names and abbreviations for the metric or SI units used in measurements of length, volume, mass, temperature, and time.

2.1 State the metric and SI units used to measure each of the following:
a. length and height
b. volume
c. temperature
d. time

2.2 Which one of the measurement in metric or SI unit may be used to clear the following statements?
a. My new trouser has a length of 46.
b. This watermelon weighs 5.
c. Water boils at 100.
d. Stadium is 300 from my house.

2.3 State the name of the unit and their use of measurement for each of the following quantities.
a. 3.4 cm **b.** 500 mg **c.** 2.0 L
d. 60 s **e.** 300 K

2.4 Arrange the following in descending order:
a. 10 cm **b.** 1 m **c.** 100 mm
d. 1 km

2.5 Give the abbreviation for each of the following
a. Milimeter **b.** Miligram **c.** Second
d. Fahrenheit **e.** Meter cube

2.6 Give the abbreviation for each of the following:
a. kilogram **b.** kelvin **c.** quart
d. meter **e.** milliliter

LEARNING GOAL

Identify a number as measured or exact; determine the number of significant figures in a measured number.

2.2 Measured Numbers and Significant Figures

When you make a measurement, you use some type of measuring device. For example, you may use a meterstick to measure your height, a scale to check your weight, or a thermometer to take your temperature.

Measured Numbers

Measured numbers are the numbers you obtain when you measure a quantity such as your height, weight, or temperature. Suppose you are going to measure the lengths of the objects in Figure 2.5. The metric ruler that you use may have lines marked in 1-cm divisions or perhaps in divisions of 0.1 cm. To report the length of the object, you observe the numerical values of the marked lines at the end of object. Then, you can *estimate* by visually dividing the space between the marked lines. This estimated value is the final digit in a measured number.

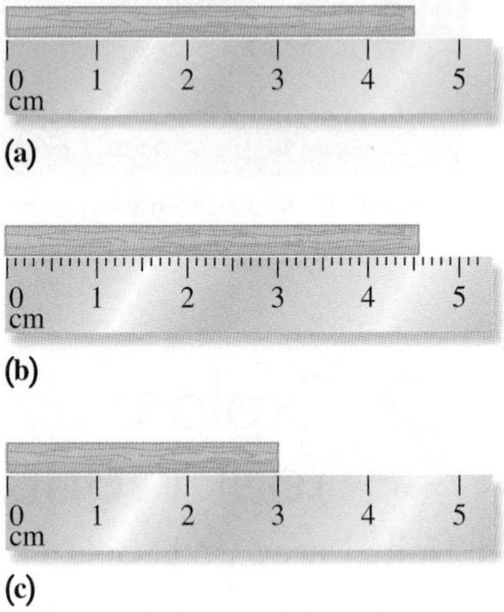

FIGURE 2.5 ▶ The lengths of the rectangular objects are measured as **(a)** 4.5 cm and **(b)** 4.55 cm.

Q What is the length of the object in **(c)**?

For example, in Figure 2.5a, the end of the object is between the marks of 4 cm and 5 cm, which means that the length is more than 4 cm but less than 5 cm. This is written as 4 cm plus an estimated digit. If you estimate that the end of the object is halfway between 4 cm and 5 cm, you would report its length as 4.5 cm. The last digit in a measured number may differ because people do not always estimate in the same way. Thus, someone else might report the length of the same object as 4.4 cm.

The metric ruler shown in Figure 2.5b is marked at every 0.1 cm. With this ruler, you can estimate to the hundredths place (0.01 cm). Now you can determine that the end of the object is between 4.5 cm and 4.6 cm. Perhaps you report its length as 4.55 cm, while another student reports its length as 4.56 cm. Both results are acceptable.

In Figure 2.5c, the end of the object appears to line up with the 3-cm mark. Because the divisions are marked in units of 1 cm, the length of the object is between 3 cm and 4 cm. Because the end of the object is on the 3-cm mark, the estimated digit is 0, which means the measurement is reported as 3.0 cm.

CORE CHEMISTRY SKILL

Counting Significant Figures

Significant Figures

In a measured number, the **significant figures (SFs)** are all the digits including the estimated digit. Nonzero numbers are always counted as significant figures. However, a zero may or may not be a significant figure depending on its position in a number. Table 2.2 gives the rules and examples of counting significant figures.

TABLE 2.2 Significant Figures in Measured Numbers

Rule	Measured Number	Number of Significant Figures
1. A number is a *significant figure* if it is		
a. not a zero	4.5 g	2
	122.35 m	5
b. a zero between nonzero digits	205 m	3
	5.008 kg	4
c. a zero at the end of a decimal number	50. L	2
	25.0 °C	3
	16.00 g	4
d. in the coefficient of a number written in scientific notation	4.8×10^5 m	2
	5.70×10^{-3} g	3
2. A zero is *not significant* if it is		
a. at the beginning of a decimal number	0.0004 s	1
	0.075 cm	2
b. used as a placeholder in a large number without a decimal point	850 000 m	2
	1 250 000 g	3

Scientific Notation and Significant Zeros

When one or more zeros in a large number are significant, they are shown clearly by writing the number in scientific notation. For example, if the first zero in the measurement 300 m is significant, but the second zero is not, the measurement is written as 3.0×10^2 m. *In this text, we will place a decimal point after a significant zero at the end of a number.* For example, if a measurement is written as 500. g, the decimal point after the second zero indicates that *both zeros* are significant. To show this more clearly, we can write it as 5.00×10^2 g. We will assume that zeros at the end of large standard numbers without a decimal point are not significant. Therefore, we write 400 000 g as 4×10^5 g, which has only one significant figure.

SAMPLE PROBLEM 2.2 Significant Zeros

Identify the significant zeros in each of the following measured numbers:

a. 0.000 250 m **b.** 70.040 g **c.** 1 020 000 L

SOLUTION

a. The zero in the last decimal place following the 5 is significant. The zeros preceding the 2 are not significant.

b. Zeros between nonzero digits or at the end of decimal numbers are significant. All the zeros in 70.040 g are significant.

c. Zeros between nonzero digits are significant. Zeros at the end of a large number with no decimal point are placeholders but not significant. The zero between 1 and 2 is significant, but the four zeros following the 2 are not significant.

STUDY CHECK 2.2

Identify the significant zeros in each of the following measured numbers:

a. 0.04008 m **b.** 6.00×10^3 g

Exact Numbers

Exact numbers *are those numbers obtained by counting items or using a definition that compares two units in the same measuring system.* Suppose a friend asks you how many classes you are taking. You would answer by counting the number of classes in your schedule. It would not use any measuring tool. Suppose you are asked to state the number of seconds in one minute. Without using any measuring device, you would give the definition: There are 60 s in 1 min. *Exact numbers are not measured, do not have a limited number of significant figures, and do not affect the number of significant figures in a calculated answer.* For more examples of exact numbers, see Table 2.3.

TABLE 2.3 Examples of Some Exact Numbers

Counted Numbers	Defined Equalities	
Items	Metric System	U.S. System
8 doughnuts	1 L = 1000 mL	1 ft = 12 in.
2 baseballs	1 m = 100 cm	1 qt = 4 cups
5 capsules	1 kg = 1000 g	1 lb = 16 oz

The number of baseballs is counted, which means 2 is an exact number.

For example, a mass of 42.2 g and a length of 5.0×10^{-3} cm are measured numbers because they are obtained using measuring tools. There are three SFs in 42.2 g because all nonzero digits are always significant. There are two SFs in 5.0×10^{-3} cm because all the digits in the coefficient of a number written in scientific notation are significant. However, a quantity of three eggs is an exact number that is obtained by counting the eggs. In the equality 1 kg = 1000 g, the masses of 1 kg and 1000 g are both exact numbers because this equality is a definition in the metric system of measurement.

QUESTIONS AND PROBLEMS

2.2 Measured Numbers and Significant Figures

LEARNING GOAL Identify a number as measured or exact; determine the number of significant figures in a measured number.

2.7 Identify the numbers in each of the following statements as measured or exact:
a. My chemistry book weight 8 lb.
b. There are 12 red roses in this bouquet.
c. In metric system, 1 m is equal to 100 cm.
d. There are 20 types of cakes in this bakery.

2.8 Identify the numbers in each of the following statements as measured or exact:
a. There are 31 students in the laboratory.
b. The oldest known flower lived 1.20×10^8 yr ago.
c. The largest gem ever found, an aquamarine, has a mass of 104 kg.
d. A laboratory test shows a blood cholesterol level of 184 mg/dL.

2.9 Identify the measured number(s), if any, in each of the following pairs of numbers:
a. 3 hamburgers and 6 oz of hamburger
b. 1 table and 4 chairs
c. 0.75 lb of grapes and 350 g of butter
d. 60 s = 1 min

2.10 Identify the exact number(s), if any, in each of the following statements:
a. Density of water is 1 g/mL
b. 1 g of water is equal to 1 cm^3
c. 1 g of water can fill the container
d. 1 qt of milk contains 700 mL water

2.11 Indicate if the zeros are significant in each of the following measurements.
a. 1.008 m **b.** 2.8700 cm
c. 3000 L **d.** 5.60×10^{-5} Kg
e. 9670000 g

2.12 Indicate if the zeros are significant in each of the following measurements:
a. 20.05 °C **b.** 5.00 m
c. 0.000 02 g **d.** 120 000 yr
e. 8.05×10^2 L

2.13 How many significant figures are there in each of the following?
a. 28.003 g **b.** 0.000057 m
c. 890000000 km **d.** 4.50×10^6 kg
e. 0.7005 L **f.** 19.0 °C

2.14 How many significant figures are in each of the following?
a. 3.1416 m **b.** 59600700 g
c. 300. K **d.** 120.5 °C
e. 10.102 g

2.15 In which of the following pairs do both numbers contain the same number of significant figures.
a. 51.00 kg and 510000 kg
b. 0.825 m and 0.00825 m
c. 0.000073 s and 7.30×10^4 s
d. 480.0 L and 0.0480 L

2.16 In which of the following pairs do both numbers contain the same number of significant figures?
a. 0.005 75 g and 5.75×10^{-3} g
b. 405 K and 405.0 K
c. 150 000 s and 1.50×10^4 s
d. 3.8×10^{-2} L and 3.0×10^5 L

2.17 Write each of the following in scientific notation with two significant figures:
a. 8537 L **b.** 31 000 g
c. 160 000 m **d.** 0.0001 20 cm

2.18 Write each of the following in scientific notation with two significant figures:
a. 5 100 000 g **b.** 26 000 s
c. 400 000 m **d.** 0.000 82 kg

LEARNING GOAL

Adjust calculated answers to give the correct number of significant figures.

2.3 Significant Figures in Calculations

In the sciences, we measure many things: the length of a bacterium, the volume of a gas sample, the temperature of a reaction mixture, or the mass of iron in a sample. The numbers obtained from these types of measurements are often used in calculations. The number of significant figures in the measured numbers determines the number of significant figures in the calculated answer.

Using a calculator will help you perform calculations faster. However, calculators cannot think for you. It is up to you to enter the numbers correctly, press the correct function keys, and give an answer with the correct number of significant figures.

Rounding Off

CORE CHEMISTRY SKILL

Rounding Off

Suppose you decide to buy carpeting for a room that measures 5.52 m by 3.58 m. Each of these measurements has three significant figures because the measuring tape limits your estimated place to 0.01 m. To determine how much carpeting you need, you would calculate the area of the room by multiplying 5.52 times 3.58 on your calculator. The calculator shows the number 19.7616 in its display. However, this is not the correct final answer because there are too many numbers, which is the result of the multiplication process. Because each of the original measurements has only three significant figures, the displayed number (19.7616) must be *rounded off* to three significant figures, 19.8. Therefore, you can order carpeting that will cover an area of 19.8 m^2.

Each time you use a calculator, it is important to look at the original measurements and determine the number of significant figures that can be used for the answer. You can use the following rules to round off the numbers shown in a calculator display.

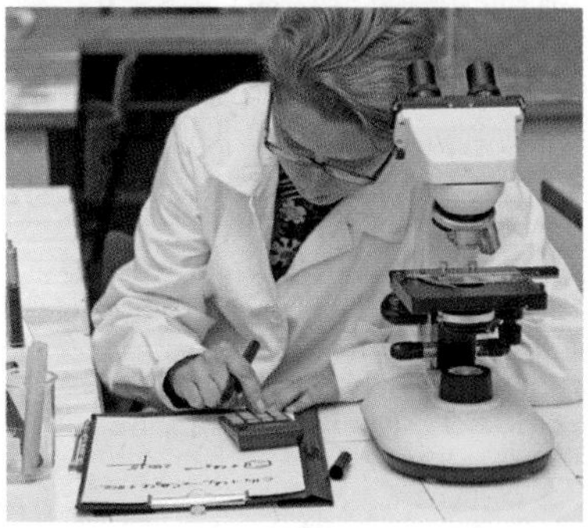

A technician uses a calculator in the laboratory.

Rules for Rounding Off

1. If the first digit to be dropped is *4 or less*, then it and all following digits are simply dropped from the number.
2. If the first digit to be dropped is *5 or greater*, then the last retained digit of the number is increased by 1.

Number to Round Off	Three Significant Figures	Two Significant Figures
8.4234	8.42 (drop 34)	8.4 (drop 234)
14.780	14.8 (drop 80, increase the last retained digit by 1)	15 (drop 780, increase the last retained digit by 1)
3256	3260* (drop 6, increase the last retained digit by 1, add 0) (3.26×10^3)	3300* (drop 56, increase the last retained digit by 1, add 00) (3.3×10^3)

*The value of a large number is retained by using placeholder zeros to replace dropped digits.

SAMPLE PROBLEM 2.3 Rounding Off

Round off each of the following numbers to three significant figures:

a. 35.7823 m **b.** 0.002 621 7 L **c.** 3.8268×10^3 g

SOLUTION

a. To round off 35.7823 m to three significant figures, drop 823 and increase the last retained digit by 1 to give 35.8 m.
b. To round off 0.002 621 7 L to three significant figures, drop 17 to give 0.002 62 L.
c. To round off 3.8268×10^3 g to three significant figures, drop 68 and increase the last retained digit by 1 to give 3.83×10^3 g.

STUDY CHECK 2.3

Round off each of the numbers in Sample Problem 2.3 to two significant figures.

CORE CHEMISTRY SKILL

Using Significant Figures in Calculations

A calculator is helpful in working problems and doing calculations faster.

Multiplication and Division with Measured Numbers

In multiplication or division, the final answer is written so that it has the same number of significant figures (SFs) as the measurement with the fewest SFs. An example of rounding off a calculator display follows:

Perform the following operations with measured numbers:

$$\frac{2.8 \times 67.40}{34.8} =$$

When the problem has multiple steps, the numbers in the numerator are multiplied and then divided by each of the numbers in the denominator.

2.8	⊠	67.40	⊟	34.8	=	5.422988506	=	5.4
Two SFs		Four SFs		Three SFs		Calculator display		Answer, rounded off to two SFs

Because the calculator display has more digits than the significant figures in the measured numbers allow, we need to round off. Using the measured number that has the fewest number (two) of significant figures, 2.8, we round off the calculator display to the answer with two SFs.

Adding Significant Zeros

Sometimes, a calculator display gives a small whole number. Then we add one or more significant zeros to the calculator display to obtain the correct number of significant figures. For example, suppose the calculator display is 4, but you used measurements that have three significant numbers. Then two significant zeros are *added* to give 4.00 as the correct answer.

Three SFs

$$\frac{8.00}{2.00} = 4. = 4.00$$

Three SFs — Calculator display — Final answer, two zeros added to give three SFs

SAMPLE PROBLEM 2.4 Significant Figures in Multiplication and Division

Perform the following calculations of measured numbers. Round off the calculator display or add zeros to give each answer with the correct number of significant figures.

a. 56.8×0.37 **b.** $\frac{(2.075)(0.585)}{(8.42)(0.0045)}$ **c.** $\frac{25.0}{5.00}$

SOLUTION

a. 21 **b.** 32 **c.** 5.00 (add two significant zeros)

STUDY CHECK 2.4

Perform the following calculations of measured numbers and give the answers with the correct number of significant figures:

a. $45.26 \times 0.010\,88$ **b.** $2.6 \div 324$ **c.** $\frac{4.0 \times 8.00}{16}$

Addition and Subtraction with Measured Numbers

In addition or subtraction, the final answer is written so that it has the same number of decimal places as the measurement having the fewest decimal places.

2.045	Thousandths place
⊞ 34.1	Tenths place
36.145	Calculator display
36.1	Answer, rounded off to the tenths place

When numbers are added or subtracted to give an answer ending in zero, the zero does not appear after the decimal point in the calculator display. For example, 14.5 g − 2.5 g = 12.0 g. However, if you do the subtraction on your calculator, the display shows 12. To write the correct answer, a significant zero is written after the decimal point.

SAMPLE PROBLEM 2.5 Significant Figures in Addition and Subtraction

Perform the following calculations and round off the calculator display or add zeros to give each answer with the correct number of decimal places:

a. 104.45 mL + 0.838 mL + 46 mL **b.** 153.247 g − 14.82 g

SOLUTION

a.	104.45 mL	Hundredths place
	0.838 mL	Thousandths place
	+ 46 mL	Ones place
	151 mL	Answer, rounded off to the ones place
b.	153.247 g	Thousandths place
	− 14.82 g	Hundredths place
	138.43 g	Answer, rounded off to the hundredths place

STUDY CHECK 2.5

Perform the following calculations and give the answers with the correct number of decimal places:

a. 82.45 mg + 1.245 mg + 0.000 56 mg **b.** 4.259 L − 3.8 L

QUESTIONS AND PROBLEMS

2.3 Significant Figures in Calculations

LEARNING GOAL Adjust calculated answers to give the correct number of significant figures.

2.19 Why do we usually need to round off calculations that use measured numbers?

2.20 Why do we sometimes add a zero to a number in a calculator display?

2.21 Round off each of the following measurements to three significant figures:
a. 310 cm **b.** 3.832×10^5 s
c. 2007 mL **d.** 0.6738265 cm
e. 28.2038 L

2.22 Round off each of the measurements in problem 2.21 to two significant figures.

2.23 Round off to three significant figures and express in scientific notation.
a. 5080 L **b.** 37 400 g
c. 104 720 m **d.** 0.000 250 82 s

2.24 Round off to three significant figures and express in scientific notation.
a. 5 100 000 L **b.** 26 711 s
c. 0.003 378 m **d.** 56.982 g

2.25 Perform each of the following calculations, and give an answer with the correct number of significant figures:
a. 45.7×0.034 **b.** $0.002\,78 \times 5$
c. $\frac{34.56}{1.25}$ **d.** $\frac{(0.2465)(25)}{1.78}$

2.26 Perform each of the following calculations, and give an answer with the correct number of significant figures:
a. 400×185 **b.** $\frac{2.40}{(4)(125)}$
c. $0.825 \times 3.6 \times 5.1$ **d.** $\frac{3.5 \times 0.261}{8.24 \times 20.0}$

2.27 Perform each of the following calculations, and give an answer with the correct number of decimal places:
a. 45.48 cm + 8.057 cm
b. 23.45 g + 104.1 g + 0.025 g
c. 145.675 mL − 24.2 mL
d. 1.08 L − 0.585 L

2.28 Perform each of the following calculations, and give an answer with the correct number of decimal places:
a. 236.02 108.01
b. 56.8×300
c. $\frac{28.7}{7}$

2.4 Prefixes and Equalities

LEARNING GOAL

Use the numerical values of prefixes to write a metric equality.

The special feature of the SI as well as the metric system is that a **prefix** can be placed in front of any unit to increase or decrease its size by some factor of 10. For example, the prefixes *milli* and *micro* are used to make the smaller units, milligram (mg) and microgram (μg). Table 2.4 lists some of the metric prefixes, their symbols, and their numerical values.

CORE CHEMISTRY SKILL

Using Prefixes

TABLE 2.4 Metric and SI Prefixes

Prefix	Symbol	Numerical Value	Scientific Notation	Equality
Prefixes That Increase the Size of the Unit				
tera	T	1 000 000 000 000	10^{12}	$1\text{ Ts} = 1 \times 10^{12}\text{ s}$ $1\text{ s} = 1 \times 10^{-12}\text{ Ts}$
giga	G	1 000 000 000	10^{9}	$1\text{ Gm} = 1 \times 10^{9}\text{ m}$ $1\text{ m} = 1 \times 10^{-9}\text{ Gm}$
mega	M	1 000 000	10^{6}	$1\text{ Mg} = 1 \times 10^{6}\text{ g}$ $1\text{ g} = 1 \times 10^{-6}\text{ Mg}$
kilo	k	1 000	10^{3}	$1\text{ km} = 1 \times 10^{3}\text{ m}$ $1\text{ m} = 1 \times 10^{-3}\text{ km}$
Prefixes That Decrease the Size of the Unit				
deci	d	0.1	10^{-1}	$1\text{ dL} = 1 \times 10^{-1}\text{ L}$ $1\text{ L} = 10\text{ dL}$
centi	c	0.01	10^{-2}	$1\text{ cm} = 1 \times 10^{-2}\text{ m}$ $1\text{ m} = 100\text{ cm}$
milli	m	0.001	10^{-3}	$1\text{ ms} = 1 \times 10^{-3}\text{ s}$ $1\text{ s} = 1 \times 10^{3}\text{ ms}$
micro	μ*	0.000 001	10^{-6}	$1\ \mu\text{g} = 1 \times 10^{-6}\text{ g}$ $1\text{ g} = 1 \times 10^{6}\ \mu\text{g}$
nano	n	0.000 000 001	10^{-9}	$1\text{ nm} = 1 \times 10^{-9}\text{ m}$ $1\text{ m} = 1 \times 10^{9}\text{ nm}$
pico	p	0.000 000 000 001	10^{-12}	$1\text{ ps} = 1 \times 10^{-12}\text{ s}$ $1\text{ s} = 1 \times 10^{12}\text{ ps}$

*In medicine and nursing, the abbreviation mc for the prefix micro is used because the symbol μ may be misread, which could result in a medication error. Thus, 1 μg would be written as 1 mcg.

TABLE 2.5 Daily Values for Selected Nutrients

Nutrient	Amount Recommended
Protein	44 g
Vitamin C	60 mg
Vitamin B_{12}	6 μg (6 mcg)
Calcium	1000 mg
Copper	2 mg
Iodine	150 μg (150 mcg)
Iron	18 mg
Magnesium	400 mg
Niacin	20 mg
Potassium	3500 mg
Selenium	55 μg (55 mcg)
Sodium	2400 mg
Zinc	15 mg

Source: U.S. Food and Drug Administration

The prefix *centi* is like cents in a dollar. One cent would be a "centidollar" or 0.01 of a dollar. That also means that one dollar is the same as 100 cents. The prefix *deci* is like dimes in a dollar. One dime would be a "decidollar" or 0.1 of a dollar. That also means that one dollar is the same as 10 dimes.

The U.S. Food and Drug Administration has determined the Daily Values (DV) for nutrients for adults and children aged 4 or older. Examples of these recommended Daily Values, some of which use prefixes, are listed in Table 2.5.

The relationship of a prefix to a unit can be expressed by replacing the prefix with its numerical value. For example, when the prefix *kilo* in kilometer is replaced with its value of 1000, we find that a kilometer is equal to 1000 meters. Other examples follow:

1 **kilo**meter (1 km) = **1000** meters ($1000\text{ m} = 10^3\text{ m}$)
1 **kilo**liter (1 kL) = **1000** liters ($1000\text{ L} = 10^3\text{ L}$)
1 **kilo**gram (1 kg) = **1000** grams ($1000\text{ g} = 10^3\text{ g}$)

SAMPLE PROBLEM 2.6 Prefixes and Equalities

An endoscopic camera has a width of 1 mm. Complete each of the following equalities involving millimeters:

a. 1 m = ______ mm **b.** 1 cm = ______ mm

SOLUTION

a. 1 m = 1000 mm **b.** 1 cm = 10 mm

STUDY CHECK 2.6

What is the relationship between millimeters and micrometers?

Measuring Length

An ophthalmologist may measure the diameter of the retina of an eye in centimeters (cm), whereas a surgeon may need to know the length of a nerve in millimeters (mm). When the prefix *centi* is used with the unit meter, it becomes *centimeter*, a length that is one-hundredth of a meter (0.01 m). When the prefix *milli* is used with the unit meter, it becomes *millimeter*, a length that is one-thousandth of a meter (0.001 m). There are 100 cm and 1000 mm in a meter.

If we compare the lengths of a millimeter and a centimeter, we find that 1 mm is 0.1 cm; there are 10 mm in 1 cm. These comparisons are examples of **equalities**, which show the relationship between two units that measure the same quantity. In every equality, each quantity has both a number and a unit. Examples of equalities between different metric units of length follow:

$$1\ \text{m} = 100\ \text{cm} = 1 \times 10^2\ \text{cm}$$
$$1\ \text{m} = 1000\ \text{mm} = 1 \times 10^3\ \text{mm}$$
$$1\ \text{cm} = 10\ \text{mm} = 1 \times 10^1\ \text{mm}$$

Some metric units for length are compared in Figure 2.6.

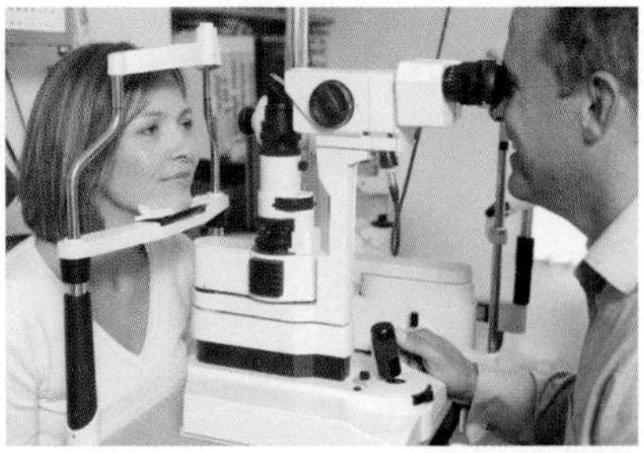

Using a retinal camera, an ophthalmologist photographs the retina of an eye.

First quantity			Second quantity	
1	m	=	100	cm
↑	↑		↑	↑
Number	+ unit		Number	+ unit

This example of an equality shows the relationship between meters and centimeters.

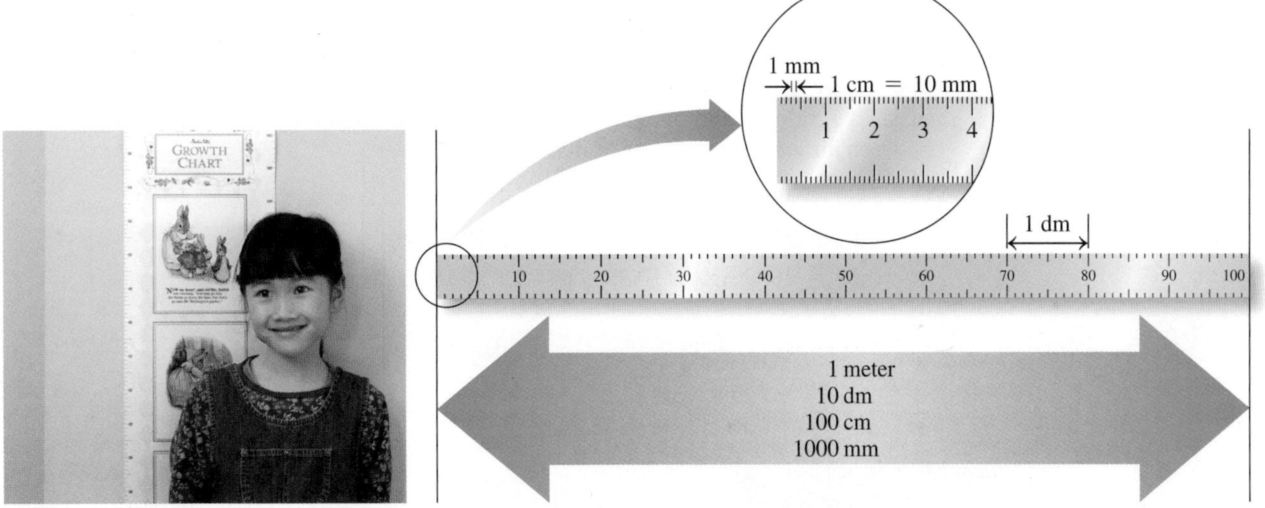

FIGURE 2.6 ▶ The metric length of 1 m is the same length as 10 dm, 100 cm, or 1000 mm.

Q How many millimeters (mm) are in 1 centimeter (cm)?

Measuring Volume

Volumes of 1 L or smaller are common in the health sciences. When a liter is divided into 10 equal portions, each portion is a deciliter (dL). There are 10 dL in 1 L. Laboratory results for blood work are often reported in mass per deciliter. Table 2.6 lists typical laboratory test values for some substances in the blood.

TABLE 2.6 Some Typical Laboratory Test Values

Substance in Blood	Typical Range
Albumin	3.5–5.0 g/dL
Ammonia	20–150 mcg/dL
Calcium	8.5–10.5 mg/dL
Cholesterol	105–250 mg/dL
Iron (male)	80–160 mcg/dL
Protein (total)	6.0–8.0 g/dL

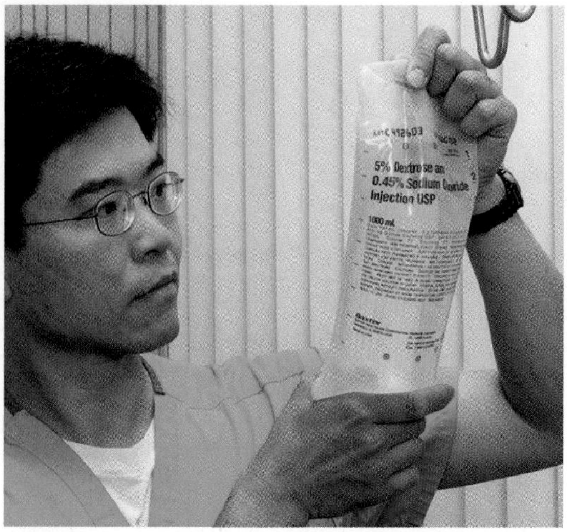

FIGURE 2.7 ▶ A plastic intravenous fluid container contains 1000 mL.

Q How many liters of solution are in the intravenous fluid container?

When a liter is divided into a thousand parts, each of the smaller volumes is called a milliliter (mL). In a 1-L container of physiological saline, there are 1000 mL of solution (see Figure 2.7). Examples of equalities between different metric units of volume follow:

$$1\text{ L} = 10\text{ dL} = 1 \times 10^1\text{ dL}$$
$$1\text{ L} = 1000\text{ mL} = 1 \times 10^3\text{ mL}$$
$$1\text{ dL} = 100\text{ mL} = 1 \times 10^2\text{ mL}$$

The **cubic centimeter** (abbreviated as **cm³** or **cc**) is the volume of a cube whose dimensions are 1 cm on each side. A cubic centimeter has the same volume as a milliliter, and the units are often used interchangeably.

$$1\text{ cm}^3 = 1\text{ cc} = 1\text{ mL}$$

When you see *1 cm*, you are reading about length; when you see *1 cc* or *1 cm³* or *1 mL*, you are reading about volume. A comparison of units of volume is illustrated in Figure 2.8.

Measuring Mass

When you go to the doctor for a physical examination, your mass is recorded in kilograms, whereas the results of your laboratory tests are reported in grams, milligrams (mg), or micrograms (μg or mcg). A kilogram is equal to 1000 g. One gram represents the same mass as 1000 mg, and one mg equals 1000 μg (or 1000 mcg). Examples of equalities between different metric units of mass follow:

$$1\text{ kg} = 1000\text{ g} = 1 \times 10^3\text{ g}$$
$$1\text{ g} = 1000\text{ mg} = 1 \times 10^3\text{ mg}$$
$$1\text{ mg} = 1000\ \mu\text{g (or 1000 mcg)} = 1 \times 10^3\ \mu\text{g (or } 1 \times 10^3\text{ mcg)}$$

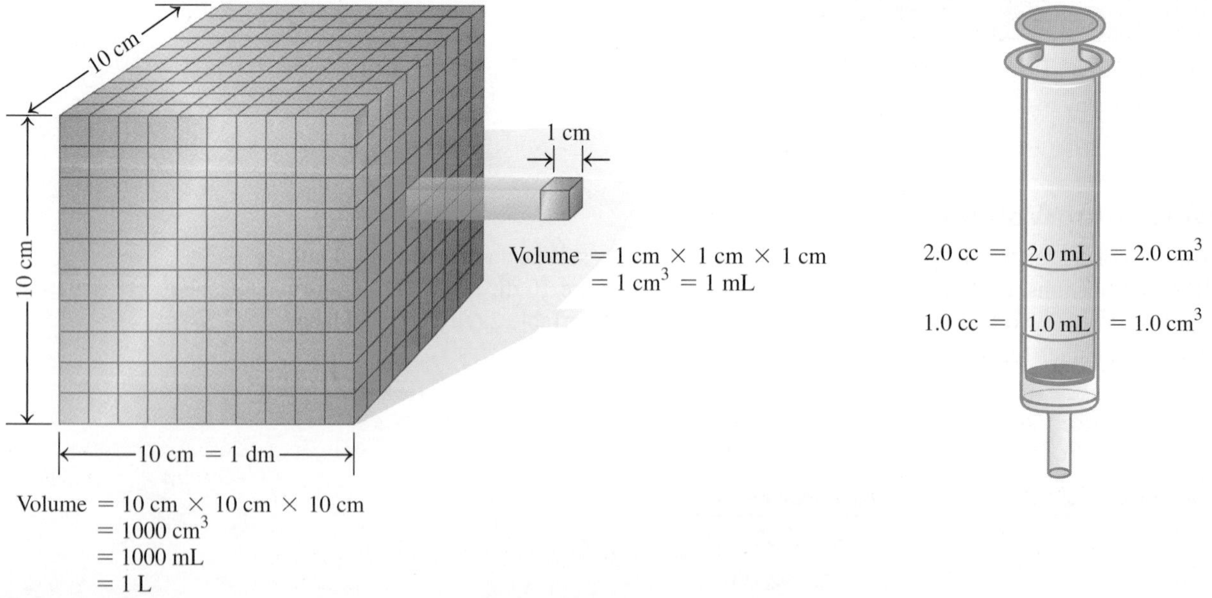

FIGURE 2.8 ▶ A cube measuring 10 cm on each side has a volume of 1000 cm³, or 1 L; a cube measuring 1 cm on each side has a volume of 1 cm³ (cc) or 1 mL.

Q What is the relationship between a milliliter (mL) and a cubic centimeter (cm³)?

QUESTIONS AND PROBLEMS

2.4 Prefixes and Equalities

LEARNING GOAL Use the numerical values of prefixes to write a metric equality.

2.29 The speedometer is marked in both km/h and mi/h, or mph. What is the meaning of each abbreviation?

2.30 In a French car, the odometer reads 2250. What units would this be? What units would it be if this were an odometer in a car made for the United States?

2.31 Write the abbreviation for each of the following units:
a. milligram **b.** deciliter
c. kilometer **d.** centigram

2.32 Write the abbreviation for each of the following units:
a. gigagram **b.** megameter
c. microliter **d.** nanosecond

2.33 Write the complete name for each of the following units.
a. ps **b.** cm **c.** nm **d.** mL

2.34 Write the complete name for each of the following units:
a. dL **b.** Ts **c.** mcg **d.** pm

2.35 Write the numerical value for each of the following prefixes:
a. centi **b.** tera **c.** milli **d.** deci

2.36 Write the numerical value for each of the following prefixes:
a. giga **b.** micro **c.** mega **d.** nano

2.37 Use a prefix to write the name for each of the following:
a. 0.1 g **b.** 10^{-6} g **c.** 1000 g **d.** 0.01 g

2.38 Use a prefix to write the name for each of the following:
a. 10^{9} m **b.** 10^{6} m **c.** 0.001 m **d.** 10^{-12} m

2.39 Complete each of the following metric relationships.
a. 1 s = ________ Ts **b.** 1 pm = ________ m
c. 1 cm = ________ m **d.** 1 m = ________ Gm

2.40 Complete each of the following metric relationships:
a. 1 Mg = ________ g **b.** 1 mL = ________ μL
c. 1 g = ________ kg **d.** 1 g = ________ mg

2.41 For each of the following pairs, which has the larger amount?
a. 1000 g or 0.1 kg **b.** 0.5 dL or 60 mL
c. 1000 nm or 10 m **d.** 2 dm or 0.2 m

2.42 For each of the following pairs, which is the smaller unit?
a. mg or g **b.** centimeter or nanometer
c. millimeters or micrometers **d.** mL or dL
e. centigrams or megagrams

2.5 Writing Conversion Factors

LEARNING GOAL

Write a conversion factor for two units that describe the same quantity.

Many problems in chemistry and the health sciences require you to change from one unit to another unit. You make changes in units every day. For example, suppose you worked 2.0 hours (h) on your homework, and someone asked you how many minutes that was. You would answer 120 minutes (min). You must have multiplied 2.0 h × 60 min/h because you knew the equality (1 h = 60 min) that related the two units. When you expressed 2.0 h as 120 min, you did not change the amount of time you spent studying. You changed only the unit of measurement used to express the time. *Any equality can be written as fractions called* **conversion factors** *with one of the quantities in the numerator and the other quantity in the denominator.* Be sure to include the units when you write the conversion factors. Two conversion factors are always possible from any equality.

CORE CHEMISTRY SKILL

Writing Conversion Factors from Equalities

Two Conversion Factors for the Equality 60 min = 1 h

$$\frac{\text{Numerator}}{\text{Denominator}} \begin{matrix}\longrightarrow \\ \longrightarrow\end{matrix} \quad \frac{60\text{ min}}{1\text{ h}} \quad \text{and} \quad \frac{1\text{ h}}{60\text{ min}}$$

These factors are read as "60 minutes per 1 hour" and "1 hour per 60 minutes." The term *per* means "divide." Some common relationships are given in Table 2.7. It is important that the equality you select to form a conversion factor is a *true* relationship.

The numbers in any equality between two metric units or between two U.S. system units are obtained by definition. Because numbers in a definition are exact, they are not

Explore Your World

SI and Metric Equalities on Product Labels

Read the labels on some food products on your kitchen shelves and in your refrigerator, or refer to the labels in Figure 2.9. List the amount of product given in different units. Write a relationship for two of the amounts for the same product and container. Look for measurements of grams and pounds or quarts and milliliters.

Questions

1. Use the stated measurement to derive a metric–U.S. conversion factor.
2. How do your results compare with the conversion factors we have described in this text?

TABLE 2.7 Some Common Equalities

Quantity	Metric (SI)	U.S.	Metric–U.S.
Length	1 km = 1000 m	1 ft = 12 in.	2.54 cm = 1 in. (exact)
	1 m = 1000 mm	1 yd = 3 ft	1 m = 39.4 in.
	1 cm = 10 mm	1 mi = 5280 ft	1 km = 0.621 mi
Volume	1 L = 1000 mL	1 qt = 4 cups	946 mL = 1 qt
	1 dL = 100 mL	1 qt = 2 pt	1 L = 1.06 qt
	1 mL = 1 cm^3	1 gal = 4 qt	473 mL = 1 pt
	1 mL = 1 cc*		1 mL = 15 drops*
			5 mL = 1 tsp*
			15 mL = 1 T (tbsp)*
Mass	1 kg = 1000 g	1 lb = 16 oz	1 kg = 2.20 lb
	1 g = 1000 mg		454 g = 1 lb
	1 mg = 1000 mcg*		
Time	1 h = 60 min	1 h = 60 min	
	1 min = 60 s	1 min = 60 s	

*Used in nursing and medicine

used to determine significant figures. For example, the equality of 1 g = 1000 mg is defined, which means that both of the numbers 1 and 1000 are exact. *However, when an equality consists of a metric unit and a U.S. unit, one of the numbers in the equality is obtained by measurement and counts toward the significant figures in the answer.* For example, the equality of 1 lb = 454 g is obtained by measuring the grams in exactly 1 lb. In this equality, the measured quantity 454 g has three significant figures, whereas the 1 is exact. An exception is the relationship of 1 in. = 2.54 cm where 2.54 has been defined as exact.

Metric Conversion Factors

We can write metric conversion factors for any of the metric relationships. For example, from the equality for meters and centimeters, we can write the following factors:

Metric Equality	Conversion Factors
1 m = 100 cm	$\frac{100\text{ cm}}{1\text{ m}}$ and $\frac{1\text{ m}}{100\text{ cm}}$

Both are proper conversion factors for the relationship; one is just the inverse of the other. *The usefulness of conversion factors is enhanced by the fact that we can turn a conversion factor over and use its inverse.* The number 100 and 1 in this equality and its conversion factors are both *exact* numbers.

Metric–U.S. System Conversion Factors

Suppose you need to convert from pounds, a unit in the U.S. system, to kilograms in the metric (or SI) system. A relationship you could use is

1 kg = 2.20 lb

The corresponding conversion factors would be

$$\frac{2.20\text{ lb}}{1\text{ kg}} \quad \text{and} \quad \frac{1\text{ kg}}{2.20\text{ lb}}$$

Figure 2.9 illustrates the contents of some packaged foods in both U.S. and metric units.

FIGURE 2.9 ▶ In the United States, the contents of many packaged foods are listed in both U.S. and metric units.

Q What are some advantages of using the metric system?

SAMPLE PROBLEM 2.7 Writing Conversion Factors

Identify the correct conversion factor(s) for the equality of 1 pint of blood contains 473 mL of blood.

a. $\frac{473\text{ pt}}{1\text{ mL}}$ **b.** $\frac{1\text{ pt}}{473\text{ mL}}$

c. $\frac{473\text{ mL}}{1\text{ pt}}$ **d.** $\frac{1\text{ mL}}{473\text{ pt}}$

SOLUTION

The equality for pint and milliliters is 1 pt = 473 mL. Answers **b** and **c** are correctly written conversion factors for the equality.

STUDY CHECK 2.7

What are the two correctly written conversion factors for the equality 1000 mm = 1 m?

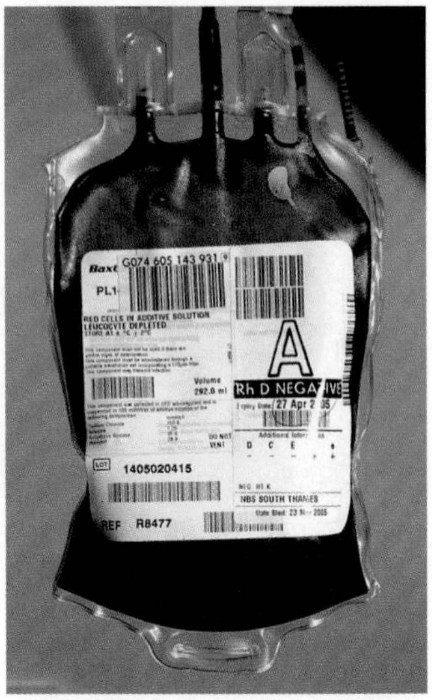

1 pint of blood contains 473 mL.

Equalities and Conversion Factors Stated Within a Problem

An equality may also be stated within a problem that applies only to that problem. For example, the speed of a car in kilometers per hour or the milligrams of vitamin C in a tablet would be specific relationships for that problem only. However, it is possible to identify these relationships within a problem and to write corresponding conversion factors.

From each of the following statements, we can write an equality, and its conversion factors, and identify each number as exact or give its significant figures.

The car was traveling at a speed of 85 km/h.

Equality	Conversion Factors	Significant Figures or Exact
85 km = 1 h	$\frac{85\text{ km}}{1\text{ h}}$ and $\frac{1\text{ h}}{85\text{ km}}$	The 85 km is measured: It has two significant figures. The 1 in 1 h is exact.

One tablet contains 500 mg of vitamin C.

Equality	Conversion Factors	Significant Figures or Exact
1 tablet = 500 mg of vitamin C	$\frac{500\text{ mg vitamin C}}{1\text{ tablet}}$ and $\frac{1\text{ tablet}}{500\text{ mg vitamin C}}$	The 500 mg is measured: It has one significant figure. The 1 in 1 tablet is exact.

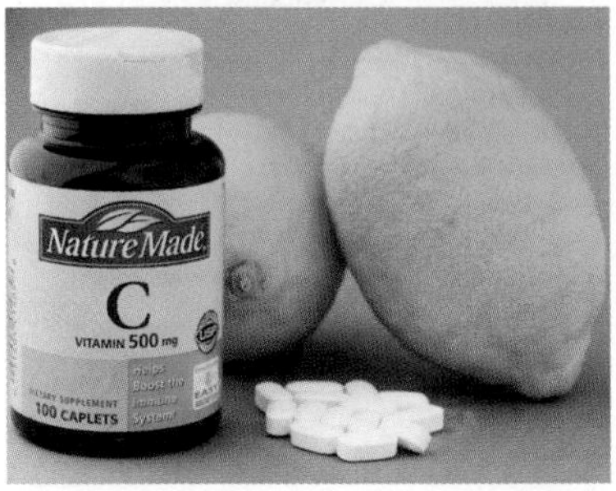

Vitamin C, an antioxidant needed by the body, is found in fruits such as lemons.

Conversion Factors from Dosage Problems

Equalities stated within dosage problems for medications can also be written as conversion factors. For example, Keflex (Cephalexin), an antibiotic used for respiratory and ear infections, is available in 250-mg capsules. The quantity of Keflex in a capsule can be written as an equality from which two conversion factors are possible.

Equality	Conversion Factors	Significant Figures or Exact
1 capsule = 250 mg of Keflex	$\frac{250\text{ mg Keflex}}{1\text{ capsule}}$ and $\frac{1\text{ capsule}}{250\text{ mg Keflex}}$	The 250 mg is measured: It has two significant figures. The 1 in 1 capsule is exact.

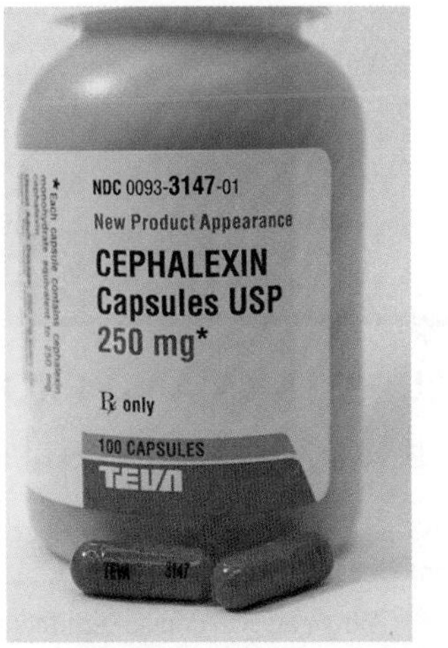

Keflex (Cephalexin) used to treat respiratory infections is available in 250-mg capsules.

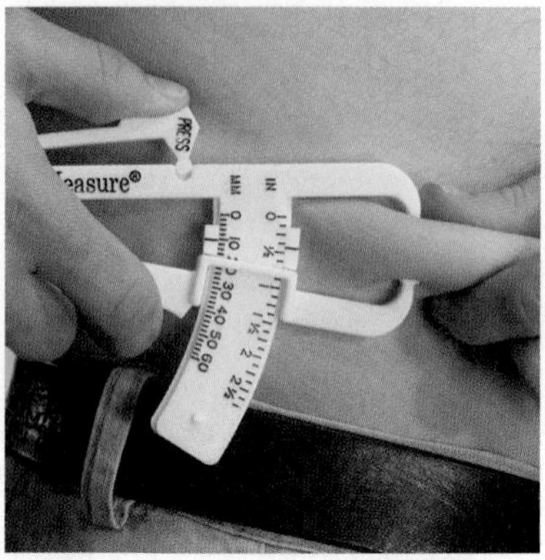

The thickness of the skin fold at the abdomen is used to determine the percentage of body fat.

Conversion Factors from a Percentage

A percentage (%) is written as a conversion factor by choosing a unit and expressing the numerical relationship of the parts of this unit to 100 parts of the whole. For example, a person might have 18% body fat by mass. The percentage quantity can be written as 18 mass units of body fat in every 100 mass units of body mass. Different mass units, such as grams (g), kilograms (kg), or pounds (lb) can be used, but both units in the factor must be the same.

Equality	Conversion Factors	Significant Figures or Exact
18 kg of body fat = 100 kg of body mass	$\frac{18\text{ kg body fat}}{100\text{ kg body mass}}$ and $\frac{100\text{ kg body mass}}{18\text{ kg body fat}}$	The 18 kg is measured: It has two significant figures. The 100 in 100 kg is exact.

Salmon contains high levels of omega-3 fatty acids.

SAMPLE PROBLEM 2.8 Equalities and Conversion Factors Stated in a Problem

Write the equality and its corresponding conversion factors, and identify each number as exact or give its significant figures for each of the following statements:

a. There are 325 mg of aspirin in 1 tablet.
b. Cold water fish such as salmon contains 1.9% omega-3 fatty acids by mass.

SOLUTION

a. There are 325 mg of aspirin in 1 tablet.

Equality	Conversion Factors	Significant Figures or Exact
1 tablet = 325 mg of aspirin	$\frac{325\text{ mg aspirin}}{1\text{ tablet}}$ and $\frac{1\text{ tablet}}{325\text{ mg aspirin}}$	The 325 mg is measured: It has three significant figures. The 1 in 1 tablet is exact.

b. 1.9 g of omega-3 fatty acids = 100 g of salmon

$$\frac{1.9\text{ g omega-3 fatty acids}}{100\text{ g salmon}} \quad \text{and} \quad \frac{100\text{ g salmon}}{1.9\text{ g omega-3 fatty acids}}$$

The 1.9 in 1.9 g is measured: It has two significant figures. The 100 in 100 g is exact.

STUDY CHECK 2.8

Levsin (hyoscyamine sulfate), used to treat stomach and bladder problems, is available as drops with 0.125 mg Levsin per 1 mL of solution. Write the equality and its corresponding conversion factors, and identify each number as exact or give its number of significant figures.

QUESTIONS AND PROBLEMS

2.5 Writing Conversion Factors

LEARNING GOAL Write a conversion factor for two units that describe the same quantity.

2.43 Why can two conversion factors be written for the equality 1 m = 100 cm?

2.44 How can you check that you have written the correct conversion factors for an equality?

2.45 Write the equality and conversion factors for each of the following pairs of units:
a. centimeters and inches
b. milligrams and grams
c. quarts and milliliters
d. deciliters and milliliters
e. days and weeks

2.46 Write the equality and conversion factors for each of the following pairs of units:
a. centimeters and meters
b. pounds and kilograms
c. pounds and grams
d. milliliters and liters
e. dimes and dollars

2.47 Write the equality and conversion factors for each of the following statements:
a. An hour has 60 mins.
b. A week has 7 days.
c. A can containing 355 mL of Coke.
d. A bottle of 20 Vitamin C tablets.

2.48 Write the equality and conversion factors for each of the following statements:
a. The label on a bottle reads 10 mg of furosemide per 1 mL.
b. The Daily Value for iodine is 150 mcg.
c. An IV of normal saline solution has a flow rate of 85 mL per hour.
d. Gold jewelry contains 58% by mass gold.
e. One capsule of fish oil contains 360 mg of omega-3 fatty acids.

Chemistry Link to **Health**

Toxicology and Risk–Benefit Assessment

Each day, we make choices about what we do or what we eat, often without thinking about the risks associated with these choices. We are aware of the risks of cancer from smoking or the risks of lead poisoning, and we know there is a greater risk of having an accident if we cross a street where there is no light or crosswalk.

A basic concept of toxicology is the statement of Paracelsus that the dose is the difference between a poison and a cure. To evaluate the level of danger from various substances, natural or synthetic, a risk assessment is made by exposing laboratory animals to the substances and monitoring the health effects. Often, doses very much greater than humans might ordinarily encounter are given to the test animals.

Many hazardous chemicals or substances have been identified by these tests. One measure of toxicity is the LD_{50}, or lethal dose, which is the concentration of the substance that causes death in 50% of the test animals. A dosage is typically measured in milligrams per kilogram (mg/kg) of body mass or micrograms per kilogram (mcg/kg) of body mass.

Other evaluations need to be made, but it is easy to compare LD_{50} values. Parathion, a pesticide, with an LD_{50} of 3 mg/kg, would be highly toxic. This means that 3 mg of parathion per kg of body mass would be fatal to half the test animals. Table salt (sodium chloride) with an LD_{50} of 3000 mg/kg would have a much lower toxicity. You would need to ingest a huge amount of salt before any toxic effect would be observed. Although the risk to animals can be evaluated in the laboratory, it is more difficult to determine the impact in the environment since there is also a difference between continued exposure and a single, large dose of the substance.

The LD_{50} of caffeine is 192 mg/kg.

Table 2.8 lists some LD_{50} values and compares substances in order of increasing toxicity.

TABLE 2.8 Some LD_{50} Values for Substances Tested in Rats

Substance	LD_{50} (mg/kg)
Table sugar	29 700
Boric acid	5140
Baking soda	4220
Table salt	3000
Ethanol	2080
Aspirin	1100
Caffeine	192
DDT	113
Dichlorvos (pesticide strips)	56
Sodium cyanide	6
Parathion	3

2.6 Problem Solving Using Unit Conversion

LEARNING GOAL
Use conversion factors to change from one unit to another.

The process of problem solving in chemistry often requires one or more conversion factors to change a given unit to the needed unit. For the problem, the unit of the given quantity and the unit of the needed quantity are identified. From there, the problem is set up with one or more conversion factors used to convert the given unit to the needed unit as seen in the following Sample Problem.

Given unit × one or more conversion factors = needed unit

SAMPLE PROBLEM 2.9 Problem Solving Using Conversion Factors

In radiological imaging such as PET or CT scans, dosages of pharmaceuticals are based on body mass. If a person weighs 164 lb, what is the body mass in kilograms?

Guide to Problem Solving Using Conversion Factors

STEP 1 State the given and needed quantities.

STEP 2 Write a plan to convert the given unit to the needed unit.

STEP 3 State the equalities and conversion factors.

STEP 4 Set up the problem to cancel units and calculate the answer.

SOLUTION

STEP 1 State the given and needed quantities.

ANALYZE THE PROBLEM	Given	Need
	164 lb	kilograms

STEP 2 Write a plan to convert the given unit to the needed unit. We see that the given unit is in the U.S. system of measurement and the needed unit in the metric system. Therefore, the conversion factor must relate the U.S. unit lb to the metric unit kg.

pounds → U.S.–Metric factor → kilograms

STEP 3 State the equalities and conversion factors.

$$1\text{ kg} = 2.20\text{ lb}$$

$$\frac{2.20\text{ lb}}{1\text{ kg}} \quad\text{and}\quad \frac{1\text{ kg}}{2.20\text{ lb}}$$

STEP 4 Set up the problem to cancel units and calculate the answer. Write the given, 164 lb, and multiply by the conversion factor that has the unit lb in the denominator (bottom number) to cancel out the given unit (lb) in the numerator.

Unit for answer goes here

$$164\ \cancel{\text{lb}} \times \frac{1\text{ kg}}{2.20\ \cancel{\text{lb}}} = 74.5\text{ kg}$$

Given — Conversion factor — Answer

Look at how the units cancel. The given unit lb cancels out and the needed unit kg is in the numerator. *The unit you want in the final answer is the one that remains after all the other units have canceled out.* This is a helpful way to check that you set up a problem properly.

$$\cancel{\text{lb}} \times \frac{\text{kg}}{\cancel{\text{lb}}} = \text{kg} \quad \text{Unit needed for answer}$$

The calculator display gives the numerical answer, which is adjusted to give a final answer with the proper number of significant figures (SFs).

$$164 \times \frac{1}{2.20} = 164 \div 2.20 = 74.54545455 = 74.5$$

164: Three SFs; 1: Exact; 2.20: Three SFs; 74.54545455: Calculator display; 74.5: Three SFs (rounded off)

The value of 74.5 combined with the unit, kg, gives the final answer of 74.5 kg. With few exceptions, answers to numerical problems contain a number and a unit.

STUDY CHECK 2.9

If 1890 mL of orange juice is prepared from orange juice concentrate, how many liters of orange juice is that?

Interactive Video

Conversion Factors

Using Two or More Conversion Factors

CORE CHEMISTRY SKILL
Using Conversion Factors

In problem solving, two or more conversion factors are often needed to complete the change of units. In setting up these problems, one factor follows the other. Each factor is arranged to cancel the preceding unit until the needed unit is obtained. Once the problem is set up to cancel units properly, the calculations can be done without writing intermediate results. The process is worth practicing until you understand unit cancelation, the steps on the calculator, and rounding off to give a final answer. In this text, when two or more conversion factors are required, the final answer will be based on obtaining a final calculator display and rounding off (or adding zeros) to give the correct number of significant figures.

SAMPLE PROBLEM 2.10 Using Two or More Conversion Factors

A doctor's order for 0.50 g of Keflex is available as 250-mg tablets. How many tablets of Keflex are needed? In the following setup, fill in the missing parts of the conversion factors, show the canceled units, and give the correct answer:

$$0.50\text{ g Keflex} \times \frac{\_\_\text{mg Keflex}}{1\text{ g Keflex}} \times \frac{1\text{ tablet}}{\_\_\text{mg Keflex}} =$$

SOLUTION

$$0.50\text{ }\cancel{\text{g Keflex}} \times \frac{1000\text{ }\cancel{\text{mg Keflex}}}{1\text{ }\cancel{\text{g Keflex}}} \times \frac{1\text{ tablet}}{250\text{ }\cancel{\text{mg Keflex}}} = 2\text{ tablets}$$

STUDY CHECK 2.10

A newborn has a length of 450 mm. In the following setup that calculates the baby's length in inches, fill in the missing parts of the conversion factors, show the canceled units, and give the correct answer:

$$450\text{ mm} \times \frac{1\text{ cm}}{\_\_\text{mm}} \times \frac{1\text{ in.}}{\_\_\text{cm}} =$$

SAMPLE PROBLEM 2.11 Problem Solving Using Two Conversion Factors

Synthroid is used as a replacement or supplemental therapy for diminished thyroid function. A doctor's order prescribed a dosage of 0.150 mg of Synthroid. If tablets in stock contain 75 mcg of Synthroid, how many tablets are required to provide the prescribed medication?

SOLUTION

STEP 1 **State the given and needed quantities.**

ANALYZE THE PROBLEM	Given	Need
	0.150 mg of Synthroid	number of tablets

STEP 2 **Write a plan to convert the given unit to the needed unit.**

milligrams → Metric factor → micrograms → Clinical factor → number of tablets

STEP 3 **State the equalities and conversion factors.** In the problem, the information for the dosage used as an equality along with the metric equality for milligrams and micrograms provides the following conversion factors:

$$1\text{ mg} = 1000\text{ mcg}$$
$$\frac{1\text{ mg}}{1000\text{ mcg}} \text{ and } \frac{1000\text{ mcg}}{1\text{ mg}}$$

$$1\text{ tablet} = 75\text{ mcg}$$
$$\frac{1\text{ tablet}}{75\text{ mcg}} \text{ and } \frac{75\text{ mcg}}{1\text{ tablet}}$$

STEP 4 **Set up the problem to cancel units and calculate the answer.** The problem can be set up using the metric factor to cancel milligrams, and then the clinical factor to obtain the number of tablets as the final unit.

$$\underset{\text{Three SFs}}{0.150\ \cancel{\text{mg}}} \times \overset{\text{Exact}}{\underset{\text{Exact}}{\frac{1000\ \cancel{\text{mcg}}}{1\ \cancel{\text{mg}}}}} \times \overset{\text{Exact}}{\underset{\text{Two SFs}}{\frac{1\ \text{tablet}}{75\ \cancel{\text{mcg}}}}} = 2\ \text{tablets}$$

STUDY CHECK 2.11

A bottle contains 120 mL of cough syrup. If one teaspoon (5 mL) is given four times a day, how many days will elapse before a refill is needed?

SAMPLE PROBLEM 2.12 Using a Percentage as a Conversion Factor

A person who exercises regularly has 16% body fat. If this person weighs 155 lb, what is the mass, in kilograms, of body fat?

SOLUTION

STEP 1 **State the given and needed quantities.**

ANALYZE THE PROBLEM	Given	Need
	155 lb body weight, 16% body fat	kilograms of body fat

STEP 2 **Write a plan to convert the given unit to the needed unit.**

pounds of body weight → U.S.–Metric factor → kilograms of body mass → Percentage factor → kilograms of body fat

STEP 3 **State the equalities and conversion factors.** One equality is the U.S.–metric factor for lb and kg. The second is the percentage factor derived from the percentage information given in the problem.

1 kg of body mass = 2.20 lb of body weight

$$\frac{2.20\ \text{lb body weight}}{1\ \text{kg body mass}} \text{ and } \frac{1\ \text{kg body mass}}{2.20\ \text{lb body weight}}$$

16 kg of body fat = 100 kg of body mass

$$\frac{16\ \text{kg body fat}}{100\ \text{kg body mass}} \text{ and } \frac{100\ \text{kg body mass}}{16\ \text{kg body fat}}$$

STEP 4 **Set up the problem to cancel units and calculate the answer.** We can set up the problem using conversion factors to cancel each unit, starting with lb of body weight, until we obtain the final unit, kg of body fat, in the numerator.

$$\underset{\text{Three SFs}}{155\ \cancel{\text{lb body weight}}} \times \overset{\text{Exact}}{\underset{\text{Three SFs}}{\frac{1\ \cancel{\text{kg body mass}}}{2.20\ \cancel{\text{lb body weight}}}}} \times \overset{\text{Two SFs}}{\underset{\text{Exact}}{\frac{16\ \text{kg body fat}}{100\ \cancel{\text{kg body mass}}}}} = \underset{\text{Two SFs}}{11\ \text{kg of body fat}}$$

STUDY CHECK 2.12

Uncooked lean ground beef can contain up to 22% fat by mass. How many grams of fat would be contained in 0.25 lb of the ground beef?

QUESTIONS AND PROBLEMS

2.6 Problem Solving Using Unit Conversion

LEARNING GOAL Use conversion factors to change from one unit to another.

2.49 When you convert one unit to another, how do you know which unit of the conversion factor to place in the denominator?

2.50 When you convert one unit to another, how do you know which unit of the conversion factor to place in the numerator?

2.51 Use metric conversion factors to solve each of the following problems:
- **a.** The height of a student is 175 cm. How tall is the student in meters?
- **b.** A cooler has a volume of 5500 mL. What is the capacity of the cooler in liters?
- **c.** A Bee Hummingbird has a mass of 0.0018 kg. What is the mass of the hummingbird in grams?

2.52 Use metric conversion factors to solve each of the following problems:
- **a.** The Daily Value for phosphorus is 800 mg. How many grams of phosphorus are recommended?
- **b.** A glass of orange juice contains 0.85 dL of juice. How many milliliters of orange juice is that?
- **c.** A package of chocolate instant pudding contains 2840 mg of sodium. How many grams of sodium is that?

2.53 Solve each of the following problems using one or more conversion factors:
- **a.** A container holds 0.500 qt of liquid. How many milliliters of lemonade will it hold?
- **b.** What is the mass, in kilograms, of a person who weighs 145 lb?
- **c.** An athlete has 15% by mass body fat. What is the weight of fat, in pounds, of a 74-kg athlete?

Agricultural fertilizers applied to a field provide nitrogen for plant growth.

- **d.** A plant fertilizer contains 15% by mass nitrogen (N). In a container of soluble plant food, there are 10.0 oz of fertilizer. How many grams of nitrogen are in the container?
- **e.** In a candy factory, the nutty chocolate bars contain 22.0% by mass pecans. If 5.0 kg of pecans were used for candy last Tuesday, how many pounds of nutty chocolate bars were made?

2.54 Using metric conversion factors, convert each of the following to the required unit:
- **a.** 2 g/L to __________ mg/dL
- **b.** 0.8 mg/cc to __________ g/L
- **c.** 0.1 kg/mL to __________ g/L
- **d.** 30 m/mL to __________ km/L

2.55 Using conversion factors, solve each of the following clinical problems:
- **a.** You have used 250 L of distilled water for a dialysis patient. How many gallons of water is that?
- **b.** A patient needs 0.024 g of a sulfa drug. There are 8-mg tablets in stock. How many tablets should be given?
- **c.** The daily dose of ampicillin for the treatment of an ear infection is 115 mg/kg of body weight. What is the daily dose for a 34-lb child?

2.56 Using conversion factors, solve each of the following clinical problems:
- **a.** The physician has ordered 1.0 g of tetracycline to be given every six hours to a patient. If your stock on hand is 500-mg tablets, how many will you need for one day's treatment?
- **b.** An intramuscular medication is given at 5.00 mg/kg of body weight. What is the dose for a 180-lb patient?
- **c.** A physician has ordered 0.50 mg of atropine, intramuscularly. If atropine were available as 0.10 mg/mL of solution, how many milliliters would you need to give?

2.7 Density

LEARNING GOAL

Calculate the density of a substance; use the density to calculate the mass or volume of a substance.

The mass and volume of any object can be measured. If we compare the mass of the object to its volume, we obtain a relationship called **density**.

$$\text{Density} = \frac{\text{mass of substance}}{\text{volume of substance}}$$

Every substance has a unique density, which distinguishes it from other substances. For example, lead has a density of 11.3 g/mL, whereas cork has a density of 0.26 g/mL. From these densities, we can predict if these substances will sink or float in water. *If an object is less dense than a liquid, the object floats when placed in the liquid.* If a substance, such as cork, is less dense than water, it will float. However, a lead object sinks because its density is greater than that of water (see Figure 2.10).

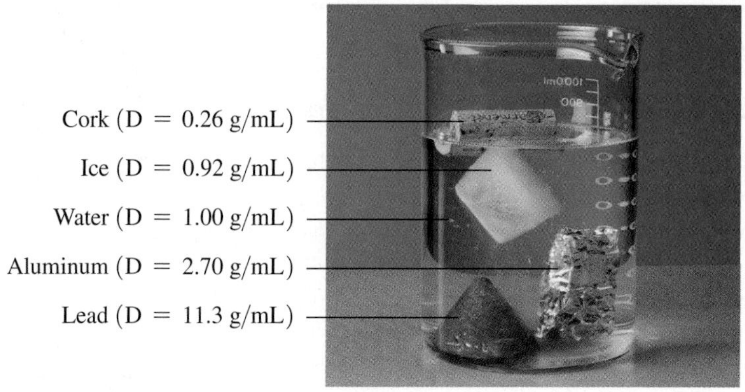

FIGURE 2.10 ▶ Objects that sink in water are more dense than water; objects that float are less dense.

Q Why does an ice cube float and a piece of aluminum sink?

Metals such as gold and lead tend to have higher densities, whereas gases have low densities. In the metric system, the densities of solids and liquids are usually expressed as grams per cubic centimeter (g/cm^3) or grams per milliliter (g/mL). The densities of gases are usually stated as grams per liter (g/L). Table 2.9 gives the densities of some common substances.

TABLE 2.9 Densities of Some Common Substances

Solids (at 25 °C)	Density (g/mL)	Liquids (at 25 °C)	Density (g/mL)	Gases (at 0 °C)	Density (g/L)
Cork	0.26	Gasoline	0.74	Hydrogen	0.090
Wood (maple)	0.75	Ethanol	0.79	Helium	0.179
Ice (at 0 °C)	0.92	Olive oil	0.92	Methane	0.714
Sugar	1.59	Water (at 4 °C)	1.00	Neon	0.902
Bone	1.80	Urine	1.003–1.030	Nitrogen	1.25
Salt (NaCl)	2.16	Plasma (blood)	1.03	Air (dry)	1.29
Aluminum	2.70	Milk	1.04	Oxygen	1.43
Diamond	3.52	Mercury	13.6	Carbon Dioxide	1.96
Iron	7.86				
Copper	8.92				
Silver	10.5				
Lead	11.3				
Gold	19.3				

Calculating Density

We can calculate the density of a substance from its mass and volume. High-density lipoprotein (HDL) is a type of cholesterol, sometimes called "good cholesterol," that is measured in a routine blood test. If we have a 0.258-g sample of HDL with a volume of 0.22 mL, we can calculate the density of the HDL as follows:

$$\text{Density} = \frac{\text{mass of HDL sample}}{\text{volume of HDL sample}} = \frac{\overset{\text{Three SFs}}{0.258\text{ g}}}{\underset{\text{Two SFs}}{0.22\text{ mL}}} = \underset{\text{Two SFs}}{1.2\text{ g/mL}}$$

Density of Solids

The volume of a solid can be determined by volume displacement. When a solid is completely submerged in water, it displaces a volume that is equal to the volume of the solid. In Figure 2.11, the water level rises from 35.5 mL to 45.0 mL after the zinc object is added. This means that 9.5 mL of water is displaced and that the volume of the object is 9.5 mL. The density of the zinc is calculated using volume displacement as follows:

$$\text{Density} = \frac{\overset{\text{Four SFs}}{68.60\text{ g zinc}}}{\underset{\text{Two SFs}}{9.5\text{ mL}}} = \underset{\text{Two SFs}}{7.2\text{ g/mL}}$$

Explore Your World

Sink or Float

1. Fill a large container or bucket with water. Place a can of diet soft drink and a can of a regular soft drink in the water. What happens? Using information on the label, how might you account for your observations?
2. Design an experiment to determine the substance that is more dense in each of the following:
   a. water and vegetable oil
   b. water and ice
   c. rubbing alcohol and ice
   d. vegetable oil, water, and ice

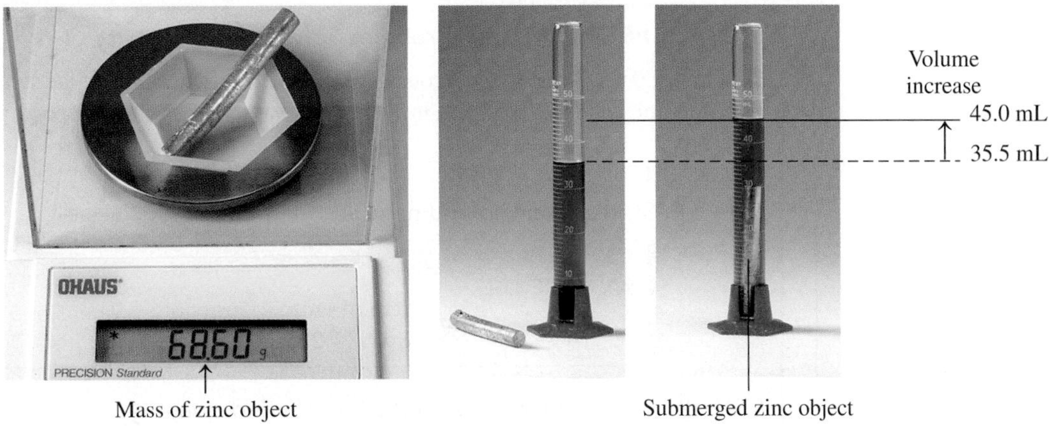

FIGURE 2.11 ▶ The density of a solid can be determined by volume displacement because a submerged object displaces a volume of water equal to its own volume.

Q How is the volume of the zinc object determined?

Chemistry Link to **Health**

Bone Density

The density of our bones determines their health and strength. Our bones are constantly gaining and losing minerals such as calcium, magnesium, and phosphate. In childhood, bones form at a faster rate than they break down. As we age, the breakdown of bone occurs more rapidly than new bone forms. As the loss of bone minerals increases, bones begin to thin, causing a decrease in mass and density. Thinner bones lack strength, which increases the risk of fracture. Hormonal changes, disease, and certain medications can also contribute to the thinning of bone. Eventually, a condition of severe thinning of bone known as *osteoporosis* may occur. *Scanning electron micrographs* (SEMs) show **(a)** normal bone and **(b)** bone with osteoporosis due to loss of bone minerals.

Bone density is often determined by passing low-dose X-rays through the narrow part at the top of the femur (hip) and the spine **(c)**. These locations are where fractures are more likely to occur, especially as we age. Bones with high density will block more of the X-rays compared to bones that are less dense. The results of a bone density test are compared to a healthy young adult as well as to other people of the same age.

Recommendations to improve bone strength include supplements of calcium and vitamin D. Weight-bearing exercise such as walking and lifting weights can also improve muscle strength, which in turn, increases bone strength.

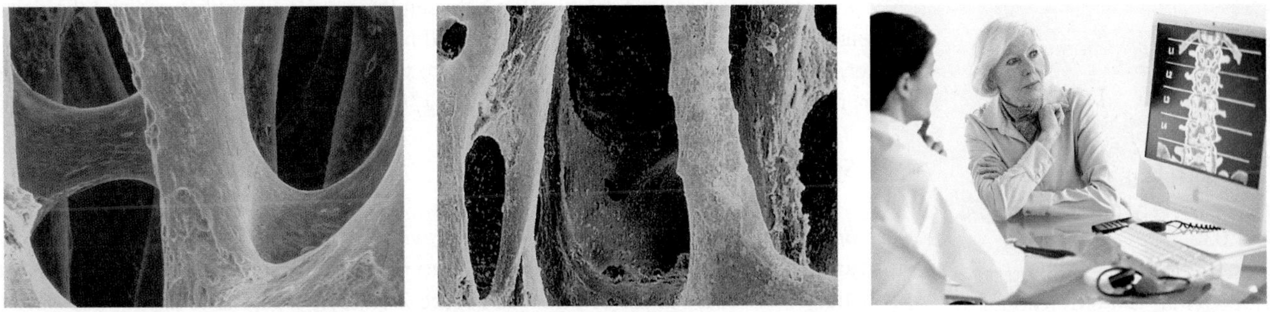

(a) Normal bone **(b)** Bone with osteoporosis **(c)** Viewing a low-dose X-ray of the spine

Problem Solving Using Density

CORE CHEMISTRY SKILL

Using Density as a Conversion Factor

Density can be used as a conversion factor. For example, if the volume and the density of a sample are known, the mass in grams of the sample can be calculated as shown in the following Sample Problem.

SAMPLE PROBLEM 2.13 Problem Solving with Density

John took 2.0 teaspoons (tsp) of cough syrup for a persistent cough. If the syrup had a density of 1.20 g/mL and there is 5.0 mL in 1 tsp, what was the mass, in grams, of the cough syrup?

SOLUTION

STEP 1 State the given and needed quantities.

ANALYZE THE PROBLEM	Given	Need
	2.0 tsp of syrup, density of syrup (1.20 g/mL)	grams of syrup

STEP 2 Write a plan to calculate the needed quantity.

teaspoons → U.S.–Metric factor → milliliters → Density factor → grams

STEP 3 Write the equalities and their conversion factors including density.

$$1\text{ tsp} = 5.0\text{ mL}\qquad \frac{5.0\text{ mL}}{1\text{ tsp}} \text{ and } \frac{1\text{ tsp}}{5.0\text{ mL}}$$

$$1\text{ mL} = 1.20\text{ g of syrup}\qquad \frac{1\text{ mL}}{1.20\text{ g syrup}} \text{ and } \frac{1.20\text{ g syrup}}{1\text{ mL}}$$

STEP 4 Set up the problem to calculate the needed quantity.

$$\underset{\text{Two SFs}}{2.0\,\cancel{\text{tsp}}} \times \overset{\text{Two SFs}}{\underset{\text{Exact}}{\frac{5.0\,\cancel{\text{mL}}}{1\,\cancel{\text{tsp}}}}} \times \overset{\text{Three SFs}}{\underset{\text{Exact}}{\frac{1.20\text{ g syrup}}{1\,\cancel{\text{mL}}}}} = \underset{\text{Two SFs}}{12\text{ g of syrup}}$$

STUDY CHECK 2.13

The density of maple syrup is 1.33 g/mL. A bottle of maple syrup contains 740 mL of syrup. What is the mass of the syrup?

Guide to Using Density

STEP 1
State the given and needed quantities.

STEP 2
Write a plan to calculate the needed quantity.

STEP 3
Write the equalities and their conversion factors including density.

STEP 4
Set up the problem to calculate the needed quantity.

Specific Gravity

Specific gravity (sp gr) is a relationship between the density of a substance and the density of water. Specific gravity is calculated by dividing the density of a sample by the density of water, which is 1.00 g/mL at 4 °C. A substance with a specific gravity of 1.00 has the same numerical value as the density of water (1.00 g/mL).

$$\text{Specific gravity} = \frac{\text{density of sample}}{\text{density of water}}$$

Specific gravity is one of the few unitless values you will encounter in chemistry. An instrument called a hydrometer is often used to measure the specific gravity of fluids such as battery fluid or a sample of urine. In Figure 2.12, a hydrometer is used to measure the specific gravity of urine.

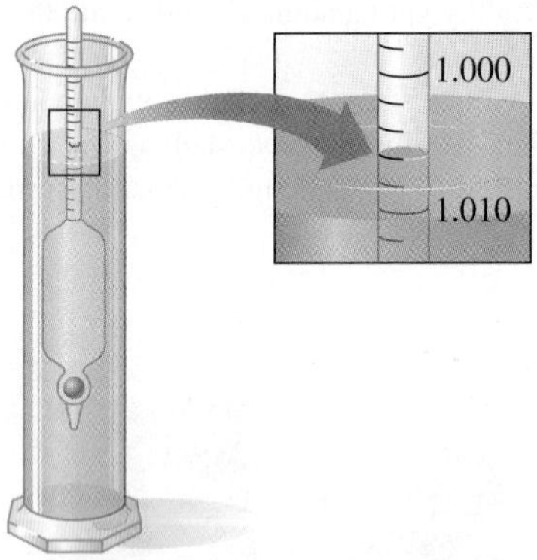

FIGURE 2.12 ▶ A hydrometer is used to measure the specific gravity of urine, which, for adults, is 1.002–1.030.

Q If the hydrometer reading is 1.006, what is the density of the urine?

QUESTIONS AND PROBLEMS

2.7 Density

LEARNING GOAL Calculate the density of a substance; use the density to calculate the mass or volume of a substance.

2.57 In an old trunk, you find a piece of metal that you think may be aluminum, silver, or lead. In lab, you find it has a mass of 217 g and a volume of 19.2 cm^3. Using Table 2.9, what is the metal you found?

2.58 Suppose you have two 100-mL graduated cylinders. In each cylinder there is 40.0 mL of water. You also have two cubes: One is lead, and the other is aluminum. Each cube measures 2.0 cm on each side. After you carefully lower each cube into the water of its own cylinder, what will the new water level be in each of the cylinders?

2.59 What is the density (g/mL) of each of the following samples?

a. A 20.0-mL sample of a salt solution that has a mass of 24.0 g.

b. A solid object with a mass of 1.65 lb and a volume of 170 mL.

c. A gem has a mass of 4.50 g. When the gem is placed in a graduated cylinder containing 20.00 mL of water, the water level rises to 21.45 mL.

d. A lightweight head on the driver of a golf club is made of titanium. If the volume of a sample of titanium is 114 cm^3 and the mass is 514.1 g, what is the density of titanium?

Lightweight heads on golf clubs are made of titanium.

2.60 Calculate the sample volume in a. and b., and the mass in c. and d. of each of the following samples:

a. What is the volume of 1.22 g of a solid with a density of 0.92 g/mL?

b. What is the volume of 0.20 kg of a solid with a density of 3.0 g/mL?

c. What is the mass, in kilograms, of 30 mL of a liquid with a density of 4.3 g/mL?

d. What is the mass, in grams, of 0.03 L of a liquid with a density of 0.83 g/mL?

2.61 Use the density value to solve the following problems:

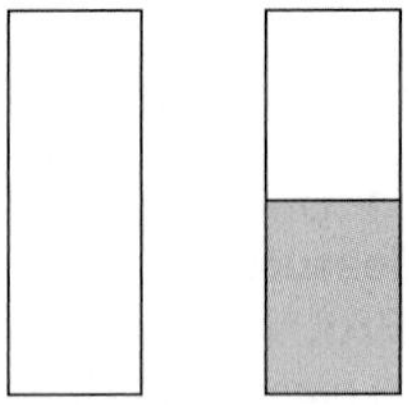

115.25 g **182.48 g**

a. What is the mass, in grams, of 150 mL of a liquid with a density of 1.4 g/mL?

b. What is the mass of a glucose solution that fills a 0.500-L intravenous bottle if the density of the glucose solution is 1.15 g/mL?

c. A sculptor has prepared a mold for casting a bronze figure. The figure has a volume of 225 mL. If bronze has a density of 7.8 g/mL, how many ounces of bronze are needed in the preparation of the bronze figure?

2.62 What is the specific gravity of a solid sample which has a density of 3.04 kg/m^3?

CONCEPT MAP

CHEMISTRY AND MEASUREMENTS

Measurements

in chemistry involve

Metric Units — for measuring — **Length (m)**, **Mass (g)**, **Volume (L)**, **Temperature (°C)**, **Time (s)**

Mass (g) and **Volume (L)** — give — **Density**

Measured Numbers — have — **Significant Figures** — that require — **Rounding Off Answers** — or — **Adding Zeros**

Prefixes — that change the size of — **Metric Units** — to give — **Equalities** — used for — **Conversion Factors** — to change units in — **Problem Solving**

CHAPTER REVIEW

2.1 Units of Measurement

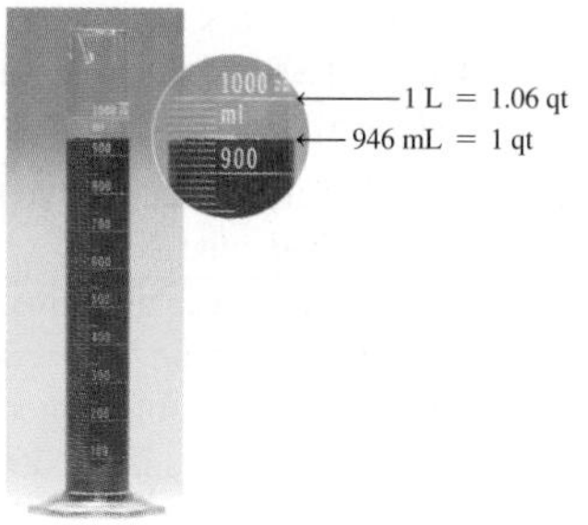

LEARNING GOAL Write the names and abbreviations for the metric or SI units used in measurements of length, volume, mass, temperature, and time.

- In science, physical quantities are described in units of the metric or International System of Units (SI).
- Some important units are meter (m) for length, liter (L) for volume, gram (g) and kilogram (kg) for mass, degree Celsius (°C) and Kelvin (K) for temperature, and second (s) for time.

2.2 Measured Numbers and Significant Figures

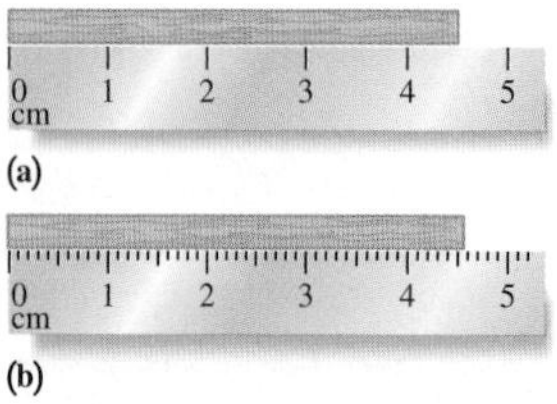

LEARNING GOAL Identify a number as measured or exact; determine the number of significant figures in a measured number.

- A measured number is any number obtained by using a measuring device.
- An exact number is obtained by counting items or from a definition; no measuring device is needed.
- Significant figures are the numbers reported in a measurement including the estimated digit.
- Zeros in front of a decimal number or at the end of a nondecimal number are not significant.

2.3 Significant Figures in Calculations

LEARNING GOAL Adjust calculated answers to give the correct number of significant figures.

- In multiplication and division, the final answer is written so that it has the same number of significant figures as the measurement with the fewest significant figures.
- In addition and subtraction, the final answer is written so that it has the same number of decimal places as the measurement with the fewest decimal places.

2.4 Prefixes and Equalities

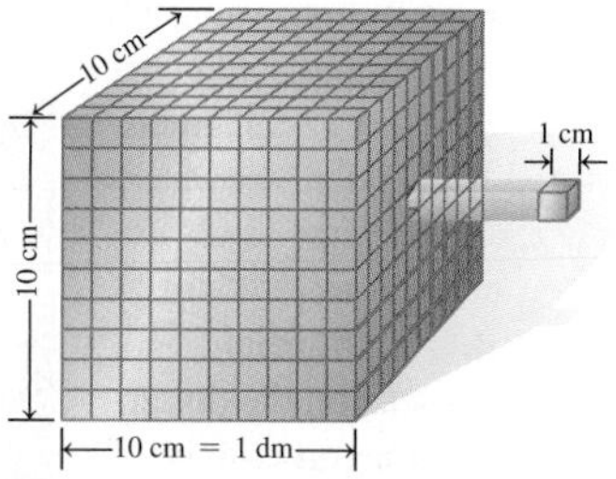

LEARNING GOAL Use the numerical values of prefixes to write a metric equality.

- A prefix placed in front of a metric or SI unit changes the size of the unit by factors of 10.
- Prefixes such as *centi*, *milli*, and *micro* provide smaller units; prefixes such as *kilo*, *mega*, and *tera* provide larger units.
- An equality shows the relationship between two units that measure the same quantity of length, volume, mass, or time.
- Examples of metric equalities are 1 m = 100 cm, 1 L = 1000 mL, 1 kg = 1000 g, and 1 min = 60 s.

2.5 Writing Conversion Factors

LEARNING GOAL Write a conversion factor for two units that describe the same quantity.

- Conversion factors are used to express a relationship in the form of a fraction.
- Two conversion factors can be written for any relationship in the metric or U.S. system.
- A percentage is written as a conversion factor by expressing matching units as the parts in 100 parts of the whole.

2.6 Problem Solving Using Unit Conversion

LEARNING GOAL Use conversion factors to change from one unit to another.

- Conversion factors are useful when changing a quantity expressed in one unit to a quantity expressed in another unit.
- In the problem-solving process, a given unit is multiplied by one or more conversion factors that cancel units until the needed answer is obtained.

2.7 Density

LEARNING GOAL Calculate the density of a substance; use the density to calculate the mass or volume of a substance.

- The density of a substance is a ratio of its mass to its volume, usually g/mL or g/cm^3.
- The units of density can be used to write conversion factors that convert between the mass and volume of a substance.

KEY TERMS

Celsius (°C) temperature scale A temperature scale on which water has a freezing point of 0 °C and a boiling point of 100 °C.

centimeter (cm) A unit of length in the metric system; there are 2.54 cm in 1 in.

conversion factor A ratio in which the numerator and denominator are quantities from an equality or given relationship. For example, the conversion factors for the relationship 1 kg = 2.20 lb are written as

$$\frac{2.20 \text{ lb}}{1 \text{ kg}} \quad \text{and} \quad \frac{1 \text{ kg}}{2.20 \text{ lb}}$$

cubic centimeter (cm^3, cc) The volume of a cube that has 1-cm sides; 1 cm^3 is equal to 1 mL.

density The relationship of the mass of an object to its volume expressed as grams per cubic centimeter (g/cm^3), grams per milliliter (g/mL), or grams per liter (g/L).

equality A relationship between two units that measure the same quantity.

exact number A number obtained by counting or by definition.

gram (g) The metric unit used in measurements of mass.

International System of Units (SI) An international system of units that modifies the metric system.

Kelvin (K) temperature scale A temperature scale on which the lowest possible temperature is 0 K.

kilogram (kg) A metric mass of 1000 g, equal to 2.20 lb. The kilogram is the SI standard unit of mass.

liter (L) The metric unit for volume that is slightly larger than a quart.

mass A measure of the quantity of material in an object.

measured number A number obtained when a quantity is determined by using a measuring device.

meter (m) The metric unit for length that is slightly longer than a yard. The meter is the SI standard unit of length.

metric system A system of measurement used by scientists and in most countries of the world.

milliliter (mL) A metric unit of volume equal to one-thousandth of a liter (0.001 L).

prefix The part of the name of a metric unit that precedes the base unit and specifies the size of the measurement. All prefixes are related on a decimal scale.

second (s) A unit of time used in both the SI and metric systems.

significant figures (SFs) The numbers recorded in a measurement.

temperature An indicator of the hotness or coldness of an object.

volume The amount of space occupied by a substance.

CORE CHEMISTRY SKILLS

The chapter section containing each Core Chemistry Skill is shown in parentheses at the end of each heading.

Counting Significant Figures (2.2)

The significant figures are all the *measured* numbers including the last, estimated digit:

- All nonzero digits
- Zeros between nonzero digits
- Zeros within a decimal number
- All digits in a coefficient of a number written in scientific notation

An *exact* number is obtained from counting or a definition and has no effect on the number of significant figures in the final answer.

Example: State the number of significant figures in each of the following:

Example		Answer
0.003 045 mm	**Answer:**	four SFs
15 000 m		two SFs
45.067 kg		five SFs
5.30×10^3 g		three SFs
2 cans of soda		exact

Rounding Off (2.3)

Calculator displays are rounded off to give the correct number of significant figures.

- If the first digit to be dropped is *4 or less*, then it and all following digits are simply dropped from the number.
- If the first digit to be dropped is *5 or greater*, then the last retained digit of the number is increased by 1.

One or more significant zeros are added when the calculator display has fewer digits than the needed number of significant figures.

Example: Round off each of the following to three significant figures:

Example		Answer
3.608 92 L	**Answer:**	3.61 L
0.003 870 298 m		0.003 87 m
6 g		6.00 g

Using Significant Figures in Calculations (2.3)

- In multiplication or division, the final answer is written so that it has the same number of significant figures as the measurement with the fewest significant figures (SFs).
- In addition or subtraction, the final answer is written so that it has the same number of decimal places as the measurement having the fewest decimal places.

Perform the following calculations using measured numbers and give answers with the correct number of SFs:

Example: $4.05 \text{ m} \times 0.6078 \text{ m}$

Answer: 2.46159 rounded off to 2.46 m^2 (three SFs)

Example: $\dfrac{4.50 \text{ g}}{3.27 \text{ mL}}$

Answer: 1.376 146 789 rounded off to 1.38 g/mL (three SFs)

Example: $0.758 \text{ g} + 3.10 \text{ g}$

Answer: 3.858 rounded off to 3.86 g (hundredths place)

Example: $13.538 \text{ km} - 8.6 \text{ km}$

Answer: 4.938 rounded off to 4.9 km (tenths place)

Using Prefixes (2.4)

- In the metric and SI systems of units, a prefix attached to any unit increases or decreases its size by some factor of 10.
- When the prefix *centi* is used with the unit meter, it becomes centimeter, a length that is one-hundredth of a meter (0.01 m).
- When the prefix *milli* is used with the unit meter, it becomes millimeter, a length that is one-thousandth of a meter (0.001 m).

Example: Complete the following statements with the correct prefix symbol:

a. 1000 m = 1 ____ m **b.** 0.01 g = 1 ____ g

Answer:

a. 1000 m = 1 km **b.** 0.01 g = 1 cg

Writing Conversion Factors from Equalities (2.5)

- A conversion factor allows you to change from one unit to another.
- Two conversion factors can be written for any equality in the metric, U.S., or metric–U.S. systems of measurement.
- Two conversion factors can be written for a relationship stated within a problem or for a percentage (%).

Example: Write two conversion factors for the equality 1 L = 1000 mL.

Answer: $\dfrac{1000 \text{ mL}}{1 \text{ L}}$ and $\dfrac{1 \text{ L}}{1000 \text{ mL}}$

Using Conversion Factors (2.6)

In problem solving, conversion factors are used to cancel the given unit and to provide the needed unit for the answer.

- State the given and needed quantities.
- Write a plan to convert the given unit to the needed unit.
- State the equalities and conversion factors.
- Set up the problem to cancel units and calculate the answer.

Example: A computer chip has a width of 0.75 in. What is that distance in millimeters?

Answer:

Equalities: 1 in. = 2.54 cm; 1 cm = 10 mm

Conversion Factors: $\frac{2.54\text{ cm}}{1\text{ in.}}$ and $\frac{1\text{ in.}}{2.54\text{ cm}}$; $\frac{10\text{ mm}}{1\text{ cm}}$ and $\frac{1\text{ cm}}{10\text{ mm}}$

Problem Setup: $0.75\ \cancel{\text{in.}} \times \frac{2.54\ \cancel{\text{cm}}}{1\ \cancel{\text{in.}}} \times \frac{10\text{ mm}}{1\ \cancel{\text{cm}}} = 19\text{ mm}$

Using Density as a Conversion Factor (2.7)

Density is an equality of mass and volume for a substance, which is written as the *density expression*.

$$\text{Density} = \frac{\text{mass of substance}}{\text{volume of substance}}$$

Density is useful as a conversion factor to convert between mass and volume.

Example: The element tungsten used in lightbulb filaments has a density of 19.3 g/cm^3. What is the volume, in cubic centimeters, of 250 g of tungsten?

Answer: *Equality*: 1 cm^3 = 19.3 g

Conversion Factors: $\frac{19.3\text{ g}}{1\text{ cm}^3}$ and $\frac{1\text{ cm}^3}{19.3\text{ g}}$

Problem Setup: $250\ \cancel{\text{g}} \times \frac{1\text{ cm}^3}{19.3\ \cancel{\text{g}}} = 13\text{ cm}^3$

UNDERSTANDING THE CONCEPTS

The chapter sections to review are shown in parentheses at the end of each question.

2.63 In which of the following pairs do both numbers contain the same number of significant figures? (2.2)

a. 2.0500 m and 0.0205 m
b. 600.0 K and 60 K
c. 0.000 75 s and 75 000 s
d. 6.240 L and 6.240×10^{-2} L

2.64 In which of the following pairs do both numbers contain the same number of significant figures? (2.2)

a. 3.44×10^{-3} g and 0.0344 g
b. 0.0098 s and 9.8×10^4 s
c. 6.8×10^3 m and 68 000 m
d. 258.000 g and 2.58×10^{-2} g

2.65 Indicate if each of the following is answered with an exact number or a measured number: (2.2)

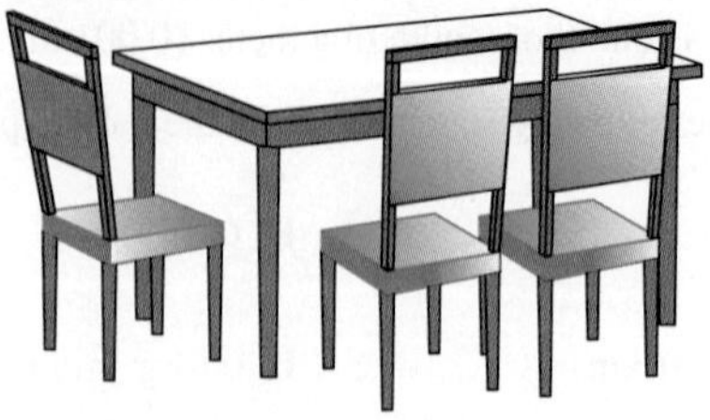

a. number of legs
b. height of table
c. number of chairs at the table
d. area of tabletop

2.66 Measure the length of each of the objects in diagrams **(a)**, **(b)**, and **(c)** using the metric ruler in the figure. Indicate the number of significant figures for each and the estimated digit for each. (2.2)

(a)

(b)

(c)

2.67 The length of this rug is 38.4 in. and the width is 24.2 in. (2.3)

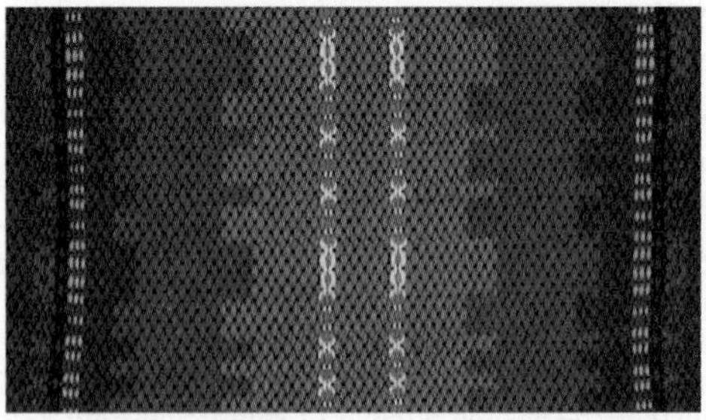

a. What is the length of this rug measured in centimeters?
b. What is the width of this rug measured in centimeters?
c. How many significant figures are in the length measurement?
d. Calculate the area of the rug, in cm^2, to the correct number of significant figures.

2.68 Each of the following diagrams represents a container of water and a cube. Some cubes float while others sink. Match diagrams **1**, **2**, **3**, or **4** with one of the following descriptions and explain your choices: (2.7)

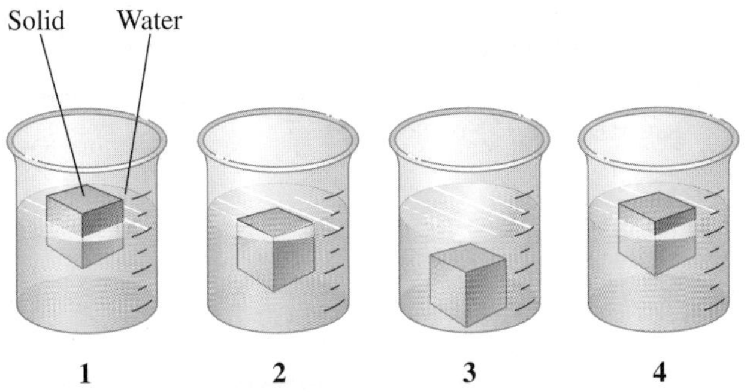

a. The cube has a greater density than water.
b. The cube has a density that is 0.60 to 0.80 g/mL.
c. The cube has a density that is one-half the density of water.
d. The cube has the same density as water.

2.69 What is the density of the solid object that is weighed and submerged in water? (2.7)

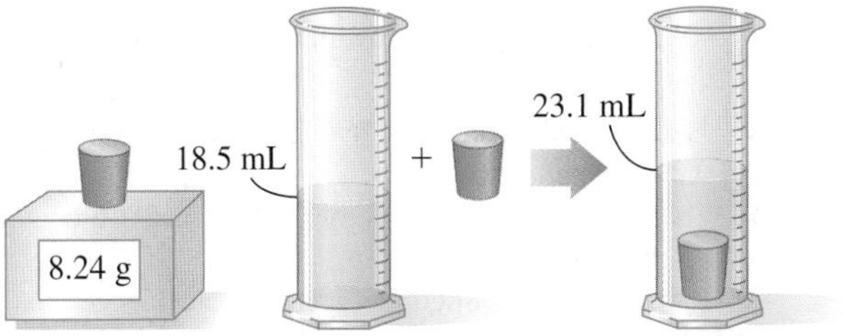

2.70 Consider the following solids. The solids **A**, **B**, and **C** represent aluminum, gold, and silver. If each has a mass of 10.0 g, what is the identity of each solid? (2.7)

Density of aluminum = 2.70 g/mL
Density of gold = 19.3 g/mL
Density of silver = 10.5 g/mL

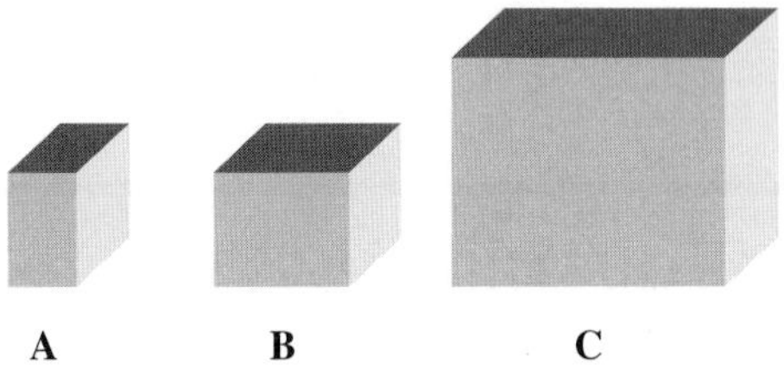

ADDITIONAL QUESTIONS AND PROBLEMS

2.71 Round off or add zeros to the following calculated answers to give a final answer with three significant figures: (2.2)
a. 0.000 012 58 L
b. 3.528×10^2 kg
c. 125 111 m

2.72 Round off or add zeros to the following calculated answers to give a final answer with three significant figures: (2.2)
a. 58.703 g
b. 3×10^{-3} s
c. 0.010 826 g

2.73 A dessert contains 137.25 g of vanilla ice cream, 84 g of fudge sauce, and 43.7 g of nuts. (2.3, 2.6)
a. What is the total mass, in grams, of the dessert?
b. What is the total weight, in pounds, of the dessert?

2.74 A fish company delivers 22 kg of salmon, 5.5 kg of crab, and 3.48 kg of oysters to your seafood restaurant. (2.3, 2.6)
a. What is the total mass, in kilograms, of the seafood?
b. What is the total number of pounds?

2.75 In France, grapes are 1.75 Euros per kilogram. What is the cost of grapes, in dollars per pound, if the exchange rate is 1.36 dollars/Euro? (2.6)

2.76 In Mexico, avocados are 48 pesos per kilogram. What is the cost, in cents, of an avocado that weighs 0.45 lb if the exchange rate is 13 pesos to the dollar? (2.6)

2.77 Bill's recipe for onion soup calls for 4.0 lb of thinly sliced onions. If an onion has an average mass of 115 g, how many onions does Bill need? (2.6)

2.78 The price of 1 lb of potatoes is \$1.75. If all the potatoes sold today at the store bring in \$1420, how many kilograms of potatoes did grocery shoppers buy? (2.6)

2.79 The following nutrition information is listed on a box of crackers: (2.6)

Serving size 0.50 oz (6 crackers)
Fat 4 g per serving Sodium 140 mg per serving

a. If the box has a net weight (contents only) of 8.0 oz, about how many crackers are in the box?
b. If you ate 10 crackers, how many ounces of fat are you consuming?
c. How many grams of sodium are used to prepare 50 boxes of crackers in part **a**?

2.80 A dialysis unit requires 75 000 mL of distilled water. How many gallons of water are needed? (1 gal = 4 qt) (2.6)

2.81 To prevent bacterial infection, a doctor orders 4 tablets of amoxicillin per day for 10 days. If each tablet contains 250 mg of amoxicillin, how many ounces of the medication are given in 10 days? (2.6)

2.82 A metal cube has a density of 3 g/mL. How can it be floated in a liquid which has a density of 2 g/mL?

2.83 What is a cholesterol level of 1.85 g/L in the standard units of mg/dL? (2.6)

2.84 An object has a mass of 3.15 oz and a volume of 0.1173 L. What is the density (g/mL) of the object? (2.7)

2.85 The density of lead is 11.3 g/mL. The water level in a graduated cylinder initially at 215 mL rises to 285 mL after a piece of lead is submerged. What is the mass, in grams, of the lead? (2.7)

2.86 A measuring cylinder contains 50 mL of oil (insoluble in water) with a density of 1.5 g/mL. What is the new level of oil in the cylinder after 40.3 mL of water is added to it?

2.87 How many cubic centimeters (cm^3) of olive oil have the same mass as 1.50 L of gasoline (see Table 2.9)? (2.7)

2.88 Ethanol has a density of 0.79 g/mL. What is the volume, in quarts, of 1.50 kg of alcohol? (2.7)

CHALLENGE QUESTIONS

The following groups of questions are related to the topics in this chapter. However, they do not all follow the chapter order, and they require you to combine concepts and skills from several sections. These questions will help you increase your critical thinking skills and prepare for your next exam.

2.89 A balance measures mass to 0.001 g. If you determine the mass of an object that weighs about 30 g, would you record the mass as 30 g, 31 g, 31.2 g, 31.16 g, or 31.075 g? Explain your choice by writing two to three complete sentences that describe your thinking. (2.3)

2.90 When three students use the same meterstick to measure the length of a paper clip, they obtain results of 5.8 cm, 5.75 cm, and 5.76 cm. If the meterstick has millimeter markings, what are some reasons for the different values? (2.3)

2.91 When a 25000 g metal block was added to a container full of water, 8.0 m^3 of water overflowed. What is the density of the metal block? (2.6)

2.92 A 50.0-g silver object and a 50.0-g gold object are both added to 75.5 mL of water contained in a graduated cylinder. What is the new water level in the cylinder? (2.7)

2.93 In the manufacturing of computer chips, cylinders of silicon are cut into thin wafers that are 3.00 in. in diameter and have a mass of 1.50 g of silicon. How thick, in millimeters, is each wafer if silicon has a density of 2.33 g/cm^3? (The volume of a cylinder is $V = \pi r^2 h$.) (2.6)

2.94 A sunscreen preparation contains 2.50% by mass benzyl salicylate. If a tube contains 4.0 ounces of sunscreen, how many kilograms of benzyl salicylate are needed to manufacture 325 tubes of sunscreen? (2.6)

2.95 Based on the list of typical test values given in Table 2.6, indicate if the following test values for substances in a blood sample are within the normal range:
a. Calcium: 2 g/L
b. Cholesterol: 0.5 mg/mL
c. Albumin: 45 g/L
d. Protein: 0.4 mg/mL

2.96 A dietary supplement contains the following components. If a person uses the dietary supplement once a day, how many milligrams of each component would that person consume in one week? (2.6)
a. calcium 0.20 g
b. iron 0.50 mg
c. iodine 53 mcg
d. chromium 45 mcg
e. threonine 0.285 g

2.97 **a.** Some athletes have as little as 3.0% body fat. If such a person has a body mass of 65 kg, how many pounds of body fat does that person have? (2.6)
b. In liposuction, a doctor removes fat deposits from a person's body. If body fat has a density of 0.94 g/mL and 3.0 L of fat is removed, how many pounds of fat were removed from the patient?

2.98 An 18-karat gold necklace is 75% gold by mass, 16% silver, and 9.0% copper. (2.6, 2.7)
a. What is the mass, in grams, of the necklace if it contains 0.24 oz of silver?
b. How many grams of copper are in the necklace?
c. If 18-karat gold has a density of 15.5 g/cm^3, what is the volume in cubic centimeters?

2.99 A graduated cylinder contains three liquids **A**, **B**, and **C**, which have different densities and do not mix: mercury (D = 13.6 g/mL), vegetable oil (D = 0.92 g/mL), and water (D = 1.00 g/mL). Identify the liquids **A**, **B**, and **C** in the cylinder. (2.7)

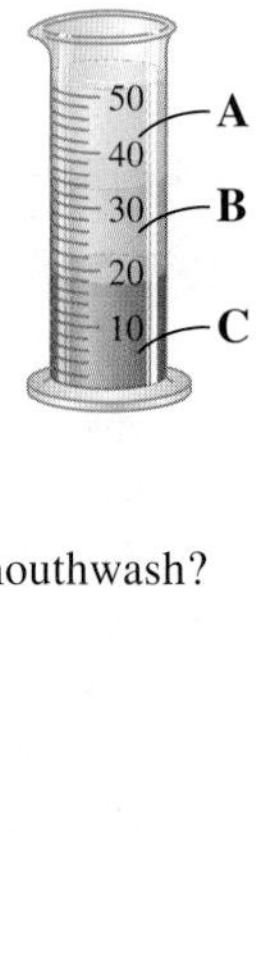

2.100 A mouthwash is 21.6% ethanol by mass. If each bottle contains 0.358 pt of mouthwash with a density of 0.876 g/mL, how many kilograms of ethanol are in 180 bottles of the mouthwash? (2.6, 2.7)

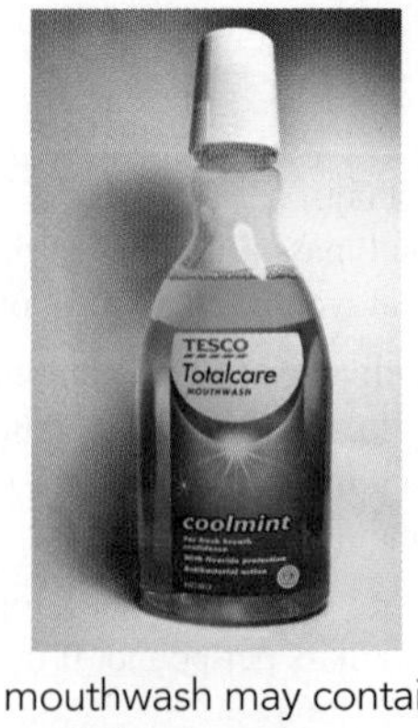

A mouthwash may contain over 20% ethanol.

ANSWERS

Answers to Study Checks

2.1 **a.** temperature **b.** time

2.2 **a.** The zeros between 4 and 8 are significant. The zeros preceding the first nonzero digit 4 are not significant.
b. All zeros in the coefficient of a number written in scientific notation are significant.

2.3 **a.** 36 m **b.** 0.0026 L **c.** 3.8×10^3 g

2.4 **a.** 0.4924 **b.** 0.0080 or 8.0×10^{-3}
c. 2.0

2.5 **a.** 83.70 mg **b.** 0.5 L

2.6 1 mm = 1000 μm (mcm)

2.7 $\frac{1000 \text{ mm}}{1 \text{ m}}$ and $\frac{1 \text{ m}}{1000 \text{ mm}}$

2.8 0.125 mg of Levsin = 1 mL of solution
$\frac{0.125 \text{ mg Levsin}}{1 \text{ mL solution}}$ and $\frac{1 \text{ mL solution}}{0.125 \text{ mg Levsin}}$
The 125 in 0.125 mg is measured: It has three SFs. The 1 in 1 mL is exact.

2.9 1.89 L

2.10 $450 \cancel{\text{mm}} \times \frac{1 \cancel{\text{cm}}}{10 \cancel{\text{mm}}} \times \frac{1 \text{ in.}}{2.54 \cancel{\text{cm}}} = 18 \text{ in.}$

2.11 6 days

2.12 25 g of fat

2.13 980 g

Answers to Selected Questions and Problems

2.1 Metric: meter, SI: meter
Metric: liter, SI: cubic meter
Metric: degree Celsius, SI: Kelvin
Metric: second (s), SI: second (s)

2.3 centimeter which is the measurement of length
milligram is the unit of mass
litre which is the unit of volume
second which is the unit of time
kelvin which is the unit of temperature

2.5 **a.** mm **b.** mg **c.** S
d. °F **e.** m^3

2.7 **a.** measured **b.** exact
c. measured **d.** exact

2.9 **a.** 6 oz of hamburger **b.** none
c. 0.75 lb, 350 g **d.** none (definitions are exact)

2.11 **a.** significant **b.** significant
c. not significant **d.** significant
e. not significant

2.13 **a.** 5 **b.** 2 **c.** 2
d. 3 **e.** 4 **f.** 3

2.15 **d.** 4 and 4

2.17 **a.** 8.5×10^3 L **b.** 3.1×10^4 g
c. 1.6×10^5 m **d.** 1.2×10^{-4} cm

2.19 A calculator often gives more digits than the number of significant figures allowed in the answer.

2.21 **a.** 310 cm **b.** 3.832×10^5 s
c. 2007 mL **d.** 0.6738265 cm
e. 28.2038

2.23 **a.** 5.08×10^3 L **b.** 3.74×10^4 g
c. 1.05×10^5 m **d.** 2.51×10^{-4} s

2.25 **a.** 1.6 **b.** 0.01
c. 27.6 **d.** 3.5

2.27 **a.** 53.54 cm **b.** 127.6 g
c. 121.5 mL **d.** 0.50 L

2.29 km/h is kilometers per hour; mi/h is miles per hour.

2.31 **a.** mg **b.** dL
c. km **d.** cg

2.33 **a.** picosecond **b.** Centimeter
c. Nanometer **d.** mililitre

2.35 **a.** 0.01 **b.** 10^{12}
c. 0.001 **d.** 0.1

2.37 **a.** decigram **b.** microgram
c. kilogram **d.** centigram

2.39 **a.** 0.1 dL **b.** 10 cc
c. 10000 L **d.** 10 cm^3

2.41 **a.** 1000 g **b.** 60 mL
c. 10 m **d.** same

2.43 A conversion factor can be inverted to give a second conversion factor.

2.45 **a.** 2.54 cm = 1 in., $\frac{2.54 \text{ cm}}{1 \text{ in.}}$ and $\frac{1 \text{ in.}}{2.54 \text{ cm}}$
b. 1000 mg = 1 g, $\frac{1000 \text{ mg}}{1 \text{ g}}$ and $\frac{1 \text{ g}}{1000 \text{ mg}}$
c. 946 mL = 1 qt, $\frac{946 \text{ mL}}{1 \text{ qt}}$ and $\frac{1 \text{ qt}}{946 \text{ mL}}$
d. 1 dL = 100 mL, $\frac{100 \text{ mL}}{1 \text{ dL}}$ and $\frac{1 \text{ dL}}{100 \text{ mL}}$
e. 7 days = 1 week, $\frac{7 \text{ days}}{1 \text{ week}}$ and $\frac{1 \text{ week}}{7 \text{ days}}$

2.47 **a.** 1 hour = 60 mins
b. 1 week = 7 days
c. 1 can of Coke = 335 mL
d. 1 bottle of vitamin C = 20 tablets

2.49 The unit in the denominator must cancel with the preceding unit in the numerator.

2.51 **a.** 1.75 m **b.** 5.5 L **c.** 1.8 g

2.53 **a.** 473 mL **b.** 65.9 kg **c.** 24 lb
d. 43 g **e.** 50. lb

2.55 **a.** 66 gal **b.** 3 tablets
c. 1800 mg

2.57 lead

2.59 **a.** 1.20 g/mL **b.** 4.4 g/mL
c. 3.10 g/mL **d.** 4.51 g/mL

2.61 **a.** 210 g **b.** 575 g **c.** 62 oz

2.63 c and **d**

2.65 **a.** exact **b.** measured
c. exact **d.** measured

2.67 **a.** length = 97.5 cm **b.** width = 61.5 cm
c. 3 SFs **d.** 6.00×10^3 cm^2

2.69 1.8 g/mL

2.71 **a.** 0.000 012 6 L **b.** 3.53×10^2 kg
c. 125 000 m

2.73 **a.** 265 g **b.** 0.584 lb

2.75 $1.08 per lb

2.77 16 onions

2.79 **a.** 96 crackers **b.** 0.2 oz of fat
c. 110 g of sodium

2.81 0.35 oz

2.83 185 mg/dL

2.85 790 g

2.87 1200 cm^3

2.89 You would record the mass as 31.075 g. Since the balance will weigh to the nearest 0.001 g, the mass value would be reported to 0.001 g.

2.91 3.1 kg/m^3

2.93 0.141 mm

2.95 **a.** abnormal **b.** abnormal
c. normal **d.** abnormal

2.97 **a.** 4.3 lb of body fat **b.** 6.2 lb

2.99 **A** is vegetable oil, **B** is water, and **C** is mercury.

3 Matter and Energy

Charles is 13 years old and overweight. His doctor is worried that Charles is at risk for type 2 diabetes and advises his mother to make an appointment with a dietitian. Daniel, a dietitian, explains to them that choosing the appropriate foods is important to living a healthy lifestyle, losing weight, and preventing or managing diabetes.

Daniel also explains that food contains potential or stored energy, and different foods contain different amounts of potential energy. For instance, carbohydrates contain 4 kcal/g while fats contain 9 kcal/g. He then explains that diets high in fat require more exercise to burn the fats, as they contain more potential energy. Daniel encourages Charles and his mother to include whole grains, fruits, and vegetables in their diet instead of foods high in fat. They also discuss food labels and the fact that smaller serving sizes of healthy foods are necessary in order to lose weight. Before leaving, Charles and his mother are given a menu for the following two weeks, as well as a diary to keep track of what, and how much, they actually consume.

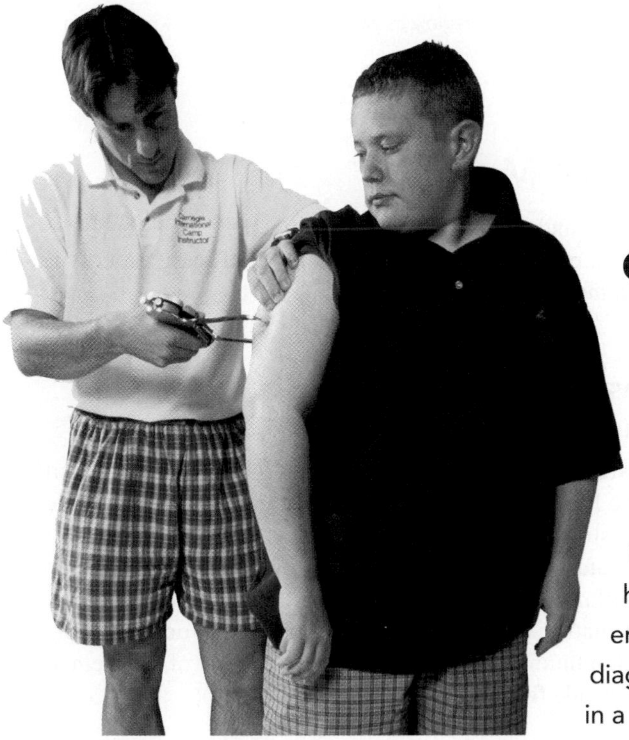

CAREER Dietitian

Dietitians specialize in helping individuals learn about good nutrition and the need for a balanced diet. This requires them to understand biochemical processes, the importance of vitamins, and food labels, as well as the differences between carbohydrates, fats, and proteins in terms of their energy value and how they are metabolized. Dietitians work in a variety of environments including hospitals, nursing homes, school cafeterias, and public health clinics. In these environments, they create specialized diets for individuals diagnosed with a specific disease, or create meal plans for those in a nursing home.

LOOKING AHEAD

Every day, we see a variety of materials with many different shapes and forms. To a scientist, all of this material is *matter*. Matter is everywhere around us: the orange juice we had for breakfast, the water we put in the coffee maker, the plastic bag we put our sandwich in, our toothbrush and toothpaste, the oxygen we inhale, and the carbon dioxide we exhale.

When we look around, we see that matter takes the physical form of a solid, a liquid, or a gas. Water is a familiar example that we routinely observe in all three states. In the solid state, water can be an ice cube or a snowflake. It is a liquid when it comes out of a faucet or fills a pool. Water forms a gas, or vapor, when it evaporates from wet clothes or boils in a pan. In these examples, water changes state by gaining or losing energy. For example, energy is added to melt ice cubes and to boil water in a teakettle. Conversely, energy is removed to freeze liquid water in an ice cube tray and to condense water vapor to liquid droplets.

Almost everything we do involves energy. We use energy when we heat water, cook food, turn on lights, walk, study, use a washing machine, or drive our cars. Of course, that energy has to come from something. In our bodies, the food we eat provides us with energy. Energy from burning fossil fuels or the Sun is used to heat a home or water for a pool.

CHAPTER READINESS*

KEY MATH SKILLS

- Using Positive and Negative Numbers in Calculations (1.4B)
- Solving Equations (1.4D)
- Interpreting a Graph (1.4E)
- Writing Numbers in Scientific Notation (1.4F)

CORE CHEMISTRY SKILLS

- Counting Significant Figures (2.2)
- Rounding Off (2.3)
- Using Significant Figures in Calculations (2.3)
- Writing Conversion Factors from Equalities (2.5)
- Using Conversion Factors (2.6)

*These Key Math Skills and Core Chemistry Skills from previous chapters are listed here for your review as you proceed to the new material in this chapter.

LEARNING GOAL

Classify examples of matter as pure substances or mixtures.

CORE CHEMISTRY SKILL

Classifying Matter

3.1 Classification of Matter

Matter is anything that has mass and occupies space. Matter makes up all the things we use such as water, wood, plates, plastic bags, clothes, and shoes. The different types of matter are classified by their composition.

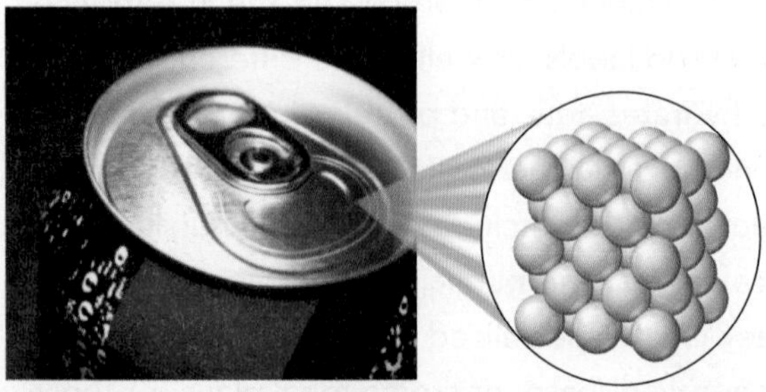

An aluminum can consists of many atoms of aluminum.

Pure Substances: Elements and Compounds

A **pure substance** is matter that has a fixed or definite composition. There are two kinds of pure substances: elements and compounds. An **element**, the simplest type of a pure substance, is composed of only one type of material such as silver, iron, or aluminum. Every element is composed of *atoms*, which are extremely tiny particles that make up each type of matter. Silver is composed of silver atoms, iron of iron atoms, and aluminum of aluminum atoms. A full list of the elements is found on the inside front cover of this text.

A **compound** is also a pure substance, but it consists of atoms of two or more elements always chemically combined in the same proportion. In compounds, the atoms are held together by attractions called *bonds*, which form small groups of atoms called *molecules*. For example, a molecule of the compound water has two hydrogen atoms for every one oxygen atom and is represented by the formula H_2O. This means that water found anywhere always has the same composition of H_2O. Another compound that consists of a chemical combination of hydrogen and oxygen is hydrogen peroxide. It has two hydrogen atoms for every two oxygen atoms, and is represented by the formula H_2O_2. Thus, water (H_2O) and hydrogen peroxide (H_2O_2) are different compounds that have different properties.

Pure substances that are compounds can be broken down by chemical processes into their elements. They cannot be broken down through physical methods such as boiling or sifting. For example, ordinary table salt consists of the compound NaCl, which can be separated by chemical processes into sodium metal and chlorine gas, as seen in Figure 3.1. Elements cannot be broken down further.

A molecule of water, H_2O, consists of two atoms of hydrogen (white) for one atom of oxygen (red).

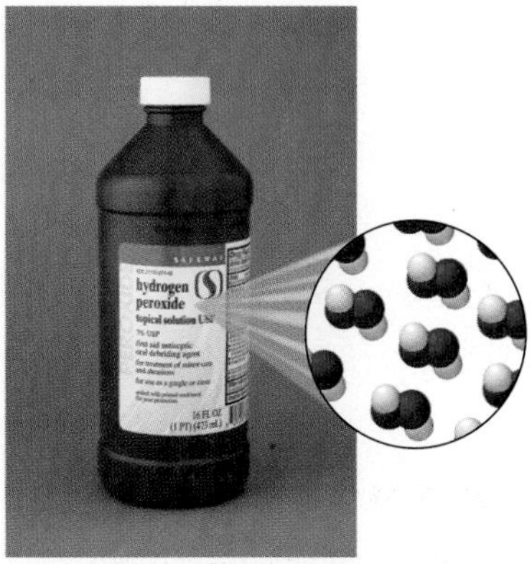

A molecule of hydrogen peroxide, H_2O_2, consists of two atoms of hydrogen (white) for every two atoms of oxygen (red).

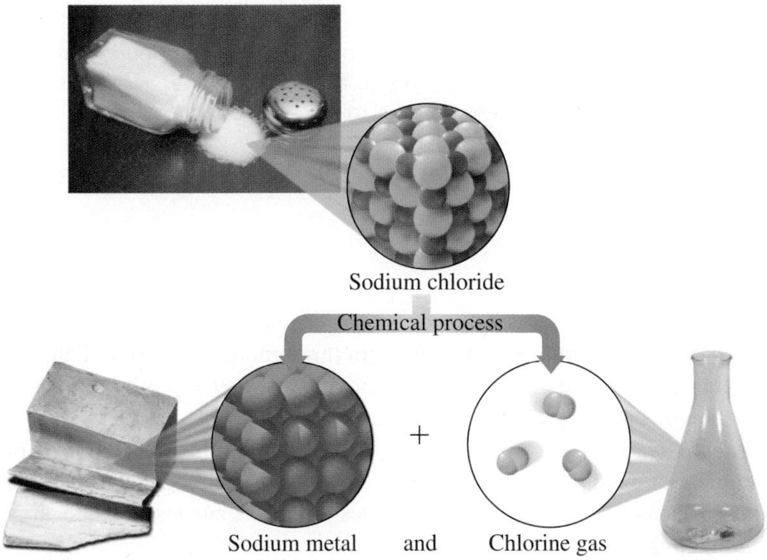

FIGURE 3.1 ▶ The decomposition of salt, NaCl, produces the elements sodium and chlorine.

Q How do elements and compounds differ?

Mixtures

In a **mixture**, two or more different substances are physically mixed, but not chemically combined. Much of the matter in our everyday lives consists of mixtures. The air we breathe is a mixture of mostly oxygen and nitrogen gases. The steel in buildings and railroad tracks is a mixture of iron, nickel, carbon, and chromium. The brass in doorknobs and fixtures is a mixture of zinc and copper. Different types of brass have different properties, depending on the ratio of copper to zinc (see Figure 3.2). Tea, coffee, and ocean water are mixtures too. Unlike compounds, the proportions of the substances in a mixture are not consistent but can vary. For example, two sugar–water mixtures may look the same, but the one with the higher ratio of sugar to water would taste sweeter.

Physical processes can be used to separate mixtures because there are no chemical interactions between the components. For example, different coins, such as nickels, dimes, and quarters, can be separated by size; iron particles mixed with sand can be picked up with a magnet; and water is separated from cooked spaghetti by using a strainer (see Figure 3.3).

FIGURE 3.3 ▶ A mixture of spaghetti and water is separated using a strainer, a physical method of separation.

Q Why can physical methods be used to separate mixtures but not compounds?

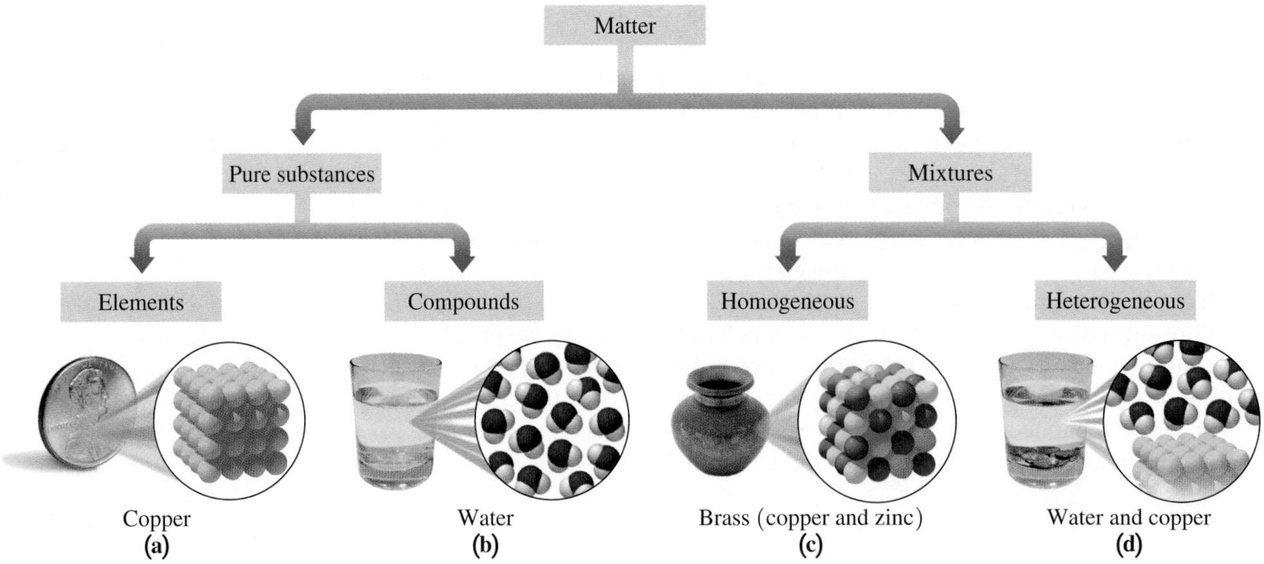

FIGURE 3.2 ▶ Matter is organized by its components: elements, compounds, and mixtures. **(a)** The element copper consists of copper atoms. **(b)** The compound water consists of H_2O molecules. **(c)** Brass is a homogeneous mixture of copper and zinc atoms. **(d)** Copper metal in water is a heterogeneous mixture of copper atoms and H_2O molecules.

Q Why are copper and water pure substances, but brass is a mixture?

Types of Mixtures

Mixtures are classified further as homogeneous or heterogeneous. In a *homogeneous mixture*, also called a *solution*, the composition is uniform throughout the sample. Familiar examples of homogeneous mixtures are air, which contains oxygen and nitrogen gases, and seawater, a solution of salt and water.

In a *heterogeneous mixture*, the components do not have a uniform composition throughout the sample. For example, a mixture of oil and water is heterogeneous because the oil floats on the surface of the water. Other examples of heterogeneous mixtures are a cookie with raisins and pulp in orange juice.

In the chemistry laboratory, mixtures are separated by various methods. Solids are separated from liquids by *filtration*, which involves pouring a mixture through a filter paper set in a funnel. In *chromatography*, different components of a liquid mixture separate as they move at different rates up the surface of a piece of chromatography paper.

(a) A mixture of a liquid and a solid is separated by filtration. **(b)** Different substances are separated as they travel at different rates up the surface of chromatography paper.

SAMPLE PROBLEM 3.1 Classifying Mixtures

Classify each of the following as a pure substance (element or compound) or a mixture (homogeneous or heterogeneous):

a. copper in copper wire
b. a chocolate-chip cookie
c. nitrox, a combination of oxygen and nitrogen used to fill scuba tanks

SOLUTION

a. Copper is an element, which is a pure substance.
b. A chocolate-chip cookie does not have a uniform composition, which makes it a heterogeneous mixture.
c. The gases oxygen and nitrogen have a uniform composition in nitrox, which makes it a homogeneous mixture.

STUDY CHECK 3.1

A salad dressing is prepared with oil, vinegar, and chunks of blue cheese. Is this a homogeneous or heterogeneous mixture?

Chemistry Link to **Health**

Breathing Mixtures

The air we breathe is composed mostly of the gases oxygen (21%) and nitrogen (79%). The homogeneous breathing mixtures used by scuba divers differ from the air we breathe depending on the depth of the dive. Nitrox is a mixture of oxygen and nitrogen, but with more oxygen gas (up to 32%) and less nitrogen gas (68%) than air. A breathing mixture with less nitrogen gas decreases the risk of *nitrogen narcosis* associated with breathing regular air while diving. Heliox contains oxygen and helium, which is typically used for diving to more than 200 ft. By replacing nitrogen with helium, nitrogen narcosis does not occur. However, at dive depths over 300 ft, helium is associated with severe shaking and body temperature drop.

A breathing mixture used for dives over 400 ft is trimix, which contains oxygen, helium, and some nitrogen. The addition of some nitrogen lessens the problem of shaking that comes with breathing high levels of helium. Heliox and trimix are used only by professional, military, or other highly trained divers.

In hospitals, heliox may be used as a treatment for respiratory disorders and lung constriction in adults and premature infants. Heliox is less dense than air, which reduces the effort of breathing, and helps distribute the oxygen gas to the tissues.

A nitrox mixture is used to fill scuba tanks.

QUESTIONS AND PROBLEMS

3.1 Classification of Matter

LEARNING GOAL Classify examples of matter as pure substances or mixtures.

3.1 Classify each of the following as a pure substance or a mixture:
a. Distilled water **b.** Dry ice
c. Fruit custard **d.** Iron
e. Trimix (hydrogen, argon, nitrogen) in a scuba tank

3.2 Classify each of the following as a pure substance or a mixture:
a. a soft drink **b.** propane (C_3H_8)
c. a cheese sandwich **d.** an iron (Fe) nail
e. salt substitute (KCl)

3.3 Classify each of the following pure substances as an element or compound:
a. Al_2O_3 (Aluminum oxide) **b.** Li (lithium)
c. NH_4OH (Ammonium hydroxide) **d.** glucose ($C_6H_{12}O_6$)
e. P (Phosphorus)

3.4 Classify each of the following pure substances as an element or a compound:
a. helium gas (He) **b.** mercury (Hg) in a thermometer
c. sugar ($C_{12}H_{22}O_{11}$) **d.** sulfur (S)
e. lye (NaOH)

3.5 Classify each of the following mixture as homogeneous or heterogeneous
a. Sugar solution **b.** Aerated drinks
c. sand in water **d.** mixed vegetable
e. mango juice with mango slices

3.6 Classify each of the following mixtures as homogeneous or heterogeneous:
a. brass **b.** chocolate-chip ice cream
c. normal saline **d.** peanut butter
e. pulp in orange juice

LEARNING GOAL

Identify the states and the physical and chemical properties of matter.

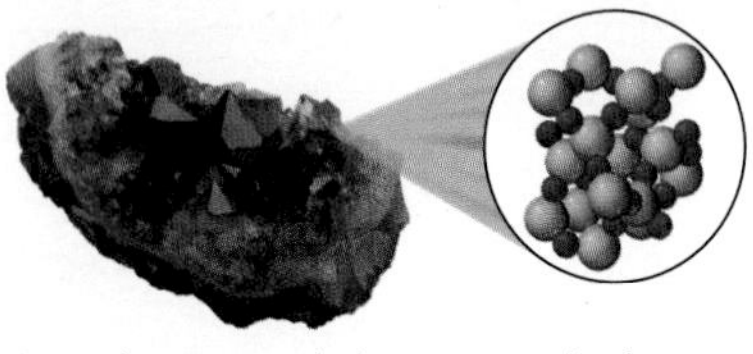

Amethyst, a solid, is a purple form of quartz (SiO_2).

3.2 States and Properties of Matter

On Earth, matter exists in one of three *physical forms* called the **states of matter**: solids, liquids, and gases. A **solid**, such as a pebble or a baseball, has a definite shape and volume. You can probably recognize several solids within your reach right now such as books, pencils, or a computer mouse. In a solid, strong attractive forces hold the particles such as atoms or molecules close together. The particles in a solid are arranged in such a rigid pattern, their only movement is to vibrate slowly in fixed positions. For many solids, this rigid structure produces a crystal such as that seen in amethyst.

A **liquid** has a definite volume, but not a definite shape. In a liquid, the particles move in random directions but are sufficiently attracted to each other to maintain a definite volume, although not a rigid structure. Thus, when water, oil, or vinegar is poured from one container to another, the liquid maintains its own volume but takes the shape of the new container.

A **gas** does not have a definite shape or volume. In a gas, the particles are far apart, have little attraction to each other, and move at high speeds, taking the shape and volume of their container. When you inflate a bicycle tire, the air, which is a gas, fills the entire volume of the tire. The propane gas in a tank fills the entire volume of the tank. Table 3.1 compares the three states of matter.

Water as a liquid takes the shape of its container.

A gas takes the shape and volume of its container.

TABLE 3.1 A Comparison of Solids, Liquids, and Gases

Characteristic	Solid	Liquid	Gas
Shape	Has a definite shape	Takes the shape of the container	Takes the shape of the container
Volume	Has a definite volume	Has a definite volume	Fills the volume of the container
Arrangement of Particles	Fixed, very close	Random, close	Random, far apart
Interaction between Particles	Very strong	Strong	Essentially none
Movement of Particles	Very slow	Moderate	Very fast
Examples	Ice, salt, iron	Water, oil, vinegar	Water vapor, helium, air

CORE CHEMISTRY SKILL

Identifying Physical and Chemical Changes

Physical Properties and Physical Changes

One way to describe matter is to observe its properties. For example, if you were asked to describe yourself, you might list your characteristics such as the color of your eyes and skin or the length, color, and texture of your hair.

Physical properties are those characteristics that can be observed or measured without affecting the identity of a substance. In chemistry, typical physical properties include the shape, color, melting point, boiling point, and physical state of a substance. For example, a penny has the physical properties of a round shape, an orange-red color, solid state, and a shiny luster. Table 3.2 gives more examples of physical properties of copper found in pennies, electrical wiring, and copper pans.

Water is a substance that is commonly found in all three states: solid, liquid, and gas. When matter undergoes a **physical change**, its state, size, or its appearance will change, but its composition remains the same. The solid state of water, snow or ice, has a different appearance than its liquid or gaseous state, but all three states are water.

The evaporation of water from seawater gives white, solid crystals of salt (sodium chloride).

Copper, used in cookware, is a good conductor of heat.

TABLE 3.2 Some Physical Properties of Copper

State at 25 °C	Solid
Color	Orange-red
Odor	Odorless
Melting Point	1083 °C
Boiling Point	2567 °C
Luster	Shiny
Conduction of Electricity	Excellent
Conduction of Heat	Excellent

The physical appearance of a substance can change in other ways too. Suppose that you dissolve some salt in water. The appearance of the salt changes, but you could re-form the salt crystals by evaporating the water. Thus in a physical change of state, no new substances are produced. Table 3.3 gives more examples of physical changes.

Chemical Properties and Chemical Changes

Chemical properties are those that describe the ability of a substance to change into a new substance. When a **chemical change** takes place, the original substance is converted into one or more new substances, which have different physical and chemical properties. For example, the rusting or corrosion of a metal such as iron is a chemical property. In the rain, an iron (Fe) nail undergoes a chemical change when it reacts with oxygen (O_2) to form rust (Fe_2O_3). A chemical change has taken place: Rust is a new substance with new physical and chemical properties. Table 3.4 gives examples of some chemical changes.

Flan has a topping of caramelized sugar.

TABLE 3.3 Examples of Some Physical Changes

Water boils to form water vapor.
Sugar dissolves in water to form a solution.
Copper is drawn into thin copper wires.
Paper is cut into tiny pieces of confetti.
Pepper is ground into flakes.

TABLE 3.4 Examples of Some Chemical Changes

Shiny, silver metal reacts in air to give a black, grainy coating.
A piece of wood burns with a bright flame, and produces heat, ashes, carbon dioxide, and water vapor.
Heating white, granular sugar forms a smooth, caramel-colored substance.
Iron, which is gray and shiny, combines with oxygen to form orange-red rust.

SAMPLE PROBLEM 3.2 Physical and Chemical Changes

Classify each of the following as a physical or chemical change:

a. A gold ingot is hammered to form gold leaf.
b. Gasoline burns in air.
c. Garlic is chopped into small pieces.

SOLUTION

a. A physical change occurs when the gold ingot changes shape.
b. A chemical change occurs when gasoline burns and forms different substances with new properties.
c. A physical change occurs when the size of an object changes.

STUDY CHECK 3.2

Classify each of the following as a physical or chemical change:

a. Water freezes on a pond.
b. Gas bubbles form when baking powder is placed in vinegar.
c. A log is chopped for firewood.

Interactive Video

Chemical vs. Physical Change

A gold ingot is hammered to form gold leaf.

QUESTIONS AND PROBLEMS

3.2 States and Properties of Matter

LEARNING GOAL Identify the states and the physical and chemical properties of matter.

3.7 Indicate whether each of the following describes a gas, a liquid, or a solid:
- **a.** The air in a scuba tank has no definite volume or shape.
- **b.** The neon atoms in a lighting display do not interact with each other.
- **c.** The particles in an ice cube are held in a rigid structure.

3.8 Indicate whether each of the following describes a gas, a liquid, or a solid:
- **a.** Lemonade has a definite volume but takes the shape of its container.
- **b.** The particles in a tank of oxygen are very far apart.
- **c.** Helium occupies the entire volume of a balloon.

3.9 Describe each of the following as physical or chemical property:
- **a.** Sodium is a soft metal.
- **b.** Sodium reacts readily with water producing flammable hydrogen gas.
- **c.** Fermentation of glucose gives alcohol.
- **d.** Aluminum can be converted into thin aluminum sheets.

3.10 Describe each of the following as a physical or chemical property:
- **a.** Neon is a colorless gas at room temperature.
- **b.** Apple slices turn brown when they are exposed to air.
- **c.** Phosphorus will ignite when exposed to air.
- **d.** At room temperature, mercury is a liquid.
- **e.** Propane gas is compressed to a liquid for placement in a small cylinder.

3.11 What type of change, physical or chemical, takes place in each of the following?
- **a.** Water vapor condenses to form rain.
- **b.** Cesium metal reacts explosively with water.
- **c.** Gold melts at 1064 °C.
- **d.** A puzzle is cut into 1000 pieces.
- **e.** Cheese is grated.

3.12 What type of change, physical or chemical, takes place in each of the following?
- **a.** NaOH(s) is added to HCl solution.
- **b.** Water is poured in ethanol.
- **c.** CO_2 is pressurized to form dry ice.
- **d.** Enzymatic hydrolysis.
- **e.** A chocolate bar melts.

3.13 What type of change, physical or chemical, takes place in each of the following?
- **a.** Bromine is dissolved in water.
- **b.** Sodium is dissolved in water.
- **c.** Sodium Chloride is dissolved in water.
- **d.** Oxygen is dissolved in water.

3.14 Describe each property of the element zirconium as physical or chemical.
- **a.** melts at 1852 °C
- **b.** is resistant to corrosion
- **c.** has a grayish white color
- **d.** ignites spontaneously in air when finely divided
- **e.** is a shiny metal

3.3 Temperature

LEARNING GOAL

Given a temperature, calculate a corresponding temperature on another scale.

CORE CHEMISTRY SKILL

Converting between Temperature Scales

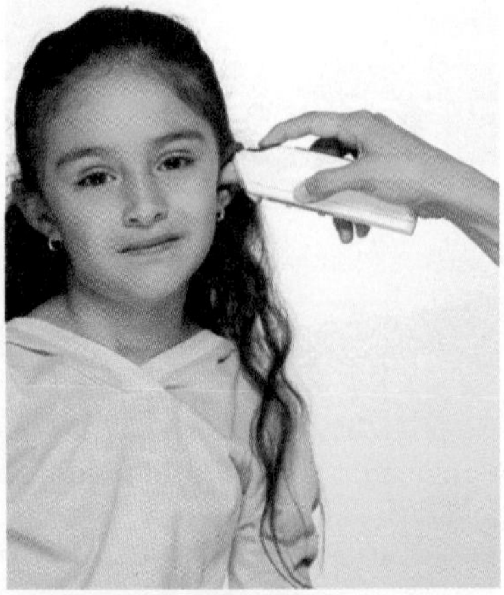

A digital ear thermometer is used to measure body temperature.

Temperatures in science, are measured and reported in *Celsius* (°C) units. On the Celsius scale, the reference points are the freezing point of water, defined as 0 °C, and the boiling point, 100 °C. In the United States, everyday temperatures are commonly reported in *Fahrenheit* (°F) units. On the Fahrenheit scale, water freezes at exactly 32 °F and boils at 212 °F. A typical room temperature of 22 °C would be the same as 72 °F. Normal human body temperature is 37.0 °C, which is the same temperature as 98.6 °C.

On the Celsius and Fahrenheit temperature scales, the temperature difference between freezing and boiling is divided into smaller units called *degrees*. On the Celsius scale, there are 100 degrees Celsius between the freezing and boiling points of water, whereas the Fahrenheit scale has 180 degrees Fahrenheit between the freezing and boiling points of water. That makes a degree Celsius almost twice the size of a degree Fahrenheit: 1 °C = 1.8 °F (see Figure 3.4).

$$180 \text{ Fahrenheit degrees} = 100 \text{ degrees Celsius}$$

$$\frac{180 \text{ Fahrenheit degrees}}{100 \text{ degrees Celsius}} = \frac{1.8\ ^\circ\text{F}}{1\ ^\circ\text{C}}$$

We can write a temperature equation that relates a Fahrenheit temperature and its corresponding Celsius temperature.

$$T_F = 1.8(T_C) + 32 \quad \text{or} \quad T_F = 1.8(T_C) + 32$$

Changes °C to °F — Adjusts freezing point

Temperature equation to obtain degrees Fahrenheit

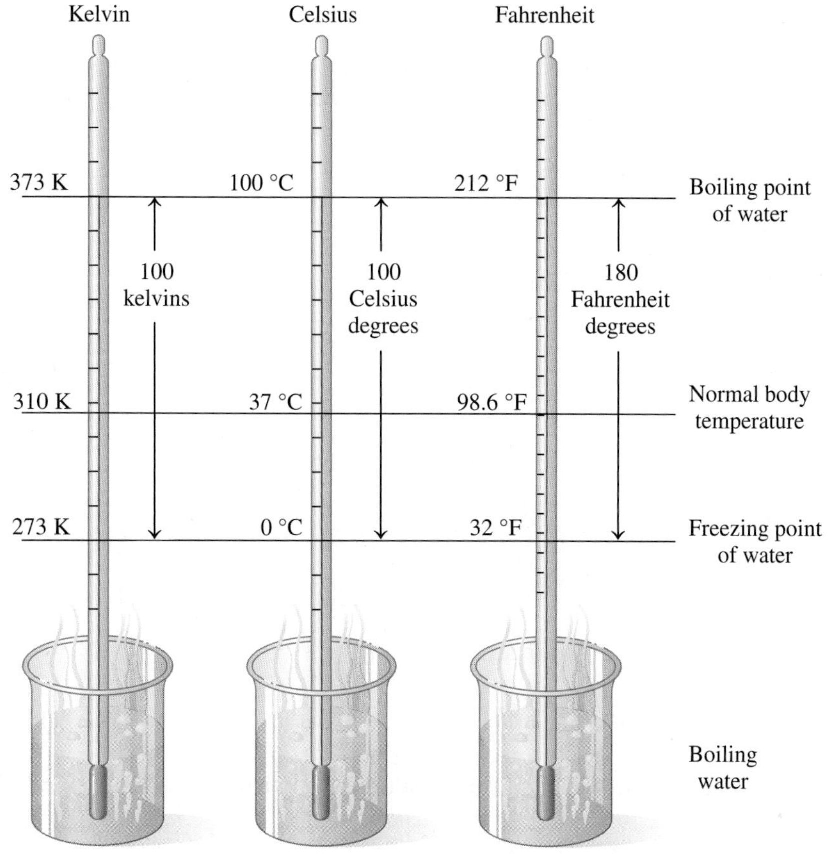

FIGURE 3.4 ▶ A comparison of the Fahrenheit, Celsius, and Kelvin temperature scales between the freezing and boiling points of water.

Q What is the difference in the freezing points of water on the Celsius and Fahrenheit temperature scales?

In this equation, the Celsius temperature is multiplied by 1.8 to change °C to °F; then 32 is added to adjust the freezing point from 0 °C, the Fahrenheit freezing point, to 32 °F. The values, 1.8 and 32, used in the temperature equation are exact numbers and are not used to determine significant figures in the answer.

To convert from Fahrenheit to Celsius, the temperature equation is rearranged to solve for T_C. First, we subtract 32 from both sides since we must apply the same operation to both sides of the equation.

$$T_F - 32 = 1.8(T_C) + \cancel{32} - \cancel{32}$$
$$T_F - 32 = 1.8(T_C)$$

Second, we solve the equation for T_C by dividing both sides by 1.8.

$$\frac{T_F - 32}{1.8} = \frac{\cancel{1.8}(T_C)}{\cancel{1.8}}$$

$$\frac{T_F - 32}{1.8} = T_C \quad \text{Temperature equation to obtain degrees Celsius}$$

Scientists have learned that the coldest temperature possible is −273 °C (more precisely, −273.15 °C). On the *Kelvin* scale, this temperature, called *absolute zero*, has the value of 0 K. Units on the Kelvin scale are called kelvins (K); *no degree symbol is used.* Because there are no lower temperatures, the Kelvin scale has no negative temperature values. Between the freezing point of water, 273 K, and the boiling point, 373 K, there are 100 kelvins, which makes a kelvin equal in size to a Celsius degree.

$$1\ \text{K} = 1\ °\text{C}$$

We can write an equation that relates a Celsius temperature to its corresponding Kelvin temperature by adding 273 to the Celsius temperature. Table 3.5 gives a comparison of some temperatures on the three scales.

TABLE 3.5 A Comparison of Temperatures

Example	Fahrenheit (°F)	Celsius (°C)	Kelvin (K)
Sun	9937	5503	5776
A hot oven	450	232	505
A desert	120	49	322
A high fever	104	40	313
Room temperature	70	21	294
Water freezes	32	0	273
A northern winter	−66	−54	219
Nitrogen liquefies	−346	−210	63
Helium boils	−452	−269	4
Absolute zero	−459	−273	0

$T_K = T_C + 273$ Temperature equation to obtain kelvins

Suppose that a dermatologist uses liquid nitrogen at −196 °C to remove skin lesions and some skin cancer. We can calculate the temperature of the liquid nitrogen in kelvins by adding 273 to the temperature in degrees Celsius.

$T_K = -196 + 273 = 77\ K$

Guide to Calculating Temperature

STEP 1
State the given and needed quantities.

STEP 2
Write a temperature equation.

STEP 3
Substitute in the known values and calculate the new temperature.

SAMPLE PROBLEM 3.3 Converting from Celsius to Fahrenheit

The typical temperature in a hospital room is set at 21 °C. What is that temperature in degrees Fahrenheit?

SOLUTION

STEP 1 State the given and needed quantities.

ANALYZE THE PROBLEM	Given	Need
	21 °C	*T* in degrees Fahrenheit

STEP 2 Write a temperature equation.

$T_F = 1.8(T_C) + 32$

STEP 3 Substitute in the known values and calculate the new temperature.

$T_F = 1.8(21) + 32$ 1.8 is exact; 32 is exact

Two SFs

$T_F = 38 + 32$

$= 70.\ °F$ Answer to the ones place

In the equation, *the values of 1.8 and 32 are exact numbers*, which do not affect the number of SFs used in the answer.

STUDY CHECK 3.3

In the process of making ice cream, rock salt is added to crushed ice to chill the ice cream mixture. If the temperature drops to −11 °C, what is it in degrees Fahrenheit?

SAMPLE PROBLEM 3.4 Converting from Fahrenheit to Celsius

In a type of cancer treatment called *thermotherapy*, temperatures as high as 113 °F are used to destroy cancer cells. What is that temperature in degrees Celsius?

SOLUTION

STEP 1 State the given and needed quantities.

ANALYZE THE PROBLEM	Given	Need
	113 °F	*T* in degrees Celsius

STEP 2 Write a temperature equation.

$$T_C = \frac{T_F - 32}{1.8}$$

STEP 3 Substitute in the known values and calculate the new temperature.

$$T_C = \frac{(113 - 32)}{1.8}$$ 32 is exact; 1.8 is exact

$$= \frac{81}{1.8} = 45\ ^\circ\text{C}$$ Answer to the ones place

STUDY CHECK 3.4

A child has a temperature of 103.6 °F. What is this temperature on a Celsius thermometer?

Chemistry Link to **Health**

Variation in Body Temperature

Normal body temperature is considered to be 37.0 °C, although it varies throughout the day and from person to person. Oral temperatures of 36.1 °C are common in the morning and climb to a high of 37.2 °C between 6 P.M. and 10 P.M. Temperatures above 37.2 °C for a person at rest are usually an indication of illness. Individuals who are involved in prolonged exercise may also experience elevated temperatures. Body temperatures of marathon runners can range from 39 °C to 41 °C as heat production during exercise exceeds the body's ability to lose heat.

Changes of more than 3.5 °C from the normal body temperature begin to interfere with bodily functions. Body temperatures above 41 °C, *hyperthermia*, can lead to convulsions, particularly in children, which may cause permanent brain damage. Heatstroke occurs above 41.1 °C. Sweat production stops, and the skin becomes hot and dry. The pulse rate is elevated, and respiration becomes weak and rapid. The person can become lethargic and lapse into a coma. Damage to internal organs is a major concern, and treatment, which must be immediate, may include immersing the person in an ice-water bath.

At the low temperature extreme of *hypothermia*, body temperature can drop as low as 28.5 °C. The person may appear cold and pale and have an irregular heartbeat. Unconsciousness can occur if the body temperature drops below 26.7 °C. Respiration becomes slow and shallow, and oxygenation of the tissues decreases. Treatment involves providing oxygen and increasing blood volume with glucose and saline fluids. Injecting warm fluids (37.0 °C) into the peritoneal cavity may restore the internal temperature.

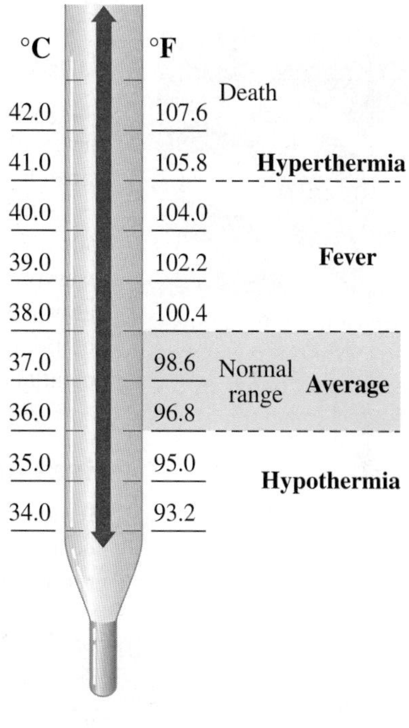

QUESTIONS AND PROBLEMS

3.3 Temperature

LEARNING GOAL Given a temperature, calculate a corresponding temperature on another scale.

3.15 Solve the following temperature conversions:
a. 50 °F = _______ °C
b. 283 K = _______ °C
c. 3000 K = _______ °C

3.16 Your partner is preparing the apparatus for an experiment in the lab. The reaction needs to be carried out at 273 K, but you notice that your partner is searching for an oil-heating bath. What would you advise him to do, and what would be the right reaction medium?

3.17 Solve the following temperature conversions:
a. 37.0 °C = _______ °F
b. 65.3 °F = _______ °C
c. −27 °C = _______ K
d. 62 °C = _______ K
e. 114 °F = _______ °C

3.18 Solve the following temperature conversions:
a. 25 °C = _______ °F
b. 155 °C = _______ °F
c. −25 °F = _______ °C
d. 224 K = _______ °C
e. 145 °C = _______ K

3.19 **a.** A patient with hyperthermia has a temperature of 106 °F. What does this read on a Celsius thermometer?
b. Because high fevers can cause convulsions in children, the doctor wants to be called if the child's temperature goes over 40.0 °C. Should the doctor be called if a child has a temperature of 103 °F?

3.20 **a.** Hot compresses for a patient are prepared with water heated to 145 °F. What is the temperature of the hot water in degrees Celsius?
b. During extreme hypothermia, a boy's temperature dropped to 20.6 °C. What was his temperature in degrees Fahrenheit?

LEARNING GOAL

Identify energy as potential or kinetic; convert between units of energy.

3.4 Energy

When you are running, walking, dancing, or thinking, you are using energy to do *work*, any activity that requires energy. In fact, **energy** is defined as the ability to do work. Suppose you are climbing a steep hill and you become too tired to go on. At that moment, you do not have the energy to do any more work. Now suppose you sit down and have lunch. In a while, you will have obtained some energy from the food, and you will be able to do more work and complete the climb.

Water at the top of the dam stores potential energy. When the water flows over the dam, potential energy is converted to kinetic energy.

Kinetic and Potential Energy

Energy can be classified as kinetic energy or potential energy. **Kinetic energy** is the energy of motion. Any object that is moving has kinetic energy. **Potential energy** is determined by the position of an object or by the chemical composition of a substance. A boulder resting on top of a mountain has potential energy because of its location. If the boulder rolls down the mountain, the potential energy becomes kinetic energy. Water stored in a reservoir has potential energy. When the water goes over the dam and falls to the stream below, its potential energy is converted to kinetic energy. Foods and fossil fuels have potential energy in their molecules. When you digest food or burn gasoline in your car, potential energy is converted to kinetic energy to do work.

Heat and Energy

Heat is the energy associated with the motion of particles. An ice cube feels cold because heat flows from your hand into the ice cube. The faster the particles move, the greater the heat or thermal energy of the substance. In the ice cube, the particles are moving very slowly. As heat is added, the motions of the particles in the ice cube increase. Eventually, the particles have enough energy to make the ice cube melt as it changes from a solid to a liquid.

TABLE 3.6 A Comparison of Energy for Various Resources

Energy in Joules	
10^{27}	Energy radiated by Sun in 1 s (10^{26})
10^{24}	World reserves of fossil fuel (10^{23})
10^{21}	Energy consumption for 1 yr in the United States (10^{20})
10^{18}	Solar energy reaching the Earth in 1 s (10^{17})
10^{15}	
10^{12}	Energy use per person in 1 yr in the United States (10^{11})
10^{9}	Energy from 1 gal of gasoline (10^{8})
10^{6}	Energy from one serving of pasta, a doughnut, or needed to bicycle for 1 h (10^{6})
10^{3}	Energy used to sleep for 1 h (10^{5})
10^{0}	

Units of Energy

The SI unit of energy and work is the **joule (J)** (pronounced "jewel"). The joule is a small amount of energy, so scientists often use the kilojoule (kJ), 1000 joules. To heat water for one cup of tea, you need about 75 000 J or 75 kJ of heat. Table 3.6 shows a comparison of energy in joules for several energy sources.

You may be more familiar with the unit **calorie (cal)**, from the Latin *caloric*, meaning "heat." The calorie was originally defined as the amount of energy (heat) needed to raise the temperature of 1 g of water by 1 °C. Now one calorie is defined as exactly 4.184 J. This equality can be written as two conversion factors:

$$1 \text{ cal} = 4.184 \text{ J (exact)} \qquad \frac{4.184 \text{ J}}{1 \text{ cal}} \text{ and } \frac{1 \text{ cal}}{4.184 \text{ J}}$$

CORE CHEMISTRY SKILL

Using Energy Units

One *kilocalorie* (kcal) is equal to 1000 calories, and one *kilojoule* (kJ) is equal to 1000 joules. The equalities and conversion factors follow:

$$1 \text{ kcal} = 1000 \text{ cal} \qquad \frac{1 \text{ kcal}}{1000 \text{ cal}} \text{ and } \frac{1000 \text{ cal}}{1 \text{ kcal}}$$

$$1 \text{ kJ} = 1000 \text{ J} \qquad \frac{1 \text{ kJ}}{1000 \text{ J}} \text{ and } \frac{1000 \text{ J}}{1 \text{ kJ}}$$

SAMPLE PROBLEM 3.5 Energy Units

A defibrillator gives a high-energy-shock output of 360 J. What is this quantity of energy in calories?

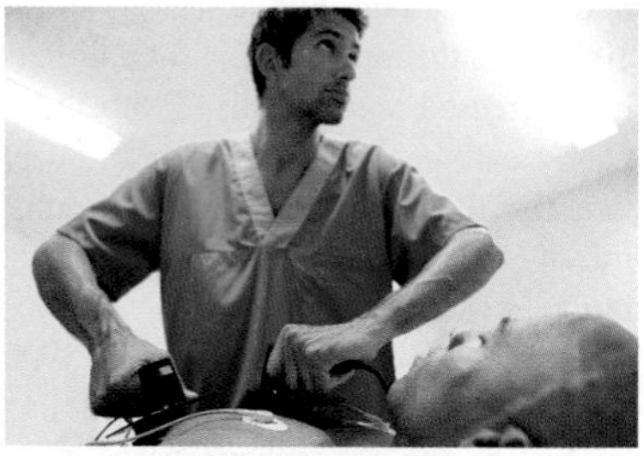

A defibrillator provides electrical energy to heart muscle to re-establish normal rhythm.

SOLUTION

STEP 1 State the given and needed quantities.

ANALYZE THE PROBLEM	Given	Need
	360 J	calories

STEP 2 Write a plan to convert the given unit to the needed unit.

joules → Energy factor → calories

STEP 3 State the equalities and conversion factors.

$$1 \text{ cal} = 4.184 \text{ J}$$

$$\frac{1 \text{ cal}}{4.184 \text{ J}} \text{ and } \frac{4.184 \text{ J}}{1 \text{ cal}}$$

STEP 4 Set up the problem to calculate the needed quantity.

$$\underset{\text{Two SFs}}{360 \, \cancel{\text{J}}} \times \underset{\text{Exact}}{\overset{\text{Exact}}{\frac{1 \text{ cal}}{4.184 \, \cancel{\text{J}}}}} = \underset{\text{Two SFs}}{86 \text{ cal}}$$

STUDY CHECK 3.5

The burning of 1.0 g of coal produces 8.4 kcal. How many kilojoules are produced?

QUESTIONS AND PROBLEMS

3.4 Energy

LEARNING GOAL Identify energy as potential or kinetic; convert between units of energy.

3.21 Discuss the changes in the potential and kinetic energy of a roller-coaster ride as the roller-coaster car climbs to the top and goes down the other side.

3.22 Calculate the SI units for the following:
- **a.** 500 kcal
- **b.** 40 cal

3.23 Indicate whether each of the following statements describes potential or kinetic energy.
- **a.** When book is placed on the top of the table.
- **b.** A car on top of the hill.
- **c.** Collision of balls while playing billiard.
- **d.** Gasoline stored in the fuel tank.

3.24 Indicate whether each of the following statements describes potential or kinetic energy:
- **a.** the energy in your food
- **b.** a tightly wound spring
- **c.** an earthquake
- **d.** a car speeding down the freeway

3.25 The energy needed to keep a 75-watt lightbulb burning for 1.0 h is 270 kJ. Calculate the energy required to keep the lightbulb burning for 3.0 h in each of the following energy units:
- **a.** joules
- **b.** kilocalories

3.26 An athlete uses 870 kcal for a long jump. Calculate the energy used for the jump in each of the following energy units:
- **a.** Joules
- **b.** Kilojoules

Chemistry Link to the Environment

Carbon Dioxide and Climate Change

The Earth's climate is a product of interactions between sunlight, the atmosphere, and the oceans. The Sun provides us with energy in the form of solar radiation. Some of this radiation is reflected back into space. The rest is absorbed by the clouds, atmospheric gases including carbon dioxide, and the Earth's surface. For millions of years, concentrations of carbon dioxide (CO_2) have fluctuated. However in the last 100 years, the amount of carbon dioxide (CO_2) gas in our atmosphere has increased significantly. From the years 1000 to 1800, the atmospheric carbon dioxide averaged 280 ppm. The concentration of ppm indicates the parts per million by volume, which for gases is the same as mL of CO_2 per kL of air. But since the beginning of the Industrial Revolution in 1800 up until 2005, the level of atmospheric carbon dioxide has risen from about 280 ppm to about 390 ppm in 2011, a 40% increase.

As the atmospheric CO_2 level increases, more solar radiation is trapped by the atmospheric gases, which raises the temperature at the surface of the Earth. Some scientists have estimated that if the carbon dioxide level doubles from its level before the Industrial Revolution, the average global temperature could increase by 2.0 to 4.4 °C. Although this seems to be a small temperature change, it could have dramatic impact worldwide. Even now, glaciers and snow cover in much of the world have diminished. Ice sheets in Antarctica and Greenland are melting faster and breaking apart. Although no one knows for sure how rapidly the ice in the polar regions is melting, this accelerating change will contribute to a rise in sea level. In the twentieth century, the sea level rose 15 to 23 cm. Some scientists predict the sea level will rise 1 m in this century. Such an increase will have a major impact on coastal areas.

Until recently, the carbon dioxide level was maintained as algae in the oceans and the trees in the forests utilized the carbon dioxide. However, the ability of these and other forms of plant life to absorb carbon dioxide is not keeping up with the increase in carbon dioxide. Most scientists agree that the primary source of the increase of carbon dioxide is the burning of fossil fuels such as gasoline, coal, and natural gas. The cutting and burning of trees in the rain forests (deforestation) also reduces the amount of carbon dioxide removed from the atmosphere.

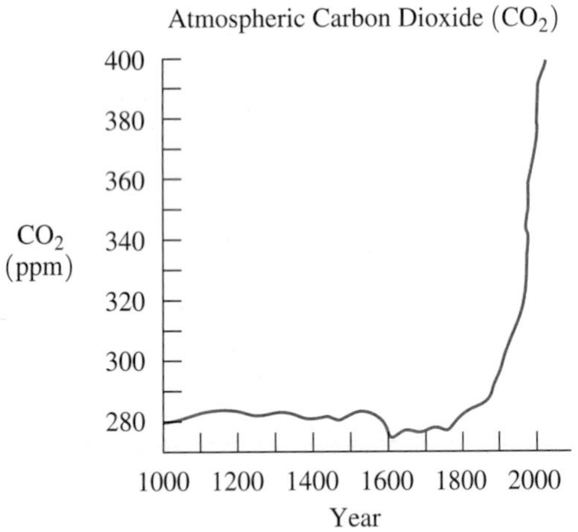

Atmospheric carbon dioxide levels are shown for the years from 1000 C.E. to 2010 C.E.

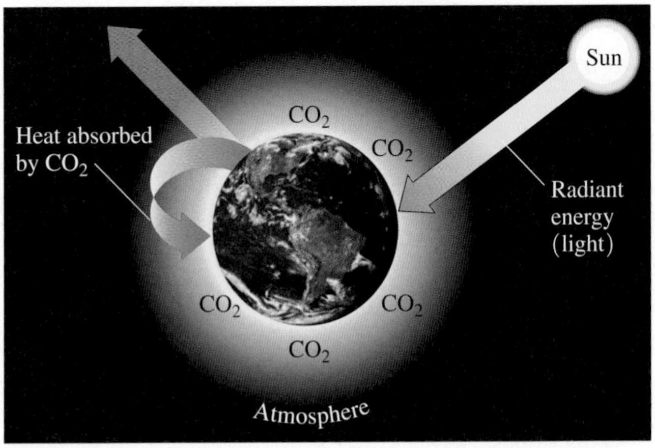

Heat from the Sun is trapped by the CO_2 layer in the atmosphere.

Worldwide efforts are being made to reduce the carbon dioxide produced by burning fossil fuels that heat our homes, run our cars, and provide energy for industries. Scientists are exploring ways to provide alternative energy sources and to reduce the effects of deforestation. Meanwhile, we can reduce energy use in our homes by using appliances that are more energy efficient such as replacing incandescent light bulbs with fluorescent lights. Such an effort worldwide will reduce the possible impact of climate change and at the same time save our fuel resources.

LEARNING GOAL

Use the energy values to calculate the kilocalories (kcal) or kilojoules (kJ) for a food.

3.5 Energy and Nutrition

The food we eat provides energy to do work in the body, which includes the growth and repair of cells. Carbohydrates are the primary fuel for the body, but if the carbohydrate reserves are exhausted, fats and then proteins are used for energy.

For many years in the field of nutrition, the energy from food was measured as Calories or kilocalories. The nutritional unit *Calorie*, *Cal* (with an uppercase C), is the same as 1000 cal, or 1 kcal. The international unit, kilojoule (kJ), is becoming more prevalent. For

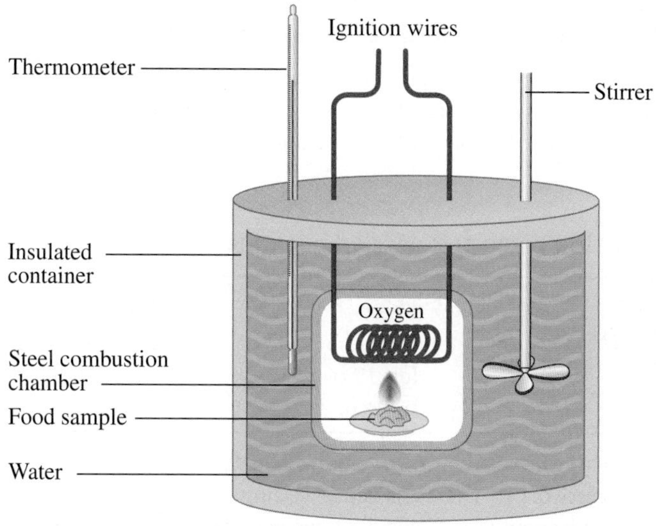

FIGURE 3.5 ▶ Heat released from burning a food sample in a calorimeter is used to determine the energy value for the food.

Q What happens to the temperature of water in a calorimeter during the combustion of a food sample?

example, a baked potato has an energy value of 100 Calories, which is 100 kcal or 440 kJ. A typical diet that provides 2100 Cal (kcal) is the same as an 8800 kJ diet.

$$1 \text{ Cal} = 1 \text{ kcal} = 1000 \text{ cal}$$
$$1 \text{ Cal} = 4.184 \text{ kJ} = 4184 \text{ J}$$

In the nutrition laboratory, foods are burned in a *calorimeter* to determine their *energy value* (kJ/g or kcal/g) (see Figure 3.5). A sample of food is placed in a steel container called a calorimeter filled with oxygen with a measured amount of water that fills the surrounding chamber. The food sample is ignited, releasing heat that increases the temperature of the water. From the mass of the food and water as well as the temperature increase, the energy value for the food is calculated. We will assume that the energy absorbed by the calorimeter is negligible.

TABLE 3.7 Typical Energy Values for the Three Food Types

Food Type	kcal/g	kJ/g
Carbohydrate	4	17
Fat	9	38
Protein	4	17

Energy Values for Foods

The **energy values** for food are the kilocalories or kilojoules obtained from burning 1 g of carbohydrate, fat, or protein, which are listed in Table 3.7.

Using these energy values, we can calculate the total energy for a food if the mass of each food type is known.

$$\text{kilocalories} = \cancel{g} \times \frac{\text{kcal}}{\cancel{g}} \qquad \text{kilojoules} = \cancel{g} \times \frac{\text{kJ}}{\cancel{g}}$$

On packaged food, the energy content is listed in the Nutrition Facts label, usually in terms of the number of Calories or kilojoules for one serving. The general composition and energy content for some foods are given in Table 3.8.

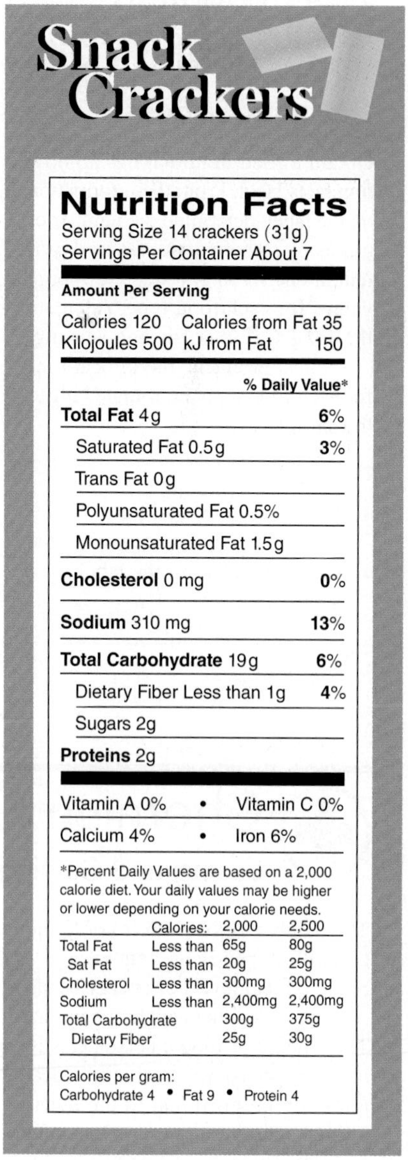

The nutrition facts include the total Calories and kilojoules, and the grams of carbohydrate, fat, and protein per serving.

TABLE 3.8 General Composition and Energy Content for Some Foods

Food	Carbohydrate (g)	Fat (g)	Protein (g)	Energy*
Banana, 1 medium	26	0	1	110 kcal (460 kJ)
Beef, ground, 3 oz	0	14	22	220 kcal (910 kJ)
Carrots, raw, 1 cup	11	0	1	50 kcal (200 kJ)
Chicken, no skin, 3 oz	0	3	20	110 kcal (460 kJ)
Egg, 1 large	0	6	6	80 kcal (330 kJ)
Milk, 4% fat, 1 cup	12	9	9	170 kcal (700 kJ)
Milk, nonfat, 1 cup	12	0	9	90 kcal (360 kJ)
Potato, baked	23	0	3	100 kcal (440 kJ)
Salmon, 3 oz	0	5	16	110 kcal (460 kJ)
Steak, 3 oz	0	27	19	320 kcal (1350 kJ)

*Energy values are rounded off to the tens place.

Guide to Calculating the Energy from a Food

STEP 1
State the given and needed quantities.

STEP 2
Use the energy value for each food type and calculate the kJ or kcal rounded off to the tens place.

STEP 3
Add the energy for each food type to give the total energy from the food.

Explore Your World

Counting Calories

Obtain a food item that has a nutrition facts label. From the information on the label, determine the number of grams of carbohydrate, fat, and protein in one serving. Using the energy values for each food type, calculate the total Calories for one serving. (For most products, the kilocalories for each food type are rounded off to the tens place.)

Question

1. How does your total for the Calories in one serving compare to the Calories stated on the label for a single serving?

SAMPLE PROBLEM 3.6 Energy Content for a Food

At a fast-food restaurant, a hamburger contains 37 g of carbohydrate, 19 g of fat, and 23 g of protein. What is the energy from each food type and the total energy, in kilocalories, for the hamburger? Round off the kilocalories for each food type to the tens place.

SOLUTION

STEP 1 **State the given and needed quantities.**

	Given	Need
ANALYZE THE PROBLEM	carbohydrate, 37 g fat, 19 g protein, 23 g	kilocalories for each food type, total number of kilocalories

STEP 2 **Use the energy value for each food type and calculate the kJ or kcal rounded off to the tens place.** Using the energy values for carbohydrate, fat, and protein (see Table 3.7), we can calculate the energy for each type of food.

Food Type	Mass		Energy Value		Energy
Carbohydrate	37 $\cancel{g}$	$\times$	$\frac{4\ \text{kcal}}{1\ \cancel{g}}$	=	150 kcal
Fat	19 $\cancel{g}$	$\times$	$\frac{9\ \text{kcal}}{1\ \cancel{g}}$	=	170 kcal
Protein	23 $\cancel{g}$	$\times$	$\frac{4\ \text{kcal}}{1\ \cancel{g}}$	=	90 kcal

STEP 3 **Add the energy for each food type to give the total energy from the food.**

Total energy = 150 kcal + 170 kcal + 90 kcal = 410 kcal

STUDY CHECK 3.6

Suppose your breakfast consisted of a 1-oz (28 g) serving of oat-bran hot cereal with half a cup of whole milk. The breakfast contains 22 g of carbohydrate, 7 g of fat, and 10 g of protein.

a. What is the energy, in kilojoules, from each food type? Round off the kilojoules for each food type to the tens place.

b. What is the total energy, in kilojoules, from your breakfast?

Chemistry Link to Health

Losing and Gaining Weight

The number of kilocalories or kilojoules needed in the daily diet of an adult depends on gender, age, and level of physical activity. Some typical levels of energy needs are given in Table 3.9.

A person gains weight when food intake exceeds energy output. The amount of food a person eats is regulated by the hunger center in the hypothalamus, which is located in the brain. Food intake is normally proportional to the nutrient stores in the body. If these nutrient stores are low, you feel hungry; if they are high, you do not feel like eating.

TABLE 3.9 Typical Energy Requirements for Adults

Gender	Age	Moderately Active kcal (kJ)	Highly Active kcal (kJ)
Female	19–30	2100 (8800)	2400 (10 000)
	31–50	2000 (8400)	2200 (9200)
Male	19–30	2700 (11 300)	3000 (12 600)
	31–50	2500 (10 500)	2900 (12 100)

A person loses weight when food intake is less than energy output. Many diet products contain cellulose, which has no nutritive value but provides bulk and makes you feel full. Some diet drugs depress the hunger center and must be used with caution, because they excite the nervous system and can elevate blood pressure. Because muscular exercise is an important way to expend energy, an increase in daily exercise aids weight loss. Table 3.10 lists some activities and the amount of energy they require.

One hour of swimming uses 2100 kJ of energy.

TABLE 3.10 Energy Expended by a 70.0-kg (154-lb) Adult

Activity	Energy (kcal/h)	Energy (kJ/h)
Sleeping	60	250
Sitting	100	420
Walking	200	840
Swimming	500	2100
Running	750	3100

QUESTIONS AND PROBLEMS

3.5 Energy and Nutrition

LEARNING GOAL Use the energy values to calculate the kilocalories (kcal) or kilojoules (kJ) for a food.

3.27 Calculate the kilocalories for each of the following:
- **a.** One potato that produces 140 kJ when burned in a calorimeter
- **b.** A sample of vegetable oil that produces 530 kJ when burned in a calorimeter

3.28 Calculate the kilocalories for each of the following:
- **a.** one cup of popcorn that produces 131 kJ when burned in a calorimeter
- **b.** a sample of butter that produces 23.4 kJ when burned in a calorimeter

3.29 Using the energy values for foods (see Table 3.7), determine each of the following (round off the answers in kilocalories or kilojoules to the tens place):
- **a.** the total kilojoules for one cup of orange juice that contains 26 g of carbohydrate, no fat, and 2 g of protein
- **b.** the grams of carbohydrate in one apple if the apple has no fat and no protein and provides 72 kcal of energy
- **c.** the kilocalories in one tablespoon of vegetable oil, which contains 14 g of fat and no carbohydrate or protein
- **d.** the grams of fat in one avocado that has 405 kcal, 13 g of carbohydrate, and 5 g of protein

3.30 Using the energy values for foods (see Table 3.7), determine each of the following (round off the answers in kilocalories or kilojoules to the tens place):
- **a.** the total kilojoules in two tablespoons of crunchy peanut butter that contains 6 g of carbohydrate, 16 g of fat, and 7 g of protein
- **b.** the grams of protein in one cup of soup that has 110 kcal with 9 g of carbohydrate and 6 g of fat
- **c.** the kilocalories in one can of cola if it has 40. g of carbohydrate and no fat or protein
- **d.** the total kilocalories for a diet that consists of 68 g of carbohydrate, 9 g of fat, and 150 g of protein

3.31 One cup of a health drink contains 20 g of carbohydrate, 15 g of fat and 12 g of protein. How much energy in kilocalories and kilojoules is in the health drink? (Round off the kilocalories or kilojoules to the tens place.)

3.32 A high-protein diet contains 70.0 g of carbohydrate, 5.0 g of fat, and 150 g of protein. How much energy, in kilocalories and kilojoules, does this diet provide? (Round off the kilocalories and kilojoules to the tens place.)

3.6 Specific Heat

LEARNING GOAL

Use specific heat to calculate heat loss or gain.

Every substance has its own characteristic ability to absorb heat. When you bake a potato, you place it in a hot oven. If you are cooking pasta, you add the pasta to boiling water. You already know that adding heat to water increases its temperature until it boils. Certain substances must absorb more heat than others to reach a certain temperature.

The energy requirements for different substances are described in terms of a physical property called *specific heat*. The **specific heat (*SH*)** for a substance is defined as the amount of heat needed to raise the temperature of exactly 1 g of a substance by exactly 1 °C. This temperature change is written as ΔT (*delta T*), where the delta symbol means "change in."

$$\text{Specific heat } (SH) = \frac{\text{heat}}{\text{grams} \times \Delta T} = \frac{\text{cal (or J)}}{\text{g °C}}$$

The specific heat for water is written using our definition of the calorie and joule.

$$SH \text{ for } H_2O(l) = \frac{1.00 \text{ cal}}{\text{g °C}} = \frac{4.184 \text{ J}}{\text{g °C}}$$

TABLE 3.11 Specific Heats for Some Substances

Substance	cal/g °C	J/g °C
Elements		
Aluminum, $Al(s)$	0.214	0.897
Copper, $Cu(s)$	0.0920	0.385
Gold, $Au(s)$	0.0308	0.129
Iron, $Fe(s)$	0.108	0.452
Silver, $Ag(s)$	0.0562	0.235
Titanium, $Ti(s)$	0.125	0.523
Compounds		
Ammonia, $NH_3(g)$	0.488	2.04
Ethanol, $C_2H_5OH(l)$	0.588	2.46
Sodium chloride, $NaCl(s)$	0.207	0.864
Water, $H_2O(l)$	1.00	4.184
Water, $H_2O(s)$	0.485	2.03

If we look at Table 3.11, we see that 1 g of water requires 1.00 cal or 4.184 J to increase its temperature by 1 °C. Water has a large specific heat that is about five times the specific heat of aluminum. Aluminum has a specific heat that is about twice that of copper. However, adding the same amount of heat (1.00 cal or 4.184 J) will raise the temperature of 1 g of aluminum by about 5 °C and 1 g of copper by about 10 °C. The low specific heats of aluminum and copper mean they transfer heat efficiently, which makes them useful in cookware.

The high specific heat of water has a major impact on the temperatures in a coastal city compared to an inland city. A large mass of water near a coastal city can absorb or release five times the energy absorbed or released by the same mass of rock near an inland city. This means that in the summer a body of water absorbs large quantities of heat, which cools a coastal city, and then in the winter that same body of water releases large quantities of heat, which provides warmer temperatures. A similar effect happens with our bodies, which contain 70% by mass water. Water in the body absorbs or releases large quantities of heat in order to maintain an almost constant body temperature.

The high specific heat of water keeps temperatures more moderate in summer and winter.

Calculations Using Specific Heat

When we know the specific heat of a substance, we can calculate the heat lost or gained by measuring the mass of the substance and the initial and final temperature. We can substitute these measurements into the specific heat expression that is rearranged to solve for heat, which we call the *heat equation.*

$$\text{Heat} = \text{mass} \times \text{temperature change} \times \text{specific heat}$$

$$\text{Heat} = m \times \Delta T \times SH$$

$$\text{cal} = \cancel{\text{g}} \times \cancel{°\text{C}} \times \frac{\text{cal}}{\cancel{\text{g}}\ \cancel{°\text{C}}}$$

$$\text{J} = \cancel{\text{g}} \times \cancel{°\text{C}} \times \frac{\text{J}}{\cancel{\text{g}}\ \cancel{°\text{C}}}$$

CORE CHEMISTRY SKILL

Using the Heat Equation

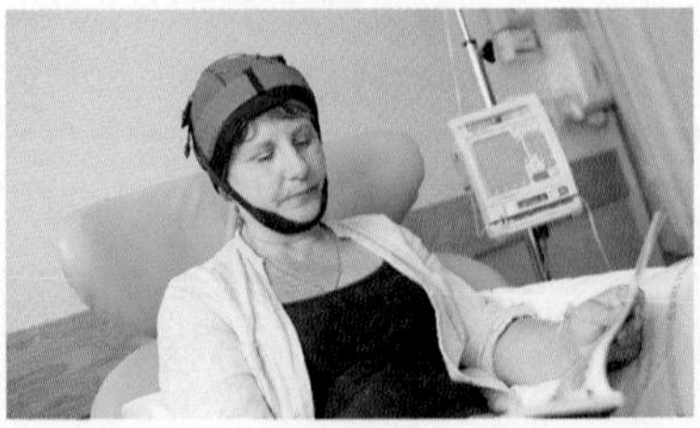

A cooling cap lowers the body temperature to reduce the oxygen required by the tissues.

SAMPLE PROBLEM 3.7 Calculating Heat Loss

During cardiac arrest, surgery, and strokes, body temperature is lowered, which reduces the amount of oxygen needed by the body. Some methods used to lower body temperature include cooled saline solution, cool water blankets, or cooling caps worn on the head. How many kilojoules are lost when the body temperature of a surgery patient with a blood volume of 5500 mL is cooled from 38.5 °C to 33.2 °C? (Assume that the specific heat and density of blood is the same as for water, 1.0 g/mL.)

SOLUTION

STEP 1 State the given and needed quantities.

ANALYZE THE PROBLEM	Given	Need
	5500 mL of blood = 5500 g of blood	kilojoules removed

STEP 2 Calculate the temperature change (ΔT).

$\Delta T = 38.5\ °C - 33.2\ °C = 5.3\ °C$

STEP 3 Write the heat equation.

$\text{Heat} = m \times \Delta T \times SH$

STEP 4 Substitute in the given values and calculate the heat, making sure units cancel.

$$\text{Heat} = \underset{\text{Two SFs}}{5500\ \cancel{g}} \times \underset{\text{Two SFs}}{5.3\ \cancel{°C}} \times \overset{\text{Exact}}{\frac{4.184\ \cancel{J}}{\cancel{g}\ \cancel{°C}}} \times \overset{\text{Exact}}{\frac{1\ \text{kJ}}{\underset{\text{Exact}}{1000\ \cancel{J}}}} = \underset{\text{Two SFs}}{120\ \text{kJ}}$$

STUDY CHECK 3.7

How many kilocalories are lost when 560 g of water cools from 67 °C to 22 °C?

Guide to Calculations Using Specific Heat

STEP 1 State the given and needed quantities.

STEP 2 Calculate the temperature change (ΔT).

STEP 3 Write the heat equation.

STEP 4 Substitute in the given values and calculate the heat, making sure units cancel.

QUESTIONS AND PROBLEMS

3.6 Specific Heat

LEARNING GOAL Use specific heat to calculate heat loss or gain.

3.33 If the same amount of heat is supplied to samples of 10.0 g each of aluminum, iron, and copper all at 15.0 °C, which sample would reach the highest temperature (see Table 3.11)?

3.34 Substances **A** and **B** are the same mass and at the same initial temperature. When they are heated, the final temperature of **A** is 55 °C higher than the temperature of **B**. What does this tell you about the specific heats of **A** and **B**?

3.35 Use the heat equation to calculate the energy for each of the following (See Table 3.11):
a. Calories of heat 10.4 g of water from 25 °C to 45 °C
b. Joules of heat 68 g of water from 12 °C to 37 °C
c. Kilocalories of heat 136 g of water from 27 °C to 81 °C
d. Kilojoules of heat 185 g of copper from 30 °C to 190 °C

3.36 Use the heat equation to calculate the energy for each of the following (see Table 3.11):
a. calories lost when 85 g of water cools from 45 °C to 25 °C
b. joules lost when 25 g of water cools from 86 °C to 61 °C
c. kilocalories to heat 5.0 kg of water from 22 °C to 28 °C
d. kilojoules to heat 224 g of gold from 18 °C to 185 °C

3.37 Use the heat equation to calculate the energy, in joules and calories, for each of the following (see Table 3.11):
a. to heat 25.0 g of water from 12.5 °C to 25.7 °C
b. to heat 38.0 g of copper from 122 °C to 246 °C
c. lost when 15.0 g of ethanol, C_2H_5OH, cools from 60.5 °C to −42.0 °C
d. lost when 125 g of iron cools from 118 °C to 55 °C

3.38 Use the heat equation to calculate the energy, in joules and calories, for each of the following (see Table 3.11):
a. to heat 5.25 g of water from 5.5 °C to 64.8 °C
b. lost when 75.0 g of water cools from 86.4 °C to 2.1 °C
c. to heat 10.0 g of silver from 112 °C to 275 °C
d. lost when 18.0 g of gold cools from 224 °C to 118 °C

3.7 Changes of State

LEARNING GOAL

Describe the changes of state between solids, liquids, and gases; calculate the energy involved.

Matter undergoes a **change of state** when it is converted from one state to another state (see Figure 3.6).

Melting and Freezing

When heat is added to a solid, the particles move faster. At a temperature called the **melting point (mp)**, the particles of a solid gain sufficient energy to overcome the attractive

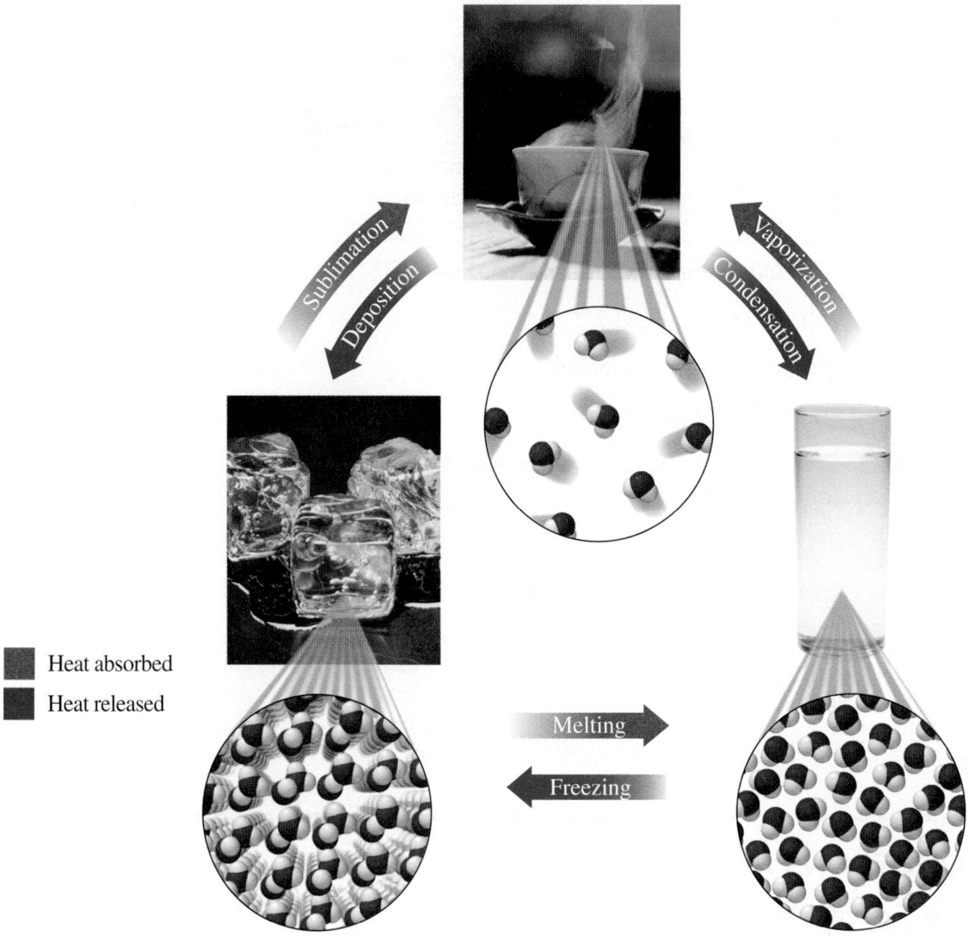

FIGURE 3.6 ▶ Changes of state include melting and freezing, vaporization and condensation, sublimation and deposition.

Q Is heat added or released when liquid water freezes?

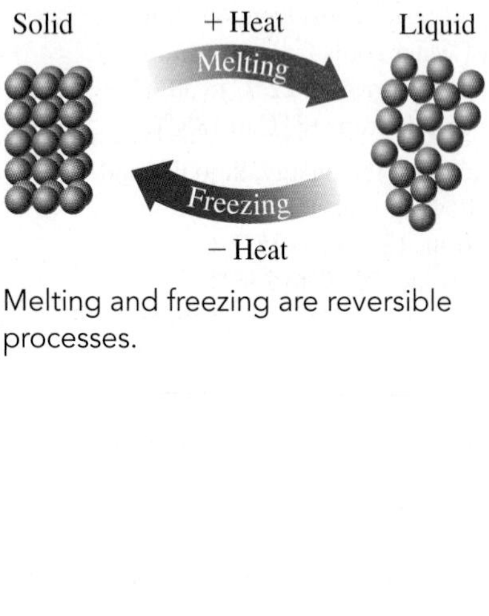

Melting and freezing are reversible processes.

forces that hold them together. The particles in the solid separate and move about in random patterns. The substance is **melting**, changing from a solid to a liquid.

If the temperature of a liquid is lowered, the reverse process takes place. Kinetic energy is lost, the particles slow down, and attractive forces pull the particles close together. The substance is **freezing**. A liquid changes to a solid at the **freezing point (fp)**, which is the same temperature as its melting point. Every substance has its own freezing (melting) point: Solid water (ice) melts at 0 °C when heat is added, and freezes at 0 °C when heat is removed. Gold melts at 1064 °C when heat is added, and freezes at 1064 °C when heat is removed.

During a change of state, the temperature of a substance remains constant. Suppose we have a glass containing ice and water. The ice melts when heat is added at 0 °C, forming more liquid. When heat is removed at 0 °C, the liquid water freezes, forming more solid. The process of melting requires heat; the process of freezing releases heat. Melting and freezing are reversible at 0 °C.

Heat of Fusion

During melting, the **heat of fusion** is the energy that must be added to convert exactly 1 g of solid to liquid at the melting point. For example, 80. cal (334 J) of heat is needed to melt exactly 1 g of ice at its melting point (0 °C).

$$H_2O(s) + 80.\ \text{cal/g (or 334 J/g)} \longrightarrow H_2O(l)$$

Heat of Fusion for Water

$$\frac{80.\ \text{cal}}{1\ \text{g}\ H_2O} \quad \text{and} \quad \frac{1\ \text{g}\ H_2O}{80.\ \text{cal}} \qquad \frac{334\ \text{J}}{1\ \text{g}\ H_2O} \quad \text{and} \quad \frac{1\ \text{g}\ H_2O}{334\ \text{J}}$$

The heat of fusion (80. cal/g or 334 J/g) is also the quantity of heat that must be removed to freeze exactly 1 g of water at its freezing point (0 °C).

Water is sometimes sprayed in fruit orchards during subfreezing weather. If the air temperature drops to 0 °C, the water begins to freeze. Because heat is released as the water molecules form solid ice, the air warms above 0 °C and protects the fruit.

$$H_2O(l) \longrightarrow H_2O(s) + 80.\ \text{cal/g (or 334 J/g)}$$

The heat of fusion can be used as a conversion factor in calculations. For example, to determine the heat needed to melt a sample of ice, the mass of ice, in grams, is multiplied by the heat of fusion. Because the temperature remains constant as long as the ice is melting, there is no temperature change given in the calculation, as shown in Sample Problem 3.8.

Calculating Heat to Melt (or Freeze) Water

Heat = mass × heat of fusion

$$\text{cal} = \cancel{g} \times \frac{80.\ \text{cal}}{\cancel{g}} \qquad \text{J} = \cancel{g} \times \frac{334\ \text{J}}{\cancel{g}}$$

SAMPLE PROBLEM 3.8 Heat of Fusion

Ice cubes at 0 °C with a mass of 26.0 g are added to your soft drink.

a. How much heat (joules) will be absorbed to melt all the ice at 0 °C?
b. What happens to the temperature of your soft drink? Why?

SOLUTION

a.

STEP 1 State the given and needed quantities.

ANALYZE THE PROBLEM	Given	Need
	26.0 g of ice at 0 °C, heat of fusion = 334 J/g	joules to melt ice at 0 °C

STEP 2 Write a plan to convert the given quantity to the needed quantity.

grams of ice → Heat of fusion → joules

STEP 3 Write the heat conversion factor and any metric factor.

$$1\ \text{g of}\ H_2O(s \rightarrow l) = 334\ \text{J}$$

$$\frac{334\ \text{J}}{1\ \text{g}\ H_2O} \quad \text{and} \quad \frac{1\ \text{g}\ H_2O}{334\ \text{J}}$$

STEP 4 Set up the problem and calculate the needed quantity.

$$26.0\ \cancel{\text{g}\ H_2O} \times \frac{334\ \text{J}}{1\ \cancel{\text{g}\ H_2O}} = 8680\ \text{J}$$

Three SFs (26.0 g), Exact (1 g), Three SFs (334 J), Three SFs (8680 J)

b. The temperature of the soft drink will decrease as heat is removed from the soft drink to melt the ice.

STUDY CHECK 3.8

In a freezer, 150 g of water at 0 °C is placed in an ice cube tray. How much heat, in kilocalories, must be removed to form ice cubes at 0 °C?

Guide to Calculations Using a Heat Conversion Factor

STEP 1 State the given and needed quantities.

STEP 2 Write a plan to convert the given quantity to the needed quantity.

STEP 3 Write the heat conversion factor and any metric factor.

STEP 4 Set up the problem and calculate the needed quantity.

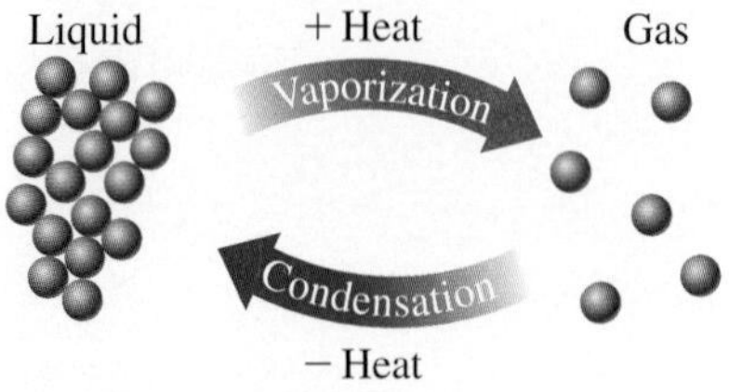

Vaporization and condensation are reversible processes.

Evaporation, Boiling, and Condensation

Water in a mud puddle disappears, unwrapped food dries out, and clothes hung on a line dry. **Evaporation** is taking place as water molecules with sufficient energy escape from the liquid surface and enter the gas phase (see Figure 3.7a). The loss of the "hot" water molecules removes heat, which cools the remaining liquid water. As heat is added, more and more water molecules gain sufficient energy to evaporate. At the **boiling point (bp)**, the molecules within a liquid have enough energy to overcome their attractive forces and become a gas. We observe the **boiling** of a liquid such as water as gas bubbles form throughout the liquid, rise to the surface, and escape (see Figure 3.7b).

When heat is removed, a reverse process takes place. In **condensation**, water vapor is converted back to liquid as the water molecules lose kinetic energy and slow down. Condensation occurs at the same temperature as boiling but differs because heat is removed. You may have noticed that condensation occurs when you take a hot shower and the water vapor forms water droplets on a mirror. Because a substance loses heat as it condenses, its surroundings become warmer. That is why, when a rainstorm is approaching, you may notice a warming of the air as gaseous water molecules condense to rain.

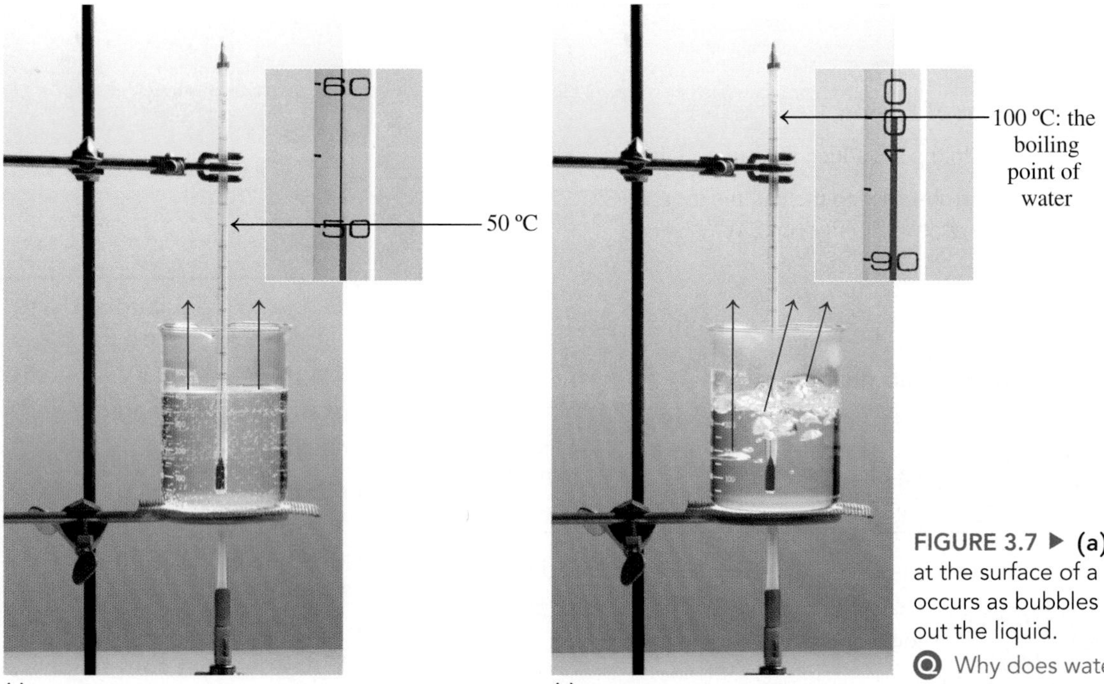

(a) (b)

FIGURE 3.7 ▶ **(a)** Evaporation occurs at the surface of a liquid. **(b)** Boiling occurs as bubbles of gas form throughout the liquid.

Q Why does water evaporate faster at 85 °C than at 15 °C?

Sublimation

In a process called **sublimation**, the particles on the surface of a solid change directly to a gas with no temperature change and without going through the liquid state. In the reverse process called **deposition**, gas particles change directly to a solid. For example, dry ice, which is solid carbon dioxide, sublimes at −78 °C. It is called "dry" because it does not form a liquid as it warms. In extremely cold areas, snow does not melt but sublimes directly to water vapor.

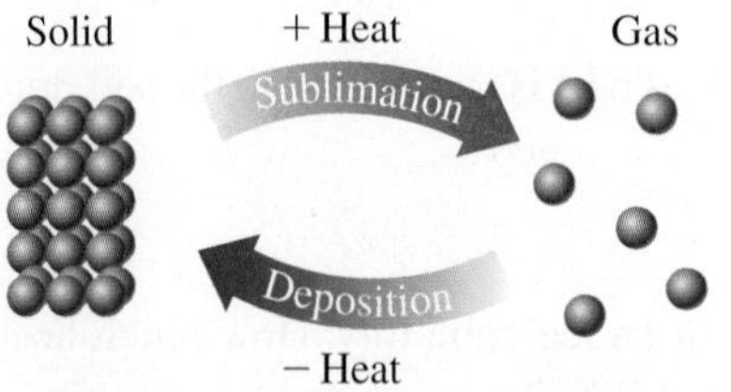

Sublimation and deposition are reversible processes.

Dry ice sublimes at −78 °C.

When frozen foods are left in the freezer for a long time, so much water sublimes that foods, especially meats, become dry and shrunken, a condition called *freezer burn*. Deposition occurs in a freezer when water vapor forms ice crystals on the surface of freezer bags and frozen food.

Freeze-dried foods prepared by sublimation are convenient for long-term storage and for camping and hiking. A food that has been frozen is placed in a vacuum chamber where it dries as the ice sublimes. The dried food retains all of its nutritional value and needs only water to be edible. A food that is freeze-dried does not need refrigeration because bacteria cannot grow without moisture.

Water vapor will change to solid on contact with a cold surface.

Freeze-dried foods have a long shelf life because they contain no water.

Heat of Vaporization

The **heat of vaporization** is the energy that must be added to convert exactly 1 g of liquid to gas at its boiling point. For water, 540 cal (2260 J) is needed to convert 1 g of water to vapor at 100 °C.

$$H_2O(l) + 540\ \text{cal/g (or 2260 J/g)} \longrightarrow H_2O(g)$$

This same amount of heat is released when 1 g of water vapor (gas) changes to liquid at 100 °C.

$$H_2O(g) \longrightarrow H_2O(l) + 540\ \text{cal/g (or 2260 J/g)}$$

Therefore, 2260 J/g or 540 cal/g is also the *heat of condensation* of water.

Heat of Vaporization for Water

$$\frac{540\ \text{cal}}{1\ \text{g}\ H_2O} \text{ and } \frac{1\ \text{g}\ H_2O}{540\ \text{cal}} \qquad \frac{2260\ \text{J}}{1\ \text{g}\ H_2O} \text{ and } \frac{1\ \text{g}\ H_2O}{2260\ \text{J}}$$

To determine the heat needed to boil a sample of water, the mass, in grams, is multiplied by the heat of vaporization. Because the temperature remains constant as long as the water is boiling, there is no temperature change given in the calculation, as shown in Sample Problem 3.9.

Calculating Heat to Vaporize (or Condense) Water

Heat = mass × heat of vaporization

$$\text{cal} = \cancel{\text{g}} \times \frac{540\ \text{cal}}{\cancel{\text{g}}} \qquad \text{J} = \cancel{\text{g}} \times \frac{2260\ \text{J}}{\cancel{\text{g}}}$$

SAMPLE PROBLEM 3.9 Using Heat of Vaporization

In a sauna, 122 g of water is converted to steam at 100 °C. How many kilojoules of heat are needed?

SOLUTION

STEP 1 State the given and needed quantities.

ANALYZE THE PROBLEM	Given	Need
	122 g of $H_2O(l)$ at 100 °C, heat of vaporization = 2260 J/g	kilojoules to convert to $H_2O(g)$ at 100 °C

STEP 2 Write a plan to convert the given quantity to the needed quantity.

grams of water → Heat of vaporization → joules → Metric factor → kilojoules

STEP 3 Write the heat conversion factor and any metric factor.

$$1\ \text{g of } H_2O(l \rightarrow g) = 2260\ \text{J} \qquad 1\ \text{kJ} = 1000\ \text{J}$$

$$\frac{2260\ \text{J}}{1\ \text{g}\ H_2O} \text{ and } \frac{1\ \text{g}\ H_2O}{2260\ \text{J}} \qquad \frac{1000\ \text{J}}{1\ \text{kJ}} \text{ and } \frac{1\ \text{kJ}}{1000\ \text{J}}$$

STEP 4 Set up the problem and calculate the needed quantity.

$$\underset{\text{Three SFs}}{122\,\cancel{g\,H_2O}} \times \overset{\text{Three SFs}}{\underset{\text{Exact}}{\frac{2260\,\cancel{J}}{1\,\cancel{g\,H_2O}}}} \times \overset{\text{Exact}}{\underset{\text{Exact}}{\frac{1\,kJ}{1000\,\cancel{J}}}} = \underset{\text{Three SFs}}{276\,kJ}$$

STUDY CHECK 3.9

When steam from a pan of boiling water reaches a cool window, it condenses. How much heat, in kilojoules, is released when 25.0 g of steam condenses at 100 °C?

Chemistry Link to **Health**

Steam Burns

Hot water at 100 °C will cause burns and damage to the skin. However, getting steam on the skin is even more dangerous. If 25 g of hot water at 100 °C falls on a person's skin, the temperature of the water will drop to body temperature, 37 °C. The heat released during cooling can cause severe burns. The amount of heat can be calculated from the mass, the temperature change, 100 °C − 37 °C = 63 °C, and the specific heat of water, 4.184 J/g °C.

$$25\,\cancel{g} \times 63\,\cancel{°C} \times \frac{4.184\,J}{\cancel{g}\,\cancel{°C}} = 6600\text{ J of heat released when water cools}$$

However, getting steam on the skin is even more damaging. The condensation of the same quantity of steam to liquid at 100 °C releases much more heat—almost ten times as much. This amount of heat can be calculated using the heat of vaporization, which is 2260 J/g for water at 100 °C.

$$25\,\cancel{g} \times \frac{2260\,J}{1\,\cancel{g}} = 57\,000\text{ J released when water (gas) condenses to water (liquid) at 100 °C}$$

The total heat released is calculated by combining the heat from the condensation at 100 °C and the heat from cooling of the steam from 100 °C to 37 °C (body temperature). We can see that most of the heat is from the condensation of steam. This large amount of heat released on the skin is what causes damage from steam burns.

Condensation (100 °C)	= 57 000 J
Cooling (100 °C to 37 °C)	= 6 600 J
Heat released	= 64 000 J (rounded off)

The amount of heat released from steam is almost ten times greater than the heat from the same amount of hot water.

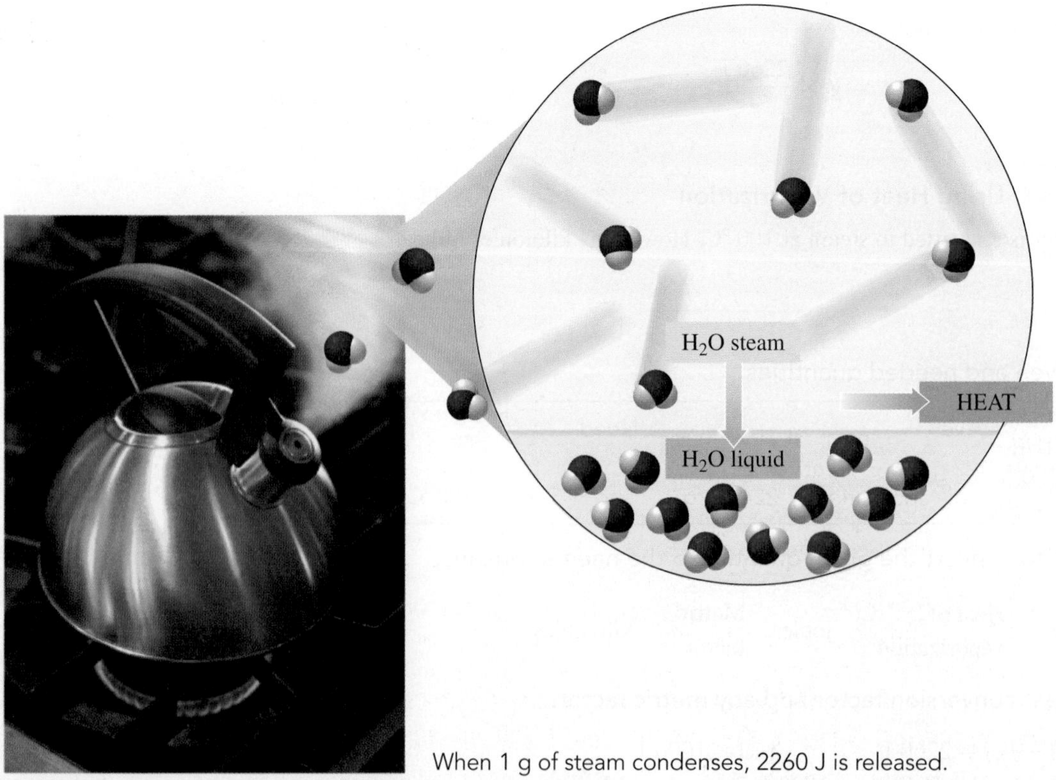

When 1 g of steam condenses, 2260 J is released.

Heating and Cooling Curves

All the changes of state during the heating or cooling of a substance can be illustrated visually. On a **heating curve** or **cooling curve**, the temperature is shown on the vertical axis and the loss or gain of heat is shown on the horizontal axis.

Steps on a Heating Curve

The first diagonal line indicates a warming of a solid as heat is added. When the melting temperature is reached, a horizontal line, or plateau, indicates that the solid is melting. As melting takes place, the solid is changing to liquid without any change in temperature (see Figure 3.8).

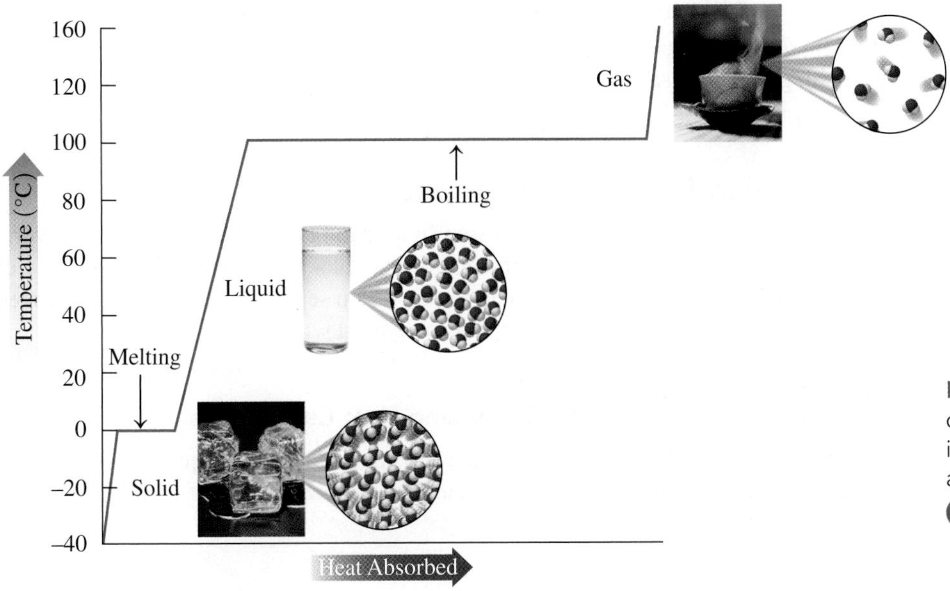

FIGURE 3.8 ▶ A heating curve diagrams the temperature increases and changes of state as heat is added.

Q What does the plateau at 100 °C represent on the heating curve for water?

Once all the particles are in the liquid state, heat that is added will increase the temperature of the liquid. This increase is drawn as a diagonal line from the melting point temperature to the boiling point temperature. Once the liquid reaches its boiling point, a horizontal line indicates that the temperature is constant as liquid changes to gas. Because the heat of vaporization is greater than the heat of fusion, the horizontal line at the boiling point is longer than the line at the melting point. Once all the liquid becomes gas, adding more heat increases the temperature of the gas.

Steps on a Cooling Curve

A cooling curve is a diagram of the cooling process in which the temperature decreases as heat is removed. Initially, a diagonal line to the boiling (condensation) point is drawn to show that heat is removed from a gas until it begins to condense. At the boiling (condensation) point, a horizontal line is drawn that indicates a change of state as gas condenses to form a liquid. After all the gas has changed into liquid, further cooling lowers the temperature. The decrease in temperature is shown as a diagonal line from the condensation temperature to the freezing temperature. At the freezing point, another horizontal line indicates that liquid is changing to solid at the freezing point temperature. Once all the substance is frozen, the removal of more heat decreases the temperature of the solid below its freezing point, which is shown as a diagonal line below its freezing point.

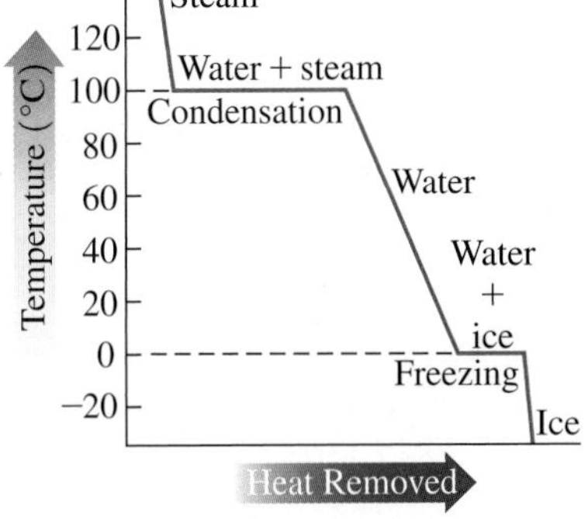

A cooling curve for water illustrates the change in temperature and changes of state as heat is removed.

SAMPLE PROBLEM 3.10 Using a Cooling Curve

Using the cooling curve for water, identify the state or change of state for water as solid, liquid, gas, condensation, or freezing.

a. at 120 °C **b.** at 100 °C **c.** at 40 °C

SOLUTION

a. A temperature of 120 °C occurs on the diagonal line above the boiling (condensation) point indicating that water is a gas.
b. A temperature of 100 °C, shown as a horizontal line, indicates that the water vapor is changing to liquid water, or condensing.
c. A temperature of 40 °C occurs on the diagonal line below the boiling point but above the freezing point, which indicates that the water is in the liquid state.

STUDY CHECK 3.10

Using the cooling curve for water, identify the state or change of state for water as solid, liquid, gas, condensation, or freezing.

a. at 0 °C **b.** at −20 °C

Combining Energy Calculations

Up to now, we have calculated one step in a heating or cooling curve. However, many problems require a combination of steps that include a temperature change as well as a change of state. The heat is calculated for each step separately and then added together to find the total energy, as seen in Sample Problem 3.11.

SAMPLE PROBLEM 3.11 Combining Heat Calculations

Calculate the total heat, in joules, needed to convert 15.0 g of liquid ethanol at 25.0 °C to gas at its boiling point of 78.0 °C. Ethanol has a specific heat of 2.46 J/g °C, and a heat of vaporization of 841 J/g.

SOLUTION

STEP 1 State the given and needed quantities.

	Given	Need
ANALYZE THE PROBLEM	15.0 g of ethanol at 25.0 °C, specific heat of ethanol = 2.46 J/g °C, heat of vaporization of ethanol = 841 J/g	total joules to raise temperature of ethanol from 25.0 °C to 78.0 °C, and convert to gas (78.0 °C)

STEP 2 Write a plan to convert the given quantity to the needed quantity. When several changes occur, draw a diagram of heating and changes of state.

Total heat = joules needed to warm ethanol from 25.0 °C to 78.0 °C (boiling point)
+ joules to change liquid to gas at 78.0 °C (boiling point)

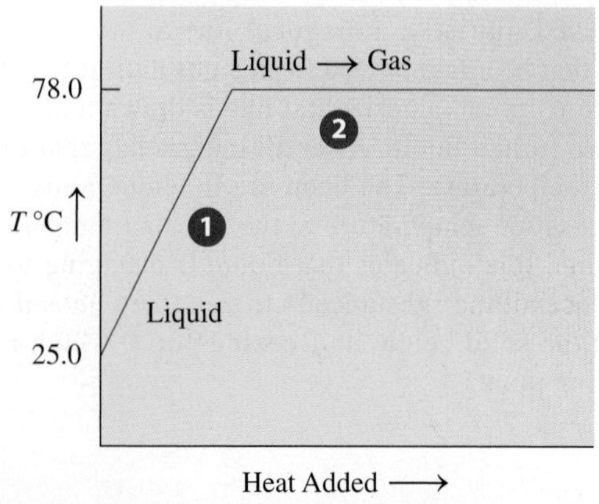

STEP 3 Write the heat conversion factor and any metric factor.

$$SH_{\text{Ethanol}} = \frac{2.46\ \text{J}}{\text{g °C}}$$

$$\frac{2.46\ \text{J}}{\text{g °C}} \text{ and } \frac{\text{g °C}}{2.46\ \text{J}}$$

$$1\ \text{g of ethanol } (l \rightarrow g) = 841\ \text{J}$$

$$\frac{841\ \text{J}}{1\ \text{g ethanol}} \text{ and } \frac{1\ \text{g ethanol}}{841\ \text{J}}$$

STEP 4 Set up the problem and calculate the needed quantity.

$\Delta T = 78.0\ °\text{C} - 25.0\ °\text{C} = 53.0\ °\text{C}$

Heat needed to warm ethanol (liquid) from 25.0 °C to ethanol (liquid) at 78.0 °C.

$$15.0\ \cancel{\text{g}} \times 53.0\ °\cancel{\text{C}} \times \frac{2.46\ \text{J}}{\cancel{\text{g}}\ °\cancel{\text{C}}} = 1960\ \text{J}$$

Three SFs (15.0 g), Three SFs (53.0 °C), Three SFs (2.46 J/g °C), Three SFs (1960 J)

Heat needed to change ethanol (liquid) to ethanol (gas) at 78.0 °C.

$$15.0\ \cancel{\text{g ethanol}} \times \frac{841\ \text{J}}{1\ \cancel{\text{g ethanol}}} = 12\ 600\ \text{J}$$

Three SFs (15.0 g ethanol), Three SFs (841 J), Exact (1 g ethanol), Three SFs (12 600 J)

Calculate the total heat:

Heating ethanol (25.0 °C to boiling point 78.0 °C)	1 960 J
Changing liquid to gas at boiling point (78.0 °C)	12 600 J
Total heat needed	14 600 J (rounded off to hundreds place)

STUDY CHECK 3.11

How many kilojoules are released when 75.0 g of steam at 100 °C condenses, cools to 0 °C, and freezes at 0 °C? (*Hint*: The solution will require three energy calculations.)

QUESTIONS AND PROBLEMS

3.7 Changes of State

LEARNING GOAL Describe the changes of state between solids, liquids, and gases; calculate the energy involved.

3.39 Identify each of the following changes of state as melting, freezing, sublimation or deposition
- **a.** The ice sculpture start breaking down as temperature rises above 0 °C.
- **b.** Ice form on the strawberry plants in a cold wintry night.
- **c.** Water becomes solid at 0 °C.
- **d.** Dry ice on heating becomes gas.

3.40 Identify each of the following changes of state as melting, freezing, sublimation, or deposition:
- **a.** Dry ice in an ice-cream cart disappears.
- **b.** Snow on the ground turns to liquid water.
- **c.** Heat is removed from 125 g of liquid water at 0 °C.
- **d.** Frost (ice) forms on the walls of a freezer unit of a refrigerator.

3.41 Calculate the heat change at 0 °C for each of the following problems. Indicate whether heat was absorbed or released:
- **a.** calories to melt 65 g of ice
- **b.** joules to melt 17.0 g of ice
- **c.** kilocalories to freeze 225 g of water
- **d.** kilojoules to freeze 50.0 g of water

3.42 Calculate the amount of H_2O involved as heat change at 0 °C for each of the following problems and indicate whether the water freezes or melts:
- **a.** 5.6 kcal of energy was released
- **b.** 45925 J of energy was released
- **c.** 33.6 kcal of energy was absorbed
- **d.** 8.35 kJ of energy was absorbed

3.43 Identify each of the following changes of state as evaporation, boiling, or condensation:
- **a.** The water vapor in the clouds changes to rain.
- **b.** Wet clothes dry on a clothesline.
- **c.** Lava flows into the ocean and steam forms.
- **d.** After a hot shower, your bathroom mirror is covered with water.

3.44 Identify each of the following changes of state as evaporation, boiling, or condensation:
- **a.** At 100 °C, the water in a pan changes to steam.
- **b.** On a cool morning, the windows in your car fog up.
- **c.** A shallow pond dries up in the summer.
- **d.** Your teakettle whistles when the water is ready for tea.

3.45 Calculate the heat change at 100 °C for each of the following and indicate whether heat was absorbed or released:
- **a.** calories to vaporize 10.0 g of water
- **b.** joules to vaporize 5.00 g of water

c. kilocalories to condense 8.0 kg of steam
d. kilojoules to condense 175 g of steam

3.46 Calculate the heat change at 100 °C for each of the following and indicate whether heat was absorbed or released:
a. calories to condense 10.0 g of steam
b. joules to condense 7.60 g of steam
c. kilocalories to vaporize 44 g of water
d. kilojoules to vaporize 5.00 kg of water

3.47 Draw a heating curve for a sample of ice that is heated from −20 °C to 150 °C. Indicate the segment of the graph that corresponds to each of the following:
a. solid b. melting c. liquid
d. boiling e. gas

3.48 Draw a cooling curve for a sample of steam that cools from 110 °C to −10 °C. Indicate the segment of the graph that corresponds to each of the following:
a. solid b. freezing c. liquid
d. condensation e. gas

3.49 Using the values for the heat of fusion, specific heat of water, and/or heat of vaporization, calculate the amount of heat energy in each of the following:
a. joules needed to melt 50.0 g of ice at 0 °C and to warm the liquid to 65.0 °C
b. kilocalories released when 15.0 g of steam condenses at 100 °C and the liquid cools to 0 °C
c. kilojoules needed to melt 24.0 g of ice at 0 °C, warm the liquid to 100 °C, and change it to steam at 100 °C

3.50 Using the values for the heat of fusion, specific heat of water, and/or heat of vaporization, calculate the amount of heat energy in each of the following:
a. joules released when 125 g of steam at 100 °C condenses and cools to liquid at 15.0 °C.
b. kilocalories needed to melt a 525-g ice sculpture at 0 °C and to warm the liquid to 15.0 °C
c. kilojoules released when 85.0 g of steam condenses at 100 °C, cools, and freezes at 0 °C

CONCEPT MAP

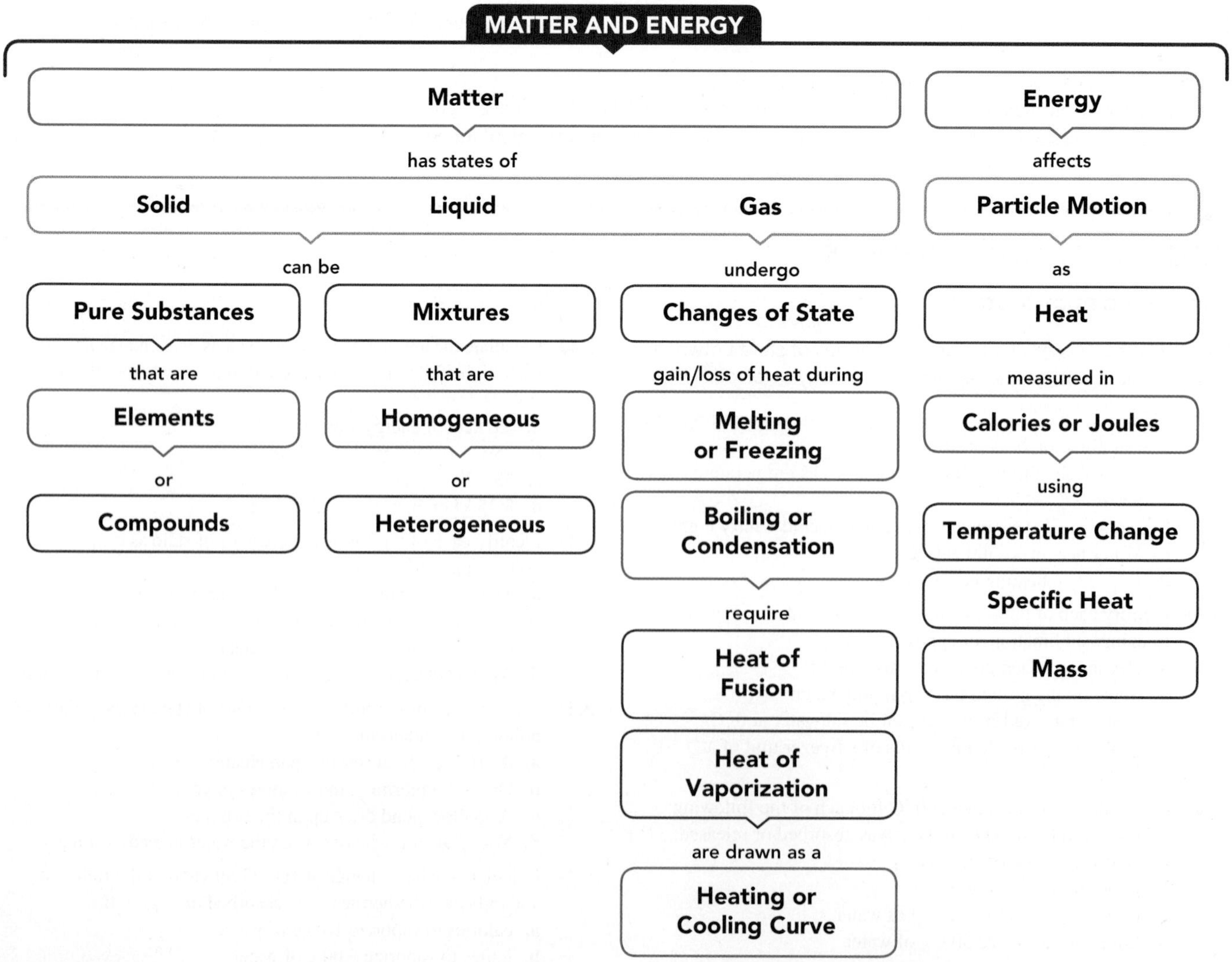

CHAPTER REVIEW

3.1 Classification of Matter

LEARNING GOAL Classify examples of matter as pure substances or mixtures.

- Matter is anything that has mass and occupies space.
- Matter is classified as pure substances or mixtures.
- Pure substances, which are elements or compounds, have fixed compositions, and mixtures have variable compositions.
- The substances in mixtures can be separated using physical methods.

3.2 States and Properties of Matter

LEARNING GOAL Identify the states and the physical and chemical properties of matter.

- The three states of matter are solid, liquid, and gas.
- A physical property is a characteristic of a substance in that it can be observed or measured without affecting the identity of the substance.
- A physical change occurs when physical properties change, but not the composition of the substance.
- A chemical property indicates the ability of a substance to change into another substance.
- A chemical change occurs when one or more substances react to form a substance with new physical and chemical properties.

3.3 Temperature

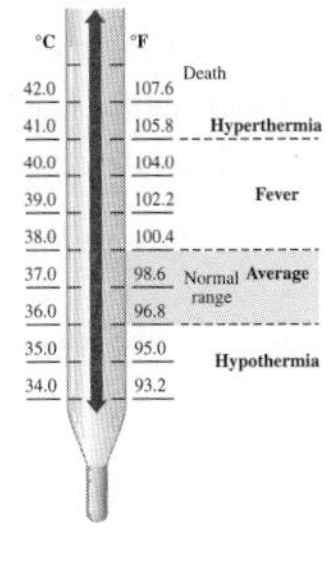

LEARNING GOAL Given a temperature, calculate a corresponding temperature on another scale.

- In science, temperature is measured in degrees Celsius (°C) or kelvins (K).
- On the Celsius scale, there are 100 units between the freezing point of water (0 °C) and the boiling point (100 °C).
- On the Fahrenheit scale, there are 180 units between the freezing point of water (32 °F) and the boiling point (212 °F). A Fahrenheit temperature is related to its Celsius temperature by the equation $T_F = 1.8(T_C) + 32$.
- The SI unit, kelvin, is related to the Celsius temperature by the equation $T_K = T_C + 273$.

3.4 Energy

LEARNING GOAL Identify energy as potential or kinetic; convert between units of energy.

- Energy is the ability to do work.
- Potential energy is stored energy; kinetic energy is the energy of motion.
- Common units of energy are the calorie (cal), kilocalorie (kcal), joule (J), and kilojoule (kJ).
- One calorie is equal to 4.184 J.

3.5 Energy and Nutrition

LEARNING GOAL Use the energy values to calculate the kilocalories (kcal) or kilojoules (kJ) for a food.

- The nutritional Calorie is the same amount of energy as 1 kcal or 1000 calories.
- The energy of a food is the sum of kilocalories or kilojoules from carbohydrate, fat, and protein.

3.6 Specific Heat

LEARNING GOAL Use specific heat to calculate heat loss or gain.

- Specific heat is the amount of energy required to raise the temperature of exactly 1 g of a substance by exactly 1 °C.
- The heat lost or gained by a substance is determined by multiplying its mass, the temperature change, and its specific heat.

3.7 Changes of State

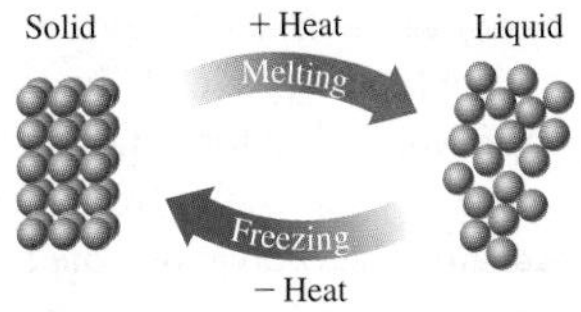

LEARNING GOAL Describe the changes of state between solids, liquids, and gases; calculate the energy involved.

- Melting occurs when the particles in a solid absorb enough energy to break apart and form a liquid.
- The amount of energy required to convert exactly 1 g of solid to liquid is called the heat of fusion.
- For water, 80. cal (334 J) is needed to melt 1 g of ice or must be removed to freeze 1 g of water.
- Evaporation occurs when particles in a liquid state absorb enough energy to break apart and form gaseous particles.
- Boiling is the vaporization of liquid at its boiling point. The heat of vaporization is the amount of heat needed to convert exactly 1 g of liquid to vapor.
- For water, 540 cal (2260 J) is needed to vaporize 1 g of water or must be removed to condense 1 g of steam.
- A heating or cooling curve illustrates the changes in temperature and state as heat is added to or removed from a substance. Plateaus on the graph indicate changes of state.
- The total heat absorbed or removed from a substance undergoing temperature changes and changes of state is the sum of energy calculations for change(s) of state and change(s) in temperature.
- Sublimation is a process whereby a solid changes directly to a gas.

KEY TERMS

boiling The formation of bubbles of gas throughout a liquid.

boiling point (bp) The temperature at which a liquid changes to gas (boils) and gas changes to liquid (condenses).

calorie (cal) The amount of heat energy that raises the temperature of exactly 1 g of water exactly 1 °C.

change of state The transformation of one state of matter to another; for example, solid to liquid, liquid to solid, liquid to gas.

chemical change A change during which the original substance is converted into a new substance that has a different composition and new physical and chemical properties.

chemical properties The properties that indicate the ability of a substance to change into a new substance.

compound A pure substance consisting of two or more elements, with a definite composition, that can be broken down into simpler substances only by chemical methods.

condensation The change of state from a gas to a liquid.

cooling curve A diagram that illustrates temperature changes and changes of state for a substance as heat is removed.

deposition The change of a gas directly into a solid; the reverse of sublimation.

element A pure substance containing only one type of matter, which cannot be broken down by chemical methods.

energy The ability to do work.

energy value The kilocalories (or kilojoules) obtained per gram of the food types: carbohydrate, fat, and protein.

evaporation The formation of a gas (vapor) by the escape of high-energy molecules from the surface of a liquid.

freezing The change of state from liquid to solid.

freezing point (fp) The temperature at which a liquid changes to a solid (freezes), a solid changes to a liquid (melts).

gas A state of matter that does not have a definite shape or volume.

heat The energy associated with the motion of particles in a substance.

heat of fusion The energy required to melt exactly 1 g of a substance at its melting point. For water, 80. cal (334 J) is needed to melt 1 g of ice; 80. cal (334 J) is released when 1 g of water freezes.

heat of vaporization The energy required to vaporize exactly 1 g of a substance at its boiling point. For water, 540 cal (2260 J) is needed to vaporize 1 g of liquid; 1 g of steam gives off 540 cal (2260 J) when it condenses.

heating curve A diagram that illustrates the temperature changes and changes of state of a substance as it is heated.

joule (J) The SI unit of heat energy; 4.184 J = 1 cal.

kinetic energy The energy of moving particles.

liquid A state of matter that takes the shape of its container but has a definite volume.

matter The material that makes up a substance and has mass and occupies space.

melting The change of state from a solid to a liquid.

melting point (mp) The temperature at which a solid becomes a liquid (melts). It is the same temperature as the freezing point.

mixture The physical combination of two or more substances that does not change the identities of the mixed substances.

physical change A change in which the physical properties of a substance change but its identity stays the same.

physical properties The properties that can be observed or measured without affecting the identity of a substance.

potential energy A type of energy related to position or composition of a substance.

pure substance A type of matter that has a definite composition.

solid A state of matter that has its own shape and volume.

specific heat (*SH*) A quantity of heat that changes the temperature of exactly 1 g of a substance by exactly 1 °C.

states of matter Three forms of matter: solid, liquid, and gas.

sublimation The change of state in which a solid is transformed directly to a gas without forming a liquid first.

CORE CHEMISTRY SKILLS

The chapter section containing each Core Chemistry Skill is shown in parentheses at the end of each heading.

Classifying Matter (3.1)

- A pure substance is matter that has a fixed or constant composition.
- An element, the simplest type of a pure substance, is composed of only one type of matter, such as silver, iron, or aluminum.
- A compound is also a pure substance, but it consists of two or more elements chemically combined in the same proportion.
- In a *homogeneous mixture*, also called a *solution*, the composition of the substances in the mixture is uniform.
- In a *heterogeneous mixture*, the components are visible and do not have a uniform composition throughout the sample.

Example: Classify each of the following as a pure substance (element or compound) or a mixture (homogeneous or heterogeneous):

a. iron in a nail
b. black coffee
c. carbon dioxide, a greenhouse gas

Answer: **a.** Iron is a pure substance, which is an element.
b. Black coffee contains different substances with uniform composition, which makes it a homogeneous mixture.
c. The gas carbon dioxide, which is a pure substance that contains two elements chemically combined, is a compound.

Identifying Physical and Chemical Changes (3.2)

- When matter undergoes a physical change, its state or its appearance changes, but its composition remains the same.
- When a chemical change takes place, the original substance is converted into a new substance, which has different physical and chemical properties.

Example: Classify each of the following as a physical or chemical property:

a. Helium in a balloon is a gas.
b. Methane, in natural gas, burns.
c. Hydrogen sulfide smells like rotten eggs.

Answer: **a.** A gas is a state of matter, which makes it a physical property.
b. When methane burns, it changes to different substances with new properties, which is a chemical property.
c. The odor of hydrogen sulfide is a physical property.

Converting between Temperature Scales (3.3)

- The temperature equation $T_F = 1.8(T_C) + 32$ is used to convert from Celsius to Fahrenheit and can be rearranged to convert from Fahrenheit to Celsius.
- The temperature equation $T_K = T_C + 273$ is used to convert from Celsius to Kelvin and can be rearranged to convert from Kelvin to Celsius.

Example: Convert 75.0 °C to degrees Fahrenheit.

Answer: $T_F = 1.8(T_C) + 32$

$T_F = 1.8(75.0) + 32 = 135 + 32$

$= 167\text{ °F}$

Example: Convert 355 K to degrees Celsius.

Answer: $T_K = T_C + 273$

To solve the equation for T_C, subtract 273 from both sides.

$T_K - 273 = T_C + \cancel{273} - \cancel{273}$

$T_C = T_K - 273$

$T_C = 355 - 273$

$= 82\text{ °C}$

Using Energy Units (3.4)

- Equalities for energy units include 1 cal = 4.184 J, 1 kcal = 1000 cal, and 1 kJ = 1000 J.
- Each equality for energy units can be written as two conversion factors:

$$\frac{4.184\text{ J}}{1\text{ cal}} \text{ and } \frac{1\text{ cal}}{4.184\text{ J}}; \quad \frac{1000\text{ cal}}{1\text{ kcal}} \text{ and } \frac{1\text{ kcal}}{1000\text{ cal}}; \quad \frac{1000\text{ J}}{1\text{ kJ}} \text{ and } \frac{1\text{ kJ}}{1000\text{ J}}$$

- The energy unit conversion factors are used to cancel given units of energy and to obtain the needed unit of energy.

Example: Convert 45 000 J to kilocalories.

Answer: Using the conversion factors above, we start with the given 45 000 J and convert it to kilocalories.

$$45\,000\text{ J} \times \frac{1\text{ cal}}{4.184\text{ J}} \times \frac{1\text{ kcal}}{1000\text{ cal}} = 11\text{ kcal}$$

Using the Heat Equation (3.6)

- The quantity of heat absorbed or lost by a substance is calculated using the heat equation.

  Heat $= m \times \Delta T \times SH$
- Heat, in joules, is obtained when the specific heat of a substance in J/g °C is used.
- To cancel, the unit grams is used for mass, and the unit °C is used for temperature change.

Example: How many joules are required to heat 5.25 g of titanium from 85.5 °C to 132.5 °C?

Answer: $m = 5.25\text{ g}$, $\Delta T = 132.5\text{ °C} - 85.5\text{ °C} = 47.0\text{ °C}$
SH for titanium $= 0.523\text{ J/g °C}$
The known values are substituted into the heat equation making sure units cancel.

$$\text{Heat} = m \times \Delta T \times SH = 5.25\text{ g} \times 47.0\text{ °C} \times \frac{0.523\text{ J}}{\text{g °C}}$$

$$= 129\text{ J}$$

UNDERSTANDING THE CONCEPTS

The chapter sections to review are shown in parentheses at the end of each question.

3.51 Identify each of the following as an element, a compound, or a mixture. Explain your choice. (3.1)

a.
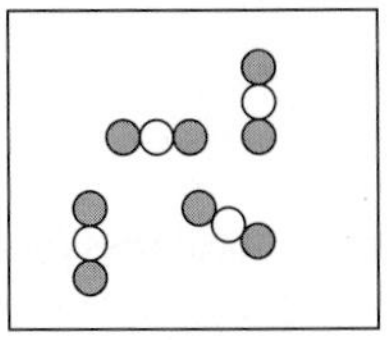

b.
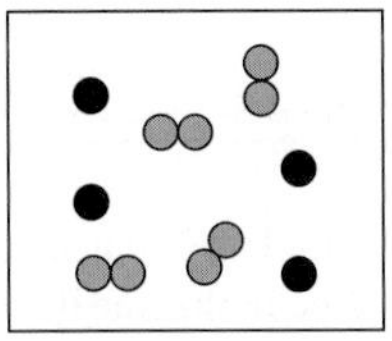

c.
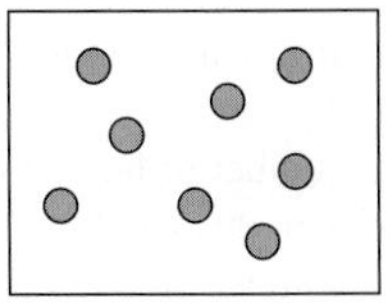

3.52 Identify each of the following as a homogeneous or heterogeneous mixture. Explain your choice. (3.1)

a.
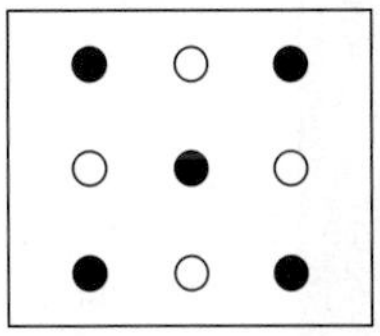

b.
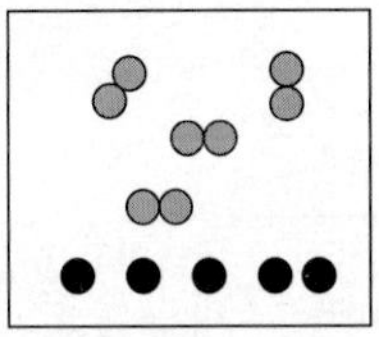

c.
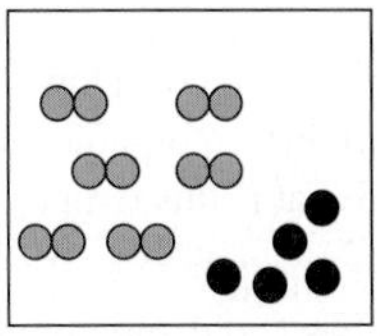

3.53 Classify each of the following as a homogeneous or heterogeneous mixture: (3.1)
a. lemon-flavored water
b. stuffed mushrooms
c. eye drops

3.54 Classify each of the following as a homogeneous or heterogeneous mixture: (3.1)

a. ketchup **b.** hard-boiled egg **c.** tortilla soup

3.55 State the temperature on the Celsius thermometer and convert to Fahrenheit. (3.3)

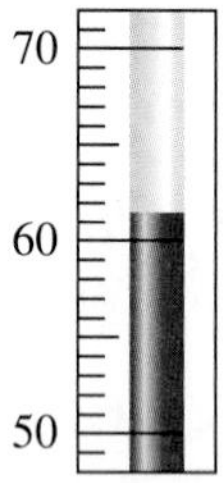

3.56 State the temperature on the Celsius thermometer and convert to Fahrenheit. (3.3)

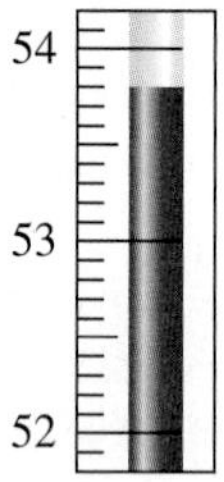

3.57 Compost can be made at home from grass clippings, some kitchen scraps, and dry leaves. As microbes break down organic matter, heat is generated and the compost can reach a temperature of 155 °F, which kills most pathogens. What is this initial temperature in degrees Celsius? In kelvins? (3.3)

Compost produced from decayed plant material is used to enrich the soil.

3.58 After a week, biochemical reactions in compost slow, and the temperature drops to 45 °C. The dark brown organic-rich mixture is ready for use in the garden. What is this temperature in degrees Fahrenheit? In kelvins? (3.3)

3.59 Calculate the energy to heat two cubes (gold and aluminum) each with a volume of 10.0 cm^3 from 15 °C to 25 °C. Refer to Tables 2.9 and 3.11. (3.6)

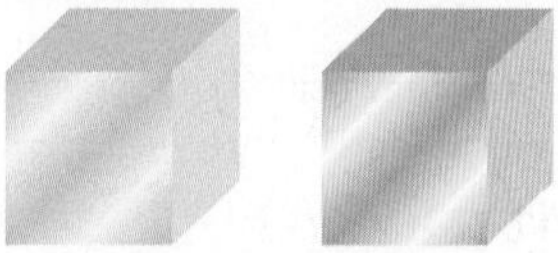

3.60 Calculate the energy to heat two cubes (silver and copper), each with a volume of 10.0 cm^3 from 15 °C to 25 °C. Refer to Tables 2.9 and 3.11. (3.6)

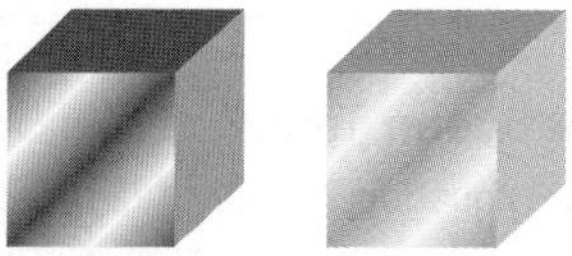

3.61 A 70.0-kg person had a quarter-pound cheeseburger, french fries, and a chocolate shake. (3.5)

Item	Carbohydrate (g)	Fat (g)	Protein (g)
Cheeseburger	46	40.	47
French fries	47	16	4
Chocolate shake	76	10.	10.

a. Using Table 3.7, calculate the total kilocalories for each food type in this meal (round off the kilocalories to the tens place).

b. Determine the total kilocalories for the meal (round off to the tens place).

c. Using Table 3.10, determine the number of hours of sleeping needed to burn off the kilocalories in this meal.

d. Using Table 3.10, determine the number of hours of running needed to burn off the kilocalories in this meal.

3.62 Your friend, who has a mass of 70.0 kg, has a slice of pizza, a cola soft drink, and ice cream. (3.5)

Item	Carbohydrate (g)	Fat (g)	Protein (g)
Pizza	29	10.	13
Cola	51	0	0
Ice cream	44	28	8

a. Using Table 3.7, calculate the total kilocalories for each food type in this meal (round off the kilocalories to the tens place).

b. Determine the total kilocalories for the meal (round off to the tens place).

c. Using Table 3.10, determine the number of hours of sitting needed to burn off the kilocalories in this meal.

d. Using Table 3.10, determine the number of hours of swimming needed to burn off the kilocalories in this meal.

ADDITIONAL QUESTIONS AND PROBLEMS

3.63 Classify each of the following as an element, a compound, or a mixture: (3.1)
a. carbon in pencils
b. carbon monoxide (CO) in automobile exhaust
c. orange juice

3.64 Classify each of the following as an element, a compound, or a mixture: (3.1)
a. neon gas in lights
b. a salad dressing of oil and vinegar
c. sodium hypochlorite (NaOCl) in bleach

3.65 Classify each of the following mixtures as homogeneous or heterogeneous: (3.1)
a. hot fudge sundae **b.** herbal tea **c.** vegetable oil

3.66 Classify each of the following mixtures as homogeneous or heterogeneous: (3.1)
a. water and sand **b.** mustard **c.** blue ink

3.67 Identify each of the following as solid, liquid, or gas: (3.2)
a. mercury metal
b. hydrogen
c. cappuccino
d. dry ice
e. charcoal briquettes on a barbecue

3.68 Identify each of the following as solid, liquid, or gas: (3.2)
a. popcorn in a bag
b. water in a garden hose
c. a computer mouse
d. air in a tire
e. hot tea in a teacup

3.69 Identify each of the following as a physical or chemical property: (3.2)
a. Gold is shiny.
b. Gold melts at 1064 °C.
c. Gold is a good conductor of electricity.
d. When gold reacts with yellow sulfur, a black sulfide compound forms.

3.70 Identify each of the following as a physical or chemical property: (3.2)
a. A candle is 10 in. high and 2 in. in diameter.
b. A candle burns.
c. The wax of a candle softens on a hot day.
d. A candle is blue.

3.71 Identify each of the following as a physical or chemical change: (3.2)
a. A plant grows a new leaf.
b. Chocolate is melted for a dessert.
c. Wood is chopped for the fireplace.
d. Wood burns in a woodstove.

3.72 Identify each of the following as a physical or chemical change: (3.2)
a. Aspirin tablets are broken in half.
b. Carrots are grated for use in a salad.
c. Malt undergoes fermentation to make beer.
d. A copper pipe reacts with air and turns green.

3.73 Calculate each of the following temperatures in degrees Celsius and kelvins: (3.3)
a. The highest recorded temperature in the continental United States was 134 °F in Death Valley, California, on July 10, 1913.
b. The lowest recorded temperature in the continental United States was −69.7 °F in Rodgers Pass, Montana, on January 20, 1954.

3.74 Calculate each of the following temperatures in kelvins and degrees Fahrenheit: (3.3)
a. The highest recorded temperature in the world was 58.0 °C in El Azizia, Libya, on September 13, 1922.
b. The lowest recorded temperature in the world was −89.2 °C in Vostok, Antarctica, on July 21, 1983.

3.75 What is −15 °F in degrees Celsius and in kelvins? (3.3)

3.76 The highest recorded body temperature that a person has survived is 46.5 °C. Calculate that temperature in degrees Fahrenheit and in kelvins. (3.3)

3.77 If you want to lose 1 lb of "body fat," which is 15% water, how many kilocalories do you need to expend? (3.5)

3.78 Calculate the Cal (kcal) in one cup of whole milk that contains 12 g of carbohydrate, 9 g of fat, and 9 g of protein. (Round off to the tens place.) (3.5)

On a sunny day, the sand gets hot but the water stays cool.

3.79 On a hot day, the beach sand gets hot but the water stays cool. Would you predict that the specific heat of sand is higher or lower than that of water? Explain. (3.6)

3.80 On a hot sunny day, you get out of the swimming pool and sit in a metal chair, which is very hot. Would you predict that the specific heat of the metal is higher or lower than that of water? Explain. (3.6)

3.81 What is freeze-drying in food storage? How is bacterial activity affected by this? (3.7)

3.82 A large bottle containing 883 g of water at 4 °C is removed from the refrigerator. How many kilojoules are absorbed to warm the water to room temperature of 22 °C? (3.6)

3.83 A 0.50-g sample of vegetable oil is placed in a calorimeter. When the sample is burned, 18.9 kJ is given off. What is the energy value (kcal/g) for the oil? (3.5)

3.84 A 1.3-g sample of rice is placed in a calorimeter. When the sample is burned, 22 kJ is given off. What is the energy value (kcal/g) for the rice? (3.5)

3.85 The following graph is a heating curve for chloroform, a solvent for fats, oils, and waxes: (3.7)

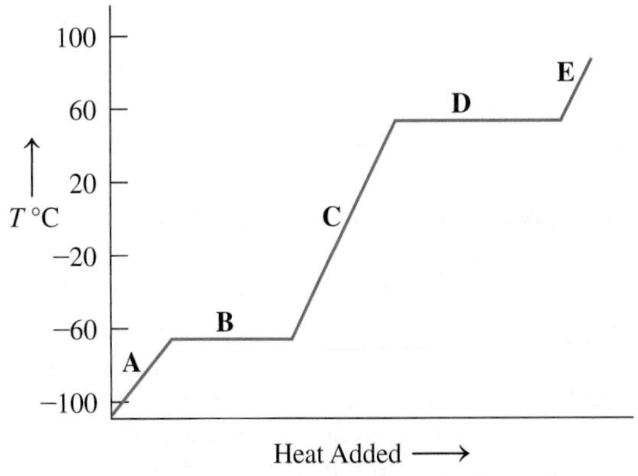

a. What is the approximate melting point of chloroform?
b. What is the approximate boiling point of chloroform?

c. On the heating curve, identify the segments **A**, **B**, **C**, **D**, and **E** as solid, liquid, gas, melting, or boiling.

d. At the following temperatures, is chloroform a solid, liquid, or gas?

−80 °C; −40 °C; 25 °C; 80 °C

3.86 Associate the contents of the beakers (**1–5**) with segments (**A–E**) on the following heating curve for water: (3.7)

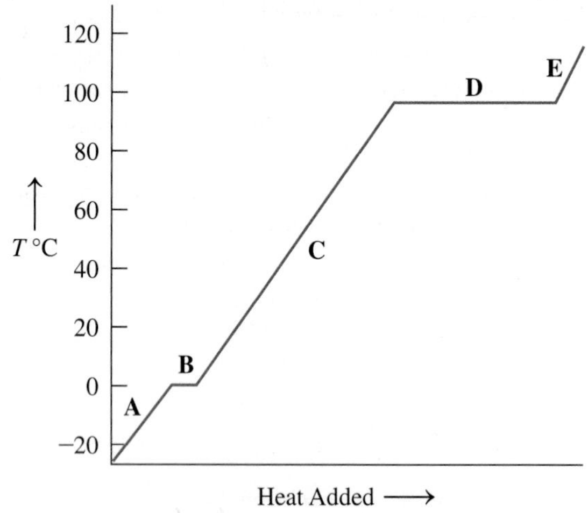

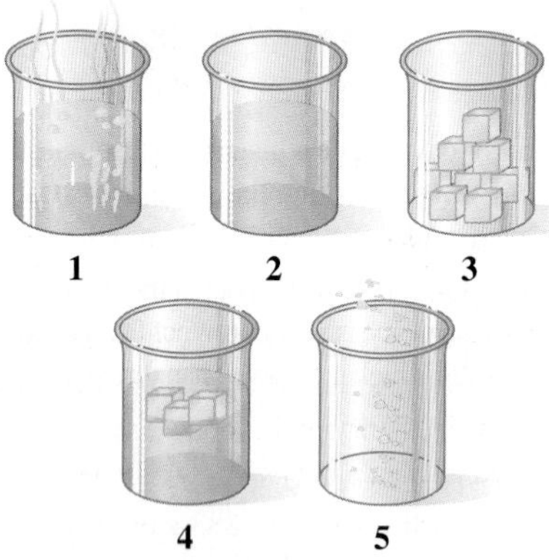

3.87 Calculate the heat of fusion of the following matters: (3.7)

a. 11.2 kcal of energy is absorbed to melt 140 g of a compound Z

b. 2.8 kJ of energy is absorbed to melt 60 g of a compound Y

3.88 The melting point of benzene is 5.5 °C and its boiling point is 80.1 °C. Sketch a heating curve for benzene from 0 °C to 100 °C. (3.7)

a. What is the state of benzene at 15 °C?

b. What happens on the curve at 5.5 °C?

c. What is the state of benzene at 63 °C?

d. What is the state of benzene at 98 °C?

e. At what temperature will both liquid and gas be present?

CHALLENGE QUESTIONS

The following groups of questions are related to the topics in this chapter. However, they do not all follow the chapter order, and they require you to combine concepts and skills from several sections. These questions will help you increase your critical thinking skills and prepare for your next exam.

3.89 When a 0.66-g sample of olive oil is burned in a calorimeter, the heat released increases the temperature of 370 g of water from 22.7 °C to 38.8 °C. What is the energy value for the olive oil in kcal/g? (3.5, 3.6)

3.90 A 45-g piece of ice at 0.0 °C is added to a sample of water at 8.0 °C. All of the ice melts and the temperature of the water decreases to 0.0 °C. How many grams of water were in the sample? (3.6, 3.7)

3.91 In a large building, oil is used in a steam boiler heating system. The combustion of 1.0 lb of oil provides 2.4×10^7 J. (3.4, 3.6)

a. How many kilograms of oil are needed to heat 150 kg of water from 22 °C to 100 °C?

b. How many kilograms of oil are needed to change 150 kg of water to steam at 100 °C?

3.92 When 1.0 g of gasoline burns, it releases 11 kcal of heat. The density of gasoline is 0.74 g/mL. (3.4, 3.6)

a. How many megajoules are released when 1.0 gal of gasoline burns?

b. If a television requires 150 kJ/h to run, how many hours can the television run on the energy provided by 1.0 gal of gasoline?

3.93 An ice bag containing 275 g of ice at 0 °C was used to treat sore muscles. When the bag was removed, the ice had melted and the liquid water had a temperature of 24.0 °C. How many kilojoules of heat were absorbed? (3.6, 3.7)

3.94 A 115-g sample of steam at 100 °C is emitted from a volcano. It condenses, cools, and falls as snow at 0 °C. How many kilojoules of heat were released? (3.6, 3.7)

ANSWERS

Answers to Study Checks

3.1 heterogeneous

3.2 a. physical change **b.** chemical change
c. physical change

3.3 12 °F

3.4 39.8 °C

3.5 35 kJ

3.6 a. carbohydrate, 370 kJ; fat, 270 kJ; protein, 170 kJ
b. 810 kJ

3.7 25 kcal

3.8 12 kcal

3.9 56.5 kJ released

3.10 a. freezing **b.** solid

3.11 226 kJ

Answers to Selected Questions and Problems

3.1 a. pure substance **b.** pure substance
c. mixture **d.** pure substance
e. mixture

3.3 a. compound **b.** element
c. compound **d.** compound
e. element

3.5 a. homogeneous **b.** homogeneous
c. heterogeneous **d.** heterogeneous
e. heterogeneous

3.7 a. gas **b.** gas
c. solid

3.9 a. physical **b.** chemical
c. chemical **d.** physical

3.11 a. physical **b.** chemical
c. physical **d.** physical
e. physical

3.13 a. chemical **b.** chemical
c. physical **d.** physical

3.15 a. 10 **b.** 10
c. 2727

3.17 a. 123.8 °F **b.** 23.22 °C
c. 266 K **d.** 321 K
e. 53.89 °C

3.19 a. 41 °C
b. No. The temperature is equivalent to 39 °C.

3.21 When the roller-coaster car is at the top of the ramp, it has its maximum potential energy. As it descends, potential energy changes to kinetic energy. At the bottom, all the energy is kinetic.

3.23 a. potential **b.** kinetic
c. potential **d.** potential

3.25 a. 8.1×10^5 J **b.** 190 kcal

3.27 a. 33.46 kcal **b.** 126.7 kcal

3.29 a. 470 kJ **b.** 18 g
c. 130 kcal **d.** 37 g

3.31 263 kcal and 1114 kJ

3.33 Copper has the lowest specific heat of the samples and will reach the highest temperature.

3.35 a. 208 cal **b.** 7112.8 J
c. 7.34 kcal **d.** 11.396 kJ

3.37 a. 1380 J, 330. cal **b.** 1810 J, 434 cal
c. 3780 J, 904 cal **d.** 3600 J, 850 cal

3.39 a. melting **b.** deposition
c. freezing **d.** sublimation

3.41 a. 5200 cal absorbed **b.** 5680 J absorbed
c. 18 kcal released **d.** 16.7 kJ released

3.43 a. condensation **b.** evaporation
c. boiling **d.** condensation

3.45 a. 5400 cal absorbed **b.** 11 300 J absorbed
c. 4300 kcal released **d.** 396 kJ released

3.47

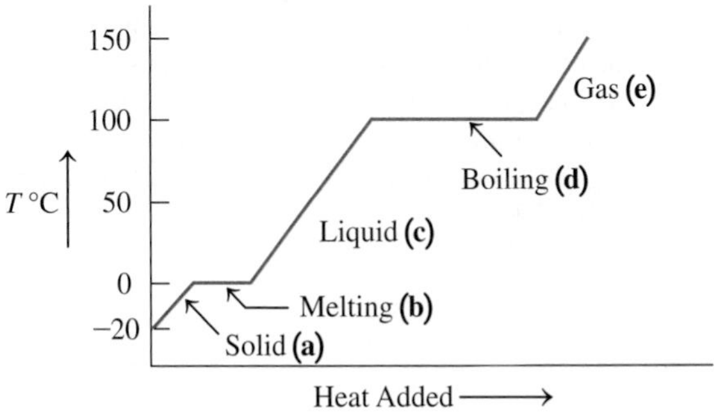

3.49 a. 30 300 J **b.** 9.6 kcal
c. 72.2 kJ

3.51 a. compound, the molecules have a definite 2:1 ratio of atoms
b. mixture, has two different kinds of atoms and molecules
c. element, has a single kind of atom

3.53 a. homogeneous **b.** heterogeneous
c. homogeneous

3.55 61.4 °C, 143 °F

3.57 68.3 °C, 341 K

3.59 gold, 250 J or 59 cal; aluminum, 240 J or 58 cal

3.61 a. carbohydrate, 680 kcal; fat, 590 kcal; protein, 240 kcal
b. 1510 kcal
c. 25 h
d. 2.0 h

3.63 a. element **b.** compound
c. mixture

3.65 a. heterogeneous **b.** homogeneous
c. homogeneous

3.67 a. liquid **b.** gas
c. liquid **d.** solid
e. solid

3.69 a. physical **b.** physical
c. physical **d.** chemical

3.71 a. chemical **b.** physical
c. physical **d.** chemical

3.73 **a.** 56.7 °C, 330. K
b. −56.5 °C, 217 K

3.75 −26 °C, 247 K

3.77 3500 kcal

3.79 The same amount of heat causes a greater temperature change in the sand than in the water, thus the sand must have a lower specific heat than that of water.

3.81 Freeze-drying removes water from the food by sublimation. This prevents bacterial activity as microbes cannot survive without moisture.

3.83 9.0 kcal/g

3.85 **a.** about −60 °C
b. about 60 °C
c. The diagonal line A represents the solid state as temperature increases. The horizontal line B represents the change from solid to liquid or melting of the substance. The diagonal line C represents the liquid state as temperature increases. The horizontal line D represents the change from liquid to gas or boiling of the liquid. The diagonal line E represents the gas state as temperature increases.
d. At −80 °C, solid; at −40 °C, liquid; at 25 °C, liquid; at 80 °C, gas

3.87

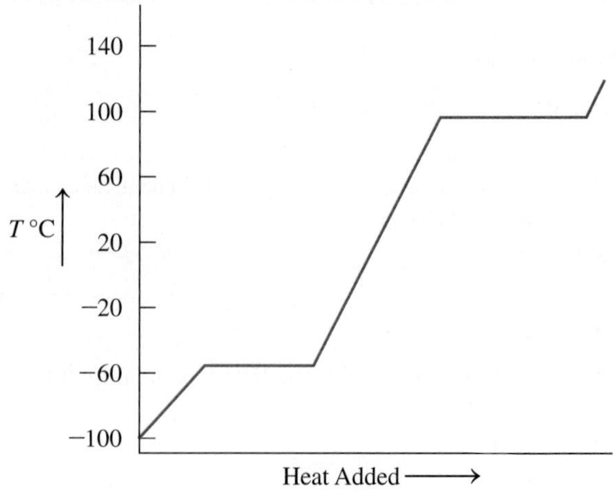

a. 80 cal/g **b.** 46.7 J/g

3.89 9.0 kcal/g

3.91 **a.** 0.93 kg **b.** 6.4 kg

3.93 119.5 kJ

COMBINING IDEAS FROM Chapters 1 to 3

CI.1 Gold, one of the most sought-after metals in the world, has a density of 19.3 g/cm^3, a melting point of 1064 °C, a specific heat of 0.129 J/g °C, and a heat of fusion of 63.6 J/g. A gold nugget found in Alaska in 1998 weighs 20.17 lb. (2.4, 2.6, 2.7, 3.3, 3.6, 3.7)

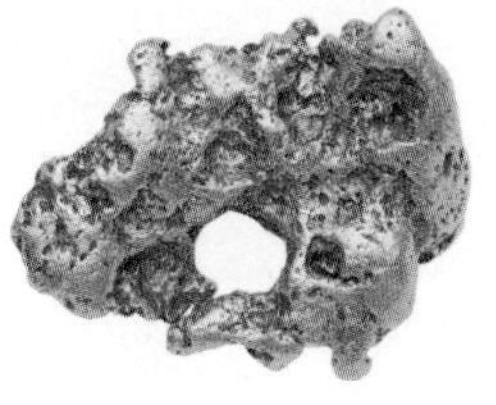

Gold nuggets, also called native gold, can be found in streams and mines.

a. How many significant figures are in the measurement of the weight of the nugget?
b. Which is the mass of the nugget in kilograms?
c. If the nugget were pure gold, what would its volume be in cm^3?
d. What is the melting point of gold in degrees Fahrenheit and kelvins?
e. How many kilocalories are required to raise the temperature of the nugget from 500. °C to 1064 °C and melt all the gold to liquid at 1064 °C?
f. If the price of gold is $55.98 per gram, what is the nugget worth, in dollars?

CI.2 The mileage for a motorcycle with a fuel-tank capacity of 22 L is 35 mi/gal. (2.5, 2.6, 2.7, 3.4)

When 1.00 g of gasoline burns, 47 kJ of energy are released.

a. How long a trip, in kilometers, can be made on one full tank of gasoline?
b. If the price of gasoline is $3.82 per gallon, what would be the cost of fuel for the trip?
c. If the average speed during the trip is 44 mi/h, how many hours will it take to reach the destination?
d. If the density of gasoline is 0.74 g/mL, what is the mass, in grams, of the fuel in the tank?
e. When 1.00 g of gasoline burns, 47 kJ of energy is released. How many kilojoules are produced when the fuel in one full tank is burned?

CI.3 Answer the following questions for the water samples **A** and **B** shown in the diagrams: (3.1, 3.2, 3.6)

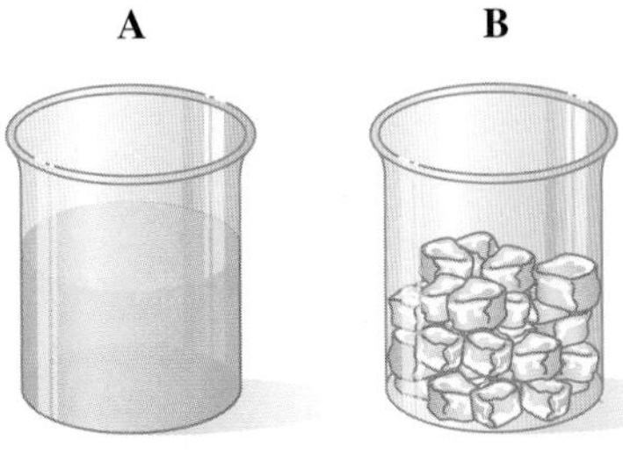

a. In which sample (**A** or **B**) does the water have its own shape?
b. Which diagram (**1** or **2** or **3**) represents the arrangement of particles in water sample **A**?
c. Which diagram (**1** or **2** or **3**) represents the arrangement of particles in water sample **B**?

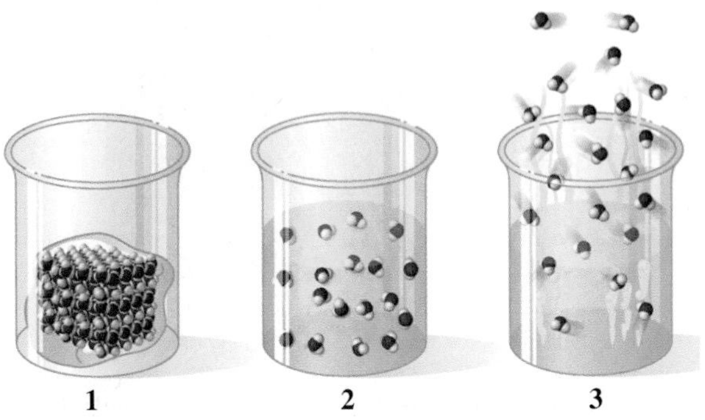

Answer the following for diagrams **1**, **2**, and **3**: (3.2, 3.3)
d. The state of matter indicated in diagram **1** is a ______; in diagram **2**, it is a ______; and in diagram **3**, it is a ______.
e. The motion of the particles is slowest in diagram ______.
f. The arrangement of particles is farthest apart in diagram ______.
g. The particles fill the volume of the container in diagram ______.
h. If the water in diagram **2** has a mass of 19 g and a temperature of 45 °C, how much heat, in kilojoules, is removed to cool the liquid to 0 °C?

CI.4 The label of a black cherry almond energy bar with a mass of 68 g lists the nutrition facts as 39 g of carbohydrate, 5 g of fat, and 10. g of protein. (2.5, 2.6, 3.4, 3.5)

An energy bar contains carbohydrate, fat, and protein.

a. Using the energy values for carbohydrates, fats, and proteins (see Table 3.7), what are the total kilocalories (Calories) listed for a black cherry almond bar? (Round off answers for each food type to the tens place.)
b. What are the kilojoules for the black cherry almond bar? (Round off answers for each food type to the tens place.)
c. If you obtain 160 kJ, how many grams of the black cherry almond bar did you eat?
d. If you are walking and using energy at a rate of 840 kJ/h, how many minutes will you need to walk to expend the energy from two black cherry almond bars?

CI.5 In one box of nails, there are 75 iron nails weighing 0.250 lb. The density of iron is 7.86 g/cm^3. The specific heat of iron is 0.452 J/g °C. The melting point of iron is 1535 °C. The heat of fusion for iron is 272 J/g. (2.5, 2.6, 2.7, 3.4, 3.6, 3.7)

Nails made of iron have a density of 7.86 g/cm^3.

a. What is the volume, in cm^3, of the iron nails in the box?
b. If 30 nails are added to a graduated cylinder containing 17.6 mL of water, what is the new level of water, in milliliters, in the cylinder?
c. How much heat, in joules, must be added to the nails in the box to raise their temperature from 16 °C to 125 °C?
d. How much heat, in joules, is required to heat one nail from 25 °C to its melting point and change it to liquid iron?

CI.6 A hot tub is filled with 450 gal of water. (2.5, 2.6, 2.7, 3.3, 3.4, 3.6)

A hot tub filled with water is heated to 105 °F.

a. What is the volume of water, in liters, in the tub?
b. What is the mass, in kilograms, of water in the tub?
c. How many kilocalories are needed to heat the water from 62 °F to 105 °F?
d. If the hot-tub heater provides 5900 kJ/min, how long, in minutes, will it take to heat the water in the hot tub from 62 °F to 105 °F?

ANSWERS

CI.1
a. 4 significant figures
b. 9.17 kg
c. 475 cm^3
d. 1947 °F; 1337 K
e. 298 kcal
f. $513 000

CI.3
a. **B**
b. **A** is represented by diagram **2**.
c. **B** is represented by diagram **1**.
d. solid; liquid; gas
e. diagram **1**
f. diagram **3**
g. diagram **3**
h. 3.6 kJ

CI.5
a. 14.4 cm^3
b. 23.4 mL
c. 5590 J
d. 1440 J

4

Atoms and Elements

John is preparing for the next growing season as he decides how much of each crop should be planted and their location on his farm. Part of this decision is determined by the quality of the soil including the pH, the amount of moisture, and the nutrient content in the soil. He begins by sampling the soil and performing a few chemical tests on the soil. John determines that several of his fields need additional fertilizer before the crops can be planted. John considers several different types of fertilizers as each supplies different nutrients to the soil to help increase crop production. Plants need three basic elements for growth. These elements are potassium, nitrogen, and phosphorus. Potassium (K on the periodic table) is a metal, while nitrogen (N) and phosphorus (P) are nonmetals. Fertilizers may also contain several other elements including calcium (Ca), magnesium (Mg), and sulfur (S).
John applies a fertilizer containing a mixture of all of these elements to his soil and plans to re-check the soil nutrient content in a few days.

CAREER Farmer

Farming involves much more than growing crops and raising animals. Farmers must understand how to perform chemical tests and how to apply fertilizer to soil and pesticides or herbicides to crops. Pesticides are chemicals used to kill insects that could destroy the crop, while herbicides are chemicals used to kill weeds that would compete for the crop's water and nutrient supply. This requires knowledge of how these chemicals work, their safety, effectiveness, and their storage. In using this information, farmers are able to grow crops that produce a higher yield, greater nutritional value, and better taste.

LOOKING AHEAD

All matter is composed of *elements*, of which there are 118 different kinds. Of these, 88 elements occur naturally and make up all the substances in our world. Many elements are already familiar to you. Perhaps you use aluminum in the form of foil or drink soft drinks from aluminum cans. You may have a ring or necklace made of gold, silver, or perhaps platinum. If you play tennis or golf, then you may have noticed that your racket or clubs may be made from the elements titanium or carbon. In our bodies, calcium and phosphorus form the structure of bones and teeth, iron and copper are needed in the formation of red blood cells, and iodine is required for the proper functioning of the thyroid.

The correct amounts of certain elements are crucial to the proper growth and function of the body. Low levels of iron can lead to anemia, while lack of iodine can cause hypothyroidism and goiter. Some elements known as microminerals, such as chromium, cobalt, and selenium, are needed in our bodies in very small amounts. Laboratory tests are used to confirm that these elements are within normal ranges in our bodies.

CHAPTER READINESS*

KEY MATH SKILLS

- Using Positive and Negative Numbers in Calculations (1.4B)
- Calculating a Percentage (1.4C)

CORE CHEMISTRY SKILLS

- Counting Significant Figures (2.2)
- Rounding Off (2.3)
- Using Significant Figures in Calculations (2.3)

*These Key Math Skills and Core Chemistry Skills from previous chapters are listed here for your review as you proceed to the new material in this chapter.

LEARNING GOAL

Given the name of an element, write its correct symbol; from the symbol, write the correct name.

4.1 Elements and Symbols

Elements are pure substances from which all other things are built. Elements cannot be broken down into simpler substances. Over the centuries, elements have been named for planets, mythological figures, colors, minerals, geographic locations, and famous people. Some sources of names of elements are listed in Table 4.1. A complete list of all the elements and their symbols are found on the inside front cover of this text.

TABLE 4.1 Some Elements, Symbols, and Source of Names

Element	Symbol	Source of Name
Uranium	U	The planet Uranus
Titanium	Ti	Titans (mythology)
Chlorine	Cl	*Chloros*: "greenish yellow" (Greek)
Iodine	I	*Ioeides*: "violet" (Greek)
Magnesium	Mg	Magnesia, a mineral
Californium	Cf	California
Curium	Cm	Marie and Pierre Curie
Copernicium	Cn	Nicolaus Copernicus

Chemical Symbols

Chemical symbols are one- or two-letter abbreviations for the names of the elements. Only the first letter of an element's symbol is capitalized. If the symbol has a second letter, it is lowercase so that we know when a different element is indicated. If two letters are capitalized, they represent the symbols of two different elements. For example, the element cobalt has the symbol Co. However, the two capital letters CO specify two elements, carbon (C) and oxygen (O).

Although most of the symbols use letters from the current names, some are derived from their ancient names. For example, Na, the symbol for sodium, comes from the Latin word *natrium*. The symbol for iron, Fe, is derived from the Latin name *ferrum*. Table 4.2 lists the names and symbols of some common elements. Learning their names and symbols will greatly help your learning of chemistry.

One-Letter Symbols		Two-Letter Symbols	
C	Carbon	Co	Cobalt
S	Sulfur	Si	Silicon
N	Nitrogen	Ne	Neon
I	Iodine	Ni	Nickel

TABLE 4.2 Names and Symbols of Some Common Elements

Name*	Symbol	Name*	Symbol	Name*	Symbol
Aluminum	Al	Gold (*aurum*)	Au	Oxygen	O
Argon	Ar	Helium	He	Phosphorus	P
Arsenic	As	Hydrogen	H	Platinum	Pt
Barium	Ba	Iodine	I	Potassium (*kalium*)	K
Boron	B	Iron (*ferrum*)	Fe	Radium	Ra
Bromine	Br	Lead (*plumbum*)	Pb	Silicon	Si
Cadmium	Cd	Lithium	Li	Silver (*argentum*)	Ag
Calcium	Ca	Magnesium	Mg	Sodium (*natrium*)	Na
Carbon	C	Manganese	Mn	Strontium	Sr
Chlorine	Cl	Mercury (*hydrargyrum*)	Hg	Sulfur	S
Chromium	Cr	Neon	Ne	Tin (*stannum*)	Sn
Cobalt	Co	Nickel	Ni	Titanium	Ti
Copper (*cuprum*)	Cu	Nitrogen	N	Uranium	U
Fluorine	F			Zinc	Zn

*Names given in parentheses are ancient Latin or Greek words from which the symbols are derived.

Aluminum

Carbon

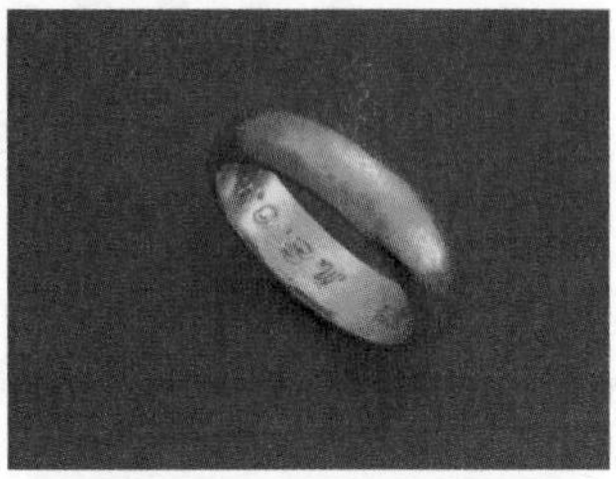
Gold

Silver

Sulfur

SAMPLE PROBLEM 4.1 Writing Chemical Symbols

What are the chemical symbols for the following elements?

a. nickel **b.** nitrogen **c.** neon

SOLUTION

a. Ni **b.** N **c.** Ne

STUDY CHECK 4.1

What are the chemical symbols for the elements silicon, sulfur, and silver?

SAMPLE PROBLEM 4.2 Names and Symbols of Chemical Elements

Give the name of the element that corresponds to each of the following chemical symbols:

a. Zn **b.** K **c.** H **d.** Fe

SOLUTION

a. zinc **b.** potassium **c.** hydrogen **d.** iron

STUDY CHECK 4.2

What are the names of the elements with the chemical symbols Mg, Al, and F?

Chemistry Link to the Environment

Many Forms of Carbon

Carbon has the symbol C. However, its atoms can be arranged in different ways to give several different substances. Two forms of carbon—diamond and graphite—have been known since prehistoric times. In diamond, carbon atoms are arranged in a rigid structure. A diamond is transparent and harder than any other substance, whereas graphite is black and soft. In graphite, carbon atoms are arranged in sheets that slide over each other. Graphite is used as pencil lead, as lubricants, and as carbon fibers for the manufacture of lightweight golf clubs and tennis rackets.

Two other forms of carbon have been discovered more recently. In the form called *Buckminsterfullerene* or *buckyball* (named after R. Buckminster Fuller, who popularized the geodesic dome), 60 carbon atoms are arranged as rings of five and six atoms to give a spherical, cage-like structure. When a fullerene structure is stretched out, it produces a cylinder with a diameter of only a few nanometers called a *nanotube*. Practical uses for buckyballs and nanotubes are not yet developed, but they are expected to find use in lightweight structural materials, heat conductors, computer parts, and medicine. Recent research has shown that carbon nanotubes (CNTs) can carry many drug molecules that can be released once the CNTs enter the targeted cells.

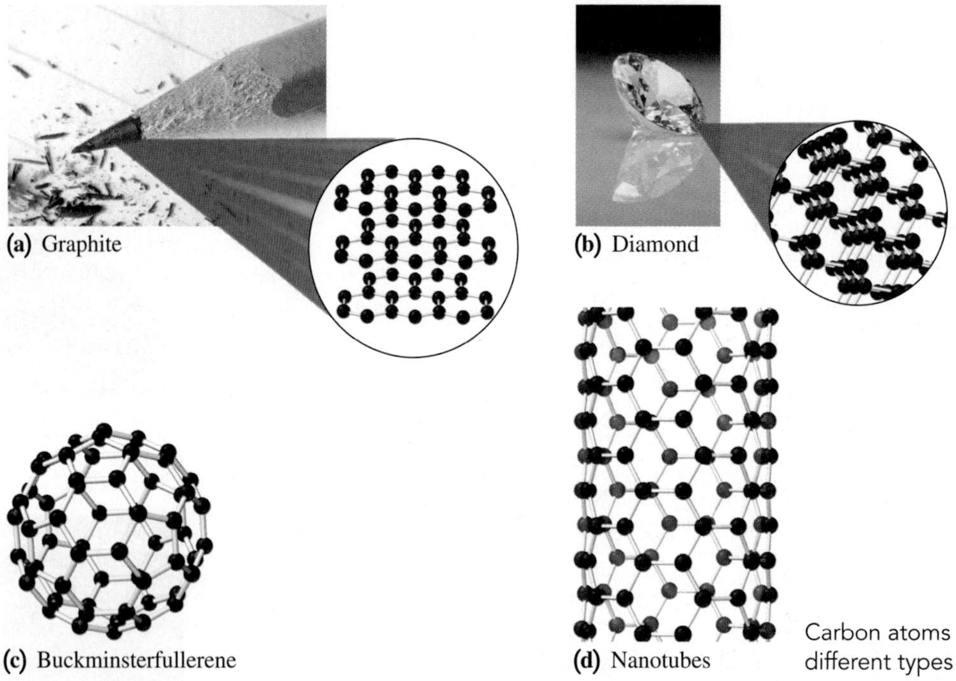

Carbon atoms can form different types of structures.

QUESTIONS AND PROBLEMS

4.1 Elements and Symbols

LEARNING GOAL Given the name of an element, write its correct symbol; from the symbol, write the correct name.

4.1 Write the symbol for the following elements:
a. sodium **b.** potassium **c.** nickel **d.** silver
e. nitrogen **f.** palladium **g.** vanadium **h.** zinc

4.2 Write the symbols for the following elements:
a. oxygen **b.** lithium **c.** uranium **d.** titanium
e. hydrogen **f.** chromium **g.** tin **h.** gold

4.3 Write the name of the element for each symbol.
a. S **b.** F **c.** Ne **d.** Au
e. Al **f.** Cu **g.** Pd **h.** Mg

4.4 Write the name of the element for each symbol.
a. He **b.** P **c.** Na **d.** As
e. Ca **f.** Br **g.** Cd **h.** Si

4.5 What elements are represented by the symbols in each of the following substances?
a. table salt, NaCl **b.** plaster casts, $CaSO_4$
c. Demerol, $C_{15}H_{22}ClNO_2$ **d.** pigment, $BaCrO_4$

4.6 What elements are represented by the symbols in each of the following substances?
a. fireworks, KNO_3 **b.** dental cement, $Zn_3(PO_4)_2$
c. antacid, $Mg(OH)_2$ **d.** manufacture of aluminum, AlF_3

Chemistry Link to Health

Toxicity of Mercury

Mercury is a silvery, shiny element that is a liquid at room temperature. Mercury can enter the body through inhaled mercury vapor, contact with the skin, or ingestion of foods or water contaminated with mercury. In the body, mercury destroys proteins and disrupts cell function. Long-term exposure to mercury can damage the brain and kidneys, cause mental retardation, and decrease physical development. Blood, urine, and hair samples are used to test for mercury.

In both freshwater and seawater, bacteria convert mercury into toxic methylmercury, which attacks the central nervous system. Because fish absorb methylmercury, we are exposed to mercury by consuming mercury-contaminated fish. As levels of mercury ingested from fish became a concern, the Food and Drug Administration set a maximum level of one part mercury per million parts seafood (1 ppm), which is the same as 1 mg of mercury in every kilogram of seafood. Fish higher in the food chain, such as swordfish and shark, can have such high levels of mercury that the Environmental Protection Agency recommends they be consumed no more than once a week.

One of the worst incidents of mercury poisoning occurred in Minamata and Niigata, Japan, in 1950. At that time, the ocean was polluted with high levels of mercury from industrial wastes. Because fish were a major food in the diet, more than 2000 people were affected with mercury poisoning and died or developed neural damage. In the United States between 1988 and 1997, the use of mercury decreased by 75% when the use of mercury was banned in paint and pesticides, and regulated in batteries and other products. Certain batteries and compact fluorescent light bulbs (CFL) contain mercury, and instructions for their safe disposal should be followed.

This mercury fountain, housed in glass, was designed by Alexander Calder for the 1937 World's Fair in Paris.

4.2 The Periodic Table

LEARNING GOAL

Use the periodic table to identify the group and the period of an element; identify the element as a metal, a nonmetal, or a metalloid.

As more elements were discovered, it became necessary to organize them into some type of classification system. By the late 1800s, scientists recognized that certain elements looked alike and behaved much the same way. In 1872, a Russian chemist, Dmitri Mendeleev, arranged the 60 elements known at that time into groups with similar properties and placed them in order of increasing atomic masses. Today, this arrangement of 118 elements is known as the **periodic table** (see Figure 4.1).

Periods and Groups

Each horizontal row in the periodic table is a **period** (see Figure 4.2). The periods are counted from the top of the table as Periods 1 to 7. The first period contains two elements: hydrogen (H) and helium (He). The second period contains eight elements: lithium (Li), beryllium (Be), boron (B), carbon (C), nitrogen (N), oxygen (O), fluorine (F), and neon (Ne). The third period also contains eight elements beginning with sodium (Na) and ending with argon (Ar). The fourth period, which begins with potassium (K), and the fifth period, which begins with rubidium (Rb), have 18 elements each. The sixth period, which begins with cesium (Cs), has 32 elements. The seventh period, as of today, contains 32 elements, for a total of 118 elements.

Each vertical column on the periodic table contains a **group** (or family) of elements that have similar properties. A **group number** is written at the top of each vertical column (group) in the periodic table. For many years, the **representative elements** have had group numbers 1A to 8A. In the center of the periodic table is a block of elements known as the **transition elements**, which have had group numbers followed by the letter "B." A newer system assigns numbers of 1 to 18 to all of the groups going left to right across the periodic table. Because both systems are currently in use, they are both shown on the periodic table in this text and are included in our discussions of elements and group numbers. The two rows of 14 elements called the *lanthanides* and *actinides* (or the inner transition elements), which are part of Periods 6 and 7, are placed at the bottom of the periodic table to allow it to fit on a page.

FIGURE 4.1 ▶ On the periodic table, groups are the elements arranged as vertical columns, and periods are the elements in each horizontal row.

Q What is the symbol of the alkali metal in Period 3?

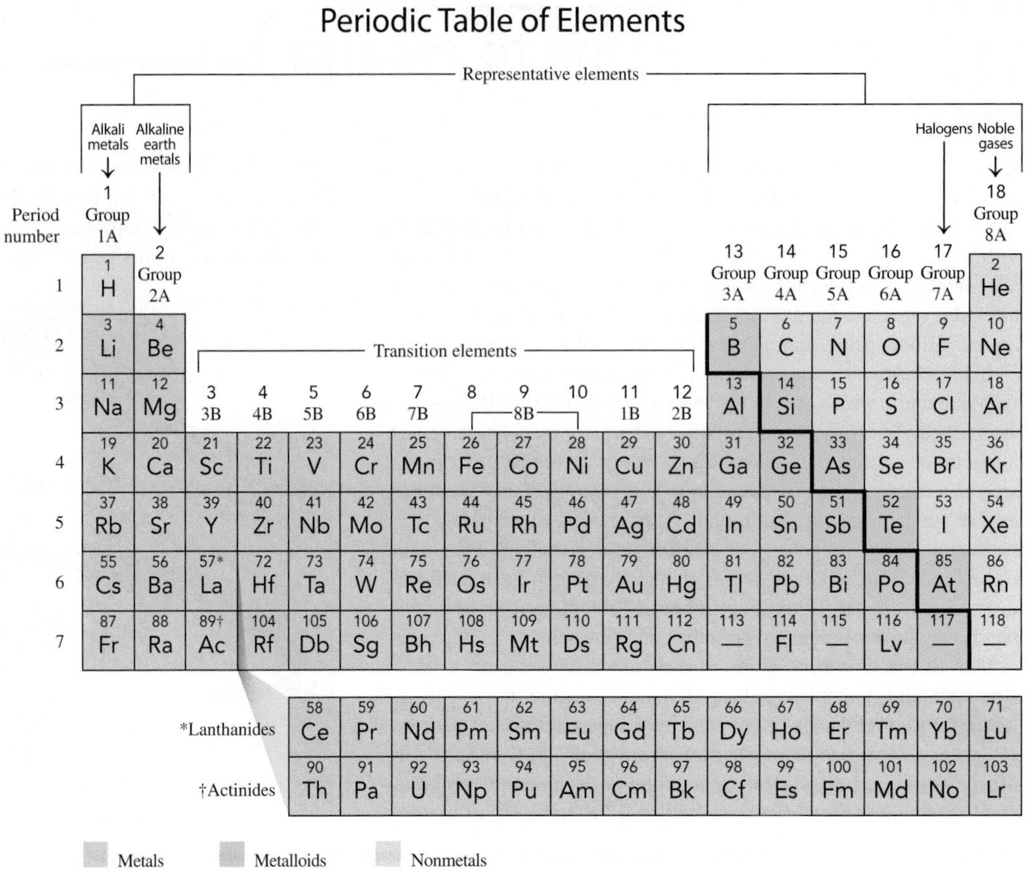

Names of Groups

Several groups in the periodic table have special names (see Figure 4.3). Group 1A (1) elements—lithium (Li), sodium (Na), potassium (K), rubidium (Rb), cesium (Cs), and francium (Fr)—are a family of elements known as the **alkali metals** (see Figure 4.4). The elements within this group are soft, shiny metals that are good conductors of heat and electricity and have relatively low melting points. Alkali metals react vigorously with water and form white products when they combine with oxygen.

FIGURE 4.2 ▶ On the periodic table, each vertical column represents a group of elements and each horizontal row of elements represents a period.

Q Are the elements Si, P, and S part of a group or a period?

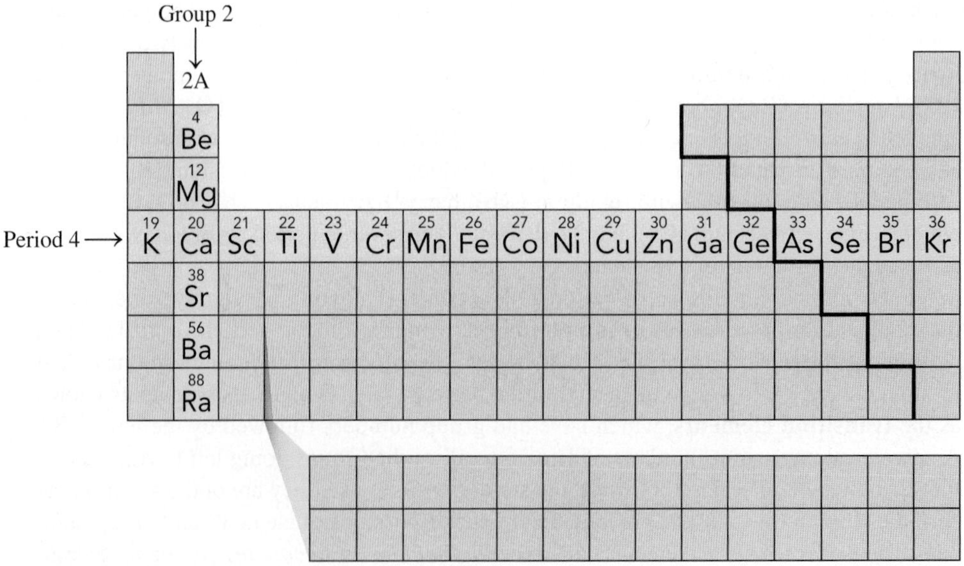

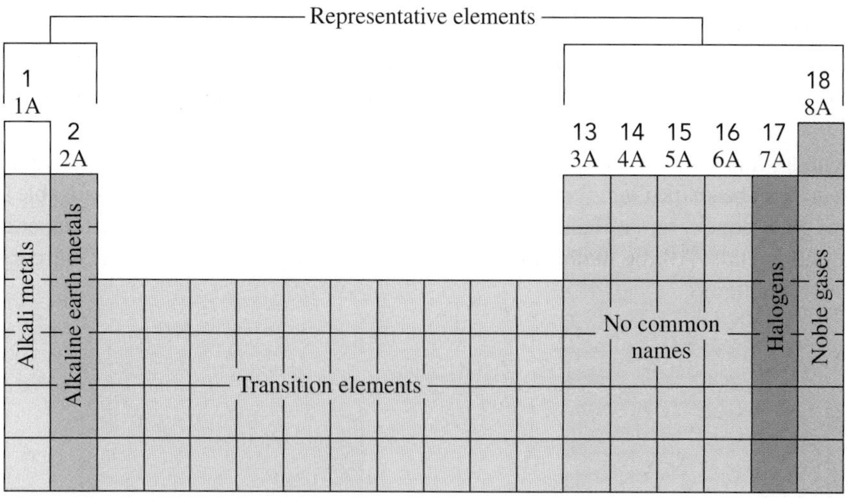

FIGURE 4.3 ▶ Certain groups on the periodic table have common names.

Q What is the common name for the group of elements that includes helium and argon?

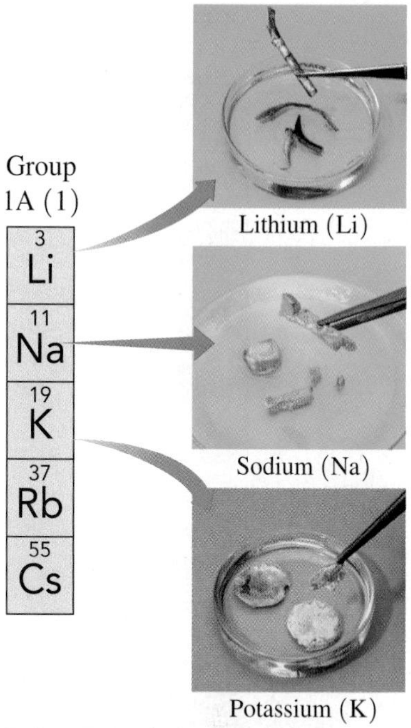

FIGURE 4.4 ▶ Lithium (Li), sodium (Na), and potassium (K) are some alkali metals from Group 1A (1).

Q What physical properties do these alkali metals have in common?

Although hydrogen (H) is at the top of Group 1A (1), it is not an alkali metal and has very different properties than the rest of the elements in this group. Thus hydrogen is not included in the alkali metals.

The **alkaline earth metals** are found in Group 2A (2). They include the elements beryllium (Be), magnesium (Mg), calcium (Ca), strontium (Sr), barium (Ba), and radium (Ra). The alkaline earth metals are shiny metals like those in Group 1A (1), but they are not as reactive.

The **halogens** are found on the right side of the periodic table in Group 7A (17). They include the elements fluorine (F), chlorine (Cl), bromine (Br), iodine (I), and astatine (At) (see Figure 4.5). The halogens, especially fluorine and chlorine, are highly reactive and form compounds with most of the elements.

The **noble gases** are found in Group 8A (18). They include helium (He), neon (Ne), argon (Ar), krypton (Kr), xenon (Xe), and radon (Rn). They are quite unreactive and are seldom found in combination with other elements.

SAMPLE PROBLEM 4.3 Group and Period Numbers of Some Elements

Give the group and period numbers for each of the following elements and identify each as a representative or a transition element:

a. manganese **b.** barium **c.** gold

SOLUTION

a. Manganese (Mn) in Group 7B (7) and Period 4 is a transition element.
b. Barium (Ba) in Group 2A (2) and Period 6 is a representative element.
c. Gold (Au) in Group 1B (11) and Period 6 is a transition element.

STUDY CHECK 4.3

Strontium is an element that gives a brilliant red color to fireworks.

a. In what group is strontium found?
b. What is the name of this chemical family?
c. In what period is strontium found?

Strontium provides the red color in fireworks.

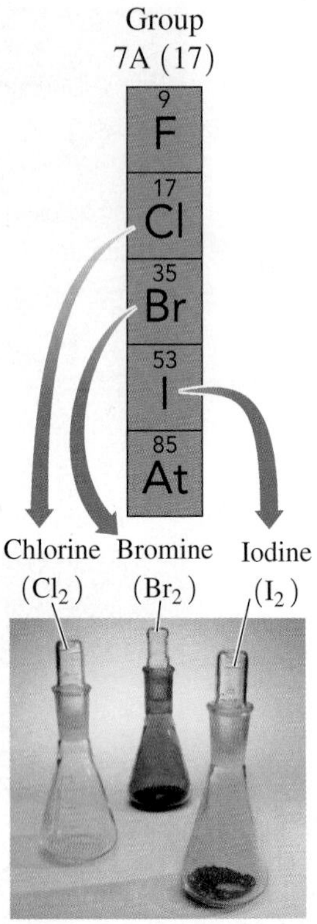

FIGURE 4.5 ▶ Chlorine (Cl_2), bromine (Br_2), and iodine (I_2) are examples of halogens from Group 7A (17).

Ⓠ What elements are in the halogen group?

Metals, Nonmetals, and Metalloids

Another feature of the periodic table is the heavy zigzag line that separates the elements into the *metals* and the *nonmetals*. *Except for hydrogen*, the metals are to the left of the line with the nonmetals to the right (see Figure 4.6).

In general, most **metals** are shiny solids, such as copper (Cu), gold (Au), and silver (Ag). Metals can be shaped into wires (ductile) or hammered into a flat sheet (malleable). Metals are good conductors of heat and electricity. They usually melt at higher temperatures than nonmetals. All the metals are solids at room temperature, except for mercury (Hg), which is a liquid.

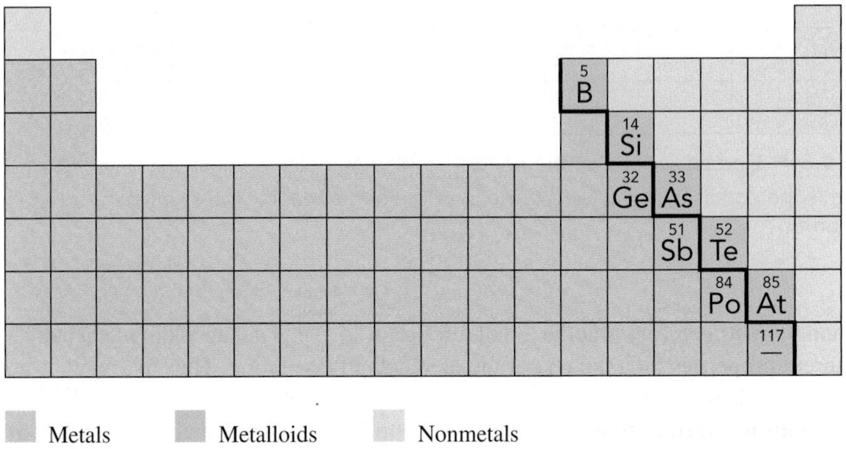

FIGURE 4.6 ▶ Along the heavy zigzag line on the periodic table that separates the metals and nonmetals are metalloids, which exhibit characteristics of both metals and nonmetals.

Ⓠ On which side of the heavy zigzag line are the nonmetals located?

Nonmetals are not especially shiny, ductile, or malleable, and they are often poor conductors of heat and electricity. They typically have low melting points and low densities. Some examples of nonmetals are hydrogen (H), carbon (C), nitrogen (N), oxygen (O), chlorine (Cl), and sulfur (S).

Except for aluminum, the elements located along the heavy line are **metalloids**: B, Si, Ge, As, Sb, Te, Po, and At. Metalloids are elements that exhibit some properties that are typical of the metals and other properties that are characteristic of the nonmetals. For example, they are better conductors of heat and electricity than the nonmetals, but not as good as the metals. The metalloids are *semiconductors* because they can be modified to function as conductors or insulators. Table 4.3 compares some characteristics of silver, a metal, with those of antimony, a metalloid, and sulfur, a nonmetal.

A silver cup is shiny, antimony is a blue-gray solid, and sulfur is a dull, yellow color.

TABLE 4.3 Some Characteristics of a Metal, a Metalloid, and a Nonmetal

Silver (Ag)	Antimony (Sb)	Sulfur (S)
Metal	Metalloid	Nonmetal
Shiny	Blue-gray, shiny	Dull, yellow
Extremely ductile	Brittle	Brittle
Can be hammered into sheets (malleable)	Shatters when hammered	Shatters when hammered
Good conductor of heat and electricity	Poor conductor of heat and electricity	Poor conductor of heat and electricity, good insulator
Used in coins, jewelry, tableware	Used to harden lead, color glass and plastics	Used in gunpowder, rubber, fungicides
Density 10.5 g/mL	Density 6.7 g/mL	Density 2.1 g/mL
Melting point 962 °C	Melting point 630 °C	Melting point 113 °C

SAMPLE PROBLEM 4.4 Metals, Nonmetals, and Metalloids

Use the periodic table to classify each of the following elements by its group and period, group name (if any), and as a metal, a nonmetal, or a metalloid:

a. Na, important in nerve impulses, regulates blood pressure
b. I, needed to produce thyroid hormones
c. Si, needed for tendons and ligaments

SOLUTION

a. Na (sodium), Group 1A (1), Period 3, is an alkali metal.
b. I (iodine), Group 7A (17), Period 5, halogen, is a nonmetal.
c. Si (silicon), Group 4A (14), Period 3, is a metalloid.

STUDY CHECK 4.4

Give the name and symbol of the element represented by each of the following:

a. Group 5A (15), Period 4
b. A noble gas in Period 6
c. A metalloid in Period 2

Chemistry Link to **Health**

Elements Essential to Health

Of all the elements, only about 20 are essential for the well-being and survival of the human body. Of those, four elements—oxygen, carbon, hydrogen, and nitrogen—which are representative elements in Period 1 and Period 2 on the periodic table, make up 96% of our body mass. Most of the food in our daily diet provides these elements to maintain a healthy body. These elements are found in carbohydrates, fats, and proteins. Most of the hydrogen and oxygen is found in water, which makes up 55 to 60% of our body mass.

The *macrominerals*—Ca, P, K, Cl, S, Na, and Mg—are representative elements located in Period 3 and Period 4 of the periodic table. They are involved in the formation of bones and teeth, maintenance of heart and blood vessels, muscle contraction, nerve impulses, acid–base balance of body fluids, and regulation of cellular metabolism. The macrominerals are present in lower amounts than the major elements, so that smaller amounts are required in our daily diets.

The other essential elements, called *microminerals* or *trace elements*, are mostly transition elements in Period 4 along with Si in Period 3 and Mo and I in Period 5. They are present in the human body in very small amounts, some less than 100 mg. In recent years, the detection of such small amounts has improved so that researchers can more easily identify the roles of trace elements. Some trace elements such as arsenic, chromium, and selenium are toxic at high levels in the body but are still required by the body. Other elements, such as tin and nickel, are thought to be essential, but their metabolic role has not yet been determined. Some examples and the amounts present in a 60-kg person are listed in Table 4.4.

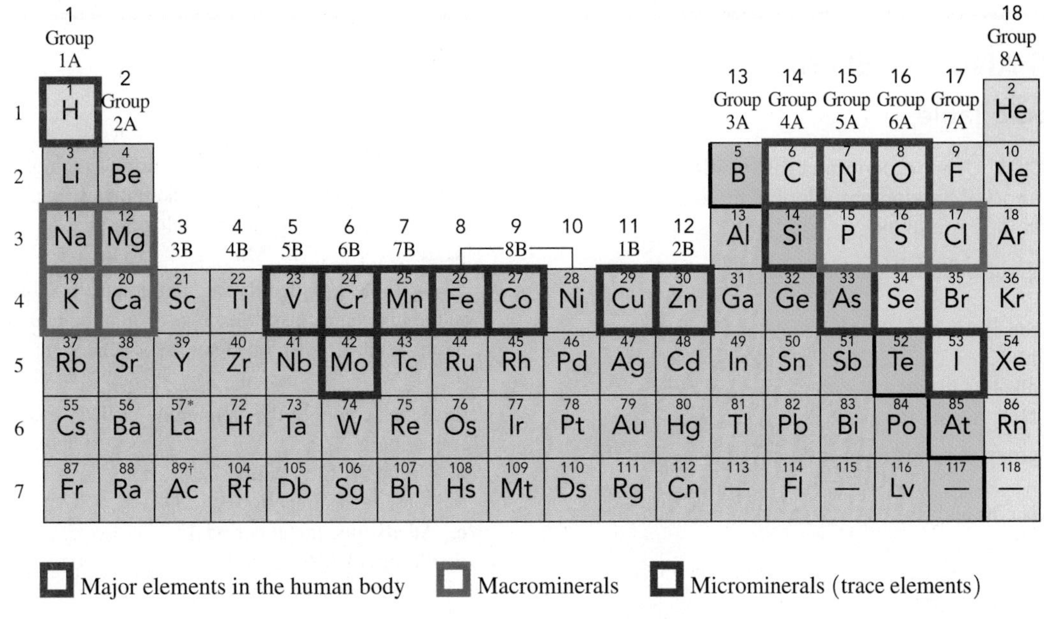

TABLE 4.4 Typical Amounts of Essential Elements in a 60-kg Adult

Element	Quantity	Function
Major Elements		
Oxygen (O)	39 kg	Building block of biomolecules and water (H_2O)
Carbon (C)	11 kg	Building block of organic and biomolecules
Hydrogen (H)	6 kg	Component of biomolecules, water (H_2O), and pH of body fluids, stomach acid (HCl)
Nitrogen (N)	1.5 kg	Component of proteins and nucleic acids
Macrominerals		
Calcium (Ca)	1000 g	Needed for bone and teeth, muscle contraction, nerve impulses
Phosphorus (P)	600 g	Needed for bone and teeth, nucleic acids
Potassium (K)	120 g	Most prevalent positive ion (K^+) in cells, muscle contraction, nerve impulses
Chlorine (Cl)	100 g	Most prevalent negative ion (Cl^-) in fluids outside cells, stomach acid (HCl)
Sulfur (S)	86 g	Component of proteins, liver, vitamin B_1, insulin
Sodium (Na)	60 g	Most prevalent positive ion (Na^+) in fluids outside cells, water balance, muscle contraction, nerve impulses
Magnesium (Mg)	36 g	Component of bone, required for metabolic reactions
Microminerals (trace elements)		
Iron (Fe)	3600 mg	Component of oxygen carrier hemoglobin
Silicon (Si)	3000 mg	Needed for growth and maintenance of bone and teeth, tendons and ligaments, hair and skin
Zinc (Zn)	2000 mg	Used in metabolic reactions in cells, DNA synthesis, growth of bone, teeth, connective tissue, immune system
Copper (Cu)	240 mg	Needed for blood vessels, blood pressure, immune system
Manganese (Mn)	60 mg	Needed for bone growth, blood clotting, necessary for metabolic reactions
Iodine (I)	20 mg	Needed for proper thyroid function
Molybdenum (Mo)	12 mg	Needed to process Fe and N from food
Arsenic (As)	3 mg	Needed for growth and reproduction
Chromium (Cr)	3 mg	Needed for maintenance of blood sugar levels, synthesis of biomolecules
Cobalt (Co)	3 mg	Component of vitamin B_{12}, red blood cells
Selenium (Se)	2 mg	Used in the immune system, health of heart and pancreas
Vanadium (V)	2 mg	Needed in the formation of bone and teeth, energy from food

QUESTIONS AND PROBLEMS

4.2 The Periodic Table

LEARNING GOAL Use the periodic table to identify the group and the period of an element; identify the element as a metal, a nonmetal, or a metalloid.

4.7 Identify the group or period number described by each of the following:
a. contains the elements O, F, Ne
b. begins with K
c. contains alkaline metals
d. contains N, P, As, Sb, Bi

4.8 Identify the group or period number described by each of the following:
a. contains Na, K, and Rb
b. begins with Li
c. contains the noble gases
d. contains F, Cl, Br, and I

4.9 Using Table 4.4 identify the function of each of the following in the body and classify each as alkali metal, an alkaline earth metal, a transition metal or a halogen:
a. K **b.** Mn **c.** Co **d.** I

4.10 Using Table 4.4, identify the function of each of the following in the body and classify each as an alkali metal, an alkaline earth metal, a transition element, or a halogen:
a. Mg **b.** Cu **c.** I **d.** Na

4.11 Give the symbol of the element described by each of the following:
a. Group 5A (15) period 2
b. A noble gas in period 3
c. An alkaline metal period 4
d. Group 6A (16) period 3
e. Group 7A (17) period 4

4.12 Give the symbol of the element described by each of the following:
- **a.** an alkaline earth metal in Period 2
- **b.** Group 5A (15), Period 3
- **c.** a noble gas in Period 4
- **d.** a halogen in Period 5
- **e.** Group 4A (14), Period 4

4.13 Identify each of the following elements as a metal, a non metal or a metalloid.
- **a.** Lithium
- **b.** Phosphorous
- **c.** Element used in coins, jewelry
- **d.** Element which is poor conductor of electricity and heat
- **e.** Silicon
- **f.** Hydrogen
- **g.** Strontium
- **h.** Cesium

4.14 Identify each of the following elements as a metal, a nonmetal, or a metalloid:
- **a.** located in Group 2A (2)
- **b.** a good conductor of electricity
- **c.** chlorine
- **d.** arsenic
- **e.** an element that is not shiny
- **f.** oxygen
- **g.** nitrogen
- **h.** tin

4.3 The Atom

LEARNING GOAL

Describe the electrical charge and location in an atom for a proton, a neutron, and an electron.

All the elements listed on the periodic table are made up of atoms. An **atom** is the smallest particle of an element that retains the characteristics of that element. Imagine that you are tearing a piece of aluminum foil into smaller and smaller pieces. Now imagine that you have a microscopic piece so small that it cannot be divided any further. Then you would have a single atom of aluminum.

The concept of the atom is relatively recent. Although the Greek philosophers in 500 B.C.E. reasoned that everything must contain minute particles they called *atomos*, the idea of atoms did not become a scientific theory until 1808. Then John Dalton (1766–1844) developed an atomic theory that proposed that atoms were responsible for the combinations of elements found in compounds.

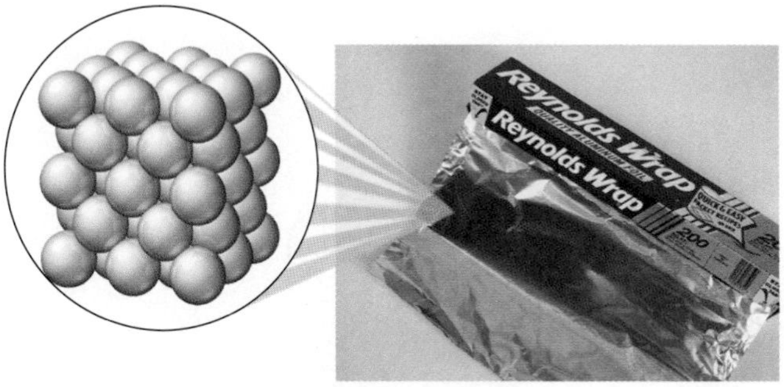

Aluminum foil consists of atoms of aluminum.

Dalton's Atomic Theory

1. All matter is made up of tiny particles called atoms.
2. All atoms of a given element are similar to one another and different from atoms of other elements.
3. Atoms of two or more different elements combine to form compounds. A particular compound is always made up of the same kinds of atoms and always has the same number of each kind of atom.
4. A chemical reaction involves the rearrangement, separation, or combination of atoms. Atoms are never created or destroyed during a chemical reaction.

Dalton's atomic theory formed the basis of current atomic theory, although we have modified some of Dalton's statements. We now know that atoms of the same element are not completely identical to each other and consist of even smaller particles. However, an atom is still the smallest particle that retains the properties of an element.

FIGURE 4.7 ▶ Images of nickel atoms are produced when nickel is magnified millions of times by a scanning tunneling microscope (STM).

Q Why is a microscope with extremely high magnification needed to see these atoms?

Although atoms are the building blocks of everything we see around us, we cannot see an atom or even a billion atoms with the naked eye. However, when billions and billions of atoms are packed together, the characteristics of each atom are added to those of the next until we can see the characteristics we associate with the element. For example, a small piece of the shiny element nickel consists of many, many nickel atoms. A special kind of microscope called a *scanning tunneling microscope* (STM) produces images of individual atoms (see Figure 4.7).

Electrical Charges in an Atom

By the end of the 1800s, experiments with electricity showed that atoms were not solid spheres but were composed of even smaller bits of matter called *subatomic particles*, three of which are the *proton*, *neutron*, and *electron*. Two of these subatomic particles were discovered because they have electrical charges.

An electrical charge can be positive or negative. Experiments show that like charges repel, or push away from each other. When you brush your hair on a dry day, electrical charges that are alike build up on the brush and in your hair. As a result, your hair flies away from the brush. However, opposite or unlike charges attract. The crackle of clothes taken from the clothes dryer indicates the presence of electrical charges. The clinginess of the clothing results from the attraction of opposite, unlike charges (see Figure 4.8).

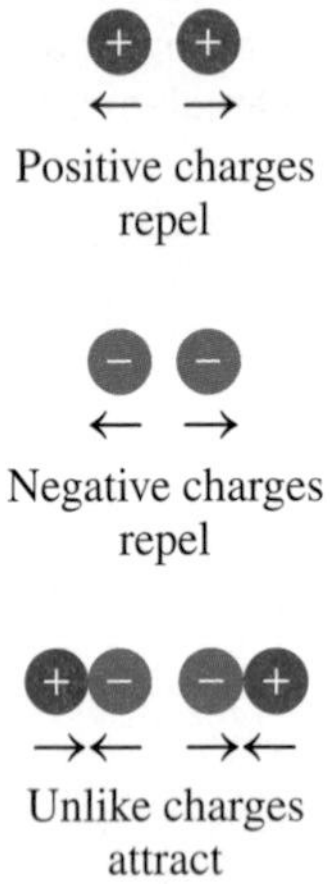

FIGURE 4.8 ▶ Like charges repel and unlike charges attract.

Q Why are the electrons attracted to the protons in the nucleus of an atom?

Structure of the Atom

In 1897, J. J. Thomson, an English physicist, applied electricity to a glass tube, which produced streams of small particles called *cathode rays*. Because these rays were attracted to a positively charged electrode, Thomson realized that the particles in the rays must be negatively charged. In further experiments, these particles called **electrons** were found to be much smaller than the atom and to have extremely small masses. Because atoms are neutral, scientists soon discovered that atoms contained positively charged particles called **protons** that were much heavier than the electrons.

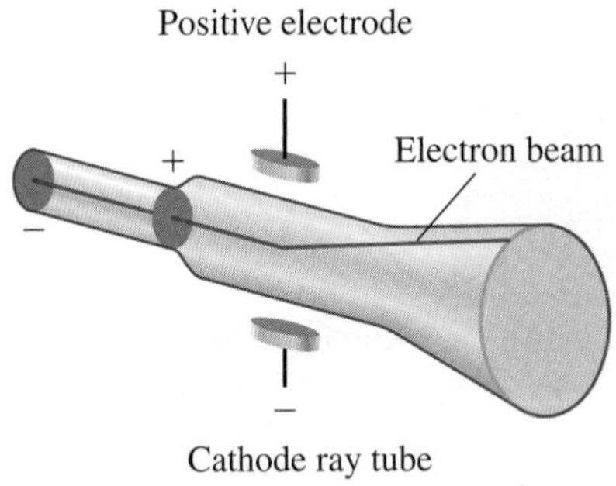

Negatively charged cathode rays (electrons) are attracted to the positive electrode.

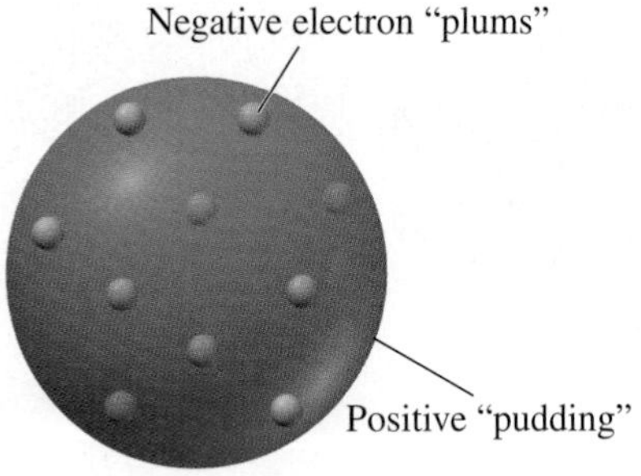

Thomson's "plum-pudding" model had protons and electrons scattered throughout the atom.

Thomson proposed a "plum-pudding" model for the atom in which the electrons and protons were randomly distributed in a positively charged cloud like "plums in a pudding" or "chocolate chips in a cookie." In 1911, Ernest Rutherford worked with Thomson to test this model. In Rutherford's experiment, positively charged particles were aimed at a thin sheet of gold foil (see Figure 4.9). If the Thomson model were correct, the particles would travel in straight paths through the gold foil. Rutherford was greatly surprised to find that some of the particles were deflected as they passed through the gold foil, and a few particles were deflected so much that they went back in the opposite direction. According to Rutherford, it was as though he had shot a cannonball at a piece of tissue paper, and it bounced back at him.

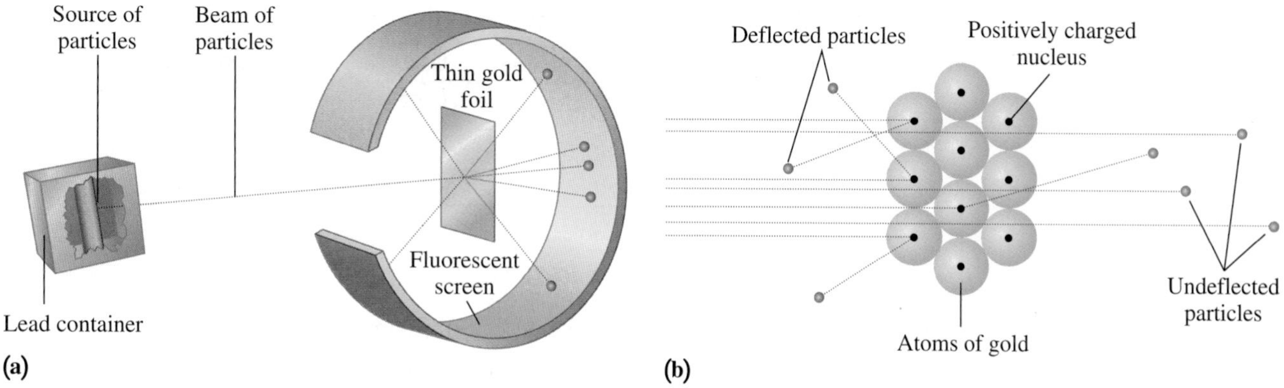

FIGURE 4.9 ▶ **(a)** Positive particles are aimed at a piece of gold foil. **(b)** Particles that come close to the atomic nuclei are deflected from their straight path.

Q Why are some particles deflected while most pass through the gold foil undeflected?

From his gold-foil experiments, Rutherford realized that the protons must be contained in a small, positively charged region at the center of the atom, which he called the **nucleus**. He proposed that the electrons in the atom occupy the space surrounding the nucleus through which most of the particles traveled undisturbed. Only the particles that came near this dense, positive center were deflected. If an atom were the size of a football stadium, the nucleus would be about the size of a golf ball placed in the center of the field.

Interactive Video

Rutherford Gold-Foil Experiment

Scientists knew that the nucleus was heavier than the mass of the protons, so they looked for another subatomic particle. Eventually, they discovered that the nucleus also contained a particle that is neutral, which they called a **neutron**. Thus, the masses of the protons and neutrons in the nucleus determine the mass of an atom (see Figure 4.10).

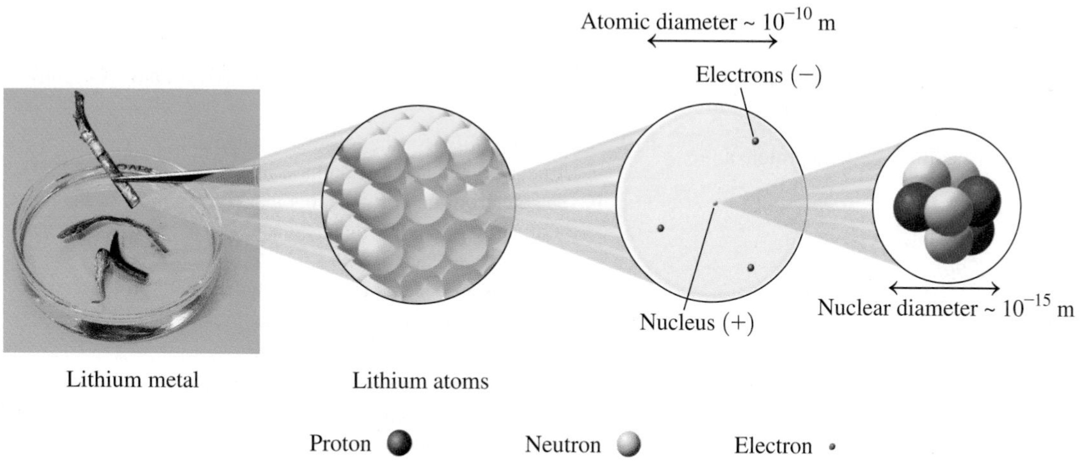

FIGURE 4.10 ▶ In an atom, the protons and neutrons that make up almost all the mass are packed into the tiny volume of the nucleus. The rapidly moving electrons (negative charge) surround the nucleus and account for the large volume of the atom.

Q Why can we say that the atom is mostly empty space?

Mass of the Atom

All the subatomic particles are extremely small compared with the things you see around you. One proton has a mass of 1.67×10^{-24} g, and the neutron is about the same. However, the electron has a mass 9.11×10^{-28} g, which is much less than the mass of either a proton or neutron. Because the masses of subatomic particles are so small, chemists use a very small unit of mass called an **atomic mass unit (amu)**. An amu is defined as one-twelfth of the mass of a carbon atom which has a nucleus containing 6 protons and 6 neutrons. In biology, the atomic mass unit is called a *Dalton* (Da) in honor of John Dalton. On the amu scale, the proton and neutron each have a mass of about 1 amu. Because the electron mass is so small, it is usually ignored in atomic mass calculations. Table 4.5 summarizes some information about the subatomic particles in an atom.

TABLE 4.5 Subatomic Particles in the Atom

Particle	Symbol	Charge	Mass (amu)	Location in Atom
Proton	p or p^+	1+	1.007	Nucleus
Neutron	n or n^0	0	1.008	Nucleus
Electron	e^-	1−	0.000 55	Outside nucleus

Explore Your World

Repulsion and Attraction

Tear a small piece of paper into bits. Brush your hair several times, and place the brush just above the bits of paper. Use your knowledge of electrical charges to give an explanation for your observations. Try the same experiment with a comb.

Questions

1. What happens when objects with like charges are placed close together?
2. What happens when objects with unlike charges are placed close together?

SAMPLE PROBLEM 4.5 Identifying Subatomic Particles

Identify the subatomic particle that has the following characteristics:

a. no charge
b. a mass of 0.000 55 amu
c. a mass about the same as a neutron

SOLUTION

a. neutron **b.** electron **c.** proton

STUDY CHECK 4.5

Is the following statement *true* or *false*?

The nucleus occupies a large volume in an atom.

QUESTIONS AND PROBLEMS

4.3 The Atom

LEARNING GOAL Describe the electrical charge and location in an atom for a proton, a neutron, and an electron.

4.15 Identify each of the following as describing either a proton, a neutron, or an electron:
a. has the smallest mass
b. has a 1+ charge
c. is found outside the nucleus
d. is electrically neutral

4.16 Identify each of the following as describing either a proton, a neutron, or an electron:
a. has a mass about the same as a proton
b. is found in the nucleus
c. is attracted to the protons
d. has a 1− charge

4.17 What did Rutherford determine about the structure of the atom from his gold-foil experiment?

4.18 How did Thomson determine that the electrons have a negative charge?

4.19 State *true* or *false* for the following statements.
a. A neutron and an electron have opposite charges.
b. The electrons and protons together contain most of the mass of an atom.
c. Protons repel each other.
d. Electrons are located inside the nucleus with protons.

4.20 Is each of the following statements *true* or *false*?
a. A proton is attracted to an electron.
b. A neutron has twice the mass of a proton.
c. Neutrons repel each other.
d. Electrons and neutrons have opposite charges.

4.21 On a dry day, your hair flies apart when you brush it. How would you explain this?

4.22 Why do cations repel each other?

LEARNING GOAL

Given the atomic number and the mass number of an atom, state the number of protons, neutrons, and electrons.

4.4 Atomic Number and Mass Number

All the atoms of the same element always have the same number of protons. This feature distinguishes atoms of one element from atoms of all the other elements.

Atomic Number

The **atomic number** of an element is equal to the number of protons in every atom of that element. The atomic number is the whole number that appears above the symbol of each element on the periodic table.

Atomic number = number of protons in an atom

CORE CHEMISTRY SKILL

Counting Protons and Neutrons

The periodic table on the inside front cover of this text shows the elements in order of atomic number from 1 to 118. We can use an atomic number to identify the number of protons in an atom of any element. For example, a lithium atom, with atomic number 3, has 3 protons. Every lithium atom has 3 and only 3 protons. Any atom with 3 protons is always a lithium atom. In the same way, we determine that a carbon atom, with atomic number 6, has 6 protons. Every carbon atom has 6 protons, and any atom with 6 protons is carbon; every copper atom, with atomic number 29, has 29 protons and any atom with 29 protons is copper.

An atom is electrically neutral. That means that the number of protons in an atom is equal to the number of electrons, which gives every atom an overall charge of zero. Thus, for every atom, the atomic number also gives the number of electrons.

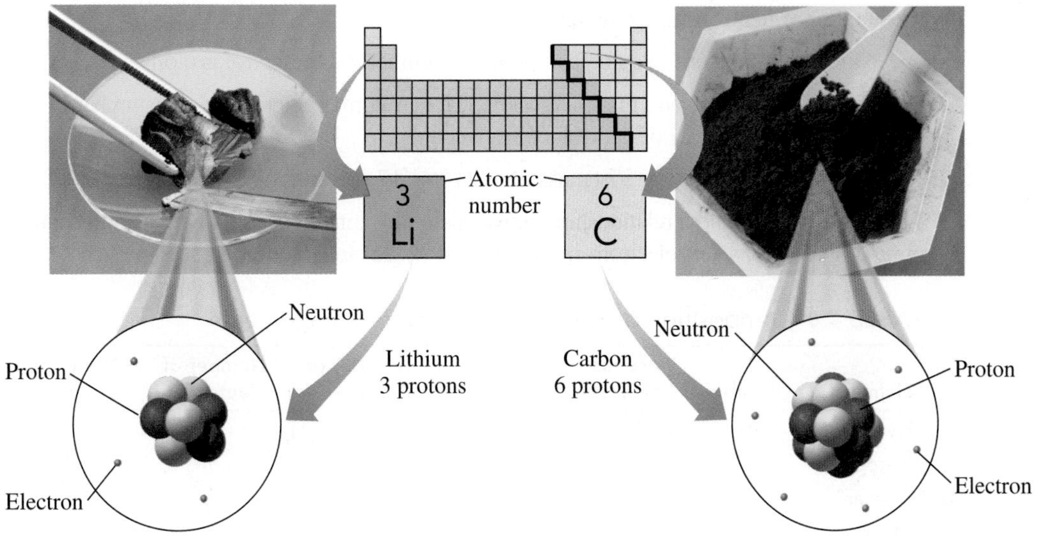

All atoms of lithium (left) contain three protons, and all atoms of carbon (right) contain six protons.

SAMPLE PROBLEM 4.6 Atomic Number, Protons, and Electrons

Using the periodic table, state the atomic number, number of protons, and number of electrons for an atom of each of the following elements:

a. nitrogen **b.** magnesium **c.** bromine

SOLUTION

ANALYZE THE PROBLEM	Atomic Number	Number of Protons	Number of Electrons
	shown above the element symbol on the periodic table	equal to the atomic number	equal to the number of protons

a. On the periodic table, the atomic number for nitrogen (N) is 7. Thus, nitrogen atoms contain 7 protons and 7 electrons.
b. On the periodic table, the atomic number for magnesium (Mg) is 12. Thus, magnesium atoms contain 12 protons and 12 electrons.
c. On the periodic table, the atomic number for bromine (Br) is 35. Thus, bromine atoms contain 35 protons and 35 electrons.

STUDY CHECK 4.6

Consider an atom that has 79 electrons.

a. How many protons are in its nucleus?
b. What is its atomic number?
c. What is its name and symbol?

Mass Number

We now know that the protons and neutrons determine the mass of the nucleus. Thus, for a single atom, we assign a **mass number**, which is the total number of protons and neutrons in its nucleus. However, the mass number does not appear on the periodic table because it applies to single atoms only.

Mass number = number of protons + number of neutrons

For example, the nucleus of a single oxygen atom that contains 8 protons and 8 neutrons has a mass number of 16. An atom of iron that contains 26 protons and 32 neutrons has a mass number of 58.

If we are given the mass number of an atom and its atomic number, we can calculate the number of neutrons in its nucleus.

Number of neutrons in a nucleus = mass number − number of protons

For example, if we are given a mass number of 37 for an atom of chlorine (atomic number 17), we can calculate the number of neutrons in its nucleus.

Number of neutrons = 37 (mass number) − 17 (protons) = 20 neutrons

Table 4.6 illustrates these relationships between atomic number, mass number, and the number of protons, neutrons, and electrons in examples of single atoms for different elements.

TABLE 4.6 Composition of Some Atoms of Different Elements

Element	Symbol	Atomic Number	Mass Number	Number of Protons	Number of Neutrons	Number of Electrons
Hydrogen	H	1	1	1	0	1
Nitrogen	N	7	14	7	7	7
Oxygen	O	8	16	8	8	8
Chlorine	Cl	17	37	17	20	17
Iron	Fe	26	58	26	32	26
Gold	Au	79	197	79	118	79

SAMPLE PROBLEM 4.7 Calculating Numbers of Protons, Neutrons, and Electrons

As a micronutrient, zinc is needed for metabolic reactions in cells, DNA synthesis, the growth of bone, teeth, and connective tissue, and the proper functioning of the immune system. For an atom of zinc that has a mass number of 68, determine the following:

a. the number of protons **b.** the number of neutrons **c.** the number of electrons

SOLUTION

ANALYZE THE PROBLEM	Element	Atomic Number	Mass Number	Number of Protons	Number of Neutrons	Number of Electrons
	Zinc (Zn)	30 (periodic table)	68	equal to atomic number	mass number − number of protons	equal to number of protons

a. Zinc (Zn), with an atomic number of 30, has 30 protons.

b. The number of neutrons in this atom is found by subtracting number of protons (atomic number) from the mass number.

Mass number	−	atomic number	=	number of neutrons
68	−	30	=	38

c. Because the zinc atom is neutral, the number of electrons is equal to the number of protons. A zinc atom has 30 electrons.

STUDY CHECK 4.7

How many neutrons are in the nucleus of a bromine atom that has a mass number of 80?

QUESTIONS AND PROBLEMS

4.4 Atomic Number and Mass Number

LEARNING GOAL Given the atomic number and the mass number of an atom, state the number of protons, neutrons, and electrons.

4.23 Would you use the atomic number, mass number, or both to determine each of the following?
a. number of protons in an atom
b. number of neutrons in an atom
c. number of particles in the nucleus
d. number of electrons in a neutral atom

4.24 Identify the type of subatomic particles described by each of the following:
a. atomic number
b. mass number
c. mass number − atomic number
d. mass number + atomic number

4.25 Write the names and symbols of the elements with the following atomic numbers:
a. 5 **b.** 29 **c.** 24 **d.** 37
e. 13 **f.** 44 **g.** 55 **h.** 6

4.26 Identify the metallic elements from the following atomic numbers and give their names and the symbols:
a. 1 **b.** 11 **c.** 19 **d.** 22
e. 35 **f.** 47 **g.** 79 **h.** 29

4.27 How many protons and electrons are there in a neutral atom of each of the following elements?
a. oxygen **b.** chromium
c. chlorine **d.** calcium

4.28 How many protons and electrons are there in a neutral atom of each of the following elements?
a. carbon **b.** fluorine
c. tin **d.** nickel

4.29 Complete the following table for a neutral atom of each element:

Name of the Element	Symbol	Atomic Number	Mass Number	Number of Protons	Number of Neutrons	Number of Electrons
	Al		27			
		12			12	
Potassium					20	
				16	15	
			56			26

4.30 Complete the following table for a neutral atom of each element:

Name of the Element	Symbol	Atomic Number	Mass Number	Number of Protons	Number of Neutrons	Number of Electrons
	N		15			
Calcium			42			
				38	50	
		14			16	
		56	138			

4.5 Isotopes and Atomic Mass

LEARNING GOAL

Determine the number of protons, electrons, and neutrons in one or more of the isotopes of an element; calculate the atomic mass of an element using the percent abundance and mass of its naturally occurring isotopes.

We have seen that all atoms of the same element have the same number of protons and electrons. However, the atoms of any one element are not entirely identical because the atoms of most elements have different numbers of neutrons. When a sample of an element consists of two or more atoms with differing numbers of neutrons, those atoms are called *isotopes*.

Atoms and Isotopes

CORE CHEMISTRY SKILL

Writing Atomic Symbols for Isotopes

Isotopes are atoms of the same element that have the same atomic number but different numbers of neutrons. For example, all atoms of the element magnesium (Mg) have an atomic number of 12. Thus every magnesium atom always has 12 protons. However, some naturally occurring magnesium atoms have 12 neutrons, others have 13 neutrons, and still others have 14 neutrons. The different numbers of neutrons give the magnesium atoms different mass numbers but do not change their chemical behavior. The three isotopes of magnesium have the same atomic number but different mass numbers.

To distinguish between the different isotopes of an element, we write an **atomic symbol** for a particular isotope that indicates the mass number in the upper left corner and the atomic number in the lower left corner.

An isotope may be referred to by its name or symbol, followed by its mass number, such as magnesium-24 or Mg-24. Magnesium has three naturally occurring isotopes, as

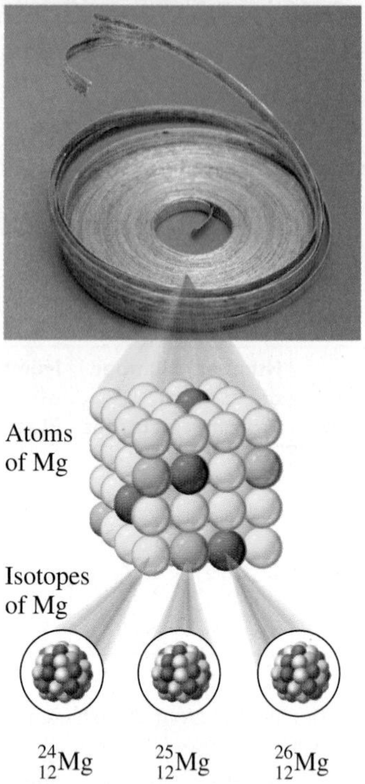

The nuclei of three naturally occurring magnesium isotopes have the same number of protons but different numbers of neutrons.

Mass number → $^{24}_{12}Mg$ ← Symbol of element
Atomic number →

Atomic symbol for an isotope of magnesium, Mg-24.

shown in Table 4.7. In a large sample of naturally occurring magnesium atoms, each type of isotope can be present as a low percentage or a high percentage. For example, the Mg-24 isotope makes up almost 80% of the total sample, whereas Mg-25 and Mg-26 each make up only about 10% of the total number of magnesium atoms.

TABLE 4.7 Isotopes of Magnesium

Atomic Symbol	$^{24}_{12}Mg$	$^{25}_{12}Mg$	$^{26}_{12}Mg$
Name	Mg-24	Mg-25	Mg-26
Number of Protons	12	12	12
Number of Electrons	12	12	12
Mass Number	**24**	**25**	**26**
Number of Neutrons	**12**	**13**	**14**
Mass of Isotope (amu)	**23.99**	**24.99**	**25.98**
Percent Abundance	78.70	10.13	11.17

SAMPLE PROBLEM 4.8 Identifying Protons and Neutrons in Isotopes

The element neon has three naturally occurring isotopes: Ne-20, Ne-21, and Ne-22. State the number of protons and neutrons in these stable isotopes of neon.

a. $^{20}_{10}Ne$ **b.** $^{21}_{10}Ne$ **c.** $^{22}_{10}Ne$

SOLUTION

ANALYZE THE PROBLEM	Atomic Number	Mass Number	Number of Protons	Number of Neutrons
	number in lower left corner	number in upper left corner	equal to atomic number	equal to mass number − number of protons

In the atomic symbol, the mass number is shown in the upper left corner of the symbol, and the atomic number is shown in the lower left corner of the symbol. Thus, each isotope of Ne, atomic number 10, has 10 protons. The number of neutrons is found by subtracting the number of protons (10) from the mass number of each isotope.

Isotope	Atomic Number	Mass Number	Number of Protons	Number of Neutrons
a. $^{20}_{10}Ne$	10	20	10	$20 - 10 = 10$
b. $^{21}_{10}Ne$	10	21	10	$21 - 10 = 11$
c. $^{22}_{10}Ne$	10	22	10	$22 - 10 = 12$

STUDY CHECK 4.8

Write an atomic symbol for each of the following:

a. a nitrogen atom with eight neutrons
b. an atom with 20 protons and 22 neutrons
c. an atom with mass number 27 and 14 neutrons

Atomic Mass

In laboratory work, a chemist generally uses samples with many atoms that contain all the different atoms or isotopes of an element. Because each isotope has a different mass, chemists have calculated an **atomic mass** for an "average atom," which is a *weighted average* of the masses of all the naturally occurring isotopes of that element. On the periodic table, the atomic mass is the number including decimal places that is given below the symbol of each element. Most elements consist of two or more isotopes, which is one reason that the atomic masses on the periodic table are seldom whole numbers.

Weighted Average Analogy

To understand how the atomic mass as a weighted average for a group of isotopes is calculated, we will use an analogy of bowling balls with different weights. Suppose that a bowling alley has ordered 5 bowling balls that weigh 8 lb each and 20 bowling balls that weigh 14 lb each. In this group of bowling balls, there are more 14-lb balls than 8-lb balls. The abundance of the 14-lb balls is 80.% (20/25), and the abundance of the 8-lb balls is 20.% (5/25). Now we can calculate a weighted average for the "average" bowling ball using the weight and percent abundance of the two types of bowling balls.

The weighted average of 8-lb and 14-lb bowling balls is calculated using their percent abundance.

	Weight		Percent Abundance		Weight from Each Type
14-lb bowling balls	14 lb	×	$\frac{80.}{100}$	=	11.2 lb
8-lb bowling balls	8 lb	×	$\frac{20.}{100}$	=	1.6 lb
	Weighted average of a bowling ball			=	12.8 lb
	"Atomic mass" of a bowling ball			=	12.8 lb

Calculating Atomic Mass

To calculate the atomic mass of an element, we need to know the percentage abundance and the mass of each isotope, which must be determined experimentally. For example, a large sample of naturally occurring chlorine atoms consists of 75.76% of ${}^{35}_{17}Cl$ atoms and 24.24% of ${}^{37}_{17}Cl$ atoms. The ${}^{35}_{17}Cl$ isotope has a mass of 34.97 amu and the ${}^{37}_{17}Cl$ isotope has a mass of 36.97 amu.

$$\text{Atomic mass of Cl} = \underbrace{\text{mass of } {}^{35}_{17}\text{Cl} \times \frac{{}^{35}_{17}\text{Cl}\%}{100\%}}_{\text{amu from } {}^{35}_{17}\text{Cl}} + \underbrace{\text{mass of } {}^{37}_{17}\text{Cl} \times \frac{{}^{37}_{17}\text{Cl}\%}{100\%}}_{\text{amu from } {}^{37}_{17}\text{Cl}}$$

Isotope	Mass (amu)	×	Abundance (%)	=	Contribution to Average Cl Atom
${}^{35}_{17}Cl$	34.97	×	$\frac{75.76}{100}$	=	26.49 amu
${}^{37}_{17}Cl$	36.97	×	$\frac{24.24}{100}$	=	8.962 amu
			Atomic mass of Cl	=	35.45 amu (weighted average mass)

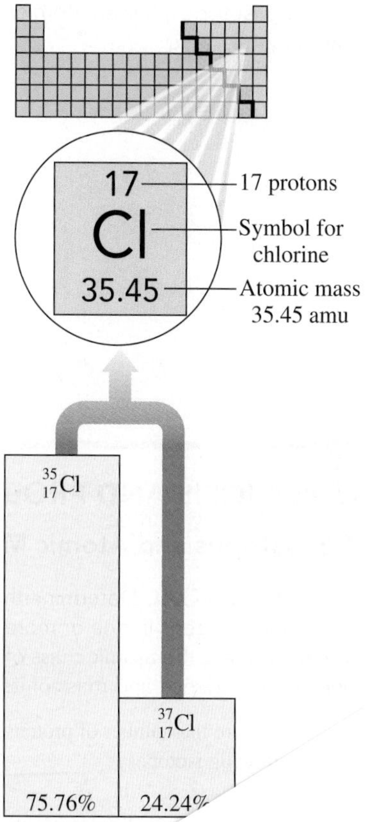

Chlorine, with tw
ring isotopes, has a
of 35.45 amu.

The atomic mass of 35.45 amu is the weighted average mass of a sample of Cl atoms, although no individual Cl atom actually has this mass. An atomic mass of 35.45, which is closer to the mass number of Cl-35, also indicates there is a higher percentage of ${}^{35}_{17}Cl$ atoms in the chlorine sample. In fact, there are about three atoms of ${}^{35}_{17}Cl$ for every one atom of ${}^{37}_{17}Cl$ in a sample of chlorine atoms.

Table 4.8 lists the naturally occurring isotopes of some selected elements and their atomic masses along with their most prevalent isotopes.

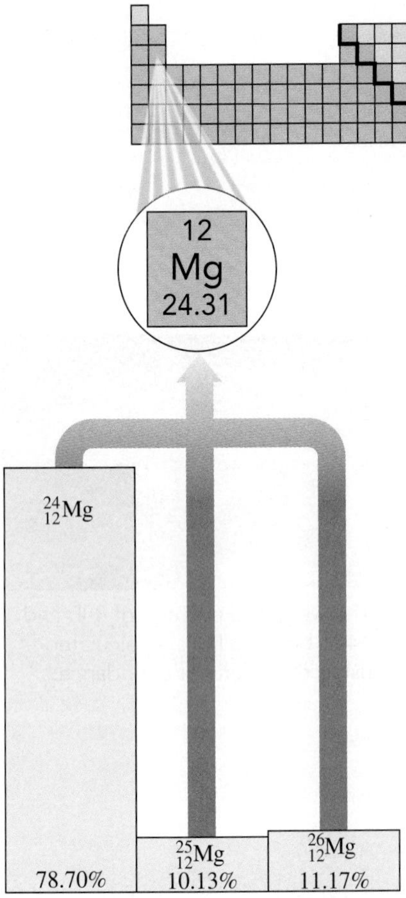

Magnesium, with three naturally occurring isotopes, has an atomic mass of 24.31 amu.

TABLE 4.8 The Atomic Mass of Some Elements

Element	Isotopes	Atomic Mass (weighted average)	Most Prevalent Isotope
Lithium	$^{6}_{3}Li$, $^{7}_{3}Li$	6.941 amu	$^{7}_{3}Li$
Carbon	$^{12}_{6}C$, $^{13}_{6}C$, $^{14}_{6}C$	12.01 amu	$^{12}_{6}C$
Oxygen	$^{16}_{8}O$, $^{17}_{8}O$, $^{18}_{8}O$	16.00 amu	$^{16}_{8}O$
Fluorine	$^{19}_{9}F$	19.00 amu	$^{19}_{9}F$
Sulfur	$^{32}_{16}S$, $^{33}_{16}S$, $^{34}_{16}S$, $^{36}_{16}S$	32.07 amu	$^{32}_{16}S$
Copper	$^{63}_{29}Cu$, $^{65}_{29}Cu$	63.55 amu	$^{63}_{29}Cu$

SAMPLE PROBLEM 4.9 Calculating Atomic Mass

Using Table 4.7, calculate the atomic mass for magnesium using the weighted average mass method.

SOLUTION

To determine atomic mass, calculate the contribution of each isotope to the total atomic mass by multiplying the mass of each of the isotopes by its percent abundance/100 and adding the results. Atomic mass = mass of isotope 1 × % abundance/100 isotope 1 + mass of isotope 2 × % abundance/100 isotope 2 + · · ·

Isotope	Mass (amu)		Abundance (%)		Contribution to the Atomic Mass
$^{24}_{12}Mg$	23.99	×	$\frac{78.70}{100}$	=	18.88 amu
$^{25}_{12}Mg$	24.99	×	$\frac{10.13}{100}$	=	2.531 amu
$^{26}_{12}Mg$	25.98	×	$\frac{11.17}{100}$	=	2.902 amu
			Atomic mass of Mg	=	24.31 amu (weighted average mass)

STUDY CHECK 4.9

There are two naturally occurring isotopes of boron. The isotope $^{10}_{5}B$ has a mass of 10.01 amu with an abundance of 19.80%, and the isotope $^{11}_{5}B$ has a mass of 11.01 amu with an abundance of 80.20%. Calculate the atomic mass for boron using the weighted average mass method.

QUESTIONS AND PROBLEMS

4.5 Isotopes and Atomic Mass

LEARNING GOAL Determine the number of protons, electrons, and neutrons in one or more of the isotopes of an element; calculate the atomic mass of an element using the percent abundance and mass of its naturally occurring isotopes.

4.31 What are the number of protons, neutrons and electrons in the following isotopes?
a. $^{4}_{2}He$ **b.** $^{14}_{6}C$ **c.** $^{235}_{92}U$ **d.** $^{35}_{17}Cl$

4.32 What are the number of protons, neutrons, and electrons in the following isotopes?
a. $^{2}_{1}H$ **b.** $^{14}_{7}N$ **c.** $^{26}_{14}Si$ **d.** $^{70}_{30}Zn$

4.33 Write the atomic symbols for the isotope with each of the following characteristics:
a. 7 protons 8 neutrons
b. 29 protons 36 neutrons
c. 56 electrons 81 neutrons
d. A calcium atom with 22 neutrons
e. A silver atom with 60 neutrons

4.34 Write the atomic symbol for the isotope with each of the following characteristics:
a. an oxygen atom with 10 neutrons
b. 4 protons and 5 neutrons

c. 25 electrons and 28 neutrons
d. a mass number of 24 and 13 neutrons
e. a nickel atom with 32 neutrons

4.35 Argon has three naturally occurring isotopes, with mass numbers 36, 38, and 40.
a. Write the atomic symbol for each of these atoms.
b. How are these isotopes alike?
c. How are they different?
d. Why is the atomic mass of argon listed on the periodic table not a whole number?
e. Which isotope is the most prevalent in a sample of argon?

4.36 Strontium has four naturally occurring isotopes, with mass numbers 84, 86, 87, and 88.
a. Write the atomic symbol for each of these atoms.
b. How are these isotopes alike?
c. How are they different?
d. Why is the atomic mass of strontium listed on the periodic table not a whole number?
e. Which isotope is the most prevalent in a sample of strontium?

4.37 Two isotopes of gallium are naturally occurring, with $^{69}_{31}Ga$ at 60.11% (68.93 amu) and $^{71}_{31}Ga$ at 39.89% (70.92 amu). Calculate the atomic mass for gallium using the weighted average mass method.

4.38 Two isotopes of rubidium occur naturally, with $^{85}_{37}Rb$ at 72.17% (84.91 amu) and $^{87}_{37}Rb$ at 27.83% (86.91 amu). Calculate the atomic mass for rubidium using the weighted average mass method.

4.6 Electron Energy Levels

LEARNING GOAL

Given the name or symbol of one of the first 20 elements in the periodic table, write the electron arrangement.

When we listen to a radio, use a microwave oven, turn on a light, see the colors of a rainbow, or have an X-ray taken, we are experiencing various forms of *electromagnetic radiation.* Light and other electromagnetic radiation consist of energy particles that move as waves of energy. In a wave, just like the waves in an ocean, the distance between the peaks is called the *wavelength.* All forms of electromagnetic radiation travel in space at the speed of light, 3.0×10^8 m/s, but differ in energy and wavelength. High-energy radiation has short wavelengths compared to low-energy radiation, which has longer wavelengths. The *electromagnetic spectrum* shows the arrangement of different types of electromagnetic radiation in order of increasing energy (see Figure 4.11).

When the light from the Sun passes through a prism, the light separates into a continuous color spectrum, which consists of the colors we see in a rainbow. In contrast, when light from a heated element passes through a prism, it separates into distinct lines of color separated by dark areas called an *atomic spectrum.* Each element has its own unique atomic spectrum.

A rainbow forms when light passes through water droplets.

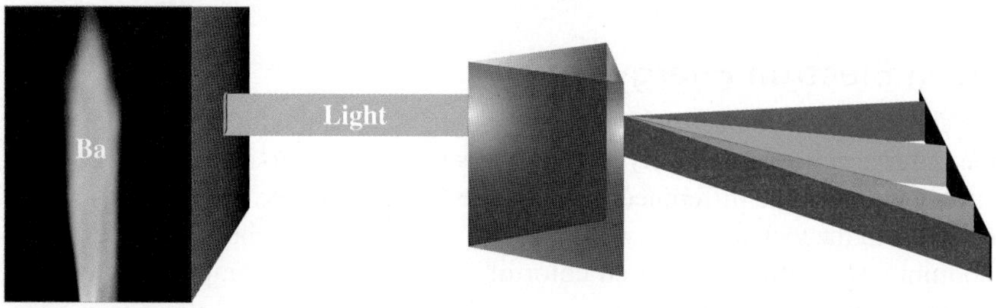

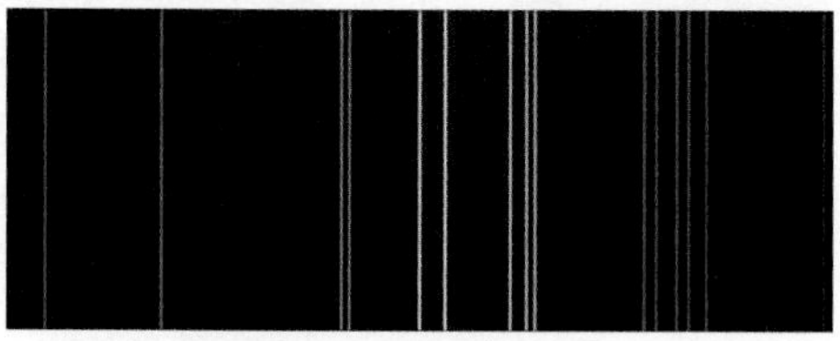

In an atomic spectrum, light from a heated element separates into distinct lines.

Barium light spectrum

Electron Energy Levels

Scientists have now determined that the lines in the atomic spectra of elements are associated with changes in the energies of the electrons. In an atom, each electron has a specific energy known as its **energy level**, which is assigned values called *principal quantum numbers* (n), ($n = 1, n = 2, \ldots$). Generally, electrons in the lower energy levels are closer to the nucleus, while electrons in the higher energy levels are farther away. The energy of an electron is *quantized*, which means that the energy of an electron can only have specific energy values, but cannot have values between them.

Principal Quantum Number (n)

$$1 < 2 < 3 < 4 < 5 < 6 < 7$$

Lowest Energy $\longrightarrow$ Highest Energy

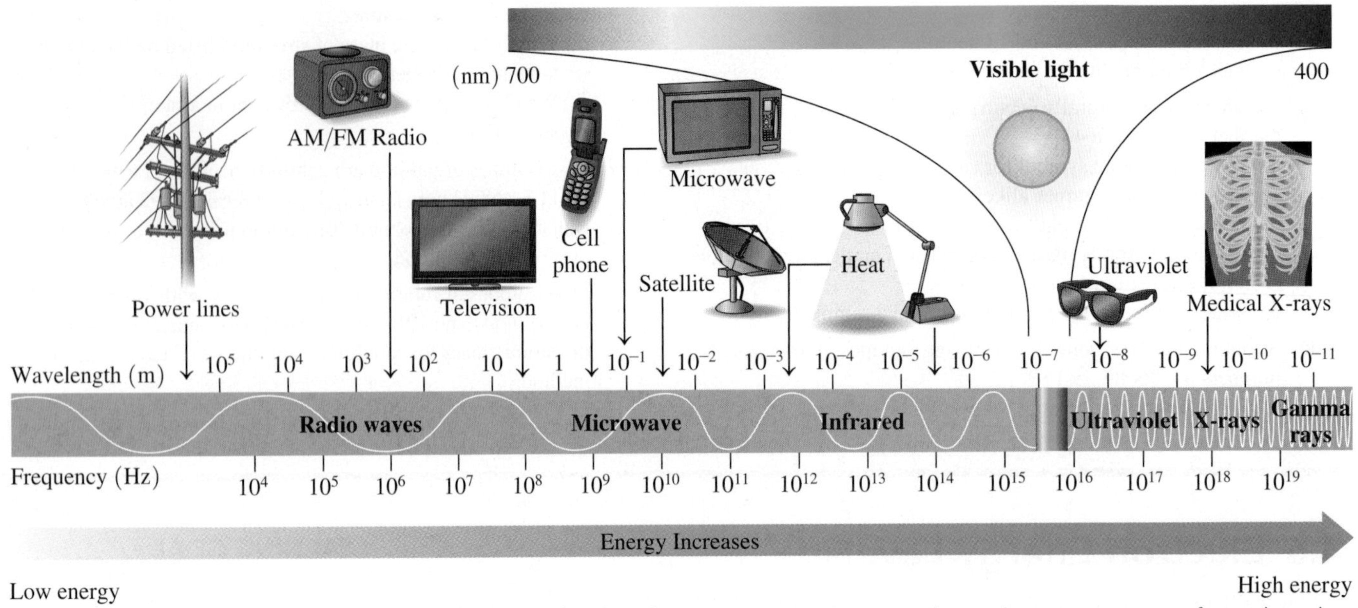

FIGURE 4.11 ▶ The electromagnetic spectrum shows the arrangement of wavelengths of electromagnetic radiation. The visible portion consists of wavelengths from 700 nm to 400 nm.

Q How does the wavelength of ultraviolet light compare to that of a microwave?

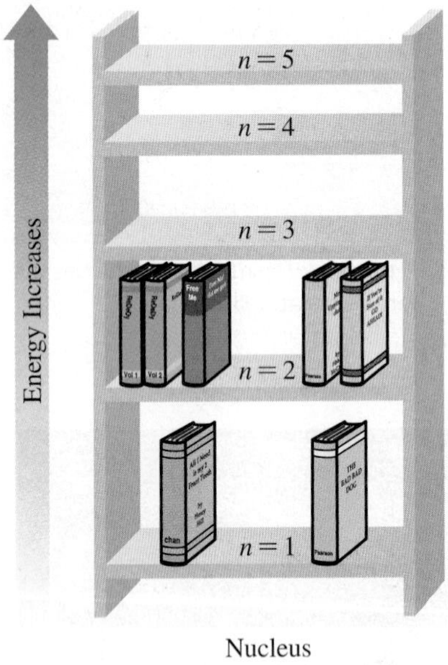

An electron can have only the energy of one of the energy levels in an atom.

All the electrons with the same energy are grouped in the same energy level. As an analogy, we can think of the energy levels of an atom as similar to the shelves in a bookcase. The first shelf is the lowest energy level; the second shelf is the second energy level, and so on. If we are arranging books on the shelves, it would take less energy to fill the bottom shelf first, and then the second shelf, and so on. However, we could never get any book to stay in the space between any of the shelves. Similarly, the energy of an electron must be at specific energy levels, and not between.

Unlike standard bookcases, however, there is a large difference between the energy of the first and second levels, but then the higher energy levels are closer together. Another difference is that the lower electron energy levels hold fewer electrons than the higher energy levels.

Changes in Electron Energy Level

An electron can change from one energy level to a higher level only if it absorbs the energy equal to the difference in energy levels. When an electron changes to a lower energy level, it emits energy equal to the difference between the two levels (see Figure 4.12). If the energy emitted is in the visible range, we see one of the colors of visible light. The yellow color of sodium streetlights and the red color of neon lights are examples of electrons emitting energy in the visible color range.

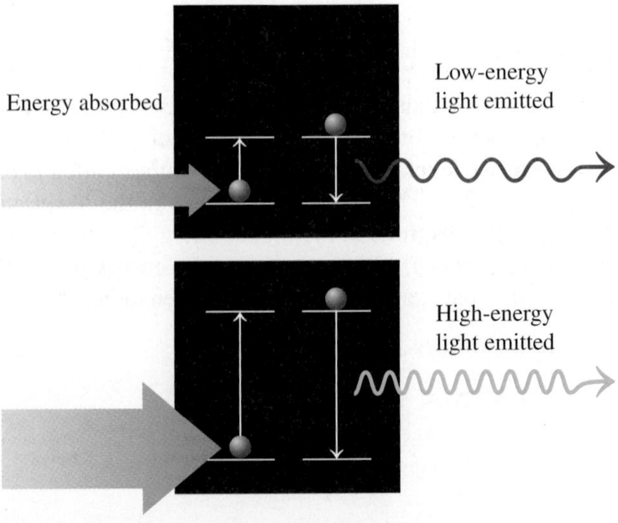

FIGURE 4.12 ▶ Electrons absorb a specific amount of energy to move to a higher energy level. When electrons lose energy, a specific quantity of energy is emitted.

Q What causes electrons to move to higher energy levels?

Chemistry Link to the Environment

Energy-Saving Fluorescent Bulbs

Compact fluorescent lights (CFL) are replacing the standard light bulb we use in our homes and workplaces. Compared to a standard light bulb, the CFL has a longer life and uses less electricity. Within about 20 days of use, the CFL saves enough money in electricity costs to pay for its higher initial cost.

A standard incandescent light bulb has a thin tungsten filament inside a sealed glass bulb. When the light is switched on, electricity flows through this filament, and electrical energy is converted to heat energy. When the filament reaches a temperature around 2300 °C, we see white light.

A fluorescent bulb produces light in a different way. When the switch is turned on, electrons move between two electrodes and collide with mercury atoms in a mixture of mercury and argon gas inside the light. When the electrons in the mercury atoms absorb energy from the collisions, they are raised to higher energy levels. As electrons fall to lower energy levels emitting ultraviolet radiation, they strike the phosphor coating inside the tube, and fluorescence occurs as visible light is emitted.

The production of light in a fluorescent bulb is more efficient than in an incandescent light bulb. A 75-watt incandescent bulb can be replaced by a 20-watt CFL that gives the same amount of light, providing a 70% reduction in electricity costs. A typical incandescent light bulb lasts for one to two months, whereas a CFL lasts from 1 to 2 yr. One drawback of the CFL is that each contains about 4 mg of mercury. As long as the bulb stays intact, no mercury is released. However, used CFL bulbs should not be disposed of in household trash but rather should be taken to a recycling center.

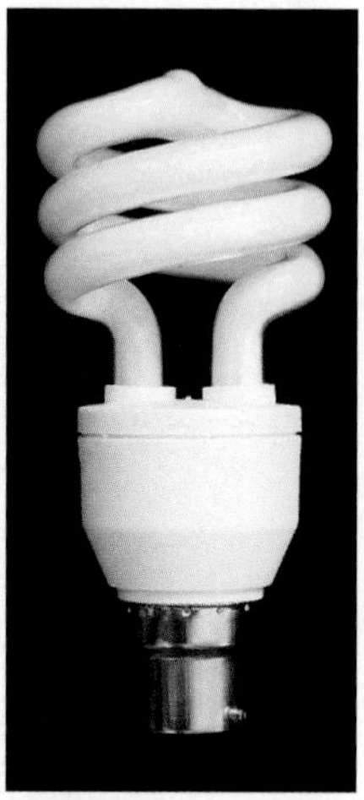

A compact fluorescent light (CFL) uses up to 70% less energy.

Electron Arrangements for the First 20 Elements

CORE CHEMISTRY SKILL

Writing Electron Arrangements

The *electron arrangement* of an atom gives the number of electrons in each energy level. We can write the electron arrangements for the first 20 elements by placing electrons in energy levels beginning with the lowest. There is a limit to the number of electrons allowed in each energy level. Only a few electrons can occupy the lower energy levels, while more electrons can be accommodated in higher energy levels. We can now look at the numbers of electrons in the first four energy levels for the first 20 elements as shown in Table 4.9.

TABLE 4.9 Electron Arrangements for the First 20 Elements

			Number of Electrons in Energy Level			
Element	**Symbol**	**Atomic Number**	**1**	**2**	**3**	**4**
Hydrogen	H	1	1			
Helium	He	2	2			
Lithium	Li	3	2	1		
Beryllium	Be	4	2	2		
Boron	B	5	2	3		
Carbon	C	6	2	4		
Nitrogen	N	7	2	5		
Oxygen	O	8	2	6		
Fluorine	F	9	2	7		
Neon	Ne	10	2	8		
Sodium	Na	11	2	8	1	
Magnesium	Mg	12	2	8	2	
Aluminum	Al	13	2	8	3	
Silicon	Si	14	2	8	4	
Phosphorus	P	15	2	8	5	
Sulfur	S	16	2	8	6	
Chlorine	Cl	17	2	8	7	
Argon	Ar	18	2	8	8	
Potassium	K	19	2	8	8	1
Calcium	Ca	20	2	8	8	2

Hydrogen 1
Helium 2

Period 1

The single electron of hydrogen goes into energy level 1, and the two electrons of helium fill energy level 1. Thus, energy level 1 can hold just two electrons. The electron arrangements for H and He are shown in the margin.

Lithium 2,1
Carbon 2,4
Neon 2,8

Period 2

For the elements of the second period (lithium, Li, to neon, Ne), we fill the first energy level with two electrons, and place the remaining electrons in the second energy level. For example, lithium has three electrons. Two of those electrons fill energy level 1. The remaining electron goes into the second energy level. We can write this electron arrangement as 2,1. Going across Period 2, more electrons are added to the second energy level. For example, an atom of carbon, with a total of six electrons, fills energy level 1, which leaves the four remaining electrons in the second energy level. The electron arrangement for carbon can be written 2,4. For neon, the last element in Period 2, both the first and second energy levels are filled to give an electron arrangement of 2,8.

Sodium 2,8,1
Sulfur 2,8,6
Argon 2,8,8

Period 3

For sodium, the first element in Period 3, 10 electrons fill the first and second energy levels, which means that the remaining electron must go into the third energy level. The electron arrangement for sodium can be written 2,8,1. The elements that follow sodium in the third period continue to add electrons, one at a time, until the third level is complete. For example, a sulfur atom with 16 electrons fills the first and second energy levels, which leaves six electrons in the third level. The electron arrangement for sulfur can be written 2,8,6. At the end of Period 3, argon has eight electrons in the third level. Once energy level 3 has eight electrons, it stops filling. However, 10 more electrons will be added later.

Potassium 2,8,8,1
Calcium 2,8,8,2

Period 4

For potassium, the first element in Period 4, electrons fill the first and second energy levels, and eight electrons are in the third energy level. The remaining electron of potassium enters energy level 4, which gives the electron arrangement of 2,8,8,1. For calcium, the process is similar, except that calcium has two electrons in the fourth energy level. The electron arrangement for calcium can be written 2,8,8,2.

Elements Beyond 20

Following calcium, the next 10 electrons (scandium to zinc) are added to the third energy level, which already has 8 electrons, until it is complete with 18 electrons. The higher energy levels 5, 6, and 7 can theoretically accommodate up to 50, 72, and 98 electrons, but they are not completely filled. Beyond the first 20 elements, the electron arrangements become more complicated and will not be covered in this text.

Energy Level (n)	1	2	3	4	5	6	7
Number of Electrons	2	8	18	32	32	18	8

SAMPLE PROBLEM 4.10 Writing Electron Arrangements

Write the electron arrangement for each of the following:

a. oxygen **b.** chlorine

SOLUTION

a. Oxygen with atomic number 8 has eight electrons, which are arranged with two electrons in energy level 1 and six electrons in energy level 2.

2,6

b. Chlorine with atomic number 17 has 17 electrons, which are arranged with two electrons in energy level 1, eight electrons in energy level 2, and seven electrons in energy level 3.

2,8,7

STUDY CHECK 4.10

What element has an electron arrangement of 2,8,5?

Chemistry Link to Health

Biological Reactions to UV Light

Our everyday life depends on sunlight, but exposure to sunlight can have damaging effects on living cells, and too much exposure can even cause their death. The light energy, especially ultraviolet (UV), excites electrons and may lead to unwanted chemical reactions. The list of damaging effects of sunlight includes sunburn; wrinkling; premature aging of the skin; changes in the DNA of the cells, which can lead to skin cancers; inflammation of the eyes; and perhaps cataracts. Some drugs, like the acne medications Accutane and Retin-A, as well as antibiotics, diuretics, sulfonamides, and estrogen, make the skin extremely sensitive to light.

However, medicine does take advantage of the beneficial effect of sunlight. *Phototherapy* can be used to treat certain skin conditions, including psoriasis, eczema, and dermatitis. In the treatment of psoriasis, for example, oral drugs are given to make the skin more photosensitive; exposure to UV follows. Low-energy light is used to break down bilirubin in neonatal jaundice. Sunlight is also a factor in stimulating the immune system.

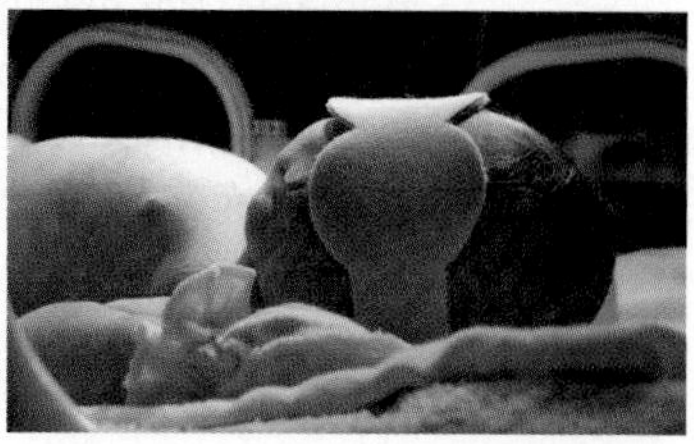

Babies with neonatal jaundice are treated with UV light.

In a disorder called *seasonal affective disorder* or SAD, people experience mood swings and depression during the winter. Some research suggests that SAD is the result of a decrease of serotonin, or an increase in melatonin, when there are fewer hours of sunlight. One treatment for SAD is therapy using bright light provided by a lamp called a light box. A daily exposure to intense light for 30 to 60 min seems to reduce symptoms of SAD.

QUESTIONS AND PROBLEMS

4.6 Electron Energy Levels

LEARNING GOAL Given the name or symbol of one of the first 20 elements in the periodic table, write the electron arrangement.

4.39 Electrons can move to higher energy levels when they _______ (absorb/emit) energy.

4.40 Electrons drop to lower energy levels when they _______ (absorb/emit) energy.

4.41 Identify the form of electromagnetic radiation in each pair that has the greater energy:
a. green light or yellow light
b. microwaves or blue light

4.42 Identify the form of electromagnetic radiation in each pair that has the shorter wavelength:
a. microwaves or radiowave
b. infrared light or ultraviolet light

4.43 Write the electron arrangement for each of the following elements (*Example*: sodium 2,8,1)
a. oxygen **b.** krypton **c.** calcium
d. chlorine **e.** boron **f.** nitrogen

4.44 Write the electron arrangement for each of the following elements: (*Example*: sodium 2,8,1)
a. phosphorus **b.** neon **c.** sulfur
d. magnesium **e.** aluminum **f.** fluorine

4.45 Identify the elements that have the following electron arrangements:

	Energy Level			
	1	**2**	**3**	**4**
a.	2	1		
b.	2	8	2	
c.	1			
d.	2	8	7	
e.	2	6		

4.46 Identify the elements that have the following electron arrangements:

	Energy Level			
	1	**2**	**3**	**4**
a.	2	5		
b.	2	8	6	
c.	2	4		
d.	2	8	8	1
e.	2	8	3	

4.7 Trends in Periodic Properties

LEARNING GOAL

Use the electron arrangement of elements to explain the trends in periodic properties.

The electron arrangements of atoms are an important factor in the physical and chemical properties of the elements. Now we will look at the *valence electrons* in atoms, the trends in *atomic size*, *ionization energy*, and *metallic character*. Known as *periodic properties*, there is a pattern of regular change going across a period, and then the trend is repeated again in each successive period.

CORE CHEMISTRY SKILL

Identifying Trends in Periodic Properties

Group Number and Valence Electrons

The chemical properties of representative elements in Groups 1A(1) to 8A (18) are mostly due to the **valence electrons**, which are the electrons in the outermost energy level. The group number gives the number of valence electrons for each group of representative elements. For example, all the elements in Group 1A (1) have one valence electron. All the elements in Group 2A (2) have two valence electrons. The halogens in Group 7A (17) all have seven valence electrons. Table 4.10 shows how the number of valence electrons for common representative elements is consistent with the group number.

TABLE 4.10 Comparison of Electron Arrangements, by Group, for Some Representative Elements

			Number of Electrons in Energy Level			
Group Number	**Element**	**Symbol**	**1**	**2**	**3**	**4**
1A (1)	Lithium	Li	2	**1**		
	Sodium	Na	2	8	**1**	
	Potassium	K	2	8	8	**1**
2A (2)	Beryllium	Be	2	**2**		
	Magnesium	Mg	2	8	**2**	
	Calcium	Ca	2	8	8	**2**
3A (13)	Boron	B	2	**3**		
	Aluminum	Al	2	8	**3**	
	Gallium	Ga	2	8	18	**3**
4A (14)	Carbon	C	2	**4**		
	Silicon	Si	2	8	**4**	
	Germanium	Ge	2	8	18	**4**
5A (15)	Nitrogen	N	2	**5**		
	Phosphorus	P	2	8	**5**	
	Arsenic	As	2	8	18	**5**
6A (16)	Oxygen	O	2	**6**		
	Sulfur	S	2	8	**6**	
	Selenium	Se	2	8	18	**6**
7A (17)	Fluorine	F	2	**7**		
	Chlorine	Cl	2	8	**7**	
	Bromine	Br	2	8	18	**7**
8A (18)	Helium	He	**2**			
	Neon	Ne	2	**8**		
	Argon	Ar	2	8	**8**	
	Krypton	Kr	2	8	18	**8**

SAMPLE PROBLEM 4.11 Using Group Numbers

Using the periodic table, write the group number and the number of valence electrons for each of the following elements:

a. cesium **b.** iodine

SOLUTION

a. Cesium (Cs) is in Group 1A (1); cesium has one valence electron.
b. Iodine (I) is in Group 7A (17); iodine has seven valence electrons.

STUDY CHECK 4.11

What is the group number of elements with atoms that have five valence electrons?

Number of Valence Electrons Increases →

TABLE 4.11 Electron-Dot Symbols for Selected Elements in Periods 1–4

	Group Number							
	1A (1)	2A (2)	3A (13)	4A (4)	5A (15)	6A (16)	7A (17)	8A (18)
Number of Valence Electrons	1	2	3	4	5	6	7	8
Electron-Dot Symbol								
	H•							He:*
	Li•	Ḃe•	•Ḃ•	•Ċ• (with dot below)	•N̈• (with dot below)	•Ö: (with dot below)	•F̈: (with pair below)	:N̈e: (with pair below)
	Na•	Ṁg•	•Ȧl•	•Ṡi• (with dot below)	•P̈• (with dot below)	•S̈: (with dot below)	•C̈l: (with pair below)	:Är: (with pair below)
	K•	Ċa•	•Ġa•	•Ġe• (with dot below)	•Äs• (with dot below)	•S̈e: (with dot below)	•B̈r: (with pair below)	:K̈r: (with pair below)

*Helium (He) is stable with two valence electrons.

Electron-Dot Symbols

CORE CHEMISTRY SKILL

Drawing Electron-Dot Symbols

An **electron-dot symbol**, also known as a Lewis structure, represents the valence electrons as dots that are placed on the sides, top, or bottom of the symbol for the element. One to four valence electrons are arranged as single dots. When an atom has five to eight valence electrons, one or more electrons are paired. Any of the following would be an acceptable electron-dot symbol for magnesium, which has two valence electrons:

Electron-dot symbols for magnesium

Ṁg· Ṁg (dot below) ·Ṁg ·Mg· Mg· (dot below) ·Mg (dot below)

Electron-dot symbols for selected elements are given in Table 4.11.

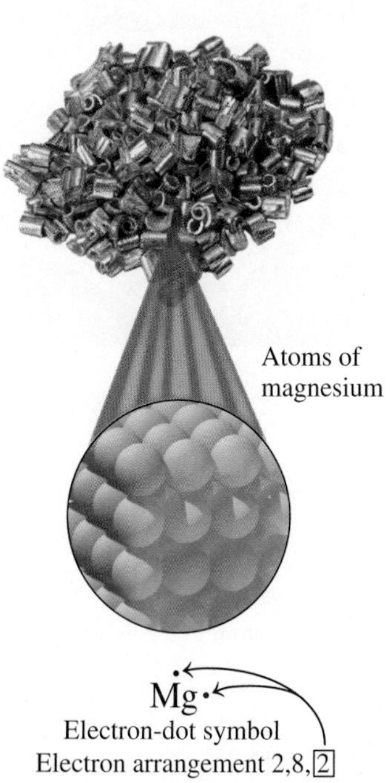

Magnesium (Mg) has two valence electrons, which give it an electron-dot symbol with two dots.

SAMPLE PROBLEM 4.12 Drawing Electron-Dot Symbols

Draw the electron-dot symbol for each of the following:

a. bromine **b.** aluminum

SOLUTION

a. Because the group number for bromine is 7A (17), bromine has seven valence electrons, which are drawn as seven dots, three pairs and one single dot, around the symbol Br.

·B̈r: (with pair below)

b. Aluminum, in Group 3A (13), has three valence electrons, which are shown as three single dots around the symbol Al.

·Ȧl·

STUDY CHECK 4.12

What is the electron-dot symbol for phosphorus?

Atomic Size

The size of an atom is determined by the distance of the valence electrons from the nucleus. For each group of representative elements, the atomic size *increases* going from the top to the bottom because the outermost electrons in each energy level are farther from the nucleus. For example, in Group 1A (1), Li has a valence electron in energy level 2; Na has a valence electron in energy level 3; and K has a valence electron in energy level 4. This means that a K atom is larger than a Na atom, and a Na atom is larger than a Li atom (see Figure 4.13).

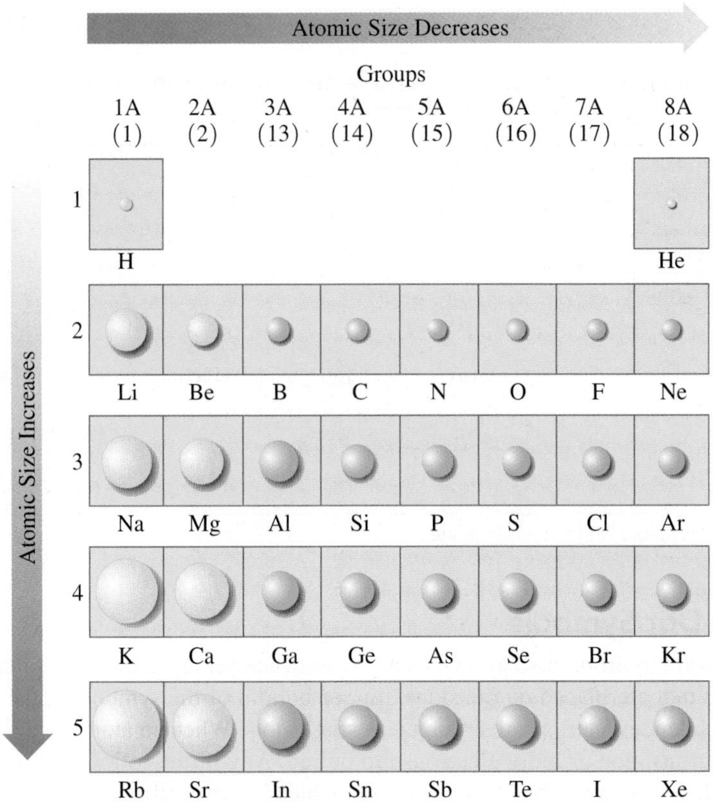

FIGURE 4.13 ▶ For representative elements, the atomic size increases going down a group but decreases going from left to right across a period.

Q Why does the atomic size increase going down a group of representative elements?

For the elements in a period, an increase in the number of protons in the nucleus increases the attraction for the outermost electrons. As a result, the outer electrons are pulled closer to the nucleus, which means that the size of representative elements *decreases* going from left to right across a period.

SAMPLE PROBLEM 4.13 Size of Atoms

Identify the smaller atom in each of the following pairs and explain your choice:

a. N or F **b.** K or Kr **c.** Ca and Sr

SOLUTION

a. The F atom has a greater positive charge on the nucleus, which pulls electrons closer, and makes the F atom smaller than the N atom. Atomic size decreases going from left to right across a period.

b. The Kr atom has a greater positive charge on the nucleus, which pulls electrons closer, and makes the Kr atom smaller than the K atom. Atomic size decreases going from left to right across a period.

c. The outer electrons in the Ca atom are closer to the nucleus than in the Sr atom, which makes the Ca atom smaller than the Sr atom. Atomic size increases going down a group.

STUDY CHECK 4.13

Which atom has the largest atomic size, Mg, Ca, or Cl?

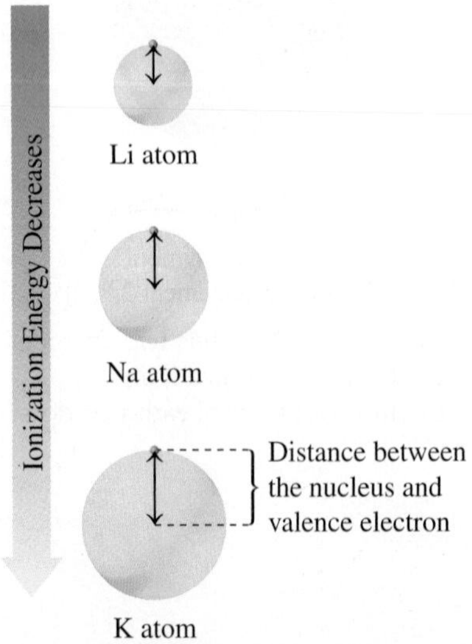

As the distance from the nucleus to the valence electrons increases in Group 1A (1), the ionization energy decreases.

Ionization Energy

In an atom, negatively charged electrons are attracted to the positive charge of the protons in the nucleus. Thus, a quantity of energy known as the **ionization energy** is required to remove one of the outermost electrons. When an electron is removed from a neutral atom, a positive particle called a *cation* with a 1+ charge is formed.

$$Na(g) + \text{energy (ionization)} \longrightarrow Na^{+}(g) + e^{-}$$

The ionization energy *decreases* going down a group. Less energy is needed to remove an electron because nuclear attraction decreases when electrons are farther from the nucleus. Going across a period from left to right, the ionization energy *increases*. As the positive charge of the nucleus increases, more energy is needed to remove an electron.

In Period 1, the valence electrons are close to the nucleus and strongly held. H and He have high ionization energies because a large amount of energy is required to remove an electron. The ionization energy for He is the highest of any element because He has a full, stable, energy level which is disrupted by removing an electron. The high ionization energies of the noble gases indicate that their electron arrangements are especially stable. In general, the ionization energy is low for the metals and high for the nonmetals (see Figure 4.14).

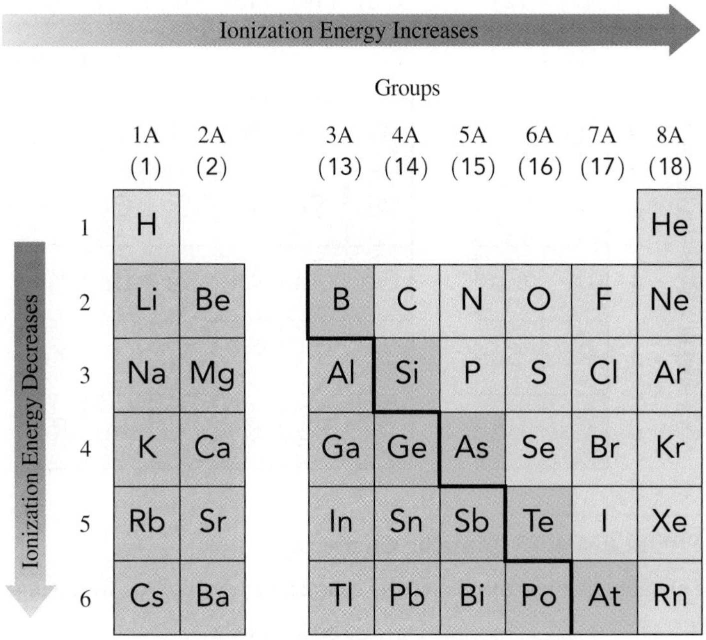

FIGURE 4.14 ▶ Ionization energies for the representative elements decrease going down a group and increase going across a period.

Q Why is the ionization energy for F greater than that for Cl?

SAMPLE PROBLEM 4.14 Ionization Energy

Indicate the element in each group that has the higher ionization energy and explain your choice.

a. K or Na **b.** Mg or Cl **c.** F, N, or C

SOLUTION

a. Na. In Na, an electron is removed from an energy level closer to the nucleus, which requires a higher ionization energy for Na compared to K.

b. Cl. The increased nuclear charge of Cl increases the attraction for the valence electrons, which requires a higher ionization energy for Cl compared to Mg.

c. F. The increased nuclear charge of F increases the attraction for the valence electrons, which requires a higher ionization energy for F compared to C or N.

STUDY CHECK 4.14

Arrange Sn, Sr, and I in order of increasing ionization energy.

Metallic Character

An element that has **metallic character** is an element that loses valence electrons easily. Metallic character is more prevalent in the elements (metals) on the left side of the periodic table and decreases going from left to right across a period. The elements (nonmetals) on the right side of the periodic table do not easily lose electrons, which means they are the least metallic. Most of the metalloids between the metals and nonmetals tend to lose electrons, but not as easily as the metals. Thus, in Period 3, sodium, which loses electrons most easily, would be the most metallic. Going across from left to right in Period 3, metallic character decreases to argon, which has the least metallic character.

For elements in the same group of representative elements, metallic character increases going from top to bottom. Atoms at the bottom of any group have more electron levels, which makes it easier to lose electrons. Thus, the elements at the bottom of a group on the periodic table have lower ionization energy and are more metallic compared to the elements at the top (see Figure 4.15).

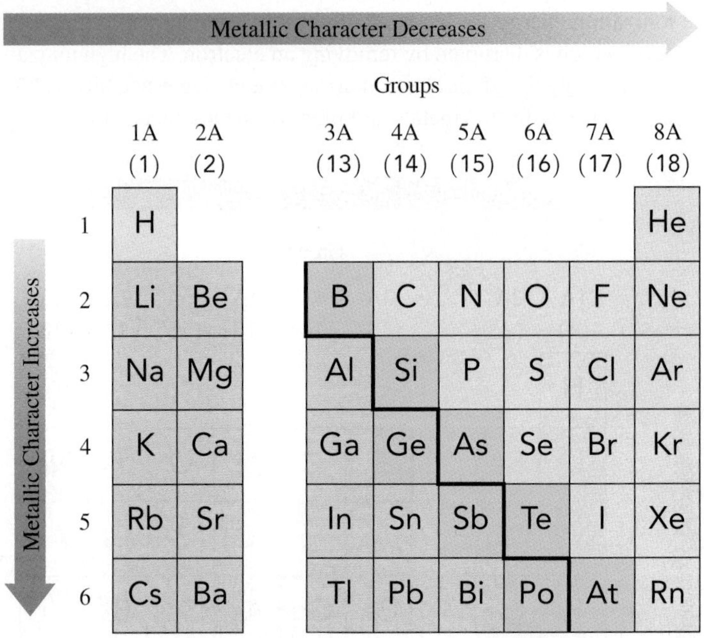

FIGURE 4.15 ▶ Metallic character of the representative elements increases going down a group and decreases going from left to right across a period.

Q Why is the metallic character greater for Rb than for Li?

SAMPLE PROBLEM 4.15 Metallic Character

Identify the element that has more metallic character in each of the following:

a. Mg or Al **b.** Mg or Ca

SOLUTION

a. Mg is more metallic than Al because metallic character decreases going from left to right across a period.

b. Ca is more metallic than Mg because metallic character increases going down a group.

STUDY CHECK 4.15

Of the elements Na, K, and S, which has the most metallic character?

A summary of the trends in periodic properties we have discussed is given in Table 4.12.

TABLE 4.12 Summary of Trends in Periodic Properties of Representative Elements

Periodic Property	Top to Bottom within a Group	Left to Right across a Period
Valence Electrons	Remains the same	Increases
Atomic Size	Increases due to the increase in number of energy levels	Decreases due to the increase of protons in the nucleus that pull electrons closer
Ionization Energy	Decreases because valence electrons are easier to remove when they are farther away from the nucleus	Increases as the attraction of the protons for outer electrons requires more energy to remove an electron
Metallic Character	Increases because valence electrons are easier to remove when they are farther from the nucleus	Decreases as the attraction of the protons makes it more difficult to remove an electron

QUESTIONS AND PROBLEMS

4.7 Trends in Periodic Properties

LEARNING GOAL Use the electron arrangement of elements to explain the trends in periodic properties.

4.47 What is the group number and number of valence electrons for each of the following elements?
a. magnesium **b.** iodine **c.** oxygen
d. phosphorus **e.** tin **f.** boron

4.48 What is the group number and number of valence electrons for each of the following elements?
a. potassium **b.** silicon **c.** neon
d. aluminum **e.** barium **f.** bromine

4.49 Write the group number and draw the electron-dot symbol for each of the following elements:
a. sulfur **b.** nitrogen **c.** calcium
d. sodium **e.** gallium

4.50 Write the group number and draw the electron-dot symbol for each of the following elements:
a. carbon **b.** oxygen **c.** argon
d. lithium **e.** chlorine

4.51 Select the larger atom in each pair.
a. Na or Cl **b.** Na or Rb **c.** Na or Mg **d.** Rb or I

4.52 Select the larger atom in each pair.
a. S or Cl **b.** S or O **c.** S or Se **d.** S or Mg

4.53 Place the elements in each set in order of increasing atomic size.
a. C, Si, Ge **b.** K, Na, Cs **c.** F, Cl, I **d.** As, N, P

4.54 Place the elements in each set in order of decreasing atomic size.
a. Cl, S, P **b.** Ge, Si, C **c.** Ba, Ca, Sr **d.** S, O, Se

4.55 Select the element in each pair with the higher ionization energy.
a. Br or I **b.** Mg or Sr **c.** Si or P **d.** I or Xe

4.56 Select the element in each pair with the higher ionization energy.
a. O or Ne **b.** K or Br **c.** Ca or Ba **d.** Ne or N

4.57 Place the elements in each set in order of increasing ionization energy.
a. F, Cl, Br **b.** Na, Cl, Al **c.** Na, K, Cs **d.** As, Ca, Br

4.58 Place the elements in each set in order of increasing ionization energy.
a. O, N, C **b.** S, P, Cl **c.** As, P, N **d.** Al, Si, P

4.59 Fill in the following blanks using *larger* or *smaller*, *more metallic* or *less metallic*. Na has a ______ atomic size and is ______ than P.

4.60 Fill in the following blanks using *larger* or *smaller*, *higher* or *lower*. Mg has a ______ atomic size and a ______ ionization energy than Cs.

4.61 Place the following in order of decreasing metallic character:
Br, Ge, Ca, Ga

4.62 Place the following in order of increasing metallic character:
Mg, P, Al, Ar

4.63 Fill in each of the following blanks using *higher* or *lower*, *more* or *less*:
Sr has a ______ ionization energy and ______ metallic character than Sb.

4.64 Fill in each of the following blanks using *higher* or *lower*, *more* or *less*:
N has a ______ ionization energy and ______ metallic character than As.

4.65 Complete each of the statements **a–d** using **1**, **2**, or **3**:
1. decreases **2.** increases **3.** remains the same
Going down Group 6A (16),
a. the ionization energy ______
b. the atomic size ______
c. the metallic character ______
d. the number of valence electrons ______

4.66 Complete each of the statements **a–d** using **1**, **2**, or **3**:
1. decreases **2.** increases **3.** remains the same
Going from left to right across Period 4,
a. the ionization energy ______
b. the atomic size ______
c. the metallic character ______
d. the number of valence electrons ______

4.67 Which statements completed with **a–e** will be *true* and which will be *false*?
An atom of N compared to an atom of Li has a larger (greater)
a. atomic size **b.** ionization energy
c. number of protons **d.** metallic character
e. number of valence electrons

4.68 Which statements completed with **a–d** will be *true* and which will be *false*?
An atom of Pb compared to an atom of C has a larger (greater)
a. number of neutrons **b.** ionization energy
c. atomic size **d.** metallic character

CONCEPT MAP

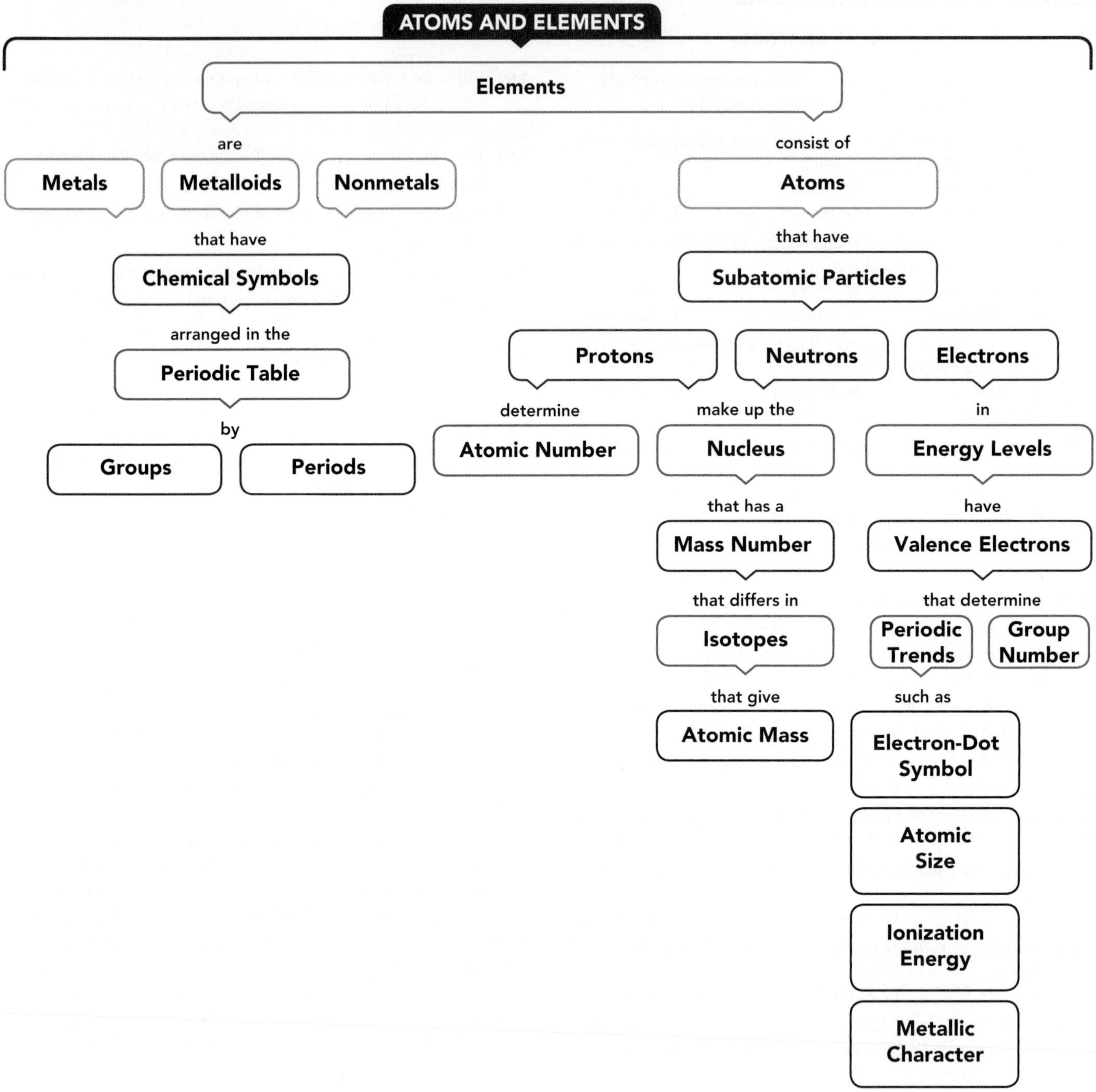

CHAPTER REVIEW

4.1 Elements and Symbols

LEARNING GOAL Given the name of an element, write its correct symbol; from the symbol, write the correct name.

- Elements are the primary substances of matter.
- Chemical symbols are one- or two-letter abbreviations of the names of the elements.

4.2 The Periodic Table

LEARNING GOAL Use the periodic table to identify the group and the period of an element; identify the element as a metal, a nonmetal, or a metalloid.

- The periodic table is an arrangement of the elements by increasing atomic number.

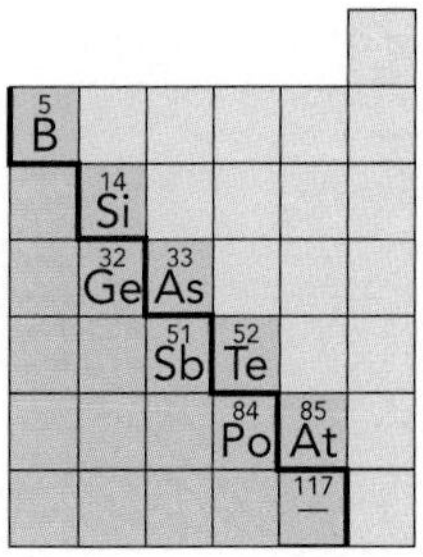

- A horizontal row is called a *period*.
  A vertical column on the periodic table containing elements with similar properties is called a *group*.
- Elements in Group 1A (1) are called the *alkali metals*; Group 2A (2), the *alkaline earth metals*; Group 7A (17), the *halogens*; and Group 8A (18), the *noble gases*.
- On the periodic table, *metals* are located on the left of the heavy zigzag line, and *nonmetals* are to the right of the heavy zigzag line.
- Except for aluminum, elements located along the heavy zigzag line are called *metalloids*.

4.3 The Atom

LEARNING GOAL Describe the electrical charge and location in an atom for a proton, a neutron, and an electron.

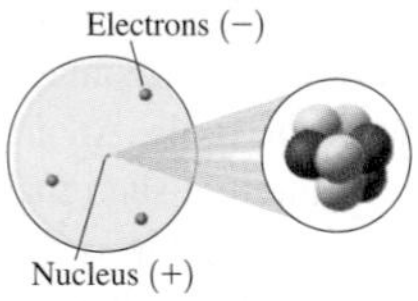

- An atom is the smallest particle that retains the characteristics of an element.
- Atoms are composed of three types of subatomic particles.
- Protons have a positive charge (+), electrons carry a negative charge (−), and neutrons are electrically neutral.
- The protons and neutrons are found in the tiny, dense nucleus; electrons are located outside the nucleus.

4.4 Atomic Number and Mass Number

LEARNING GOAL Given the atomic number and the mass number of an atom, state the number of protons, neutrons, and electrons.

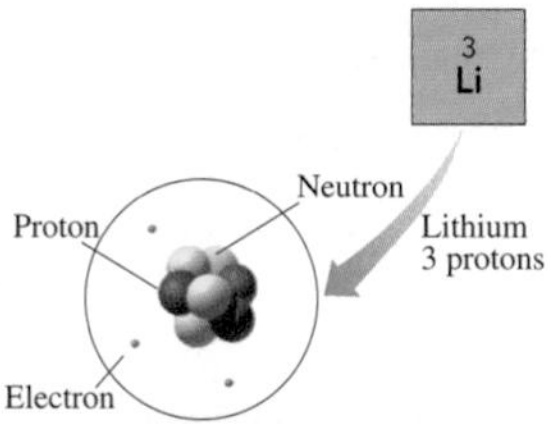

- The atomic number gives the number of protons in all the atoms of the same element.
- In a neutral atom, the number of protons and electrons is equal.
- The mass number is the total number of protons and neutrons in an atom.

4.5 Isotopes and Atomic Mass

LEARNING GOAL Determine the number of protons, electrons, and neutrons in one or more of the isotopes of an element; calculate the atomic mass of an element using the percent abundance and mass of its naturally occurring isotopes.

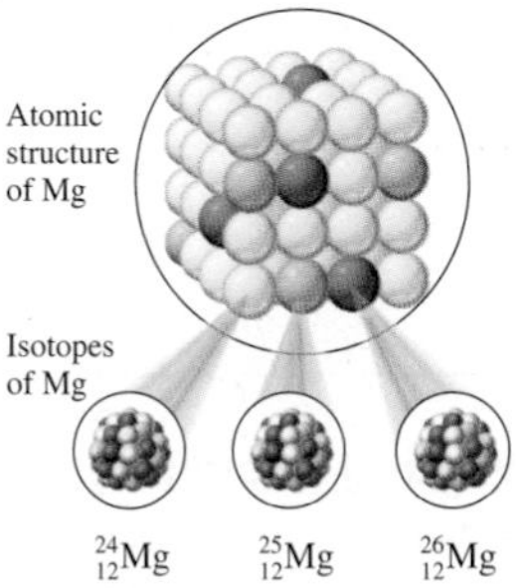

- Atoms that have the same number of protons but different numbers of neutrons are called *isotopes*.
- The atomic mass of an element is the weighted average mass of all the isotopes in a naturally occurring sample of that element.

4.6 Electron Energy Levels

LEARNING GOAL Given the name or symbol of one of the first 20 elements in the periodic table, write the electron arrangement.

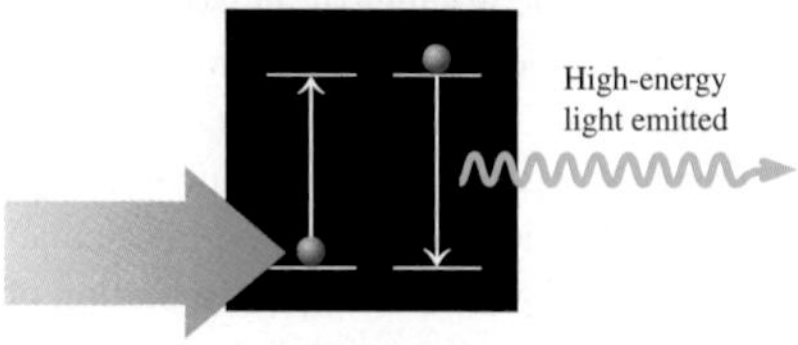

- Every electron has a specific amount of energy.
- In an atom, the electrons of similar energy are grouped in specific energy levels.
- The first level nearest the nucleus can hold 2 electrons, the second level can hold 8 electrons, and the third level will take up to 18 electrons.
- The electron arrangement is written by placing the number of electrons in that atom in order from the lowest energy levels and filling to higher levels.

4.7 Trends in Periodic Properties

LEARNING GOAL Use the electron arrangement of elements to explain the trends in periodic properties.

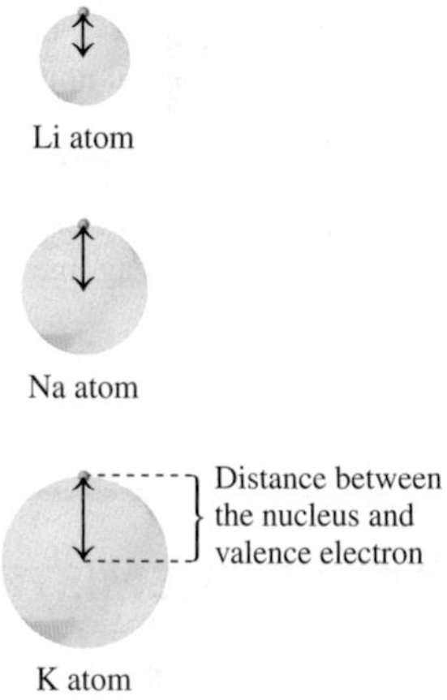

- The properties of elements are related to the valence electrons of the atoms.
- With only a few exceptions, each group of elements has the same arrangement of valence electrons differing only in the energy level.
- Valence electrons are represented as dots around the symbol of the element.
- The size of an atom increases going down a group and decreases going from left to right across a period.
- The energy required to remove a valence electron is the ionization energy, which decreases going down a group, and increases going from left to right across a period.
- The metallic character of an element increases going down a group and decreases going from left to right across a period.

KEY TERMS

alkali metals Elements of Group 1A (1) except hydrogen; these are soft, shiny metals with one valence electron.

alkaline earth metals Group 2A (2) elements, which have two valence electrons.

atom The smallest particle of an element that retains the characteristics of the element.

atomic mass The weighted average mass of all the naturally occurring isotopes of an element.

atomic mass unit (amu) A small mass unit used to describe the mass of very small particles such as atoms and subatomic particles; 1 amu is equal to one-twelfth the mass of a $^{12}_{6}C$ atom.

atomic number A number that is equal to the number of protons in an atom.

atomic symbol An abbreviation used to indicate the mass number and atomic number of an isotope.

chemical symbol An abbreviation that represents the name of an element.

electron A negatively charged subatomic particle having a very small mass that is usually ignored in calculations; its symbol is e^-.

electron-dot symbol The representation of an atom that shows valence electrons as dots around the symbol of the element.

energy level A group of electrons with similar energy.

group A vertical column in the periodic table that contains elements having similar physical and chemical properties.

group number A number that appears at the top of each vertical column (group) in the periodic table and indicates the number of electrons in the outermost energy level.

halogens Group 7A (17) elements—fluorine, chlorine, bromine, iodine, and astatine—which have seven valence electrons.

ionization energy The energy needed to remove the least tightly bound electron from the outermost energy level of an atom.

isotope An atom that differs only in mass number from another atom of the same element. Isotopes have the same atomic number (number of protons), but different numbers of neutrons.

mass number The total number of protons and neutrons in the nucleus of an atom.

metal An element that is shiny, malleable, ductile, and a good conductor of heat and electricity. The metals are located to the left of the heavy zigzag line on the periodic table.

metallic character A measure of how easily an element loses a valence electron.

metalloid Elements with properties of both metals and nonmetals, located along the heavy zigzag line on the periodic table.

neutron A neutral subatomic particle having a mass of about 1 amu and found in the nucleus of an atom; its symbol is n or n^0.

noble gas An element in Group 8A (18) of the periodic table, generally unreactive and seldom found in combination with other elements.

nonmetal An element with little or no luster that is a poor conductor of heat and electricity. The nonmetals are located to the right of the heavy zigzag line on the periodic table.

nucleus The compact, very dense center of an atom, containing the protons and neutrons of the atom.

period A horizontal row of elements in the periodic table.

periodic table An arrangement of elements by increasing atomic number such that elements having similar chemical behavior are grouped in vertical columns.

proton A positively charged subatomic particle having a mass of about 1 amu and found in the nucleus of an atom; its symbol is p or p^+.

representative elements Elements found in Groups 1A (1) through 8A (18) excluding B groups (3–12) of the periodic table.

transition elements Elements located between Groups 2A (2) and 3A (13) on the periodic table.

valence electrons Electrons in the outermost energy level of an atom.

CORE CHEMISTRY SKILLS

The chapter section containing each Core Chemistry Skill is shown in parentheses at the end of each heading.

Counting Protons and Neutrons (4.4)

- The atomic number of an element is equal to the number of protons in every atom of that element. The atomic number is the whole number that appears above the symbol of each element on the periodic table.

  Atomic number = number of protons in an atom

- Because atoms are neutral, the number of electrons is equal to the number of protons. Thus, the atomic number gives the number of electrons.
- The mass number is the total number of protons and neutrons in the nucleus of an atom.

  Mass number = number of protons + number of neutrons

- The number of neutrons is calculated from the mass number and atomic number.

  Number of neutrons = mass number − number of protons

Example: Calculate the number of protons, neutrons, and electrons in a krypton atom with a mass number of 80.

Answer:

Element	Atomic Number	Mass Number	Number of Protons	Number of Neutrons	Number of Electrons
Kr	36	80	equal to atomic number (36)	equal to mass number − number of protons 80 − 36 = 44	equal to number of protons (36)

Writing Atomic Symbols for Isotopes (4.5)

- Isotopes are atoms of the same element that have the same atomic number but different numbers of neutrons.
- An atomic symbol is written for a particular isotope, with its mass number (protons and neutrons) shown in the upper left corner and its atomic number (protons) shown in the lower left corner.

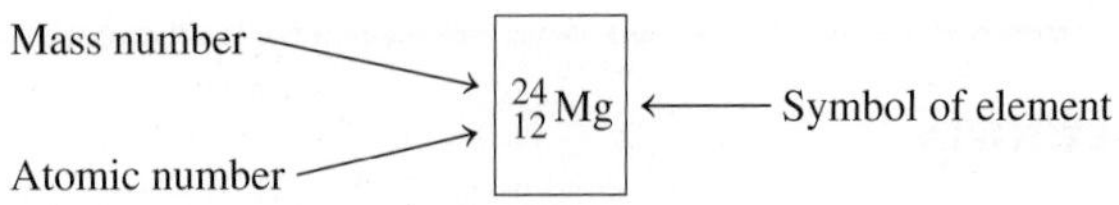

Example: Calculate the number of protons and neutrons in the cadmium isotope $^{112}_{48}\text{Cd}$.

Answer:

Isotope	Atomic Number	Mass Number	Number of Protons	Number of Neutrons
$^{112}_{48}\text{Cd}$	number (48) in lower left corner	number (112) in upper left corner	equal to atomic number (48)	equal to mass number − number of protons 112 − 48 = 64

Writing Electron Arrangements (4.6)

- The electron arrangement for an atom specifies the energy levels occupied by the electrons of an atom.
- An electron arrangement is written starting with the lowest energy level, followed by the next lowest energy level.

Example: Write the electron arrangement for phosphorus.

Answer: Phosphorus has atomic number 15, which means it has 15 protons and 15 electrons.

2,8,5

Identifying Trends in Periodic Properties (4.7)

- The size of an atom increases going down a group and decreases going from left to right across a period.
- The ionization energy decreases going down a group and increases going from left to right across a period.
- The metallic character of an element increases going down a group and decreases going from left to right across a period.

Example: For Mg, P, and Cl, identify which has the

a. largest atomic size
b. highest ionization energy
c. most metallic character

Answer: **a.** Mg **b.** Cl **c.** Mg

Drawing Electron-Dot Symbols (4.7)

- The valence electrons are the electrons in the outermost energy level.
- The number of valence electrons is the same as the group number for the representative elements.
- An electron-dot symbol represents the number of valence electrons shown as dots placed around the symbol for the element.

Example: Give the group number and number of valence electrons, and draw the electron-dot symbol for each of the following:

a. Rb **b.** Se **c.** Xe

Answer: **a.** Group 1A (1), one valence electron, Rb·
b. Group 6A (16), six valence electrons, ·S̈e:
c. Group 8A (18), eight valence electrons, :Ẍ̤e:

UNDERSTANDING THE CONCEPTS

The chapter sections to review are shown in parentheses at the end of the question.

4.69 According to Dalton's atomic theory, which of the following are *true* or *false*? If false, correct the statement to make it true. (4.3)

a. Atoms of an element are identical to atoms of other elements.
b. Every element is made of atoms.
c. Atoms of different elements combine to form compounds.
d. In a chemical reaction, some atoms disappear and new atoms appear.

4.70 Use Rutherford's gold-foil experiment to answer each of the following: (4.3)

a. What did Rutherford expect to happen when he aimed particles at the gold foil?
b. How did the results differ from what he expected?
c. How did he use the results to propose a model of the atom?

4.71 Match the subatomic particles (**1–3**) to each of the descriptions below: (4.4)

1. protons **2.** neutrons **3.** electrons

a. atomic mass
b. atomic number
c. positive charge
d. negative charge
e. mass number – atomic number

4.72 Match the subatomic particles (**1–3**) to each of the descriptions below: (4.4)

1. protons **2.** neutrons **3.** electrons

a. mass number
b. surround the nucleus
c. in the nucleus
d. charge of 0
e. equal to number of electrons

4.73 Consider the following atoms in which X represents the chemical symbol of the element: (4.4, 4.5)

$^{16}_{8}\text{X}$ $^{16}_{9}\text{X}$ $^{18}_{10}\text{X}$ $^{17}_{8}\text{X}$ $^{18}_{8}\text{X}$

a. What atoms have the same number of protons?
b. Which atoms are isotopes? Of what element?
c. Which atoms have the same mass number?
d. What atoms have the same number of neutrons?

4.74 For each of the following, write the symbol and name for X and the number of protons and neutrons. Which are isotopes of each other? (4.4, 4.5)

a. $^{124}_{47}\text{X}$ **b.** $^{116}_{49}\text{X}$ **c.** $^{116}_{50}\text{X}$
d. $^{124}_{50}\text{X}$ **e.** $^{116}_{48}\text{X}$

4.75 For each of the following, find out the number of protons and neutrons: (4.4, 4.5)

a. $^{235}_{92}\text{U}$ **b.** $^{239}_{92}\text{U}$ **c.** $^{37}_{17}\text{Cl}$

4.76 Complete the following table for the three naturally occurring isotopes of silicon, the major component in computer chips: (4.4, 4.5)

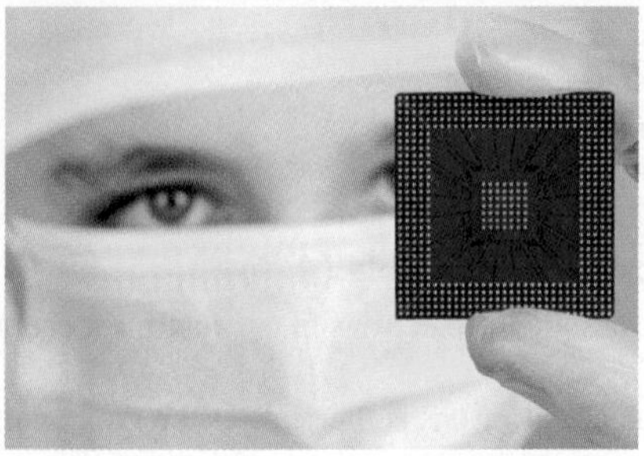

Computer chips consist primarily of the element silicon.

	Isotope		
	${}^{28}_{14}\text{Si}$	${}^{29}_{14}\text{Si}$	${}^{30}_{14}\text{Si}$
Atomic Number			
Mass Number			
Number of Protons			
Number of Neutrons			
Number of Electrons			

4.77 For each representation of a nucleus **A** through **E**, write the atomic symbol, and identify which are isotopes. (4.4, 4.5)

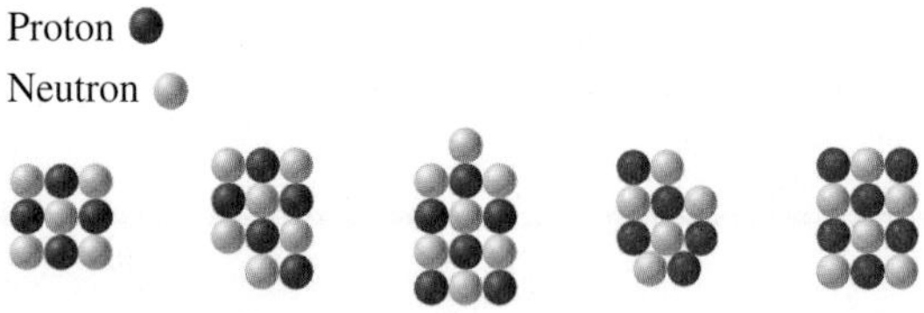

A B C D E

4.78 Identify the element represented by each nucleus **A** through **E** in problem 4.77 as a metal, a nonmetal, or a metalloid. (4.3)

4.79 Match the spheres **A** through **D** with atoms of Li, Na, K, and Rb. (4.7)

A B C D

4.80 Match the spheres **A** through **D** with atoms of K, Ge, Ca, and Kr. (4.7)

A B C D

4.81 Of the elements Si, F, Sb, Sn, Ar, He, and I, identify one that fits each of the following: (4.2, 4.6, 4.7)
a. halogen
b. in Group 4A
c. metalloid
d. a gas (molecule)
e. period 3 (16)

4.82 Of the elements Sn, Xe, Te, Sr, I, and Rb, identify one that fits each of the following: (4.2, 4.6, 4.7)
a. smallest atomic size
b. an alkaline earth metal
c. a metalloid
d. lowest ionization energy
e. in Group 4A (14)
f. least metallic character
g. seven valence electrons

ADDITIONAL QUESTIONS AND PROBLEMS

4.83 Give the group and period numbers for each of the following elements: (4.2)
a. bromine **b.** argon **c.** potassium **d.** radium

4.84 Give the group and period numbers for each of the following elements: (4.2)
a. radon **b.** arsenic **c.** carbon **d.** neon

4.85 Indicate if each of the following statements is *true* or *false*: (4.3)
a. The proton is a negatively charged particle.
b. The neutron is 2000 times as heavy as a proton.
c. The atomic mass unit is based on a carbon atom with six protons and six neutrons.
d. The nucleus is the largest part of the atom.
e. The electrons are located outside the nucleus.

4.86 Indicate if each of the following statements is *true* or *false*: (4.3)
a. The neutron has no net charges.
b. Most of the mass of an atom is due to the protons and electrons.
c. The total charge of the electrons is equal, but opposite, to the charge of the nucleus.
d. The proton and the electron have about the same mass.
e. The mass number is the number of protons.

4.87 Complete the following statements: (4.2, 4.3)
a. The atomic number gives the number of _______ in the nucleus.
b. In an atom, the number of electrons is equal to the number of _______.
c. Sodium and potassium are examples of elements called _______.

4.88 Complete the following statements: (4.2, 4.3)
a. Elements that have properties of both metals and nonmetals are called _______.
b. _______ lose electrons more easily than metalloids.

4.89 Complete the following statements: (4.2, 4.3)
a. 28 **b.** 56 **c.** 88 **d.** 33
e. 50 **f.** 55 **g.** 79 **h.** 80

4.90 Write the name and symbol of the element with the following atomic number: (4.1, 4.2)
a. 36 **b.** 22 **c.** 48 **d.** 26
e. 54 **f.** 78 **g.** 83 **h.** 92

4.91 For the following atoms, determine the number of protons, neutrons, and electrons: (4.4, 4.5)
a. $^{114}_{48}Cd$ **b.** $^{98}_{43}Tc$ **c.** $^{199}_{79}Au$
d. $^{222}_{86}Rn$ **e.** $^{136}_{54}Xe$

4.92 For the following atoms, determine the number of protons, neutrons, and electrons: (4.4, 4.5)
a. $^{202}_{80}Hg$ **b.** $^{127}_{53}I$ **c.** $^{75}_{35}Br$
d. $^{133}_{55}Cs$ **e.** $^{195}_{78}Pt$

4.93 Complete the following table: (4.4, 4.5)

Name	Atomic Symbol	Number of Protons	Number of Neutrons	Number of Electrons
	$^{34}_{16}S$			
		28	34	
Magnesium			14	
	$^{220}_{86}Rn$			

4.94 Complete the following table: (4.4, 4.5)

Name	Atomic Symbol	Number of Protons	Number of Neutrons	Number of Electrons
Potassium			22	
	$^{51}_{23}V$			
		48	64	
Barium			82	

4.95 Write the atomic symbol for each of the following: (4.4, 4.5)
a. an atom with four protons and five neutrons
b. an atom with 12 protons and 14 neutrons
c. a calcium atom with a mass number of 46
d. an atom with 30 electrons and 40 neutrons

4.96 Write the atomic symbol for each of the following: (4.4, 4.5)
a. an aluminum atom with 14 neutrons
b. an atom with atomic number 26 and 32 neutrons
c. a strontium atom with 50 neutrons
d. an atom with a mass number of 72 and atomic number 33

4.97 The most prevalent isotope of lead is $^{208}_{82}Pb$. (4.4, 4.5)
a. How many protons, neutrons, and electrons are in $^{208}_{82}Pb$?
b. What is the atomic symbol of another isotope of lead with 132 neutrons?
c. What is the atomic symbol and name of an atom with the same mass number as in part **b** and 131 neutrons?

4.98 The most prevalent isotope of silver is $^{107}_{47}Ag$. (4.4, 4.5)
a. How many protons, neutrons, and electrons are in $^{107}_{47}Ag$?
b. What is the atomic symbol of another isotope of silver with 62 neutrons?
c. What is the atomic symbol and name of an atom with the same mass number as in part **b** and 61 neutrons?

4.99 Write the period number for each of the following and state if they are metal, nonmetal or metalloid:
a. P **b.** Ge
c. F **d.** W
e. Cs

4.100 Write the group number and the electron arrangement for each of the following: (4.6)
a. magnesium **b.** chlorine
c. beryllium **d.** argon

4.101 Why is the ionization energy of Ca higher than K, but lower than that of Mg? (4.7)

4.102 Why is the ionization energy of Cl lower than F, but higher than that of S? (4.7)

4.103 Of the elements Li, Be, N, and F, which (4.7)
a. is an alkaline earth metal?
b. has the largest atomic size?
c. has the highest ionization energy?
d. is found in Group 5A (15)?
e. has the most metallic character?

4.104 Of the elements C, Si, Sn, Pb, and Ge, which (4.7)
a. has the largest atomic size?
b. has the smallest atomic size?
c. has the lowest ionization energy?
d. requires the most energy to remove an electron?
e. is found in Period 4?

CHALLENGE QUESTIONS

The following groups of questions are related to the topics in this chapter. However, they do not all follow the chapter order, and they require you to combine concepts and skills from several sections. These questions will help you increase your critical thinking skills and prepare for your next exam.

4.105 There are four naturally occurring isotopes of strontium: $^{84}_{38}Sr$, $^{86}_{38}Sr$, $^{87}_{38}Sr$, $^{88}_{38}Sr$. (4.4, 4.5)
a. How many protons, neutrons, and electrons are in Sr-87?
b. What is the most prevalent isotope in a strontium sample?
c. How many neutrons are in Sr-84?
d. Why don't any of the isotopes of strontium have the atomic mass of 87.62 amu listed on the periodic table?

4.106 There are four naturally occurring isotopes of iron: $^{54}_{26}Fe$, $^{56}_{26}Fe$, $^{57}_{26}Fe$, and $^{58}_{26}Fe$. (4.4, 4.5)
a. How many protons, neutrons, and electrons are in $^{58}_{26}Fe$?
b. What is the most prevalent isotope in an iron sample?
c. How many neutrons are in $^{57}_{26}Fe$?
d. Why don't any of the isotopes of iron have the atomic mass of 55.85 amu listed on the periodic table?

4.107 Give the symbol of the element that has the (4.7)
a. smallest atomic size in Group 6A (16)
b. smallest atomic size in Period 3
c. highest ionization energy in Group 5A (15)
d. lowest ionization energy in Period 3
e. most metallic character in Group 2A (2)

4.108 Give the symbol of the element that has
a. the second largest atomic size in Group 1A (1)
b. the largest atomic size in Period 4
c. the second highest ionization energy in Group 2A (2)
d. the third highest ionization energy in Group 6A (16)
e. least metallic character in Group 4A (14)

4.109 A lead atom has a mass of 3.4×10^{-22} g. How many lead atoms are in a cube of lead that has a volume of 2.00 cm^3 if the density of lead is 11.3 g/cm^3? (4.1)

4.110 If the diameter of an atom is 3.14×10^{-8} cm, how many atoms would fit along a line exactly 1 cm long? (4.1)

4.111 Silicon has three naturally occurring isotopes: Si-28 (27.977 amu) with an abundance of 92.23%, Si-29 (28.976 amu) with a 4.68% abundance, and Si-30 (29.974 amu) with a 3.09% abundance. Calculate the atomic mass for silicon using the weighted average mass method. (4.5)

4.112 Antimony (Sb) has two naturally occurring isotopes: Sb-121 and Sb-123. If Sb-121 has a 57.21% abundance and a mass of 120.9 amu, and Sb-123 has a 42.79% abundance and a mass of 122.9 amu, calculate the atomic mass for antimony using the weighted average mass method. (4.5)

ANSWERS

Answers to Study Checks

4.1 Si, S, and Ag

4.2 magnesium, aluminum, and fluorine

4.3 **a.** Strontium is in Group 2A (2).
b. Strontium is an alkaline earth metal.
c. Strontium is in Period 5.

4.4 **a.** arsenic, As **b.** radon, Rn **c.** boron, B

4.5 False; the nucleus occupies a very small volume in an atom.

4.6 **a.** 79 **b.** 79 **c.** gold, Au

4.7 45 neutrons

4.8 **a.** $^{15}_{7}\text{N}$ **b.** $^{42}_{20}\text{Ca}$ **c.** $^{27}_{13}\text{Al}$

4.9 10.81 amu

4.10 phosphorus

4.11 Group 5A (15)

4.12 $\cdot\ddot{\underset{\cdot}{\text{P}}}\cdot$

4.13 Ca

4.14 Ionization energy increases going left to right across a period: Sr is lowest, Sn is higher, and I is the highest of this set.

4.15 K

Answers to Selected Questions and Problems

4.1 **a.** Na **b.** K **c.** Ni **d.** Ag
e. N **f.** Pd **g.** V **h.** Zn

4.3 **a.** sulfur **b.** fluorine **c.** neon **d.** gold
e. aluminum **f.** copper **g.** palladium **h.** magnesium

4.5 **a.** sodium, chlorine
b. calcium, sulfur, oxygen
c. carbon, hydrogen, chlorine, nitrogen, oxygen
d. barium, chromium, oxygen

4.7 **a.** 2 Period **b.** 4 Period
c. Group 2A **d.** Group 15 or Group 5A

4.9 **a.** K:—Alkali metal needed for muscle contraction and nerve impulses.
b. Mn:—Transition metal needed for bone growth, blood clotting and necessary for metabolic reactions.
c. Co:—Transition metal which is the component of Vitamin B12 and red blood cells.
d. I:—Halogen needed for proper thyroid function.

4.11 **a.** N **b.** Ar **c.** Ca
d. S **e.** Br

4.13 **a.** Metal **b.** Nonmetal **c.** Au—metal
d. Sb—metalloid **e.** Si—metalloid **f.** Nonmetal
g. Metal **h.** Metal

4.15 **a.** electron **b.** proton
c. electron **d.** neutron

4.17 Rutherford determined that an atom contains a small, compact nucleus that is positively charged.

4.19 **a.** False **b.** False
c. True **d.** False

4.21 In the process of brushing hair, strands of hair become charged with like charges that repel each other.

4.23 **a.** atomic number **b.** both
c. mass number **d.** atomic number

4.25 **a.** lithium, Li **b.** Copper Cu
c. Chromium Cr **d.** Rubidium Rb
e. Aluminum Al **f.** Ruthenium Ru
g. Cesium Cs **h.** Carbon C

4.27 **a.** 8 Protons 8 electrons
b. 24 Protons 24 electrons
c. 17 Protons 17 electrons
d. 20 Protons 20 electrons

4.29

Name of the Element	Symbol	Atomic Number	Mass Number	Number of Protons	Number of Neutrons	Number of Electrons
Aluminum	Al	13	27	13	14	13
Magnesium	Mg	12	24	12	12	12
Potassium	K	19	39	19	20	19
Sulfur	S	16	31	16	15	16
Iron	Fe	26	56	26	30	26

4.31 **a.** Protons = 2 neutrons = 2 electrons = 2
b. Protons = 6 neutrons = 8 electrons = 6
c. Protons = 92 neutrons = 143 electrons = 92
d. Protons = 17 neutrons = 18 electrons = 17

4.33 **a.** $^{15}_{7}\text{N}$ **b.** $^{65}_{29}\text{Cu}$ **c.** $^{137}_{56}\text{Ba}$
d. $^{42}_{20}\text{Ca}$ **e.** $^{107}_{47}\text{Ag}$

4.35 **a.** $^{36}_{18}\text{Ar}$ $^{38}_{18}\text{Ar}$ $^{40}_{18}\text{Ar}$
b. They all have the same number of protons and electrons.
c. They have different numbers of neutrons, which gives them different mass numbers.
d. The atomic mass of Ar listed on the periodic table is the weighted average atomic mass of all the naturally occurring isotopes.
e. The isotope Ar-40 is most prevalent because its mass is closest to the atomic mass of Ar on the periodic table.

4.37 69.72 amu

4.39 absorb

4.41 **a.** green light **b.** blue light

4.43 **a.** 2,4 **b.** 2,8,8 **c.** 2,8,8,1
d. 2,8,4 **e.** 2 **f.** 2,5

4.45 **a.** Li **b.** Mg **c.** H
d. Cl **e.** O

4.47 **a.** Group 2A (2), 2 e^- **b.** Group 7A (17), 7 e^-
c. Group 6A (16), 6 e^- **d.** Group 5A (15), 5 e^-
e. Group 4A (14), 4 e^- **f.** Group 3A (13), 3 e^-

4.49 **a.** Group 6A (16)

·$\ddot{\text{S}}$:

b. Group 5A (15)

·$\ddot{\text{N}}$·

c. Group 2A (2)

$\dot{\text{C}}$a·

d. Group 1A (1)

Na·

e. Group 3A (13)

·$\dot{\text{G}}$a·

4.51 **a.** Na **b.** Rb **c.** Na **d.** Rb

4.53 **a.** C < Si < Ge **b.** Na < K < Cs **c.** F < Cl < I **d.** N < P < As

4.55 **a.** Br **b.** Mg **c.** P **d.** Xe

4.57 **a.** Br, Cl, F **b.** Na, Al, Cl **c.** Cs, K, Na **d.** Ca, As, Br

4.59 larger, more metallic

4.61 Ca, Ga, Ge, Br

4.63 lower, more

4.65 **a.** decreases **b.** increases **c.** increases **d.** remains the same

4.67 **b**, **c**, and **e** are true, **a** and **d** are false.

4.69 **a.** False; All atoms of a given element are different from atoms of other elements.
b. True
c. True
d. False; In a chemical reaction atoms are never created or destroyed.

4.71 **a.** 1 + 2 **b.** 1 **c.** 1 **d.** 3 **e.** 2

4.73 **a.** $^{16}_{8}X$, $^{17}_{8}X$, $^{18}_{8}X$ all have eight protons.
b. $^{16}_{8}X$, $^{17}_{8}X$, $^{18}_{8}X$ are all isotopes of oxygen.
c. $^{16}_{8}X$ and $^{16}_{9}X$ have mass numbers of 16, and $^{18}_{10}X$ and $^{18}_{8}X$ have mass numbers of 18.
d. $^{16}_{8}X$ and $^{18}_{10}X$ have eight neutrons each.

4.75 **a.** 92 protons, 143 neutrons
b. 92 protons, 147 neutrons
c. 17 protons, 20 neutrons

4.77 **a.** $^{9}_{4}Be$ **b.** $^{11}_{5}B$ **c.** $^{13}_{6}C$ **d.** $^{10}_{5}B$ **e.** $^{12}_{6}C$

Representations **B** and **D** are isotopes of boron; **C** and **E** are isotopes of carbon.

4.79 Li is **D**, Na is **A**, K is **C**, and Rb is **B**.

4.81 **a.** F, I **b.** Si, Sn **c.** Si, Sb **d.** F_2, Ar, He, I_2 **e.** Si, Ar

4.83 **a.** Group 7A (17), Period 4
b. Group 8A (18), Period 3
c. Group 1A (1), Period 4
d. Group 2A (2), Period 7

4.85 **a.** False **b.** False **c.** True **d.** False **e.** True

4.87 **a.** protons **b.** protons **c.** alkali metals

4.89 **a.** nickel, Ni **b.** barium, Ba **c.** radium, Ra **d.** arsenic, As **e.** tin, Sn **f.** cesium, Cs **g.** gold, Au **h.** mercury, Hg

4.91 **a.** 48 protons, 66 neutrons, 48 electrons
b. 43 protons, 55 neutrons, 43 electrons
c. 79 protons, 120 neutrons, 79 electrons
d. 86 protons, 136 neutrons, 86 electrons
e. 54 protons, 82 neutrons, 54 electrons

4.93

Name	Atomic Symbol	Number of Protons	Number of Neutrons	Number of Electrons
Sulfur	$^{34}_{16}S$	16	18	16
Nickel	$^{62}_{28}Ni$	28	34	28
Magnesium	$^{26}_{12}Mg$	12	14	12
Radon	$^{220}_{86}Rn$	86	134	86

4.95 **a.** $^{9}_{4}Be$ **b.** $^{26}_{12}Mg$ **c.** $^{46}_{20}Ca$ **d.** $^{70}_{30}Zn$

4.97 **a.** 82 protons, 126 neutrons, 82 electrons
b. $^{214}_{82}Pb$
c. $^{214}_{83}Bi$, bismuth

4.99 **a.** Period 3; nonmetal
b. Period 4; metalloid
c. Period 2; nonmetal
d. Period 6; Metal
e. Period 6; Metal

4.101 Calcium has a greater number of protons than K. The increase in positive charge increases the attraction for electrons, which means that more energy is required to remove the valence electrons. The valence electrons are farther from the nucleus in Ca than in Mg and less energy is needed to remove them.

4.103 **a.** Be **b.** Li **c.** F **d.** N **e.** Li

4.105 **a.** 38 protons, 49 neutrons, 38 electrons
b. $^{88}_{38}Sr$
c. 46 neutrons
d. The atomic mass on the periodic table is the weighted average of all the naturally occurring isotopes.

4.107 **a.** O **b.** Ar **c.** N **d.** Na **e.** Ra

4.109 6.6×10^{22} atoms of Pb

4.111 28.09 amu

5 Nuclear Chemistry

Luke has been experiencing weight loss, loss of appetite, vomiting, and some abdominal pain. Luke visits his doctor, who suspects a problem with his liver. The doctor orders blood work to be completed, and also refers Luke to a radiation technologist for a CT scan. Computed tomography, more commonly known as CT, uses X-rays to obtain a series of two-dimensional images and is commonly used for the diagnosis of abdominal diseases.

Fred, a radiation technologist, begins by explaining the procedure for the CT scan to Luke and asking him if he has any allergies. Luke indicates that he has none, and Fred places an IV in Luke's arm. Fred then positions Luke into the scanner, and an initial scan of Luke's liver is taken. Fred then injects the first dose of contrast into Luke's bloodstream and a contrast scan is taken. From the CT scans, it is clear that a change in the liver tissue has occurred, possibly due to hepatitis or an infection.

The liver has many functions and is critical to metabolism. The liver produces bile, consisting of bile salts and other chemicals, which is required for the digestion of lipids. The liver is also responsible for the conversion of waste products from protein metabolism into urea, which is eliminated in the urine.

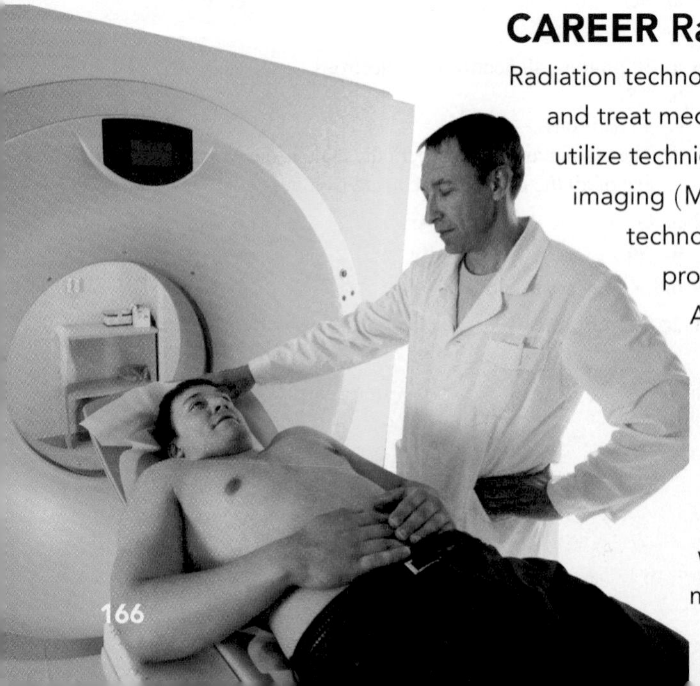

CAREER Radiation Technologist

Radiation technologists specialize in the use of imaging techniques to diagnose and treat medical problems. With additional education, they are also able to utilize techniques such as computed tomography (CT), magnetic resonance imaging (MRI), and positron emission tomography (PET). Radiation technologists prepare a patient by explaining the procedure and properly positioning the patient within the scanning apparatus. A patient may be given radioactive tracers such as technetium-99m, iodine-131, and gallium-67 that emit gamma radiation, which is detected and used to develop an image of the kidneys or thyroid or to follow the blood flow in the heart muscle. Radiation technologists must be knowledgeable about radiation exposure to limit the amount of radiation which patients are exposed to, as well as themselves. They may also prepare a contrast solution for a patient to drink, and maintain patient records and equipment.

With the production of artificial radioactive substances in 1934, the field of nuclear medicine was established. In 1937, the first radioactive isotope was used to treat a person with leukemia at the University of California at Berkeley. Major strides in the use of radioactivity in medicine occurred in 1946, when a radioactive iodine isotope was successfully used to diagnose thyroid function and to treat hyperthyroidism and thyroid cancer. Radioactive substances are now used to produce images of organs, such as liver, spleen, thyroid gland, kidneys, and the brain, and to detect heart disease. Today, procedures in nuclear medicine provide information about the function and structure of every organ in the body, which allows the nuclear physician to diagnose and treat diseases early.

LOOKING AHEAD

CHAPTER READINESS*

KEY MATH SKILLS

- Using Positive and Negative Numbers in Calculations (1.4B)
- Solving Equations (1.4D)
- Interpreting a Graph (1.4E)

CORE CHEMISTRY SKILLS

- Using Conversion Factors (2.6)
- Counting Protons and Neutrons (4.4)
- Writing Atomic Symbols for Isotopes (4.5)

*These Key Math Skills and Core Chemistry Skills from previous chapters are listed here for your review as you proceed to the new material in this chapter.

5.1 Natural Radioactivity

LEARNING GOAL

Describe alpha, beta, positron, and gamma radiation.

Most naturally occurring isotopes of elements up to atomic number 19 have stable nuclei. Elements with atomic numbers 20 and higher usually have one or more isotopes that have unstable nuclei in which the nuclear forces cannot offset the repulsions between the protons. An unstable nucleus is *radioactive*, which means that it spontaneously emits small particles of energy called **radiation** to become more stable. Radiation may take the form of alpha (α) and beta (β) particles, positrons (β^+), or pure energy such as gamma (γ) rays. An isotope of an element that emits radiation is called a **radioisotope**. For most types of radiation, there is a change in the number of protons in the nucleus, which means that an atom is converted into an atom of a different element. This kind of nuclear change was not evident to Dalton when he made his predictions about atoms. Elements with atomic numbers of 93 and higher are produced artificially in nuclear laboratories and consist only of radioactive isotopes.

The atomic symbols for the different isotopes are written with the mass number in the upper left corner and the atomic number in the lower left corner. The mass number is the sum of the number of protons and neutrons in the nucleus, and the atomic number is equal to the number of protons. For example, a radioactive isotope of iodine used in the diagnosis and treatment of thyroid conditions has a mass number of 131 and an atomic number of 53.

Mass number (protons and neutrons) → $^{131}_{53}\text{I}$ ← Symbol of element

Atomic number (protons) → $^{131}_{53}\text{I}$

Radioactive isotopes are identified by writing the mass number after the element's name or symbol. Thus, in this example, the isotope is called iodine-131 or I-131. Table 5.1 compares some stable, nonradioactive isotopes with some radioactive isotopes.

TABLE 5.1 Stable and Radioactive Isotopes of Some Elements

Magnesium	Iodine	Uranium
Stable Isotopes		
$^{24}_{12}Mg$	$^{127}_{53}I$	None
Magnesium-24	Iodine-127	
Radioactive Isotopes		
$^{23}_{12}Mg$	$^{125}_{53}I$	$^{235}_{92}U$
Magnesium-23	Iodine-125	Uranium-235
$^{27}_{12}Mg$	$^{131}_{53}I$	$^{238}_{92}U$
Magnesium-27	Iodine-131	Uranium-238

$^{4}_{2}He$

Alpha (α) particle

$^{0}_{-1}e$

Beta (β) particle

$$^{1}_{0}n \longrightarrow {}^{1}_{1}H + {}^{0}_{-1}e \text{ (or } \beta)$$

Neutron in the nucleus | New proton remains in the nucleus | New electron emitted as a beta particle

$$^{1}_{1}H \longrightarrow {}^{1}_{0}n + {}^{0}_{+1}e \text{ (or } \beta^{+})$$

Proton in the nucleus | New neutron remains in the nucleus | Positron emitted

$^{0}_{0}\gamma$

Gamma (γ) ray

Types of Radiation

By emitting radiation, an unstable nucleus forms a more stable, lower energy nucleus. One type of radiation consists of *alpha particles*. An **alpha particle** is identical to a helium (He) nucleus, which has two protons and two neutrons. An alpha particle has a mass number of 4, an atomic number of 2, and a charge of 2+. The symbol for an alpha particle is the Greek letter alpha (α) or the symbol of a helium nucleus except that the 2+ charge is omitted.

Another type of radiation occurs when a radioisotope emits a *beta particle*. A **beta particle**, which is a high-energy electron, has a charge of 1−, and because its mass is so much less than the mass of a proton, it has a mass number of 0. It is represented by the Greek letter beta (β) or by the symbol for the electron including the mass number and the charge, ($^{0}_{-1}e$). A beta particle is formed when a neutron in an unstable nucleus changes into a proton.

A **positron**, similar to a beta particle, has a positive (1+) charge with a mass number of 0. It is represented by the Greek letter beta with a 1+ charge, β^{+}, or by the symbol of an electron, which includes the mass number and the charge, $^{0}_{+1}e$. A positron is produced by an unstable nucleus when a proton is transformed into a neutron and a positron.

A positron is an example of *antimatter*, a term physicists use to describe a particle that is the opposite of another particle, in this case, an electron. When an electron and a positron collide, their minute masses are completely converted to energy in the form of *gamma rays*.

$$^{0}_{-1}e + {}^{0}_{+1}e \longrightarrow 2\,{}^{0}_{0}\gamma$$

Gamma rays are high-energy radiation, released when an unstable nucleus undergoes a rearrangement of its particles to give a more stable, lower energy nucleus. Gamma rays are often emitted along with other types of radiation. A gamma ray is written as the Greek letter gamma ($^{0}_{0}\gamma$). Because gamma rays are energy only, zeros are used to show that a gamma ray has no mass or charge.

Table 5.2 summarizes the types of radiation we will use in nuclear equations.

TABLE 5.2 Some Forms of Radiation

Type of Radiation	Symbol		Change in Nucleus	Mass Number	Charge
Alpha Particle	α	$^{4}_{2}He$	Two protons and two neutrons are emitted as an alpha particle.	4	2+
Beta Particle	β	$^{0}_{-1}e$	A neutron changes to a proton and an electron is emitted.	0	1−
Positron	β^{+}	$^{0}_{+1}e$	A proton changes to a neutron and a positron is emitted.	0	1+
Gamma Ray	γ	$^{0}_{0}\gamma$	Energy is lost to stabilize the nucleus.	0	0
Proton	p	$^{1}_{1}H$	A proton is emitted.	1	1+
Neutron	n	$^{1}_{0}n$	A neutron is emitted.	1	0

SAMPLE PROBLEM 5.1 Radiation Particles

Identify and write the symbol for each of the following types of radiation:

a. contains two protons and two neutrons
b. has a mass number of 0 and a 1− charge

SOLUTION

a. An alpha (α) particle, ${}^{4}_{2}\text{He}$, has two protons and two neutrons.
b. A beta (β) particle, ${}^{0}_{-1}e$, is like an electron with a mass number of 0 and a 1− charge.

STUDY CHECK 5.1

Identify and write the symbol for the type of radiation that has a mass number of zero and a 1+ charge.

Biological Effects of Radiation

When radiation strikes molecules in its path, electrons may be knocked away, forming unstable ions. If this *ionizing radiation* passes through the human body, it may interact with water molecules, removing electrons and producing H_2O^+, which can cause undesirable chemical reactions.

The cells most sensitive to radiation are the ones undergoing rapid division—those of the bone marrow, skin, reproductive organs, and intestinal lining, as well as all cells of growing children. Damaged cells may lose their ability to produce necessary materials. For example, if radiation damages cells of the bone marrow, red blood cells may no longer be produced. If sperm cells, ova, or the cells of a fetus are damaged, birth defects may result. In contrast, cells of the nerves, muscles, liver, and adult bones are much less sensitive to radiation because they undergo little or no cellular division.

Cancer cells are another example of rapidly dividing cells. Because cancer cells are highly sensitive to radiation, large doses of radiation are used to destroy them. The normal tissue that surrounds cancer cells divides at a slower rate and suffers less damage from radiation. However, radiation may cause malignant tumors, leukemia, anemia, and genetic mutations.

Radiation Protection

Radiation technologists, chemists, doctors, and nurses who work with radioactive isotopes must use proper radiation protection. Proper *shielding* is necessary to prevent exposure. Alpha particles, which have the largest mass and charge of the radiation particles, travel only a few centimeters in the air before they collide with air molecules, acquire electrons, and become helium atoms. A piece of paper, clothing, and our skin are protection against alpha particles. Lab coats and gloves will also provide sufficient shielding. However, if alpha emitters are ingested or inhaled, the alpha particles they give off can cause serious internal damage.

Beta particles have a very small mass and move much faster and farther than alpha particles, traveling as much as several meters through air. They can pass through paper and penetrate as far as 4 to 5 mm into body tissue. External exposure to beta particles can burn the surface of the skin, but they do not travel far enough to reach the internal organs. Heavy clothing such as lab coats and gloves are needed to protect the skin from beta particles.

Gamma rays travel great distances through the air and pass through many materials, including body tissues. Because gamma rays penetrate so deeply, exposure to gamma rays can be extremely hazardous. Only very dense shielding, such as lead or concrete, will stop them. Syringes used for injections of radioactive materials use shielding made of lead or heavy-weight materials such as tungsten and plastic composites.

When working with radioactive materials, medical personnel wear protective clothing and gloves and stand behind a shield (see Figure 5.1). Long tongs may be used to pick up vials of radioactive material, keeping them away from the hands and body. Table 5.3 summarizes the shielding materials required for the various types of radiation.

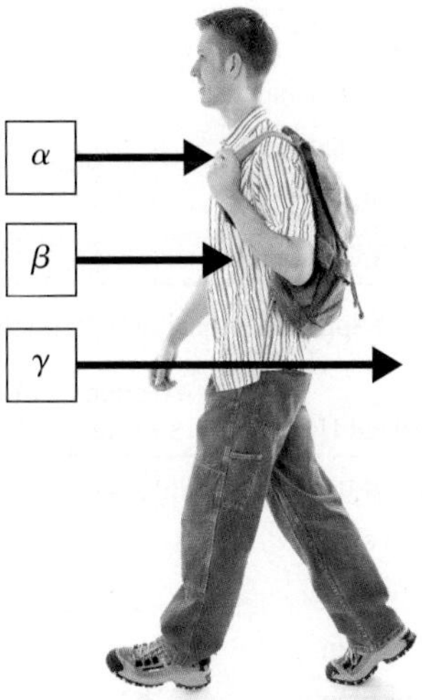

Different types of radiation penetrate the body to different depths.

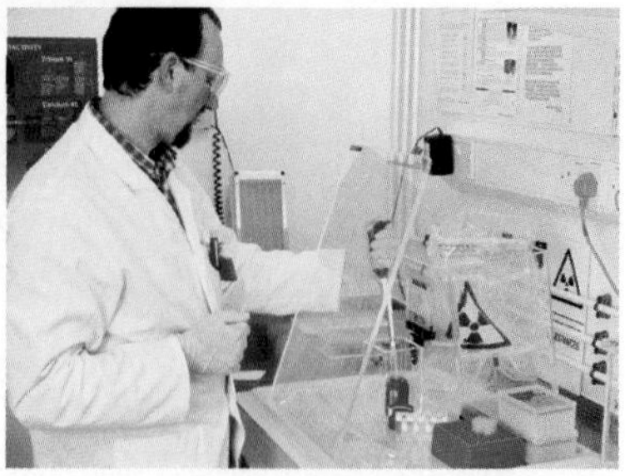

FIGURE 5.1 ▶ A person working with radioisotopes wears protective clothing and gloves and stands behind a shield.

Q What types of radiation does a lead shield block?

TABLE 5.3 Properties of Radiation and Shielding Required

Property	Alpha (α) Particle	Beta (β) Particle	Gamma (γ) Ray
Travel Distance in Air	2 to 4 cm	200 to 300 cm	500 m
Tissue Depth	0.05 mm	4 to 5 mm	50 cm or more
Shielding	Paper, clothing	Heavy clothing, lab coats, gloves	Lead, thick concrete
Typical Source	Radium-226	Carbon-14	Technetium-99m

If you work in an environment where radioactive materials are present, such as a nuclear medicine facility, try to keep the time you spend in a radioactive area to a minimum. Remaining in a radioactive area twice as long exposes you to twice as much radiation.

Keep your distance! The greater the distance from the radioactive source, the lower the intensity of radiation received. By doubling your distance from the radiation source, the intensity of radiation drops to $(\frac{1}{2})^2$, or one-fourth of its previous value.

QUESTIONS AND PROBLEMS

5.1 Natural Radioactivity

LEARNING GOAL Describe alpha, beta, positron, and gamma radiation.

5.1 Identify the type of particle or radiation for each of the following:
a. $^{4}_{2}\text{He}$ **b.** $^{0}_{+1}e$ **c.** $^{0}_{0}\gamma$

5.2 Identify the type of particle or radiation for each of the following:
a. $^{0}_{-1}e$ **b.** $^{1}_{1}\text{H}$ **c.** $^{1}_{0}n$

5.3 Naturally occurring potassium consists of three isotopes: potassium-39, potassium-40, and potassium-41.
a. Write the atomic symbol for each isotope.
b. In what ways are the isotopes similar, and in what ways do they differ?

5.4 Naturally occurring iodine is iodine-127. Medically, radioactive isotopes of iodine-125 and iodine-131 are used.
a. Write the atomic symbol for each isotope.
b. In what ways are the isotopes similar, and in what ways do they differ?

5.5 Supply the missing information in the following table:

Medical Use	Atomic Symbol	Mass Number	Number of Protons	Number of Neutrons
Heart imaging	$^{201}_{81}\text{Tl}$			
Radiation therapy		60	27	
Abdominal scan			31	36
Hyperthyroidism	$^{131}_{53}\text{I}$			
Leukemia treatment		32		17

5.6 Supply the missing information in the following table:

Medical Use	Atomic Symbol	Mass Number	Number of Protons	Number of Neutrons
Cancer treatment	$^{131}_{55}\text{Cs}$			
Brain scan			43	56
Blood flow		141	58	
Bone scan		85		47
Lung function	$^{133}_{54}\text{Xe}$			

5.7 Write the symbol for each of the following isotopes used in nuclear medicine:
a. Iodine-131
b. cobalt-60
c. strontium-89
d. Ruthenium-106

5.8 Write the symbol for each of the following isotopes:
a. U-235
b. U-239
c. C-14
d. He-4

5.9 Identify each of the following:
a. $^{13}_{7}\text{X}$ **b.** $^{18}_{9}\text{X}$
c. $^{0}_{+1}\text{X}$ **d.** $^{128}_{53}\text{X}$

5.10 Identify each of the following:
a. $^{1}_{1}\text{X}$ **b.** $^{81}_{35}\text{X}$ **c.** $^{0}_{0}\text{X}$
d. $^{59}_{26}\text{X}$ **e.** $^{0}_{+1}\text{X}$

5.11 Match the type of radiation (1–3) with each of the following statements:
1. alpha particle
2. beta particle
3. gamma radiation

a. does not penetrate skin
b. shielding protection includes lead or thick concrete
c. can be very harmful if ingested

5.12 Match the type of radiation (1–3) with each of the following statements:
1. alpha particle
2. beta particle
3. gamma radiation

a. penetrates farthest into skin and body tissues
b. shielding protection includes lab coats and gloves
c. travels only a short distance in air

5.2 Nuclear Reactions

LEARNING GOAL

Write a balanced nuclear equation showing mass numbers and atomic numbers for radioactive decay.

In a process called **radioactive decay**, a nucleus spontaneously breaks down by emitting radiation. This process is shown by writing a *nuclear equation* with the atomic symbols of the original radioactive nucleus on the left, an arrow, and the new nucleus and the type of radiation emitted on the right.

$$\text{Radioactive nucleus} \longrightarrow \text{new nucleus} + \text{radiation } (\alpha, \beta, \beta^+, \gamma)$$

In a nuclear equation, the sum of the mass numbers and the sum of the atomic numbers on one side of the arrow must equal the sum of the mass numbers and the sum of the atomic numbers on the other side.

CORE CHEMISTRY SKILL

Writing Nuclear Equations

Alpha Decay

An unstable nucleus may emit an alpha particle, which consists of two protons and two neutrons. Thus, the mass number of the radioactive nucleus decreases by 4, and its atomic number decreases by 2. For example, when uranium-238 emits an alpha particle, the new nucleus that forms has a mass number of 234. Compared to uranium with 92 protons, the new nucleus has 90 protons, which is thorium.

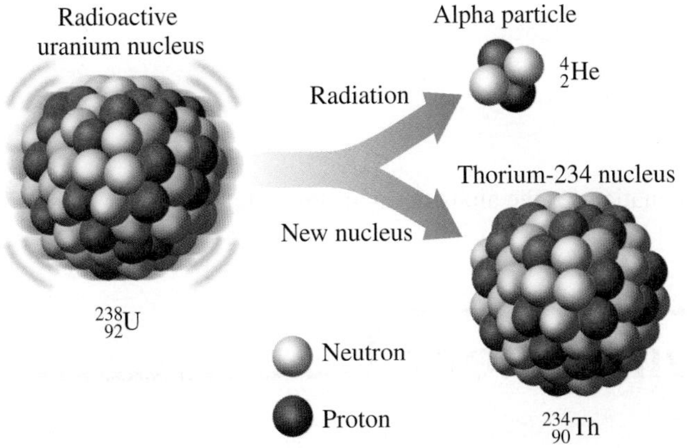

146 neutrons, 92 protons		144 neutrons, 90 protons		2 neutrons, 2 protons
$^{238}_{92}U$	$\longrightarrow$	$^{234}_{90}Th$	+	$^{4}_{2}He$
Radioactive nucleus		New nucleus		Alpha particle

In the nuclear equation for alpha decay, the mass number of the new nucleus decreases by 4 and its atomic number decreases by 2.

We can look at writing a balanced nuclear equation for americium-241, which undergoes alpha decay as shown in Sample Problem 5.2.

SAMPLE PROBLEM 5.2 Writing an Equation for Alpha Decay

Smoke detectors that are used in homes and apartments contain americium-241, which undergoes alpha decay. When alpha particles collide with air molecules, charged particles are produced that generate an electrical current. If smoke particles enter the detector, they interfere with the formation of charged particles in the air, and the electrical current is interrupted. This causes the alarm to sound and warns the occupants of the danger of fire. Complete the following nuclear equation for the decay of americium-241:

$$^{241}_{95}Am \longrightarrow ? + ^{4}_{2}He$$

SOLUTION

ANALYZE THE PROBLEM	Given	Need
	Am-241; alpha decay	balanced nuclear equation

A smoke detector sounds an alarm when smoke enters its ionization chamber.

Guide to Completing a Nuclear Equation

STEP 1 Write the incomplete nuclear equation.

STEP 2 Determine the missing mass number.

STEP 3 Determine the missing atomic number.

STEP 4 Determine the symbol of the new nucleus.

STEP 5 Complete the nuclear equation.

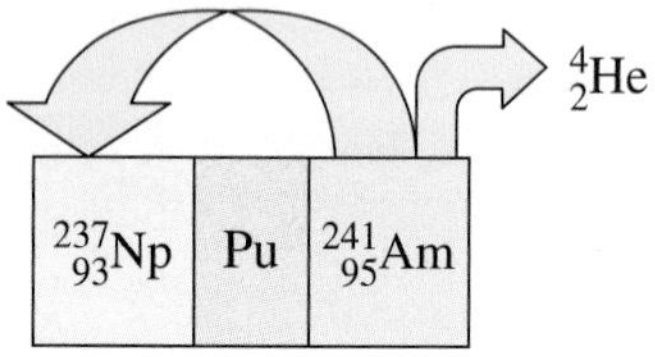

STEP 1 **Write the incomplete nuclear equation.**

$$^{241}_{95}Am \longrightarrow ? + ^{4}_{2}He$$

STEP 2 **Determine the missing mass number.** In the equation, the mass number of the americium, 241, is equal to the sum of the mass numbers of the new nucleus and the alpha particle.

$$241 = ? + 4$$
$$241 - 4 = ?$$
$$241 - 4 = 237 \text{ (mass number of new nucleus)}$$

STEP 3 **Determine the missing atomic number.** The atomic number of americium, 95, must equal the sum of the atomic numbers of the new nucleus and the alpha particle.

$$95 = ? + 2$$
$$95 - 2 = ?$$
$$95 - 2 = 93 \text{ (atomic number of new nucleus)}$$

STEP 4 **Determine the symbol of the new nucleus.** On the periodic table, the element that has atomic number 93 is neptunium, Np. The nucleus of this isotope of Np is written as $^{237}_{93}Np$.

STEP 5 **Complete the nuclear equation.**

$$^{241}_{95}Am \longrightarrow ^{237}_{93}Np + ^{4}_{2}He$$

STUDY CHECK 5.2

Write the balanced nuclear equation for the alpha decay of Po-214.

Chemistry Link to the Environment

Radon in Our Homes

The presence of radon gas has become a much publicized environmental and health issue because of the radiation danger it poses. Radioactive isotopes such as radium-226 are naturally present in many types of rocks and soils. Radium-226 emits an alpha particle and is converted into radon gas, which diffuses out of the rocks and soil.

$$^{226}_{88}Ra \longrightarrow ^{222}_{86}Rn + ^{4}_{2}He$$

Outdoors, radon gas poses little danger because it disperses in the air. However, if the radioactive source is under a house or building, the radon gas can enter the house through cracks in the foundation or other openings. Those who live or work there may inhale the radon. Inside the lungs, radon-222 emits alpha particles to form polonium-218, which is known to cause lung cancer.

$$^{222}_{86}Rn \longrightarrow ^{218}_{84}Po + ^{4}_{2}He$$

The Environmental Protection Agency (EPA) recommends that the maximum level of radon not exceed 4 picocuries (pCi) per liter of air in a home. One picocurie (pCi) is equal to 10^{-12} curies (Ci); curies are described in section 5.3. In California, 1% of all the houses surveyed exceeded the EPA's recommended maximum radon level.

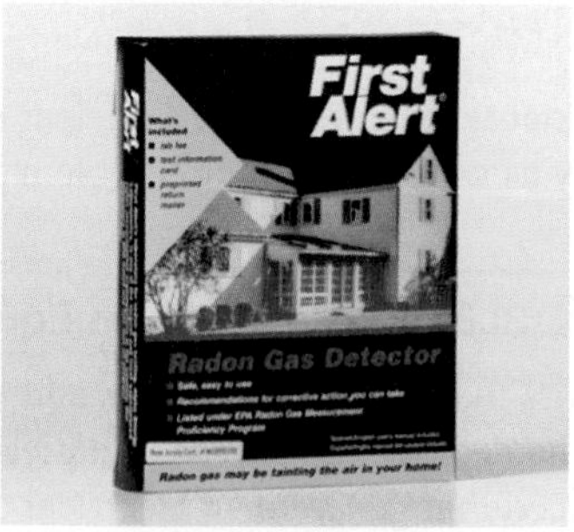

A radon gas detector is used to determine radon levels in buildings.

Beta Decay

The formation of a beta particle is the result of the breakdown of a neutron into a proton and an electron (beta particle). Because the proton remains in the nucleus, the number of protons increases by one, while the number of neutrons decreases by one. Thus, in a nuclear equation for beta decay, the mass number of the radioactive nucleus and the mass number of the new nucleus are the same. However, the atomic number of the new nucleus increases by one, which makes it a nucleus of a different element (*transmutation*). For example, the beta decay of a carbon-14 nucleus produces a nitrogen-14 nucleus.

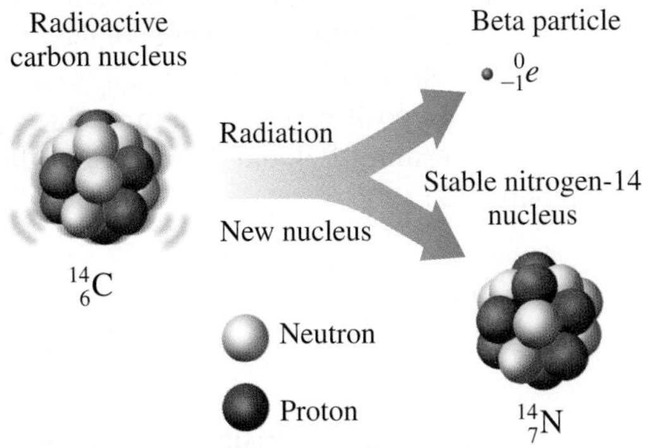

8 neutrons, 6 protons; 7 neutrons, 7 protons; 0 neutrons, 0 protons, 1− charge

$$^{14}_{6}C \longrightarrow {}^{14}_{7}N + {}^{0}_{-1}e$$

Radioactive nucleus — New nucleus — Beta particle

In the nuclear equation for beta decay, the mass number of the new nucleus remains the same and its atomic number increases by 1.

SAMPLE PROBLEM 5.3 Writing a Nuclear Equation for Beta Decay

The radioactive isotope yttrium-90, a beta emitter, is used in cancer treatment and as a colloidal injection into large joints to relieve the pain of arthritis. Write the balanced nuclear equation for the beta decay of yttrium-90.

SOLUTION

ANALYZE THE PROBLEM	Given	Need
	Y-90; beta decay	balanced nuclear equation

STEP 1 **Write the incomplete nuclear equation.**

$$^{90}_{39}Y \longrightarrow ? + {}^{0}_{-1}e$$

STEP 2 **Determine the missing mass number.** In the equation, the mass number of the yttrium, 90, is equal to the sum of the mass numbers of the new nucleus and the beta particle.

$90 = ? + 0$
$90 - 0 = ?$
$90 - 0 = 90$ (mass number of new nucleus)

STEP 3 **Determine the missing atomic number.** The atomic number of yttrium, 39, must equal the sum of the atomic numbers of the new nucleus and the beta particle.

$39 = ? - 1$
$39 + 1 = ?$
$39 + 1 = 40$ (atomic number of new nucleus)

STEP 4 **Determine the symbol of the new nucleus.** On the periodic table, the element that has atomic number 40 is zirconium, Zr. The nucleus of this isotope of Zr is written as $^{90}_{40}Zr$.

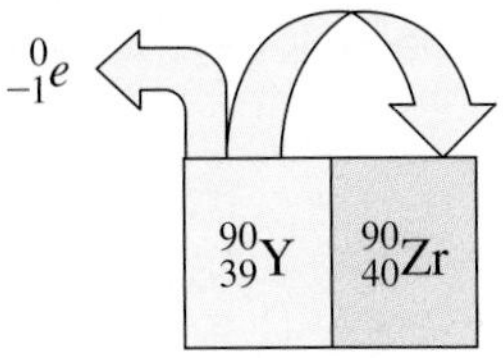

STEP 5 **Complete the nuclear equation.**

$$^{90}_{39}Y \longrightarrow {}^{90}_{40}Zr + {}^{0}_{-1}e$$

STUDY CHECK 5.3

Write the balanced nuclear equation for the beta decay of chromium-51.

Positron Emission

In positron emission, a proton in an unstable nucleus is converted to a neutron and a positron. The neutron remains in the nucleus, but the positron is emitted from the nucleus. In a nuclear equation for positron emission, the mass number of the radioactive nucleus and the mass number of the new nucleus are the same. However, the atomic number of the new nucleus decreases by one, indicating a change of one element into another (*transmutation*). For example, an aluminum-24 nucleus undergoes positron emission to produce a magnesium-24 nucleus. The atomic number of magnesium (12) and the charge of the positron (1+) give the atomic number of aluminum (13).

$${}^{24}_{13}\text{Al} \longrightarrow {}^{24}_{12}\text{Mg} + {}^{0}_{+1}e$$

SAMPLE PROBLEM 5.4 Writing a Nuclear Equation for Positron Emission

Write the balanced nuclear equation for manganese-49, which decays by emitting a positron.

SOLUTION

ANALYZE THE PROBLEM	Given	Need
	Mn-49; positron emission	balanced nuclear equation

STEP 1 Write the incomplete nuclear equation.

$${}^{49}_{25}\text{Mn} \longrightarrow ? + {}^{0}_{+1}e$$

STEP 2 Determine the missing mass number. In the equation, the mass number of manganese, 49, is equal to the sum of the mass numbers of the new nucleus and the positron.

$49 = ? + 0$
$49 - 0 = ?$
$49 - 0 = 49$ (mass number of new nucleus)

STEP 3 Determine the missing atomic number. The atomic number of manganese, 25, must equal the sum of the atomic numbers of the new nucleus and the positron.

$25 = ? + 1$
$25 - 1 = ?$
$25 - 1 = 24$ (atomic number of new nucleus)

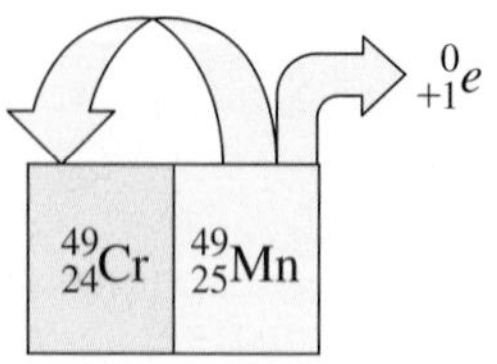

STEP 4 Determine the symbol of the new nucleus. On the periodic table, the element that has atomic number 24 is chromium (Cr). The nucleus of this isotope of Cr is written as ${}^{49}_{24}\text{Cr}$.

STEP 5 Complete the nuclear equation.

$${}^{49}_{25}\text{Mn} \longrightarrow {}^{49}_{24}\text{Cr} + {}^{0}_{+1}e$$

STUDY CHECK 5.4

Write the balanced nuclear equation for xenon-118, which undergoes positron emission.

Gamma Emission

Pure gamma emitters are rare, although gamma radiation accompanies most alpha and beta radiation. In radiology, one of the most commonly used gamma emitters is technetium (Tc). The unstable isotope of technetium is written as the *metastable* (symbol m) isotope technetium-99m, Tc-99m, or ${}^{99m}_{43}Tc$. By emitting energy in the form of gamma rays, the nucleus becomes more stable.

$${}^{99m}_{43}Tc \longrightarrow {}^{99}_{43}Tc + {}^{0}_{0}\gamma$$

Figure 5.2 summarizes the changes in the nucleus for alpha, beta, positron, and gamma radiation.

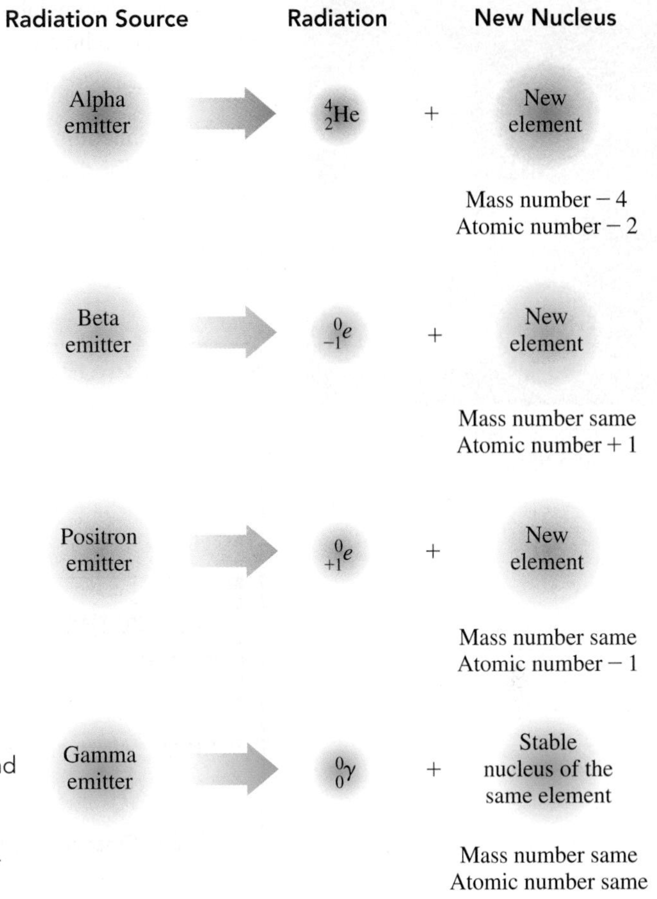

FIGURE 5.2 ▶ When the nuclei of alpha, beta, positron, and gamma emitters emit radiation, new, more stable nuclei are produced.

Q What changes occur in the number of protons and neutrons when an alpha emitter gives off radiation?

Producing Radioactive Isotopes

Today, many radioisotopes are produced in small amounts by converting stable, nonradioactive isotopes into radioactive ones. In a process called *transmutation*, a stable nucleus is bombarded by high-speed particles such as alpha particles, protons, neutrons, and small nuclei. When one of these particles is absorbed, the stable nucleus is converted to a radioactive isotope and usually some type of radiation particle.

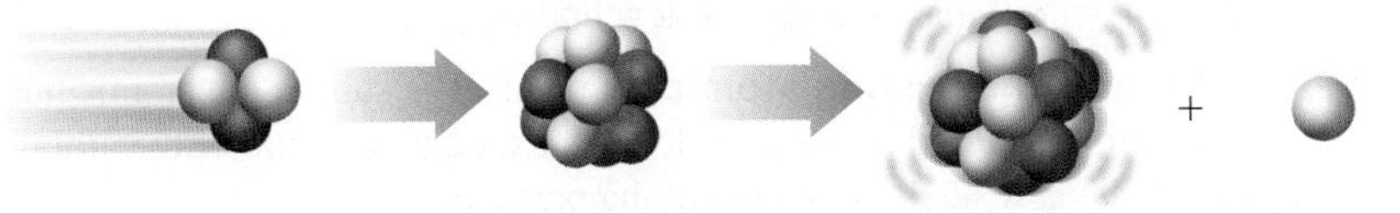

$${}^{4}_{2}He + {}^{10}_{5}B \longrightarrow {}^{13}_{7}N + {}^{1}_{0}n$$

Bombarding particle | Stable nucleus | New radioactive nucleus | Neutron

When nonradioactive B-10 is bombarded by an alpha particle, the products are radioactive N-13 and a neutron.

All elements that have an atomic number greater than 92 have been produced by bombardment. Most have been produced in only small amounts and exist for only a short time, making it difficult to study their properties. For example, when californium-249 is bombarded with nitrogen-15, the radioactive element dubnium-260 and four neutrons are produced.

$${}^{15}_{7}N + {}^{249}_{98}Cf \longrightarrow {}^{260}_{105}Db + 4\,{}^{1}_{0}n$$

Technetium-99m is a radioisotope used in nuclear medicine for several diagnostic procedures, including the detection of brain tumors and examinations of the liver and spleen. The source of technetium-99m is molybdenum-99, which is produced in a nuclear reactor by neutron bombardment of molybdenum-98.

$${}^{1}_{0}n + {}^{98}_{42}Mo \longrightarrow {}^{99}_{42}Mo$$

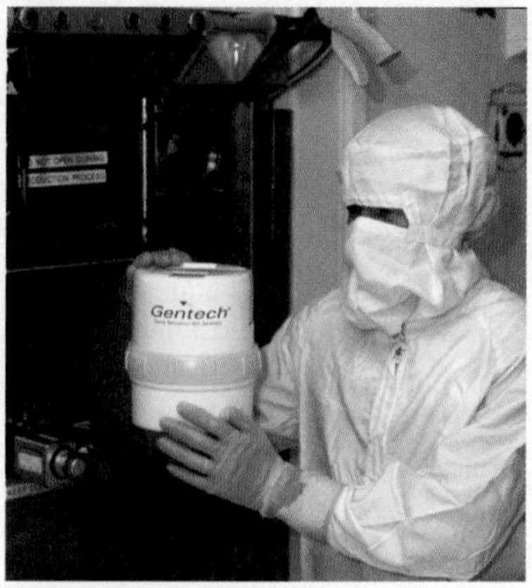

A generator is used to prepare technetium-99m.

Many radiology laboratories have small generators containing molybdenum-99, which decays to the technetium-99m radioisotope.

$$^{99}_{42}\text{Mo} \longrightarrow {}^{99m}_{43}\text{Tc} + {}^{0}_{-1}e$$

The technetium-99m radioisotope decays by emitting gamma rays. Gamma emission is desirable for diagnostic work because the gamma rays pass through the body to the detection equipment.

$$^{99m}_{43}\text{Tc} \longrightarrow {}^{99}_{43}\text{Tc} + {}^{0}_{0}\gamma$$

SAMPLE PROBLEM 5.5 Writing an Equation for an Isotope Produced by Bombardment

Write the balanced nuclear equation for the bombardment of nickel-58 by a proton, $^{1}_{1}\text{H}$, which produces a radioactive isotope and an alpha particle.

SOLUTION

ANALYZE THE PROBLEM	Given	Need
	proton bombardment of Ni-58	balanced nuclear equation

STEP 1 **Write the incomplete nuclear equation.**

$$^{1}_{1}\text{H} + {}^{58}_{28}\text{Ni} \longrightarrow ? + {}^{4}_{2}\text{He}$$

STEP 2 **Determine the missing mass number.** In the equation, the sum of the mass numbers of the proton, 1, and the nickel, 58, must equal the sum of the mass numbers of the new nucleus and the alpha particle.

$1 + 58 = ? + 4$
$59 - 4 = ?$
$59 - 4 = 55$ (mass number of new nucleus)

STEP 3 **Determine the missing atomic number.** The sum of the atomic numbers of the proton, 1, and of the nickel, 28, must equal the sum of the atomic numbers of the new nucleus and the alpha particle.

$1 + 28 = ? + 2$
$29 - 2 = ?$
$29 - 2 = 27$ (atomic number of new nucleus)

STEP 4 **Determine the symbol of the new nucleus.** On the periodic table, the element that has atomic number 27 is cobalt, Co. The nucleus of this isotope of Co is written as $^{55}_{27}\text{Co}$.

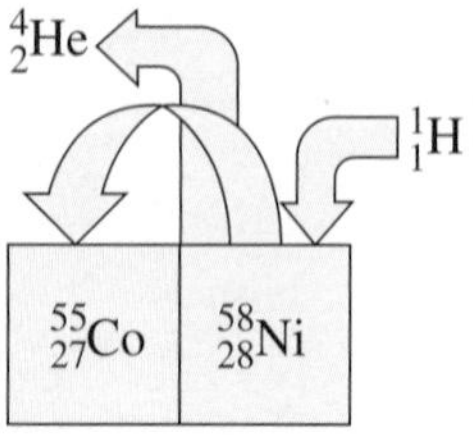

STEP 5 **Complete the nuclear equation.**

$$^{1}_{1}\text{H} + {}^{58}_{28}\text{Ni} \longrightarrow {}^{55}_{27}\text{Co} + {}^{4}_{2}\text{He}$$

STUDY CHECK 5.5

The first radioactive isotope was produced in 1934 by the bombardment of aluminum-27 by an alpha particle to produce a radioactive isotope and one neutron. What is the balanced nuclear equation for this transmutation?

QUESTIONS AND PROBLEMS

5.2 Nuclear Reactions

LEARNING GOAL Write a balanced nuclear equation showing mass numbers and atomic numbers for radioactive decay.

5.13 Write a balanced nuclear equation for the alpha decay of each of the following radioactive isotopes:

a. ${}^{238}_{92}\text{U}$ **b.** ${}^{240}_{94}\text{Pu}$

c. ${}^{60}_{27}\text{Co}$ **d.** ${}^{226}_{88}\text{Ra}$

5.14 Write a balanced nuclear equation for the alpha decay of each of the following radioactive isotopes:

a. curium-243 **b.** ${}^{252}_{99}\text{Es}$

c. ${}^{251}_{98}\text{Cf}$ **d.** ${}^{261}_{107}\text{Bh}$

5.15 Write a balanced nuclear equation for the beta decay of each of the following radioactive isotopes:

a. ${}^{14}_{6}\text{C}$ **b.** ${}^{18}_{9}\text{F}$

c. ${}^{12}_{5}\text{B}$ **d.** ${}^{137}_{55}\text{Cs}$

5.16 Write a balanced nuclear equation for the beta decay of each of the following radioactive isotopes:

a. ${}^{44}_{19}\text{K}$ **b.** iron-59

c. potassium-42 **d.** ${}^{141}_{56}\text{Ba}$

5.17 Write a balanced nuclear equation for the positron emission of each of the following radioactive isotopes:

a. ${}^{10}_{6}\text{C}$ **b.** ${}^{18}_{9}\text{F}$

c. ${}^{22}_{11}\text{Na}$ **d.** ${}^{19}_{10}\text{Ne}$

5.18 Write a balanced nuclear equation for the positron emission of each of the following radioactive isotopes:

a. boron-8 **b.** ${}^{15}_{8}\text{O}$

c. ${}^{40}_{19}\text{K}$ **d.** nitrogen-13

5.19 Complete each of the following nuclear equations and describe the type of radiation:

a. ${}^{28}_{13}\text{Al} \longrightarrow ? + {}^{0}_{-1}e$ **b.** ${}^{180m}_{73}\text{Ta} \longrightarrow {}^{180}_{73}\text{Ta} + ?$

c. ${}^{66}_{29}\text{Cu} \longrightarrow {}^{66}_{30}\text{Zn} + ?$ **d.** $? \longrightarrow {}^{234}_{90}\text{Th} + {}^{4}_{2}\text{He}$

e. ${}^{188}_{80}\text{Hg} \longrightarrow ? + {}^{0}_{+1}e$

5.20 Complete each of the following nuclear equations and describe the type of radiation:

a. ${}^{11}_{6}\text{C} \longrightarrow {}^{11}_{5}\text{B} + ?$ **b.** ${}^{35}_{16}\text{S} \longrightarrow ? + {}^{0}_{-1}e$

c. $? \longrightarrow {}^{90}_{39}\text{Y} + {}^{0}_{-1}e$ **d.** ${}^{210}_{83}\text{Bi} \longrightarrow ? + {}^{4}_{2}\text{He}$

e. $? \longrightarrow {}^{89}_{39}\text{Y} + {}^{0}_{+1}e$

5.21 Complete each of the following bombardment reactions:

a. ${}^{1}_{0}n + {}^{19}_{5}\text{Be} \longrightarrow ?$

b. ${}^{1}_{0}n + {}^{19}_{9}\text{F} \longrightarrow ? + {}^{0}_{+1}\text{e}$

c. ${}^{1}_{0}n + ? \longrightarrow {}^{231}_{89}\text{Ac} + {}^{4}_{2}\text{He}$

d. ${}^{4}_{2}\text{He} + {}^{40}_{19}\text{N} \longrightarrow ? + {}^{1}_{1}\text{H}$

5.22 Complete each of the following bombardment reactions:

a. $? + {}^{40}_{18}\text{Ar} \longrightarrow {}^{43}_{19}\text{K} + {}^{1}_{1}\text{H}$

b. ${}^{1}_{0}n + {}^{238}_{92}\text{U} \longrightarrow ?$

c. ${}^{1}_{0}n + ? \longrightarrow {}^{14}_{6}\text{C} + {}^{1}_{1}\text{H}$

d. $? + {}^{64}_{28}\text{Ni} \longrightarrow {}^{272}_{111}\text{Rg} + {}^{1}_{0}n$

5.3 Radiation Measurement

LEARNING GOAL

Describe the detection and measurement of radiation.

One of the most common instruments for detecting beta and gamma radiation is the Geiger counter. It consists of a metal tube filled with a gas such as argon. When radiation enters a window on the end of the tube, it forms charged particles in the gas, which produce an electrical current. Each burst of current is amplified to give a click and a reading on a meter.

$$\text{Ar} + \text{radiation} \longrightarrow \text{Ar}^{+} + e^{-}$$

Measuring Radiation

Radiation is measured in several different ways. When a radiology laboratory obtains a radioisotope, the *activity* of the sample is measured in terms of the number of nuclear disintegrations per second. The **curie (Ci)**, the original unit of activity, was defined as the number of disintegrations that occurs in 1 s for 1 g of radium, which is equal to 3.7×10^{10} disintegrations/s. The unit was named for the Polish scientist Marie Curie, who along with her husband, Pierre, discovered the radioactive elements radium and polonium. The SI unit of radiation activity is the **becquerel (Bq)**, which is 1 disintegration/s.

The **rad (radiation absorbed dose)** is a unit that measures the amount of radiation absorbed by a gram of material such as body tissue. The SI unit for absorbed dose is the **gray (Gy)**, which is defined as the joules of energy absorbed by 1 kg of body tissue. The gray is equal to 100 rad.

A radiation technician uses a Geiger counter to check radiation levels.

The **rem (radiation equivalent in humans)** is a unit that measures the biological effects of different kinds of radiation. Although alpha particles do not penetrate the skin, if they should enter the body by some other route, they can cause extensive damage within a short distance in tissue. High-energy radiation, such as beta particles, high-energy protons, and neutrons that travel into tissue, cause more damage. Gamma rays are damaging because they travel a long way through body tissue.

To determine the **equivalent dose** or rem dose, the absorbed dose (rad) is multiplied by a factor that adjusts for biological damage caused by a particular form of radiation. For beta and gamma radiation the factor is 1, so the biological damage in rems is the same as the absorbed radiation (rad). For high-energy protons and neutrons, the factor is about 10, and for alpha particles it is 20.

$$\text{Biological damage (rem)} = \text{Absorbed dose (rad)} \times \text{Factor}$$

Often, the measurement for an equivalent dose will be in units of millirems (mrem). One rem is equal to 1000 mrem. The SI unit is the **sievert (Sv)**. One sievert is equal to 100 rem. Table 5.4 summarizes the units used to measure radiation.

TABLE 5.4 Units of Radiation Measurement

Measurement	Common Unit	SI Unit	Relationship
Activity	curie (Ci) 1 Ci = 3.7×10^{10} disintegrations/s	becquerel (Bq) 1 Bq = 1 disintegration/s	1 Ci = 3.7×10^{10} Bq
Absorbed Dose	rad	gray (Gy) 1 Gy = 1 J/kg of tissue	1 Gy = 100 rad
Biological Damage	rem	sievert (Sv)	1 Sv = 100 rem

Chemistry Link to Health

Radiation and Food

Foodborne illnesses caused by pathogenic bacteria such as *Salmonella*, *Listeria*, and *Escherichia coli* have become a major health concern in the United States. The Centers for Disease Control and Prevention estimates that each year, *E. coli* in contaminated foods infects 20 000 people in the United States and that 500 people die. *E. coli* has been responsible for outbreaks of illness from contaminated ground beef, fruit juices, lettuce, and alfalfa sprouts.

The U.S. Food and Drug Administration (FDA) has approved the use of 0.3 kGy to 1 kGy of radiation produced by cobalt-60 or cesium-137 for the treatment of foods. The irradiation technology is much like that used to sterilize medical supplies. Cobalt pellets are placed in stainless steel tubes, which are arranged in racks. When food moves through the series of racks, the gamma rays pass through the food and kill the bacteria.

It is important for consumers to understand that when food is irradiated, it never comes in contact with the radioactive source. The gamma rays pass through the food to kill bacteria, but that does not make the food radioactive. The radiation kills bacteria because it stops their ability to divide and grow. We cook or heat food thoroughly for the same purpose. Radiation, as well as heat, has little effect on the food itself because its cells are no longer dividing or growing. Thus irradiated food is not harmed, although a small amount of vitamin B_1 and C may be lost.

Currently, tomatoes, blueberries, strawberries, and mushrooms are being irradiated to allow them to be harvested when completely ripe and extend their shelf life (see Figure 5.3). The FDA has also approved the irradiation of pork, poultry, and beef in order to decrease potential infections and to extend shelf life. Currently, irradiated vegetable and meat products are available in retail markets in more than 40 countries. In the United States, irradiated foods such as tropical fruits, spinach, and ground meats are found in some stores. *Apollo 17* astronauts ate irradiated foods on the moon, and some U.S. hospitals and nursing homes now use irradiated poultry to reduce the possibility of salmonella infections among residents. The extended shelf life of irradiated food also makes it useful for campers and military personnel. Soon, consumers concerned about food safety will have a choice of irradiated meats, fruits, and vegetables at the market.

FIGURE 5.3 ▶ **(a)** The FDA requires this symbol to appear on irradiated retail foods. **(b)** After two weeks, the irradiated strawberries on the right show no spoilage. Mold is growing on the nonirradiated ones on the left.

Q Why are irradiated foods used on spaceships and in nursing homes?

People who work in radiology laboratories wear dosimeters attached to their clothing to determine any exposure to radiation such as X-rays, gamma ray, or beta particles. A dosimeter can be thermoluminescent (TLD), optically stimulated luminescence (OSL), or electronic personal (EPD). Dosimeters provide real time radiation levels measured by monitors in the work area.

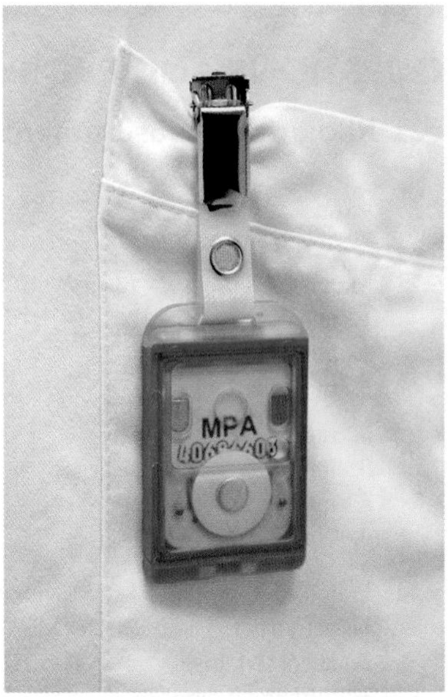

A dosimeter measures radiation exposure.

SAMPLE PROBLEM 5.6 Radiation Measurement

One treatment for bone pain involves intravenous administration of the radioisotope phosphorus-32, which is incorporated into bone. A typical dose of 7 mCi can produce up to 450 rad in the bone. What is the difference between the units of mCi and rad?

SOLUTION

The millicuries (mCi) indicate the activity of the P-32 in terms of nuclei that break down in 1 s. The radiation absorbed dose (rad) is a measure of amount of radiation absorbed by the bone.

STUDY CHECK 5.6

For Sample Problem 5.6, what is the absorbed dose of radiation in grays (Gy)?

Exposure to Radiation

Every day, we are exposed to low levels of radiation from naturally occurring radioactive isotopes in the buildings where we live and work, in our food and water, and in the air we breathe. For example, potassium-40, a naturally occurring radioactive isotope, is present in any potassium-containing food. Other naturally occurring radioisotopes in air and food are carbon-14, radon-222, strontium-90, and iodine-131. The average person in the United States is exposed to about 360 mrem of radiation annually. Medical sources of radiation, including dental, hip, spine, and chest X-rays and mammograms, add to our radiation exposure. Table 5.5 lists some common sources of radiation.

Another source of background radiation is cosmic radiation produced in space by the Sun. People who live at high altitudes or travel by airplane receive a greater amount of cosmic radiation because there are fewer molecules in the atmosphere to absorb the radiation. For example, a person living in Denver receives about twice the cosmic radiation as a person living in Los Angeles. A person living close to a nuclear power plant normally does not receive much additional radiation, perhaps 0.1 mrem in 1 yr. (One rem equals 1000 mrem.) However, in the accident at the Chernobyl nuclear power plant in 1986 in Ukraine, it is estimated that people in a nearby town received as much as 1 rem/h.

TABLE 5.5 Average Annual Radiation Received by a Person in the United States

Source	Dose (mrem)
Natural	
Ground	20
Air, water, food	30
Cosmic rays	40
Wood, concrete, brick	50
Medical	
Chest X-ray	20
Dental X-ray	20
Mammogram	40
Hip X-ray	60
Lumbar spine X-ray	70
Upper gastrointestinal tract X-ray	200
Other	
Nuclear power plants	0.1
Television	20
Air travel	10
Radon	200*

*Varies widely.
Source: U.S. Environmental Protection Agency

Radiation Sickness

The larger the dose of radiation received at one time, the greater the effect on the body. Exposure to radiation of less than 25 rem usually cannot be detected. Whole-body exposure of 100 rem produces a temporary decrease in the number of white blood cells. If the exposure to radiation is greater than 100 rem, a person may suffer the symptoms of radiation sickness: nausea, vomiting, fatigue, and a reduction in white-cell count. A whole-body dosage greater than 300 rem can decrease the white-cell count to zero. The person suffers diarrhea, hair loss, and infection. Exposure to radiation of 500 rem is expected to cause death in 50% of the people receiving that dose. This amount of radiation to the whole body is called the *lethal dose for one-half the population*, or the LD_{50}. The LD_{50} varies for different life forms, as Table 5.6 shows. Whole-body radiation of 600 rem or greater would be fatal to all humans within a few weeks.

TABLE 5.6 Lethal Doses of Radiation for Some Life Forms

Life Form	LD_{50} (rem)
Insect	100 000
Bacterium	50 000
Rat	800
Human	500
Dog	300

QUESTIONS AND PROBLEMS

5.3 Radiation Measurement

LEARNING GOAL Describe the detection and measurement of radiation.

5.23 Match each property (**1–3**) with its unit of measurement.
1. activity
2. absorbed dose
3. biological damage

a. rad
b. mrem
c. mCi
d. Gy

5.24 Match each property (**1–3**) with its unit of measurement.
1. activity
2. absorbed dose
3. biological damage

a. mrad
b. gray
c. becquerel
d. Sv

5.25 Two patients in a nuclear laboratory were accidently exposed to radiation. If one was exposed to 12mGy and the other to 9 rad, which patient received more radiation?

5.26 Two samples of a radioisotope were spilled in a nuclear laboratory. The activity of one sample was 10 kBq and the other 15 mCi. Which sample produced the higher amount of radiation?

5.27 **a.** The recommended dosage of iodine-131 is 4.20 μCi/kg of body mass for treatment of hyperthyroidism. How many microcuries of iodine-131 are needed for a 75.0-kg person?
b. A person receives 50 rad of gamma radiation. What is that amount in grays?

5.28 **a.** The dosage of technetium-99m for a lung scan is 20. μCi/kg of body mass. How many millicuries of technetium-99m should be given to a 50.0-kg person $(1 \text{ mCi} = 1000\ \mu\text{Ci})$?
b. Suppose a person absorbed 50 mrad of alpha radiation. What would be the equivalent dose in millirems?

LEARNING GOAL

Given the half-life of a radioisotope, calculate the amount of radioisotope remaining after one or more half-lives.

5.4 Half-Life of a Radioisotope

The **half-life** of a radioisotope is the amount of time it takes for one-half of a sample to decay. For example, $^{131}_{53}\text{I}$ has a half-life of 8.0 days. As $^{131}_{53}\text{I}$ decays, it produces the nonradioactive isotope $^{131}_{54}\text{Xe}$ and a beta particle.

$$^{131}_{53}\text{I} \longrightarrow {}^{131}_{54}\text{Xe} + {}^{\ 0}_{-1}e$$

Suppose we have a sample that initially contains 20. mg of $^{131}_{53}\text{I}$. In 8.0 days, one-half (10. mg) of all the I-131 nuclei in the sample will decay, which leaves 10. mg of I-131. After 16 days (two half-lives), 5.0 mg of the remaining I-131 decays, which leaves 5.0 mg of I-131. After 24 days (three half-lives), 2.5 mg of the remaining I-131 decays, which leaves 2.5 mg of I-131 nuclei still capable of producing radiation.

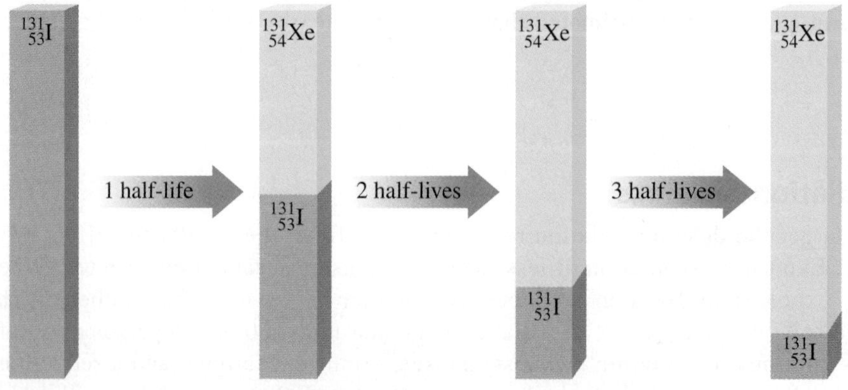

$$20.\text{ mg of } ^{131}_{53}\text{I} \xrightarrow{\text{1 half-life}} 10.\text{ mg of } ^{131}_{53}\text{I} \xrightarrow{\text{2 half-lives}} 5.0\text{ mg of } ^{131}_{53}\text{I} \xrightarrow{\text{3 half-lives}} 2.5\text{ mg of } ^{131}_{53}\text{I}$$

A **decay curve** is a diagram of the decay of a radioactive isotope. Figure 5.4 shows such a curve for the $^{131}_{53}\text{I}$ we have discussed.

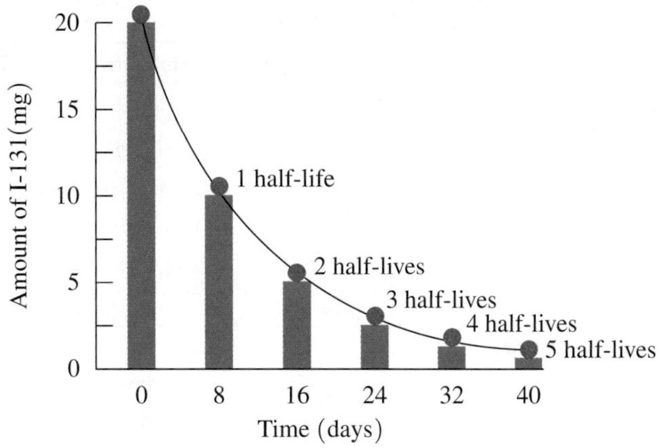

FIGURE 5.4 ▶ The decay curve for iodine-131 shows that one-half of the radioactive sample decays and one-half remains radioactive after each half-life of 8.0 days.

Q How many milligrams of the 20.-mg sample remain radioactive after two half-lives?

SAMPLE PROBLEM 5.7 Using Half-Lives of a Radioisotope

CORE CHEMISTRY SKILL
Using Half-Lives

Phosphorus-32, a radioisotope used in the treatment of leukemia, has a half-life of 14.3 days. If a sample contains 8.0 mg of phosphorus-32, how many milligrams of phosphorus-32 remain after 42.9 days?

SOLUTION

STEP 1 State the given and needed quantities.

	Given	Need
ANALYZE THE PROBLEM	8.0 mg of P-32, 42.9 days elapsed, half-life = 14.3 days	milligrams of P-32 remaining

STEP 2 Write a plan to calculate the unknown quantity.

days → **Half-life** → number of half-lives

milligrams of $^{32}_{15}P$ → **Number of half-lives** → milligrams of $^{32}_{15}P$ remaining

STEP 3 Write the half-life equality and conversion factors.

1 half-life = 14.3 days

$$\frac{14.3\text{ days}}{1\text{ half-life}} \text{ and } \frac{1\text{ half-life}}{14.3\text{ days}}$$

STEP 4 Set up the problem to calculate the needed quantity. First we determine the number of half-lives in the amount of time that has elapsed.

$$\text{number of half-lives} = 42.9\ \cancel{\text{days}} \times \frac{1\text{ half-life}}{14.3\ \cancel{\text{days}}} = 3.00\text{ half-lives}$$

Now we can determine how much of the sample decays in 3 half-lives and how many grams of the phosphorus remain.

$$8.0\text{ mg of }^{32}_{15}P \xrightarrow{\text{1 half-life}} 4.0\text{ mg of }^{32}_{15}P \xrightarrow{\text{2 half-lives}} 2.0\text{ mg of }^{32}_{15}P \xrightarrow{\text{3 half-lives}} 1.0\text{ mg of }^{32}_{15}P$$

Guide to Using Half-Lives

STEP 1 State the given and needed quantities.

STEP 2 Write a plan to calculate the unknown quantity.

STEP 3 Write the half-life equality and conversion factors.

STEP 4 Set up the problem to calculate the needed quantity.

STUDY CHECK 5.7

Iron-59 has a half-life of 44 days. If a nuclear laboratory receives a sample of 8.0 μg of iron-59, how many micrograms of iron-59 are still active after 176 days?

Interactive Video

Half-Lives

TABLE 5.7 Half-Lives of Some Radioisotopes

Element	Radioisotope	Half-Life	Type of Radiation
Naturally Occurring Radioisotopes			
Carbon-14	${}^{14}_{6}C$	5730 yr	Beta
Potassium-40	${}^{40}_{19}K$	1.3×10^9 yr	Beta, gamma
Radium-226	${}^{226}_{88}Ra$	1600 yr	Alpha
Strontium-90	${}^{90}_{38}Sr$	38.1 yr	Alpha
Uranium-238	${}^{238}_{92}U$	4.5×10^9 yr	Alpha
Some Medical Radioisotopes			
Chromium-51	${}^{51}_{24}Cr$	28 days	Gamma
Iodine-131	${}^{131}_{53}I$	8.0 days	Gamma
Iridium-192	${}^{192}_{77}Ir$	74 days	Beta, gamma
Iron-59	${}^{59}_{26}Fe$	44 days	Beta, gamma
Radon-222	${}^{222}_{86}Rn$	3.8 days	Alpha
Technetium-99m	${}^{99m}_{43}Tc$	6.0 h	Beta, gamma

Naturally occurring isotopes of the elements usually have long half-lives, as shown in Table 5.7. They disintegrate slowly and produce radiation over a long period of time, even hundreds or millions of years. In contrast, the radioisotopes used in nuclear medicine have much shorter half-lives. They disintegrate rapidly and produce almost all their radiation in a short period of time. For example, technetium-99m emits half of its radiation in the first 6 h. This means that a small amount of the radioisotope given to a patient is essentially gone within two days. The decay products of technetium-99m are totally eliminated by the body.

Chemistry Link to the **Environment**

Dating Ancient Objects

Radiological dating is a technique used by geologists, archaeologists, and historians to determine the age of ancient objects. The age of an object derived from plants or animals (such as wood, fiber, natural pigments, bone, and cotton or woolen clothing) is determined by measuring the amount of carbon-14, a naturally occurring radioactive form of carbon. In 1960, Willard Libby received the Nobel Prize for the work he did developing carbon-14 dating techniques during the 1940s.[1] Carbon-14 is produced in the upper atmosphere by the bombardment of ${}^{14}_{7}N$ by high-energy neutrons from cosmic rays.

$${}^{1}_{0}n + {}^{14}_{7}N \longrightarrow {}^{14}_{6}C + {}^{1}_{1}H$$

Neutron from cosmic rays; Nitrogen in atmosphere; Radioactive carbon-14; Proton

The carbon-14 reacts with oxygen to form radioactive carbon dioxide, ${}^{14}_{6}CO_2$. Living plants continuously absorb carbon dioxide, which incorporates carbon-14 into the plant material. The uptake of carbon-14 stops when the plant dies.

$${}^{14}_{6}C \longrightarrow {}^{14}_{7}N + {}^{0}_{-1}e$$

As the carbon-14 decays, the amount of radioactive carbon-14 in the plant material steadily decreases. In a process called *carbon dating*, scientists use the half-life of carbon-14 (5730 yr) to calculate the length of time since the plant died. For example, a wooden beam found in an ancient dwelling might have one-half of the carbon-14 found in living plants today. Because one half-life of carbon-14 is 5730 yr, the dwelling was constructed about 5730 yr ago. Carbon-14 dating was used to determine that the Dead Sea Scrolls are about 2000 yr old.

The age of the Dead Sea Scrolls was determined using carbon-14.

A radiological dating method used for determining the age of much older items is based on the radioisotope uranium-238, which decays through a series of reactions to lead-206. The uranium-238 isotope has an incredibly long half-life, about 4×10^9 (4 billion) years. Measurements of the amounts of uranium-238 and lead-206 enable geologists to determine the age of rock samples. The older rocks will have a higher percentage of lead-206 because more of the uranium-238 has decayed. The age of rocks brought back from the Moon by the *Apollo* missions, for example, was determined using uranium-238. They were found to be about 4×10^9 yr old, approximately the same age calculated for the Earth.

[1]Nobel Media AB 2013.

SAMPLE PROBLEM 5.8 Dating Using Half-Lives

Carbon material in the bones of humans and animals assimilates carbon until death. Using radiocarbon dating, the number of half-lives of carbon-14 from a bone sample determines the age of the bone. Suppose a sample is obtained from a prehistoric animal and used for radiocarbon dating. We can calculate the age of the bone or the years elapsed since the animal died by using the half-life of carbon-14, which is 5730 yr. If the sample shows that four half-lives have passed, how much time has elapsed since the animal died?

The age of a bone sample from a skeleton can be determined by carbon dating.

SOLUTION

STEP 1 State the given and needed quantities.

ANALYZE THE PROBLEM	Given	Need
	1 half-life of C-14 = 5730 yr, 4 half-lives elapsed	years elapsed

STEP 2 Write a plan to calculate the unknown quantity.

4 half-lives → Half-life → years elapsed

STEP 3 Write the half-life equality and conversion factors.

$$1 \text{ half-life} = 5730 \text{ yr}$$
$$\frac{5730 \text{ yr}}{1 \text{ half-life}} \quad \text{and} \quad \frac{1 \text{ half-life}}{5730 \text{ yr}}$$

STEP 4 Set up the problem to calculate the needed quantity.

$$\text{Years elapsed} = 4.0 \cancel{\text{half-lives}} \times \frac{5730 \text{ yr}}{1 \cancel{\text{half-life}}} = 23\,000 \text{ yr}$$

We would estimate that the animal lived 23 000 yr ago.

STUDY CHECK 5.8

Suppose that a piece of wood found in a tomb had one-eighth of its original carbon-14 activity. About how many years ago was the wood part of a living tree?

QUESTIONS AND PROBLEMS

5.4 Half-Life of a Radioisotope

LEARNING GOAL Given the half-life of a radioisotope, calculate the amount of radioisotope remaining after one or more half-lives.

5.29 For each of the following, indicate if the number of half-lives elapsed is:

1. one half-life
2. two half-lives
3. three half-lives

a. a sample of Pd-103 with a half-life of 17 days after 34 days
b. a sample of C-11 with a half-life of 20 min after 20 min
c. a sample of At-211 with a half-life of 7 h after 21 h

5.30 For each of the following, indicate if the number of half-lives elapsed is:

1. one half-life
2. two half-lives
3. three half-lives

a. a sample of Ce-141 with a half-life of 32.5 days after 32.5 days
b. a sample of F-18 with a half-life of 110 min after 330 min
c. a sample of Au-198 with a half-life of 2.7 days after 5.4 days

5.31 Technetium-99m is an ideal radioisotope for scanning organs because it has a half-life of 6.0 h and is a pure gamma emitter. Suppose that 80.0 mg were prepared in the technetium generator this morning. How many milligrams of technetium-99m would remain after the following intervals?

a. one half-life
b. two half-lives
c. 18 h
d. 24 h

5.32 A sample of sodium-24 with an activity of 12 mCi is used to study the rate of blood flow in the circulatory system. If sodium-24 has a half-life of 15 h, what is the activity after each of the following intervals?
a. one half-life **b.** 30 h
c. three half-lives **d.** 2.5 days

5.33 Strontium-85, used for bone scans, has a half-life of 65 days.
a. How long will it take for the radiation level of strontium-85 to drop to one-fourth of its original level?
b. How long will it take for the radiation level of strontium-85 to drop to one-eighth of its original level?

5.34 Fluorine-18, which has a half-life of 110 min, is used in PET scans.
a. If 100. mg of fluorine-18 is shipped at 8:00 A.M., how many milligrams of the radioisotope are still active after 110 min?
b. If 100. mg of fluorine-18 is shipped at 8:00 A.M., how many milligrams of the radioisotope are still active when the sample arrives at the radiology laboratory at 1:30 P.M.?

LEARNING GOAL

Describe the use of radioisotopes in medicine.

5.5 Medical Applications Using Radioactivity

To determine the condition of an organ in the body, a radiation technologist may use a radioisotope that concentrates in that organ. The cells in the body do not differentiate between a nonradioactive atom and a radioactive one, so these radioisotopes are easily incorporated. Then the radioactive atoms are detected because they emit radiation. Some radioisotopes used in nuclear medicine are listed in Table 5.8.

TABLE 5.8 Medical Applications of Radioisotopes

Isotope	Half-Life	Radiation	Medical Application
Au-198	2.7 days	Beta	Liver imaging; treatment of abdominal carcinoma
Ce-141	32.5 days	Beta	Gastrointestinal tract diagnosis; measuring blood flow to the heart
Cs-131	9.7 days	Gamma	Prostate brachytherapy
F-18	110 min	Positron	Positron emission tomography (PET)
Ga-67	78 h	Gamma	Abdominal imaging; tumor detection
Ga-68	68 min	Gamma	Detection of pancreatic cancer
I-123	13.2 h	Gamma	Treatment of thyroid, brain, and prostate cancer
I-131	8.0 days	Beta	Treatment of Graves' disease, goiter, hyperthyroidism, thyroid and prostate cancer
Ir-192	74 days	Gamma	Treatment of breast and prostate cancer
P-32	14.3 days	Beta	Treatment of leukemia, excess red blood cells, pancreatic cancer
Pd-103	17 days	Gamma	Prostate brachytherapy
Sr-85	65 days	Gamma	Detection of bone lesions; brain scans
Tc-99m	6.0 h	Gamma	Imaging of skeleton and heart muscle, brain, liver, heart, lungs, bone, spleen, kidney, and thyroid; most widely used radioisotope in nuclear medicine
Y-90	2.7 days	Beta	Treatment of liver cancer

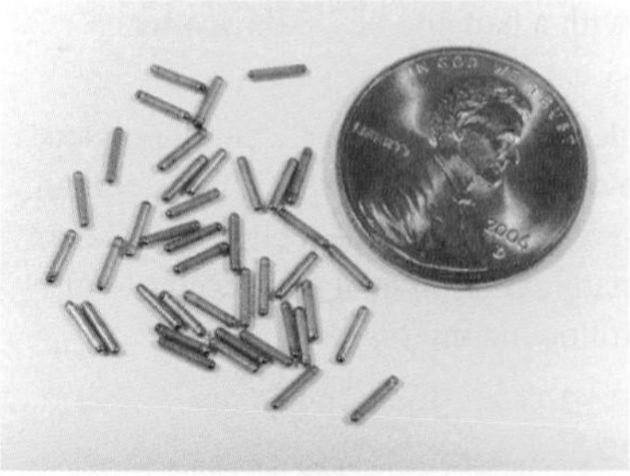

Titanium "seeds" filled with a radioactive isotope are implanted in the body to treat cancer.

SAMPLE PROBLEM 5.9 Medical Applications of Radioactivity

In the treatment of abdominal carcinoma, a person is treated with "seeds" of gold-198, which is a beta emitter. Write the balanced nuclear equation for the beta decay of gold-198.

SOLUTION

ANALYZE THE PROBLEM	Given	Need
	gold-198; beta decay	balanced nuclear equation

We can write the incomplete nuclear equation starting with gold-198.

$$^{198}_{79}Au \longrightarrow ? + ^{0}_{-1}e$$

In beta decay, the mass number, 198, does not change, but the atomic number of the new nucleus increases by one. The new atomic number is 80, which is mercury, Hg.

$$^{198}_{79}Au \longrightarrow ^{198}_{80}Hg + ^{0}_{-1}e$$

STUDY CHECK 5.9

In an experimental treatment, a person is given boron-10, which is taken up by malignant tumors. When bombarded with neutrons, boron-10 decays by emitting alpha particles that destroy the surrounding tumor cells. Write the balanced nuclear equation for the reaction for this experimental procedure.

Scans with Radioisotopes

After a person receives a radioisotope, the radiologist determines the level and location of radioactivity emitted by the radioisotope. An apparatus called a *scanner* is used to produce an image of the organ. The scanner moves slowly across the body above the region where the organ containing the radioisotope is located. The gamma rays emitted from the radioisotope in the organ can be used to expose a photographic plate, producing a *scan* of the organ. On a scan, an area of decreased or increased radiation can indicate conditions such as a disease of the organ, a tumor, a blood clot, or edema.

A common method of determining thyroid function is the use of *radioactive iodine uptake*. Taken orally, the radioisotope iodine-131 mixes with the iodine already present in the thyroid. Twenty-four hours later, the amount of iodine taken up by the thyroid is determined. A detection tube held up to the area of the thyroid gland detects the radiation coming from the iodine-131 that has located there (see Figure 5.5).

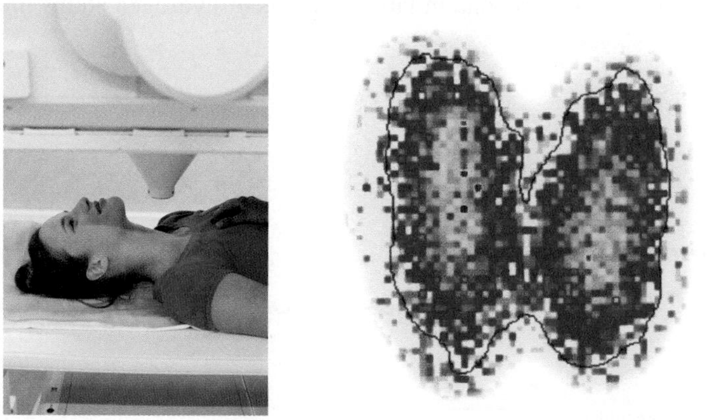

(a) (b)

FIGURE 5.5 ▶ **(a)** A scanner is used to detect radiation from a radioisotope that has accumulated in an organ. **(b)** A scan of the thyroid shows the accumulation of radioactive iodine-131 in the thyroid.

Q What type of radiation would move through body tissues to create a scan?

A person with a hyperactive thyroid will have a higher than normal level of radioactive iodine, whereas a person with a hypoactive thyroid will have lower values. If a person has hyperthyroidism, treatment is begun to lower the activity of the thyroid. One treatment involves giving a therapeutic dosage of radioactive iodine, which has a higher radiation level than the diagnostic dose. The radioactive iodine goes to the thyroid where its radiation destroys some of the thyroid cells. The thyroid produces less thyroid hormone, bringing the hyperthyroid condition under control.

Positron Emission Tomography (PET)

Positron emitters with short half-lives such as carbon-11, oxygen-15, nitrogen-13, and fluorine-18 are used in an imaging method called *positron emission tomography* (PET). A positron-emitting isotope such as fluorine-18 combined with substances in the body such as glucose is used to study brain function, metabolism, and blood flow.

$$^{18}_{9}F \longrightarrow ^{18}_{8}O + ^{0}_{+1}e$$

As positrons are emitted, they combine with electrons to produce gamma rays that are detected by computerized equipment to create a three-dimensional image of the organ (see Figure 5.6).

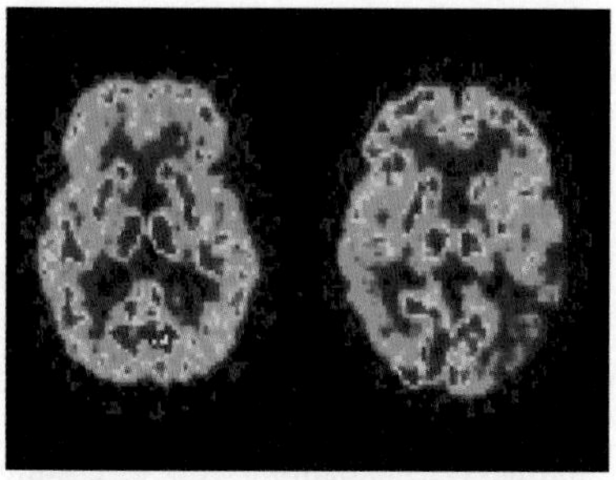

FIGURE 5.6 ▶ These PET scans of the brain show a normal brain on the left and a brain affected by Alzheimer's disease on the right.

Q When positrons collide with electrons, what type of radiation is produced that gives an image of an organ?

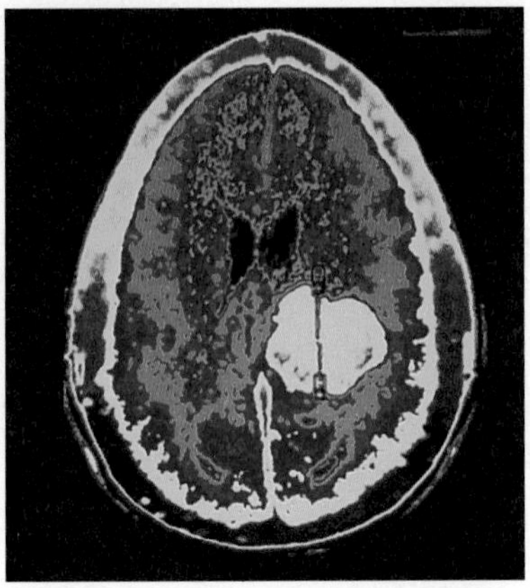
A CT scan shows a tumor (yellow) in the brain.

Computed Tomography (CT)

Another imaging method used to scan organs such as the brain, lungs, and heart is *computed tomography* (CT). A computer monitors the absorption of 30 000 X-ray beams directed at successive layers of the target organ. Based on the densities of the tissues and fluids in the organ, the differences in absorption of the X-rays provide a series of images of the organ. This technique is successful in the identification of hemorrhages, tumors, and atrophy.

Magnetic Resonance Imaging (MRI)

Magnetic resonance imaging (MRI) is a powerful imaging technique that does not involve X-ray radiation. It is the least invasive imaging method available. MRI is based on the absorption of energy when the protons in hydrogen atoms are excited by a strong magnetic field. Hydrogen atoms make up 63% of all the atoms in the body. In the hydrogen nuclei, the protons act like tiny bar magnets. With no external field, the protons have random orientations. However, when placed within a strong magnetic field, the protons align with the field. A proton aligned with the field has a lower energy than one that is aligned against the field. As the MRI scan proceeds, radiofrequency pulses of energy are applied and the hydrogen nuclei resonate at a certain frequency. Then the radio waves are quickly turned off and the protons slowly return to their natural alignment within the magnetic field, and resonate at a different frequency. They release the energy absorbed from the radio wave pulses. The difference in energy between the two states is released, which produces the electromagnetic signal that the scanner detects. These signals are sent to a computer system, where a color image of the body is generated. Because hydrogen atoms in the body are in different chemical environments, different energies are absorbed. MRI is particularly useful in obtaining images of soft tissues, which contain large amounts of hydrogen atoms in the form of water.

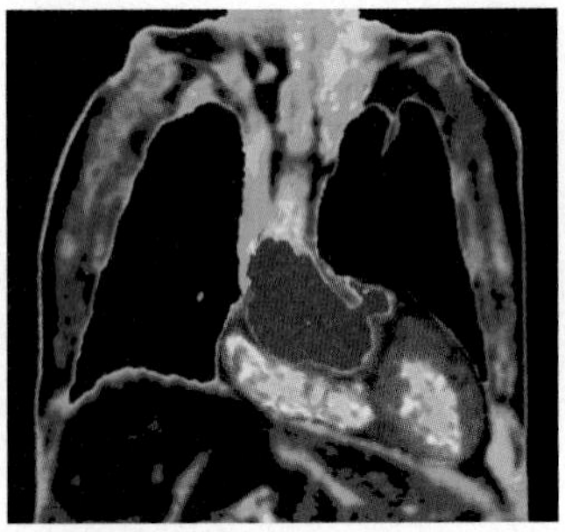
An MRI scan provides images of the heart and lungs.

Chemistry Link to Health

Brachytherapy

The process called *brachytherapy*, or seed implantation, is an internal form of radiation therapy. The prefix *brachy* is from the Greek word for short distance. With internal radiation, a high dose of radiation is delivered to a cancerous area, while normal tissue sustains minimal damage. Because higher doses are used, fewer treatments of shorter duration are needed. Conventional external treatment delivers a lower dose per treatment, but requires six to eight weeks of treatments.

Permanent Brachytherapy

One of the most common forms of cancer in males is prostate cancer. In addition to surgery and chemotherapy, one treatment option is to place 40 or more titanium capsules, or "seeds," in the malignant area. Each seed, which is the size of a grain of rice, contains radioactive iodine-125, palladium-103, or cesium-131, which decay by gamma emission. The radiation from the seeds destroys the cancer by interfering with the reproduction of cancer cells with minimal damage to adjacent normal tissues. Ninety percent (90%) of the radioisotopes decay within a few months because they have short half-lives.

Isotope	I-125	Pd-103	Cs-131
Radiation	Gamma	Gamma	Gamma
Half-Life	60 days	17 days	10 days
Time Required to Deliver 90% of Radiation	7 months	2 months	1 month

Almost no radiation passes out of the patient's body. The amount of radiation received by a family member is no greater than that received on a long plane flight. Because the radioisotopes decay to products that are not radioactive, the inert titanium capsules can be left in the body.

Temporary Brachytherapy

In another type of treatment for prostate cancer, long needles containing iridium-192 are placed in the tumor. However, the needles are removed after 5 to 10 min, depending on the activity of the iridium isotope. Compared to permanent brachytherapy, temporary brachytherapy can deliver a higher dose of radiation over a shorter time. The procedure may be repeated in a few days.

Brachytherapy is also used following breast cancer lumpectomy. An iridium-192 isotope is inserted into the catheter implanted in the space left by the removal of the tumor. The isotope is removed after 5 to 10 min, depending on the activity of the iridium source. Radiation is delivered primarily to the tissue surrounding the cavity that contained the tumor and where the cancer is most likely to recur. The procedure is repeated twice a day for five days to give an absorbed dose of 34 Gy (3400 rad). The catheter is removed, and no radioactive material remains in the body.

In conventional external beam therapy for breast cancer, a patient is given 2 Gy once a day for six to seven weeks, which gives a total absorbed dose of about 80 Gy or 8000 rad. The external beam therapy irradiates the entire breast, including the tumor cavity.

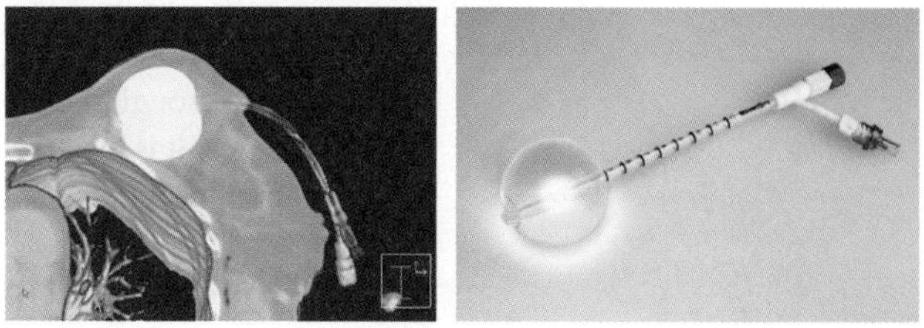

A catheter placed temporarily in the breast for radiation from Ir-192.

QUESTIONS AND PROBLEMS

5.5 Medical Applications Using Radioactivity

LEARNING GOAL Describe the use of radioisotopes in medicine.

5.35 Bone and bony structures contain calcium and phosphorus.
- **a.** Why would the radioisotopes calcium-47 and phosphorus-32 be used in the diagnosis and treatment of bone diseases?
- **b.** During nuclear tests, scientists were concerned that strontium-85, a radioactive product, would be harmful to the growth of bone in children. Explain.

5.36
- **a.** Technetium-99m emits only gamma radiation. Why would this type of radiation be used in diagnostic imaging rather than an isotope that also emits beta or alpha radiation?
- **b.** A person with *polycythemia vera* (excess production of red blood cells) receives radioactive phosphorus-32. Why would this treatment reduce the production of red blood cells in the bone marrow of the patient?

5.37 In a diagnostic test for leukemia, a person receives 4.4 mL of a solution containing selenium-75. If the activity of the selenium-75 is 45 mCi/mL, what dose, in microcuries, does the patient receive?

5.38 A vial contains radioactive iodine-131 with an activity of 2.0 mCi/mL. If a thyroid test requires 3.0 mCi in an "atomic cocktail," how many milliliters are used to prepare the iodine-131 solution?

5.6 Nuclear Fission and Fusion

LEARNING GOAL

Describe the processes of nuclear fission and fusion.

During the 1930s, scientists bombarding uranium-235 with neutrons discovered that the U-235 nucleus splits into two smaller nuclei and produces a great amount of energy. This was the discovery of nuclear **fission**. The energy generated by splitting the atom was called *atomic energy*. A typical equation for nuclear fission is:

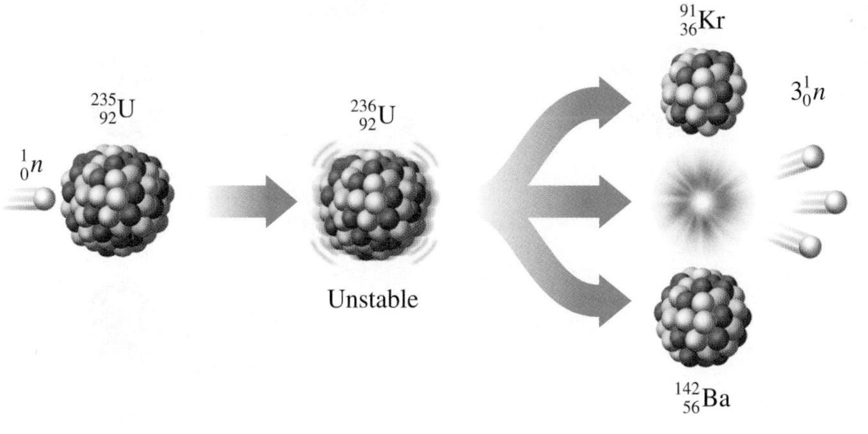

$$^{1}_{0}n + ^{235}_{92}U \longrightarrow ^{236}_{92}U \longrightarrow ^{91}_{36}Kr + ^{142}_{56}Ba + 3^{1}_{0}n + \text{energy}$$

If we could determine the mass of the products krypton, barium, and three neutrons, with great accuracy, we would find that their total mass is slightly less than the mass of the starting materials. The missing mass has been converted into an enormous amount of energy, consistent with the famous equation derived by Albert Einstein.

$$E = mc^2$$

where E is the energy released, m is the mass lost, and c is the speed of light, 3×10^8 m/s. Even though the mass loss is very small, when it is multiplied by the speed of light squared, the result is a large value for the energy released. The fission of 1 g of uranium-235 produces about as much energy as the burning of 3 tons of coal.

Chain Reaction

Fission begins when a neutron collides with the nucleus of a uranium atom. The resulting nucleus is unstable and splits into smaller nuclei. This fission process also releases several neutrons and large amounts of gamma radiation and energy. The neutrons emitted have high energies and bombard other uranium-235 nuclei. In a **chain reaction**, there is a rapid increase in the number of high-energy neutrons available to react with more uranium. To sustain a nuclear chain reaction, sufficient quantities of uranium-235 must be brought together to provide a *critical mass* in which almost all the neutrons immediately collide with more uranium-235 nuclei. So much heat and energy build up that an atomic explosion can occur (see Figure 5.7).

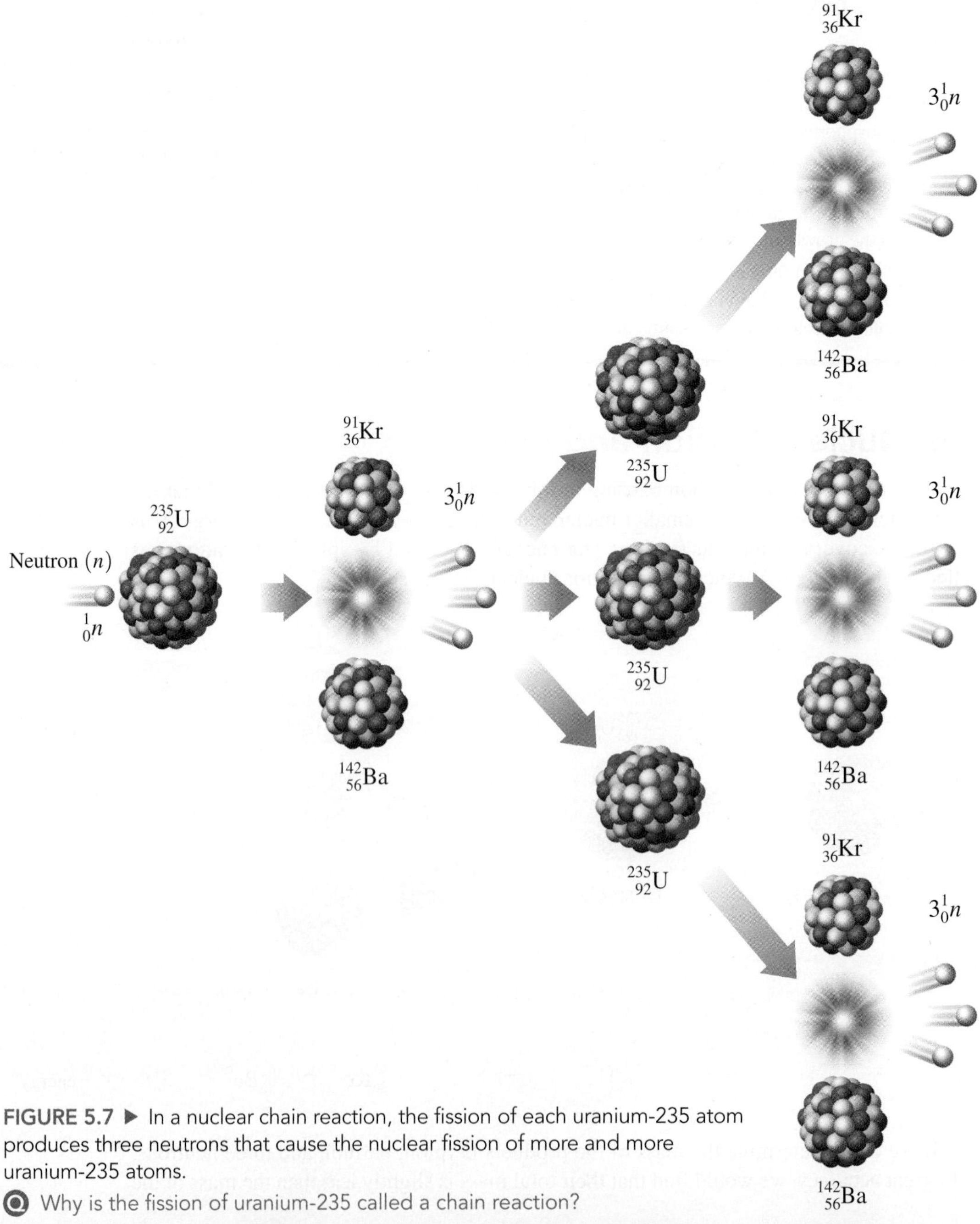

FIGURE 5.7 ▶ In a nuclear chain reaction, the fission of each uranium-235 atom produces three neutrons that cause the nuclear fission of more and more uranium-235 atoms.

Q Why is the fission of uranium-235 called a chain reaction?

Nuclear Fusion

In **fusion**, two small nuclei combine to form a larger nucleus. Mass is lost, and a tremendous amount of energy is released, even more than the energy released from nuclear fission. However, a fusion reaction requires a temperature of 100 000 000 °C to overcome the repulsion of the hydrogen nuclei and cause them to undergo fusion. Fusion reactions occur continuously in the Sun and other stars, providing us with heat and light. The huge amounts of energy produced by our sun come from the fusion of 6×10^{11} kg of hydrogen every second. In a fusion reaction, isotopes of hydrogen combine to form helium and large amounts of energy.

Scientists expect less radioactive waste with shorter half-lives from fusion reactors. However, fusion is still in the experimental stage because the extremely high temperatures needed have been difficult to reach and even more difficult to maintain. Research groups around the world are attempting to develop the technology needed to make the harnessing of the fusion reaction for energy a reality in our lifetime.

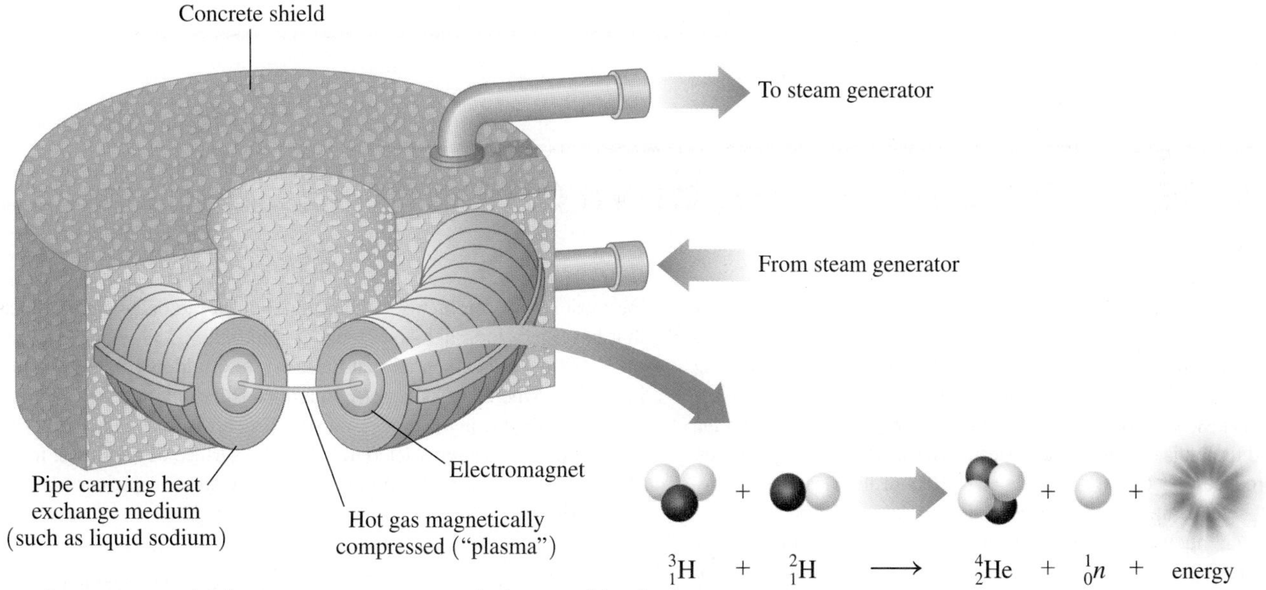

In a fusion reactor, high temperatures are needed to combine hydrogen atoms.

SAMPLE PROBLEM 5.10 Identifying Fission and Fusion

Classify the following as pertaining to fission, fusion, or both:

a. A large nucleus breaks apart to produce smaller nuclei.
b. Large amounts of energy are released.
c. Extremely high temperatures are needed for reaction.

SOLUTION

a. When a large nucleus breaks apart to produce smaller nuclei, the process is fission.
b. Large amounts of energy are generated in both the fusion and fission processes.
c. An extremely high temperature is required for fusion.

STUDY CHECK 5.10

Classify the following nuclear equation as pertaining to fission, fusion, or both:

$$^{3}_{1}H + ^{2}_{1}H \longrightarrow ^{4}_{2}He + ^{1}_{0}n + \text{energy}$$

QUESTIONS AND PROBLEMS

5.6 Nuclear Fission and Fusion

LEARNING GOAL Describe the processes of nuclear fission and fusion.

5.39 What is nuclear fission?

5.40 How does a chain reaction occur in nuclear fission?

5.41 Complete the following fission reaction:

$$^{1}_{0}n + ^{235}_{92}U \longrightarrow ^{94}_{38}Sr + ^{140}_{54}Xe + ?^{1}_{0}n$$

5.42 In another fission reaction, uranium-235 bombarded with a neutron produces strontium-94, another small nucleus, and three neutrons. Write the balanced nuclear equation for the fission reaction.

5.43 Indicate whether each of the following is characteristic of the fission or fusion process, or both:
a. Neutrons bombard a nucleus.
b. The nuclear process occurs in the Sun.
c. A large nucleus splits into smaller nuclei.
d. Small nuclei combine to form larger nuclei.

5.44 Indicate whether each of the following is characteristic of the fission or fusion process, or both:
a. Very high temperatures are required to initiate the reaction.
b. Less radioactive waste is produced.
c. Hydrogen nuclei are the reactants.
d. Large amounts of energy are released when the nuclear reaction occurs.

Chemistry Link to the **Environment**

Nuclear Power Plants

In a nuclear power plant, the quantity of uranium-235 is held below a critical mass, so it cannot sustain a chain reaction. The fission reactions are slowed by placing control rods, which absorb some of the fast-moving neutrons, among the uranium samples. In this way, less fission occurs, and there is a slower, controlled production of energy. The heat from the controlled fission is used to produce steam. The steam drives a generator, which produces electricity. Approximately 10% of the electrical energy produced in the United States is generated in nuclear power plants.

Although nuclear power plants help meet some of our energy needs, there are some problems associated with nuclear power. One of the most serious problems is the production of radioactive byproducts that have very long half-lives, such as plutonium-239 with a half-life of 24 000 yr. It is essential that these waste products be stored safely in a place where they do not contaminate the environment. Several countries are now in the process of selecting areas where nuclear waste can be placed in caverns 1000 m below the surface of the Earth. In the United States, a current proposed repository site for nuclear waste is Yucca Mountain, Nevada.

Nuclear power plants supply about 10% of electricity in the United States.

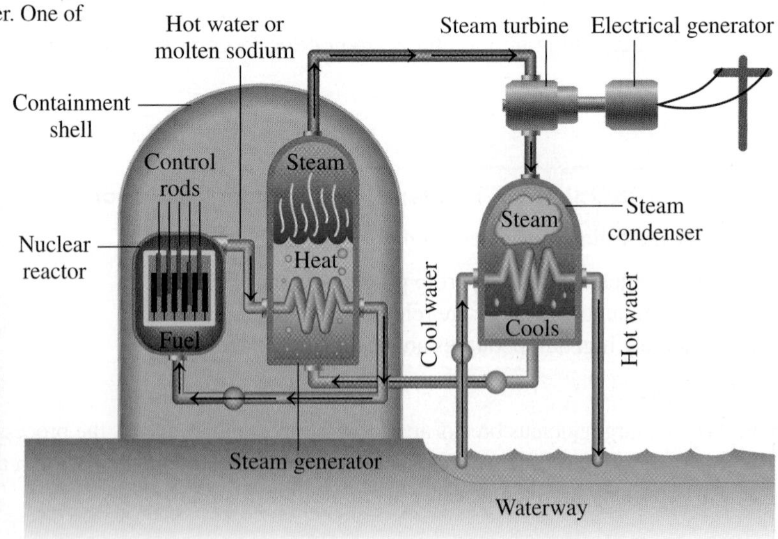

Heat from nuclear fission is used to generate electricity.

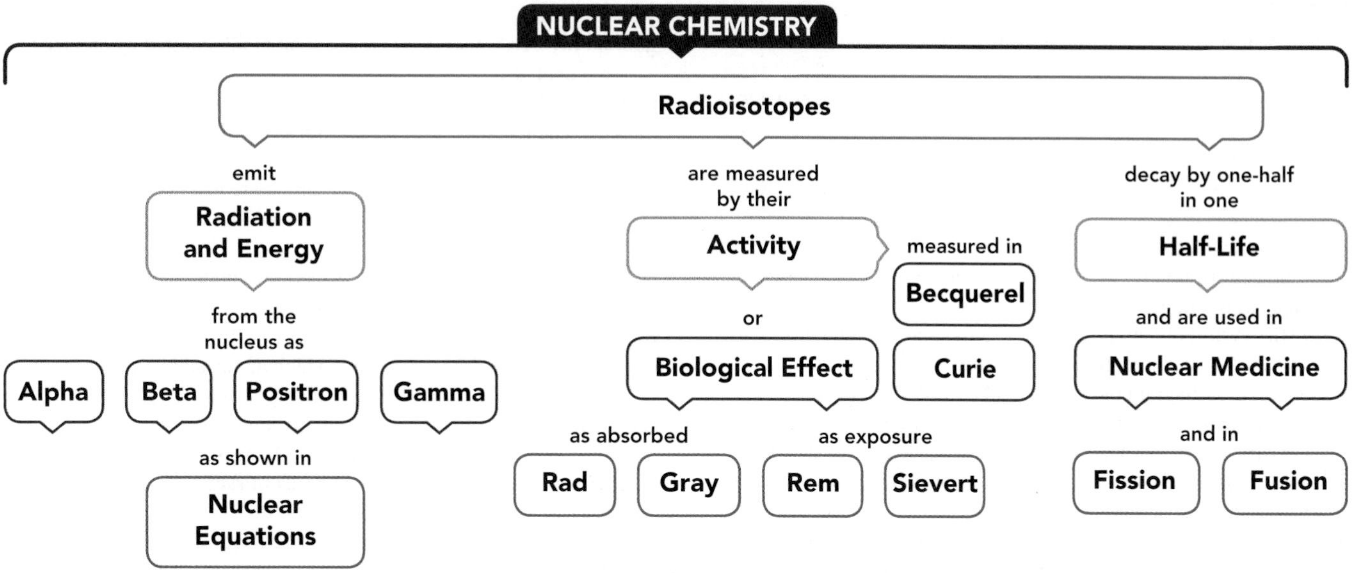

CHAPTER REVIEW

5.1 Natural Radioactivity

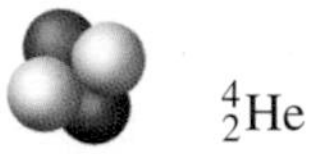

Alpha (α) particle

LEARNING GOAL Describe alpha, beta, positron, and gamma radiation.

- Radioactive isotopes have unstable nuclei that break down (decay), spontaneously emitting alpha (α), beta (β), positron (β^+), and gamma (γ) radiation.
- Because radiation can damage the cells in the body, proper protection must be used: shielding, limiting the time of exposure, and distance.

5.2 Nuclear Reactions

LEARNING GOAL Write a balanced nuclear equation showing mass numbers and atomic numbers for radioactive decay.

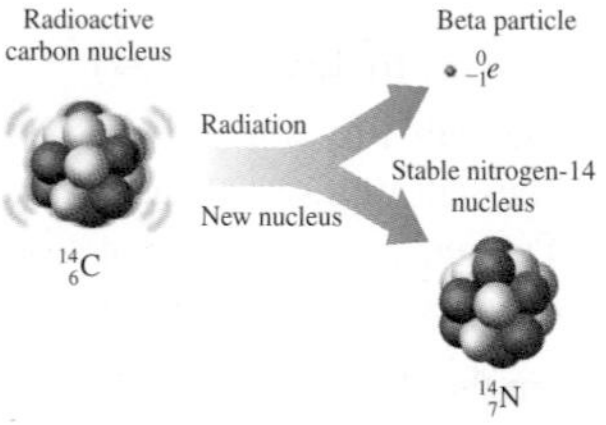

- A balanced nuclear equation is used to represent the changes that take place in the nuclei of the reactants and products.
- The new isotopes and the type of radiation emitted can be determined from the symbols that show the mass numbers and atomic numbers of the isotopes in the nuclear equation.
- A radioisotope is produced artificially when a nonradioactive isotope is bombarded by a small particle.

5.3 Radiation Measurement

LEARNING GOAL Describe the detection and measurement of radiation.

- In a Geiger counter, radiation produces charged particles in the gas contained in a tube, which generates an electrical current.
- The curie (Ci) and the becquerel (Bq) measure the activity, which is the number of nuclear transformations per second.
- The amount of radiation absorbed by a substance is measured in rads or the gray (Gy).
- The rem and the sievert (Sv) are units used to determine the biological damage from the different types of radiation.

5.4 Half-Life of a Radioisotope

LEARNING GOAL Given the half-life of a radioisotope, calculate the amount of radioisotope remaining after one or more half-lives.

- Every radioisotope has its own rate of emitting radiation.
- The time it takes for one-half of a radioactive sample to decay is called its half-life.
- For many medical radioisotopes, such as Tc-99m and I-131, half-lives are short.
- For other isotopes, usually naturally occurring ones such as C-14, Ra-226, and U-238, half-lives are extremely long.

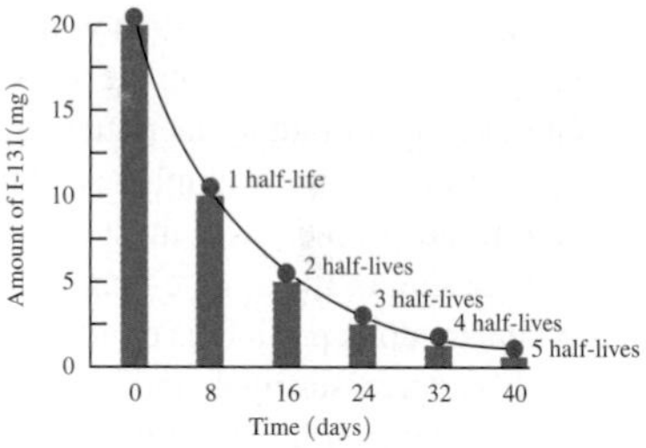

5.5 Medical Applications Using Radioactivity

LEARNING GOAL Describe the use of radioisotopes in medicine.

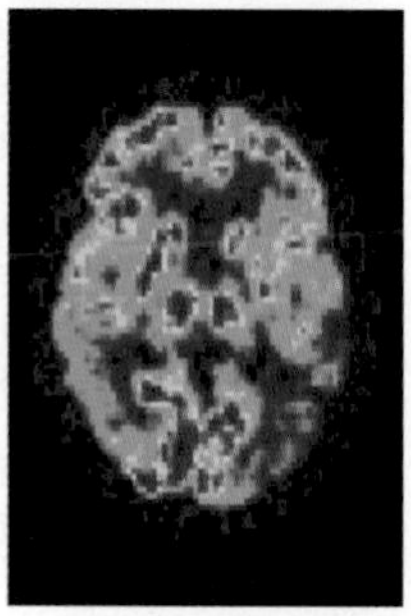

- In nuclear medicine, radioisotopes that go to specific sites in the body are given to the patient.
- By detecting the radiation they emit, an evaluation can be made about the location and extent of an injury, disease, tumor, or the level of function of a particular organ.
- Higher levels of radiation are used to treat or destroy tumors.

5.6 Nuclear Fission and Fusion

LEARNING GOAL Describe the processes of nuclear fission and fusion.

- In fission, the bombardment of a large nucleus breaks it apart into smaller nuclei, releasing one or more types of radiation and a great amount of energy.
- In fusion, small nuclei combine to form larger nuclei while great amounts of energy are released.

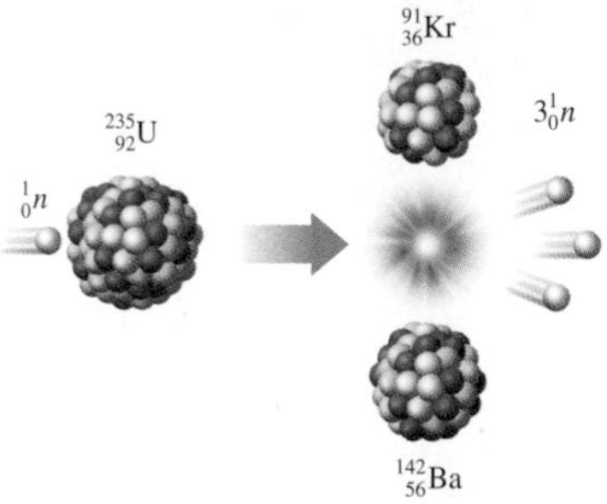

KEY TERMS

alpha particle A nuclear particle identical to a helium nucleus, symbol α or ${}^{4}_{2}He$.

becquerel (Bq) A unit of activity of a radioactive sample equal to one disintegration per second.

beta particle A particle identical to an electron, symbol ${}^{0}_{-1}e$ or β, that forms in the nucleus when a neutron changes to a proton and an electron.

chain reaction A fission reaction that will continue once it has been initiated by a high-energy neutron bombarding a heavy nucleus such as uranium-235.

curie (Ci) A unit of activity of a radioactive sample equal to 3.7×10^{10} disintegrations/s.

decay curve A diagram of the decay of a radioactive element.

equivalent dose The measure of biological damage from an absorbed dose that has been adjusted for the type of radiation.

fission A process in which large nuclei are split into smaller pieces, releasing large amounts of energy.

fusion A reaction in which large amounts of energy are released when small nuclei combine to form larger nuclei.

gamma ray High-energy radiation, symbol ${}^{0}_{0}\gamma$, emitted by an unstable nucleus.

gray (Gy) A unit of absorbed dose equal to 100 rad.

half-life The length of time it takes for one-half of a radioactive sample to decay.

positron A particle of radiation with no mass and a positive charge, symbol β^+ or ${}^{0}_{+1}e$, produced when a proton is transformed into a neutron and a positron.

rad (radiation absorbed dose) A measure of an amount of radiation absorbed by the body.

radiation Energy or particles released by radioactive atoms.

radioactive decay The process by which an unstable nucleus breaks down with the release of high-energy radiation.

radioisotope A radioactive atom of an element.

rem (radiation equivalent in humans) A measure of the biological damage caused by the various kinds of radiation (rad $\times$ radiation biological factor).

sievert (Sv) A unit of biological damage (equivalent dose) equal to 100 rem.

CORE CHEMISTRY SKILLS

The chapter section containing each Core Chemistry Skill is shown in parentheses at the end of each heading.

Writing Nuclear Equations (5.2)

- A nuclear equation is written with the atomic symbols of the original radioactive nucleus on the left, an arrow, and the new nucleus and the type of radiation emitted on the right.
- The sum of the mass numbers and the sum of the atomic numbers on one side of the arrow must equal the sum of the mass numbers and the sum of the atomic numbers on the other side.
- When an alpha particle is emitted, the mass number of the new nucleus decreases by 4, and its atomic number decreases by 2.
- When a beta particle is emitted, there is no change in the mass number of the new nucleus, but its atomic number increases by one.
- When a positron is emitted, there is no change in the mass number of the new nucleus, but its atomic number decreases by one.
- In gamma emission, there is no change in the mass number or the atomic number of the new nucleus.

Example: **a.** Write a balanced nuclear equation for the alpha decay of Po-210.

b. Write a balanced nuclear equation for the beta decay of Co-60.

Answer: **a.** When an alpha particle is emitted, we calculate the decrease of 4 in the mass number (210) of polonium, and a decrease of 2 in its atomic number.

$${}^{210}_{84}Po \longrightarrow {}^{206}_{82}? + {}^{4}_{2}He$$

Because lead has atomic number 82, the new nucleus must be an isotope of lead.

$${}^{210}_{84}Po \longrightarrow {}^{206}_{82}Pb + {}^{4}_{2}He$$

b. When a beta particle is emitted, there is no change in the mass number (60) of cobalt, but there is an increase of 1 in its atomic number.

$${}^{60}_{27}Co \longrightarrow {}^{60}_{28}? + {}^{0}_{-1}e$$

Because nickel has atomic number 28, the new nucleus must be an isotope of nickel.

$${}^{60}_{27}Co \longrightarrow {}^{60}_{28}Ni + {}^{0}_{-1}e$$

Using Half-Lives (5.4)

- The half-life of a radioisotope is the amount of time it takes for one-half of a sample to decay.
- The remaining amount of a radioisotope is calculated by dividing its quantity or activity by one-half for each half-life that has elapsed.

Example: Co-60 has a half-life of 5.3 yr. If the initial sample of Co-60 has an activity of 1200 Ci, what is its activity after 15.9 yr?

Answer:

Years	Number of Half-Lives	Co-60 (Activity)
0	0	1200 Ci
5.3	1	600 Ci
10.6	2	300 Ci
15.9	3	150 Ci

In 15.9 yr, three half-lives have passed. Thus, the activity was reduced from 1200 Ci to 150 Ci.

$$1200\text{ Ci} \xrightarrow{\text{1 half-life}} 600\text{ Ci} \xrightarrow{\text{2 half-lives}} 300\text{ Ci} \xrightarrow{\text{3 half-lives}} 150\text{ Ci}$$

UNDERSTANDING THE CONCEPTS

The chapter sections to review are shown in parentheses at the end of each question.

In problems 5.45 to 5.48, a nucleus is shown with protons and neutrons.

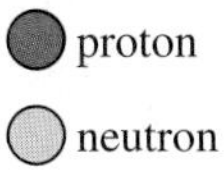

5.45 Draw the new nucleus when this isotope emits a positron to complete the following figure: (5.2)

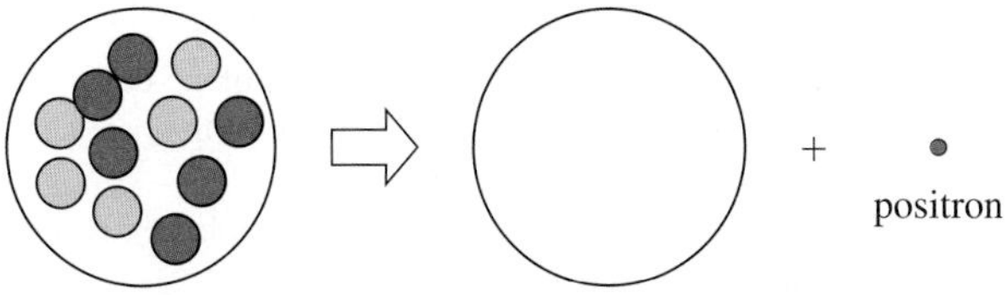

5.46 Draw the nucleus that emits a beta particle to complete the following figure: (5.2)

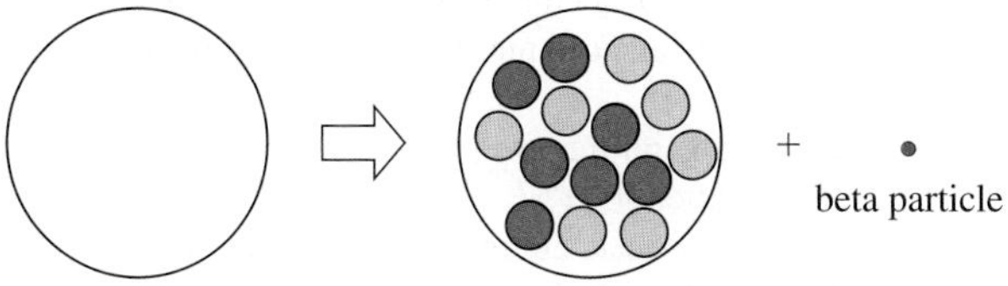

5.47 Draw the nucleus of the isotope that is bombarded in the following figure: (5.2)

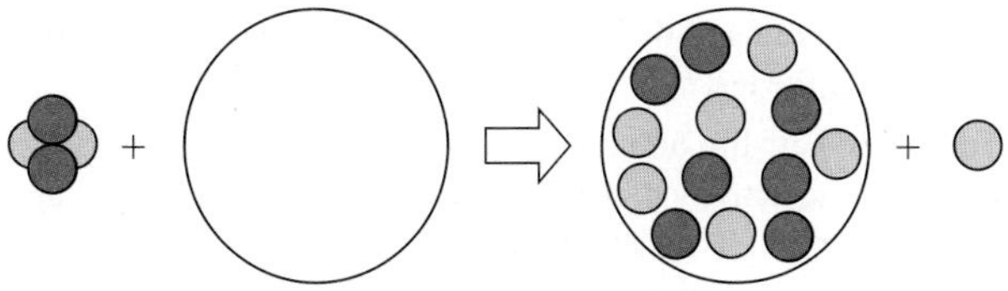

5.48 Complete the following bombardment reaction by drawing the nucleus of the new isotope that is produced in the following figure: (5.2)

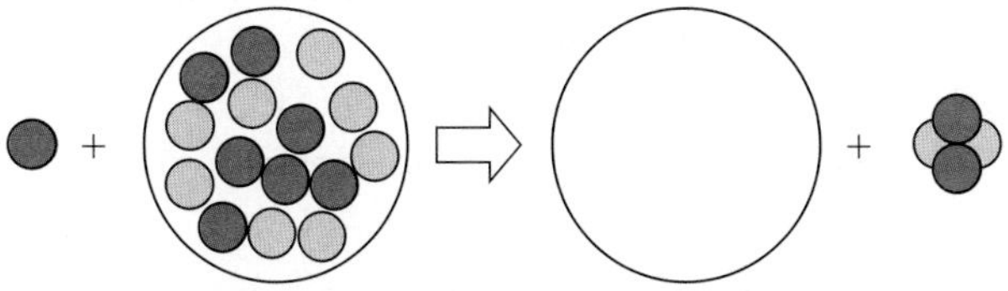

5.49 Carbon dating of small bits of charcoal used in cave paintings has determined that some of the paintings are from 10 000 to 30 000 yr old. Carbon-14 has a half-life of 5730 yr. In a 1 μg-sample of carbon from a live tree, the activity of carbon-14 is 6.4 μCi. If researchers determine that 1 μg of charcoal from a prehistoric cave painting in France has an activity of 0.80 μCi, what is the age of the painting? (5.4)

The technique of carbon dating is used to determine the age of ancient cave paintings.

5.50 Use the following decay curve for iodine-131 to answer questions **a–c**: (5.4)

a. Complete the values for the mass of radioactive iodine-131 on the vertical axis.

b. Complete the number of days on the horizontal axis.

c. What is the half-life, in days, of iodine-131?

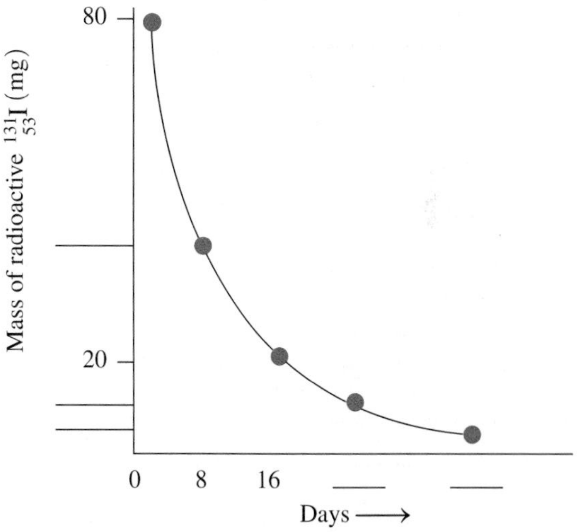

ADDITIONAL QUESTIONS AND PROBLEMS

5.51 Determine the number of protons and neutrons in the nucleus of each of the following:

1. oxygen-18
2. Potassium-42
3. Nitrogen-15
4. Samarium-153

5.52 Determine the number of protons and number of neutrons in the nucleus of each of the following: (5.1)

a. boron-10
b. zinc-72
c. iron-59
d. gold-198

5.53 Identify each of the following as alpha decay, beta decay, positron decay or gamma decay.
a. $^{131}_{53}I \longrightarrow ^{131}_{53}I + ^{0}_{0}\gamma$
b. $^{42}_{19}K \longrightarrow ^{42}_{20}Ca + ^{0}_{-1}e$
c. $^{236}_{92}U \longrightarrow ^{232}_{90}Th + ^{4}_{2}He$

5.54 Identify each of the following as alpha decay, beta decay, positron emission, or gamma emission: (5.1, 5.2)
a. $^{127}_{55}Cs \longrightarrow ^{127}_{54}Xe + ^{0}_{+1}e$
b. $^{90}_{38}Sr \longrightarrow ^{90}_{39}Y + ^{0}_{-1}e$
c. $^{218}_{85}At \longrightarrow ^{214}_{83}Bi + ^{4}_{2}He$

5.55 Write the balanced nuclear equation for each of the following: (5.1, 5.2)
a. Th-225 (α decay)
b. Bi-210 (α decay)
c. cesium-137 (β decay)
d. tin-126 (β decay)
e. F-18 (β^{+} emission)

5.56 Write the balanced nuclear equation for each of the following: (5.1, 5.2)
a. potassium-40 (β decay)
b. sulfur-35 (β decay)
c. platinum-190 (α decay)
d. Ra-210 (α decay)
e. In-113m (γ emission)

5.57 Complete each of the following nuclear equations: (5.2)
a. $^{4}_{2}He + ^{14}_{7}N \longrightarrow ? + ^{1}_{1}H$
b. $^{4}_{2}He + ^{27}_{13}Al \longrightarrow ^{30}_{14}Si + ?$
c. $^{1}_{0}n + ^{235}_{92}U \longrightarrow ^{90}_{38}Sr + 3^{1}_{0}n + ?$
d. $^{23m}_{12}Mg \longrightarrow ? + ^{0}_{0}\gamma$

5.58 Complete each of the following nuclear equations: (5.2)
a. $? + ^{59}_{27}Co \longrightarrow ^{56}_{25}Mn + ^{4}_{2}He$
b. $? \longrightarrow ^{14}_{7}N + ^{0}_{-1}e$
c. $^{0}_{-1}e + ^{76}_{36}Kr \longrightarrow ?$
d. $^{4}_{2}He + ^{241}_{95}Am \longrightarrow ? + 2^{1}_{0}n$

5.59 Write the balanced nuclear equation for each of the following: (5.2)
a. When two oxygen-16 atoms collide, one of the products is an alpha particle.
b. When californium-249 is bombarded by oxygen-18, a new element, seaborgium-263, and four neutrons are produced.
c. Radon-222 undergoes alpha decay.
d. An atom of strontium-80 emits a positron.

5.60 Write the balanced nuclear equation for each of the following: (5.2)
a. Polonium-210 decays to give lead-206.
b. Bismuth-211 emits an alpha particle.
c. A radioisotope emits a positron to form titanium-48.
d. An atom of germanium-69 emits a positron.

5.61 The activity of K-40 in a 50 kg human body is estimated to be 110nCi. What is the activity in becquerels? (5.3)

5.62 The activity of C-14 in a 70.-kg human body is estimated to be 3.7 kBq. What is this activity in microcuries? (5.3)

5.63 If the amount of radioactive phosphorus-32 in a sample decreases from 1.2 mg to 0.30 mg in 28.6 days, what is the half-life of phosphorus-32? (5.4)

5.64 If the amount of radioactive iodine-123 in a sample decreases from 0.4 mg to 0.1 mg in 26.4 h, what is the half-life of iodine-123? (5.4)

5.65 Calcium-47, a beta emitter, has a half-life of 4.5 days. (5.2, 5.4)
a. Write the balanced nuclear equation for the beta decay of calcium-47.
b. How many milligrams of a 16-mg sample of calcium-47 remain after 18 days?
c. How many days have passed if 4.8 mg of calcium-47 decayed to 1.2 mg of calcium-47?

5.66 Cesium-137, a beta emitter, has a half-life of 30 yr. (5.2, 5.4)
a. Write the balanced nuclear equation for the beta decay of cesium-137.
b. How many milligrams of a 16-mg sample of cesium-137 remain after 90 yr?
c. How many years are required for 28 mg of cesium-137 to decay to 3.5 mg of cesium-137?

5.67 A 120-mg sample of technetium-99m is used for a diagnostic test. If technetium-99m has a half-life of 6.0 h, how many milligrams of the technetium-99m sample remains active 24 h after the test? (5.4)

5.68 The half-life of oxygen-15 is 124 s. If a sample of oxygen-15 has an activity of 4000 Bq, how many minutes will elapse before it has an activity of 500 Bq? (5.4)

5.69 What is the difference between fission and fusion? (5.6)

5.70 **a.** What are the products in the fission of uranium-235 that make possible a nuclear chain reaction? (5.6)
b. What is the purpose of placing control rods among uranium samples in a nuclear reactor?

5.71 Where does fusion occur naturally? (5.6)

5.72 What are the advantages of a fusion reactor? (5.6)

CHALLENGE QUESTIONS

The following groups of questions are related to the topics in this chapter. However, they do not all follow the chapter order, and they require you to combine concepts and skills from several sections. These questions will help you increase your critical thinking skills and prepare for your next exam.

5.73 Write the balanced nuclear equation for each of the following radioactive emissions: (5.2)
a. an alpha particle from Hg-180
b. a beta particle from Au-198
c. a positron from Rb-82

5.74 Write the balanced nuclear equation for each of the following radioactive emissions: (5.2)
a. an alpha particle from Gd-148
b. a beta particle from Sr-90
c. a positron from Al-25

5.75 All the elements beyond uranium, the transuranium elements, have been prepared by bombardment and are not naturally occurring elements. The first transuranium element neptunium, Np, was prepared by bombarding U-238 with neutrons to form a neptunium atom and a beta particle. Complete the following equation: (5.2)

$$^{1}_{0}n + ^{238}_{92}U \longrightarrow ? + ?$$

5.76 One of the most recent transuranium elements ununoctium-294 (Uuo-294), atomic number 118, was prepared by bombarding californium-249 with another isotope. Complete the following equation for the preparation of this new element: (5.2)

$$? + {}^{249}_{98}\text{Cf} \longrightarrow {}^{294}_{118}\text{Uuo} + 3{}^{1}_{0}n$$

5.77 A nuclear technician was accidentally exposed to potassium-42 while doing brain scans for possible tumors. The error was not discovered until 36 h later when the activity of the potassium-42 sample was 2.0 μCi. If potassium-42 has a half-life of 12 h, what was the activity of the sample at the time the technician was exposed? (5.3, 5.4)

5.78 A wooden object from the site of an ancient temple has a carbon-14 activity of 5 counts/min compared with a reference piece of wood cut today that has an activity of 40 counts/min. If the half-life for carbon-14 is 5730 yr, what is the age of the ancient wood object? (5.3, 5.4)

5.79 The half-life for the radioactive decay of calcium-47 is 4.5 days. If a sample has an activity of 1.0 μCi after 27 days, what was the initial activity, in microcuries, of the sample? (5.3, 5.4)

5.80 The half-life for the radioactive decay of Ce-141 is 32.5 days. If a sample has an activity of 4.0 μCi after 130 days have elapsed, what was the initial activity, in microcuries, of the sample? (5.3, 5.4)

5.81 A 64-μCi sample of Tl-201 decays to 4.0 μCi in 12 days. What is the half-life, in days, of Tl-201? (5.3, 5.4)

5.82 A 16-μg sample of sodium-24 decays to 2.0 μg in 45 h. What is the half-life, in hours, of sodium-24? (5.4)

5.83 Element 114 was recently named flerovium, symbol Fl. The reaction for its synthesis involves bombarding Pu-244 with Ca-48. Write the balanced nuclear equation for the synthesis of flerovium. (5.2)

5.84 Element 116 was recently named livermorium, symbol Lv. The reaction for its synthesis involves bombarding Cm-248 with Ca-48. Write the balanced nuclear equation for the synthesis of livermorium. (5.2)

ANSWERS

Answers to Study Checks

5.1 A positron, ${}^{0}_{+1}e$ has a mass number of 0 and a 1+ charge.

5.2 ${}^{214}_{84}\text{Po} \longrightarrow {}^{210}_{82}\text{Pb} + {}^{4}_{2}\text{He}$

5.3 ${}^{51}_{24}\text{Cr} \longrightarrow {}^{51}_{25}\text{Mn} + {}^{0}_{-1}e$

5.4 ${}^{118}_{54}\text{Xe} \longrightarrow {}^{118}_{53}\text{I} + {}^{0}_{+1}e$

5.5 ${}^{4}_{2}\text{He} + {}^{27}_{13}\text{Al} \longrightarrow {}^{30}_{15}\text{P} + {}^{1}_{0}n$

5.6 4.5 Gy

5.7 0.50 μg of iron-59

5.8 17 200 yr

5.9 ${}^{1}_{0}n + {}^{10}_{5}\text{B} \longrightarrow {}^{7}_{3}\text{Li} + {}^{4}_{2}\text{He}$

5.10 When small nuclei combine to release energy, the process is fusion.

Answers to Selected Questions and Problems

5.1 **a.** alpha particle
b. positron
c. gamma radiation

5.3 **a.** ${}^{39}_{19}\text{K}$, ${}^{40}_{19}\text{K}$, ${}^{41}_{19}\text{K}$
b. They all have 19 protons and 19 electrons, but they differ in the number of neutrons.

5.5

Medical Use	Atomic Symbol	Mass Number	Number of Protons	Number of Neutrons
Heart imaging	${}^{201}_{81}\text{Tl}$	201	81	120
Radiation therapy	${}^{60}_{27}\text{Co}$	60	27	33
Abdominal scan	${}^{67}_{31}\text{Ga}$	67	31	36
Hyperthyroidism	${}^{131}_{53}\text{I}$	131	53	78
Leukemia treatment	${}^{32}_{15}\text{P}$	32	15	17

5.7 **a.** ${}^{131}_{53}\text{I}$ **b.** ${}^{60}_{27}\text{Co}$
c. ${}^{80}_{38}\text{Sr}$ **d.** ${}^{106}_{44}\text{Ru}$

5.9 **a.** ${}^{13}_{7}\text{N}$ **b.** ${}^{18}_{9}\text{F}$
c. ${}^{0}_{+1}e$ or β^+ **d.** ${}^{123}_{53}\text{I}$

5.11 **a.** **1.** alpha particle
b. **3.** gamma radiation
c. **1.** alpha particle

5.13 **a.** ${}^{238}_{92}\text{U} \longrightarrow {}^{234}_{90}\text{Th} + {}^{4}_{2}\text{He}$
b. ${}^{240}_{94}\text{Pu} \longrightarrow {}^{236}_{92}\text{U} + {}^{4}_{2}\text{He}$
c. ${}^{60}_{27}\text{Co} \longrightarrow {}^{56}_{25}\text{Mn} + {}^{4}_{2}\text{He}$
d. ${}^{226}_{88}\text{Ra} \longrightarrow {}^{222}_{86}\text{Rn} + {}^{4}_{2}\text{He}$

5.15 **a.** ${}^{25}_{11}\text{Na} \longrightarrow {}^{25}_{12}\text{Mg} + {}^{0}_{-1}e$
b. ${}^{20}_{8}\text{O} \longrightarrow {}^{20}_{9}\text{F} + {}^{0}_{-1}e$
c. ${}^{92}_{38}\text{Sr} \longrightarrow {}^{92}_{39}\text{Y} + {}^{0}_{-1}e$
d. ${}^{60}_{26}\text{Fe} \longrightarrow {}^{60}_{27}\text{Co} + {}^{0}_{-1}e$

5.17 **a.** ${}^{14}_{6}\text{C} \longrightarrow {}^{14}_{7}\text{N} + {}^{0}_{-1}e$
b. ${}^{18}_{9}\text{F} \longrightarrow {}^{18}_{10}\text{Ne} + {}^{0}_{-1}e$
c. ${}^{12}_{5}\text{B} \longrightarrow {}^{12}_{6}\text{C} + {}^{0}_{-1}e$
d. ${}^{137}_{55}\text{Cs} \longrightarrow {}^{137}_{56}\text{Ba} + {}^{0}_{-1}e$

5.19 **a.** ${}^{28}_{14}\text{Si}$, beta decay
b. ${}^{0}_{0}\gamma$, gamma emission
c. ${}^{0}_{-1}e$, beta decay
d. ${}^{238}_{92}\text{U}$, alpha decay
e. ${}^{188}_{79}\text{Au}$, positron emission

5.21 **a.** ${}^{13}_{5}\text{B}$ **b.** ${}^{20}_{8}\text{O}$
c. ${}^{234}_{91}\text{Pa}$ **d.** ${}^{43}_{20}\text{Ca}$

5.23 **a.** **2.** absorbed dose
b. **3.** biological damage
c. **1.** activity
d. **2.** absorbed dose

5.25 0.012 Gy = 1.2 rad
Patient with 9 rad received more radiation

5.27 a. 315 mCi **b.** 0.5 Gy

5.29 a. two half-lives **b.** one half-life
c. three half-lives

5.31 a. 40.0 mg **b.** 20.0 mg
c. 10.0 mg **d.** 5.00 mg

5.33 a. 130 days **b.** 195 days

5.35 a. Since the elements Ca and P are part of the bone, the radioactive isotopes of Ca and P will become part of the bony structures of the body, where their radiation can be used to diagnose or treat bone diseases.
b. Strontium (Sr) acts much like calcium (Ca) because both are Group 2A (2) elements. The body will accumulate radioactive strontium in bones in the same way that it incorporates calcium. Radioactive strontium is harmful to children because the radiation it produces causes more damage in cells that are dividing rapidly.

5.37 180 μCi

5.39 Nuclear fission is the splitting of a large atom into smaller fragments with the release of large amounts of energy.

5.41 2 neutrons

5.43 a. fission **b.** fusion
c. fission **d.** fusion

5.45

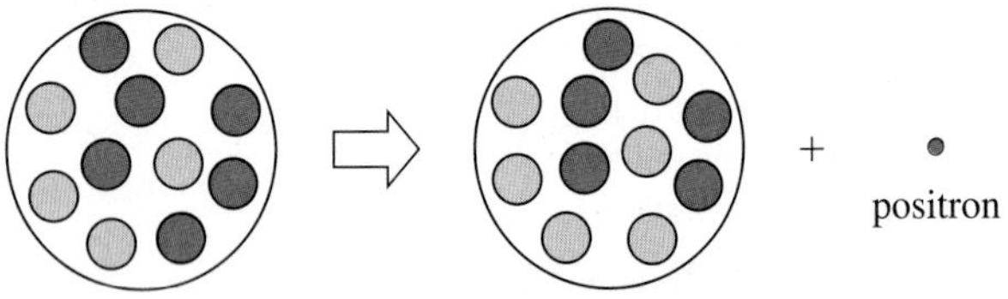

5.47

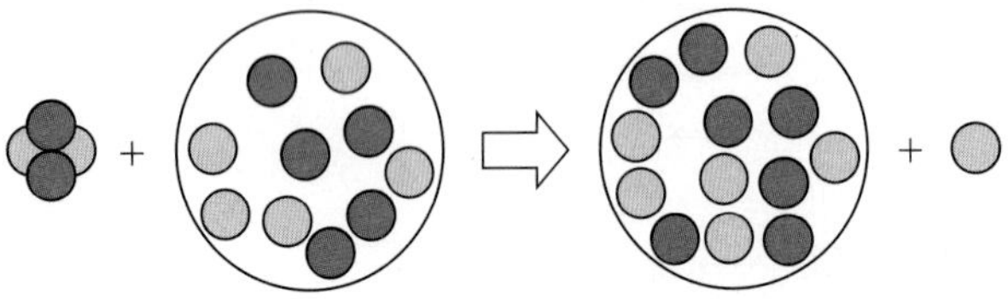

5.49 17 200 yr old

5.51 a. Proton = 8, neutron = 10
b. Proton = 19, neutron = 23
c. proton = 7, neutron = 8
d. Proton = 62, neutron = 91

5.53 a. gamma decay
b. beta decay
c. alpha decay

5.55 a. ${}^{225}_{90}Th \longrightarrow {}^{221}_{88}Ra + {}^{4}_{2}He$
b. ${}^{210}_{83}Bi \longrightarrow {}^{206}_{81}Tl + {}^{4}_{2}He$
c. ${}^{137}_{55}Cs \longrightarrow {}^{137}_{56}Ba + {}^{0}_{-1}e$
d. ${}^{126}_{50}Sn \longrightarrow {}^{126}_{51}Sb + {}^{0}_{-1}e$
e. ${}^{18}_{9}F \longrightarrow {}^{18}_{8}O + {}^{0}_{+1}e$

5.57 a. ${}^{17}_{8}O$ **b.** ${}^{1}_{1}H$
c. ${}^{143}_{54}Xe$ **d.** ${}^{23}_{12}Mg$

5.59 a. ${}^{16}_{8}O + {}^{16}_{8}O \longrightarrow {}^{28}_{14}Si + {}^{4}_{2}He$
b. ${}^{18}_{8}O + {}^{249}_{98}Cf \longrightarrow {}^{263}_{106}Sg + 4{}^{1}_{0}n$
c. ${}^{222}_{86}Rn \longrightarrow {}^{218}_{84}Po + {}^{4}_{2}He$
d. ${}^{80}_{38}Sr \longrightarrow {}^{80}_{37}Rb + {}^{0}_{+1}e$

5.61 4.4×10^3 Bq

5.63 14.3 days

5.65 a. ${}^{47}_{20}Ca \longrightarrow {}^{47}_{21}Sc + {}^{0}_{-1}e$
b. 1.0 mg of Ca-47
c. 9.0 days

5.67 7.5 mg of Tc-99m

5.69 In the fission process, an atom splits into smaller nuclei. In fusion, small nuclei combine (fuse) to form a larger nucleus.

5.71 Fusion occurs naturally in the Sun and other stars.

5.73 a. ${}^{180}_{80}Hg \longrightarrow {}^{176}_{78}Pt + {}^{4}_{2}He$
b. ${}^{198}_{79}Au \longrightarrow {}^{198}_{80}Hg + {}^{0}_{-1}e$
c. ${}^{82}_{37}Rb \longrightarrow {}^{82}_{36}Kr + {}^{0}_{+1}e$

5.75 ${}^{1}_{0}n + {}^{238}_{92}U \longrightarrow {}^{239}_{93}Np + {}^{0}_{-1}e$

5.77 16 μCi

5.79 64 μCi

5.81 3.0 days

5.83 ${}^{48}_{20}Ca + {}^{244}_{94}Pu \longrightarrow {}^{292}_{114}Fl$

6

Ionic and Molecular Compounds

Sarah, a pharmacy technician, is working at a local drug store. A customer asks Sarah about the effects of aspirin, as his doctor has recommended that he take a low-dose aspirin (81 mg) every day to prevent a heart attack or stroke.

Sarah informs her customer that aspirin is acetylsalicylic acid, and has the chemical formula, $C_9H_8O_4$. Aspirin is molecular compound, often referred to as an organic molecule because it contains the nonmetals carbon (C), hydrogen (H), and oxygen (O). Sarah explains to her customer that aspirin is used to relieve minor pains, to reduce inflammation and fever, and to slow blood clotting. Some potential side effects of aspirin may include heartburn, upset stomach, nausea, and an increased risk of a stomach ulcer. Sarah then refers the customer to the licensed pharmacist on duty for additional information.

CAREER Pharmacy Technician

Pharmacy technicians work under the supervision of a pharmacist, and their main responsibility is to fill prescriptions by preparing pharmaceutical medications. They obtain the proper medication, calculate, measure, and label the patient's medication, which is then approved by the pharmacist. After the prescription is filled, the technician prices and files the prescription. Pharmacy technicians also provide customer service by receiving prescription requests, interacting with customers, and answering any questions they may have about the drugs and their health condition. Pharmacy technicians also prepare insurance claims, and create and maintain patient profiles.

LOOKING AHEAD

Small amounts of metals cause the different colors of gemstones.

In nature, atoms of almost all the elements on the periodic table are found in combination with other atoms. Only the atoms of the noble gases—He, Ne, Ar, Kr, Xe, and Rn—do not combine in nature with other atoms. A *compound* is composed of two or more elements, with a definite composition.

Compounds may be ionic or molecular. In an *ionic compound*, one or more electrons are transferred from metals to nonmetals, which form positive and negative ions. The attraction between these ions is called an *ionic bond*.

We utilize ionic compounds every day, including salt, NaCl, and baking soda, $NaHCO_3$. Milk of magnesia, $Mg(OH)_2$, or calcium carbonate, $CaCO_3$, may be taken to settle an upset stomach. In a mineral supplement, iron may be present as iron(II) sulfate, $FeSO_4$, iodine as potassium iodide, KI, and manganese as manganese(II) sulfate, $MnSO_4$. Some sunscreens contain zinc oxide, ZnO, and tin(II) fluoride, SnF_2, in toothpaste provides fluoride to help prevent tooth decay. Precious and semiprecious gemstones are examples of ionic compounds called minerals that are cut and polished to make jewelry. For example, sapphires and rubies are made of aluminum oxide, Al_2O_3. Impurities of chromium make rubies red, and iron and titanium make sapphires blue.

A *molecular compound* consists of two or more nonmetals that share one or more valence electrons. The resulting molecules are held together by *covalent bonds*. There are many more molecular compounds than there are ionic ones. For example, water (H_2O) and carbon dioxide (CO_2) are both molecular compounds. Molecular compounds consist of molecules, which are discrete groups of atoms in a definite proportion. A molecule of water (H_2O) consists of two atoms of hydrogen and one atom of oxygen. When you have iced tea, perhaps you add molecules of sugar ($C_{12}H_{22}O_{11}$), which is a molecular compound. Other molecular compounds include propane (C_3H_8), alcohol (C_2H_6O), the antibiotic amoxicillin ($C_{16}H_{19}N_3O_5S$), and the antidepressant Prozac ($C_{17}H_{18}F_3NO$).

CHAPTER READINESS*

KEY MATH SKILLS

- Using Positive and Negative Numbers in Calculations (1.4B)
- Solving Equations (1.4D)

CORE CHEMISTRY SKILLS

- Writing Electron Arrangements (4.6)
- Drawing Electron-Dot Symbols (4.7)

*These Key Math Skills and Core Chemistry Skills from previous chapters are listed here for your review as you proceed to the new material in this chapter.

LEARNING GOAL

Write the symbols for the simple ions of the representative elements.

6.1 Ions: Transfer of Electrons

Most of the elements, except the noble gases, are found in nature combined as compounds. The noble gases are so stable that they form compounds only under extreme conditions. One explanation for the stability of noble gases is that they have a filled valence electron energy level.

Compounds form when electrons are transferred or shared to give stable electron arrangement to the atoms. In the formation of either an *ionic bond* or a *covalent bond*, atoms lose, gain, or share valence electrons to acquire an **octet** of eight valence electrons. This tendency of atoms to attain a stable electron arrangement is known as the **octet rule** and provides a key to our understanding of the ways in which atoms bond and form compounds. A few elements achieve the stability of helium with two valence electrons. However, we do not use the octet rule with transition elements.

Ionic bonds occur when the valence electrons of atoms of a metal are transferred to atoms of nonmetals. For example, sodium atoms lose electrons and chlorine atoms gain electrons to form the ionic compound NaCl. *Covalent bonds* form when atoms of nonmetals share valence electrons. In the molecular compounds H_2O and C_3H_8, atoms share electrons.

Type	Ionic Compounds	Molecular Compounds	
Particles	Ions	Molecules	
Bonding	Ionic bonds	Covalent bonds	
Examples		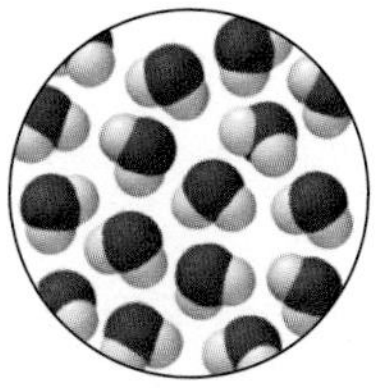	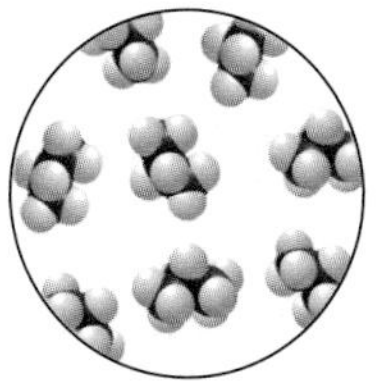
	Na^+ Cl^- ions	H_2O molecules	C_3H_8 molecules

Positive Ions: Loss of Electrons

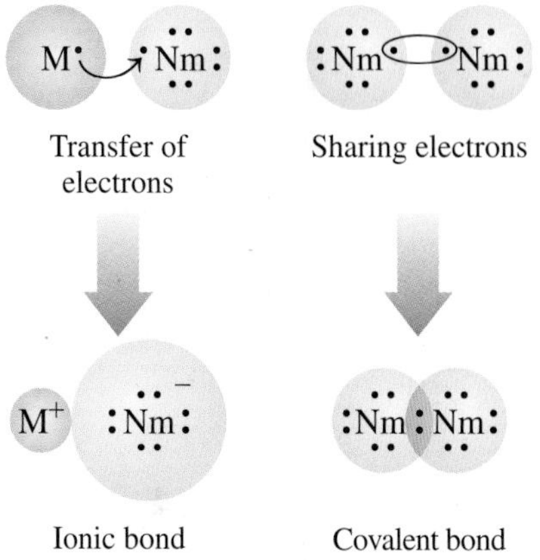

M is a metal
Nm is a nonmetal

In ionic bonding, **ions**, which have electrical charges, form when atoms lose or gain electrons to form a stable electron arrangement. Because the ionization energies of metals of Groups 1A (1), 2A (2), and 3A (13) are low, metal atoms readily lose their valence electrons. In doing so, they form ions with positive charges. A metal atom obtains the same electron arrangement as its nearest noble gas (usually eight valence electrons). For example, when a sodium atom loses its single valence electron, the remaining electrons have a stable electron configuration. By losing an electron, sodium has 10 negatively charged electrons instead of 11. Because there are still 11 positively charged protons in its nucleus, the atom is no longer neutral. It is now a sodium ion with a positive electrical charge, called an **ionic charge**, of 1+. In the symbol for the sodium ion, the ionic charge of 1+ is written in the upper right-hand corner, Na^+, where the 1 is understood. The sodium ion is smaller than the sodium atom because the ion has lost its outermost electron from the third energy level. The positively charged ions of metals are called **cations** (pronounced *cat-eye-uns*) and use the name of the element.

$$\text{Ionic charge} = \text{Charge of protons} + \text{Charge of electrons}$$
$$1+ = (11+) \qquad + (10-)$$

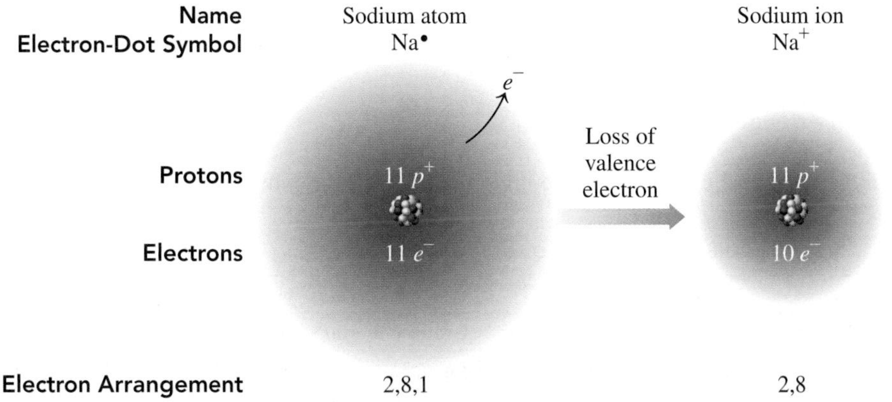

Magnesium, a metal in Group 2A (2), obtains a stable electron arrangement by losing two valence electrons to form a magnesium ion with a 2+ ionic charge, Mg^{2+}. A metal ion is named by its element name. Thus, Mg^{2+} is named the *magnesium ion*. The magnesium ion is smaller than the magnesium atom because the outermost electrons in the third energy level were removed. The octet in the magnesium ion is made up of electrons that fill its second energy level.

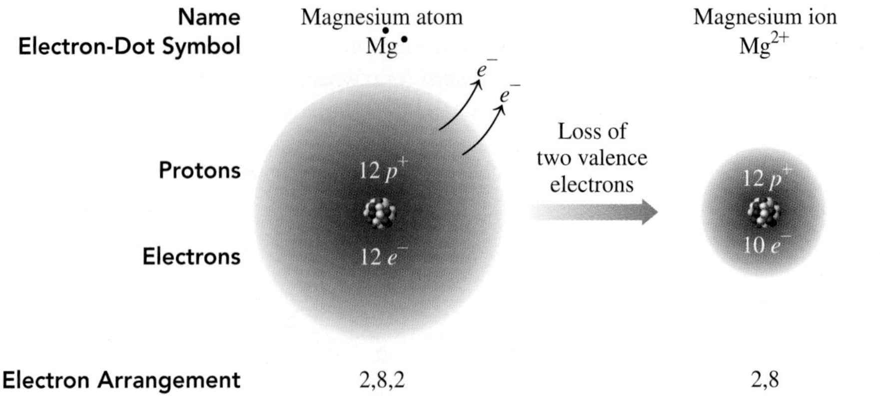

CORE CHEMISTRY SKILL

Writing Positive and Negative Ions

Negative Ions: Gain of Electrons

The ionization energy of a nonmetal atom in Groups 5A (15), 6A (16), or 7A (17) is high. In an ionic compound, a nonmetal atom gains one or more valence electrons to obtain a stable electron configuration. By gaining electrons, a nonmetal atom forms a negatively charged ion. For example, an atom of chlorine with seven valence electrons gains one electron to form an octet. Because it now has 18 electrons and 17 protons in its nucleus, the chlorine atom is no longer neutral. It is a *chloride ion* with an ionic charge of 1−, which is written as Cl^-, with the 1 understood. A negatively charged ion, called an **anion** (pronounced *an-eye-un*), is named by using the first syllable of its element name followed by *ide*. The chloride ion is larger than the chlorine atom because the ion has an additional electron, which completes its outermost energy level.

$$\text{Ionic charge} = \text{Charge of protons} + \text{Charge of electrons}$$
$$1- = (17+) + (18-)$$

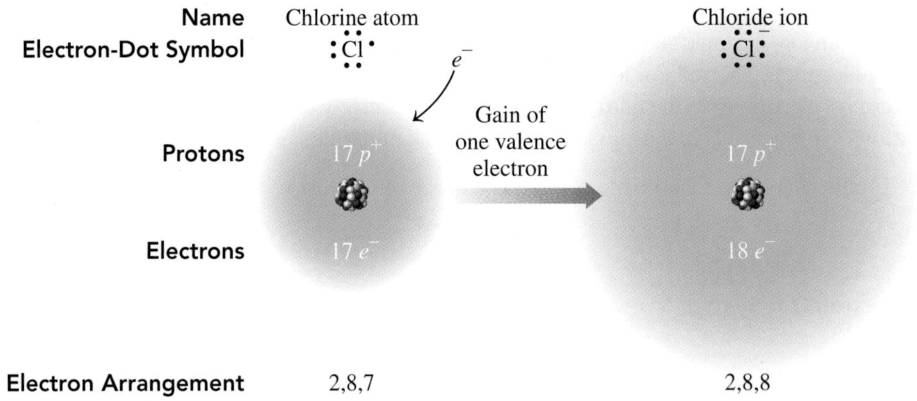

Table 6.1 lists the names of some important metal and nonmetal ions.

TABLE 6.1 Formulas and Names of Some Common Ions

Metals			Nonmetals		
Group Number	Cation	Name of Cation	Group Number	Anion	Name of Anion
1A (1)	Li^+	Lithium	5A (15)	N^{3-}	Nitride
	Na^+	Sodium		P^{3-}	Phosphide
	K^+	Potassium	6A (16)	O^{2-}	Oxide
2A (2)	Mg^{2+}	Magnesium		S^{2-}	Sulfide
	Ca^{2+}	Calcium	7A (17)	F^-	Fluoride
	Ba^{2+}	Barium		Cl^-	Chloride
3A (13)	Al^{3+}	Aluminum		Br^-	Bromide
				I^-	Iodide

SAMPLE PROBLEM 6.1 Ions

a. Write the symbol and name for the ion that has 7 protons and 10 electrons.
b. Write the symbol and name for the ion that has 20 protons and 18 electrons.

SOLUTION

a. The element with seven protons is nitrogen. In an ion of nitrogen with 10 electrons, the ionic charge would be 3−, $[(7+) + (10-) = 3-]$. The ion, written as N^{3-}, is the *nitride* ion.
b. The element with 20 protons is calcium. In an ion of calcium with 18 electrons, the ionic charge would be 2+, $[(20+) + (18-) = 2+]$. The ion, written as Ca^{2+}, is the *calcium* ion.

STUDY CHECK 6.1

How many protons and electrons are in each of the following ions?

a. S^{2-} **b.** Al^{3+}

Ionic Charges from Group Numbers

In ionic compounds, representative elements usually lose or gain electrons to give eight valence electrons like their nearest noble gas (or two for helium). We can use the group numbers in the periodic table to determine the charges for the ions of the representative elements. The elements in Group 1A (1) lose one electron to form ions with a 1+ charge. The elements in Group 2A (2) lose two electrons to form ions with a 2+ charge. The elements in Group 3A (13) lose three electrons to form ions with a 3+ charge. In this text, we do not use the group numbers of the transition elements to determine their ionic charges.

In ionic compounds, the elements in Group 7A (17) gain one electron to form ions with a 1− charge. The elements in Group 6A (16) gain two electrons to form ions with a 2− charge. The elements in Group 5A (15) gain three electrons to form ions with a 3− charge.

The nonmetals of Group 4A (14) do not typically form ions. However, the metals Sn and Pb in Group 4A (14) lose electrons to form positive ions. Table 6.2 lists the ionic charges for some common monatomic ions of representative elements.

TABLE 6.2 Examples of Monatomic Ions and Their Nearest Noble Gases

		Metals Lose Valence Electrons			Nonmetals Gain Valence Electrons				
Noble Gases		1A (1)	2A (2)	3A (13)	5A (15)	6A (16)	7A (17)		Noble Gases
He	⇦	Li^+							
Ne	⇦	Na^+	Mg^{2+}	Al^{3+}	N^{3-}	O^{2-}	F^-	⇨	**Ne**
Ar	⇦	K^+	Ca^{2+}		P^{3-}	S^{2-}	Cl^-	⇨	**Ar**
Kr	⇦	Rb^+	Sr^{2+}				Br^-	⇨	**Kr**
Xe	⇦	Cs^+	Ba^{2+}				I^-	⇨	**Xe**

SAMPLE PROBLEM 6.2 Writing Symbols for Ions

Consider the elements aluminum and oxygen.

a. Identify each as a metal or a nonmetal.
b. State the number of valence electrons for each.
c. State the number of electrons that must be lost or gained for each to achieve an octet.
d. Write the symbol, including its ionic charge, and name for each resulting ion.

SOLUTION

Aluminum	Oxygen
a. metal	nonmetal
b. three valence electrons	six valence electrons
c. loses 3 e^-	gains 2 e^-
d. Al^{3+}, $[(13+) + (10-) = 3+]$, aluminum ion	O^{2-}, $[(8+) + (10-) = 2-]$, oxide ion

STUDY CHECK 6.2

What are the symbols for the ions formed by potassium and sulfur?

Chemistry Link to **Health**

Some Important Ions in the Body

Several ions in body fluids have important physiological and metabolic functions. Some are listed in Table 6.3.

Foods such as bananas, milk, cheese, and potatoes provide the body with ions that are important in regulating body functions.

Milk, cheese, bananas, cereal, and potatoes provide ions for the body.

TABLE 6.3 Ions in the Body

Ion	Occurrence	Function	Source	Result of Too Little	Result of Too Much
Na^+	Principal cation outside the cell	Regulation and control of body fluids	Salt, cheese, pickles	Hyponatremia, anxiety, diarrhea, circulatory failure, decrease in body fluid	Hypernatremia, little urine, thirst, edema
K^+	Principal cation inside the cell	Regulation of body fluids and cellular functions	Bananas, orange juice, milk, prunes, potatoes	Hypokalemia (hypopotassemia), lethargy, muscle weakness, failure of neurological impulses	Hyperkalemia (hyperpotassemia), irritability, nausea, little urine, cardiac arrest
Ca^{2+}	Cation outside the cell; 90% of calcium in the body in bone as $Ca_3(PO_4)_2$ or $CaCO_3$	Major cation of bone; needed for muscle contraction	Milk, yogurt, cheese, greens, spinach	Hypocalcemia, tingling fingertips, muscle cramps, osteoporosis	Hypercalcemia, relaxed muscles, kidney stones, deep bone pain
Mg^{2+}	Cation outside the cell; 50% of magnesium in the body in bone structure	Essential for certain enzymes, muscles, nerve control	Widely distributed (part of chlorophyll of all green plants), nuts, whole grains	Disorientation, hypertension, tremors, slow pulse	Drowsiness
Cl^-	Principal anion outside the cell	Gastric juice, regulation of body fluids	Salt	Same as for Na^+	Same as for Na^+

QUESTIONS AND PROBLEMS

6.1 Ions: Transfer of Electrons

LEARNING GOAL Write the symbols for the simple ions of the representative elements.

6.1 State the number of electrons that must be lost by atoms of each of the following to achieve a noble gas configuration.
a. Na **b.** Mg **c.** Al **d.** Sr **e.** K

6.2 State the number of electrons that must be gained by atoms of each of the following to achieve a noble gas configuration.
a. F **b.** O **c.** P **d.** As **e.** S

6.3 State the number of electrons lost or gained when the following elements forms ions:
a. Be **b.** Sb **c.** Cl
d. Ba **e.** I

6.4 State the number of electrons lost or gained when the following elements form ions:
a. O **b.** Group 2A (2) **c.** F
d. K **e.** Rb

6.5 Write the symbols for the ions with the following number of protons and electrons:
a. 3 protons, 3 electrons **b.** 17 protons, 18 electrons
c. 22 protons, 18 electrons **d.** 44 protons, 41 electrons

6.6 Write the symbols for the ions with the following number of protons and electrons:
a. 8 protons, 10 electrons **b.** 19 protons, 18 electrons
c. 35 protons, 36 electrons **d.** 50 protons, 46 electrons

6.7 Write the symbol for the ion of each of the following:
a. Bromine **b.** Lithium
c. Selenium **d.** Indium

6.8 Write the symbol for the ion of each of the following:
a. fluorine **b.** calcium
c. sodium **d.** iodine

6.2 Writing Formulas for Ionic Compounds

LEARNING GOAL

Using charge balance, write the correct formula for an ionic compound.

Ionic compounds consist of positive and negative ions. The ions are held together by strong electrical attractions between the oppositely charged ions, called *ionic bonds*. The positive ions are formed by *metals* losing electrons, and the negative ions are formed when *nonmetals* gain electrons. Even though noble gases are nonmetals, they already have stable electron configurations and do not form compounds.

Properties of Ionic Compounds

The physical and chemical properties of an ionic compound such as NaCl are very different from those of the original elements. For example, the original elements of NaCl were sodium, which is a soft, shiny metal, and chlorine, which is a yellow-green poisonous gas. However, when they react and form positive and negative ions, they produce NaCl, which is ordinary table salt, a hard, white, crystalline substance that is important in our diet.

In a crystal of NaCl, the larger Cl^- ions are arranged in a three-dimensional structure in which the smaller Na^+ ions occupy the spaces between the Cl^- ions (see Figure 6.1). In this crystal, every Na^+ ion is surrounded by six Cl^- ions, and every Cl^- ion is surrounded by six Na^+ ions. Thus, there are many strong attractions between the positive and negative ions, which account for the high melting points of ionic compounds. For example, the melting point of NaCl is 801 °C. At room temperature, ionic compounds are solids.

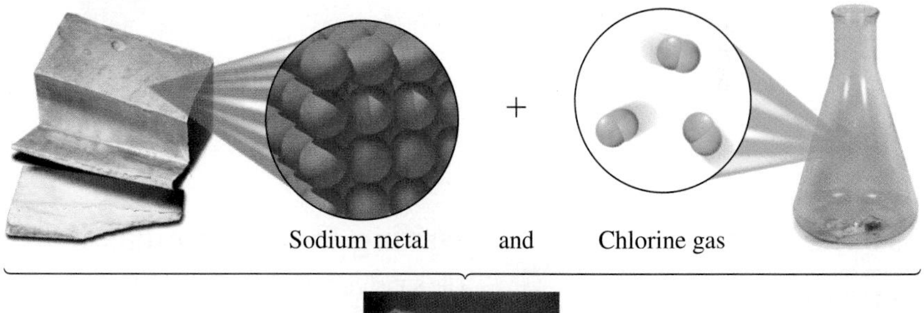

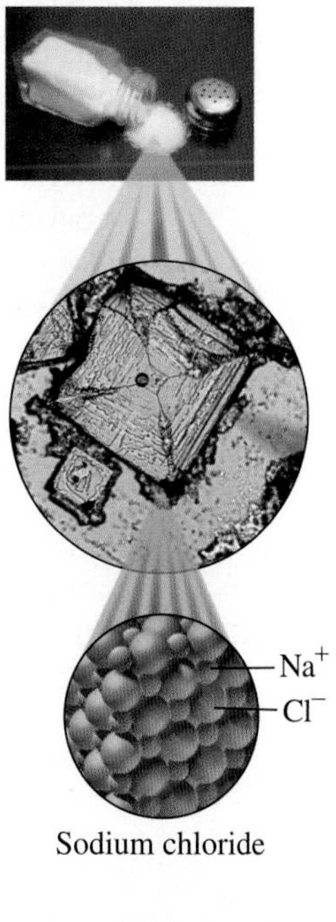

FIGURE 6.1 ▶ The elements sodium and chlorine react to form the ionic compound sodium chloride, the compound that makes up table salt. The magnification of NaCl crystals shows the arrangement of Na^+ and Cl^- ions in a NaCl crystal.

Q What is the type of bonding between Na^+ and Cl^- ions in NaCl?

Chemical Formulas of Ionic Compounds

The **chemical formula** of a compound represents the symbols and subscripts in the lowest whole-number ratio of the atoms or ions. In the formula of an ionic compound, the sum of the ionic charges in the formula is always zero. *Thus, the total amount of positive charge is equal to the total amount of negative charge.* For example, to achieve a stable electron configuration, a Na atom (metal) loses its one valence electron to form Na^+, and one Cl atom (nonmetal) gains one electron to form a Cl^- ion. The formula NaCl indicates that the compound has charge balance because there is one sodium ion, Na^+, for every chloride ion, Cl^-. Although the ions are positively or negatively charged, they are not shown in the formula of the compound.

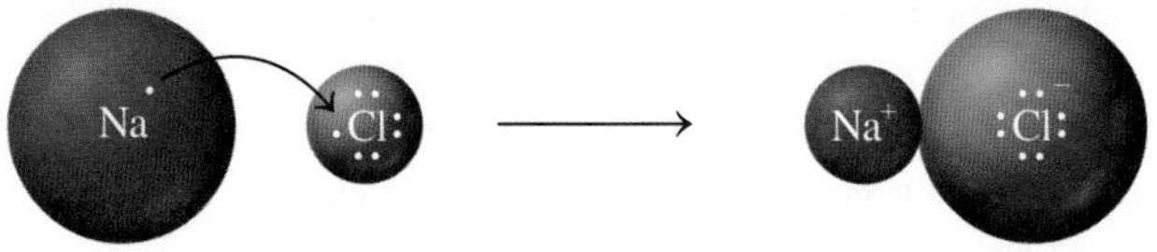

Loses 1 e^- Gains 1 e^-

Na^+ Cl^-
$1(1+) + 1(1-) = 0$
NaCl, sodium chloride

Subscripts in Formulas

Consider a compound of magnesium and chlorine. To achieve a stable electron arrangement, a Mg atom (metal) loses its two valence electrons to form Mg^{2+}. Two Cl atoms (nonmetals) each gain one electron to form two Cl^- ions. The two Cl^- ions are needed to balance the positive charge of Mg^{2+}. This gives the formula $MgCl_2$, magnesium chloride, in which the subscript 2 shows that two Cl^- ions are needed for charge balance.

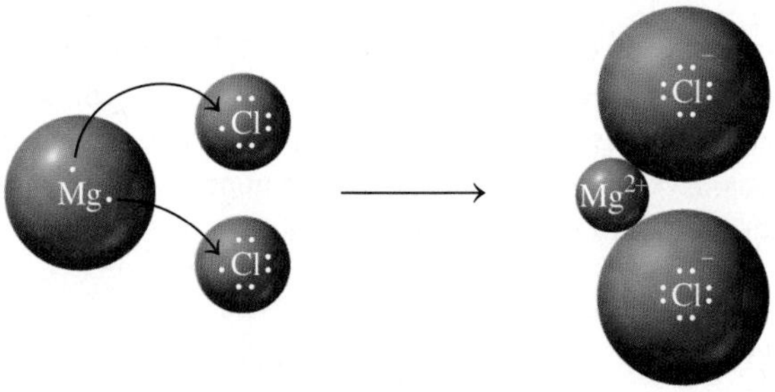

Loses 2 e^- Each gains 1 e^-

Mg^{2+} $2Cl^-$
$1(2+) + 2(1-) = 0$
$MgCl_2$, magnesium chloride

CORE CHEMISTRY SKILL
Writing Ionic Formulas

Writing Ionic Formulas from Ionic Charges

The subscripts in the formula of an ionic compound represent the number of positive and negative ions that give an overall charge of zero. Thus, we can now write a formula directly from the ionic charges of the positive and negative ions. Suppose we wish to write the formula for the ionic compound containing Na^+ and S^{2-} ions. To balance the ionic charge of the S^{2-} ion, we will need to place two Na^+ ions in the formula. This gives the formula Na_2S, which has an overall charge of zero. In the formula of an ionic compound, the cation is written first followed by the anion. Appropriate subscripts are used to show the number of each of the ions. This formula, which is the lowest ratio of the ions in an ionic compound, is called a *formula unit*.

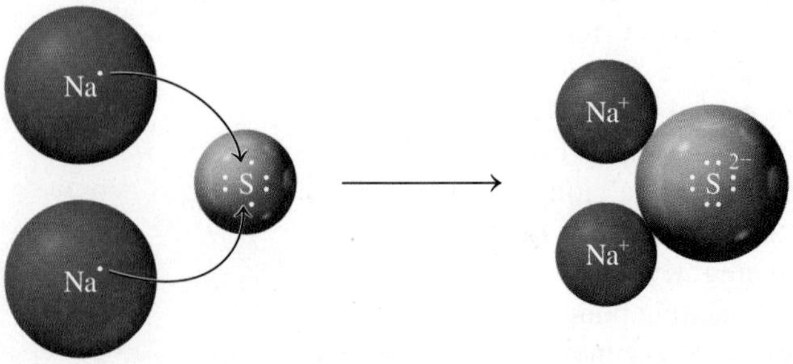

Each loses 1 e^- Gains 2 e^-

$2Na^+$ S^{2-}
$2(1+) + 1(2-) = 0$
Na_2S, sodium sulfide

SAMPLE PROBLEM 6.3 Writing Formulas from Ionic Charges

Write the symbols for the ions, and the correct formula for the ionic compound formed when lithium and nitrogen react.

SOLUTION

Lithium, which is a metal in Group 1A (1), forms Li^+; nitrogen, which is a nonmetal in Group 5A (15), forms N^{3-}. The charge of 3− is balanced by three Li^+ ions.

$$3(1+) + 1(3-) = 0$$

Writing the cation (positive ion) first and the anion (negative ion) second gives the formula Li_3N.

STUDY CHECK 6.3

Write the symbol for the ions, and the correct formula for the ionic compound that would form when calcium and oxygen react.

QUESTIONS AND PROBLEMS

6.2 Writing Formulas for Ionic Compounds

LEARNING GOAL Using charge balance, write the correct formula for an ionic compound.

6.9 Which of the following pairs of elements are likely to form an ionic compound?
a. Sodium and chloride **b.** Nitrogen and fluorine
c. Lithium and oxygen **d.** Barium and sodium
e. Phosphorus and sulfur **f.** Potassium and bromine

6.10 Which of the following pairs of elements are likely to form an ionic compound?
a. helium and oxygen **b.** magnesium and chlorine
c. chlorine and bromine **d.** potassium and sulfur
e. sodium and potassium **f.** nitrogen and iodine

6.11 Write the correct ionic formula for the compound formed between each of the following pairs of ions:
a. Li^+ and O^{2-} **b.** Ca^{+2} and Cl^- **c.** Be^{+2} and F^-
d. B^{+3} and N^{3-} **e.** Al^{+3} and F^-

6.12 Write the correct ionic formula for the compound formed between each of the following pairs of ions:
a. Al^{3+} and Cl^- **b.** Ca^{2+} and S^{2-} **c.** Li^+ and S^{2-}
d. Rb^+ and P^{3-} **e.** Cs^+ and I^-

6.13 Write the symbols for the ions, and the correct formula for the ionic compound formed by each of the following:
a. potassium and sulfur **b.** phosphorus and bromide
c. aluminum and chloride **d.** silicon and oxygen

6.14 Write the symbols for the ions, and the correct formula for the ionic compound formed by each of the following:
a. calcium and chlorine **b.** rubidium and bromine
c. sodium and phosphorus **d.** magnesium and oxygen

6.3 Naming Ionic Compounds

LEARNING GOAL

Given the formula of an ionic compound, write the correct name; given the name of an ionic compound, write the correct formula.

CORE CHEMISTRY SKILL

Naming Ionic Compounds

In the name of an ionic compound made up of two elements, the name of the metal ion, which is written first, is the same as its element name. The name of the nonmetal ion is obtained by using the first syllable of its element name followed by *ide*. In the name of any ionic compound, a space separates the name of the cation from the name of the anion. Subscripts are not used; they are understood because of the charge balance of the ions in the compound (see Table 6.4).

Iodized salt contains KI to prevent iodine deficiency.

TABLE 6.4 Names of Some Ionic Compounds

Compound	Metal Ion	Nonmetal Ion	Name
KI	K^+ Potassium	I^- Iodide	Potassium iodide
$MgBr_2$	Mg^{2+} Magnesium	Br^- Bromide	Magnesium bromide
Al_2O_3	Al^{3+} Aluminum	O^{2-} Oxide	Aluminum oxide

SAMPLE PROBLEM 6.4 Naming Ionic Compounds

Write the name for the ionic compound Mg_3N_2.

SOLUTION

STEP 1 Identify the cation and anion. The cation, Mg^{2+}, is from Group 2A (2), and the anion, N^{3-}, is from Group 5A (15).

STEP 2 Name the cation by its element name. The cation Mg^{2+} is magnesium.

STEP 3 Name the anion by using the first syllable of its element name followed by *ide*. The anion N^{3-} is nitride.

STEP 4 Write the name for the cation first and the name for the anion second. Mg_3N_2 magnesium nitride

STUDY CHECK 6.4

Name the compound Ga_2S_3.

Guide to Naming Ionic Compounds with Metals That Form a Single Ion

STEP 1
Identify the cation and anion.

STEP 2
Name the cation by its element name.

STEP 3
Name the anion by using the first syllable of its element name followed by *ide*.

STEP 4
Write the name for the cation first and the name for the anion second.

Metals with Variable Charges

We have seen that the charge of an ion of a representative element can be obtained from its group number. However, we cannot determine the charge of a transition element because it typically forms two or more positive ions. The transition elements lose electrons, but they are lost from the highest energy level and sometimes from a lower energy level as well. This is also true for metals of representative elements in Groups 4A (14) and 5A (15), such as Pb, Sn, and Bi.

In some ionic compounds, iron is in the Fe^{2+} form, but in other compounds, it has the Fe^{3+} form. Copper also forms two different ions, Cu^{+} and Cu^{2+}. When a metal can form two or more types of ions, it has *variable charge*. Then we cannot predict the ionic charge from the group number.

For metals that form two or more ions, a naming system is used to identify the particular cation. To do this, a Roman numeral that is equal to the ionic charge is placed in parentheses immediately after the name of the metal. For example, Fe^{2+} is iron(II), and Fe^{3+} is iron(III). Table 6.5 lists the ions of some metals that produce more than one ion.

Figure 6.2 shows some ions and their location on the periodic table. The transition elements form more than one positive ion except for zinc (Zn^{2+}), cadmium (Cd^{2+}), and silver (Ag^{+}), which form only one ion. Thus, no Roman numerals are used with zinc, cadmium, and silver when naming their cations in ionic compounds. Metals in Groups 4A (14) and 5A (15) also form more than one type of positive ion. For example, lead and tin in Group 4A (14) form cations with charges of 2+ and 4+ and bismuth in Group 5A (15) forms cations with charges of 3+ and 5+.

TABLE 6.5 Some Metals That Form More Than One Positive Ion

Element	Possible Ions	Name of Ion
Bismuth	Bi^{3+}	Bismuth(III)
	Bi^{5+}	Bismuth(V)
Chromium	Cr^{2+}	Chromium(II)
	Cr^{3+}	Chromium(III)
Cobalt	Co^{2+}	Cobalt(II)
	Co^{3+}	Cobalt(III)
Copper	Cu^{+}	Copper(I)
	Cu^{2+}	Copper(II)
Gold	Au^{+}	Gold(I)
	Au^{3+}	Gold(III)
Iron	Fe^{2+}	Iron(II)
	Fe^{3+}	Iron(III)
Lead	Pb^{2+}	Lead(II)
	Pb^{4+}	Lead(IV)
Manganese	Mn^{2+}	Manganese(II)
	Mn^{3+}	Manganese(III)
Mercury	Hg_2^{2+}	Mercury(I)*
	Hg^{2+}	Mercury(II)
Nickel	Ni^{2+}	Nickel(II)
	Ni^{3+}	Nickel(III)
Tin	Sn^{2+}	Tin(II)
	Sn^{4+}	Tin(IV)

*Mercury(I) ions form an ion pair with a 2+ charge.

Determination of Variable Charge

When you name an ionic compound, you need to determine if the metal is a representative element or a transition element. If it is a transition element, except for zinc, cadmium, or silver, you will need to use its ionic charge as a Roman numeral as part of its name. The calculation of ionic charge depends on the negative charge of the anions in the formula. For example, we use charge balance to determine the charge of a copper cation in the ionic compound $CuCl_2$. Because there are two chloride ions, each with a 1− charge, the total negative charge is 2−. To balance this 2− charge, the copper ion must have a charge of 2+, or Cu^{2+}:

$\mathbf{CuCl_2}$

$$\begin{aligned} &\text{Cu charge} + 2\,\text{Cl}^- \text{ charge} = 0 \\ &? \quad + 2(1-) = 0 \\ &2+ \quad + 2- = 0 \end{aligned}$$

To indicate the 2+ charge for the copper ion Cu^{2+}, we place the Roman numeral (II) immediately after copper when naming this compound: copper(II) chloride.

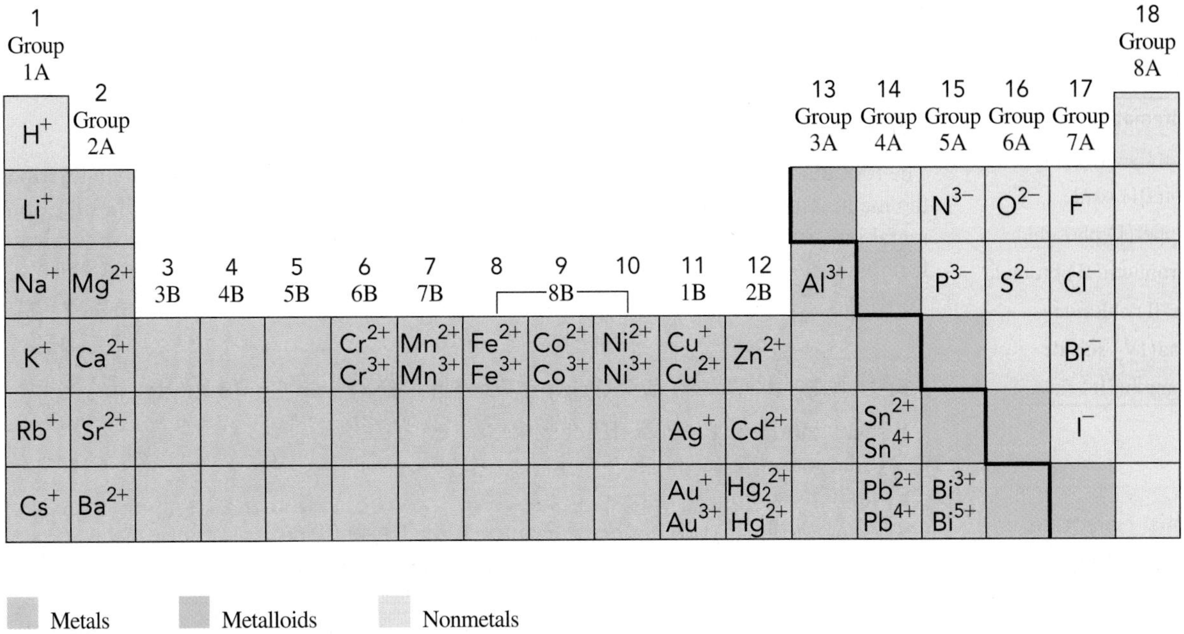

FIGURE 6.2 ▶ In ionic compounds, metals form positive ions, and nonmetals form negative ions.

Q What are the typical ions of calcium, copper, and oxygen in ionic compounds?

SAMPLE PROBLEM 6.5 Naming Ionic Compounds with Variable Charge Metal Ions

Antifouling paint contains Cu_2O, which prevents the growth of barnacles and algae on the bottoms of boats. What is the name of Cu_2O?

SOLUTION

STEP 1 Determine the charge of the cation from the anion.

ANALYZE THE PROBLEM		Metal		Nonmetal
	Formula		Cu_2O	
	Elements	copper (Cu)		oxygen (O)
	Groups	transition element		6A (16)
	Ions	Cu?		O^{2-}
	Charge Balance	2Cu?	+	$2- = 0$
		$\frac{2Cu?}{2} = \frac{2+}{2} = 1+$		
	Ions	Cu^{+}		O^{2-}

STEP 2 Name the cation by its element name and use a Roman numeral in parentheses for the charge. copper(I)

STEP 3 Name the anion by using the first syllable of its element name followed by *ide*. oxide

STEP 4 Write the name for the cation first and the name for the anion second. copper(I) oxide

STUDY CHECK 6.5

Write the name for the compound with the formula Mn_2S_3.

The growth of barnacles is prevented by using a paint with Cu_2O on the bottom of a boat.

Guide to Naming Ionic Compounds with Variable Charge Metals

STEP 1 Determine the charge of the cation from the anion.

STEP 2 Name the cation by its element name, and use a Roman numeral in parentheses for the charge.

STEP 3 Name the anion by using the first syllable of its element name followed by *ide*.

STEP 4 Write the name for the cation first and the name for the anion second.

TABLE 6.6 Some Ionic Compounds of Metals That Form Two Kinds of Positive Ions

Compound	Systematic Name
$FeCl_2$	Iron(II) chloride
Fe_2O_3	Iron(III) oxide
Cu_3P	Copper(I) phosphide
$CrBr_2$	Chromium(II) bromide
$SnCl_2$	Tin(II) chloride
PbS_2	Lead(IV) sulfide
BiF_3	Bismuth(III) fluoride

Table 6.6 lists the names of some ionic compounds in which the transition elements and metals from Groups 4A (14) and 5A (15) have more than one positive ion.

Writing Formulas from the Name of an Ionic Compound

The formula for an ionic compound is written from the first part of the name that describes the metal ion including its charge, and the second part of the name, that specifies the nonmetal ion. Subscripts are added, as needed, to balance the charge. The steps for writing a formula from the name of an ionic compound are shown in Sample Problem 6.6.

Guide to Writing Formulas from the Name of an Ionic Compound

STEP 1 Identify the cation and anion.

STEP 2 Balance the charges.

STEP 3 Write the formula, cation first, using subscripts from the charge balance.

SAMPLE PROBLEM 6.6 Writing Formulas for Ionic Compounds

Write the formula for iron(III) chloride.

SOLUTION

STEP 1 **Identify the cation and anion.** Iron(III) chloride consists of a metal and a nonmetal, which means that it as an ionic compound.

ANALYZE THE PROBLEM	Type of Ion	Cation	Anion
	Name	iron(III)	chloride
	Group	transition element	7A (17)
	Symbol of Ion	Fe^{3+}	Cl^-

STEP 2 **Balance the charges.**

$$\begin{array}{ll} Fe^{3+} & Cl^- \\ & Cl^- \\ & Cl^- \\ \hline \end{array}$$

$$\mathbf{1}(3+) + \mathbf{3}(1-) = 0$$

↑ Becomes a subscript in the formula

STEP 3 **Write the formula, cation first, using subscripts from the charge balance.**

$FeCl_3$

STUDY CHECK 6.6

Write the correct formula for chromium(III) oxide.

The pigment chrome oxide green contains chromium(III) oxide.

QUESTIONS AND PROBLEMS

6.3 Naming Ionic Compounds

LEARNING GOAL Given the formula of an ionic compound, write the correct name; given the name of an ionic compound, write the correct formula.

6.15 Write the names of each of the following ionic compound:
a. K_2O **b.** $BeCl_2$ **c.** $AlBr_3$
d. $CsCl$ **e.** CaO **f.** Ca_3P_2

6.16 Write the name for each of the following ionic compounds:
a. $MgCl_2$ **b.** K_3P **c.** Li_2S
d. CsF **e.** MgO **f.** $SrBr_2$

6.17 Write the name of each of the following ions (include the Roman numeral when necessary):
a. Fe^{+3} **b.** Ti^{+2} **c.** Co^{+3}
d. Mn^{+2} **e.** Hg^{+2} **f.** Cu^{+2}

6.18 Write the name for each of the following ions (include the Roman numeral when necessary):
a. Ag^+ **b.** Cu^+ **c.** Bi^{3+}
d. Sn^{2+} **e.** Au^{3+} **f.** Ni^{2+}

6.19 Write the name for each of the following ionic compounds:
a. Fe_2O_3 **b.** FeO **c.** Cu_2S
d. CuS **e.** $CdBr_2$ **f.** $PdBr_2$

6.20 Write the name for each of the following ionic compounds:
a. Ag_3P **b.** PbS **c.** SnO_2
d. $MnCl_3$ **e.** Bi_2O_3 **f.** $CoCl_2$

6.21 Write the symbol for the cation in each of the following ionic compounds:
a. $AgCl$ **b.** HgI_2 **c.** MnO **d.** $ZnCl_2$

6.22 Write the symbol for the cation in each of the following ionic compounds:
a. $FeCl_2$ **b.** CrO **c.** Ni_2S_3 **d.** AlP

6.23 Write the formula for each of the following ionic compounds:
a. magnesium chloride **b.** sodium sulfide
c. copper(I) oxide **d.** zinc phosphide
e. gold(III) nitride **f.** cobalt(III) fluoride

6.24 Write the formula for each of the following ionic compounds:
a. nickel(III) oxide **b.** barium fluoride
c. tin(IV) chloride **d.** silver sulfide
e. bismuth(V) chloride **f.** potassium nitride

6.25 Write the formula for each of the following ionic compounds:
a. cobalt(III) chloride **b.** lead(IV) oxide
c. silver iodide **d.** calcium nitride
e. copper(I) phosphide **f.** chromium(II) chloride

6.26 Write the chemical formula for each of the following compounds:
a. zinc bromide **b.** iron(III) sulfide
c. manganese(IV) oxide **d.** potassium permanganate
e. lithium nitride **f.** lithium aluminum hydride(Hydride: H^-)

6.4 Polyatomic Ions

LEARNING GOAL

Write the name and formula for an ionic compound containing a polyatomic ion.

An ionic compound may also contain a *polyatomic ion* as one of its cations or anions. A **polyatomic ion** is a group of covalently bonded atoms that has an overall ionic charge. Most polyatomic ions consist of a nonmetal such as phosphorus, sulfur, carbon, or nitrogen covalently bonded to oxygen atoms.

Almost all the polyatomic ions are anions with charges of $1-$, $2-$, or $3-$. Only one common polyatomic ion, NH_4^+, has a positive charge. Some models of common polyatomic ions are shown in Figure 6.3.

Names of Polyatomic Ions

The names of the most common polyatomic ions end in *ate*, such as nitrate and sulfate. When a related ion has one less oxygen atom, the *ite* ending is used for its name such as nitrite and sulfite. Recognizing these endings will help you identify polyatomic ions in the name of a compound. The hydroxide ion (OH^-) and cyanide ion (CN^-) are exceptions to this naming pattern.

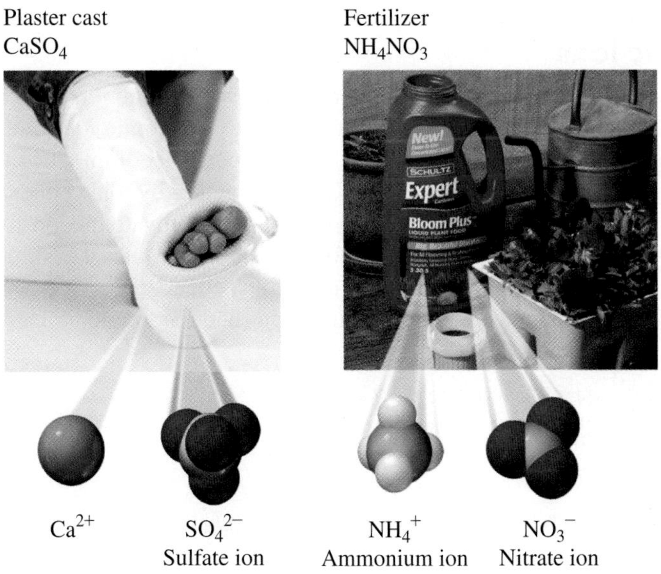

FIGURE 6.3 ▶ Many products contain polyatomic ions, which are groups of atoms that have an ionic charge.

Q What is the charge of a sulfate ion?

TABLE 6.7 Names and Formulas of Some Common Polyatomic Ions

Nonmetal	Formula of Ion*	Name of Ion
Hydrogen	OH^-	Hydroxide
Nitrogen	NH_4^+	Ammonium
	$\mathbf{NO_3^-}$	**Nitrate**
	NO_2^-	Nitrite
Chlorine	ClO_4^-	Perchlorate
	$\mathbf{ClO_3^-}$	**Chlorate**
	ClO_2^-	Chlorite
	ClO^-	Hypochlorite
Carbon	$\mathbf{CO_3^{2-}}$	**Carbonate**
	HCO_3^-	Hydrogen carbonate (or bicarbonate)
	CN^-	Cyanide
	$C_2H_3O_2^-$	Acetate
Sulfur	$\mathbf{SO_4^{2-}}$	**Sulfate**
	HSO_4^-	Hydrogen sulfate (or bisulfate)
	SO_3^{2-}	Sulfite
	HSO_3^-	Hydrogen sulfite (or bisulfite)
Phosphorus	$\mathbf{PO_4^{3-}}$	**Phosphate**
	HPO_4^{2-}	Hydrogen phosphate
	$H_2PO_4^-$	Dihydrogen phosphate
	PO_3^{3-}	Phosphite

*Formulas and names in bold type indicate the most common polyatomic ion for that element.

By learning the formulas, charges, and names of the polyatomic ions shown in bold type in Table 6.7, you can derive the related ions. Note that both the *ate* ion and *ite* ion of a particular nonmetal have the same ionic charge. For example, the sulfate ion is SO_4^{2-}, and the sulfite ion, which has one less oxygen atom, is SO_3^{2-}. Phosphate and phosphite ions each have a 3− charge; nitrate and nitrite each have a 1− charge; and perchlorate, chlorate, chlorite, and hypochlorite all have a 1− charge. The halogens form four different polyatomic ions with oxygen.

The formula of hydrogen carbonate, or *bicarbonate*, is written by placing a hydrogen in front of the polyatomic ion formula for carbonate (CO_3^{2-}), and the charge is decreased from 2− to 1− to give HCO_3^-.

$$H^+ + CO_3^{2-} = HCO_3^-$$

Sodium chlorite is used in the processing and bleaching of pulp from wood fibers and recycled cardboard.

Writing Formulas for Compounds Containing Polyatomic Ions

No polyatomic ion exists by itself. Like any ion, a polyatomic ion must be associated with ions of opposite charge. The bonding between polyatomic ions and other ions is one of electrical attraction. For example, the compound sodium chlorite consists of sodium ions (Na^+) and chlorite ions (ClO_2^-) held together by ionic bonds.

To write correct formulas for compounds containing polyatomic ions, we follow the same rules of charge balance that we used for writing the formulas for simple ionic compounds. The total negative and positive charges must equal zero. For example, consider the formula for a compound containing sodium ions and chlorite ions. The ions are written as

	Na^+	ClO_2^-
	Sodium ion	Chlorite ion
Ionic charge:	(1+) +	(1−) = 0

Because one ion of each balances the charge, the formula is written as

$$NaClO_2$$

Sodium chlorite

When more than one polyatomic ion is needed for charge balance, parentheses are used to enclose the formula of the ion. A subscript is written outside the right parenthesis of the polyatomic ion to indicate the number needed for charge balance. Consider the formula for magnesium nitrate. The ions in this compound are the magnesium ion and the nitrate ion, a polyatomic ion.

Mg^{2+}	NO_3^-
Magnesium ion	Nitrate ion

To balance the positive charge of 2+ on the magnesium ion, two nitrate ions are needed. In the formula of the compound, parentheses are placed around the nitrate ion, and the subscript 2 is written outside the right parenthesis.

$$Mg^{2+} \quad NO_3^- \quad NO_3^-$$

$$(2+) + 2(1-) = 0$$

Magnesium nitrate

$Mg(NO_3)_2$

Parentheses enclose the formula of the nitrate ion

Subscript outside the parenthesis indicates the use of two nitrate ions

SAMPLE PROBLEM 6.7 Writing Formulas Containing Polyatomic Ions

Write the formula for aluminum bicarbonate.

SOLUTION

STEP 1 Identify the cation and polyatomic ion (anion).

Cation	Polyatomic Ion (anion)
aluminum	bicarbonate
Al^{3+}	HCO_3^-

STEP 2 Balance the charges.

Al^{3+} $\quad HCO_3^-$ $\quad HCO_3^-$ $\quad HCO_3^-$

$$\mathbf{1}(3+) + \mathbf{3}(1-) = 0$$

↑ Becomes a subscript in the formula

STEP 3 **Write the formula, cation first, using the subscripts from charge balance.** The formula for the compound is written by enclosing the formula of the bicarbonate ion, HCO_3^-, in parentheses and writing the subscript 3 outside the right parenthesis.

$Al(HCO_3)_3$

STUDY CHECK 6.7

Write the formula for a compound containing ammonium ions and phosphate ions.

Guide to Writing Formulas with Polyatomic Ions

STEP 1 Identify the cation and polyatomic ion (anion).

STEP 2 Balance the charges.

STEP 3 Write the formula, cation first, using the subscripts from charge balance.

Naming Ionic Compounds Containing Polyatomic Ions

When naming ionic compounds containing polyatomic ions, we first write the positive ion, usually a metal, and then we write the name for the polyatomic ion. It is important that you learn to recognize the polyatomic ion in the formula and name it correctly. As with other ionic compounds, no prefixes are used.

Na_2SO_4	$FePO_4$	$Al_2(CO_3)_3$
$Na_2\boxed{SO_4}$	$Fe\boxed{PO_4}$	$Al_2(\boxed{CO_3})_3$
Sodium sulfate	Iron(III) phosphate	Aluminum carbonate

Table 6.8 lists the formulas and names of some ionic compounds that include polyatomic ions and also gives their uses in medicine and industry.

TABLE 6.8 Some Compounds That Contain Polyatomic Ions

Formula	Name	Use
$BaSO_4$	Barium sulfate	Contrast medium for X-rays
$CaCO_3$	Calcium carbonate	Antacid, calcium supplement
$Ca_3(PO_4)_2$	Calcium phosphate	Calcium dietary supplement
$CaSO_3$	Calcium sulfite	Preservative in cider and fruit juices
$CaSO_4$	Calcium sulfate	Plaster casts
$AgNO_3$	Silver nitrate	Topical anti-infective
$NaHCO_3$	Sodium bicarbonate *or* sodium hydrogen carbonate	Antacid
$Zn_3(PO_4)_2$	Zinc phosphate	Dental cement
$FePO_4$	Iron(III) phosphate	Food additive
K_2CO_3	Potassium carbonate	Alkalizer, diuretic
$Al_2(SO_4)_3$	Aluminum sulfate	Antiperspirant, anti-infective
$AlPO_4$	Aluminum phosphate	Antacid
$MgSO_4$	Magnesium sulfate	Cathartic, Epsom salts

SAMPLE PROBLEM 6.8 Naming Compounds Containing Polyatomic Ions

Name the following ionic compounds:

a. $Cu(NO_2)_2$ **b.** $KClO_3$

SOLUTION

Formula	STEP 1 Cation	Anion	STEP 2 Name of Cation	STEP 3 Name of Anion	STEP 4 Name of Compound
a. $Cu(NO_2)_2$	Cu^{2+}	NO_2^-	Copper(II) ion	Nitrite ion	Copper(II) nitrite
b. $KClO_3$	K^+	ClO_3^-	Potassium ion	Chlorate ion	Potassium chlorate

STUDY CHECK 6.8

What is the name of $Co_3(PO_4)_2$?

Guide to Naming Ionic Compounds with Polyatomic Ions

STEP 1 Identify the cation and polyatomic ion (anion).

STEP 2 Name the cation using a Roman numeral, if needed.

STEP 3 Name the polyatomic ion.

STEP 4 Write the name for the compound, cation first and the polyatomic ion second.

Summary of Naming Ionic Compounds

In these sections, we have examined strategies for naming ionic compounds. Now we can summarize the rules, as illustrated in Figure 6.4. In general, ionic compounds that have two elements are named by stating the first element, followed by the second element with an *ide* ending. For ionic compounds, it is necessary to determine whether the metal can form more than one positive ion; if so, a Roman numeral following the name of the metal indicates the particular ionic charge. Ionic compounds with three or more elements include some type of polyatomic ion. They are named by ionic rules but usually have an *ate* or *ite* ending when the polyatomic ion has a negative charge.

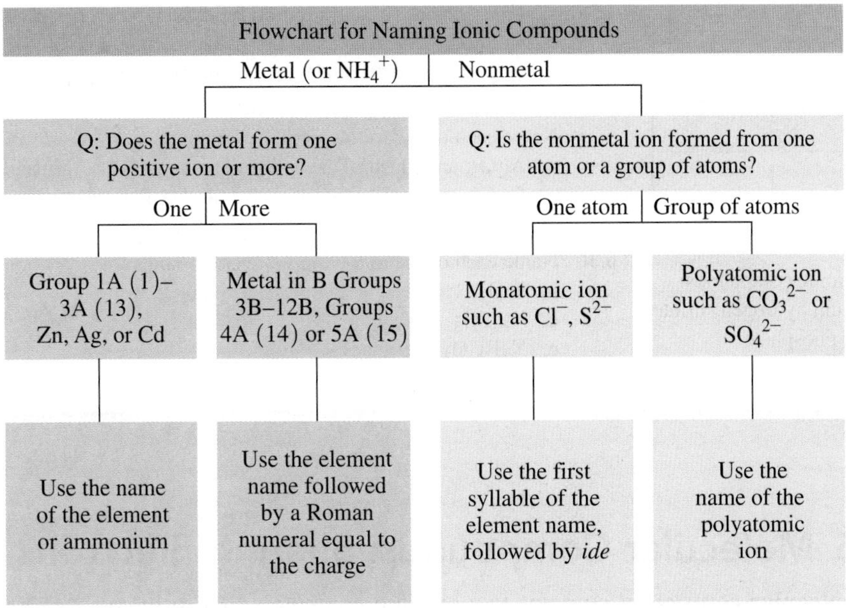

FIGURE 6.4 ▶ A flowchart illustrates the naming for ionic compounds.

Q Why are the names of some metal ions followed by a Roman numeral in the name of a compound?

SAMPLE PROBLEM 6.9 Naming Ionic Compounds

Name the following ionic compounds:

a. Na_3P **b.** $CuSO_4$ **c.** $Cr(ClO)_3$

SOLUTION

	STEP 1		STEP 2	STEP 3	STEP 4
Formula	Cation	Anion	Name of Cation	Name of Anion	Name of Compound
a. Na_3P	Na^+	P^{3-}	Sodium ion	Phosphide ion	Sodium phosphide
b. $CuSO_4$	Cu^{2+}	SO_4^{2-}	Copper(II) ion	Sulfate ion	Copper(II) sulfate
c. $Cr(ClO)_3$	Cr^{3+}	ClO^-	Chromium(III) ion	Hypochlorite ion	Chromium(III) hypochlorite

STUDY CHECK 6.9

What is the name of $Fe(NO_3)_2$?

QUESTIONS AND PROBLEMS

6.4 Polyatomic Ions

LEARNING GOAL Write the name and formula for an ionic compound containing a polyatomic ion.

6.27 Write the formula including the charge for each of the following polyatomic ions:
a. hydrogen carbonate (bicarbonate) **b.** ammonium
c. phosphite **d.** chlorate

6.28 Write the formula including the charge for each of the following polyatomic ions:
a. nitrite **b.** sulfite
c. hydroxide **d.** acetate

6.29 Name the following polyatomic ions:
a. NO_3^- **b.** ClO_4^- **c.** HSO_4^- **d.** PO_3^{3-}

6.30 Name the following polyatomic ions:
a. OH^- **b.** PO_4^{3-} **c.** CN^- **d.** NO_2^-

6.31 Complete the following table with the formula and name of the compound that forms between each pair of ions:

	NO_2^-	CO_3^{2-}	HSO_4^-	PO_4^{3-}
Li^+				
Cu^{2+}				
Ba^{2+}				

6.32 Complete the following table with the formula and name of the compound that forms between each pair of ions:

	NO_3^-	HCO_3^-	SO_3^{2-}	HPO_4^{2-}
NH_4^+				
Al^{3+}				
Pb^{4+}				

6.33 Write the correct formula for the following ionic compounds:
a. barium hydroxide
b. sodium hydrogen sulfate
c. iron(II) nitrite
d. zinc phosphate
e. iron(III) carbonate

6.34 Write the correct formula for the following ionic compounds:
a. aluminum chlorate
b. ammonium oxide
c. magnesium bicarbonate
d. sodium nitrite
e. copper(I) sulfate

6.35 Name each of the following ionic compounds:
a. Al_2O_3, an industrial catalyst
b. $Mg_3(PO_4)_2$, antacid
c. Cr_2O_3, green pigment
d. Na_3PO_4, laxative
e. Bi_2S_3, found in bismuth ore

6.36 Name each of the following ionic compounds:
a. SeO_2
b. K_2SO_4
c. Na_2SO_3
d. $SnCl_4$
e. $NaHCO_3$

LEARNING GOAL

Given the formula of a molecular compound, write its correct name; given the name of a molecular compound, write its formula.

6.5 Molecular Compounds: Sharing Electrons

A **molecular compound** contains two or more nonmetals that form *covalent bonds*. Because nonmetals have high ionization energies, valence electrons are shared by nonmetal atoms to achieve stability. When atoms share electrons, the bond is a **covalent bond**. When two or more atoms share electrons, they form a **molecule**.

Formation of a Hydrogen Molecule

The simplest molecule is hydrogen, H_2. When two H atoms are far apart, there is no attraction between them. As the H atoms move closer, the positive charge of each nucleus attracts the electron of the other atom. This attraction, which is greater than the repulsion between the valence electrons, pulls the H atoms closer until they share a pair of valence electrons (see Figure 6.5). The result is called a *covalent bond*, in which the shared electrons give the noble gas arrangement of He to *each* of the H atoms. When the H atoms form H_2, they are more stable than two individual H atoms.

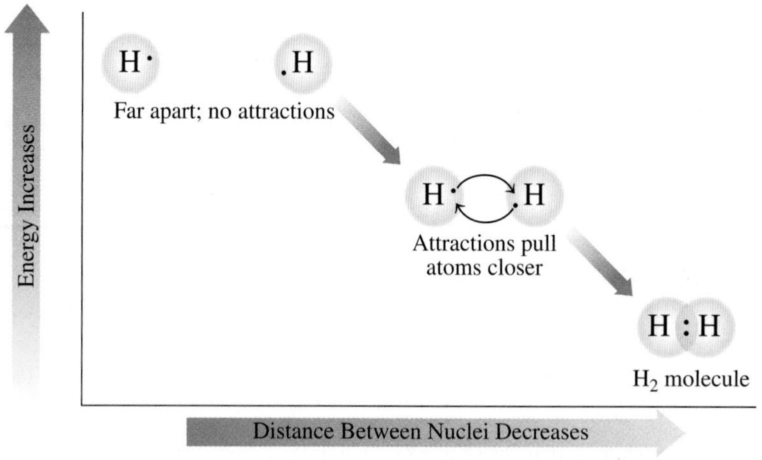

FIGURE 6.5 ▶ A covalent bond forms as H atoms move close together to share electrons.

Q What determines the attraction between two H atoms?

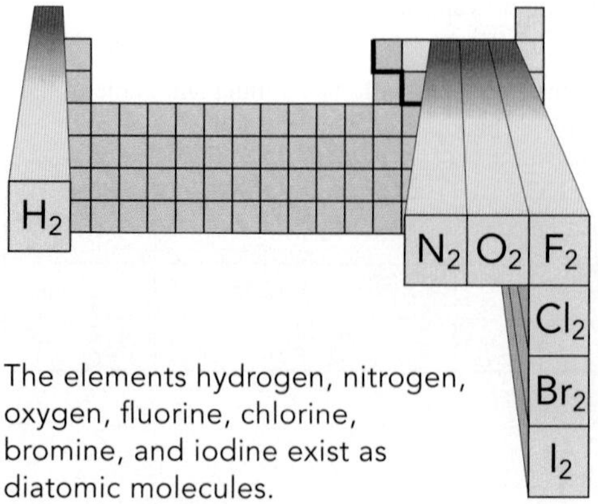

The elements hydrogen, nitrogen, oxygen, fluorine, chlorine, bromine, and iodine exist as diatomic molecules.

Electron-Dot Formulas of Molecular Compounds

The valence electrons in molecules are shown using an *electron-dot formula*, also called a Lewis structure. The shared electrons, or *bonding pairs*, are shown as two dots or a single line between atoms. The nonbonding pairs of electrons, or *lone pairs*, are placed on the outside. For example, a fluorine molecule, F_2, consists of two fluorine atoms, which are in Group 7A (17), each with seven valence electrons. In the F_2 molecule, each F atom achieves an octet by sharing its unpaired valence electron.

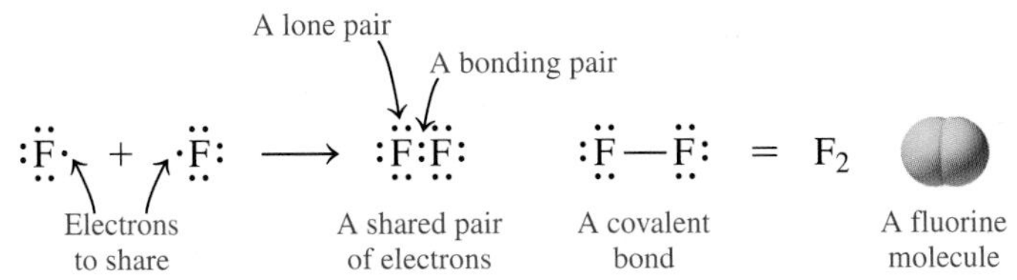

Hydrogen (H_2) and fluorine (F_2) are examples of nonmetal elements whose natural state is diatomic; that is, they contain two like atoms. The elements that exist as diatomic molecules are listed in Table 6.9.

TABLE 6.9 Elements That Exist as Diatomic Molecules

Diatomic Molecule	Name
H_2	Hydrogen
N_2	Nitrogen
O_2	Oxygen
F_2	Fluorine
Cl_2	Chlorine
Br_2	Bromine
I_2	Iodine

Sharing Electrons Between Atoms of Different Elements

The number of electrons that a nonmetal atom shares and the number of covalent bonds it forms are usually equal to the number of electrons it needs to achieve a noble gas arrangement. Table 6.10 gives the most typical bonding patterns for some nonmetals. For example, the element carbon combines with hydrogen to form a molecular compound, CH_4, methane, which is a component of natural gas.

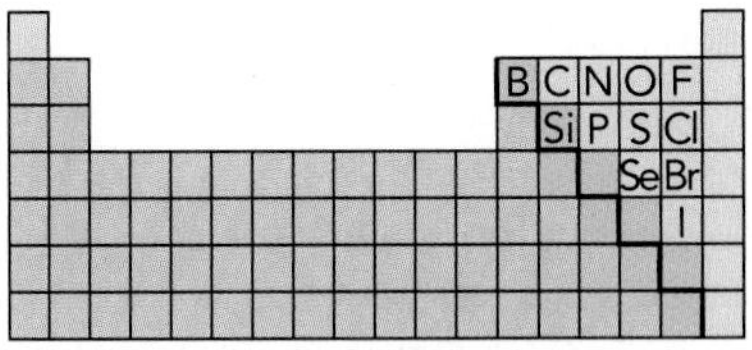

Examples: CO_2, SF_2, PCl_3, CH_4

The number of covalent bonds that a nonmetal atom forms is usually equal to the number of electrons it needs to achieve a stable electron arrangement.

TABLE 6.10 Typical Bonding Patterns of Some Nonmetals

1A (1)	3A (13)	4A (14)	5A (15)	6A (16)	7A (17)
*H 1 bond					
	*B 3 bonds	C 4 bonds	N 3 bonds	O 2 bonds	F 1 bond
		Si 4 bonds	P 3 bonds	S 2 bonds	Cl, Br, I 1 bond

*H and B do not form octets. H atoms share one electron pair; B atoms share three electron pairs for a set of six electrons.

CORE CHEMISTRY SKILL

Drawing Electron-Dot Formulas

Drawing Electron-Dot Formulas

To draw the electron-dot formula for CH_4, we first draw the electron-dot symbols for carbon and hydrogen.

Then we can determine the number of valence electrons needed for carbon and hydrogen. When a carbon atom shares its four electrons with four hydrogen atoms, carbon obtains an octet and each hydrogen atom is complete with two shared electrons. The electron-dot formula is drawn with the carbon atom as the central atom with the hydrogen atoms on each of the sides. The bonding pairs of electrons, which are single covalent bonds, may also be shown as single lines between the carbon atom and each of the hydrogen atoms.

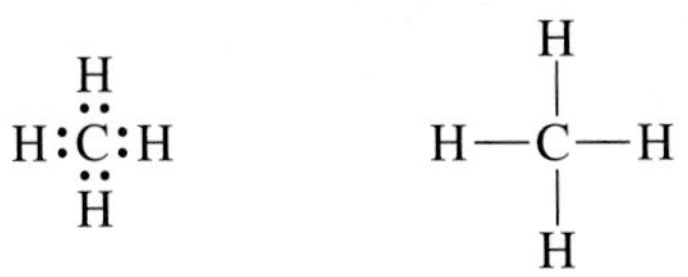

Table 6.11 gives the formulas and three-dimensional models of some molecules.

TABLE 6.11 Electron-Dot Formulas for Some Molecular Compounds

CH_4	NH_3	H_2O

Electron-Dot Formulas

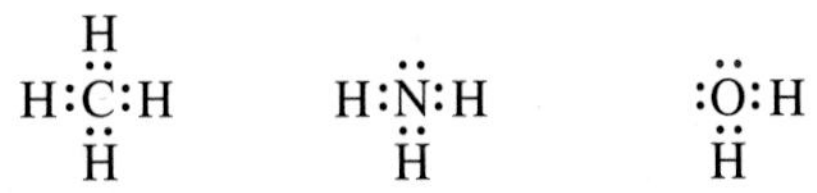

Formulas Using Bonds and Electron Dots

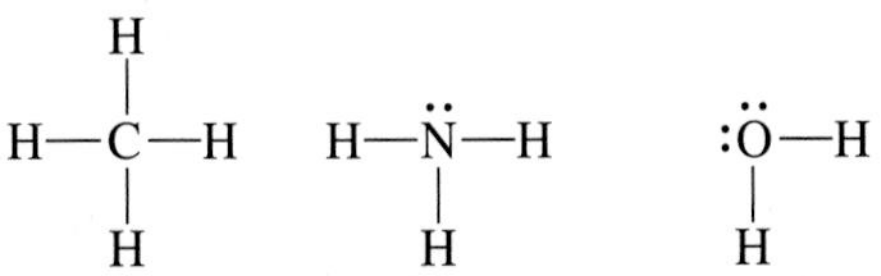

Molecular Models

Methane molecule	Ammonia molecule	Water molecule

SAMPLE PROBLEM 6.10 Drawing Electron-Dot Formulas

Use the electron-dot symbols of S and F to draw the electron-dot formula for SF_2, in which sulfur is the central atom.

SOLUTION

To draw the electron-dot formula for SF_2, we first draw the electron-dot symbols for sulfur with six valence electrons and fluorine with seven valence electrons.

Then we can determine the number of valence electrons needed for sulfur and fluorine. A sulfur atom will share each of its two unpaired electrons with the unpaired electrons of two fluorine atoms. In this way, both the S atom and the two F atoms achieve octets. The electron-dot formula for SF_2 is drawn with the sulfur atom with two fluorine atoms attached with paired electrons.

STUDY CHECK 6.10

Draw the electron-dot formula for PH_3.

Exceptions to the Octet Rule

While the octet rule is useful, there are exceptions. We have already seen that a hydrogen (H_2) molecule requires just two electrons or a single bond to achieve the stability of the nearest noble gas, helium. Although the nonmetals typically form octets, atoms such as P, S, Cl, Br, and I can form compounds with 10 or 12 valence electrons. For example, in PCl_3, the P atom has an octet, but in PCl_5, the P atom has 10 valence electrons or five bonds. In H_2S, the S atom has an octet, but in SF_6, there are 12 valence electrons or six bonds to the sulfur atom. In this text, we will encounter formulas with expanded octets, but we do not represent them with electron-dot formulas.

Double and Triple Bonds

Up to now, we have looked at bonding in molecules having only single bonds. In many molecular compounds, atoms share two or three pairs of electrons to complete their octets. A **double bond** occurs when two pairs of electrons are shared; in a **triple bond**, three pairs of electrons are shared. Atoms of carbon, oxygen, nitrogen, and sulfur are most likely to form multiple bonds.

Double and triple bonds form when the number of valence electrons is not enough to complete the octets of all the atoms in the molecule. Then one or more lone pairs of electrons from the atoms attached to the central atom are shared with the central atom.

For example, there are double bonds in CO_2 because two pairs of electrons are shared between the carbon atom and each oxygen atom to give octets. Atoms of hydrogen and the halogens do not form double or triple bonds. The process of drawing an electron-dot formula with multiple bonds is shown in Sample Problem 6.11.

SAMPLE PROBLEM 6.11 Drawing Electron-Dot Formulas with Multiple Bonds

Draw the electron-dot formula for carbon dioxide, CO_2, in which the central atom is C.

SOLUTION

STEP 1 Determine the arrangement of atoms. O C O

STEP 2 Determine the total number of valence electrons. Using the group number to determine valence electrons, the carbon atom has four valence electrons, and each oxygen atom has six valence electrons, which gives a total of 16 valence electrons.

Element	Group	Atoms	Valence Electrons	=	Total
C	4A (14)	1 C	$\times 4\ e^-$	=	$4\ e^-$
O	6A (16)	2 O	$\times 6\ e^-$	=	$12\ e^-$
	Total valence electrons for CO_2			=	$16\ e^-$

STEP 3 **Attach each bonded atom to the central atom with a pair of electrons.**

O:C:O or O—C—O

STEP 4 **Place the remaining electrons to complete octets.** Because we used four valence electrons to attach the C atom to two O atoms, there are 12 valence electrons remaining.

16 valence $e^- - 4$ bonding $e^- = 12\ e^-$ remaining

The 12 remaining electrons are placed as six lone pairs of electrons on the outside O atoms. However, this does not complete the octet of the C atom.

:Ö:C:Ö: or :Ö—C—Ö:

To obtain an octet, the C atom must share lone pairs of electrons from each of the O atoms. When two bonding pairs occur between atoms, it is known as a double bond.

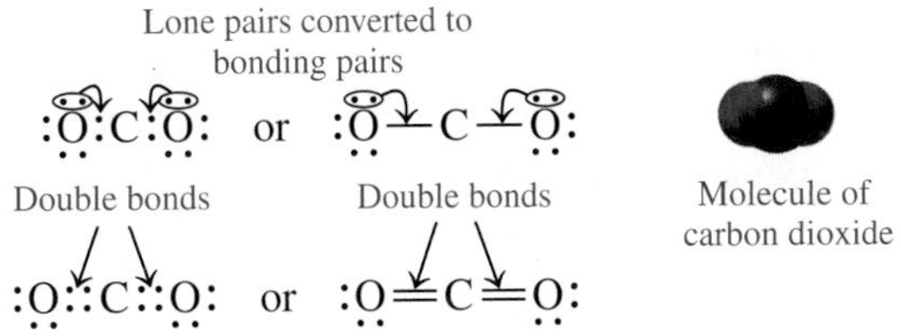

STUDY CHECK 6.11

Draw the electron-dot formula for SO_2. The atoms are arranged as O S O.

Forming Triple Bonds

The molecule N_2 contains a triple bond. Because nitrogen is in Group 5A (15), each N atom has five valence electrons. An octet for each N atom cannot be achieved by sharing only one or two pairs of electrons. However, each N atom achieves an octet by sharing three bonding pairs of electrons to form a triple bond.

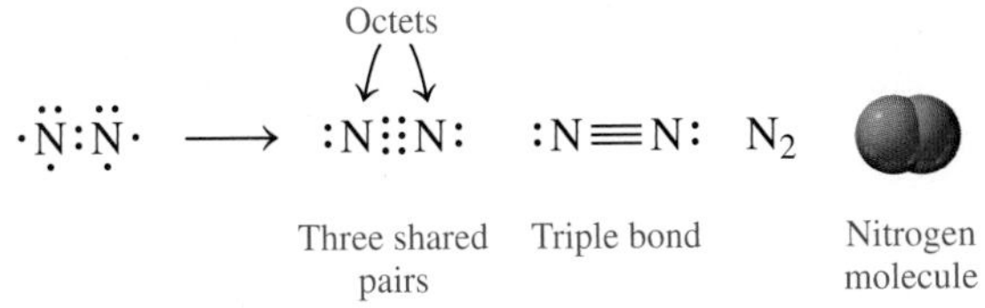

Names and Formulas of Molecular Compounds

When naming a molecular compound, the first nonmetal in the formula is named by its element name; the second nonmetal is named using the first syllable of its element name, followed by *ide*. When a subscript indicates two or more atoms of an element, a prefix is shown in front of its name. Table 6.12 lists prefixes used in naming molecular compounds.

Guide to Drawing Electron-Dot Formulas

STEP 1 Determine the arrangement of atoms.

STEP 2 Determine the total number of valence electrons.

STEP 3 Attach each bonded atom to the central atom with a pair of electrons.

STEP 4 Place the remaining electrons to complete octets (two for H).

CORE CHEMISTRY SKILL

Writing the Names and Formulas for Molecular Compounds

TABLE 6.12 Prefixes Used in Naming Molecular Compounds

1 mono	6 hexa
2 di	7 hepta
3 tri	8 octa
4 tetra	9 nona
5 penta	10 deca

The names of molecular compounds need prefixes because several different compounds can be formed from the same two nonmetals. For example, carbon and oxygen can form two different compounds, carbon monoxide, CO, and carbon dioxide, CO_2, in which the number of atoms of oxygen in each compound is indicated by the prefixes *mono* or *di* in their names.

When the vowels *o* and *o* or *a* and *o* appear together, the first vowel is omitted, as in carbon monoxide. In the name of a molecular compound, the prefix *mono* is usually omitted, as in NO, nitrogen oxide. Traditionally, however, CO is named carbon monoxide. Table 6.13 lists the formulas, names, and commercial uses of some molecular compounds.

TABLE 6.13 Some Common Molecular Compounds

Formula	Name	Commercial Uses
CS_2	Carbon disulfide	Manufacture of rayon
CO_2	Carbon dioxide	Fire extinguishers, dry ice, propellant in aerosols, carbonation of beverages
NO	Nitrogen oxide	Stabilizer
N_2O	Dinitrogen oxide	Inhalation anesthetic, "laughing gas"
SO_2	Sulfur dioxide	Preserving fruits, vegetables; disinfectant in breweries; bleaching textiles
SO_3	Sulfur trioxide	Manufacture of explosives
SF_6	Sulfur hexafluoride	Electrical circuits

Guide to Naming Molecular Compounds

STEP 1
Name the first nonmetal by its element name.

STEP 2
Name the second nonmetal by using the first syllable of its element name followed by *ide*.

STEP 3
Add prefixes to indicate the number of atoms (subscripts).

SAMPLE PROBLEM 6.12 Naming Molecular Compounds

Name the molecular compound NCl_3.

SOLUTION

ANALYZE THE PROBLEM	Symbol of Element	N	Cl
	Name	nitrogen	chloride
	Subscript	1	3
	Prefix	none	tri

STEP 1 **Name the first nonmetal by its element name.** In NCl_3, the first nonmetal (N) is nitrogen.

STEP 2 **Name the second nonmetal by using the first syllable of its element name followed by *ide*.** The second nonmetal (Cl) is named chloride.

STEP 3 **Add prefixes to indicate the number of atoms (subscripts).** Because there is one nitrogen atom, no prefix is needed. The subscript 3 for the Cl atoms is shown as the prefix *tri*. The name of NCl_3 is nitrogen trichloride.

STUDY CHECK 6.12

Write the name for each of the following molecular compounds:

a. $SiBr_4$ **b.** Br_2O

Writing Formulas from the Names of Molecular Compounds

In the name of a molecular compound, the names of two nonmetals are given along with prefixes for the number of atoms of each. To write the formula from the name, we use the symbol for each element and a subscript if a prefix indicates two or more atoms.

SAMPLE PROBLEM 6.13 Writing Formulas for Molecular Compounds

Write the formula for the molecular compound diboron trioxide.

SOLUTION

STEP 1 **Write the symbols in the order of the elements in the name.**

ANALYZE THE PROBLEM	Name of Element	boron	oxygen
	Symbol of element	B	O
	Subscript	2 (di)	3 (tri)

STEP 2 **Write any prefixes as subscripts.** The prefix *di* in *di*boron indicates that there are two atoms of boron, shown as a subscript 2 in the formula. The prefix *tri* in *tri*oxide indicates that there are three atoms of oxygen, shown as a subscript 3 in the formula.

B_2O_3

STUDY CHECK 6.13

What is the formula of the molecular compound iodine pentafluoride?

Guide to Writing Formulas for Molecular Compounds

STEP 1
Write the symbols in the order of the elements in the name.

STEP 2
Write any prefixes as subscripts.

Summary of Naming Ionic and Molecular Compounds

We have now examined strategies for naming ionic and molecular compounds. In general, compounds having two elements are named by stating the first element name, followed by the name of the second element with an *ide* ending. If the first element is a metal, the compound is usually ionic; if the first element is a nonmetal, the compound is usually molecular. For ionic compounds, it is necessary to determine whether the metal can form more than one type of positive ion; if so, a Roman numeral following the name of the metal indicates the particular ionic charge. One exception is the ammonium ion, NH_4^+, which is also written first as a positively charged polyatomic ion. Ionic compounds having three or more elements include some type of polyatomic ion. They are named by ionic rules but have an *ate* or *ite* ending when the polyatomic ion has a negative charge.

In naming molecular compounds having two elements, prefixes are necessary to indicate two or more atoms of each nonmetal as shown in that particular formula (see Figure 6.6).

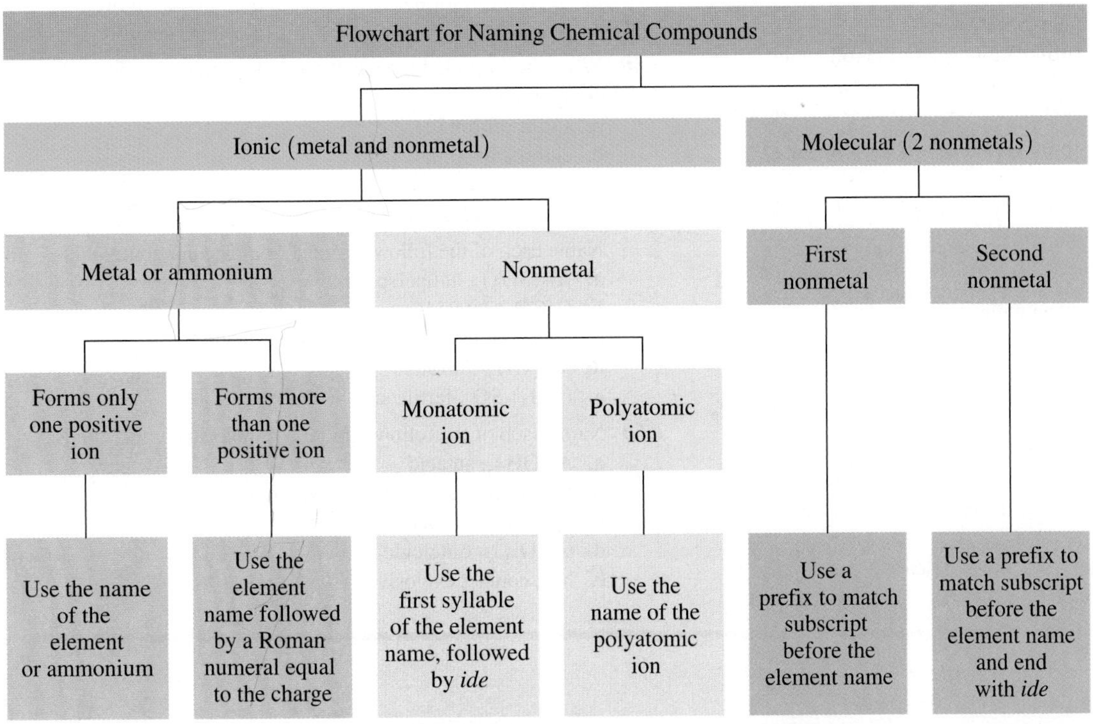

FIGURE 6.6 ▶ A flowchart illustrates naming for ionic and molecular compounds.

Q Why are the names of some metal ions followed by a Roman numeral in the name of a compound?

SAMPLE PROBLEM 6.14 Naming Ionic and Molecular Compounds

Identify each of the following compounds as ionic or molecular and give its name:

a. K_3P **b.** $NiSO_4$ **c.** SO_3

SOLUTION

a. K_3P, consisting of a metal and nonmetal, is an ionic compound. As a representative element in Group 1A (1), K forms the sodium ion, K^+. Phosphorus, as a representative element in Group 5A (15), forms a phosphide ion, P^{3-}. Writing the name for the cation followed by the name of the anion gives the name potassium phosphide.

b. $NiSO_4$, consisting of a cation of a transition element and a polyatomic ion SO_4^{2-}, is an ionic compound. As a transition element, Ni forms more than one type of ion. In this formula, the 2− charge of SO_4^{2-} is balanced by one nickel ion, Ni^{2+}. In the name, a Roman numeral written after the metal name, nickel(II), specifies the 2+ charge. The anion SO_4^{2-} is a polyatomic ion named sulfate. The compound is named nickel(II) sulfate.

c. SO_3 consists of two nonmetals, which indicates that it is a molecular compound. The first element S is *sulfur* (no prefix is needed). The second element O, *oxide*, has subscript 3, which requires a prefix *tri* in the name. The compound is named sulfur trioxide.

STUDY CHECK 6.14

What is the name of $Fe(NO_3)_3$?

QUESTIONS AND PROBLEMS

6.5 Molecular Compounds: Sharing Electrons

LEARNING GOAL Given the formula of a molecular compound, write its correct name; given the name of a molecular compound, write its formula.

6.37 What elements on the periodic table are most likely to form molecular compounds?

6.38 How does the bond that forms between Na and Cl differ from a bond that forms between N and Cl?

6.39 Draw the electron-dot formula for each of the following molecules:
a. Br_2 **b.** H_2 **c.** HF **d.** OF_2

6.40 Draw the electron-dot formula for each of the following molecules:
a. NI_3 **b.** CCl_4 **c.** Cl_2 **d.** SiF_4

6.41 Name each of the following molecular compounds:
a. PBr_3 **b.** Cl_2O **c.** CBr_4 **d.** HF **e.** NF_3

6.42 Name each of the following molecular compounds:
a. CS_2 **b.** P_2O_5 **c.** SiO_2 **d.** PCl_3 **e.** CO

6.43 Name each of the following molecular compounds:
a. N_2 **b.** Si_2Br_6 **c.** TiO_2 **d.** PCl_5 **e.** SeF_6

6.44 Name each of the following molecular compounds:
a. SiF_4 **b.** IBr_3 **c.** CO_2 **d.** N_2F_2 **e.** N_2S_3

6.45 Write the formula for each of the following molecular compounds:
a. carbon tetrachloride **b.** carbon monoxide
c. phosphorus trichloride **d.** dinitrogen tetroxide

6.46 Write the formula for each of the following molecular compounds:
a. sulfur dioxide **b.** silicon tetrachloride
c. iodine trifluoride **d.** dinitrogen oxide

6.47 Write the formula for each of the following molecular compounds:
a. oxygen difluoride **b.** boron trichloride
c. dinitrogen trioxide **d.** sulfur hexafluoride

6.48 Write the formula for each of the following molecular compounds:
a. sulfur dibromide
b. carbon disulfide
c. tetraphosphorus hexoxide
d. dinitrogen pentoxide

6.49 Name each of the following ionic or molecular compounds:
a. $Al_2(SO_4)_3$, antiperspirant
b. $CaCO_3$, antacid
c. N_2O, "laughing gas," inhaled anesthetic
d. $Mg(OH)_2$, laxative
e. $(NH_4)_2SO_4$, fertilizer

6.50 Name each of the following ionic or molecular compounds:
a. $Al(OH)_3$, antacid
b. $FeSO_4$, iron supplement in vitamins
c. N_2O_4, rocket fuel
d. $Cu(OH)_2$, fungicide
e. NI_3, contact explosive

6.6 Electronegativity and Bond Polarity

LEARNING GOAL

Use electronegativity to determine the polarity of a bond.

We can learn more about the chemistry of compounds by looking at how bonding electrons are shared between atoms. The bonding electrons are shared equally in a bond between identical nonmetal atoms. However, when a bond is between atoms of different elements, the electron pairs are usually shared unequally. Then the shared pairs of electrons are attracted to one atom in the bond more than the other.

The **electronegativity** of an atom is its ability to attract the shared electrons in a chemical bond. Nonmetals have higher electronegativities than do metals, because nonmetals have a greater attraction for electrons than metals. On the electronegativity scale, fluorine was assigned a value of 4.0, and the electronegativities for all other elements were determined relative to the attraction of fluorine for shared electrons. The nonmetal fluorine, which has the highest electronegativity (4.0), is located in the upper right corner of the periodic table. The metal cesium, which has the lowest electronegativity (0.7), is located in the lower left corner of the periodic table. The electronegativities for the representative elements are shown in Figure 6.7. Note that there are no electronegativity values for the noble gases because they do not typically form bonds. The electronegativity values for transition elements are also low, but we have not included them in our discussion.

CORE CHEMISTRY SKILL

Using Electronegativity

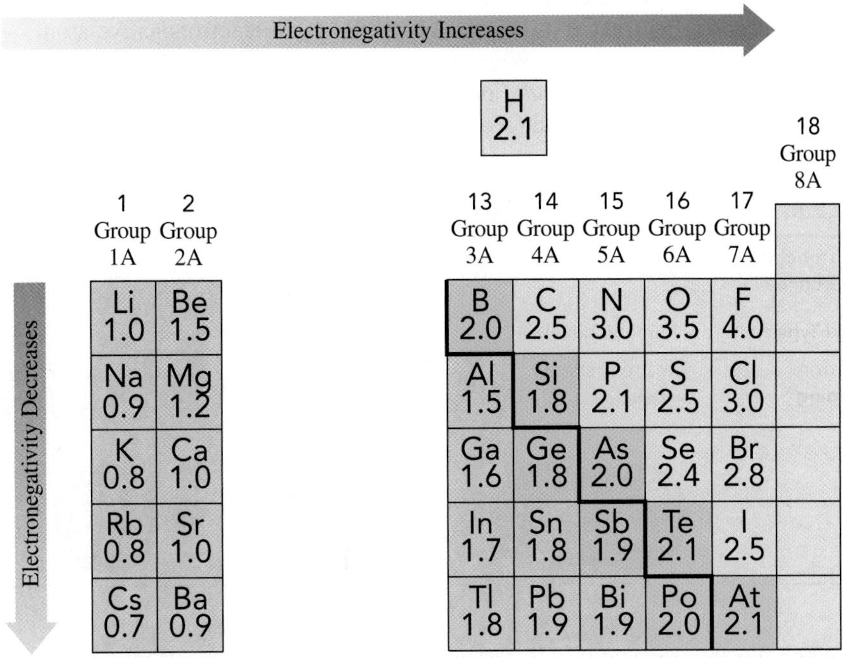

FIGURE 6.7 ▶ The electronegativity values of representative elements in Group 1A (1) to Group 7A (17), which indicate the ability of atoms to attract shared electrons, increase going across a period from left to right and decrease going down a group.

Q What element on the periodic table has the strongest attraction for shared electrons?

Polarity of Bonds

The difference in the electronegativity values of two atoms can be used to predict the type of chemical bond, ionic or covalent, that forms. For the H—H bond, the electronegativity difference is zero ($2.1 - 2.1 = 0$), which means the bonding electrons are shared equally. A bond between atoms with identical or very similar electronegativity values is a **nonpolar covalent bond**. However, when bonds are between atoms with different electronegativity values, the electrons are shared unequally; the bond is a **polar covalent bond**. For the H—Cl bond, there is an electronegativity difference of $3.0\,(\text{Cl}) - 2.1(\text{H}) = 0.9$, which means that the H—Cl bond is polar covalent (see Figure 6.8).

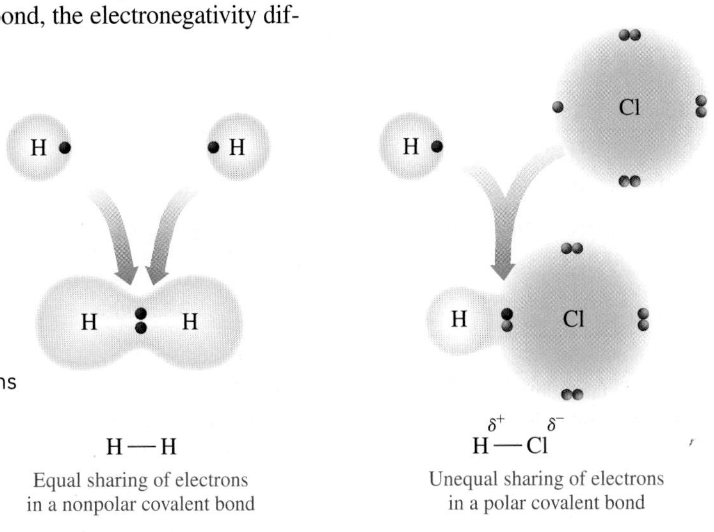

FIGURE 6.8 ▶ In the nonpolar covalent bond of H_2, electrons are shared equally. In the polar covalent bond of HCl, electrons are shared unequally.

Q H_2 has a nonpolar covalent bond, but HCl has a polar covalent bond. Explain.

Dipoles and Bond Polarity

The **polarity** of a bond depends on the difference in the electronegativity values of its atoms. In a polar covalent bond, the shared electrons are attracted to the more electronegative atom, which makes it partially negative, due to the negatively charged electrons around that atom. At the other end of the bond, the atom with the lower electronegativity becomes partially positive due to the lack of the electrons at that atom.

A bond becomes more *polar* as the electronegativity difference increases. A polar covalent bond that has a separation of charges is called a **dipole**. The positive and negative ends of the dipole are indicated by the lowercase Greek letter delta with a positive or negative sign, δ^+ and δ^-. Sometimes we use an arrow that points from the positive charge to the negative charge $\mapsto$ to indicate the dipole.

Examples of Dipoles in Polar Covalent Bonds

$\overset{\delta^+}{C}—\overset{\delta^-}{O}$ $\quad$ $\overset{\delta^+}{N}—\overset{\delta^-}{O}$ $\quad$ $\overset{\delta^+}{Cl}—\overset{\delta^-}{F}$

$\mapsto$ $\quad$ $\mapsto$ $\quad$ $\mapsto$

Variations in Bonding

The variations in bonding are continuous; there is no definite point at which one type of bond stops and the next starts. When the electronegativity difference is between 0.0 and 0.4, the electrons are considered to be shared equally in a *nonpolar covalent bond*. For example, the C—C bond $(2.5 - 2.5 = 0.0)$ and the C—H bond $(2.5 - 2.1 = 0.4)$ are classified as nonpolar covalent bonds. As the electronegativity difference increases, the shared electrons are attracted more strongly to the more electronegative atom, which increases the polarity of the bond. When the electronegativity difference is from 0.5 to 1.8, the bond is a *polar covalent bond*. For example, the O—H bond $(3.5 - 2.1 = 1.4)$ is classified as a polar covalent bond (see Table 6.14).

TABLE 6.14 Electronegativity Differences and Types of Bonds

Electronegativity Difference	0	0.4	1.8 → 3.3
Bond Type	Nonpolar covalent	Polar covalent	Ionic
Electron Bonding	Electrons shared equally	Electrons shared unequally	Electrons transferred
		δ^+ δ^-	$+$ $-$

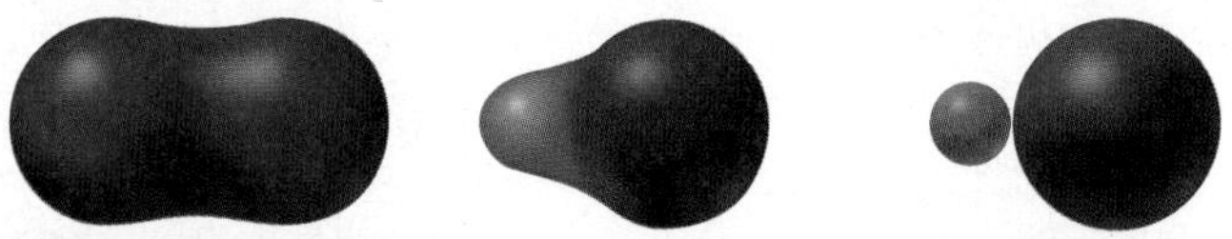

When the electronegativity difference is greater than 1.8, electrons are transferred from one atom to another, which results in an *ionic bond*. For example, the electronegativity difference for the ionic compound NaCl is $3.0 - 0.9 = 2.1$. Thus for large differences in electronegativity, we would predict an ionic bond (see Table 6.15).

TABLE 6.15 Predicting Bond Type from Electronegativity Differences

Molecule	Bond	Type of Electron Sharing	Electronegativity Difference*	Type of Bond	Reason
H_2	H—H	Shared equally	$2.1 - 2.1 = 0.0$	Nonpolar covalent	Between 0.0 and 0.4
BrCl	Br—Cl	Shared about equally	$3.0 - 2.8 = 0.2$	Nonpolar covalent	Between 0.0 and 0.4
HBr	$\overset{\delta^+}{H}—\overset{\delta^-}{Br}$	Shared unequally	$2.8 - 2.1 = 0.7$	Polar covalent	Greater than 0.4, but less than 1.8
HCl	$\overset{\delta^+}{H}—\overset{\delta^-}{Cl}$	Shared unequally	$3.0 - 2.1 = 0.9$	Polar covalent	Greater than 0.4, but less than 1.8
NaCl	Na^+Cl^-	Electron transfer	$3.0 - 0.9 = 2.1$	Ionic	Greater than 1.8
MgO	$Mg^{2+}O^{2-}$	Electron transfer	$3.5 - 1.2 = 2.3$	Ionic	Greater than 1.8

*Values are taken from Figure 6.7.

SAMPLE PROBLEM 6.15 Bond Polarity

Using electronegativity values, classify each of the following bonds as nonpolar covalent, polar covalent, or ionic:

O—K, Cl—As, N—N, P—Br

SOLUTION

For each bond, we obtain the electronegativity values and calculate the difference in electronegativity.

Bond	Electronegativity Difference	Type of Bond
O—K	3.5 − 0.8 = 2.7	Ionic
Cl—As	3.0 − 2.0 = 1.0	Polar covalent
N—N	3.0 − 3.0 = 0.0	Nonpolar covalent
P—Br	2.8 − 2.1 = 0.7	Polar covalent

STUDY CHECK 6.15

Using electronegativity values, classify each of the following bonds as nonpolar covalent, polar covalent, or ionic:

a. P—Cl **b.** Br—Br **c.** Na—O

QUESTIONS AND PROBLEMS

6.6 Electronegativity and Bond Polarity

LEARNING GOAL Use electronegativity to determine the polarity of a bond.

6.51 Describe the trend in electronegativity as *increases* or *decreases* for each of the following:
a. from B to F **b.** from Mg to Ba **c.** from F to I

6.52 Describe the trend in electronegativity as *increases* or *decreases* for each of the following:
a. from Al to Cl **b.** from Br to K **c.** from Li to Cs

6.53 Using the periodic table, arrange the atoms in each of the following sets in order of increasing electronegativity:
a. Li, Na, K **b.** Na, Cl, P **c.** Se, Ca, O

6.54 Using the periodic table, arrange the molecules in each of the following in order of increasing bond polarity:
a. NaCl **b.** NaI **c.** NaBr **d.** Cl-Cl

6.55 Predict whether each of the following bonds is nonpolar covalent, polar covalent, or ionic:
a. Si—Br **b.** Li—F **c.** Br—F
d. I—I **e.** N—P **f.** C—P

6.56 Predict whether each of the following bonds is nonpolar covalent, polar covalent, or ionic:
a. Si—O **b.** K—Cl **c.** S—F
d. P—Br **e.** Li—O **f.** N—S

6.57 For each of the following bonds, indicate the positive end with δ^+ and the negative end with δ^-. Draw an arrow to show the dipole for each.
a. N—F **b.** Si—Br **c.** C—O
d. P—Br **e.** N—P

6.58 For each of the following bonds, indicate the positive end with δ^+ and the negative end with δ^-. Draw an arrow to show the dipole for each.
a. P—Cl **b.** Se—F **c.** Br—F
d. N—H **e.** B—Cl

6.7 Shapes and Polarity of Molecules

LEARNING GOAL

Predict the three-dimensional structure of a molecule, and classify it as polar or nonpolar.

Using information about electron-dot formulas, we can predict the three-dimensional shapes of many molecules. The shape is important in our understanding of how molecules interact with enzymes or certain antibiotics or produce our sense of taste and smell.

CORE CHEMISTRY SKILL

Predicting Shape

The three-dimensional shape of a molecule is determined by drawing an electron-dot formula and identifying the number of electron groups (one or more electron pairs) around the central atom. We count lone pairs of electrons, single, double, or triple bonds as *one* electron group. In the **valence shell electron-pair repulsion (VSEPR) theory**, the electron groups are arranged as far apart as possible around the central atom to minimize the repulsion between their negative charges. Once we have counted the number of electron groups surrounding the central atom, we can determine its specific shape from the number of atoms bonded to the central atom.

Central Atoms with Two Electron Groups

In the electron-dot formula for CO_2, there are two electron groups (two double bonds) attached to the central atom. According to VSEPR theory, minimal repulsion occurs when two electron groups are on opposite sides of the central C atom. This gives the CO_2 molecule a *linear* electron-group geometry and a *linear shape* with a bond angle of 180°.

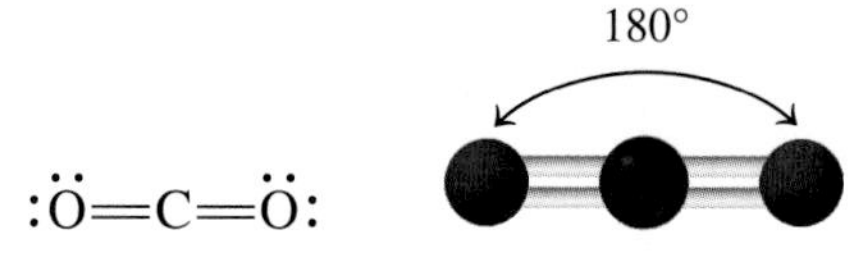

Linear electron-group geometry

Linear shape

Central Atoms with Three Electron Groups

In the electron-dot formula for formaldehyde, H_2CO, the central atom C is attached to two H atoms by single bonds and to the O atom by a double bond. Minimal repulsion occurs when three electron groups are as far apart as possible around the central C atom, which gives 120° bond angles. This type of electron-group geometry is *trigonal planar* and gives a shape for H_2CO called *trigonal planar.*

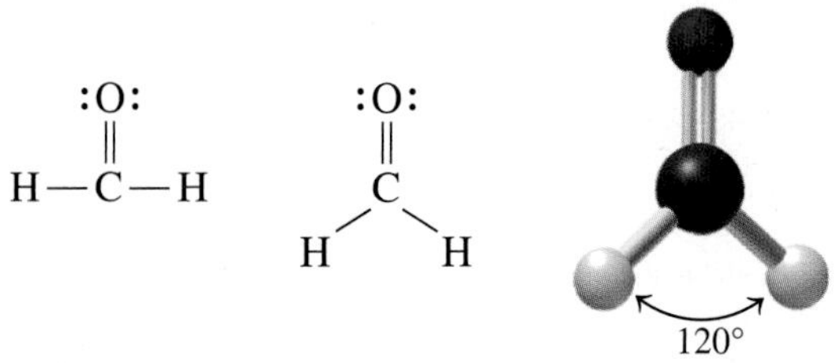

Electron-dot formula

Trigonal planar electron-group geometry

Trigonal planar shape

In the electron-dot formula for SO_2, there are also three electron groups around the central S atom: a single bond to an O atom, a double bond to another O atom, and a lone pair of electrons. As in H_2CO, three electron groups have minimal repulsion when they form trigonal planar electron-group geometry. However, in SO_2 one of the electron groups is a lone pair of electrons. Therefore, the shape of the SO_2 molecule is determined by the two O atoms bonded to the central S atom, which gives the SO_2 molecule a *bent shape* with a bond angle of 120°. When there are one or more lone pairs on the central atom, the shape has a different name than that of the electron-group geometry.

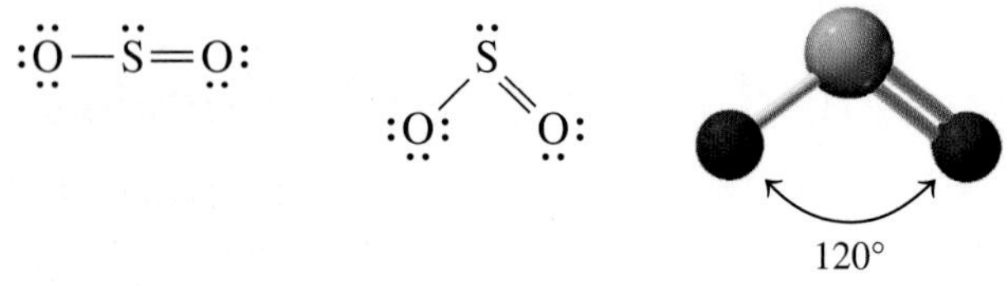

Electron-dot formula

Trigonal planar electron-group geometry

Bent shape

Central Atom with Four Electron Groups

In a molecule of methane, CH_4, the central C atom is bonded to four H atoms. From the electron-dot formula, you may think that CH_4 is planar with 90° bond angles. However, the best geometry for minimal repulsion is *tetrahedral*, giving bond angles of 109°. When there are four atoms attached to four electron groups, the shape of the molecule is *tetrahedral*.

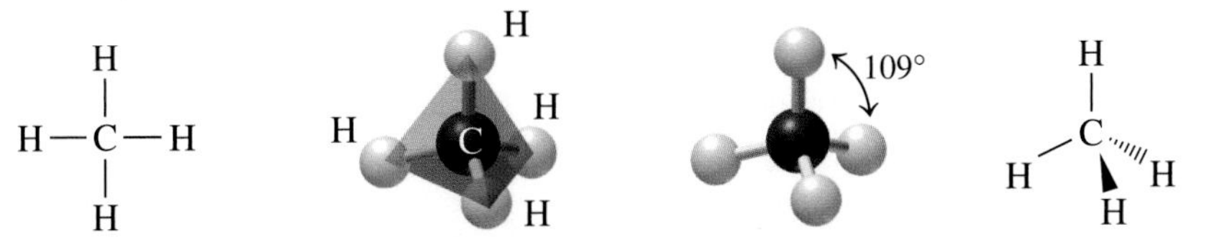

Electron-dot formula | Tetrahedral electron-group geometry | Tetrahedral shape | Tetrahedral wedge–dash notation

A way to represent the three-dimensional structure of methane is to use the wedge–dash notation. In this representation, the two bonds connecting carbon to hydrogen by solid lines are in the plane of the paper. The wedge represents a carbon-to-hydrogen bond coming out of the page toward us, whereas the dash represents a carbon-to-hydrogen bond going into the page away from us.

Now we can look at molecules that have four electron groups, of which one or more are lone pairs of electrons. Then the central atom is attached to only two or three atoms. For example, in the electron-dot formula for ammonia, NH_3, four electron groups have a tetrahedral electron-group geometry. However, in NH_3 one of the electron groups is a lone pair of electrons. Therefore, the shape of NH_3 is determined by the three H atoms bonded to the central N atom. Therefore, the shape of the NH_3 molecule is *trigonal pyramidal*, with a bond angle of 109°. The wedge–dash notation can also represent this three-dimensional structure of ammonia with one N—H bond in the plane, one N—H bond coming toward us, and one N—H bond going away from us.

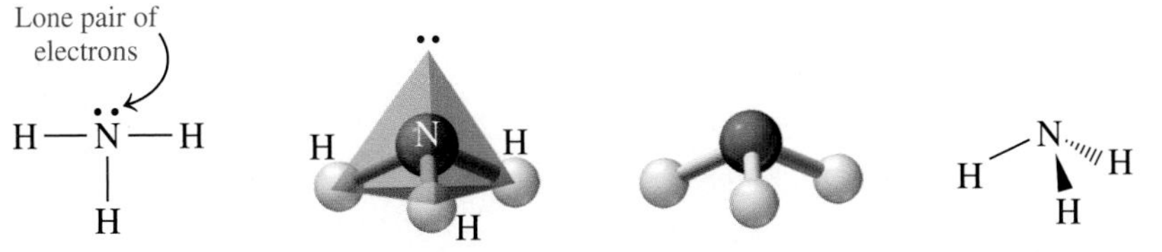

Electron-dot formula | Tetrahedral electron-group geometry | Trigonal pyramidal shape | Trigonal pyramidal wedge–dash notation

In the electron-dot formula for water, H_2O, there are also four electron groups, which have minimal repulsion when the electron-group geometry is tetrahedral. However, in H_2O, two of the electron groups are lone pairs of electrons. Because the shape of H_2O is determined by the two H atoms bonded to the central O atom, the H_2O molecule has a *bent shape* with a bond angle of 109°. Table 6.16 gives the shapes of molecules with two, three, or four bonded atoms.

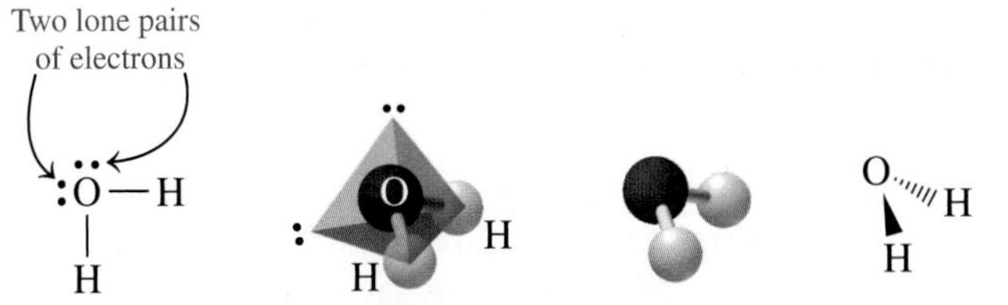

Electron-dot formula | Tetrahedral electron-group geometry | Bent shape | Bent wedge–dash notation

TABLE 6.16 Molecular Shapes for a Central Atom with Two, Three, and Four Bonded Atoms

Electron Groups	Electron-Group Geometry	Bonded Atoms	Lone Pairs	Bond Angle*	Molecular Shape	Example	Three-Dimensional Model
2	Linear	2	0	180°	Linear	CO_2	
3	Trigonal planar	3	0	120°	Trigonal planar	H_2CO	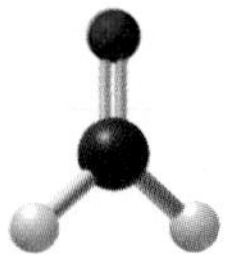
3	Trigonal planar	2	1	120°	Bent	SO_2	
4	Tetrahedral	4	0	109°	Tetrahedral	CH_4	
4	Tetrahedral	3	1	109°	Trigonal pyramidal	NH_3	
4	Tetrahedral	2	2	109°	Bent	H_2O	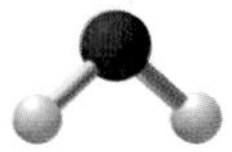

*The bond angles in actual molecules may vary slightly.

Guide to Predicting Molecular Shape (VSEPR Theory)

STEP 1
Draw the electron-dot formula.

STEP 2
Arrange the electron groups around the central atom to minimize repulsion.

STEP 3
Use the atoms bonded to the central atom to determine the molecular shape.

SAMPLE PROBLEM 6.16 Shapes of Molecules

Predict the shape of a molecule of $SiCl_4$.

SOLUTION

	Name of Element	Silicon		Chlorine
ANALYZE THE PROBLEM	Symbol of element	Si		Cl
	Atoms of element	1		4
	Valence electrons	4 e^-		7 e^-
	Total electrons	1(4 e^-)	+	4(7 e^-) = 32 e^-

STEP 1 **Draw the electron-dot formula.** Using 32 e^-, we draw the bonds and lone pairs for the electron-dot formula of $SiCl_4$.

:Cl:
:Cl:Si:Cl:
:Cl:

STEP 2 **Arrange the electron groups around the central atom to minimize repulsion.** To minimize repulsion, the electron-group geometry would be tetrahedral.

STEP 3 **Use the atoms bonded to the central atom to determine the molecular shape.** Because the central Si atom has four bonded pairs and no lone pairs of electrons, the $SiCl_4$ molecule has a tetrahedral shape.

STUDY CHECK 6.16

Predict the shape of SCl_2.

Polarity of Molecules

CORE CHEMISTRY SKILL

Identifying Polarity of Molecules

We have seen that covalent bonds in molecules can be polar or nonpolar. Now we will look at how the bonds in a molecule and its shape determine whether that molecule is classified as polar or nonpolar.

Nonpolar Molecules

In a **nonpolar molecule**, all the bonds are nonpolar or the polar bonds cancel each other out. Molecules such as H_2, Cl_2, and CH_4 are nonpolar because they contain only nonpolar covalent bonds.

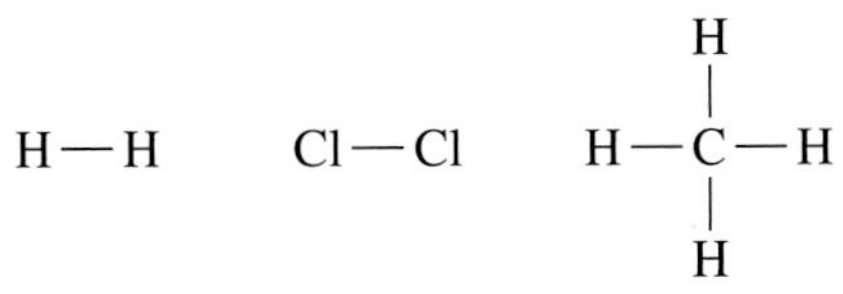

Nonpolar molecules

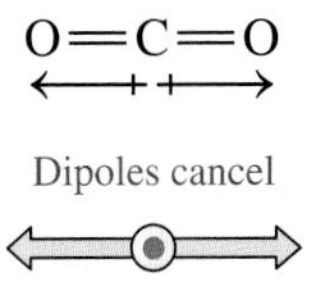

CO_2 is a nonpolar molecule.

A *nonpolar molecule* also occurs when polar bonds (dipoles) cancel each other because they are in a symmetrical arrangement. For example, CO_2, a linear molecule, contains two equal polar covalent bonds whose dipoles point in opposite directions. As a result, the dipoles cancel out, which makes a CO_2 molecule nonpolar.

Another example of a nonpolar molecule is the CCl_4 molecule, which has four polar bonds symmetrically arranged around the central C atom. Each of the C—Cl bonds has the same polarity, but because they have a tetrahedral arrangement, their opposing dipoles cancel out. As a result, a molecule of CCl_4 is nonpolar.

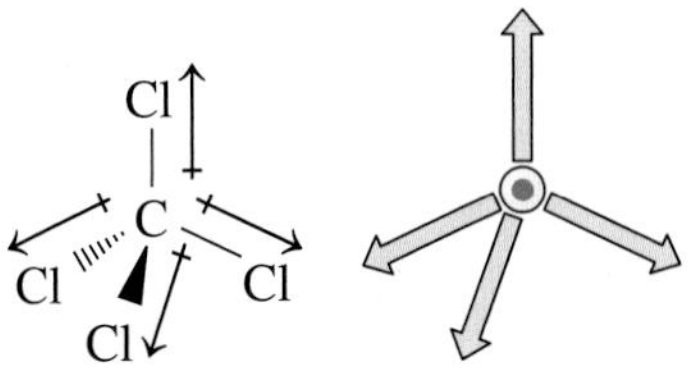

The four C—Cl dipoles cancel out.

Polar Molecules

In a **polar molecule**, one end of the molecule is more negatively charged than the other end. Polarity in a molecule occurs when the dipoles from the individual polar bonds do not cancel each other. For example, HCl is a polar molecule because it has one covalent bond that is polar.

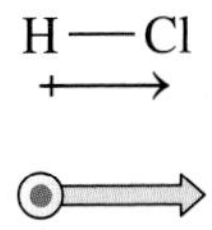

A single dipole does not cancel.

In molecules with two or more electron-groups, the shape, such as bent or trigonal pyramidal, determines whether or not the dipoles cancel. For example, we have seen that H_2O has a bent shape. Thus, a water molecule is polar because the individual dipoles do not cancel.

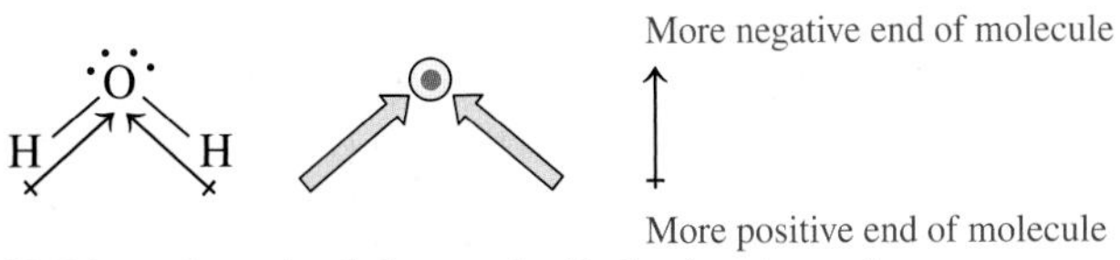

H_2O is a polar molecule because its dipoles do not cancel.

The NH_3 molecule has a tetrahedral electron-group geometry with three bonded atoms, which gives it a trigonal pyramidal shape. Thus, a NH_3 molecule is polar because the individual N—H dipoles do not cancel.

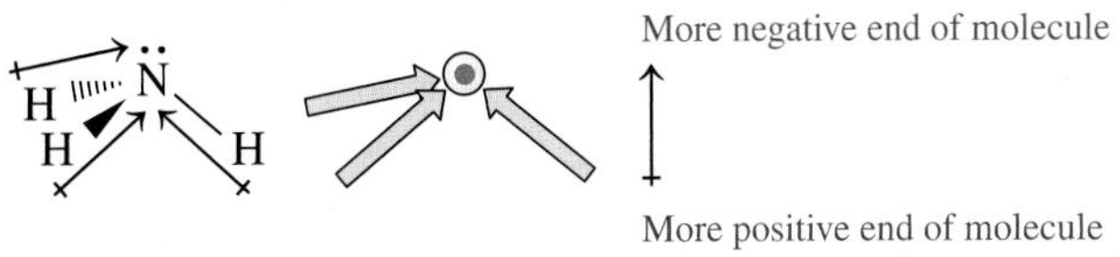

NH_3 is a polar molecule because its dipoles do not cancel.

In the molecule CH_3F, the C—F bond is polar covalent, but the three C—H bonds are nonpolar covalent. Because there is only one dipole in CH_3F, which does not cancel, CH_3F is a polar molecule.

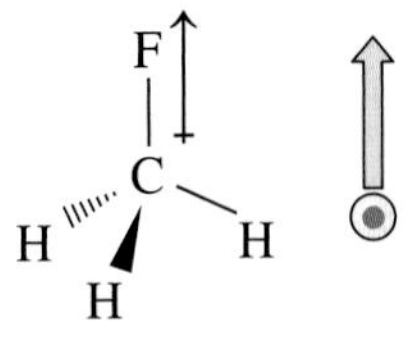

CH_3F is a polar molecule.

SAMPLE PROBLEM 6.17 Polarity of Molecules

Determine whether a molecule of OF_2 is polar or nonpolar.

Guide to Determination of Polarity of a Molecule

STEP 1
Determine if the bonds are polar covalent or nonpolar covalent.

STEP 2
If the bonds are polar covalent, draw the electron-dot formula and determine if the dipoles cancel.

SOLUTION

STEP 1 Determine if the bonds are polar covalent or nonpolar covalent. From Figure 6.7, F 4.0 and O 3.5 give an electronegativity difference of 0.5, which makes each of the O—F bonds polar covalent.

STEP 2 If the bonds are polar covalent, draw the electron-dot formula and determine if the dipoles cancel. The electron-dot formula for OF_2 has four electron groups and two bonded atoms. The molecule has a bent shape in which the dipoles of the O—F bonds do not cancel. The OF_2 molecule would be polar.

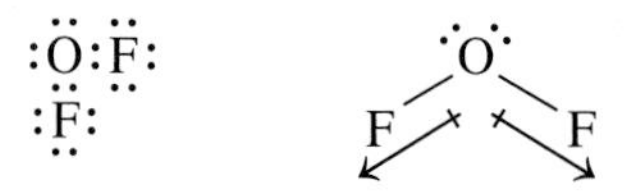

OF_2 is a polar molecule.

STUDY CHECK 6.17

Would a PCl_3 molecule be polar or nonpolar?

QUESTIONS AND PROBLEMS

6.7 Shapes and Polarity of Molecules

LEARNING GOAL Predict the three-dimensional structure of a molecule, and classify it as polar or nonpolar.

6.59 Choose the shape (**1** to **6**) that matches each of the following descriptions (**a** to **c**):

1. linear **2.** bent (109°) **3.** trigonal planar
4. bent (120°) **5.** trigonal pyramidal **6.** tetrahedral

a. a molecule with a central atom that has four electron groups and four bonded atoms
b. a molecule with a central atom that has four electron groups and three bonded atoms
c. a molecule with a central atom that has three electron groups and three bonded atoms

6.60 Choose the shape (**1** to **6**) that matches each of the following descriptions (**a** to **c**):

1. linear **2.** bent (109°) **3.** trigonal planar
4. bent (120°) **5.** trigonal pyramidal **6.** tetrahedral

a. a molecule with a central atom that has four electron groups and two bonded atoms
b. a molecule with a central atom that has two electron groups and two bonded atoms
c. a molecule with a central atom that has three electron groups and two bonded atoms

6.61 Predict the three-dimensional structure of the following compounds:
a. $Ni(CO)_4$
b. CS_2
c. NEt_3
d. SeO_3:

6.62 Complete each of the following statements for a molecule of H_2S:
a. There are _______ electron groups around the central S atom.
b. The electron-group geometry is _______.
c. The number of atoms attached to the central S atom is _______.
d. The shape of the molecule is _______.

6.63 Compare the electron-dot formulas of PH_3 and NH_3. Why do these molecules have the same shape?

6.64 Compare the electron-dot formulas of CH_4 and H_2O. Why do these molecules have approximately the same bond angles but different molecular shapes?

6.65 Use VSEPR theory to predict the shape of each of the following:
a. $SeBr_2$ **b.** CCl_4 **c.** OBr_2

6.66 Use VSEPR theory to predict the shape of each of the following:
a. NCl_3 **b.** SeO_2 **c.** SiF_2Cl_2

6.67 The gas molecule O_2 is nonpolar, but HOH is polar. Explain.

6.68 The molecules CH_4 and CH_3Cl both contain four bonds. Why is CH_4 nonpolar whereas CH_3Cl is polar?

6.69 Identify each of the following molecules as polar or nonpolar:
a. HBr **b.** NF_3 **c.** CHF_3

6.70 Identify each of the following molecules as polar or nonpolar:
a. SeF_2 **b.** PBr_3 **c.** $SiCl_4$

6.8 Attractive Forces in Compounds

In gases, the interactions between particles are minimal, which allows gas molecules to move far apart from each other. In solids and liquids, there are sufficient interactions between the particles to hold them close together, although some solids have low melting points while others have very high melting points. Such differences in properties are explained by looking at the various kinds of attractive forces between particles including *dipole–dipole attractions*, *hydrogen bonding*, *dispersion forces*, as well as *ionic bonds*.

Ionic compounds have high melting points. For example, solid NaCl melts at 801 °C. Large amounts of energy are needed to overcome the strong attractive forces between positive and negative ions. In solids containing molecules with covalent bonds, there are attractive forces too, but they are weaker than those of an ionic compound.

LEARNING GOAL

Describe the attractive forces between ions, polar covalent molecules, and nonpolar covalent molecules.

CORE CHEMISTRY SKILL

Identifying Attractive Forces

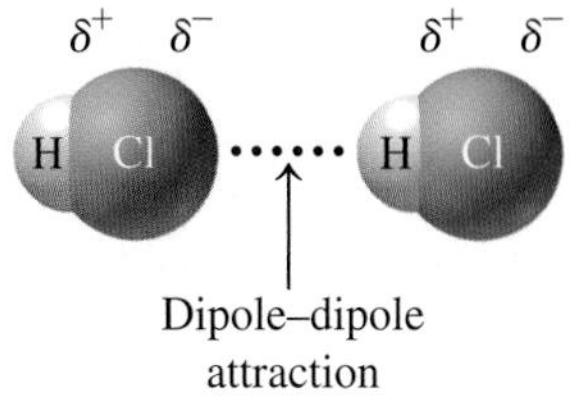

Dipole–Dipole Attractions

For polar molecules, attractive forces called **dipole–dipole attractions** occur between the positive end of one molecule and the negative end of another. For a polar molecule with a dipole such as HCl, the partially positive H atom of one HCl molecule attracts the partially negative Cl atom in another HCl molecule.

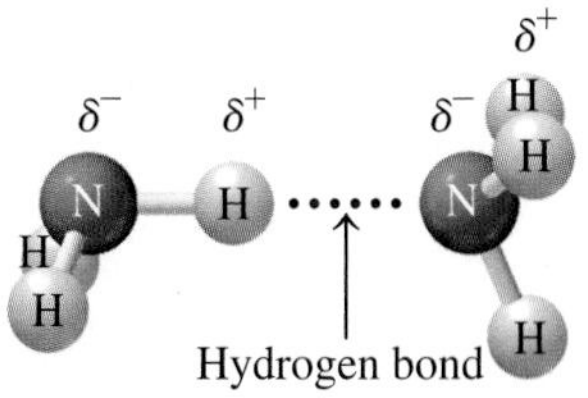

Hydrogen Bonds

Polar molecules containing hydrogen atoms bonded to highly electronegative atoms of nitrogen, oxygen, or fluorine form especially strong dipole–dipole attractions. This type of attraction, called a **hydrogen bond**, occurs between the partially positive hydrogen atom in one molecule and the partially negative nitrogen, oxygen, or fluorine atom in another molecule. Hydrogen bonds are the strongest type of attractive forces between polar covalent molecules. They are a major factor in the formation and structure of biological molecules such as proteins and DNA.

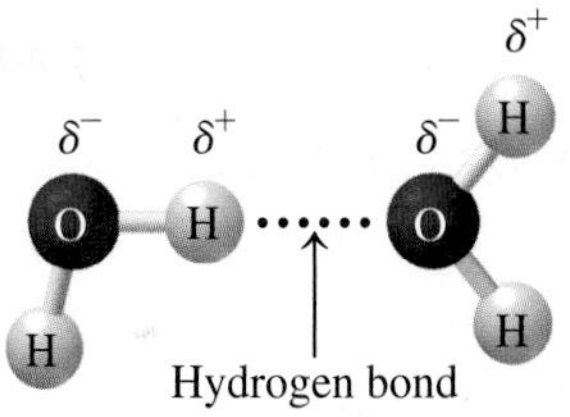

Dispersion Forces

Nonpolar compounds do form solids, but at low temperatures. Very weak attractions called **dispersion forces** occur between nonpolar molecules. Usually, the electrons in a nonpolar covalent molecule are distributed symmetrically. However, electrons may accumulate more in one part of the molecule than another, which forms a *temporary dipole*. Although dispersion forces are very weak, they make it possible for nonpolar molecules to form liquids and solids.

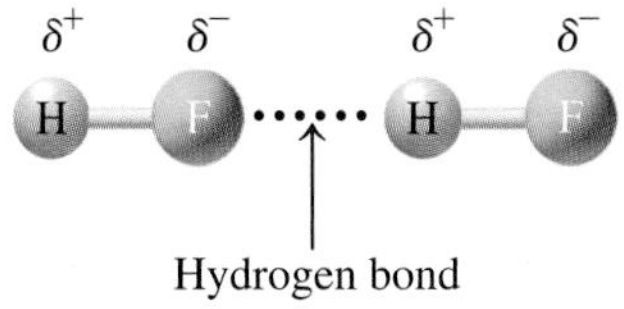

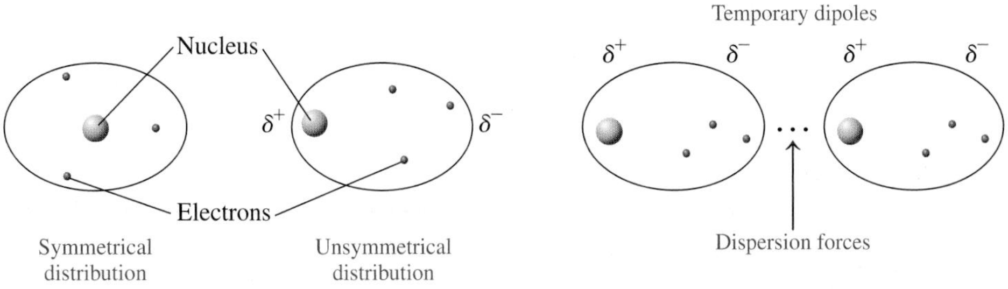

Nonpolar covalent molecules have weak attractions when they form temporary dipoles.

The melting points of substances are related to the strength of the attractive forces. Compounds with weak attractive forces such as dispersion forces have low melting points because only a small amount of energy is needed to separate the molecules and form a liquid. Compounds with hydrogen bonds and dipole–dipole attractions require more energy to break the attractive forces between the molecules. The highest melting points are seen with ionic compounds that have very strong attractions between ions. Table 6.17 compares the melting points of some substances with various kinds of attractive forces. The various types of attractions between particles in solids and liquids are summarized in Table 6.18.

TABLE 6.17 Melting Points of Selected Substances

Substance	Melting Point (°C)
Ionic Bonds	
MgF_2	1248
NaCl	801
Hydrogen Bonds	
H_2O	0
NH_3	−78
Dipole–Dipole Attractions	
HBr	−89
HCl	−115
Dispersion Forces	
Cl_2	−101
F_2	−220

TABLE 6.18 Comparison of Bonding and Attractive Forces

Type of Force	Particle Arrangement	Example	Strength
Between Atoms or Ions			Strong
Ionic bond	+ − + / − + −	Na^+Cl^-	↑
Covalent bond (X = nonmetal)	X : X	Cl—Cl	
Between Molecules			
Hydrogen bond (X = F, O, or N)	$\delta^+ \delta^-$ H X · · · $\delta^+ \delta^-$ H X	$\overset{\delta^+}{H}—\overset{\delta^-}{F} \cdots \overset{\delta^+}{H}—\overset{\delta^-}{F}$	
Dipole–dipole attractions (X and Y = nonmetals)	$\delta^+ \delta^-$ Y X · · · $\delta^+ \delta^-$ Y X	$\overset{\delta^+}{Br}—\overset{\delta^-}{Cl} \cdots \overset{\delta^+}{Br}—\overset{\delta^-}{Cl}$	
Dispersion forces (temporary shift of electrons in nonpolar bonds)	$\delta^+ \delta^-$ X :X · · · $\delta^+ \delta^-$ X :X (temporary dipoles)	$\overset{\delta^+}{F}—\overset{\delta^-}{F} \cdots \overset{\delta^+}{F}—\overset{\delta^-}{F}$	Weak

SAMPLE PROBLEM 6.18 Attractive Forces Between Particles

Indicate the major type of molecular interaction—dipole–dipole attractions, hydrogen bonding, or dispersion forces—expected of each of the following:

a. HF **b.** Br_2 **c.** PCl_3

SOLUTION

a. HF is a polar molecule that interacts with other HF molecules by hydrogen bonding.
b. Br_2 is nonpolar; only dispersion forces provide attractive forces.
c. The polarity of the PCl_3 molecules provides dipole–dipole attractions.

STUDY CHECK 6.18

Why is the melting point of H_2S lower than that of H_2O?

Chemistry Link to **Health**

Attractive Forces in Biological Compounds

The proteins in our bodies are biological molecules that have many different functions. They are needed for structural components such as cartilage, muscles, hair, and nails, and for the formation of enzymes that regulate biological reactions. Other proteins, such as hemoglobin and myoglobin, transport oxygen in the blood and muscle.

Proteins are composed of building blocks called *amino acids.* Every amino acid has a central carbon atom bonded to $—NH_3^+$ from an amine, a $—COO^-$ from a carboxylic acid, an H atom, and a side chain called an R group, which is unique for each amino acid. For example, the R group for the amino acid alanine is a methyl group ($—CH_3$).

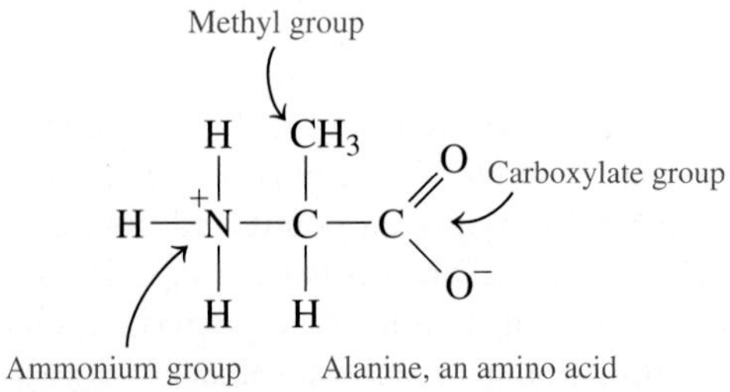

The ionized form of alanine contains $—NH_3^+$, $—COO^-$, H, and a $—CH_3$ group.

Several of the amino acids have side chains that contain groups such as hydroxyl —OH, amide $—CONH_2$, amine $—NH_2$, which is ionized as ammonium $—NH_3^+$, and carboxyl —COOH, which is ionized as carboxylate $—COO^-$.

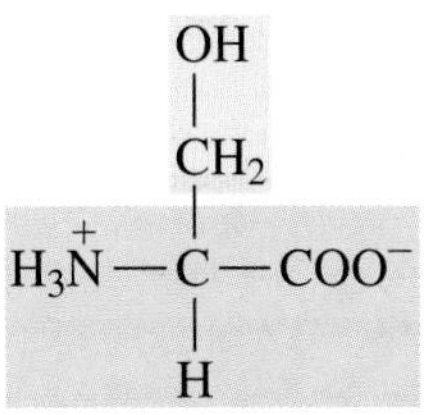

Serine

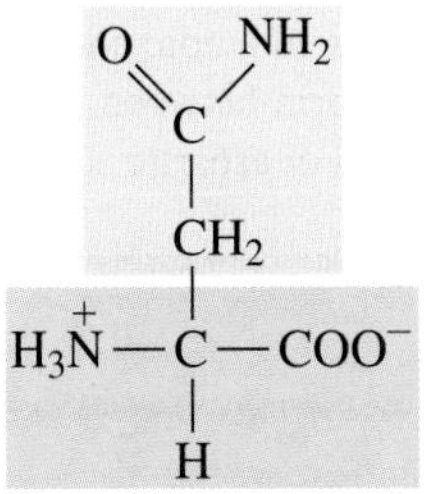

Asparagine

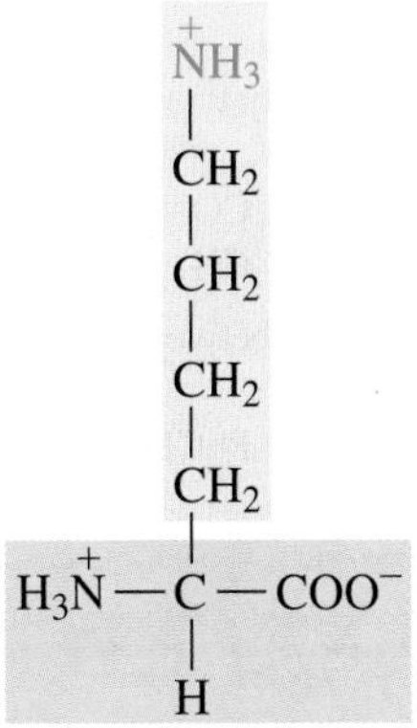

Lysine

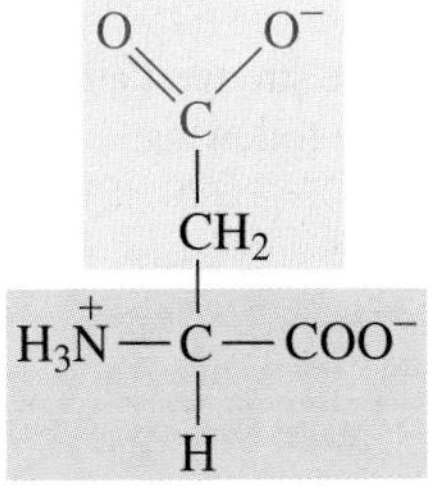

Aspartic acid

In the primary structure of proteins, amino acids are linked together by peptide bonds, which are amide bonds that form between the $—COO^-$ group of one amino acid and the $—NH_3^+$ group of the next amino acid. When there are more than 50 amino acids in a chain, it is called a protein. Every protein in our bodies has a unique sequence of amino acids in a primary structure that determines its biological function. In addition, proteins have higher levels of structure. In an *alpha helix*, hydrogen bonds form between the hydrogen atom of the N—H groups and the oxygen atom of a C=O group in the next turn. Because many hydrogen bonds form, the backbone of a protein takes a helical shape similar to a corkscrew or a spiral staircase.

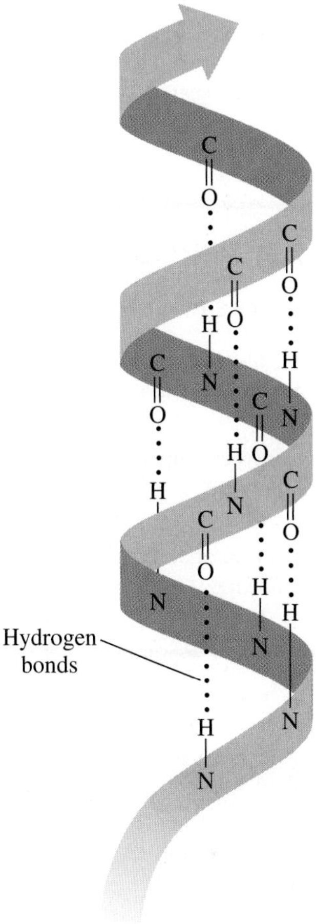

The shape of an alpha helix is stabilized by hydrogen bonds.

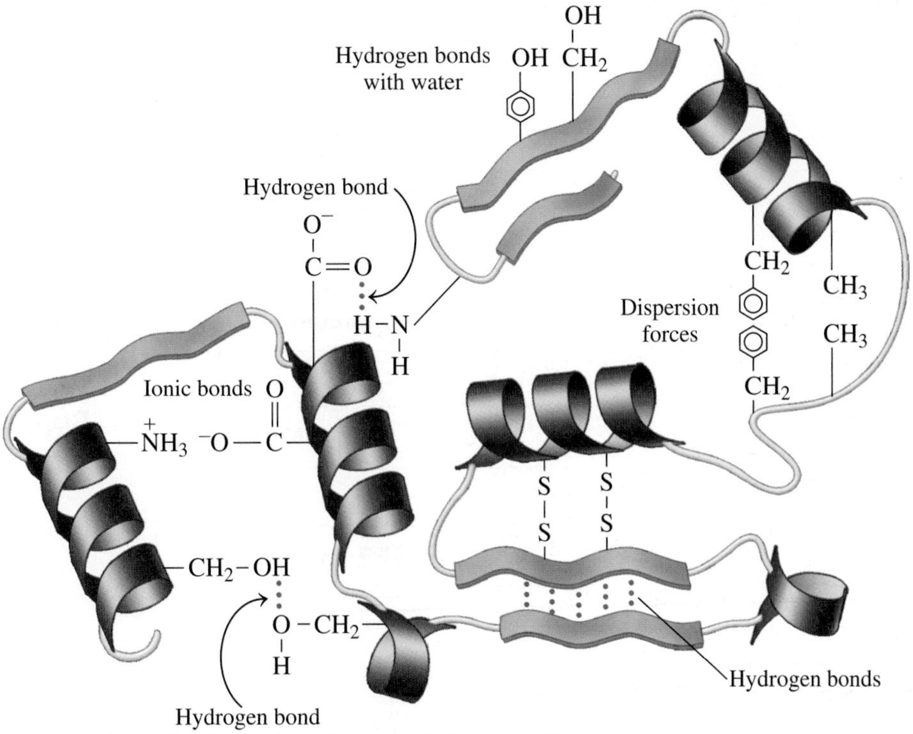

Attractive forces hold the protein in a specific shape.

Interactive Video

Interactive Activity on Viewing Molecules

Shapes of Biologically Active Proteins

Many proteins have compact, spherical shapes because sections of the chain fold over on top of other sections of the chain. This shape is stabilized by attractive forces between functional groups of side chains (R groups), causing the protein to twist and bend into a specific three-dimensional shape.

Hydrogen bonds form between the side chains within the protein. For example, a hydrogen bond can occur between the —OH groups on two serines or between the —OH of a serine and the $—NH_2$ of asparagine. Hydrogen bonding also occurs between the polar side chains of the amino acids on the outer surface of the protein and —OH or —H of polar water molecules in the external aqueous environment.

Ionic bonds can form between the positively and negatively charged R groups of acidic and basic amino acids. For example, an ionic bond can form between the $—NH_3^+$ in the R group of lysine and the $—COO^-$ of aspartic acid.

QUESTIONS AND PROBLEMS

6.8 Attractive Forces in Compounds

LEARNING GOAL Describe the attractive forces between ions, polar covalent molecules, and nonpolar covalent molecules.

6.71 Identify the major type of attractive force between the particles of each of the following:
a. BrF **b.** KCl **c.** NF_3 **d.** Cl_2

6.72 Identify the major type of attractive force between the particles of each of the following:
a. HCl **b.** MgF_2 **c.** PBr_3 **d.** NH_3

6.73 Identify the strongest attractive forces between the particles of each of the following:
a. CH_3OH **b.** CO
c. CF_4 **d.** $CH_3—CH_3$

6.74 Identify the strongest attractive forces between the particles of each of the following:
a. O_2 **b.** ICl **c.** $PhCH_2Cl$ **d.** H_2O_2

CONCEPT MAP

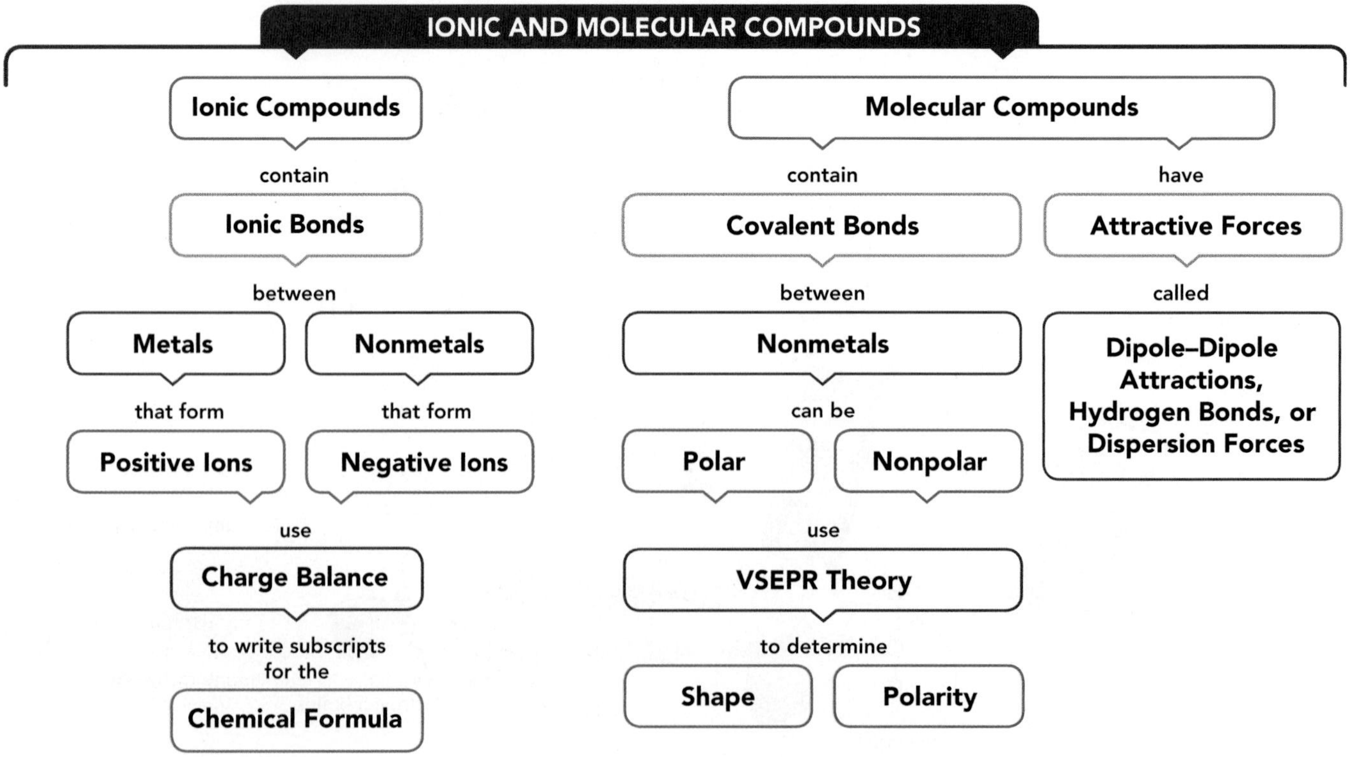

CHAPTER REVIEW

6.1 Ions: Transfer of Electrons

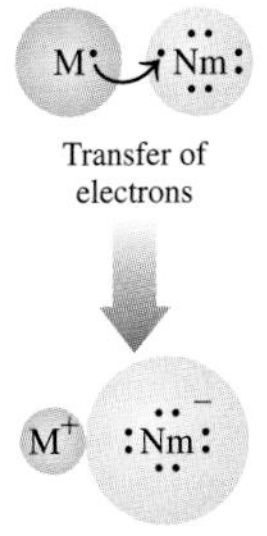

LEARNING GOAL Write the symbols for the simple ions of the representative elements.

- The stability of the noble gases is associated with a complete electron arrangement in the outermost energy level.
- With the exception of helium, which has two electrons, noble gases have eight valence electrons, which is an octet.
- Atoms of elements in Groups 1A–7A (1, 2, 13–17) achieve stability by losing, gaining, or sharing their valence electrons in the formation of compounds.
- Metals of the representative elements lose valence electrons to form positively charged ions (cations): Group 1A (1), 1+, Group 2A (2), 2+, and Group 3A (13), 3+.
- When reacting with metals, nonmetals gain electrons to form octets and form negatively charged ions (anions): Groups 5A (15), 3−, 6A (16), 2−, and 7A (17), 1−.

6.2 Writing Formulas for Ionic Compounds

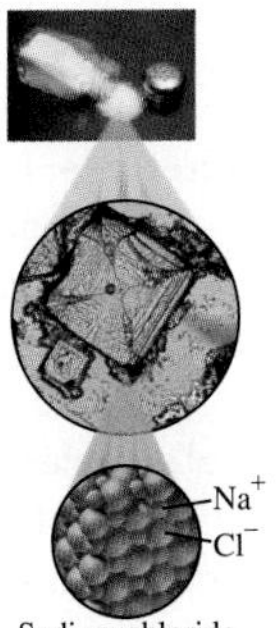

Sodium chloride

LEARNING GOAL Using charge balance, write the correct formula for an ionic compound.

- The total positive and negative ionic charge is balanced in the formula of an ionic compound.
- Charge balance in a formula is achieved by using subscripts after each symbol so that the overall charge is zero.

6.3 Naming Ionic Compounds

LEARNING GOAL Given the formula of an ionic compound, write the correct name; given the name of an ionic compound, write the correct formula.

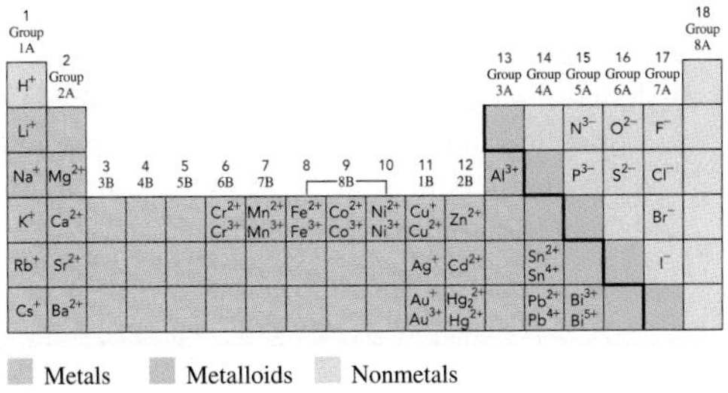

- In naming ionic compounds, the positive ion is given first, followed by the name of the negative ion.
- The names of ionic compounds containing two elements end with *ide*.
- Except for Ag, Cd, and Zn, transition elements form cations with two or more ionic charges.
- The charge of the cation is determined from the total negative charge in the formula and included as a Roman numeral immediately following the name of the metal that has a variable charge.

6.4 Polyatomic Ions

LEARNING GOAL Write the name and formula for an ionic compound containing a polyatomic ion.

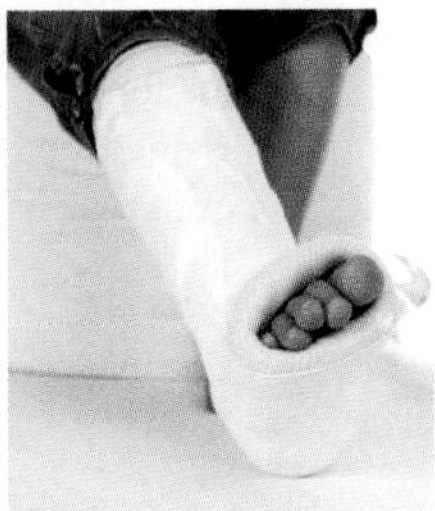

- A polyatomic ion is a covalently bonded group of atoms with an electrical charge; for example, the carbonate ion has the formula CO_3^{2-}.
- Most polyatomic ions have names that end with *ate* or *ite*.
- Most polyatomic ions contain a nonmetal and one or more oxygen atoms.
- The ammonium ion, NH_4^+, is a positive polyatomic ion.
- When more than one polyatomic ion is used for charge balance, parentheses enclose the formula of the polyatomic ion.

6.5 Molecular Compounds: Sharing Electrons

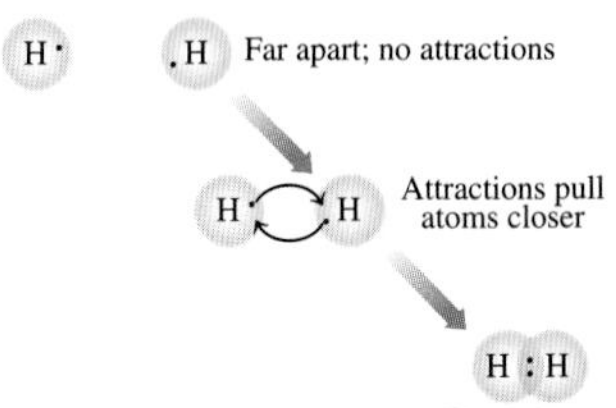

H_2 molecule

LEARNING GOAL Given the formula of a molecular compound, write its correct name; given the name of a molecular compound, write its formula.

- In a covalent bond, atoms of nonmetals share valence electrons such that each atom has a stable electron arrangement.
- The first nonmetal in a molecular compound uses its element name; the second nonmetal uses the first syllable of its element name followed by *ide*.
- The name of a molecular compound with two different atoms uses prefixes to indicate the subscripts in the formula.

6.6 Electronegativity and Bond Polarity

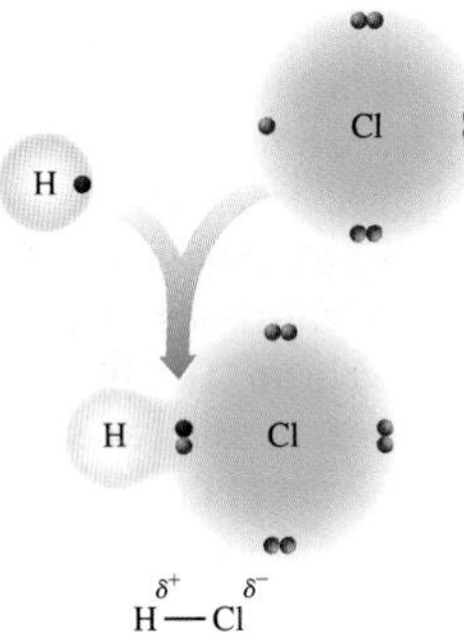

Unequal sharing of electrons in a polar covalent bond

LEARNING GOAL Use electronegativity to determine the polarity of a bond.

- Electronegativity is the ability of an atom to attract the electrons it shares with another atom. In general, the electronegativities of metals are low, while nonmetals have high electronegativities.
- In a nonpolar covalent bond, atoms share electrons equally.
- In a polar covalent bond, the electrons are unequally shared because they are attracted to the more electronegative atom.
- The atom in a polar bond with the lower electronegativity is partially positive (δ^+), and the atom with the higher electronegativity is partially negative (δ^-).
- Atoms that form ionic bonds have large differences in electronegativities.

6.7 Shapes and Polarity of Molecules

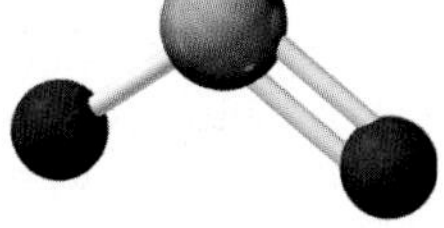

LEARNING GOAL Predict the three-dimensional structure of a molecule, and classify it as polar or nonpolar.

- The shape of a molecule is determined from the electron-dot formula, the electron-group geometry, and the number of bonded atoms.
- The electron-group geometry around a central atom with two electron groups is linear; with three electron groups, the geometry is trigonal planar; and with four electron groups, the geometry is tetrahedral.

- When all the electron groups are bonded to atoms, the shape has the same name as the electron arrangement.
- A central atom with three electron groups and two bonded atoms has a bent shape, 120°.
- A central atom with four electron groups and three bonded atoms has a trigonal pyramidal shape.
- A central atom with four electron groups and two bonded atoms has a bent shape, 109°.
- Nonpolar molecules contain nonpolar covalent bonds or have an arrangement of bonded atoms that causes the dipoles to cancel out.
- In polar molecules, the dipoles do not cancel.

6.8 Attractive Forces in Compounds

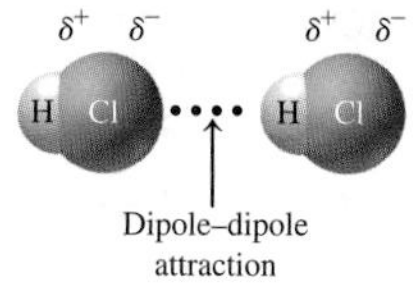

LEARNING GOAL Describe the attractive forces between ions, polar covalent molecules, and nonpolar covalent molecules.

- In ionic solids, oppositely charged ions are held in a rigid structure by ionic bonds.
- Attractive forces called dipole–dipole attractions and hydrogen bonds hold the solid and liquid states of polar molecular compounds together.
- Nonpolar compounds form solids and liquids by weak attractions between temporary dipoles called dispersion forces.

KEY TERMS

anion A negatively charged ion such as Cl^-, O^{2-}, or SO_4^{2-}.
cation A positively charged ion such as Na^+, Mg^{2+}, Al^{3+}, or NH_4^+.
chemical formula The group of symbols and subscripts that represents the atoms or ions in a compound.
covalent bond A sharing of valence electrons by atoms.
dipole The separation of positive and negative charges in a polar bond indicated by an arrow that is drawn from the more positive atom to the more negative atom.
dipole–dipole attractions Attractive forces between oppositely charged ends of polar molecules.
dispersion forces Weak dipole bonding that results from a momentary polarization of nonpolar molecules.
double bond A sharing of two pairs of electrons by two atoms.
electronegativity The relative ability of an element to attract electrons in a bond.
hydrogen bond The attraction between a partially positive H atom and a strongly electronegative atom of F, O, or N.
ion An atom or group of atoms having an electrical charge because of a loss or gain of electrons.
ionic charge The difference between the number of protons (positive) and the number of electrons (negative) written in the upper right corner of the symbol for the element or polyatomic ion.
ionic compound A compound of positive and negative ions held together by ionic bonds.
molecular compound A combination of atoms in which noble gas arrangements are attained by sharing electrons.
molecule The smallest unit of two or more atoms held together by covalent bonds.
nonpolar covalent bond A covalent bond in which the electrons are shared equally between atoms.
nonpolar molecule A molecule that has only nonpolar bonds or in which the bond dipoles cancel.
octet A set of eight valence electrons.
octet rule Elements in Groups 1A–7A (1, 2, 13–17) react with other elements by forming ionic or covalent bonds to produce a noble gas arrangement, usually eight electrons in the outer shell.
polar covalent bond A covalent bond in which the electrons are shared unequally between atoms.
polar molecule A molecule containing bond dipoles that do not cancel.
polarity A measure of the unequal sharing of electrons indicated by the difference in electronegativities.
polyatomic ion A group of covalently bonded nonmetal atoms that has an overall electrical charge.
triple bond A sharing of three pairs of electrons by two atoms.
valence shell electron-pair repulsion (VSEPR) theory A theory that predicts the shape of a molecule by placing the electron pairs on a central atom as far apart as possible to minimize the mutual repulsion of the electrons.

CORE CHEMISTRY SKILLS

The chapter section containing each Core Chemistry Skill is shown in parentheses at the end of each heading.

Writing Positive and Negative Ions (6.1)

- In the formation of an ionic bond, atoms of a metal lose and atoms of a nonmetal gain valence electrons to acquire a stable electron arrangement, usually eight valence electrons.
- This tendency of atoms to attain a stable electron arrangement is known as the octet rule.

Example: State the number of electrons lost or gained by atoms and the ion formed for each of the following to obtain a stable electron arrangement:

a. Br **b.** Ca **c.** S

Answer: **a.** Br atoms gain one electron to achieve a stable electron arrangement, Br^-.
b. Ca atoms lose two electrons to achieve a stable electron arrangement, Ca^{2+}.
c. S atoms gain two electrons to achieve a stable electron arrangement, S^{2-}.

Writing Ionic Formulas (6.2)

- The chemical formula of a compound represents the lowest whole-number ratio of the atoms or ions.
- In the chemical formula of an ionic compound, the sum of the positive and negative charges is always zero.
- Thus, in a chemical formula of an ionic compound, the total positive charge is equal to the total negative charge.

Example: Write the formula for magnesium phosphate.

Answer: Magnesium phosphate is an ionic compound that contains the ions Mg^{2+} and PO_4^{3-}.
Using charge balance, we determine the number(s) of each type of ion.

$$3(2+) + 2(3-) = 0$$

$3Mg^{2+}$ and $2PO_4^{3-}$ gives the formula $Mg_3(PO_4)_2$.

Naming Ionic Compounds (6.3)

- In the name of an ionic compound made up of two elements, the name of the metal ion, which is written first, is the same as its element name.
- For metals that form two or more different ions, a Roman numeral that is equal to the ionic charge is placed in parentheses immediately after the name of the metal.
- The name of a nonmetal ion is obtained by using the first syllable of its element name followed by *ide*.
- The name of a polyatomic ion usually ends in *ate* or *ite*.

Example: What is the name of $PbSO_4$?

Answer: This compound contains the polyatomic ion SO_4^{2-}, which is sulfate with a 2− charge.
For charge balance, the positive ion must have a charge of 2+.

$$Pb? + (2-) = 0; \quad Pb = 2+$$

Because lead can form two different positive ions, a Roman numeral II is used in the name of the compound: lead(II) sulfate.

Drawing Electron-Dot Formulas (6.5)

- The electron-dot formula for a molecule shows the sequence of atoms, the bonding pairs of electrons shared between atoms, and the nonbonding or *lone pairs* of electrons.
- Double or triple bonds result when a second or third electron pair is shared between the same atoms to complete octets.

Example: Draw the electron-dot formula for CS_2.

Answer: The central atom in the atom arrangement is C.

S C S

Determine the total number of valence electrons.

$$1\,C \times 4\,e^- = 4\,e^-$$
$$2\,S \times 6\,e^- = 12\,e^-$$
$$\text{Total} = 16\,e^-$$

Attach each bonded atom to the central atom using a pair of electrons. Two bonding pairs use four electrons.

S:C:S

Place the 12 remaining electrons as lone pairs around the S atoms.

:S̤̈:C:S̤̈:

To complete the octet for C, a lone pair of electrons from each of the S atoms is shared with C, which forms two double bonds.

:S̈::C::S̈: or :S̈=C=S̈:

Writing the Names and Formulas for Molecular Compounds (6.5)

- When naming a molecular compound, the first nonmetal in the formula is named by its element name; the second nonmetal is named using the first syllable of its element name, followed by *ide*.

Example: Name the molecular compound BrF_5.

Answer: Two nonmetals share electrons and form a molecular compound. Br (first nonmetal) is bromine; F (second nonmetal) is fluoride. In the name for a molecular compound, prefixes indicate the subscripts in the formulas. The subscript 1 is understood for Br. The subscript 5 for fluoride is written with the prefix *penta*. The name is bromine pentafluoride.

Using Electronegativity (6.6)

- The electronegativity values indicate the ability of atoms to attract shared electrons.
- Electronegativity values increase going across a period from left to right, and decrease going down a group.
- A nonpolar covalent bond occurs between atoms with identical or very similar electronegativity values such that the electronegativity difference is 0.0 to 0.4.
- A polar covalent bond occurs when electrons are shared unequally between atoms with electronegativity differences from 0.5 to 1.8.
- An ionic bond typically occurs when the difference in electronegativity for two atoms is greater than 1.8.

Example: Use electronegativity values to classify each of the following bonds as nonpolar covalent, polar covalent, or ionic:

a. Sr—Cl **b.** C—S **c.** O—Br

Answer: **a.** An electronegativity difference of 2.0 (Cl 3.0 − Sr 1.0) makes this an ionic bond.
b. An electronegativity difference of 0.0 (C 2.5 − S 2.5) makes this a nonpolar covalent bond.
c. An electronegativity difference of 0.7 (O 3.5 − Br 2.8) makes this a polar covalent bond.

Predicting Shape (6.7)

- The three-dimensional shape of a molecule or polyatomic ion is determined by drawing an electron-dot formula and identifying the number of electron groups (one or more electron pairs) around the central atom and the number of bonded atoms.
- In the valence shell electron-pair repulsion (VSEPR) theory, the electron groups are arranged as far apart as possible around a central atom to minimize the repulsion.
- A central atom with two electron groups bonded to two atoms is linear. A central atom with three electron groups bonded to three atoms is trigonal planar, and to two atoms is bent (120°). A central atom with four electron groups bonded to four atoms is tetrahedral, to three atoms is trigonal pyramidal, and to two atoms is bent (109°).

Identifying Polarity of Molecules (6.7)

- A molecule is nonpolar if all of its bonds are nonpolar or it has polar bonds that cancel out. CCl_4 is a nonpolar molecule that consists of four polar bonds that cancel out.

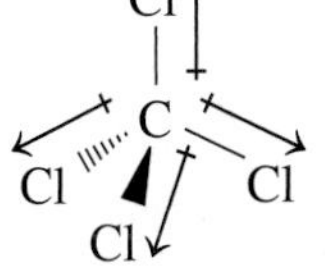

- A molecule is polar if it contains polar bonds that do not cancel out. H_2O is a polar molecule that consists of polar bonds that do not cancel out.

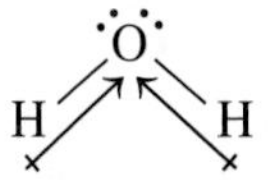

Example: Predict whether $AsCl_3$ is polar or nonpolar.

Answer: From its electron-dot formula, we see that $AsCl_3$ has four electron groups with three bonded atoms.

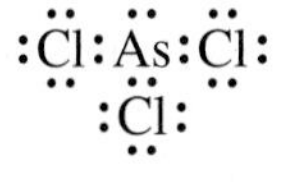

The shape of a molecule of $AsCl_3$ would be trigonal pyramidal with three polar bonds (As—Cl = 3.0 − 2.0 = 1.0) that do not cancel. Thus, it is a polar molecule.

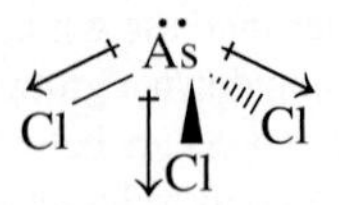

Identifying Attractive Forces (6.8)

- Dipole–dipole attractions occur between the dipoles in polar compounds because the positively charged end of one molecule is attracted to the negatively charged end of another molecule.

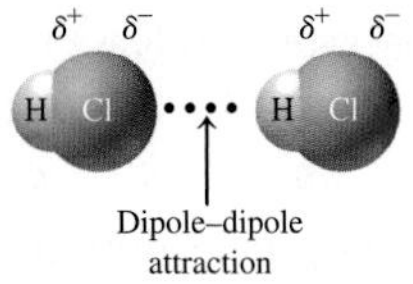

- Strong dipole–dipole attractions called hydrogen bonds occur in compounds in which H is bonded to F, O, or N. The partially positive H atom in one molecule has a strong attraction to the partially negative F, O, or N in another molecule.
- Dispersion forces are very weak attractive forces between nonpolar molecules that occur when *temporary dipoles* form as electrons are unequally distributed.

Example: Identify the strongest type of attractive forces in each of the following:

a. HF **b.** F_2 **c.** NF_3

Answer: **a.** HF molecules, which are polar with H bonded to F, have hydrogen bonding.
b. Nonpolar F_2 molecules have only dispersion forces.
c. NF_3 molecules, which are polar, have dipole–dipole attractions.

UNDERSTANDING THE CONCEPTS

The chapter sections to review are shown in parentheses at the end of each question.

6.75 **a.** How does the octet rule explain the formation of a magnesium ion? (6.1)
b. What noble gas has the same electron arrangement as the magnesium ion?
c. Why are Group 1A (1) and Group 2A (2) elements found in many compounds, but not Group 8A (18) elements?

6.76 **a.** How does the octet rule explain the formation of a chloride ion? (6.1)
b. What noble gas has the same electron arrangement as the chloride ion?
c. Why are Group 7A (17) elements found in many compounds, but not Group 8A (18) elements?

6.77 Identify each of the following atoms or ions: (6.1)

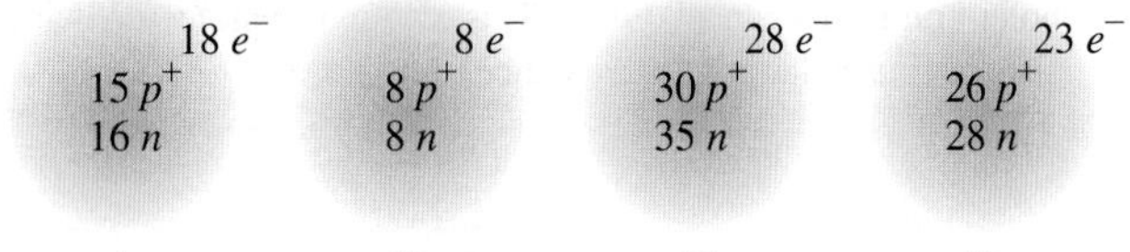

A **B** **C** **D**

6.78 Identify each of the following atoms or ions: (6.1)

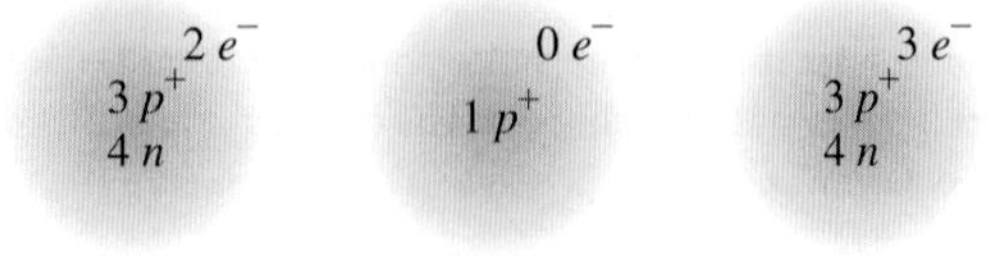

A **B** **C**

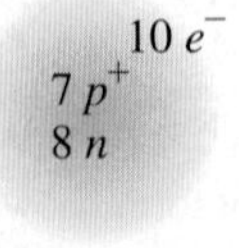

D

6.79 Consider the following electron-dot symbols for elements X and Y: (6.1, 6.2, 6.5)

X· ·Ÿ·

a. What are the group numbers of X and Y?
b. Will a compound of X and Y be ionic or molecular?
c. What ions would be formed by X and Y?
d. What would be the formula of a compound of X and Y?
e. What would be the formula of a compound of X and sulfur?
f. What would be the formula of a compound of Y and chlorine?
g. Is the compound in part **f** ionic or molecular?

6.80 Consider the following electron-dot symbols for elements X and Y: (6.1, 6.5)

Ẋ· ·Ÿ·

a. What are the group numbers of X and Y?
b. Will a compound of X and Y be ionic or molecular?
c. What ions would be formed by X and Y?
d. What would be the formula of a compound of X and Y?
e. What would be the formula of a compound of X and sulfur?
f. What would be the formula of a compound of Y and chlorine?
g. Is the compound in part **f** ionic or molecular?

6.81 Match each of the electron-dot formulas (**a** to **c**) with the correct diagram (**1** to **3**) of its shape, and name the shape; indicate if each molecule is polar or nonpolar. (Assume X and Y are nonmetals and all bonds are polar covalent.) (6.7)

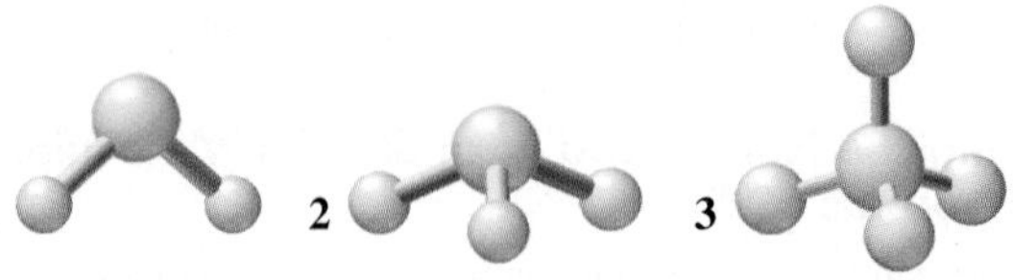

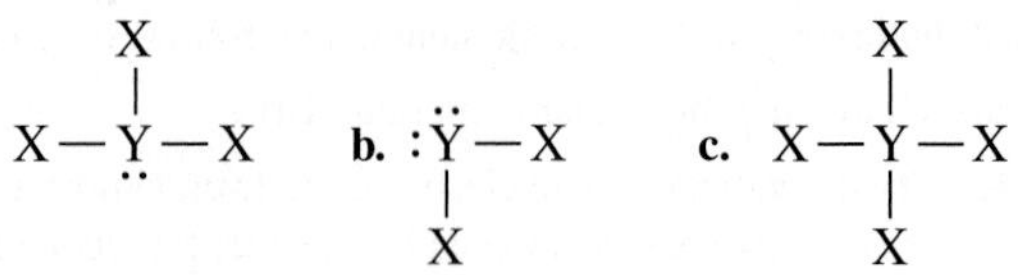

6.82 Match each of the formulas (**a** to **c**) with the correct diagram (**1** to **3**) of its shape, and name the shape; indicate if each molecule is polar or nonpolar. (6.7)

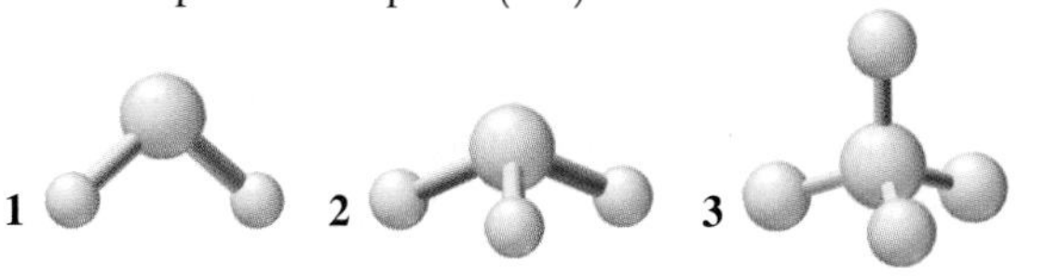

a. PBr_3 **b.** $SiCl_4$ **c.** OF_2

6.83 Using each of the following electron arrangements, give the formulas of the cation and anion that form, the formula for the compound they form, and its name. (6.2, 6.3)

Electron Arrangements		Cation	Anion	Formula of Compound	Name of Compound
2,8,2	2,5				
2,8,8,1	2,6				
2,8,3	2,8,7				

6.84 Using each of the following electron arrangements, give the formulas of the cation and anion that form, the formula of the compound they form, and its name. (6.2, 6.3)

Electron Arrangements		Cation	Anion	Formula of Compound	Name of Compound
2,8,1	2,7				
2,8,8,2	2,8,6				
2,8,3	2,8,5				

6.85 State the number of valence electrons, bonding pairs, and lone pairs in each of the following electron-dot formulas: (6.5)

a. $\mathrm{H:H}$ **b.** $\mathrm{H:\underset{\cdot\cdot}{\overset{\cdot\cdot}{Br}}:}$ **c.** $\mathrm{:\underset{\cdot\cdot}{\overset{\cdot\cdot}{Br}}:\underset{\cdot\cdot}{\overset{\cdot\cdot}{Br}}:}$

6.86 State the number of valence electrons, bonding pairs, and lone pairs in each of the following electron-dot formulas: (6.5)

a. $\mathrm{H:\underset{\cdot\cdot}{\overset{H}{O}}:}$ **b.** $\mathrm{H:\underset{H}{\overset{\cdot\cdot}{N}}:H}$ **c.** $\mathrm{:\underset{\cdot\cdot}{\overset{\cdot\cdot}{Br}}:\underset{\cdot\cdot}{\overset{\cdot\cdot}{O}}:\underset{\cdot\cdot}{\overset{\cdot\cdot}{Br}}:}$

ADDITIONAL QUESTIONS AND PROBLEMS

6.87 Write the electron arrangement for the following: (6.1)
a. N^{3-} **b.** Mg^{2+} **c.** P^{3-} **d.** Al^{3+} **e.** Li^{+}

6.88 Write the electron arrangement for the following: (6.1)
a. K^{+} **b.** Na^{+} **c.** S^{2-} **d.** Cl^{-} **e.** Ca^{2+}

6.89 Consider a set of trigonal pyramidal molecules with the general formula AX_3.
a. How will the A-X bond polarity change down the group of X?
b. How will the X-A-X bond angle change when X becomes larger?
c. What will be the change in the molecular polarity when X becomes larger?

6.90 Consider an ion with the symbol Y^{3-} formed from a representative element. (6.1, 6.2, 6.3)
a. What is the group number of the element?
b. What is the electron-dot symbol of the element?
c. If Y is in Period 3, what is the element?
d. What is the formula of the compound formed from the barium ion and Y?

6.91 One of the ions of tin is tin(IV). (6.1, 6.2, 6.3, 6.4)
a. What is the symbol for this ion?
b. How many protons and electrons are in the ion?
c. What is the formula of tin(IV) oxide?
d. What is the formula of tin(IV) phosphate?

6.92 There are three common oxidation states of cobalt complexes: cobalt (I), (II), and (III). (6.1, 6.2, 6.3, 6.4)
a. What are the symbols for those ions?
b. What are the formulas of cobalt (II) and (III) iodides?
c. What are the formulas of cobalt (II) and (III) oxides?

6.93 Write the formula for each of the following ionic compounds: (6.2, 6.3)
a. tin(II) sulfide **b.** lead(IV) oxide
c. silver chloride **d.** calcium nitride
e. copper(I) phosphide **f.** chromium(II) bromide

6.94 Write the formula for each of the following ionic compounds: (6.2, 6.3)
a. nickel(III) oxide **b.** iron(III) sulfide
c. lead(II) sulfate **d.** chromium(III) iodide
e. lithium nitride **f.** gold(I) oxide

6.95 Name each of the following molecular compounds: (6.5)
a. NCl_3 **b.** N_2S_3 **c.** N_2O
d. IF **e.** BF_3 **f.** P_2O_5

6.96 Name each of the following molecular compounds: (6.5)
a. CBr_4 **b.** SF_6 **c.** BrCl
d. N_2O_4 **e.** SO_2 **f.** CS_2

6.97 Write the formula for each of the following molecular compounds: (6.5)
a. carbon sulfide **b.** diphosphorus pentoxide
c. dihydrogen sulfide **d.** sulfur dichloride

6.98 Write the formula for each of the following molecular compounds: (6.5)
a. silicon dioxide **b.** carbon tetrabromide
c. diphosphorus tetraiodide **d.** dinitrogen trioxide

6.99 Classify each of the following as ionic or molecular, and give its name: (6.3, 6.5)
a. $FeCl_3$ **b.** Na_2SO_4 **c.** NO_2
d. N_2 **e.** PF_5 **f.** CF_4

6.100 Classify each of the following as ionic or molecular, and give its name: (6.3, 6.5)
a. $Al_2(CO_3)_3$ **b.** ClF_5 **c.** H_2
d. Mg_3N_2 **e.** ClO_2 **f.** $CrPO_4$

6.101 Write the formula for each of the following: (6.3, 6.4, 6.5)
a. tin(II) carbonate **b.** lithium phosphide
c. silicon tetrachloride **d.** manganese(III) oxide
e. bromine **f.** calcium bromide

6.102 Write the formula for each of the following: (6.3, 6.4, 6.5)
a. sodium carbonate **b.** nitrogen dioxide
c. aluminum nitrate **d.** copper(I) nitride
e. potassium phosphate **f.** cobalt(III) sulfate

6.103 Select the more polar bond in each of the following pairs: (6.6)
a. C—N or C—O **b.** N—F or N—Br
c. Br—Cl or S—Cl **d.** Br—Cl or Br—I
e. N—F or N—O

6.104 Select the more polar bond in each of the following pairs: (6.6)
a. C—C or C—O **b.** P—Cl or P—Br
c. Si—S or Si—Cl **d.** F—Cl or F—Br
e. P—O or P—S

6.105 Calculate the electronegativity difference and classify each of the following bonds as nonpolar covalent, polar covalent, or ionic: (6.6)
a. Si—Cl **b.** C—C
c. Na—Cl **d.** C—H
e. F—F

6.106 Calculate the electronegativity difference and classify each of the following bonds as nonpolar covalent, polar covalent, or ionic: (6.6)
a. C—N **b.** Cl—Cl
c. K—Br **d.** H—H
e. N—F

6.107 Predict the shape and polarity of each of the following molecules: (6.7)
a. A central atom with three identical bonded atoms and one lone pair.
b. A central atom with two bonded atoms and two lone pairs.

6.108 Predict the shape and polarity of each of the following molecules: (6.7)
a. A main group atom with four identical bonded atoms and no lone pairs.
b. A main group atom with three bonded atoms that are not identical and has lone pairs.

6.109 Classify the following molecules as polar or nonpolar: (6.7)
a. NH_3 **b.** CH_3Cl **c.** SiF_4

6.110 Classify the following molecules as polar or nonpolar: (6.7)
a. GeH_4 **b.** SeO_2 **c.** SCl_2

6.111 Predict the shape and polarity of each of the following molecules: (6.7)
a. SI_2 **b.** PF_3

6.112 Predict the shape and polarity of each of the following molecules: (6.7)
a. H_2O **b.** CF_4

6.113 Indicate the major type of attractive force—(1) ionic bonds, (2) dipole–dipole attractions, (3) hydrogen bonds, (4) dispersion forces—that occurs between particles of the following substances: (6.8)
a. NH_3 **b.** HI **c.** I_2 **d.** Cs_2O

6.114 Indicate the major type of attractive force—(1) ionic bonds, (2) dipole–dipole attractions, (3) hydrogen bonds, (4) dispersion forces—that occurs between particles of the following substances: (6.8)
a. CHF_3 **b.** H_2O **c.** LiCl **d.** N_2

CHALLENGE QUESTIONS

The following groups of questions are related to the topics in this chapter. However, they do not all follow the chapter order, and they require you to combine concepts and skills from several sections. These questions will help you increase your critical thinking skills and prepare for your next exam.

6.115 Complete the following table for atoms or ions: (6.1)

Atom or Ion	Number of Protons	Number of Electrons	Electrons Lost/Gained
K^+			
	$12\ p^+$	$10\ e^-$	
	$8\ p^+$		$2\ e^-$ gained
		$10\ e^-$	$3\ e^-$ lost

6.116 Complete the following table for atoms or ions: (6.1)

Atom or Ion	Number of Protons	Number of Electrons	Electrons Lost/Gained
	$30\ p^+$		$2\ e^-$ lost
	$36\ p^+$	$36\ e^-$	
	$16\ p^+$		$2\ e^-$ gained
		$46\ e^-$	$4\ e^-$ lost

6.117 Identify the group number in the periodic table of X, a representative element, in each of the following ionic compounds: (6.2)
a. XCl_3 **b.** Al_2X_3 **c.** XCO_3

6.118 Identify the group number in the periodic table of X, a representative element, in each of the following ionic compounds: (6.2)
a. X_2O_3 **b.** X_2SO_3 **c.** Na_3X

6.119 Classify each of the following as ionic or molecular, and name each: (6.2, 6.3, 6.4, 6.5)
a. Na_2O **b.** WO_2 **c.** $BaCO_3$
d. NF_3 **e.** CS_2 **f.** Cs_3PO_4
g. $K(AuCl_4)$ **h.** Cl_2

6.120 Classify each of the following as ionic or molecular, and name each: (6.2, 6.3, 6.4, 6.5)
a. $FeCl_2$ **b.** Cl_2O_7 **c.** Br_2
d. $Ca_3(PO_4)_2$ **e.** PCl_3 **f.** $Al(NO_3)_3$
g. $PbCl_4$ **h.** $MgCO_3$

ANSWERS

Answers to Study Checks

6.1 **a.** 16 protons, 18 electrons
b. 13 protons, 10 electrons

6.2 K^+ and S^{2-}

6.3 Ca^{2+} and O^{2-}, CaO

6.4 gallium sulfide

6.5 manganese(III) sulfide

6.6 Cr_2O_3

6.7 $(NH_4)_3PO_4$

6.8 cobalt(II) phosphate

6.9 iron(II) nitrate

6.10 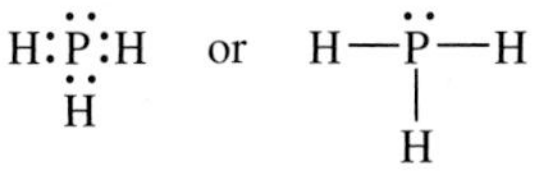

6.11 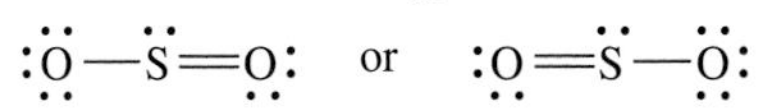

6.12 **a.** silicon tetrabromide **b.** dibromine oxide

6.13 IF_5

6.14 iron(III) nitrate

6.15 **a.** polar covalent (0.9) **b.** nonpolar covalent (0.0)
c. ionic (2.6)

6.16 The central atom sulfur has four electron groups: two bonded atoms and two lone pairs of electrons. The geometry of SCl_2 would be bent, 109°.

6.17 polar

6.18 The attractive forces between H_2S molecules are dipole–dipole attractions, whereas H_2O molecules have hydrogen bonds, which are stronger and require more energy to break.

Answers to Selected Questions and Problems

6.1 **a.** Na **b.** Mg **c.** Al **d.** Sr **e.** K

6.3 **a.** 2 *e*− lost **b.** 3 *e*− gained **c.** 1 *e*− gained
d. 2 *e*− lost **e.** 1 *e*− gained

6.5 **a.** Li, **b.** Cl^-, **c.** Ti^{4+} **d.** Ru^{3+}

6.7 **a.** Br^- **b.** Li^+ **c.** Se^{2-} **d.** In^{+3}

6.9 (a), (c) and (f)

6.11 **a.** Li_2O **b.** $CaCl_2$ **c.** BeF_2
d. BN **e.** AlF_3

6.13 **a.** K^+, S^{2-}, K_2S **b.** P^{3+}, Br^-, PBr_3
c. Al^{3+}, Cl^-, $AlCl_3$ **d.** Si^{4+}, O^{2-}, SiO_2

6.15 **a.** Potassium oxide **b.** Berilium chloride
c. Aluminium bromide **d.** Cesium chloride
e. Calcium oxide **f.** Calcium phosphide

6.17 **a.** iron (III) **b.** titanium (II)
c. Cobalt (III) **d.** Manganese (II)
e. Mercury (II) **f.** Copper

6.19 **a.** Iron (III)oxide **b.** Nickel (II) oxide;
c. copper (I) sulfate **d.** gold (III) chloride
e. cadmium (II) iodide **f.** palladium (II) bromide

6.21 **a.** Ag^+ **b.** Hg^{+2} **c.** Mn^{+2} **d.** Zn^{+2}

6.23 **a.** $MgCl_2$ **b.** Na_2S **c.** Cu_2O **d.** Zn_3P_2
e. AuN **f.** CoF_3

6.25 **a.** $CoCl_3$ **b.** PbO_2 **c.** AgI **d.** Ca_3N_2
e. Cu_3P **f.** $CrCl_2$

6.27 **a.** HCO_3^- **b.** NH_4^+ **c.** PO_3^{3-} **d.** ClO_3^-

6.29 **a.** Nitrate **b.** Perchlorate
c. Hydrogen sulfate **d.** Phosphite

6.31

	NO_2^-	CO_3^{2-}	HSO_4^-	PO_4^{3-}
Li^+	$LiNO_2$ Lithium nitrite	Li_2CO_3 Lithium carbonate	$LiHSO_4$ Lithium hydrogen sulfate	Li_3PO_4 Lithium phosphate
Cu^{2+}	$Cu(NO_2)_2$ Copper(II) nitrite	$CuCO_3$ Copper(II) carbonate	$Cu(HSO_4)_2$ Copper(II) hydrogen sulfate	$Cu_3(PO_4)_2$ Copper(II) phosphate
Ba^{2+}	$Ba(NO_2)_2$ Barium nitrite	$BaCO_3$ Barium carbonate	$Ba(HSO_4)_2$ Barium hydrogen sulfate	$Ba_3(PO_4)_2$ Barium phosphate

6.33 **a.** $Ba(OH)_2$ **b.** $NaHSO_4$ **c.** $Fe(NO_2)_2$
d. $Zn_3(PO_4)_2$ **e.** $Fe_2(CO_3)_3$

6.35 **a.** aluminum oxide **b.** magnesium phosphate
c. chromium(III) oxide **d.** sodium phosphate
e. bismuth(III) sulfide

6.37 Nonmetals share electrons to form molecular compounds.

6.39 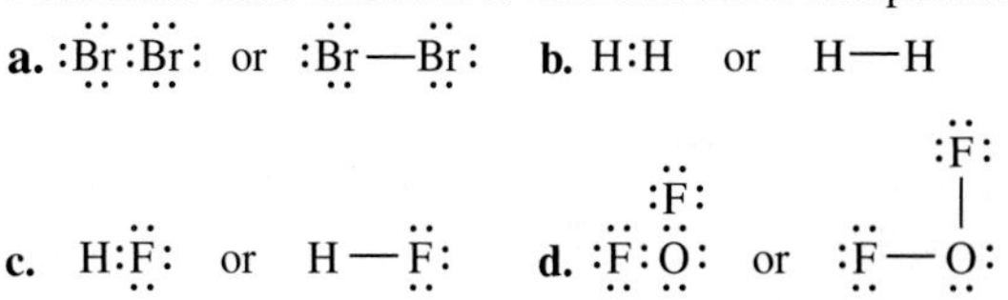

6.41 **a.** phosphorus tribromide **b.** dichlorine oxide
c. carbon tetrabromide **d.** hydrogen fluoride
e. nitrogen trifluoride

6.43 **a.** dinitrogen (nitrogen gas) **b.** disilicon hexabromide
c. titanium dioxide **d.** phosphorus pentachloride
e. selenium hexafluoride

6.45 **a.** CCl_4 **b.** CO **c.** PCl_3 **d.** N_2O_4

6.47 **a.** OF_2 **b.** BCl_3 **c.** N_2O_3 **d.** SF_6

6.49 **a.** dinitrogen (nitrogen gas); **b.** disilicon hexabromide;
c. titanium dioxide; **d.** phosphorus pentachloride;
e. selenium hexafluoride

6.51 **a.** increases **b.** decreases **c.** decreases

6.53 **a.** K, Na, Li **b.** Na, P, Cl **c.** Ca, Se, O

6.55 **a.** polar covalent **b.** ionic
c. polar covalent **d.** nonpolar covalent
e. polar covalent **f.** nonpolar covalent

6.57 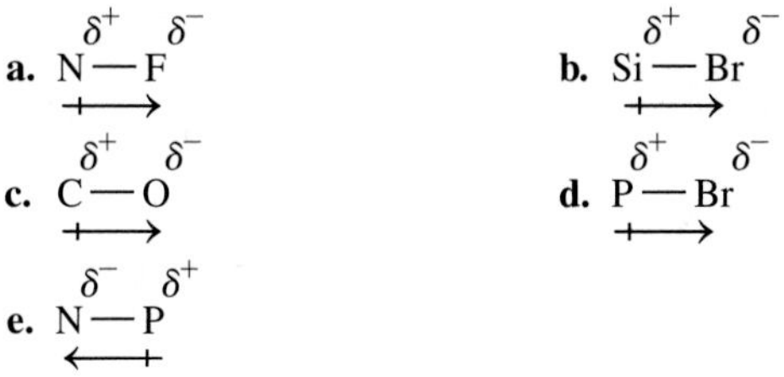

6.59 **a.** **6.** tetrahedral **b.** **5.** trigonal pyramidal
c. **3.** trigonal planar

6.61 **a.** tetrahedral **b.** linear
c. trigonal pyramidal **d.** trigonal planar

6.63 In both PH_3 and NH_3, there are three bonded atoms and one lone pair of electrons on the central atoms. The shapes of both are trigonal pyramidal.

6.65 **a.** bent, 109° **b.** tetrahedral **c.** bent, 120°

6.67 O_2 is a nonpolar molecule because there is a nonpolar covalent bond between O atoms, which have identical electronegativity values.

6.69 **a.** polar **b.** polar **c.** polar

6.71 **a.** dipole–dipole attractions
b. ionic bonds
c. dipole–dipole attractions
d. dispersion forces

6.73 **a.** hydrogen bonds
b. dipole–dipole attractions
c. dispersion forces
d. dispersion forces

6.75 **a.** By losing two valence electrons from the third energy level, magnesium achieves an octet in the second energy level.
b. The magnesium ion Mg^{2+} has the same electron arrangement as Ne (2,8).
c. Group 1A (1) and 2A (2) elements achieve octets by losing electrons to form compounds. Group 8A (18) elements are stable with octets (or two electrons for helium).

6.77 **a.** P^{3-} ion **b.** O atom
c. Zn^{2+} ion **d.** Fe^{3+} ion

6.79 **a.** X = Group 1A (1), Y = Group 6A (16)
b. ionic **c.** X^+ and Y^{2-} **d.** X_2Y
e. X_2S **f.** YCl_2 **g.** molecular

6.81 **a.** **2.** trigonal pyramidal, polar
b. **1.** bent, 109°, polar **c.** **3.** tetrahedral, nonpolar

6.83

Electron Arrangements		Cation	Anion	Formula of Compound	Name of Compound
2,8,2	2,5	Mg^{2+}	N^{3-}	Mg_3N_2	Magnesium nitride
2,8,8,1	2,6	K^+	O^{2-}	K_2O	Potassium oxide
2,8,3	2,8,7	Al^{3+}	Cl^-	$AlCl_3$	Aluminum chloride

6.85 **a.** two valence electrons, one bonding pair, no lone pairs
b. eight valence electrons, one bonding pair, three lone pairs
c. 14 valence electrons, one bonding pair, six lone pairs

6.87 **a.** 2,8 **b.** 2,8 **c.** 2,8,8 **d.** 2,8 **e.** 2

6.89 **a.** 2A (2) **b.** $\dot{\text{X}}\cdot$ **c.** Mg **d.** X_3N_2

6.91 **a.** Sn^{4+} **b.** 50 protons and 46 electrons
c. SnO_2 **d.** $Sn_3(PO_4)_4$

6.93 **a.** SnS **b.** PbO_2 **c.** AgCl
d. Ca_3N_2 **e.** Cu_3P **f.** $CrBr_2$

6.95 **a.** nitrogen trichloride **b.** dinitrogen trisulfide
c. dinitrogen oxide **d.** iodine fluoride
e. boron trifluoride **f.** diphosphorus pentoxide

6.97 **a.** CS **b.** P_2O_5 **c.** H_2S **d.** SCl_2

6.99 **a.** ionic, iron(III) chloride
b. ionic, sodium sulfate
c. molecular, nitrogen dioxide
d. molecular, nitrogen
e. molecular, phosphorus pentafluoride
f. molecular, carbon tetrafluoride

6.101 **a.** $SnCO_3$ **b.** Li_3P **c.** $SiCl_4$
d. Mn_2O_3 **e.** Br_2 **f.** $CaBr_2$

6.103 **a.** C—O **b.** N—F **c.** S—Cl
d. Br—I **e.** N—F

6.105 **a.** polar covalent (Cl 3.0 − Si 1.8 = 1.2)
b. nonpolar covalent (C 2.5 − C 2.5 = 0.0)
c. ionic (Cl 3.0 − Na 0.9 = 2.1)
d. nonpolar covalent (C 2.5 − H 2.1 = 0.4)
e. nonpolar covalent (F 4.0 − F 4.0 = 0.0)

6.107 **a.** trigonal pyramidal, polar
b. bent 109°, polar

6.109 **a.** polar **b.** polar **c.** nonpolar

6.111 **a.** bent 109°, nonpolar
b. trigonal pyramidal, polar

6.113 **a.** (3) hydrogen bonds **b.** (2) dipole–dipole attractions
c. (4) dispersion forces **d.** (1) ionic bonds

6.115

Atom or Ion	Number of Protons	Number of Electrons	Electrons Lost/Gained
K^+	19 p^+	18 e^-	1 e^- lost
Mg^{2+}	12 p^+	10 e^-	2 e^- lost
O^{2-}	8 p^+	10 e^-	2 e^- gained
Al^{3+}	13 p^+	10 e^-	3 e^- lost

6.117 **a.** Group 3A (13) **b.** Group 6A (16) **c.** Group 2A (2)

6.119 **a.** ionic, lithium oxide
b. ionic, chromium(II) nitrate
c. ionic, magnesium bicarbonate or magnesium hydrogen carbonate
d. molecular, nitrogen trifluoride
e. ionic, calcium chloride
f. ionic, potassium phosphate
g. ionic, gold(III) sulfite
h. molecular, iodine

CI.7 For parts **a** to **f**, consider the loss of electrons by atoms of the element X, and a gain of electrons by atoms of the element Y, if X is in Group 2A (2), Period 3, and Y is in Group 7A (17), Period 3. (4.6, 6.1, 6.2, 6.3, 6.6)

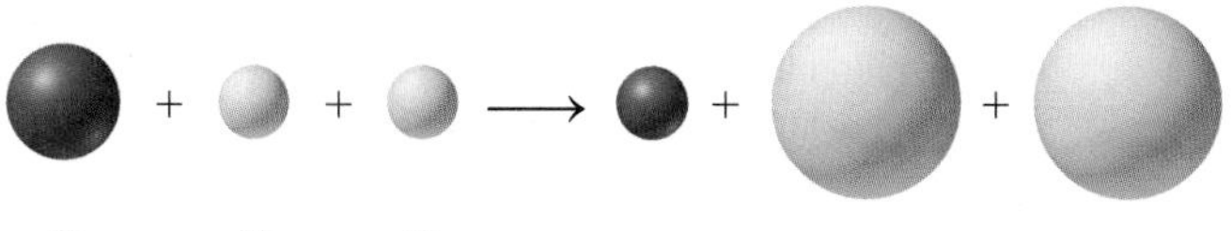

X Y Y

a. Which reactant has the higher electronegativity?
b. What are the ionic charges of **X** and **Y** in the product?
c. Write the electron arrangements for the atoms **X** and **Y**.
d. Write the electron arrangements for the ions of **X** and **Y**.
e. Give the names for the noble gases with the same electron arrangements as each of these ions.
f. Write the formula and name for the ionic compound formed by the ions of **X** and **Y**.

CI.8 A sterling silver bracelet, which is 92.5% silver by mass, has a volume of 25.6 cm^3 and a density of 10.2 g/cm^3. (2.10, 4.5, 5.2, 5.4)

Sterling silver is 92.5% silver by mass.

a. What is the mass, in kilograms, of the bracelet?
b. Determine the number of protons and neutrons in each of the two stable isotopes of silver:

$^{107}_{47}Ag$ $^{109}_{47}Ag$

c. Silver (Ag) has two naturally occurring isotopes: Ag-107 (106.905 amu) with an abundance of 51.84% and Ag-109 (108.905 amu) with an abundance of 48.16%. Calculate the atomic mass for silver.
d. Ag-112 decays by beta emission. Write the balanced nuclear equation for the decay of Ag-112.
e. A 64.0-μCi sample of Ag-112 decays to 8.00 μCi in 9.3 h. What is the half-life of Ag-112?

CI.9 Some of the isotopes of silicon are listed in the following table: (4.5, 4.6, 5.2, 5.4, 6.6, 6.7)

Isotope	% Natural Abundance	Atomic Mass	Half-Life (radioactive)	Radiation
$^{27}_{14}Si$		26.987	4.9 s	Positron
$^{28}_{14}Si$	92.23	27.977	Stable	None
$^{29}_{14}Si$	4.683	28.976	Stable	None
$^{30}_{14}Si$	3.087	29.974	Stable	None
$^{31}_{14}Si$		30.975	2.6 h	Beta

a. In the following table, indicate the number of protons, neutrons, and electrons for each isotope listed:

Isotope	Number of Protons	Number of Neutrons	Number of Electrons
$^{27}_{14}Si$			
$^{28}_{14}Si$			
$^{29}_{14}Si$			
$^{30}_{14}Si$			
$^{31}_{14}Si$			

b. What is the electron arrangement of silicon?
c. Calculate the atomic mass for silicon using the isotopes that have a natural abundance.
d. Write the balanced nuclear equations for the positron emission of Si-27 and the beta decay of Si-31.
e. Draw the electron-dot formula and predict the shape of $SiCl_4$.
f. How many hours are needed for a sample of Si-31 with an activity of 16 μCi to decay to 2.0 μCi?

CI.10 The iceman known as Ötzi was discovered in a high mountain pass on the Austrian–Italian border. Samples of his hair and bones had carbon-14 activity that was 50% of that present in new hair or bone. Carbon-14 undergoes beta decay and has a half-life of 5730 yr. (5.2, 5.4)

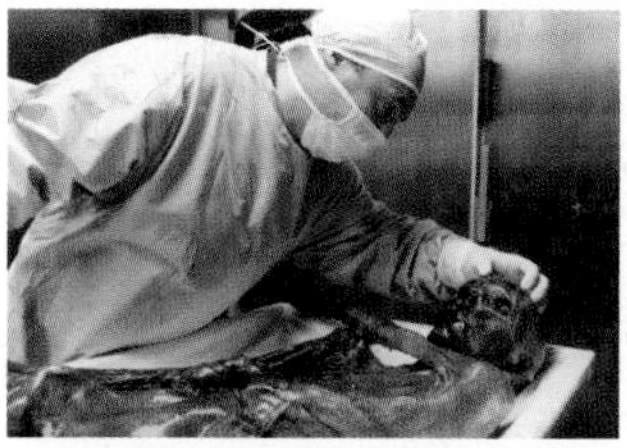

The mummified remains of Ötzi were discovered in 1991.

a. How long ago did Ötzi live?
b. Write a balanced nuclear equation for the decay of carbon-14.

CI.11 K^+ is an electrolyte required by the human body and found in many foods as well as salt substitutes. One of the isotopes of potassium is potassium-40, which has a natural abundance of 0.012% and a half-life of 1.30×10^9 yr. The isotope potassium-40 decays to calcium-40 or to argon-40. A typical activity for potassium-40 is 7.0 μCi per gram. (5.2, 5.3, 5.4)

Potassium chloride is used as a salt substitute.

a. Write a balanced nuclear equation for each type of decay and identify the particle emitted.
b. A shaker of salt substitute contains 1.6 oz of K. What is the activity, in milliCuries and becquerels, of the potassium in the shaker?

CI.12 Of much concern to environmentalists is radon-222, which is a radioactive noble gas that can seep from the ground into basements of homes and buildings. Radon-222 is a product of the decay of radium-226 that occurs naturally in rocks and soil in much of the United States. Radon-222, which has a half-life of 3.8 days, decays by emitting an alpha particle. Radon-222, which is a gas, can be inhaled into the lungs where it is strongly associated with lung cancer. Radon levels in a home can be measured with a home radon-detection kit. Environmental agencies have set the maximum level of radon-222 in a home at 4 picocuries per liter (pCi/L) of air. (5.2, 5.3, 5.4)

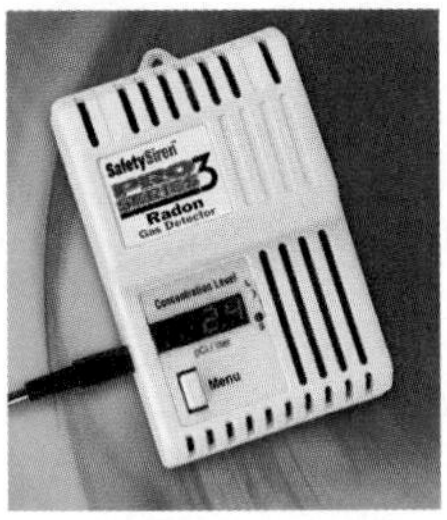

A home detection kit is used to measure the level of radon-222.

a. Write the balanced nuclear equation for the decay of Ra-226.

b. Write the balanced nuclear equation for the decay of Rn-222.

c. If a room contains 24 000 atoms of radon-222, how many atoms of radon-222 remain after 15.2 days?

d. Suppose a room has a volume of 72 000 L (7.2×10^4 L). If the radon level is the maximum allowed (4 pCi/L), how many alpha particles are emitted from Rn-222 in 1 day? (1 Ci = 3.7×10^{10} disintegrations/s)

ANSWERS

CI.7 **a.** Y has the higher electronegativity.

b. X^{2+}, Y^{-}

c. X = 2,8,2 Y = 2,8,7

d. X^{2+} = 2,8 Y^{-} = 2,8,8

e. X^{2+} has the same electron arrangement as Ne. Y^{-} has the same electron arrangement as Ar.

f. $MgCl_2$, magnesium chloride

CI.9 **a.**

Isotope	Number of Protons	Number of Neutrons	Number of Electrons
$^{27}_{14}Si$	14	13	14
$^{28}_{14}Si$	14	14	14
$^{29}_{14}Si$	14	15	14
$^{30}_{14}Si$	14	16	14
$^{31}_{14}Si$	14	17	14

b. 2,8,4

c. 28.09 amu

d. $^{27}_{14}Si \longrightarrow {}^{27}_{13}Al + {}^{0}_{+1}e$

$^{31}_{14}Si \longrightarrow {}^{31}_{15}P + {}^{0}_{-1}e$

e. 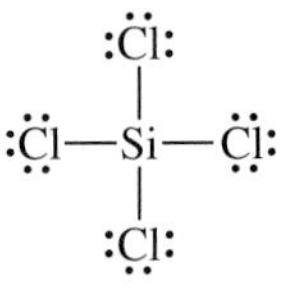 Tetrahedral

f. 7.8 h

CI.11 **a.** $^{40}_{19}K \longrightarrow {}^{40}_{20}Ca + {}^{0}_{-1}e$ Beta particle

$^{40}_{19}K \longrightarrow {}^{40}_{18}Ar + {}^{0}_{+1}e$ Positron

b. 3.8×10^{-5} mCi; 1.4×103 Bq

7

Chemical Quantities and Reactions

Natalie was recently diagnosed with mild emphysema due to secondhand cigarette smoke. She has been referred to Angela, an exercise physiologist, who begins to assess Natalie's condition by connecting her to an EKG, a pulse oximeter, and a blood pressure cuff. The EKG tracks Natalie's heart rate and rhythm, the pulse oximeter tracks the oxygen levels in her blood, while the blood pressure cuff determines the pressure exerted by the heart in pumping her blood. Natalie then walks on a treadmill to determine her overall physical condition.

Based on Natalie's results, Angela creates a workout regime. They begin with low-intensity exercises that utilize smaller muscles instead of larger muscles, which require more O_2 and can deplete a significant amount of the O_2 in her blood. During the exercises, Angela continues to monitor Natalie's heart rate, blood O_2 level, and blood pressure to ensure that Natalie is exercising at a level that will enable her to become stronger without breaking down muscle due to a lack of oxygen.

CAREER Exercise Physiologist

Exercise physiologists work with athletes as well as patients who have been diagnosed with diabetes, heart disease, pulmonary disease, or other chronic disability or disease. Patients who have been diagnosed with one of these diseases are often prescribed exercise as a form of treatment, and they are referred to an exercise physiologist. The exercise physiologist evaluates the patient's overall health and then creates a customized exercise program for that individual.

A program for an athlete might focus on reducing the number of injuries, while a program for a cardiac patient would focus on strengthening the heart muscles. The exercise physiologist also monitors the patient for improvement and notes if the exercise is reducing or reversing the progression of the disease.

LOOKING AHEAD

In chemistry, we calculate and measure the amounts of substances to use in lab. Actually, measuring the amount of a substance is something you do every day. When you cook, you measure out the proper amounts of ingredients so you don't have too much of one or too little of another. At the gas station, you pump a certain amount of fuel into your gas tank. If you paint the walls of a room, you measure the area and purchase the amount of paint that will cover the walls. In the lab, the chemical formula of a substance tells us the number and kinds of atoms it has, which we then use to determine the mass of the substance to use in an experiment.

Chemical reactions occur everywhere. The fuel in our cars burns with oxygen to make the car move and run the air conditioner. When we cook our food or bleach our hair, chemical reactions take place. In our bodies, chemical reactions convert food into molecules that build muscles and move them. In the leaves of trees and plants, carbon dioxide and water are converted into carbohydrates.

Some chemical reactions are simple, whereas others are quite complex. However, they can all be written with chemical equations that chemists use to describe chemical reactions. In every chemical reaction, the atoms in the reacting substances, called *reactants*, are rearranged to give new substances called *products*.

We do the same thing at home when we use a recipe to make cookies. At the automotive repair shop, a mechanic goes through a similar process when adjusting the fuel system of an engine to allow for the correct amounts of fuel and oxygen. In the body, a certain amount of O_2 must reach the tissues for efficient metabolic reactions. If the oxygenation of the blood is low, the respiratory therapist will oxygenate the patient and recheck the blood levels.

CHAPTER READINESS*

KEY MATH SKILLS

- Calculating a Percentage (1.4C)
- Solving Equations (1.4D)
- Writing Numbers in Scientific Notation (1.4F)

CORE CHEMISTRY SKILLS

- Counting Significant Figures (2.2)
- Using Significant Figures in Calculations (2.3)
- Writing Conversion Factors from Equalities (2.5)
- Using Conversion Factors (2.6)
- Using Energy Units (3.4)
- Writing Ionic Formulas (6.2)
- Naming Ionic Compounds (6.3)
- Writing the Names and Formulas for Molecular Compounds (6.5)

*These Key Math Skills and Core Chemistry Skills from previous chapters are listed here for your review as you proceed to the new material in this chapter.

LEARNING GOAL

Use Avogadro's number to determine the number of particles in a given number of moles.

7.1 The Mole

At the grocery store, you buy eggs by the dozen or soda by the case. In an office-supply store, pencils are ordered by the gross and paper by the ream. The terms *dozen*, *case*, *gross*, and *ream* are used to count the number of items present. For example, when you buy a dozen eggs, you know that there will be 12 eggs in the carton.

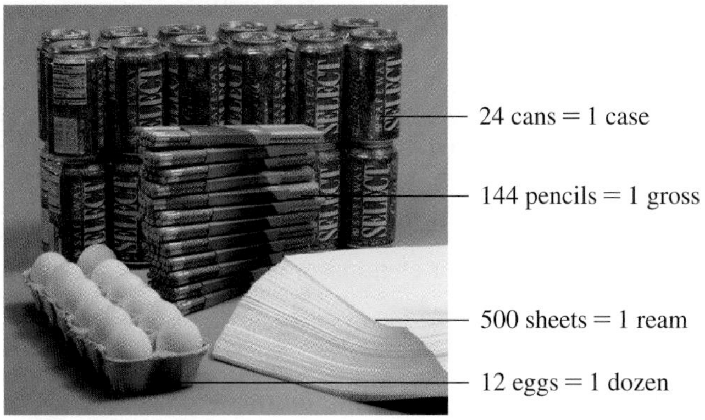

Collections of items include dozen, gross, ream, and mole.

Avogadro's Number

In chemistry, particles such as atoms, molecules, and ions are counted by the **mole**, which contains 6.02×10^{23} items. This value, known as **Avogadro's number**, is a very big number because atoms are so small that it takes an extremely large number of atoms to provide a sufficient amount to weigh and use in chemical reactions. Avogadro's number is named for Amedeo Avogadro (1776–1856), an Italian physicist.

CORE CHEMISTRY SKILL

Converting Particles to Moles

Avogadro's number

$$602\,000\,000\,000\,000\,000\,000\,000 = 6.02 \times 10^{23}$$

One mole of any element always contains Avogadro's number of atoms. For example, 1 mole of carbon contains 6.02×10^{23} carbon atoms; 1 mole of aluminum contains 6.02×10^{23} aluminum atoms; 1 mole of sulfur contains 6.02×10^{23} sulfur atoms.

$$1 \text{ mole of an element} = 6.02 \times 10^{23} \text{ atoms of that element}$$

One mole of sulfur contains 6.02×10^{23} sulfur atoms.

Avogadro's number tells us that one mole of a compound contains 6.02×10^{23} of the particular type of particles that make up that compound. One mole of a molecular compound contains Avogadro's number of molecules. For example, 1 mole of CO_2 contains 6.02×10^{23} molecules of CO_2. One mole of an ionic compound contains Avogadro's number of **formula units**, which are the groups of ions represented by the formula of an ionic compound. One mole of NaCl contains 6.02×10^{23} formula units of NaCl (Na^+, Cl^-). Table 7.1 gives examples of the number of particles in some 1-mole quantities.

TABLE 7.1 Number of Particles in 1-Mole Samples

Substance	Number and Type of Particles
1 mole of Al	6.02×10^{23} atoms of Al
1 mole of S	6.02×10^{23} atoms of S
1 mole of water (H_2O)	6.02×10^{23} molecules of H_2O
1 mole of vitamin C ($C_6H_8O_6$)	6.02×10^{23} molecules of vitamin C
1 mole of NaCl	6.02×10^{23} formula units of NaCl

We use Avogadro's number as a conversion factor to convert between the moles of a substance and the number of particles it contains.

$$\frac{6.02 \times 10^{23} \text{ particles}}{1 \text{ mole}} \quad \text{and} \quad \frac{1 \text{ mole}}{6.02 \times 10^{23} \text{ particles}}$$

For example, we use Avogadro's number to convert 4.00 moles of sulfur to atoms of sulfur.

$$4.00 \cancel{\text{moles S}} \times \underbrace{\frac{6.02 \times 10^{23} \text{ atoms S}}{1 \cancel{\text{mole S}}}}_{\text{Avogadro's number as a conversion factor}} = 2.41 \times 10^{24} \text{ atoms of S}$$

We also use Avogadro's number to convert 3.01×10^{24} molecules of CO_2 to moles of CO_2.

$$3.01 \times 10^{24}\ \cancel{\text{molecules } CO_2} \times \underbrace{\frac{1 \text{ mole } CO_2}{6.02 \times 10^{23}\ \cancel{\text{molecules } CO_2}}}_{\text{Avogadro's number as a conversion factor}} = 5.00 \text{ moles of } CO_2$$

In calculations that convert between moles and particles, the number of moles will be a small number compared to the number of atoms or molecules, which will be large.

SAMPLE PROBLEM 7.1 Calculating the Number of Molecules

How many molecules are present in 1.75 moles of carbon dioxide, CO_2?

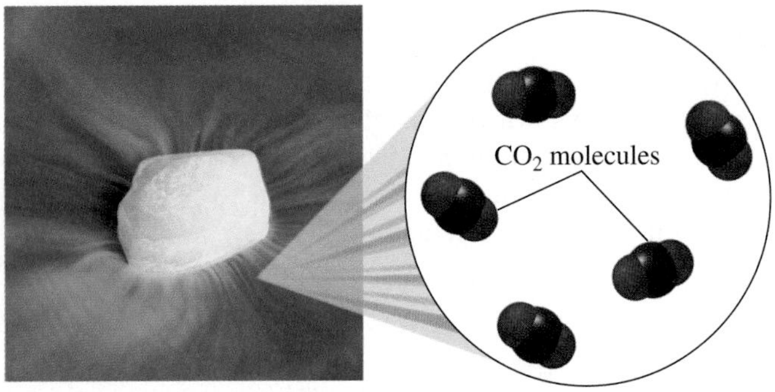

The solid form of carbon dioxide is known as "dry ice."

Guide to Calculating the Atoms or Molecules of a Substance

STEP 1 State the given and needed quantities.

STEP 2 Write a plan to convert moles to atoms or molecules.

STEP 3 Use Avogadro's number to write conversion factors.

STEP 4 Set up the problem to calculate the number of particles.

SOLUTION

STEP 1 State the given and needed quantities.

ANALYZE THE PROBLEM	Given	Need
	1.75 moles of CO_2	molecules of CO_2

STEP 2 Write a plan to convert moles to atoms or molecules.

moles of CO_2 $\xrightarrow{\text{Avogadro's number}}$ molecules of CO_2

STEP 3 Use Avogadro's number to write conversion factors.

$$1 \text{ mole of } CO_2 = 6.02 \times 10^{23} \text{ molecules of } CO_2$$

$$\frac{6.02 \times 10^{23} \text{ molecules } CO_2}{1 \text{ mole } CO_2} \quad \text{and} \quad \frac{1 \text{ mole } CO_2}{6.02 \times 10^{23} \text{ molecules } CO_2}$$

STEP 4 Set up the problem to calculate the number of particles.

$$1.75\ \cancel{\text{moles } CO_2} \times \frac{6.02 \times 10^{23} \text{ molecules } CO_2}{1\ \cancel{\text{mole } CO_2}} = 1.05 \times 10^{24} \text{ molecules of } CO_2$$

STUDY CHECK 7.1

How many moles of water, H_2O, contain 2.60×10^{23} molecules of water?

Moles of Elements in a Chemical Formula

We have seen that the subscripts in the chemical formula of a compound indicate the number of atoms of each type of element in the compound. For example, aspirin, $C_9H_8O_4$, is a drug used to reduce pain and inflammation in the body. Using the subscripts in the chemical formula of aspirin shows that there are 9 carbon atoms, 8 hydrogen atoms, and 4 oxygen atoms in each molecule of aspirin. The subscripts also tell us the number of moles of each element in 1 mole of aspirin: 9 moles of C atoms, 8 moles of H atoms, and 4 moles of O atoms.

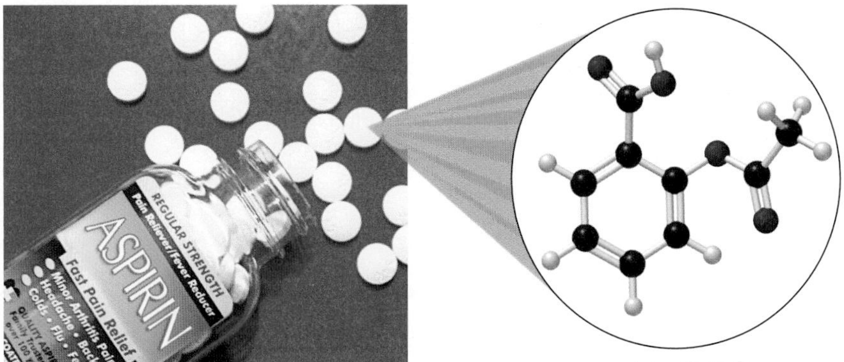

Aspirin $C_9H_8O_4$

Number of atoms in 1 molecule
Carbon (C) Hydrogen (H) Oxygen (O)

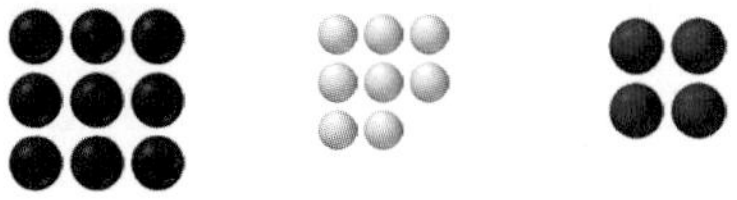

The chemical formula subscripts specify the $C_9H_8O_4$

	Carbon	**Hydrogen**	**Oxygen**
Atoms in 1 molecule	9 atoms of C	8 atoms of H	4 atoms of O
Moles of each element in 1 mole	9 moles of C	8 moles of H	4 moles of O

Using the subscripts from the formula, $C_9H_8O_4$, we can write the conversion factors for each of the elements in 1 mole of aspirin.

$$\frac{9 \text{ moles C}}{1 \text{ mole } C_9H_8O_4} \quad \frac{8 \text{ moles H}}{1 \text{ mole } C_9H_8O_4} \quad \frac{4 \text{ moles O}}{1 \text{ mole } C_9H_8O_4}$$

$$\frac{1 \text{ mole } C_9H_8O_4}{9 \text{ moles C}} \quad \frac{1 \text{ mole } C_9H_8O_4}{8 \text{ moles H}} \quad \frac{1 \text{ mole } C_9H_8O_4}{4 \text{ moles O}}$$

SAMPLE PROBLEM 7.2 Calculating the Moles of an Element in a Compound

Propyl acetate, $C_5H_{10}O_2$, gives the odor and taste to pears. How many moles of C are present in 1.50 moles of propyl acetate?

SOLUTION

STEP 1 State the given and needed quantities.

ANALYZE THE PROBLEM	Given	Need
	1.50 moles of propyl acetate, chemical formula $C_5H_{10}O_2$	moles of C

STEP 2 Write a plan to convert moles of compound to moles of an element.

moles of $C_5H_{10}O_2$ → Subscript → moles of C atoms

Guide to Calculating the Moles of an Element in a Compound

STEP 1 State the given and needed quantities.

STEP 2 Write a plan to convert moles of compound to moles of an element.

STEP 3 Write equalities and conversion factors using subscripts.

STEP 4 Set up the problem to calculate the moles of an element.

The compound propyl acetate provides the odor and taste of pears.

STEP 3 Write equalities and conversion factors using subscripts.

$$1 \text{ mole of } C_5H_{10}O_2 = 5 \text{ moles of C}$$

$$\frac{5 \text{ moles C}}{1 \text{ mole } C_5H_{10}O_2} \quad \text{and} \quad \frac{1 \text{ mole } C_5H_{10}O_2}{5 \text{ moles C}}$$

STEP 4 Set up the problem to calculate the moles of an element.

$$1.50 \cancel{\text{moles } C_5H_{10}O_2} \times \frac{5 \text{ moles C}}{1 \cancel{\text{mole } C_5H_{10}O_2}} = 7.50 \text{ moles of C}$$

STUDY CHECK 7.2

How many moles of propyl acetate, $C_5H_{10}O_2$, contain 0.480 mole of O?

QUESTIONS AND PROBLEMS

7.1 The Mole

LEARNING GOAL Use Avogadro's number to determine the number of particles in a given number of moles.

7.1 What is a mole?

7.2 What is Avogadro's number?

7.3 Calculate each of the following:
- **a.** Number of N atoms in 0.700mole of N_2
- **b.** Number of CO molecules in 2.36moles of CO
- **c.** Moles of Co in 7.84×10^{22} atoms of Co
- **d.** Moles of CH_4 in 5.62×10^{24} molecules of CH_4

7.4 Calculate each of the following:
- **a.** number of O atoms in 0.0180 mole of dry ice
- **b.** number of K in 14.5 mole of K_2CO_3
- **c.** mole of O in 8.60×10^{21} molecules of H_2SO_4
- **d.** mole of C_6H_6 in 6.90×10^{23} molecules of C_6H_6

7.5 Calculate each of the following quantities in 5.00 moles of H_2SO_4:
- **a.** moles of H
- **b.** moles of O
- **c.** atoms of S
- **d.** atoms of O

7.6 Calculate each of the following quantities in 0.365 moles of C_3H_6O:
- **a.** moles of C
- **b.** moles of O
- **c.** atoms of H
- **d.** atoms of C

7.7 Oxepine ($C_6H_{12}O$) is a seven-membered ring oxacycle.
- **a.** How many moles of carbon are in 1.0 millimoles of oxepine?
- **b.** How many moles of H are 1.0 moles of oxepine?
- **c.** How many moles of O are in 5 moles of oxepine?

7.8 Aluminum sulfate, $Al_2(SO_4)_3$, is used in some antiperspirants.
- **a.** How many moles of S are present in 3.0 moles of $Al_2(SO_4)_3$?
- **b.** How many moles of aluminum ions (Al^{3+}) are present in 0.40 mole of $Al_2(SO_4)_3$?
- **c.** How many moles of sulfate ions (SO_4^{2-}) are present in 1.5 moles of $Al_2(SO_4)_3$?

LEARNING GOAL

Calculate the molar mass for a substance given its chemical formula; use molar mass to convert between grams and moles.

7.2 Molar Mass and Calculations

A single atom or molecule is much too small to weigh, even on the most accurate balance. In fact, it takes a huge number of atoms or molecules to make enough of a substance for you to see. An amount of water that contains Avogadro's number of water molecules is only a few sips. In the laboratory, we can use a balance to weigh out Avogadro's number of particles or 1 mole of a substance.

For any element, the quantity called **molar mass** is the quantity in grams that equals the atomic mass of that element. We are counting 6.02×10^{23} atoms of an element when we weigh out the number of grams equal to its molar mass. For example, carbon has an atomic mass of 12.01 on the periodic table. This means 1 mole of carbon atoms has a mass of 12.01 g. Then to obtain 1 mole of carbon atoms, we would weigh out 12.01 g of carbon. The molar mass of carbon is found by looking at its atomic mass on the periodic table.

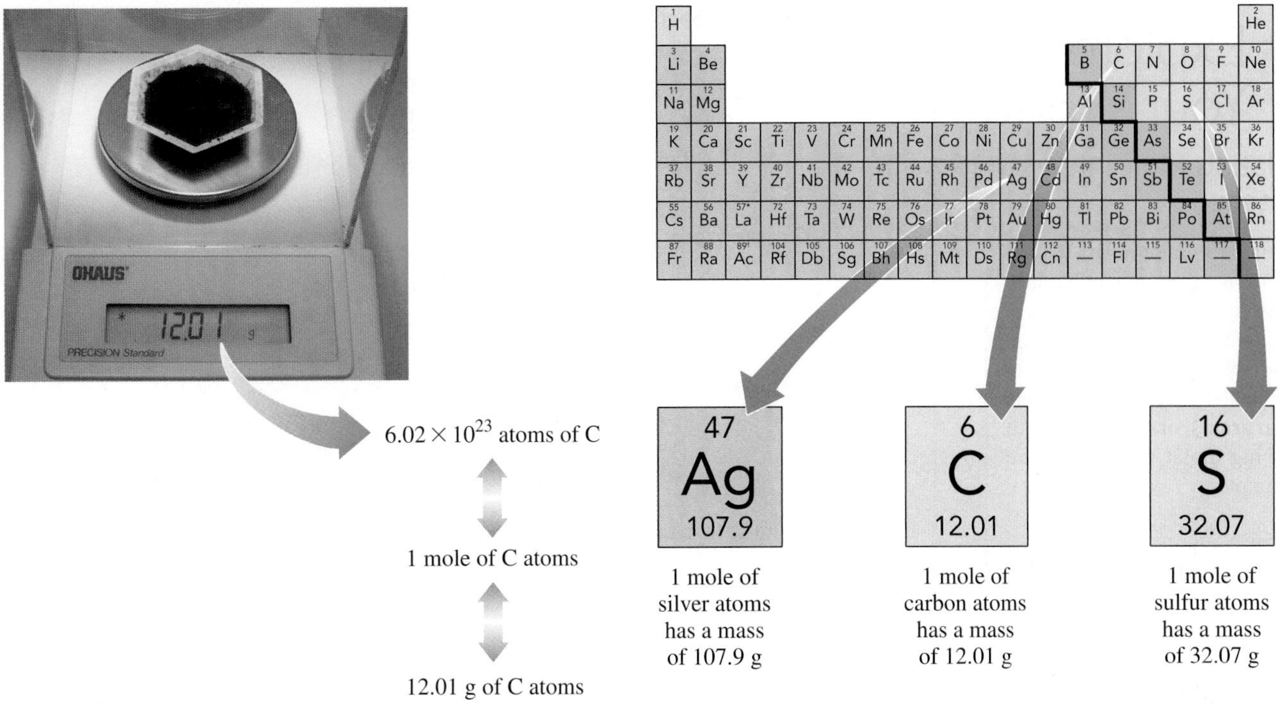

The molar mass of an element is useful to convert moles of an element to grams, or grams to moles. For example, 1 mole of sulfur has a mass of 32.07 g. We can express the molar mass of sulfur as an equality

1 mole of S = 32.07 g of S

From this equality, two conversion factors can be written as

$$\frac{32.07 \text{ g S}}{1 \text{ mole S}} \quad \text{and} \quad \frac{1 \text{ mole S}}{32.07 \text{ g S}}$$

CORE CHEMISTRY SKILL

Calculating Molar Mass

Lithium carbonate produces a red color in fireworks.

Molar Mass of a Compound

To calculate the molar mass for a compound, we multiply the molar mass of each element by its subscript in the formula, and add the results as shown in Sample Problem 7.3.

SAMPLE PROBLEM 7.3 Calculating the Molar Mass for a Compound

Calculate the molar mass for lithium carbonate, Li_2CO_3, used to produce red color in fireworks.

SOLUTION

ANALYZE THE PROBLEM	Given	Need
	formula Li_2CO_3	molar mass of Li_2CO_3

STEP 1 Obtain the molar mass of each element.

$$\frac{6.941 \text{ g Li}}{1 \text{ mole Li}} \qquad \frac{12.01 \text{ g C}}{1 \text{ mole C}} \qquad \frac{16.00 \text{ g O}}{1 \text{ mole O}}$$

Guide to Calculating Molar Mass

STEP 1
Obtain the molar mass of each element.

STEP 2
Multiply each molar mass by the number of moles (subscript) in the formula.

STEP 3
Calculate the molar mass by adding the masses of the elements.

Explore Your World

Calculating Moles in the Kitchen

The labels on food products list the components in grams and milligrams. It is possible to convert those values to moles, which is also interesting. Read the labels of some products in the kitchen and convert the amounts given in grams or milligrams to moles using molar mass. Here are some examples.

Questions

1. How many moles of NaCl are in a box of salt?
2. How many moles of sugar are contained in a 5-lb bag of sugar if sugar has the formula $C_{12}H_{22}O_{11}$?
3. A serving of cereal contains 90 mg of potassium. If there are 11 servings of cereal in the box, how many moles of K^+ are present in the cereal in the box?

STEP 2 Multiply each molar mass by the number of moles (subscript) in the formula.

Grams from 2 moles of Li

$$2\ \cancel{\text{moles Li}} \times \frac{6.941\text{ g Li}}{1\ \cancel{\text{mole Li}}} = 13.88\text{ g of Li}$$

Grams from 1 mole of C

$$1\ \cancel{\text{mole C}} \times \frac{12.01\text{ g C}}{1\ \cancel{\text{mole C}}} = 12.01\text{ g of C}$$

Grams from 3 moles of O

$$3\ \cancel{\text{moles O}} \times \frac{16.00\text{ g O}}{1\ \cancel{\text{mole O}}} = 48.00\text{ g of O}$$

STEP 3 Calculate the molar mass by adding the masses of the elements.

2 moles of Li	=	13.88 g of Li
1 mole of C	=	12.01 g of C
3 moles of O	=	+48.00 g of O
Molar mass of Li_2CO_3	=	73.89 g

STUDY CHECK 7.3

Calculate the molar mass for salicylic acid, $C_7H_6O_3$.

Figure 7.1 shows some 1-mole quantities of substances. Table 7.2 lists the molar mass for several 1-mole samples.

1-Mole Quantities

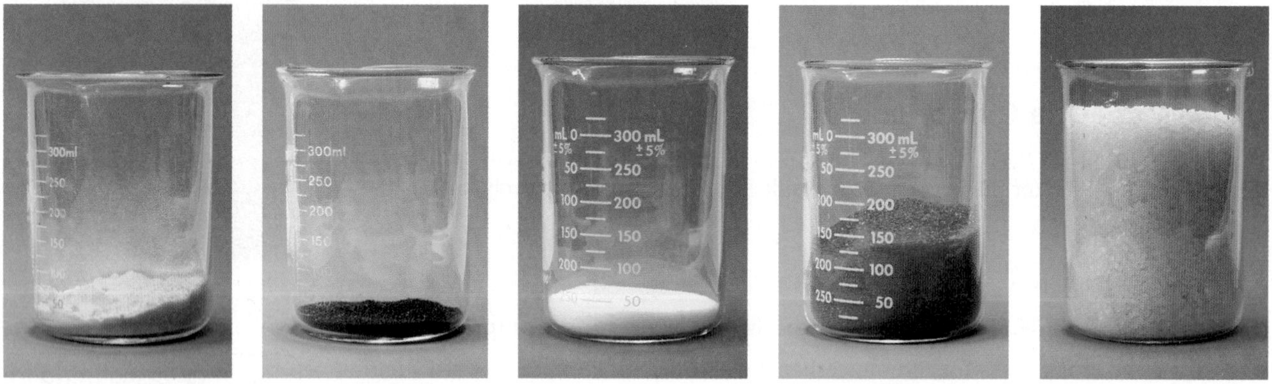

S Fe NaCl $K_2Cr_2O_7$ $C_{12}H_{22}O_{11}$

FIGURE 7.1 ▶ 1-mole samples: sulfur, S (32.07 g); iron, Fe (55.85 g); salt, NaCl (58.44 g); potassium dichromate, $K_2Cr_2O_7$ (294.2 g); and sucrose, $C_{12}H_{22}O_{11}$ (342.3 g).

Q How is the molar mass for $K_2Cr_2O_7$ obtained?

TABLE 7.2 The Molar Mass of Selected Elements and Compounds

Substance	Molar Mass
1 mole of C	12.01 g
1 mole of Na	22.99 g
1 mole of Fe	55.85 g
1 mole of NaF	41.99 g
1 mole of $C_6H_{12}O_6$ (glucose)	180.16 g
1 mole of $C_8H_{10}N_4O_2$ (caffeine)	194.20 g

Calculations Using Molar Mass

We can now change from moles to grams, or grams to moles, using the conversion factors derived from the molar mass as shown in Sample Problem 7.4. (Remember, you must determine the molar mass of the substance first.)

SAMPLE PROBLEM 7.4 Converting Moles of an Element to Grams

Silver metal is used in the manufacture of tableware, mirrors, jewelry, and dental alloys. If the design for a piece of jewelry requires 0.750 mole of silver, how many grams of silver are needed?

SOLUTION

STEP 1 State the given and needed quantities.

ANALYZE THE PROBLEM	Given	Need
	0.750 mole of Ag	grams of Ag

STEP 2 Write a plan to convert moles to grams.

moles of Ag → Molar mass factor → grams of Ag

STEP 3 Determine the molar mass and write conversion factors.

$$1 \text{ mole of Ag} = 107.9 \text{ g of Ag}$$
$$\frac{107.9 \text{ g Ag}}{1 \text{ mole Ag}} \text{ and } \frac{1 \text{ mole Ag}}{107.9 \text{ g Ag}}$$

STEP 4 Set up the problem to convert moles to grams.
Calculate the grams of silver using the molar mass factor that cancels mole Ag.

$$0.750 \cancel{\text{mole Ag}} \times \frac{107.9 \text{ g Ag}}{1 \cancel{\text{mole Ag}}} = 80.9 \text{ g of Ag}$$

STUDY CHECK 7.4

Calculate the number of grams of gold (Au) present in 0.124 mole of gold.

CORE CHEMISTRY SKILL

Using Molar Mass as a Conversion Factor

Silver metal is used to make jewelry.

Guide to Calculating the Moles (or Grams) of a Substance from Grams (or Moles)

STEP 1 State the given and needed quantities.

STEP 2 Write a plan to convert moles to grams (or grams to moles).

STEP 3 Determine the molar mass and write conversion factors.

STEP 4 Set up the problem to convert moles to grams (or grams to moles).

The conversion factors for a compound are also written from the molar mass. For example, the molar mass of the compound H_2O is written

$$1 \text{ mole of } H_2O = 18.02 \text{ g of } H_2O$$

From this equality, conversion factors for the molar mass of H_2O are written as

$$\frac{18.02 \text{ g } H_2O}{1 \text{ mole } H_2O} \text{ and } \frac{1 \text{ mole } H_2O}{18.02 \text{ g } H_2O}$$

SAMPLE PROBLEM 7.5 Converting the Mass of a Compound to Moles

A box of salt contains 737 g of NaCl. How many moles of NaCl are present?

SOLUTION

STEP 1 State the given and needed quantities.

ANALYZE THE PROBLEM	Given	Need
	737 g of NaCl	moles of NaCl

STEP 2 Write a plan to convert grams to moles.

grams of NaCl → Molar mass factor → moles of NaCl

Table salt is sodium chloride, NaCl.

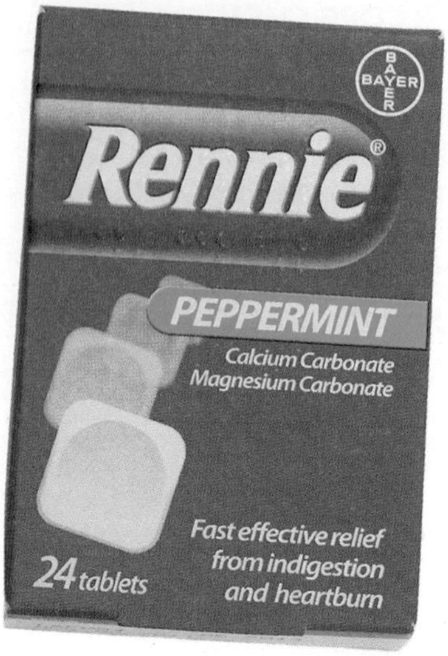

STEP 3 Determine the molar mass and write conversion factors.

$$(1 \times 22.99) + (1 \times 35.45) = 58.44 \text{ g/mole}$$

$$1 \text{ mole of NaCl} = 58.44 \text{ g of NaCl}$$

$$\frac{58.44 \text{ g NaCl}}{1 \text{ mole NaCl}} \quad \text{and} \quad \frac{1 \text{ mole NaCl}}{58.44 \text{ g NaCl}}$$

STEP 4 Set up the problem to convert grams to moles.

$$737 \cancel{\text{g NaCl}} \times \frac{1 \text{ mole NaCl}}{58.44 \cancel{\text{g NaCl}}} = 12.6 \text{ moles of NaCl}$$

STUDY CHECK 7.5

One tablet of an antacid contains 680. mg of $CaCO_3$. How many moles of $CaCO_3$ are present?

We can summarize the calculations to show the connections between the moles of a compound, its mass in grams, number of molecules (or formula units if ionic), and the moles and atoms of each element in that compound in the following flowchart:

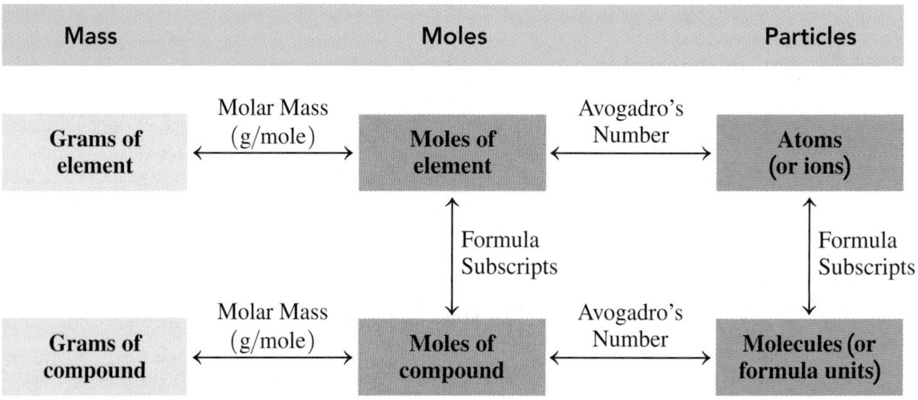

QUESTIONS AND PROBLEMS

7.2 Molar Mass and Calculations

LEARNING GOAL Calculate the molar mass for a substance given its chemical formula; use molar mass to convert between grams and moles.

7.9 Calculate the molar mass for each of the following:
- **a.** $KC_4H_5O_6$ (cream of tartar)
- **b.** Fe_2O_3 (rust)
- **c.** $C_{19}H_{20}FNO_3$ (Paxil, an antidepressant)
- **d.** $Al_2(SO_4)_3$ (antiperspirant)
- **e.** $Mg(OH)_2$ (antacid)

7.10 Calculate the molar mass for each of the following:
- **a.** $FeSO_4$ (iron supplement)
- **b.** Al_2O_3 (absorbent and abrasive)
- **c.** $C_7H_5NO_3S$ (saccharin)
- **d.** C_3H_8O (rubbing alcohol)
- **e.** $(NH_4)_2CO_3$ (baking powder)

7.11 Calculate the molar mass for each of the following:
a. Br_2 **b.** C_3H_8O **c.** $CuSO_4$
d. BF_3 **e.** C_6H_5Cl

7.12 Calculate the molar mass for each of the following:
a. O_2 **b.** KH_2PO_4 **c.** $Fe(ClO_4)_3$
d. $C_4H_8O_4$ **e.** $Ga_2(CO_3)_3$

7.13 Calculate the mass, in grams, for each of the following:
a. 3.00 moles of Li **b.** 1.90 moles of Mg
c. 0.245 mole of Fe **d.** 2.30 moles of Ag

7.14 Calculate the mass, in grams, for each of the following:
a. 1.50 moles of K **b.** 2.5 moles of C
c. 0.25 mole of P **d.** 12.5 moles of He

7.15 Calculate the number of grams in each of the following:
a. 2.0 moles of $MgCl_2$ **b.** 3.5 moles of benzene C_6H_6
c. 5.00s mole of NaOMe **d.** 0.5 mole of $C_6H_{12}O_6$

7.16 Calculate the number of grams in each of the following:
a. 4.0 mole of $MgCl_2$ **b.** 7.0 mole of benzene (C_6H_6)
c. 5.00 mole of $NaOCH_3$ **d.** 1.5 mole of glucose $(C_6H_{12}O_6)$

7.17 **a.** The compound $MgSO_4$, Epsom salts, is used to soothe sore feet and muscles. How many grams will you need to prepare a bath containing 5.00 moles of Epsom salts?
b. In a bottle of soda, there is 0.25 mole of CO_2. How many grams of CO_2 are in the bottle?

7.18 **a.** Cyclopropane, C_3H_6, is an anesthetic given by inhalation. How many grams are in 0.25 mole of cyclopropane?
b. The sedative Demerol hydrochloride has the formula $C_{15}H_{22}ClNO_2$. How many grams are in 0.025 mole of Demerol hydrochloride?

7.19 How many moles are contained in each of the following?
a. 25.0g of Cu **b.** 0.600g of P
c. 12.0g of CO **d.** 60.5g of CH_4

7.20 How many moles are contained in each of the following?
a. 25.0 g of Ca **b.** 5.00 g of S
c. 40.0 g of H_2O **d.** 100.0 g of O_2

7.21 Calculate the number of moles in 15.0g of each of the following:
a. Ar **b.** N_2
c. Fe_2O_3 **d.** $CaCl_2$

7.22 Calculate the number of moles in 4.00 g of each of the following:
a. He **b.** SnO_2
c. $Cr(OH)_3$ **d.** Ca_3N_2

7.23 How many moles of O are in each of the following quantities?
a. 23g of O **b.** 110g of CO_2 **c.** 4.0 moles of Al_2O_3

7.24 How many moles of O are in each of the following quantities?
a. 88 g of CO_2 **b.** 0.25 kg of NO_2 **c.** 100 g ice

7.25 Methyl salicylate or Oil of wintergreen ($C_8H_8O_3$) is an organic ester widely used as a fragrance in foods & beverages and in leniments.
a. How many grams of the compound are in 1.75 moles of methyl salicylate?
b. How many moles of the compound are in 42.0g of methyl salicylate?
c. How many grams of O are in 42.0g of methyl salicylate?

7.26 Allyl sulfide, $C_6H_{10}S$, is the substance that gives garlic, onions, and leeks their characteristic odor. Some studies indicate that garlic may be beneficial for the heart and in lowering cholesterol.
a. How many moles of H are in 0.75 mole of $C_6H_{10}S$?
b. How many moles of S are in 23.2 g of $C_6H_{10}S$?
c. How many grams of C are in 44.0 g of $C_6H_{10}S$?

The characteristic odor of garlic is a sulfur-containing compound.

7.3 Equations for Chemical Reactions

LEARNING GOAL

Write a balanced chemical equation from the formulas of the reactants and products for a reaction; determine the number of atoms in the reactants and products.

A *chemical change* occurs when a substance is converted into one or more new substances that have different formulas and different properties. For example, when silver tarnishes, the shiny silver metal (Ag) reacts with sulfur (S) to become the dull, black substance we call tarnish (Ag_2S) (see Figure 7.2).

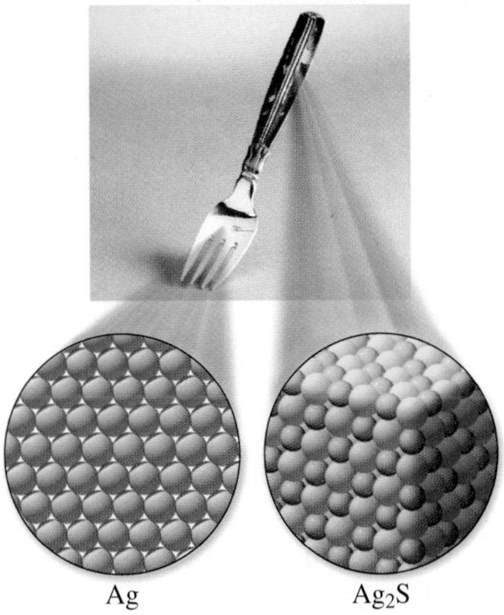

FIGURE 7.2 ▶ A chemical change produces new substances with new properties.

Q Why is the formation of tarnish a chemical change?

FIGURE 7.3 ▶ Chemical reactions involve chemical changes. When iron (Fe) reacts with oxygen (O_2), the product is rust (Fe_2O_3).

Q What is the evidence for chemical change in this chemical reaction?

A *chemical reaction* always involves chemical change because atoms of the reacting substances form new combinations with new properties. For example, a chemical reaction takes place when a piece of iron (Fe) combines with oxygen (O_2) in the air to produce a new substance, rust (Fe_2O_3), which has a reddish-brown color (see Figure 7.3). During a chemical change, new properties become visible, which are an indication that a chemical reaction has taken place (see Table 7.3).

When you build a model airplane, prepare a new recipe, or mix a medication, you follow a set of directions. These directions tell you what materials to use and the products you will obtain. In chemistry, a *chemical equation* tells us the materials we need and the products that will form.

TABLE 7.3 Types of Visible Evidence of a Chemical Reaction

1.	Change in color
2.	Formation of a gas (bubbles)
3.	Formation of a solid (precipitate)
4.	Heat (or a flame) produced or heat absorbed

Writing a Chemical Equation

Suppose you work in a bicycle shop, assembling wheels and frames into bicycles. You could represent this process by a simple equation:

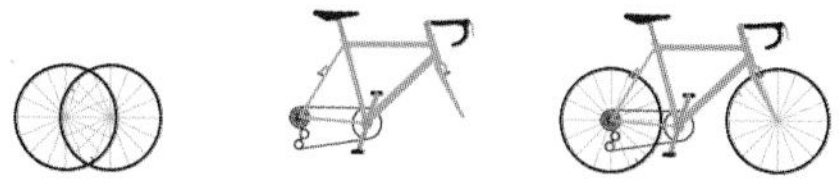

Equation: 2 Wheels + 1 Frame ⟶ 1 Bicycle

Reactants (2 Wheels + 1 Frame) — Product (1 Bicycle)

When you burn charcoal in a grill, the carbon in the charcoal combines with oxygen to form carbon dioxide. We can represent this reaction by a chemical equation.

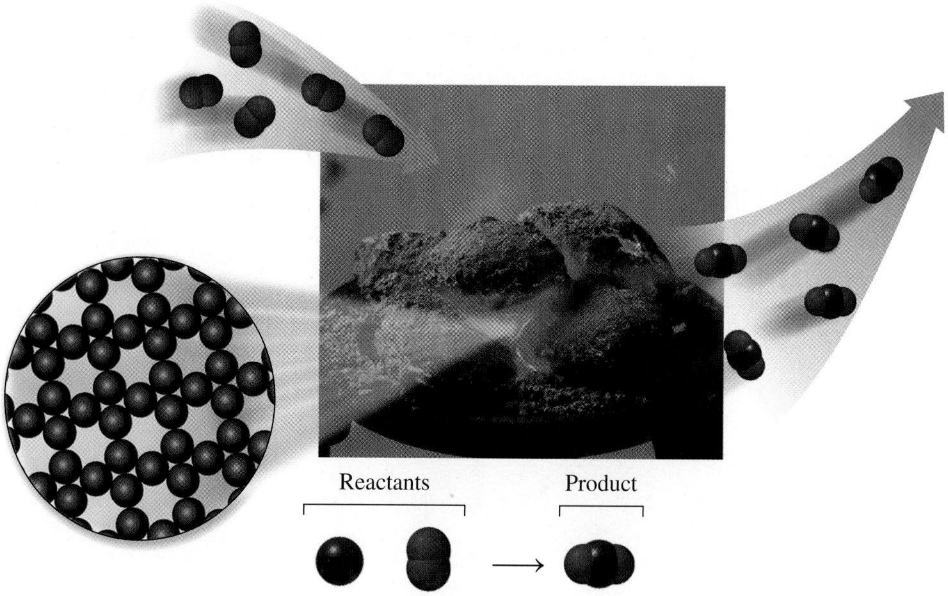

Equation: $C(s) + O_2(g) \xrightarrow{\Delta} CO_2(g)$

In a **chemical equation**, the formulas of the **reactants** are written on the left of the arrow and the formulas of the **products** on the right. When there are two or more formulas on the same side, they are separated by plus (+) signs. The delta sign (Δ) indicates that heat was used to start the reaction.

Generally, each formula in an equation is followed by an abbreviation, in parentheses, that gives the physical state of the substance: solid (*s*), liquid (*l*), or gas (*g*). If a substance is dissolved in water, it is an aqueous (*aq*) solution. Table 7.4 summarizes some of the symbols used in equations.

TABLE 7.4 Some Symbols Used in Writing Equations

Symbol	Meaning
+	Separates two or more formulas
$\longrightarrow$	Reacts to form products
$\xrightarrow{\Delta}$	Reactants are heated
(*s*)	Solid
(*l*)	Liquid
(*g*)	Gas
(*aq*)	Aqueous

Identifying a Balanced Chemical Equation

When a chemical reaction takes place, the bonds between the atoms of the reactants are broken and new bonds are formed to give the products. All atoms are conserved, which means that atoms cannot be gained, lost, or changed into other types of atoms during a chemical reaction. Every chemical reaction must be written as a **balanced equation**, which shows the same number of atoms for each element in the reactants as well as in the products. For example, the chemical equation we wrote previously for burning carbon is balanced because there is one carbon atom and two oxygen atoms in both the reactants and the products:

$$C(s) + O_2(g) \xrightarrow{\Delta} CO_2(g)$$

C O C O

Reactant atoms = Product atoms

Now consider the reaction in which hydrogen reacts with oxygen to form water written as follows:

$$H_2(g) + O_2(g) \longrightarrow H_2O(g) \quad \text{Unbalanced}$$

In the *balanced* equation, there are whole numbers called **coefficients** in front of the formulas. On the reactant side, the coefficient of 2 in front of the H_2 formula represents two molecules of hydrogen, which is 4 atoms of H. A coefficient of 1 is understood for O_2, which gives 2 atoms of O. On the product side, the coefficient of 2 in front of the H_2O formula represents 2 molecules of water. Because the coefficient of 2 multiplies all the atoms in H_2O, there are 4 hydrogen atoms and 2 oxygen atoms in the products. Because there are the same number of hydrogen atoms and oxygen atoms in the reactants as in the products, we know that the equation is *balanced*. This illustrates the *Law of Conservation of Matter*, which states that matter cannot be created or destroyed during a chemical reaction.

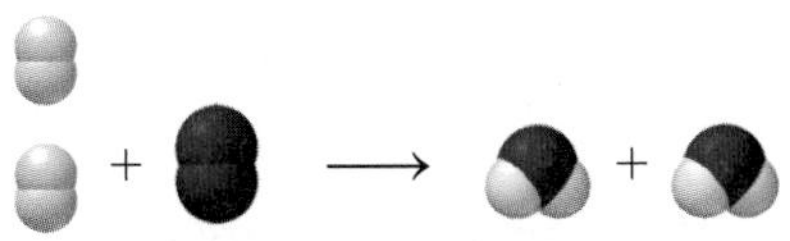

$$2H_2(g) + O_2(g) \longrightarrow 2H_2O(g)$$

Balanced

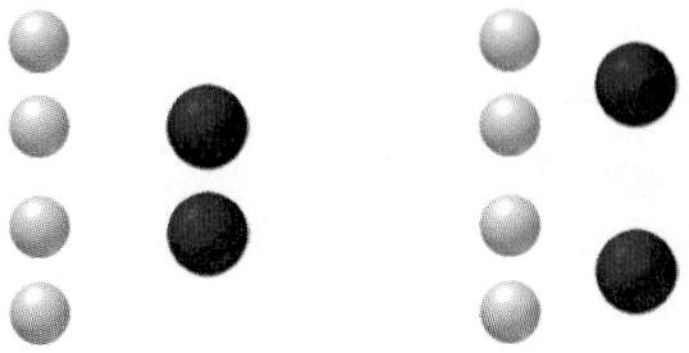

H O H O

Reactant atoms = Product atoms

SAMPLE PROBLEM 7.6 Number of Atoms in Balanced Chemical Equations

Indicate the number of each type of atom in the following balanced chemical equation:

$$Fe_2S_3(s) + 6HCl(aq) \longrightarrow 2FeCl_3(aq) + 3H_2S(g)$$

	Reactants	Products
Fe		
S		
H		
Cl		

SOLUTION

The total number of atoms in each formula is obtained by multiplying the coefficient by each subscript in a chemical formula.

	Reactants	Products
Fe	2 (1 × 2)	2 (2 × 1)
S	3 (1 × 3)	3 (3 × 1)
H	6 (6 × 1)	6 (3 × 2)
Cl	6 (6 × 1)	6 (2 × 3)

STUDY CHECK 7.6

When ethane, C_2H_6, burns in oxygen, the products are carbon dioxide and water. The balanced chemical equation is written as

$$2C_2H_6(g) + 7O_2(g) \xrightarrow{\Delta} 4CO_2(g) + 6H_2O(g)$$

Calculate the number of each type of atom in the reactants and in the products.

CORE CHEMISTRY SKILL

Balancing a Chemical Equation

Balancing a Chemical Equation

The chemical reaction that occurs in the flame of a gas burner you use in the laboratory or a gas cooktop is the reaction of methane gas, CH_4, and oxygen to produce carbon dioxide and water. We now show the process of balancing a chemical equation in Sample Problem 7.7.

SAMPLE PROBLEM 7.7 Balancing a Chemical Equation

The chemical reaction of methane, CH_4, and oxygen gas, O_2, produces carbon dioxide (CO_2) and water (H_2O). Write a balanced chemical equation for this reaction.

Guide to Balancing a Chemical Equation

STEP 1
Write an equation using the correct formulas for the reactants and products.

STEP 2
Count the atoms of each element in the reactants and products.

STEP 3
Use coefficients to balance each element.

STEP 4
Check the final equation to confirm it is balanced.

SOLUTION

STEP 1 Write an equation using the correct formulas for the reactants and products.

$$CH_4(g) + O_2(g) \xrightarrow{\Delta} CO_2(g) + H_2O(g)$$

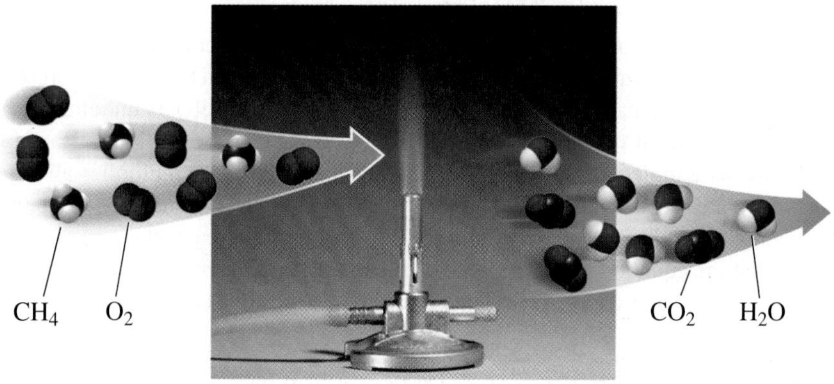

STEP 2 Count the atoms of each element in the reactants and products. When we compare the atoms on the reactant side and the atoms on the product side, we see that there are more H atoms in the reactants and more O atoms in the products.

$$CH_4(g) + O_2(g) \xrightarrow{\Delta} CO_2(g) + H_2O(g)$$

Reactants	Products	
1 C atom	1 C atom	Balanced
4 H atoms	2 H atoms	Not balanced
2 O atoms	3 O atoms	Not balanced

STEP 3 Use coefficients to balance each element. We will start by balancing the H atoms in CH_4 because it has the most atoms. By placing a coefficient of 2 in front of the formula for H_2O, a total of 4 H atoms in the products is obtained. *Only use coefficients to balance an equation. Do not change any of the subscripts: This would alter the chemical formula of a reactant or product.*

$$CH_4(g) + O_2(g) \xrightarrow{\Delta} CO_2(g) + 2H_2O(g)$$

Reactants	Products	
1 C atom	1 C atom	Balanced
4 H atoms	4 H atoms	Balanced
2 O atoms	4 O atoms	Not balanced

We can balance the O atoms on the reactant side by placing a coefficient of 2 in front of the formula O_2. There are now 4 O atoms in both the reactants and products.

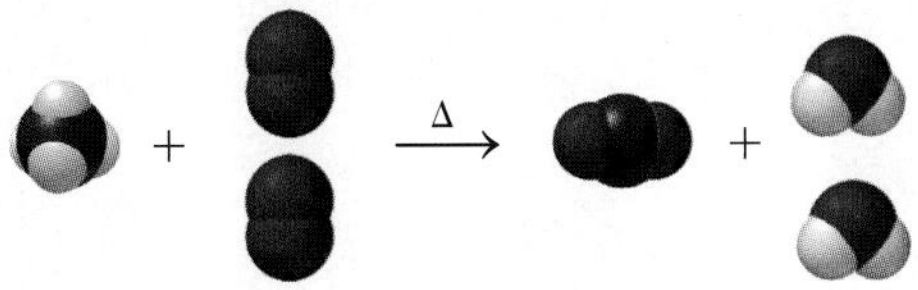

$$CH_4(g) + 2O_2(g) \xrightarrow{\Delta} CO_2(g) + 2H_2O(g)$$ Balanced

STEP 4 **Check the final equation to confirm it is balanced.** In the final equation, the numbers of atoms of C, H, and O are the same in both the reactants and the products. The equation is balanced.

$$CH_4(g) + 2O_2(g) \xrightarrow{\Delta} CO_2(g) + 2H_2O(g)$$

Reactants	Products	
1 C atom	1 C atom	Balanced
4 H atoms	4 H atoms	Balanced
4 O atoms	4 O atoms	Balanced

In a balanced chemical equation, the coefficients must be the *lowest possible whole numbers*. Suppose you had obtained the following for the balanced equation:

$$2CH_4(g) + 4O_2(g) \xrightarrow{\Delta} 2CO_2(g) + 4H_2O(g)$$ Incorrect

Although there are equal numbers of atoms on both sides of the equation, this is not written correctly. To obtain coefficients that are the lowest whole numbers, we divide all the coefficients by 2.

STUDY CHECK 7.7

Balance the following chemical equation:

$$Al(s) + Cl_2(g) \longrightarrow AlCl_3(s)$$

SAMPLE PROBLEM 7.8 Balancing Chemical Equations with Polyatomic Ions

Balance the following chemical equation:

$$Na_3PO_4(aq) + MgCl_2(aq) \longrightarrow Mg_3(PO_4)_2(s) + NaCl(aq)$$

SOLUTION

STEP 1 **Write an equation using the correct formulas for the reactants and products.**

$$Na_3PO_4(aq) + MgCl_2(aq) \longrightarrow Mg_3(PO_4)_2(s) + NaCl(aq)$$ Unbalanced

STEP 2 **Count the atoms of each element in the reactants and products.** When we compare the number of ions in the reactants and products, we find that the equation is not balanced. In this equation, we can balance the phosphate ion as a group of atoms because it appears on both sides of the equation.

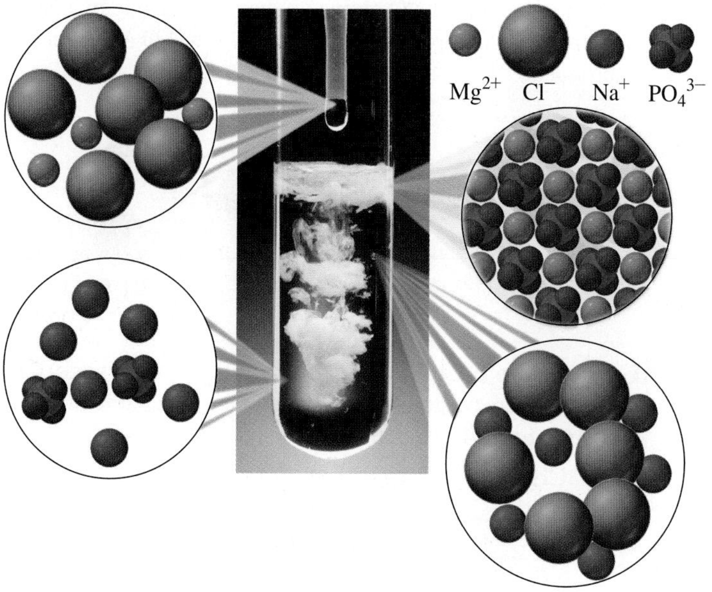

$$Na_3PO_4(aq) + MgCl_2(aq) \longrightarrow Mg_3(PO_4)_2(s) + NaCl(aq)$$

Reactants	Products	
3 Na^+	1 Na^+	Not balanced
1 PO_4^{3-}	2 PO_4^{3-}	Not balanced
1 Mg^{2+}	3 Mg^{2+}	Not balanced
2 Cl^-	1 Cl^-	Not balanced

STEP 3 **Use coefficients to balance each element.** We begin with the formula that has the highest subscript values, which in this equation is $Mg_3(PO_4)_2$. The subscript 3 in $Mg_3(PO_4)_2$ is used as a coefficient for $MgCl_2$ to balance magnesium. The subscript 2 in $Mg_3(PO_4)_2$ is used as a coefficient for Na_3PO_4 to balance the phosphate ion.

$$2Na_3PO_4(aq) + 3MgCl_2(aq) \longrightarrow Mg_3(PO_4)_2(s) + NaCl(aq)$$

Reactants	Products	
6 Na^+	1 Na^+	Not balanced
2 PO_4^{3-}	2 PO_4^{3-}	Balanced
3 Mg^{2+}	3 Mg^{2+}	Balanced
6 Cl^-	1 Cl^-	Not balanced

In the reactants and products, we see that the sodium and chloride ions are not yet balanced. A coefficient of 6 is placed in front of the NaCl to balance the equation.

$$2Na_3PO_4(aq) + 3MgCl_2(aq) \longrightarrow Mg_3(PO_4)_2(s) + 6NaCl(aq)$$

STEP 4 **Check the final equation to confirm it is balanced.** A check of the total number of ions confirms the equation is balanced. A coefficient of 1 is understood and not usually written.

$$2Na_3PO_4(aq) + 3MgCl_2(aq) \longrightarrow Mg_3(PO_4)_2(s) + 6NaCl(aq) \text{ Balanced}$$

Reactants	Products	
6 Na^+	6 Na^+	Balanced
2 PO_4^{3-}	2 PO_4^{3-}	Balanced
3 Mg^{2+}	3 Mg^{2+}	Balanced
6 Cl^-	6 Cl^-	Balanced

STUDY CHECK 7.8

Balance the following chemical equation:

$Sb_2S_3(s) + HCl(aq) \longrightarrow SbCl_3(s) + H_2S(g)$

QUESTIONS AND PROBLEMS

7.3 Equations for Chemical Reactions

LEARNING GOAL Write a balanced chemical equation from the formulas of the reactants and products for a reaction; determine the number of atoms in the reactants and products.

7.27 Determine whether each of the following chemical equations is balanced or not balanced:

a. $S(s) + O_2(g) \longrightarrow SO_3(g)$

b. $2Al(s) + 3Cl_2(g) \longrightarrow 2AlCl_3(s)$

c. $H_2(g) + O_2(g) \longrightarrow H_2O(g)$

d. $C_3H_8(g) + 5O_2(g) \xrightarrow{\Delta} 3CO_2(g) + 4H_2O(g)$

7.28 Determine whether each of the following chemical equations is balanced or not balanced:

a. $PCl_3(s) + Cl_2(g) \longrightarrow PCl_5(s)$

b. $CO(g) + 2H_2(g) \longrightarrow CH_3OH(g)$

c. $2KClO_3(s) \longrightarrow 2KCl(s) + O_2(g)$

d. $Mg(s) + N_2(g) \longrightarrow Mg_3N_2(s)$

7.29 Balance each of the following chemical equations:

a. $N_2(g) + O_2(g) \longrightarrow NO(g)$

b. $HgO(s) \xrightarrow{\Delta} Hg(l) + O_2(g)$

c. $Fe(s) + O_2(g) \longrightarrow Fe_2O_3(s)$

d. $Na(s) + Cl_2(g) \longrightarrow NaCl(s)$

7.30 Balance each of the following chemical equations:

a. $Mg(s) + Br_2(l)$

b. $O_2(g) + P_4(s)$

c. $K_2CO_3(s) + HCl(aq)$

d. $H_2O(l) + NaCl(s)$

7.31 Balance each of the following chemical equations:

a. $Mg(s) + AgNO_3(aq) \longrightarrow Mg(NO_3)_2(aq) + Ag(s)$

b. $Al(s) + CuSO_4(aq) \longrightarrow Al_2(SO_4)_3(aq) + Cu(s)$

c. $Pb(NO_3)_2(aq) + NaCl(aq) \longrightarrow PbCl_2(s) + NaNO_3(aq)$

d. $Al(s) + HCl(aq) \longrightarrow AlCl_3(aq) + H_2(g)$

7.32 Balance each of the following chemical equations:

a. $Zn(s) + HNO_3(aq) \longrightarrow Zn(NO_3)_2(aq) + H_2(g)$

b. $Al(s) + H_2SO_4(aq) \longrightarrow Al_2(SO_4)_3(aq) + H_2(g)$

c. $K_2SO_4(aq) + BaCl_2(aq) \longrightarrow KCl(aq) + BaSO_4(s)$

d. $CaCO_3(s) \longrightarrow CaO(s) + CO_2(g)$

7.4 Types of Reactions

LEARNING GOAL

Identify a reaction as a combination, decomposition, single replacement, double replacement, or combustion.

A great number of reactions occur in nature, in biological systems, and in the laboratory. However, there are some general patterns that help us classify most reactions into five general types.

Combination Reactions

In a **combination reaction**, two or more elements or compounds bond to form one product. For example, sulfur and oxygen combine to form the product sulfur dioxide.

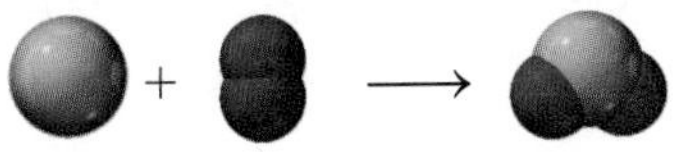

$$S(s) + O_2(g) \longrightarrow SO_2(g)$$

Combination

Two or more reactants		combine to yield	a single product
A	+ B	$\longrightarrow$	A B

In Figure 7.4, the elements magnesium and oxygen combine to form a single product, magnesium oxide.

$$2Mg(s) + O_2(g) \longrightarrow 2MgO(s)$$

In other examples of combination reactions, elements or compounds combine to form a single product.

$$N_2(g) + 3H_2(g) \longrightarrow 2NH_3(g)$$
$$Cu(s) + S(s) \longrightarrow CuS(s)$$
$$MgO(s) + CO_2(g) \longrightarrow MgCO_3(s)$$

CORE CHEMISTRY SKILL

Classifying Types of Chemical Reactions

$$\underset{\text{Magnesium}}{2Mg(s)} + \underset{\text{Oxygen}}{O_2(g)} \xrightarrow{\Delta} \underset{\text{Magnesium oxide}}{2MgO(s)}$$

FIGURE 7.4 ▶ In a combination reaction, two or more substances combine to form one substance as product.

Q What happens to the atoms in the reactants in a combination reaction?

Decomposition

A reactant	splits into	two or more products
A B	⟶	A + B

Decomposition Reactions

In a **decomposition reaction**, a reactant splits into two or more simpler products. For example, when mercury(II) oxide is heated, the compound breaks apart into mercury atoms and oxygen (see Figure 7.5).

$$2HgO(s) \xrightarrow{\Delta} 2Hg(l) + O_2(g)$$

In another example of a decomposition reaction, calcium carbonate breaks apart into simpler compounds of calcium oxide and carbon dioxide.

$$CaCO_3(s) \xrightarrow{\Delta} CaO(s) + CO_2(g)$$

Replacement Reactions

In a replacement reaction, elements in a compound are replaced by other elements. In a **single replacement reaction**, a reacting element switches place with an element in the other reacting compound.

Single replacement

One element	replaces	another element
A + B C	⟶	A C + B

In the single replacement reaction shown in Figure 7.6, zinc replaces hydrogen in hydrochloric acid, HCl(*aq*).

$$Zn(s) + 2HCl(aq) \longrightarrow H_2(g) + ZnCl_2(aq)$$

In another single replacement reaction, chlorine replaces bromine in the compound potassium bromide.

$$Cl_2(g) + 2KBr(s) \longrightarrow 2KCl(s) + Br_2(l)$$

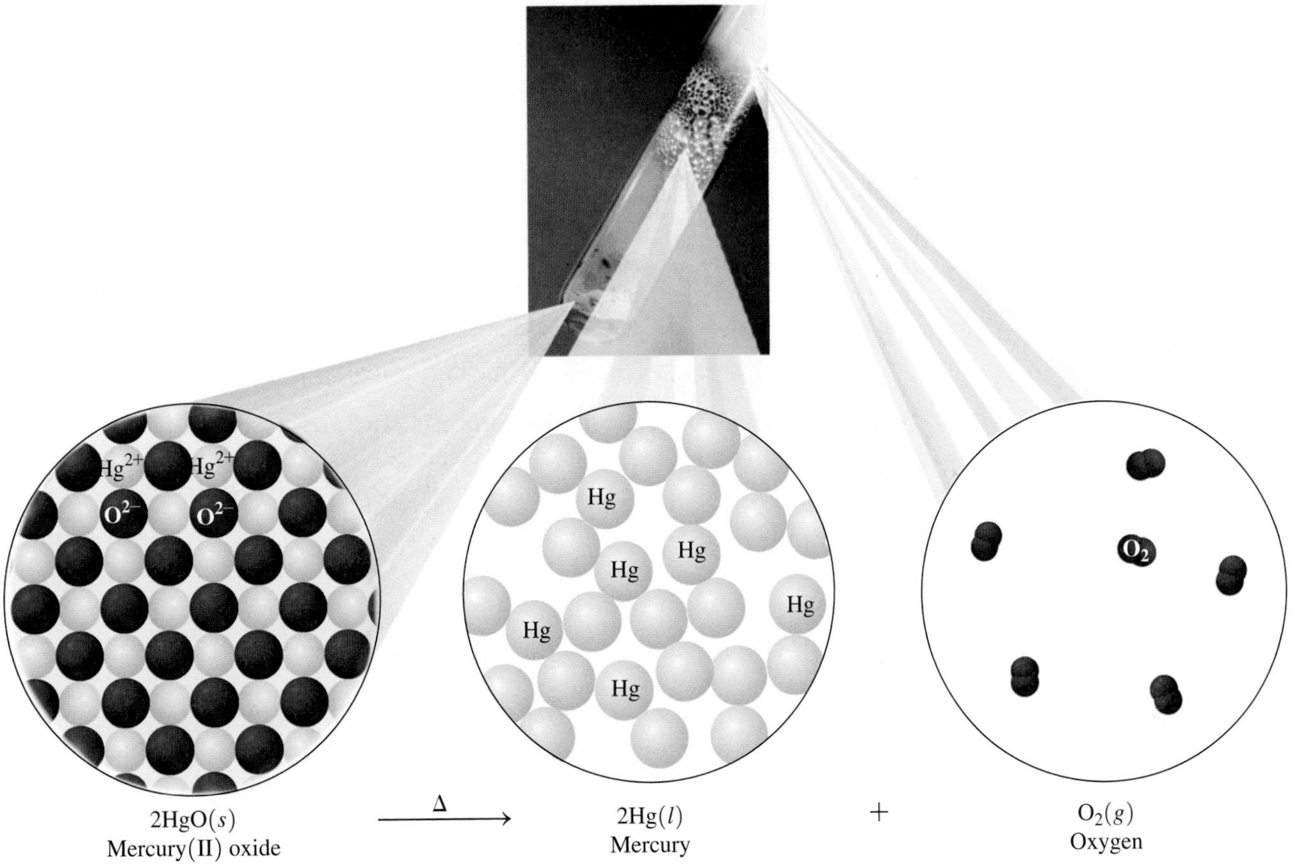

FIGURE 7.5 ▶ In a decomposition reaction, one reactant breaks down into two or more products.

Q How do the differences in the reactant and products classify this as a decomposition reaction?

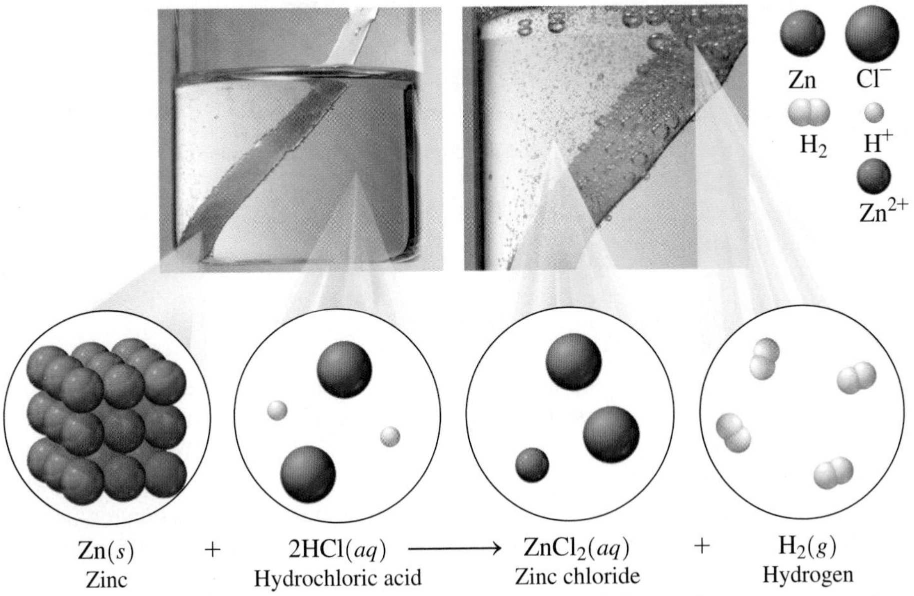

FIGURE 7.6 ▶ In a single replacement reaction, an atom or ion replaces an atom or ion in a compound.

Q What changes in the formulas of the reactants identify this equation as a single replacement?

In a **double replacement reaction**, the positive ions in the reacting compounds switch places.

Double replacement

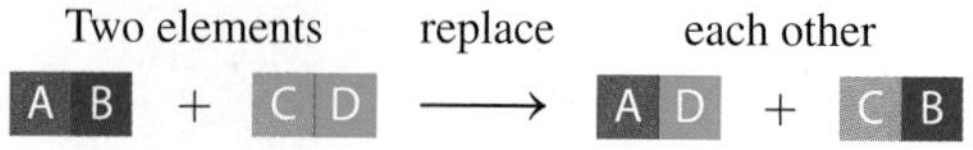

In the reaction shown in Figure 7.7, barium ions change places with sodium ions in the reactants to form sodium chloride and a white solid precipitate of barium sulfate. The formulas of the products depend on the charges of the ions.

$$BaCl_2(aq) + Na_2SO_4(aq) \longrightarrow BaSO_4(s) + 2NaCl(aq)$$

When sodium hydroxide and hydrochloric acid (HCl) react, sodium and hydrogen ions switch places, forming water and sodium chloride.

$$NaOH(aq) + HCl(aq) \longrightarrow HOH(l) + NaCl(aq)$$

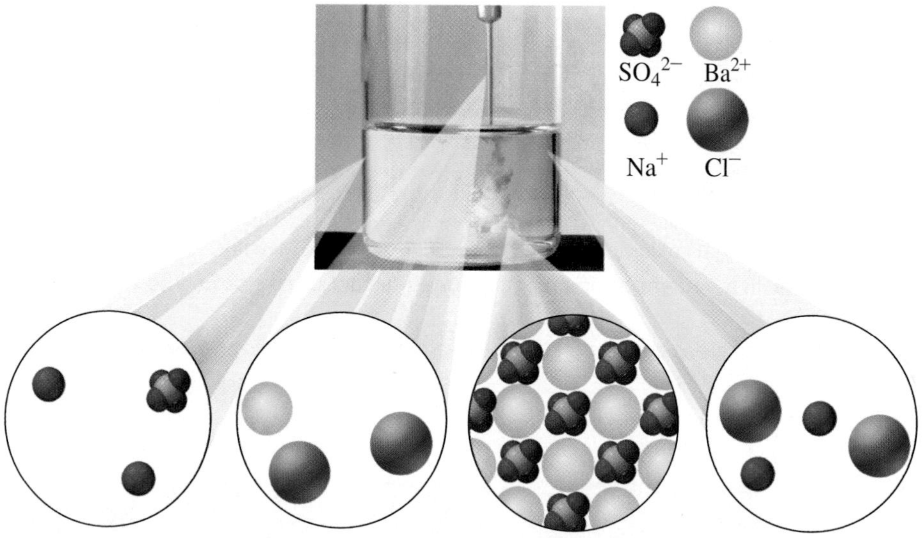

$$Na_2SO_4(aq) + BaCl_2(aq) \longrightarrow BaSO_4(s) + 2NaCl(aq)$$

Sodium sulfate | Barium chloride | Barium sulfate | Sodium chloride

FIGURE 7.7 ▶ In a double replacement reaction, the positive ions in the reactants replace each other.

Q How do the changes in the formulas of the reactants identify this equation as a double replacement reaction?

In a combustion reaction, a candle burns using the oxygen in the air.

Combustion Reactions

The burning of a candle and the burning of fuel in the engine of a car are examples of combustion reactions. In a **combustion reaction**, a carbon-containing compound, usually a fuel, burns in oxygen from the air to produce carbon dioxide (CO_2), water (H_2O), and energy in the form of heat or a flame. For example, methane gas (CH_4) undergoes combustion when used to cook our food on a gas cooktop and to heat our homes. In the equation for the combustion of methane, each element in the fuel (CH_4) forms a compound with oxygen.

$$\underset{\text{Methane}}{CH_4(g)} + 2O_2(g) \xrightarrow{\Delta} CO_2(g) + 2H_2O(g) + \text{energy}$$

The balanced equation for the combustion of propane (C_3H_8) is

$$C_3H_8(g) + 5O_2(g) \xrightarrow{\Delta} 3CO_2(g) + 4H_2O(g) + \text{energy}$$

Propane is the fuel used in portable heaters and gas barbecues. Gasoline, a mixture of liquid hydrocarbons, is the fuel that powers our cars, lawn mowers, and snow blowers.

Table 7.5 summarizes the reaction types and gives examples.

TABLE 7.5 Summary of Reaction Types

Reaction Type	Example
Combination	
$A + B \longrightarrow AB$	$Ca(s) + Cl_2(g) \longrightarrow CaCl_2(s)$
Decomposition	
$AB \longrightarrow A + B$	$Fe_2S_3(s) \longrightarrow 2Fe(s) + 3S(s)$
Single Replacement	
$A + BC \longrightarrow AC + B$	$Cu(s) + 2AgNO_3(aq) \longrightarrow Cu(NO_3)_2(aq) + 2Ag(s)$
Double Replacement	
$AB + CD \longrightarrow AD + CB$	$BaCl_2(aq) + K_2SO_4(aq) \longrightarrow BaSO_4(s) + 2KCl(aq)$
Combustion	
$C_XH_Y + ZO_2(g) \xrightarrow{\Delta} XCO_2(g) + \frac{Y}{2}H_2O(g) + \text{energy}$	$CH_4(g) + 2O_2(g) \xrightarrow{\Delta} CO_2(g) + 2H_2O(g) + \text{energy}$

SAMPLE PROBLEM 7.9 Identifying Reactions and Predicting Products

Classify each of the following as a combination, decomposition, single replacement, double replacement, or combustion reaction:

a. $2Fe_2O_3(s) + 3C(s) \longrightarrow 3CO_2(g) + 4Fe(s)$

b. $2KClO_3(s) \xrightarrow{\Delta} 2KCl(s) + 3O_2(g)$

c. $C_2H_4(g) + 3O_2(g) \xrightarrow{\Delta} 2CO_2(g) + 2H_2O(g) + \text{energy}$

SOLUTION

a. In this single replacement reaction, a C atom replaces Fe in Fe_2O_3 to form the compound CO_2 and Fe atoms.

b. When one reactant breaks down to produce two products, the reaction is decomposition.

c. The reaction of a carbon compound with oxygen to produce carbon dioxide, water, and energy makes this a combustion reaction.

STUDY CHECK 7.9

Nitrogen gas (N_2) and oxygen gas (O_2) react to form nitrogen dioxide gas. Write the balanced chemical equation using the correct chemical formulas of the reactants and product, and identify the reaction type.

Chemistry Link to **Health**

Incomplete Combustion: Toxicity of Carbon Monoxide

When a propane heater, fireplace, or woodstove is used in a closed room, there must be adequate ventilation. If the supply of oxygen is limited, incomplete combustion from burning gas, oil, or wood produces carbon monoxide. The incomplete combustion of methane in natural gas is written

$$\underset{\text{Limited oxygen supply}}{2CH_4(g) + 3O_2(g)} \xrightarrow{\Delta} \underset{\text{Carbon monoxide}}{2CO(g)} + 4H_2O(g) + \text{heat}$$

Carbon monoxide (CO) is a colorless, odorless, poisonous gas. When inhaled, CO passes into the bloodstream, where it attaches to hemoglobin, which reduces the amount of oxygen (O_2) reaching the cells. As a result, a person can experience a reduction in exercise capability, visual perception, and manual dexterity.

Hemoglobin is the protein that transports O_2 in the blood. When the amount of hemoglobin bound to CO (COHb) is about 10%, a person may experience shortness of breath, mild headache, and drowsiness. Heavy smokers can have levels of COHb in their blood as high as 9%. When as much as 30% of the hemoglobin is bound to CO, a person may experience more severe symptoms, including dizziness, mental confusion, severe headache, and nausea. If 50% or more of the hemoglobin is bound to CO, a person could become unconscious and die if not treated immediately with oxygen.

QUESTIONS AND PROBLEMS

7.4 Types of Reactions

LEARNING GOAL Identify a reaction as a combination, decomposition, single replacement, double replacement, or combustion.

7.33 Classify each of the following as a combination, decomposition, single replacement, double replacement, or combustion reaction:

a. $2Al_2O_3(s) \xrightarrow{\Delta} 4Al(s) + 3O_2(g)$
b. $Br_2(g) + BaI_2(s) \longrightarrow BaBr_2(s) + I_2(g)$
c. $2C_2H_2(g) + 5O_2(g) \xrightarrow{\Delta} 4CO_2(g) + 2H_2O(g)$
d. $BaCl_2(aq) + K_2CO_3(aq) \longrightarrow BaCO_3(s) + 2KCl(aq)$
e. $Pb(s) + O_2(g) \longrightarrow PbO_2(s)$

7.34 Classify each of the following as a combination, decomposition, single replacement, double replacement, or combustion reaction:

a. $H_2(g) + Br_2(g) \longrightarrow 2HBr(g)$
b. $AgNO_3(aq) + NaCl(aq) \longrightarrow AgCl(s) + NaNO_3(aq)$
c. $2H_2O_2(aq) \longrightarrow 2H_2O(l) + O_2(g)$
d. $Zn(s) + CuCl_2(aq) \longrightarrow Cu(s) + ZnCl_2(aq)$
e. $C_5H_8(g) + 7O_2(g) \xrightarrow{\Delta} 5CO_2(g) + 4H_2O(g)$

7.35 Classify each of the following as a combination, decomposition, single replacement, double replacement, or combustion reaction:

a. $4Fe(s) + 3O_2(g) \longrightarrow 2Fe_2O_3(s)$
b. $Mg(s) + 2AgNO_3(aq) \longrightarrow Mg(NO_3)_2(aq) + 2Ag(s)$
c. $CuCO_3(s) \xrightarrow{\Delta} CuO(s) + CO_2(g)$
d. $Al_2(SO_4)_3(aq) + 6KOH(aq) \longrightarrow 2Al(OH)_3(s) + 3K_2SO_4(aq)$
e. $C_4H_8(g) + 6O_2(g) \xrightarrow{\Delta} 4CO_2(g) + 4H_2O(g)$

7.36 Classify each of the following as a combination, decomposition, single replacement, double replacement, or combustion reaction:

a. $HCl(aq) + NaOH(aq) \longrightarrow NaCl(aq) + H_2O(l)$
b. $2Na(s) + H_2O(l) \longrightarrow H_2(g) + 2NaOH(aq)$
c. $Ca(s) + 2AuNO_3(aq) \longrightarrow Ca(NO_3)_2(aq) + 2Au(s)$
d. $4Al(s) + 3O_2(g) \longrightarrow 2Al_2O_3(s)$
e. $C_2H_6(g) + 2O_2(g) \longrightarrow CO_2(g) + CO(g) + 3H_2O + \text{energy}$

7.37 Using Table 7.5, predict the products that would result from each of the following reactions and balance:

a. combination: $Mg(s) + Cl_2(g) \longrightarrow$
b. decomposition: $HBr(g) \longrightarrow$
c. single replacement: $Mg(s) + Zn(NO_3)_2(aq) \longrightarrow$
d. double replacement: $K_2S(aq) + Pb(NO_3)_2(aq) \longrightarrow$
e. combustion: $C_2H_6(g) + O_2(g) \xrightarrow{\Delta}$

7.38 Using Table 7.5, predict the products that would result from each of the following reactions and balance:

a. combination: $Ca(s) + O_2(g) \longrightarrow$
b. combustion: $C_6H_6(g) + O_2(g) \xrightarrow{\Delta}$
c. decomposition: $PbO_2(s) \xrightarrow{\Delta}$
d. single replacement: $KI(s) + Cl_2(g) \longrightarrow$
e. double replacement: $CuCl_2(aq) + Na_2S(aq) \longrightarrow$

LEARNING GOAL

Define the terms oxidation and reduction; identify the reactants oxidized and reduced.

7.5 Oxidation–Reduction Reactions

Perhaps you have never heard of an oxidation and reduction reaction. However, this type of reaction has many important applications in your everyday life. When you see a rusty nail, tarnish on a silver spoon, or corrosion on metal, you are observing oxidation.

$$4Fe(s) + 3O_2(g) \longrightarrow \underset{\text{Rust}}{2Fe_2O_3(s)} \quad \text{Fe is oxidized}$$

Rust forms when the oxygen in the air reacts with iron.

When we turn the lights on in our automobiles, an oxidation–reduction reaction within the car battery provides the electricity. On a cold, wintry day, we might build a fire. As the wood burns, oxygen combines with carbon and hydrogen to produce carbon dioxide, water, and heat. In the previous section, we called this a combustion reaction, but it is also an *oxidation–reduction reaction*. When we eat foods with starches in them, the starches break down to give glucose, which is oxidized in our cells to give us energy along with carbon dioxide and water. Every breath we take provides oxygen to carry out oxidation in our cells.

$$\underset{\text{Glucose}}{C_6H_{12}O_6(aq)} + 6O_2(g) \longrightarrow 6CO_2(g) + 6H_2O(l) + \text{energy}$$

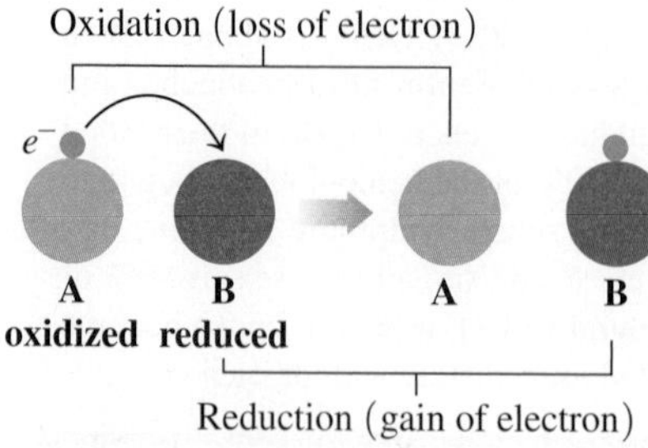

Oxidation–Reduction Reactions

In an **oxidation–reduction reaction** (*redox*), electrons are transferred from one substance to another. If one substance loses electrons, another substance must gain electrons. **Oxidation** is defined as the *loss* of electrons; **reduction** is the *gain* of electrons.

One way to remember these definitions is to use the following acronym:

OIL RIG

Oxidation **I**s **L**oss of electrons

Reduction **I**s **G**ain of electrons

Oxidation–Reduction

In general, atoms of metals lose electrons to form positive ions, whereas nonmetals gain electrons to form negative ions. Now we can say that metals are oxidized and nonmetals are reduced.

The green color that appears on copper surfaces from weathering, known as *patina*, is a mixture of $CuCO_3$ and CuO. We can now look at the oxidation and reduction reactions that take place when copper metal reacts with oxygen in the air to produce copper(II) oxide.

$$2Cu(s) + O_2(g) \longrightarrow 2CuO(s)$$

The element Cu in the reactants has a charge of 0, but in the CuO product, it is present as Cu^{2+}, which has a 2+ charge. Because the Cu atom lost two electrons, the charge is more positive. This means that Cu was oxidized in the reaction.

$$Cu^0(s) \longrightarrow Cu^{2+}(s) + 2\,e^- \quad \text{Oxidation: loss of electrons by Cu}$$

At the same time, the element O in the reactants has a charge of 0, but in the CuO product, it is present as O^{2-}, which has a 2− charge. Because the O atom has gained two electrons, the charge is more negative. This means that O was reduced in the reaction.

$$O_2^0(g) + 4\,e^- \longrightarrow 2O^{2-}(s) \quad \text{Reduction: gain of electrons by O}$$

Thus, the overall equation for the formation of CuO involves an oxidation and a reduction that occur simultaneously. In every oxidation and reduction, the number of electrons lost must be equal to the number of electrons gained. Therefore, we multiply the oxidation reaction of Cu by 2. Canceling the 4 e^- on each side, we obtain the overall oxidation–reduction equation for the formation of CuO.

$$2Cu(s) \longrightarrow 2Cu^{2+}(s) + \cancel{4e^-} \quad \text{Oxidation}$$
$$O_2(g) + \cancel{4e^-} \longrightarrow 2O^{2-}(s) \quad \text{Reduction}$$
$$2Cu(s) + O_2(g) \longrightarrow 2CuO(s) \quad \text{Oxidation–reduction equation}$$

As we see in the next reaction between zinc and copper(II) sulfate, there is always an oxidation with every reduction (see Figure 7.8).

$$Zn(s) + CuSO_4(aq) \longrightarrow ZnSO_4(aq) + Cu(s)$$

We rewrite the equation to show the atoms and ions.

$$\mathbf{Zn}(s) + \mathbf{Cu^{2+}}(aq) + SO_4^{2-}(aq) \longrightarrow \mathbf{Zn^{2+}}(aq) + SO_4^{2-}(aq) + \mathbf{Cu}(s)$$

FIGURE 7.8 ▶ In this single replacement reaction, Zn(s) is oxidized to $Zn^{2+}(aq)$ when it provides two electrons to reduce $Cu^{2+}(aq)$ to Cu(s):

$$Zn(s) + Cu^{2+}(aq) \longrightarrow Cu(s) + Zn^{2+}(aq)$$

Q In the oxidation, does Zn(s) lose or gain electrons?

The green patina on copper is due to oxidation.

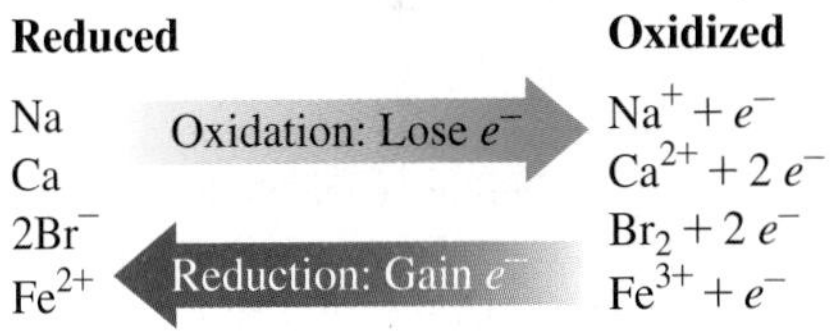

Oxidation is a loss of electrons; reduction is a gain of electrons.

CORE CHEMISTRY SKILL

Identifying Oxidized and Reduced Substances

Explore Your World

Oxidation of Fruits and Vegetables

Freshly cut surfaces of fruits and vegetables discolor when exposed to oxygen in the air. Cut three slices of a fruit or vegetable such as apple, potato, avocado, or banana. Leave one piece on the kitchen counter (uncovered). Wrap one piece in plastic wrap and leave it on the kitchen counter. Dip one piece in lemon juice and leave it uncovered.

Questions

1. What changes take place in each sample after 1 to 2 h?
2. Why would wrapping fruits and vegetables slow the rate of discoloration?
3. If lemon juice contains vitamin C (an antioxidant), why would dipping a fruit or vegetable in lemon juice affect the oxidation reaction on the surface of the fruit or vegetable?
4. Other kinds of antioxidants are vitamin E, citric acid, and BHT. Look for these antioxidants on the labels of cereals, potato chips, and other foods in your kitchen. Why are antioxidants added to food products that will be stored on our kitchen shelves?

In this reaction, Zn atoms lose two electrons to form Zn^{2+}. The increase in positive charge indicates that Zn is oxidized. At the same time, Cu^{2+} gains two electrons. The decrease in charge indicates that Cu is reduced. The SO_4^{2-} ions are *spectator ions*, which are present in both the reactants and products and do not change.

$$Zn(s) \longrightarrow Zn^{2+}(aq) + 2\,e^- \quad \text{Oxidation of Zn}$$

$$Cu^{2+}(aq) + 2\,e^- \longrightarrow Cu(s) \quad \text{Reduction of } Cu^{2+}$$

In this single replacement reaction, zinc was oxidized and copper(II) ion was reduced.

Oxidation and Reduction in Biological Systems

Oxidation may also involve the addition of oxygen or the loss of hydrogen, and reduction may involve the loss of oxygen or the gain of hydrogen. In the cells of the body, oxidation of organic (carbon) compounds involves the transfer of hydrogen atoms (H), which are composed of electrons and protons. For example, the oxidation of a typical biochemical molecule can involve the transfer of two hydrogen atoms (or $2H^+$ and $2\,e^-$) to a hydrogen ion acceptor such as the coenzyme FAD (flavin adenine dinucleotide). The coenzyme is reduced to $FADH_2$.

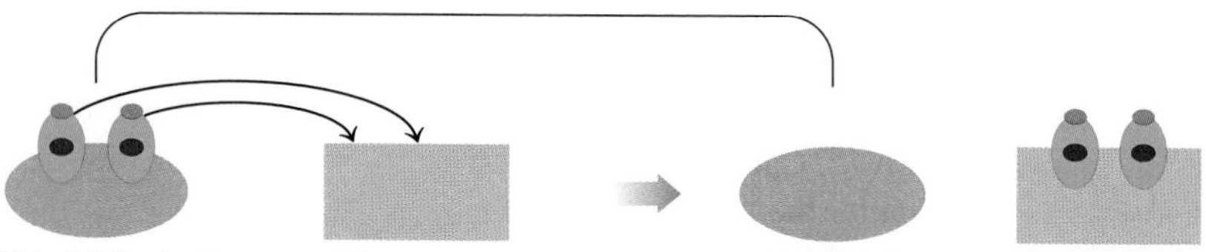

In many biochemical oxidation–reduction reactions, the transfer of hydrogen atoms is necessary for the production of energy in the cells. For example, methyl alcohol (CH_3OH), a poisonous substance, is metabolized in the body by the following reactions:

$$\underset{\text{Methyl alcohol}}{CH_3OH} \longrightarrow \underset{\text{Formaldehyde}}{H_2CO} + 2H \quad \text{Oxidation: loss of H atoms}$$

The formaldehyde can be oxidized further, this time by the addition of oxygen, to produce formic acid.

$$\underset{\text{Formaldehyde}}{2H_2CO} + O_2 \longrightarrow \underset{\text{Formic acid}}{2H_2CO_2} \quad \text{Oxidation: addition of O atoms}$$

Finally, formic acid is oxidized to carbon dioxide and water.

$$\underset{\text{Formic acid}}{2H_2CO_2} + O_2 \longrightarrow 2CO_2 + 2H_2O \quad \text{Oxidation: addition of O atoms}$$

The intermediate products of the oxidation of methyl alcohol are quite toxic, causing blindness and possibly death as they interfere with key reactions in the cells of the body.

In summary, we find that the particular definition of oxidation and reduction we use depends on the process that occurs in the reaction. All these definitions are summarized in Table 7.6. Oxidation always involves a loss of electrons, but it may also be seen as an addition of oxygen or the loss of hydrogen atoms. A reduction always involves a gain of electrons and may also be seen as the loss of oxygen or the gain of hydrogen.

TABLE 7.6 Characteristics of Oxidation and Reduction

Always Involves	May Involve
Oxidation	
Loss of electrons	Addition of oxygen
	Loss of hydrogen
Reduction	
Gain of electrons	Loss of oxygen
	Gain of hydrogen

QUESTIONS AND PROBLEMS

7.5 Oxidation–Reduction Reactions

LEARNING GOAL Define the terms oxidation and reduction; identify the reactants oxidized and reduced.

7.39 Identify each of the following as an oxidation or a reduction:
a. $Na^+(aq) + e^- \longrightarrow Na(s)$
b. $Ni(s) \longrightarrow Ni^{2+}(aq) + 2\,e^-$
c. $Cr^{3+}(aq) + 3\,e^- \longrightarrow Cr(s)$
d. $2H^+(aq) + 2\,e^- \longrightarrow H_2(g)$

7.40 Identify each of the following as an oxidation or a reduction:
a. $O_2(g) + 4\,e^- \longrightarrow 2O^{2-}(aq)$
b. $Ag(s) \longrightarrow Ag^+(aq) + e^-$
c. $Fe^{3+}(aq) + e^- \longrightarrow Fe^{2+}(aq)$
d. $2Br^-(aq) \longrightarrow Br_2(l) + 2\,e^-$

7.41 In each of the following reactions, identify the reactant that is oxidized and the reactant that is reduced:
a. $Zn(s) + Cl_2(g) \longrightarrow ZnCl_2(s)$
b. $Cl_2(g) + 2NaBr(aq) \longrightarrow 2NaCl(aq) + Br_2(l)$
c. $2PbO(s) \longrightarrow 2Pb(s) + O_2(g)$
d. $2Fe^{3+}(aq) + Sn^{2+}(aq) \longrightarrow 2Fe^{2+}(aq) + Sn^{4+}(aq)$

7.42 In each of the following reactions, identify the reactant that is oxidized and the reactant that is reduced:
a. $2Li(s) + F_2(g) \longrightarrow 2LiF(s)$
b. $Cl_2(g) + 2KI(aq) \longrightarrow 2KCl(aq) + I_2(s)$
c. $2Al(s) + 3Sn^{2+}(aq) \longrightarrow 2Al^{3+}(aq) + 3Sn(s)$
d. $Fe(s) + CuSO_4(aq) \longrightarrow FeSO_4(aq) + Cu(s)$

7.43 In the mitochondria of human cells, energy is provided by the oxidation and reduction reactions of the iron ions in the cytochromes in electron transport. Identify each of the following reactions as an oxidation or reduction:
a. $Fe^{3+} + e^- \longrightarrow Fe^{2+}$ **b.** $Fe^{2+} \longrightarrow Fe^{3+} + e^-$

7.44 Chlorine (Cl_2) is a strong germicide used to disinfect drinking water and to kill microbes in swimming pools. If the product is Cl^-, was the elemental chlorine oxidized or reduced?

7.45 When olefin, an unsaturated hydrocarbon, reacts with hydrogen, it forms a saturated hydrocarbon. Is olefin oxidized or reduced in the hydrogenation reaction?

$$C_8H_{16} + 2H_2 \longrightarrow C_8H_{18}$$
Linoleic acid

7.46 In one of the reactions in the citric acid cycle, which provides energy, succinic acid is converted to fumaric acid.

$$C_4H_6O_4 \longrightarrow C_4H_4O_4 + 2H$$
Succinic acid Fumaric acid

The reaction is accompanied by a coenzyme, flavin adenine dinucleotide (FAD).

$$FAD + 2H \longrightarrow FADH_2$$

a. Is succinic acid oxidized or reduced?
b. Is FAD oxidized or reduced?
c. Why would the two reactions occur together?

7.6 Mole Relationships in Chemical Equations

LEARNING GOAL

Given a quantity in moles of reactant or product, use a mole–mole factor from the balanced chemical equation to calculate the number of moles of another substance in the reaction.

In any chemical reaction, the total amount of matter in the reactants is equal to the total amount of matter in the products. Thus, the total mass of all the reactants must be equal to the total mass of all the products. This is known as the *law of conservation of mass*, which states that there is no change in the total mass of the substances reacting in a chemical reaction. Thus, no material is lost or gained as original substances are changed to new substances.

For example, tarnish (Ag_2S) forms when silver reacts with sulfur to form silver sulfide.

$$2Ag(s) + S(s) \longrightarrow Ag_2S(s)$$

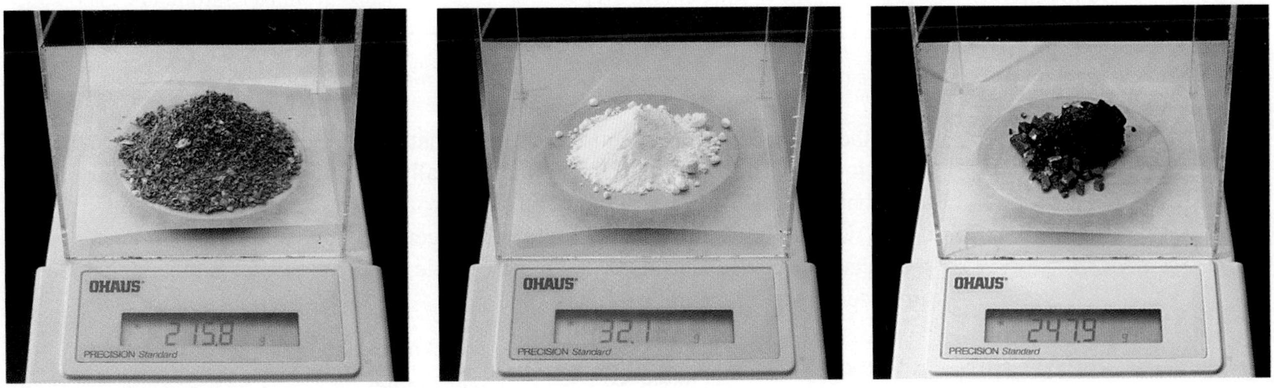

$2Ag(s)$ + $S(s)$ $\longrightarrow$ $Ag_2S(s)$

Mass of reactants = Mass of product

In the chemical reaction of Ag and S, the mass of the reactants is the same as the mass of the product, Ag_2S.

In this reaction, the number of silver atoms that reacts is twice the number of sulfur atoms. When 200 silver atoms react, 100 sulfur atoms are required. However, in the actual chemical reaction, many more atoms of both silver and sulfur would react. If we are dealing with moles of silver and sulfur, then the coefficients in the equation can be interpreted in terms of moles. Thus, 2 moles of silver react with 1 mole of sulfur to form 1 mole of Ag_2S. Because the molar mass of each can be determined, the moles of Ag, S, and Ag_2S can also be stated in terms of mass in grams of each. Thus, 215.8 g of Ag and 32.1 g of S react to form 247.9 g of Ag_2S. The total mass of the reactants (247.9 g) is equal to the mass of product (247.9 g). The various ways in which a chemical equation can be interpreted are seen in Table 7.7.

TABLE 7.7 Information Available from a Balanced Equation

	Reactants		Products
Equation	$2Ag(s)$	$+ S(s)$	$\longrightarrow Ag_2S(s)$
Atoms	2 Ag atoms	+ 1 S atom	$\longrightarrow$ 1 Ag_2S formula unit
	200 Ag atoms	+ 100 S atoms	$\longrightarrow$ 100 Ag_2S formula units
Avogadro's Number of Atoms	$2(6.02 \times 10^{23})$ Ag atoms	$+ 1(6.02 \times 10^{23})$ S atoms	$\longrightarrow 1(6.02 \times 10^{23})$ Ag_2S formula units
Moles	2 moles of Ag	+ 1 mole of S	$\longrightarrow$ 1 mole of Ag_2S
Mass (g)	2(107.9 g) of Ag	+ 1(32.07 g) of S	$\longrightarrow$ 1(247.9 g) of Ag_2S
Total Mass (g)	247.9 g		$\longrightarrow$ 247.9 g

Mole–Mole Factors from an Equation

When iron reacts with sulfur, the product is iron(III) sulfide.

$$2Fe(s) + 3S(s) \longrightarrow Fe_2S_3(s)$$

Iron (Fe) $2Fe(s)$ + Sulfur (S) $3S(s)$ $\longrightarrow$ Iron(III) sulfide (Fe_2S_3) $Fe_2S_3(s)$

In the chemical reaction of Fe and S, the mass of the reactants is the same as the mass of the product, Fe_2S_3.

From the balanced chemical equation, we see that 2 moles of iron reacts with 3 moles of sulfur to form 1 mole of iron(III) sulfide. Actually, any amount of iron or sulfur may be used, but the *ratio* of iron reacting with sulfur will always be the same. From the coefficients, we can write **mole–mole factors** between reactants and between reactants and products. The coefficients used in the mole–mole factors are *exact* numbers; they do not limit the number of significant figures.

Fe and S: $\dfrac{2 \text{ moles Fe}}{3 \text{ moles S}}$ and $\dfrac{3 \text{ moles S}}{2 \text{ moles Fe}}$

Fe and Fe_2S_3: $\dfrac{2 \text{ moles Fe}}{1 \text{ mole } Fe_2S_3}$ and $\dfrac{1 \text{ mole } Fe_2S_3}{2 \text{ moles Fe}}$

S and Fe_2S_3: $\dfrac{3 \text{ moles S}}{1 \text{ mole } Fe_2S_3}$ and $\dfrac{1 \text{ mole } Fe_2S_3}{3 \text{ moles S}}$

Using Mole–Mole Factors in Calculations

Whenever you prepare a recipe, adjust an engine for the proper mixture of fuel and air, or prepare medicines in a pharmaceutical laboratory, you need to know the proper amounts of reactants to use and how much of the product will form. Now that we have written all the possible conversion factors for the balanced equation $2Fe(s) + 3S(s) \longrightarrow Fe_2S_3(s)$, we will use those mole–mole factors in a chemical calculation in Sample Problem 7.10.

CORE CHEMISTRY SKILL

Using Mole–Mole Factors

SAMPLE PROBLEM 7.10 Calculating Moles of a Reactant

In the chemical reaction of iron and sulfur, how many moles of sulfur are needed to react with 1.42 moles of iron?

$$2Fe(s) + 3S(s) \longrightarrow Fe_2S_3(s)$$

SOLUTION

STEP 1 State the given and needed quantities (moles).

ANALYZE THE PROBLEM	Given	Need
	1.42 moles of Fe	moles of S
	Equation	
	$2Fe(s) + 3S(s) \longrightarrow Fe_2S_3(s)$	

STEP 2 Write a plan to convert the given to the needed quantity (moles).

moles of Fe → Mole–mole factor → moles of S

STEP 3 Use coefficients to write mole–mole factors.

$$2 \text{ moles of Fe} = 3 \text{ moles of S}$$

$$\frac{2 \text{ moles Fe}}{3 \text{ moles S}} \quad \text{and} \quad \frac{3 \text{ moles S}}{2 \text{ moles Fe}}$$

STEP 4 Set up the problem to give the needed quantity (moles).

$$1.42 \cancel{\text{moles Fe}} \times \frac{3 \text{ moles S}}{2 \cancel{\text{moles Fe}}} = 2.13 \text{ moles of S}$$

Three SFs — Exact (numerator), Exact (denominator) — Three SFs

STUDY CHECK 7.10

Using the equation in Sample Problem 7.10, calculate the number of moles of iron needed to react with 2.75 moles of sulfur.

Guide to Calculating the Quantities of Reactants and Products in a Chemical Reaction

STEP 1 State the given and needed quantities (moles or grams).

STEP 2 Write a plan to convert the given to the needed quantity (moles or grams).

STEP 3 Use coefficients to write mole–mole factors; write molar mass factors if needed.

STEP 4 Set up the problem to give the needed quantity (moles or grams).

QUESTIONS AND PROBLEMS

7.6 Mole Relationships in Chemical Equations

LEARNING GOAL Given a quantity in moles of reactant or product, use a mole–mole factor from the balanced chemical equation to calculate the number of moles of another substance in the reaction.

7.47 Write all the mole–mole factors for each of the following chemical equations:
a. $2SO_2(g) + O_2(g) \longrightarrow 2SO_3(g)$
b. $4P(s) + 5O_2(g) \longrightarrow 2P_2O_5(s)$

7.48 Write all the mole–mole factors for each of the following chemical equations:
a. $2Al(s) + 3Cl_2(g) \longrightarrow 2AlCl_3(s)$
b. $4HCl(g) + O_2(g) \longrightarrow 2Cl_2(g) + 2H_2O(g)$

7.49 The chemical reaction of hydrogen with oxygen produces water.

$$2H_2(g) + O_2(g) \longrightarrow 2H_2O(g)$$

a. How many moles of O_2 are required to react with 2.0 moles of H_2?
b. How many moles of H_2 are needed to react with 5.0 moles of O_2?
c. How many moles of H_2O form when 2.5 moles of O_2 reacts?

7.50 Ammonia is produced by the chemical reaction of hydrogen and nitrogen.

$$N_2(g) + 3H_2(g) \longrightarrow \underset{\text{Ammonia}}{2NH_3(g)}$$

a. How many moles of H_2 are needed to react with 1.0 mole of N_2?
b. How many moles of N_2 reacted if 0.60 mole of NH_3 is produced?
c. How many moles of NH_3 are produced when 1.4 moles of H_2 reacts?

7.51 Carbon disulfide and carbon monoxide are produced when carbon is heated with sulfur dioxide.

$$5C(s) + 2SO_2(g) \xrightarrow{\Delta} CS_2(l) + 4CO(g)$$

a. How many moles of C are needed to react with 0.500 mole of SO_2?
b. How many moles of CO are produced when 1.2 moles of C reacts?
c. How many moles of SO_2 are needed to produce 0.50 mole of CS_2?
d. How many moles of CS_2 are produced when 2.5 moles of C reacts?

7.52 Write a balanced equation for the complete combustion of ethane (C_2H_6).
a. How many moles of water are produced when 10 mmol of ethane reacts in the above combustion?
b. What if the amount of oxygen is not enough for the ethane to combust?
c. Give a balanced equation if there are only 2 equivalents of oxygen available for the combustion.

LEARNING GOAL

Given the mass in grams of a substance in a reaction, calculate the mass in grams of another substance in the reaction.

CORE CHEMISTRY SKILL

Converting Grams to Grams

7.7 Mass Calculations for Reactions

When we have the balanced chemical equation for a reaction, we can use the mass of one of the substances (A) in the reaction to calculate the mass of another substance (B) in the reaction. However, the calculations require us to convert the mass of A to moles of A using the molar mass factor for A. Then we use the mole–mole factor that links substance A to substance B, which we obtain from the coefficients in the balanced equation. This mole–mole factor (B/A) will convert the moles of A to moles of B. Then the molar mass factor of B is used to calculate the grams of substance B.

Substance A | **Substance B**

grams of A → Molar mass factor A → moles of A → Mole–mole factor B/A → moles of B → Molar mass factor B → grams of B

A mixture of acetylene and oxygen undergoes combustion during the welding of metals.

SAMPLE PROBLEM 7.11 Mass of Product

When acetylene, C_2H_2, burns in oxygen, high temperatures are produced that are used for welding metals.

$$2C_2H_2(g) + 5O_2(g) \xrightarrow{\Delta} 4CO_2(g) + 2H_2O(g)$$

How many grams of CO_2 are produced when 54.6 g of C_2H_2 is burned?

SOLUTION

STEP 1 State the given and needed quantities (grams).

	Given	Need
ANALYZE THE PROBLEM	54.6 g of C_2H_2	grams of CO_2
	Equation	
	$2C_2H_2(g) + 5O_2(g) \xrightarrow{\Delta} 4CO_2(g) + 2H_2O(g)$ + energy	

STEP 2 Write a plan to convert the given to the needed quantity (grams).

grams of C_2H_2 → Molar mass → moles of C_2H_2 → Mole–mole factor → moles of CO_2 → Molar mass → grams of CO_2

STEP 3 Use coefficients to write mole–mole factors; write molar mass factors if needed.

$$1 \text{ mole of } C_2H_2 = 26.04 \text{ g of } C_2H_2$$

$$\frac{26.04 \text{ g } C_2H_2}{1 \text{ mole } C_2H_2} \quad \text{and} \quad \frac{1 \text{ mole } C_2H_2}{26.04 \text{ g } C_2H_2}$$

$$2 \text{ moles of } C_2H_2 = 4 \text{ moles of } CO_2$$

$$\frac{2 \text{ moles } C_2H_2}{4 \text{ moles } CO_2} \quad \text{and} \quad \frac{4 \text{ moles } CO_2}{2 \text{ moles } C_2H_2}$$

$$1 \text{ mole of } CO_2 = 44.01 \text{ g of } CO_2$$

$$\frac{44.01 \text{ g } CO_2}{1 \text{ mole } CO_2} \quad \text{and} \quad \frac{1 \text{ mole } CO_2}{44.01 \text{ g } CO_2}$$

STEP 4 Set up the problem to give the needed quantity (grams).

$$\underset{\text{Three SFs}}{54.6 \; \cancel{\text{g } C_2H_2}} \times \overset{\text{Exact}}{\underset{\text{Four SFs}}{\frac{1 \; \cancel{\text{mole } C_2H_2}}{26.04 \; \cancel{\text{g } C_2H_2}}}} \times \overset{\text{Exact}}{\underset{\text{Exact}}{\frac{4 \; \cancel{\text{moles } CO_2}}{2 \; \cancel{\text{moles } C_2H_2}}}} \times \overset{\text{Four SFs}}{\underset{\text{Exact}}{\frac{44.01 \text{ g } CO_2}{1 \; \cancel{\text{mole } CO_2}}}} = \underset{\text{Three SFs}}{185 \text{ g of } CO_2}$$

STUDY CHECK 7.11

Using the equation in Sample Problem 7.11, calculate the grams of CO_2 that can be produced when 25.0 g of O_2 reacts.

QUESTIONS AND PROBLEMS

7.7 Mass Calculations for Reactions

LEARNING GOAL Given the mass in grams of a substance in a reaction, calculate the mass in grams of another substance in the reaction.

7.53 Sodium reacts with oxygen to produce sodium oxide.

$$4Na(s) + O_2(g) \longrightarrow 2Na_2O(s)$$

a. How many grams of Na_2O are produced when 57.5 g of Na reacts?
b. If you have 18.0 g of Na, how many grams of O_2 are needed for the reaction?
c. How many grams of O_2 are needed in a reaction that produces 75.0 g of Na_2O?

7.54 Nitrogen gas reacts with hydrogen gas to produce ammonia.

$$N_2(g) + 3H_2(g) \longrightarrow 2NH_3(g)$$

a. If you have 3.64 g of H_2, how many grams of NH_3 can be produced?
b. How many grams of H_2 are needed to react with 2.80 g of N_2?
c. How many grams of NH_3 can be produced from 12.0 g of H_2?

7.55 Ammonia and oxygen react to form nitrogen and water.

$$4NH_3(g) + 3O_2(g) \longrightarrow 2N_2(g) + 6H_2O(g)$$

a. How many grams of O_2 are needed to react with 13.6 g of NH_3?
b. How many grams of N_2 can be produced when 6.50 g of O_2 reacts?
c. How many grams of H_2O are formed from the reaction of 34.0 g of NH_3?

7.56 Iron(III) oxide reacts with carbon to give iron and carbon monoxide.

$$Fe_2O_3(s) + 3C(s) \longrightarrow 2Fe(s) + 3CO(g)$$

a. How many grams of C are required to react with 16.5 g of Fe_2O_3?
b. How many grams of CO are produced when 36.0 g of C reacts?
c. How many grams of Fe can be produced when 6.00 g of Fe_2O_3 reacts?

7.57 Nitrogen dioxide and water react to produce nitric acid, HNO_3, and nitrogen oxide.

$$3NO_2(g) + H_2O(l) \longrightarrow 2HNO_3(aq) + NO(g)$$

a. How many grams of H_2O are needed to react with 28.0 g of NO_2?
b. How many grams of NO are produced from 15.8 g of NO_2?
c. How many grams of HNO_3 are produced from 8.25 g of NO_2?

7.58 Calcium cyanamide reacts with water to form calcium carbonate and ammonia.

$$CaCN_2(s) + 3H_2O(l) \longrightarrow CaCO_3(s) + 2NH_3(g)$$

a. How many grams of H_2O are needed to react with 75.0 g of $CaCN_2$?
b. How many grams of NH_3 are produced from 5.24 g of $CaCN_2$?
c. How many grams of $CaCO_3$ form if 155 g of H_2O reacts?

7.59 When solid lead(II) sulfide ore burns in oxygen gas, the products are solid lead(II) oxide and sulfur dioxide gas.
a. Write the balanced chemical equation for the reaction.
b. How many grams of oxygen are required to react with 30.0 g of lead(II) sulfide?

Interactive Video

Problem 7.59

c. How many grams of sulfur dioxide can be produced when 65.0 g of lead(II) sulfide reacts?

d. How many grams of lead(II) sulfide are used to produce 128 g of lead(II) oxide?

7.60 Aluminum oxidizes to form aluminum oxide.

a. Write the balanced chemical equation for the reaction.

b. How many mmol and grams of aluminum are needed to react with 6 mmol of oxygen?

c. How many moles and grams of aluminum oxide are produced with 4 mole of aluminum?

LEARNING GOAL

Describe exothermic and endothermic reactions and factors that affect the rate of a reaction.

7.8 Energy in Chemical Reactions

For a chemical reaction to take place, the molecules of the reactants must collide with each other and have the proper orientation and energy. Even when a collision has the proper orientation, there still must be sufficient energy to break the bonds of the reactants. The **activation energy** is the amount of energy required to break the bonds between atoms of the reactants. If the energy of a collision is less than the activation energy, the molecules bounce apart without reacting. Many collisions occur, but only a few actually lead to the formation of product.

The concept of activation energy is analogous to climbing over a hill. To reach a destination on the other side, we must expend energy to climb to the top of the hill. Once we are at the top, we can easily run down the other side. The energy needed to get us from our starting point to the top of the hill would be the activation energy.

Three Conditions Required for a Reaction to Occur

1. **Collision** The reactants must collide.
2. **Orientation** The reactants must align properly to break and form bonds.
3. **Energy** The collision must provide the energy of activation.

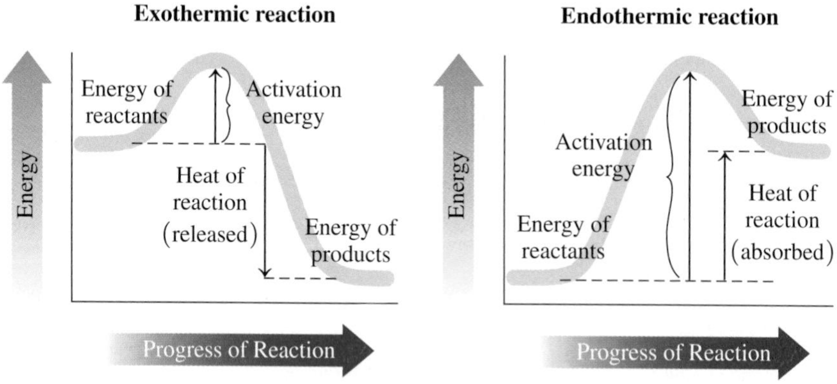

The activation energy is the energy needed to convert reacting molecules into products.

Exothermic Reactions

In every chemical reaction, heat is absorbed or released as reactants are converted to products. The *heat of reaction* is the difference between the energy of the reactants and the energy of the products. In an **exothermic reaction** (*exo* means "out"), the energy of the products is lower than the energy of the reactants. Thus, heat is released in exothermic reactions. For example, in the thermite reaction, the reaction of aluminum and iron(III) oxide produces so much heat that temperatures of 2500 °C can be reached. The thermite reaction has been used to cut or weld railroad tracks. In the equation for an exothermic reaction, the heat of reaction is written on the same side as the products.

The high temperature of the thermite reaction has been used to cut or weld railroad tracks.

Exothermic, Heat Released

$$2Al(s) + Fe_2O_3(s) \longrightarrow 2Fe(s) + Al_2O_3(s) + 850\ kJ \quad \text{Heat is a product}$$

Endothermic Reactions

In **endothermic reactions** (*endo* means "within"), the energy of the products is higher than that of the reactants. Thus, heat is absorbed in endothermic reactions. For example, when hydrogen and iodine react to form hydrogen iodide, heat must be absorbed. In the equation for an endothermic reaction, the heat of reaction is written on the same side as the reactants.

Endothermic, Heat Absorbed

$H_2(g) + I_2(g) + 55\text{ kJ} \longrightarrow 2HI(g)$ Heat is a reactant

Reaction	Energy Change	Heat in the Equation
Exothermic	Heat released	Product side
Endothermic	Heat absorbed	Reactant side

Chemistry Link to **Health**

Cold Packs and Hot Packs

In a hospital, at a first-aid station, or at an athletic event, an instant *cold pack* may be used to reduce swelling from an injury, remove heat from inflammation, or decrease capillary size to lessen the effect of hemorrhaging. Inside the plastic container of a cold pack, there is a compartment containing solid ammonium nitrate (NH_4NO_3) that is separated from a compartment containing water. The pack is activated when it is hit or squeezed hard enough to break the walls between the compartments and cause the ammonium nitrate to mix with the water (shown as H_2O over the reaction arrow). In an endothermic process, 1 mole of NH_4NO_3 that dissolves in water absorbs 26 kJ. The temperature drops to about 4 to 5 °C to give a cold pack that is ready to use.

Endothermic Reaction in a Cold Pack

$$NH_4NO_3(s) + 26\text{ kJ} \xrightarrow{H_2O} NH_4NO_3(aq)$$

Hot packs are used to relax muscles, lessen aches and cramps, and increase circulation by expanding capillary size. Constructed in the same way as cold packs, a hot pack contains a salt such as $CaCl_2$. When 1 mole of $CaCl_2$ dissolves in water, 82 kJ is released as heat. The temperature rises to as much as 66 °C to give a hot pack that is ready to use.

Exothermic Reaction in a Hot Pack

$$CaCl_2(s) \xrightarrow{H_2O} CaCl_2(aq) + 82\text{ kJ}$$

Cold packs use an endothermic reaction.

Rate of Reaction

The *rate* (or speed) *of reaction* is measured by the amount of reactant used up, or the amount of product formed, in a certain period of time. Reactions with low activation energies go faster than reactions with high activation energies. Some reactions go very fast, while others are very slow. For any reaction, the rate is affected by changes in temperature, changes in the concentration of the reactants, and the addition of catalysts.

Temperature

At higher temperatures, the increase in kinetic energy of the reactants makes them move faster and collide more often, and it provides more collisions with the required energy of activation. Reactions almost always go faster at higher temperatures. For every 10 °C increase in temperature, most reaction rates approximately double. If we want food to cook faster, we raise the temperature. When body temperature rises, there is an increase in the pulse rate, rate of breathing, and metabolic rate. On the other hand, we slow down a reaction by lowering the temperature. For example, we refrigerate perishable foods to make them last longer. In some cardiac surgeries, body temperature is lowered to 28 °C so the heart can be stopped and less oxygen is required by the brain. This is also the reason why some people have survived submersion in icy lakes for long periods of time.

Concentrations of Reactants

The rate of a reaction also increases when reactants are added. Then there are more collisions between the reactants and the reaction goes faster (see Table 7.8). For example, a

TABLE 7.8 Factors That Increase Reaction Rate

Factor	Reason
Increase reactant concentration	More collisions
Increase temperature	More collisions, more collisions with energy of activation
Add a catalyst	Lowers energy of activation

patient having difficulty breathing may be given a breathing mixture with a higher oxygen content than the atmosphere. The increase in the number of oxygen molecules in the lungs increases the rate at which oxygen combines with hemoglobin. The increased rate of oxygenation of the blood means that the patient can breathe more easily.

$$\underset{\text{Hemoglobin}}{Hb(aq)} + \underset{\text{Oxygen}}{O_2(g)} \longrightarrow \underset{\text{Oxyhemoglobin}}{HbO_2(aq)}$$

Catalysts

Another way to speed up a reaction is to lower the energy of activation. This can be done by adding a **catalyst**. Earlier, we discussed the energy required to climb a hill. If instead we find a tunnel through the hill, we do not need as much energy to get to the other side. A catalyst acts by providing an alternate pathway with a lower energy requirement. As a result, more collisions form product successfully. Catalysts have found many uses in industry. In the production of margarine, the reaction of hydrogen with vegetable oils is normally very slow. However, when finely divided platinum is present as a catalyst, the reaction occurs rapidly. In the body, biocatalysts called enzymes make most metabolic reactions go at the rates necessary for proper cellular activity.

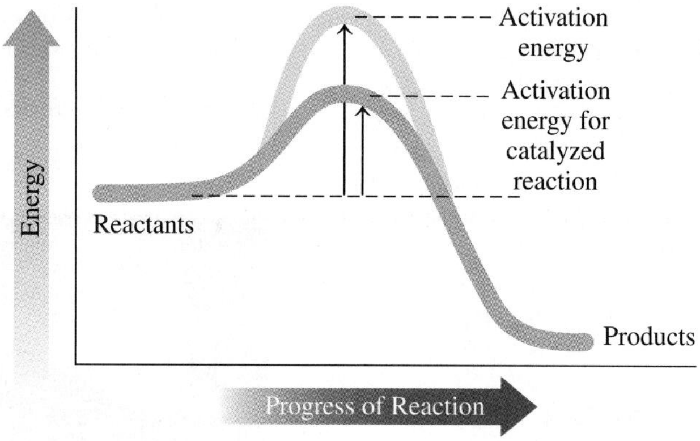

A catalyst lowers the activation energy of a chemical reaction.

SAMPLE PROBLEM 7.12 Reactions and Rates

In the reaction of 1 mole of solid carbon with oxygen gas, the energy of the carbon dioxide gas produced is 393 kJ lower than the energy of the reactants.

a. Is the reaction exothermic or endothermic?
b. Write the balanced equation for the reaction, including the heat of reaction.
c. Indicate how an increase in the number of O_2 molecules will change the rate of the reaction.

SOLUTION

a. When the products have a lower energy than the reactants, the reaction is exothermic.
b. $C(s) + O_2(g) \longrightarrow CO_2(g) + 393\text{ kJ}$
c. An increase in the number of O_2 molecules will increase the rate of the reaction.

STUDY CHECK 7.12

How does lowering the temperature affect the rate of reaction?

QUESTIONS AND PROBLEMS

7.8 Energy in Chemical Reactions

LEARNING GOAL Describe exothermic and endothermic reactions and factors that affect the rate of a reaction.

7.61 **a.** Why do chemical reactions require energy of activation?
b. What is the function of a catalyst?
c. In an exothermic reaction, is the energy of the products higher or lower than that of the reactants?
d. Draw an energy diagram for an exothermic reaction.

7.62 **a.** What is measured by the heat of reaction?
b. How does the heat of reaction differ in exothermic and endothermic reactions?

c. In an endothermic reaction, is the energy of the products higher or lower than that of the reactants?
d. Draw an energy diagram for an endothermic reaction.

7.63 Classify each of the following as an exothermic or endothermic reaction:
a. A reaction releases 550 kJ.
b. The energy level of the products is higher than that of the reactants.
c. The metabolism of glucose in the body provides energy.

7.64 Classify each of the following as an exothermic or endothermic reaction:
a. The energy level of the products is lower than that of the reactants.
b. In the body, the synthesis of proteins requires energy.
c. A reaction absorbs 125 kJ.

7.65 Classify each of the following as an exothermic or endothermic reaction:
a. $CH_4(g) + 2O_2(g) \xrightarrow{\Delta} CO_2(g) + 2H_2O(g) + 802\text{ kJ}$
b. $Ca(OH)_2(s) + 65.3\text{ kJ} \longrightarrow CaO(s) + H_2O(l)$
c. $2Al(s) + Fe_2O_3(s) \longrightarrow Al_2O_3(s) + 2Fe(s) + 850\text{ kJ}$

7.66 Classify each of the following as an exothermic or endothermic reaction:
a. $C_3H_8(g) + 5O_2(g) \xrightarrow{\Delta} 3CO_2(g) + 4H_2O(g) + 2220\text{ kJ}$
b. $2Na(s) + Cl_2(g) \longrightarrow 2NaCl(s) + 819\text{ kJ}$
c. $PCl_5(g) + 67\text{ kJ} \longrightarrow PCl_3(g) + Cl_2(g)$

7.67 **a.** What is meant by the rate of a reaction?
b. Why does bread grow mold more quickly at room temperature than in the refrigerator?

7.68 **a.** How does a catalyst affect the activation energy?
b. Why is pure oxygen used in respiratory distress?

7.69 How would each of the following change the rate of the reaction shown here?

$$2SO_2(g) + O_2(g) \longrightarrow 2SO_3(g)$$

a. adding some $SO_2(g)$
b. increasing the temperature
c. adding a catalyst
d. removing some $O_2(g)$

7.70 How would each of the following change the rate of the reaction shown here?

$$2NO(g) + 2H_2(g) \longrightarrow N_2(g) + 2H_2O(g)$$

a. adding some $NO(g)$
b. lowering the temperature
c. removing some $H_2(g)$
d. adding a catalyst

CONCEPT MAP

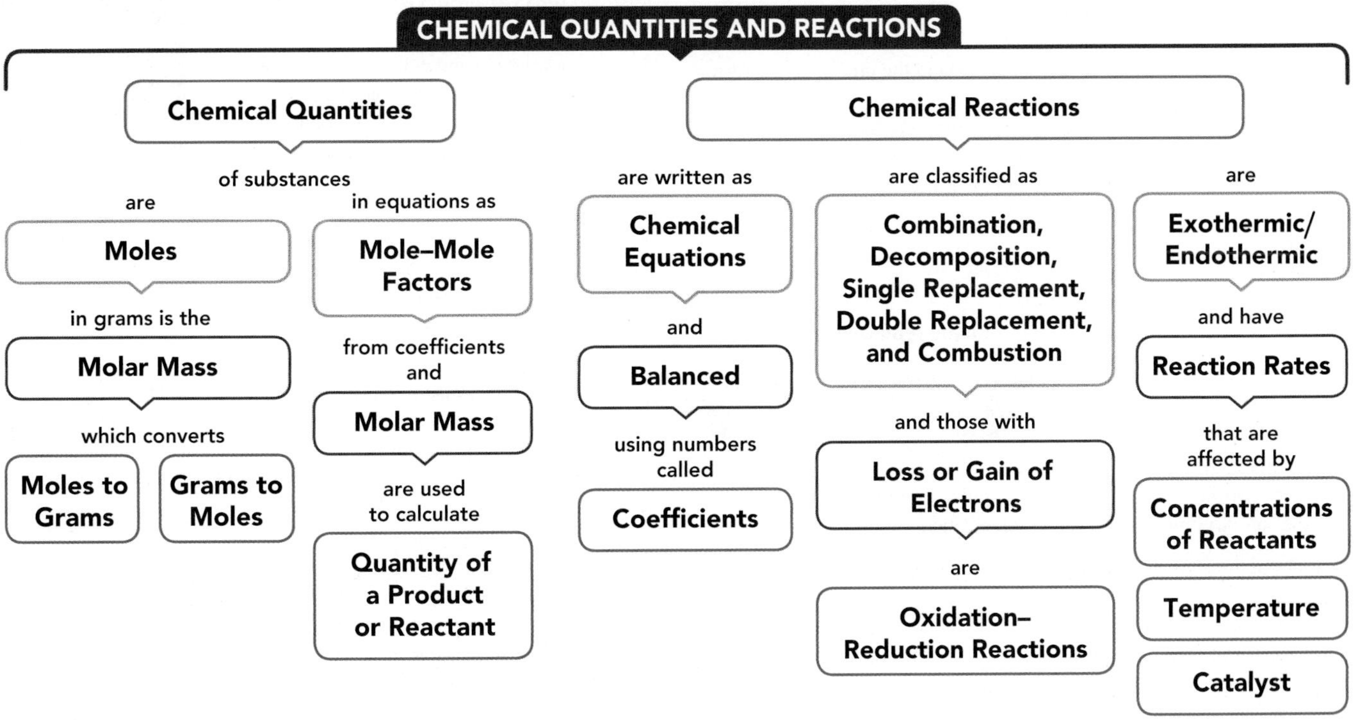

CHAPTER REVIEW

7.1 The Mole

LEARNING GOAL Use Avogadro's number to determine the number of particles in a given number of moles.

- One mole of an element contains 6.02×10^{23} atoms; 1 mole of a compound contains 6.02×10^{23} molecules or formula units.

7.2 Molar Mass and Calculations

LEARNING GOAL Calculate the molar mass for a substance given its chemical formula; use molar mass to convert between grams and moles.

- The molar mass (g/mole) of any substance is the mass in grams equal numerically to its atomic mass, or the sum of the atomic masses, which have been multiplied by their subscripts in a formula.
- The molar mass is used as a conversion factor to change a quantity in grams to moles or to change a given number of moles to grams.

7.3 Equations for Chemical Reactions

LEARNING GOAL Write a balanced chemical equation from the formulas of the reactants and products for a reaction; determine the number of atoms in the reactants and products.

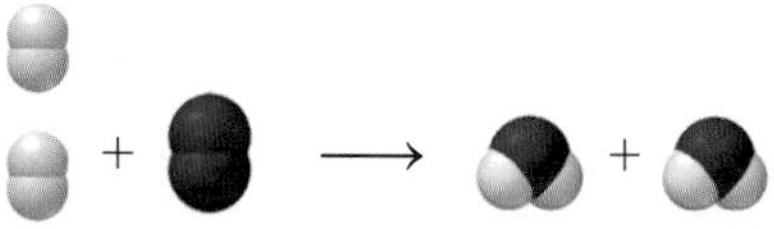

$$2H_2(g) + O_2(g) \longrightarrow 2H_2O(g)$$

Balanced

- A chemical change occurs when the atoms of the initial substances rearrange to form new substances.
- A chemical equation shows the formulas of the substances that react on the left side of a reaction arrow and the products that form on the right side of the reaction arrow.
- A chemical equation is balanced by writing coefficients, small whole numbers, in front of formulas to equalize the atoms of each of the elements in the reactants and the products.

7.4 Types of Reactions

LEARNING GOAL Identify a reaction as a combination, decomposition, single replacement, double replacement, or combustion.

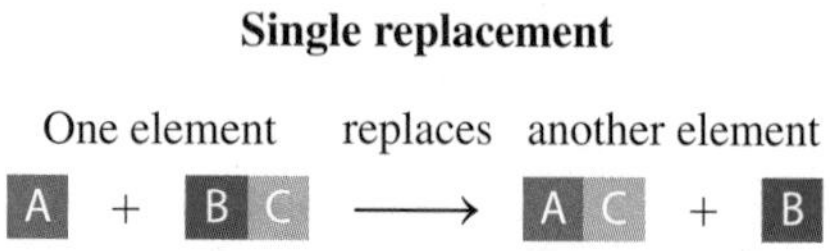

- Many chemical reactions can be organized by reaction type: combination, decomposition, single replacement, double replacement, or combustion.

7.5 Oxidation–Reduction Reactions

LEARNING GOAL Define the terms oxidation and reduction; identify the reactants oxidized and reduced.

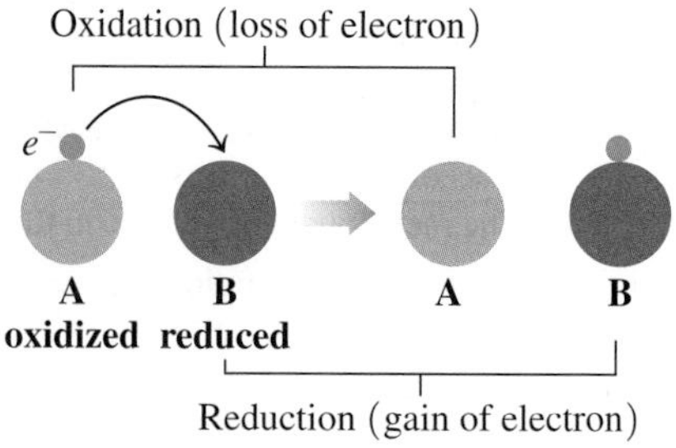

- When electrons are transferred in a reaction, it is an oxidation–reduction reaction.
- One reactant loses electrons, and another reactant gains electrons.
- Overall, the number of electrons lost and gained is equal.

7.6 Mole Relationships in Chemical Equations

LEARNING GOAL Given a quantity in moles of reactant or product, use a mole–mole factor from the balanced chemical equation to calculate the number of moles of another substance in the reaction.

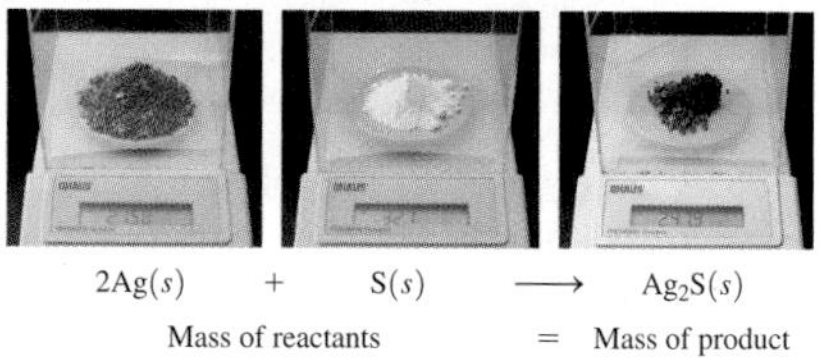

- In a balanced equation, the total mass of the reactants is equal to the total mass of the products.
- The coefficients in an equation describing the relationship between the moles of any two components are used to write mole–mole factors.
- When the number of moles for one substance is known, a mole–mole factor is used to find the moles of a different substance in the reaction.

7.7 Mass Calculations for Reactions

LEARNING GOAL Given the mass in grams of a substance in a reaction, calculate the mass in grams of another substance in the reaction.

- In calculations using equations, the molar masses of the substances and their mole–mole factors are used to change the number of grams of one substance to the corresponding grams of a different substance.

7.8 Energy in Chemical Reactions

LEARNING GOAL Describe exothermic and endothermic reactions and factors that affect the rate of a reaction.

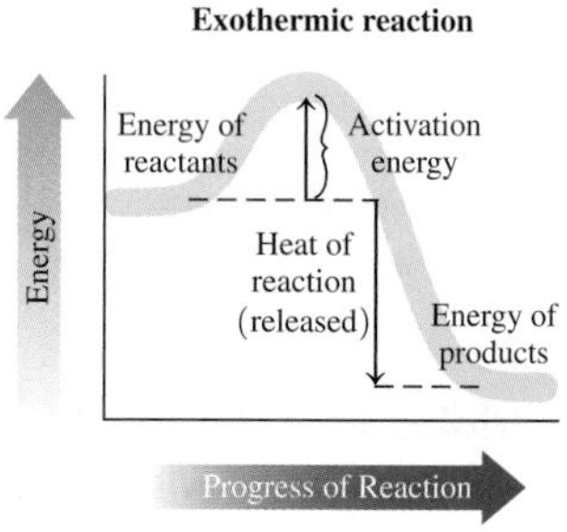

- In chemical reactions, the heat of reaction is the energy difference between the reactants and the products.
- In an exothermic reaction, heat is released because the energy of the products is lower than that of the reactants.
- In an endothermic reaction, heat is absorbed because the energy of the products is higher than that of the reactants.
- The rate of a reaction is the speed at which the reactants are converted to products.
- Increasing the concentrations of reactants, raising the temperature, or adding a catalyst can increase the rate of a reaction.

KEY TERMS

activation energy The energy needed upon collision to break apart the bonds of the reacting molecules.

Avogadro's number The number of items in a mole, equal to 6.02×10^{23}.

balanced equation The final form of a chemical equation that shows the same number of atoms of each element in the reactants and products.

catalyst A substance that increases the rate of reaction by lowering the activation energy.

chemical equation A shorthand way to represent a chemical reaction using chemical formulas to indicate the reactants and products and coefficients to show reacting ratios.

coefficients Whole numbers placed in front of the formulas to balance the number of atoms or moles of atoms of each element on both sides of an equation.

combination reaction A chemical reaction in which reactants combine to form a single product.

combustion reaction A chemical reaction in which a fuel containing carbon and hydrogen reacts with oxygen to produce CO_2, H_2O, and energy.

decomposition reaction A reaction in which a single reactant splits into two or more simpler substances.

double replacement reaction A reaction in which parts of two different reactants exchange places.

endothermic reaction A reaction that requires heat; the energy of the products is higher than the energy of the reactants.

exothermic reaction A reaction that releases heat; the energy of the products is lower than the energy of the reactants.

formula unit The group of ions represented by the formula of an ionic compound.

molar mass The mass in grams of 1 mole of an element equal numerically to its atomic mass. The molar mass of a compound is equal to the sum of the masses of the elements multiplied by their subscripts in the formula.

mole A group of atoms, molecules, or formula units that contains 6.02×10^{23} of these items.

mole–mole factor A conversion factor that relates the number of moles of two compounds derived from the coefficients in an equation.

oxidation The loss of electrons by a substance. Biological oxidation may involve the addition of oxygen or the loss of hydrogen.

oxidation–reduction reaction A reaction in which the oxidation of one reactant is always accompanied by the reduction of another reactant.

products The substances formed as a result of a chemical reaction.

reactants The initial substances that undergo change in a chemical reaction.

reduction The gain of electrons by a substance. Biological reduction may involve the loss of oxygen or the gain of hydrogen.

single replacement reaction A reaction in which an element replaces a different element in a compound.

CORE CHEMISTRY SKILLS

The chapter section containing each Core Chemistry Skill is shown in parentheses at the end of each heading.

Converting Particles to Moles (7.1)

- In chemistry, atoms, molecules, and ions are counted by the mole, a unit that contains 6.02×10^{23} items, which is Avogadro's number.
- For example, 1 mole of carbon contains 6.02×10^{23} atoms of carbon and 1 mole of H_2O contains 6.02×10^{23} molecules of H_2O.
- Avogadro's number is used to convert between particles and moles.

Example: How many moles of nickel contain 2.45×10^{24} Ni atoms?

Answer:

$$\underset{\text{Three SFs}}{2.45 \times 10^{24}\ \cancel{\text{Ni atoms}}} \times \frac{\overset{\text{Exact}}{1\ \text{mole Ni}}}{\underset{\text{Three SFs}}{6.02 \times 10^{23}\ \cancel{\text{Ni atoms}}}} = \underset{\text{Three SFs}}{4.07\ \text{moles of Ni}}$$

Calculating Molar Mass (7.2)

- The molar mass of a compound is the sum of the molar mass of each element in its chemical formula multiplied by its subscript in the formula.

Pinene is a component of pine sap.

Example: Pinene, $C_{10}H_{16}$, which is found in pine tree sap and essential oils, has anti-inflammatory properties. Calculate the molar mass for pinene.

Answer:

$$10\ \cancel{\text{moles C}} \times \frac{12.01\ \text{g C}}{1\ \cancel{\text{mole C}}} = 120.1\ \text{g of C}$$

$$16\ \cancel{\text{moles H}} \times \frac{1.008\ \text{g H}}{1\ \cancel{\text{mole H}}} = 16.13\ \text{g of H}$$

$$\text{Molar mass of } C_{10}H_{16} = 136.2\ \text{g of } C_{10}H_{16}$$

Using Molar Mass as a Conversion Factor (7.2)

- The molar mass of an element is its mass in grams equal numerically to its atomic mass.
- The molar mass of a compound is its mass in grams equal numerically to the sum of the masses of its elements.
- Molar mass is used as a conversion factor to convert between the moles and grams of a substance.

A bicycle with an aluminum frame.

Equality: 1 mole of Al = 26.98 g of Al

Conversion Factors: $\frac{26.98\ \text{g Al}}{1\ \text{mole Al}}$ and $\frac{1\ \text{mole Al}}{26.98\ \text{g Al}}$

Example: The frame of a bicycle contains 6500 g of aluminum. How many moles of aluminum are in the bicycle frame?

Answer:

$$\underset{\text{Two SFs}}{6500\ \cancel{\text{g Al}}} \times \frac{\overset{\text{Exact}}{1\ \text{mole Al}}}{\underset{\text{Four SFs}}{26.98\ \cancel{\text{g Al}}}} = \underset{\text{Two SFs}}{240\ \text{moles of Al}}$$

Balancing a Chemical Equation (7.3)

- In a *balanced* equation, whole numbers called coefficients multiply each of the atoms in the chemical formulas so that the number of each type of atom in the reactants is equal to the number of the same type of atom in the products.

Example: Balance the following chemical equation:

$$SnCl_4(s) + H_2O(l) \longrightarrow Sn(OH)_4(s) + HCl(aq) \quad \text{Unbalanced}$$

Answer: When we compare the atoms on the reactant side and the product side, we see that there are more Cl atoms in the reactants and more O and H atoms in the products.

To balance the equation, we need to use coefficients in front of the formulas containing the Cl atoms, H atoms, and O atoms.

- Place a 4 in front of the formula HCl to give 8 H atoms and 4 Cl atoms in the products.

$$SnCl_4(s) + H_2O(l) \longrightarrow Sn(OH)_4(s) + 4HCl(aq)$$

- Place a 4 in front of the formula H_2O to give 8 H atoms and 4 O atoms in the reactants.

$$SnCl_4(s) + 4H_2O(l) \longrightarrow Sn(OH)_4(s) + 4HCl(aq)$$

- The total number of Sn (1), Cl (4), H (8), and O (4) atoms is now equal on both sides of the equation.

$$SnCl_4(s) + 4H_2O(l) \longrightarrow Sn(OH)_4(s) + 4HCl(aq) \ \text{Balanced}$$

Classifying Types of Chemical Reactions (7.4)

- Chemical reactions are classified by identifying general patterns in their equations.
- In a combination reaction, two or more elements or compounds bond to form one product.
- In a decomposition reaction, a single reactant splits into two or more products.
- In a single replacement reaction, an uncombined element takes the place of an element in a compound.
- In a double replacement reaction, the positive ions in the reacting compounds switch places.
- In combustion reaction, a carbon-containing compound that is the fuel burns in oxygen from the air to produce carbon dioxide (CO_2), water (H_2O), and energy.

Example: Classify the type of the following reaction:

$$2Al(s) + Fe_2O_3(s) \xrightarrow{\Delta} Al_2O_3(s) + 2Fe(l)$$

Answer: The iron in iron(III) oxide is replaced by aluminum, which makes this a single replacement reaction.

Identifying Oxidized and Reduced Substances (7.5)

- In an oxidation–reduction reaction (abbreviated *redox*), one reactant is oxidized when it loses electrons, and another reactant is reduced when it gains electrons.
- Oxidation is the *loss* of electrons; reduction is the *gain* of electrons.

Example: For the following redox reaction, identify the reactant that is oxidized, and the reactant that is reduced:

$$Fe(s) + Cu^{2+}(aq) \longrightarrow Fe^{2+}(aq) + Cu(s)$$

Answer: $Fe^0(s) \longrightarrow Fe^{2+}(aq) + 2\,e^-$

Fe loses electrons; it is oxidized.

$Cu^{2+}(aq) + 2\,e^- \longrightarrow Cu^0(s)$

Cu^{2+} gains electrons; it is reduced.

Using Mole–Mole Factors (7.6)

Consider the balanced chemical equation

$$4Na(s) + O_2(g) \longrightarrow 2Na_2O(s)$$

- The coefficients in a balanced chemical equation represent the moles of reactants and the moles of products. Thus, 4 moles of Na react with 1 mole of O_2 to form 2 moles of Na_2O.
- From the coefficients, mole–mole factors can be written for any two substances as follows:

Na and O_2	$\frac{4 \text{ moles Na}}{1 \text{ mole } O_2}$	and	$\frac{1 \text{ mole } O_2}{4 \text{ moles Na}}$
Na and Na_2O	$\frac{4 \text{ moles Na}}{2 \text{ moles } Na_2O}$	and	$\frac{2 \text{ moles } Na_2O}{4 \text{ moles Na}}$
O_2 and Na_2O	$\frac{1 \text{ mole } O_2}{2 \text{ moles } Na_2O}$	and	$\frac{2 \text{ moles } Na_2O}{1 \text{ mole } O_2}$

- A mole–mole factor is used to convert the number of moles of one substance in the reaction to the number of moles of another substance in the reaction.

Example: How many moles of sodium are needed to produce 3.5 moles of sodium oxide?

Answer: **Given:** 3.5 moles of Na_2O **Need:** moles of Na

$$\underset{\text{Two SFs}}{3.5 \ \cancel{\text{moles } Na_2O}} \times \overset{\text{Exact}}{\frac{4 \text{ moles Na}}{2 \ \cancel{\text{moles } Na_2O}}}_{\text{Exact}} = \underset{\text{Two SFs}}{7.0 \text{ moles of Na}}$$

Converting Grams to Grams (7.7)

- When we have the balanced chemical equation for a reaction, we can use the mass of substance A and then calculate the mass of substance B.

$$A \longrightarrow B$$

- The process is as follows:

(1) Use the molar mass factor of A to convert the mass, in grams, of A to moles of A.
(2) Use the mole–mole factor that converts moles of A to moles of B.
(3) Use the molar mass factor of B to calculate the mass, in grams, of B.

$$\text{grams of A} \xrightarrow{\textit{molar mass A}} \text{moles of A} \xrightarrow{\textit{mole–mole factor}} \text{moles of B} \xrightarrow{\textit{molar mass B}} \text{grams of B}$$

Example: How many grams of O_2 are needed to completely react with 14.6 g of Na?

$$4Na(s) + O_2(g) \longrightarrow 2Na_2O(s)$$

Answer:

	Molar mass of Na	Mole–mole factor	Molar mass of O_2	
	Exact	Exact	Four SFs	
$14.6 \ \cancel{g\ Na} \times$	$\frac{1 \ \cancel{\text{mole Na}}}{22.99 \ \cancel{g\ Na}} \times$	$\frac{1 \ \cancel{\text{mole } O_2}}{4 \ \cancel{\text{moles Na}}} \times$	$\frac{32.00 \text{ g } O_2}{1 \ \cancel{\text{mole } O_2}}$	$= 5.08$ g of O_2
Three SFs	Four SFs	Exact	Exact	Three SFs

UNDERSTANDING THE CONCEPTS

The chapter sections to review are shown in parentheses at the end of each question.

7.71 Using the models of the molecules (black = C, white = H, yellow = S, green = Cl), determine each of the following for models of compounds **1** and **2**: (7.1, 7.2)

1.

2.

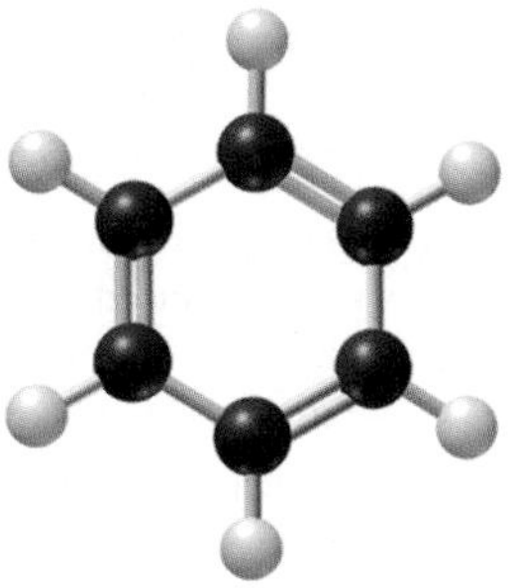

a. molecular formula
b. molar mass
c. number of moles in 10.0 g

7.72 Using the models of the molecules (black = C, white = H, yellow = S, red = O), determine each of the following for models of compounds **1** and **2**: (7.1, 7.2)

1.

2.

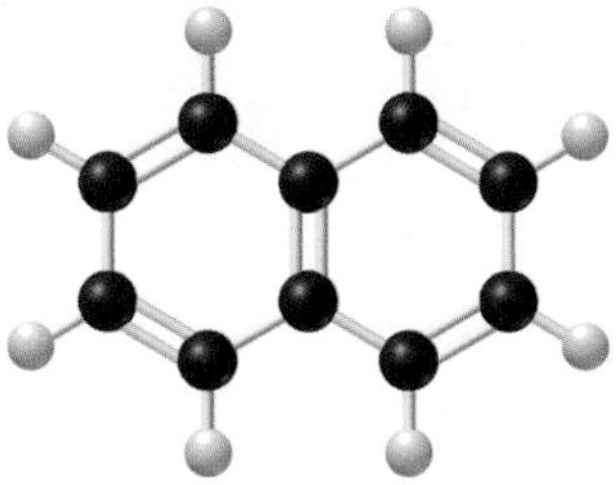

a. molecular formula
b. molar mass
c. number of moles in 10.0 g

7.73 A dandruff shampoo contains dipyrithione, $C_{10}H_8N_2O_2S_2$, which acts as an antibacterial and antifungal agent. (7.1, 7.2)
a. What is the molar mass of dipyrithione?
b. How many moles of dipyrithione are in 25.0 g?
c. How many moles of C are in 25.0 g of dipyrithione?
d. How many moles of dipyrithione contain 8.2×10^{24} atoms of N?

Dandruff shampoo contains dipyrithione.

7.74 Ibuprofen, an anti-inflammatory, has the formula $C_{13}H_{18}O_2$. (7.1, 7.2)

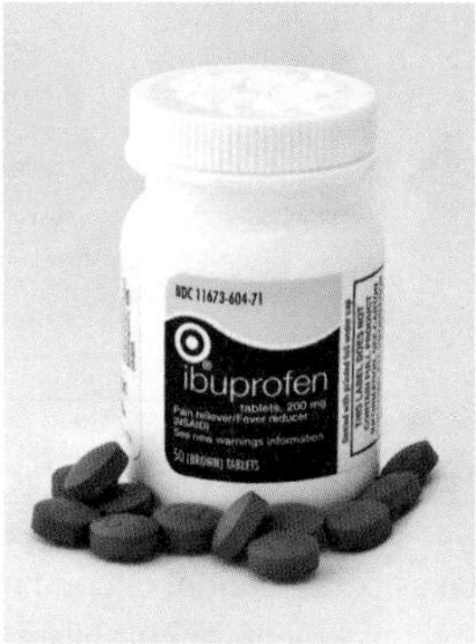

Ibuprofen is an anti-inflammatory.

a. What is the molar mass of ibuprofen?
b. How many grams of ibuprofen are in 0.525 mole?
c. How many moles of C are in 12.0 g of ibuprofen?
d. How many moles of ibuprofen contain 1.22×10^{23} atoms of C?

7.75 Balance each of the following by adding coefficients, and identify the type of reaction for each: (7.3, 7.4)

a.

b.

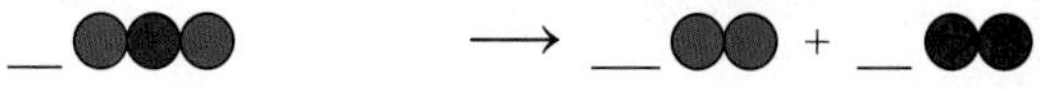

7.76 Balance each of the following by adding coefficients, and identify the type of reaction for each: (7.3, 7.4)

a.

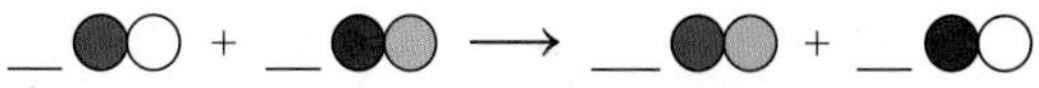

b.

7.77 If red spheres represent oxygen atoms, blue spheres represent nitrogen atoms, and all the molecules are gases, (7.3, 7.4)

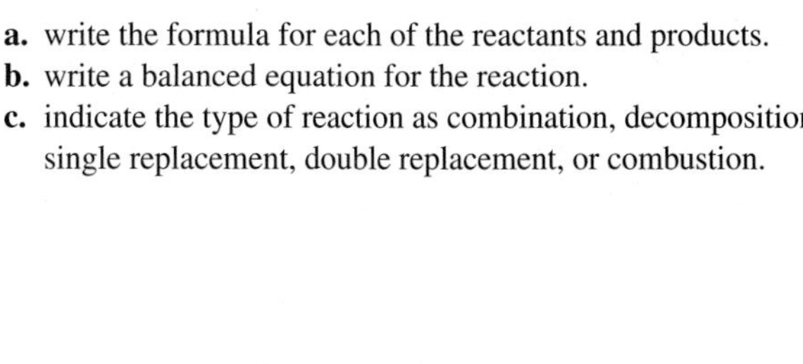

a. write the formula for each of the reactants and products.
b. write a balanced equation for the reaction.
c. indicate the type of reaction as combination, decomposition, single replacement, double replacement, or combustion.

7.78 If purple spheres represent iodine atoms, white spheres represent hydrogen atoms, and all the molecules are gases, (7.3, 7.4)

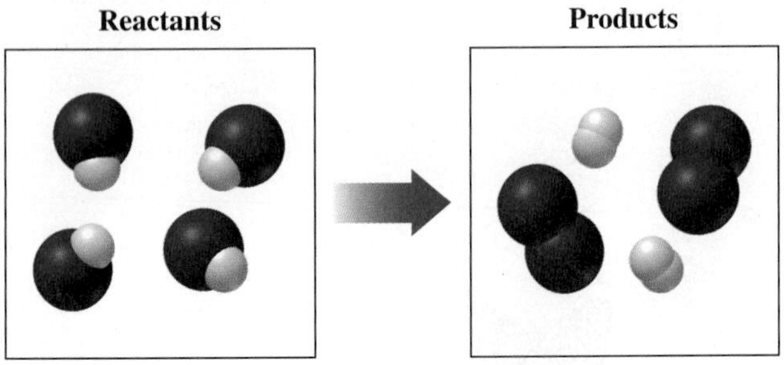

a. write the formula for each of the reactants and products.
b. write a balanced equation for the reaction.
c. indicate the type of reaction as combination, decomposition, single replacement, double replacement, or combustion.

7.79 If blue spheres represent nitrogen atoms, purple spheres represent iodine atoms, the reacting molecules are solid, and the products are gases, (7.3, 7.4)

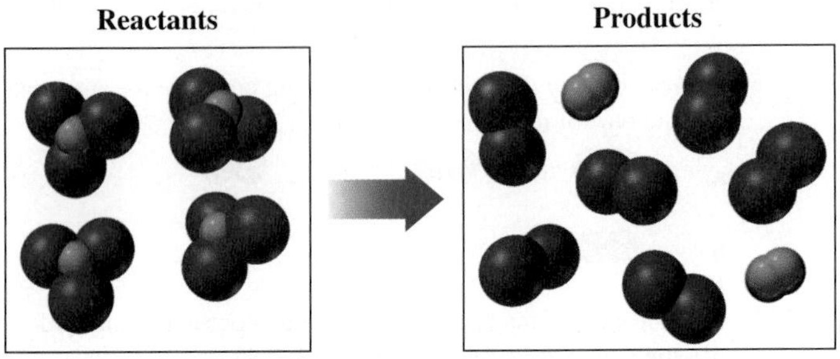

a. write the formula for each of the reactants and products.
b. write a balanced equation for the reaction.
c. indicate the type of reaction as combination, decomposition, single replacement, double replacement, or combustion.

7.80 If green spheres represent chlorine atoms, yellow-green spheres represent fluorine atoms, white spheres represent hydrogen atoms, and all the molecules are gases, (7.3, 7.4)

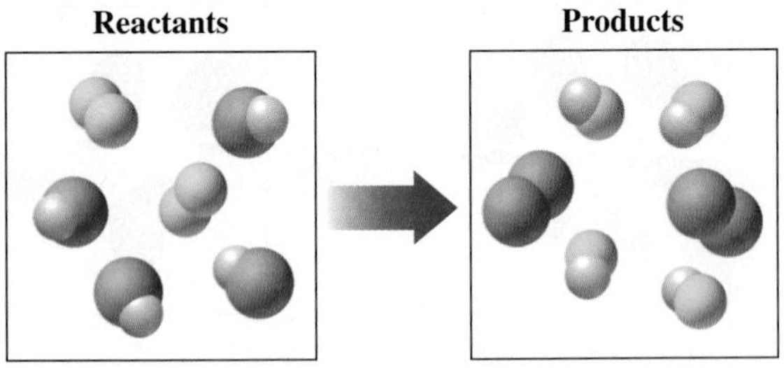

a. write the formula for each of the reactants and products.
b. write a balanced equation for the reaction.
c. indicate the type of reaction as combination, decomposition, single replacement, double replacement, or combustion.

7.81 If green spheres represent chlorine atoms, red spheres represent oxygen atoms, and all the molecules are gases, (7.3, 7.4)

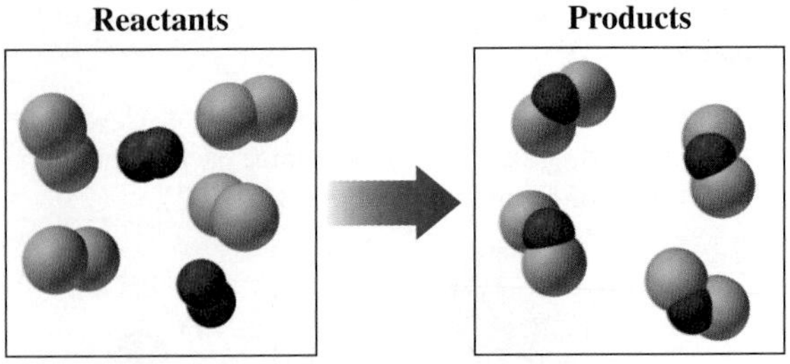

a. write the formula for each of the reactants and products.
b. write a balanced equation for the reaction.
c. indicate the type of reaction as combination, decomposition, single replacement, double replacement, or combustion.

7.82 If blue spheres represent nitrogen atoms, purple spheres represent iodine atoms, the reacting molecules are gases, and the products are solid, (7.3, 7.4)

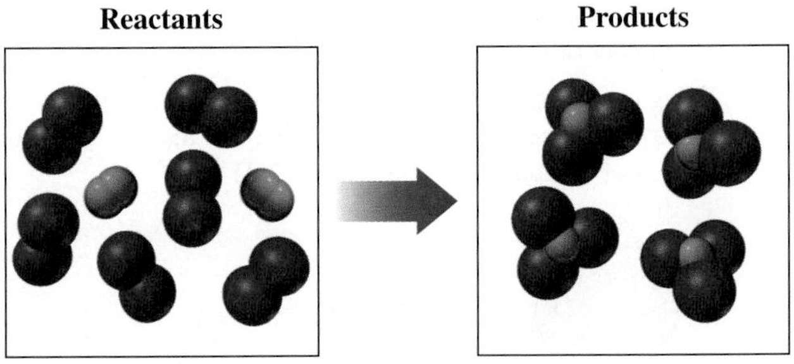

a. write the formula for each of the reactants and products.
b. write a balanced equation for the reaction.
c. indicate the type of reaction as combination, decomposition, single replacement, double replacement, or combustion.

ADDITIONAL QUESTIONS AND PROBLEMS

7.83 Calculate the molar mass for each of the following: (7.2)
a. $CaSO_4$
b. H_2SO_4
c. NaCl
d. C_6H_6O

7.84 Calculate the molar mass for each of the following: (7.2)
a. $MgCO_3$, magnesium carbonate, used in antacids
b. $Au(OH)_3$, gold(III) hydroxide, used in gold plating
c. $C_{18}H_{34}O_2$, oleic acid, from olive oil
d. $C_{21}H_{26}O_5$, prednisone, anti-inflammatory

7.85 How many grams are in 0.150 mole of each of the following? (7.2)
a. K **b.** Cl_2 **c.** Na_2CO_3

7.86 How many grams are in 2.25 moles of each of the following? (7.2)
a. N_2 **b.** NaBr **c.** C_6H_{14}

7.87 How many moles of O are in 25.0 g of each of the following compounds? (7.2)
a. CO_2 **b.** Al_2O_3 **c.** $NiCl_2.6H_2O$

7.88 How many moles of O are in 25.0 g of each of the following compounds? (7.2)
a. NH_3 **b.** $Ca(NO_3)_2$ **c.** SO_3

7.89 Identify the type of reaction for each of the following as combination, decomposition, single replacement, double replacement, or combustion: (7.4)
a. A metal and a nonmetal form an ionic compound.
b. A compound of hydrogen and carbon reacts with oxygen to produce carbon dioxide and water.
c. Heating calcium carbonate produces calcium oxide and carbon dioxide.
d. Zinc replaces copper in $Cu(NO_3)_2$.

7.90 Identify the type of reaction for each of the following as combination, decomposition, single replacement, double replacement, or combustion: (7.4)
a. A compound breaks apart into its elements.
b. Copper and bromine form copper(II) bromide.
c. Iron(II) sulfite breaks down to iron(II) oxide and sulfur dioxide.
d. Silver ion from $AgNO_3(aq)$ forms a solid with bromide ion from $KBr(aq)$.

7.91 Balance each of the following chemical equations, and identify the type of reaction: (7.3, 7.4)
a. $NH_3(g) + HCl(g) \longrightarrow NH_4Cl(s)$
b. $C_4H_8(g) + O_2(g) \xrightarrow{\Delta} CO_2(g) + H_2O(g)$
c. $Sb(s) + Cl_2(g) \longrightarrow SbCl_3(s)$
d. $NI_3(s) \longrightarrow N_2(g) + I_2(g)$
e. $KBr(aq) + Cl_2(aq) \longrightarrow KCl(aq) + Br_2(l)$
f. $Fe(s) + H_2SO_4(aq) \longrightarrow Fe_2(SO_4)_3(aq) + H_2(g)$
g. $Al_2(SO_4)_3(aq) + NaOH(aq) \longrightarrow Na_2SO_4(aq) + Al(OH)_3(s)$

7.92 Balance each of the following chemical equations, and identify the type of reaction: (7.3, 7.4)
a. $Si_3N_4(s) \longrightarrow Si(s) + N_2(g)$
b. $Mg(s) + N_2(g) \longrightarrow Mg_3N_2(s)$
c. $Al(s) + H_3PO_4(aq) \longrightarrow AlPO_4(aq) + H_2(g)$
d. $C_3H_4(g) + O_2(g) \xrightarrow{\Delta} CO_2(g) + H_2O(g)$
e. $Cr_2O_3(s) + H_2(g) \longrightarrow Cr(s) + H_2O(g)$
f. $Al(s) + Cl_2(g) \longrightarrow AlCl_3(s)$
g. $MgCl_2(aq) + AgNO_3(aq) \longrightarrow Mg(NO_3)_2(aq) + AgCl(s)$

7.93 Predict the products and write a balanced equation for each of the following: (7.3, 7.4)
a. single replacement:
$Zn(s) + HCl(aq) \longrightarrow$ ______ + ______
b. decomposition: $BaCO_3(s) \xrightarrow{\Delta}$ ______ + ______
c. double replacement:
$NaOH(aq) + HCl(aq) \longrightarrow$ ______ + ______
d. combination: $Al(s) + F_2(g) \longrightarrow$ ______

7.94 Predict the products and write a balanced equation for each of the following: (7.3, 7.4)
a. decomposition: $NaCl(s) \xrightarrow{\text{Electricity}}$ ______ + ______
b. combination: $Ca(s) + Br_2(g) \longrightarrow$ ______
c. combustion:
$C_2H_4(g) + O_2(g) \xrightarrow{\Delta}$ ______ + ______
d. double replacement:
$NiCl_2(aq) + NaOH(aq) \longrightarrow$ ______ + ______

7.95 Identify each of the following as an oxidation or a reduction reaction: (7.5)
a. $Zn^{2+}(aq) + 2\,e^- \longrightarrow Zn(s)$
b. $Al(s) \longrightarrow Al^{3+}(aq) + 3\,e^-$
c. $Pb(s) \longrightarrow Pb^{2+}(aq) + 2\,e^-$
d. $Cl_2(g) + 2\,e^- \longrightarrow 2Cl^-(aq)$

7.96 Write a balanced chemical equation for each of the following oxidation–reduction reactions: (7.3, 7.5)
a. Sulfur reacts with molecular chlorine to form sulfur dichloride.
b. Molecular chlorine and sodium bromide react to form molecular bromine and sodium chloride.
c. Aluminum metal and iron(III) oxide react to produce aluminum oxide and elemental iron.
d. Copper(II) oxide reacts with carbon to form elemental copper and carbon dioxide.

7.97 When ammonia (NH_3) reacts with fluorine, the products are dinitrogen tetrafluoride (N_2F_4) and hydrogen fluoride (HF) (7.3, 7.6, 7.7).
a. Write the balanced chemical equation.
b. How many moles of each reactant are needed to produce 4.00 moles of HF?
c. How many grams of F_2 are needed to react with 25.5 g of NH_3?
d. How many grams of N_2F_4 can be produced when 3.40 g of NH_3 reacts?

7.98 When nitrogen dioxide (NO_2) from car exhaust combines with water in the air, it forms nitric acid (HNO_3), which causes acid rain, and nitrogen oxide. (7.3, 7.6, 7.7)
a. Write the balanced chemical equation.
b. How many moles of each product are produced from 0.250 mole of H_2O?
c. How many grams of HNO_3 are produced when 60.0 g of NO_2 completely reacts?
d. How many grams of NO_2 are needed to form 75.0 g of HNO_3?

7.99 Pentane gas, C_5H_{12}, undergoes combustion with oxygen to produce carbon dioxide and water. (7.3, 7.6, 7.7)

a. Write the balanced chemical equation.

b. How many grams of C_5H_{12} are needed to produce 72 g of water?

c. How many grams of CO_2 are produced from 32.0 g of O_2?

7.100 Propane gas, C_3H_8, reacts with oxygen to produce water and carbon dioxide. Propane has a density of 2.02 g/L at room temperature. (7.3, 7.6, 7.7)

a. Write the balanced chemical equation.

b. How many grams of H_2O form when 5.00 L of C_3H_8 reacts?

c. How many grams of CO_2 are produced from 18.5 g of O_2?

d. How many grams of H_2O can be produced from the reaction of 100. g of C_3H_8?

7.101 The equation for the formation of silicon tetrachloride from silicon and chlorine is (7.8)

$$Si(s) + 2Cl_2(g) \longrightarrow SiCl_4(g) + 157 \text{ kcal}$$

a. Is the formation of $SiCl_4$ an endothermic or exothermic reaction?

b. Is the energy of the product higher or lower than the energy of the reactants?

7.102 The equation for the formation of nitrogen oxide is (7.8)

$$N_2(g) + O_2(g) + 90.2 \text{ kJ} \longrightarrow 2NO(g)$$

a. Is the formation of NO an endothermic or exothermic reaction?

b. Is the energy of the product higher or lower than the energy of the reactants?

CHALLENGE QUESTIONS

The following groups of questions are related to the topics in this chapter. However, they do not all follow the chapter order, and they require you to combine concepts and skills from several sections. These questions will help you increase your critical thinking skills and prepare for your next exam.

7.103 At a winery, glucose ($C_6H_{12}O_6$) in grapes undergoes fermentation to produce ethanol (C_2H_6O) and carbon dioxide. (7.7)

$$\underset{\text{Glucose}}{C_6H_{12}O_6} \longrightarrow \underset{\text{Ethanol}}{2C_2H_6O} + 2CO_2$$

Glucose in grapes ferments to produce ethanol.

a. How many grams of glucose are required to form 124 g of ethanol?

b. How many grams of ethanol would be formed from the reaction of 0.240 kg of glucose?

7.104 Benzyl alcohol ($C_6H_5CH_2OH$) is a fuel that reacts with sodium to produce hydrogen and the corresponding salt.

a. Write the balanced chemical equation.

b. Classify the reaction as a combustion, double displacement, single displacement, decomposition or combination.

c. Calculate the amount of sodium required to react with 10 mmol of benzyl alcohol.

7.105 Consider the following *unbalanced* equation: (7.3, 7.4, 7.6, 7.7)

$$Al(s) + O_2(g) \longrightarrow Al_2O_3(s)$$

a. Write the balanced chemical equation.

b. Identify the type of reaction.

c. How many moles of O_2 are needed to react with 4.50 moles of Al?

d. How many grams of Al_2O_3 are produced when 50.2 g of Al reacts?

e. When Al is reacted in a closed container with 8.00 g of O_2, how many grams of Al_2O_3 can form?

7.106 A toothpaste contains 0.240% by mass sodium fluoride used to prevent dental caries and 0.30% by mass triclosan $C_{12}H_7Cl_3O_2$, an antigingivitis agent. One tube contains 119 g of toothpaste. (7.1, 7.2)

Components in toothpaste include triclosan and NaF.

a. How many moles of NaF are in the tube of toothpaste?

b. How many fluoride ions, F^-, are in the tube of toothpaste?

c. How many grams of sodium ion, Na^+, are in 1.50 g of toothpaste?

d. How many molecules of triclosan are in the tube of toothpaste?

7.107 During heavy exercise and workouts, lactic acid, $C_3H_6O_3$, accumulates in the muscles where it can cause pain and soreness. (7.1, 7.2)

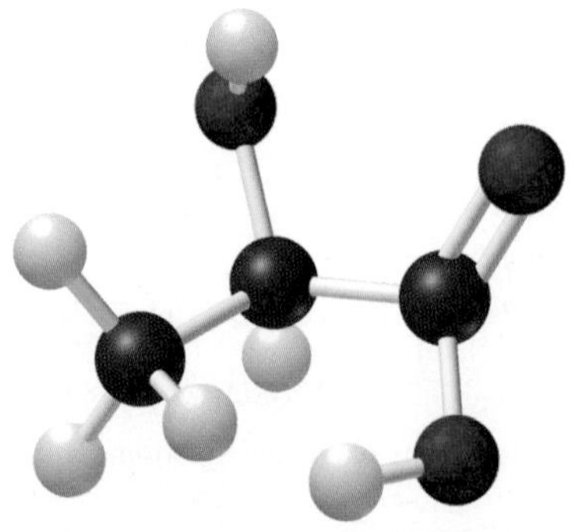

In the ball-and-stick model of lactic acid, black spheres = C, white spheres = H, and red spheres = O.

a. How many molecules are in 0.500 mole of lactic acid?
b. How many atoms of C are in 1.50 moles of lactic acid?
c. How many moles of lactic acid contain 4.5×10^{24} atoms of O?
d. What is the molar mass of lactic acid?

7.108 Ammonium sulfate, $(NH_4)_2SO_4$, is used in fertilizers to provide nitrogen for the soil. (7.1, 7.2)
a. How many formula units are in 0.200 mole of ammonium sulfate?
b. How many H atoms are in 0.100 mole of ammonium sulfate?
c. How many moles of ammonium sulfate contain 7.4×10^{25} atoms of N?
d. What is the molar mass of ammonium sulfate?

7.109 The gaseous hydrocarbon acetylene, C_2H_2, used in welders' torches, releases 1300 kJ when 1 mole of C_2H_2 undergoes combustion. (7.1, 7.3, 7.6, 7.7, 7.8)
a. Write a balanced equation for the reaction, including the heat of reaction.
b. Is the reaction endothermic or exothermic?
c. How many moles of H_2O are produced when 2.00 moles of O_2 reacts?
d. How many grams of O_2 are needed to react with 9.80 g of C_2H_2?

7.110 Methanol (CH_3OH), which is used as a cooking fuel, burns with oxygen gas (O_2) to produce the gases carbon dioxide (CO_2) and water (H_2O). The reaction produces 363 kJ of heat per mole of methanol. (7.1, 7.3, 7.6, 7.7, 7.8)
a. Write a balanced equation for the reaction, including the heat of reaction.
b. Is the reaction endothermic or exothermic?
c. How many moles of O_2 must react with 0.450 mole of CH_3OH?
d. How many grams of CO_2 are produced when 78.0 g of CH_3OH reacts?

ANSWERS

Answers to Study Checks

7.1 0.432 mole of H_2O

7.2 0.240 mole of propyl acetate

7.3 138.12 g

7.4 24.4 g of Au

7.5 0.00679 mole of $CaCO_3$

7.6 In both the reactants and products, there are 4 C atoms, 12 H atoms, and 14 O atoms.

7.7 $2Al(s) + 3Cl_2(g) \longrightarrow 2AlCl_3(s)$

7.8 $Sb_2S_3(s) + 6HCl(aq) \longrightarrow 2SbCl_3(s) + 3H_2S(g)$

7.9 $N_2(g) + 2O_2(g) \longrightarrow 2NO_2(g)$ combination

7.10 1.83 moles of iron

7.11 27.5 g of CO_2

7.12 The rate of reaction will decrease because the number of collisions between reacting particles will be fewer, and a smaller number of the collisions that do occur will have sufficient activation energy.

Answers to Selected Questions and Problems

7.1 One mole contains 6.02×10^{23} atoms of an element, molecules of a molecular substance, or formula units of an ionic substance.

7.3 **a.** 4.3×10^{23} atoms
b. 14.2×10^{23} molecules
c. 0.13 mole
d. 9.3 moles

7.5 **a.** 10.00 moles of H
b. 20.00 moles of O
c. 30.1×10^{23} atoms of S
d. 1.20×10^{25} atoms of O

7.7 **a.** 24 moles of H
b. 1.0×10^2 moles of C
c. 0.040 mole of N

7.9 **a.** 188.18 g **b.** 159.7 g **c.** 329.4 g
d. 342.2 g **e.** 58.33 g

7.11 **a.** 159.81 $gmol^{-1}$ **b.** 60.1 $gmol^{-1}$ **c.** 159.61 $gmol^{-1}$
d. 67.81 $gmol^{-1}$ **e.** 112.56 $gmol^{-1}$

7.13 **a.** 20.82 g **b.** 46.19 g
c. 13.68 g **d.** 248.1 g

7.15 **a.** 89.46 g **b.** 30.23 g
c. 9.54 g **d.** 403 g

7.17 **a.** 602 g **b.** 11 g

7.19 **a.** 0.393 mole **b.** 0.02 mole
c. 0.428 mole **d.** 3.77 moles

7.21 **a.** 0.376 mole **b.** 0.536 mole
c. 0.094 mole **d.** 0.135 mole

7.23 **a.** 1.44 moles **b.** 5 moles
c. 12 moles

7.25 **a.** 266.26 g **b.** 0.276 mole
c. 13.25 g

7.27 **a.** not balanced **b.** balanced
c. not balanced **d.** balanced

7.29 **a.** $N_2(g) + O_2(g) \longrightarrow 2NO(g)$
b. $2HgO(s) \longrightarrow 2Hg(l) + O_2(g)$
c. $4Fe(s) + 3O_2(g) \longrightarrow 2Fe_2O_3(s)$
d. $2Na(s) + Cl_2(g) \longrightarrow 2NaCl(s)$

7.31 **a.** $Mg(s) + 2AgNO_3(aq) \longrightarrow Mg(NO_3)_2(aq) + 2Ag(s)$
b. $2Al(s) + 3CuSO_4(aq) \longrightarrow 3Cu(s) + Al_2(SO_4)_3(aq)$
c. $Pb(NO_3)_2(aq) + 2NaCl(aq) \longrightarrow PbCl_2(s) + 2NaNO_3(aq)$
d. $2Al(s) + 6HCl(aq) \longrightarrow 2AlCl_3(aq) + 3H_2(g)$

7.33 **a.** decomposition **b.** single replacement
c. combustion **d.** double replacement
e. combination

7.35 **a.** combination **b.** single replacement
c. decomposition **d.** double replacement
e. combustion

7.37 **a.** $Mg(s) + Cl_2(g) \longrightarrow MgCl_2(s)$
b. $2HBr(g) \longrightarrow H_2(g) + Br_2(g)$
c. $Mg(s) + Zn(NO_3)_2(aq) \longrightarrow Zn(s) + Mg(NO_3)_2(aq)$
d. $K_2S(aq) + Pb(NO_3)_2(aq) \longrightarrow 2KNO_3(aq) + PbS(s)$
e. $2C_2H_6(g) + 7O_2(g) \xrightarrow{\Delta} 4CO_2(g) + 6H_2O(g)$

7.39 **a.** reduction **b.** oxidation
c. reduction **d.** reduction

7.41 **a.** Zn is oxidized, Cl_2 is reduced.
b. The Br^- in NaBr is oxidized, Cl_2 is reduced.
c. The O^{2-} in PbO is oxidized, the Pb^{2+} in PbO is reduced.
d. Sn^{2+} is oxidized, Fe^{3+} is reduced.

7.43 **a.** reduction **b.** oxidation

7.47 **a.** $\frac{2 \text{ moles } SO_2}{1 \text{ mole } O_2}$ and $\frac{1 \text{ mole } O_2}{2 \text{ moles } SO_2}$

$\frac{2 \text{ moles } SO_2}{2 \text{ moles } SO_3}$ and $\frac{2 \text{ moles } SO_3}{2 \text{ moles } SO_2}$

$\frac{2 \text{ moles } SO_3}{1 \text{ mole } O_2}$ and $\frac{1 \text{ mole } O_2}{2 \text{ moles } SO_3}$

b. $\frac{4 \text{ moles } P}{5 \text{ moles } O_2}$ and $\frac{5 \text{ moles } O_2}{4 \text{ moles } P}$

$\frac{4 \text{ moles } P}{2 \text{ moles } P_2O_5}$ and $\frac{2 \text{ moles } P_2O_5}{4 \text{ moles } P}$

$\frac{5 \text{ moles } O_2}{2 \text{ moles } P_2O_5}$ and $\frac{2 \text{ moles } P_2O_5}{5 \text{ moles } O_2}$

7.49 **a.** 1.0 mole of O_2 **b.** 10. moles of H_2
c. 5.0 moles of H_2O

7.51 **a.** 1.25 moles of C **b.** 0.96 mole of CO
c. 1.0 mole of SO_2 **d.** 0.50 mole of CS_2

7.53 **a.** 77.5 g of Na_2O **b.** 6.26 g of O_2
c. 19.4 g of O_2

7.55 **a.** 19.2 g of O_2 **b.** 3.79 g of N_2
c. 54.0 g of H_2O

7.57 **a.** 3.66 g of H_2O **b.** 3.44 g of NO
c. 7.53 g of HNO_3

7.59 **a.** $2PbS(s) + 3O_2(g) \longrightarrow 2PbO(s) + 2SO_2(g)$
b. 6.02 g of O_2
c. 17.4 g of SO_2
d. 137 g of PbS

7.61 **a.** The energy of activation is the energy required to break the bonds of the reacting molecules.
b. A catalyst provides a pathway that lowers the activation energy and speeds up a reaction.
c. In exothermic reactions, the energy of the products is lower than the energy of the reactants.
d.

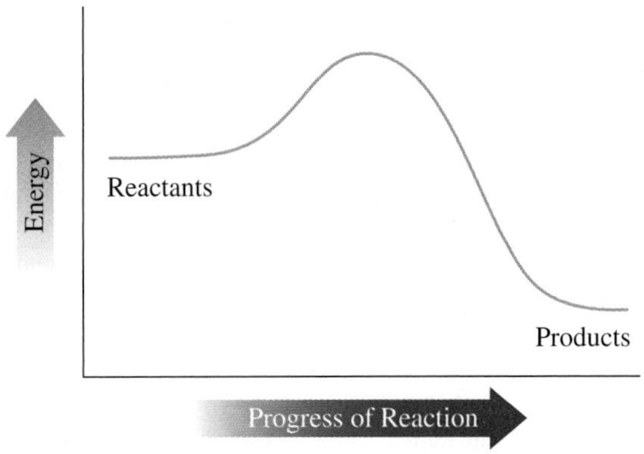

7.63 **a.** exothermic **b.** endothermic
c. exothermic

7.65 **a.** exothermic **b.** endothermic
c. exothermic

7.67 **a.** The rate of a reaction tells how fast the products are formed or how fast the reactants are consumed.
b. Reactions go faster at higher temperatures.

7.69 **a.** increase **b.** increase
c. increase **d.** decrease

7.71 **1. a.** S_2Cl_2 **b.** 135.04 g/mole
c. 0.0741 mole

2. a. C_6H_6 **b.** 78.11 g/mole
c. 0.128 mole

7.73 **a.** 252.3 g/mole
b. 0.0991 mole of dipyrithione
c. 0.991 mole of C
d. 6.8 moles of dipyrithione

7.75 **a.** 1,1,2 combination **b.** 2,2,1 decomposition

7.77 **a.** reactants NO and O_2; product NO_2
b. $2NO(g) + O_2(g) \longrightarrow 2NO_2(g)$
c. combination

7.79 **a.** reactant NI_3; product N_2 and I_2
b. $2NI_3(s) \longrightarrow N_2(g) + 3I_2(g)$
c. decomposition

7.81 **a.** reactants Cl_2 and O_2; product OCl_2
b. $2Cl_2(g) + O_2(g) \longrightarrow 2OCl_2(g)$
c. combination

7.85 **a.** 5.87 g **b.** 10.6 g **c.** 15.9 g

7.89 **a.** combination **b.** combustion
c. decomposition **d.** single replacement

7.91 **a.** $NH_3(g) + HCl(g) \longrightarrow NH_4Cl(s)$ combination
b. $C_4H_8(g) + 6O_2(g) \xrightarrow{\Delta} 4CO_2(g) + 4H_2O(g)$ combustion
c. $2Sb(s) + 3Cl_2(g) \longrightarrow 2SbCl_3(s)$ combination
d. $2NI_3(s) \longrightarrow N_2(g) + 3I_2(g)$ decomposition
e. $2KBr(aq) + Cl_2(aq) \longrightarrow 2KCl(aq) + Br_2(l)$ single replacement
f. $2Fe(s) + 3H_2SO_4(aq) \longrightarrow Fe_2(SO_4)_3(aq) + 3H_2(g)$ single replacement
g. $Al_2(SO_4)_3(aq) + 6NaOH(aq) \longrightarrow 3Na_2SO_4(aq) + 2Al(OH)_3(s)$ double replacement

7.93 **a.** $Zn(s) + 2HCl(aq) \longrightarrow ZnCl_2(aq) + H_2(g)$
b. $BaCO_3(s) \xrightarrow{\Delta} BaO(s) + CO_2(g)$
c. $NaOH(aq) + HCl(aq) \longrightarrow NaCl(aq) + H_2O(l)$
d. $2Al(s) + 3F_2(g) \longrightarrow 2AlF_3(s)$

7.95 **a.** reduction **b.** oxidation
c. oxidation **d.** reduction

7.97 **a.** $2NH_3(g) + 5F_2(g) \longrightarrow N_2F_4(g) + 6HF(g)$
b. 1.33 moles of NH_3 and 3.33 moles of F_2
c. 142 g of F_2
d. 10.4 g of N_2F_4

7.99 **a.** $C_5H_{12}(g) + 8O_2(g) \xrightarrow{\Delta} 5CO_2(g) + 6H_2O(g) + \text{energy}$
b. 48 g of pentane
c. 27.5 g of CO_2

7.101 **a.** exothermic **b.** lower

7.103 **a.** 242 g of glucose **b.** 123 g of ethanol

7.105 **a.** $4Al(s) + 3O_2(g) \longrightarrow 2Al_2O_3(s)$
b. combination
c. 3.38 moles of oxygen
d. 94.9 g of Al_2O_3
e. 17.0 g of Al_2O_3

7.107 **a.** 3.01×10^{23} molecules
b. 2.71×10^{24} atoms of C
c. 2.5 moles of lactic acid
d. 90.08 g

7.109 **a.** $2C_2H_2(g) + 5O_2(g) \xrightarrow{\Delta} 4CO_2(g) + 2H_2O(g) + 2600\text{ kJ}$
b. exothermic
c. 0.800 mole of H_2O
d. 30.1 g of O_2

8 Gases

After soccer practice, Whitney complained that she was having difficulty breathing. Her father quickly took her to the emergency room where she was seen by a respiratory therapist who listened to Whitney's chest and then tested her breathing capacity using a spirometer. Based on her limited breathing capacity and the wheezing noise in her chest, Whitney was diagnosed as having asthma.

The therapist gave Whitney a nebulizer containing a bronchodilator that opens the airways and allows more air to go into the lungs. During the breathing treatment, the respiratory therapist measured the amount of oxygen (O_2) in her blood and explained to Whitney and her father that air is a mixture of gases containing 78% nitrogen (N_2) gas and 21% oxygen (O_2) gas. Because Whitney had difficulty obtaining sufficient oxygen breathing air, the respiratory therapist gave her supplemental oxygen through an oxygen mask. Within a short period of time, Whitney's breathing returned to normal. The therapist then explained that the lungs work according to Boyle's law: The volume of the lungs increases upon inhalation, and the pressure decreases to make air flow in. However, during an asthma attack, the airways become restricted, and it becomes more difficult to expand the volume of the lungs.

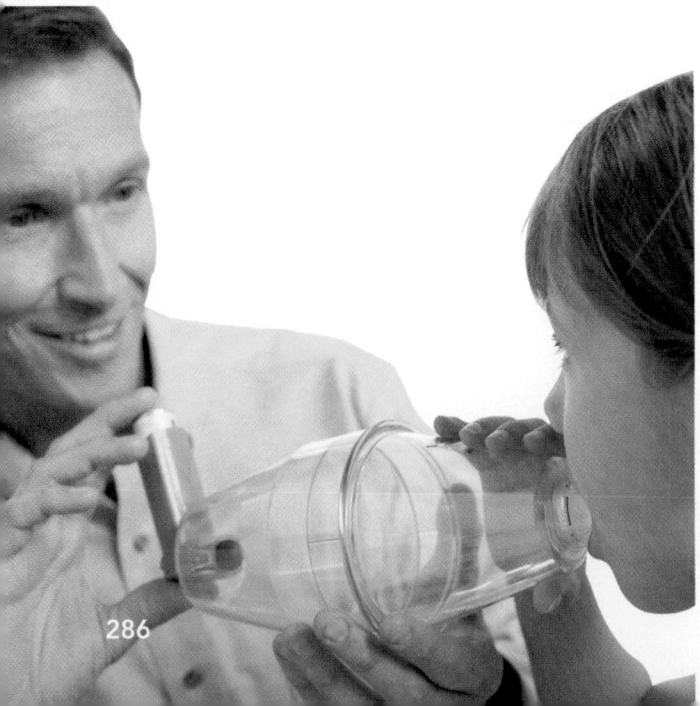

CAREER Respiratory Therapist

Respiratory therapists assess and treat a range of patients, including premature infants whose lungs have not developed and asthmatics or patients with emphysema or cystic fibrosis. In assessing patients, they perform a variety of diagnostic tests including breathing capacity, concentrations of oxygen and carbon dioxide in a patient's blood, as well as blood pH.

In order to treat patients, therapists provide oxygen or aerosol medications to the patient, as well as chest physiotherapy to remove mucus from their lungs. Respiratory therapists also educate patients on how to correctly use their inhalers.

We all live at the bottom of a sea of gases called the atmosphere. The most important of these gases is oxygen, which constitutes about 21% of the atmosphere. Without oxygen, life on this planet would be impossible: Oxygen is vital to all life processes of plants and animals. Ozone (O_3), formed in the upper atmosphere by the interaction of oxygen with ultraviolet light, absorbs some of the harmful radiation before it can strike the Earth's surface. The other gases in the atmosphere include nitrogen (78% of the atmosphere), argon, carbon dioxide (CO_2), and water vapor. Carbon dioxide gas, a product of combustion and metabolism, is used by plants in photosynthesis, a process that produces the oxygen that is essential for humans and animals.

The atmosphere has become a dumping ground for other gases, such as methane, chlorofluorocarbons, and nitrogen oxides as well as volatile organic compounds, which are gases from paints, paint thinners, and cleaning supplies. The chemical reactions of these gases with sunlight and oxygen in the air are contributing to air pollution, ozone depletion, climate change, and acid rain. Such chemical changes can seriously affect our health and our lifestyle. An understanding of gases and some of the laws that govern gas behavior can help us understand the nature of matter and allow us to make decisions concerning important environmental and health issues.

LOOKING AHEAD

CHAPTER READINESS*

KEY MATH SKILLS

- Solving Equations (1.4D)

CORE CHEMISTRY SKILLS

- Using Significant Figures in Calculations (2.3)
- Writing Conversion Factors from Equalities (2.5)
- Using Conversion Factors (2.6)
- Using Molar Mass as a Conversion Factor (7.2)
- Using Mole–Mole Factors (7.6)

*These Key Math Skills and Core Chemistry Skills from previous chapters are listed here for your review as you proceed to the new material in this chapter.

8.1 Properties of Gases

LEARNING GOAL

Describe the kinetic molecular theory of gases and the units of measurement used for gases.

We are surrounded by gases, but are not often aware of their presence. Of the elements on the periodic table, only a handful exist as gases at room temperature: H_2, N_2, O_2, F_2, Cl_2, and the noble gases. Another group of gases includes the oxides of the nonmetals on the upper right corner of the periodic table, such as CO, CO_2, NO, NO_2, SO_2, and SO_3. Generally molecules that are gases at room temperature have fewer than five atoms from the first or second period.

The behavior of gases is quite different from that of liquids and solids. Gas particles are far apart, whereas particles of both liquids and solids are held close together. A gas has no definite shape or volume and will completely fill any container. Because, there are great distances between gas particles, a gas is less dense than a solid or liquid, and easy to compress. A model for the behavior of a gas, called the **kinetic molecular theory of gases**, helps us understand gas behavior.

Kinetic Molecular Theory of Gases

1. **A gas consists of small particles (atoms or molecules) that move randomly with high velocities.** Gas molecules moving in random directions at high speeds cause a gas to fill the entire volume of a container.
2. **The attractive forces between the particles of a gas are usually very small.** Gas particles are far apart and fill a container of any size and shape.
3. **The actual volume occupied by gas molecules is extremely small compared with the volume that the gas occupies.** The volume of the gas is considered equal to the volume of the container. Most of the volume of a gas is empty space, which allows gases to be easily compressed.
4. **Gas particles are in constant motion, moving rapidly in straight paths.** When gas particles collide, they rebound and travel in new directions. Every time they hit the walls of the container, they exert pressure. An increase in the number or force of collisions against the walls of the container causes an increase in the pressure of the gas.
5. **The average kinetic energy of gas molecules is proportional to the Kelvin temperature.** Gas particles move faster as the temperature increases. At higher temperatures, gas particles hit the walls of the container more often and with more force, producing higher pressures.

Interactive Video

Kinetic Molecular Theory

The kinetic molecular theory helps explain some of the characteristics of gases. For example, we can quickly smell perfume when a bottle is opened on the other side of a room, because its particles move rapidly in all directions. At room temperature, the molecules of air are moving at about 450 m/s, which is 1000 mi/h. They move faster at higher temperatures and more slowly at lower temperatures. Sometimes tires or gas-filled containers explode when temperatures are too high. From the kinetic molecular theory, we know that gas particles move faster when heated, hit the walls of a container with more force, and cause a buildup of pressure inside a container.

When we talk about a gas, we describe it in terms of four properties: pressure, volume, temperature, and the amount of gas.

Pressure (*P*)

Gas particles are extremely small and move rapidly. When they hit the walls of a container, they exert **pressure** (see Figure 8.1). If we heat the container, the molecules move faster and smash into the walls more often and with increased force, thus increasing the pressure. The gas particles in the air, mostly oxygen and nitrogen, exert a pressure on us called **atmospheric pressure** (see Figure 8.2). As you go to higher altitudes, the atmospheric pressure is less because there are fewer particles in the air. The most common units used for gas pressure measurement are the *atmosphere* (atm) and *millimeters of mercury* (mmHg). On the TV weather report, you may hear or see the atmospheric pressure given in inches of mercury, or kilopascals in countries other than the United States. In a hospital, the unit torr or psi (pounds per square inch) may be used.

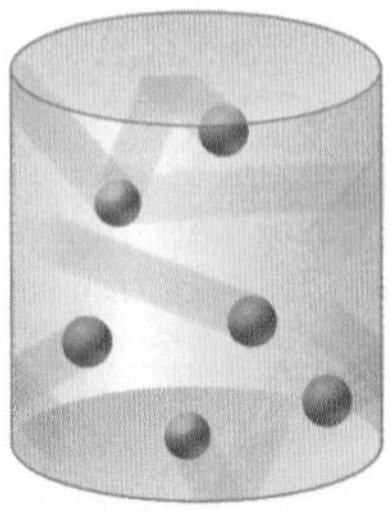

FIGURE 8.1 ▶ Gas particles move in straight lines within a container. The gas particles exert pressure when they collide with the walls of the container.

Q Why does heating the container increase the pressure of the gas within it?

Volume (*V*)

The volume of gas equals the size of the container in which the gas is placed. When you inflate a tire or a basketball, you are adding more gas particles. The increase in the number of particles hitting the walls of the tire or basketball increases the volume. Sometimes, on a cold morning, a tire looks flat. The volume of the tire has decreased because a lower temperature decreases the speed of the molecules, which in turn reduces the force of their impacts on the walls of the tire. The most common units for volume measurement are liters (L) and milliliters (mL).

Temperature (*T*)

The temperature of a gas is related to the kinetic energy of its particles. For example, if we have a gas at 200 K in a rigid container and heat it to a temperature of 400 K, the gas particles will have twice the kinetic energy that they did at 200 K. This also means that the gas at 400 K exerts twice the pressure of the gas at 200 K. Although you measure gas temperature

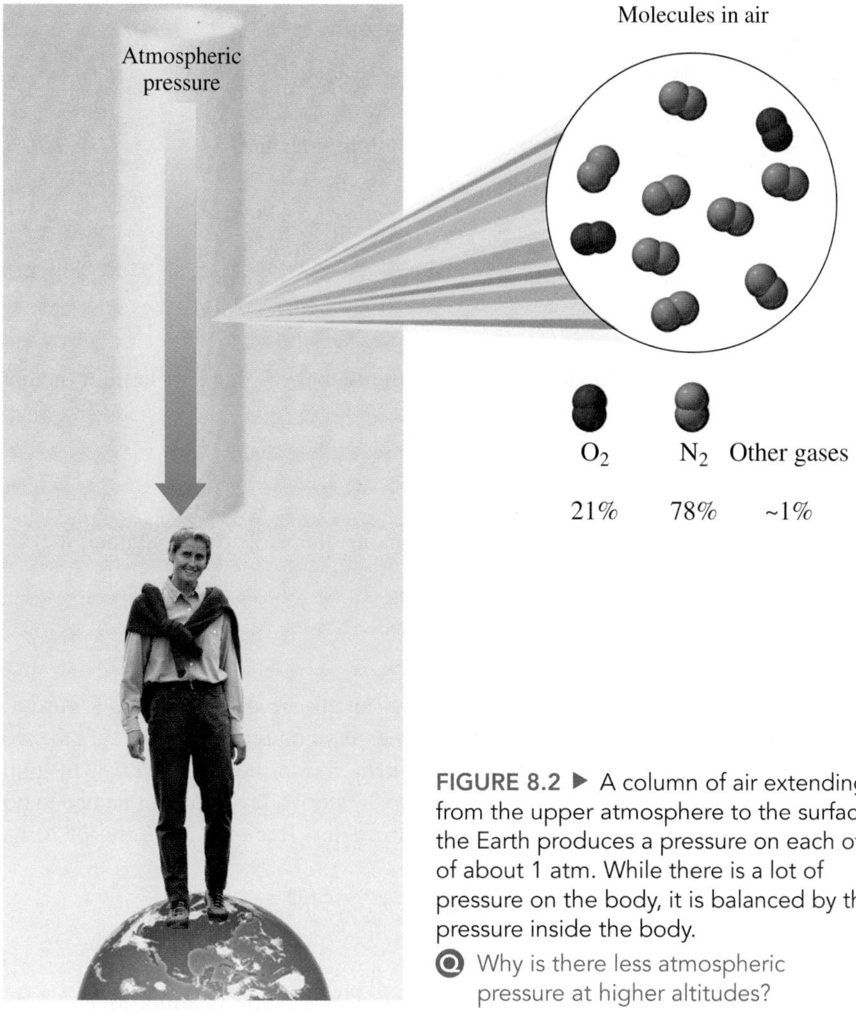

FIGURE 8.2 ▶ A column of air extending from the upper atmosphere to the surface of the Earth produces a pressure on each of us of about 1 atm. While there is a lot of pressure on the body, it is balanced by the pressure inside the body.

Q Why is there less atmospheric pressure at higher altitudes?

using a Celsius thermometer, all comparisons of gas behavior and all calculations related to temperature must use the Kelvin temperature scale. No one has quite created the conditions for absolute zero (0 K), but we predict that the particles will have zero kinetic energy and the gas will exert zero pressure at absolute zero.

Amount of Gas (*n*)

When you add air to a bicycle tire, you increase the amount of gas, which results in a higher pressure in the tire. Usually we measure the amount of gas by its mass in grams. In gas law calculations, we need to change the grams of gas to moles.

A summary of the four properties of a gas are given in Table 8.1.

TABLE 8.1 Properties That Describe a Gas

Property	Description	Units of Measurement
Pressure (*P*)	The force exerted by a gas against the walls of the container	atmosphere (atm); millimeters of mercury (mmHg); torr; pascal (Pa)
Volume (*V*)	The space occupied by a gas	liter (L); milliliter (mL)
Temperature (*T*)	The determining factor of the kinetic energy and rate of motion of gas particles	degree Celsius (°C); Kelvin (K) *is required in calculations*
Amount (*n*)	The quantity of gas present in a container	grams (g); moles (*n*) *is required in calculations*

Explore Your World

Forming a Gas

Obtain baking soda and a jar or a plastic bottle. You will also need an elastic glove that fits over the mouth of the jar or a balloon that fits snugly over the top of the plastic bottle. Place a cup of vinegar in the jar or bottle. Sprinkle some baking soda into the fingertips of the glove or into the balloon. Carefully fit the glove or balloon over the top of the jar or bottle. Slowly lift the fingers of the glove or the balloon so that the baking soda falls into the vinegar. Watch what happens. Squeeze the glove or balloon.

Questions

1. Describe the properties of gas that you observe as the reaction takes place between vinegar and baking soda.
2. How do you know that a gas was formed?

SAMPLE PROBLEM 8.1 Properties of Gases

Identify the property of a gas that is described by each of the following:

a. increases the kinetic energy of gas particles
b. the force of the gas particles hitting the walls of the container
c. the space that is occupied by a gas

SOLUTION

a. temperature **b.** pressure **c.** volume

STUDY CHECK 8.1

As more helium gas is added to a balloon, the number of grams of helium increases. What property of a gas is described?

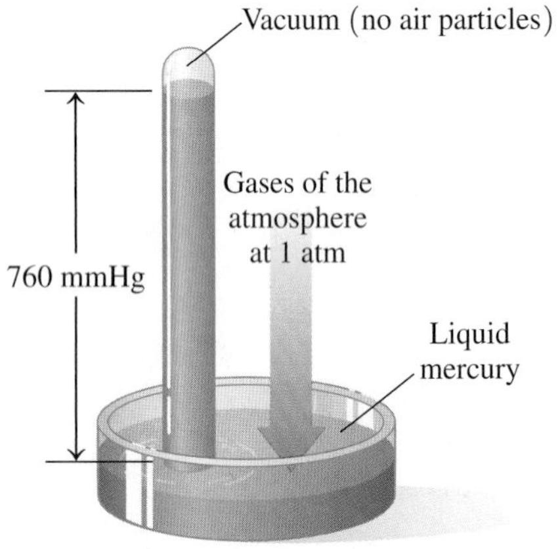

FIGURE 8.3 ▶ A barometer: The pressure exerted by the gases in the atmosphere is equal to the downward pressure of a mercury column in a closed glass tube. The height of the mercury column measured in mmHg is called atmospheric pressure.

Q Why does the height of the mercury column change from day to day?

Measurement of Gas Pressure

When billions and billions of gas particles hit against the walls of a container, they exert *pressure*, which is defined as a force acting on a certain area.

$$\text{Pressure } (P) = \frac{\text{force}}{\text{area}}$$

The atmospheric pressure can be measured using a barometer (see Figure 8.3). At a pressure of exactly 1 atmosphere (atm), a mercury column in an inverted tube would be *exactly* 760 mm high. One **atmosphere (atm)** is defined as *exactly* 760 mmHg (millimeters of mercury). One atmosphere is also 760 torr, a pressure unit named to honor Evangelista Torricelli, the inventor of the barometer. Because the units of torr and mmHg are equal, they are used interchangeably.

$$1 \text{ atm} = 760 \text{ mmHg} = 760 \text{ torr (exact)}$$
$$1 \text{ mmHg} = 1 \text{ torr (exact)}$$

In SI units, pressure is measured in pascals (Pa); 1 atm is equal to 101 325 Pa. Because a pascal is a very small unit, pressures are usually reported in kilopascals.

$$1 \text{ atm} = 1.01325 \times 10^5 \text{ Pa} = 101.325 \text{ kPa}$$

The U.S. equivalent of 1 atm is 14.7 lb/in.^2 (psi). When you use a pressure gauge to check the air pressure in the tires of a car, it may read 30 to 35 psi. This measurement is actually 30 to 35 psi above the pressure that the atmosphere exerts on the outside of the tire.

$$1 \text{ atm} = 14.7 \text{ lb/in.}^2$$

Table 8.2 summarizes the various units used in the measurement of pressure.

TABLE 8.2 Units for Measuring Pressure

Unit	Abbreviation	Unit Equivalent to 1 atm
Atmosphere	atm	1 atm (exact)
Millimeters of Hg	mmHg	760 mmHg (exact)
Torr	torr	760 torr (exact)
Inches of Hg	in. Hg	29.9 in. Hg
Pounds per square inch	lb/in.^2 (psi)	14.7 lb/in.^2
Pascal	Pa	101 325 Pa
Kilopascal	kPa	101.325 kPa

If you have a barometer in your home, it probably gives pressure in inches of mercury. One atmosphere is equal to the pressure of a column of mercury that is 29.9 in. high. Atmospheric pressure changes with variations in weather and altitude. On a hot, sunny day, a column of air has more particles, which increases the pressure on the mercury surface. The mercury column rises, indicating a higher atmospheric pressure. On a rainy day, the atmosphere exerts less pressure, which causes the mercury column to fall. In the weather report, this type of weather is called a *low-pressure system*. Above sea level, the density of the gases in the air decreases, which causes lower atmospheric pressures; the atmospheric pressure is greater than 760 mmHg at the Dead Sea because it is below sea level (see Table 8.3).

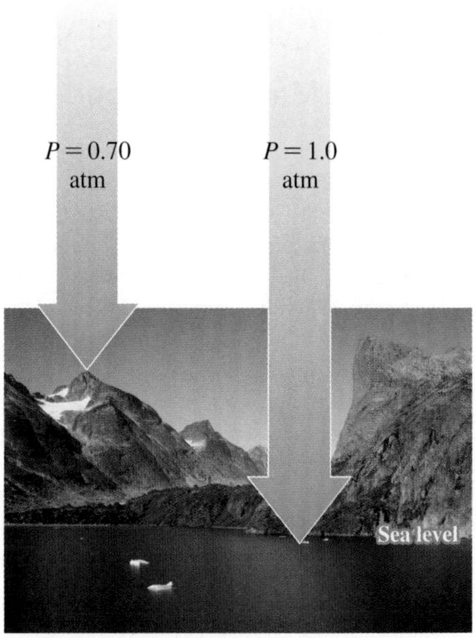

The atmospheric pressure decreases as the altitude increases.

TABLE 8.3 Altitude and Atmospheric Pressure

Location	Altitude (km)	Atmospheric Pressure (mmHg)
Dead Sea	−0.40	800
Sea level	0.00	760
Los Angeles	0.09	752
Las Vegas	0.70	700
Denver	1.60	630
Mount Whitney	4.50	440
Mount Everest	8.90	253

Divers must be concerned about increasing pressures on their ears and lungs when they dive below the surface of the ocean. Because water is more dense than air, the pressure on a diver increases rapidly as the diver descends. At a depth of 33 ft below the surface of the ocean, an additional 1 atm of pressure is exerted by the water on a diver, for a total of 2 atm. At 100 ft, there is a total pressure of 4 atm on a diver. The regulator that a diver uses continuously adjusts the pressure of the breathing mixture to match the increase in pressure.

SAMPLE PROBLEM 8.2 Units of Pressure

A sample of neon gas has a pressure of 0.50 atm. Calculate the pressure, in millimeters of mercury, of the neon (see Table 8.2).

SOLUTION

The equality 1 atm = 760 mmHg can be written as two conversion factors:

$$\frac{760\text{ mmHg}}{1\text{ atm}} \quad \text{or} \quad \frac{1\text{ atm}}{760\text{ mmHg}}$$

Using the conversion factor that cancels atm and gives mmHg, we can set up the problem as

$$0.50\,\cancel{\text{atm}} \times \frac{760\text{ mmHg}}{1\,\cancel{\text{atm}}} = 380\text{ mmHg}$$

STUDY CHECK 8.2

A tank of nitrous oxide (N_2O) used in a hospital has a pressure of 48 psi. What is that pressure in atmospheres (see Table 8.2)?

Chemistry Link to Health

Measuring Blood Pressure

Your blood pressure is one of the vital signs a doctor or nurse checks during a physical examination. It actually consists of two separate measurements. Acting like a pump, the heart contracts to create the pressure that pushes blood through the circulatory system. During contraction, the blood pressure is at its highest; this is your *systolic* pressure. When the heart muscles relax, the blood pressure falls; this is your *diastolic* pressure. The normal range for systolic pressure is 100 to 120 mmHg. For diastolic pressure, it is 60 to 80 mmHg. These two measurements are usually expressed as a ratio such as 100/80. These values are somewhat higher in older people. When blood pressures are elevated, such as 140/90, there is a greater risk of stroke, heart attack, or kidney damage. Low blood pressure prevents the brain from receiving adequate oxygen, causing dizziness and fainting.

The blood pressures are measured by a sphygmomanometer, an instrument consisting of a stethoscope and an inflatable cuff connected to a tube of mercury called a manometer. After the cuff is wrapped around the upper arm, it is pumped up with air until it cuts off the flow of blood through the arm. With the stethoscope over the artery, the air is slowly released from the cuff, decreasing the pressure on the artery. When the blood flow first starts again in the artery, a noise can be heard through the stethoscope signifying the systolic blood pressure as the pressure shown on the manometer. As air continues to be released, the cuff deflates until no sound is heard in the artery. A second pressure reading is taken at the moment of silence and denotes the diastolic pressure, the pressure when the heart is not contracting.

The use of digital blood pressure monitors is becoming more common. However, they have not been validated for use in all situations and can sometimes give inaccurate readings.

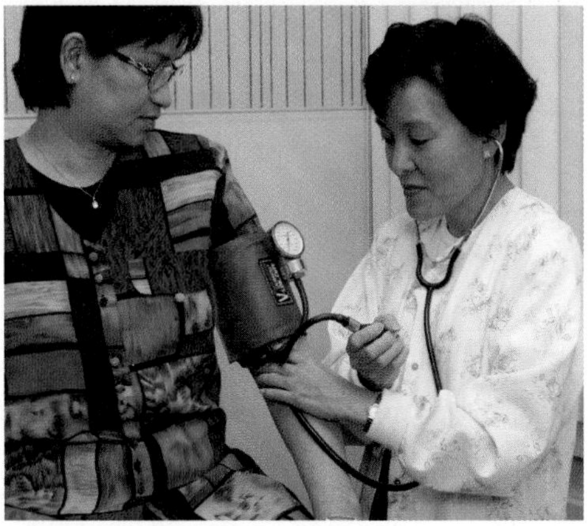

The measurement of blood pressure is part of a routine physical exam.

QUESTIONS AND PROBLEMS

8.1 Properties of Gases

LEARNING GOAL Describe the kinetic molecular theory of gases and the units of measurement used for gases.

8.1 Use the kinetic molecular theory of gases to explain each of the following:
- **a.** Gases move faster at higher temperatures.
- **b.** Gases can be compressed much more than liquids or solids.
- **c.** Gases have low densities.

8.2 Use the kinetic molecular theory of gases to explain each of the following:
- **a.** A container of nonstick cooking spray explodes when thrown into a fire.
- **b.** The air in a hot-air balloon is heated to make the balloon rise.
- **c.** You can smell the odor of cooking onions from far away.

8.3 Identify the property of a gas that is measured in each of the following:
- **a.** 350 K **b.** 125 mL
- **c.** 2.00 g of O_2 **d.** 755 mmHg

8.4 Identify the property of a gas that is measured in each of the following:
- **a.** 425 K **b.** 1.0 atm
- **c.** 10.0 L **d.** 0.50 mole of He

8.5 Which of the following statement(s) describes the pressure of a gas?
- **a.** the force of the gas particles on the walls of the container
- **b.** the number of gas particles in a container
- **c.** 4.5 L of helium gas
- **d.** 750 torr
- **e.** 28.8 $lb/in.^2$

8.6 Which of the following statement(s) describes the pressure of a gas?
- **a.** 320 °C
- **b.** 2500 mmHg
- **c.** 670 torr
- **d.** 101 Pa
- **e.** 201 kPa

8.7 A Helium tank contains helium (He) at a pressure of 4.00 atm. What is the pressure in the tank in terms of the following units?
- **a.** torr **b.** mmHg
- **c.** kPa **d.** psi

8.8 On a climb up Mt. Whitney, the atmospheric pressure drops to 467 mmHg. What is the pressure in terms of the following units?
- **a.** atm **b.** torr

8.2 Pressure and Volume (Boyle's Law)

Imagine that you can see air particles hitting the walls inside a bicycle tire pump. What happens to the pressure inside the pump as you push down on the handle? As the volume decreases, there is a decrease in the surface area of the container. The air particles are crowded together, more collisions occur, and the pressure increases within the container.

When a change in one property (in this case, volume) causes a change in another property (in this case, pressure), the properties are related. If the changes occur in opposite directions, the properties have an *inverse relationship*. The inverse relationship between the pressure and volume of a gas is known as **Boyle's law**. The law states that the volume (V) of a sample of gas changes inversely with the pressure (P) of the gas as long as there is no change in the temperature (T) or amount of gas (n) (see Figure 8.4).

If the volume or pressure of a gas sample changes without any change in the temperature or in the amount of the gas, then the final volume will give the same PV product as the initial pressure and volume. Then we can set the initial and final PV products equal to each other.

LEARNING GOAL

Use the pressure–volume relationship (Boyle's law) to determine the final pressure or volume when the temperature and amount of gas are constant.

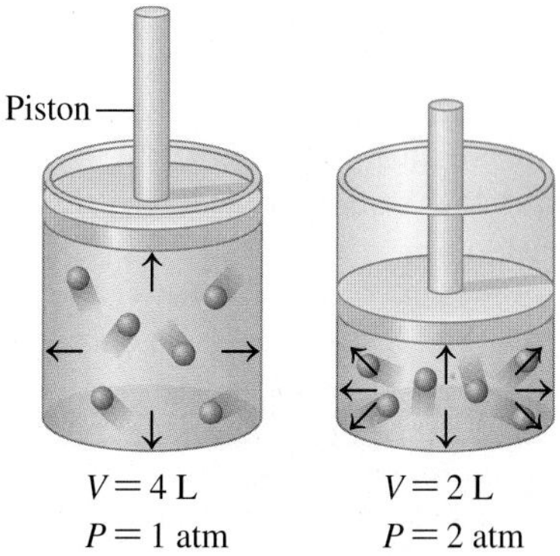

FIGURE 8.4 ▶ **Boyle's law:** As volume decreases, gas molecules become more crowded, which causes the pressure to increase. Pressure and volume are inversely related.

Q If the volume of a gas increases, what will happen to its pressure?

Boyle's Law

$P_1V_1 = P_2V_2$ No change in number of moles and temperature

SAMPLE PROBLEM 8.3 Calculating Volume When Pressure Changes

Oxygen gas is delivered through a face mask during oxygen therapy. The gauge on a 12-L tank of compressed oxygen reads 3800 mmHg. How many liters would this same gas occupy at a pressure of 0.75 atm when temperature and amount of gas do not change?

SOLUTION

STEP 1 Organize the data in a table of initial and final conditions. To match the units for the initial and final pressures, we can convert either atm to mmHg or mmHg to atm.

$$0.75 \text{ atm} \times \frac{760 \text{ mmHg}}{1 \text{ atm}} = 570 \text{ mmHg}$$

or

$$3800 \text{ mmHg} \times \frac{1 \text{ atm}}{760 \text{ mmHg}} = 5.0 \text{ atm}$$

We place the gas data using units of mmHg for pressure and liters for volume in a table. (We could have both pressures in units of atm as well.) The properties that do not change, which are temperature and amount of gas, are shown below the table. We know that pressure decreases, and can predict that the volume increases.

	Conditions 1	Conditions 2	Know	Predict
ANALYZE THE PROBLEM	$P_1 = 3800$ mmHg (5.0 atm)	$P_2 = 570$ mmHg (0.75 atm)	P decreases	
	$V_1 = 12$ L	$V_2 = ?$ L		V increases

Factors that remain constant: T and n

STEP 2 Rearrange the gas law equation to solve for the unknown quantity. For a PV relationship, we use Boyle's law and solve for V_2 by dividing both sides by P_2. According to Boyle's law, a decrease in the pressure will cause an increase in the volume when T and n remain constant.

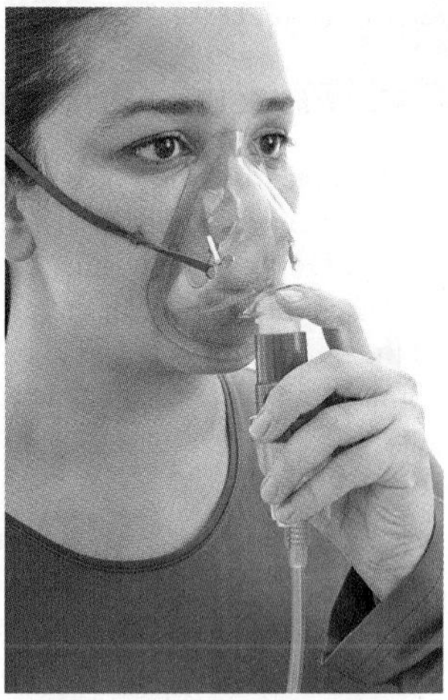

Oxygen therapy increases the oxygen available to the tissues of the body.

Guide to Using the Gas Laws

STEP 1
Organize the data in a table of initial and final conditions.

STEP 2
Rearrange the gas law equation to solve for the unknown quantity.

STEP 3
Substitute values into the gas law equation and calculate.

CORE CHEMISTRY SKILL

Using the Gas Laws

$$P_1V_1 = P_2V_2$$

$$\frac{P_1V_1}{P_2} = \frac{\cancel{P_2}V_2}{\cancel{P_2}}$$

$$V_2 = V_1 \times \frac{P_1}{P_2}$$

STEP 3 **Substitute values into the gas law equation and calculate.** When we substitute in the values with pressures in units of mmHg or atm, the ratio of pressures (pressure factor) is greater than 1, which increases the volume as predicted in Step 1.

$$V_2 = 12\ \text{L} \times \underbrace{\frac{3800\ \cancel{\text{mmHg}}}{570\ \cancel{\text{mmHg}}}}_{\text{Pressure factor increases volume}} = 80.\ \text{L}$$

or

$$V_2 = 12\ \text{L} \times \underbrace{\frac{5.0\ \cancel{\text{atm}}}{0.75\ \cancel{\text{atm}}}}_{\text{Pressure factor increases volume}} = 80.\ \text{L}$$

STUDY CHECK 8.3

In an underground natural gas reserve, a bubble of methane gas, CH_4, has a volume of 45.0 mL at 1.60 atm. What volume, in milliliters, will the gas bubble occupy when it reaches the surface where the atmospheric pressure is 744 mmHg, if there is no change in the temperature or amount of gas?

Chemistry Link to Health

Pressure–Volume Relationship in Breathing

The importance of Boyle's law becomes more apparent when you consider the mechanics of breathing. Our lungs are elastic, balloon-like structures contained within an airtight chamber called the thoracic cavity. The diaphragm, a muscle, forms the flexible floor of the cavity.

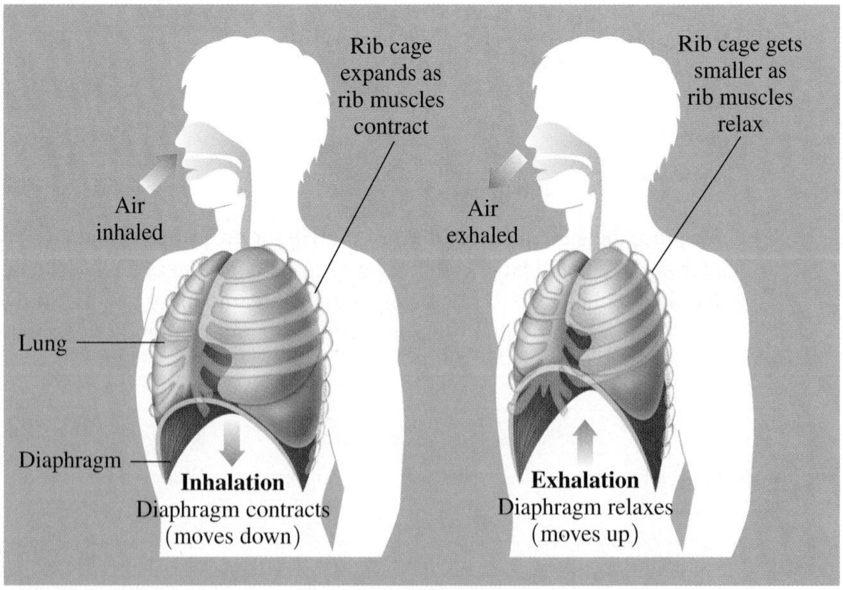

Inspiration

The process of taking a breath of air begins when the diaphragm contracts and the rib cage expands, causing an increase in the volume of the thoracic cavity. The elasticity of the lungs allows them to expand when the thoracic cavity expands. According to Boyle's law, the pressure inside the lungs decreases when their volume increases, causing the pressure inside the lungs to fall below the pressure of the atmosphere. This difference in pressures produces a *pressure gradient* between the lungs and the atmosphere. In a pressure gradient, molecules flow from an area of higher pressure to an area of lower pressure. During the inhalation phase of breathing, air flows into the lungs (*inspiration*), until the pressure within the lungs becomes equal to the pressure of the atmosphere.

Expiration

Expiration, or the exhalation phase of breathing, occurs when the diaphragm relaxes and moves back up into the thoracic cavity to its resting position. The volume of the thoracic cavity decreases, which squeezes the lungs and decreases their volume. Now the pressure in the lungs is higher than the pressure of the atmosphere, so air flows out of the lungs. Thus, breathing is a process in which pressure gradients are continuously created between the lungs and the environment because of the changes in the volume and pressure.

QUESTIONS AND PROBLEMS

8.2 Pressure and Volume (Boyle's Law)

LEARNING GOAL Use the pressure–volume relationship (Boyle's law) to determine the final pressure or volume when the temperature and amount of gas are constant.

8.9 Why do scuba divers need to exhale air when they ascend to the surface of the water?

8.10 Why does a sealed bag of chips expand when you take it to a higher altitude?

8.11 The air in a cylinder with a piston has a volume of 220 mL and a pressure of 650 mmHg.

a. To obtain a higher pressure inside the cylinder at constant temperature and amount of gas, would the cylinder change as shown in **A** or **B**? Explain your choice.

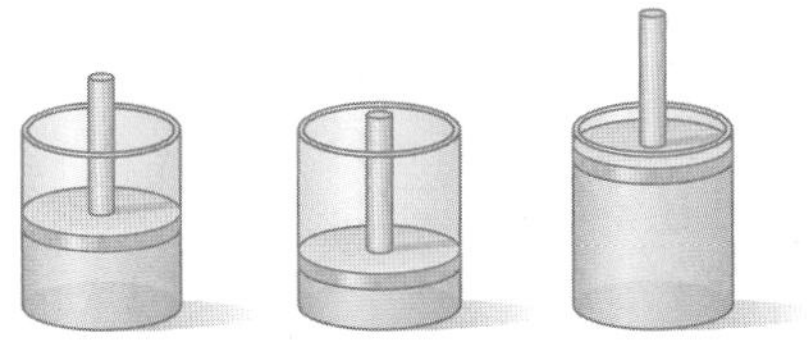

b. If the pressure inside the cylinder increases to 1.2 atm, what is the final volume, in milliliters, of the cylinder? Complete the following data table:

Property	Conditions 1	Conditions 2	Know	Predict
Pressure (P)				
Volume (V)				

8.12 A balloon is filled with helium gas. When each of the following changes are made at constant temperature, which of the following diagrams (**A**, **B**, or **C**) shows the final volume of the balloon?

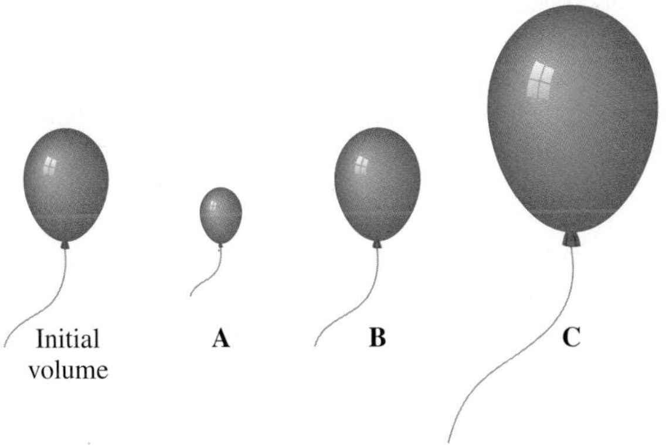

a. The balloon floats to a higher altitude where the outside pressure is lower.

b. The balloon is taken inside the house, but the atmospheric pressure does not change.

c. The balloon is put in a hyperbaric chamber in which the pressure is increased.

8.13 A gas with a volume of 4.0 L is in a closed container. Indicate the changes (*increases, decreases, does not change*) in pressure when the volume undergoes the following changes at constant temperature and amount of gas:

a. The volume is compressed to 2.0 L.

b. The volume expands to 12 L.

c. The volume is compressed to 0.40 L.

8.14 A gas at 0.395 atm is in a closed container. Indicate the changes (*increases, decreases, no change*) in its volume when the pressure undergoes the following changes at constant temperature and amount of gas:

a. The pressure increases to 6.0 atm.

b. The pressure remains at 2.0 atm.

c. The pressure drops to 0.40 atm.

8.15 A 25.0-L balloon contains hydrogen gas at a pressure of 570 mmHg. What is the final pressure, in millimeters of mercury, of the hydrogen gas at each of the following volumes, if there is no change in temperature and amount of gas?

a. 10.0 L **b.** 5.60 L

c. 16400 mL **d.** 1580 mL

8.16 The air in a 5.00-L tank has a pressure of 1.20 atm. What is the final pressure, in atmospheres, when the air is placed in tanks that have the following volumes, if there is no change in temperature and amount of gas?

a. 1.00 L **b.** 2500. mL

c. 750. mL **d.** 8.00 L

8.17 A sample of oxygen (O_2) has a volume of 30.0 L at a pressure of 760. mmHg. What is the final volume, in liters, of the gas at each of the following pressures, if there is no change in temperature and amount of gas?

a. 625 mmHg **b.** 3.0 atm

c. 0.800 atm **d.** 350 torr

8.18 A sample of methane (CH_4) has a volume of 25 mL at a pressure of 0.80 atm. What is the final volume, in milliliters, of the gas at each of the following pressures, if there is no change in temperature and amount of gas?

a. 0.40 atm **b.** 2.00 atm

c. 2500 mmHg **d.** 80.0 torr

8.19 At a constant temperature, what will be the final volume of an amount of gas at 20 atm if the initial volume is 100 L and pressure is 0.5 atm?

8.20 A tank of oxygen holds 20.0 L of oxygen (O_2) at a pressure of 15.0 atm. What is the volume, in liters, of this gas when it is released at a pressure of 1.00 atm with no change in temperature and amount of gas?

8.21 Use the word *inspiration* or *expiration* to describe the part of the breathing cycle that occurs as a result of each of the following:

a. The diaphragm contracts (flattens out).

b. The volume of the lungs decreases.

c. The pressure within the lungs is less than that of the atmosphere.

8.22 Use the word *inspiration* or *expiration* to describe the part of the breathing cycle that occurs as a result of each of the following:

a. The diaphragm relaxes, moving up into the thoracic cavity.

b. The volume of the lungs expands.

c. The pressure within the lungs is greater than that of the atmosphere.

LEARNING GOAL

Use the temperature–volume relationship (Charles's law) to determine the final temperature or volume when the pressure and amount of gas are constant.

As the gas in a hot-air balloon is heated, it expands.

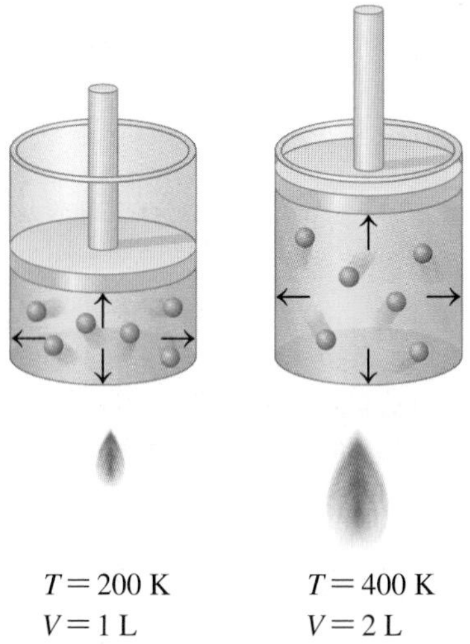

FIGURE 8.5 ▶ **Charles's law:** The Kelvin temperature of a gas is directly related to the volume of the gas when there is no change in the pressure and amount. When the temperature increases, making the molecules move faster, the volume must increase to maintain constant pressure.

Q If the temperature of a gas decreases at constant pressure and amount of gas, how will the volume change?

8.3 Temperature and Volume (Charles's Law)

Suppose that you are going to take a ride in a hot-air balloon. The captain turns on a propane burner to heat the air inside the balloon. As the air is heated, it expands and becomes less dense than the air outside, causing the balloon and its passengers to lift off. In 1787, Jacques Charles, a balloonist as well as a physicist, proposed that the volume of a gas is related to the temperature. This became **Charles's law**, which states that the volume (V) of a gas is directly related to the temperature (T) when there is no change in the pressure (P) or amount (n) of gas (see Figure 8.5). A *direct relationship* is one in which the related properties increase or decrease together. For two conditions, initial and final, we can write Charles's law as follows:

Charles's Law

$$\frac{V_1}{T_1} = \frac{V_2}{T_2} \quad \text{No change in number of moles and pressure}$$

All temperatures used in gas law calculations must be converted to their corresponding Kelvin (K) temperatures.

To determine the effect of changing temperature on the volume of a gas, the pressure and the amount of gas are kept constant. If we increase the temperature of a gas sample, we know from the kinetic molecular theory that the motion (kinetic energy) of the gas particles will also increase. To keep the pressure constant, the volume of the container must increase (see Figure 8.5). If the temperature of the gas decreases, the volume of the container must decrease to maintain the same pressure when the amount of gas is constant.

SAMPLE PROBLEM 8.4 Calculating Volume When Temperature Changes

A sample of argon gas has a volume of 5.40 L and a temperature of 15 °C. What is the final volume, in liters, of the gas if the temperature has been increased to 42 °C at constant pressure and amount of gas?

SOLUTION

STEP 1 **Organize the data in a table of initial and final conditions.** The properties that change, which are the temperature and volume, are listed in the following table. The properties that do not change, which are pressure and amount of gas, are shown below the table. When the temperature is given in degrees Celsius, it must be changed to kelvins. Because we know the initial and final temperatures of the gas, we know that the temperature increases. Thus, we can predict that the volume increases.

$$T_1 = 15\text{ °C} + 273 = 288\text{ K}$$
$$T_2 = 42\text{ °C} + 273 = 315\text{ K}$$

ANALYZE THE PROBLEM	Conditions 1	Conditions 2	Know	Predict
	$T_1 = 288$ K	$T_2 = 315$ K	T increases	
	$V_1 = 5.40$ L	$V_2 = ?$ L		V increases

Factors that remain constant: P and n

STEP 2 **Rearrange the gas law equation to solve for the unknown quantity.** In this problem, we want to know the final volume (V_2) when the temperature increases. Using Charles's law, we solve for V_2 by multiplying both sides by T_2.

$$\frac{V_1}{T_1} = \frac{V_2}{T_2}$$

$$\frac{V_1}{T_1} \times T_2 = \frac{V_2}{\cancel{T_2}} \times \cancel{T_2}$$

$$V_2 = V_1 \times \frac{T_2}{T_1}$$

STEP 3 **Substitute values into the gas law equation and calculate.** From the table, we see that the temperature has increased. Because temperature is directly related to volume, the volume must increase. When we substitute in the values, we see that the ratio of the temperatures (temperature factor) is greater than 1, which increases the volume as predicted in Step 1.

$$V_2 = 5.40 \text{ L} \times \frac{315 \cancel{\text{K}}}{288 \cancel{\text{K}}} = 5.91 \text{ L}$$

Temperature factor increases volume

STUDY CHECK 8.4

A mountain climber with a body temperature of 37 °C inhales 486 mL of air at a temperature of −8 °C. What volume, in milliliters, will the air occupy in the lungs, if the pressure and amount of gas do not change?

QUESTIONS AND PROBLEMS

8.3 Temperature and Volume (Charles's Law)

LEARNING GOAL Use the temperature–volume relationship (Charles's law) to determine the final temperature or volume when the pressure and amount of gas are constant.

8.23 Select the diagram that shows the final volume of a balloon when the following changes are made at constant pressure and amount of gas:

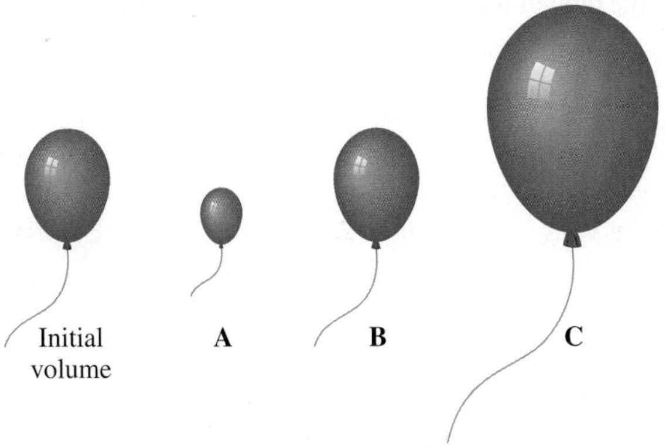

a. The temperature changes from 100 K to 300 K.
b. The balloon is placed in a freezer.
c. The balloon is first warmed and then returned to its starting temperature.

8.24 Indicate whether the final volume of gas in each of the following is the *same*, *larger*, or *smaller* than the initial volume, if pressure and amount of gas do not change:
a. A volume of 505 mL of air on a cold winter day at −15 °C is breathed into the lungs, where body temperature is 37 °C.
b. The heater used to heat the air in a hot-air balloon is turned off.
c. A balloon filled with helium at the amusement park is left in a car on a hot day.

8.25 A sample of argon initially has a volume of 3.80 L at 25 °C. What final temperature, in degrees Celsius, is needed to change the volume of the gas to each of the following, if n and P do not change?
a. 6.00 L **b.** 2100 mL
c. 8.50 L **d.** 4250 mL

8.26 A gas has a volume of 4.00 L at 23 °C. What final temperature, in degrees Celsius, is needed to change the volume of the gas to each of the following, if n and P do not change?
a. 1.50 L **b.** 1200 mL
c. 10.0 L **d.** 50.0 mL

8.27 A balloon contains 4300 mL of neon gas at 45 °C. What is the final volume, in milliliters, of the gas when the temperature changes to each of the following, if n and P do not change?
a. 35 °C **b.** 560 K
c. −15 °C **d.** 640 K

8.28 An air bubble has a volume of 0.500 L at 18 °C. What is the final volume, in liters, of the gas when the temperature changes to each of the following, if n and P do not change?
a. 0 °C **b.** 425 K
c. −12 °C **d.** 575 K

LEARNING GOAL

Use the temperature–pressure relationship (Gay-Lussac's law) to determine the final temperature or pressure when the volume and amount of gas are constant.

8.4 Temperature and Pressure (Gay-Lussac's Law)

If we could observe the molecules of a gas as the temperature rises, we would notice that they move faster and hit the sides of the container more often and with greater force. If we maintain a constant volume and amount of gas, the pressure will increase. In the temperature–pressure relationship known as **Gay-Lussac's law**, the pressure of a gas is directly related to its Kelvin temperature. This means that an increase in temperature increases the pressure of a gas and a decrease in temperature decreases the pressure of the gas, as long as the volume and amount of gas do not change (see Figure 8.6).

$T = 200$ K $\quad$ $T = 400$ K

$P = 1$ atm $\quad$ $P = 2$ atm

FIGURE 8.6 ▶ **Gay-Lussac's law:** When the Kelvin temperature of a gas doubles at constant volume and amount of gas, the pressure also doubles.

Q How does a decrease in the temperature of a gas affect its pressure at constant volume and amount of gas?

Gay-Lussac's Law

$$\frac{P_1}{T_1} = \frac{P_2}{T_2} \quad \text{No change in number of moles and volume}$$

All temperatures used in gas law calculations must be converted to their corresponding Kelvin (K) temperatures.

SAMPLE PROBLEM 8.5 Calculating Pressure When Temperature Changes

Aerosol containers can be dangerous if they are heated, because they can explode. Suppose a container of hair spray with a pressure of 4.0 atm at a room temperature of 25 °C is thrown into a fire. If the temperature of the gas inside the aerosol can reach 402 °C, what will be its pressure in atmospheres? The aerosol container may explode if the pressure inside exceeds 8.0 atm. Would you expect it to explode?

SOLUTION

STEP 1 **Organize the data in a table of initial and final conditions.** We list the properties that change, which are the pressure and temperature, in a table. The properties that do not change, which are volume and amount of gas, are shown below the table. The temperatures given in degrees Celsius must be changed to kelvins. Because we know the initial and final temperatures of the gas, we know that the temperature increases. Thus, we can predict that the pressure increases.

$$T_1 = 25\ ^\circ\text{C} + 273 = 298\ \text{K}$$

$$T_2 = 402\ ^\circ\text{C} + 273 = 675\ \text{K}$$

ANALYZE THE PROBLEM	Conditions 1	Conditions 2	Know	Predict
	$P_1 = 4.0$ atm	$P_2 = ?$ atm		P increases
	$T_1 = 298$ K	$T_2 = 675$ K	T increases	

Factors that remain constant: V and n

STEP 2 **Rearrange the gas law equation to solve for the unknown quantity.** Using Gay-Lussac's law, we can solve for P_2 by multiplying both sides by T_2.

$$\frac{P_1}{T_1} = \frac{P_2}{T_2}$$

$$\frac{P_1}{T_1} \times T_2 = \frac{P_2}{\cancel{T_2}} \times \cancel{T_2}$$

$$P_2 = P_1 \times \frac{T_2}{T_1}$$

STEP 3 Substitute values into the gas law equation and calculate. When we substitute in the values, we see that the ratio of the temperatures (temperature factor), is greater than 1, which increases pressure as predicted in Step 1.

$$P_2 = 4.0 \text{ atm} \times \frac{675 \cancel{\text{K}}}{298 \cancel{\text{K}}} = 9.1 \text{ atm}$$

Temperature factor increases pressure

Because the calculated pressure of 9.1 atm of the gas exceeds the limit of 8.0 atm for the can, we would expect the aerosol can to explode.

STUDY CHECK 8.5

In a storage area where the temperature has reached 55 °C, the pressure of oxygen gas in a 15.0-L steel cylinder is 965 torr. To what temperature, in degrees Celsius, would the gas have to be cooled to reduce the pressure to 850. torr when the volume and amount of the gas do not change?

QUESTIONS AND PROBLEMS

8.4 Temperature and Pressure (Gay-Lussac's Law)

LEARNING GOAL Use the temperature–pressure relationship (Gay-Lussac's law) to determine the final temperature or pressure when the volume and amount of gas are constant.

8.29 Calculate the final pressure, in torr, for each of the following, if n and V are constant:

a. A gas with an initial pressure of 1500 torr at 175 °C is cooled to 20 °C.

b. A gas in an aerosol can with an initial pressure of 3.10 atm at 24 °C is heated to 45 °C.

8.30 Calculate the final pressure, in atmospheres, for each of the following, if n and V are constant:

a. A gas with an initial pressure of 1.20 atm at 75 °C is cooled to −32 °C.

b. A sample of N_2 gas with an initial pressure of 780. mmHg at −75 °C is heated to 28 °C.

8.31 Calculate the final temperature, in degrees Celsius, for each of the following, if n and V are constant:

a. A sample of helium gas at 55 °C and 780. mmHg is cooled to give a pressure of 520. mmHg.

b. A tank of a neon gas with a pressure of 0.730 atm at −12 °C is heated to give a pressure of 1650 torr.

8.32 Calculate the final temperature, in degrees Celsius, for each of the following, if n and V are constant:

a. A tank of helium gas with a pressure of 250 torr at 0 °C is heated to give a pressure of 1500 torr.

b. A sample of air at 40. °C and 740. mmHg is cooled to give a pressure of 680. mmHg.

8.5 The Combined Gas Law

LEARNING GOAL

Use the combined gas law to calculate the final pressure, volume, or temperature of a gas when changes in two of these properties are given and the amount of gas is constant.

All of the pressure–volume–temperature relationships for gases that we have studied may be combined into a single relationship called the **combined gas law**. This expression is useful for studying the effect of changes in two of these variables on the third as long as the amount of gas (number of moles) remains constant.

Combined Gas Law

$$\frac{P_1V_1}{T_1} = \frac{P_2V_2}{T_2} \quad \text{No change in number of moles of gas}$$

By using the combined gas law, we can derive any of the gas laws by omitting those properties that do not change as seen in Table 8.4.

TABLE 8.4 Summary of Gas Laws

Combined Gas Law	Properties Held Constant	Relationship	Name of Gas Law
$\frac{P_1V_1}{\cancel{T_1}} = \frac{P_2V_2}{\cancel{T_2}}$	T, n	$P_1V_1 = P_2V_2$	Boyle's law
$\frac{\cancel{P_1}V_1}{T_1} = \frac{\cancel{P_2}V_2}{T_2}$	P, n	$\frac{V_1}{T_1} = \frac{V_2}{T_2}$	Charles's law
$\frac{P_1\cancel{V_1}}{T_1} = \frac{P_2\cancel{V_2}}{T_2}$	V, n	$\frac{P_1}{T_1} = \frac{P_2}{T_2}$	Gay-Lussac's law

Under water, the pressure on a diver is greater than the atmospheric pressure.

SAMPLE PROBLEM 8.6 Using the Combined Gas Law

A 25.0-mL bubble is released from a diver's air tank at a pressure of 4.00 atm and a temperature of 11 °C. What is the volume, in milliliters, of the bubble when it reaches the ocean surface, where the pressure is 1.00 atm and the temperature is 18 °C? (Assume the amount of gas in the bubble does not change.)

SOLUTION

STEP 1 Organize the data in a table of initial and final conditions. We list the properties that change, which are the pressure, volume, and temperature, in a table. The property that remains constant, which is the amount of gas, is shown below the table. The temperatures in degrees Celsius must be changed to kelvins.

$$T_1 = 11\text{ °C} + 273 = 284\text{ K}$$
$$T_2 = 18\text{ °C} + 273 = 291\text{ K}$$

	Conditions 1	Conditions 2
ANALYZE THE PROBLEM	$P_1 = 4.00$ atm	$P_2 = 1.00$ atm
	$V_1 = 25.0$ mL	$V_2 = ?$ mL
	$T_1 = 284$ K	$T_2 = 291$ K

Factor that remains constant: n

STEP 2 Rearrange the gas law equation to solve for the unknown quantity. For changes in two conditions, we rearrange the combined gas law to solve for V_2.

$$\frac{P_1V_1}{T_1} = \frac{P_2V_2}{T_2}$$

$$\frac{P_1V_1}{T_1} \times \frac{T_2}{P_2} = \frac{\cancel{P_2}V_2}{\cancel{T_2}} \times \frac{\cancel{T_2}}{\cancel{P_2}}$$

$$V_2 = V_1 \times \frac{P_1}{P_2} \times \frac{T_2}{T_1}$$

STEP 3 Substitute values into the gas law equation and calculate. From the data table, we determine that both the pressure decrease and the temperature increase will increase the volume.

$$V_2 = 25.0\text{ mL} \times \frac{4.00\ \cancel{\text{atm}}}{1.00\ \cancel{\text{atm}}} \times \frac{291\ \cancel{\text{K}}}{284\ \cancel{\text{K}}} = 102\text{ mL}$$

Pressure factor increases volume; Temperature factor increases volume

However, when the unknown value is decreased by one change but increased by the second change, it is difficult to predict the overall change for the unknown.

STUDY CHECK 8.6

A weather balloon is filled with 15.0 L of helium at a temperature of 25 °C at a pressure of 685 mmHg. What is the pressure, in millimeters of mercury, of the helium in the balloon in the upper atmosphere when the temperature is −35 °C and the final volume becomes 34.0 L, if the amount of He does not change?

QUESTIONS AND PROBLEMS

8.5 The Combined Gas Law

LEARNING GOAL Use the combined gas law to calculate the final pressure, volume, or temperature of a gas when changes in two of these properties are given and the amount of gas is constant.

8.33 A sample of xenon gas has a volume of 7.50 L at a pressure of 675 mmHg and a temperature of 35 °C. What is the final pressure of the gas, in atmospheres, when the volume and temperature of the gas sample are changed to the following, if the amount of gas does not change?
a. 1560 mL and 373 K **b.** 4.35 L and 20 °C
c. 18.5 L and 53 °C

8.34 A sample of argon gas has a volume of 735 mL at a pressure of 1.20 atm and a temperature of 112 °C. What is the final volume of the gas, in milliliters, when the pressure and temperature of the gas sample are changed to the following, if the amount of gas does not change?
a. 658 mmHg and 281 K **b.** 0.55 atm and 75 °C
c. 15.4 atm and −15 °C

8.35 A 124-mL bubble of hot gases at 212 °C and 1.80 atm is emitted from an active volcano. What is the final temperature, in degrees Celsius, of the gas in the bubble outside the volcano if the final volume of the bubble is 138 mL and the pressure is 0.800 atm, if the amount of gas does not change?

8.36 A scuba diver 60 ft below the ocean surface inhales 50.0 mL of compressed air from a scuba tank at a pressure of 3.00 atm and a temperature of 8 °C. What is the final pressure of air, in atmospheres, in the lungs when the gas expands to 150.0 mL at a body temperature of 37 °C, if the amount of gas does not change?

8.6 Volume and Moles (Avogadro's Law)

LEARNING GOAL

Use Avogadro's law to calculate the amount or volume of a gas when the pressure and temperature are constant.

In our study of the gas laws, we have looked at changes in properties for a specified amount (n) of gas. Now we will consider how the properties of a gas change when there is a change in the number of moles or grams of the gas.

When you blow up a balloon, its volume increases because you add more air molecules. If the balloon has a hole in it, air leaks out, causing its volume to decrease. In 1811, Amedeo Avogadro formulated **Avogadro's law**, which states that the volume of a gas is directly related to the number of moles of a gas when temperature and pressure do not change. For example, if the number of moles of a gas is doubled, then the volume will also double as long as we do not change the pressure or the temperature (see Figure 8.7). At constant pressure and temperature, we can write Avogadro's law:

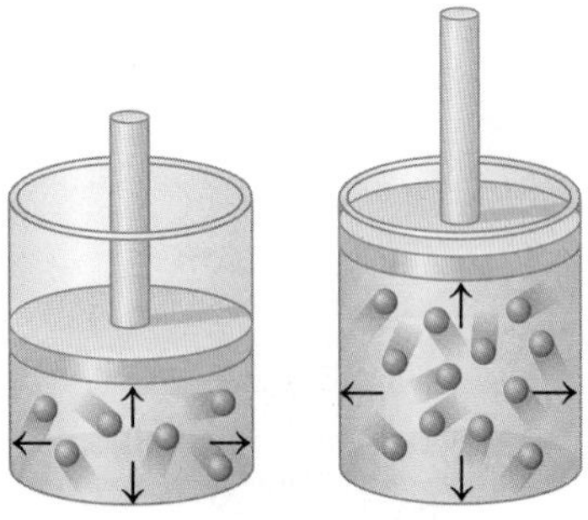

$n = 1$ mole $\quad$ $n = 2$ moles
$V = 1$ L $\quad$ $V = 2$ L

FIGURE 8.7 ▶ Avogadro's law: The volume of a gas is directly related to the number of moles of the gas. If the number of moles is doubled, the volume must double at constant temperature and pressure.

Q If a balloon has a leak, what happens to its volume?

Avogadro's Law

$$\frac{V_1}{n_1} = \frac{V_2}{n_2} \quad \text{No change in pressure and temperature}$$

SAMPLE PROBLEM 8.7 Calculating Volume for a Change in Moles

A weather balloon with a volume of 44 L is filled with 2.0 moles of helium. What is the final volume, in liters, if 3.0 moles of helium gas are added, to give a total of 5.0 moles of helium gas, if the pressure and temperature do not change?

SOLUTION

STEP 1 **Organize the data in a table of initial and final conditions.** We list those properties that change, which are volume and amount (moles) of gas, in a table. The properties that do not change, which are pressure and temperature, are shown below the table. Because there is an increase in the number of moles of gas, we can predict that the volume increases.

ANALYZE THE PROBLEM	Conditions 1	Conditions 2	Know	Predict
	$V_1 = 44$ L	$V_2 = ?$ L		V increases
	$n_1 = 2.0$ moles	$n_2 = 5.0$ moles	n increases	

Factors that remain constant: P and T

STEP 2 Rearrange the gas law equation to solve for the unknown quantity. Using Avogadro's law, we can solve for V_2 by multiplying both sides by n_2.

$$\frac{V_1}{n_1} = \frac{V_2}{n_2}$$

$$\frac{V_1}{n_1} \times n_2 = \frac{V_2}{\cancel{n_2}} \times \cancel{n_2}$$

$$V_2 = V_1 \times \frac{n_2}{n_1}$$

STEP 3 Substitute values into the gas law equation and calculate. When we substitute in the values, we see that the ratio of the moles (mole factor) is greater than 1, which increases the volume as predicted in Step 1.

$$V_2 = 44\text{ L} \times \frac{5.0\text{ }\cancel{\text{moles}}}{2.0\text{ }\cancel{\text{moles}}} = 110\text{ L}$$

Mole factor increases volume

STUDY CHECK 8.7

A sample containing 8.00 g of oxygen gas has a volume of 5.00 L. What is the final volume, in liters, after 4.00 g of oxygen gas is added to the 8.00 g of oxygen in the balloon, if the temperature and pressure do not change?

STP and Molar Volume

Using Avogadro's law, we can say that any two gases will have equal volumes if they contain the same number of moles of gas at the same temperature and pressure. To help us make comparisons between different gases, arbitrary conditions called *standard temperature* (273 K) and *standard pressure* (1 atm) together abbreviated **STP**, were selected by scientists:

STP Conditions

Standard temperature is *exactly* 0 °C (273 K).

Standard pressure is *exactly* 1 atm (760 mmHg).

At STP, one mole of any gas occupies a volume of 22.4 L, which is about the same as the volume of three basketballs. This volume, 22.4 L, of any gas at STP is called its **molar volume** (see Figure 8.8).

When a gas is at STP conditions (0 °C and 1 atm), its molar volume can be used to write conversion factors between the number of moles of gas and its volume, in liters.

The molar volume of a gas at STP is about the same as the volume of three basketballs.

Molar Volume Conversion Factors

1 mole of gas = 22.4 L of gas (STP)

$$\frac{1\text{ mole gas}}{22.4\text{ L gas (STP)}} \quad \text{and} \quad \frac{22.4\text{ L gas (STP)}}{1\text{ mole gas}}$$

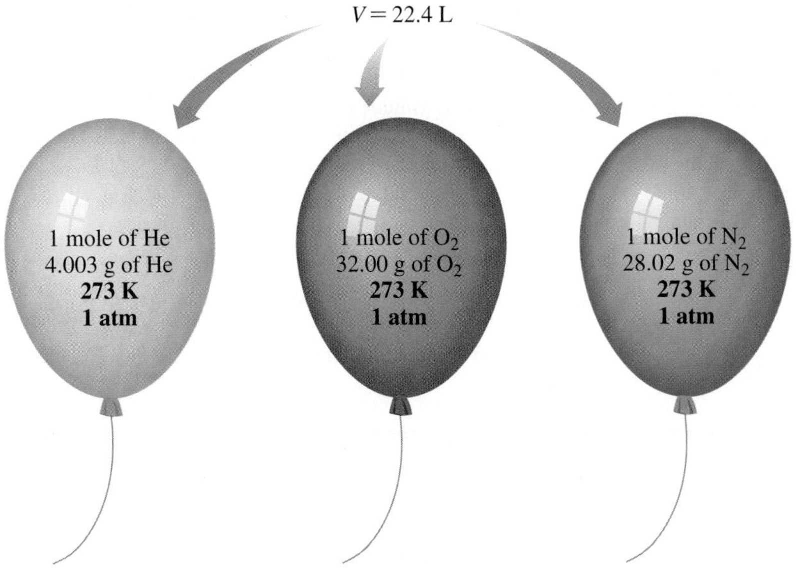

FIGURE 8.8 ▶ Avogadro's law indicates that 1 mole of any gas at STP has a volume of 22.4 L.

Q What volume of gas, in liters, is occupied by 16.0 g of methane gas, CH_4, at STP?

SAMPLE PROBLEM 8.8 Using Molar Volume to Find Volume at STP

What is the volume, in liters, of 64.0 g of O_2 gas at STP?

SOLUTION

STEP 1 State the given and needed quantities.

	Given	Need
ANALYZE THE PROBLEM	64.0 g of $O_2(g)$ at STP	liters of O_2 gas at STP

STEP 2 Write a plan to calculate the needed quantity. The grams of O_2 are converted to moles using molar mass. Then a molar volume conversion factor is used to convert the number of moles to volume (L).

grams of O_2 → Molar mass → moles of O_2 → Molar volume → liters of O_2

STEP 3 Write the equalities and conversion factors including 22.4 L/mole at STP.

1 mole of O_2 = 32.00 g of O_2

$$\frac{32.00 \text{ g } O_2}{1 \text{ mole } O_2} \text{ and } \frac{1 \text{ mole } O_2}{32.00 \text{ g } O_2}$$

1 mole of O_2 = 22.4 L of O_2 (STP)

$$\frac{22.4 \text{ L } O_2 \text{ (STP)}}{1 \text{ mole } O_2} \text{ and } \frac{1 \text{ mole } O_2}{22.4 \text{ L } O_2 \text{ (STP)}}$$

STEP 4 Set up the problem with factors to cancel units.

$$64.0 \cancel{\text{g } O_2} \times \frac{1 \cancel{\text{mole } O_2}}{32.00 \cancel{\text{g } O_2}} \times \frac{22.4 \text{ L } O_2 \text{ (STP)}}{1 \cancel{\text{mole } O_2}} = 44.8 \text{ L of } O_2 \text{ (STP)}$$

STUDY CHECK 8.8

How many grams of $N_2(g)$ are in 5.6 L of $N_2(g)$ at STP?

Guide to Using Molar Volume

STEP 1 State the given and needed quantities.

STEP 2 Write a plan to calculate the needed quantity.

STEP 3 Write the equalities and conversion factors including 22.4 L/mole at STP.

STEP 4 Set up the problem with factors to cancel units.

QUESTIONS AND PROBLEMS

8.6 Volume and Moles (Avogadro's Law)

LEARNING GOAL Use Avogadro's law to calculate the amount or volume of a gas when the pressure and temperature are constant.

8.37 What is the difference in volume between 1 mole of CO_2 and 1 mole of H_2 at STP?

8.38 Why do bubbles form when a can of soda is opened?

8.39 A sample containing 1.50 moles of Ne gas has an initial volume of 8.00 L. What is the final volume of the gas, in liters, when each of the following changes occurs in the quantity of the gas at constant pressure and temperature?

a. A leak allows one-half of the Ne atoms to escape.
b. A sample of 3.50 moles of Ne is added to the 1.50 moles of Ne gas in the container.
c. A sample of 25.0 g of Ne is added to the 1.50 moles of Ne gas in the container.

8.40 A sample containing 4.80 g of O_2 gas has an initial volume of 15.0 L. What is the final volume, in liters, when each of the following changes occurs in the quantity of the gas at constant pressure and temperature?
a. A sample of 0.500 mole of O_2 is added to the 4.80 g of O_2 in the container.
b. A sample of 2.00 g of O_2 is removed.
c. A sample of 4.00 g of O_2 is added to the 4.80 g of O_2 gas in the container.

8.41 Use molar volume to calculate each of the following at STP:
a. the number of moles of O_2 in 44.8 L of O_2 gas
b. the volume, in liters, occupied by 0.420 mole of He gas

8.42 Use molar volume to calculate each of the following at STP:
a. the volume, in liters, occupied by 2.50 moles of N_2 gas
b. the number of moles of CO_2 in 4.00 L of CO_2 gas

8.43 Use molar volume to calculate each of the following at STP:
a. the volume, in liters, of 6.40 g of O_2 gas
b. the number of grams of H_2 in 1620 mL of H_2 gas

8.44 Use molar volume to calculate each of the following at STP:
a. the number of grams of Ne contained in 11.2 L of Ne gas
b. the volume, in liters, occupied by 50.0 g of Ar gas

LEARNING GOAL

Use Dalton's law of partial pressures to calculate the total pressure of a mixture of gases.

8.7 Partial Pressures (Dalton's Law)

Many gas samples are a mixture of gases. For example, the air you breathe is a mixture of mostly oxygen and nitrogen gases. In gas mixtures, scientists observed that all gas particles behave in the same way. Therefore, the total pressure of the gases in a mixture is a result of the collisions of the gas particles regardless of what type of gas they are.

In a gas mixture, each gas exerts its **partial pressure**, which is the pressure it would exert if it were the only gas in the container. **Dalton's law** states that the total pressure of a gas mixture is the sum of the partial pressures of the gases in the mixture.

Dalton's Law

$$P_{total} = P_1 + P_2 + P_3 + \cdots$$

Total pressure of a gas mixture = Sum of the partial pressures of the gases in the mixture

CORE CHEMISTRY SKILL

Calculating Partial Pressure

Suppose we have two separate tanks, one filled with helium at 2.0 atm and the other filled with argon at 4.0 atm. When the gases are combined in a single tank at the same volume and temperature, the number of gas molecules, not the type of gas, determines the pressure in a container. Then the pressure of the gases in the gas mixture would be 6.0 atm, which is the sum of their individual or partial pressures.

$P_{He} = 2.0$ atm $P_{Ar} = 4.0$ atm

$$P_{total} = P_{He} + P_{Ar} = 2.0\text{ atm} + 4.0\text{ atm} = 6.0\text{ atm}$$

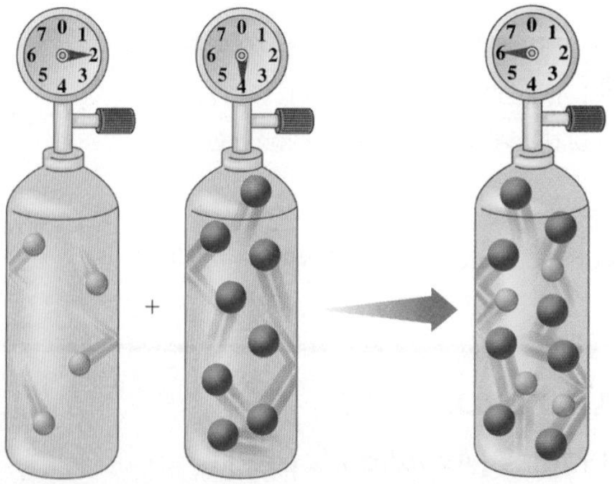

The total pressure of two gases is the sum of their partial pressures.

SAMPLE PROBLEM 8.9 Partial Pressure of a Gas in a Mixture

A heliox breathing mixture of oxygen and helium is prepared for a scuba diver who is going to descend 200 ft below the ocean surface. At that depth, the diver breathes a gas mixture that has a total pressure of 7.00 atm. If the partial pressure of the oxygen in the tank at that depth is 1140 mmHg, what is the partial pressure, in atmospheres, of the helium in the breathing mixture?

SOLUTION

STEP 1 Write the equation for the sum of the partial pressures.

$$P_{total} = P_{O_2} + P_{He}$$

STEP 2 Rearrange the equation to solve for the unknown pressure. To solve for the partial pressure of helium (P_{He}), we rearrange the equation to give the following:

$$P_{He} = P_{total} - P_{O_2}$$

Convert units to match.

$$P_{O_2} = 1140\ \cancel{\text{mmHg}} \times \frac{1\ \text{atm}}{760\ \cancel{\text{mmHg}}} = 1.50\ \text{atm}$$

STEP 3 Substitute known pressures into the equation and calculate the unknown partial pressure.

$$P_{He} = P_{total} - P_{O_2}$$

$$P_{He} = 7.00\ \text{atm} - 1.50\ \text{atm} = 5.50\ \text{atm}$$

Guide to Calculating Partial Pressure

STEP 1 Write the equation for the sum of the partial pressures.

STEP 2 Rearrange the equation to solve for the unknown pressure.

STEP 3 Substitute known pressures into the equation and calculate the unknown partial pressure.

STUDY CHECK 8.9

A common anesthetic in dentistry known as Nitronox consists of a mixture of dinitrogen oxide, N_2O, and oxygen gas, O_2, which is administered through an inhaler over the nose. If the anesthetic mixture has a total pressure of 740 mmHg, and the partial pressure of the oxygen is 370 mmHg, what is the partial pressure, in torr, of the dinitrogen oxide?

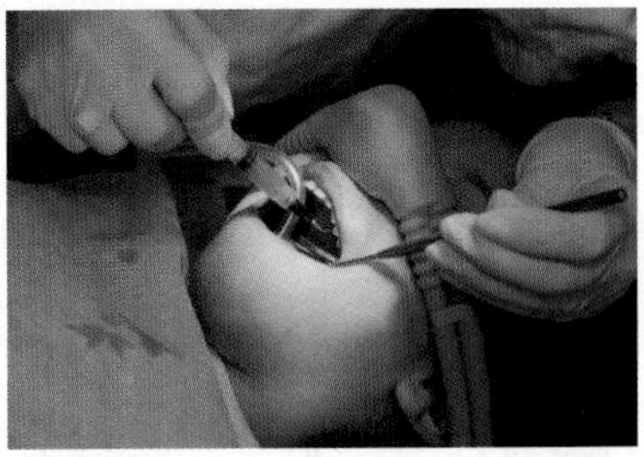

Dinitrogen oxide is used as an anesthetic in dentistry.

Chemistry Link to **Health**

Blood Gases

Our cells continuously use oxygen and produce carbon dioxide. Both gases move in and out of the lungs through the membranes of the alveoli, the tiny air sacs at the ends of the airways in the lungs. An exchange of gases occurs in which oxygen from the air diffuses into the lungs and into the blood, while carbon dioxide produced in the cells is carried to the lungs to be exhaled. In Table 8.5, partial pressures are given for the gases in air that we inhale (inspired air), air in the alveoli, and air that we exhale (expired air).

At sea level, oxygen normally has a partial pressure of 100 mmHg in the alveoli of the lungs. Because the partial pressure of oxygen in venous blood is 40 mmHg, oxygen diffuses from the alveoli into the bloodstream. The oxygen combines with hemoglobin, which carries it to the tissues of the body where the partial pressure of oxygen can be very low, less than 30 mmHg. Oxygen diffuses from the blood, where the partial pressure of O_2 is high, into the tissues where O_2 pressure is low.

TABLE 8.5 Partial Pressures of Gases During Breathing

	Partial Pressure (mmHg)		
Gas	**Inspired Air**	**Alveolar Air**	**Expired Air**
Nitrogen, N_2	594	573	569
Oxygen, O_2	160.	100.	116
Carbon dioxide, CO_2	0.3	40.	28
Water vapor, H_2O	5.7	47	47
Total	760.	760.	760.

As oxygen is used in the cells of the body during metabolic processes, carbon dioxide is produced, so the partial pressure of CO_2 may be as high as 50 mmHg or more. Carbon dioxide diffuses from the tissues into the bloodstream and is carried to the lungs. There it diffuses out of the blood, where CO_2 has a partial pressure of 46 mmHg, into the alveoli, where the CO_2 is at 40 mmHg and is exhaled. Table 8.6 gives the partial pressures of blood gases in the tissues and in oxygenated and deoxygenated blood.

TABLE 8.6 Partial Pressures of Oxygen and Carbon Dioxide in Blood and Tissues

	Partial Pressure (mmHg)		
Gas	Oxygenated Blood	Deoxygenated Blood	Tissues
O_2	100	40	30 or less
CO_2	40	46	50 or greater

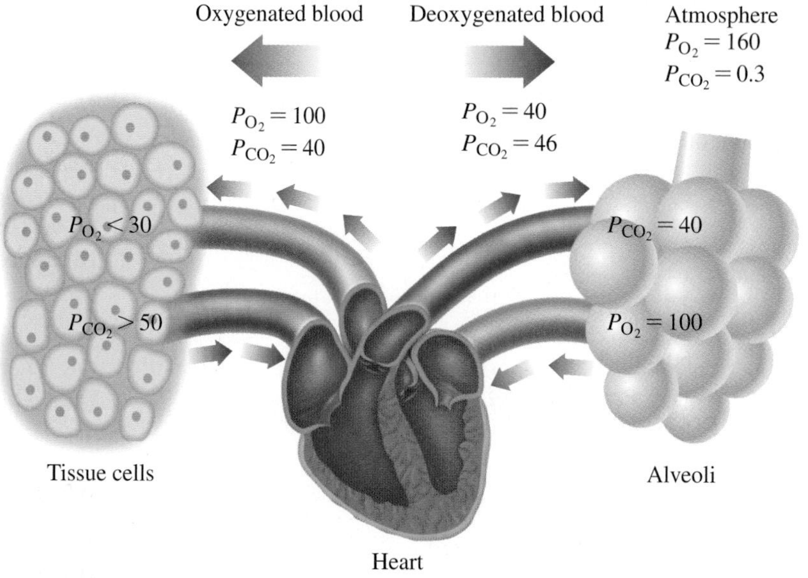

Chemistry Link to Health

Hyperbaric Chambers

A burn patient may undergo treatment for burns and infections in a hyperbaric chamber, a device in which pressures can be obtained that are two to three times greater than atmospheric pressure. A greater oxygen pressure increases the level of dissolved oxygen in the blood and tissues, where it fights bacterial infections. High levels of oxygen are toxic to many strains of bacteria. The hyperbaric chamber may also be used during surgery, to help counteract carbon monoxide (CO) poisoning, and to treat some cancers.

The blood is normally capable of dissolving up to 95% of the oxygen. Thus, if the partial pressure of the oxygen in the hyperbaric chamber is 2280 mmHg (3 atm), about 2170 mmHg of oxygen can dissolve in the blood where it saturates the tissues. In the treatment for carbon monoxide poisoning, oxygen at high pressure is used to displace the CO from the hemoglobin faster than breathing pure oxygen at 1 atm.

A patient undergoing treatment in a hyperbaric chamber must also undergo decompression (reduction of pressure) at a rate that slowly reduces the concentration of dissolved oxygen in the blood. If decompression is too rapid, the oxygen dissolved in the blood may form gas bubbles in the circulatory system.

Similarly, if a scuba diver does not decompress slowly, a condition called the "bends" may occur. While below the surface of the ocean, a diver uses a breathing mixture with a higher pressure. If there is nitrogen in the mixture, higher quantities of nitrogen gas will dissolve in the blood. If the diver ascends to the surface too quickly, the dissolved nitrogen forms gas bubbles that can block a blood vessel and cut off the flow of blood in the joints and tissues of the body and be quite painful. A diver suffering from the bends is placed immediately into a hyperbaric chamber where pressure is first increased and then slowly decreased. The dissolved nitrogen can then diffuse through the lungs until atmospheric pressure is reached.

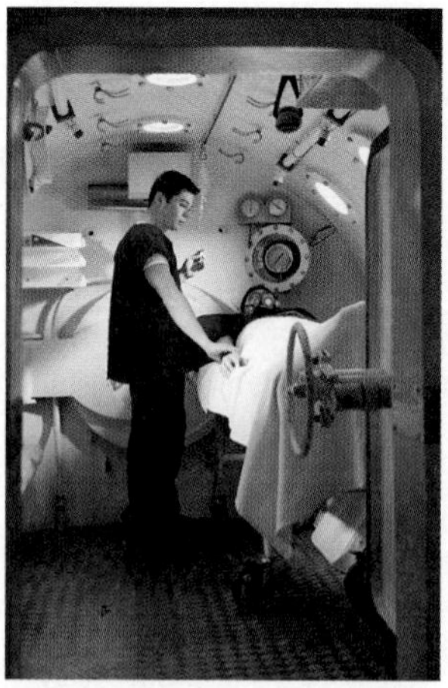

A hyperbaric chamber is used in the treatment of certain diseases.

QUESTIONS AND PROBLEMS

8.7 Partial Pressures (Dalton's Law)

LEARNING GOAL Use Dalton's law of partial pressures to calculate the total pressure of a mixture of gases.

8.45 A typical air sample in the lungs contains oxygen at 100 mmHg, nitrogen at 573 mmHg, carbon dioxide at 40 mmHg, and water vapor at 47 mmHg. Why are these pressures called partial pressures?

8.46 Suppose a mixture contains helium and oxygen gases. If the partial pressure of helium is the same as the partial pressure of oxygen, what do you know about the number of helium atoms compared to the number of oxygen molecules? Explain.

8.47 In a gas mixture, the partial pressures are nitrogen 425 torr, oxygen 115 torr, and helium 225 torr. What is the total pressure, in torr, exerted by the gas mixture?

8.48 In a gas mixture, the partial pressures are argon 415 mmHg, neon 75 mmHg, and nitrogen 125 mmHg. What is the total pressure, in mmHg, exerted by the gas mixture?

8.49 A gas mixture containing oxygen, nitrogen, and helium exerts a total pressure of 925 torr. If the partial pressures are oxygen 425 torr and helium 75 torr, what is the partial pressure, in torr, of the nitrogen in the mixture?

8.50 A gas mixture containing oxygen, nitrogen, and neon exerts a total pressure of 1.20 atm. If helium added to the mixture increases the pressure to 1.50 atm, what is the partial pressure, in atmospheres, of the helium?

8.51 In certain lung ailments such as emphysema, there is a decrease in the ability of oxygen to diffuse into the blood.

a. How would the partial pressure of oxygen in the blood change?

b. Why does a person with severe emphysema sometimes use a portable oxygen tank?

8.52 An accident to the head can affect the ability of a person to ventilate (breathe in and out).

a. What would happen to the partial pressures of oxygen and carbon dioxide in the blood if a person cannot properly ventilate?

b. When a person who cannot breathe properly is placed on a ventilator, an air mixture is delivered at pressures that are alternately above the air pressure in the person's lung, and then below. How will this move oxygen gas into the lungs, and carbon dioxide out?

CONCEPT MAP

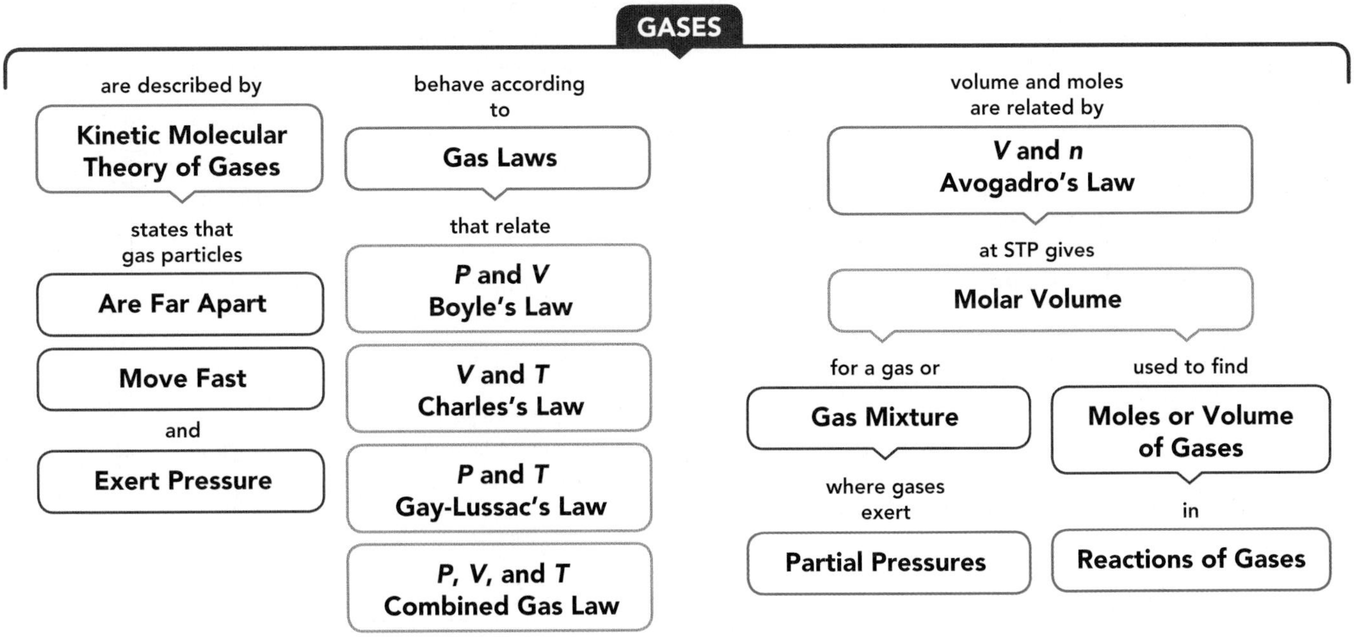

CHAPTER REVIEW

8.1 Properties of Gases

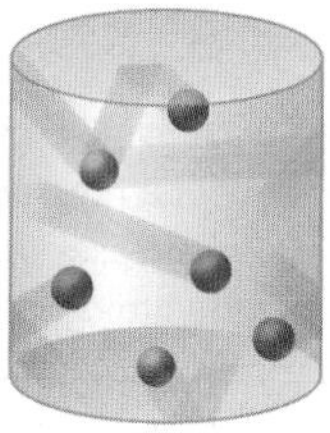

LEARNING GOAL Describe the kinetic molecular theory of gases and the units of measurement used for gases.

- In a gas, particles are so far apart and moving so fast that their attractions are negligible.
- A gas is described by the physical properties of pressure (P), volume (V), temperature (T) in kelvins (K), and the amount in moles (n).
- A gas exerts pressure, the force of the gas particles striking the walls of a container.
- Gas pressure is measured in units such as torr, mmHg, atm, and Pa.

8.2 Pressure and Volume (Boyle's Law)

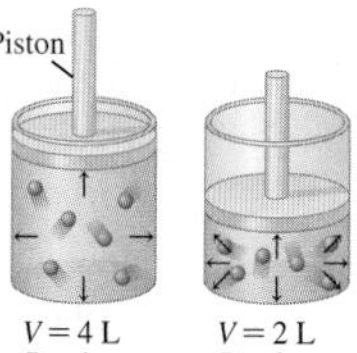

LEARNING GOAL Use the pressure–volume relationship (Boyle's law) to determine the final pressure or volume when the temperature and amount of gas are constant.

- The volume (V) of a gas changes inversely with the pressure (P) of the gas if there is no change in the temperature and the amount of gas.

$$P_1V_1 = P_2V_2$$

- The pressure increases if volume decreases; pressure decreases if volume increases.

8.3 Temperature and Volume (Charles's Law)

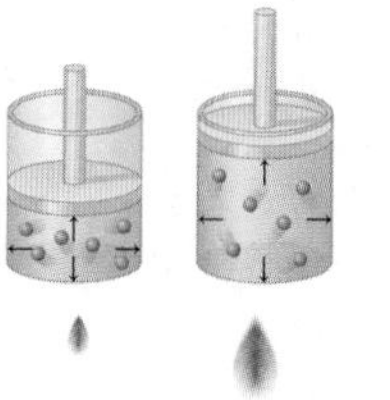

LEARNING GOAL Use the temperature–volume relationship (Charles's law) to determine the final temperature or volume when the pressure and amount of gas are constant.

- The volume (V) of a gas is directly related to its Kelvin temperature (T) when there is no change in the pressure and amount of the gas.

$$\frac{V_1}{T_1} = \frac{V_2}{T_2}$$

- As temperature of a gas increases, its volume increases; if its temperature decreases, volume decreases.

8.4 Temperature and Pressure (Gay-Lussac's Law)

LEARNING GOAL Use the temperature–pressure relationship (Gay-Lussac's law) to determine the final temperature or pressure when the volume and amount of gas are constant.

- The pressure (P) of a gas is directly related to its Kelvin temperature (T) when there is no change in the volume and amount of the gas.

$$\frac{P_1}{T_1} = \frac{P_2}{T_2}$$

- As temperature of a gas increases, its pressure increases; if its temperature decreases, pressure decreases.

8.5 The Combined Gas Law

LEARNING GOAL Use the combined gas law to calculate the final pressure, volume, or temperature of a gas when changes in two of these properties are given and the amount of gas is constant.

- The combined gas law is the relationship of pressure (P), volume (V), and temperature (T) for a constant amount (n) of gas.

$$\frac{P_1V_1}{T_1} = \frac{P_2V_2}{T_2}$$

- The combined gas law is used to determine the effect of changes in two of the variables on the third.

8.6 Volume and Moles (Avogadro's Law)

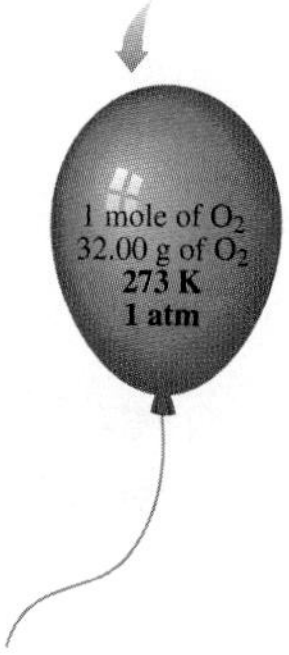

LEARNING GOAL Use Avogadro's law to calculate the amount or volume of a gas when the pressure and temperature are constant.

- The volume (V) of a gas is directly related to the number of moles (n) of the gas when the pressure and temperature of the gas do not change.

$$\frac{V_1}{n_1} = \frac{V_2}{n_2}$$

- If the moles of gas are increased, the volume must increase; if the moles of gas are decreased, the volume must decrease.
- At standard temperature (273 K) and standard pressure (1 atm), abbreviated STP, one mole of any gas has a volume of 22.4 L.

8.7 Partial Pressures (Dalton's Law)

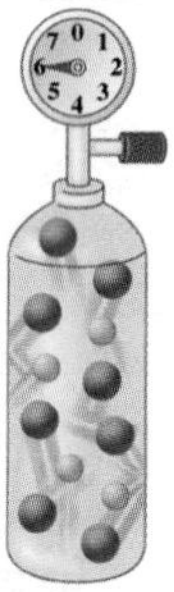

LEARNING GOAL Use Dalton's law of partial pressures to calculate the total pressure of a mixture of gases.

- In a mixture of two or more gases, the total pressure is the sum of the partial pressures of the individual gases.

$$P_{total} = P_1 + P_2 + P_3 + \ldots$$

- The partial pressure of a gas in a mixture is the pressure it would exert if it were the only gas in the container.

KEY TERMS

atmosphere (atm) A unit equal to pressure exerted by a column of mercury 760 mm high.

atmospheric pressure The pressure exerted by the atmosphere.

Avogadro's law A gas law stating that the volume of a gas is directly related to the number of moles of gas when pressure and temperature do not change.

Boyle's law A gas law stating that the pressure of a gas is inversely related to the volume when temperature and moles of the gas do not change.

Charles's law A gas law stating that the volume of a gas changes directly with a change in Kelvin temperature when pressure and moles of the gas do not change.

combined gas law A relationship that combines several gas laws relating pressure, volume, and temperature when the amount of gas does not change.

$$\frac{P_1V_1}{T_1} = \frac{P_2V_2}{T_2}$$

Dalton's law A gas law stating that the total pressure exerted by a mixture of gases in a container is the sum of the partial pressures that each gas would exert alone.

Gay-Lussac's law A gas law stating that the pressure of a gas changes directly with a change in the Kelvin temperature when the number of moles of a gas and its volume do not change.

kinetic molecular theory of gases A model used to explain the behavior of gases.

molar volume A volume of 22.4 L occupied by 1 mole of a gas at STP conditions of 0 °C (273 K) and 1 atm.

partial pressure The pressure exerted by a single gas in a gas mixture.

pressure The force exerted by gas particles that hit the walls of a container.

STP Standard conditions of exactly 0 °C (273 K) temperature and 1 atm pressure used for the comparison of gases.

CORE CHEMISTRY SKILLS

The chapter section containing each Core Chemistry Skill is shown in parentheses at the end of each heading.

Using the Gas Laws (8.2, 8.3, 8.4, 8.5, 8.6)

- Boyle's, Charles's, Gay-Lussac's, and Avogadro's laws show the relationships between two properties of a gas.

$P_1V_1 = P_2V_2$ Boyle's law

$\frac{V_1}{T_1} = \frac{V_2}{T_2}$ Charles's law

$\frac{P_1}{T_1} = \frac{P_2}{T_2}$ Gay-Lussac's law

$\frac{V_1}{n_1} = \frac{V_2}{n_2}$ Avogadro's law

- The combined gas law shows the relationship between *P*, *V*, and *T* for a gas.

$$\frac{P_1V_1}{T_1} = \frac{P_2V_2}{T_2}$$

- When two properties of a gas vary and the other two are constant, we list the initial and final conditions of each property in a table.

Example: A sample of helium gas (He) has a volume of 6.8 L and a pressure of 2.5 atm. What is the final volume, in liters, if it has a final pressure of 1.2 atm with no change in temperature and amount of gas?

Answer:

ANALYZE THE PROBLEM	Conditions 1	Conditions 2	Know	Predict
	$V_1 = 6.8$ L	$V_2 = ?$ L		*V* increases
	$P_1 = 2.5$ atm	$P_2 = 1.2$ atm	*P* decreases	

Factors that remain constant: *T* and *n*

Using Boyle's law, we can write the relationship for V_2, which we predict will increase.

$$V_2 = V_1 \times \frac{P_1}{P_2}$$

$$V_2 = 6.8\text{ L} \times \frac{2.5\ \cancel{\text{atm}}}{1.2\ \cancel{\text{atm}}} = 14\text{ L}$$

Calculating Partial Pressure (8.7)

- In a gas mixture, each gas exerts its partial pressure, which is the pressure it would exert if it were the only gas in the container.
- Dalton's law states that the total pressure of a gas mixture is the sum of the partial pressures of the gases in the mixture.

$$P_{total} = P_1 + P_2 + P_3 + \cdots$$

Example: A gas mixture with a total pressure of 1.18 atm contains helium gas at a partial pressure of 465 mmHg and nitrogen gas. What is the partial pressure, in atmospheres, of the nitrogen gas?

Answer: Initially, we convert the partial pressure of helium gas from mmHg to atm.

$$465\ \cancel{\text{mmHg}} \times \frac{1\text{ atm}}{760\ \cancel{\text{mmHg}}} = 0.612\text{ atm of He gas}$$

Using Dalton's law, we solve for the needed quantity, P_{N_2} in atm.

$$P_{total} = P_{N_2} + P_{He}$$

$$P_{N_2} = P_{total} - P_{He}$$

$$P_{N_2} = 1.18\text{ atm} - 0.612\text{ atm} = 0.57\text{ atm}$$

UNDERSTANDING THE CONCEPTS

The chapter sections to review are shown in parentheses at the end of each question.

8.53 Two flasks of equal volume and at the same temperature contain different gases. One flask contains 10.0 g of Ne, and the other flask contains 10.0 g of He. Is each of the following statements *true* or *false*? Explain. (8.1)

a. The flask that contains He has a higher pressure than the flask that contains Ne.

b. The densities of the gases are the same.

8.54 Two flasks of equal volume and at the same temperature contain different gases. One flask contains 5.0 g of O_2, and the other flask contains 5.0 g of H_2. Is each of the following statements *true* or *false*? Explain. (8.1)

a. Both flasks contain the same number of molecules.

b. The pressures in the flasks are the same.

8.55 At 100 °C, which of the following diagrams (**1**, **2**, or **3**) represents a gas sample that exerts the (8.1)

a. lowest pressure? **b.** highest pressure?

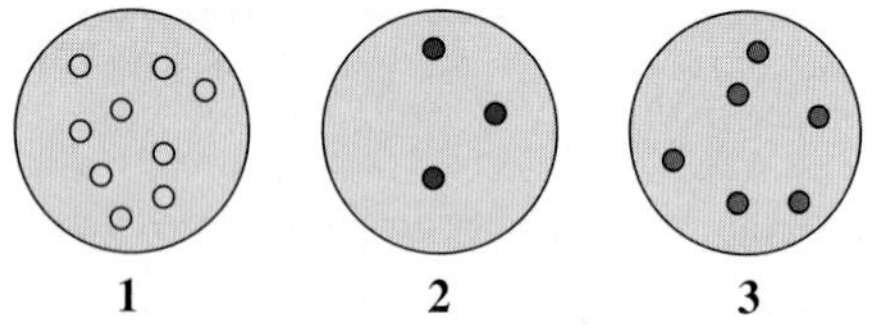

8.56 Indicate which diagram (**1**, **2**, or **3**) represents the volume of the gas sample in a flexible container when each of the following changes (**a** to **e**) takes place: (8.2, 8.3, 8.4)

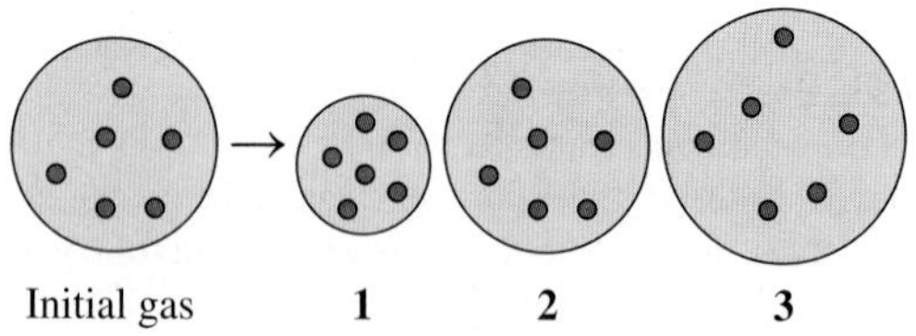

a. Temperature increases at constant pressure.

b. Temperature decreases at constant pressure.

c. Atmospheric pressure increases at constant temperature.

d. Atmospheric pressure decreases at constant temperature.

e. Doubling the atmospheric pressure and doubling the Kelvin temperature.

8.57 A balloon is filled with helium gas with a partial pressure of 1.00 atm and neon gas with a partial pressure of 0.50 atm. For each of the following changes of the initial balloon, select the diagram (**A**, **B**, or **C**) that shows the final volume of the balloon: (8.2, 8.3, 8.4, 8.6)

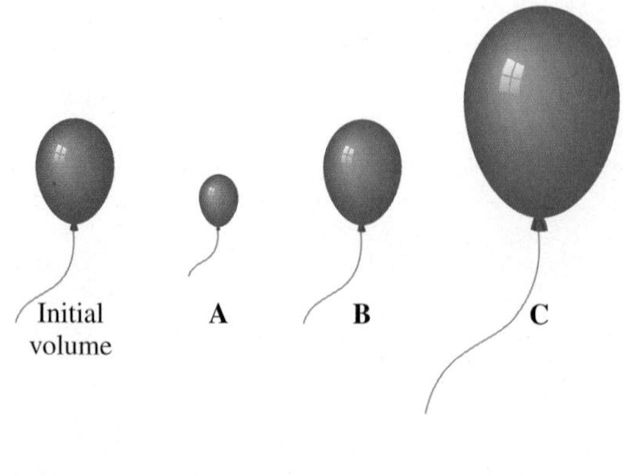

a. The balloon is put in a cold storage unit (*P* and *n* constant).

b. The balloon floats to a higher altitude where the pressure is less (*n* and *T* constant).

c. All of the neon gas is removed (*T* and *P* constant).

d. The Kelvin temperature doubles and half of the gas atoms leak out (*P* constant).

e. 2.0 moles of O_2 gas is added (*T* and *P* constant).

8.58 Indicate if pressure *increases*, *decreases*, or *stays the same* in each of the following: (8.2, 8.4, 8.6)

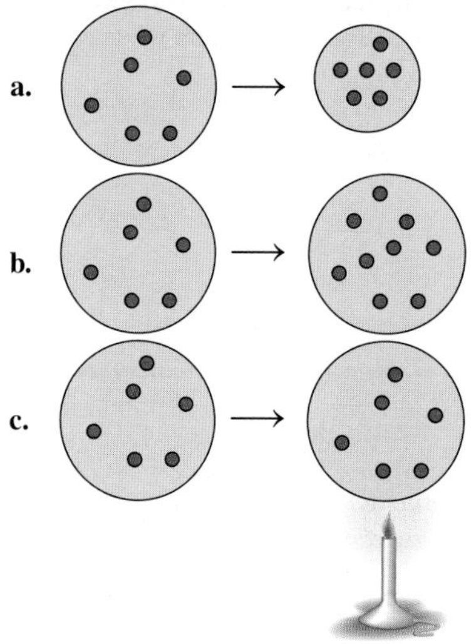

8.59 At a restaurant, a customer chokes on a piece of food. You put your arms around the person's waist and use your fists to push up on the person's abdomen, an action called the Heimlich maneuver. (8.2)

a. How would this action change the volume of the chest and lungs?

b. Why does it cause the person to expel the food item from the airway?

8.60 The partial pressure of O_2 is the double of H_2 in a container. If the total pressure of O_2 is 10 atm after adding 1 atm of N_2, what is the partial pressure of H_2?

ADDITIONAL QUESTIONS AND PROBLEMS

8.61 In 1783, Jacques Charles launched his first balloon filled with hydrogen gas, which he chose because it was lighter than air. If the balloon had a volume of 31 000 L, how many kilograms of hydrogen were needed to fill the balloon at STP? (8.6)

Jacques Charles used hydrogen to launch his balloon in 1783.

8.62 In problem 8.61, the balloon reached an altitude of 1000 m, where the pressure was 658 mmHg and the temperature was −8 °C. What was the volume, in liters, of the balloon at these conditions? (8.5)

8.63 A sample of hydrogen (H_2) gas at 127 °C has a pressure of 2.00 atm. At what final temperature, in degrees Celsius, will the pressure of the H_2 decrease to 0.25 atm, if V and n are constant? (8.4)

8.64 A fire extinguisher has a pressure of 10. atm at 25 °C. What is the final pressure, in atmospheres, when the fire extinguisher is used at a temperature of 75 °C, if V and n are constant? (8.4)

8.65 A weather balloon has a volume of 750 L when filled with helium at 8 °C at a pressure of 380 torr. What is the final volume, in liters, of the balloon, where the pressure is 0.20 atm and the temperature is −45 °C, when n remains constant? (8.5)

8.66 During laparoscopic surgery, carbon dioxide gas is used to expand the abdomen to help create a larger working space. If 4.80 L of CO_2 gas at 18 °C at 785 mmHg is used, what is the final volume, in liters, of the gas at 37 °C and a pressure of 745 mmHg, if the amount of CO_2 remains the same? (8.5)

8.67 A weather balloon is partially filled with helium to allow for expansion at high altitudes. At STP, a weather balloon is filled with enough helium to give a volume of 25.0 L. At an altitude of 30.0 km and −35 °C, it has expanded to 2460 L. The increase in volume causes it to burst and a small parachute returns the instruments to the Earth. (8.5, 8.6)

a. How many grams of helium are added to the balloon?

b. What is the final pressure, in millimeters of mercury, of the helium inside the balloon when it bursts?

8.68 A mixture of nitrogen (N_2) and helium has a volume of 250 mL at 30 °C and a total pressure of 745 mmHg. (8.5, 8.6, 8.7)

a. If the partial pressure of helium is 32 mmHg, what is the partial pressure of the nitrogen?

b. What is the final volume in liters, of the nitrogen at STP?

8.69 Two gas containers of equal volume X are partitioned with a valve. One contains H_2 at a pressure of 1 atm and the other contains O_2 at a pressure of 2 atm. What is the final pressure of the gases when the partition is removed?

8.70 What is the total pressure, in millimeters of mercury, of a gas mixture containing argon gas at 0.25 atm, helium gas at 350 mmHg, and nitrogen gas at 360 torr? (8.7)

CHALLENGE QUESTIONS

The following groups of questions are related to the topics in this chapter. However, they do not all follow the chapter order, and they require you to combine concepts and skills from several sections. These questions will help you increase your critical thinking skills and prepare for your next exam.

8.71 Your spaceship has docked at a space station above Mars. The temperature inside the space station is a carefully controlled 24 °C at a pressure of 745 mmHg. A balloon with a volume of 425 mL drifts into the airlock where the temperature is −95 °C and the pressure is 0.115 atm. What is the final volume, in milliliters, of the balloon if n remains constant and the balloon is very elastic? (8.5)

8.72 You are doing research on planet X. The temperature inside the space station is a carefully controlled 24 °C and the pressure is 755 mmHg. Suppose that a balloon, which has a volume of 850. mL inside the space station, is placed into the airlock, and floats out to planet X. If planet X has an atmospheric pressure of 0.150 atm and the volume of the balloon changes to 3.22 L, what is the temperature, in degrees Celsius, on planet X (n does not change)? (8.5)

8.73 A gas sample has a volume of 4250 mL at 15 °C and 745 mmHg. What is the final temperature, in degrees Celsius, after the sample is transferred to a different container with a volume of 2.50 L and a pressure of 1.20 atm when n is constant? (8.5)

8.74 In the fermentation of glucose (wine making), 780 mL of CO_2 gas was produced at 37 °C and 1.00 atm. What is the final volume, in liters, of the gas when measured at 22 °C and 675 mmHg, when n is constant? (8.5)

8.75 If the pressure of O_2 inside a chamber is 3.0 atm, what is the volume, in liters, of the chamber containing 2.4 kg of O_2 at 28 °C? (8.5, 8.6)

8.76 Hyperbaric therapy uses 100% oxygen at pressure to help heal wounds and infections, and to treat carbon monoxide poisoning. A hyperbaric chamber has a volume of 1510 L. How many kilograms of O_2 gas are needed to give an oxygen pressure of 2.04 atm at 25 °C? (8.5, 8.6)

ANSWERS

Answers to Study Checks

8.1 The mass, in grams, gives the amount of gas.

8.2 3.3 atm

8.3 73.5 mL

8.4 569 mL

8.5 16 °C

8.6 241 mmHg

8.7 7.50 L

8.8 7.0 g of N_2

8.9 370 torr

Answers to Selected Questions and Problems

8.1 a. At a higher temperature, gas particles have greater kinetic energy, which makes them move faster.
b. Because there are great distances between the particles of a gas, they can be pushed closer together and still remain a gas.
c. Gas particles are very far apart, which means that the mass of a gas in a certain volume is very small, resulting in a low density.

8.3 a. temperature **b.** volume
c. amount **d.** pressure

8.5 a, **d**, and **e** describe the pressure of a gas.

8.7 a. 3040 torr **b.** 3040 mmHg
c. 405.3kPa **d.** 58.8psi

8.9 As a diver ascends to the surface, external pressure decreases. If the air in the lungs were not exhaled, its volume would expand and severely damage the lungs. The pressure in the lungs must adjust to changes in the external pressure.

8.11 a. The pressure is greater in cylinder A. According to Boyle's law, a decrease in volume pushes the gas particles closer together, which will cause an increase in the pressure.
b.

Property	Conditions 1	Conditions 2	Know	Predict
Pressure (P)	650 mmHg	1.2 atm (910 mmHg)	P increases	
Volume (V)	220 mL	160 mL		V decreases

8.13 a. increases **b.** decreases **c.** increases

8.15 a. 1425 mmHg **b.** 2544.64 mmHg
c. 868.90 mmHg **d.** 9019 mmHg

8.17 a. 36.48 L **b.** 10 L
c. 37.5 L **d.** 65.14 L

8.19 24.5 L

8.21 a. inspiration **b.** expiration **c.** inspiration

8.23 a. C **b.** A **c.** B

8.25 a. 197.53 °C **b.** −108.32 °C
c. 393.58 °C **d.** 60.29 °C

8.27 a. 4165 mL **b.** 7572 mL
c. 3489 mL **d.** 8654 mL

8.29 a. 981,03 torr **b.** 2522.59 torr

8.31 a. −54.33 °C **b.** 503.23 °C

8.33 a. 3930.05 mmHg **b.** 1107.12 mmHg **c.** 289.64 mmHg

8.35 −33 °C

8.39 a. 4.00 L **b.** 26.7 L **c.** 14.6 L

8.41 a. 2.00 moles of O_2 **b.** 9.41 L of He

8.43 a. 4.48 L of O_2 **b.** 0.146 g of H_2

8.45 In a gas mixture, the pressure that each gas exerts as part of the total pressure is called the partial pressure of that gas. Because the air sample is a mixture of gases, the total pressure is the sum of the partial pressures of each gas in the sample.

8.47 765 torr

8.49 425 torr

8.51 a. The partial pressure of oxygen will be lower than normal.
b. Breathing a higher concentration of oxygen will help to increase the supply of oxygen in the lungs and blood and raise the partial pressure of oxygen in the blood.

8.53 a. True. The flask containing helium has more moles of helium and thus more helium atoms.
b. True. The mass and volume of each are the same, which means the mass/volume ratio or density is the same in both flasks.

8.55 a. 2 **b.** 1

8.57 a. A **b.** C **c.** A
d. B **e.** C

8.59 a. The volume of the chest and lungs is decreased.
b. The decrease in volume increases the pressure, which can dislodge the food in the trachea.

8.61 2.8 kg of H_2

8.63 −223 °C

8.65 1500 L of He

8.67 a. 4.46 g of helium **b.** 6.73 mmHg

8.71 2170 mL

8.73 −66 °C

8.75 620 L

9 Solutions

When Michelle's kidneys stopped functioning, she was placed on dialysis three times a week. As she enters the dialysis unit, her dialysis nurse, Amanda, asks Michelle how she is feeling. Michelle indicates that she feels tired today and has considerable swelling around her ankles. The dialysis nurse informs Michelle that her side effects are due to her body's inability to regulate the amount of water in her cells. She explains that the amount of water is regulated by the concentration of electrolytes in her body fluids and the rate at which waste products are removed from her body. Amanda explains that although water is essential for the many chemical reactions that occur in the body, the amount of water can become too high or too low, due to various diseases and conditions. Because Michelle's kidneys no longer perform dialysis, she cannot regulate the amount of electrolytes or waste in her body fluids. As a result, she has an electrolyte imbalance and a buildup of waste products, so her body is retaining water. Amanda then explains that the dialysis machine does the work of her kidneys to reduce the high levels of electrolytes and waste products.

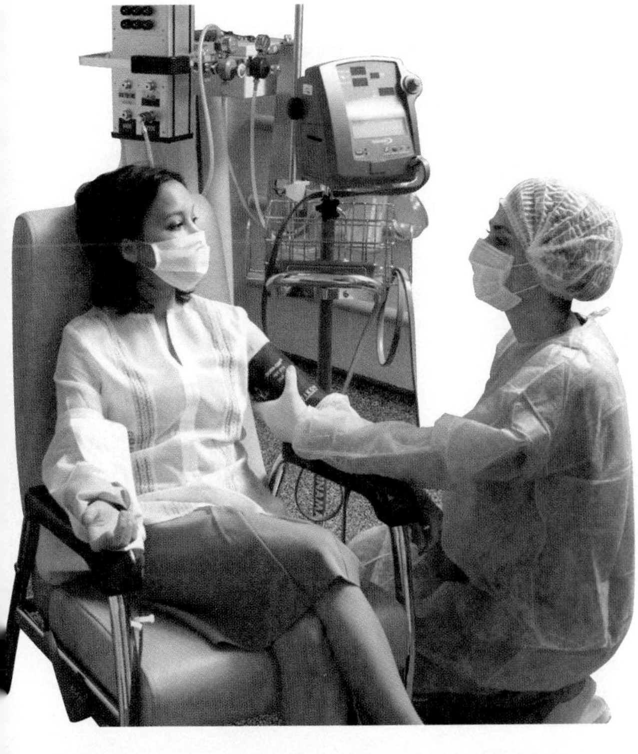

CAREER Dialysis Nurse

A dialysis nurse specializes in assisting patients with kidney disease undergoing dialysis. This requires monitoring the patient before, during, and after dialysis for any complications such as a drop in blood pressure or cramping. The dialysis nurse connects the patient to the dialysis unit via a dialysis catheter that is inserted into the neck or chest, which must be kept clean to prevent infection. A dialysis nurse must have considerable knowledge about how the dialysis machine functions to ensure that it is operating correctly at all times.

LOOKING AHEAD

Solutions are everywhere around us. Most of the gases, liquids, and solids we see are mixtures of at least one substance dissolved in another. There are different types of solutions. The air we breathe is a solution that is primarily oxygen and nitrogen gases. Carbon dioxide gas dissolved in water makes carbonated drinks. When we make solutions of coffee or tea, we use hot water to dissolve substances from coffee beans or tea leaves. The ocean is also a solution, consisting of many salts such as sodium chloride dissolved in water. In your medicine cabinet, the antiseptic tincture of iodine is a solution of iodine dissolved in ethanol.

Because the individual components in any mixture are not bonded to each other, the composition of those components can vary. Also, some of the physical properties of the individual components are still noticeable. For example, in ocean water, we detect the dissolved sodium chloride by the salty taste. The flavor we associate with coffee is due to the dissolved components. In a solution, the components cannot be distinguished one from the other. Syrup is a solution of sugar and water: The sugar cannot be distinguished from the water.

Our body fluids contain water and dissolved substances such as glucose and urea and electrolytes such as K^+, Na^+, Cl^-, Mg^{2+}, HCO_3^-, and HPO_4^{2-}. Proper amounts of each of these dissolved substances and water must be maintained in the body fluids. Small changes in electrolyte levels can seriously disrupt cellular processes and endanger our health. Solutions can be described by their concentration, which is the amount of solute in a specific amount of that solution. These relationships, which include mass % (m/m), volume % (v/v), mass/volume % (m/v), and molarity (M), can be used to convert between the amount of a solute and the quantity of its solution. Solutions are also diluted by adding a specific amount of solvent to a solution.

In the processes of osmosis and dialysis, water, essential nutrients, and waste products enter and leave the cells of the body. The kidneys utilize osmosis and dialysis to regulate the amount of water and electrolytes that are excreted.

CHAPTER READINESS*

KEY MATH SKILLS

- Calculating a Percentage (1.4C)
- Solving Equations (1.4D)
- Interpreting a Graph (1.4E)

CORE CHEMISTRY SKILLS

- Writing Conversion Factors from Equalities (2.5)
- Using Conversion Factors (2.6)
- Identifying Attractive Forces (6.8)
- Using Mole–Mole Factors (7.6)

*These Key Math Skills and Core Chemistry Skills from previous chapters are listed here for your review as you proceed to the new material in this chapter.

LEARNING GOAL

Identify the solute and solvent in a solution; describe the formation of a solution.

9.1 Solutions

A **solution** is a homogeneous mixture in which one substance, called the **solute**, is uniformly dispersed in another substance called the **solvent**. Because the solute and the solvent do not react with each other, they can be mixed in varying proportions. A solution of a little salt dissolved in water tastes slightly salty. When a large amount of salt is dissolved in water, the solution tastes very salty. Usually, the solute (in this case, salt) is the substance present in the lesser amount, whereas the solvent (in this case, water) is present in the greater amount. For example, in a solution composed of 5.0 g of salt and 50. g of water, salt is the solute and water is the solvent. In a solution, the particles of the solute are evenly dispersed among the molecules within the solvent (see Figure 9.1).

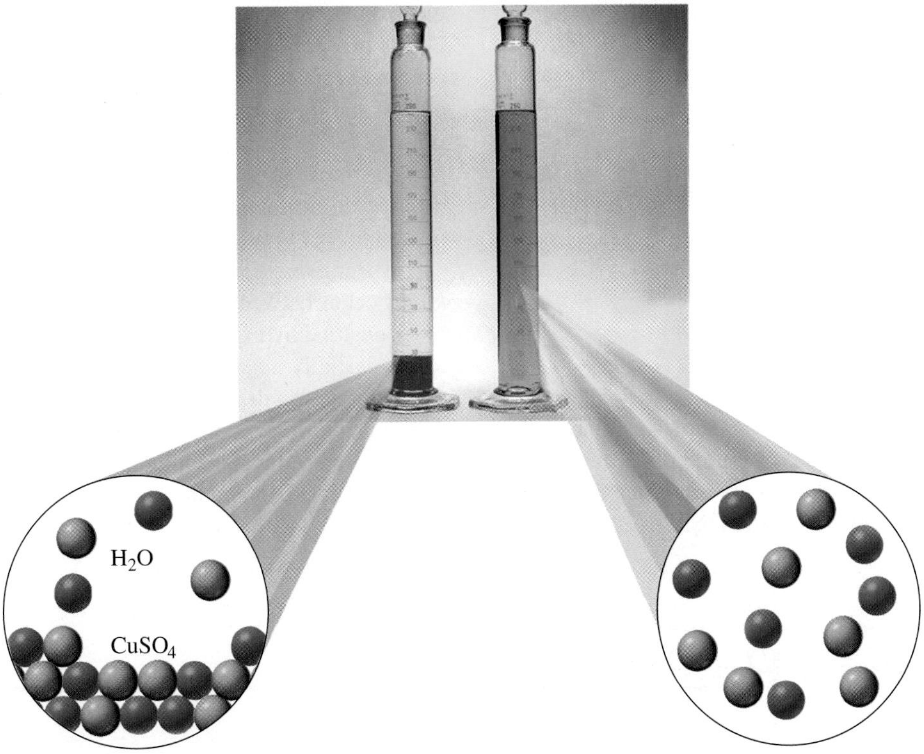

FIGURE 9.1 ▶ A solution of copper(II) sulfate ($CuSO_4$) forms as particles of solute dissolve, move away from the crystal, and become evenly dispersed among the solvent (water) molecules.

Q What does the uniform blue color in the graduated cylinder on the right indicate?

A solution consists of at least one solute dispersed in a solvent.

Types of Solutes and Solvents

Solutes and solvents may be solids, liquids, or gases. The solution that forms has the same physical state as the solvent. When sugar crystals are dissolved in water, the resulting sugar solution is liquid. Sugar is the solute, and water is the solvent. Soda water and soft drinks are prepared by dissolving carbon dioxide gas in water. The carbon dioxide gas is the solute, and water is the solvent. Table 9.1 lists some solutes and solvents and their solutions.

TABLE 9.1 Some Examples of Solutions

Type	Example	Primary Solute	Solvent
Gas Solutions			
Gas in a gas	Air	Oxygen (gas)	Nitrogen (gas)
Liquid Solutions			
Gas in a liquid	Soda water	Carbon dioxide (gas)	Water (liquid)
	Household ammonia	Ammonia (gas)	Water (liquid)
Liquid in a liquid	Vinegar	Acetic acid (liquid)	Water (liquid)
Solid in a liquid	Seawater	Sodium chloride (solid)	Water (liquid)
	Tincture of iodine	Iodine (solid)	Ethanol (liquid)
Solid Solutions			
Solid in a solid	Brass	Zinc (solid)	Copper (solid)
	Steel	Carbon (solid)	Iron (solid)

Water as a Solvent

Water is one of the most common solvents in nature. In the H_2O molecule, an oxygen atom shares electrons with two hydrogen atoms. Because oxygen is much more

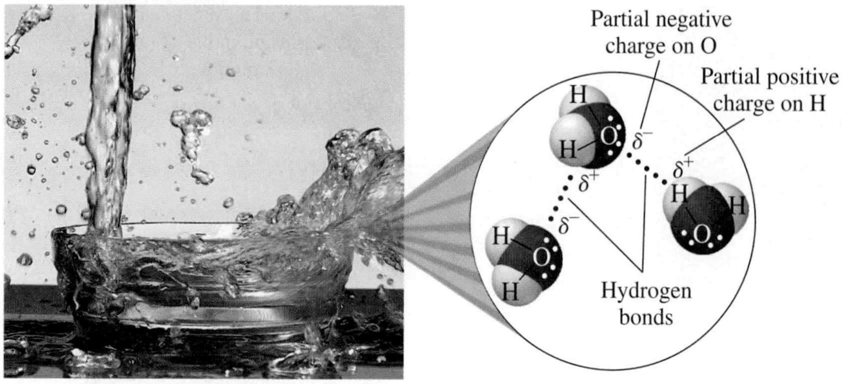

In water, hydrogen bonds form between an oxygen atom in one water molecule and the hydrogen atom in another.

electronegative than hydrogen, the O—H bonds are polar. In each polar bond, the oxygen atom has a partial negative (δ^-) charge and the hydrogen atom has a partial positive (δ^+) charge. Because the shape of a water molecule is bent, not linear, its dipoles do not cancel out. Thus, water is polar, and is a *polar solvent*.

Attractive forces known as *hydrogen bonds* occur between molecules where partially positive hydrogen atoms are attracted to the partially negative atoms N, O, or F. As seen in the diagram on the left, the hydrogen bonds are shown as a series of dots. Although hydrogen bonds are much weaker than covalent or ionic bonds, there are many of them linking water molecules together. Hydrogen bonds are important in the properties of biological compounds such as proteins, carbohydrates, and DNA.

Chemistry Link to **Health**

Water in the Body

The average adult is about 60% water by mass, and the average infant about 75%. About 60% of the body's water is contained within the cells as intracellular fluids; the other 40% makes up extracellular fluids, which include the interstitial fluid in tissue and the plasma in the blood. These external fluids carry nutrients and waste materials between the cells and the circulatory system.

Typical water gain and loss during 24 hours

Water gain	
Liquid	1000 mL
Food	1200 mL
Metabolism	300 mL
Total	2500 mL

Water loss	
Urine	1500 mL
Perspiration	300 mL
Breath	600 mL
Feces	100 mL
Total	2500 mL

Every day you lose between 1500 and 3000 mL of water from the kidneys as urine, from the skin as perspiration, from the lungs as you exhale, and from the gastrointestinal tract. Serious dehydration can occur in an adult if there is a 10% net loss in total body fluid; a 20% loss of fluid can be fatal. An infant suffers severe dehydration with only a 5 to 10% loss in body fluid.

Water loss is continually replaced by the liquids and foods in the diet and from metabolic processes that produce water in the cells of the body. Table 9.2 lists the percentage by mass of water contained in some foods.

The water lost from the body is replaced by the intake of fluids.

TABLE 9.2 Percentage of Water in Some Foods

Food	Water (% by mass)	Food	Water (% by mass)
Vegetables		**Meats/Fish**	
Carrot	88	Chicken, cooked	71
Celery	94	Hamburger, broiled	60
Cucumber	96	Salmon	71
Tomato	94		
Fruits		**Milk Products**	
Apple	85	Cottage cheese	78
Cantaloupe	91	Milk, whole	87
Orange	86	Yogurt	88
Strawberry	90		
Watermelon	93		

Formation of Solutions

The interactions between solute and solvent will determine whether a solution will form. Initially, energy is needed to separate the particles in the solute and the solvent particles. Then energy is released as solute particles move between the solvent particles to form a solution. However, there must be attractions between the solute and the solvent particles to provide the energy for the initial separation. These attractions occur when the solute and the solvent have similar polarities. The expression "like dissolves like" is a way of saying that the polarities of a solute and a solvent must be similar in order for a solution to form (see Figure 9.2). In the absence of attractions between a solute and a solvent, there is insufficient energy to form a solution (see Table 9.3).

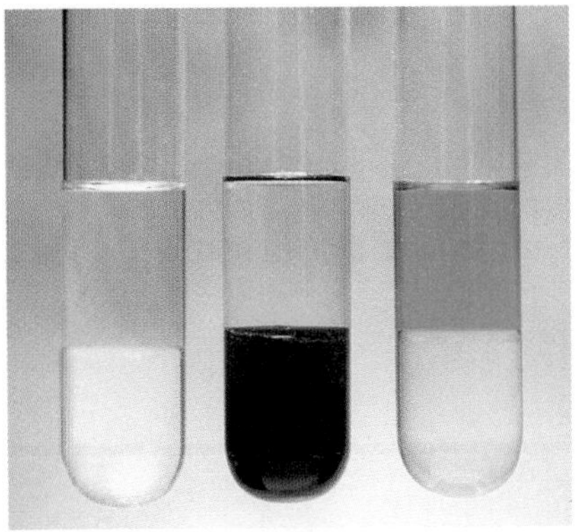
(a) (b) (c)

FIGURE 9.2 ▶ Like dissolves like. In each test tube, the lower layer is CH_2Cl_2 (more dense), and the upper layer is water (less dense). **(a)** CH_2Cl_2 is nonpolar and water is polar; the two layers do not mix. **(b)** The nonpolar solute I_2 (purple) is soluble in the nonpolar solvent CH_2Cl_2. **(c)** The ionic solute $Ni(NO_3)_2$ (green) is soluble in the polar solvent water.

Q In which layer would polar molecules of sucrose ($C_{12}H_{22}O_{11}$) be soluble?

TABLE 9.3 Possible Combinations of Solutes and Solvents

Solutions Will Form		Solutions Will Not Form	
Solute	**Solvent**	**Solute**	**Solvent**
Polar	Polar	Polar	Nonpolar
Nonpolar	Nonpolar	Nonpolar	Polar

Solutions with Ionic and Polar Solutes

In ionic solutes such as sodium chloride, NaCl, there are strong ionic bonds between positively charged Na^+ ions and negatively charged Cl^- ions. In water, a polar solvent, the hydrogen bonds provide strong solvent–solvent attractions. When NaCl crystals are placed in water, partially negative oxygen atoms in water molecules attract positive Na^+ ions, and the partially positive hydrogen atoms in other water molecules attract negative Cl^- ions (see Figure 9.3). As soon as the Na^+ ions and the Cl^- ions form a solution, they undergo **hydration** as water molecules surround each ion. Hydration of the ions diminishes their attraction to other ions and keeps them in solution. The strong solute–solvent attractions between Na^+ and Cl^- ions and the polar water molecules provide energy needed to form the solution.

In the equation for the formation of the NaCl solution, the solid and aqueous NaCl are shown with the formula H_2O over the arrow, which indicates that water is needed for the dissociation process but is not a reactant.

$$NaCl(s) \xrightarrow[\text{Dissociation}]{H_2O} Na^+(aq) + Cl^-(aq)$$

In another example, we find that a polar molecular compound such as methanol, $CH_3—OH$, is soluble in water because methanol has a polar —OH group that forms hydrogen bonds with water (see Figure 9.4). Polar solutes require polar solvents for a solution to form.

Explore Your World

Like Dissolves Like

Obtain small samples of vegetable oil, water, vinegar, salt, and sugar. Then combine and mix the substances as listed in **a** through **e**.

a. oil and water
b. water and vinegar
c. salt and water
d. sugar and water
e. salt and oil

Questions

1. Which of the mixtures formed a solution?
2. Which of the mixtures did not form a solution?
3. Why do some mixtures form solutions, but others do not?

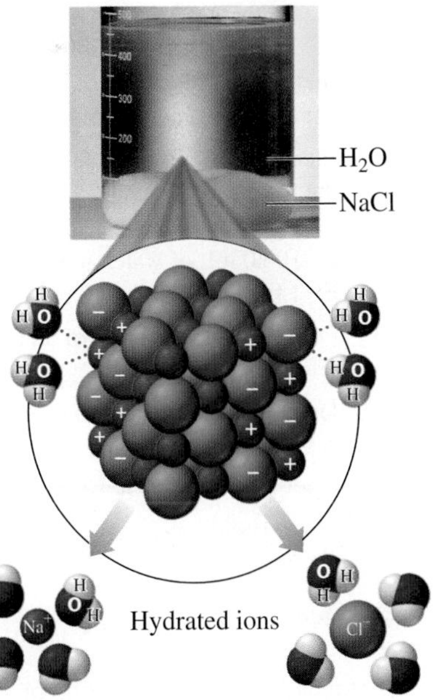

FIGURE 9.3 ▶ Ions on the surface of a crystal of NaCl dissolve in water as they are attracted to the polar water molecules that pull the ions into solution and surround them.

Q What helps keep Na^+ and Cl^- ions in solution?

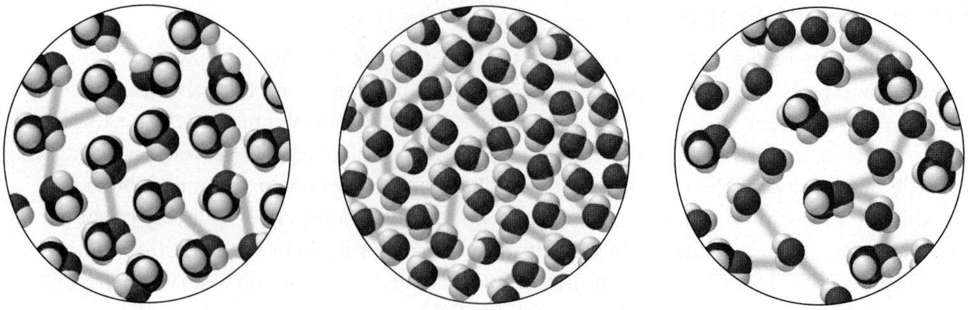

FIGURE 9.4 ▶ Molecules of methanol, $CH_3—OH$, form hydrogen bonds with polar water molecules to form a methanol–water solution.

Q Why are there attractions between the solute methanol and the solvent water?

Solutions with Nonpolar Solutes

Compounds containing nonpolar molecules, such as iodine (I_2), oil, or grease, do not dissolve in water because there are essentially no attractions between the particles of a nonpolar solute and the polar solvent. Nonpolar solutes require nonpolar solvents for a solution to form.

QUESTIONS AND PROBLEMS

9.1 Solutions

LEARNING GOAL Identify the solute and solvent in a solution; describe the formation of a solution.

9.1 Identify the solute and the solvent in each solution composed of the following:
- **a.** 20.0 g of KCl and 200.0 g of H_2O
- **b.** 60.0 mL of methanol, CH_3OH, and 20.0 mL of H_2O
- **c.** 1.5g of C and 50g of Fe

9.2 Identify the solute and the solvent in each solution composed of the following:
- **a.** 10.0 mL of acetic acid and 200. mL of water
- **b.** 100.0 mL of water and 5.0 g of sugar
- **c.** 1.0 g of Br_2 and 50.0 mL of methylene chloride(l)

9.3 Describe the formation of an aqueous KI solution, when solid KI dissolves in water.

9.4 Describe the formation of an aqueous LiBr solution, when solid LiBr dissolves in water.

9.5 Water is a polar solvent and benzene is a nonpolar solvent. In which solvent is each of the following more likely to be soluble?
- **a.** Hexane, nonpolar
- **b.** gasoline, nonpolar
- **c.** KCl, ionic
- **d.** $Ni(NO_3)_2$, ionic

9.6 Water is a polar solvent and hexane is a nonpolar solvent. In which solvent is each of the following more likely to be soluble?
- **a.** vegetable oil, nonpolar
- **b.** benzene, nonpolar
- **c.** LiCl, ionic
- **d.** Na_2SO_4, ionic

LEARNING GOAL

Identify solutes as electrolytes or nonelectrolytes.

9.2 Electrolytes and Nonelectrolytes

Solutes can be classified by their ability to conduct an electrical current. When **electrolytes** dissolve in water, the process of *dissociation* separates them into ions forming solutions that conduct electricity. When **nonelectrolytes** dissolve in water, they do not separate into ions and their solutions do not conduct electricity.

To test solutions for the presence of ions, we can use an apparatus that consists of a battery and a pair of electrodes connected by wires to a light bulb. The light bulb glows when electricity can flow, which can only happen when electrolytes provide ions that move between the electrodes to complete the circuit.

Types of Electrolytes

Electrolytes can be further classified as *strong electrolytes* or *weak electrolytes*. For a **strong electrolyte**, such as sodium chloride (NaCl), there is 100% dissociation of the solute into ions. When the electrodes from the light bulb apparatus are placed in the NaCl solution, the light bulb glows very brightly.

In an equation for dissociation of a compound in water, the charges must balance. For example, magnesium nitrate dissociates to give one magnesium ion for every two nitrate ions. However, only the ionic bonds between Mg^{2+} and NO_3^- are broken, not the covalent bonds within the polyatomic ion. The equation for the dissociation of $Mg(NO_3)_2$ is written as follows:

$$Mg(NO_3)_2(s) \xrightarrow[\text{Dissociation}]{H_2O} Mg^{2+}(aq) + 2NO_3^-(aq)$$

A **weak electrolyte** is a compound that dissolves in water mostly as molecules. Only a few of the dissolved solute molecules undergo *ionization*, producing a small number of ions in solution. Thus solutions of weak electrolytes do not conduct electrical current as well as solutions of strong electrolytes. When the electrodes are placed in a solution of a weak electrolyte, the glow of the light bulb is very dim. In an aqueous solution of the weak electrolyte HF, a few HF molecules ionize to produce H^+ and F^- ions. As more H^+ and F^- ions form, some recombine to give HF molecules. These forward and reverse reactions of molecules to ions and back again are indicated by two arrows between reactant and products that point in opposite directions:

$$HF(aq) \underset{\text{Recombination}}{\overset{\text{Ionization}}{\rightleftarrows}} H^+(aq) + F^-(aq)$$

A nonelectrolyte such as sucrose (sugar) dissolves in water only as molecules, which do not ionize. When electrodes of the light bulb apparatus are placed in a solution of a nonelectrolyte, the light bulb does not glow, because the solution does not contain ions and cannot conduct electricity.

$$C_{12}H_{22}O_{11}(s) \xrightarrow{H_2O} C_{12}H_{22}O_{11}(aq)$$

Table 9.4 summarizes the classification of solutes in aqueous solutions.

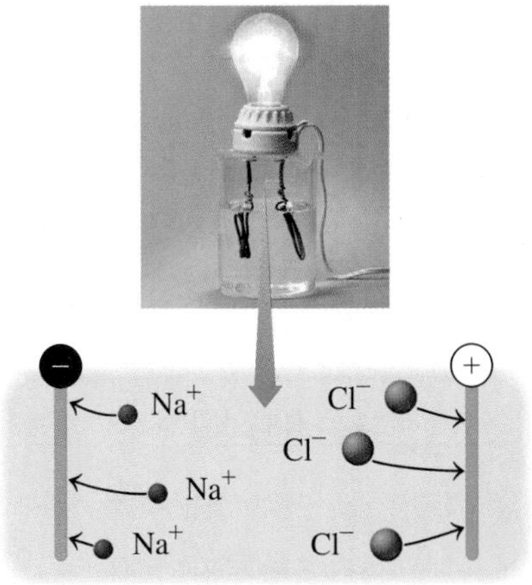

Strong electrolyte

(a)
A strong electrolyte completely dissociates into ions in an aqueous solution.

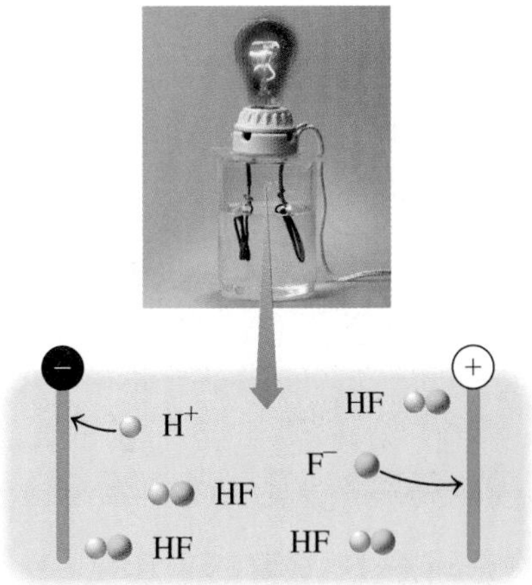

Weak electrolyte

(b)
A weak electrolyte forms mostly molecules and a few ions in an aqueous solution.

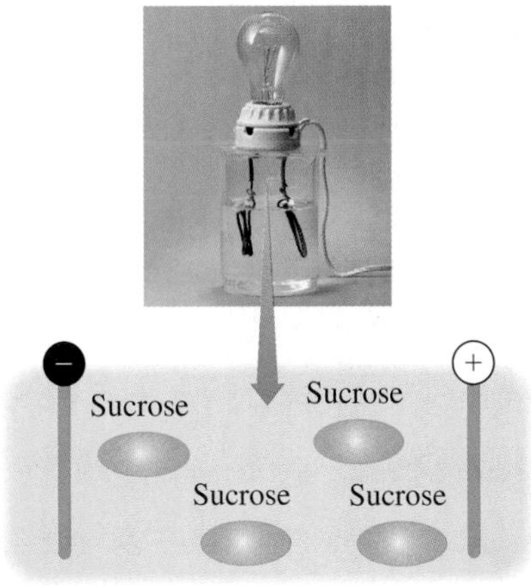

Nonelectrolyte

(c)
A nonelectrolyte dissolves as molecules in an aqueous solution.

SAMPLE PROBLEM 9.1 Solutions of Electrolytes and Nonelectrolytes

Indicate whether solutions of each of the following contain only ions, only molecules, or mostly molecules and a few ions. Write the equation for the formation of a solution for each of the following:

a. $Na_2SO_4(s)$, a strong electrolyte
b. $CH_3OH(l)$, a nonelectrolyte

SOLUTION

a. An aqueous solution of $Na_2SO_4(s)$ contains only the ions Na^+ and SO_4^{2-}.

$$Na_2SO_4(s) \xrightarrow{H_2O} 2Na^+(aq) + SO_4^{2-}(aq)$$

b. A nonelectrolyte such as $CH_3OH(l)$ produces only molecules when it dissolves in water.

$$CH_3OH(l) \xrightarrow{H_2O} CH_3OH(aq)$$

STUDY CHECK 9.1

Boric acid, $H_3BO_3(s)$, is a weak electrolyte. Would you expect a boric acid solution to contain only ions, only molecules, or mostly molecules and a few ions?

TABLE 9.4 Classification of Solutes in Aqueous Solutions

Type of Solute	In Solution	Type(s) of Particles in Solution	Conducts Electricity?	Examples
Strong electrolyte	Dissociates completely	Ions only	Yes	Ionic compounds such as NaCl, KBr, $MgCl_2$, $NaNO_3$; bases such as NaOH, KOH; acids such as HCl, HBr, HI, HNO_3, $HClO_4$, H_2SO_4
Weak electrolyte	Ionizes partially	Mostly molecules and a few ions	Weakly	HF, H_2O, NH_3, $HC_2H_3O_2$ (acetic acid)
Nonelectrolyte	No ionization	Molecules only	No	Carbon compounds such as CH_3OH (methanol), C_2H_5OH (ethanol), $C_{12}H_{22}O_{11}$ (sucrose), CH_4N_2O (urea)

Equivalents

Body fluids contain a mixture of electrolytes, such as Na^+, Cl^-, K^+, and Ca^{2+}. We measure each individual ion in terms of an **equivalent (Eq)**, which is the amount of that ion equal to 1 mole of positive or negative electrical charge. For example, 1 mole of Na^+ ions and 1 mole of Cl^- ions are each 1 equivalent because they each contain 1 mole of charge. For an ion with a charge of 2+ or 2−, there are 2 equivalents for each mole. Some examples of ions and equivalents are shown in Table 9.5.

TABLE 9.5 Equivalents of Electrolytes

Ion	Ionic Charge	Number of Equivalents in 1 Mole
Na^+	1+	1 Eq
Ca^{2+}	2+	2 Eq
Fe^{3+}	3+	3 Eq
Cl^-	1−	1 Eq
SO_4^{2-}	2−	2 Eq

In any solution, the charge of the positive ions is always balanced by the charge of the negative ions. The concentrations of electrolytes in intravenous fluids are expressed in milliequivalents per liter (mEq/L); 1 Eq = 1000 mEq. For example, a solution containing 25 mEq/L of Na^+ and 4 mEq/L of K^+ has a total positive charge of 29 mEq/L. If Cl^- is the only anion, its concentration must be 29 mEq/L.

SAMPLE PROBLEM 9.2 Electrolyte Concentration

The laboratory tests for a patient indicate a blood calcium level of 8.8 mEq/L.

a. How many moles of calcium ion are in 0.50 L of blood?
b. If chloride ion is the only other ion present, what is its concentration in mEq/L?

SOLUTION

a. Using the volume and the electrolyte concentration in mEq/L, we can find the number of equivalents in 0.50 L of blood.

$$0.50\,\cancel{L} \times \frac{8.8\,\cancel{mEq\,Ca^{2+}}}{1\,\cancel{L}} \times \frac{1\,Eq\,Ca^{2+}}{1000\,\cancel{mEq\,Ca^{2+}}} = 0.0044\,Eq\text{ of }Ca^{2+}$$

We can then convert equivalents to moles (for Ca^{2+} there are 2 Eq per mole).

$$0.0044\,\cancel{Eq\,Ca^{2+}} \times \frac{1\,mole\,Ca^{2+}}{2\,\cancel{Eq\,Ca^{2+}}} = 0.0022\,mole\text{ of }Ca^{2+}$$

b. If the concentration of Ca^{2+} is 8.8 mEq/L, then the concentration of Cl^- must be 8.8 mEq/L to balance the charge.

STUDY CHECK 9.2

A Ringer's solution for intravenous fluid replacement contains 155 mEq of Cl^- per liter of solution. If a patient receives 1250 mL of Ringer's solution, how many moles of chloride ion were given?

Chemistry Link to **Health**

Electrolytes in Body Fluids

Electrolytes in the body play an important role in maintaining the proper function of the cells and organs in the body. Typically, the electrolytes sodium, potassium, chloride, and bicarbonate are measured in a blood test. Sodium ions regulate the water content in the body and are important in carrying electrical impulses through the nervous system. Potassium ions are also involved in the transmission of electrical impulses and play a role in the maintenance of a regular heartbeat. Chloride ions balance the charges of the positive ions and also control the balance of fluids in the body. Bicarbonate is important in maintaining the proper pH of the blood. Sometimes when vomiting, diarrhea, or sweating is excessive, the concentrations of certain electrolytes may decrease. Then fluids such as Pedialyte may be given to return electrolyte levels to normal.

The concentrations of electrolytes present in body fluids and in intravenous fluids given to a patient are often expressed in milliequivalents per liter (mEq/L) of solution. For example, one liter of Pedialyte contains the following electrolytes: Na^+ 45 mEq, Cl^- 35 mEq, and citrate^{3-} 30 mEq.

Table 9.6 gives the concentrations of some typical electrolytes in blood plasma. There is a charge balance because the total number of positive charges is equal to the total number of negative charges. The use of a specific intravenous solution depends on the nutritional, electrolyte, and fluid needs of the individual patient. Examples of various types of solutions are given in Table 9.7.

TABLE 9.6 Some Typical Concentrations of Electrolytes in Blood Plasma

Electrolyte	Concentration (mEq/L)
Cations	
Na^+	138
K^+	5
Mg^{2+}	3
Ca^{2+}	4
Total	150
Anions	
Cl^-	110
HCO_3^-	30
HPO_4^{2-}	4
Proteins	6
Total	150

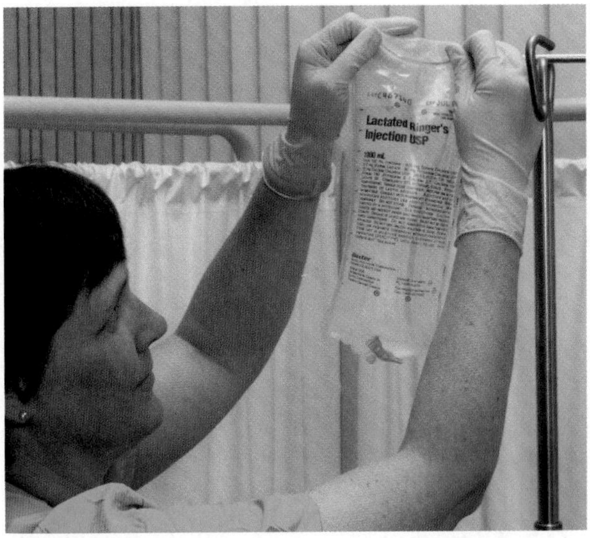

An intravenous solution is used to replace electrolytes in the body.

TABLE 9.7 Electrolyte Concentrations in Intravenous Replacement Solutions

Solution	Electrolytes (mEq/L)	Use
Sodium chloride (0.9%)	Na^+ 154, Cl^- 154	Replacement of fluid loss
Potassium chloride with 5.0% dextrose	K^+ 40, Cl^- 40	Treatment of malnutrition (low potassium levels)
Ringer's solution	Na^+ 147, K^+ 4, Ca^{2+} 4, Cl^- 155	Replacement of fluids and electrolytes lost through dehydration
Maintenance solution with 5.0% dextrose	Na^+ 40, K^+ 35, Cl^- 40, lactate$^-$ 20, HPO_4^{2-} 15	Maintenance of fluid and electrolyte levels
Replacement solution (extracellular)	Na^+ 140, K^+ 10, Ca^{2+} 5, Mg^{2+} 3, Cl^- 103, acetate$^-$ 47, citrate^{3-} 8	Replacement of electrolytes in extracellular fluids

QUESTIONS AND PROBLEMS

9.2 Electrolytes and Nonelectrolytes

LEARNING GOAL Identify solutes as electrolytes or nonelectrolytes.

9.7 KF is a strong electrolyte, and HF is a weak electrolyte. How is the solution of KF different from that of HF?

9.8 NaOH is a strong electrolyte, and CH_3OH is a nonelectrolyte. How is the solution of NaOH different from that of CH_3OH?

9.9 Write a balanced equation for the dissociation of each of the following strong electrolytes in water:
a. NaCl **b.** $MgCl_2$
c. HF **d.** CH_3COOH

9.10 Write a balanced equation for the dissociation of each of the following strong electrolytes in water:
a. LiBr **b.** $NaNO_3$
c. $CuCl_2$ **d.** K_2CO_3

9.11 The nature of the compound determines the form in which it exists in water. Rate the following solutes in terms of their degree of ionization in water:
a. vinegar, a weak electrolyte
b. K_2CO_3, a strong electrolyte
c. cellulose, a nonelectrolyte

9.12 Indicate whether aqueous solutions of each of the following solutes contain only ions, only molecules, or mostly molecules and a few ions:
a. NH_4Cl, a strong electrolyte
b. ethanol, C_2H_5OH, a nonelectrolyte
c. HCN, hydrocyanic acid, a weak electrolyte

9.13 Classify the solute represented in each of the following equations as a strong, weak, or nonelectrolyte:
a. $K_2SO_4(s) \xrightarrow{H_2O} 2K^+(aq) + SO_4^{2-}(aq)$
b. $NH_3(g) + H_2O(l) \rightleftarrows NH_4^+(aq) + OH^-(aq)$
c. $C_6H_{12}O_6(s) \xrightarrow{H_2O} C_6H_{12}O_6(aq)$

9.14 Classify the solute represented in each of the following equations as a strong, weak, or nonelectrolyte:
a. $C_2H_5OH(aq) \xrightarrow{H_2O} C_2H_5OH(aq)$
b. $NaCl \xrightarrow{H_2O} Na^+ + Cl^-$
c. $HF \xrightarrow{H_2O} H_3O^+ + F^-$

9.15 Indicate the number of equivalents in each of the following:
a. 2 mole of Li^+
b. 1 moles of Br^-
c. 2 mole of Mg^{2+}
d. 3 moles of CO^{3+}

9.16 Indicate the number of equivalents in each of the following:
a. 1 mole of Mg^{2+}
b. 0.5 mole of H^+
c. 4 moles of Cl^-
d. 2 moles of Fe^{3+}

9.17 A normal KCl solution contains 175 mEq/L each of K^+ and Cl^-. How many moles each of K^+ and Cl^- are in 2.00 L of the KCl solution?

9.18 A solution to replace potassium loss contains 40. mEq/L each of K^+ and Cl^-. How many moles each of K^+ and Cl^- are in 1.5 L of the solution?

9.19 A solution contains 40. mEq/L of Cl^- and 15 mEq/L of HPO_4^{2-}. If Na^+ is the only cation in the solution, what is the Na^+ concentration, in milliequivalents per liter?

9.20 A sample of Ringer's solution contains the following concentrations (mEq/L) of cations: Na^+ 147, K^+ 4, and Ca^{2+} 4. If Cl^- is the only anion in the solution, what is the Cl^- concentration, in milliequivalents per liter?

LEARNING GOAL

Define solubility; distinguish between an unsaturated and a saturated solution. Identify a salt as soluble or insoluble.

9.3 Solubility

The term *solubility* is used to describe the amount of a solute that can dissolve in a given amount of solvent. Many factors, such as the type of solute, the type of solvent, and the temperature, affect the solubility of a solute. **Solubility**, usually expressed in grams of solute in 100 g of solvent, is the maximum amount of solute that can be dissolved at a certain temperature. If a solute readily dissolves when added to the solvent, the solution does not contain the maximum amount of solute. We call this solution an **unsaturated solution**.

A solution that contains all the solute that can dissolve is a **saturated solution**. When a solution is saturated, the rate at which the solute dissolves becomes equal to the rate at which solid forms, a process known as *recrystallization*. Then there is no further change in the amount of dissolved solute in solution.

$$\text{Solute} + \text{solvent} \underset{\text{Solute recrystallizes}}{\overset{\text{Solute dissolves}}{\rightleftarrows}} \text{Saturated solution}$$

We can prepare a saturated solution by adding an amount of solute greater than that needed to reach solubility. Stirring the solution will dissolve the maximum amount of solute and leave the excess on the bottom of the container. Once we have a saturated solution, the addition of more solute will only increase the amount of undissolved solute.

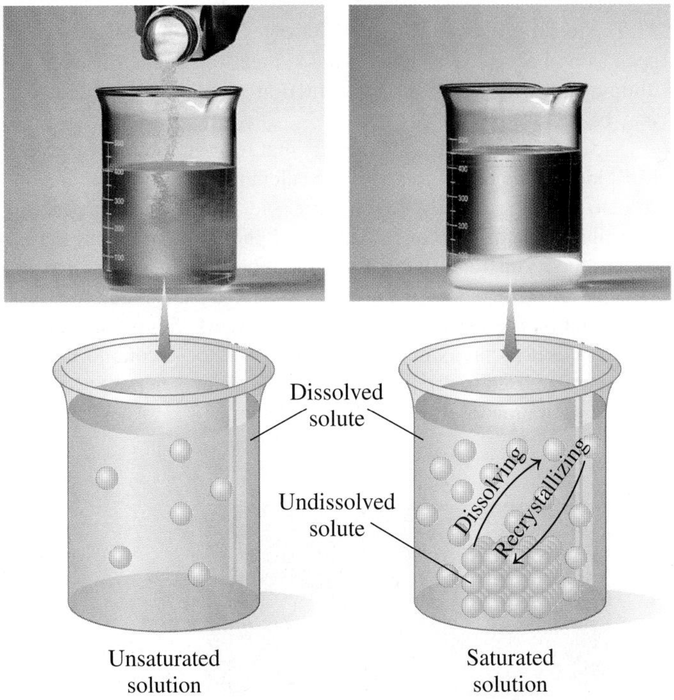

More solute can dissolve in an unsaturated solution, but not in a saturated solution.

SAMPLE PROBLEM 9.3 Saturated Solutions

At 20 °C, the solubility of KCl is 34 g/100 g of H_2O. In the laboratory, a student mixes 75 g of KCl with 200. g of H_2O at a temperature of 20 °C.

a. How much of the KCl will dissolve?
b. Is the solution saturated or unsaturated?
c. What is the mass, in grams, of any solid KCl left undissolved on the bottom of the container?

SOLUTION

a. KCl has a solubility of 34 g of KCl in 100 g of water. Using the solubility as a conversion factor, we can calculate the maximum amount of KCl that can dissolve in 200. g of water as follows:

$$200.\ \text{g}\ \cancel{H_2O} \times \frac{34\ \text{g KCl}}{100\ \text{g}\ \cancel{H_2O}} = 68\ \text{g of KCl}$$

b. Because 75 g of KCl exceeds the maximum amount (68 g) that can dissolve in 200. g of water, the KCl solution is saturated.
c. If we add 75 g of KCl to 200. g of water and only 68 g of KCl can dissolve, there is 7 g (75 g − 68 g) of solid (undissolved) KCl on the bottom of the container.

STUDY CHECK 9.3

At 40 °C, the solubility of KNO_3 is 65 g/100 g of H_2O. How many grams of KNO_3 will dissolve in 120 g of H_2O at 40 °C?

Chemistry Link to Health

Gout and Kidney Stones: A Problem of Saturation in Body Fluids

The conditions of gout and kidney stones involve compounds in the body that exceed their solubility levels and form solid products. Gout affects adults, primarily men, over the age of 40. Attacks of gout may occur when the concentration of uric acid in blood plasma exceeds its solubility, which is 7 mg/100 mL of plasma at 37 °C. Insoluble deposits of needle-like crystals of uric acid can form in the cartilage, tendons, and soft tissues where they cause painful gout attacks. They may also form in the tissues of the kidneys, where they can cause renal damage. High levels of uric acid in the body can be caused by an increase in uric acid production, failure of the kidneys to remove uric acid, or a diet with an overabundance of foods containing purines, which are metabolized to uric acid in the body. Foods in the diet that contribute to high levels of uric acid include certain meats, sardines, mushrooms, asparagus, and beans. Drinking alcoholic beverages may also significantly increase uric acid levels and bring about gout attacks.

Treatment for gout involves diet changes and drugs. Medications, such as probenecid, which helps the kidneys eliminate uric acid, or allopurinol, which blocks the production of uric acid by the body may be useful.

Kidney stones are solid materials that form in the urinary tract. Most kidney stones are composed of calcium phosphate and calcium oxalate, although they can be solid uric acid. Insufficient water intake and high levels of calcium, oxalate, and phosphate in the urine can lead to the formation of kidney stones. When a kidney stone passes through the urinary tract, it causes considerable pain and discomfort, necessitating the use of painkillers and surgery. Sometimes ultrasound is used to break up kidney stones. Persons prone to kidney stones are advised to drink six to eight glasses of water every day to prevent saturation levels of minerals in the urine.

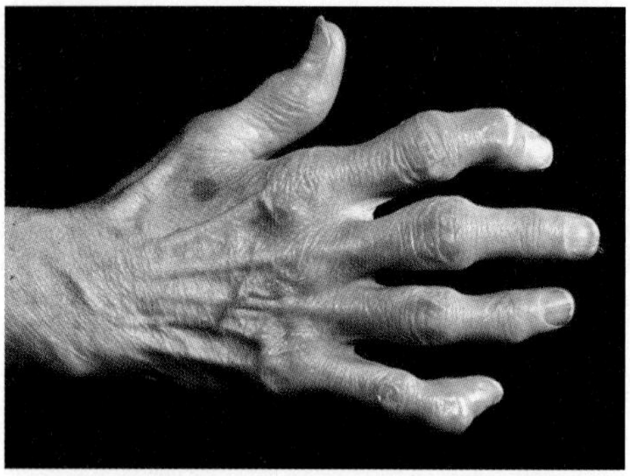

Gout occurs when uric acid exceeds its solubility.

Kidney stones form when calcium phosphate exceeds its solubility.

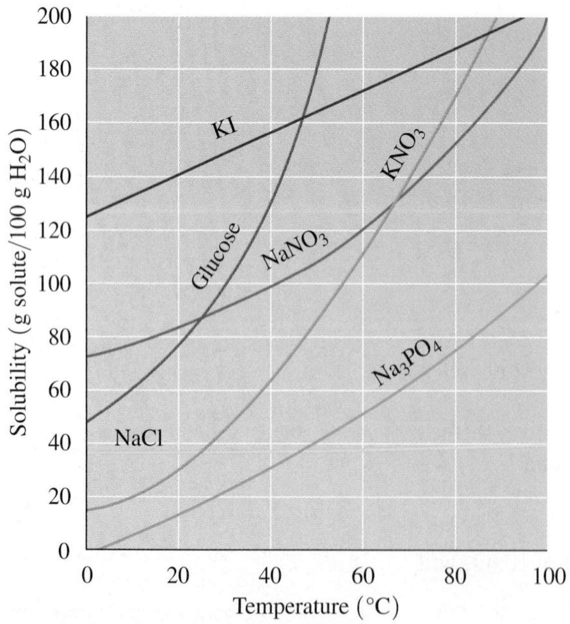

FIGURE 9.5 ▶ In water, most common solids are more soluble as the temperature increases.

Q Compare the solubility of $NaNO_3$ at 20 °C and 60 °C.

Effect of Temperature on Solubility

The solubility of most solids is greater as temperature increases, which means that solutions usually contain more dissolved solute at higher temperature. A few substances show little change in solubility at higher temperatures, and a few are less soluble (see Figure 9.5). For example, when you add sugar to iced tea, some undissolved sugar may form on the bottom of the glass. But if you add sugar to hot tea, many teaspoons of sugar are needed before solid sugar appears. Hot tea dissolves more sugar than does cold tea because the solubility of sugar is much greater at a higher temperature.

When a saturated solution is carefully cooled, it becomes a *supersaturated solution* because it contains more solute than the solubility allows. Such a solution is unstable, and if the solution is agitated or if a solute crystal is added, the excess solute will recrystallize to give a saturated solution again.

Conversely, the solubility of a gas in water decreases as the temperature increases. At higher temperatures, more gas molecules have the energy to escape from the solution. Perhaps you have observed the bubbles escaping from a cold carbonated soft drink as it warms. At high temperatures, bottles containing carbonated solutions may burst as more gas molecules leave the solution and increase the gas pressure inside the bottle. Biologists have found that increased temperatures in rivers and lakes cause the amount of dissolved oxygen to decrease until the warm water can no longer support a biological community. Electricity-generating plants are required to have their own ponds to use with their cooling towers to lessen the threat of thermal pollution in surrounding waterways.

Explore Your World

Preparing Rock Candy

Need: one cup of water, three cups of granulated sugar, clean narrow glass or jar, wooden stick (skewer) or thick string that is the height of the glass, pencil, and food coloring (optional)

Rock candy can be made from a saturated solution of sugar (sucrose).

Process:

1. Place the water and two cups of sugar in a pan and begin heating and stirring. The sugar should all dissolve. Continue heating, but not boiling, adding small amounts of the remaining sugar and stirring thoroughly each time, until some of the sugar no longer dissolves. There may be some sugar crystals on the bottom of the pan. Carefully pour the sugar solution into the glass or jar. Add two to three drops of food coloring, if desired.
2. Wet the wooden stick and roll it in granulated sugar to provide crystals for the sugar solution to attach to. Place the stick in the sugar solution. If using string, tape the string to a pencil so it will hang slightly above the bottom of the glass. Wet the lower half of the string, roll it in the granulated sugar, and place the pencil across the top of the glass with the string in the sugar solution. Place the glass and wooden stick or string in a place where it will not be disturbed. You should see sugar crystals grow over the next several days.

Questions

1. Why did more sugar dissolve as the solution was heated?
2. How did you know when you obtained a saturated solution?
3. Why did you see crystals forming on the stick or string over time?

Henry's Law

Henry's law states that the solubility of gas in a liquid is directly related to the pressure of that gas above the liquid. At higher pressures, there are more gas molecules available to enter and dissolve in the liquid. A can of soda is carbonated by using CO_2 gas at high pressure to increase the solubility of the CO_2 in the beverage. When you open the can at atmospheric pressure, the pressure on the CO_2 drops, which decreases the solubility of CO_2. As a result, bubbles of CO_2 rapidly escape from the solution. The burst of bubbles is even more noticeable when you open a warm can of soda.

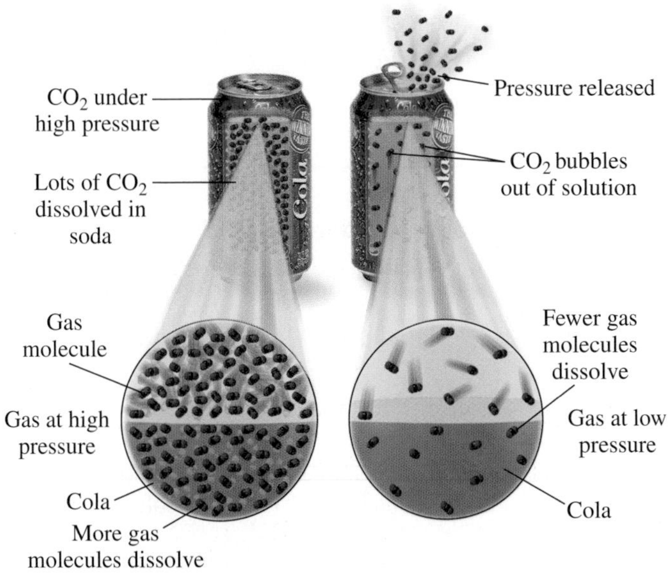

When the pressure of a gas above a solution decreases, the solubility of that gas in the solution also decreases.

Explore Your World

Preparing Solutions

Obtain a drinking glass and water and place one-fourth or half cup of cold water in a glass. Begin adding one tablespoon of sugar at a time and stir thoroughly. Take a sip of the liquid in the glass as you proceed. As the sugar solution becomes more concentrated, you may need to stir for a few minutes until all the sugar dissolves. Each time, observe the solution after several minutes to determine when it is saturated.

Repeat the above activity with one-fourth or half cup of warm water. Count the number of tablespoons of sugar you need to form a saturated solution.

Questions

1. What do you notice about how sweet each sugar solution tastes?
2. How did you know when you obtained a saturated solution?
3. How much sugar dissolved in the warm water compared to the cold water?

CORE CHEMISTRY SKILL

Using Solubility Rules

TABLE 9.8 Solubility Rules for Ionic Solids in Water

An Ionic Solid Is:	
Soluble If It Contains:	**Insoluble If It Contains:**
Li^+, Na^+, K^+	None
NH_4^+	None
NO_3^-, $C_2H_3O_2^-$	None
Cl^-, Br^-, I^-	Ag^+, Pb^{2+}, or Hg_2^{2+}
SO_4^{2-}	Ba^{2+}, Pb^{2+}, Ca^{2+}, Sr^{2+}, CO_3^{2-}, S^{2-}, PO_4^{3-}, OH^-

Soluble and Insoluble Salts

In our discussion up to now, we have considered ionic compounds that dissolve in water; they are **soluble salts**. However, some ionic compounds do not dissociate into ions in water. They are **insoluble salts** that remain as solids even in contact with water.

Salts that are soluble in water typically contain at least one of the following ions: Li^+, Na^+, K^+, NH_4^+, NO_3^-, or $C_2H_3O_2^-$ (acetate). *Only a salt containing a soluble cation or anion will dissolve in water.* Most salts containing Cl^- are soluble, but AgCl, $PbCl_2$, and Hg_2Cl_2 are insoluble chloride salts. Similarly, most salts containing SO_4^{2-} are soluble, but a few are insoluble, as shown in Table 9.8. Most other salts are insoluble and do not dissolve in water (see Figure 9.6). In an insoluble salt, the ionic bonds between its positive and negative ions are too strong for the polar water molecules to break. We can use the solubility rules to predict whether a salt (a solid ionic compound) would be expected to dissolve in water. Table 9.9 illustrates the use of these rules.

In medicine, the insoluble salt $BaSO_4$ is used as an opaque substance to enhance X-rays of the gastrointestinal tract. $BaSO_4$ is so insoluble that it does not dissolve in gastric fluids (see Figure 9.7). Other barium salts cannot be used because they would dissolve in water, releasing Ba^{2+} which is poisonous.

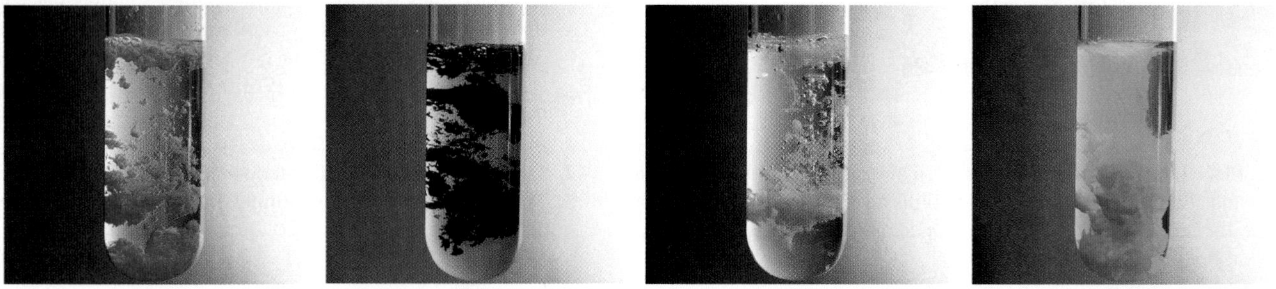

CdS FeS PbI_2 $Ni(OH)_2$

FIGURE 9.6 ▶ If a salt contains a combination of a cation and an anion that are not soluble, that salt is insoluble. For example, combinations of cadmium and sulfide, iron and sulfide, lead and iodide, and nickel and hydroxide do not contain any soluble ions. Thus, they form insoluble salts.

Q What ions make each of these salts insoluble in water?

TABLE 9.9 Using Solubility Rules

Ionic Compound	Solubility in Water	Reasoning
K_2S	Soluble	Contains K^+
$Ca(NO_3)_2$	Soluble	Contains NO_3^-
$PbCl_2$	Insoluble	Is an insoluble chloride
NaOH	Soluble	Contains Na^+
$AlPO_4$	Insoluble	Contains no soluble ions

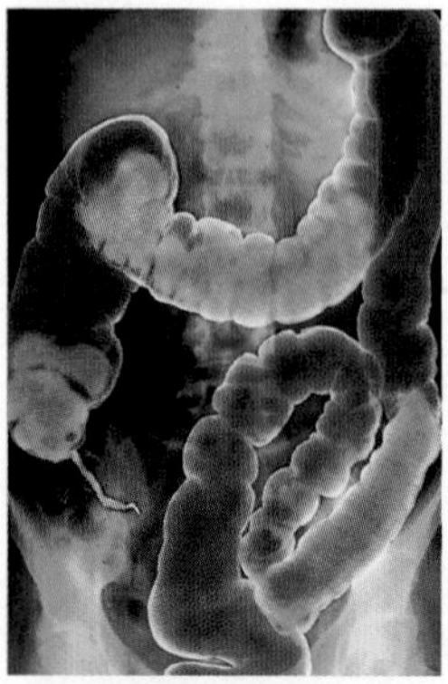

FIGURE 9.7 ▶ A barium sulfate-enhanced X-ray of the abdomen shows the large intestine.

Q Is $BaSO_4$ a soluble or an insoluble substance?

SAMPLE PROBLEM 9.4 Soluble and Insoluble Salts

Predict whether each of the following salts is soluble in water and explain why:

a. Na_3PO_4 **b.** $CaCO_3$

SOLUTION

a. The salt Na_3PO_4 is soluble in water because any compound that contains Na^+ is soluble.

b. The salt $CaCO_3$ is not soluble. The compound does not contain a soluble positive ion, which means that a calcium salt containing CO_3^{2-} is not soluble.

STUDY CHECK 9.4

Predict whether $MgCl_2$ is soluble in water and explain why.

QUESTIONS AND PROBLEMS

9.3 Solubility

LEARNING GOAL Define solubility; distinguish between an unsaturated and a saturated solution. Identify a salt as soluble or insoluble.

9.21 State whether each of the following refers to a saturated or an unsaturated solution:
a. A crystal added to a solution does not change in size.
b. A sugar cube completely dissolves when added to a cup of coffee.

9.22 State whether each of the following refers to a saturated or an unsaturated solution:
a. A spoonful of salt added to boiling water dissolves.
b. A layer of sugar forms on the bottom of a glass of tea as ice is added.

Use the following table for problems 9.23 to 9.26:

	Solubility (g/100 g H_2O)	
Substance	20 °C	50 °C
KCl	34	43
$NaNO_3$	88	110
$C_{12}H_{22}O_{11}$ (sugar)	204	260

9.23 Determine whether each of the following solutions will be saturated or unsaturated at 20 °C:
a. adding 37 g of $NaNO_3$ to 100. g of H_2O
b. adding 50 g of KCl to 80 g of H_2O
c. adding 250 g of sugar to 150 g of H_2O

9.24 Determine whether each of the following solutions will be saturated or unsaturated at 50 °C:
a. adding 25 g of KCl to 50. g of H_2O
b. adding 150. g of $NaNO_3$ to 75 g of H_2O
c. adding 80. g of sugar to 25 g of H_2O

9.25 A solution containing 430g of sugar ($C_{12}H_{22}O_{11}$)in 200g of H_2O at 50 °C is cooled to 20 °C.
a. How many grams of sugar remain in solution at 20 °C?
b. How many grams of solid sugar crystallized after cooling?

9.26 A solution containing 80. g of $NaNO_3$ in 75 g of H_2O at 50 °C is cooled to 20 °C.
a. How many grams of $NaNO_3$ remain in solution at 20 °C?
b. How many grams of solid $NaNO_3$ crystallized after cooling?

9.27 Explain the following observations:
a. More sugar dissolves in hot tea than in iced tea.
b. Champagne in a warm room goes flat.
c. A warm can of soda has more spray when opened than a cold one.

9.28 Explain the following observations:
a. An open can of soda loses its "fizz" faster at room temperature than in the refrigerator.
b. Chlorine gas in tap water escapes as the sample warms to room temperature.
c. Less sugar dissolves in iced coffee than in hot coffee.

9.29 Predict whether each of the following ionic compounds is soluble in water:
a. NaCl **b.** $PbCl_2$
c. $Mg(OH)_2$ **d.** NH_4NO_3
e. KBr

9.30 Predict whether each of the following ionic compounds is soluble in water:
a. PbS **b.** KI
c. Na_2S **d.** Ag_2O
e. $CaSO_4$

9.4 Concentrations of Solutions

LEARNING GOAL

Calculate the concentration of a solute in a solution; use concentration units to calculate the amount of solute or solution.

The amount of solute dissolved in a certain amount of solution is called the **concentration** of the solution. We will look at ways to express a concentration as a ratio of a certain amount of solute in a given amount of solution. The amount of a solute may be expressed in units of grams, milliliters, or moles. The amount of a solution may be expressed in units of grams, milliliters, or liters.

$$\text{Concentration of a solution} = \frac{\text{amount of solute}}{\text{amount of solution}}$$

Mass Percent (m/m) Concentration

Mass percent (m/m) describes the mass of the solute in grams for exactly 100 g of solution. In the calculation of mass percent (m/m), the units of mass of the solute and solution must be the same. If the mass of the solute is given as grams, then the mass of the solution must also be grams. The mass of the solution is the sum of the mass of the solute and the mass of the solvent.

Add 8.00 g of KCl

Add water until the solution has a mass of 50.00 g

When water is added to 8.00 g of KCl to form 50.00 g of a KCl solution, the mass percent concentration is 16.0% (m/m).

$$\text{Mass percent (m/m)} = \frac{\text{mass of solute (g)}}{\text{mass of solute (g)} + \text{mass of solvent (g)}} \times 100\%$$
$$= \frac{\text{mass of solute (g)}}{\text{mass of solution (g)}} \times 100\%$$

Suppose we prepared a solution by mixing 8.00 g of KCl (solute) with 42.00 g of water (solvent). Together, the mass of the solute and mass of solvent give the mass of the solution (8.00 g + 42.00 g = 50.00 g). Mass percent is calculated by substituting the mass of the solute and the mass of the solution into the mass percent expression.

$$\frac{8.00\text{ g KCl}}{\underbrace{50.00\text{ g solution}}_{\substack{8.00\text{ g KCl} + 42.00\text{ g H}_2\text{O} \\ \text{Solute} + \text{Solvent}}}} \times 100\% = 16.0\%\text{ (m/m) KCl solution}$$

SAMPLE PROBLEM 9.5 Calculating Mass Percent (m/m) Concentration

What is the mass percent of NaOH in a solution prepared by dissolving 30.0 g of NaOH in 120.0 g of H_2O?

SOLUTION

STEP 1 State the given and needed quantities.

	Given	Need
ANALYZE THE PROBLEM	30.0 g of NaOH solute, 30.0 g NaOH + 120.0 g H_2O = 150.0 g of NaOH solution	mass percent (m/m)

STEP 2 Write the concentration expression.

$$\text{Mass percent (m/m)} = \frac{\text{grams of solute}}{\text{grams of solution}} \times 100\%$$

STEP 3 Substitute solute and solution quantities into the expression and calculate.

$$\text{Mass percent (m/m)} = \frac{30.0\text{ g NaOH}}{150.0\text{ g solution}} \times 100\%$$
$$= 20.0\%\text{ (m/m) NaOH solution}$$

STUDY CHECK 9.5

What is the mass percent (m/m) of NaCl in a solution made by dissolving 2.0 g of NaCl in 56.0 g of H_2O?

Guide to Calculating Solution Concentration

STEP 1
State the given and needed quantities.

STEP 2
Write the concentration expression.

STEP 3
Substitute solute and solution quantities into the expression and calculate.

Volume Percent (v/v) Concentration

Because the volumes of liquids or gases are easily measured, the concentrations of their solutions are often expressed as **volume percent (v/v)**. The units of volume used in the ratio must be the same, for example, both in milliliters or both in liters.

$$\text{Volume percent (v/v)} = \frac{\text{volume of solute}}{\text{volume of solution}} \times 100\%$$

We interpret a volume percent as the volume of solute in exactly 100 mL of solution. On a bottle of extract of vanilla, a label that reads alcohol 35% (v/v) means 35 mL of ethanol solute in exactly 100 mL of vanilla solution.

The label indicates that vanilla extract contains 35% (v/v) alcohol.

SAMPLE PROBLEM 9.6 Calculating Volume Percent (v/v) Concentration

A bottle contains 59 mL of lemon extract solution. If the extract contains 49 mL of alcohol, what is the volume percent (v/v) of the alcohol in the solution?

SOLUTION

STEP 1 State the given and needed quantities.

ANALYZE THE PROBLEM	Given	Need
	49 mL of alcohol, 59 mL of solution	volume percent (v/v)

STEP 2 Write the concentration expression.

$$\text{Volume percent (v/v)} = \frac{\text{volume of solute}}{\text{volume of solution}} \times 100\%$$

STEP 3 Substitute solute and solution quantities into the expression and calculate.

$$\text{Volume percent (v/v)} = \frac{49 \text{ mL alcohol}}{59 \text{ mL solution}} \times 100\%$$

$$= 83\% \text{ (v/v) alcohol solution}$$

STUDY CHECK 9.6

What is the volume percent (v/v) of Br_2 in a solution prepared by dissolving 12 mL of liquid bromine (Br_2) in the solvent carbon tetrachloride (CCl_4) to make 250 mL of solution?

CORE CHEMISTRY SKILL

Calculating Concentration

Lemon extract is a solution of lemon flavor and alcohol.

Mass/Volume Percent (m/v) Concentration

Mass/volume percent (m/v) describes the mass of the solute in grams for exactly 100 mL of solution. In the calculation of mass/volume percent, the unit of mass of the solute is grams and the unit of the solution volume is milliliters.

$$\text{Mass/volume percent (m/v)} = \frac{\text{grams of solute}}{\text{milliliters of solution}} \times 100\%$$

The mass/volume percent is widely used in hospitals and pharmacies for the preparation of intravenous solutions and medicines. For example, a 5% (m/v) glucose solution contains 5 g of glucose in 100 mL of solution. The volume of solution represents the combined volumes of the glucose and H_2O.

SAMPLE PROBLEM 9.7 Calculating Mass/Volume Percent (m/v) Concentration

A student prepared a solution by dissolving 5.0 g of KI in enough water to give a final volume of 250 mL. What is the mass/volume percent (m/v) of the KI solution?

SOLUTION

STEP 1 State the given and needed quantities.

ANALYZE THE PROBLEM	Given	Need
	5.0 g of KI solute, 250 mL of KI solution	mass/volume percent (m/v)

STEP 2 Write the concentration expression.

$$\text{Mass/volume percent (m/v)} = \frac{\text{mass of solute}}{\text{volume of solution}} \times 100\%$$

STEP 3 Substitute solute and solution quantities into the expression and calculate.

$$\text{Mass/volume percent (m/v)} = \frac{5.0 \text{ g KI}}{250 \text{ mL solution}} \times 100\%$$
$$= 2.0\% \text{ (m/v) KI solution}$$

STUDY CHECK 9.7

What is the mass/volume percent (m/v) of NaOH in a solution prepared by dissolving 12 g of NaOH in enough water to make 220 mL of solution?

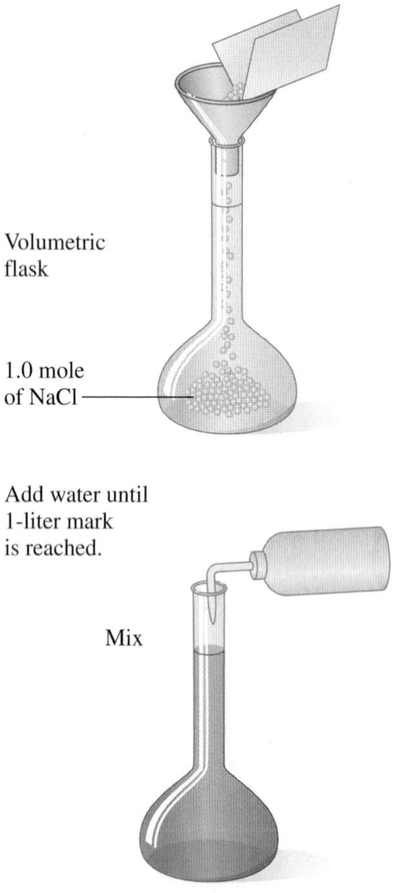

A 1.0 molar (M) NaCl solution

Molarity (M) Concentration

When chemists work with solutions, they often use **molarity (M)**, a concentration that states the number of moles of solute in exactly 1 L of solution.

$$\text{Molarity (M)} = \frac{\text{moles of solute}}{\text{liters of solution}}$$

The molarity of a solution can be calculated knowing the moles of solute and the volume of solution in liters. For example, if 1.0 mole of NaCl were dissolved in enough water to prepare 1.0 L of solution, the resulting NaCl solution has a molarity of 1.0 M. The abbreviation M indicates the units of mole per liter (mole/L).

$$\text{M} = \frac{\text{moles of solute}}{\text{liters of solution}} = \frac{1.0 \text{ mole NaCl}}{1 \text{ L solution}} = 1.0 \text{ M NaCl solution}$$

SAMPLE PROBLEM 9.8 Calculating Molarity

What is the molarity (M) of 60.0 g of NaOH in 0.250 L of NaOH solution?

SOLUTION

STEP 1 State the given and needed quantities.

ANALYZE THE PROBLEM	Given	Need
	60.0 g of NaOH, 0.250 L of NaCl solution	molarity (mole/L)

To calculate the moles of NaOH, we need to write the equality and conversion factors for the molar mass of NaOH. Then the moles in 60.0 g of NaOH can be determined.

$$1 \text{ mole of NaOH} = 40.00 \text{ g of NaOH}$$
$$\frac{1 \text{ mole NaOH}}{40.00 \text{ g NaOH}} \text{ and } \frac{40.00 \text{ g NaOH}}{1 \text{ mole NaOH}}$$

$$\text{moles of NaOH} = 60.0 \text{ g } \cancel{\text{NaOH}} \times \frac{1 \text{ mole NaOH}}{40.00 \text{ g } \cancel{\text{NaOH}}}$$
$$= 1.50 \text{ moles of NaOH}$$

$$\text{volume of solution} = 0.250 \text{ L of NaOH solution}$$

STEP 2 Write the concentration expression.

$$\text{Molarity (M)} = \frac{\text{moles of solute}}{\text{liters of solution}}$$

STEP 3 Substitute solute and solution quantities into the expression and calculate.

$$M = \frac{1.50 \text{ moles NaOH}}{0.250 \text{ L solution}} = \frac{6.00 \text{ moles NaOH}}{1 \text{ L solution}} = 6.00 \text{ M NaOH solution}$$

STUDY CHECK 9.8

What is the molarity of a solution that contains 75.0 g of KNO_3 dissolved in 0.350 L of solution?

Table 9.10 summarizes the types of units used in the various types of concentration expressions for solutions.

TABLE 9.10 Summary of Types of Concentration Expressions and Their Units

Concentration Units	Mass Percent (m/m)	Volume Percent (v/v)	Mass/Volume Percent (m/v)	Molarity (M)
Solute	g	mL	g	mole
Solution	g	mL	mL	L

Using Concentration as a Conversion Factor

CORE CHEMISTRY SKILL
Using Concentration as a Conversion Factor

In the preparation of solutions, we often need to calculate the amount of solute or solution. Then the concentration is useful as a conversion factor. Some examples of percent concentrations and molarity, their meanings, and conversion factors are given in Table 9.11. Some examples of using percent concentration or molarity as conversion factors are given in Sample Problems 9.9 and 9.10.

TABLE 9.11 Conversion Factors from Concentrations

Percent Concentration	Meaning	Conversion Factors
10% (m/m) KCl solution	10 g of KCl in 100 g of KCl solution	$\frac{10 \text{ g KCl}}{100 \text{ g solution}}$ and $\frac{100 \text{ g solution}}{10 \text{ g KCl}}$
12% (v/v) ethanol solution	12 mL of ethanol in 100 mL of ethanol solution	$\frac{12 \text{ mL ethanol}}{100 \text{ mL solution}}$ and $\frac{100 \text{ mL solution}}{12 \text{ mL ethanol}}$
5% (m/v) glucose solution	5 g of glucose in 100 mL of glucose solution	$\frac{5 \text{ g glucose}}{100 \text{ mL solution}}$ and $\frac{100 \text{ mL solution}}{5 \text{ g glucose}}$
Molarity		
6.0 M HCl solution	6.0 moles of HCl in 1 liter of HCl solution	$\frac{6.0 \text{ moles HCl}}{1 \text{ L solution}}$ and $\frac{1 \text{ L solution}}{6.0 \text{ moles HCl}}$

SAMPLE PROBLEM 9.9 Using Mass/Volume Percent to Find Mass of Solute

A topical antibiotic is 1.0% (m/v) clindamycin. How many grams of clindamycin are in 60. mL of the 1.0% (m/v) solution?

SOLUTION

STEP 1 State the given and needed quantities.

	Given	Need
ANALYZE THE PROBLEM	60. mL of 1.0% (m/v) clindamycin solution	grams of clindamycin

Guide to Using Concentration to Calculate Mass or Volume

STEP 1 State the given and needed quantities.

STEP 2 Write a plan to calculate the mass or volume.

STEP 3 Write equalities and conversion factors.

STEP 4 Set up the problem to calculate the mass or volume.

STEP 2 Write a plan to calculate the mass.

milliliters of solution → % (m/v) factor → grams of clindamycin

STEP 3 Write equalities and conversion factors. The percent (m/v) indicates the grams of a solute in every 100 mL of a solution. The 1.0% (m/v) can be written as two conversion factors.

$$100\text{ mL of solution} = 1.0\text{ g of clindamycin}$$
$$\frac{1.0\text{ g clindamycin}}{100\text{ mL solution}} \quad \text{and} \quad \frac{100\text{ mL solution}}{1.0\text{ g clindamycin}}$$

STEP 4 Set up the problem to calculate the mass. The volume of the solution is converted to mass of solute using the conversion factor that cancels mL.

$$60.\ \cancel{\text{mL solution}} \times \frac{1.0\text{ g clindamycin}}{100\ \cancel{\text{mL solution}}} = 0.60\text{ g of clindamycin}$$

STUDY CHECK 9.9

Calculate the grams of KCl in 300. mL of a 6.00% (m/v) KCl solution.

SAMPLE PROBLEM 9.10 Using Molarity to Calculate Volume of Solution

How many liters of a 2.00 M NaCl solution are needed to provide 67.3 g of NaCl?

SOLUTION

STEP 1 State the given and needed quantities.

ANALYZE THE PROBLEM	Given	Need
	67.3 g of NaCl, 2.00 M NaCl solution	liters of NaCl solution

STEP 2 Write a plan to calculate the volume.

grams of NaCl → Molar mass → moles of NaCl → Molarity → liters of NaCl solution

STEP 3 Write equalities and conversion factors.

$$1\text{ mole of NaCl} = 58.44\text{ g of NaCl}$$
$$\frac{1\text{ mole NaCl}}{58.44\text{ g NaCl}} \quad \text{and} \quad \frac{58.44\text{ g NaCl}}{1\text{ mole NaCl}}$$

$$1\text{ L of NaCl solution} = 2.00\text{ moles of NaCl}$$
$$\frac{1\text{ L NaCl solution}}{2.00\text{ moles NaCl}} \quad \text{and} \quad \frac{2.00\text{ moles NaCl}}{1\text{ L NaCl solution}}$$

STEP 4 Set up the problem to calculate the volume.

$$\text{liters of NaCl solution} = 67.3\ \cancel{\text{g NaCl}} \times \frac{1\ \cancel{\text{mole NaCl}}}{58.44\ \cancel{\text{g NaCl}}} \times \frac{1\text{ L NaCl solution}}{2.00\ \cancel{\text{moles NaCl}}}$$
$$= 0.576\text{ L of NaCl solution}$$

STUDY CHECK 9.10

How many milliliters of a 6.0 M HCl solution will provide 164 g of HCl?

QUESTIONS AND PROBLEMS

9.4 Concentrations of Solutions

LEARNING GOAL Calculate the concentration of a solute in a solution; use concentration units to calculate the amount of solute or solution.

9.31 What is the difference between a 5% (m/m) glucose solution and a 5% (m/v) glucose solution?

9.32 What is the difference between a 10% (v/v) methanol (CH_3OH) solution and a 10% (m/m) methanol solution?

9.33 Calculate the mass percent (m/m) for the solute in each of the following:
a. 40 g of LiCl and150 g of H_2O
b. 25 g of sugar in 200 g of sugar solution
c. 6.0 g of $MgCl_2$ in 50.0 g of $MgCl_2$ solution

9.34 Calculate the mass percent (m/m) for the solute in each of the following:
a. 75 g of NaOH in 325 g of NaOH solution
b. 2.0 g of KOH and 20.0 g of H_2O
c. 48.5 g of Na_2CO_3 in 250.0 g of Na_2CO_3 solution

9.35 Calculate the mass/volume percent (m/v) for the solute in each of the following:
a. 0.02 kg of caffeine in 200 mL of coffee
b. 17 g of KCl in 250 mL of KCl solution

9.36 Calculate the mass/volume percent (m/v) for the solute in each of the following:
a. 2.50 g of LiCl in 40.0 mL of LiCl solution
b. 7.5 g of casein in 120 mL of low-fat milk

9.37 Calculate the grams or milliliters of solute needed to prepare the following:
a. 50. g of a 5.0% (m/m) KCl solution
b. 1250 mL of a 4.0% (m/v) NH_4Cl solution
c. 250. mL of a 10.0% (v/v) acetic acid solution

9.38 Calculate the grams or milliliters of solute needed to prepare the following:
a. 150. g of a 40.0% (m/m) $LiNO_3$ solution
b. 450 mL of a 2.0% (m/v) KOH solution
c. 225 mL of a 15% (v/v) isopropyl alcohol solution

9.39 A mouthwash contains 22.5% (v/v) alcohol. If the bottle of mouthwash contains 355 mL, what is the volume, in milliliters, of alcohol?

9.40 A bottle of champagne is 11% (v/v) alcohol. If there are 750 mL of champagne in the bottle, what is the volume, in milliliters, of alcohol?

Interactive Video

Problem 9.40

9.41 A patient receives 100. mL of 20.% (m/v) mannitol solution every hour.
a. How many grams of mannitol are given in 1 h?
b. How many grams of mannitol does the patient receive in 12 h?

9.42 A patient receives 250 mL of a 4.0% (m/v) amino acid solution twice a day.
a. How many grams of amino acids are in 250 mL of solution?
b. How many grams of amino acids does the patient receive in 1 day?

9.43 A patient needs 50. g of glucose in the next 12 h. How many liters of a 3% (m/v) glucose solution must be given?

9.44 A patient received 2.0 g of NaCl in 8 h. How many milliliters of a 0.90% (m/v) NaCl (saline) solution were delivered?

9.45 Calculate the amount of formic acid in each of the following solutions, given that the density of formic acid is 1.22 g/mL at 25 °C:
a. 10% formic acid (m/v) in 10 mL formic acid solution
b. 10% formic acid (v/v) in 10 mL formic acid solution

9.46 For each of the following solutions, calculate the:
a. grams of 2.0% (m/m) NaCl solution that contains 7.50 g of NaCl
b. milliliters of 25% (m/v) NaF solution that contains 4.0 g of NaF
c. milliliters of 8.0% (v/v) ethanol solution that contains 20.0 mL of ethanol

9.47 Calculate the molarity of each of the following:
a. 1.00 moles of NaCl in 10.00 L of a NaCl solution
b. 7.00 g of glucose in 10.00 L of a glucose solution
c. 2.95 g of KOH in 700. mL of a KOH solution

9.48 Calculate the molarity of each of the following:
a. 0.500 mole of glucose in 0.200 L of a glucose solution
b. 73.0 g of HCl in 2.00 L of a HCl solution
c. 30.0 g of NaOH in 350. mL of a NaOH solution

9.49 Calculate the grams of solute needed to prepare each of the following:
a. 4.00 L of a 2.0 M KCl solution
b. 7.00 L of a 0.300 M $MgCl_2$ solution
c. 145.0 mL of a 4.00 M HCl solution

9.50 Calculate the grams of solute needed to prepare each of the following:
a. 2.00 L of a 6.00 M NaOH solution
b. 5.00 L of a 0.100 M $CaCl_2$ solution
c. 175 mL of a 3.00 M $NaNO_3$ solution

9.51 For each of the following solutions, calculate the:
a. liters of a 2.00 M KBr solution to obtain 3.00 moles of KBr
b. liters of a 1.50 M NaCl solution to obtain 15.0 moles of NaCl
c. milliliters of a 0.800 M $Ca(NO_3)_2$ solution to obtain 0.0500 mole of $Ca(NO_3)_2$

9.52 For each of the following solutions, calculate the:
a. liters of a 4.00 M KCl solution to obtain 0.100 mole of KCl
b. liters of a 6.00 M HCl solution to obtain 5.00 moles of HCl
c. milliliters of a 2.50 M K_2SO_4 solution to obtain 1.20 moles of K_2SO_4

LEARNING GOAL

Describe the dilution of a solution; calculate the final concentration or volume of a diluted solution.

9.5 Dilution of Solutions

In chemistry and biology, we often prepare diluted solutions from more concentrated solutions. In a process called **dilution**, a solvent, usually water, is added to a solution, which increases the volume. As a result, the concentration of the solution decreases. In an everyday example, you are making a dilution when you add three cans of water to a can of concentrated orange juice.

1 can of orange juice concentrate + 3 cans of water = 4 cans of orange juice

Although the addition of solvent increases the volume, the amount of solute does not change; it is the same in the concentrated solution and the diluted solution (see Figure 9.8).

$$\underset{\text{Concentrated solution}}{\text{Grams or moles of solute}} = \underset{\text{Diluted solution}}{\text{grams or moles of solute}}$$

We can write this equality in terms of the concentration, C, and the volume, V. The concentration, C, may be percent concentration or molarity.

$$\underset{\text{Concentrated solution}}{C_1V_1} = \underset{\text{Diluted solution}}{C_2V_2}$$

If we are given any three of the four variables (C_1, C_2, V_1, or V_2), we can rearrange the dilution expression to solve for the unknown quantity as seen in Sample Problem 9.11.

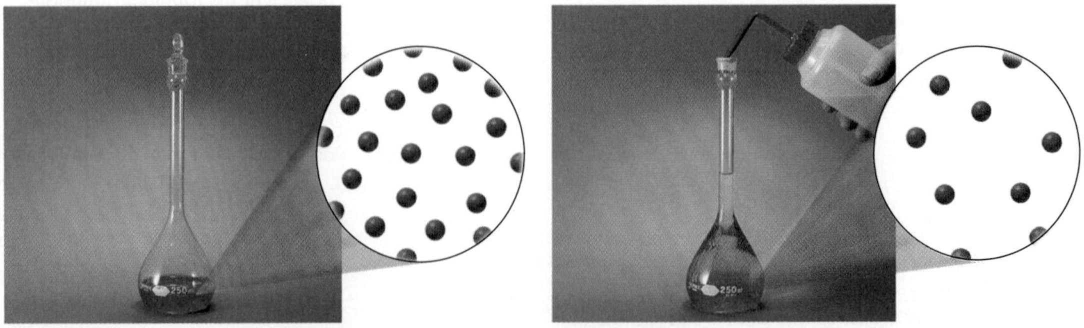

FIGURE 9.8 ▶ When water is added to a concentrated solution, there is no change in the number of particles. However, the solute particles spread out as the volume of the diluted solution increases.

Q What is the concentration of the diluted solution after an equal volume of water is added to a sample of a 6 M HCl solution?

SAMPLE PROBLEM 9.11 Volume of a Diluted Solution

What volume, in milliliters, of a 2.5% (m/v) KOH solution can be prepared by diluting 50.0 mL of a 12% (m/v) KOH solution?

SOLUTION

STEP 1 Prepare a table of the concentrations and volumes of the solutions. For our problem analysis, we organize the solution data in a table, making sure that the units of concentration and volume are the same.

ANALYZE THE PROBLEM	Concentrated Solution	Diluted Solution	Know	Predict
	$C_1 = 12\%$ (m/v) KOH	$C_2 = 2.5\%$ (m/v) KOH	C decreases	
	$V_1 = 50.0$ mL	$V_2 = ?$ mL		V increases

STEP 2 Rearrange the dilution expression to solve for the unknown quantity.

$$C_1V_1 = C_2V_2$$

$$\frac{C_1V_1}{C_2} = \frac{\cancel{C_2}V_2}{\cancel{C_2}} \quad \text{Divide both sides by } C_2$$

$$V_2 = V_1 \times \frac{C_1}{C_2}$$

STEP 3 Substitute the known quantities into the dilution expression and calculate.

$$V_2 = 50.0\text{ mL} \times \underbrace{\frac{12\%}{2.5\%}}_{\text{Concentration factor increases volume}} = 240\text{ mL of diluted KOH solution}$$

When the initial volume (V_1) is multiplied by a ratio of the percent concentrations (concentration factor) that is greater than 1, the volume of the solution increases as predicted in Step 1.

STUDY CHECK 9.11

What is the final volume, in milliliters, when 25.0 mL of a 15% (m/v) KCl solution is diluted to a 3.0% (m/v) KCl solution?

Guide to Calculating Dilution Quantities

STEP 1 Prepare a table of the concentrations and volumes of the solutions.

STEP 2 Rearrange the dilution expression to solve for the unknown quantity.

STEP 3 Substitute the known quantities into the dilution expression and calculate.

SAMPLE PROBLEM 9.12 Molarity of a Diluted Solution

What is the molarity of a solution when 75.0 mL of a 4.00 M KCl solution is diluted to a volume of 500. mL?

SOLUTION

STEP 1 Prepare a table of the concentrations and volumes of the solutions.

ANALYZE THE PROBLEM	Concentrated Solution	Diluted Solution	Know	Predict
	$C_1 = 4.00$ M KCl	$C_2 = ?$ M KCl		C (molarity) decreases
	$V_1 = 75.0$ mL $= 0.0750$ L	$V_2 = 500.$ mL $= 0.500$ L	V increases	

STEP 2 **Rearrange the dilution expression to solve for the unknown quantity.**

$$C_1V_1 = C_2V_2$$

$$\frac{C_1V_1}{V_2} = \frac{C_2\cancel{V_2}}{\cancel{V_2}} \quad \text{Divide both sides by } V_2$$

$$C_2 = C_1 \times \frac{V_1}{V_2}$$

STEP 3 **Substitute the known quantities into the dilution expression and calculate.**

$$C_2 = 4.00\ \text{M} \times \frac{75.0\ \cancel{\text{mL}}}{500.\ \cancel{\text{mL}}} = 0.600\ \text{M (diluted KCl solution)}$$

Volume factor decreases concentration

When the initial molarity (C_1) is multiplied by a ratio of the volumes (volume factor) that is less than 1, the molarity of the diluted solution decreases as predicted in Step 1.

STUDY CHECK 9.12

What is the molarity of a solution when 50.0 mL of 4.00 M KOH is diluted to 200. mL?

QUESTIONS AND PROBLEMS

9.5 Dilution of Solutions

LEARNING GOAL Describe the dilution of a solution; calculate the final concentration or volume of a diluted solution.

9.53 To make tomato soup, you add one can of water to the condensed soup. Why is this a dilution?

9.54 What is the meaning of dilution? Is dilution a physical change or a chemical change?

9.55 Calculate the final concentration of each of the following:

- **a.** 2.0 L of a 6.0 M HCl solution is added to water so that the final volume is 6.0 L.
- **b.** Water is added to 0.50 L of a 12 M NaOH solution to make 3.0 L of a diluted NaOH solution.
- **c.** A 10.0-mL sample of a 25% (m/v) KOH solution is diluted with water so that the final volume is 100.0 mL.
- **d.** A 50.0-mL sample of a 15% (m/v) H_2SO_4 solution is added to water to give a final volume of 250 mL.

9.56 Calculate the final concentration of each of the following:

- **a.** 1.0 L of a 4.0 M HNO_3 solution is added to water so that the final volume is 8.0 L.
- **b.** Water is added to 0.25 L of a 6.0 M NaF solution to make 2.0 L of a diluted NaF solution.
- **c.** A 50.0-mL sample of an 8.0% (m/v) KBr solution is diluted with water so that the final volume is 200.0 mL.
- **d.** A 5.0-mL sample of a 50.0% (m/v) acetic acid ($HC_2H_3O_2$) solution is added to water to give a final volume of 25 mL.

9.57 Determine the final volume, in milliliters, of each of the following:

- **a.** a 1.5 M HCl solution prepared from 20.0 mL of a 6.0 M HCl solution
- **b.** a 2.0% (m/v) LiCl solution prepared from 50.0 mL of a 10.0% (m/v) LiCl solution
- **c.** a 0.500 M H_3PO_4 solution prepared from 50.0 mL of a 6.00 M H_3PO_4 solution
- **d.** a 5.0% (m/v) glucose solution prepared from 75 mL of a 12% (m/v) glucose solution

9.58 Determine the final volume, in milliliters, of each of the following:

- **a.** a 1.00% (m/v) H_2SO_4 solution prepared from 10.0 mL of a 20.0% H_2SO_4 solution
- **b.** a 0.10 M HCl solution prepared from 25 mL of a 6.0 M HCl solution
- **c.** a 1.0 M NaOH solution prepared from 50.0 mL of a 12 M NaOH solution
- **d.** a 1.0% (m/v) $CaCl_2$ solution prepared from 18 mL of a 4.0% (m/v) $CaCl_2$ solution

9.59 Determine the volume, in milliliters, required to prepare each of the following:

- **a.** 255 mL of a 0.200 M HNO_3 solution using a 4.00 M HNO_3 solution
- **b.** 715 mL of a 0.100 M $MgCl_2$ solution using a 6.00 M $MgCl_2$ solution
- **c.** 0.100 L of a 0.150 M KCl solution using an 8.00 M KCl solution

9.60 Determine the volume, in milliliters, required to prepare each of the following:

- **a.** 20.0 mL of a 0.250 M KNO_3 solution using a 6.00 M KNO_3 solution
- **b.** 25.0 mL of a 2.50 M H_2SO_4 solution using a 12.0 M H_2SO_4 solution
- **c.** 0.500 L of a 1.50 M NH_4Cl solution using a 10.0 M NH_4Cl solution

9.6 Properties of Solutions

LEARNING GOAL

Identify a mixture as a solution, a colloid, or a suspension. Describe how the number of particles in a solution affects the osmotic pressure of a solution.

The size and number of solute particles in different types of mixtures play an important role in determining the properties of that mixture.

Solutions

In the solutions discussed up to now, the solute was dissolved as small particles that are uniformly dispersed throughout the solvent to give a homogeneous solution. When you observe a solution, such as salt water, you cannot visually distinguish the solute from the solvent. The solution appears transparent, although it may have a color. The particles are so small that they go through filters and through *semipermeable membranes*. A semipermeable membrane allows solvent molecules such as water and very small solute particles to pass through, but does allow the passage of large solute molecules.

Colloids

The particles in a **colloid** are much larger than solute particles in a solution. Colloidal particles are large molecules, such as proteins, or groups of molecules or ions. Colloids, similar to solutions, are homogeneous mixtures that do not separate or settle out. Colloidal particles are small enough to pass through filters, but too large to pass through semipermeable membranes. Table 9.12 lists several examples of colloids.

TABLE 9.12 Examples of Colloids

Colloid	Substance Dispersed	Dispersing Medium
Fog, clouds, hair sprays	Liquid	Gas
Dust, smoke	Solid	Gas
Shaving cream, whipped cream, soapsuds	Gas	Liquid
Styrofoam, marshmallows	Gas	Solid
Mayonnaise, homogenized milk	Liquid	Liquid
Cheese, butter	Liquid	Solid
Blood plasma, paints (latex), gelatin	Solid	Liquid

Suspensions

Suspensions are heterogeneous, nonuniform mixtures that are very different from solutions or colloids. The particles of a suspension are so large that they can often be seen with the naked eye. They are trapped by filters and semipermeable membranes.

The weight of the suspended solute particles causes them to settle out soon after mixing. If you stir muddy water, it mixes but then quickly separates as the suspended particles settle to the bottom and leave clear liquid at the top. You can find suspensions among the medications in a hospital or in your medicine cabinet. These include Kaopectate, calamine lotion, antacid mixtures, and liquid penicillin. It is important to follow the instructions on the label that states "shake well before using" so that the particles form a suspension.

Water-treatment plants make use of the properties of suspensions to purify water. When chemicals such as aluminum sulfate or iron(III) sulfate are added to untreated water, they react with impurities to form large suspended particles called *floc*. In the water-treatment plant, a system of filters traps the suspended particles, but clean water passes through.

Table 9.13 compares the different types of mixtures and Figure 9.9 illustrates some properties of solutions, colloids, and suspensions.

TABLE 9.13 Comparison of Solutions, Colloids, and Suspensions

Type of Mixture	Type of Particle	Settling	Separation
Solution	Small particles such as atoms, ions, or small molecules	Particles do not settle	Particles cannot be separated by filters or semipermeable membranes
Colloid	Larger molecules or groups of molecules or ions	Particles do not settle	Particles can be separated by semipermeable membranes but not by filters
Suspension	Very large particles that may be visible	Particles settle rapidly	Particles can be separated by filters

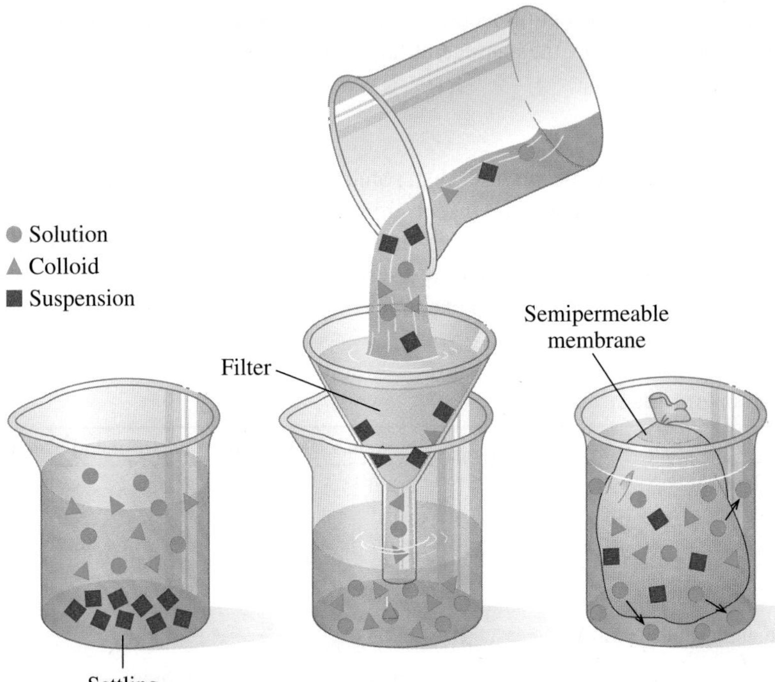

FIGURE 9.9 ▶ Properties of different types of mixtures: **(a)** suspensions settle out; **(b)** suspensions are separated by a filter; **(c)** solution particles go through a semipermeable membrane, but colloids and suspensions do not.

Q A filter can be used to separate suspension particles from a solution, but a semipermeable membrane is needed to separate colloids from a solution. Explain.

Interactive Video

Properties of Solutions

Chemistry Link to **Health**

Colloids and Solutions in the Body

In the body, colloids are retained by semipermeable membranes. For example, the intestinal lining allows solution particles to pass into the blood and lymph circulatory systems. However, the colloids from foods are too large to pass through the membrane, and they remain in the intestinal tract. Digestion breaks down large colloidal particles, such as starch and protein, into smaller particles, such as glucose and amino acids that can pass through the intestinal membrane and enter the circulatory system. Certain foods, such as bran, a fiber, cannot be broken down by human digestive processes, and they move through the intestine intact.

Because large proteins, such as enzymes, are colloids, they remain inside cells. However, many of the substances that must be obtained by cells, such as oxygen, amino acids, electrolytes, glucose, and minerals, can pass through cellular membranes. Waste products, such as urea and carbon dioxide, pass out of the cell to be excreted.

Osmotic Pressure

The movement of water into and out of the cells of plants as well as the cells of our bodies is an important biological process that also depends on the solute concentration. In a process called **osmosis**, water molecules move through a semipermeable membrane from the solution with the lower concentration of solute into a solution with the higher solute concentration. In an osmosis apparatus, water is placed on one side of a semipermeable membrane and a sucrose (sugar) solution on the other side. The semipermeable membrane allows water molecules to flow back and forth but blocks the sucrose molecules because they cannot pass through the membrane. Because the sucrose solution has a higher solute concentration, more water molecules flow into the sucrose solution than out of the sucrose solution. The volume level of the sucrose solution rises as the volume level on the water side falls. The increase of water dilutes the sucrose solution to equalize (or attempt to equalize) the concentrations on both sides of the membrane.

Eventually the height of the sucrose solution creates sufficient pressure to equalize the flow of water between the two compartments. This pressure, called **osmotic pressure**, prevents the flow of additional water into the more concentrated solution. Then there is no further change in the volumes of the two solutions. The osmotic pressure depends on the concentration of solute particles in the solution. The greater the number of particles dissolved, the higher its osmotic pressure. In this example, the sucrose solution has a higher osmotic pressure than pure water, which has an osmotic pressure of zero.

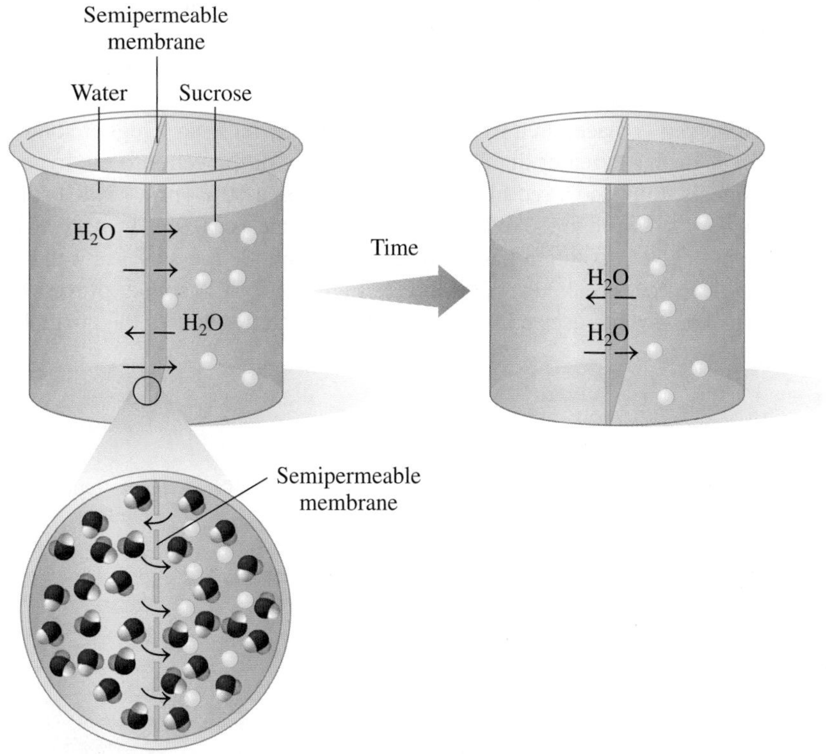

Water flows into the solution with a higher solute concentration until the flow of water becomes equal in both directions.

Explore Your World

Everyday Osmosis

1. Place a few pieces of dried fruit such as raisins, prunes, or banana chips in water. Observe them after 1 hour or more. Look at them again the next day.
2. Place some grapes in a concentrated salt-water solution. Observe them after 1 hour or more. Look at them again the next day.
3. Place one potato slice in water and another slice in a concentrated salt-water solution. After 1 to 2 hours, observe the shapes and size of the slices. Look at them again the next day.

Questions

1. How did the shape of the dried fruit change after being in water? Explain.
2. How did the appearance of the grapes change after being in a concentrated salt solution? Explain.
3. How does the appearance of the potato slice that was placed in water compare to the appearance of the potato slice placed in salt water? Explain.
4. At the grocery store, why are sprinklers used to spray water on fresh produce such as lettuce, carrots, and cucumbers?

In a process called *reverse osmosis*, a pressure greater than the osmotic pressure is applied to a solution so that it is forced through a purification membrane. The flow of water is reversed because water flows from an area of lower water concentration to an area of higher water concentration. The molecules and ions in solution stay behind, trapped by the membrane, while water passes through the membrane. This process of reverse osmosis is used in a few desalination plants to obtain pure water from sea (salt) water. However, the pressure that must be applied requires so much energy that reverse osmosis is not yet an economical method for obtaining pure water in most parts of the world.

Isotonic Solutions

Because the cell membranes in biological systems are semipermeable, osmosis is an ongoing process. The solutes in body solutions such as blood, tissue fluids, lymph, and plasma all exert osmotic pressure. Most intravenous (IV) solutions used in a hospital are **isotonic solutions**,

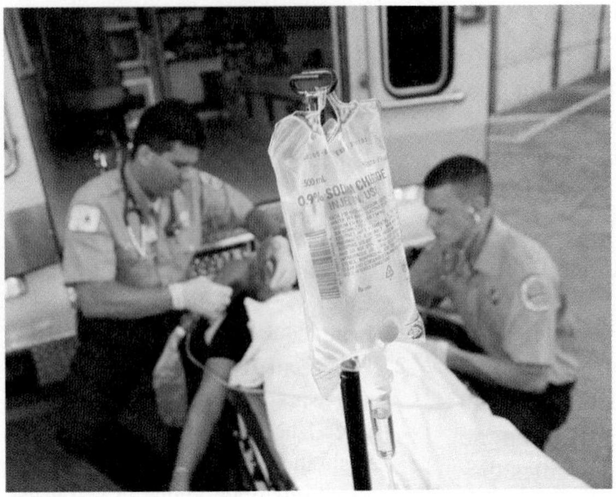

A 0.9% NaCl solution is isotonic with the solute concentration of the blood cells of the body.

which exert the same osmotic pressure as body fluids such as blood. The percent concentration typically used in IV solutions is mass/volume percent (m/v), which is a type of percent concentration we have already discussed. The most typical isotonic solutions are 0.9% (m/v) NaCl solution, or 0.9 g of NaCl/100 mL of solution, and 5% (m/v) glucose, or 5 g of glucose/100 mL of solution. Although they do not contain the same kinds of particles, a 0.9% (m/v) NaCl solution as well as a 5% (m/v) glucose solution are both 0.3 M (Na^+ and Cl^- ions or glucose molecules). A red blood cell placed in an isotonic solution retains its volume because there is an equal flow of water into and out of the cell (see Figure 9.10a).

Hypotonic and Hypertonic Solutions

If a red blood cell is placed in a solution that is not isotonic, the differences in osmotic pressure inside and outside the cell can drastically alter the volume of the cell. When a red blood cell is placed in a **hypotonic solution**, which has a lower solute concentration (*hypo* means "lower than"), water flows into the cell by osmosis. The increase in fluid causes the cell to swell, and possibly burst—a process called **hemolysis** (see Figure 9.10b). A similar process occurs when you place dehydrated food, such as raisins or dried fruit, in water. The water enters the cells, and the food becomes plump and smooth.

If a red blood cell is placed in a **hypertonic solution**, which has a higher solute concentration (*hyper* means "greater than"), water flows out of the cell into the hypertonic solution by osmosis. Suppose a red blood cell is placed in a 10% (m/v) NaCl solution. Because the osmotic pressure in the red blood cell is the same as a 0.9% (m/v) NaCl solution, the 10% (m/v) NaCl solution has a much greater osmotic pressure. As water leaves the cell, it shrinks, a process called **crenation** (see Figure 9.10c). A similar process occurs when making pickles, which uses a hypertonic salt solution that causes the cucumbers to shrivel as they lose water.

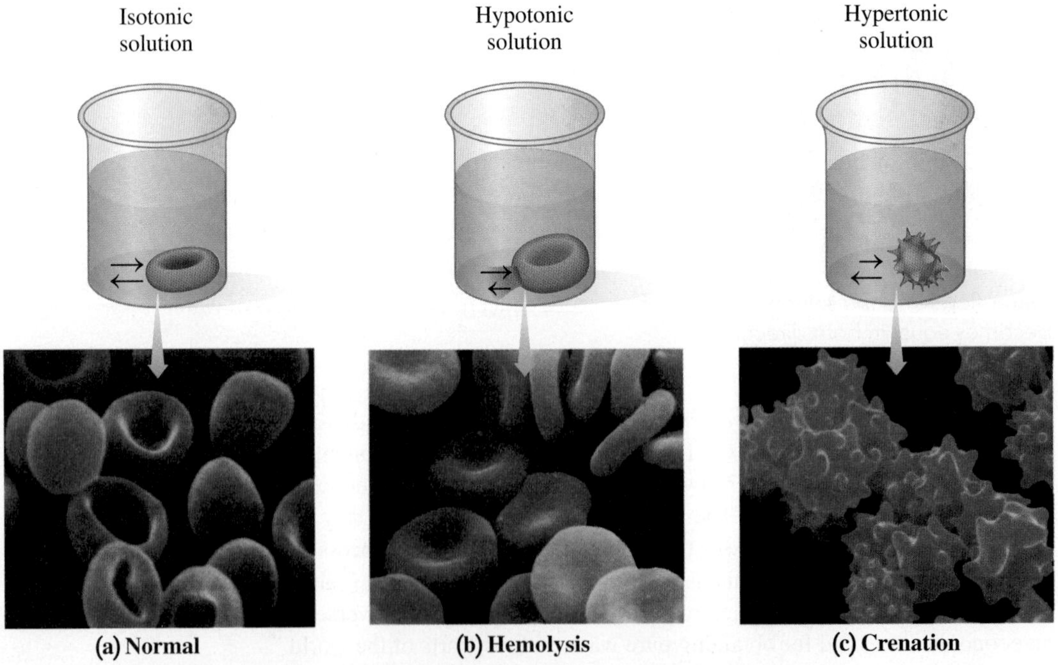

FIGURE 9.10 ▶ **(a)** In an isotonic solution, a red blood cell retains its normal volume. **(b)** Hemolysis: In a hypotonic solution, water flows into a red blood cell, causing it to swell and burst. **(c)** Crenation: In a hypertonic solution, water leaves the red blood cell, causing it to shrink.

Q What happens to a red blood cell placed in a 4% NaCl solution?

SAMPLE PROBLEM 9.13 Isotonic, Hypotonic, and Hypertonic Solutions

Describe each of the following solutions as isotonic, hypotonic, or hypertonic. Indicate whether a red blood cell placed in each solution will undergo hemolysis, crenation, or no change.

a. a 5% (m/v) glucose solution
b. a 0.2% (m/v) NaCl solution

SOLUTION

a. A 5% (m/v) glucose solution is isotonic. A red blood cell will not undergo any change.
b. A 0.2% (m/v) NaCl solution is hypotonic. A red blood cell will undergo hemolysis.

STUDY CHECK 9.13

What will happen to a red blood cell placed in a 10% (m/v) glucose solution?

Dialysis

Dialysis is a process that is similar to osmosis. In dialysis, a semipermeable membrane, called a dialyzing membrane, permits small solute molecules and ions as well as solvent water molecules to pass through, but it retains large particles, such as colloids. Dialysis is a way to separate solution particles from colloids.

Suppose we fill a cellophane bag with a solution containing NaCl, glucose, starch, and protein and place it in pure water. Cellophane is a dialyzing membrane, and the sodium ions, chloride ions, and glucose molecules will pass through it into the surrounding water. However, large colloidal particles, like starch and protein, remain inside. Water molecules will flow into the cellophane bag. Eventually the concentrations of sodium ions, chloride ions, and glucose molecules inside and outside the dialysis bag become equal. To remove more NaCl or glucose, the cellophane bag must be placed in a fresh sample of pure water.

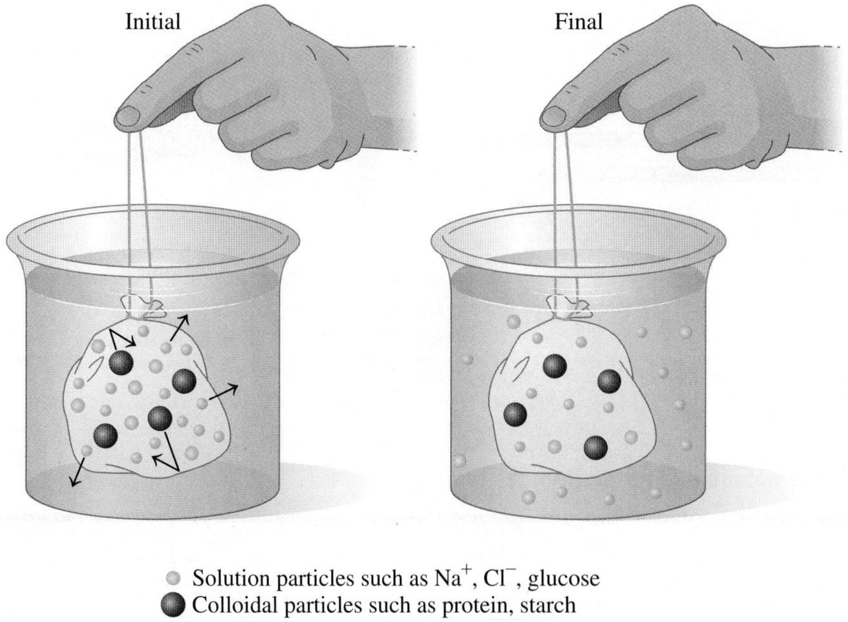

Solution particles pass through a dialyzing membrane but colloidal particles are retained.

Chemistry Link to Health

Dialysis by the Kidneys and the Artificial Kidney

The fluids of the body undergo dialysis by the membranes of the kidneys, which remove waste materials, excess salts, and water. In an adult, each kidney contains about 2 million nephrons. At the top of each nephron, there is a network of arterial capillaries called the *glomerulus*.

As blood flows into the glomerulus, small particles, such as amino acids, glucose, urea, water, and certain ions, will move through the capillary membranes into the nephron. As this solution moves through the nephron, substances still of value to the body (such as amino acids, glucose, certain ions, and 99% of the water) are reabsorbed. The major waste product, urea, is excreted in the urine.

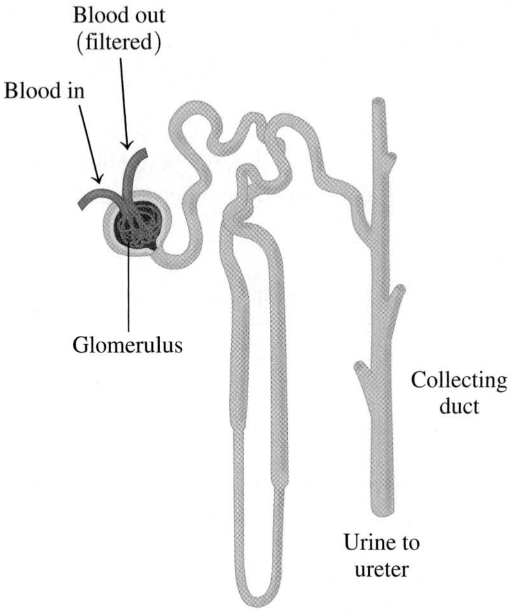

In the kidneys, each nephron contains a glomerulus where urea and waste products are removed to form urine.

Hemodialysis

If the kidneys fail to dialyze waste products, increased levels of urea can become life-threatening in a relatively short time. A person with kidney failure must use an artificial kidney, which cleanses the blood by **hemodialysis**.

A typical artificial kidney machine contains a large tank filled with water containing selected electrolytes. In the center of this dialyzing bath (dialysate), there is a dialyzing coil or membrane made of cellulose tubing. As the patient's blood flows through the dialyzing coil, the highly concentrated waste products dialyze out of the blood. No blood is lost because the membrane is not permeable to large particles such as red blood cells.

Dialysis patients do not produce much urine. As a result, they retain large amounts of water between dialysis treatments, which produces a strain on the heart. The intake of fluids for a dialysis patient may be restricted to as little as a few teaspoons of water a day. In the dialysis procedure, the pressure of the blood is increased as it circulates through the dialyzing coil so water can be squeezed out of the blood. For some dialysis patients, 2 to 10 L of water may be removed during one treatment. Dialysis patients have from two to three treatments a week, each treatment requiring about 5 to 7 h. Some of the newer treatments require less time. For many patients, dialysis is done at home with a home dialysis unit.

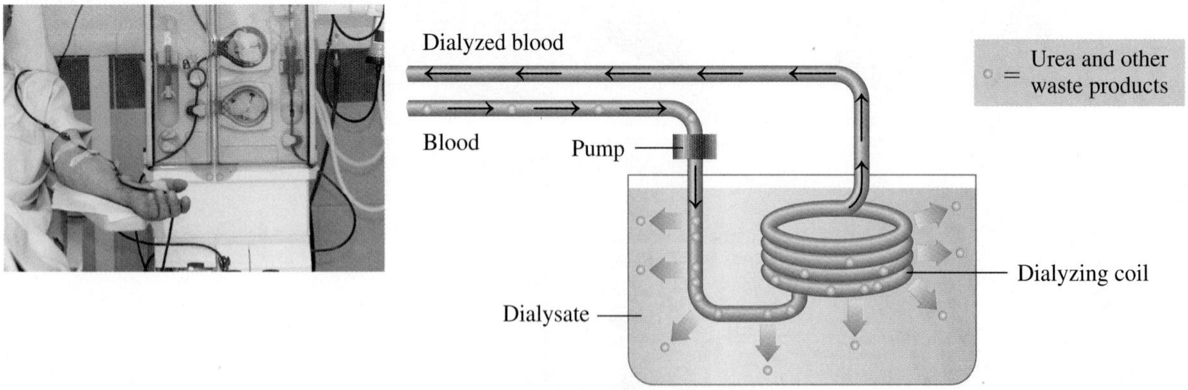

During dialysis, waste products and excess water are removed from the blood.

QUESTIONS AND PROBLEMS

9.6 Properties of Solutions

LEARNING GOAL Identify a mixture as a solution, a colloid, or a suspension. Describe how the number of particles in a solution affects the osmotic pressure of a solution.

9.61 Identify the following as characteristic of a solution, a colloid, or a suspension:
a. a mixture that cannot be separated by a semipermeable membrane
b. a mixture that settles out upon standing

9.62 Identify the following as characteristic of a solution, a colloid, or a suspension:
a. Particles of this mixture do not settle.
b. Particles in this mixture can be separated by simple filtration.

9.63 A 20% (m/v) starch solution is separated from a 1% (m/v) starch solution by a semipermeable membrane. (Starch is a colloid.)
a. Which compartment has the lower osmotic pressure?
b. In which direction will water flow initially?
c. In which compartment will the volume level rise?

9.64 A 0.1% (m/v) albumin solution is separated from a 2% (m/v) albumin solution by a semipermeable membrane. (Albumin is a colloid.)
a. Which compartment has the higher osmotic pressure?
b. In which direction will water flow initially?
c. In which compartment will the volume level rise?

9.65 Indicate the compartment (**A** or **B**) that will increase in volume for each of the following pairs of solutions separated by a semipermeable membrane:

A	B
a. 5% (m/v) sucrose	10% (m/v) sucrose
b. 8% (m/v) albumin	4% (m/v) albumin
c. 0.1% (m/v) starch	10% (m/v) starch

9.66 Indicate the compartment (**A** or **B**) that will increase in volume for each of the following pairs of solutions separated by a semipermeable membrane:

A	B
a. 20% (m/v) starch	10% (m/v) starch
b. 10% (m/v) albumin	2% (m/v) albumin
c. 0.5% (m/v) sucrose	5% (m/v) sucrose

9.67 Are the following solutions isotonic, hypotonic, or hypertonic compared with a red blood cell?
a. distilled H_2O
b. 1% (m/v) glucose
c. 0.9% (m/v) NaCl
d. 15% (m/v) glucose

9.68 Will a red blood cell undergo crenation, hemolysis, or no change in each of the following solutions?
a. 1% (m/v) glucose
b. 2% (m/v) NaCl
c. 5% (m/v) glucose
d. 0.1% (m/v) NaCl

9.69 Each of the following mixtures is placed in a dialyzing bag and immersed in distilled water. Which substances will be found outside the bag in the distilled water?
a. NaCl solution
b. starch solution (colloid) and alanine, an amino acid, solution
c. NaCl solution and starch solution (colloid)
d. urea solution

9.70 Each of the following mixtures is placed in a dialyzing bag and immersed in distilled water. Which substances will be found outside the bag in the distilled water?
a. KCl solution and glucose solution
b. albumin solution (colloid)
c. an albumin solution (colloid), KCl solution, and glucose solution
d. urea solution and NaCl solution

CONCEPT MAP

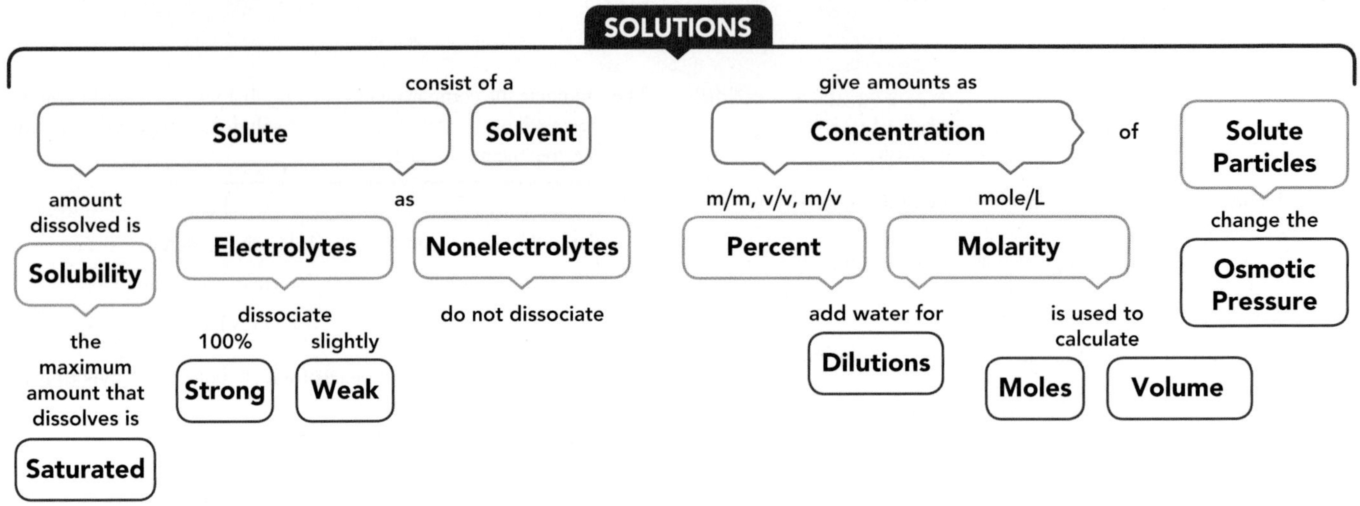

CHAPTER REVIEW

9.1 Solutions

LEARNING GOAL Identify the solute and solvent in a solution; describe the formation of a solution.

- A solution forms when a solute dissolves in a solvent.
- In a solution, the particles of solute are evenly dispersed in the solvent.
- The solute and solvent may be solid, liquid, or gas.
- The polar O—H bond leads to hydrogen bonding between water molecules.
- An ionic solute dissolves in water, a polar solvent, because the polar water molecules attract and pull the ions into solution, where they become hydrated.
- The expression *like dissolves like* means that a polar or an ionic solute dissolves in a polar solvent while a nonpolar solute dissolves in a nonpolar solvent.

9.2 Electrolytes and Nonelectrolytes

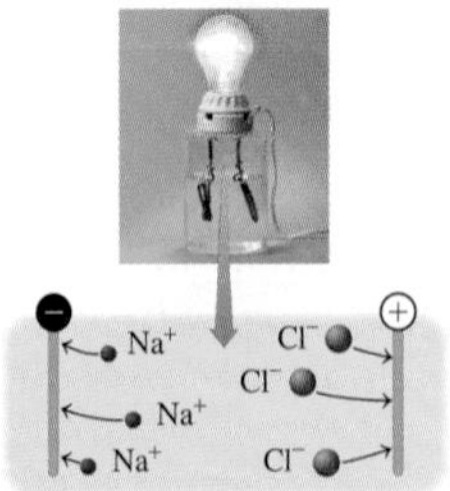

Strong electrolyte

LEARNING GOAL Identify solutes as electrolytes or nonelectrolytes.

- Substances that produce ions in water are called electrolytes because their solutions will conduct an electrical current.
- Strong electrolytes are completely dissociated, whereas weak electrolytes are only partially ionized.
- Nonelectrolytes are substances that dissolve in water to produce only molecules and cannot conduct electrical currents.
- An equivalent (Eq) is the amount of an electrolyte that carries one mole of positive or negative charge.
- One mole of Na^+ is 1 Eq. One mole of Ca^{2+} has 2 Eq.

9.3 Solubility

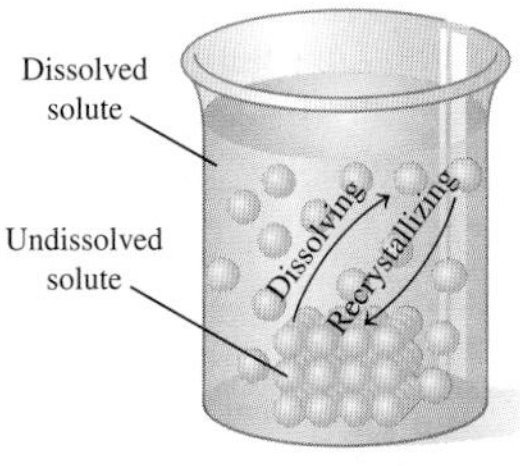

Saturated solution

LEARNING GOAL Define solubility; distinguish between an unsaturated and a saturated solution. Identify a salt as soluble or insoluble.

- The solubility of a solute is the maximum amount of a solute that can dissolve in 100 g of solvent.
- A solution that contains the maximum amount of dissolved solute is a saturated solution.
- A solution containing less than the maximum amount of dissolved solute is unsaturated.
- An increase in temperature increases the solubility of most solids in water, but decreases the solubility of gases in water.
- Salts that are soluble in water usually contain Li^+, Na^+, K^+, NH_4^+, NO_3^-, or acetate, $C_2H_3O_2^-$.

9.4 Concentrations of Solutions

LEARNING GOAL Calculate the concentration of a solute in a solution; use concentration units to calculate the amount of solute or solution.

- Mass percent expresses the mass/mass (m/m) ratio of the mass of solute to the mass of solution multiplied by 100%.
- Percent concentration can also be expressed as volume/volume (v/v) and mass/volume (m/v) ratios.
- Molarity is the moles of solute per liter of solution.
- In calculations of grams or milliliters of solute or solution, the concentration is used as a conversion factor.
- Molarity (mole/L) is written as conversion factors to solve for moles of solute or volume of solution.

9.5 Dilution of Solutions

LEARNING GOAL Describe the dilution of a solution; calculate the final concentration or volume of a diluted solution.

- In dilution, a solvent such as water is added to a solution, which increases its volume and decreases its concentration.

9.6 Properties of Solutions

LEARNING GOAL Identify a mixture as a solution, a colloid, or a suspension. Describe how the number of particles in a solution affects the osmotic pressure of a solution.

- Colloids contain particles that pass through most filters but do not settle out or pass through permeable membranes.
- Suspensions have very large particles that settle out of solution.

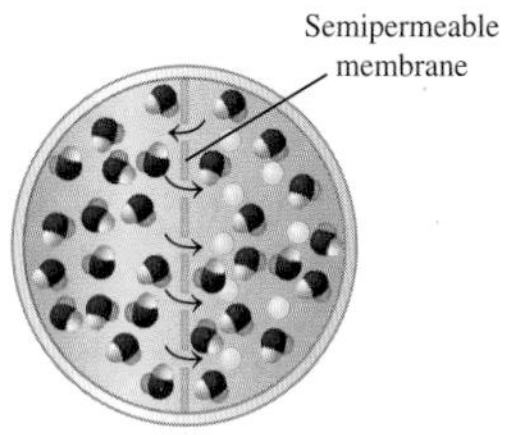

- In osmosis, solvent (water) passes through a selectively permeable membrane from a solution with a lower solute concentration to a solution with a higher concentration.
- Isotonic solutions have osmotic pressures equal to that of body fluids.
- A red blood cell maintains its volume in an isotonic solution but swells in a hypotonic solution, and shrinks in a hypertonic solution.
- In dialysis, water and small solute particles pass through a dialyzing membrane, while larger particles are retained.

KEY TERMS

colloid A mixture having particles that are moderately large. Colloids pass through filters but cannot pass through semipermeable membranes.

concentration A measure of the amount of solute that is dissolved in a specified amount of solution.

crenation The shriveling of a cell because water leaves the cell when the cell is placed in a hypertonic solution.

dialysis A process in which water and small solute particles pass through a semipermeable membrane.

dilution A process by which water (solvent) is added to a solution to increase the volume and decrease (dilute) the concentration of the solute.

electrolyte A substance that produces ions when dissolved in water; its solution conducts electricity.

equivalent (Eq) The amount of a positive or negative ion that supplies 1 mole of electrical charge.

hemodialysis A mechanical cleansing of the blood by an artificial kidney using the principle of dialysis.

hemolysis A swelling and bursting of red blood cells in a hypotonic solution because of an increase in fluid volume.

Henry's law The solubility of a gas in a liquid is directly related to the pressure of that gas above the liquid.

hydration The process of surrounding dissolved ions by water molecules.

hypertonic solution A solution that has a higher particle concentration and higher osmotic pressure than the cells of the body.

hypotonic solution A solution that has a lower particle concentration and lower osmotic pressure than the cells of the body.

insoluble salt An ionic compound that does not dissolve in water.

isotonic solution A solution that has the same particle concentration and osmotic pressure as that of the cells of the body.

mass percent (m/m) The grams of solute in exactly 100 g of solution.

mass/volume percent (m/v) The grams of solute in exactly 100 mL of solution.

molarity (M) The number of moles of solute in exactly 1 L of solution.

nonelectrolyte A substance that dissolves in water as molecules; its solution does not conduct an electrical current.

osmosis The flow of a solvent, usually water, through a semipermeable membrane into a solution of higher solute concentration.

osmotic pressure The pressure that prevents the flow of water into the more concentrated solution.

saturated solution A solution containing the maximum amount of solute that can dissolve at a given temperature. Any additional solute will remain undissolved in the container.

soluble salt An ionic compound that dissolves in water.

solubility The maximum amount of solute that can dissolve in exactly 100 g of solvent, usually water, at a given temperature.

solute The component in a solution that is present in the lesser amount.

solution A homogeneous mixture in which the solute is made up of small particles (ions or molecules) that can pass through filters and semipermeable membranes.

solvent The substance in which the solute dissolves; usually the component present in greater amount.

strong electrolyte A compound that ionizes completely when it dissolves in water. Its solution is a good conductor of electricity.

suspension A mixture in which the solute particles are large enough and heavy enough to settle out and be retained by both filters and semipermeable membranes.

unsaturated solution A solution that contains less solute than can be dissolved.

volume percent (v/v) A percent concentration that relates the volume of the solute in exactly 100 mL of solution.

weak electrolyte A substance that produces only a few ions along with many molecules when it dissolves in water. Its solution is a weak conductor of electricity.

CORE CHEMISTRY SKILLS

The chapter section containing each Core Chemistry Skill is shown in parentheses at the end of each heading.

Using Solubility Rules (9.3)

- Soluble salts contain Li^+, Na^+, K^+, NH_4^+, NO_3^-, or $C_2H_3O_2^-$ (acetate).
- Salts containing Cl^-, Br^-, or I^- are soluble, but if they are combined with Ag^+, Pb^{2+}, or Hg_2^{2+}, the salts are insoluble.
- Most salts containing SO_4^{2-} are soluble, but if they are combined with Ba^{2+}, Pb^{2+}, Ca^{2+}, or Sr^{2+}, the salt is insoluble.
- Most salts including those containing the anions CO_3^{2-}, S^{2-}, PO_4^{3-}, or OH^- are insoluble.

Example: Is K_2SO_4 soluble or insoluble in water?
Answer: K_2SO_4 is soluble in water because it contains K^+.

Calculating Concentration (9.4)

The amount of solute dissolved in a certain amount of solution is called the concentration of the solution.

- $\text{Mass percent (m/m)} = \dfrac{\text{mass of solute}}{\text{mass of solution}} \times 100\%$
- $\text{Volume percent (v/v)} = \dfrac{\text{volume of solute}}{\text{volume of solution}} \times 100\%$
- $\text{Mass/volume percent (m/v)} = \dfrac{\text{grams of solute}}{\text{milliliters of solution}} \times 100\%$
- $\text{Molarity (M)} = \dfrac{\text{moles of solute}}{\text{liters of solution}}$

Example: What is the mass/volume percent (m/v) and the molarity (M) of 225 mL (0.225 L) of a LiCl solution that contains 17.1 g of LiCl?

Answer:

$$\text{Mass/volume \% (m/v)} = \frac{\text{grams of solute}}{\text{milliliters of solution}} \times 100\%$$
$$= \frac{17.1 \text{ g LiCl}}{225 \text{ mL solution}} \times 100\%$$
$$= 7.60\% \text{ (m/v) LiCl solution}$$
$$\text{moles of LiCl} = 17.1 \cancel{\text{ g LiCl}} \times \frac{1 \text{ mole LiCl}}{42.39 \cancel{\text{ g LiCl}}}$$
$$= 0.403 \text{ mole of LiCl}$$
$$\text{Molarity (M)} = \frac{\text{moles of solute}}{\text{liters of solution}} = \frac{0.403 \text{ mole LiCl}}{0.225 \text{ L solution}}$$
$$= 1.79 \text{ M LiCl solution}$$

Using Concentration as a Conversion Factor (9.4)

- When we need to calculate the amount of solute or solution, we use the mass/volume percent (m/v) or the molarity (M) as a conversion factor.
- For example, the concentration of a 4.50 M HCl solution means there are 4.50 moles of HCl in 1 L of HCl solution, which gives two conversion factors written as

$$\frac{4.50 \text{ moles HCl}}{1 \text{ L solution}} \quad \text{and} \quad \frac{1 \text{ L solution}}{4.50 \text{ moles HCl}}$$

Example: How many milliliters of a 4.50 M HCl solution will provide 1.13 moles of HCl?

Answer:

$$1.13 \cancel{\text{ moles HCl}} \times \frac{1 \cancel{\text{ L solution}}}{4.50 \cancel{\text{ moles HCl}}} \times \frac{1000 \text{ mL solution}}{1 \cancel{\text{ L solution}}}$$
$$= 251 \text{ mL of HCl solution}$$

UNDERSTANDING THE CONCEPTS

The chapter sections to review are shown in parentheses at the end of each question.

9.71 Match the diagrams with the following: (9.1)

a. a polar solute and a polar solvent
b. a nonpolar solute and a polar solvent
c. a nonpolar solute and a nonpolar solvent

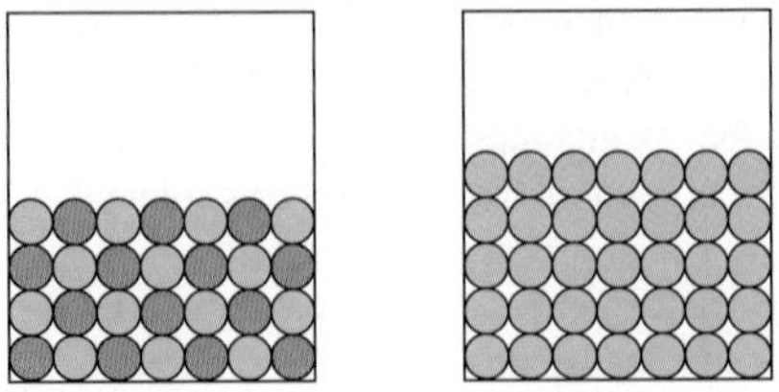

1 2

9.72 If all the solute is dissolved in diagram **1**, how would heating or cooling the solution cause each of the following changes? (9.3)

a. 2 to 3 **b.** 2 to 1

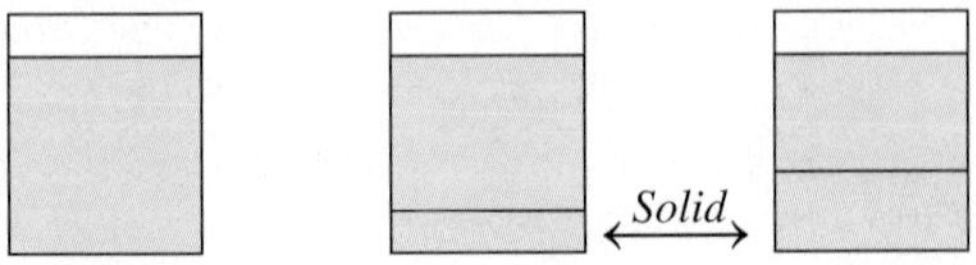

1 2 3

9.73 Select the diagram that represents the solution formed by a solute  that is a (9.2)

a. nonelectrolyte
b. weak electrolyte
c. strong electrolyte

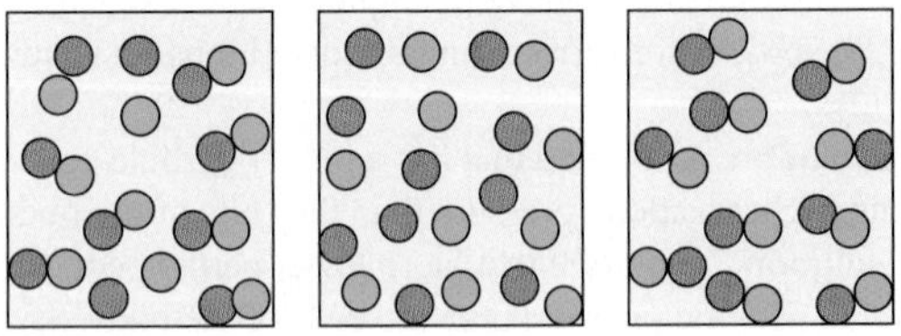

1 2 3

9.74 Why do lettuce leaves in a salad wilt after a vinaigrette dressing containing salt is added? (9.6)

9.75 A pickle is made by soaking a cucumber in brine, a salt-water solution. What makes the smooth cucumber become wrinkled like a prune? (9.6)

9.76 Select the container that represents the dilution of a 4% (m/v) KCl solution to give each of the following: (9.5)

a. a 2% (m/v) KCl solution
b. a 1% (m/v) KCl solution

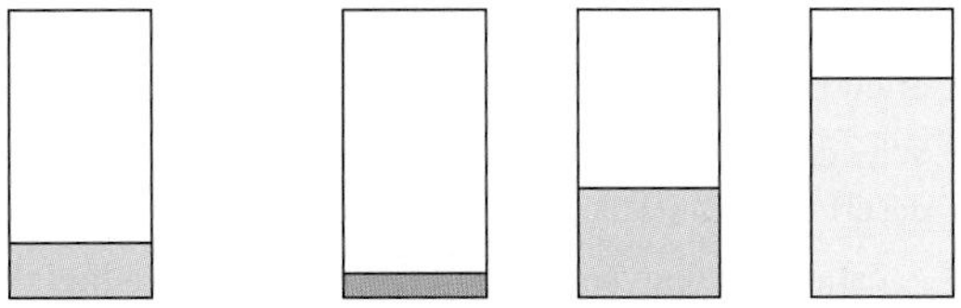

4% (m/v) KCl 1 2 3

9.77 A semipermeable membrane separates two compartments, **A** and **B**. If the levels in **A** and **B** are equal initially, select the diagram that illustrates the final levels in **a** to **d**: (9.6)

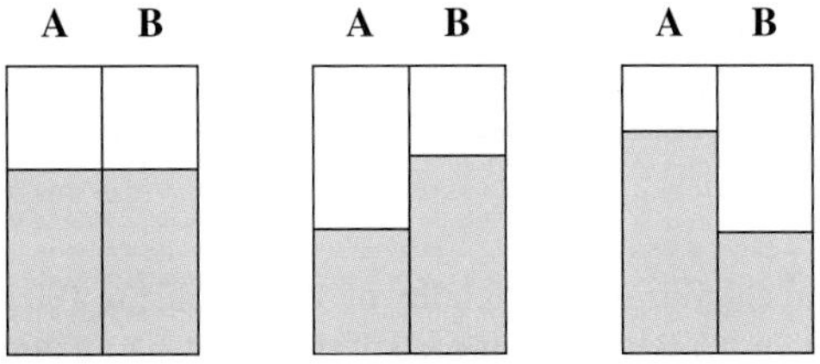

1 2 3

Solution in A	Solution in B
a. 2% (m/v) starch	8% (m/v) starch
b. 1% (m/v) starch	1% (m/v) starch
c. 5% (m/v) sucrose	1% (m/v) sucrose
d. 0.1% (m/v) sucrose	1% (m/v) sucrose

9.78 Select the diagram that represents the shape of a red blood cell when placed in each of the following **a** to **e**: (9.6)

1 2 3
Normal red blood cell

a. 0.9% (m/v) NaCl solution
b. 10% (m/v) glucose solution
c. 0.01% (m/v) NaCl solution
d. 5% (m/v) glucose solution
e. 1% (m/v) glucose solution

ADDITIONAL QUESTIONS AND PROBLEMS

9.79 Why does iodine dissolve in hexane, but not in water? (9.1)

9.80 How do temperature and pressure affect the solubility of solids and gases in water? (9.3)

9.81 Potassium nitrate has a solubility of 32 g of KNO_3 in 100 g of H_2O at 20 °C. Determine if each of the following forms an unsaturated or saturated solution at 20 °C: (9.3)

a. adding 32 g of KNO_3 to 200. g of H_2O
b. adding 19 g of KNO_3 to 50. g of H_2O
c. adding 68 g of KNO_3 to 150. g of H_2O

9.82 Potassium chloride has a solubility of 43 g of KCl in 100 g of H_2O at 50 °C. Determine if each of the following forms an unsaturated or saturated solution at 50 °C: (9.3)

a. adding 25 g of KCl to 100. g of H_2O
b. adding 15 g of KCl to 25 g of H_2O
c. adding 86 g of KCl to 150. g of H_2O

9.83 Indicate whether each of the following ionic compounds is soluble or insoluble in water: (9.3)

a. KCl **b.** $MgSO_4$ **c.** CuS
d. $AgNO_3$ **e.** $Ca(OH)_2$

9.84 Indicate whether each of the following ionic compounds is soluble or insoluble in water: (9.3)

a. $CuCO_3$ **b.** FeO **c.** $Mg_3(PO_4)_2$
d. $(NH_4)_2SO_4$ **e.** $NaHCO_3$

9.85 Calculate the mass percent (m/m) of solutions containing 15.5 g or 20.0 g of KBr and 75.5 g of H_2O. (9.4)

9.86 Calculate the mass percent (m/m) of a solution containing 26 g of K_2CO_3 and 724 g of H_2O. (9.4)

9.87 How many milliliters of a 12% (v/v) propyl alcohol solution would you need to obtain 4.5 mL of propyl alcohol? (9.4)

9.88 An 80-proof brandy is a 40.% (v/v) ethanol solution. The "proof" is twice the percent concentration of alcohol in the beverage. How many milliliters of alcohol are present in 750 mL of brandy? (9.4)

9.89 How many liters of a 12% (m/v) KOH solution would you need to obtain 86.0 g of KOH? (9.4)

9.90 How many liters of a 5.0% (m/v) glucose solution would you need to obtain 75 g of glucose? (9.4)

9.91 What is the molarity of a 500 mL KOH solution containing 8.0 g of KOH?

9.92 What is the molarity of a solution containing 15.6 g of KCl in 274 mL of KCl solution? (9.4)

9.93 What is the molarity of a 4.50% (m/v) KF solution? (9.4)

9.94 What is the molarity of a 15% (m/v) NaOH solution? (9.4)

9.95 How many grams of solute are in each of the following solutions? (9.4)
a. 2.5 L of a 3.0 M $Al(NO_3)_3$ solution
b. 75 mL of a 0.50 M $C_6H_{12}O_6$ solution
c. 235 mL of a 1.80 M LiCl solution

9.96 How many grams of solute are in each of the following solutions? (9.4)
a. 1.0 mL of a 0.450 M Na_2CO_3 solution
b. 10.5 mL of a 2.50 M $AgNO_3$ solution
c. 28.4 mL of a 6.00 mM HCl solution

9.97 Calculate the final concentration of the solution when water is added to prepare each of the following: (9.5)
a. 25.0 mL of a 0.200 M NaBr solution is diluted to 50.0 mL
b. 15.0 mL of a 12.0% (m/v) K_2SO_4 solution is diluted to 40.0 mL
c. 75.0 mL of a 6.00 M NaOH solution is diluted to 255 mL

9.98 Calculate the final concentration of the solution when water is added to prepare each of the following: (9.5)
a. 25.0 mL of a 18.0 M HCl solution is diluted to 500. mL
b. 50.0 mL of a 15.0% (m/v) NH_4Cl solution is diluted to 125 mL
c. 4.50 mL of a 8.50 M KOH solution is diluted to 75.0 mL

9.99 What is the final volume, in milliliters, when 25.0 mL of each of the following solutions is diluted to provide the given concentration? (9.5)
a. 10.0% (m/v) HCl solution to give a 2.50% (m/v) HCl solution
b. 5.00 M HCl solution to give a 1.00 M HCl solution
c. 6.00 M HCl solution to give a 0.500 M HCl solution

9.100 What is the final volume, in milliliters, when 5.00 mL of each of the following solutions is diluted to provide the given concentration? (9.5)
a. 20.0% (m/v) NaOH solution to give a 4.00% (m/v) NaOH solution
b. 0.600 M NaOH solution to give a 0.100 M NaOH solution
c. 16.0% (m/v) NaOH solution to give a 2.00% (m/v) NaOH solution

9.101 Why would solutions with high salt content be used to prepare dried flowers? (9.6)

9.102 Why can't you drink seawater even if you are stranded on a desert island? (9.6)

9.103 A patient on dialysis has a high level of urea, a high level of sodium, and a low level of potassium in the blood. Why is the dialyzing solution prepared with a high level of potassium but no sodium or urea? (9.6)

9.104 Why would a dialysis unit (artificial kidney) use isotonic concentrations of NaCl, KCl, $NaHCO_3$, and glucose in the dialysate? (9.6)

CHALLENGE QUESTIONS

The following groups of questions are related to the topics in this chapter. However, they do not all follow the chapter order, and they require you to combine concepts and skills from several sections. These questions will help you increase your critical thinking skills and prepare for your next exam.

9.105 In a laboratory experiment, a 10.0-mL sample of NaCl solution is poured into an evaporating dish with a mass of 24.10 g. The combined mass of the evaporating dish and NaCl solution is 36.15 g. After heating, the evaporating dish and dry NaCl have a combined mass of 25.50 g. (9.4)
a. What is the mass percent (m/m) of the NaCl solution?
b. What is the molarity (M) of the NaCl solution?
c. If water is added to 10.0 mL of the initial NaCl solution to give a final volume of 60.0 mL, what is the molarity of the diluted NaCl solution?

9.106 In a laboratory experiment, a 15.0-mL sample of KCl solution is poured into an evaporating dish with a mass of 24.10 g. The combined mass of the evaporating dish and KCl solution is 41.50 g. After heating, the evaporating dish and dry KCl have a combined mass of 28.28 g. (9.4)
a. What is the mass percent (m/m) of the KCl solution?
b. What is the molarity (M) of the KCl solution?
c. If water is added to 10.0 mL of the initial KCl solution to give a final volume of 60.0 mL, what is the molarity of the diluted KCl solution?

9.107 Potassium fluoride has a solubility of 92 g of KF in 100 g of H_2O at 18 °C. Determine if each of the following mixtures forms an unsaturated or saturated solution at 18 °C: (9.3)
a. adding 35 g of KF to 25 g of H_2O
b. adding 42 g of KF to 50. g of H_2O
c. adding 145 g of KF to 150. g of H_2O

9.108 Lithium chloride has a solubility of 55 g of LiCl in 100 g of H_2O at 25 °C. Determine if each of the following mixtures forms an unsaturated or saturated solution at 25 °C: (9.3)
a. adding 10 g of LiCl to 15 g of H_2O
b. adding 25 g of LiCl to 50. g of H_2O
c. adding 75 g of LiCl to 150. g of H_2O

9.109 A solution is prepared with 70.0 g of HNO_3 and 130.0 g of H_2O. The HNO_3 solution has a density of 1.21 g/mL. (9.4)
a. What is the mass percent (m/m) of the HNO_3 solution?
b. What is the total volume, in milliliters, of the solution?
c. What is the mass/volume percent (m/v) of the solution?
d. What is the molarity (M) of the solution?

Why would solutions with high salt content be used as food preservatives? (9.6)

9.110 A solution is prepared by dissolving 20 g of KCl (MW 74.55) in 100 mL 1 M NaCl solution. (9.4)
a. What is the molar ratio of KCl to NaCl?
b. What is the concentration of chloride ion in the solution?

9.111 How many milliliters of a 1.75 M LiCl solution contain 15.2 g of LiCl? (9.4)

9.112 How many milliliters of a 1.50 M NaBr solution contain 75.0 g of NaBr? (9.4)

ANSWERS

Answers to Study Checks

9.1 A solution of a weak electrolyte would contain mostly molecules and a few ions.

9.2 0.194 mole of Cl^-

9.3 78 g of KNO_3

9.4 $MgCl_2$ is soluble in water because chloride salts are soluble unless they contain Ag^+, Pb^{2+}, or Hg_2^{2+}.

9.5 3.4% (m/m) NaCl solution

9.6 4.8% (v/v) Br_2 in CCl_4

9.7 5.5% (m/v) NaOH solution

9.8 2.12 M KNO_3 solution

9.9 18.0 g of KCl

9.10 750 mL of HCl solution

9.11 125 mL

9.12 1.00 M KOH solution

9.13 The red blood cell will shrink (crenate).

Answers to Selected Questions and Problems

9.1 **a.** KCl solute and H2O solvent
b. Water solute and methanol solvent
c. carbon solute and iron solvent

9.3 The polar water molecules pull the K^+ and I^- ions away from the solid and into solution, where they are hydrated.

9.5 **a.** benzene **b.** benzene
c. water **d.** water

9.7 In a solution of KF, only the ions of K^+ and F^- are present in the solvent. In an HF solution, there are a few ions of H^+ and F^- present but mostly dissolved HF molecules.

9.9 **a.** $NaCl \longrightarrow Na^+ + Cl^-$
b. $MgCl \longrightarrow Mg^{2+} + 2Cl^-$
c. $HF \longrightarrow H^+ + F^-$
d. $CH_3COOH \longrightarrow H^+ + CH_3COO^-$

9.11 **a.** only ions
b. only ions
c. only molecules

9.13 **a.** strong electrolyte
b. weak electrolyte
c. nonelectrolyte

9.15 **a.** 2 Eq Li^+ **b.** 1 Eq Br^-
c. 4 Eq Mg^{2+} **d.** 9 Eq Co^{3+}

9.17 0.35mole of K^+, 0.35 mole of Cl^-

9.19 55 mEq/L

9.21 **a.** saturated **b.** unsaturated

9.23 **a.** unsaturated **b.** saturated
c. unsaturated

9.25 **a.** 408 g sugar **b.** 22 g crystallized sugar

9.27 **a.** The solubility of solid solutes typically increases as temperature increases.
b. The solubility of a gas is less at a higher temperature.
c. Gas solubility is less at a higher temperature and the CO_2 pressure in the can is increased.

9.29 **a.** soluble **b.** insoluble
c. insoluble **d.** soluble
e. soluble

9.31 5% (m/m) is 5 g of glucose in 100 g of solution, whereas 5% (m/v) is 5 g of glucose in 100 mL of solution.

9.33 **a.** 21.05%
b. 12.5%
c. 12%

9.35 **a.** 20%
b. 9.25%

9.37 **a.** 2.5 g of KCl
b. 50. g of NH_4Cl
c. 25.0 mL of acetic acid

9.39 79.9 mL of alcohol

9.41 **a.** 20. g of mannitol **b.** 240 g of mannitol

9.43 1.667 L sugar solution

9.45 **a.** 20. g of $LiNO_3$ solution
b. 400. mL of KOH solution
c. 20. mL of formic acid solution

9.47 **a.** 0.1M
b. 0.00389 M
c. 0.075 M

9.49 **a.** 596.4 g
b. 199.94 g or 200.0 g
c. 21.147 g

9.51 **a.** 1.50 L of KBr solution
b. 10.0 L of NaCl solution
c. 62.5 mL of $Ca(NO_3)_2$ solution

9.53 Adding water (solvent) to the soup increases the volume and dilutes the tomato soup concentration.

9.55 **a.** 2.0 M HCl solution
b. 2.0 M NaOH solution
c. 2.5% (m/v) KOH solution
d. 3.0% (m/v) H_2SO_4 solution

9.57 **a.** 80. mL of HCl solution
b. 250 mL of LiCl solution
c. 600. mL of H_3PO_4 solution
d. 180 mL of glucose solution

9.59 **a.** 12.8 mL of the HNO_3 solution
b. 11.9 mL of the $MgCl_2$ solution
c. 1.88 mL of the KCl solution

9.61 **a.** solution **b.** suspension

9.65 **a.** B 10% (m/v) sucrose solution
b. A 8% (m/v) albumin solution
c. B 10% (m/v) starch solution

9.67 **a.** hypotonic **b.** hypotonic
c. isotonic **d.** hypertonic

9.69 **a.** NaCl **b.** alanine
c. NaCl **d.** urea

9.71 **a.** 1 **b.** 2 **c.** 1

9.73 **a.** 3 **b.** 1 **c.** 2

9.75 The skin of the cucumber acts like a semipermeable membrane, and the more dilute solution inside flows into the brine solution.

9.77 **a.** 2 **b.** 1
c. 3 **d.** 2

9.79 Because iodine is a nonpolar molecule, it will dissolve in hexane, a nonpolar solvent. Iodine does not dissolve in water because water is a polar solvent.

9.81 **a.** unsaturated solution
b. saturated solution
c. saturated solution

9.83 **a.** soluble **b.** soluble
c. insoluble **d.** soluble
e. insoluble

9.87 38 mL of propyl alcohol solution

9.89 0.717 L of KOH solution

9.93 0.775 M KF solution

9.95 **a.** 1600 g of $Al(NO_3)_3$
b. 6.8 g of $C_6H_{12}O_6$
c. 17.9 g of LiCl

9.97 **a.** 0.100 M NaBr solution
b. 4.50% (m/v) K_2SO_4 solution
c. 1.76 M NaOH solution

9.99 **a.** 100. mL
b. 125 mL
c. 300. mL

9.103 A dialysis solution contains no sodium or urea so that these substances will flow out of the blood into the solution. High levels of potassium are maintained in the dialyzing solution so that the potassium will dialyze into the blood.

9.105 **a.** 11.6% (m/m) NaCl solution
b. 2.40 M NaCl solution
c. 0.400 M NaCl solution

9.107 **a.** saturated
b. unsaturated
c. saturated

9.109 **a.** 35.0% (m/m) HNO_3 solution
b. 165 mL
c. 42.4% (m/v) HNO_3 solution
d. 6.73 M HNO_3 solution

9.111 205 mL of LiCl solution

COMBINING IDEAS FROM Chapters 7 to 9

CI.13 In the following diagram, blue spheres represent the element **A** and yellow spheres represent the element **B**: (6.5, 7.3, 7.4)

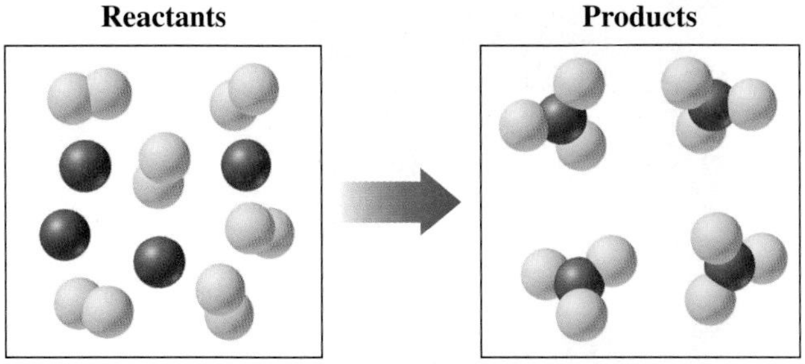

a. Write the formulas for each of the reactants and products.
b. Write a balanced chemical equation for the reaction.
c. Indicate the type of reaction as combination, decomposition, single replacement, double replacement, or combustion.

CI.14 The active ingredient in Tums is calcium carbonate. One Tums tablet contains 500. mg of calcium carbonate. (6.3, 6.4, 7.2)

The active ingredient in Tums neutralizes excess stomach acid.

a. What is the formula of calcium carbonate?
b. What is the molar mass of calcium carbonate?
c. How many moles of calcium carbonate are in one roll of Tums that contains 12 tablets?
d. If a person takes two Tums tablets, how many grams of calcium are obtained?
e. If the daily recommended quantity of Ca^{2+} to maintain bone strength in older women is 1500 mg, how many Tums tablets are needed each day to supply the needed calcium?

CI.15 Tamiflu (Oseltamivir), $C_{16}H_{28}N_2O_4$, is an antiviral drug used to treat influenza. The preparation of Tamiflu begins with the extraction of shikimic acid from the seedpods of star anise. From 2.6 g of star anise, 0.13 g of shikimic acid can be obtained and used to produce one capsule containing 75 mg of Tamiflu. The usual adult dosage for treatment of influenza is two capsules of Tamiflu daily for 5 days. (7.1, 7.2)

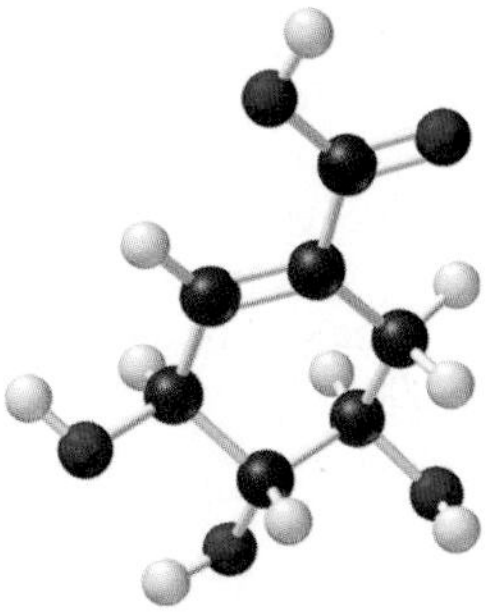

Shikimic acid is the basis for the antiviral drug in Tamiflu.

The spice called star anise is a plant source of shikimic acid.

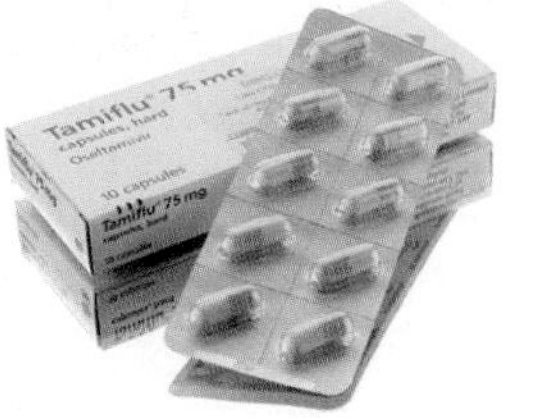

Each capsule contains 75 mg of Tamiflu.

a. If black spheres are carbon atoms, white spheres are hydrogen atoms, and red spheres are oxygen atoms, what is the formula of shikimic acid?
b. What is the molar mass of shikimic acid?
c. How many moles of shikimic acid are contained in 130 g of shikimic acid?
d. How many capsules containing 75 mg of Tamiflu could be produced from 155 g of star anise?
e. What is the molar mass of Tamiflu?
f. How many kilograms of Tamiflu would be needed to treat all the people in a city with a population of 500 000 if each person consumes two Tamiflu capsules a day for 5 days?

CI.16 The compound butyric acid gives rancid butter its characteristic odor. (7.1, 7.2)

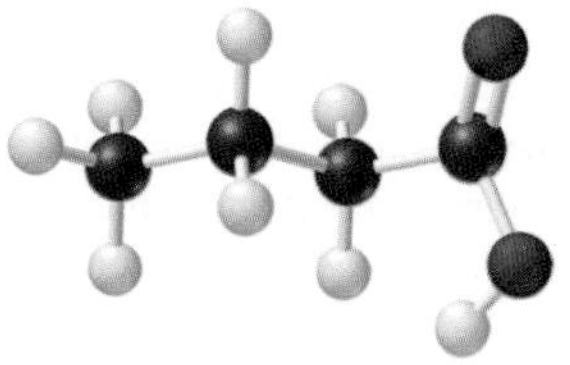

Butyric acid produces the characteristic odor of rancid butter.

a. If black spheres are carbon atoms, white spheres are hydrogen atoms, and red spheres are oxygen atoms, what is the molecular formula of butyric acid?
b. What is the molar mass of butyric acid?
c. How many grams of butyric acid contain 3.28×10^{23} atoms of oxygen?
d. How many grams of carbon are in 5.28 g of butyric acid?
e. Butyric acid has a density of 0.959 g/mL at 20 °C. How many moles of butyric acid are contained in 1.56 mL of butyric acid?
f. Identify the bonds C—C, C—H, and C—O in a molecule of butyric acid as polar covalent or nonpolar covalent.

CI.17 Methane is a major component of purified natural gas used for heating and cooking. When 1.0 mole of methane gas burns with oxygen to produce carbon dioxide and water, 883 kJ is produced. Methane gas has a density of 0.715 g/L at STP. For transport, the natural gas is cooled to −163 °C to form liquefied natural gas (LNG) with a density of 0.45 g/mL. A tank on a ship can hold 7.0 million gallons of LNG. (2.1, 2.7, 3.4, 6.7, 7.6, 7.7, 7.8, 8.6)

An LNG carrier transports liquefied natural gas.

a. Draw the electron-dot formula for methane, which has the formula CH_4.

b. What is the mass, in kilograms, of LNG (assume that LNG is all methane) transported in one tank on a ship?

c. What is the volume, in liters, of LNG (methane) from one tank when the LNG is converted to gas at STP?

d. Write a balanced equation for the combustion of methane in a gas burner, including the heat of reaction.

Methane is the fuel burned in a gas cooktop.

e. How many kilograms of oxygen are needed to react with all of the methane provided by one tank of LNG?

f. How much heat, in kilojoules, is released from burning all of the methane in one tank of LNG?

CI.18 Automobile exhaust is a major cause of air pollution. One pollutant is nitrogen oxide, which forms from nitrogen and oxygen gases in the air at the high temperatures in an automobile engine. Once emitted into the air, nitrogen oxide reacts with oxygen to produce nitrogen dioxide, a reddish brown gas with a sharp, pungent odor that makes up smog. One component of gasoline is heptane, C_7H_{16}, which has a density of 0.684 g/mL. In one year, a typical automobile uses 550 gal of gasoline and produces 41 lb of nitrogen oxide. (7.3, 7.7, 8.6)

Two gases found in automobile exhaust are carbon dioxide and nitrogen oxide.

a. Write balanced equations for the production of nitrogen oxide and nitrogen dioxide.

b. If all the nitrogen oxide emitted by one automobile is converted to nitrogen dioxide in the atmosphere, how many kilograms of nitrogen dioxide are produced in one year by a single automobile?

c. Write a balanced equation for the combustion of heptane.

d. How many moles of CO_2 are produced from the gasoline used by the typical automobile in one year, assuming the gasoline is all heptane?

e. How many liters of carbon dioxide at STP are produced in one year from the gasoline used by the typical automobile?

CI.19 When clothes have stains, bleach is often added to the wash to react with the soil and make the stains colorless. The active ingredient in bleach is sodium hypochlorite (NaClO). A bleach solution can be prepared by bubbling chlorine gas into a solution of sodium hydroxide to produce sodium hypochlorite, sodium chloride, and water. A typical bottle of bleach has a volume of 1.42 gal, with a density of 1.08 g/mL and contains 282 g of NaClO. (6.2, 6.3, 7.3, 7.7, 8.6, 9.4)

The active component of bleach is sodium hypochlorite.

a. Is sodium hypochlorite an ionic or a molecular compound?

b. What is the mass/volume percent (m/v) of sodium hypochlorite in the bleach solution?

c. Write a balanced equation for the preparation of a bleach solution.

d. How many liters of chlorine gas at STP are required to produce one bottle of bleach?

CI.20 In wine making, glucose ($C_6H_{12}O_6$) from grapes undergoes fermentation in the absence of oxygen to produce ethanol and carbon dioxide. A bottle of vintage port wine has a volume of 750 mL and contains 135 mL of ethanol (C_2H_6O). Ethanol has a density of 0.789 g/mL. In 1.5 lb of grapes, there is 26 g of glucose. (7.3, 7.7, 9.4)

When the sugar in grapes is fermented, ethanol is produced.

Port is a type of fortified wine that is produced in Portugal.

a. Calculate the volume percent (v/v) of ethanol in the port wine.

b. What is the molarity (M) of ethanol in the port wine?

c. Write a balanced equation for the fermentation reaction of glucose.

d. How many grams of glucose are required to produce one bottle of port wine?

e. How many bottles of port wine can be produced from 1.0 ton of grapes (1 ton = 2000 lb)?

ANSWERS

CI.13 **a.** Reactants: A and B_2; Products: AB_3
b. $2A + 3B_2 \longrightarrow 2AB_3$
c. combination

CI.15 **a.** $C_7H_{10}O_5$
b. 174.15 g/mole
c. 0.75 mole
d. 59 capsules
e. 312.4 g/mole
f. 400 kg

CI.17 **a.**
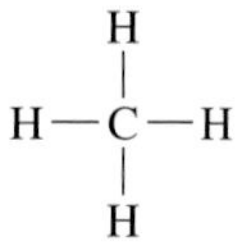

b. 1.2×10^7 kg of LNG (methane)
c. 1.7×10^{10} L of LNG (methane) at STP
d. $CH_4(g) + 2O_2(g) \xrightarrow{\Delta} CO_2(g) + 2H_2O(g) + 883\ kJ$
e. 4.8×10^7 kg of O_2
f. 6.6×10^{11} kJ

CI.19 **a.** ionic
b. 5.25 % (m/v)
c. $2NaOH(aq) + Cl_2(g) \longrightarrow NaClO(aq) + NaCl(aq) + H_2O(l)$
d. 84.9 L of chlorine gas

10 Acids and Bases and Equilibrium

A 30-year-old man has been brought to the emergency room after an automobile accident. The emergency room nurses are tending to the patient, who is unresponsive. A blood sample is taken, which is sent to Brianna, a clinical laboratory technician, who begins the process of analyzing the pH, the partial pressures of O_2 and CO_2, and the concentrations of glucose and electrolytes.

Within minutes, Brianna determines that the patient's blood pH is 7.30 and the partial pressure of CO_2 gas is above the desired level. Blood pH is typically in the range of 7.35 to 7.45, and a value less than 7.35 indicates a state of acidosis. Respiratory acidosis occurs when there is an increase in the partial pressure of CO_2 gas in the bloodstream.

Brianna recognizes these signs and immediately contacts the emergency room to inform them that the patient's airway may be blocked. In the emergency room, the staff provide the patient with an IV containing bicarbonate to increase the blood pH and begin the process of unblocking the patient's airway. Shortly afterward, the patient's airway is cleared, and his blood pH and partial pressure of CO_2 gas return to normal.

CAREER Clinical Laboratory Technician

Clinical laboratory technicians, also known as medical laboratory technicians, perform a wide variety of tests on body fluids and cells that help in the diagnosis and treatment of patients. These tests range from determining blood concentrations of glucose and cholesterol to determining drug levels in the blood for transplant patients or a patient undergoing treatment. Clinical laboratory technicians also prepare specimens in the detection of cancerous tumors, and type blood samples for transfusions. Clinical laboratory technicians must also interpret and analyze the test results, which are then passed on to the physician.

Acids and bases are important substances in health, industry, and the environment. One of the most common characteristics of acids is their sour taste. Lemons and grapefruits are sour because they contain organic acids such as citric and ascorbic acid (vitamin C). Vinegar tastes sour because it contains acetic acid. We produce lactic acid in our muscles when we exercise. Acid from bacteria turns milk sour in the production of yogurt or cottage cheese. We have hydrochloric acid in our stomachs that helps us digest food. Sometimes we take antacids, which are bases such as sodium bicarbonate or milk of magnesia, to neutralize the effects of too much stomach acid.

In the environment, the acidity, or pH, of rain, water, and soil can have significant effects. When rain becomes too acidic, it can dissolve marble statues and accelerate the corrosion of metals. In lakes and ponds, the acidity of water can affect the ability of plants and fish to survive. The acidity of soil around plants affects their growth. If the soil pH is too acidic or too basic, the roots of the plant cannot take up some nutrients. Most plants thrive in soil with a nearly neutral pH, although certain plants such as orchids, camellias, and blueberries require a more acidic soil. Major changes in the pH of the body fluids can severely affect biological activities within the cells. Buffers are present to prevent large fluctuations in pH.

LOOKING AHEAD

Citrus fruits are sour because of the presence of acids.

CHAPTER READINESS*

KEY MATH SKILLS

- Solving Equations (1.4D)
- Writing Numbers in Scientific Notation (1.4F)

CORE CHEMISTRY SKILLS

- Writing Ionic Formulas (6.2)
- Balancing a Chemical Equation (7.3)
- Using Mole–Mole Factors (7.6)
- Converting Grams to Grams (7.7)
- Using Concentration as a Conversion Factor (9.4)

*These Key Math Skills and Core Chemistry Skills from previous chapters are listed here for your review as you proceed to the new material in this chapter.

10.1 Acids and Bases

LEARNING GOAL

Describe and name acids and bases; identify Brønsted–Lowry acids and bases.

The term *acid* comes from the Latin word *acidus*, which means "sour." We are familiar with the sour tastes of vinegar and lemons and other common acids in foods.

In 1887, the Swedish chemist Svante Arrhenius was the first to describe **acids** as substances that produce hydrogen ions (H^+) when they dissolve in water. Because acids produce ions in water, they are also electrolytes. For example, hydrogen chloride ionizes completely in water to give hydrogen ions, H^+, and chloride ions, Cl^-. It is the hydrogen ions that give acids a sour taste, change blue litmus indicator to red, and corrode some metals.

$$\underset{\text{Polar molecular compound}}{HCl(g)} \xrightarrow[\text{Ionization}]{H_2O} \underset{\text{Hydrogen ion}}{H^+(aq)} + Cl^-(aq)$$

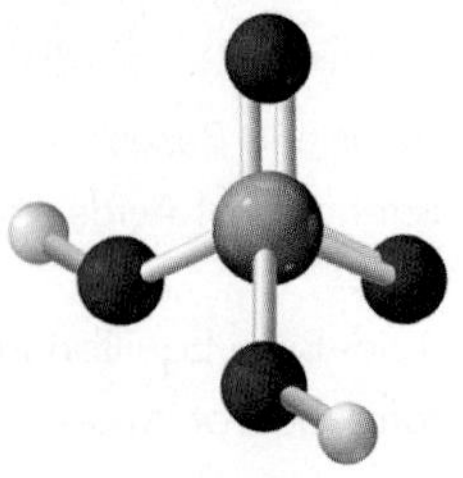

Sulfuric acid dissolves in water to produce one or two H^+ and an anion.

A soft drink contains H_3PO_4 and H_2CO_3.

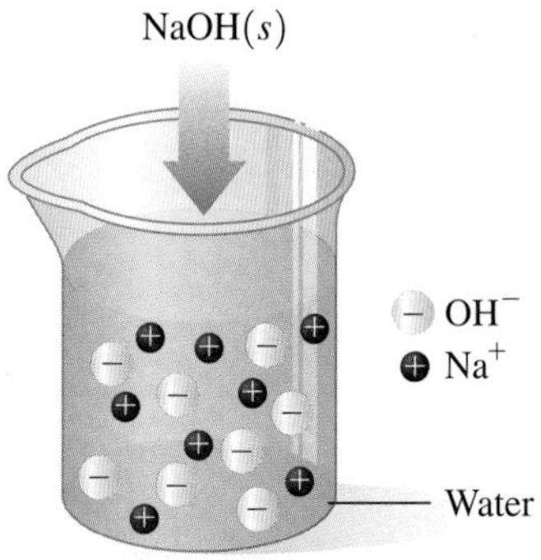

$$NaOH(s) \xrightarrow{H_2O} Na^+(aq) + OH^-(aq)$$

Ionic compound — Dissociation — Hydroxide ion

An Arrhenius base produces a cation and an OH^- anion in an aqueous solution.

Naming Acids

Acids dissolve in water to produce hydrogen ions, along with a negative ion that may be a simple nonmetal anion or a polyatomic ion. When an acid dissolves in water to produce a hydrogen ion and a simple nonmetal anion, the prefix *hydro* is used before the name of the nonmetal, and its *ide* ending is changed to *ic acid*. For example, hydrogen chloride (HCl) dissolves in water to form HCl(aq), which is named hydrochloric acid. An exception is hydrogen cyanide (HCN), which dissolves in water to form hydrocyanic acid, HCN(aq).

When an acid contains oxygen, it dissolves in water to produce a hydrogen ion and an oxygen-containing polyatomic anion. The most common form of an oxygen-containing acid has a name that ends with *ic acid*. The name of its polyatomic anion ends in *ate*. An acid that contains one less oxygen atom than the common form is named as an *ous acid*. The name of its polyatomic anion ends with *ite* (see Table 10.1). By learning the names of the most common acids, we can derive the names of the corresponding *ous acids* and their polyatomic anions.

TABLE 10.1 Names of Common Acids and Their Anions

Acid	Name of Acid	Anion	Name of Anion
HCl	**Hydro**chlor**ic acid**	Cl^-	Chlor**ide**
HBr	**Hydro**brom**ic acid**	Br^-	Brom**ide**
HI	**Hydro**iod**ic acid**	I^-	Iod**ide**
HCN	**Hydro**cyan**ic acid**	CN^-	Cyan**ide**
HNO_3	Nitr**ic acid**	NO_3^-	Nitr**ate**
HNO_2	Nitr**ous acid**	NO_2^-	Nitr**ite**
H_2SO_4	Sulfur**ic acid**	SO_4^{2-}	Sulf**ate**
H_2SO_3	Sulfur**ous acid**	SO_3^{2-}	Sulf**ite**
H_2CO_3	Carbon**ic acid**	CO_3^{2-}	Carbon**ate**
$HC_2H_3O_2$	Acet**ic acid**	$C_2H_3O_2^-$	Acet**ate**
H_3PO_4	Phosphor**ic acid**	PO_4^{3-}	Phosph**ate**
H_3PO_3	Phosphor**ous acid**	PO_3^{3-}	Phosph**ite**
$HClO_3$	Chlor**ic acid**	ClO_3^-	Chlor**ate**
$HClO_2$	Chlor**ous acid**	ClO_2^-	Chlor**ite**

Bases

You may be familiar with some household bases such as antacids, drain openers, and oven cleaners. According to the Arrhenius theory, **bases** are ionic compounds that dissociate into a metal ion and hydroxide ions (OH^-) when they dissolve in water. Thus, Arrhenius bases are also electrolytes. For example, sodium hydroxide is an Arrhenius base that dissociates in water to give sodium ions, Na^+, and hydroxide ions, OH^-.

Most Arrhenius bases are formed from Groups 1A (1) and 2A (2) metals, such as NaOH, KOH, LiOH, and $Ca(OH)_2$. The hydroxide ions (OH^-) give Arrhenius bases common characteristics such as a bitter taste and soapy, slippery feel. A base turns litmus indicator blue and phenolphthalein indicator pink.

Naming Bases

Typical Arrhenius bases are named as hydroxides.

Base	Name
LiOH	Lithium **hydroxide**
NaOH	Sodium **hydroxide**
KOH	Potassium **hydroxide**
$Ca(OH)_2$	Calcium **hydroxide**
$Al(OH)_3$	Aluminum **hydroxide**

SAMPLE PROBLEM 10.1 Dissociation of an Arrhenius Base

Write the equation for the dissociation of solid calcium hydroxide in water.

SOLUTION

When calcium hydroxide, which has the formula $Ca(OH)_2$, dissolves in water, the solution contains calcium ions (Ca^{2+}) and twice as many hydroxide ions (OH^-). The equation is written as

$$Ca(OH)_2(s) \xrightarrow{H_2O} Ca^{2+}(aq) + 2OH^-(aq)$$

STUDY CHECK 10.1

Write the equation for the dissociation of solid lithium hydroxide in water.

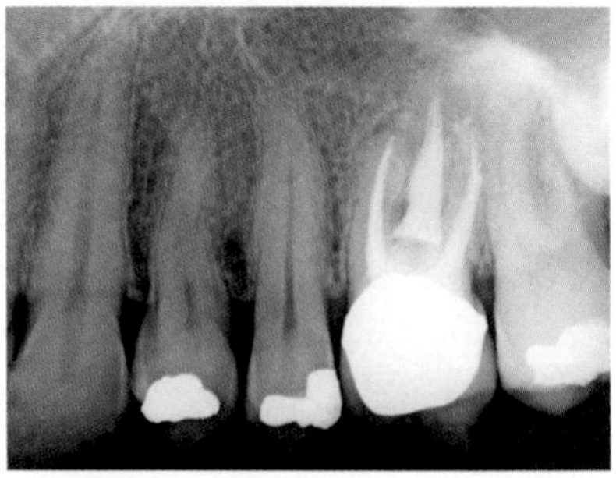

Calcium hydroxide, $Ca(OH)_2$, also called slaked lime, is used in the food industry to produce beverages, and in dentistry as a filler for root canals.

Brønsted–Lowry Acids and Bases

In 1923, J. N. Brønsted in Denmark and T. M. Lowry in Great Britain expanded the definition of acids and bases to include bases that do not contain OH^- ions. A **Brønsted–Lowry acid** can donate a hydrogen ion (H^+) to another substance, and a **Brønsted–Lowry base** can accept a hydrogen ion (H^+).

A Brønsted–Lowry acid is a substance that donates H^+.

A Brønsted–Lowry base is a substance that accepts H^+.

A free hydrogen ion, H^+, does not actually exist in water. Its attraction to polar water molecules is so strong that the H^+ bonds to a water molecule and forms a **hydronium ion**, H_3O^+.

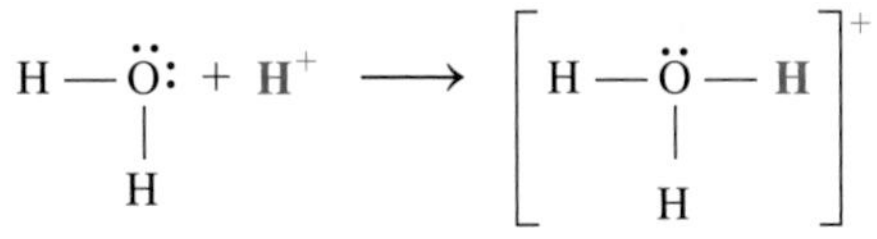

Water Hydrogen ion Hydronium ion

We can write the formation of a hydrochloric acid solution as a transfer of H^+ from hydrogen chloride to water. By accepting H^+ in the reaction, water is acting as a base according to the Brønsted–Lowry concept.

$$HCl + H_2O \longrightarrow H_3O^+ + Cl^-$$

HCl	H₂O		H₃O⁺	Cl⁻
Hydrogen chloride	Water		Hydronium ion	Chloride ion
Acid (H^+ donor)	Base (H^+ acceptor)		Acidic solution	

Ammonia, NH_3, acts as a base by accepting H^+ when it reacts with water. Because the nitrogen atom of NH_3 has a stronger attraction for H^+ than the oxygen of water, water acts as an acid by donating H^+.

$$NH_3 + H_2O \rightleftarrows NH_4^+ + OH^-$$

NH₃	H₂O		NH₄⁺	OH⁻
Ammonia	Water		Ammonium ion	Hydroxide ion
Base (H^+ acceptor)	Acid (H^+ donor)		Basic solution	

Table 10.2 compares some characteristics of acids and bases.

TABLE 10.2 Some Characteristics of Acids and Bases

Characteristic	Acids	Bases
Arrhenius	Produce H^+	Produce OH^-
Brønsted–Lowry	Donate H^+	Accept H^+
Electrolytes	Yes	Yes
Taste	Sour	Bitter, chalky
Feel	May sting	Soapy, slippery
Litmus	Red	Blue
Phenolphthalein	Colorless	Pink
Neutralization	Neutralize bases	Neutralize acids

SAMPLE PROBLEM 10.2 Acids and Bases

In each of the following equations, identify the reactant that is a Brønsted–Lowry acid and the reactant that is a Brønsted–Lowry base:

a. $HBr(aq) + H_2O(l) \longrightarrow Br^-(aq) + H_3O^+(aq)$
b. $CN^-(aq) + H_2O(l) \rightleftarrows HCN(aq) + OH^-(aq)$

SOLUTION

a. HBr, acid; H_2O, base
b. H_2O, acid; CN^-, base

STUDY CHECK 10.2

When HNO_3 reacts with water, water acts as a base (H^+ acceptor). Write the equation for the reaction.

CORE CHEMISTRY SKILL

Identifying Conjugate Acid–Base Pairs

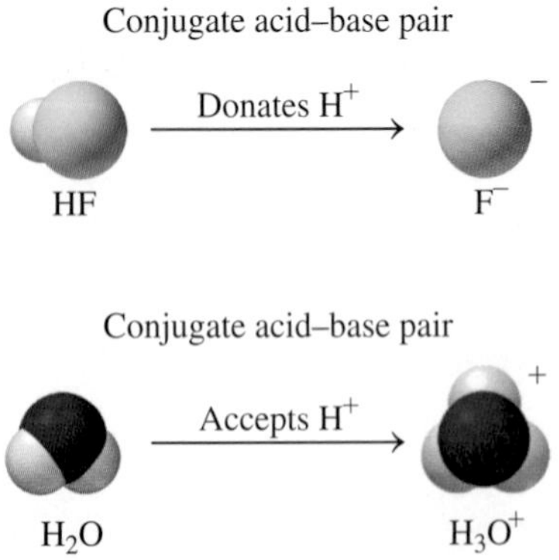

Conjugate Acid–Base Pairs

According to the Brønsted–Lowry theory, a **conjugate acid–base pair** consists of molecules or ions related by the loss of one H^+ by an acid, and the gain of one H^+ by a base. Every acid–base reaction contains two conjugate acid–base pairs because an H^+ is transferred in both the forward and the reverse directions. When an acid such as HF loses an H^+, the conjugate base F^- is formed. When the base H_2O gains an H^+, its conjugate acid, H_3O^+, is formed.

Because the overall reaction of HF is reversible, the conjugate acid H_3O^+ can donate H^+ to the conjugate base F^- and re-form the acid HF and the base H_2O. Using the relationship of loss and gain of one H^+, we can now identify the conjugate acid–base pairs as HF/F^- along with H_3O^+/H_2O.

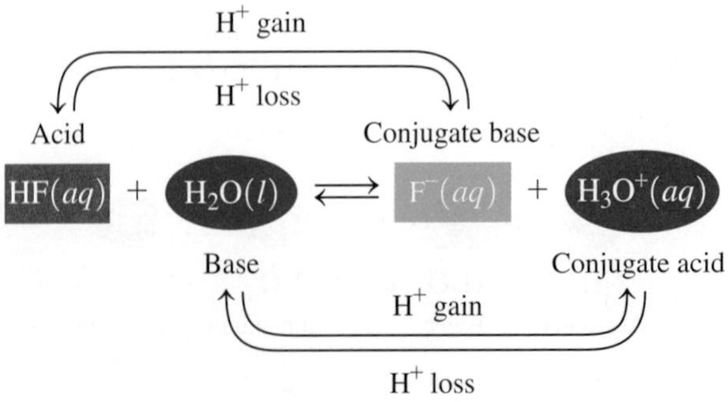

In another reaction, ammonia, NH_3, accepts H^+ from H_2O to form the conjugate acid NH_4^+ and conjugate base OH^-. Each of these conjugate acid–base pairs, NH_4^+/NH_3 and H_2O/OH^-, is related by the loss and gain of one H^+.

Conjugate acid–base pair

$$NH_3(g) + H_2O(l) \rightleftarrows NH_4^+(aq) + OH^-(aq)$$

Conjugate acid–base pair

In these two examples, we see that water can act as an acid when it donates H^+ or as a base when it accepts H^+. Substances that can act as both acids and bases are **amphoteric**. For water, the most common amphoteric substance, the acidic or basic behavior depends on the other reactant. Water donates H^+ when it reacts with a stronger base, and it accepts H^+ when it reacts with a stronger acid. Another example of an amphoteric substance is bicarbonate, HCO_3^-. With a base, HCO_3^- acts as an acid and donates one H^+ to give CO_3^{2-}. However, when HCO_3^- reacts with an acid, it acts as a base and accepts one H^+ to form H_2CO_3.

H_3O^+ / H_2CO_3	← Acts as a Base	H_2O / HCO_3^-	Acts as an Acid →	OH^- / CO_3^{2-}

Amphoteric substances act as both acids and bases.

SAMPLE PROBLEM 10.3 Identifying Conjugate Acid–Base Pairs

Identify the conjugate acid–base pairs in the following reaction:

$$HBr(aq) + NH_3(aq) \longrightarrow Br^-(aq) + NH_4^+(aq)$$

SOLUTION

STEP 1 **Identify the reactant that loses H^+ as the acid.** In the reaction, HBr donates H^+ to NH_3. Thus HBr is the acid and Br^- is its conjugate base.

STEP 2 **Identify the reactant that gains H^+ as the base.** In the reaction, NH_3 gains H^+ to form NH_4^+. Thus, NH_3 is the base and NH_4^+ is its conjugate acid.

STEP 3 **Write the conjugate acid–base pairs.** The conjugate acid–base pairs are HBr/Br^- and NH_4^+/NH_3.

STUDY CHECK 10.3

In the following reaction, identify the conjugate acid–base pairs:

$$HCN(aq) + SO_4^{2-}(aq) \rightleftarrows CN^-(aq) + HSO_4^-(aq)$$

Guide to Writing Conjugate Acid–Base Pairs

STEP 1
Identify the reactant that loses H^+ as the acid.

STEP 2
Identify the reactant that gains H^+ as the base.

STEP 3
Write the conjugate acid–base pairs.

QUESTIONS AND PROBLEMS

10.1 Acids and Bases

LEARNING GOAL Describe and name acids and bases; identify Brønsted–Lowry acids and bases.

10.1 Indicate whether each of the following statements is characteristic of an acid, a base, or both:
a. has a sour taste
b. neutralizes bases
c. produces H^+ ions in water
d. is named barium hydroxide
e. is an electrolyte

10.2 Indicate whether each of the following statements is characteristic of an acid, a base, or both:
a. neutralizes acids
b. produces OH^- ions in water
c. has a slippery feel
d. turns litmus red
e. conducts an electrical current in solution

10.3 Name each of the following acids or bases:
a. H_3PO_4 **b.** LiOH **c.** HI
d. HCN **e.** HNO_2 **f.** $Ca(OH)_2$

10.4 Name each of the following acids or bases:
a. $Al(OH)_3$ **b.** HBr **c.** H_2SO_4
d. KOH **e.** HNO_2 **f.** $HClO_2$

10.5 Write the formula for each of the following inorganic acids and bases:
a. cesium hydroxide **b.** hydrofluoric acid
c. phosphoric acid **d.** hydrobromic acid
e. sulfurous acid **f.** nitrous acid

10.6 Write the formula for each of the following acids and bases:
a. barium hydroxide **b.** hydroiodic acid
c. nitric acid **d.** strontium hydroxide
e. sodium hydroxide **f.** chloric acid

10.7 Identify the reactant that is a Brønsted–Lowry acid and the reactant that is a Brønsted–Lowry base in each of the following:
a. $HI(aq) + H_2O(l) \longrightarrow I^-(aq) + H_3O^+(aq)$
b. $F^-(aq) + H_2O(l) \rightleftarrows HF(aq) + OH^-(aq)$

10.8 Identify the reactant that is a Brønsted–Lowry acid and the reactant that is a Brønsted–Lowry base in each of the following:
a. $CO_3^{2-}(aq) + H_2O(l) \rightleftarrows HCO_3^-(aq) + OH^-(aq)$
b. $H_2SO_4(aq) + H_2O(l) \longrightarrow HSO_4^-(aq) + H_3O^+(aq)$

10.9 Write the formula for the conjugate base for each of the following acids:
a. HCl
b. HI
c. HNO_3
d. $HClO_3$

10.10 Write the formula for the conjugate base for each of the following acids:
a. HSO_3^-
b. H_3O^+
c. HPO_4^{2-}
d. HNO_2

10.11 Write the formula for the conjugate acid for each of the following bases:
a. CO_3^{2-}
b. H_2O
c. HPO_4^{2-}
d. CN^-

10.12 Write the formula for the conjugate acid for each of the following bases:
a. SO_4^{2-}
b. CN^-
c. OH^-
d. ClO_2^-

10.13 Identify the Brønsted–Lowry acid–base pairs in each of the following equations:
a. $H_2CO_3(aq) + H_2O(l) \rightleftarrows HCO_3^-(aq) + H_3O^+(aq)$
b. $NH_4^+(aq) + H_2O(l) \rightleftarrows NH_3(aq) + H_3O^+(aq)$
c. $HCN(aq) + NO_2^-(aq) \rightleftarrows CN^-(aq) + HNO_2(aq)$

10.14 Identify the Brønsted–Lowry acid–base pairs in each of the following equations:
a. $H_3PO_4(aq) + H_2O(l) \rightleftarrows H_2PO_4^-(aq) + H_3O^+(aq)$
b. $CO_3^{2-}(aq) + H_2O(l) \rightleftarrows HCO_3^-(aq) + OH^-(aq)$
c. $H_3PO_4(aq) + NH_3(aq) \rightleftarrows H_2PO_4^-(aq) + NH_4^+(aq)$

LEARNING GOAL

Write equations for the ionization of acids and bases.

10.2 Strengths of Acids and Bases

The *strength* of an acid is determined by the moles of H_3O^+ that are produced for each mole of acid that dissolves. The *strength* of a base is determined by the moles of OH^- that are produced for each mole of base that dissolves. Strong acids and strong bases ionize completely in water whereas weak acids and weak bases ionize only slightly in water, leaving most of the initial acid or base as molecules.

Strong and Weak Acids

Strong acids are examples of strong electrolytes because they donate H^+ so easily that their ionization in water is virtually complete. For example, when HCl, a strong acid, ionizes in water, H^+ is transferred to H_2O; the resulting solution contains only the ions H_3O^+ and Cl^-. We consider the reaction of HCl in H_2O as going 100% to products. Thus, the equation for a strong acid such as HCl is written with a single arrow to the products.

$$HCl(g) + H_2O(l) \longrightarrow H_3O^+(aq) + Cl^-(aq)$$

There are only six common strong acids. All other acids are weak (see Table 10.3). **Weak acids** are weak electrolytes because they ionize slightly in water, which produces only a few ions. A solution of a strong acid contains all ions, whereas a solution of a weak acid contains mostly molecules and few ions (see Figure 10.1).

TABLE 10.3 Common Strong and Weak Acids

Strong Acids	
Hydroiodic acid	HI
Hydrobromic acid	HBr
Perchloric acid	$HClO_4$
Hydrochloric acid	HCl
Sulfuric acid	H_2SO_4
Nitric acid	HNO_3
Weak Acids	
Hydronium ion	H_3O^+
Hydrogen sulfate ion	HSO_4^-
Phosphoric acid	H_3PO_4
Hydrofluoric acid	HF
Nitrous acid	HNO_2
Acetic acid	$HC_2H_3O_2$
Carbonic acid	H_2CO_3
Hydrosulfuric acid	H_2S
Dihydrogen phosphate	$H_2PO_4^{2-}$
Ammonium ion	NH_4^+
Hydrocyanic acid	HCN
Bicarbonate ion	HCO_3^-
Hydrogen sulfide ion	HS^-
Water	H_2O

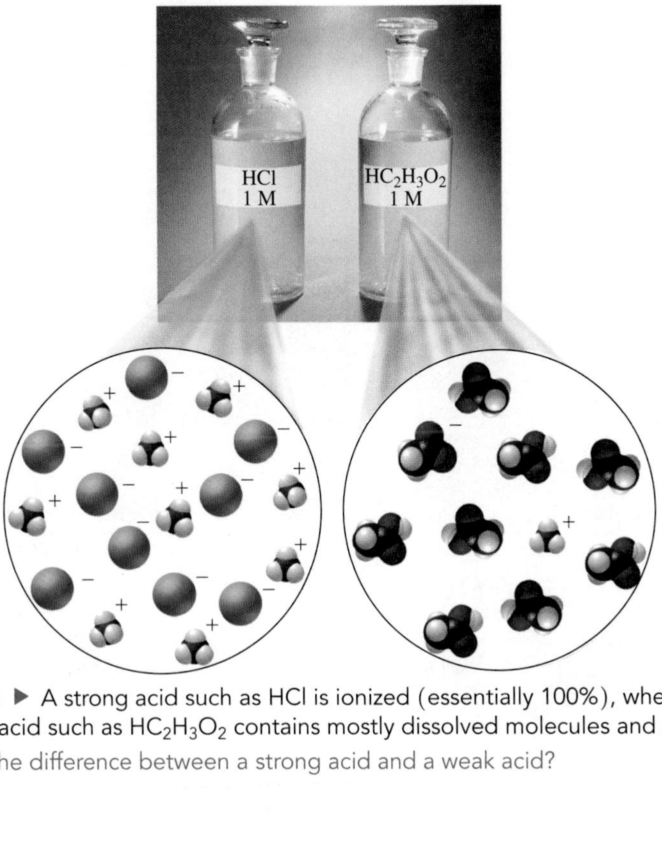

FIGURE 10.1 ▶ A strong acid such as HCl is ionized (essentially 100%), whereas a solution of weak acid such as $HC_2H_3O_2$ contains mostly dissolved molecules and a few ions.

Q What is the difference between a strong acid and a weak acid?

Acids produce hydrogen ions in aqueous solutions.

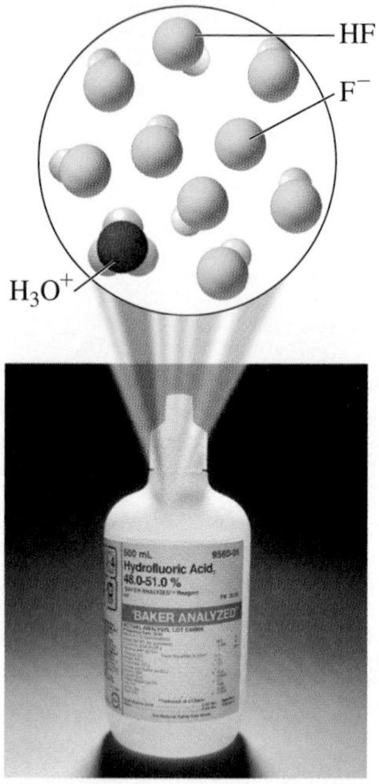

Hydrofluoric acid is the only halogen acid that is a weak acid.

Many of the products you use at home contain weak acids. In carbonated soft drinks, CO_2 dissolves in water to form carbonic acid, H_2CO_3, a weak acid. A weak acid such as H_2CO_3 contains mostly H_2CO_3 molecules and a few H_3O^+ and HCO_3^- ions. A weak acid such as carbonic acid, H_2CO_3, is written with a double arrow. Carbonic acid is a *diprotic acid* that has two H^+, which ionize one at a time. Because HCO_3^- is also a weak acid, a second ionization produces another hydronium ion and the carbonate ion, CO_3^{2-}.

$$H_2CO_3(aq) + H_2O(l) \rightleftarrows H_3O^+(aq) + HCO_3^-(aq)$$

$$\underset{\text{Bicarbonate ion (hydrogen carbonate)}}{HCO_3^-(aq)} + H_2O(l) \rightleftarrows H_3O^+(aq) + \underset{\text{Carbonate ion}}{CO_3^{2-}(aq)}$$

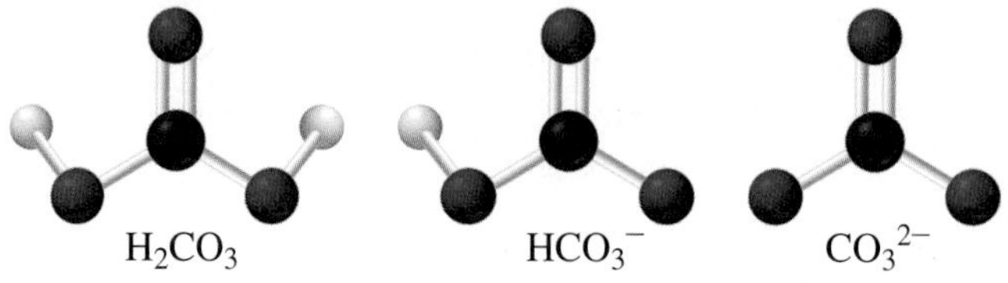

Citric acid is a weak acid found in fruits and fruit juices such as lemons, oranges, and grapefruit. In the vinegar used in salad dressings, the weak acid, acetic acid, $HC_2H_3O_2$, is present as a 5% (m/v) acetic acid solution.

$$\underset{\text{Acetic acid}}{HC_2H_3O_2(l)} + H_2O(l) \rightleftarrows H_3O^+(aq) + \underset{\text{Acetate ion}}{C_2H_3O_2^-(aq)}$$

In summary, a strong acid such as HI in water ionizes completely to form an aqueous solution of the ions H_3O^+ and I^-. A weak acid such as HF ionizes only slightly in water to form an aqueous solution that consists mostly of HF molecules and a few H_3O^+ and F^- ions (see Figure 10.2).

$$\text{Strong acid: } HI(aq) + H_2O(l) \longrightarrow H_3O^+(aq) + I^-(aq) \quad \text{Completely ionized}$$

$$\text{Weak acid: } HF(aq) + H_2O(l) \rightleftarrows H_3O^+(aq) + F^-(aq) \quad \text{Slightly ionized}$$

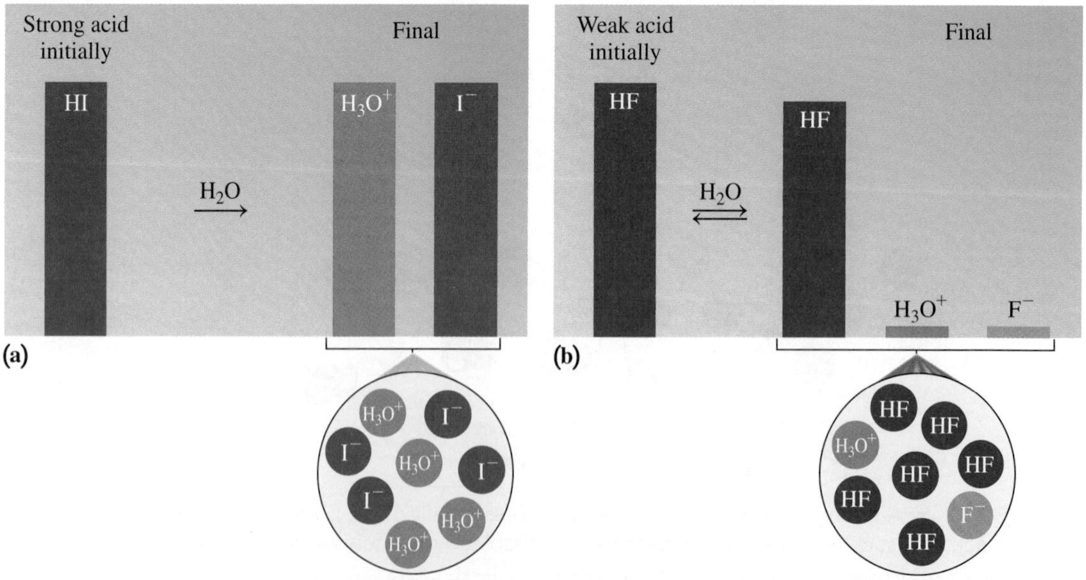

FIGURE 10.2 ▶ After ionization in water, **(a)** the strong acid (HI) has a high concentration of H_3O^+ and I^-, whereas **(b)** the weak acid (HF) has a high concentration of HF and low concentrations of H_3O^+ and F^-.

Q How do the heights of H_3O^+ and F^- compare to the height of the weak acid HF in the bar diagram?

Strong and Weak Bases

As strong electrolytes, **strong bases** dissociate completely in water. Because these strong bases are ionic compounds, they dissociate in water to give an aqueous solution of metal ions and hydroxide ions. The Group 1A (1) hydroxides are very soluble in water, which can give high concentrations of OH^- ions. For example, when KOH forms a KOH solution, it contains only the ions K^+ and OH^-.

$$KOH(s) \xrightarrow{H_2O} K^+(aq) + OH^-(aq)$$

A few strong bases are less soluble in water, but what does dissolve dissociates completely as ions.

Strong Bases

Lithium hydroxide LiOH

Sodium hydroxide NaOH

Potassium hydroxide KOH

Strontium hydroxide $Sr(OH)_2$*

Calcium hydroxide $Ca(OH)_2$*

Barium hydroxide $Ba(OH)_2$*

*Low solubility, but they dissociate completely.

Bases in household products are used to remove grease and to open drains.

Sodium hydroxide, NaOH (also known as lye), is used in household products to remove grease in ovens and to clean drains. Because high concentrations of hydroxide ions cause severe damage to the skin and eyes, directions must be followed carefully when such products are used in the home, and use in the chemistry laboratory should be carefully supervised. If you spill an acid or a base on your skin or get some in your eyes, be sure to flood the area immediately with water for at least 10 minutes and seek medical attention.

Weak bases are weak electrolytes that are poor acceptors of hydrogen ions and produce very few ions in solution. A typical weak base, ammonia, NH_3, is found in window cleaners. In an aqueous solution, only a few ammonia molecules accept H^+ to form NH_4^+ and OH^-.

$$\underset{\text{Ammonia}}{NH_3(g)} + H_2O(l) \rightleftarrows \underset{\text{Ammonium hydroxide}}{NH_4^+(aq) + OH^-(aq)}$$

Bases in Household Products

Weak Bases

Window cleaner, ammonia, NH_3

Bleach, NaOCl

Laundry detergent, Na_2CO_3, Na_3PO_4

Toothpaste and baking soda, $NaHCO_3$

Baking powder, scouring powder, Na_2CO_3

Lime for lawns and agriculture, $CaCO_3$

Laxatives, antacids, $Mg(OH)_2$, $Al(OH)_3$

Strong Bases

Drain cleaner, oven cleaner, NaOH

QUESTIONS AND PROBLEMS

10.2 Strengths of Acids and Bases

LEARNING GOAL Write equations for the ionization of acids and bases.

10.15 Using Table 10.3, identify the stronger acid in each of the following pairs:
- **a.** HI or NH_4^+
- **b.** $HClO_4$ or H_2S
- **c.** HNO_3 or H_3O^+

10.16 Using Table 10.3, identify the stronger acid in each of the following pairs:
- **a.** NH_4^+ or H_3O^+
- **b.** H_2SO_4 or HCN
- **c.** H_2O or H_2CO_3

10.17 Using Table 10.3, identify the weaker acid in each of the following pairs:
- **a.** HCN or HF
- **b.** HBr or HCO_3^-
- **c.** H_2SO_4 or H_3PO_4

10.18 Using Table 10.3, identify the weaker acid in each of the following pairs:
- **a.** HNO_3 or HCO_3^-
- **b.** HSO_4^- or H_2O
- **c.** H_2SO_4 or H_2CO_3

LEARNING GOAL

Use the concept of reversible reactions to explain acid–base equilibrium. Use Le Châtelier's principle to determine the effect on equilibrium concentrations when reaction conditions change.

10.3 Acid–Base Equilibrium

As we have seen, reactants in acid–base reactions are not always completely converted to products because a reverse reaction takes place in which products form reactants. A reversible reaction proceeds in both the forward and reverse directions. That means there are two reactions taking place: One is the reaction in the forward direction, while the other is the reaction in the reverse direction. Initially, the forward reaction occurs at a faster rate than the reverse reaction to form products. As the initial reactants are consumed, products accumulate. Then the forward reaction slows, and the rate of the reverse reaction increases.

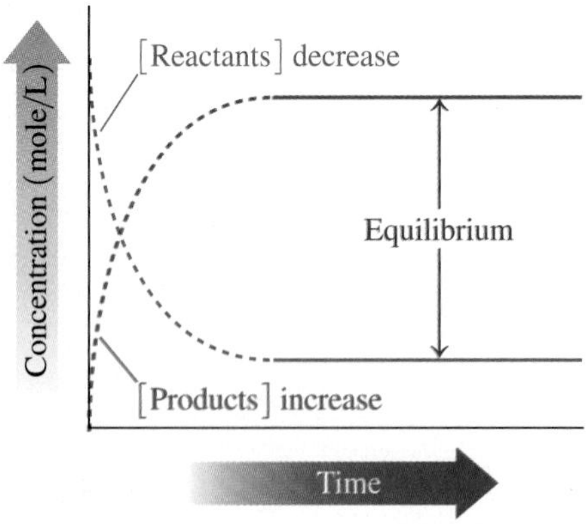

Equilibrium is reached when there are no further changes in the concentrations of reactants and products.

Equilibrium

Eventually, the rates of the forward and reverse reactions become equal. This means that the reactants form products at the same rate that the products form reactants. **Equilibrium** has been reached when no further change takes place in the concentrations of the reactants and products. However, the forward and reverse reactions continue at equal rates.

Let us look at the reaction of the weak acid HF and H_2O as it proceeds to equilibrium. Initially, only the reactants HF and H_2O are present.

$$\text{Forward reaction:}\quad HF(aq) + H_2O(l) \longrightarrow F^-(aq) + H_3O^+(aq)$$

As F^- and H_3O^+ products build up, the rate of the reverse reaction increases, while the rate of the forward reaction decreases.

$$\text{Reverse reaction:}\quad F^-(aq) + H_3O^+(aq) \longrightarrow HF(aq) + H_2O(l)$$

Eventually, the rates of the forward and reverse reactions become equal, which means that equilibrium has been reached. Then the concentrations of the reactants and products remain constant even though the forward and reverse reactions continue. The forward and reverse reactions are usually shown together in a single equation, using a double arrow.

$$HF(aq) + H_2O(l) \underset{\text{Reverse reaction}}{\overset{\text{Forward reaction}}{\rightleftarrows}} F^-(aq) + H_3O^+(aq)$$

SAMPLE PROBLEM 10.4 Reversible Reactions and Equilibrium

Complete each of the following with *equal* or *not equal*, *faster* or *slower*, *change* or *do not change*:

a. Before equilibrium is reached, the concentrations of the reactants and products ____________.

b. Initially, reactants have a ____________ rate of reaction than the rate of reaction of the products.

c. At equilibrium, the rate of the forward reaction is ____________ to the rate of the reverse reaction.

d. At equilibrium, the concentrations of the reactants and products ____________.

SOLUTION

a. change
b. faster
c. equal
d. do not change

STUDY CHECK 10.4

What does the double arrow indicate about the following reaction at equilibrium?

$$\underset{\text{Ammonia}}{NH_3(g)} + H_2O(l) \rightleftarrows \underset{\text{Ammonium hydroxide}}{NH_4^+(aq) + OH^-(aq)}$$

Le Châtelier's Principle

CORE CHEMISTRY SKILL

Using Le Châtelier's Principle

When we alter the concentration of a reactant or product of a system at equilibrium, the rates of the forward and reverse reactions will no longer be equal. We say that a *stress* is placed on the equilibrium. **Le Châtelier's principle** states that when equilibrium is disturbed, the rates of the forward and reverse reactions change to relieve that stress and reestablish equilibrium.

Suppose we have two water tanks connected by a pipe. When the water levels in the tanks are equal, water flows in the forward direction from Tank A to Tank B at the same rate as it flows in the reverse direction from Tank B to Tank A. If we add more water to Tank A, there is an increase in the rate at which water flows from Tank A into Tank B, which is shown with a longer arrow. Equilibrium is reestablished when the water levels in both tanks become equal. The water levels are higher than before, but the water flows equally between Tank A and Tank B.

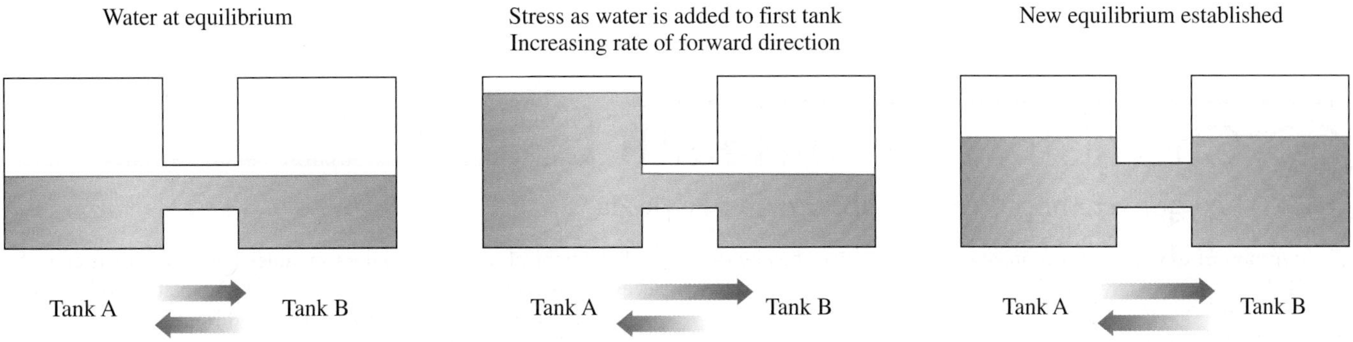

When water is added to one tank, the levels readjust to equalize.

Effect of Concentration Changes on Equilibrium

We will use the reaction of HF and H_2O to illustrate how a change in concentration disturbs a reaction at equilibrium and how the system responds to that stress.

$$HF(aq) + H_2O(l) \rightleftarrows F^-(aq) + H_3O^+(aq)$$

Suppose that more HF is added to the equilibrium mixture, which increases the concentration of HF. The system relieves this stress by increasing the rate of the forward reaction. According to Le Châtelier's principle, adding more reactant causes the system to *shift* in the direction of the products until equilibrium is reestablished.

Add HF

$$HF(aq) + H_2O(l) \rightleftarrows F^-(aq) + H_3O^+(aq)$$

In another example, suppose that some HF is removed from the reaction mixture at equilibrium. The decrease in HF concentration slows the rate of the forward reaction. According to Le Châtelier's principle, the stress of removing some of the reactant causes the system to *shift* in the direction of the reactants until equilibrium is reestablished.

Remove HF

$$HF(aq) + H_2O(l) \rightleftarrows F^-(aq) + H_3O^+(aq)$$

Stress is also placed on a system at equilibrium by changing the concentration of a product. For example, if more F^- is added to the equilibrium mixture, the rate of the reverse reaction increases as products are converted to reactants until equilibrium is reestablished. According to Le Châtelier's principle, the addition of more product causes the system to *shift* in the direction of the reactants.

Add F^-

$$HF(aq) + H_2O(l) \rightleftarrows F^-(aq) + H_3O^+(aq)$$

In summary, Le Châtelier's principle indicates that a stress caused by adding a substance at equilibrium is relieved when the system shifts the reaction away from that substance. When some of a substance is removed, the equilibrium system shifts in the direction of that substance. These features of Le Châtelier's principle are summarized in Table 10.4.

TABLE 10.4 Effect of Concentration Changes on Equilibrium

Stress	Change	Shift in the Direction of
Add reactant	Increases forward reaction rate	Products
Remove reactant	Decreases forward reaction rate	Reactants
Add product	Increases reverse reaction rate	Reactants
Remove product	Decreases reverse reaction rate	Products

Chemistry Link to **Health**

Oxygen–Hemoglobin Equilibrium and Hypoxia

The transport of oxygen involves an equilibrium between hemoglobin (Hb), oxygen, and oxyhemoglobin (HbO_2).

$$Hb(aq) + O_2(g) \rightleftarrows HbO_2(aq)$$

When the O_2 level is high in the alveoli of the lung, the reaction shifts in the direction of the product HbO_2. In the tissues where O_2 concentration is low, the reverse reaction releases the oxygen from the hemoglobin.

At normal atmospheric pressure, oxygen diffuses into the blood because the partial pressure of oxygen in the alveoli is higher than that in the blood. At an altitude above 8000 ft, a decrease in the atmospheric pressure results in a significant reduction in the partial pressure of oxygen, which means that less oxygen is available for the blood and body tissues. The fall in atmospheric pressure at higher altitudes decreases the partial pressure of inhaled oxygen. At an altitude of 18 000 ft, a person will obtain 29% less oxygen. When oxygen levels are lowered, a person may experience *hypoxia*, characterized by increased respiratory rate, headache, decreased mental acuteness, fatigue, decreased physical coordination, nausea, vomiting, and cyanosis. A similar problem occurs in persons with a history of lung disease that impairs gas diffusion in the alveoli or in persons with a reduced number of red blood cells, such as smokers.

According to Le Châtelier's principle, we see that a decrease in oxygen will shift the system in the direction of the reactants to reestablish equilibrium. Such a shift depletes the concentration of HbO_2 and causes the hypoxia.

$$Hb(aq) + O_2(g) \longleftarrow HbO_2(aq)$$

Treatment of altitude sickness includes hydration, rest, and, if necessary, descending to a lower altitude. The adaptation to lowered oxygen levels requires about 10 days. During this time, the bone marrow increases red blood cell production, providing more red blood cells and more hemoglobin. A person living at a high altitude can have 50% more red blood cells than someone at sea level. This increase in hemoglobin causes a shift in the equilibrium back in the direction of HbO_2 product. Eventually, the higher concentration of HbO_2 will provide more oxygen to the tissues and the symptoms of hypoxia will lessen.

$$Hb(aq) + O_2(g) \longrightarrow HbO_2(aq)$$

For those who climb high mountains, it is important to stop and acclimatize for several days at increasing altitudes. At very high altitudes, it may be necessary to use an oxygen tank.

Hypoxia may occur at high altitudes where the oxygen concentration is lower.

SAMPLE PROBLEM 10.5 Effect of Changes in Concentration on Equilibrium

An important reaction in the body fluids is

$$H_2CO_3(aq) + H_2O(l) \rightleftharpoons HCO_3^-(aq) + H_3O^+(aq)$$

Use Le Châtelier's principle to predict whether the system shifts in the direction of products or reactants for each of the following:

a. adding some $H_2CO_3(aq)$
b. removing some $HCO_3^-(aq)$
c. adding some $H_3O^+(aq)$

SOLUTION

According to Le Châtelier's principle, when stress is applied to a reaction at equilibrium, the system shifts to relieve that stress.

a. When the concentration of the reactant H_2CO_3 increases, the rate of the forward reaction increases to shift the system in the direction of the products until equilibrium is reestablished.
b. When the concentration of the product HCO_3^- decreases, the rate of the reverse reaction decreases. The system shifts in the direction of the products until equilibrium is reestablished.
c. When the concentration of the product H_3O^+ increases, the rate of the reverse reaction increases. The system shifts in the direction of the reactants until equilibrium is reestablished.

STUDY CHECK 10.5

Using the reaction in Sample Problem 10.5, predict whether removing H_2CO_3 causes the system to shift in the direction of products or reactants.

QUESTIONS AND PROBLEMS

10.3 Acid–Base Equilibrium

LEARNING GOAL Use the concept of reversible reactions to explain acid–base equilibrium. Use Le Châtelier's principle to determine the effect on equilibrium concentrations when reaction conditions change.

10.19 What is meant by the term *reversible reaction*?

10.20 When does a reversible reaction reach equilibrium?

10.21 Which of the following are at equilibrium?
a. The rate of the forward reaction is twice as fast as the rate of the reverse reaction.
b. The concentrations of the reactants and the products do not change.
c. The rate of the reverse reaction does not change.

10.22 Which of the following are *not* at equilibrium?
a. The rates of the forward and reverse reactions are equal.
b. The rate of the forward reaction does not change.
c. The concentrations of reactants and the products are not constant.

10.23 Use Le Châtelier's principle to predict whether each the following changes causes the system to shift in the direction of products or reactants:

$$HCHO_2(aq) + H_2O(l) \rightleftharpoons CHO_2^-(aq) + H_3O^+(aq)$$

a. adding some $CHO_2^-(aq)$
b. removing some $HCHO_2(aq)$
c. removing some $H_3O^+(aq)$
d. adding some $HCHO_2(aq)$

10.24 Use Le Châtelier's principle to predict whether each the following changes causes the system to shift in the direction of products or reactants:

$$HNO_2(aq) + H_2O(l) \rightleftharpoons NO_2^-(aq) + H_3O^+(aq)$$

a. adding some $HNO_2(aq)$
b. removing some $NO_2^-(aq)$
c. adding some $H_3O^+(aq)$
d. removing some $HNO_2(aq)$

LEARNING GOAL

Use the ion product for water to calculate the $[H_3O^+]$ and $[OH^-]$ in an aqueous solution.

10.4 Ionization of Water

We have seen that in some acid–base reactions, water is *amphoteric*, which means that it can act either as an acid or as a base. In pure water, there is a forward reaction between two water molecules that transfers H^+ from one water molecule to the other. One molecule acts as an acid by losing H^+, and the water molecule that gains H^+ acts as a base. Every time H^+ is transferred between two water molecules, the products are one H_3O^+ and one OH^-, which react in the reverse direction to re-form two water molecules. Thus, equilibrium is reached between the conjugate acid–base pairs of water.

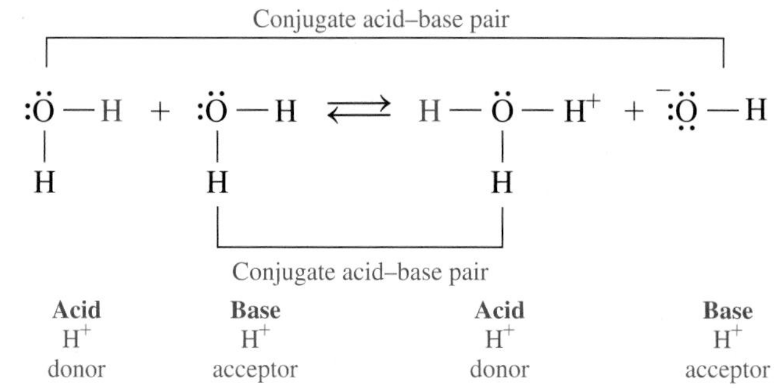

Writing the Ion Product Constant for Water, K_w

In the equation for the ionization of water, there is both a forward and a reverse reaction:

$$\underset{\text{Base}}{H_2O(l)} + \underset{\text{Acid}}{H_2O(l)} \rightleftarrows \underset{\text{Conjugate acid}}{H_3O^+(aq)} + \underset{\text{Conjugate base}}{OH^-(aq)}$$

Experiments have determined that, in pure water, the concentrations of H_3O^+ and OH^- at 25 °C are each 1.0×10^{-7} M. Square brackets are used to indicate the concentrations in moles per liter (M).

$$\text{Pure water} \quad [H_3O^+] = [OH^-] = 1.0 \times 10^{-7}\ \text{M}$$

When these concentrations are multiplied, we obtain the expression and value called the **ion product constant of water, K_w**. The concentration units are omitted in the K_w value.

$$K_w = [H_3O^+][OH^-]$$
$$= (1.0 \times 10^{-7}\ \text{M})(1.0 \times 10^{-7}\ \text{M}) = 1.0 \times 10^{-14}$$

Neutral, Acidic, and Basic Solutions

The K_w value (1.0×10^{-14}) applies to any aqueous solution at 25 °C because all aqueous solutions contain both H_3O^+ and OH^- (see Figure 10.3).

When the $[H_3O^+]$ and $[OH^-]$ in a solution are equal, the solution is **neutral**. However, most solutions are not neutral; they have different concentrations of $[H_3O^+]$ and $[OH^-]$. If acid is added to water, there is an increase in $[H_3O^+]$ and a decrease in $[OH^-]$, which makes an acidic solution. If base is added, $[OH^-]$ increases and $[H_3O^+]$ decreases, which gives a basic solution. However, for any aqueous solution, whether it is neutral, acidic, or basic, the product $[H_3O^+][OH^-]$ is equal to K_w $(1.0 \times 10^{-14}$ at 25 °C) (see Table 10.5).

Using the K_w to Calculate $[H_3O^+]$ and $[OH^-]$ in a Solution

If we know the $[H_3O^+]$ of a solution, we can use the K_w to calculate $[OH^-]$. If we know the $[OH^-]$ of a solution, we can calculate $[H_3O^+]$ as seen in Sample Problem 10.6.

$$K_w = [H_3O^+][OH^-]$$

$$[OH^-] = \frac{K_w}{[H_3O^+]} \qquad [H_3O^+] = \frac{K_w}{[OH^-]}$$

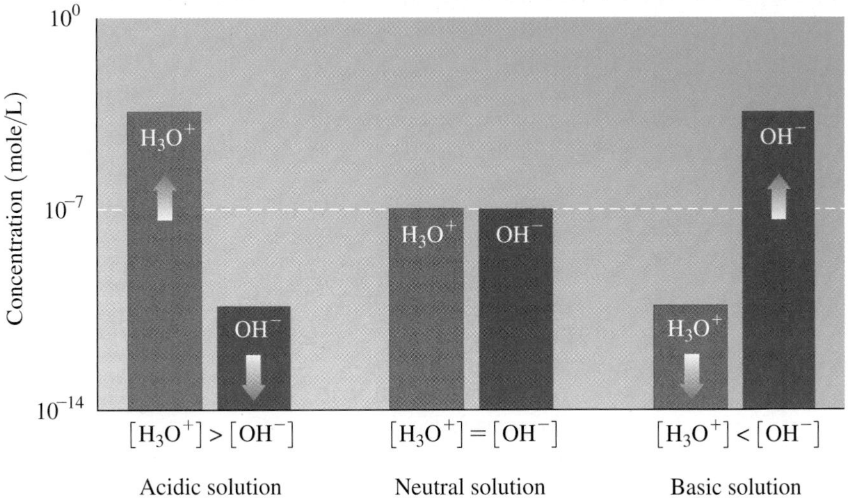

FIGURE 10.3 ▶ In a neutral solution, $[H_3O^+]$ and $[OH^-]$ are equal. In acidic solutions, the $[H_3O^+]$ is greater than the $[OH^-]$. In basic solutions, the $[OH^-]$ is greater than the $[H_3O^+]$.

Q Is a solution that has a $[H_3O^+]$ of 1.0×10^{-3} M acidic, basic, or neutral?

TABLE 10.5 Examples of $[H_3O^+]$ and $[OH^-]$ in Neutral, Acidic, and Basic Solutions

Type of Solution	$[H_3O^+]$	$[OH^-]$	K_w (25 °C)
Neutral	1.0×10^{-7} M	1.0×10^{-7} M	1.0×10^{-14}
Acidic	1.0×10^{-2} M	1.0×10^{-12} M	1.0×10^{-14}
Acidic	2.5×10^{-5} M	4.0×10^{-10} M	1.0×10^{-14}
Basic	1.0×10^{-8} M	1.0×10^{-6} M	1.0×10^{-14}
Basic	5.0×10^{-11} M	2.0×10^{-4} M	1.0×10^{-14}

SAMPLE PROBLEM 10.6 Calculating the $[H_3O^+]$ of a Solution

A vinegar solution has a $[OH^-] = 5.0 \times 10^{-12}$ M at 25 °C. What is the $[H_3O^+]$ of the vinegar solution? Is the solution acidic, basic, or neutral?

SOLUTION

STEP 1 State the given and needed quantities.

ANALYZE THE PROBLEM	Given	Need	Know
	$[OH^-] = 5.0 \times 10^{-12}$ M	$[H_3O^+]$	$K_w = [H_3O^+][OH^-] = 1.0 \times 10^{-14}$

STEP 2 Write the K_w for water and solve for the unknown $[H_3O^+]$.

Rearrange the ion product expression by dividing both sides by the $[OH^-]$.

$$K_w = [H_3O^+][OH^-] = 1.0 \times 10^{-14}$$

$$\frac{K_w}{[OH^-]} = \frac{[H_3O^+][OH^-]}{[OH^-]}$$

$$[H_3O^+] = \frac{1.0 \times 10^{-14}}{[OH^-]}$$

STEP 3 Substitute the known $[OH^-]$ into the equation and calculate.

$$[H_3O^+] = \frac{1.0 \times 10^{-14}}{[5.0 \times 10^{-12}]} = 2.0 \times 10^{-3}\text{ M}$$

Because the $[H_3O^+]$ of 2.0×10^{-3} M is larger than the $[OH^-]$ of 5.0×10^{-12} M, the solution is acidic.

STUDY CHECK 10.6

What is the $[H_3O^+]$ of an ammonia cleaning solution with $[OH^-] = 4.0 \times 10^{-4}$ M? Is the solution acidic, basic, or neutral?

CORE CHEMISTRY SKILL

Calculating $[H_3O^+]$ and $[OH^-]$ in Solutions

Guide to Calculating $[H_3O^+]$ and $[OH^-]$ in Aqueous Solutions

STEP 1 State the given and needed quantities.

STEP 2 Write the K_w for water and solve for the unknown $[H_3O^+]$ or $[OH^-]$.

STEP 3 Substitute the known $[H_3O^+]$ or $[OH^-]$ into the equation and calculate.

QUESTIONS AND PROBLEMS

10.4 Ionization of Water

LEARNING GOAL Use the ion product for water to calculate the $[H_3O^+]$ and $[OH^-]$ in an aqueous solution.

10.25 Why are the concentrations of H_3O^+ and OH^- equal in pure water?

10.26 What is the meaning and value of K_w?

10.27 In an acidic solution, how does the concentration of H_3O^+ compare to the concentration of OH^-?

10.28 If a base is added to pure water, why does the $[H_3O^+]$ decrease?

10.29 Indicate whether each of the following solutions is acidic, basic, or neutral:
- **a.** $[H_3O^+] = 3.4 \times 10^{-2}$ M
- **b.** $[H_3O^+] = 1.9 \times 10^{-7}$ M
- **c.** $[OH^-] = 6.3 \times 10^{-4}$ M
- **d.** $[OH^-] = 2.5 \times 10^{-10}$ M

10.30 Indicate whether each of the following solutions is acidic, basic, or neutral:
- **a.** $[H_3O^+] = 6.0 \times 10^{-12}$ M
- **b.** $[H_3O^+] = 1.4 \times 10^{-4}$ M
- **c.** $[OH^-] = 5.0 \times 10^{-12}$ M
- **d.** $[OH^-] = 4.5 \times 10^{-2}$ M

10.31 Calculate the $[H_3O^+]$ of each aqueous solution with the following $[OH^-]$:
- **a.** lime juice, 3.0×10^{-10} M
- **b.** tamarind, 6.0×10^{-11} M
- **c.** ammonia, 7.8×10^{-4} M
- **d.** KOH solution, 4.5×10^{-3} M

10.32 Calculate the $[H_3O^+]$ of each aqueous solution with the following $[OH^-]$:
- **a.** baking soda, 1.0×10^{-6} M
- **b.** orange juice, 5.0×10^{-11} M
- **c.** milk, 2.0×10^{-8} M
- **d.** bleach, 2.1×10^{-3} M

10.33 Calculate the $[OH^-]$ of each aqueous solution with the following $[H_3O^+]$:
- **a.** Tea, 2.0×10^{-4} M
- **b.** detergent, 4.0×10^{-9} M
- **c.** cleanser, 6.0×10^{-11} M
- **d.** sweetlime juice, 3.5×10^{-3} M

10.34 Calculate the $[OH^-]$ of each aqueous solution with the following $[H_3O^+]$:
- **a.** NaOH solution, 1.0×10^{-12} M
- **b.** aspirin, 6.0×10^{-4} M
- **c.** milk of magnesia, 1.0×10^{-9} M
- **d.** stomach acid, 5.2×10^{-2} M

LEARNING GOAL

Calculate the pH of a solution from $[H_3O^+]$; given the pH, calculate $[H_3O^+]$.

10.5 The pH Scale

Personnel working in food processing, medicine, agriculture, spa and pool maintenance, soap manufacturing, and wine making measure the $[H_3O^+]$ and $[OH^-]$ of solutions. The proper level of acidity is necessary to evaluate the functioning of the lungs and kidneys, to control bacterial growth in foods, and to prevent the growth of pests in food crops.

Although we have expressed H_3O^+ and OH^- as molar concentrations, it is more convenient to describe the acidity of solutions using the *pH scale*. On this scale, a number between 0 and 14 represents the H_3O^+ concentration for common solutions. A neutral solution has a pH of 7.0 at 25 °C. An acidic solution has a pH less than 7.0; a basic solution has a pH greater than 7.0 (see Figure 10.4).

Acidic solution	$pH < 7.0$	$[H_3O^+] > 1 \times 10^{-7}$ M
Neutral solution	$pH = 7.0$	$[H_3O^+] = 1 \times 10^{-7}$ M
Basic solution	$pH > 7.0$	$[H_3O^+] < 1 \times 10^{-7}$ M

In the laboratory, a pH meter is commonly used to determine the pH of a solution. There are also indicators and pH papers that turn specific colors when placed in solutions of different pH values. The pH is found by comparing the colors to a chart (see Figure 10.5).

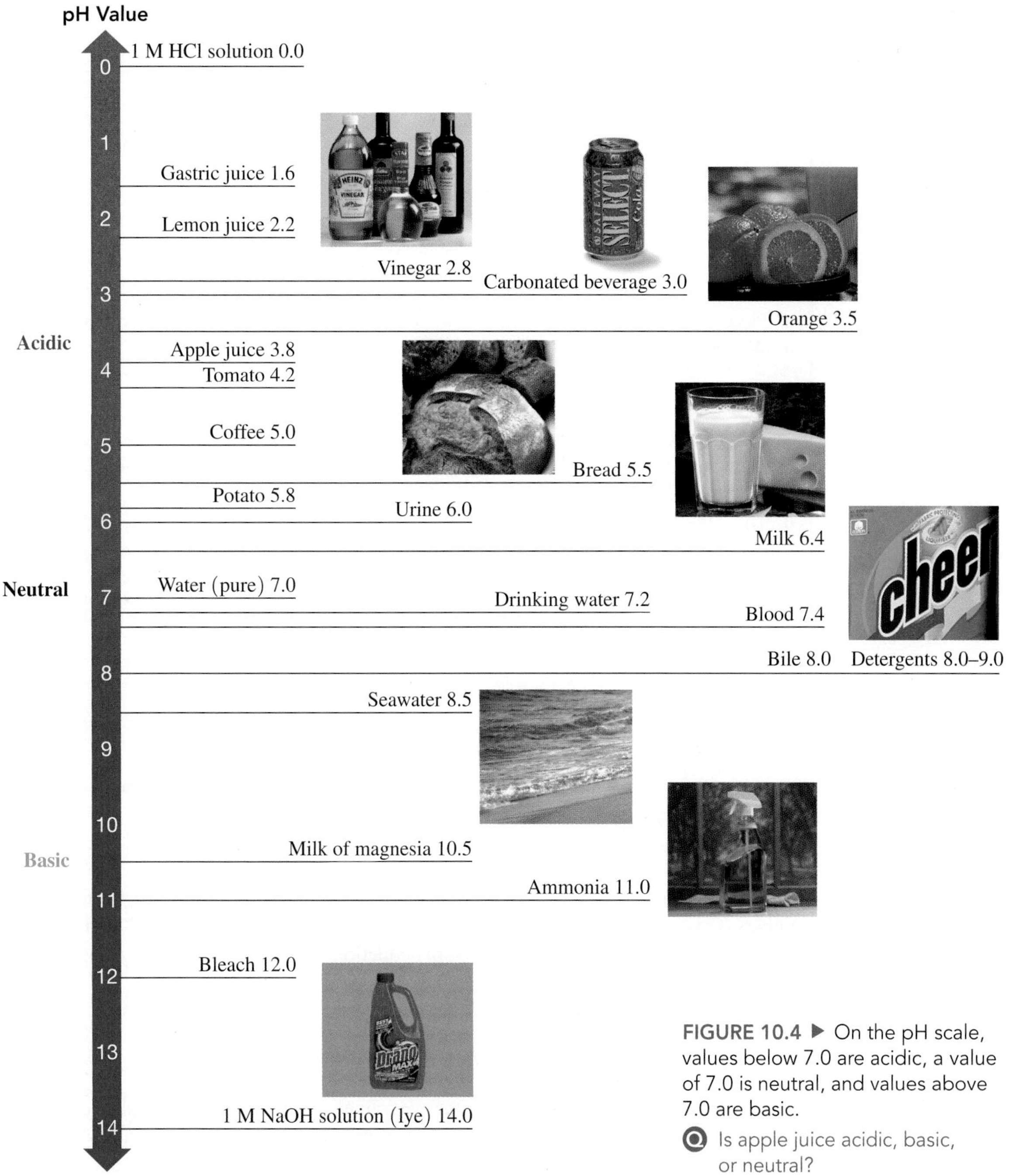

FIGURE 10.4 ▶ On the pH scale, values below 7.0 are acidic, a value of 7.0 is neutral, and values above 7.0 are basic.

Q Is apple juice acidic, basic, or neutral?

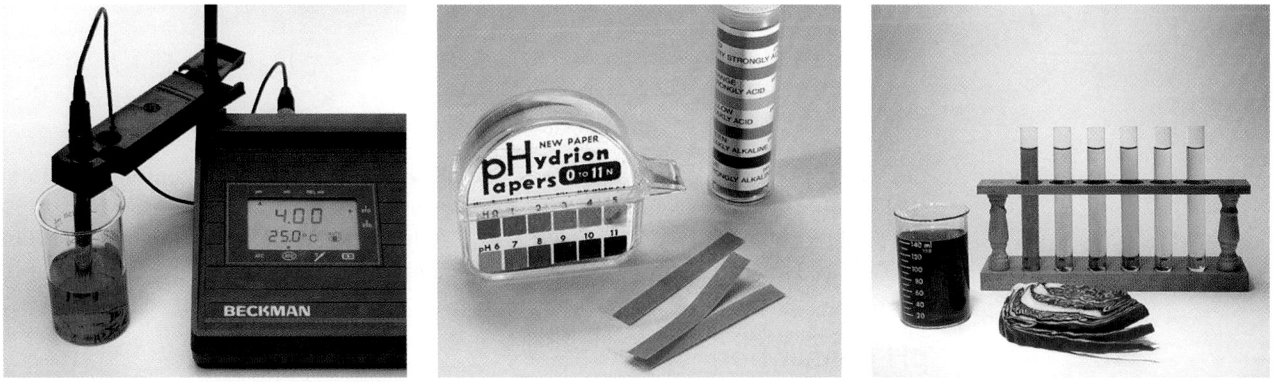

(a) (b) (c)

FIGURE 10.5 ▶ The pH of a solution can be determined using **(a)** a pH meter, **(b)** pH paper, and **(c)** indicators that turn different colors corresponding to different pH values.

Q If a pH meter reads 4.00, is the solution acidic, basic, or neutral?

Cranberry juice has a pH of 2.9.

SAMPLE PROBLEM 10.7 pH of Solutions

Consider the pH of the following items:

Item	pH
Root beer	5.8
Kitchen cleaner	10.9
Pickles	3.5
Glass cleaner	7.6
Cranberry juice	2.9

a. Place the pH values of the items on the list in order of most acidic to most basic.
b. Which item has the highest $[H_3O^+]$?

SOLUTION

a. The most acidic item is the one with the lowest pH, and the most basic is the item with the highest pH: cranberry juice (2.9), pickles (3.5), root beer (5.8), glass cleaner (7.6), kitchen cleaner (10.9).
b. The item with the highest $[H_3O^+]$ would have the lowest pH value, which is cranberry juice.

STUDY CHECK 10.7

Which item in the list in Sample Problem 10.7 has the highest $[OH^-]$?

KEY MATH SKILL

Calculating pH from $[H_3O^+]$

Calculating the pH of Solutions

The pH scale is a logarithmic scale that corresponds to the $[H_3O^+]$ of aqueous solutions. Mathematically, **pH** is the negative logarithm (base 10) of the $[H_3O^+]$.

$$pH = -\log[H_3O^+]$$

Essentially, the negative powers of 10 in the molar concentrations are converted to positive numbers. For example, a lemon juice solution with the $[H_3O^+] = 1.0 \times 10^{-2}$ M has a pH of 2.00. This can be calculated using the pH equation:

$$pH = -\log[1.0 \times 10^{-2}]$$
$$pH = -(-2.00)$$
$$= 2.00$$

The number of *decimal places* in the pH value is the same as the number of significant figures in the coefficient of $[H_3O^+]$. The number to the left of the decimal point in the pH value is the power of 10.

$$[H_3O^+] = \mathbf{1.0} \times 10^{-2}\,\text{M} \qquad pH = 2.\mathbf{00}$$

Two SFs   Two SFs

Because pH is a log scale, a change of one pH unit corresponds to a tenfold change in $[H_3O^+]$. It is important to note that the pH decreases as the $[H_3O^+]$ increases. For example, a solution with a pH of 2.00 has a $[H_3O^+]$ 10 times higher than a solution with a pH of 3.00 and 100 times higher than a solution with a pH of 4.00.

pH Calculation

The pH of a solution is calculated from the $[H_3O^+]$ by using the *log* key and changing the sign as shown in Sample Problem 10.8.

SAMPLE PROBLEM 10.8 Calculating pH from $[H_3O^+]$

Aspirin, which is acetylsalicylic acid, was the first nonsteroidal anti-inflammatory drug used to alleviate pain and fever. If a solution of aspirin has a $[H_3O^+] = 1.7 \times 10^{-3}$ M, what is the pH of the solution?

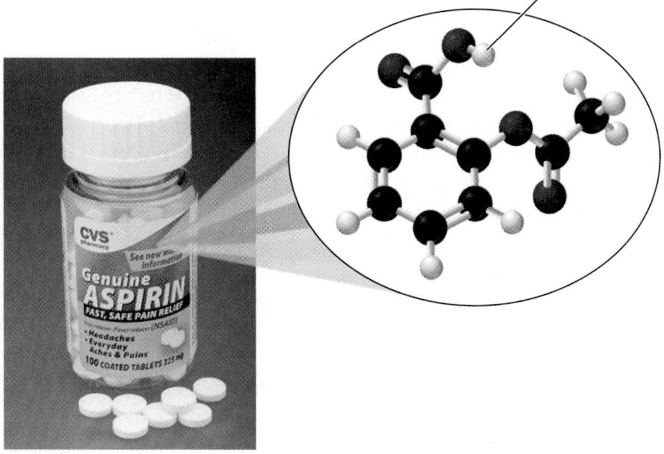

Aspirin, acetylsalicylic acid, is a weak acid.

SOLUTION

STEP 1 State the given and needed quantities.

ANALYZE THE PROBLEM	Given	Need	Know
	$[H_3O^+] = 1.7 \times 10^{-3}$ M	pH	$pH = -\log[H_3O^+]$

STEP 2 Enter the $[H_3O^+]$ into the pH equation.

$$pH = -\log[H_3O^+] = -\log[1.7 \times 10^{-3}]$$

Procedure	**Calculator Display**
Enter 1.7 and press EE or EXP	1.7^{00} or 1.7 00 or 1.7E00
Enter 3 and press +/– to change the sign.	1.7^{-03} or 1.7–03 or 1.7E–03

STEP 3 Press the *log* key and change the sign. Adjust the number of SFs on the *right* of the decimal point.

Procedure	**Calculator Display**
log +/–	2.769551079

The steps can be combined to give the calculator sequence as follows:

$$pH = -\log[1.7 \times 10^{-3}] = 1.7 \text{ EE or EXP } 3 \text{ +/– log +/–}$$
$$= 2.769551079$$

Be sure to check the instructions for your calculator. On some calculators, the log key is used first, followed by the concentration.

In a pH value, the number to the *left* of the decimal point is an *exact* number derived from the power of 10. Thus, the two SFs in the coefficient determine that there are two SFs after the decimal point in the pH value.

Coefficient **Power of ten**

1.7×10^{-3} M $\qquad pH = -\log[1.7 \times 10^{-3}] = 2.77$

Two SFs (1.7) — Exact (10^{-3}) — Exact (2) — Two SFs (.77)

STUDY CHECK 10.8

What is the pH of bleach with $[H_3O^+] = 4.2 \times 10^{-12}$ M?

Guide to Calculating pH of an Aqueous Solution

STEP 1 State the given and needed quantities.

STEP 2 Enter the $[H_3O^+]$ into the pH equation.

STEP 3 Press the *log* key and change the sign. Adjust the number of SFs on the *right* of the decimal point.

Explore Your World

Using Vegetables and Flowers as pH Indicators

Many flowers and vegetables with strong color, especially reds and purples, contain compounds that change color with changes in pH. Some examples are red cabbage, cranberry juice, and other highly colored drinks.

Materials Needed

Red cabbage or cranberry juice or drink, water, and a saucepan

Several glasses or small glass containers and some tape and a pen or pencil to mark the containers

Several colorless household solutions such as vinegar, lemon juice, other fruit juices, window cleaners, soaps, shampoos, and detergents and common household products such as baking soda, antacids, aspirin, salt, and sugar

Procedure

1. Obtain a bottle of cranberry juice or cranberry drink, or use a red cabbage to prepare the pH indicator, as follows: Tear up several red cabbage leaves and place them in a saucepan and cover with water. Boil for about 5 minutes. Cool and collect the purple solution. Alternatively, place red cabbage leaves in a blender and cover with water. Blend until thoroughly mixed, remove, and pour off the red liquid.
2. Place small amounts of each household solution into separate clear glass containers and mark what each one is. If the sample is a solid or a thick liquid, add a small amount of water. Add some cranberry juice or red cabbage indicator until you obtain a color.
3. Observe the colors of the various samples. The colors that indicate acidic solutions are the pink and orange colors (pH 1 to 4) and the pink to lavender colors (pH 5 to 6). A neutral solution has about the same purple color as the indicator. Bases will give blue to green color (pH 8 to 11) or a yellow color (pH 12 to 13).
4. Arrange your samples by color and pH. Classify each of the solutions as acidic (1 to 6), neutral (7), or basic (8 to 13).
5. Try to make an indicator using other colorful fruits or flowers.

Questions

1. Which products that tested acidic listed an acid on their labels?
2. Which products that tested basic listed a base on their labels?
3. Which products were neutral?
4. Which flowers or vegetables behaved as indicators?

KEY MATH SKILL

Calculating $[H_3O^+]$ from pH

Calculating $[H_3O^+]$ from pH

In another type of calculation, we are given the pH of a solution and asked to determine the $[H_3O^+]$. This is a reverse of the pH calculation. For whole number pH values, the negative pH value is the power of 10 in the H_3O^+ concentration.

$$[H_3O^+] = 10^{-pH}$$

For pH values that are not whole numbers, the calculation requires the use of the *10^x* key, which is usually a *2nd function* key. On some calculators, this operation is done using the *inverse* key and the *log* key as shown in Sample Problem 10.9.

Guide to Calculating $[H_3O^+]$ from pH

STEP 1 State the given and needed quantities.

STEP 2 Enter the pH value into the inverse log equation and change the sign.

STEP 3 Adjust the SFs in the coefficient.

SAMPLE PROBLEM 10.9 Calculating $[H_3O^+]$ from pH

Determine the $[H_3O^+]$ for solutions having each of the following pH values:

a. pH = 3.0
b. pH = 8.25

SOLUTION

a. For pH values that are whole numbers, the $[H_3O^+]$ can be written 1×10^{-pH}.

$$[H_3O^+] = 1 \times 10^{-3}\,M$$

b. For pH values that are not whole numbers, the $[H_3O^+]$ is calculated as follows:

STEP 1 State the given and needed quantities.

ANALYZE THE PROBLEM	Given	Need	Know
	pH = 8.25	$[H_3O^+]$	$[H_3O^+] = 10^{-pH}$

STEP 2 Enter the pH value into the inverse log equation and change the sign.

Procedure	Calculator Display
8.25 (+/−)	−8.25

Press the *2nd function* key and then the *10ˣ* key. Or press the *inverse* key and then the *log* key.

(2nd)(10ˣ) or (inv)(log) 5.623413252⁻⁰⁹ or 5.623413252−09 or 5.623413252E−09

STEP 3 **Adjust the SFs in the coefficient.** Because the pH value of 8.25 has two digits on the right of the decimal point, the $[H_3O^+]$ is written with two significant figures.

$[H_3O^+] = 5.6 \times 10^{-9}$ M

Two SFs

STUDY CHECK 10.9

What is the $[H_3O^+]$ of a beer that has a pH of 4.5?

A comparison of $[H_3O^+]$, $[OH^-]$, and their corresponding pH values is given in Table 10.6.

TABLE 10.6 A Comparison of $[H_3O^+]$, $[OH^-]$, and Corresponding pH Values

$[H_3O^+]$	pH	$[OH^-]$	
10^0	0	10^{-14}	
10^{-1}	1	10^{-13}	
10^{-2}	2	10^{-12}	
10^{-3}	3	10^{-11}	
10^{-4}	4	10^{-10}	Acidic
10^{-5}	5	10^{-9}	
10^{-6}	6	10^{-8}	
10^{-7}	7	10^{-7}	Neutral
10^{-8}	8	10^{-6}	
10^{-9}	9	10^{-5}	
10^{-10}	10	10^{-4}	Basic
10^{-11}	11	10^{-3}	
10^{-12}	12	10^{-2}	
10^{-13}	13	10^{-1}	
10^{-14}	14	10^0	

QUESTIONS AND PROBLEMS

10.5 The pH Scale

LEARNING GOAL Calculate the pH of a solution from $[H_3O^+]$; given the pH, calculate $[H_3O^+]$.

10.35 Why does a neutral solution have a pH of 7.0?

10.36 If you know the $[OH^-]$, how can you determine the pH of a solution?

10.37 State whether each of the following solutions is acidic, basic, or neutral:

a. Lime juice, pH 2.2
b. potato, pH 5.8
c. drinking water, pH 7.2
d. soft drinks, pH 3.5
e. milk of magnesia, pH 10.5
f. bleach, pH 12.0

10.38 State whether each of the following solutions is acidic, basic, or neutral:

a. soda, pH 3.22
b. shampoo, pH 5.7
c. laundry detergent, pH 9.4
d. rain, pH 5.83
e. honey, pH 3.9
f. cheese, pH 5.2

10.39 Calculate the pH of each solution given the following $[H_3O^+]$ or $[OH^-]$ values:

a. $[H_3O^+] = 1 \times 10^{-4}$ M
b. $[H_3O^+] = 3 \times 10^{-9}$ M
c. $[OH^-] = 1 \times 10^{-5}$ M
d. $[OH^-] = 2.5 \times 10^{-11}$ M
e. $[H_3O^+] = 6.7 \times 10^{-8}$ M
f. $[OH^-] = 8.2 \times 10^{-4}$ M

10.40 Calculate the pH of each solution given the following $[H_3O^+]$ or $[OH^-]$ values:

a. $[H_3O^+] = 1 \times 10^{-6}$ M
b. $[H_3O^+] = 5 \times 10^{-6}$ M
c. $[OH^-] = 4.0 \times 10^{-2}$ M
d. $[OH^-] = 8.0 \times 10^{-3}$ M
e. $[H_3O^+] = 4.7 \times 10^{-2}$ M
f. $[OH^-] = 3.9 \times 10^{-6}$ M

10.41 Complete the following table:

$[H_3O^+]$	$[OH^-]$	pH	Acidic, Basic, or Neutral?
	1×10^{-6} M		
		3.0	
2×10^{-5} M			
1×10^{-12} M			
		4.62	

10.42 Complete the following table:

$[H_3O^+]$	$[OH^-]$	pH	Acidic, Basic, or Neutral?
		10.0	
			Neutral
	1×10^{-5} M		
1×10^{-2} M			
		11.3	

Chemistry Link to **Health**

Stomach Acid, HCl

Gastric acid, which contains HCl, is produced by parietal cells that line the stomach. When the stomach expands with the intake of food, the gastric glands begin to secrete a strongly acidic solution of HCl. In a single day, a person may secrete 2000 mL of gastric juice, which contains hydrochloric acid, mucins, and the enzymes pepsin and lipase.

The HCl in the gastric juice activates a digestive enzyme from the chief cells called *pepsinogen* to form *pepsin*, which breaks down proteins in food entering the stomach. The secretion of HCl continues until the stomach has a pH of about 2, which is the optimum for activating the digestive enzymes without ulcerating the stomach lining. In addition, the low pH destroys bacteria that reach the stomach. Normally, large quantities of viscous mucus are secreted within the stomach to protect its lining from acid and enzyme damage. Gastric acid may also form under conditions of stress when the nervous system activates the production of HCl. As the contents of the stomach move into the small intestine, cells produce bicarbonate that neutralizes the gastric acid until the pH is about 5.

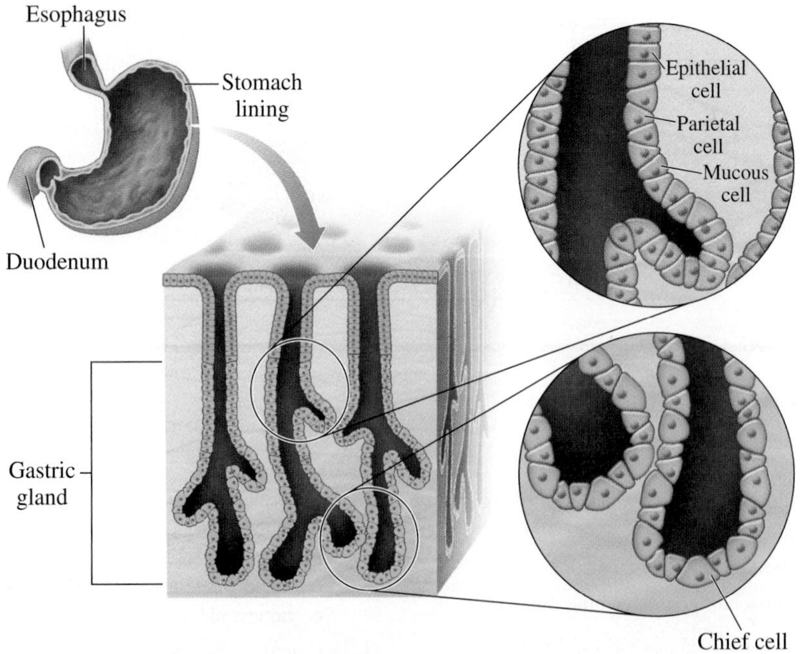

Parietal cells in the lining of the stomach secrete gastric acid HCl.

10.6 Reactions of Acids and Bases

LEARNING GOAL

Write balanced equations for reactions of acids and bases; calculate the molarity or volume of an acid from titration information.

Typical reactions of acids and bases include the reactions of acids with metals, bases, and carbonate or bicarbonate ions. For example, when you drop an antacid tablet in water, the bicarbonate ion and citric acid in the tablet react to produce carbon dioxide bubbles, water, and a salt. A *salt* is an ionic compound that does not have H^+ as the cation or OH^- as the anion.

CORE CHEMISTRY SKILL

Writing Equations for Reactions of Acids and Bases

Acids and Metals

Acids react with certain metals to produce hydrogen gas (H_2) and a salt. Metals that react with acids include potassium, sodium, calcium, magnesium, aluminum, zinc, iron, and tin. In reactions that are single replacement reactions, the metal ion replaces the hydrogen in the acid.

$$\underset{\text{Metal}}{Mg(s)} + \underset{\text{Acid}}{2HCl(aq)} \longrightarrow \underset{\text{Hydrogen}}{H_2(g)} + \underset{\text{Salt}}{MgCl_2(aq)}$$

$$\underset{\text{Metal}}{Zn(s)} + \underset{\text{Acid}}{2HNO_3(aq)} \longrightarrow \underset{\text{Hydrogen}}{H_2(g)} + \underset{\text{Salt}}{Zn(NO_3)_2(aq)}$$

Magnesium reacts rapidly with acid and forms H_2 gas and a salt of magnesium.

SAMPLE PROBLEM 10.10 Equations for Metals and Acids

Write a balanced equation for the reaction of $Al(s)$ with $HCl(aq)$.

SOLUTION

Write the reactants and products. When a metal reacts with an acid, the products are hydrogen gas and a salt. The unbalanced equation is written as

$$Al(s) + HCl(aq) \longrightarrow H_2(g) + \text{salt}(aq)$$

Determine the formula of the salt. When $Al(s)$ reacts, it forms Al^{3+}, which is balanced by $3Cl^-$ from HCl.

$$Al(s) + HCl(aq) \longrightarrow H_2(g) + AlCl_3(aq)$$

Balance the equation.

$$2Al(s) + 6HCl(aq) \longrightarrow 3H_2(g) + 2AlCl_3(aq)$$

STUDY CHECK 10.10

Write the balanced equation when $Ca(s)$ reacts with $HBr(aq)$.

Acids, Carbonates, and Bicarbonates

When an acid is added to a carbonate or bicarbonate (hydrogen carbonate), the products are carbon dioxide gas, water, and an ionic compound (salt). The acid reacts with CO_3^{2-} or HCO_3^- to produce carbonic acid, H_2CO_3, which breaks down rapidly to CO_2 and H_2O.

$$\underset{\text{Acid}}{2HCl(aq)} + \underset{\text{Carbonate}}{Na_2CO_3(aq)} \longrightarrow \underset{\text{Carbon dioxide}}{CO_2(g)} + \underset{\text{Water}}{H_2O(l)} + \underset{\text{Salt}}{2NaCl(aq)}$$

$$\underset{\text{Acid}}{HBr(aq)} + \underset{\text{Bicarbonate}}{NaHCO_3(aq)} \longrightarrow \underset{\text{Carbon dioxide}}{CO_2(g)} + \underset{\text{Water}}{H_2O(l)} + \underset{\text{Salt}}{NaBr(aq)}$$

When sodium bicarbonate (baking soda) reacts with an acid (vinegar), the products are carbon dioxide gas, water, and a salt.

Acids and Hydroxides: Neutralization

Neutralization is a reaction between an acid and a base to produce water and a salt. The H^+ of an acid that can be strong or weak and the OH^- of a strong base combine to form water as one product. The salt is the cation from the base and the anion from the

If soil is too acidic, nutrients are not absorbed by crops. Then lime ($CaCO_3$), which acts as a base, may be added to increase the soil pH.

acid. We can write the following equation for the neutralization reaction between HCl and NaOH:

$$\underset{\text{Acid}}{HCl(aq)} + \underset{\text{Base}}{NaOH(aq)} \longrightarrow \underset{\text{Water}}{H_2O(l)} + \underset{\text{Salt}}{NaCl(aq)}$$

If we write the strong acid HCl and the strong base NaOH as ions, we see that H^+ reacts with OH^- to form water, leaving the ions Na^+ and Cl^- in solution.

$$H^+(aq) + Cl^-(aq) + Na^+(aq) + OH^-(aq) \longrightarrow H_2O(l) + Na^+(aq) + Cl^-(aq)$$

When we omit the ions that do not change during the reaction (Na^+ and Cl^-), we see that the net ionic equation for neutralization is the formation of H_2O from H^+ and OH^-.

$$H^+(aq) + \cancel{Cl^-(aq)} + \cancel{Na^+(aq)} + OH^-(aq) \longrightarrow H_2O(l) + \cancel{Na^+(aq)} + \cancel{Cl^-(aq)}$$

$$H^+(aq) + OH^-(aq) \longrightarrow H_2O(l) \quad \text{Net ionic equation}$$

In a neutralization reaction, one H^+ always combines with one OH^-. Therefore, a neutralization equation uses coefficients to balance H^+ in the acid with the OH^- in the base. Sample Problem 10.11 shows how to balance the equation for the neutralization of HCl and $Ba(OH)_2$.

SAMPLE PROBLEM 10.11 Balancing Equations of Acids

Write the balanced equation for the neutralization of $HCl(aq)$ and $Ba(OH)_2(s)$.

Guide to Balancing an Equation for Neutralization

STEP 1 Write the reactants and products.

STEP 2 Balance the H^+ in the acid with the OH^- in the base.

STEP 3 Balance the H_2O with the H^+ and the OH^-.

STEP 4 Write the salt from the remaining ions.

SOLUTION

STEP 1 **Write the reactants and products.**

$$HCl(aq) + Ba(OH)_2(s) \longrightarrow H_2O(l) + \text{salt}$$

STEP 2 **Balance the H^+ in the acid with the OH^- in the base.** Placing a coefficient of 2 in front of HCl provides $2H^+$ for the $2OH^-$ in $Ba(OH)_2$.

$$2HCl(aq) + Ba(OH)_2(s) \longrightarrow H_2O(l) + \text{salt}$$

STEP 3 **Balance the H_2O with the H^+ and the OH^-.** Use a coefficient of 2 in front of H_2O to balance $2H^+$ and $2OH^-$.

$$2HCl(aq) + Ba(OH)_2(s) \longrightarrow 2H_2O(l) + \text{salt}$$

STEP 4 **Write the salt from the remaining ions.** Use the ions Ba^{2+} and $2Cl^-$ to write the formula for the salt, $BaCl_2$.

$$2H\mathbf{Cl}(aq) + \mathbf{Ba}(OH)_2(s) \longrightarrow 2H_2O(l) + \mathbf{BaCl_2}(aq)$$

STUDY CHECK 10.11

Write the balanced equation for the reaction between $H_2SO_4(aq)$ and $NaHCO_3(aq)$.

CORE CHEMISTRY SKILL

Calculating Molarity or Volume of an Acid or Base in a Titration

Acid–Base Titration

Suppose we need to find the molarity of a solution of HCl, which has an unknown concentration. We can do this by a laboratory procedure called **titration** in which we neutralize an acid sample with a known amount of base. In a titration, we place a measured volume of the acid in a flask and add a few drops of an *indicator* such as phenolphthalein. In an acidic solution, phenolphthalein is colorless. Then we fill a buret with a NaOH solution of known molarity and carefully add NaOH solution to the acid in the flask (see Figure 10.6).

In the titration, we neutralize the acid by adding a volume of base that contains a matching number of moles of OH^-. We know that neutralization has taken place when the phenolphthalein in the solution changes from colorless to pink. This is called the neutralization *endpoint*. From the volume added and molarity of the NaOH solution, we can calculate the number of moles of NaOH, the moles of acid, and then the concentration of the acid.

FIGURE 10.6 ▶ The titration of an acid. A known volume of an acid solution is placed in a flask with an indicator and titrated with a measured volume of a base solution, such as NaOH, to the neutralization endpoint.

Q What data is needed to determine the molarity of the acid in the flask?

SAMPLE PROBLEM 10.12 Titration of an Acid

A 25.0-mL (0.025 L) sample of an HCl solution is placed in a flask with a few drops of phenolphthalein (indicator). If 32.6 mL of a 0.185 M NaOH solution is needed to reach the endpoint, what is the molarity of the HCl solution?

$$NaOH(aq) + HCl(aq) \longrightarrow H_2O(l) + NaCl(aq)$$

SOLUTION

STEP 1 State the given and needed quantities.

ANALYZE THE PROBLEM	Given	Need
	25.0 mL (0.0250 L) of HCl solution, 32.6 mL of 0.185 M NaOH solution	molarity of the HCl solution
	Neutralization Equation	
	$NaOH(aq) + HCl(aq) \longrightarrow H_2O(l) + NaCl(aq)$	

STEP 2 Write a plan to calculate the molarity.

mL of NaOH solution → Metric factor → L of NaOH solution → Molarity → moles of NaOH → Mole–mole factor → moles of HCl → Divide by liters → molarity of HCl solution

STEP 3 State equalities and conversion factors, including concentrations.

$$1 \text{ L of NaOH solution} = 1000 \text{ mL of NaOH solution}$$
$$\frac{1 \text{ L NaOH solution}}{1000 \text{ mL NaOH solution}} \text{ and } \frac{1000 \text{ mL NaOH solution}}{1 \text{ L NaOH solution}}$$

$$1 \text{ L of NaOH solution} = 0.185 \text{ mole of NaOH}$$
$$\frac{1 \text{ L NaOH solution}}{0.185 \text{ mole NaOH}} \text{ and } \frac{0.185 \text{ mole NaOH}}{1 \text{ L NaOH solution}}$$

$$1 \text{ mole of HCl} = 1 \text{ mole of NaOH}$$
$$\frac{1 \text{ mole HCl}}{1 \text{ mole NaOH}} \text{ and } \frac{1 \text{ mole NaOH}}{1 \text{ mole HCl}}$$

STEP 4 Set up the problem to calculate the needed quantity.

$$32.6 \text{ }\cancel{\text{mL NaOH solution}} \times \frac{1 \text{ }\cancel{\text{L NaOH solution}}}{1000 \text{ }\cancel{\text{mL NaOH solution}}} \times \frac{0.185 \text{ }\cancel{\text{mole NaOH}}}{1 \text{ }\cancel{\text{L NaOH solution}}} \times \frac{1 \text{ mole HCl}}{1 \text{ }\cancel{\text{mole NaOH}}} = 0.00603 \text{ mole of HCl}$$

$$\text{molarity of HCl} = \frac{0.00603 \text{ mole HCl}}{0.0250 \text{ L solution}} = 0.241 \text{ M HCl solution}$$

STUDY CHECK 10.12

What is the molarity of an HCl solution, if 28.6 mL of a 0.175 M NaOH solution is needed to neutralize a 25.0-mL sample of the HCl solution?

Guide to Calculations for an Acid–Base Titration

STEP 1 State the given and needed quantities.

STEP 2 Write a plan to calculate the molarity or volume.

STEP 3 State equalities and conversion factors, including concentrations.

STEP 4 Set up the problem to calculate the needed quantity.

Interactive Video

Titration of an Acid

Chemistry Link to Health

Antacids

Antacids are substances used to neutralize excess stomach acid (HCl). Some antacids are mixtures of aluminum hydroxide and magnesium hydroxide. These hydroxides are not very soluble in water, so the levels of available OH^- are not damaging to the intestinal tract. However, aluminum hydroxide has the side effects of producing constipation and binding phosphate in the intestinal tract, which may cause weakness and loss of appetite. Magnesium hydroxide has a laxative effect. These side effects are less likely when a combination of the antacids is used.

$$Al(OH)_3(s) + 3HCl(aq) \longrightarrow 3H_2O(l) + AlCl_3(aq)$$

$$Mg(OH)_2(s) + 2HCl(aq) \longrightarrow 2H_2O(l) + MgCl_2(aq)$$

Some antacids use calcium carbonate to neutralize excess stomach acid. About 10% of the calcium is absorbed into the bloodstream, where it elevates the levels of serum calcium. Calcium carbonate is not recommended for patients who have peptic ulcers or a tendency to form kidney stones.

$$CaCO_3(s) + 2HCl(aq) \longrightarrow CO_2(g) + H_2O(l) + CaCl_2(aq)$$

Still other antacids contain sodium bicarbonate. This type of antacid has a tendency to increase blood pH and elevate sodium levels in the body fluids. It also is not recommended in the treatment of peptic ulcers.

$$NaHCO_3(s) + HCl(aq) \longrightarrow CO_2(g) + H_2O(l) + NaCl(aq)$$

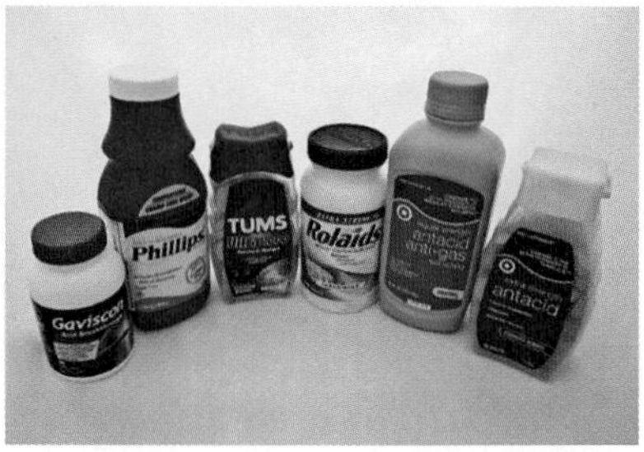

Antacids neutralize excess stomach acid.

The neutralizing substances in some antacid preparations are given in Table 10.7.

TABLE 10.7 Basic Compounds in Some Antacids

Antacid	Base(s)
Amphojel	$Al(OH)_3$
Milk of magnesia	$Mg(OH)_2$
Mylanta, Maalox, Di-Gel, Gelusil, Riopan	$Mg(OH)_2$, $Al(OH)_3$
Bisodol, Rolaids	$CaCO_3$, $Mg(OH)_2$
Titralac, Tums, Pepto-Bismol	$CaCO_3$
Alka-Seltzer	$NaHCO_3$, $KHCO_3$

QUESTIONS AND PROBLEMS

10.6 Reactions of Acids and Bases

LEARNING GOAL Write balanced equations for reactions of acids and bases; calculate the molarity or volume of an acid from titration information.

10.43 Complete and balance the equation for each of the following reactions:

a. $ZnCO_3(s) + HBr(aq) \longrightarrow$
b. $Zn(s) + HCl(aq) \longrightarrow$
c. $HCl(aq) + NaHCO_3(s) \longrightarrow$
d. $H_2SO_4(aq) + Mg(OH)_2(s) \longrightarrow$

10.44 Complete and balance the equation for each of the following reactions:

a. $KHCO_3(s) + HCl(aq) \longrightarrow$
b. $Ca(s) + H_2SO_4(aq) \longrightarrow$
c. $H_2SO_4(aq) + Ca(OH)_2(s) \longrightarrow$
d. $Na_2CO_3(s) + H_2SO_4(aq) \longrightarrow$

10.45 Balance each of the following neutralization equations:

a. $HCl(aq) + Mg(OH)_2(s) \longrightarrow H_2O(l) + MgCl_2(aq)$
b. $H_3PO_4(aq) + LiOH(aq) \longrightarrow H_2O(l) + Li_3PO_4(aq)$

10.46 Balance each of the following neutralization equations:

a. $HNO_3(aq) + Ba(OH)_2(s) \longrightarrow H_2O(l) + Ba(NO_3)_2(aq)$
b. $H_2SO_4(aq) + Al(OH)_3(aq) \longrightarrow H_2O(l) + Al_2(SO_4)_3(aq)$

10.47 Write a balanced equation for the neutralization of each of the following:

a. $H_2SO_4(aq)$ and $NaOH(aq)$
b. $HCl(aq)$ and $Fe(OH)_3(s)$
c. $H_2CO_3(aq)$ and $Mg(OH)_2(s)$

10.48 Write a balanced equation for the neutralization of each of the following:

a. $H_3PO_4(aq)$ and $NaOH(aq)$
b. $HI(aq)$ and $LiOH(aq)$
c. $HNO_3(aq)$ and $Ca(OH)_2(s)$

10.49 What is the molarity of a solution of HCl if 5.00 mL is titrated with 28.6 mL of a 0.145 M NaOH solution?

$$HCl(aq) + NaOH(aq) \longrightarrow H_2O(l) + NaCl(aq)$$

10.50 What is the molarity of the acetic acid solution if 29.7 mL of a 0.205 M KOH solution is required to neutralize 25.0 mL of a solution of $HC_2H_3O_2$?

$$HC_2H_3O_2(aq) + KOH(aq) \longrightarrow H_2O(l) + KC_2H_3O_2(aq)$$

10.51 If 38.2 mL of a 0.163 M KOH solution is required to titrate 25.0 mL of a solution of H_2SO_4, what is the molarity of the H_2SO_4 solution?

$$H_2SO_4(aq) + 2KOH(aq) \longrightarrow 2H_2O(l) + K_2SO_4(aq)$$

10.52 If 32.8 mL of a 0.162 M NaOH solution is required to titrate 25.0 mL of a solution of H_2SO_4, what is the molarity of the H_2SO_4 solution?

$$H_2SO_4(aq) + 2NaOH(aq) \longrightarrow 2H_2O(l) + Na_2SO_4(aq)$$

10.7 Buffers

LEARNING GOAL

Describe the role of buffers in maintaining the pH of a solution.

The pH of water and most solutions changes drastically when a small amount of acid or base is added. However, when an acid or base is added to a *buffer solution*, there is little change in pH. A **buffer solution** maintains pH by neutralizing small amounts of added acid or base. For example, blood contains buffers that maintain a consistent pH of about 7.4. If the pH of the blood goes slightly above or below 7.4, changes in oxygen levels and metabolic processes can be drastic enough to cause death. Even though we obtain acids and bases from foods and cellular reactions, the buffers in the body absorb those compounds so effectively that the pH of the blood remains essentially unchanged (see Figure 10.7).

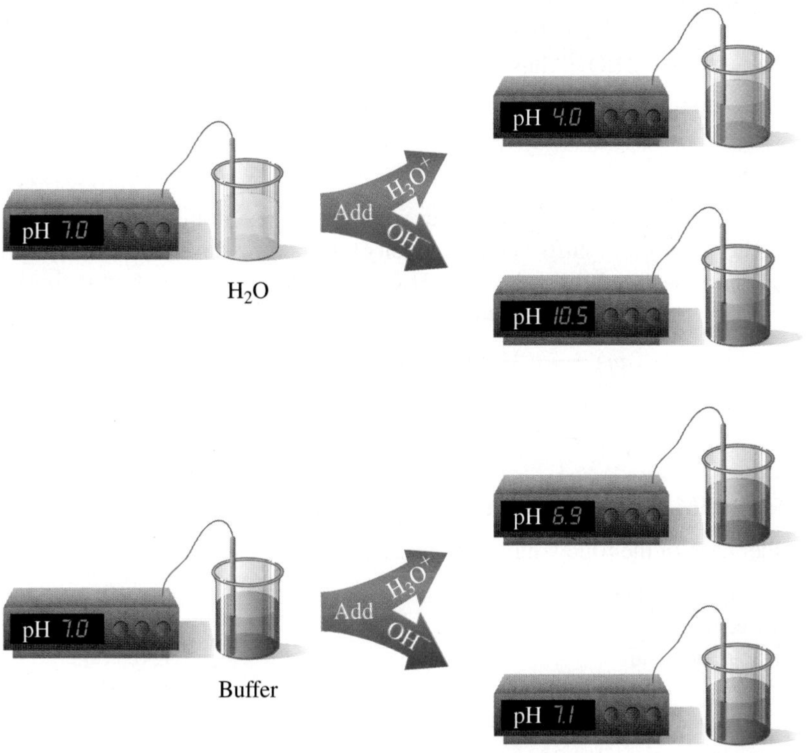

FIGURE 10.7 ▶ Adding an acid or a base to water changes the pH drastically, but a buffer resists pH change when small amounts of acid or base are added.

Q Why does the pH change several pH units when acid is added to water, but not when acid is added to a buffer?

In a buffer, an acid must be present to react with any OH^- that is added, and a base must be available to react with any added H_3O^+. However, that acid and base must not neutralize each other. Therefore, a combination of an acid–base conjugate pair is used to prepare a buffer. Most buffer solutions consist of nearly equal concentrations of a weak acid and a salt containing its conjugate base (see Figure 10.8). Buffers may also contain a weak base and a salt containing its conjugate acid.

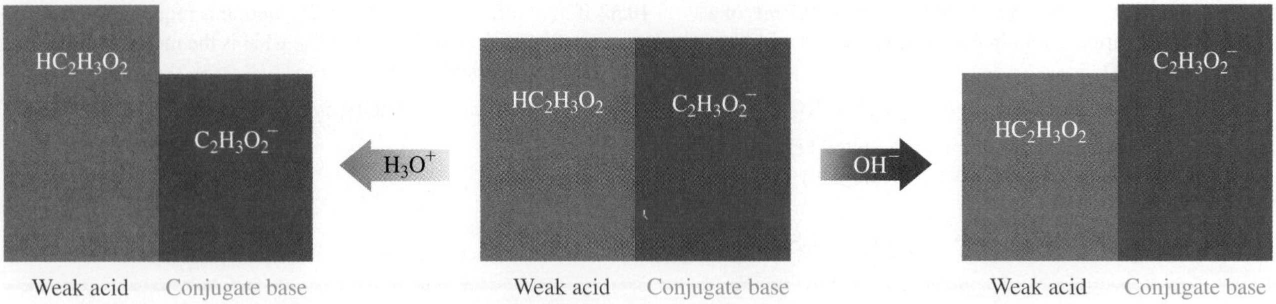

FIGURE 10.8 ▶ The buffer described here consists of about equal concentrations of acetic acid ($HC_2H_3O_2$) and its conjugate base, acetate ion ($C_2H_3O_2^-$). Adding H_3O^+ to the buffer reacts with $C_2H_3O_2^-$, whereas adding OH^- neutralizes $HC_2H_3O_2$. The pH of the solution is maintained as long as the added amounts of acid or base are small compared to the concentrations of the buffer components.

Q How does this acetic acid–acetate ion buffer maintain pH?

For example, a typical buffer can be made from the weak acid, acetic acid ($HC_2H_3O_2$), and its salt, sodium acetate ($NaC_2H_3O_2$). As a weak acid, acetic acid ionizes slightly in water to form H_3O^+ and a small amount of $C_2H_3O_2^-$. The addition of its salt provides a much larger concentration of the acetate ion ($C_2H_3O_2^-$), which is necessary for its buffering capability.

$$\underset{\text{Large amount}}{HC_2H_3O_2(aq)} + H_2O(l) \rightleftharpoons H_3O^+(aq) + \underset{\text{Large amount}}{C_2H_3O_2^-(aq)}$$

We can now describe how this buffer solution maintains the $[H_3O^+]$. When a small amount of acid is added, it combines with the acetate ion, $C_2H_3O_2^-$, causing the equilibrium to shift in the direction of $HC_2H_3O_2$. There will be a slight decrease in $[C_2H_3O_2^-]$ and a slight increase in $[HC_2H_3O_2]$; thus both $[H_3O^+]$ and pH are maintained.

$$\underset{\substack{\text{Equilibrium shifts in the}\\\text{direction of the reactants}}}{HC_2H_3O_2(aq) + H_2O(l)} \longleftarrow H_3O^+(aq) + C_2H_3O_2^-(aq)$$

If a small amount of base is added to this buffer solution, it is neutralized by the acetic acid, $HC_2H_3O_2$. The equilibrium shifts in the direction of the products, water and $C_2H_3O_2^-$. The $[HC_2H_3O_2]$ decreases slightly and the $[C_2H_3O_2^-]$ increases slightly, but again the $[H_3O^+]$ and the pH of the solution are maintained (see Figure 10.8).

$$HC_2H_3O_2(aq) + OH^-(aq) \longrightarrow \underset{\substack{\text{Equilibrium shifts in the}\\\text{direction of the products}}}{H_2O(l) + C_2H_3O_2^-(aq)}$$

SAMPLE PROBLEM 10.13 Identifying Buffer Solutions

Indicate whether each of the following would make a buffer solution:

a. HCl, a strong acid, and NaCl
b. H_3PO_4, a weak acid
c. HF, a weak acid, and NaF

SOLUTION

a. No. A buffer requires a weak acid and a salt containing its conjugate base.
b. No. A weak acid is part of a buffer, but the salt containing the conjugate base of the weak acid is also needed.
c. Yes. This mixture would be a buffer since it contains a weak acid and a salt containing its conjugate base.

STUDY CHECK 10.13

In a buffer made from the weak acid $HCHO_2$ and its salt, $KCHO_2$, when H_3O^+ is added, is it neutralized by (1) the salt, (2) H_2O, (3) OH^-, or (4) the acid?

Chemistry Link to Health

Buffers in the Blood

The arterial blood has a normal pH of 7.35 to 7.45. If changes in H_3O^+ lower the pH below 6.8 or raise it above 8.0, cells cannot function properly and death may result. In our cells, CO_2 is continually produced as an end product of cellular metabolism. Some CO_2 is carried to the lungs for elimination, and the rest dissolves in body fluids such as plasma and saliva, forming carbonic acid. As a weak acid, carbonic acid ionizes to give bicarbonate, HCO_3^-, and H_3O^+. More of the anion HCO_3^- is supplied by the kidneys to give an important buffer system in the body fluid: the H_2CO_3/HCO_3^- buffer.

$$CO_2(g) + H_2O(l) \rightleftarrows H_2CO_3(aq) + H_2O(l) \rightleftarrows H_3O^+(aq) + HCO_3^-(aq)$$

Excess H_3O^+ entering the body fluids reacts with the HCO_3^- and excess OH^- reacts with the carbonic acid.

$$H_2CO_3(aq) + H_2O(l) \longleftarrow H_3O^+(aq) + HCO_3^-(aq)$$

Equilibrium shifts in the direction of the reactants

$$H_2CO_3(aq) + OH^-(aq) \longrightarrow H_2O(l) + HCO_3^-(aq)$$

Equilibrium shifts in the direction of the products

In the body, the concentration of carbonic acid is closely associated with the partial pressure of CO_2. Table 10.8 lists the normal values for arterial blood. If the CO_2 level increases, it produces more H_2CO_3 and more H_3O^+, lowering the pH. This condition is called **acidosis**. Difficulty with ventilation or gas diffusion can lead to respiratory acidosis, which can happen in emphysema or when the medulla of the brain is affected by an accident or depressive drugs.

A decrease in the CO_2 level leads to a high blood pH, a condition called **alkalosis**. Excitement, trauma, or a high temperature may cause a person to hyperventilate, which expels large amounts of CO_2. As the partial pressure of CO_2 in the blood falls below normal, H_2CO_3 forms CO_2 and H_2O, decreasing the $[H_3O^+]$ and raising the pH. Table 10.9 lists some of the conditions that lead to changes in the blood pH and some possible treatments. The kidneys also regulate H_3O^+ and HCO_3^- components, but more slowly than the adjustment made by the lungs through ventilation.

TABLE 10.8 Normal Values for Blood Buffer in Arterial Blood

P_{CO_2}	40 mmHg
H_2CO_3	2.4 mmoles/L of plasma
HCO_3^-	24 mmoles/L of plasma
pH	7.35 to 7.45

TABLE 10.9 Acidosis and Alkalosis: Symptoms, Causes, and Treatments

Respiratory Acidosis: $CO_2\uparrow$ pH$\downarrow$	
Symptoms:	Failure to ventilate, suppression of breathing, disorientation, weakness, coma
Causes:	Lung disease blocking gas diffusion (e.g., emphysema, pneumonia, bronchitis, asthma); depression of respiratory center by drugs, cardiopulmonary arrest, stroke, poliomyelitis, or nervous system disorders
Treatment:	Correction of disorder, infusion of bicarbonate
Metabolic Acidosis: $H^+\uparrow$ pH$\downarrow$	
Symptoms:	Increased ventilation, fatigue, confusion
Causes:	Renal disease, including hepatitis and cirrhosis; increased acid production in diabetes mellitus, hyperthyroidism, alcoholism, and starvation; loss of alkali in diarrhea; acid retention in renal failure
Treatment:	Sodium bicarbonate given orally, dialysis for renal failure, insulin treatment for diabetic ketosis
Respiratory Alkalosis: $CO_2\downarrow$ pH$\uparrow$	
Symptoms:	Increased rate and depth of breathing, numbness, light-headedness, tetany
Causes:	Hyperventilation because of anxiety, hysteria, fever, exercise; reaction to drugs such as salicylate, quinine, and antihistamines; conditions causing hypoxia (e.g., pneumonia, pulmonary edema, heart disease)
Treatment:	Elimination of anxiety-producing state, rebreathing into a paper bag
Metabolic Alkalosis: $H^+\downarrow$ pH$\uparrow$	
Symptoms:	Depressed breathing, apathy, confusion
Causes:	Vomiting, diseases of the adrenal glands, ingestion of excess alkali
Treatment:	Infusion of saline solution, treatment of underlying diseases

QUESTIONS AND PROBLEMS

10.7 Buffers

LEARNING GOAL Describe the role of buffers in maintaining the pH of a solution.

10.53 Which of the following represents a buffer system? Explain.
- **a.** NaOH and NaCl
- **b.** H_2CO_3 and $NaHCO_3$
- **c.** HF and KF
- **d.** KCl and NaCl

10.54 Which of the following represents a buffer system? Explain.
- **a.** H_3PO_3
- **b.** $NaNO_3$
- **c.** $HC_2H_3O_2$ and $NaC_2H_3O_2$
- **d.** HCl and NaOH

10.55 Consider the buffer system of hydrofluoric acid, HF, and its salt, NaF.

$$HF(aq) + H_2O(l) \rightleftarrows H_3O^+(aq) + F^-(aq)$$

a. The purpose of this buffer system is to:
1. maintain [HF]
2. maintain $[F^-]$
3. maintain pH

b. The salt of the weak acid is needed to:
1. provide the conjugate base
2. neutralize added H_3O^+
3. provide the conjugate acid

c. If OH^- is added, it is neutralized by:
1. the salt
2. H_2O
3. H_3O^+

d. When H_3O^+ is added, the equilibrium shifts in the direction of the:
1. reactants
2. products
3. does not change

10.56 Consider the buffer system of nitrous acid, HNO_2, and its salt, $NaNO_2$.

$$HNO_2(aq) + H_2O(l) \rightleftarrows H_3O^+(aq) + NO_2^-(aq)$$

a. The purpose of this buffer system is to:
1. maintain $[HNO_2]$
2. maintain $[NO_2^-]$
3. maintain pH

b. The weak acid is needed to:
1. provide the conjugate base
2. neutralize added OH^-
3. provide the conjugate acid

c. If H_3O^+ is added, it is neutralized by:
1. the salt
2. H_2O
3. OH^-

d. When OH^- is added, the equilibrium shifts in the direction of the:
1. reactants
2. products
3. does not change

CONCEPT MAP

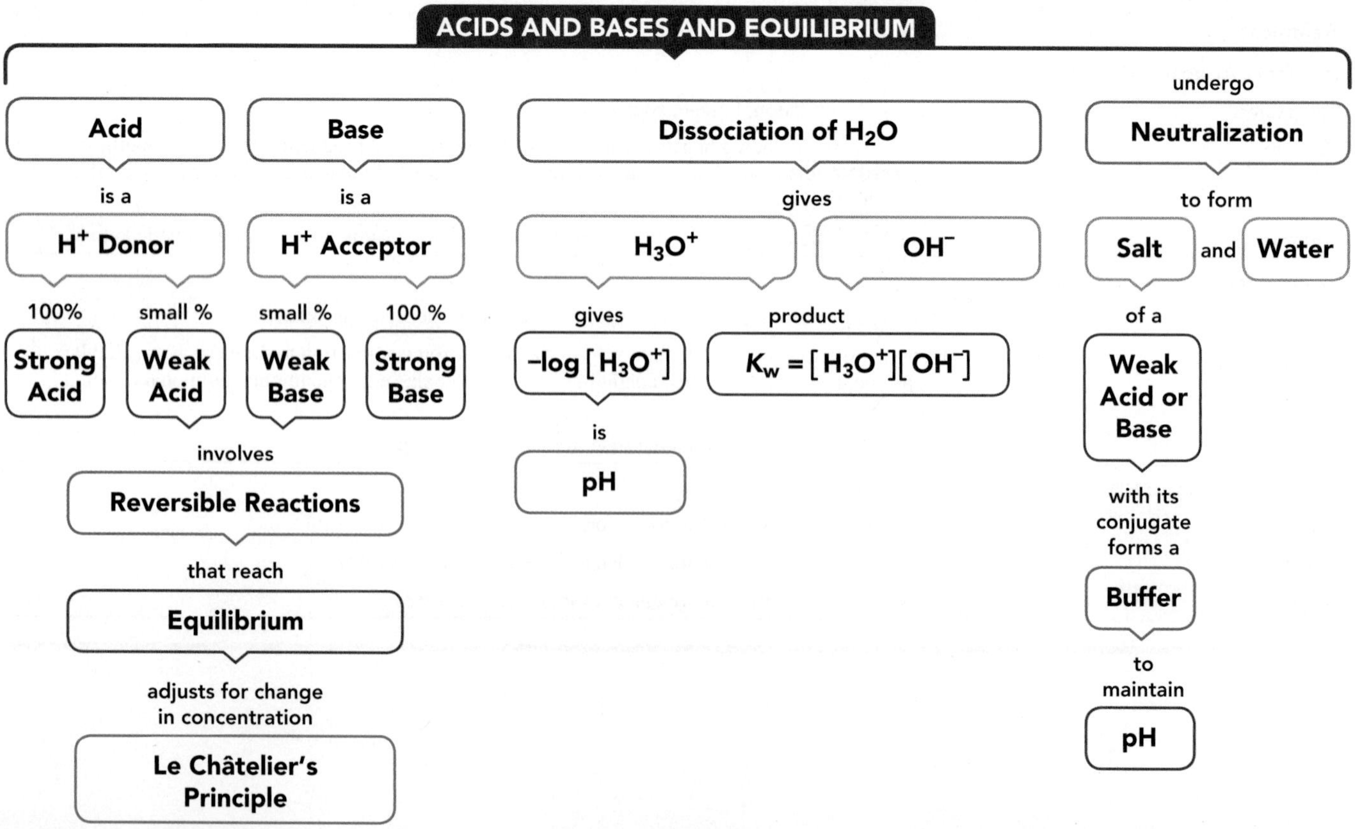

CHAPTER REVIEW

10.1 Acids and Bases

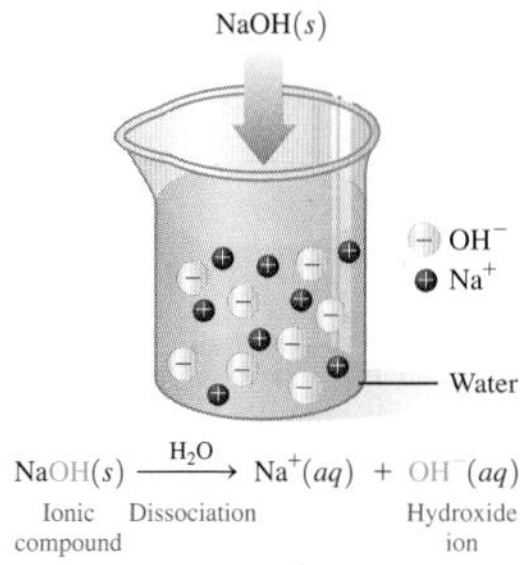

LEARNING GOAL Describe and name acids and bases; identify Brønsted–Lowry acids and bases.

- An Arrhenius acid produces H^+ and an Arrhenius base produces OH^- in aqueous solutions.
- Acids taste sour, may sting, and neutralize bases.
- Bases taste bitter, feel slippery, and neutralize acids.
- Acids containing a simple anion use a *hydro* prefix, whereas acids with oxygen-containing polyatomic anions are named as *ic* or *ous acids*.
- According to the Brønsted–Lowry theory, acids are H^+ donors and bases are H^+ acceptors.
- A conjugate acid–base pair is related by the loss or gain of one H^+.
- For example, when the acid HF donates H^+, the F^- is its conjugate base. The other acid–base pair would be H_3O^+/H_2O.

$$HF(aq) + H_2O(l) \rightleftharpoons H_3O^+(aq) + F^-(aq)$$

10.2 Strengths of Acids and Bases

LEARNING GOAL Write equations for the ionization of acids and bases.

- Strong acids ionize completely in water, and the H^+ is accepted by H_2O acting as a base.
- A weak acid ionizes slightly in water, producing only a small percentage of H_3O^+.
- Strong bases are hydroxides of Groups 1A (1) and 2A (2) that dissociate completely in water.
- An important weak base is ammonia, NH_3.

10.3 Acid–Base Equilibrium

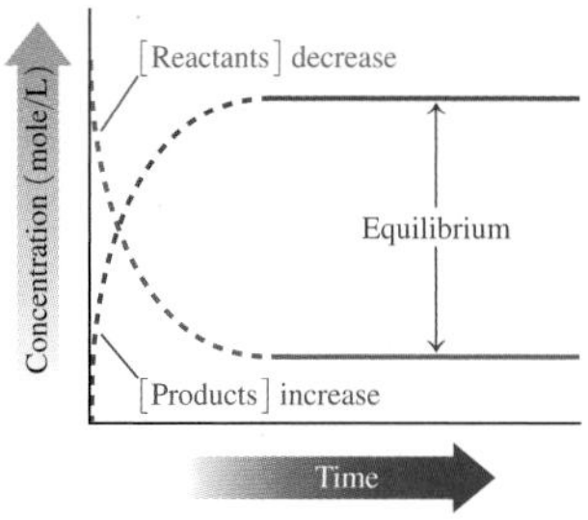

LEARNING GOAL Use the concept of reversible reactions to explain acid–base equilibrium. Use Le Châtelier's principle to determine the effect on equilibrium concentrations when reaction conditions change.

- Chemical equilibrium occurs in a reversible reaction when the rate of the forward reaction becomes equal to the rate of the reverse reaction.
- At equilibrium, no further change occurs in the concentrations of the reactants and products as the forward and reverse reactions continue.
- When reactants are removed or products are added to an equilibrium mixture, the system shifts in the direction of the reactants to reestablish equilibrium.
- When reactants are added or products are removed from an equilibrium mixture, the system shifts in the direction of the products to reestablish equilibrium.

10.4 Ionization of Water

LEARNING GOAL Use the ion product for water to calculate the $[H_3O^+]$ and $[OH^-]$ in an aqueous solution.

- In pure water, a few water molecules transfer H^+ to other water molecules, producing small, but equal, amounts of H_3O^+ and OH^-.
- In pure water, the molar concentrations of H_3O^+ and OH^- are each 1.0×10^{-7} M.
- The ion product constant for water, $K_w = [H_3O^+][OH^-] = 1.0 \times 10^{-14}$ at 25 °C.
- In acidic solutions, the $[H_3O^+]$ is greater than the $[OH^-]$.
- In basic solutions, the $[OH^-]$ is greater than the $[H_3O^+]$.

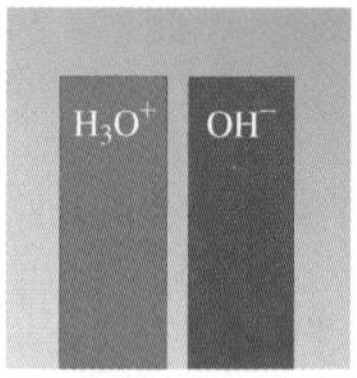

$[H_3O^+] = [OH^-]$

10.5 The pH Scale

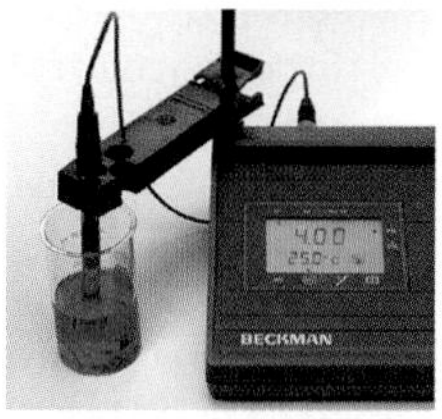

LEARNING GOAL Calculate the pH of a solution from $[H_3O^+]$; given the pH, calculate $[H_3O^+]$.

- The pH scale is a range of numbers typically from 0 to 14, which represents the $[H_3O^+]$ of the solution.
- A neutral solution has a pH of 7.0. In acidic solutions, the pH is below 7.0; in basic solutions, the pH is above 7.0.
- Mathematically, pH is the negative logarithm of the hydronium ion concentration, $pH = -\log[H_3O^+]$.

10.6 Reactions of Acids and Bases

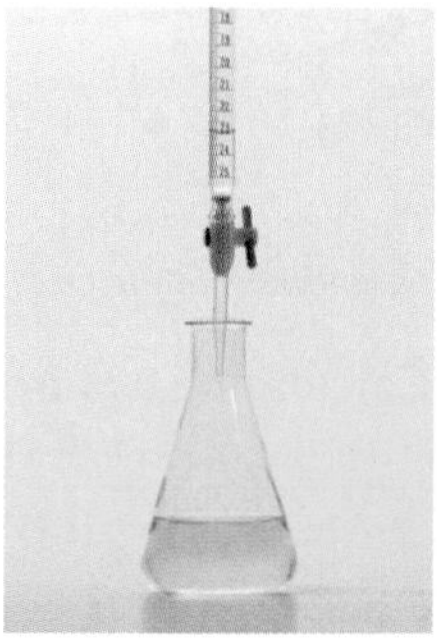

LEARNING GOAL Write balanced equations for reactions of acids and bases; calculate the molarity or volume of an acid from titration information.

- An acid reacts with a metal to produce hydrogen gas and a salt.
- The reaction of an acid with a carbonate or bicarbonate produces carbon dioxide, water, and a salt.
- In neutralization, an acid reacts with a base to produce water and a salt.
- In a titration, an acid sample is neutralized with a known amount of a base.
- From the volume and molarity of the base, the concentration of the acid is calculated.

10.7 Buffers

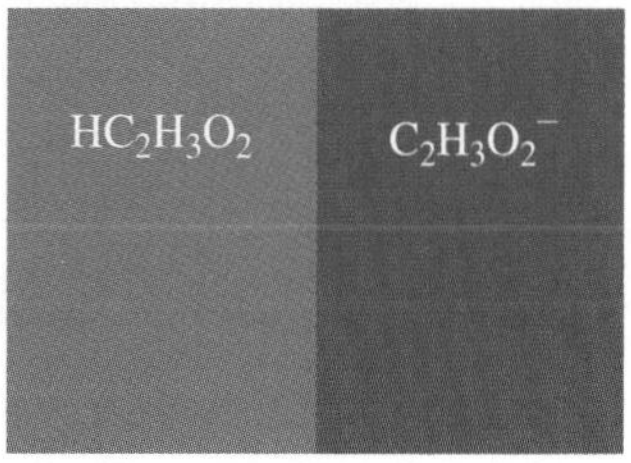

LEARNING GOAL Describe the role of buffers in maintaining the pH of a solution.

- A buffer solution resists changes in pH when small amounts of an acid or a base are added.
- A buffer contains either a weak acid and its salt or a weak base and its salt.
- In a buffer, the weak acid reacts with added OH^-, and the anion of the salt reacts with added H_3O^+.

KEY TERMS

acid A substance that dissolves in water and produces hydrogen ions (H^+), according to the Arrhenius theory. All acids are hydrogen ion (H^+) donors, according to the Brønsted–Lowry theory.

acidosis A physiological condition in which the blood pH is lower than 7.35.

alkalosis A physiological condition in which the blood pH is higher than 7.45.

amphoteric Substances that can act as either an acid or a base in water.

base A substance that dissolves in water and produces hydroxide ions (OH^-) according to the Arrhenius theory. All bases are hydrogen ion (H^+) acceptors, according to the Brønsted–Lowry theory.

Brønsted–Lowry acids and bases An acid is a hydrogen ion (H^+) donor, and a base is a hydrogen ion (H^+) acceptor.

buffer solution A solution of a weak acid and its conjugate base or a weak base and its conjugate acid that maintains the pH by neutralizing added acid or base.

conjugate acid–base pair An acid and base that differ by one H^+. When an acid donates a hydrogen ion (H^+), the product is its conjugate base, which is capable of accepting a hydrogen ion (H^+) in the reverse reaction.

equilibrium The point at which the rate of forward and reverse reactions are equal so that no further change in concentrations of reactants and products takes place.

hydronium ion The ion formed by the attraction of H^+ to a H_2O molecule, written as H_3O^+.

ion product constant of water, K_w The product of $[H_3O^+]$ and $[OH^-]$ in solution; $K_w = [H_3O^+][OH^-]$.

Le Châtelier's principle When a stress is placed on a system at equilibrium, the system shifts to relieve that stress.

neutral The term that describes a solution with equal concentrations of H_3O^+ and OH^-.

neutralization A reaction between an acid and a base to form water and a salt.

pH A measure of the $[H_3O^+]$ in a solution; $pH = -\log[H_3O^+]$.

strong acid An acid that completely ionizes in water.

strong base A base that completely dissociates in water.

titration The addition of base to an acid sample to determine the concentration of the acid.

weak acid An acid that is a poor donor of H^+ and ionizes only slightly in water.

weak base A base that is a poor acceptor of H^+ and ionizes only slightly in water.

KEY MATH SKILLS

The chapter section containing each Key Math Skill is shown in parentheses at the end of each heading.

Calculating pH from $[H_3O^+]$ (10.5)

- The pH of a solution is calculated from the negative log of the $[H_3O^+]$.

$$pH = -\log[H_3O^+]$$

Example: What is the pH of a solution that has $[H_3O^+] = 2.4 \times 10^{-11}$ M?

Answer: We substitute the given $[H_3O^+]$ into the pH equation and calculate the pH.

$$pH = -\log[H_3O^+]$$
$$pH = -\log[2.4 \times 10^{-11}]$$
$$pH = -(-10.62)$$
$$pH = 10.62$$

(Two decimal places equal the 2 SFs in the $[H_3O^+]$ coefficient.)

Calculating $[H_3O^+]$ from pH (10.5)

- The calculation of $[H_3O^+]$ from the pH is done by reversing the pH calculation using the negative pH or the inverse log (*2nd* 10^x).

$$[H_3O^+] = 10^{-pH}$$

Example: What is the $[H_3O^+]$ of a solution with a pH of 4.80?

Answer:
$$[H_3O^+] = 10^{-pH}$$
$$= 10^{-4.80}$$
= [inv] [log] −4.80

or

= [2nd] [10^x] −4.80

$$= 1.6 \times 10^{-5}\ M$$

CORE CHEMISTRY SKILLS

The chapter section containing each Core Chemistry Skill is shown in parentheses at the end of each heading.

Identifying Conjugate Acid–Base Pairs (10.1)

- According to the Brønsted–Lowry theory, a conjugate acid–base pair consists of molecules or ions related by the loss of one H^+ by an acid, and the gain of one H^+ by a base.
- Every acid–base reaction contains two conjugate acid–base pairs because an H^+ is transferred in both the forward and reverse directions.
- When an acid such as HF loses one H^+, the conjugate base F^- is formed. When H_2O acts as a base, it gains an H^+, which forms its conjugate acid, H_3O^+.

Example: Identify the conjugate acid–base pairs in the following reaction:

$$HSO_4^-(aq) + H_2O(l) \rightleftharpoons SO_4^{2-}(aq) + H_3O^+(aq)$$

Answer:
$$HSO_4^-(aq) + H_2O(l) \rightleftharpoons SO_4^{2-}(aq) + H_3O^+(aq)$$
Acid — Base — Conjugate base — Conjugate acid

Conjugate acid–base pairs: HSO_4^-/SO_4^{2-}, H_3O^+/H_2O

Using Le Châtelier's Principle (10.3)

Le Châtelier's principle states that when a system at equilibrium is disturbed by changes in concentration, the system will shift in the direction that will reduce that stress.

$$H_2S(aq) + H_2O(l) \rightleftarrows H_3O^+(aq) + HS^-(aq)$$

Example: For each of the following changes at equilibrium, indicate whether the system shifts in the direction of products or reactants:

a. removing some $H_2S(aq)$
b. adding some $H_3O^+(aq)$

Answer: **a.** Removing a reactant shifts the system in the direction of the reactants.
b. Adding a product shifts the system in the direction of the reactants.

Calculating $[H_3O^+]$ and $[OH^-]$ in Solutions (10.4)

- For all aqueous solutions, the product of $[H_3O^+]$ and $[OH^-]$ is equal to the ion product constant of water, K_w.

$$K_w = [H_3O^+][OH^-]$$

- Because pure water contains equal numbers of OH^- ions and H_3O^+ ions each with molar concentrations of 1.0×10^{-7} M, the numerical value of K_w is 1.0×10^{-14} at 25 °C.

$$K_w = [H_3O^+][OH^-] = [1.0 \times 10^{-7}][1.0 \times 10^{-7}]$$
$$= 1.0 \times 10^{-14}$$

- If we know the $[H_3O^+]$ of a solution, we can use the K_w to calculate the $[OH^-]$. If we know the $[OH^-]$ of a solution, we can calculate $[H_3O^+]$ using the K_w.

$$[OH^-] = \frac{K_w}{[H_3O^+]} \qquad [H_3O^+] = \frac{K_w}{[OH^-]}$$

Example: What is the $[OH^-]$ in a solution that has $[H_3O^+] = 2.4 \times 10^{-11}$ M? Is the solution acidic or basic?

Answer: We solve the K_w expression for $[OH^-]$ and substitute in the known values of K_w and $[H_3O^+]$.

$$[OH^-] = \frac{K_w}{[H_3O^+]} \qquad [OH^-] = \frac{1.0 \times 10^{-14}}{[2.4 \times 10^{-11}]} = 4.2 \times 10^{-4}\ \text{M}$$

Because the $[OH^-]$ is greater than the $[H_3O^+]$, this is a basic solution.

Writing Equations for Reactions of Acids and Bases (10.6)

- Acids react with certain metals to produce hydrogen gas (H_2) and a salt.

$$\underset{\text{Metal}}{Mg(s)} + \underset{\text{Acid}}{2HCl(aq)} \longrightarrow \underset{\text{Hydrogen}}{H_2(g)} + \underset{\text{Salt}}{MgCl_2(aq)}$$

- When an acid is added to a carbonate or bicarbonate, the products are carbon dioxide gas, water, and a salt.

$$\underset{\text{Acid}}{2HCl(aq)} + \underset{\text{Carbonate}}{Na_2CO_3(aq)} \longrightarrow \underset{\text{Carbon dioxide}}{CO_2(g)} + \underset{\text{Water}}{H_2O(l)} + \underset{\text{Salt}}{2NaCl(aq)}$$

- Neutralization is a reaction between a strong or a weak acid with a strong base to produce water and a salt.

$$\underset{\text{Acid}}{HCl(aq)} + \underset{\text{Base}}{NaOH(aq)} \longrightarrow \underset{\text{Water}}{H_2O(l)} + \underset{\text{Salt}}{NaCl(aq)}$$

Example: Write the balanced chemical equation for the reaction of $ZnCO_3(s)$ and hydrobromic acid $HBr(aq)$.

Answer: $ZnCO_3(s) + 2HBr(aq) \longrightarrow CO_2(g) + H_2O(l) + ZnBr_2(aq)$

Calculating Molarity or Volume of an Acid or Base in a Titration (10.6)

- In a titration, a measured volume of acid is neutralized by a basic solution of known molarity.
- From the measured volume of the strong base solution required for titration and its molarity, the number of moles of base, the moles of acid, and the concentration of the acid are calculated.

Example: A 15.0-mL (0.0150 L) sample of an H_2SO_4 solution is titrated with 24.0 mL of a 0.245 M NaOH solution. What is the molarity of the H_2SO_4 solution?

$$H_2SO_4(aq) + 2NaOH(aq) \longrightarrow 2H_2O(l) + Na_2SO_4(aq)$$

Answer: $24.0\ \cancel{\text{mL NaOH solution}} \times \frac{1\ \cancel{\text{L NaOH solution}}}{1000\ \cancel{\text{mL NaOH solution}}}$

$$\times \frac{0.245\ \cancel{\text{mole NaOH}}}{1\ \cancel{\text{L NaOH solution}}} \times \frac{1\ \text{mole } H_2SO_4}{2\ \cancel{\text{moles NaOH}}} = 0.00294\ \text{mole of } H_2SO_4$$

$$\frac{0.00294\ \text{mole } H_2SO_4}{0.0150\ \text{L } H_2SO_4\ \text{solution}} = 0.196\ \text{M } H_2SO_4\ \text{solution}$$

UNDERSTANDING THE CONCEPTS

The chapter sections to review are shown in parentheses at the end of each question.

10.57 Identify each of the following as an acid or a base: (10.1)

a. H_2SO_4 **b.** RbOH
c. $Ca(OH)_2$ **d.** HI

10.58 Identify each of the following as an acid or a base: (10.1)

a. $Sr(OH)_2$ **b.** H_2SO_3
c. $HC_2H_3O_2$ **d.** CsOH

10.59 Complete the following table: (10.1)

Acid	Conjugate Base
H_2O	
	CN^-
HNO_2	
	$H_2PO_4^-$

10.60 Complete the following table: (10.1)

Base	Conjugate Acid
	HS^-
	H_3O^+
NH_3	
HCO_3^-	

10.61 State whether each of the following solutions is acidic, basic, or neutral: (10.5)

a. rain, pH 5.2 **b.** tears, pH 7.5
c. tea, pH 3.8 **d.** cola, pH 2.5
e. photo developer, pH 12.0

10.62 State whether each of the following solutions is acidic, basic, or neutral: (10.5)
a. saliva, pH 6.8
b. urine, pH 5.9
c. pancreatic juice, pH 8.0
d. bile, pH 8.4
e. blood, pH 7.45

10.63 Determine if each of the following diagrams represents a strong acid or a weak acid. The acid has the formula HX. (10.2)

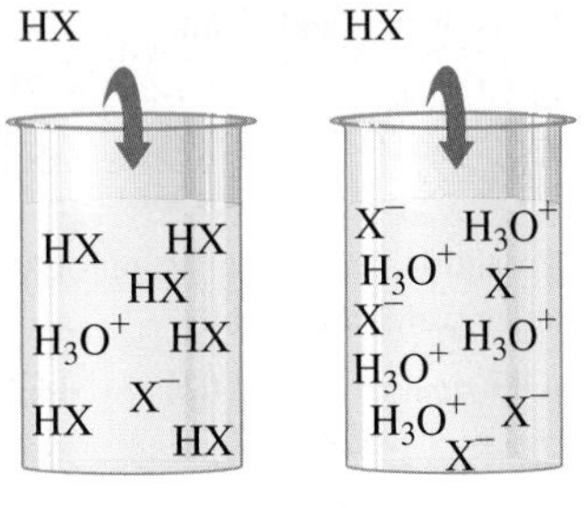

A **B**

10.64 Adding a few drops of a strong acid to water will lower the pH appreciably. However, adding the same number of drops to a buffer does not appreciably alter the pH. Explain. (10.7)

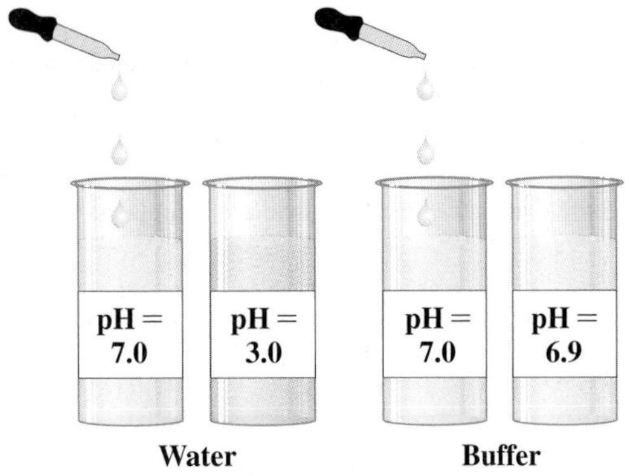

10.65 Sometimes, during stress or trauma, a person can start to hyperventilate. To avoid fainting, a person might breathe into a paper bag to avoid fainting. (10.7)
a. What changes occur in the blood pH during hyperventilation?
b. How does breathing into a paper bag help return blood pH to normal?

Breathing into a paper bag can help a person who is hyperventilating.

10.66 In the blood plasma, pH is maintained by the carbonic acid–bicarbonate buffer system. (10.7)
a. How is pH maintained when acid is added to the buffer system?
b. How is pH maintained when base is added to the buffer system?

ADDITIONAL QUESTIONS AND PROBLEMS

10.67 Identify each of the following as an acid, a base, or a salt, and give its name: (10.1)
a. LiOH
b. $Ca(NO_3)_2$
c. HBr
d. $Ba(OH)_2$
e. H_2CO_3

10.68 Identify each of the following as an acid, a base, or a salt, and give its name: (10.1)
a. Milk of magnesium
b. $MgBr_2$
c. NH_3
d. $HClO_2$
e. Vitamin C

10.69 Using Table 10.3, identify the stronger acid in each of the following pairs: (10.2)
a. HF or HCN
b. H_3O^+ or H_2S
c. HNO_2 or $HC_2H_3O_2$
d. H_2O or HCO_3^-

10.70 Using Table 10.3, identify the weaker acid in each of the following pairs: (10.2)
a. HNO_2 or H_2CO_3
b. HF or HCO_3^-
c. HBr or HSO_4^-
d. NH_4^+ or H_3PO_4

10.71 Use Le Châtelier's principle to predict if each the following changes causes the system to shift in the direction of products or reactants: (10.3)

$$H_2S(aq) + H_2O(l) \rightleftharpoons H_3O^+(aq) + HS^-(aq)$$

a. adding some $H_2S(aq)$
b. removing some $HS^-(aq)$
c. adding some $H_3O^+(aq)$
d. removing some $H_2S(aq)$

10.72 Use Le Châtelier's principle to predict if each the following changes causes the system to shift in the direction of products or reactants: (10.3)

$$HClO(aq) + H_2O(l) \rightleftharpoons H_3O^+(aq) + ClO^-(aq)$$

a. removing some $HClO(aq)$
b. adding some $ClO^-(aq)$
c. removing some $H_3O^+(aq)$
d. adding some $HClO(aq)$

10.73 Determine the pH for the following solutions: (10.5)
a. $[H_3O^+] = 1.0 \times 10^{-8}$ M
b. $[H_3O^+] = 5.0 \times 10^{-2}$ M
c. $[OH^-] = 3.5 \times 10^{-4}$ M
d. $[OH^-] = 0.005$ M

10.74 Determine the pH for the following solutions: (10.5)
a. $[OH^-] = 1.0 \times 10^{-7}$ M
b. $[H_3O^+] = 4.2 \times 10^{-3}$ M
c. $[H_3O^+] = 0.0001$ M
d. $[OH^-] = 8.5 \times 10^{-9}$ M

10.75 Identify each of the solutions in problem 10.73 as acidic, basic, or neutral. (10.5)

10.76 Identify each of the solutions in problem 10.74 as acidic, basic, or neutral. (10.5)

10.77 Calculate the $[H_3O^+]$ and $[OH^-]$ for a solution with the following pH values: (10.5)
a. 3.0 b. 6.00 c. 8.0
d. 11.0 e. 9.20

10.78 Write a balanced equation for the following reaction. What will be the effect if the atmospheric temperature is reduced? (10.5)

$$Na_3CO_3(s) + H_2SO_4(aq)$$

10.79 Solution **A** has a pH of 4.0, and solution **B** has a pH of 6.0. (10.5)
a. Which solution is more acidic?
b. What is the $[H_3O^+]$ in each?
c. What is the $[OH^-]$ in each?

10.80 Solution **X** has a pH of 9.0, and solution **Y** has a pH of 7.0. (10.5)
a. Which solution is more acidic?
b. What is the $[H_3O^+]$ in each?
c. What is the $[OH^-]$ in each?

10.81 A buffer is made by dissolving H_3PO_4 and NaH_2PO_4 in water. (10.7)
a. Write an equation that shows how this buffer neutralizes added acid.
b. Write an equation that shows how this buffer neutralizes added base.

10.82 A buffer is made by dissolving $HC_2H_3O_2$ and $NaC_2H_3O_2$ in water. (10.7)
a. Write an equation that shows how this buffer neutralizes added acid.
b. Write an equation that shows how this buffer neutralizes added base.

10.83 Calculate the volume, in milliliters, of a 0.150 M NaOH solution needed to neutralize each of the following: (10.6)
a. 25.0 mL of a 0.288 M HCl solution
b. 10.0 mL of a 0.560 M H_2SO_4 solution
c. 5.00 mL of a 0.618 M HBr solution

10.84 Calculate the volume, in milliliters, of a 0.215 M KOH solution needed to neutralize each of the following: (10.6)
a. 2.50 mL of a 0.825 M H_2SO_4 solution
b. 18.5 mL of a 0.560 M HNO_3 solution
c. 5.00 mL of a 3.18 M HCl solution

10.85 A 10.0-mL sample of vinegar, which is an aqueous solution of acetic acid, CH_3COOH, requires 16.5 mL of a 0.500 M NaOH solution to reach the endpoint in a titration. What is the molarity of the acetic acid solution? (10.6)

$$CH_3COOH(aq) + NaOH(aq) \longrightarrow H_2O(l) + CH_3COONA(aq)$$

10.86 A 10.0-mL sample of vinegar, which is an aqueous solution of acetic acid, $HC_2H_3O_2$, requires 16.5 mL of a 0.500 M NaOH solution to reach the endpoint in a titration. What is the molarity of the acetic acid solution? (10.6)

$$HC_2H_3O_2(aq) + NaOH(aq) \longrightarrow H_2O(l) + NaC_2H_3O_2(aq)$$

CHALLENGE QUESTIONS

The following groups of questions are related to the topics in this chapter. However, they do not all follow the chapter order, and they require you to combine concepts and skills from several sections. These questions will help you increase your critical thinking skills and prepare for your next exam.

10.87 For each of the following: (10.2)
1. H_2S 2. H_3PO_4
a. Write the formula for the conjugate base.
b. Which is the weaker acid (**1** or **2**)?

10.88 For each of the following: (10.2)
1. HCO_3^- 2. H_2O
a. Write the formula for the conjugate base.
b. Which is the weaker acid (**1** or **2**)?

10.89 Identify the conjugate acid–base pairs in each of the following equations: (10.1)
a. $NH_3(aq) + HNO_3(aq) \longrightarrow NH_4^+(aq) + NO_3^-(aq)$
b. $H_2O(l) + HBr(aq) \longrightarrow H_3O^+(aq) + Br^-(aq)$

10.90 Identify the conjugate acid–base pairs in each of the following equations: (10.1)
a. $HNO_2(aq) + HS^-(aq) \rightleftarrows NO_2^-(aq) + H_2S(g)$
b. $HCl(aq) + OH^-(aq) \longrightarrow Cl^-(aq) + H_2O(l)$

10.91 Complete and balance each of the following: (10.6)
a. $ZnCO_3(s) + H_2SO_4(aq) \longrightarrow$
b. $Al(s) + HCl(aq) \longrightarrow$

10.92 Complete and balance each of the following: (10.6)
a. $H_3PO_4(aq) + Ca(OH)_2(s) \longrightarrow$
b. $KHCO_3(s) + HNO_3(aq) \longrightarrow$

10.93 Determine each of the following for a 0.050 M KOH solution: (10.5, 10.6)
a. $[H_3O^+]$ b. pH
c. the balanced chemical equation for the reaction with H_2SO_4
d. milliliters of the KOH solution required to neutralize 40.0 mL of a 0.035 M H_2SO_4 solution

10.94 Complete and balance the following reaction: (10.5, 10.6)

$$Na_2CO_3(s) + H_2CO_3(aq)$$

Carbon dioxide can be dissolved in water. How will the amount of dissolved carbon dioxide in water change in an atmosphere with high pressure of carbon dioxide? What will be the effect if the above neutralization is carried out in an atmosphere with high pressure of carbon dioxide?

10.95 A 0.204 M NaOH solution is used to titrate 50.0 mL of an H_3PO_4 solution. (10.6)
a. Write the balanced chemical equation.
b. What is the molarity of the H_3PO_4 solution if 16.4 mL of the NaOH solution is required?

10.96 A 0.312 M KOH solution is used to titrate 15.0 mL of an H_2SO_4 solution. (10.6)
a. Write the balanced chemical equation.
b. What is the molarity of the H_2SO_4 solution if 28.2 mL of the KOH solution is required?

10.97 Calculate the volume, in milliliters, of a 0.250 M LiOH solution that will completely neutralize each of the following: (10.6)
a. 38.5 mL of a 0.150 M H_3PO_4 solution
b. 15.0 mL of a 0.420 M H_2SO_4 solution

10.98 How many moles of acid can 25 mL of 0.43 M $Mg(OH)_2$ solution neutralize? If it is mixed with 8.50 mL of a 0.825 M H_3PO_4 solution, will it result to an acidic or basic solution? (10.6)

10.99 One of the most acidic lakes in the United States is Little Echo Pond in the Adirondacks in New York. Recently, this lake had a pH of 4.2, well below the recommended pH of 6.5. (10.5, 10.6)

A helicopter drops calcium carbonate on an acidic lake to increase its pH.

a. What are the $[H_3O^+]$ and $[OH^-]$ of Little Echo Pond?
b. What are the $[H_3O^+]$ and $[OH^-]$ of a lake that has a pH of 6.5?
c. One way to raise the pH (and restore aquatic life) is to add limestone ($CaCO_3$). How many grams of $CaCO_3$ are needed to neutralize 1.0 kL of the acidic water from Little Echo Pond if the acid is sulfuric acid?

$$H_2SO_4(aq) + CaCO_3(s) \longrightarrow CO_2(g) + H_2O(l) + CaSO_4(aq)$$

10.100 The daily output of stomach acid (gastric juice) is 1000 mL to 2000 mL. Prior to a meal, stomach acid (HCl) typically has a pH of 1.42. (10.5, 10.6)
a. What is the $[H_3O^+]$ of stomach acid?
b. One chewable tablet of the antacid Maalox contains 600. mg of $CaCO_3$. Write the neutralization equation, and calculate the milliliters of stomach acid neutralized by two tablets of Maalox.
c. The antacid milk of magnesia contains 400. mg of $Mg(OH)_2$ per teaspoon. Write the neutralization equation, and calculate the number of milliliters of stomach acid neutralized by one tablespoon of milk of magnesia (one tablespoon = three teaspoons).

ANSWERS

Answers to Study Checks

10.1 $LiOH(s) \xrightarrow{H_2O} Li^+(aq) + OH^-(aq)$

10.2 $HNO_3(aq) + H_2O(l) \longrightarrow H_3O^+(aq) + NO_3^-(aq)$

10.3 The conjugate acid–base pairs are HCN/CN^- and HSO_4^-/SO_4^{2-}.

10.4 The double arrow indicates that the reaction is reversible and that at equilibrium the rates of forward and reverse reactions are equal.

10.5 Removing H_2CO_3 causes the equilibrium to shift in the direction of the reactants.

10.6 $[H_3O^+] = 2.5 \times 10^{-11}$ M, basic

10.7 kitchen cleaner

10.8 11.38

10.9 $[H_3O^+] = 3 \times 10^{-5}$ M

10.10 $Ca(s) + 2HBr(aq) \longrightarrow H_2(g) + CaBr_2(aq)$

10.11 $H_2SO_4(aq) + 2NaHCO_3(s) \longrightarrow 2CO_2(g) + 2H_2O(l) + Na_2SO_4(aq)$

10.12 0.200 M HCl solution

10.13 **1.** the salt

Answers to Selected Questions and Problems

10.1 **a.** acid **b.** acid **c.** acid
d. base **e.** both

10.3 **a.** Phosphoric acid **b.** Lithium hydroxide
c. Hydroiodic acid **d.** Hydrocyanic acid
e. Nitrous acid **f.** Calcium hydroxide

10.5 **a.** KOH **b.** HBr
c. H_2SO_3 **d.** NaOH
e. $Be(OH)_2$ **f.** H_2CO_3

10.7 **a.** HI is the acid (H^+ donor) and H_2O is the base (H^+ acceptor).
b. H_2O is the acid (H^+ donor) and F^- is the base (H^+ acceptor).

10.9 **a.** Cl^- **b.** I^-
c. NO_3^- **d.** ClO_3^-

10.11 **a.** H_3PO_4 **b.** HI
c. HNO_2 **d.** H_2CO_3

10.13 **a.** The conjugate acid–base pairs are H_2CO_3/HCO_3^- and H_3O^+/H_2O.
b. The conjugate acid–base pairs are NH_4^+/NH_3 and H_3O^+/H_2O.
c. The conjugate acid–base pairs are HCN/CN^- and HNO_2/NO_2^-.

10.15 **a.** HI **b.** $HClO_4$ **c.** HNO_3

10.17 **a.** HCN **b.** HCO_3^- **c.** H_3PO_4

10.19 A reversible reaction is one in which a forward reaction converts reactants to products, whereas a reverse reaction converts products to reactants.

10.21 **a.** not at equilibrium **b.** at equilibrium
c. at equilibrium

10.23 **a.** Equilibrium shifts in the direction of the reactants.
b. Equilibrium shifts in the direction of the reactants.
c. Equilibrium shifts in the direction of the products.
d. Equilibrium shifts in the direction of the products.

10.25 In pure water, $[H_3O^+] = [OH^-]$ because one of each is produced every time an H^+ transfers from one water molecule to another.

10.27 In an acidic solution, the $[H_3O^+]$ is greater than the $[OH^-]$.

10.29 **a.** acidic **b.** neutral
c. basic **d.** acidic

10.31 **a.** 3.0×10^{-5} M **b.** 1.6×10^{-4} M
c. 1.2×10^{-11} M **d.** 2.2×10^{-12} M

10.33 **a.** 5.0×10^{-11} M **b.** 2.5×10^{-6} M
c. 1.6×10^{-4} M **d.** 2.8×10^{-12} M

10.35 In a neutral solution, the $[H_3O^+]$ is 1×10^{-7} M and the pH is 7.0, which is the negative value of the power of 10.

10.37 **a.** acidic **b.** acidic **c.** neutral
d. acidic **e.** basic **f.** basic

10.39 **a.** 4.0 **b.** 8.5 **c.** 9.0
d. 3.40 **e.** 7.17 **f.** 10.92

10.41

$[H_3O^+]$	$[OH^-]$	pH	Acidic, Basic, or Neutral?
1×10^{-8} M	1×10^{-6} M	8.0	Basic
1×10^{-3} M	1×10^{-11} M	3.0	Acidic
2×10^{-5} M	5×10^{-10} M	4.7	Acidic
1×10^{-12} M	1×10^{-2} M	12.0	Basic
2.4×10^{-5} M	4.2×10^{-10} M	4.62	Acidic

10.43 **a.** $ZnCO_3(s) + 2HBr(aq) \longrightarrow CO_2(g) + H_2O(l) + ZnBr_2(aq)$
b. $Zn(s) + 2HCl(aq) \longrightarrow H_2(g) + ZnCl_2(aq)$
c. $HCl(g) + NaHCO_3(s) \longrightarrow CO_2(g) + H_2O(l) + NaCl(aq)$
d. $H_2SO_4(aq) + Mg(OH)_2(s) \longrightarrow 2H_2O(l) + MgSO_4(aq)$

10.45 **a.** $2HCl(aq) + Mg(OH)_2(s) \longrightarrow 2H_2O(l) + MgCl_2(aq)$
b. $H_3PO_4(aq) + 3LiOH(aq) \longrightarrow 3H_2O(l) + Li_3PO_4(aq)$

10.47 **a.** $H_2SO_4(aq) + 2NaOH(aq) \longrightarrow 2H_2O(l) + Na_2SO_4(aq)$
b. $3HCl(aq) + Fe(OH)_3(aq) \longrightarrow 3H_2O(l) + FeCl_3(aq)$
c. $H_2CO_3(aq) + Mg(OH)_2(s) \longrightarrow 2H_2O(l) + MgCO_3(s)$

10.49 2M HCl solution

10.51 0.124 M H_2SO_4 solution

10.53 **b** and **c** are buffer systems; **b** contains the weak acid H_2CO_3 and its salt $NaHCO_3$; **c** contains HF, a weak acid, and its salt KF.

10.55 **a.** **3** **b.** **1** and **2**
c. **3** **d.** **1**

10.57 **a.** acid **b.** base
c. base **d.** acid

10.59

Acid	Conjugate Base
H_2O	OH^-
HCN	CN^-
HNO_2	NO_2^-
H_3PO_4	$H_2PO_4^-$

10.61 **a.** acidic **b.** basic
c. acidic **d.** acidic
e. basic

10.63 **a.** weak acid **b.** strong acid

10.65 **a.** Hyperventilation will lower the CO_2 level in the blood, which lowers the $[H_2CO_3]$, which decreases the $[H_3O^+]$ and increases the blood pH.
b. Breathing into a bag increases the CO_2 level, increases the $[H_2CO_3]$, and increases the $[H_3O^+]$, which lowers the blood pH.

10.67 **a.** base, lithium hydroxide **b.** salt, calcium nitrate
c. acid, hydrobromic acid **d.** base, barium hydroxide
e. acid, carbonic acid

10.69 **a.** HF **b.** H_3O^+
c. HNO_2 **d.** HCO_3^-

10.71 **a.** System shifts in the direction of the products.
b. System shifts in the direction of the products.
c. System shifts in the direction of the reactants.
d. System shifts in the direction of the reactants.

10.73 **a.** 8.00 **b.** 1.30
c. 10.54 **d.** 11.7

10.75 **a.** basic **b.** acidic
c. basic **d.** basic

10.77 **a.** $[H_3O^+] = 1 \times 10^{-3}$ M, $[OH^-] = 1 \times 10^{-11}$ M
b. $[H_3O^+] = 1.0 \times 10^{-6}$ M, $[OH^-] = 1.0 \times 10^{-8}$ M
c. $[H_3O^+] = 1 \times 10^{-8}$ M, $[OH^-] = 1 \times 10^{-6}$ M
d. $[H_3O^+] = 1 \times 10^{-11}$ M, $[OH^-] = 1 \times 10^{-3}$ M
e. $[H_3O^+] = 6.3 \times 10^{-10}$ M, $[OH^-] = 1.6 \times 10^{-5}$ M

10.79 **a.** Solution A with a pH of 4.0 is more acidic than solution B.
b. Solution A: $[H_3O^+] = 1 \times 10^{-4}$ M,
Solution B: $[H_3O^+] = 1 \times 10^{-6}$ M
c. Solution A: $[OH^-] = 1 \times 10^{-10}$ M,
Solution B: $[OH^-] = 1 \times 10^{-8}$ M

10.81 **a.** Acid added:
$H_2PO_4^-(aq) + H_3O^+(aq) \longrightarrow H_2O(l) + H_3PO_4(aq)$
b. Base added:
$H_3PO_4(aq) + OH^-(aq) \longrightarrow H_2O(l) + H_2PO_4^-(aq)$

10.83 **a.** 48.0 mL
b. 74.7 mL
c. 20.6 mL

10.85 0.234 M H_2SO_4 solution

10.87 **a.** **1.** HS^- **2.** $H_2PO_4^-$
b. H_2S

10.89 **a.** NH_4^+/NH_3 and HNO_3/NO_3^-
b. H_3O^+/H_2O and HBr/Br^-

10.91 **a.** $ZnCO_3(s) + H_2SO_4(aq) \longrightarrow CO_2(g) + H_2O(l) + ZnSO_4(aq)$
b. $2Al(s) + 6HCl(aq) \longrightarrow 3H_2(g) + 2AlCl_3(aq)$

10.93 **a.** 2.0×10^{-13} M
b. 12.70
c. $H_2SO_4(aq) + 2KOH(aq) \longrightarrow 2H_2O(l) + K_2SO_4(aq)$
d. 56 mL

10.95 **a.** $H_3PO_4(aq) + 3NaOH(aq) \longrightarrow 3H_2O(l) + Na_3PO_4(aq)$
b. 0.0224 M H_3PO_4 solution

10.97 **a.** 69.3 mL of a LiOH solution
b. 50.4 mL of a LiOH solution

10.99 **a.** $[H_3O^+] = 6 \times 10^{-5}$ M; $[OH^-] = 2 \times 10^{-10}$ M
b. $[H_3O^+] = 3 \times 10^{-7}$ M; $[OH^-] = 3 \times 10^{-8}$ M
c. 3 g of $CaCO_3$

11 Introduction to Organic Chemistry: Hydrocarbons

A wildfire started near Rocky Mountain National Park from a lightning strike. Due to a lack of rain, there was considerable dry timber and grasses to provide fuel for the fire. The local firefighters decided to begin a backfire, which is deliberately set in the path of the wildfire so it consumes the dry timber and grasses and, thus, stops or confines the fire. Jack put on his fire-resistant pants, jacket, and face protection. Jack then started the backfire by using a drip torch containing a mixture of diesel and gasoline. Gasoline and diesel consist of organic molecules called alkanes. Alkanes, or hydrocarbons, are chains of carbon and hydrogen atoms. The alkanes present in gasoline consist of a mixture of 5 to 8 carbon atoms in a chain, while diesel typically consists of a mixture of 11 to 20 carbon atoms. Because alkanes undergo combustion reactions, they can be used to start the backfire.

CAREER Firefighter

Firefighters are first responders to fires, accidents, and other emergency situations. They are required to have an emergency medical technician or paramedic certification in order to be able to treat seriously injured people. By combining the skills of a firefighter and a paramedic, they increase the survival rates of the injured. The physical demands of firefighters are extremely high as they fight, extinguish, and prevent fires while wearing heavy protective clothing and gear. They also train for and participate in firefighting drills, and maintain fire equipment so that it is always working and ready. Firefighters must also be knowledgeable about fire codes, arson, and the handling and disposal of hazardous materials. Since firefighters also provide emergency care for sick and injured people, they need to be aware of emergency medical and rescue procedures, as well as the proper methods for controlling the spread of infectious disease.

LOOKING AHEAD

At the beginning of the nineteenth century, scientists classified chemical compounds as inorganic and organic. An inorganic compound was a substance that was composed of minerals, and an organic compound was a substance that came from an organism, thus the use of the word *organic*. Early scientists thought that some type of "vital force," which could be found only in living cells, was required to synthesize an organic compound. This idea was shown to be incorrect in 1828 when the German chemist Friedrich Wöhler synthesized urea, a product of protein metabolism, by heating an inorganic compound, ammonium cyanate.

$$\underset{\text{Ammonium cyanate (inorganic)}}{NH_4CNO} \xrightarrow{\text{Heat}} \underset{\text{Urea (organic)}}{H_2N-\overset{\overset{\displaystyle O}{\|}}{C}-NH_2}$$

The simplest organic compounds are the hydrocarbons that contain only the elements of carbon and hydrogen. All the carbon-to-carbon bonds in a hydrocarbon known as an alkane are single C—C bonds. Some common hydrocarbons include methane (CH_4), propane (C_3H_8), and butane (C_4H_{10}).

Alkenes are hydrocarbons that contain one or more carbon-to-carbon double bonds (C=C). Because double bonds are very reactive, they easily add hydrogen atoms (hydrogenation) or water (hydration) to the carbon atoms in the double bond. One important compound containing a double bond is ethene (ethylene), which is used to ripen fruit when ready for market. A common compound with a triple bond is ethyne (acetylene), which burns at high temperatures and is used in welding metals.

Learning about the structures and reactions of organic molecules will provide you with a foundation for understanding the more complex molecules of biochemistry.

CHAPTER READINESS*

CORE CHEMISTRY SKILLS

- Drawing Electron-Dot Formulas (6.5)
- Predicting Shape (6.7)
- Balancing a Chemical Equation (7.3)

*These Core Chemistry Skills from previous chapters are listed here for your review as you proceed to the new material in this chapter.

11.1 Organic Compounds

LEARNING GOAL

Identify properties characteristic of organic or inorganic compounds.

Organic chemistry is the study of carbon compounds. The element carbon has a special role because many carbon atoms can bond together to give a vast array of molecular compounds. **Organic compounds** always contain carbon and hydrogen, and sometimes other nonmetals such as oxygen, sulfur, nitrogen, phosphorus, or a halogen. We find organic compounds in many common products we use every day, such as gasoline, medicines,

Vegetable oil, an organic compound, is not soluble in water.

shampoos, plastics, and perfumes. The food we eat is composed of organic compounds such as carbohydrates, fats, and proteins that supply us with fuel for energy and the carbon atoms needed to build and repair the cells of our bodies.

The formulas of organic compounds are written with carbon first, followed by hydrogen, and then any other elements. Organic compounds typically have low melting and boiling points, are not soluble in water, and are less dense than water. For example, vegetable oil, which is a mixture of organic compounds, does not dissolve in water but floats on top. Many organic compounds undergo combustion and burn vigorously in air. By contrast, many inorganic compounds have high melting and boiling points. Inorganic compounds that are ionic are usually soluble in water, and most do not burn in air. Table 11.1 contrasts some of the properties associated with organic and inorganic compounds, such as propane, C_3H_8, and sodium chloride, NaCl (see Figure 11.1).

TABLE 11.1 Some Properties of Organic and Inorganic Compounds

Property	Organic	Example: C_3H_8	Inorganic	Example: NaCl
Elements Present	C and H, sometimes O, S, N, P, or Cl (F, Br, I)	C and H	Most metals and nonmetals	Na and Cl
Particles	Molecules	C_3H_8	Mostly ions	Na^+ and Cl^-
Bonding	Mostly covalent	Covalent	Many are ionic, some covalent	Ionic
Polarity of Bonds	Nonpolar, unless a strongly electronegative atom is present	Nonpolar	Most are ionic or polar covalent, a few are nonpolar covalent	Ionic
Melting Point	Usually low	−188 °C	Usually high	801 °C
Boiling Point	Usually low	−42 °C	Usually high	1413 °C
Flammability	High	Burns in air	Low	Does not burn
Solubility in Water	Not soluble, unless a polar group is present	No	Most are soluble, unless nonpolar	Yes

FIGURE 11.1 ▶ Propane, C_3H_8, is an organic compound, whereas sodium chloride, NaCl, is an inorganic compound.

Q Why is propane used as a fuel?

SAMPLE PROBLEM 11.1 Properties of Organic Compounds

Indicate whether the following properties are more typical of organic or inorganic compounds:

a. is not soluble in water
b. has a high melting point
c. burns in air

SOLUTION

a. Many organic compounds are not soluble in water.
b. Inorganic compounds are more likely to have high melting points.
c. Organic compounds are more likely to be flammable.

STUDY CHECK 11.1

What elements are always found in hydrocarbons?

Bonding in Organic Compounds

Hydrocarbons are organic compounds that consist of only carbon and hydrogen. In the simplest hydrocarbon, methane (CH_4), the carbon atom forms an octet by sharing its four valence electrons with four hydrogen atoms. In organic molecules, every carbon atom has four bonds. A hydrocarbon is referred to as a *saturated hydrocarbon* when all the bonds in the molecule are single bonds. We can draw an **expanded structural formula** for a compound by showing all the bonds between atoms.

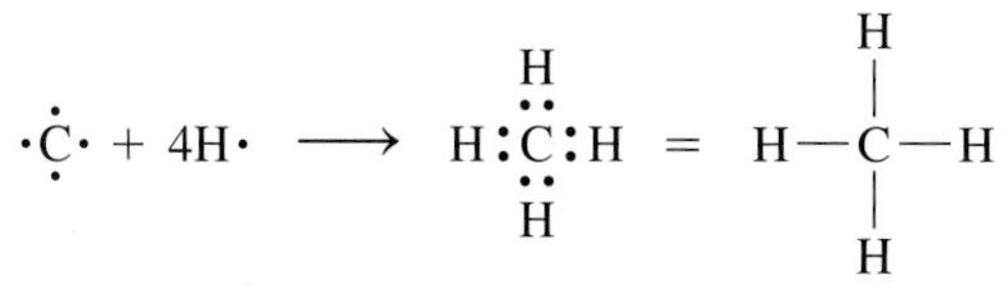

Methane

Representations of Carbon Compounds

The electron-dot formula and the expanded structural formula are two-dimensional representations in which the electron pairs and the bonds from carbon to each hydrogen atom are shown. In the three-dimensional structure of methane, CH_4, the covalent bonds from the carbon atom to each hydrogen atom are directed to the corners of a tetrahedron with bond angles of 109°. This tetrahedral shape for methane can be illustrated as a ball-and-stick model.

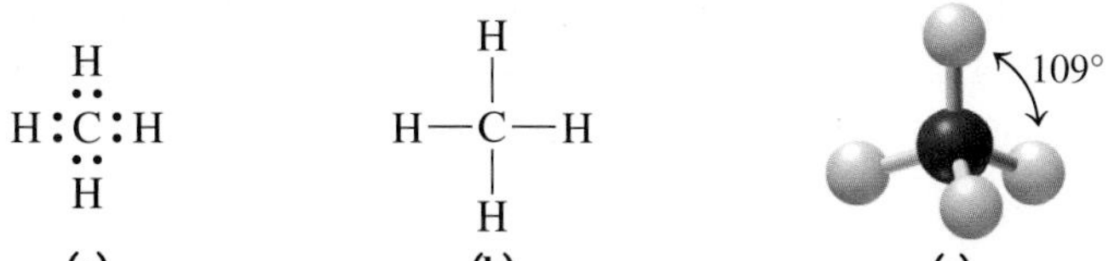

(a) (b) (c)

Two-dimensional and three-dimensional representations of methane, CH_4: **(a)** electron-dot formula, **(b)** expanded structural formula, and **(c)** ball-and-stick model.

In ethane, C_2H_6, each carbon atom is bonded to another carbon and three hydrogen atoms. As in methane, each carbon atom retains its tetrahedral shape.

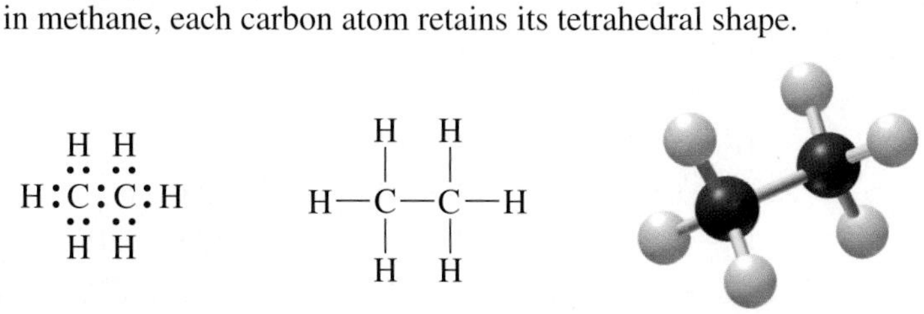

(a) (b) (c)

Two-dimensional and three-dimensional representations of ethane, C_2H_6: **(a)** electron-dot formula, **(b)** expanded structural formula, and **(c)** ball-and-stick model.

QUESTIONS AND PROBLEMS

11.1 Organic Compounds

LEARNING GOAL Identify properties characteristic of organic or inorganic compounds.

11.1 Identify each of the following as a formula of an organic or inorganic compound:

a. $MgCl_2$ **b.** C_3H_8
c. CH_4O **d.** H_3PO_4
e. C_6H_6 **f.** NaCl

11.2 Identify each of the following as a formula of an organic or inorganic compound:

a. $C_6H_{12}O_6$ **b.** K_3PO_4
c. I_2 **d.** C_2H_6S
e. $C_{10}H_{22}$ **f.** C_4H_9Br

11.3 Identify each of the following properties as more typical of an organic or inorganic compound:

a. contains no carbon
b. contains oxygen atoms
c. gives carbon dioxide and carbon monoxide upon combustion
d. contains a large amount of metal ions

11.4 Identify each of the following properties as more typical of an organic or inorganic compound:

a. dissolves easily in polarized mediums
b. has a low boiling point
c. contains carbon and hydrogen
d. forms crystals easily

11.5 Match each of the following physical and chemical properties with propane, C_3H_8, Potassium bromide, KBr:

a. Dissolves in water **b.** liquid at room temperature
c. melts at 734 °C **d.** contains carbon and hydrogen

11.6 Match each of the following physical and chemical properties with cyclohexane, C_6H_{12}, or calcium nitrate, $Ca(NO_3)_2$:

a. melts at 500 °C
b. is insoluble in water
c. does not burn in air
d. is a liquid at room temperature

LEARNING GOAL

Write the IUPAC names and draw the condensed structural formulas and skeletal formulas for alkanes and cycloalkanes.

11.2 Alkanes

More than 90% of the compounds in the world are organic compounds. The large number of carbon compounds is possible because the covalent bond between carbon atoms (C—C) is very strong, allowing carbon atoms to form long, stable chains.

The **alkanes** are a type of hydrocarbon in which the carbon atoms are connected only by single bonds. One of the most common uses of alkanes is as fuels. Methane, used in gas heaters and gas cooktops, is an alkane with one carbon atom. The alkanes ethane, propane, and butane contain two, three, and four carbon atoms, respectively, connected in a row or a *continuous chain*. As we can see, the names for alkanes end in *ane*. Such names are part of the **IUPAC (International Union of Pure and Applied Chemistry) system** used by chemists to name organic compounds. Alkanes with five or more carbon atoms in a chain are named using Greek prefixes: *pent* (5), *hex* (6), *hept* (7), *oct* (8), *non* (9), and *dec* (10) (see Table 11.2).

TABLE 11.2 IUPAC Names of the First 10 Alkanes

Number of Carbon Atoms	Prefix	Name	Molecular Formula	Condensed Structural Formula
1	Meth	Methane	CH_4	CH_4
2	Eth	Ethane	C_2H_6	$CH_3—CH_3$
3	Prop	Propane	C_3H_8	$CH_3—CH_2—CH_3$
4	But	Butane	C_4H_{10}	$CH_3—CH_2—CH_2—CH_3$
5	Pent	Pentane	C_5H_{12}	$CH_3—CH_2—CH_2—CH_2—CH_3$
6	Hex	Hexane	C_6H_{14}	$CH_3—CH_2—CH_2—CH_2—CH_2—CH_3$
7	Hept	Heptane	C_7H_{16}	$CH_3—CH_2—CH_2—CH_2—CH_2—CH_2—CH_3$
8	Oct	Octane	C_8H_{18}	$CH_3—CH_2—CH_2—CH_2—CH_2—CH_2—CH_2—CH_3$
9	Non	Nonane	C_9H_{20}	$CH_3—CH_2—CH_2—CH_2—CH_2—CH_2—CH_2—CH_2—CH_3$
10	Dec	Decane	$C_{10}H_{22}$	$CH_3—CH_2—CH_2—CH_2—CH_2—CH_2—CH_2—CH_2—CH_2—CH_3$

Condensed Structural Formulas

In a **condensed structural formula**, each carbon atom and its attached hydrogen atoms are written as a group. A subscript indicates the number of hydrogen atoms bonded to each carbon atom.

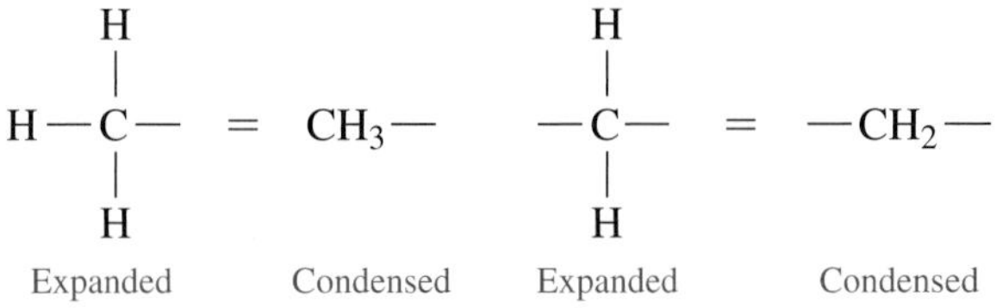

By contrast, the *molecular formula* gives the total number of carbon and hydrogen atoms, but does not indicate their arrangement in the molecule.

When an organic molecule consists of a chain of three or more carbon atoms, the carbon atoms do not lie in a straight line. Rather, they are arranged in a zigzag pattern.

A simplified structure called the *skeletal formula* shows the carbon skeleton in which carbon atoms are represented as the end of each line or as corners. The hydrogen atoms are not shown, but each carbon is understood to have bonds to four atoms. For example, in the skeletal formula of hexane, each line in the zigzag drawing represents a single bond. The carbon atoms on the ends are bonded to three hydrogen atoms. However the carbon atoms in the middle of the carbon chain are each bonded to two carbons and two hydrogen atoms (see Figure 11.2).

Conformations of Alkanes

Because an alkane has only single carbon–carbon bonds, the groups attached to each C are not in fixed positions. They can rotate freely about the bond connecting the carbon atoms. This motion is analogous to the independent rotation of the wheels of a toy car. Thus, different arrangements, known as *conformations*, occur during the rotation about a single bond.

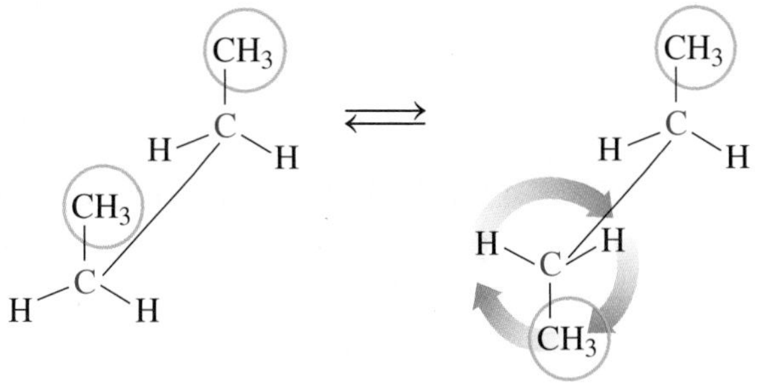

Suppose we could look at butane, C_4H_{10}, as it rotates. Sometimes the —CH_3 groups line up in front of each other, and at other times they are opposite each other. As the —CH_3 groups turn around the single bond, the carbon chain in the condensed structural formulas may appear at different angles. For example, butane can be drawn using a variety of two-dimensional condensed structural formulas as shown in Table 11.3. All these condensed structural formulas represent the same compound with four carbon atoms.

CORE CHEMISTRY SKILL

Naming and Drawing Alkanes

Alkane Name Hexane
Molecular Formula C_6H_{14}
Ball-and-Stick Model

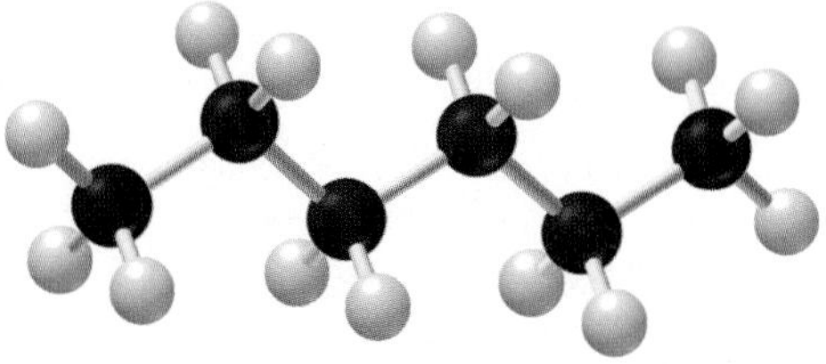

Expanded Structural Formula

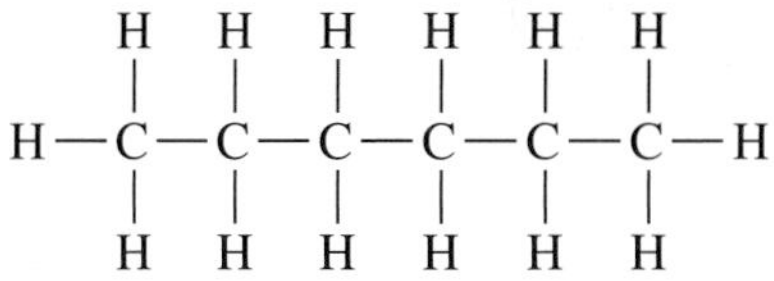

Condensed Structural Formulas

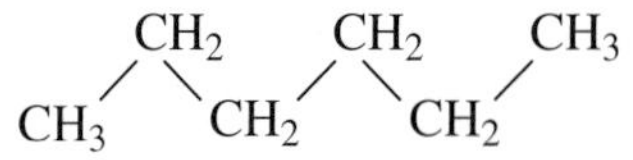

$CH_3—CH_2—CH_2—CH_2—CH_2—CH_3$

Skeletal Formula

FIGURE 11.2 ▶ A hexane molecule can be represented in several ways: molecular formula, ball-and-stick model, expanded structural formula, condensed structural formula, and skeletal formula.

Q Why do the carbon atoms in hexane appear to be arranged in a zigzag chain?

SAMPLE PROBLEM 11.2 Drawing Expanded and Condensed Structural Formulas for Alkanes

Draw the expanded structural formula, condensed structural formula, and skeletal formula for pentane.

SOLUTION

A molecule of pentane, C_5H_{12}, has five carbon atoms in a continuous chain. In the expanded structural formula, the five carbon atoms are connected to each other and to hydrogen atoms by single bonds to give each carbon atom a total of four bonds. In the condensed structural formula, the carbon atoms and their hydrogen atoms on the ends of the chain are written as $—CH_3$ and the carbon and hydrogen atoms in the middle are written as $—CH_2—$. The skeletal formula shows the carbon skeleton as a zigzag line where the ends and corners represent C atoms.

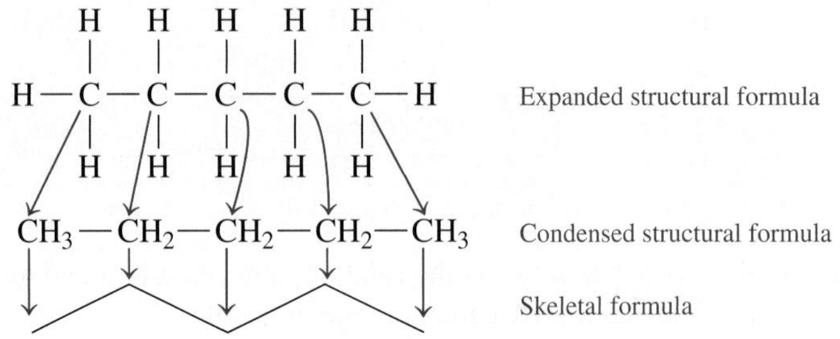

STUDY CHECK 11.2

Draw the condensed structural formula and give the name for the following skeletal formula:

TABLE 11.3 Some Structural Formulas for Butane, C_4H_{10}

Expanded Structural Formula	Ball-and-Stick Model
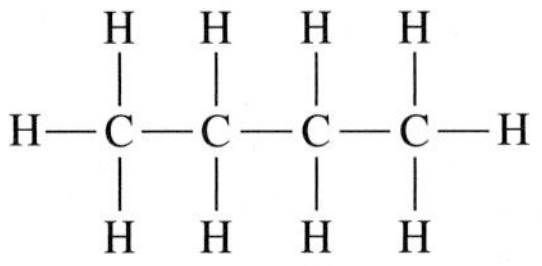	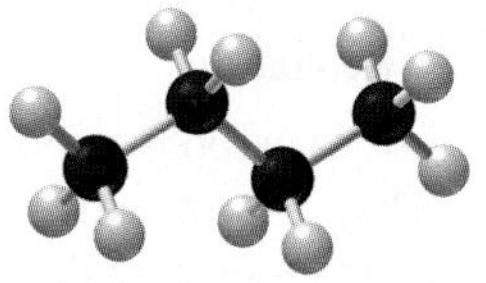

Condensed Structural Formulas

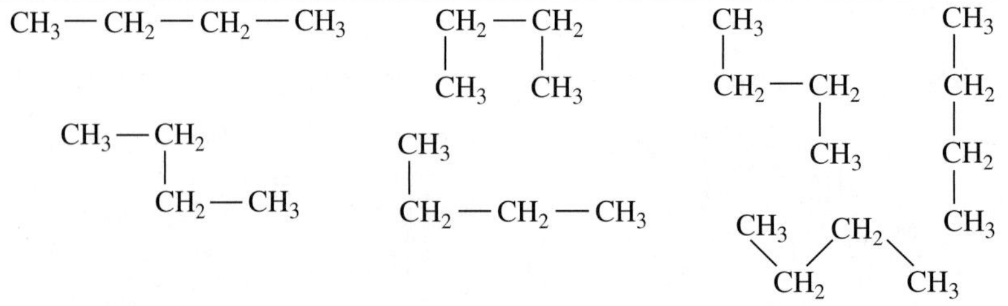

Skeletal Formulas

Cycloalkanes

Hydrocarbons can also form cyclic or ring structures called **cycloalkanes**, which have two fewer hydrogen atoms than the corresponding alkanes. The simplest cycloalkane, cyclopropane (C_3H_6), has a ring of three carbon atoms bonded to six hydrogen atoms. Most often cycloalkanes are drawn using their skeletal formulas, which appear as simple geometric figures. As seen for alkanes, each corner of the skeletal formula for a cycloalkane represents a carbon atom.

The ball-and-stick models, condensed structural formulas, and skeletal formulas for several cycloalkanes are shown in Table 11.4. A cycloalkane is named by adding the prefix *cyclo* to the name of the alkane with the same number of carbon atoms.

TABLE 11.4 Formulas of Some Common Cycloalkanes

Name			
Cyclopropane	Cyclobutane	Cyclopentane	Cyclohexane
Ball-and-Stick Model			
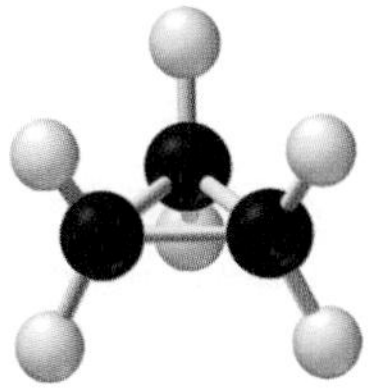	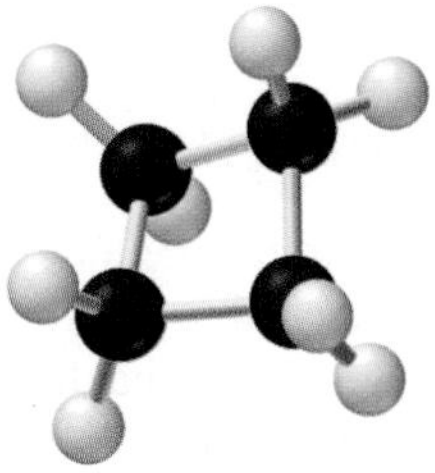	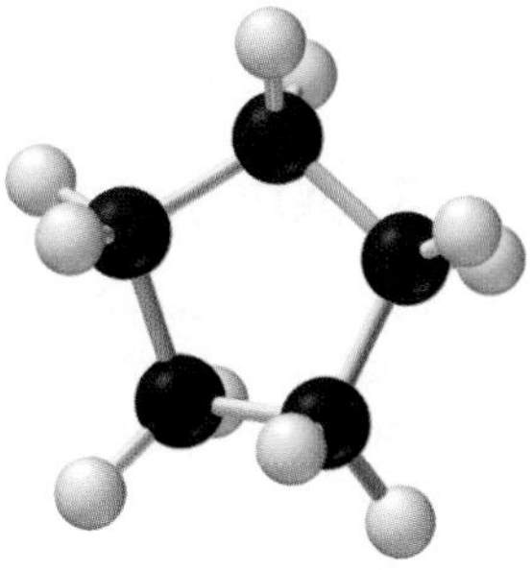	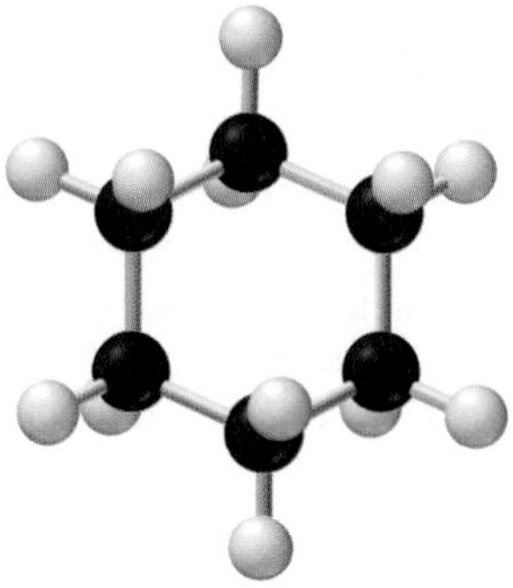
Condensed Structural Formula			
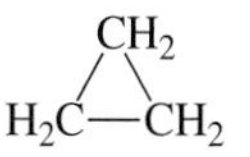	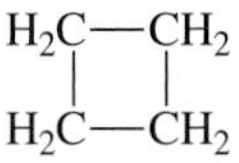	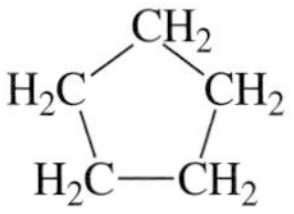	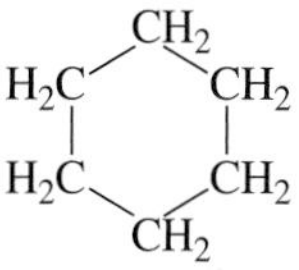
Skeletal Formula			
		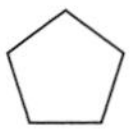	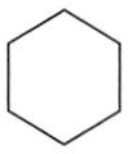

SAMPLE PROBLEM 11.3 Naming Alkanes and Cycloalkanes

Give the IUPAC name for each of the following:

a.    **b.** 

SOLUTION

a. A chain with eight carbon atoms is octane.
b. The ring of six carbon atoms is named cyclohexane.

STUDY CHECK 11.3

What is the IUPAC name of the following compound?

QUESTIONS AND PROBLEMS

11.2 Alkanes

LEARNING GOAL Write the IUPAC names and draw the condensed structural formulas and skeletal formulas for alkanes and cycloalkanes.

11.7 Give the IUPAC name for each of the following alkanes and cycloalkanes:

a. 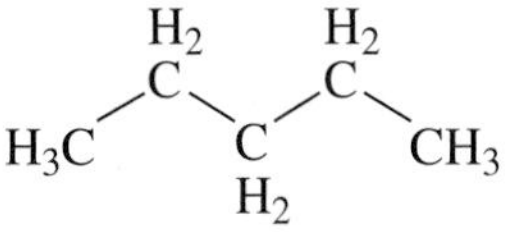   **b.** 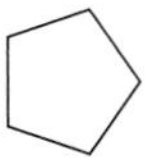

c. 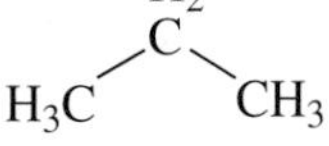   **d.** 

11.8 Give the IUPAC name for each of the following alkanes and cycloalkanes:

a. 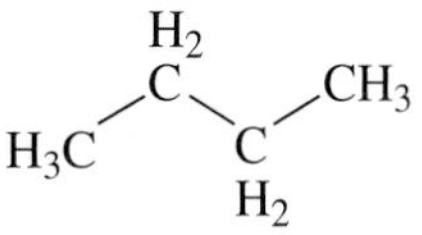   **b.** 

c. 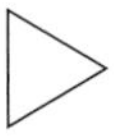   **d.** 

11.9 Draw the condensed structural formula for alkanes or the skeletal formula for cycloalkanes for each of the following:
a. butane
b. cyclohexane
c. nonane
d. ethane

11.10 Draw the condensed structural formula for alkanes or the skeletal formula for cycloalkanes for each of the following:
a. chloroethane
b. octane
c. 3-hexyne
d. 1,3-cyclohexadiene

LEARNING GOAL

Write the IUPAC names for alkanes with substituents and draw their condensed structural formulas and skeletal formulas.

11.3 Alkanes with Substituents

When an alkane has four or more carbon atoms, the atoms can be arranged so that a side group called a *branch* or **substituent** is attached to a carbon chain. For example, there are different ball-and-stick models for two compounds that have the molecular formula C_4H_{10}. One model is shown as a chain of four carbon atoms. In the other model, a carbon atom is attached as a branch or substituent to a carbon in a chain of three atoms (see Figure 11.3). An alkane with at least one branch is called a *branched alkane*. When the two compounds have the same molecular formula but different arrangements of atoms, they are called **structural isomers**.

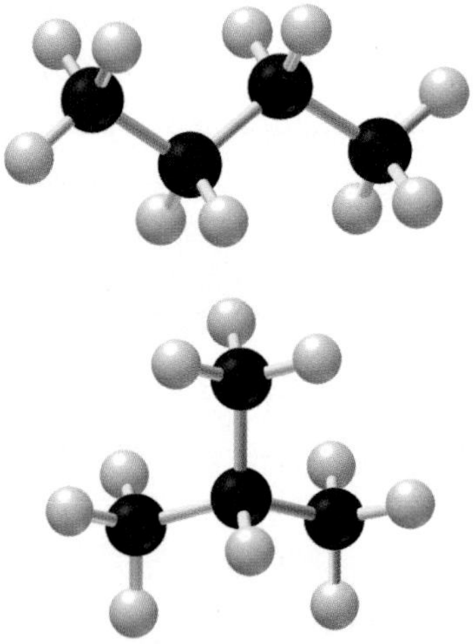

$CH_3—CH_2—CH_2—CH_3$

CH_3
|
$CH_3—CH—CH_3$

FIGURE 11.3 ▶ The structural isomers of C_4H_{10} have the same number and type of atoms but are bonded in a different order.

Q What makes these molecules structural isomers?

In another example, we can draw the condensed structural formulas for three different structural isomers with the molecular formula C_5H_{12} as follows:

Structural Isomers of C_5H_{12}

$CH_3—CH_2—CH_2—CH_2—CH_3$

CH_3
|
$CH_3—CH—CH_2—CH_3$

CH_3
|
$CH_3—C—CH_3$
|
CH_3

SAMPLE PROBLEM 11.4 Structural Isomers

Identify each pair of formulas as structural isomers or the same molecule.

a.
CH_3 CH_3
| |
$CH_2—CH_2$ and $CH_2—CH_2—CH_3$
|
CH_3

b.
CH_3
|
$CH_3—CH—CH_2—CH_2—CH_3$ and

CH_3 CH_3
| |
$CH_3—CH—CH—CH_3$

SOLUTION

a. When we add up the number of C atoms and H atoms, they give the same molecular formula C_4H_{10}. The structural formula on the left consists of a chain of four C atoms. Even though the $-CH_3$ ends are drawn up, they are not branches on the chain but part of the four-carbon chain. The structural formula on the right also consists of a four-carbon chain even though one $-CH_3$ end is drawn down. Thus both condensed structural formulas represent the same molecule and are not structural isomers.

b. When we add up the number of C atoms and H atoms, they give the same molecular formula C_6H_{14}. The structural formula on the left consists of a five-carbon chain with a $-CH_3$ substituent on the second carbon of the chain. The structural formula on the right consists of a four-carbon chain with two $-CH_3$ substituents, one bonded to the second carbon and one bonded to the third carbon. Thus there is a different order of bonding of atoms in these two structural formulas, which represent structural isomers.

STUDY CHECK 11.4

Why does the following formula represent a different structural isomer of the molecules in Sample Problem 11.4, part **b**?

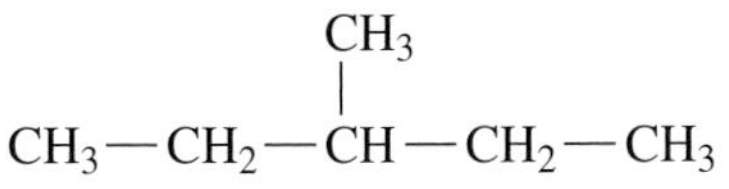

Substituents in Alkanes

In the IUPAC names for alkanes, a carbon branch is named as an **alkyl group**, which is an alkane that is missing one hydrogen atom. The alkyl group is named by replacing the *ane* ending of the corresponding alkane name with *yl*. Alkyl groups cannot exist on their own: They must be attached to a carbon chain. When a halogen atom is attached to a carbon chain, it is named as a *halo* group: *fluoro*, *chloro*, *bromo*, or *iodo*. Some of the common groups attached to carbon chains are illustrated in Table 11.5.

TABLE 11.5 Names and Formulas for Some Common Substituents

Substituent	Name
CH_3-	Methyl
CH_3-CH_2-	Ethyl
$CH_3-CH_2-CH_2-$	Propyl
$CH_3-\overset{\vert}{C}H-CH_3$	Isopropyl
$F-, Cl-, Br-, I-$	Fluoro, chloro, bromo, iodo

Naming Alkanes with Substituents

In the IUPAC system of naming, a carbon chain is numbered to give the location of the substituents. Let's take a look at how we use the IUPAC system to name the alkane shown in Sample Problem 11.5.

SAMPLE PROBLEM 11.5 Writing IUPAC Names for Alkanes with Substituents

Give the IUPAC name for the following alkane:

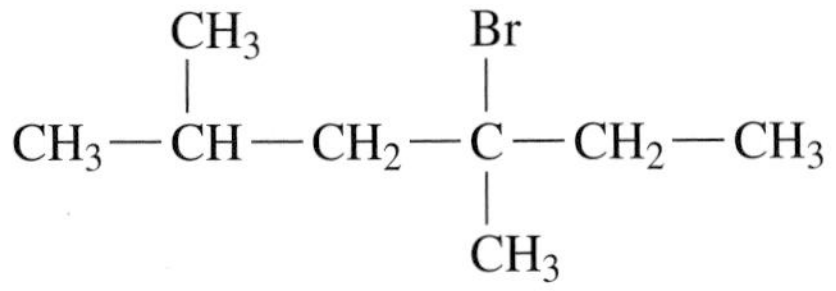

Guide to Naming Alkanes

STEP 1
Write the alkane name for the longest chain of carbon atoms.

STEP 2
Number the carbon atoms starting from the end nearer a substituent.

STEP 3
Give the location and name for each substituent (alphabetical order) as a prefix to the name of the main chain.

SOLUTION

ANALYZE THE PROBLEM	Type of Compound	Substituent	IUPAC Naming
	alkane	two methyl groups, one bromo group	number chain for substituents

STEP 1 **Write the alkane name for the longest chain of carbon atoms.** In this alkane, the longest chain has six carbon atoms, which is hexane.

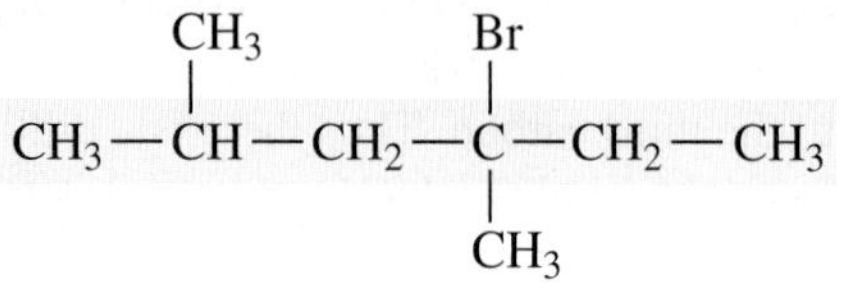

hexane

STEP 2 **Number the carbon atoms starting from the end nearer a substituent.** Carbon 1 will be the carbon atom nearer the methyl group, $-CH_3$, on the left.

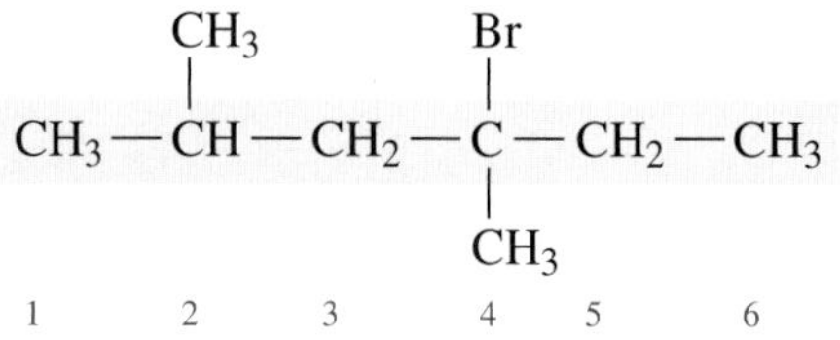

hexane

STEP 3 **Give the location and name for each substituent (alphabetical order) as a prefix to the name of the main chain.** The substituents, which are bromo and methyl groups, are listed in alphabetical order (bromo first, then methyl). A hyphen is placed between the number on the carbon chain and the substituent name. When there are two or more of the same substituent, a prefix (*di*, *tri*, *tetra*) is used in front of the name. Then commas are used to separate the numbers for the locations of the substituents.

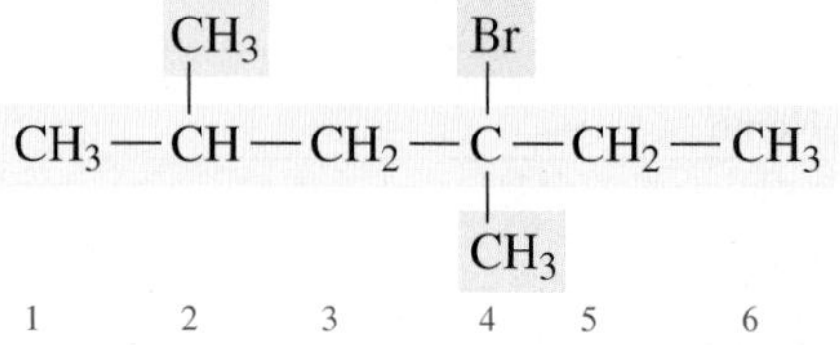

4-bromo-2,4-dimethylhexane

STUDY CHECK 11.5

Give the IUPAC name for the following compound:

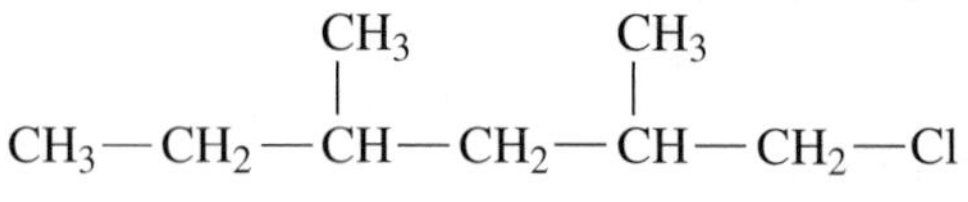

Interactive Video

Naming Alkanes

Naming Cycloalkanes with Substituents

When one substituent is attached to a carbon atom in a cycloalkane, the name of the substituent is placed in front of the cycloalkane name. No number is needed for a single alkyl group or halogen atom because the carbon atoms in the cycloalkane are equivalent.

Methylcyclopentane

Haloalkanes

In a *haloalkane*, halogen atoms replace hydrogen atoms in an alkane. The halo substituents are numbered and arranged alphabetically, just as we did with the alkyl groups. Many times chemists use the common, traditional name for these compounds rather than the systematic IUPAC name. Simple haloalkanes are commonly named as alkyl halides; the carbon group is named as an alkyl group followed by the halide name.

	$CH_3—Cl$	$CH_3—CH_2—Br$	$CH_3—CH_2—CH_2—F$
IUPAC	Chloromethane	Bromoethane	1-Fluoropropane
Common	Methyl chloride	Ethyl bromide	Propyl fluoride

Drawing Condensed Structural Formulas for Alkanes

The IUPAC name gives all the information needed to draw the condensed structural formula for an alkane. Suppose you are asked to draw the condensed structural formula for 2,3-dimethylbutane. The alkane name gives the number of carbon atoms in the longest chain. The names in the beginning indicate the substituents and where they are attached. We can break down the name in the following way:

2,3-Dimethylbutane

2,3-	Di	methyl	but	ane
Substituents on carbons 2 and 3	Two identical groups	$—CH_3$ alkyl groups	Four C atoms in the main chain	Single (C—C) bonds

SAMPLE PROBLEM 11.6 Drawing Condensed Structures from IUPAC Names

Draw the condensed structural formula and skeletal formula for 2,3-dimethylbutane.

SOLUTION

STEP 1 **Draw the main chain of carbon atoms.** For butane, we draw a chain of four carbon atoms.

C—C—C—C

STEP 2 **Number the chain and place the substituents on the carbons indicated by the numbers.** The first part of the name indicates two methyl groups $CH_3—$: one on carbon 2 and one on carbon 3.

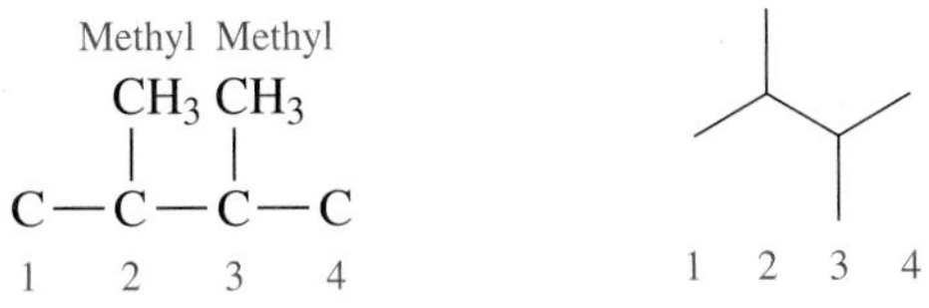

STEP 3 **Add the correct number of hydrogen atoms to give four bonds to each C atom.**

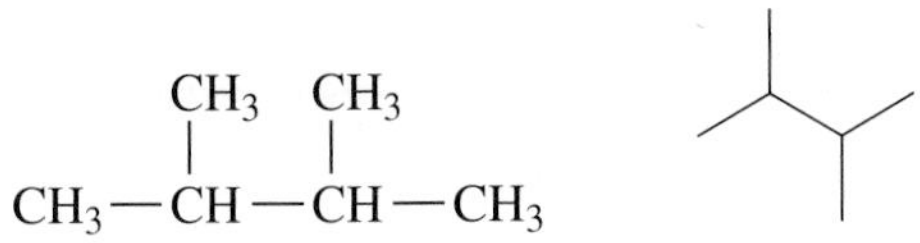

2,3-Dimethylbutane

STUDY CHECK 11.6

Draw the condensed structural formula and skeletal formula for 2-bromo-4-methylpentane.

Guide to Drawing Alkane Formulas

STEP 1 Draw the main chain of carbon atoms.

STEP 2 Number the chain and place the substituents on the carbons indicated by the numbers.

STEP 3 Add the correct number of hydrogen atoms to give four bonds to each C atom.

QUESTIONS AND PROBLEMS

11.3 Alkanes with Substituents

LEARNING GOAL Write the IUPAC names for alkanes with substituents and draw their condensed structural formulas and skeletal formulas.

11.11 Indicate whether each of the following pairs represent structural isomers or the same molecule:

a. 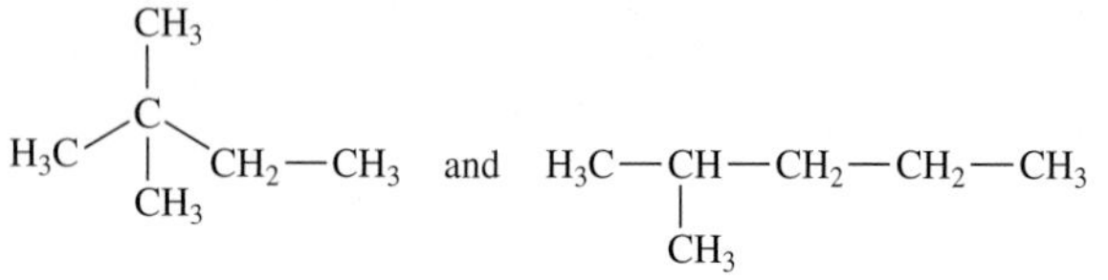

b. 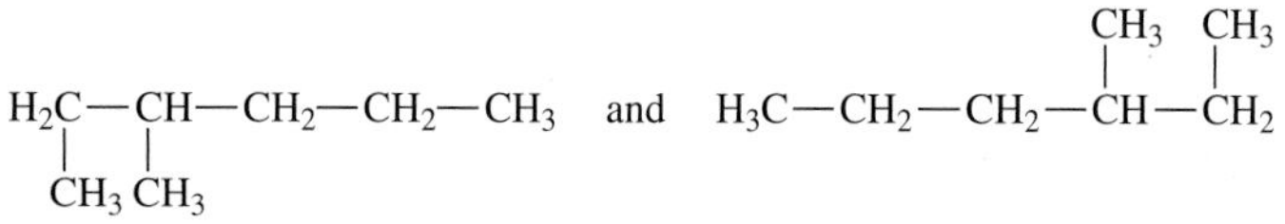

c. 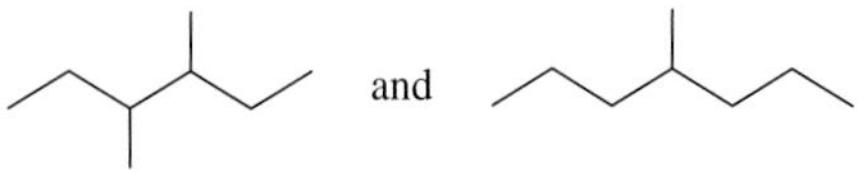

11.12 Indicate whether each of the following pairs represent structural isomers or the same molecule:

a. 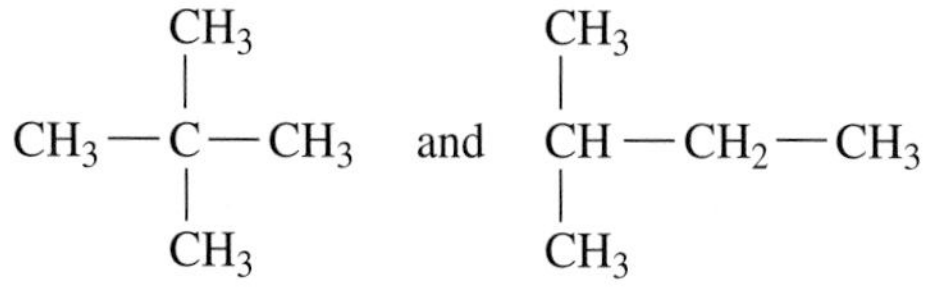

b. 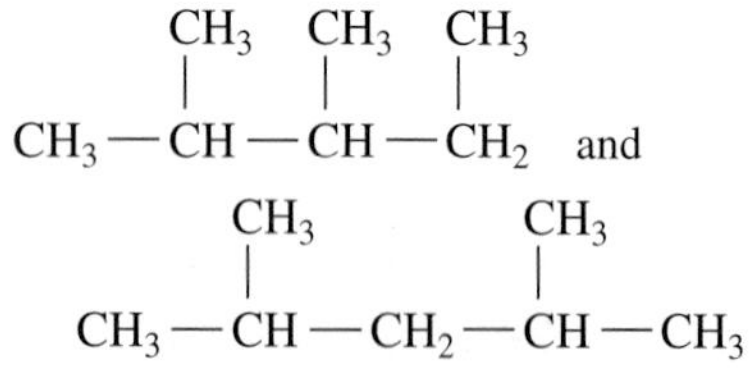

c. 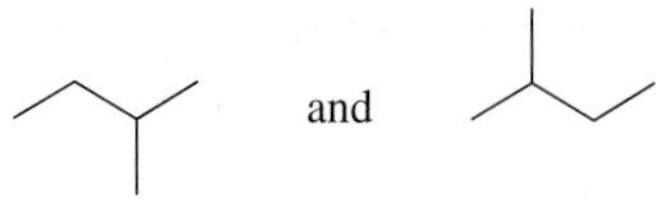

11.13 Give the IUPAC name for each of the following:

a. 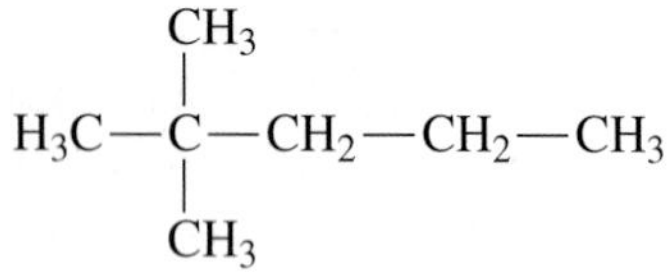

b. 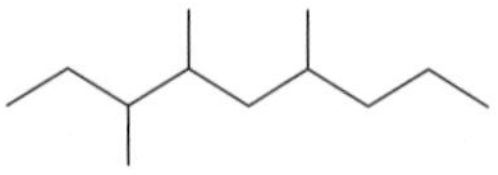

c. 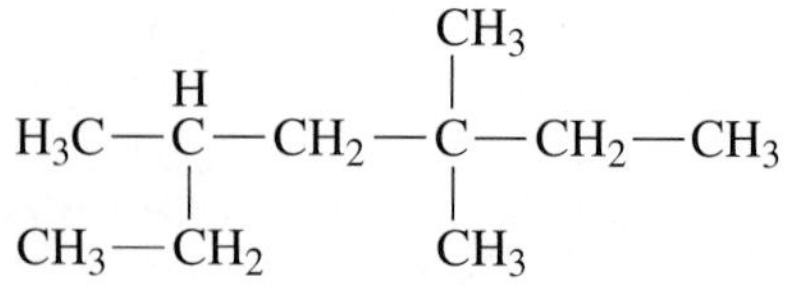

d. 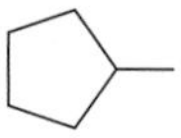

11.14 Give the IUPAC name for each of the following:

a. 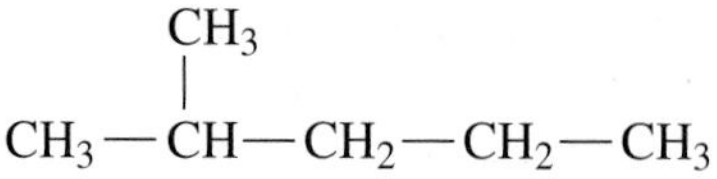

b. 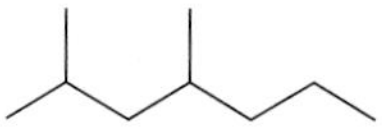

c. 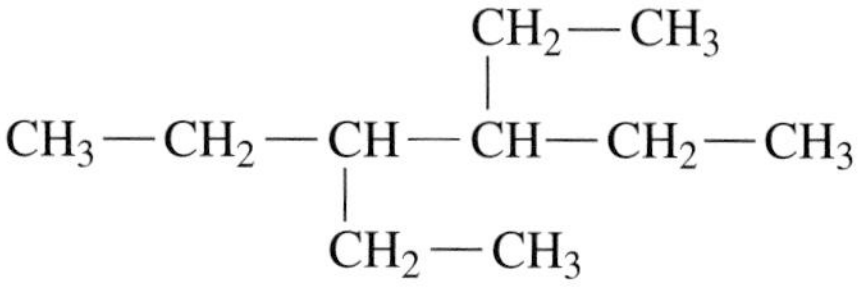

d. 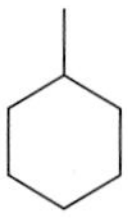

11.15 Draw the condense structural formula for each of the following alkanes:

a. 2,3-dimethylbutane
b. 2,2-dimethylpropane
c. 4-ethyl-2,3-dimethylhexane
d. 2-bromo-4-chloropentane

11.16 Draw the condensed structural formula for each of the following alkanes:

a. 3-ethylpentane
b. 3-ethyl-2-methylheptane
c. 4-ethyl-2,2-dimethyloctane
d. 2-bromopropane

11.17 Draw the skeletal formula for each of the following:

a. 3-methylhexane
b. 1-chloro-3-methylbutane
c. ethylcyclopropane
d. 3-chloroheptane

11.18 Draw the expanded structural formula for each of the following:

a. cyclopropane
b. butadiene
c. bromocyclobutane

Chemistry Link to Health

Common Uses of Halogenated Alkanes

Some common uses of haloalkanes include solvents and anesthetics. For many years, carbon tetrachloride was widely used in dry cleaners, and in home spot removers to take oils and grease out of clothes. However, its use was discontinued when carbon tetrachloride was found to be toxic to the liver, where it can cause cancer. Today, dry cleaners use other halogenated compounds such as dichloromethane, 1,1,1-trichloroethane, and 1,1,2-trichloro-1,2,2-trifluoroethane.

CH_2Cl_2 — Dichloromethane

$Cl_3C{-}CH_3$ — 1,1,1-Trichloroethane

$FCl_2C{-}CClF_2$ — 1,1,2-Trichloro-1,2,2-trifluoroethane

General anesthetics are compounds that are inhaled or injected to cause a loss of sensation so that surgery or other procedures can be done without causing pain to the patient. As nonpolar compounds, they are soluble in the nonpolar nerve membranes, where they decrease the ability of the nerve cells to conduct the sensation of pain. Trichloromethane, commonly called chloroform ($CHCl_3$), was once used as an anesthetic, but it is toxic and may be carcinogenic. For minor surgeries, a local anesthetic such as chloroethane (ethyl chloride), $CH_3{-}CH_2{-}Cl$, is applied to an area of the skin. Chloroethane evaporates quickly, cooling the skin and causing a loss of sensation.

One of the most widely used general anesthetics is halothane (2-bromo-2-chloro-1,1,1-trifluoroethane), also called fluothane. It has a pleasant odor, is nonexplosive, has few side effects, undergoes few reactions within the body, and is eliminated quickly from the body.

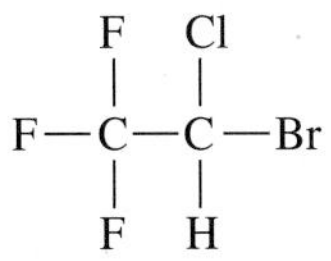

Halothane (Fluothane)

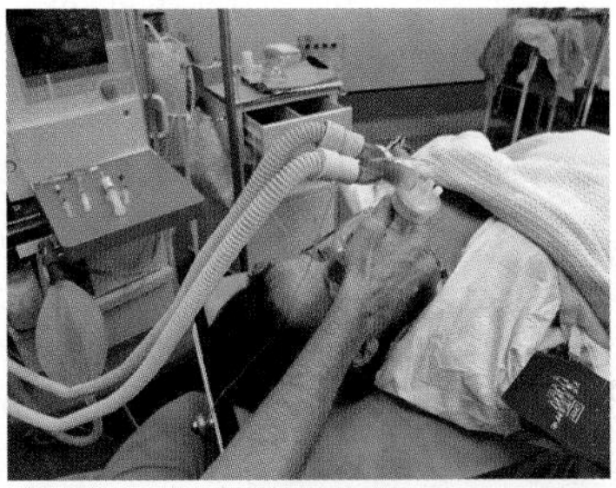

Anesthetics, such as halothane, decrease the sensation of pain.

11.4 Properties of Alkanes

LEARNING GOAL

Identify the properties of alkanes and write a balanced chemical equation for combustion.

Many types of alkanes are the components of fuels that power our cars and oil that heats our homes. You may have used a mixture of hydrocarbons such as mineral oil as a laxative or petrolatum to soften your skin. The differences in uses of many of the alkanes result from their physical properties, including solubility and density.

Some Uses of Alkanes

The first four alkanes—methane, ethane, propane, and butane—are gases at room temperature and are widely used as heating fuels.

Alkanes having five to eight carbon atoms (pentane, hexane, heptane, and octane) are liquids at room temperature. They are highly volatile, which makes them useful in fuels such as gasoline.

Liquid alkanes with 9 to 17 carbon atoms have higher boiling points and are found in kerosene, diesel, and jet fuels. Motor oil is a mixture of high-molecular-weight liquid hydrocarbons and is used to lubricate the internal components of engines. Mineral oil is a mixture of liquid hydrocarbons and is used as a laxative and a lubricant. Alkanes with 18 or more carbon atoms are waxy solids at room temperature. Known as paraffins, they are used in waxy coatings added to fruits and vegetables to retain moisture, inhibit mold, and enhance appearance. Petrolatum, or Vaseline, is a semisolid mixture of hydrocarbons with more than 25 carbon atoms used in ointments and cosmetics and as a lubricant.

The solid alkanes that make up waxy coatings on fruits and vegetables help retain moisture, inhibit mold, and enhance appearance.

Solubility and Density

Alkanes are nonpolar, which makes them insoluble in water. However, they are soluble in nonpolar solvents such as other alkanes. Alkanes have densities from 0.62 g/mL to about 0.79 g/mL, which is less than the density of water (1.0 g/mL).

FIGURE 11.4 ▶ In oil spills, large quantities of oil spread out to form a thin layer on top of the ocean surface.

Q What physical properties cause oil to remain on the surface of water?

If there is an oil spill in the ocean, the alkanes in the oil, which do not mix with water, form a thin layer on the surface that spreads over a large area. In April 2010, an explosion on an oil-drilling rig in the Gulf of Mexico caused the largest oil spill in U.S. history. At its maximum, an estimated 10 million liters of oil was leaked every day. Other major oil spills occurred in Queensland, Australia (2009), the coast of Wales (1996), the Shetland Islands (1993), and Alaska, from the *Exxon Valdez*, in 1989 (see Figure 11.4). If the crude oil reaches land, there can be considerable damage to beaches, shellfish, birds, and wildlife habitats. When animals such as birds are covered with oil, they must be cleaned quickly because ingestion of the hydrocarbons when they try to clean themselves is fatal.

Cleanup of oil spills includes mechanical, chemical, and microbiological methods. A boom may be placed around the leaking oil until it can be removed. Boats called skimmers then scoop up the oil and place it in tanks. In a chemical method, a substance that attracts oil is used to pick up oil, which is then scraped off into recovery tanks. Certain bacteria that ingest oil are also used to break oil down into less harmful products.

Butane in a portable burner undergoes combustion.

Combustion of Alkanes

The carbon–carbon single bonds in alkanes are difficult to break, which makes them the least reactive family of organic compounds. However, alkanes burn readily in oxygen to produce carbon dioxide, water, and energy. For example, methane is the natural gas we use to cook our food on a gas cooktop and heat our homes.

$$CH_4(g) + 2O_2(g) \xrightarrow{\Delta} CO_2(g) + 2H_2O(g) + \text{energy}$$

Butane is used in cooking, camping, and torches.

$$2C_4H_{10}(g) + 13O_2(g) \xrightarrow{\Delta} 8CO_2(g) + 10H_2O(g) + \text{energy}$$

QUESTIONS AND PROBLEMS

11.4 Properties of Alkanes

LEARNING GOAL Identify the properties of alkanes and write a balanced chemical equation for combustion.

11.19 Heptane, used as a solvent for rubber cement, has a density of 0.68 g/mL and boils at 98 °C.

a. Draw the condensed structural formula and skeletal formula for heptane.
b. Is heptane a solid, liquid, or gas at room temperature?
c. Is heptane soluble in water?
d. Will heptane float on water or sink?
e. Write the balanced chemical equation for the complete combustion of heptane.

11.20 Nonane has a density of 0.79 g/mL and boils at 151 °C.

a. Draw the condensed structural formula and skeletal formula for nonane.
b. Is nonane a solid, liquid, or gas at room temperature?
c. Is nonane soluble in water?
d. Will nonane float on water or sink?
e. Write the balanced chemical equation for the complete combustion of nonane.

11.21 Write the balanced chemical equation for the complete combustion of each of the following compounds:

a. ethene
b. methane
c. cyclopropane

11.22 Write the balanced chemical equation for the complete combustion of each of the following compounds:

a. hexane
b. cyclopentane
c. 2-methylbutane

11.5 Alkenes and Alkynes

LEARNING GOAL

Identify structural formulas as alkenes, cycloalkenes, and alkynes, and write their IUPAC names.

Alkenes and *alkynes* are families of hydrocarbons that contain double and triple bonds, respectively. They are called *unsaturated hydrocarbons* because they do not contain the maximum number of hydrogen atoms, as do alkanes. They react with hydrogen gas to increase the number of hydrogen atoms to become alkanes, which are *saturated hydrocarbons*.

Identifying Alkenes and Alkynes

Alkenes contain one or more carbon–carbon double bonds that form when adjacent carbon atoms share two pairs of valence electrons. Recall that *a carbon atom always forms four covalent bonds*. In the simplest alkene, ethene, C_2H_4, two carbon atoms are connected by a double bond and each is also attached to two H atoms. This gives each carbon atom in the double bond a trigonal planar arrangement with bond angles of 120°. As a result, the ethene molecule is flat because the carbon and hydrogen atoms all lie in the same plane (see Figure 11.5).

Ethene, more commonly called ethylene, is an important plant hormone involved in promoting the ripening of fruit. Commercially grown fruit, such as avocados, bananas, and tomatoes, are often picked before they are ripe. Before the fruit is brought to market, it is exposed to ethylene to accelerate the ripening process. Ethylene also accelerates the breakdown of cellulose in plants, which causes flowers to wilt and leaves to fall from trees.

In an **alkyne**, a triple bond forms when two carbon atoms share three pairs of valence electrons. In the simplest alkyne, ethyne (C_2H_2), the two carbon atoms of the triple bond are each attached to one hydrogen atom, which gives a triple bond a linear geometry. Ethyne, commonly called acetylene, is used in welding where it reacts with oxygen to produce flames with temperatures above 3300 °C.

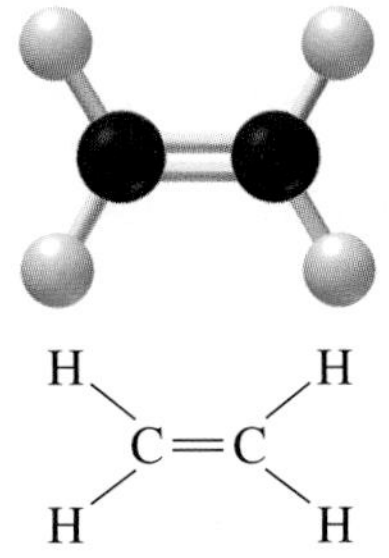

Bond angles = 120°
Ethene

H—C≡C—H

Bond angles = 180°
Ethyne

FIGURE 11.5 ▶ Ball-and-stick models of ethene and ethyne show the functional groups of double or triple bonds and the bond angles.

Q Why are these compounds called unsaturated hydrocarbons?

Fruit is ripened with ethene, a plant hormone.

A mixture of acetylene and oxygen undergoes combustion during the welding of metals.

Naming Alkenes and Alkynes

The IUPAC names for alkenes and alkynes are similar to those of alkanes. Using the alkane name with the same number of carbon atoms, the *ane* ending is replaced with *ene* for an alkene and *yne* for an alkyne (see Table 11.6). Cyclic alkenes are named as *cycloalkenes*.

Explore Your World

Ripening Fruit

Obtain two unripe green bananas. Place one in a plastic bag and seal the bag. Leave the banana and the bag with a banana on the counter. Check the bananas twice a day to observe any difference in the ripening process.

Questions

1. What compound is used to ripen the bananas?
2. What are some possible reasons for any difference in the ripening rate?
3. If you wish to ripen an avocado, what procedure might you use?

TABLE 11.6 Comparison of Names for Alkanes, Alkenes, and Alkynes

Alkane	Alkene	Alkyne
$CH_3—CH_3$	$H_2C{=}CH_2$	$HC{\equiv}CH$
Ethane	Ethene (ethylene)	Ethyne (acetylene)
$CH_3—CH_2—CH_3$	$CH_3—CH{=}CH_2$	$CH_3—C{\equiv}CH$
	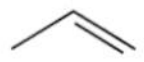	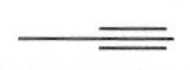
Propane	Propene	Propyne

Examples of naming an alkene and an alkyne are seen in Sample Problem 11.7.

SAMPLE PROBLEM 11.7 Naming Alkenes and Alkynes

Write the IUPAC name for each of the following:

a. $CH_3—\underset{}{CH}(CH_3)—CH{=}CH—CH_3$ (with CH_3 attached to the second carbon)

b. $CH_3—CH_2—C{\equiv}C—CH_2—CH_3$

SOLUTION

a.

ANALYZE THE PROBLEM	Feature	Type	IUPAC Naming
	double bond	alkene	replace the *ane* of the alkane name with *ene*

STEP 1 **Name the longest carbon chain that contains the double bond.** There are five carbon atoms in the longest carbon chain containing the double bond. Replace the *ane* in the corresponding alkane name with *ene* to give pentene.

$CH_3—CH(CH_3)—CH{=}CH—CH_3$ pentene

STEP 2 **Number the carbon chain starting from the end nearer the double bond.** Place the number of the first carbon in the double bond in front of the alkene name.

$\underset{5}{CH_3}—\underset{4}{CH}(CH_3)—\underset{3}{CH}{=}\underset{2}{CH}—\underset{1}{CH_3}$ 2-pentene

Alkenes or alkynes with two or three carbons do not need numbers. For example, the double bond in ethene or propene must be between carbon 1 and carbon 2.

STEP 3 **Give the location and name for each substituent (alphabetical order) as a prefix to the alkene name.** The methyl group is located on carbon 4.

$\underset{5}{CH_3}—\underset{4}{CH}(CH_3)—\underset{3}{CH}{=}\underset{2}{CH}—\underset{1}{CH_3}$ 4-methyl-2-pentene

b.

ANALYZE THE PROBLEM	Feature	Type	IUPAC Naming
	triple bond	alkyne	replace the *ane* of the alkane name with *yne*

Guide to Naming Alkenes and Alkynes

STEP 1 Name the longest carbon chain that contains the double or triple bond.

STEP 2 Number the carbon chain starting from the end nearer the double or triple bond.

STEP 3 Give the location and name for each substituent (alphabetical order) as a prefix to the alkene or alkyne name.

STEP 1 **Name the longest carbon chain that contains the triple bond.** There are six carbon atoms in the longest chain containing the triple bond. Replace the *ane* in the corresponding alkane name with *yne* to give hexyne.

$CH_3-CH_2-C{\equiv}C-CH_2-CH_3$ hexyne

STEP 2 **Number the carbon chain starting from the end nearer the triple bond.** Place the number of the first carbon in the triple bond in front of the alkyne name.

$\underset{1}{CH_3}-\underset{2}{CH_2}-\underset{3}{C}{\equiv}\underset{4}{C}-\underset{5}{CH_2}-\underset{6}{CH_3}$ 3-hexyne

STEP 3 **Give the location and name for each substituent (alphabetical order) as a prefix to the alkyne name.** There are no substituents in this formula.

STUDY CHECK 11.7

Draw the condensed structural formula for each of the following:

a. 2-pentyne
b. 2-chloro-1-hexene

Naming Cycloalkenes

Some alkenes called *cycloalkenes* have a double bond within a ring structure. If there is no substituent, the double bond does not need a number. If there is a substituent, the carbons in the double bond are numbered as 1 and 2, and the ring is numbered in the direction that will give the lower number to the substituent.

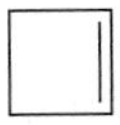

Cyclobutene

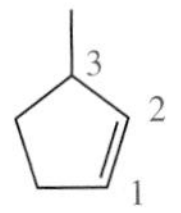

3-Methylcyclopentene

QUESTIONS AND PROBLEMS

11.5 Alkenes and Alkynes

LEARNING GOAL Identify structural formulas as alkenes, cycloalkenes, and alkynes, and write their IUPAC names.

11.23 Identify the following as alkanes, alkenes, cycloalkenes, or alkynes:

a. 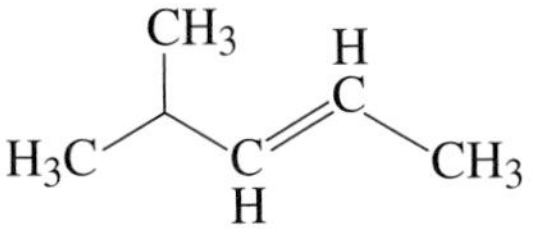

b. 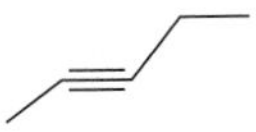

c. 

d. 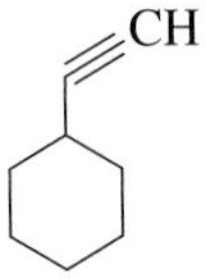

11.24 Identify the following as alkanes, alkenes, cycloalkenes, or alkynes:

a. 

b. 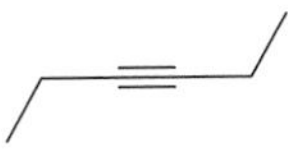

c. 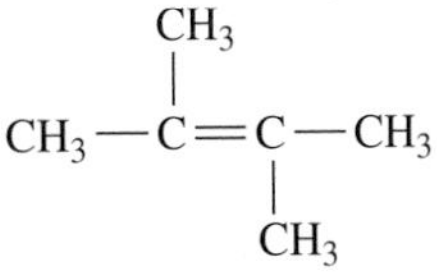

d. 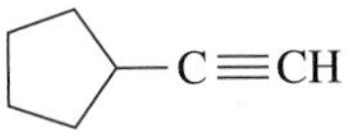

11.25 Give the IUPAC name for each of the following:

a. 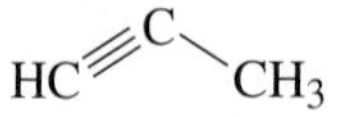

b. 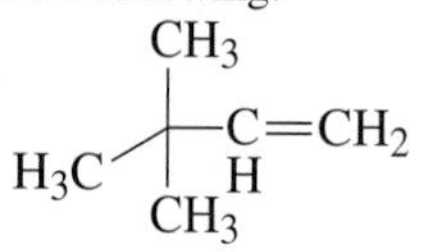

c. 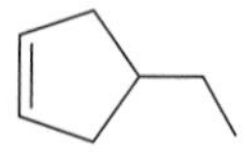

d. 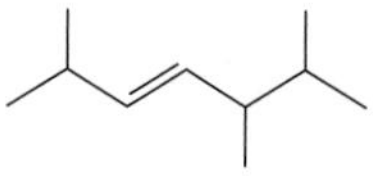

11.26 Give the IUPAC name for each of the following:

a. $H_2C{=}CH{-}CH_2{-}CH_3$

b. $CH_3{-}C{\equiv}C{-}CH_2{-}CH_2{-}\underset{}{CH}(CH_3){-}CH_3$

c. 

d. $CH_3{-}CH_2{-}CH{=}CH{-}CH_3$

11.27 Draw the condensed structural formula, or skeletal formula, for each of the following:

a. hex-1-ene

b. 4-methylpent-2-yne

c. 4-bromobut-1-ene

d. cyclohept-1-ene

11.28 Draw the condensed structural formula, or skeletal formula, if cyclic, for each of the following:

a. 1-methylcyclopentene

b. 3-methyl-1-butyne

c. 3,4-dimethyl-1-pentene

d. 2-methyl-2-hexene

LEARNING GOAL

Draw the condensed structural formulas and give the names for the cis–trans isomers of alkenes.

11.6 Cis–Trans Isomers

In any alkene, the double bond is rigid, which means there is no rotation around the double bond (see *Explore Your World* "Modeling Cis–Trans Isomers"). As a result, the atoms or groups are attached to the carbon atoms in the double bond on one side or the other, which gives two different structures called *geometric isomers* or *cis–trans isomers*.

For example, the formula for 1,2-dichloroethene can be drawn as two different molecules, which are cis–trans isomers. In the expanded structural formulas, the atoms bonded to the carbon atoms in the double bond have bond angles of 120°. When we draw 1,2-dichloroethene, we add the prefix *cis* or *trans* to denote whether the atoms bonded to the carbon atoms are on the same side or opposite sides of the molecule. In the **cis isomer**, the chlorine atoms are on the same side of the double bond. In the **trans isomer**, the chlorine atoms are on opposite sides of the double bond. *Trans* means "across," as in transcontinental; *cis* means "on this side."

$Cl{-}CH{=}CH{-}Cl$

1,2-Dichloroethene

Chlorine atoms are on the same side of the double bond.

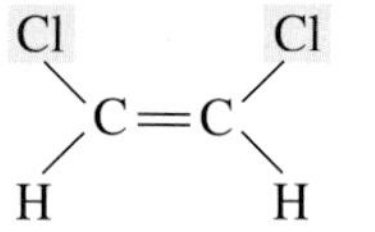

cis-1,2-Dichloroethene

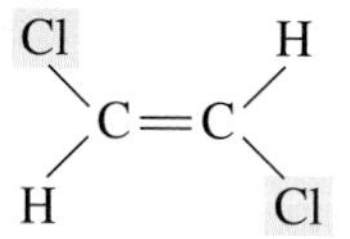

trans-1,2-Dichloroethene

Chlorine atoms are on opposite sides of the double bond.

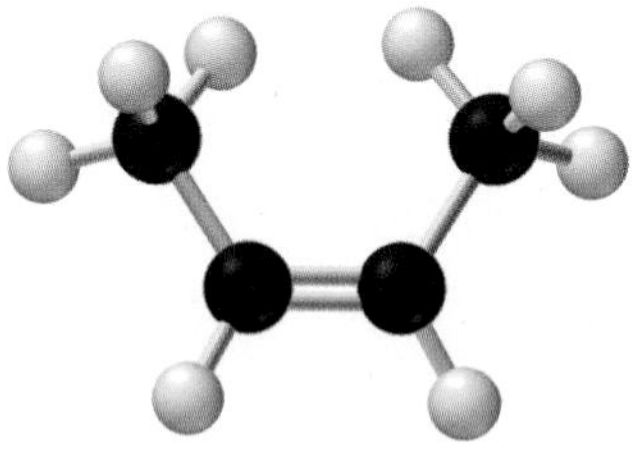

cis-2-Butene

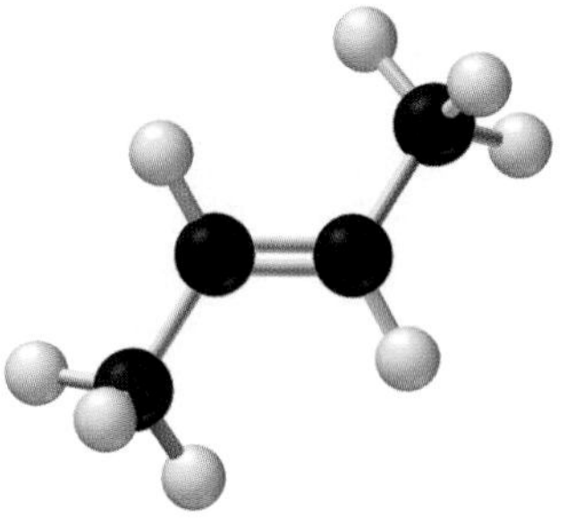

trans-2-Butene

FIGURE 11.6 ▶ Ball-and-stick models of the cis and trans isomers of 2-butene.

Q What feature in 2-butene accounts for the cis and trans isomers?

Another example of a cis–trans isomer is 2-butene. If we look closely at 2-butene, we find that each carbon in the double bond is bonded to a $-CH_3$ group and a hydrogen atom. We can draw cis–trans isomers for 2-butene. In the *cis*-2-butene isomer, the $-CH_3$ groups are attached on the same side of the double bond. In the *trans*-2-butene isomer, the $-CH_3$ groups are attached to the double bond on opposite sides, as shown in their ball-and-stick-models (see Figure 11.6).

As with any pair of cis–trans isomers, *cis*-2-butene and *trans*-2-butene are different compounds with different physical properties and chemical properties. In general, trans isomers are more stable than their cis counterparts because the groups that are bigger than hydrogen atoms on the double bond are farther apart.

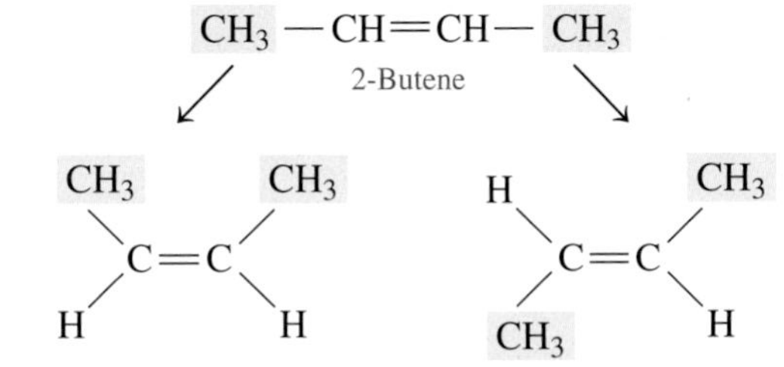

cis-2-Butene (mp −139 °C; bp 3.7 °C)   *trans*-2-Butene (mp −106 °C; bp 0.3 °C)

When the carbon atoms in the double bond are attached to two different atoms or groups of atoms, an alkene can have cis–trans isomers. For example, 3-hexene can be drawn with cis and trans isomers because there is one H atom and a CH_3-CH_2- group attached to each carbon atom in the double bond. When you are asked to draw the formula for an alkene, it is important to consider the possibility of cis and trans isomers.

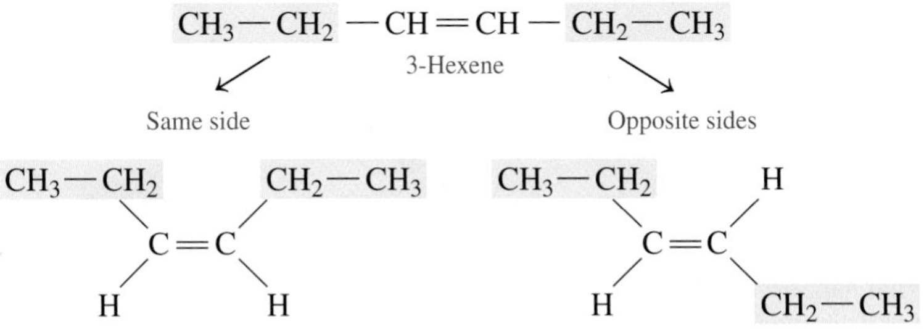

cis-3-Hexene   *trans*-3-Hexene

Same side   Opposite sides

When an alkene has identical groups on the same carbon atom of the double bond, cis and trans isomers cannot be drawn. For example, 1,1-dichloropropene has just one condensed structural formula without cis and trans isomers.

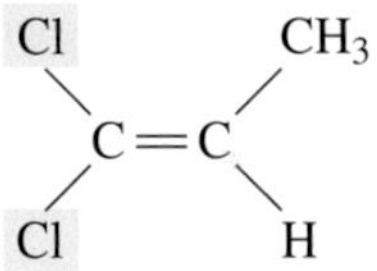

1,1-Dichloropropene

SAMPLE PROBLEM 11.8 Identifying Cis–Trans Isomers

Identify each of the following as the cis or trans isomer and give its name:

a. 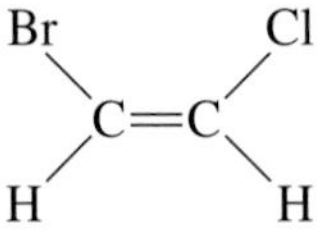

b. 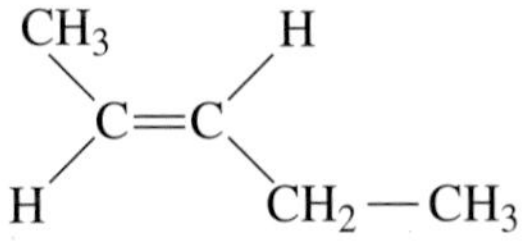

SOLUTION

a. This is a cis isomer because the two halogen atoms attached to the carbon atoms of the double bond are on the same side. The name of the two-carbon alkene, starting with the bromo group on carbon 1, is *cis*-1-bromo-2-chloroethene.

b. This is a trans isomer because the two alkyl groups attached to the carbon atoms of the double bond are on opposite sides of the double bond. This isomer of the five-carbon alkene, 2-pentene, is named *trans*-2-pentene.

STUDY CHECK 11.8

Give the name for the following compound, including cis or trans:

Explore Your World

Modeling Cis–Trans Isomers

Because cis–trans isomerism is not easy to visualize, here are some things you can do to understand the difference in rotation around a single bond compared to a double bond and how it affects groups that are attached to the carbon atoms in the double bond.

Put the tips of your index fingers together. This is a model of a single bond. Consider the index fingers as a pair of carbon atoms, and think of your thumbs and other fingers as other parts of a carbon chain. While your index fingers are touching, twist your hands and change the position of your thumbs relative to each other. Notice how the relationship of your other fingers changes.

Now place the tips of your index fingers and middle fingers together in a model of a double bond. As you did before, twist your hands to move your thumbs away from each other. What happens? Can you change the location of your thumbs relative to each other without breaking the double bond? The difficulty of moving your hands with two fingers touching represents the lack of rotation about a double bond. You have made a model of a cis isomer when both thumbs point in the same direction. If you turn one hand over so one thumb points down and the other thumb points up, you have made a model of a trans isomer.

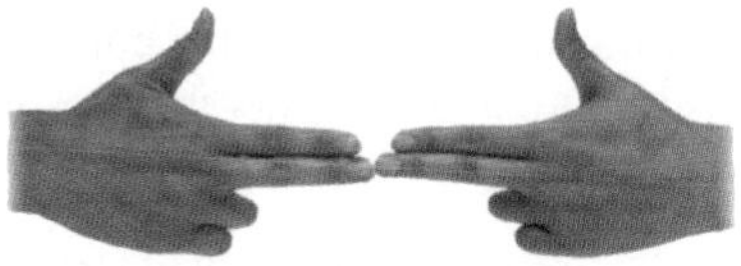

Cis-hands (cis-thumbs/fingers)

Trans-hands (trans-thumbs/fingers)

Using Gumdrops and Toothpicks to Model Cis–Trans Isomers

Obtain some toothpicks and yellow, green, and black gumdrops. The black gumdrops represent C atoms, the yellow gumdrops represent H atoms, and the green gumdrops represent Cl atoms. Place a toothpick between two black gumdrops. Use three more toothpicks to attach two yellow gumdrops and one green gumdrop to each black gumdrop carbon atom. Rotate one of the gumdrop carbon atoms to show the conformations of the attached H and Cl atoms.

Remove a toothpick and yellow gumdrop from each black gumdrop. Place a second toothpick between the carbon atoms, which makes a double bond. Try to twist the double bond of toothpicks. Can you do it? When you observe the location of the green gumdrops, does the model you made represent a cis or trans isomer? Why? If your model is a cis isomer, how would you change it to a trans isomer? If your model is a trans isomer, how would you change it to a cis isomer?

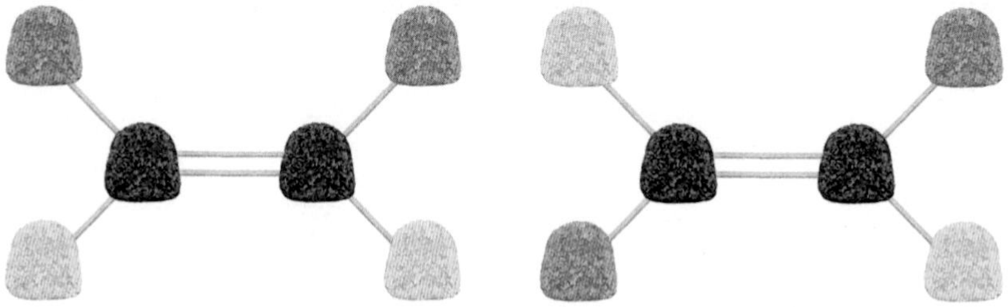

Models from gumdrops represent cis and trans isomers.

QUESTIONS AND PROBLEMS

11.6 Cis–Trans Isomers

LEARNING GOAL Draw the condensed structural formulas and give the names for the cis–trans isomers of alkenes.

11.29 Give the IUPAC name for each of the following, using cis or trans prefixes:

a. 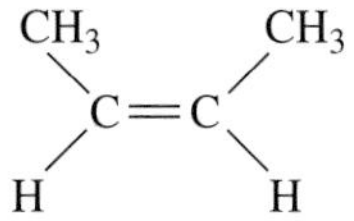

b. 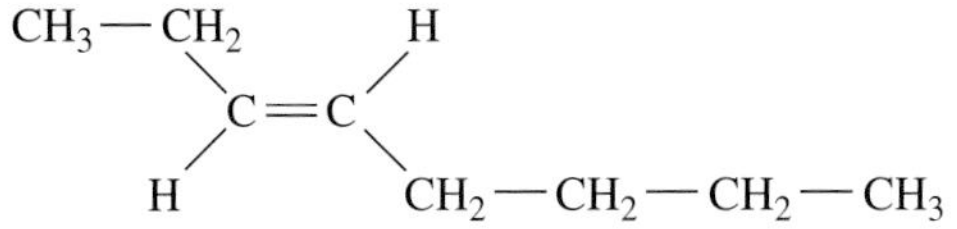

c. 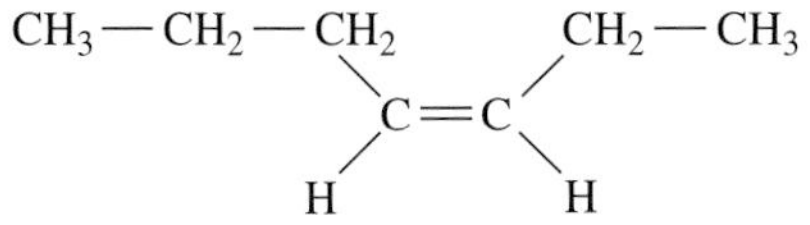

11.30 Give the IUPAC name for each of the following, using cis or trans prefixes:

a. 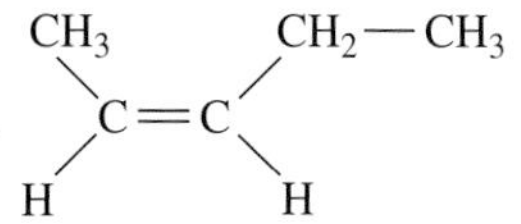

b. 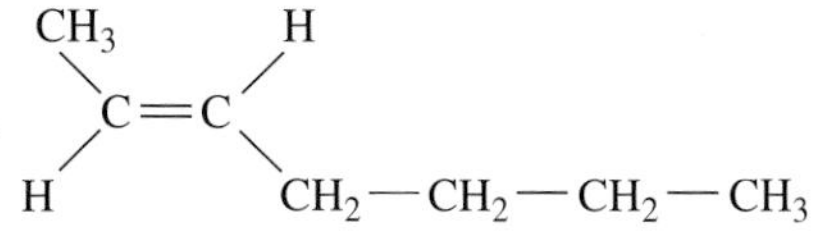

c. 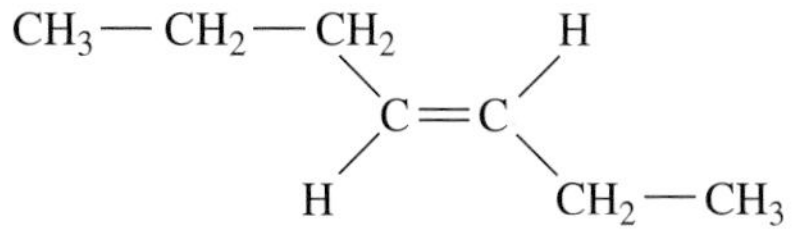

11.31 Draw the condensed structural formula for each of the following:

a. *trans*-2-bromo-3-chlorobut-2-ene

b. *trans*-hept-3-ene

c. *cis*-pent-2-ene

11.32 Draw the condensed structural formula for each of the following:

a. *cis*-1,2-difluoroethene

b. *trans*-2-pentene

c. *cis*-4-octene

Chemistry Link to the **Environment**

Pheromones in Insect Communication

Many insects emit minute quantities of chemicals called *pheromones* to send messages to individuals of the same species. Some pheromones warn of danger, others call for defense, mark a trail, or attract the opposite sex. In the past 40 yr, the structures of many pheromones have been chemically determined. One of the most studied is bombykol, the sex pheromone produced by the female silkworm moth. The bombykol molecule contains one cis double bond and one trans double bond. Even a few nanograms of bombykol will attract male silkworm moths from distances of over 1 km. The effectiveness of many of these pheromones depends on the cis or trans configuration of the double bonds in the molecules. A certain species will respond to one isomer but not the other.

Pheromones allow insects to attract mates from a great distance.

Scientists are interested in synthesizing pheromones to use as nontoxic alternatives to pesticides. When placed in a trap, bombykol can be used to capture male silkworm moths. When a synthetic pheromone is released in a field, the males cannot locate the females, which disrupts the reproductive cycle. This technique has been successful with controlling the oriental fruit moth, the grapevine moth, and the pink bollworm.

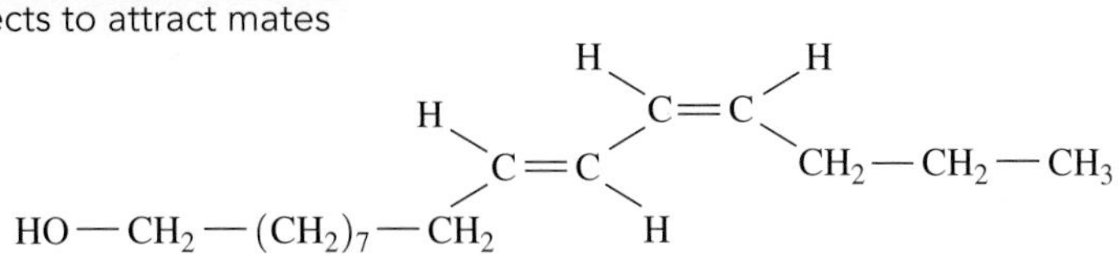

Bombykol, sex attractant for the silkworm moth

Chemistry Link to Health

Cis–Trans Isomers for Night Vision

The retinas of the eyes consist of two types of cells: rods and cones. The rods on the edge of the retina allow us to see in dim light, and the cones, in the center, produce our vision in bright light. In the rods, there is a substance called *rhodopsin* that absorbs light. Rhodopsin is composed of *cis*-11-retinal, an unsaturated compound, attached to a protein. When rhodopsin absorbs light, the *cis*-11-retinal isomer is converted to its trans isomer, which changes its shape. The trans form no longer fits the protein, and it separates from the protein. The change from the cis to trans isomer and its separation from the protein generate an electrical signal that the brain converts into an image.

An enzyme (isomerase) converts the trans isomer back to the *cis*-11-retinal isomer and the rhodopsin re-forms. If there is a deficiency of rhodopsin in the rods of the retina, *night blindness* may occur. One common cause is a lack of vitamin A in the diet. In our diet, we obtain vitamin A from plant pigments containing β-carotene, which is found in foods such as carrots, squash, and spinach. In the small intestine, the β-carotene is converted to vitamin A, which can be converted to *cis*-11-retinal or stored in the liver for future use. Without a sufficient quantity of retinal, not enough rhodopsin is produced to enable us to see adequately in dim light.

Cis–Trans Isomers of Retinal

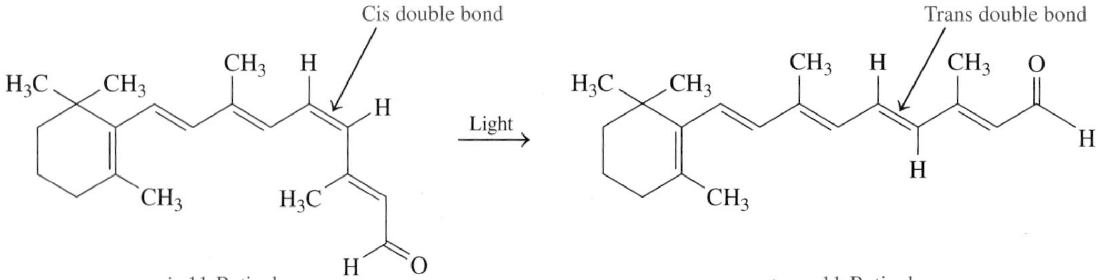

cis-11-Retinal   *trans*-11-Retinal

11.7 Addition Reactions

LEARNING GOAL

Draw the condensed structural formulas and give the names for the organic products of addition reactions of hydrogenation and hydration of alkenes.

The most characteristic reaction of alkenes is the **addition** of atoms or groups of atoms to the carbon atoms in a double bond. Addition occurs because double bonds are easily broken, providing electrons to form new single bonds.

The addition reactions have different names that depend on the type of reactant we add to the alkene, as Table 11.7 shows.

TABLE 11.7 Summary of Addition Reactions

Name of Addition Reaction	Reactants	Catalysts	Product
Hydrogenation	Alkene + H_2	Pt, Ni, or Pd	Alkane
Hydration	Alkene + H_2O	H^+ (strong acid)	Alcohol

CORE CHEMISTRY SKILL

Writing Equations for Hydrogenation and Hydration

Hydrogenation

In a reaction called **hydrogenation**, H atoms add to each of the carbon atoms in a double bond of an alkene. During hydrogenation, the double bonds are converted to single bonds in alkanes. A catalyst such as finely divided platinum (Pt), nickel (Ni), or palladium (Pd) is used to speed up the reaction. The general equation for hydrogenation can be written as follows:

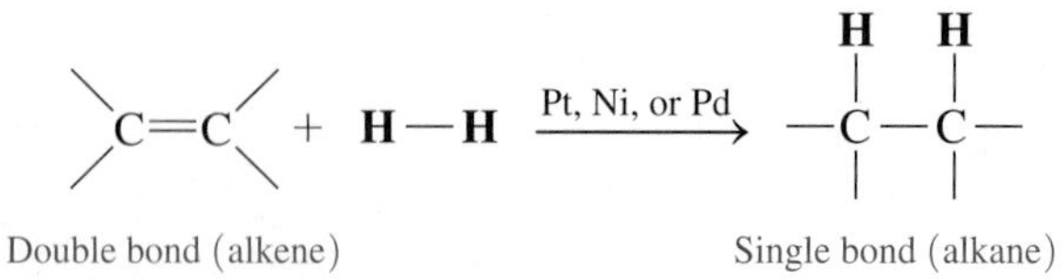

Some examples of the hydrogenation of alkenes follow:

$$CH_3-CH{=}CH-CH_3 + H-H \xrightarrow{Pt} CH_3-\underset{|}{\overset{H}{C}}H-\underset{|}{\overset{H}{C}}H-CH_3$$

2-Butene → Butane

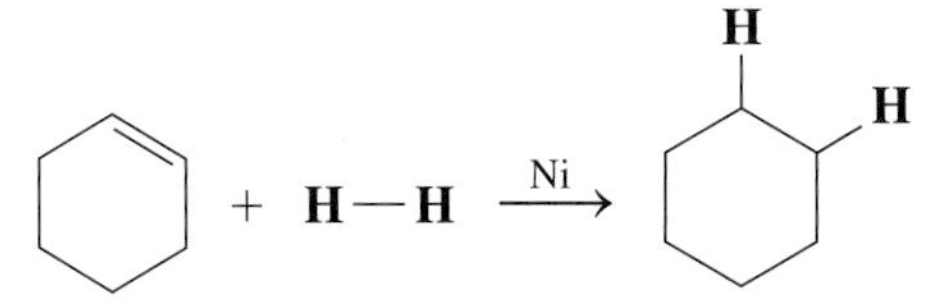

Cyclohexene → Cyclohexane

Explore Your World

Unsaturation in Fats and Oils

Read the labels on some containers of vegetable oils, margarine, peanut butter, and shortenings.

Questions

1. What terms on the label tell you that the compounds contain double bonds?
2. A label on a bottle of canola oil lists saturated, polyunsaturated, and monounsaturated fats. What do these terms tell you about the type of bonding in the fats?
3. A peanut butter label states that it contains partially hydrogenated vegetable oils or completely hydrogenated vegetable oils. What does this tell you about the type of reaction that took place in preparing the peanut butter?

SAMPLE PROBLEM 11.9 Writing Equations for Hydrogenation

Draw the condensed structural formula, or skeletal formula if cyclic, for the product of each of the following hydrogenation reactions:

a. $CH_3-CH{=}CH_2 + H_2 \xrightarrow{Pt} ?$ **b.**  $+ H_2 \xrightarrow{Pt} ?$

SOLUTION

In an addition reaction, hydrogen adds to the double bond to give an alkane.

ANALYZE THE PROBLEM	Reactant	Type of Reaction	Product
	alkene	hydrogenation (addition of H atoms to a double bond)	alkane

a. $CH_3-CH_2-CH_3$ **b.** 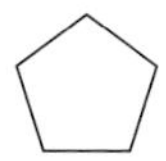

STUDY CHECK 11.9

Draw the condensed structural formula for the product of the hydrogenation of 2-methyl-1-butene, using a platinum catalyst.

Chemistry Link to Health

Hydrogenation of Unsaturated Fats

Vegetable oils such as corn oil or safflower oil are unsaturated fats composed of fatty acids that contain double bonds. The process of hydrogenation is used commercially to convert the double bonds in the unsaturated fats in vegetable oils to saturated fats such as margarine, which are more solid. Adjusting the amount of added hydrogen produces partially hydrogenated fats such as soft margarine, solid margarine in sticks, and shortenings, which are used in cooking. For example, oleic acid is a typical unsaturated fatty acid in olive oil and has a cis double bond at carbon 9. When oleic acid is hydrogenated, it is converted to stearic acid, a saturated fatty acid.

The unsaturated fats in vegetable oils are converted to saturated fats to make a more solid product.

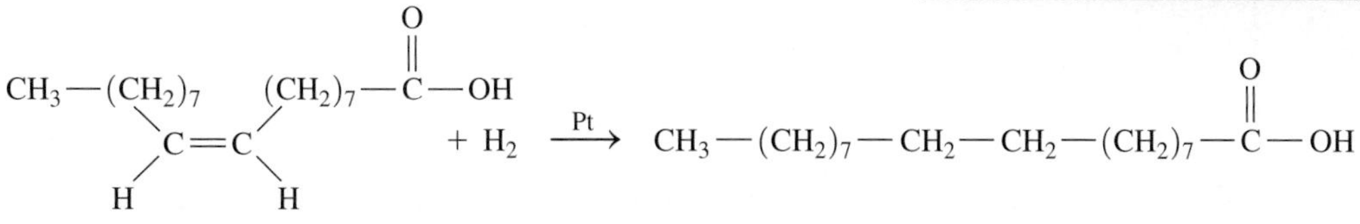

Oleic acid (found in olive oil and other unsaturated fats) → Stearic acid (found in saturated fats)

Hydration

In **hydration**, an alkene reacts with water (H—OH). A hydrogen atom (H—) from water forms a bond with one carbon atom in the double bond, and the oxygen atom in —OH forms a bond with the other carbon. The reaction is catalyzed by a strong acid such as H_2SO_4. Hydration is used to prepare alcohols, which have the hydroxyl (—OH) functional group. In the general equation for hydration, water is written as H—OH and the acid catalyst is represented by H^+.

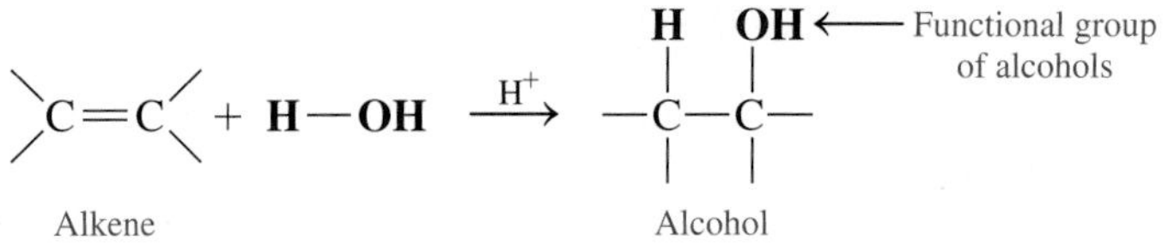

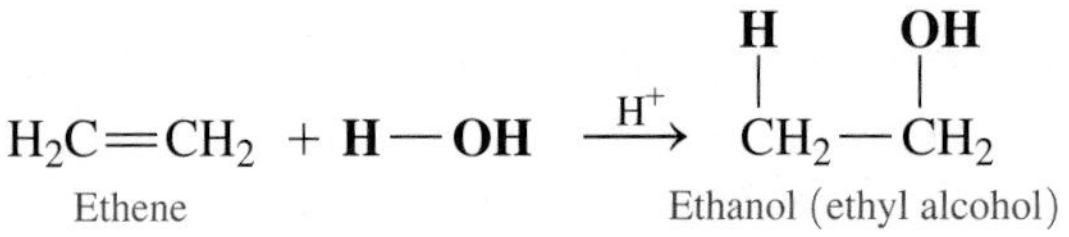

When water adds to a double bond in which the carbon atoms are attached to different numbers of H atoms (an asymmetrical double bond), the H— from H—OH attaches to the carbon that has the *greater* number of H atoms and the —OH from H—OH adds to the other carbon atom from the double bond. In the following example, the H— from H—OH attaches to the end carbon of the double bond, which has more hydrogen atoms, and the —OH adds to the middle carbon atom:

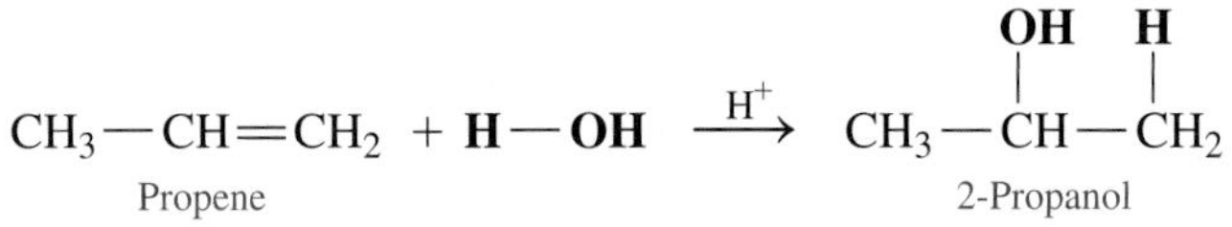

SAMPLE PROBLEM 11.10 Hydration

Draw the condensed structural formula for the product that forms in the following hydration reaction:

$$CH_3—CH_2—CH_2—CH{=}CH_2 + HOH \xrightarrow{H^+}$$

SOLUTION

ANALYZE THE PROBLEM	Reactant	Type of Reaction	Product
	alkene	hydration (asymmetrical double bond)	alcohol

The H— and —OH from water (H—OH) add to the carbon atoms in the double bond. The H— adds to the carbon with the *greater* number of hydrogen atoms, and the —OH bonds to the carbon with fewer H atoms.

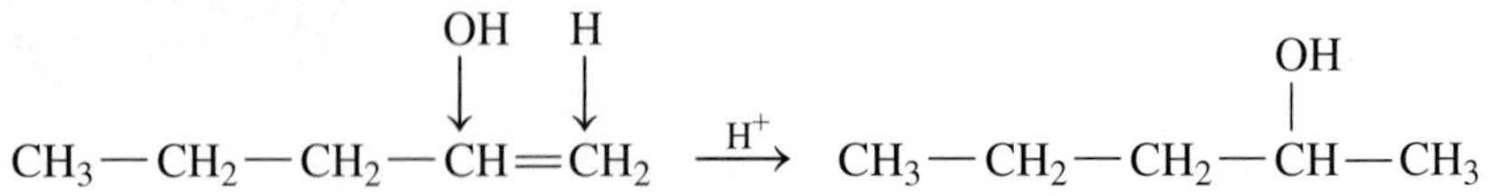

STUDY CHECK 11.10

Draw the condensed structural formula for the product obtained by the hydration of 2-methyl-2-butene.

Interactive Video

Addition to an Asymmetric Bond

QUESTIONS AND PROBLEMS

11.7 Addition Reactions

LEARNING GOAL Draw the condensed structural formulas and give the names for the organic products of addition reactions of hydrogenation and hydration of alkenes.

11.33 Draw the condensed structural formula, or skeletal formula if cyclic, for the product in each of the following reactions:

a. $CH_3—CH_2—CH_2—CH{=}CH_2 + H_2 \xrightarrow{Pt}$

b. $CH_3—CH{=}CH—CH_3 + HOH \xrightarrow{H^+}$

c. 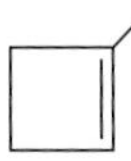 $+ HOH \xrightarrow{H^+}$

11.34 Draw the condensed structural formula, or skeletal formula if cyclic, for the product in each of the following reactions:

a. $CH_3—CH_2—CH{=}CH_2 + HOH \xrightarrow{H^+}$

b. cyclohexene $+ H_2 \xrightarrow{Pt}$

c. $CH_3—\overset{\overset{\Large CH_3}{|}}{C}{=}CH—CH_2—CH_3 + HOH \xrightarrow{H^+}$

11.8 Aromatic Compounds

LEARNING GOAL

Describe the bonding in benzene; name aromatic compounds and draw their skeletal formulas.

In 1825, Michael Faraday isolated a hydrocarbon called *benzene*, which had the molecular formula C_6H_6. A molecule of **benzene** consists of a ring of six carbon atoms with one hydrogen atom attached to each carbon. Because many compounds containing benzene have fragrant odors, the family of benzene compounds became known as **aromatic compounds**.

In benzene, each carbon atom uses three valence electrons to bond to the hydrogen atom and two adjacent carbons. That leaves one valence electron, which scientists first thought was shared in a double bond with an adjacent carbon. In 1865, August Kekulé proposed that the carbon atoms in benzene were arranged in a flat ring with alternating single and double bonds between the carbon atoms. There are two possible structural representations of benzene in which the double bonds can form between two different carbon atoms.

Today, we know that the six electrons are shared equally among the six carbon atoms and that all of the carbon–carbon bonds in benzene are identical due to resonance. This unique feature of benzene makes it especially stable. Benzene is most often represented as its skeletal formula, which shows a hexagon with a circle in the center. Some of the ways to represent benzene are shown as follows:

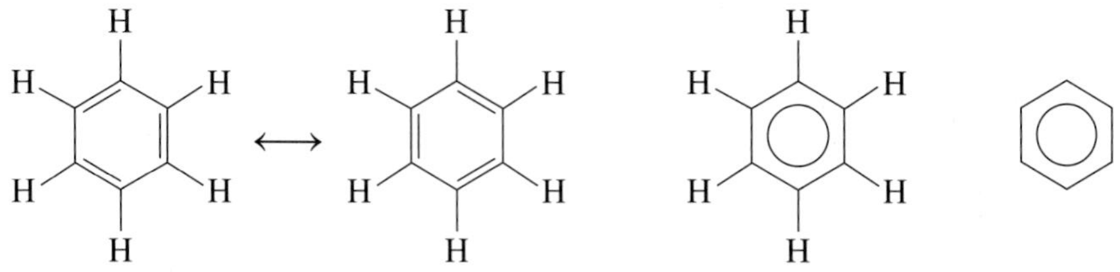

Structures for benzene

Skeletal formula for benzene ring

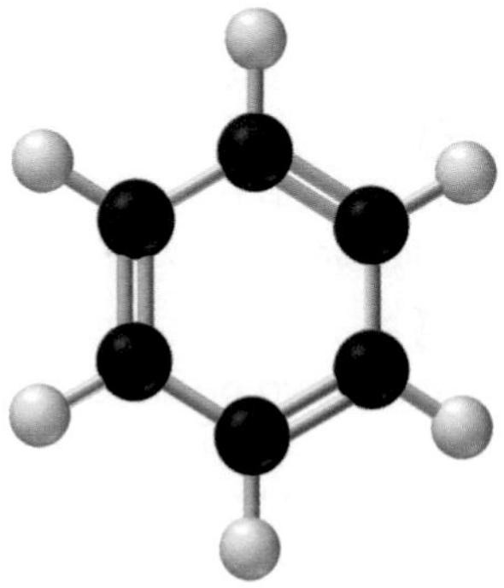

Naming Aromatic Compounds

Aromatic compounds that contain a benzene ring with a single substituent are usually named as benzene derivatives. However, many of these compounds have been important in chemistry for many years and still use their common names. Some widely used names such as toluene, aniline, and phenol are allowed by IUPAC rules.

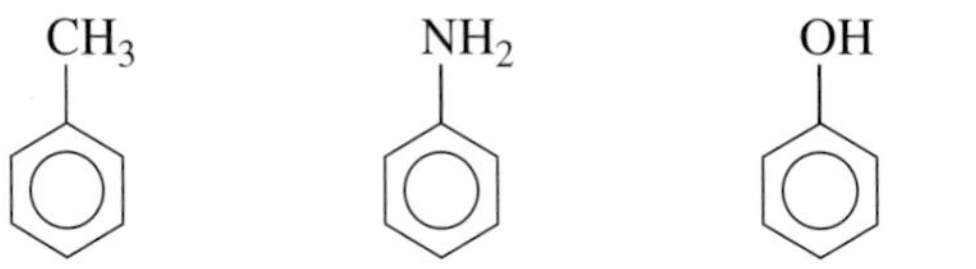

Toluene (methylbenzene)

Aniline (aminobenzene)

Phenol (hydroxybenzene)

When benzene has only one substituent, the ring is not numbered. When there are two or more substituents, the benzene ring is numbered to give the lowest numbers to the substituents.

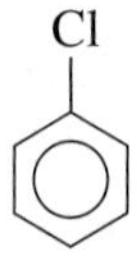
Chlorobenzene

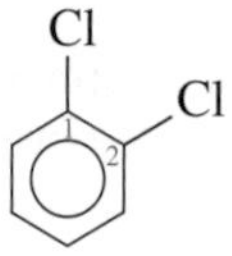
1,2-Dichlorobenzene

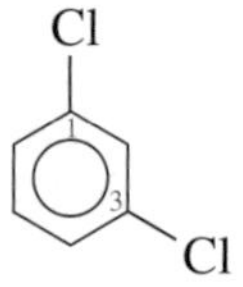
1,3-Dichlorobenzene

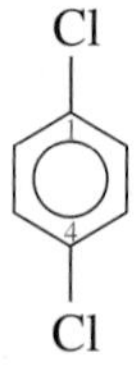
1,4-Dichlorobenzene

When a common name such as toluene, phenol, or aniline can be used, the carbon atom attached to the methyl, hydroxyl, or amine group is numbered as carbon 1. Then the substituents are named alphabetically.

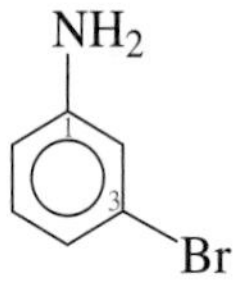
3-Bromoaniline

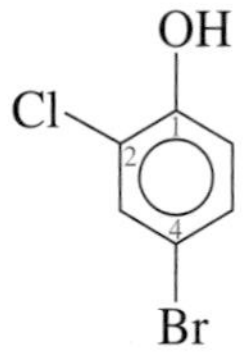
4-Bromo-2-chlorophenol

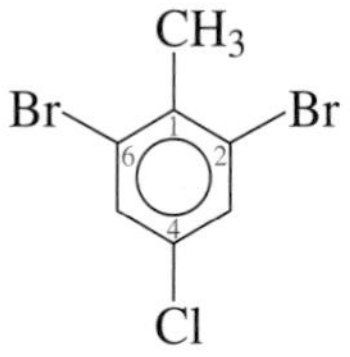
2,6-Dibromo-4-chlorotoluene

SAMPLE PROBLEM 11.11 Naming Aromatic Compounds

Give the IUPAC name for the following:

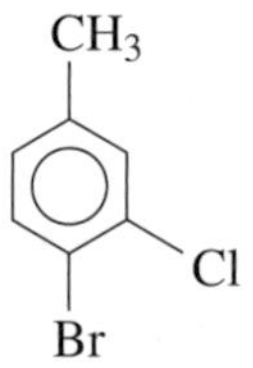

Guide to Naming Aromatic Compounds

STEP 1
Write the name for the aromatic compound.

STEP 2
If there are two or more substituents, number the aromatic ring starting from the substituent.

STEP 3
Name each substituent as a prefix.

SOLUTION

STEP 1 **Write the name for the aromatic compound.** A benzene ring with a methyl group is named toluene.

STEP 2 **If there are two or more substituents, number the aromatic ring starting from the substituent.** The methyl group of toluene is attached to carbon 1, and the ring is numbered to give the lower numbers.

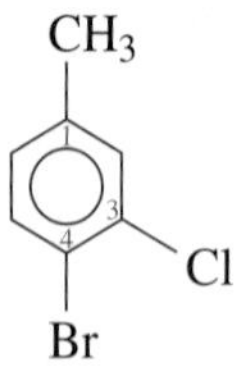

STEP 3 **Name each substituent as a prefix.** Naming the substituents in alphabetical order, this aromatic compound is 4-bromo-3-chlorotoluene.

STUDY CHECK 11.11

Give the IUPAC name for the following:

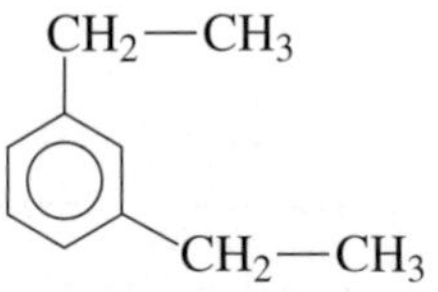

Chemistry Link to Health

Some Common Aromatic Compounds

Aromatic compounds are common in nature and in medicine. Toluene is used as a reactant to make drugs, dyes, and explosives such as TNT (trinitrotoluene). The benzene ring is found in some amino acids (the building blocks of proteins); in pain relievers such as aspirin, acetaminophen, and ibuprofen; and in flavorings such as vanillin.

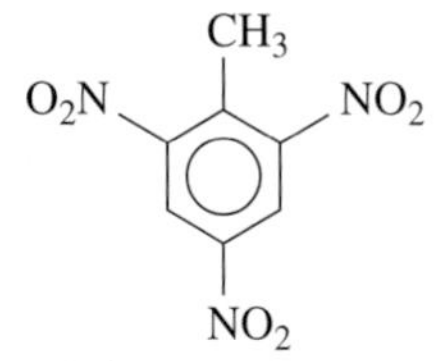
TNT (2,4,6-trinitrotoluene)

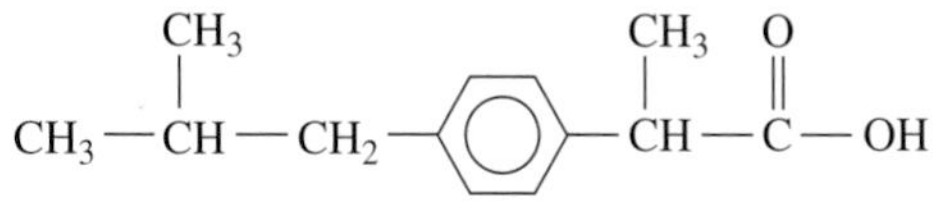
Ibuprofen

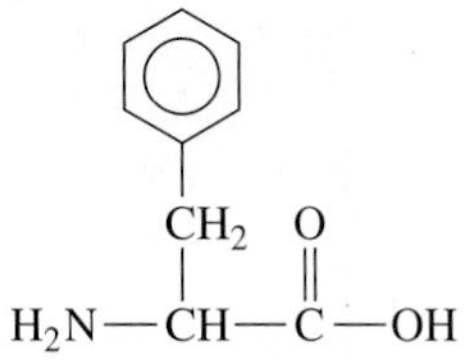
Phenylalanine (amino acid)

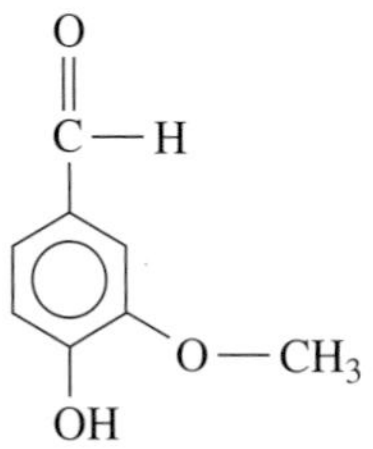
Vanillin

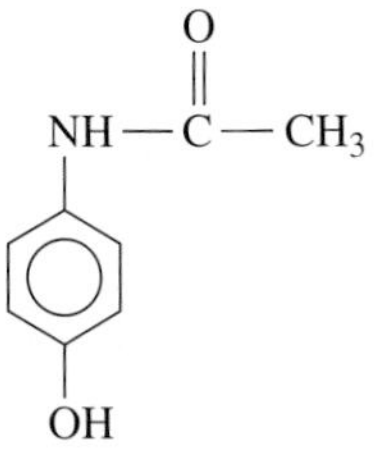
Acetaminophen

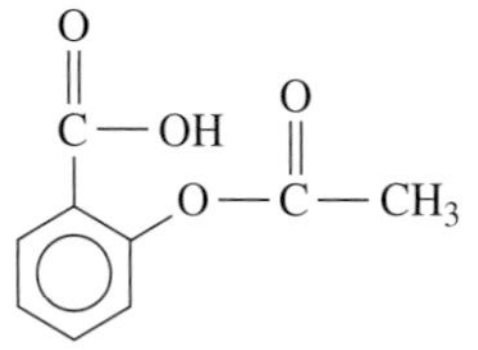
Aspirin

QUESTIONS AND PROBLEMS

11.8 Aromatic Compounds

LEARNING GOAL Describe the bonding in benzene; name aromatic compounds and draw their skeletal formulas.

11.35 Give the IUPAC name for each of the following:

a. 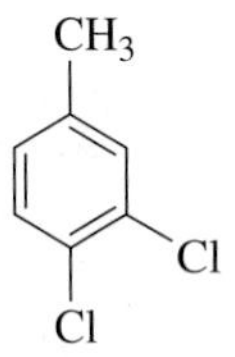

b. 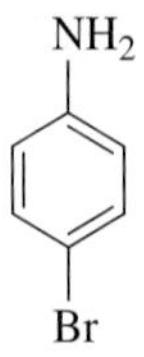

c. 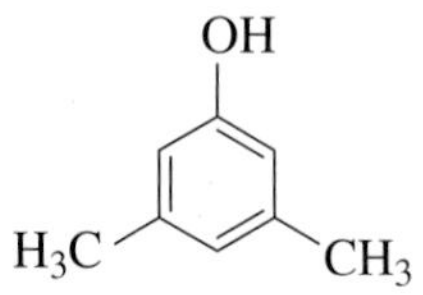

11.36 Give the IUPAC name for each of the following:

a. 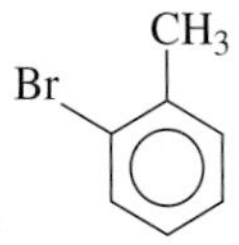

b. 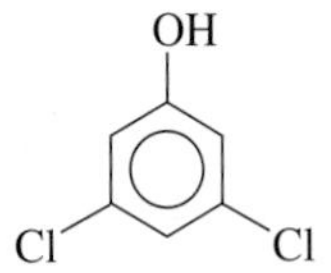

c. 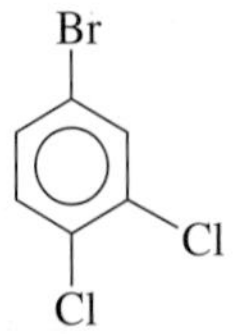

11.37 Draw the skeletal formula for each of the following compounds:

a. toluene
b. 1,3-dichlorobenzene
c. 4-ethyltoluene

11.38 Draw the skeletal formula for each of the following compounds:

a. propylbenzene
b. 4-bromoaniline
c. 1,2,4-trichlorobenzene

Chemistry Link to Health

Polycyclic Aromatic Hydrocarbons (PAHs)

Large aromatic compounds known as *polycyclic aromatic hydrocarbons* (PAHs) are formed by fusing together two or more benzene rings edge to edge. In a fused-ring compound, neighboring benzene rings share two carbon atoms. Naphthalene, with two benzene rings, is well known for its use in mothballs; anthracene, with three rings, is used in the manufacture of dyes.

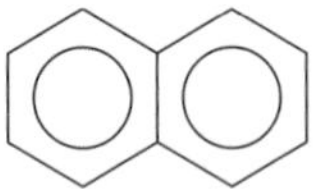

Naphthalene

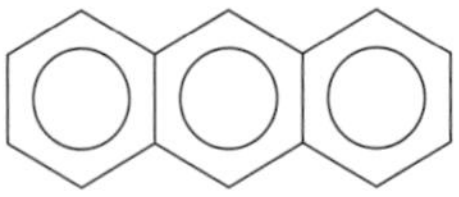

Anthracene

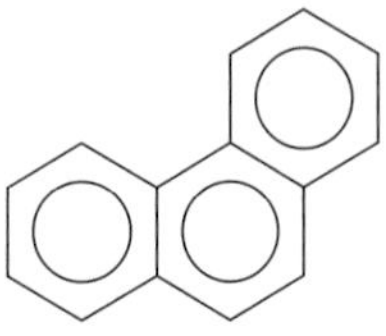

Phenanthrene

When a polycyclic compound contains phenanthrene, it may act as a carcinogen, a substance known to cause cancer.

Compounds containing five or more fused benzene rings such as benzo[*a*]pyrene are potent carcinogens. The molecules interact with the DNA in the cells, causing abnormal cell growth and cancer. Increased exposure to carcinogens increases the chance of DNA alterations in the cells. Benzo[*a*]pyrene, a product of combustion, has been identified in coal tar, tobacco smoke, barbecued meats, and automobile exhaust.

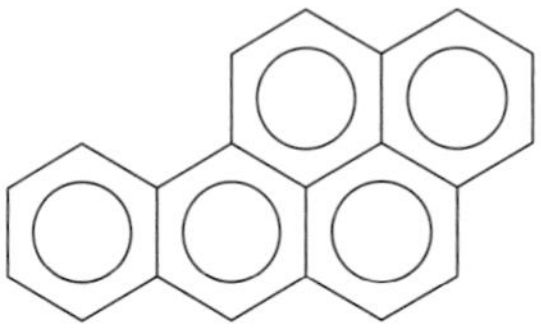

Benzo[*a*]pyrene

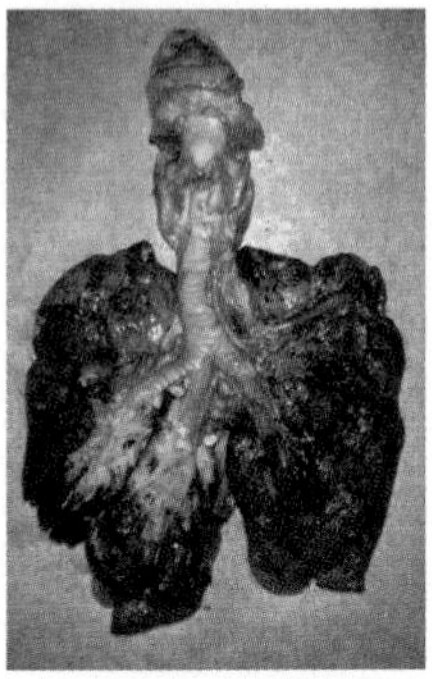

Aromatic compounds such as benzo[*a*]pyrene are strongly associated with lung cancers.

CONCEPT MAP

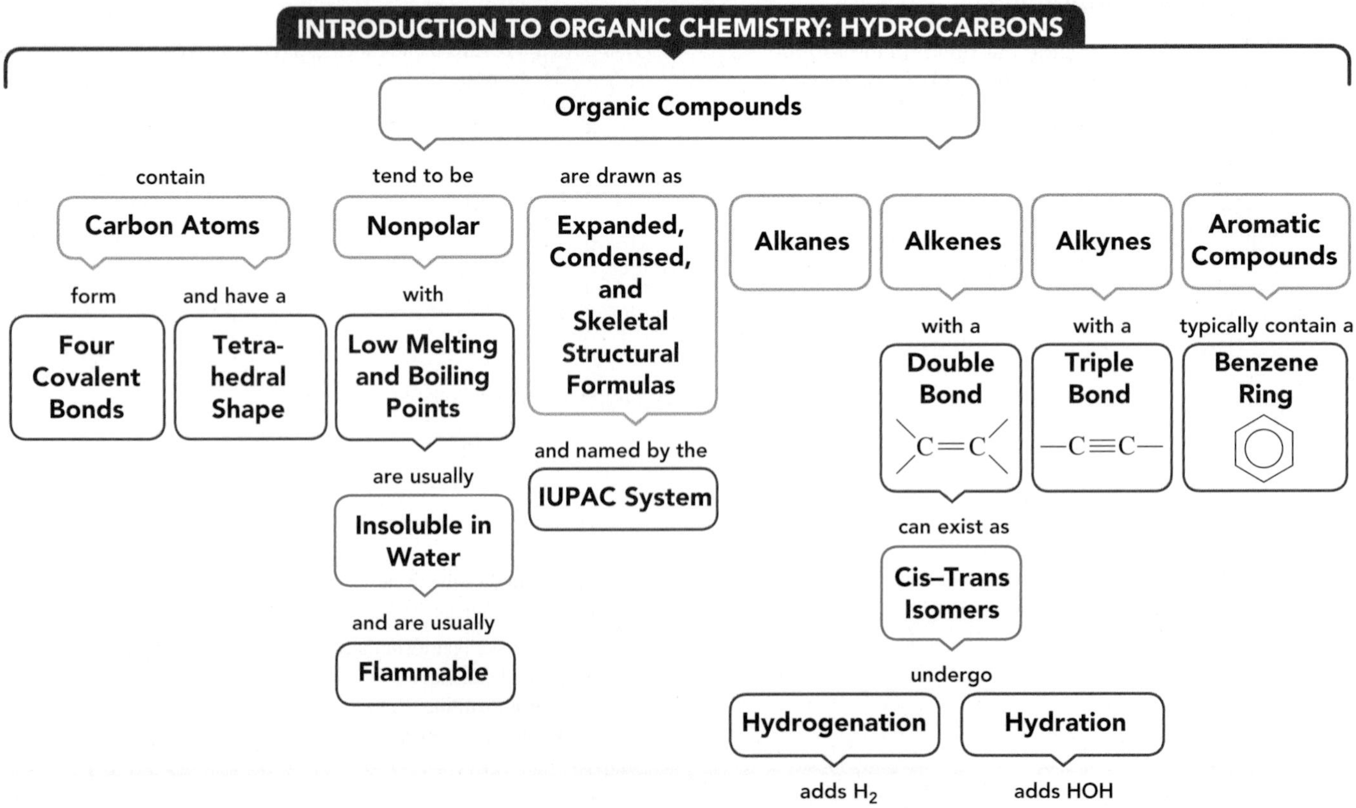

CHAPTER REVIEW

11.1 Organic Compounds

LEARNING GOAL Identify properties characteristic of organic or inorganic compounds.

- Organic compounds have covalent bonds, most form nonpolar molecules, have low melting points and low boiling points, are not very soluble in water, dissolve as molecules in solutions, and burn vigorously in air.
- Inorganic compounds are often ionic or contain polar covalent bonds and form polar molecules, have high melting and boiling points, are usually soluble in water, produce ions in water, and do not burn in air.
- In the simplest organic molecule, methane, CH_4, the C—H bonds that attach four hydrogen atoms to the carbon atom are directed to the corners of a tetrahedron with bond angles of 109°.
- In the expanded structural formula, a separate line is drawn for every bond.

11.2 Alkanes

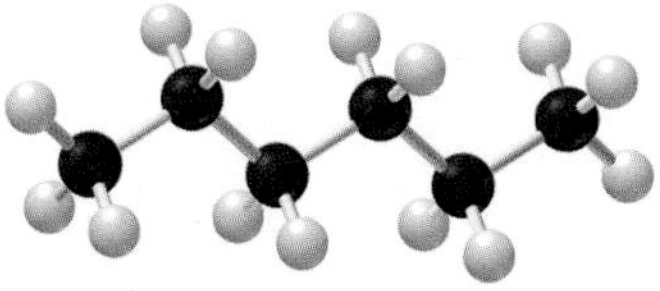

LEARNING GOAL Write the IUPAC names and draw the condensed structural formulas and skeletal formulas for alkanes and cycloalkanes.

- Alkanes are hydrocarbons that have only C—C single bonds.
- A condensed structural formula depicts groups composed of each carbon atom and its attached hydrogen atoms.
- A skeletal formula represents the carbon skeleton as ends and corners of a zigzag line or geometric figure.
- The IUPAC system is used to name organic compounds by indicating the number of carbon atoms.
- The name of a cycloalkane is written by placing the prefix *cyclo* before the alkane name with the same number of carbon atoms.

11.3 Alkanes with Substituents

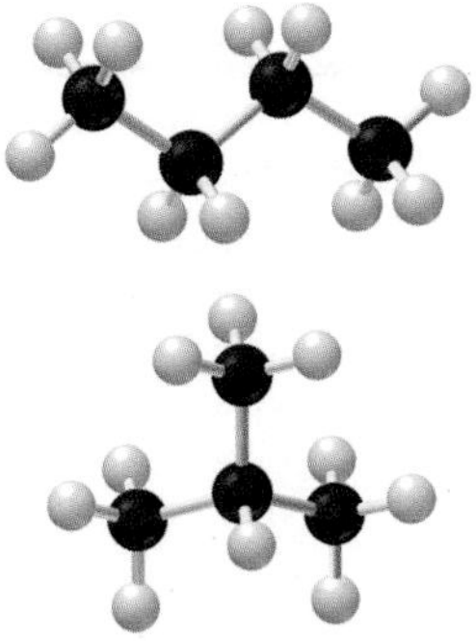

LEARNING GOAL Write the IUPAC names for alkanes with substituents and draw their condensed structural formulas and skeletal formulas.

- Structural isomers are compounds with the same molecular formulas that differ in the order in which their atoms are bonded.
- Substituents, which are attached to an alkane chain, include alkyl groups and halogen atoms (F, Cl, Br, or I).
- In the IUPAC system, alkyl substituents have names such as methyl, propyl, and isopropyl; halogen atoms are named as fluoro, chloro, bromo, or iodo.

11.4 Properties of Alkanes

LEARNING GOAL Identify the properties of alkanes and write a balanced chemical equation for combustion.

- Alkanes, which are nonpolar molecules, are not soluble in water, and usually less dense than water.
- Alkanes undergo combustion in which they react with oxygen to produce carbon dioxide, water, and energy.

11.5 Alkenes and Alkynes

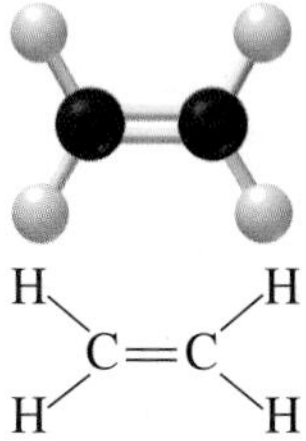

Bond angles = 120°
Ethene

LEARNING GOAL Identify structural formulas as alkenes, cycloalkenes, and alkynes, and write their IUPAC names.

- Alkenes are unsaturated hydrocarbons that contain carbon–carbon double bonds.
- Alkynes contain a carbon–carbon triple bond.
- The IUPAC names of alkenes end with *ene*, while alkyne names end with *yne*. The main chain is numbered from the end nearer the double or triple bond.

11.6 Cis–Trans Isomers

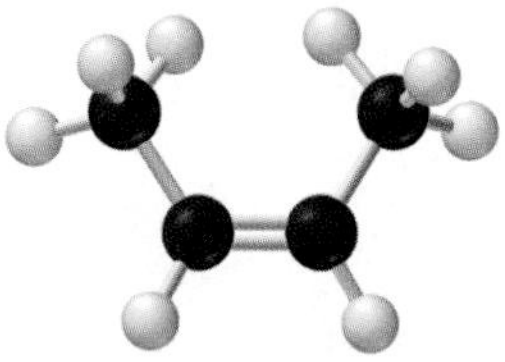

cis-2-Butene

LEARNING GOAL Draw the condensed structural formulas and give the names for the cis–trans isomers of alkenes.

- Geometric or cis–trans isomers of alkenes occur when the carbon atoms in the double bond are connected to different atoms or groups.
- In the cis isomer, the similar groups are on the same side of the double bond, whereas in the trans isomer they are connected on opposite sides of the double bond.

11.7 Addition Reactions

LEARNING GOAL Draw the condensed structural formulas and give the names for the organic products of addition reactions of hydrogenation and hydration of alkenes.

- The addition of small molecules to the double bond is a characteristic reaction of alkenes.
- Hydrogenation adds hydrogen atoms to the double bond of an alkene to yield an alkane.
- Hydration adds water (HOH) to a double bond of an alkene to form an alcohol.
- When a different number of hydrogen atoms are attached to the carbons in a double bond, the H from HOH adds to the carbon with the greater number of H atoms.

11.8 Aromatic Compounds

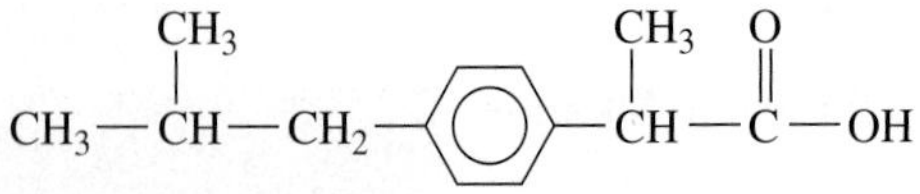

Ibuprofen

LEARNING GOAL Describe the bonding in benzene; name aromatic compounds and draw their skeletal formulas.

- Most aromatic compounds contain benzene, C_6H_6, a cyclic structure containing six carbon atoms and six hydrogen atoms.
- The structure of benzene is represented as a hexagon with a circle in the center.
- Aromatic compounds are named using the IUPAC name benzene, although common names such as toluene, aniline, and phenol are retained.
- The benzene ring is numbered, and two or more substituents are listed in alphabetical order.

SUMMARY OF NAMING

Type	Structure	Example
Alkane	$CH_3—CH_2—CH_3$	Propane
	$CH_3—CH(CH_3)—CH_2—CH_3$	2-Methylbutane
Haloalkane	$CH_3—CH_2—CH_2—Cl$	1-Chloropropane
Cycloalkane	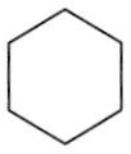	Cyclohexane
Alkene	$CH_3—CH{=}CH_2$	Propene
Cycloalkene		Cyclopropene
Alkyne	$CH_3—C{\equiv}CH$	Propyne
Aromatic	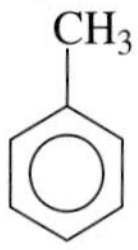	Methylbenzene, or toluene

SUMMARY OF REACTIONS

The chapter sections to review are shown after the name of the reaction.

Combustion (11.4)

$$\underset{\text{Propane}}{CH_3—CH_2—CH_3(g)} + 5O_2(g) \xrightarrow{\Delta} 3CO_2(g) + 4H_2O(g) + \text{energy}$$

Hydrogenation (11.7)

$$\underset{\text{Propene}}{H_2C{=}CH—CH_3} + H_2 \xrightarrow{Pt} \underset{\text{Propane}}{CH_3—CH_2—CH_3}$$

Hydration (11.7)

$$\underset{\text{Propene}}{H_2C{=}CH—CH_3} + H—OH \xrightarrow{H^+} \underset{\text{2-Propanol}}{CH_3—\overset{OH}{CH}—CH_3}$$

KEY TERMS

addition A reaction in which atoms or groups of atoms bond to a carbon–carbon double bond. Addition reactions include the addition of hydrogen (hydrogenation), and water (hydration).

alkane A hydrocarbon that contains only single bonds between carbon atoms.

alkene A hydrocarbon that contains one or more carbon–carbon double bonds (C=C).

alkyl group An alkane minus one hydrogen atom. Alkyl groups are named like the alkanes except a *yl* ending replaces *ane*.

alkyne A hydrocarbon that contains one or more carbon–carbon triple bonds C≡C.

aromatic compound A compound that contains the ring structure of benzene.

benzene A ring of six carbon atoms, each of which is attached to a hydrogen atom, C_6H_6.

cis isomer An isomer of an alkene in which similar groups in the double bond are on the same side.

condensed structural formula A structural formula that shows the arrangement of the carbon atoms in a molecule but groups each carbon atom with its bonded hydrogen atoms $-CH_3$, $-CH_2-$, or $-\overset{|}{C}H-$.

cycloalkane An alkane that is a ring or cyclic structure.

expanded structural formula A type of structural formula that shows the arrangement of the atoms by drawing each bond in the hydrocarbon.

hydration An addition reaction in which the components of water, H— and —OH, bond to the carbon–carbon double bond to form an alcohol.

hydrocarbon An organic compound that contains only carbon and hydrogen.

hydrogenation The addition of hydrogen (H_2) to the double bond of alkenes to yield alkanes.

IUPAC system A system for naming organic compounds devised by the International Union of Pure and Applied Chemistry.

organic compound A compound made of carbon that typically has covalent bonds, is nonpolar, has low melting and boiling points, is insoluble in water, and is flammable.

structural isomers Organic compounds in which identical molecular formulas have different arrangements of atoms.

substituent Groups of atoms such as an alkyl group or a halogen bonded to the main carbon chain or ring of carbon atoms.

trans isomer An isomer of an alkene in which similar groups in the double bond are on opposite sides.

CORE CHEMISTRY SKILLS

The chapter section containing each Core Chemistry Skill is shown in parentheses at the end of each heading.

Naming and Drawing Alkanes (11.2)

- The alkanes ethane, propane, and butane contain two, three, and four carbon atoms, respectively, connected in a row or a *continuous* chain.
- Alkanes with five or more carbon atoms in a chain are named using the prefixes *pent* (5), *hex* (6), *hept* (7), *oct* (8), *non* (9), and *dec* (10).
- In the condensed structural formula, the carbon and hydrogen atoms on the ends are written as $-CH_3$ and the carbon and hydrogen atoms in the middle are written as $-CH_2-$.

Example: **a.** What is the name of $CH_3-CH_2-CH_2-CH_3$?
b. Draw the condensed structural formula for pentane.

Answer: **a.** An alkane with a four-carbon chain is named with the prefix *but* followed by *ane*, which is butane.
b. Pentane is an alkane with a five-carbon chain. The carbon atoms on the ends are attached to three H atoms each, and the carbon atoms in the middle are attached to two H each. $CH_3-CH_2-CH_2-CH_2-CH_3$

Writing Equations for Hydrogenation and Hydration (11.7)

- The addition of small molecules to the double bond is a characteristic reaction of alkenes.
- Hydrogenation adds hydrogen atoms to the double bond of an alkene to form an alkane.
- Hydration adds water (HOH) to a double bond of an alkene to form an alcohol.
- When a different number of hydrogen atoms are attached to the carbons in a double bond, the H— from HOH adds to the carbon atom with the greater number of H atoms, and the —OH adds to the other carbon atom from the double bond.

Example: **a.** Draw the condensed structural formula for 2-methyl-2-butene.
b. Draw the condensed structural formula for the product of the hydrogenation of 2-methyl-2-butene.
c. Draw the condensed structural formula for the product of the hydration of 2-methyl-2-butene.

Answer: **a.**

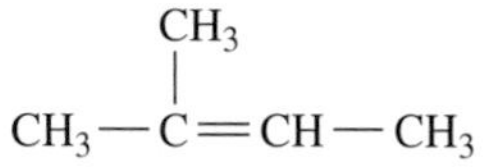

b. When H_2 adds to a double bond, the product is an alkane with the same number of carbon atoms.

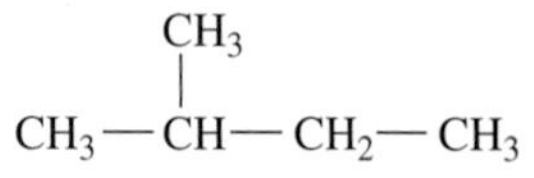

c. When HOH adds to a double bond, the product is an alcohol, which has an —OH group. The H— from HOH adds to the carbon atom in the double bond that has the greater number of H atoms.

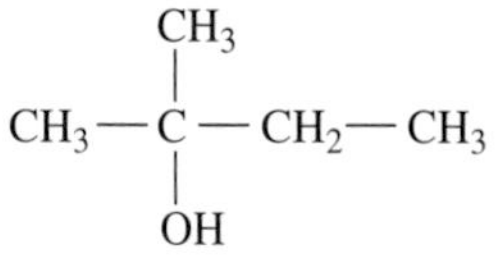

UNDERSTANDING THE CONCEPTS

The chapter sections to review are shown in parentheses at the end of each question.

11.39 Match the following physical and chemical properties with potassium chloride, KCl, used in salt substitutes, or butane, C_4H_{10}, used in lighters: (11.1)

a. melts at −138 °C
b. burns vigorously in air
c. melts at 770 °C
d. contains ionic bonds
e. is a gas at room temperature

11.40 Match the following physical and chemical properties with octane, C_8H_{18}, found in gasoline, or magnesium sulfate, $MgSO_4$, also called Epsom salts: (11.1)
a. contains only covalent bonds
b. melts at 1124 °C
c. is insoluble in water
d. is a liquid at room temperature
e. is a strong electrolyte

11.41 Identify the compounds in each of the following pairs as structural isomers or not structural isomers: (11.3)

a. 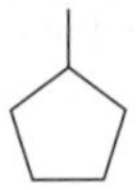 and 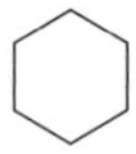

b.  and 

11.42 Identify the compounds in each of the following pairs as structural isomers or not structural isomers: (11.3)

a. 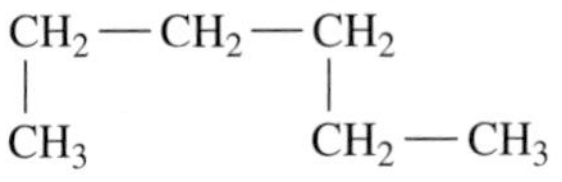 and 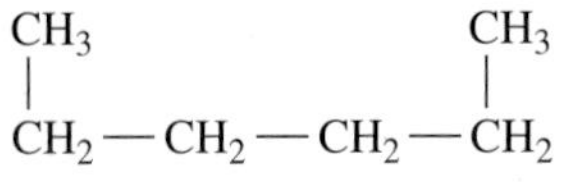

b. 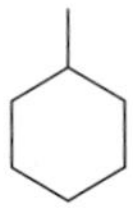 and 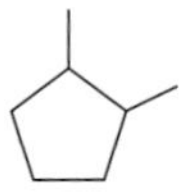

11.43 Convert each of the following skeletal formulas to a condensed structural formula and give its IUPAC name: (11.3)

a. 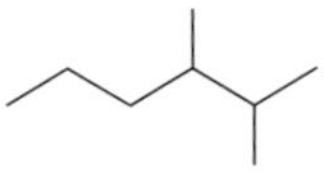 **b.** 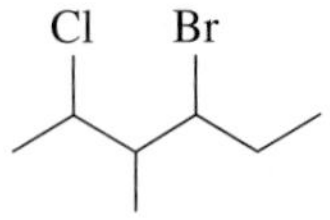

11.44 Convert each of the following skeletal formulas to a condensed structural formula and give its IUPAC name: (11.3)

a. 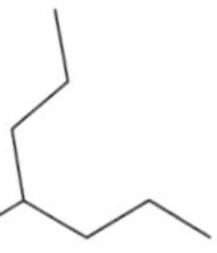 **b.** 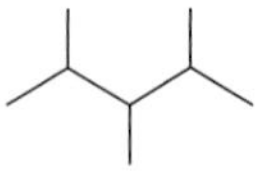

ADDITIONAL QUESTIONS AND PROBLEMS

11.45 Give the IUPAC name for each of the following: (11.3)

a. 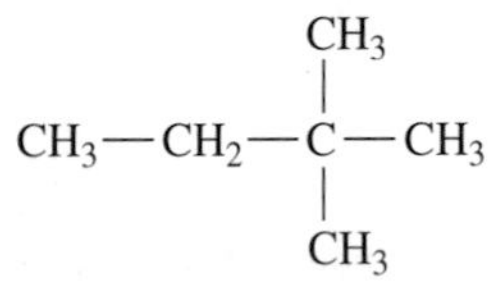

b. CH_3-CH_2-Cl

c. 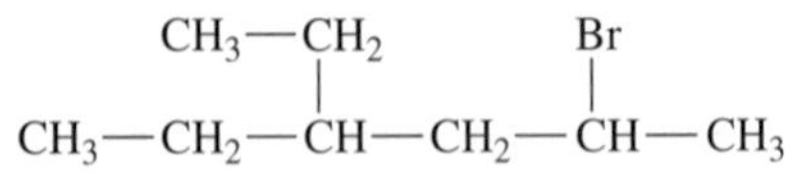

d. 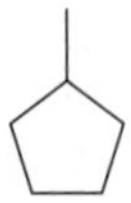

11.46 Give the IUPAC name for each of the following: (11.3)

a. 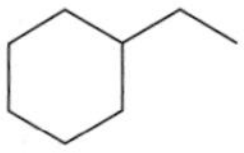

b. 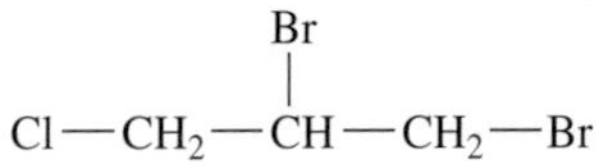

c. 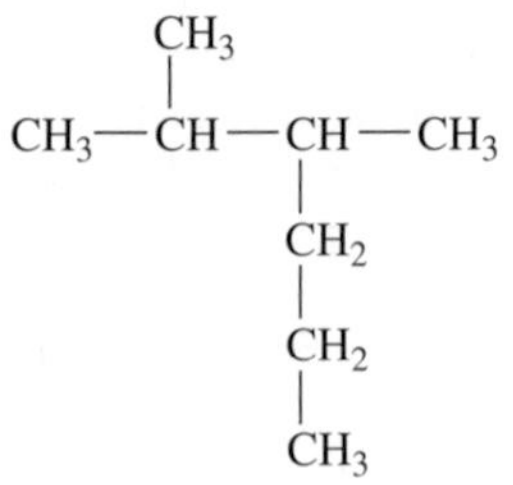

d. 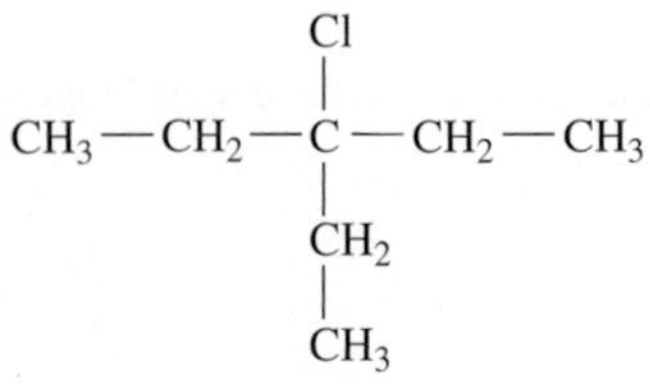

11.47 Give the IUPAC name for each of the following: (11.5, 11.6)

a. 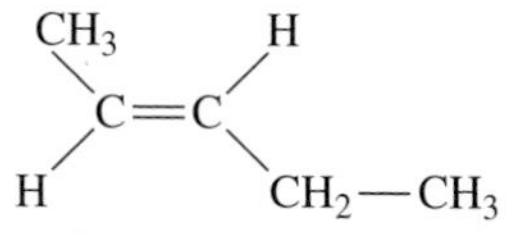

b. 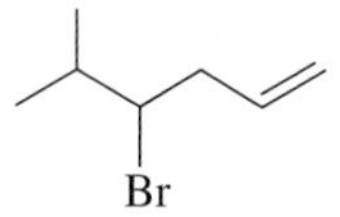

c. 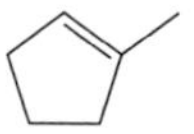

11.48 Give the IUPAC name for each of the following: (11.5, 11.6)

a. 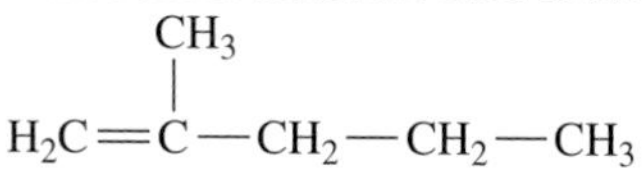

b. 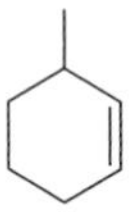

c. 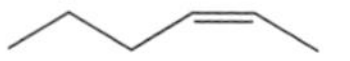

11.49 Indicate if the following pairs of structures represent structural isomers, cis–trans isomers, or the same molecule: (11.5, 11.6)

a.  and 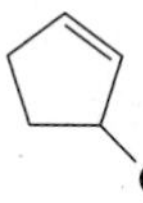

b. 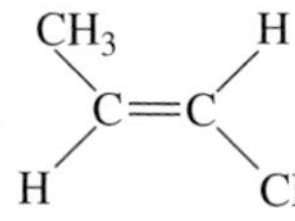 and 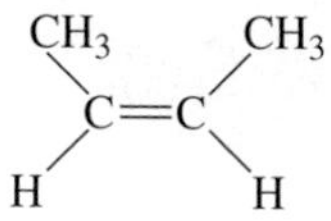

11.50 Indicate if the following pairs of structures represent structural isomers, cis–trans isomers, or the same molecule: (11.5, 11.6)

a. 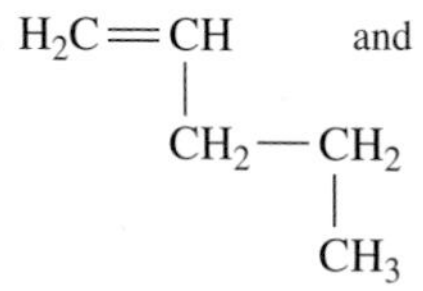 and $CH_3{-}CH_2{-}CH_2{-}CH{=}CH_2$

b. 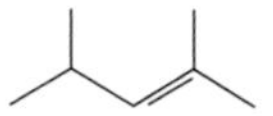 and 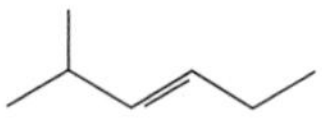

11.51 Name each of the following aromatic compounds: (11.8)

a. 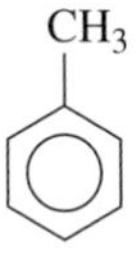

b. 

11.52 Name each of the following aromatic compounds: (11.8)

a. 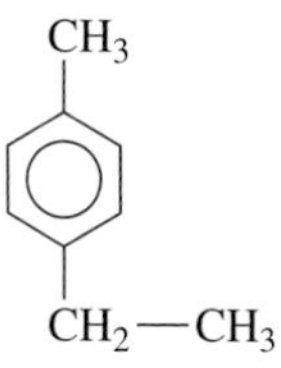

b. 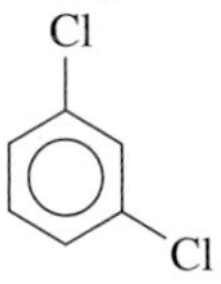

11.53 Draw the condensed structural formula, or skeletal formula if cyclic, for each of the following: (11.3, 11.5, 11.6)

a. bromocyclopropane

b. 1,1-dibromo-2-pentyne

c. *cis*-2-heptene

11.54 Draw the skeletal formula for each of the following: (11.3, 11.5, 11.6)

a. Toluene

b. Cyclopentene

c. 2,3-dichloro-1-butene

11.55 Draw the cis and trans isomers for each of the following: (11.6)

a. 2-pentene

b. 3-hexene

11.56 Draw the cis and trans isomers for each of the following: (11.6)

a. 2-butene

b. 2-hexene

11.57 Draw the condensed structural formula, or skeletal formula if cyclic, for each of the following: (11.8)

a. ethylbenzene

b. 2,5-dibromophenol

c. 1,2,4-trimethylbenzene

11.58 Draw the condensed structural formula, or skeletal formula if cyclic, for each of the following: (11.8)

a. phenol

b. 1,4-dichlorobenzene

c. 2-bromotoluene

11.59 Write a balanced chemical equation for the complete combustion of each of the following: (11.4)

a. 2,2-dimethylpropane

b. cyclobutane

c. 2-hexene

11.60 Write a balanced chemical equation for the complete combustion of each of the following: (11.4)

a. heptane

b. 2-methyl-1-pentene

c. $CH_3{-}CH_2{-}CH_2{-}CH_3$, C_4H_{10}

11.61 Give the name for the product from the hydrogenation of each of the following: (11.7)

a. 3-methyl-2-pentene

b. cyclohexene

c. propene

11.62 Give the name for the product from the complete hydrogenation of each of the following: (11.7)

a. Benzene

b. 2-Butene

c. Acetylene

11.63 Draw the condensed structural formula or skeletal formula for the product of each of the following: (11.7)

a. 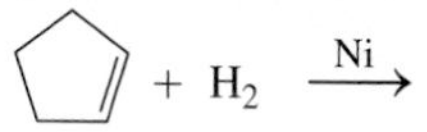

b. 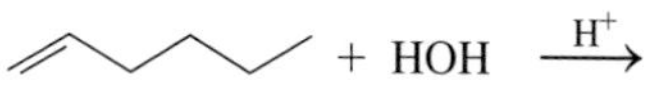

11.64 Draw the condensed structural formula or skeletal formula for the product of each of the following: (11.7)

a. 

b. $CH_3{-}CH_2{-}CH_2{-}CH{=}CH_2 + H_2 \xrightarrow{Ni}$

CHALLENGE QUESTIONS

The following groups of questions are related to the topics in this chapter. However, they do not all follow the chapter order, and they require you to combine concepts and skills from several sections. These questions will help you increase your critical thinking skills and prepare for your next exam.

11.65 The density of pentane, a component of gasoline, is 0.63 g/mL. The heat of combustion for pentane is 845 kcal/mole. (7.2, 7.4, 7.7, 7.8, 8.6, 11.2, 11.4)

a. Write the balanced chemical equation for the complete combustion of pentane. Is the reaction endothermic or exothermic?.

b. How many moles of water can be produced by the combustion of 1 mole of pentane?

c. How much heat is produced when 10 moles of pentane is burned?

d. How many moles of CO_2 are produced from the complete combustion of 10 moles of pentane?

11.66 Acetylene (ethyne) gas reacts with oxygen and burns at 3300 °C in an acetylene torch. (7.4, 7.7, 8.6, 11.4, 11.5)

a. Write the balanced chemical equation for the complete combustion of acetylene.

b. What is the molar mass of acetylene?

c. How many grams of oxygen are needed to react with 8.5 L of acetylene gas at STP?

d. How many liters of CO_2 gas at STP are produced when 30.0 g of acetylene undergoes combustion?

Acetylene and oxygen burn at 3300 °C.

11.67 Draw the condensed structural formulas for all the possible alkane isomers that have a total of six carbon atoms and a four-carbon chain. (11.3)

11.68 Draw the condensed structural formulas for all the possible haloalkane isomers that have four carbon atoms and a bromine. (11.3)

11.69 Consider the compound propane. (7.4, 7.7, 11.2)

a. Draw the condensed structural formula for propane.

b. Write the balanced chemical equation for the complete combustion of propane.

c. How many grams of O_2 are needed to react with 12.0 L of propane gas at STP?

d. How many grams of CO_2 would be produced from the reaction in part **c**?

11.70 Consider the compound ethylcyclopentane. (7.4, 7.7, 8.6, 11.2, 11.4)

a. Draw the skeletal formula for ethylcyclopentane.

b. Write the balanced chemical equation for the complete combustion of ethylcyclopentane.

c. Calculate the grams of O_2 required for the combustion of 25.0 g of ethylcyclopentane.

d. How many liters of CO_2 would be produced at STP from the reaction in part **c**?

11.71 Explosives used in mining contain TNT, or trinitrotoluene. (11.8)

The TNT in explosives is used in mining.

a. If the functional group *nitro* is $-NO_2$, what is the condensed structural formula of 2,4,6-trinitrotoluene, one isomer of TNT?

b. TNT is actually a mixture of isomers of trinitrotoluene. Draw two other possible isomers.

11.72 Margarines are produced from the hydrogenation of vegetable oils, which contain unsaturated fatty acids. How many grams of hydrogen are required to completely saturate 75.0 g of oleic acid, $C_{18}H_{34}O_2$, which has one double bond? (7.7, 11.8)

Margarines are produced by hydrogenation of unsaturated fats.

ANSWERS

Answers to Study Checks

11.1 Carbon and hydrogen are always found in hydrocarbons.

11.2 $CH_3-CH_2-CH_2-CH_2-CH_2-CH_2-CH_3$, heptane

11.3 cyclobutane

11.4 This condensed structural formula represents another isomer because it has the same molecular formula (C_6H_{14}), but it has a different arrangement of carbon atoms with one methyl group ($-CH_3$) attached to the middle (third) carbon of a five-carbon chain.

11.5 1-chloro-2,4-dimethylhexane

11.6 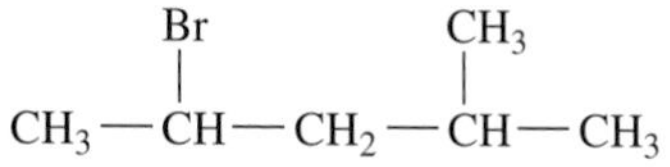 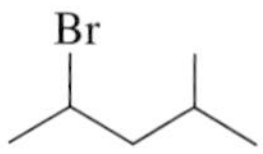

11.7 **a.** $CH_3—C≡C—CH_2—CH_3$

b. 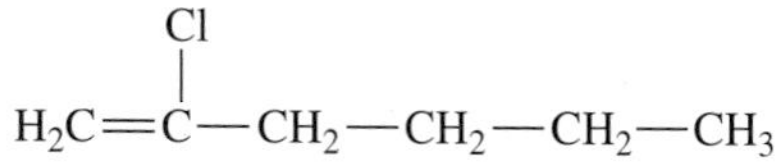

11.8 *trans*-3-hexene

11.9 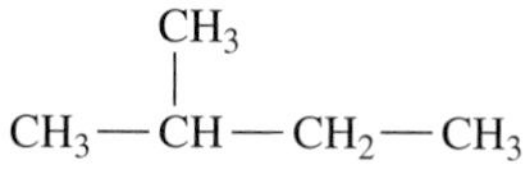

11.10 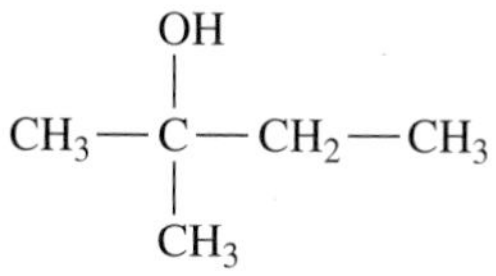

11.11 1,3-diethylbenzene

Answers to Selected Questions and Problems

11.1 **a.** inorganic **b.** organic
c. organic **d.** inorganic
e. organic **f.** inorganic

11.3 **a.** inorganic
b. organic
c. organic
d. inorganic

11.5 For KBr: (a), (c)
For propane: (b), (d)

11.7 **a.** n-pentane
b. cyclopentane
c. n-propane
d. n-octane

11.8 **a.** n-butane
b. cyclobutane
c. cyclopropane
d. n-heptane

11.9 **a.** 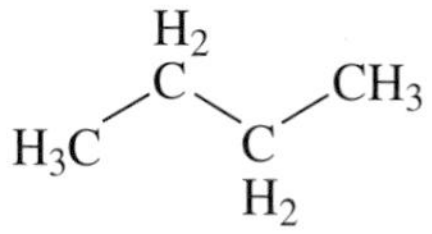 **b.** 

c. 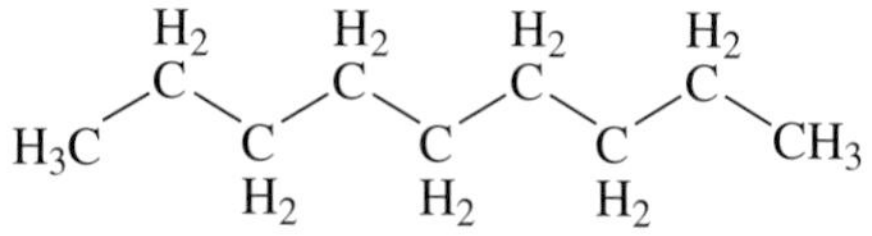

d. 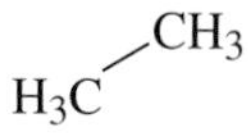

11.11 **a.** structural isomers
b. same molecule
c. structural isomers

11.13 **a.** 2,2-dimethylpentane
b. 3,4,6-trimethylnonane
c. 3,3-dimethyl-5-methylheptane
d. methylcyclopentane

11.15 **a.** 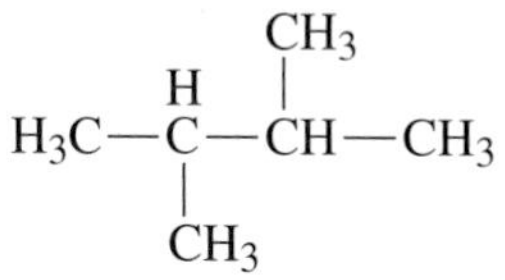 **b.** 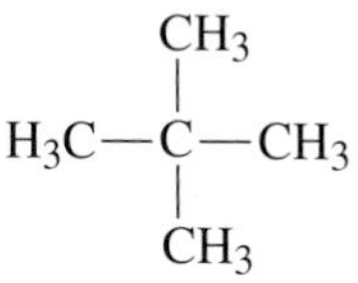

c. 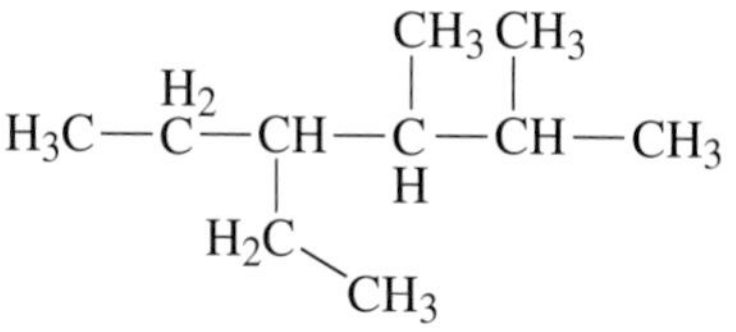

d. 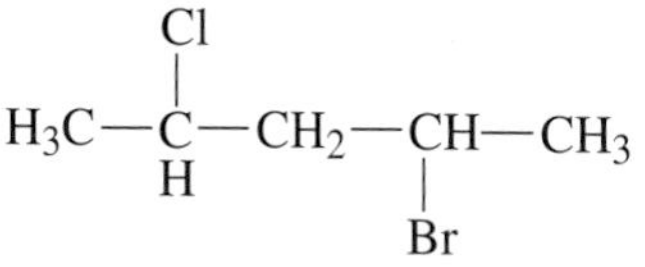

11.17 **a.** 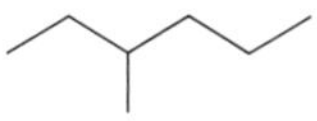 **b.** 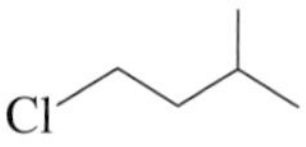

c. 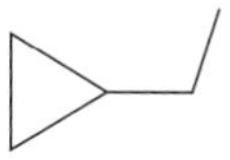 **d.** 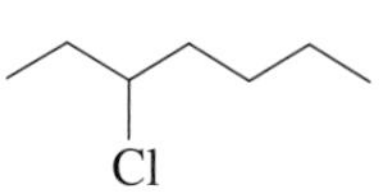

11.19 **a.** $CH_3—CH_2—CH_2—CH_2—CH_2—CH_2—CH_3$

b. liquid
c. no
d. float
e. $C_7H_{16}(g) + 11O_2(g) \xrightarrow{\Delta} 7CO_2(g) + 8H_2O(g) + \text{energy}$

11.21 **a.** $C_5H_{12} + 8O_2 \longrightarrow 5CO_2 + 6H_2O + \text{Energy}$
b. $C_4H_8 + 6O_2 \longrightarrow 4CO_2 + 4H_2O + \text{Energy}$
c. $C_7H_{16} + 11O_2 \longrightarrow 7CO_2 + 8H_2O + \text{Energy}$

11.23 **a.** alkene
b. alkyne
c. cycloalkene
d. alkyne

11.25 **a.** propyne
b. 3,3-dimethylbut-1-ene
c. 4-ethylcyclopent-1-ene
d. 2,5,6-trimethylhept-3-ene

11.27 **a.** 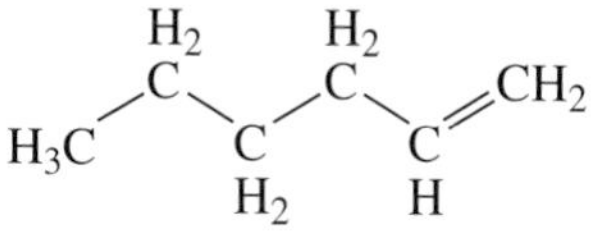

b. 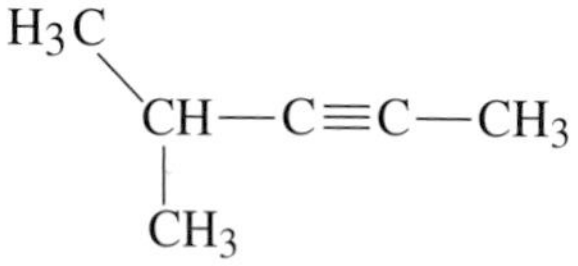

c. 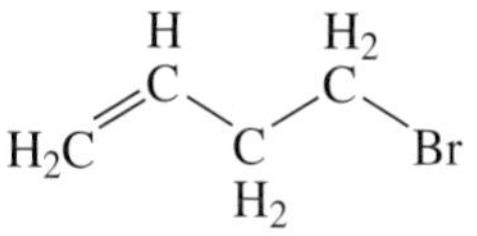 **d.** 

11.29 **a.** *cis*-2-butene
b. *trans*-3-octene
c. *cis*-3-heptene

11.31 **a.** 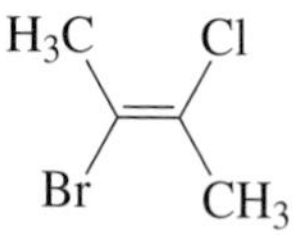

b. 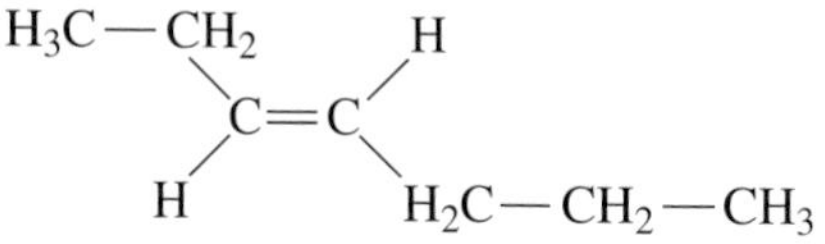

c. 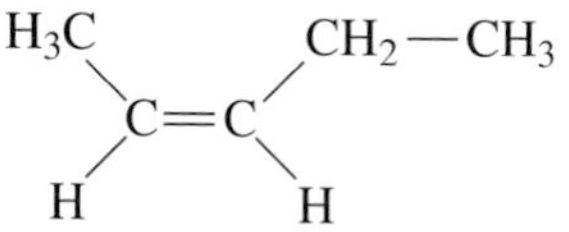

11.33 a. $CH_3—CH_2—CH_2—CH_2—CH_3$

b. 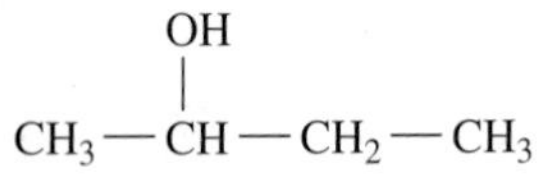

c. 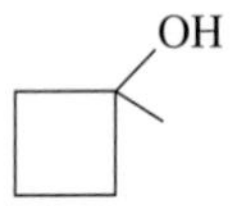

11.35 a. 3,4-dichlorotoluene
b. 4-bromoaniline
c. 3,5-dimethylphenol

11.37 a. 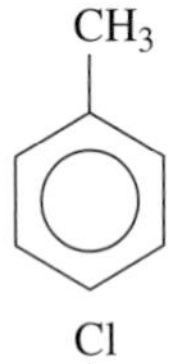

b. 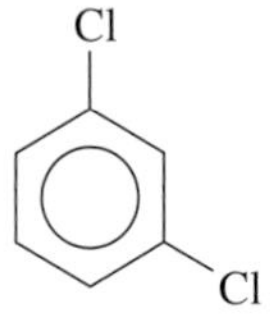

c. 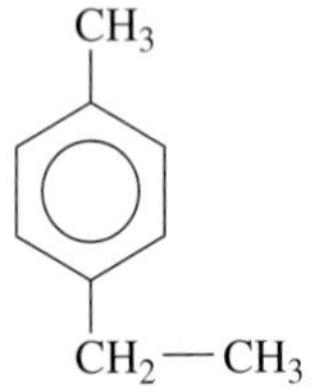

11.39 a. butane
b. butane
c. potassium chloride
d. potassium chloride
e. butane

11.41 a. structural isomers
b. not structural isomers

11.43 a. 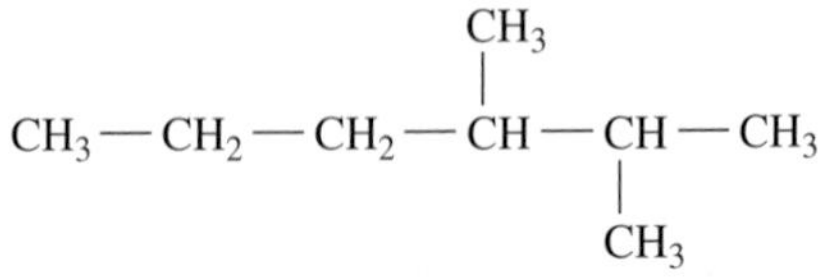

2,3-Dimethylhexane

b. 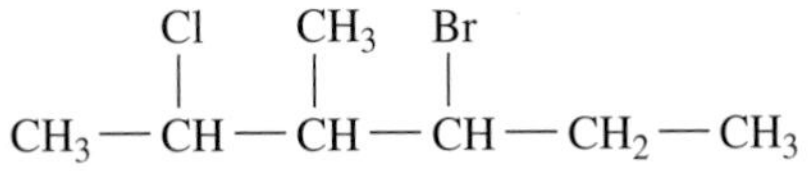

4-Bromo-2-chloro-3-methylhexane

11.45 a. 2,2-dimethylbutane
b. chloroethane
c. 2-bromo-4-ethylhexane
d. methylcyclopentane

11.47 a. *trans*-2-pentene
b. 4-bromo-5-methyl-1-hexene
c. 1-methylcyclopentene

11.49 a. structural isomers
b. cis–trans isomers

11.51 a. toluene
b. 4-bromoaniline

11.53 a. 

b. 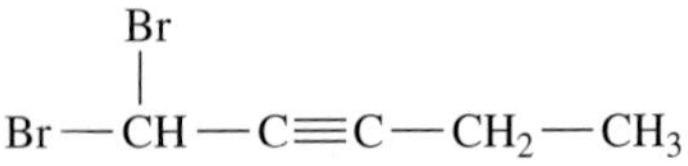

c. 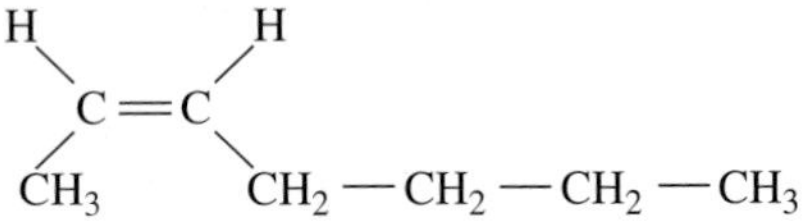

11.55 a. 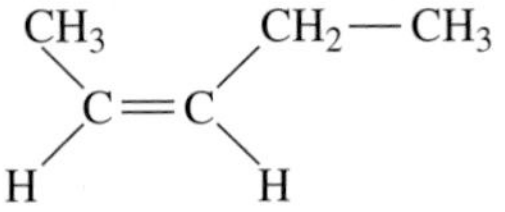

cis-2-Pentene

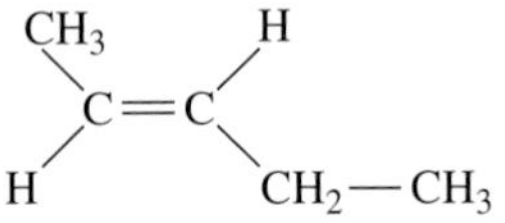

trans-2-Pentene

b. 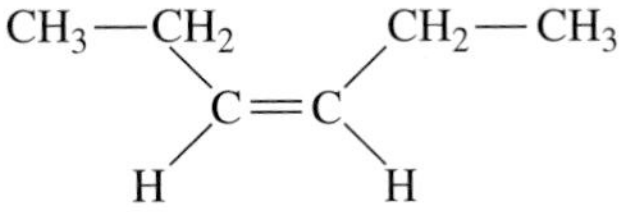

cis-3-Hexene

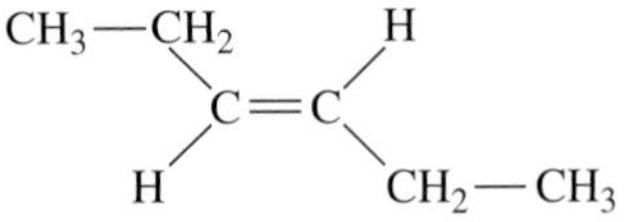

trans-3-Hexene

11.57 a. 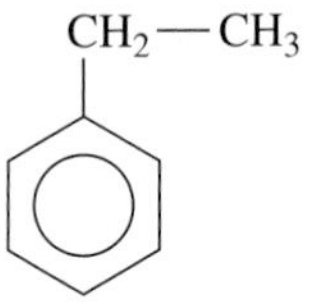

b. 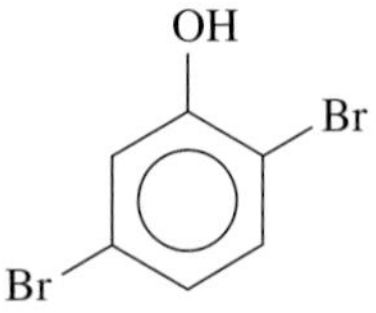

c. 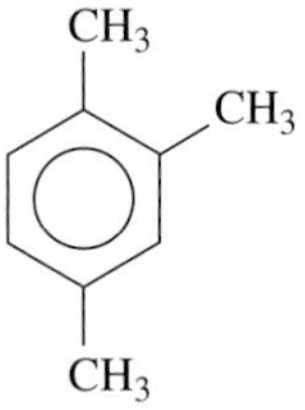

11.59 a. $C_5H_{12}(g) + 8O_2(g) \xrightarrow{\Delta} 5CO_2(g) + 6H_2O(g) + \text{energy}$
b. $C_4H_8(g) + 6O_2(g) \xrightarrow{\Delta} 4CO_2(g) + 4H_2O(g) + \text{energy}$
c. $C_6H_{12}(g) + 9O_2(g) \xrightarrow{\Delta} 6CO_2(g) + 6H_2O(g) + \text{energy}$

11.61 a. 3-methylpentane
b. cyclohexane
c. propane

11.63 a. 

b. 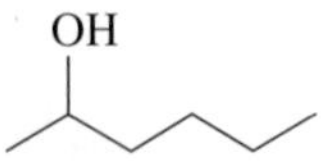

11.65 **a.** $C_5H_{12}(g) + 8O_2(g) \xrightarrow{\Delta} 5CO_2(g) + 6H_2O(g) + \text{energy}$
b. 72.15 g/mole
c. 2.8×10^4 kcal
d. 3.7×10^3 L of CO_2 at STP

11.67 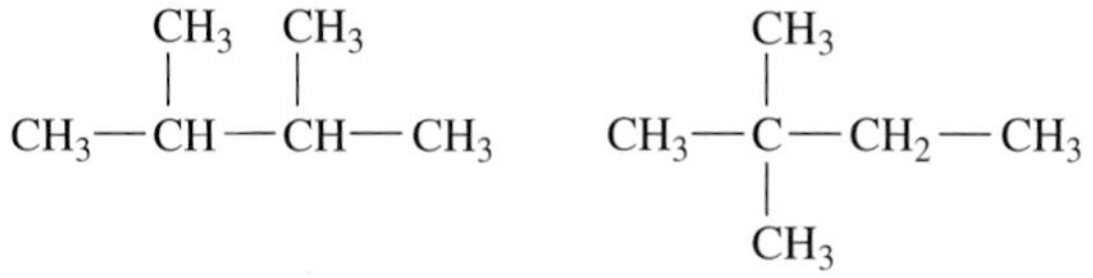

11.69 **a.** $CH_3—CH_2—CH_3$
b. $C_3H_8(g) + 5O_2(g) \xrightarrow{\Delta} 3CO_2(g) + 4H_2O(g) + \text{energy}$
c. 85.7 g of O_2
d. 70.7 g of CO_2

11.71 **a.** 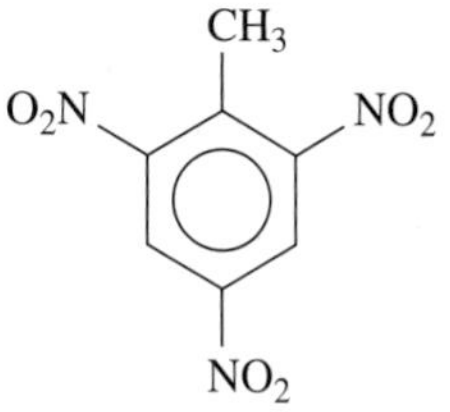

b.

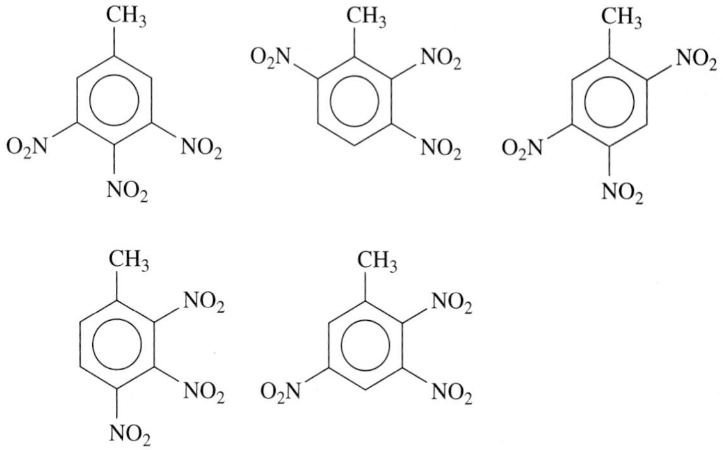

12 Alcohols, Thiols, Ethers, Aldehydes, and Ketones

An epidural is a regional anesthesia that will provide pain relief for Janet during her labor and delivery by blocking pain in the lower half of her body. Tom, a nurse anesthetist, prepares the medication for the epidural injection, which consists of a local anesthetic, chloroprocaine, and a small amount of fentanyl, a common opioid. Tom sterilizes the skin for injection using a 90% ethanol antiseptic solution. He administers the anesthetic through a catheter into Janet's lower spine and within moments, her pain subsides.

One of the earliest anesthetics used in medicine was diethyl ether, more commonly referred to as ether. Diethyl ether contains an ether functional group, which is an oxygen atom bonded by single bonds to two carbon groups. Today, ether is rarely used because it is extremely flammable, and produces undesirable side effects such as postanesthetic nausea and vomiting.

CH_3-CH_2-OH (Ethanol) $CH_3-CH_2-O-CH_2-CH_3$ (Diethyl ether)

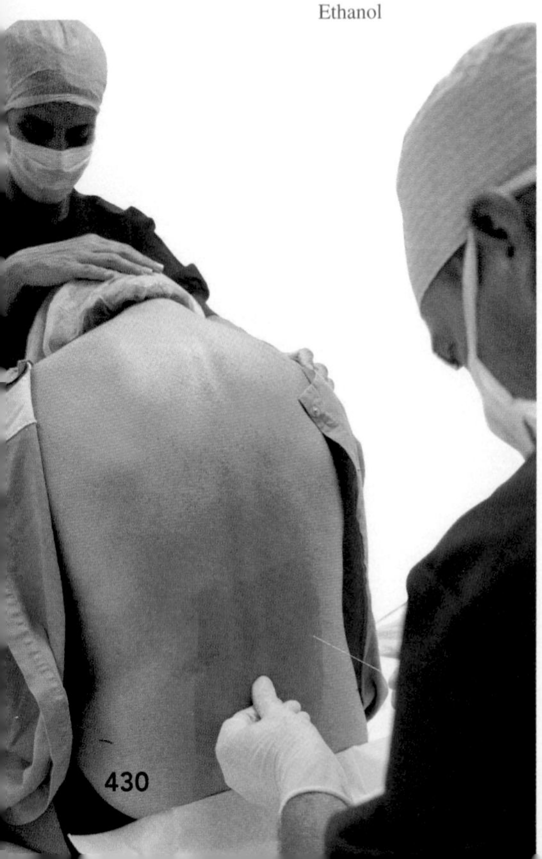

CAREER Nurse Anesthetist

More than 26 million people in the United States each year require anesthesia associated with a medical or dental procedure. Anesthesia is typically administered by a nurse anesthetist who provides care before, during, and after the procedure by giving medications to keep a patient asleep and pain-free while monitoring the patient's vital signs. A nurse anesthetist also obtains supplies, equipment, and an ample blood supply for a potential emergency, and interprets pre-surgical tests to determine how the anesthesia will affect the patient. A nurse anesthetist may also be required to insert artificial airways, administer oxygen, or work to prevent surgical shock during a procedure, working under the direction of the attending surgeon, dentist, or anesthesiologist.

In addition to carbon and hydrogen, many organic compounds contain oxygen and sulfur atoms, which are organized as *functional groups*. In this chapter, we will look at the functional groups that characterize the alcohols, phenols, thiols, ethers, aldehydes, and ketones. Alcohols, which contain the *hydroxyl group* (—OH), are commonly found in nature and are used in industry and at home. For centuries, grains, vegetables, and fruits have been fermented to produce the ethanol present in alcoholic beverages. The hydroxyl group is important in biomolecules such as sugars and starches as well as in steroids such as cholesterol and estradiol. Menthol is a cyclic alcohol with a minty odor and flavor that is used in cough drops, shaving creams, and ointments. The phenols contain the hydroxyl group attached to a benzene ring. Thiols, which contain the —SH group, give the strong odors we associate with garlic and onions. Ethers are compounds that contain an oxygen atom connected to two carbon atoms (—O—). Ethers are important solvents in chemistry and medical laboratories. Beginning in 1842, diethyl ether was used for about 100 years as a general anesthetic.

We will also study two other families of organic compounds: aldehydes and ketones. Many of the odors and flavors associated with flavorings and perfumes are due to a carbon–oxygen double bond called a *carbonyl group* (C=O). Aldehydes in foods and perfumes provide the odors and flavors of vanilla, almond, and cinnamon. In biology, you may have seen specimens preserved in a solution of formaldehyde. You probably notice the odor of a ketone when you use paint or nail polish remover.

LOOKING AHEAD

CHAPTER READINESS*

CORE CHEMISTRY SKILL

- Naming and Drawing Alkanes (11.2)

*This Core Chemistry Skill from the previous chapter is listed here for your review as you proceed to the new material in this chapter.

12.1 Alcohols, Phenols, Thiols, and Ethers

LEARNING GOAL

Give the IUPAC and common names for alcohols and phenols; give the common names for thiols and ethers. Draw their condensed structural formulas or skeletal formulas.

In an **alcohol**, the functional group known as a hydroxyl group (—OH) replaces a hydrogen atom in a hydrocarbon. Oxygen (O) atoms are shown in red in the ball-and-stick models (see Figure 12.1). In a **phenol**, the hydroxyl group replaces a hydrogen atom attached to a benzene ring. A **thiol** contains a sulfur atom, shown in yellow-green in the ball-and-stick model, which makes a thiol similar to an alcohol except that —OH is replaced by an —SH group. In an **ether**, the functional group consists of an oxygen atom, which is attached to two carbon atoms (—O—). Molecules of alcohols, phenols, thiols, and ethers have bent shapes around the oxygen or sulfur atom, similar to water.

Naming Alcohols

CORE CHEMISTRY SKILL

Identifying Functional Groups

In the IUPAC system, an alcohol is named by replacing the *e* of the corresponding alkane name with *ol*. The common name of a simple alcohol uses the name of the alkyl group followed by *alcohol*.

$CH_3—H$	$CH_3—OH$	$CH_3—CH_2—H$	$CH_3—CH_2—OH$
Methane	Methanol (methyl alcohol)	Ethane	Ethanol (ethyl alcohol)

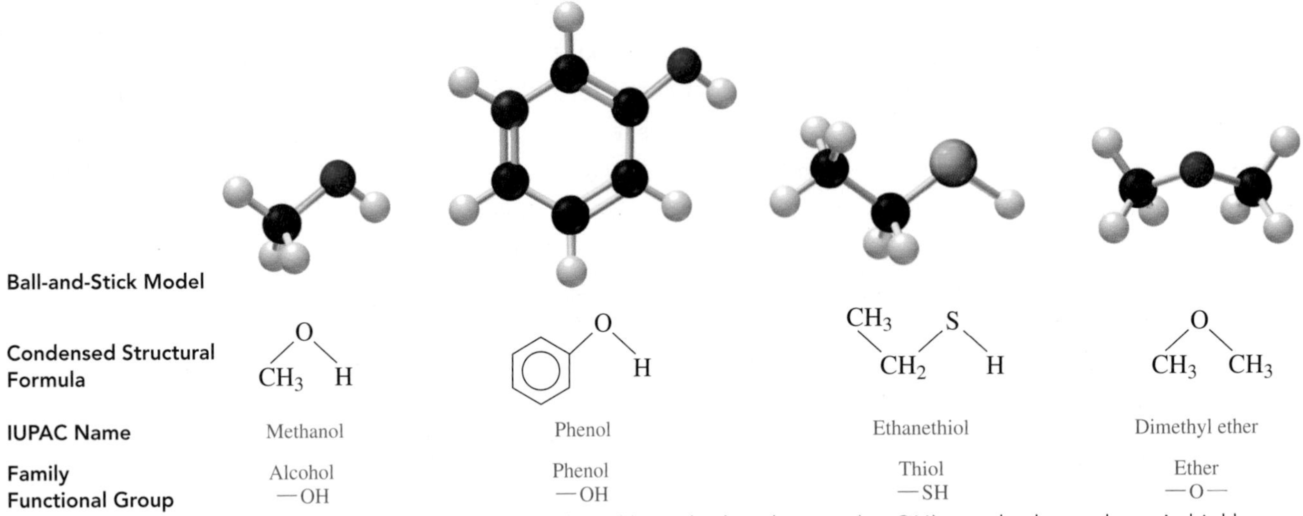

FIGURE 12.1 ▶ An alcohol or a phenol has a hydroxyl group (—OH) attached to carbon. A thiol has a thiol group (—SH) attached to carbon. An ether contains an oxygen atom (—O—) bonded to two carbon groups.

Q How is an alcohol different from a thiol?

CORE CHEMISTRY SKILL

Naming Alcohols and Phenols

When an alcohol consists of a chain with three or more carbon atoms, the chain is numbered to give the position for the —OH group and any substituents on the chain.

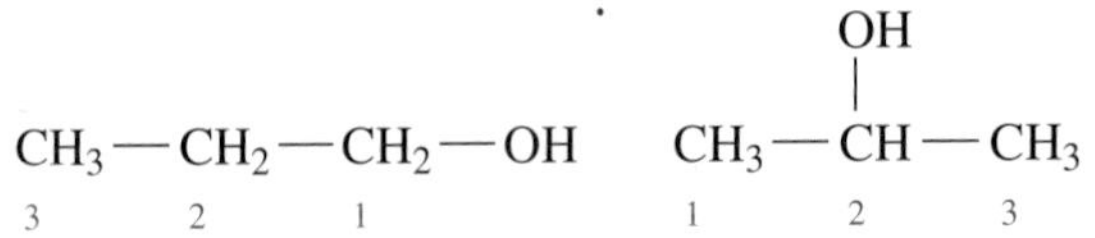

1-Propanol (propyl alcohol) 2-Propanol (isopropyl alcohol)

We can also draw the skeletal formulas for alcohols as shown for 2-propanol and 2-butanol.

2-Propanol 2-Butanol

A cyclic alcohol is named as a *cycloalkanol*. If there are substituents, the ring is numbered from carbon 1, which is the carbon attached to the —OH group. Compounds with no substituents on the ring do not require a number for the hydroxyl group.

Cyclohexanol 2-Methylcyclopentanol

Naming Phenols

The term *phenol* is the IUPAC name for a benzene ring bonded to a hydroxyl group (—OH), which is used in the name of the family of organic compounds derived from phenol. When there is a second substituent, the benzene ring is numbered starting from carbon 1, which is the carbon bonded to the —OH group.

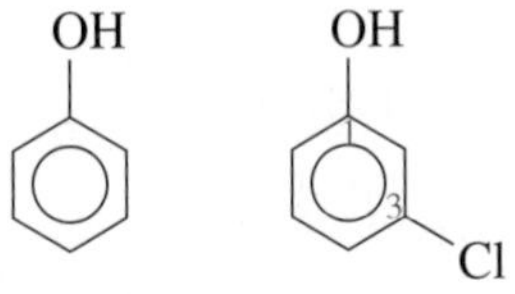

Phenol 3-Chlorophenol

SAMPLE PROBLEM 12.1 Naming Alcohols

Give the IUPAC name for the following:

$$\begin{array}{ccccccccc} & & CH_3 & & & & OH & & \\ & & | & & & & | & & \\ CH_3 & - & CH & - & CH_2 & - & CH & - & CH_3 \end{array}$$

SOLUTION

ANALYZE THE PROBLEM	Functional Group	Family	IUPAC Naming
	hydroxyl group	alcohol	replace the *e* in the alkane name with *ol*

STEP 1 **Name the longest carbon chain attached to the —OH group by replacing the e in the corresponding alkane name with *ol*.** To name the alcohol, the *e* in alkane name pentane is replaced with *ol*.

$$\begin{array}{ccccccccc} & & CH_3 & & & & OH & & \\ & & | & & & & | & & \\ CH_3 & - & CH & - & CH_2 & - & CH & - & CH_3 \end{array} \qquad \text{pentanol}$$

STEP 2 **Number the chain starting at the end nearer to the —OH group.** This carbon chain is numbered from right to left to give the position for the —OH group as carbon 2, which is shown as a prefix in the name 2-pentanol.

$$\begin{array}{ccccccccc} & & CH_3 & & & & OH & & \\ & & | & & & & | & & \\ CH_3 & - & CH & - & CH_2 & - & CH & - & CH_3 \\ 5 & & 4 & & 3 & & 2 & & 1 \end{array} \qquad \text{2-pentanol}$$

STEP 3 **Give the location and name for each substituent relative to the —OH group.** With a methyl group on carbon 4, the compound is named 4-methyl-2-pentanol.

$$\begin{array}{ccccccccc} & & CH_3 & & & & OH & & \\ & & | & & & & | & & \\ CH_3 & - & CH & - & CH_2 & - & CH & - & CH_3 \\ 5 & & 4 & & 3 & & 2 & & 1 \end{array} \qquad \text{4-methyl-2-pentanol}$$

STUDY CHECK 12.1

Give the IUPAC name for the following:

$$\begin{array}{ccccccccc} & & Cl & & & & & & \\ & & | & & & & & & \\ CH_3 & - & CH & - & CH_2 & - & CH_2 & - & OH \end{array}$$

SAMPLE PROBLEM 12.2 Naming Phenols

Give the IUPAC name for the following:

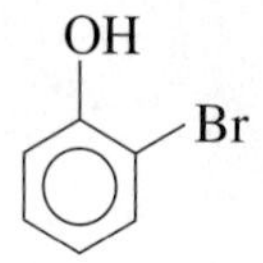

Guide to Naming Alcohols

STEP 1
Name the longest carbon chain attached to the —OH group by replacing the *e* in the corresponding alkane name with *ol*. Name an aromatic alcohol as a *phenol*.

STEP 2
Number the chain starting at the end nearer to the —OH group.

STEP 3
Give the location and name for each substituent relative to the —OH group.

Explore Your World

Alcohols in Household Products

Read the labels on household products such as sanitizers, mouthwashes, cold remedies, rubbing alcohol, and flavoring extracts. Look for names of alcohols, such as ethyl alcohol, isopropyl alcohol, thymol, and menthol.

Questions

1. What part of the name tells you that it is an alcohol?
2. What alcohol is usually meant by the term "alcohol"?
3. What is the percentage of alcohol in the products?
4. Draw the condensed structural formulas for the alcohols you find listed on the labels. You may need to use the Internet or a reference book for some structures.

SOLUTION

	Functional Group	Family	IUPAC Naming
ANALYZE THE PROBLEM	hydroxyl group on a benzene ring	phenol	phenol with substituent; number the ring to give lower numbers

STEP 1 Name an aromatic alcohol as a *phenol*. The compound is a *phenol* because it contains an —OH group attached to a benzene ring.

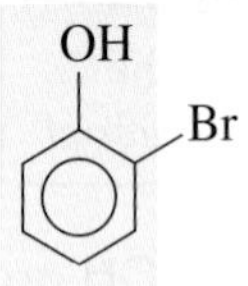

phenol

STEP 2 Number the chain starting at the end nearer to the —OH group. For a phenol, the carbon atom attached to the —OH group is carbon 1.

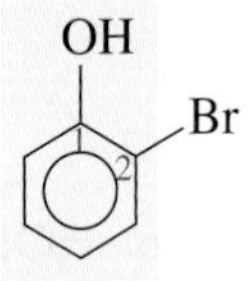

phenol

STEP 3 Give the location and name for each substituent relative to the —OH group. The IUPAC name for the compound is 2-bromophenol.

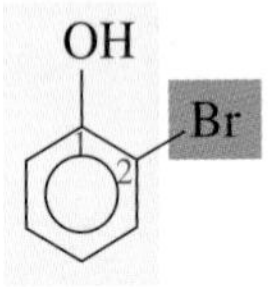

2-bromophenol

STUDY CHECK 12.2

Give the IUPAC name for the following:

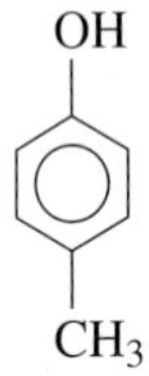

Chemistry Link to Health

Some Important Alcohols and Phenols

Methanol (*methyl alcohol*), the simplest alcohol, is found in many solvents and paint removers. If ingested, methanol is oxidized to formaldehyde, which can cause headaches, blindness, and death. Methanol is used to make plastics, medicines, and fuels. In car racing, it is used as a fuel because it is less flammable and has a higher octane rating than does gasoline.

Ethanol (*ethyl alcohol*) has been known since prehistoric times as an intoxicating product formed by the fermentation of grains, sugars, and starches.

$$C_6H_{12}O_6 \xrightarrow{\text{Fermentation}} 2CH_3{-}CH_2{-}OH + 2CO_2$$

Today, ethanol for commercial use is produced by reacting ethene and water at high temperatures and pressures. Ethanol is used as a solvent for perfumes, varnishes, and some medicines, such as tincture of iodine. Recent interest in alternative fuels has led to increased production of ethanol by the fermentation of sugars from grains such as corn, wheat, and rice. "Gasohol" is a mixture of ethanol and gasoline used as a fuel.

$$H_2C{=}CH_2 + H_2O \xrightarrow{300\ °C,\ 200\ atm,\ H^+} CH_3{-}CH_2{-}OH$$

1,2,3-Propanetriol (*glycerol* or *glycerin*), a trihydroxy alcohol, is a viscous liquid obtained from oils and fats during the production of soaps. The presence of several polar —OH groups makes it strongly attracted to water, a feature that makes glycerol useful as a skin softener in products such as skin lotions, cosmetics, shaving creams, and liquid soaps.

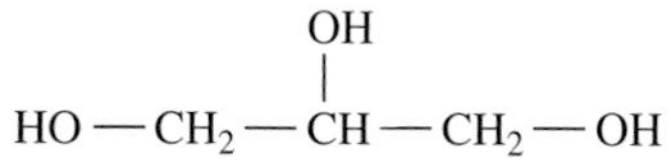

1,2,3-Propanetriol
(glycerol)

1,2-Ethanediol (*ethylene glycol*) is used as an antifreeze in heating and cooling systems. It is also a solvent for paints, inks, and plastics, and it is used in the production of synthetic fibers such as Dacron. If ingested, it is extremely toxic. In the body, it is oxidized to oxalic acid, which forms insoluble salts in the kidneys that cause renal damage, convulsions, and death. Because its sweet taste is attractive to pets and children, ethylene glycol solutions must be carefully stored.

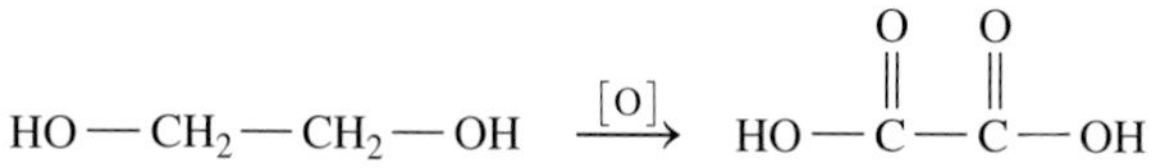

1,2-Ethanediol (ethylene glycol) — Oxalic acid

Antifreeze raises the boiling point and decreases the freezing point of water in a radiator.

Phenols are found in several of the essential oils of plants, which produce the odor or flavor of the plant. Eugenol is found in cloves, vanillin in vanilla bean, isoeugenol in nutmeg, and thymol in thyme and mint. Thymol has a pleasant, minty taste and is used in mouthwashes and by dentists to disinfect a cavity before adding a filling compound.

Bisphenol A (*BPA*) is used to make polycarbonate, a clear plastic that is used to manufacture beverage bottles, including baby bottles. Washing polycarbonate bottles with certain detergents or at high temperatures disrupts the polymer, causing small amounts of BPA to leach from the bottles. Because BPA is an estrogen mimic, there are concerns about the harmful effects from low levels of BPA. In 2008, Canada banned the use of polycarbonate baby bottles, which are now labeled "BPA free."

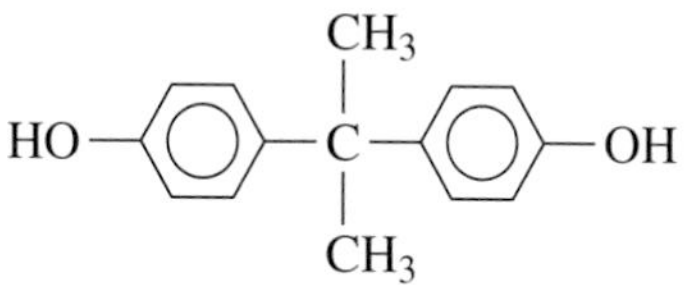

Bisphenol A (BPA)

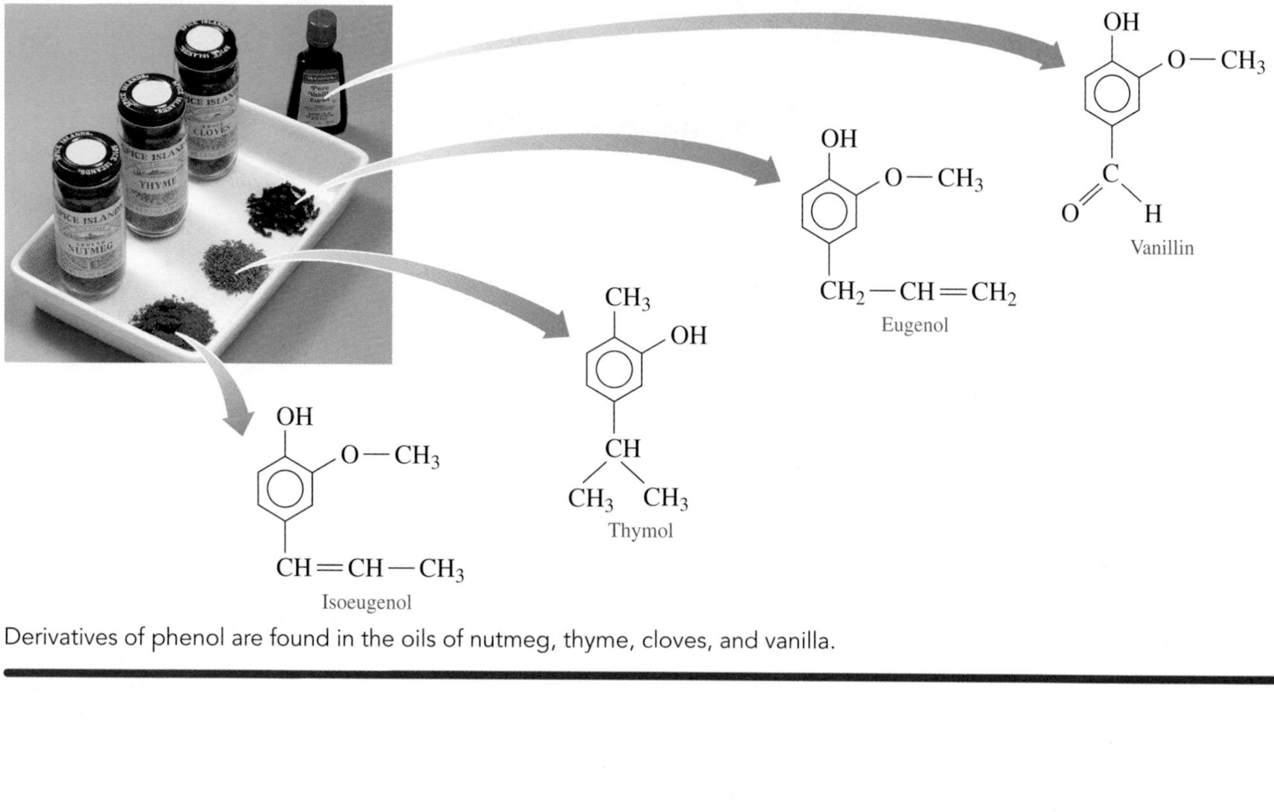

Derivatives of phenol are found in the oils of nutmeg, thyme, cloves, and vanilla.

The spray of a skunk contains a mixture of thiols.

Thiols

Thiols are a family of sulfur-containing organic compounds that have a *thiol* group ($-SH$). In the IUPAC system, thiols are named by adding *thiol* to the alkane name of the longest carbon chain and numbering the carbon chain from the end nearer the $-SH$ group.

CH_3-OH	CH_3-SH	CH_3-CH_2-SH	$CH_3-\underset{}{\overset{SH}{\overset{\vert}{C}H}}-CH_2-CH_3$
Methanol	Methanethiol	Ethanethiol	2-Butanethiol

An important property of thiols is a strong, sometimes disagreeable, odor. To help us detect natural gas (methane) leaks, a small amount of ethanethiol is added to the gas supply, which is normally odorless. There are thiols such as *trans*-2-butene-1-thiol in the spray emitted when a skunk senses danger.

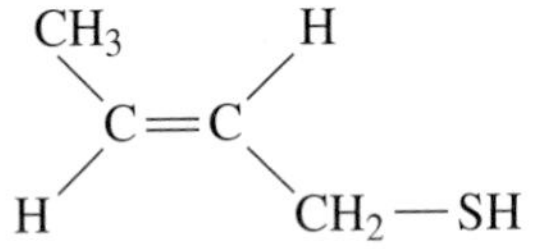

trans-2-Butene-1-thiol
(in skunk spray)

Methanethiol is the characteristic odor of oysters, cheddar cheese, onions, and garlic. Garlic also contains 2-propene-1-thiol. The odor of onions is due to 1-propanethiol, which is also a lachrymator, a substance that makes eyes tear.

Thiols are sulfur-containing compounds with $-SH$ groups, and often have strong odors.

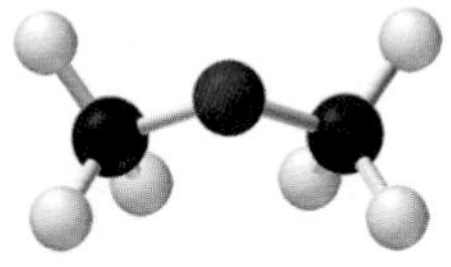

Dimethyl ether

Ethers

An ether contains an oxygen atom that is attached by single bonds to two carbon groups that are alkyls or aromatic rings. Ethers have a bent structure like water and alcohols, except both hydrogen atoms are replaced by carbon groups.

Naming Ethers

Most ethers have common names. The name of each alkyl or aromatic group attached to the oxygen atom is written in alphabetical order, followed by the word *ether*. In this text, we will use only the common names of ethers.

Methyl group ↓ Propyl group ↓

CH_3—O—CH_2—CH_2—CH_3

Common name: methyl propyl ether

Chemistry Link to Health

Ethers as Anesthetics

Anesthesia is the loss of sensation and consciousness. A general anesthetic is a substance that blocks signals to the awareness centers in the brain so the person has a loss of memory, a loss of feeling pain, and an artificial sleep. The term *ether* has been associated with anesthesia because diethyl ether was the most widely used anesthetic for more than a hundred years. Although it is easy to administer, ether is very volatile and highly flammable. A small spark in the operating room could cause an explosion. Since the 1950s, anesthetics such as Forane (isoflurane), Ethrane (enflurane), and Penthrane (methoxyflurane) have been developed that are not as flammable. Most of these anesthetics retain the ether group, but the addition of halogen atoms reduces the volatility and flammability of the ethers. More recently, they have been replaced by Fluothane (halothane) (2-bromo-2-chloro-1,1,1-trifluoroethane) and Suprane (desflurane) (1,2,2,2-tetrafluoroethyl difluoromethyl ether) because of the negative side effects of the ether-type inhalation anesthetics, such as toxicity to the liver and kidneys, and arrhythmia.

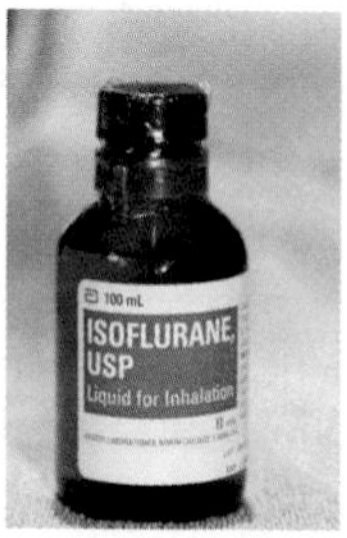

Isoflurane (Forane) is an inhaled anesthetic.

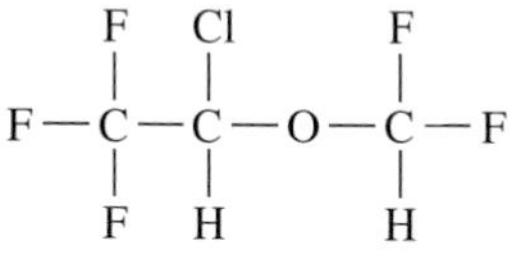

Forane (isoflurane)

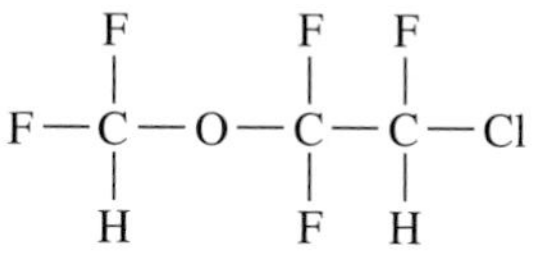

Ethrane (enflurane)

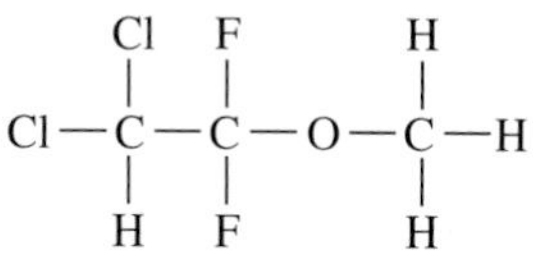

Penthrane (methoxyflurane)

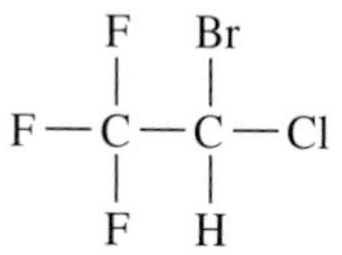

Fluothane (halothane)

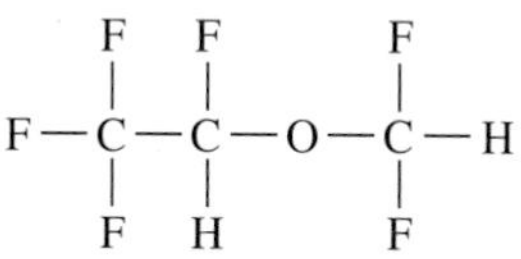

Suprane (desflurane)

QUESTIONS AND PROBLEMS

12.1 Alcohols, Phenols, Thiols, and Ethers

LEARNING GOAL Give the IUPAC and common names for alcohols and phenols; give the common names for thiols and ethers. Draw their condensed structural formulas or skeletal formulas.

12.1 State the IUPAC name for each of the following:

a. 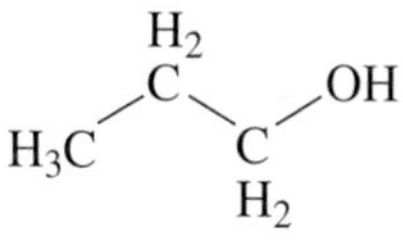

b. 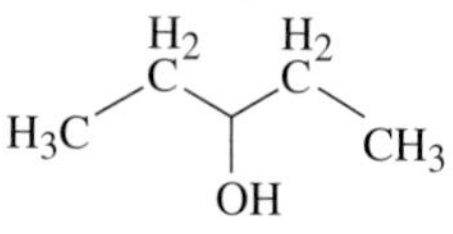

c. 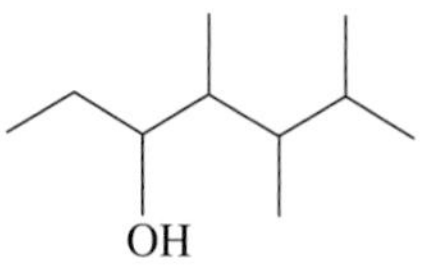

d. 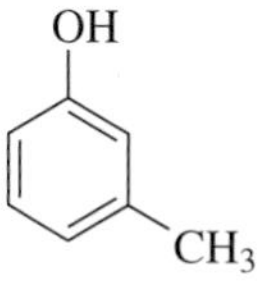

12.2 Give the IUPAC name for each of the following:

a. 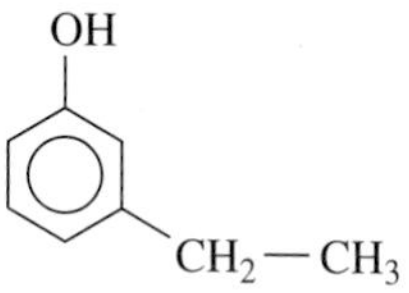

b. 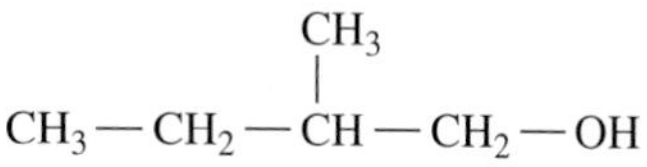

c. 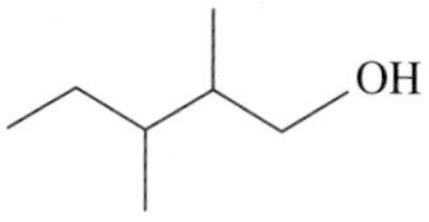

d. 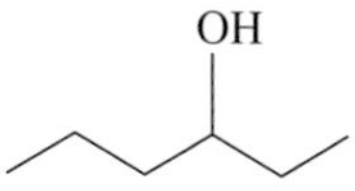

12.3 Draw the condensed structural formula for each of the following:
a. butyl alcohol
b. 3-methyl-1-pentanol
c. 3-hexanol
d. 2-bromophenol

12.4 Draw the condensed structural formula for each of the following:
a. ethanol
b. 3-methyl-2-iodo-1-butanol
c. 2-propyn-1-ol (propargyl alcohol)
d. 2-fluorophenol

12.5 Give a common name for each of the following:
a. $H_3C-O-CH_3$
b.

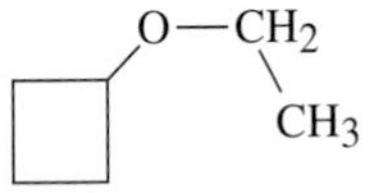

12.6 Give a common name for each of the following:
a. $CH_3-CH_2-O-CH_2-CH_2-CH_3$
b.

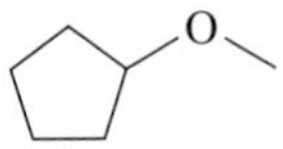

12.7 Draw the skeletal formula for each of the following:
a. ethyl methyl thioether
b. tetrahydrofuran (THF)

12.8 Draw the condensed structural formula, or skeletal formula if cyclic, for each of the following:
a. diethyl ether
b. cyclobutyl methyl ether

LEARNING GOAL

Describe the classification of alcohols; describe the solubility of alcohols in water.

12.2 Properties of Alcohols

Alcohols are classified by the number of alkyl groups attached to the carbon atom bonded to the hydroxyl group (—OH). A **primary (1°) alcohol** has one alkyl group attached to the carbon atom bonded to the —OH group, a **secondary (2°) alcohol** has two alkyl groups, and a **tertiary (3°) alcohol** has three alkyl groups.

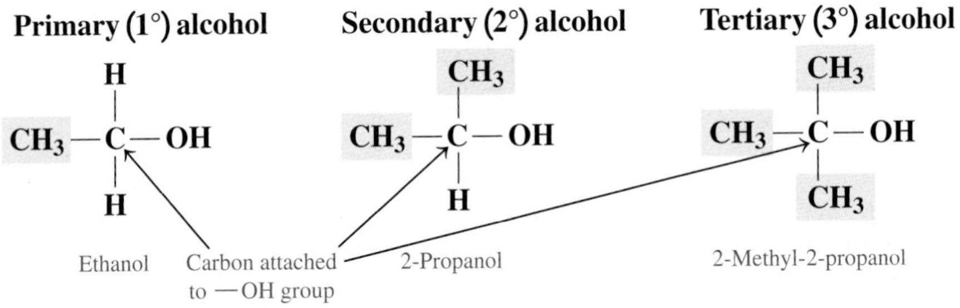

SAMPLE PROBLEM 12.3 Classifying Alcohols

Classify each of the following alcohols as primary (1°), secondary (2°), or tertiary (3°):

a. $CH_3-CH_2-CH_2-OH$

b.

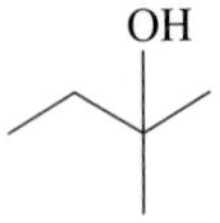

SOLUTION

a. The carbon atom bonded to the —OH group is attached to one alkyl group, which makes this a primary (1°) alcohol.
b. The carbon atom bonded to the —OH group is attached to three alkyl groups, which makes this a tertiary (3°) alcohol.

STUDY CHECK 12.3

Classify the following alcohol as primary (1°), secondary (2°), or tertiary (3°):

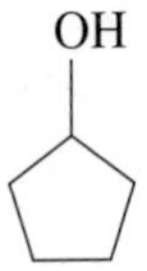

Solubility of Alcohols in Water

Hydrocarbons, which are composed of only carbon and hydrogen, are nonpolar and insoluble in water. However, the polar —OH group in an alcohol can form hydrogen bonds with the H and O atoms of water, which makes alcohols more soluble in water (see Figure 12.2).

Alcohols with one to three carbon atoms are miscible with water, which means any amount of the alcohol is completely soluble in water. However, the solubility provided by the —OH group decreases as the number of carbon atoms increases. Alcohols with four carbon atoms are slightly soluble in water and alcohols with five or more carbon atoms are insoluble. Table 12.1 compares the solubility of some alcohols.

Nonpolar carbon chain → $CH_3—CH_2$ —OH
Soluble in water

$CH_3—CH_2—CH_2—CH_2—CH_2—CH_2—CH_2—CH_2$ —OH
Insoluble in water

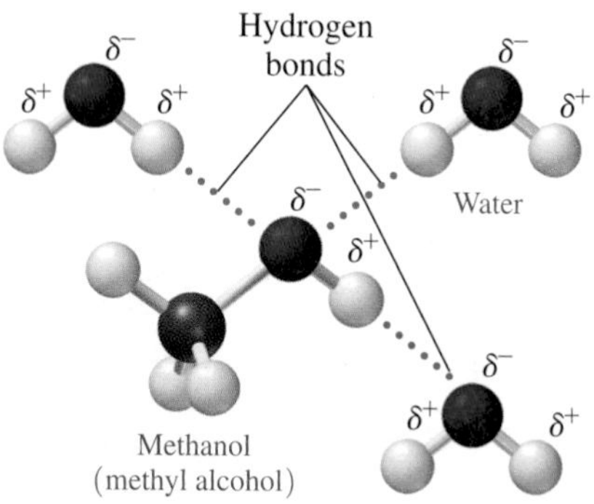

FIGURE 12.2 ▶ Methanol forms hydrogen bonds with water molecules.

Q Why is methanol more soluble in water than 1-pentanol?

TABLE 12.1 Solubility of Some Alcohols

Compound	Condensed Structural Formula	Number of Carbon Atoms	Solubility in Water
Methanol	$CH_3—OH$	1	Soluble
Ethanol	$CH_3—CH_2—OH$	2	Soluble
1-Propanol	$CH_3—CH_2—CH_2—OH$	3	Soluble
1-Butanol	$CH_3—CH_2—CH_2—CH_2—OH$	4	Slightly soluble
1-Pentanol	$CH_3—CH_2—CH_2—CH_2—CH_2—OH$	5	Insoluble

Chemistry Link to **Health**

Hand Sanitizers and Ethanol

Hand sanitizers are used as an alternative to washing hands to kill most bacteria and viruses that spread colds and flu. Many hand sanitizers use ethanol as their active ingredient. While safe for most adults, supervision is recommended when used by children because there is some concern about their risk to the health of children. After using an alcohol-based hand sanitizer, children might ingest enough alcohol to make them sick.

In an alcohol-containing sanitizer, the amount of ethanol is typically 60% (v/v) but can be as high as 85%. This amount of ethanol can make hand sanitizers a fire hazard in the home because they are highly flammable. When ethanol undergoes combustion, it produces a transparent blue flame. When using an ethanol-containing sanitizer, it is important to rub hands until they are completely dry. It is also recommended that sanitizers containing ethanol be placed in storage areas that are away from heat sources in the home.

Some sanitizers are alcohol-free, but often the active ingredient is triclosan, which contains aromatic, ether, and phenol functional groups. However, the Food and Drug Administration is considering banning triclosan in personal-care products because when mixed with tap water for disposal, the triclosan that accumulates in the environment may promote growth of antibiotic-resistant bacteria.

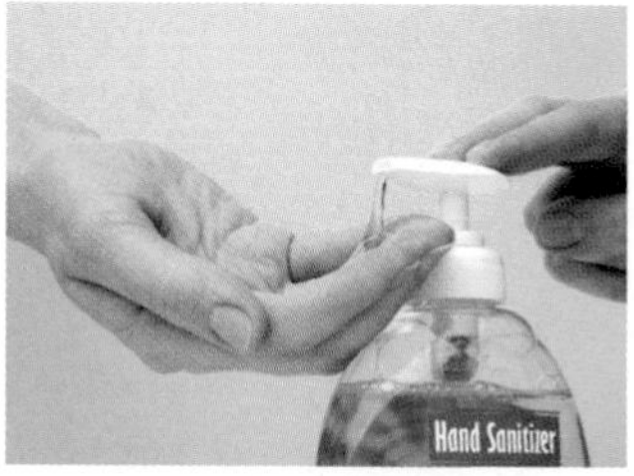

Hand sanitizers that contain ethanol are used to kill bacteria on the hands.

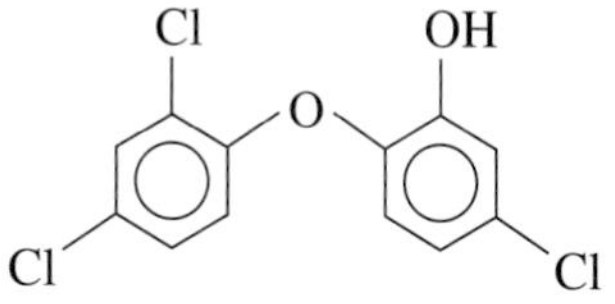

Triclosan is an antibacterial compound used in personal-care products.

Solubility of Phenols

Phenol is slightly soluble in water because the —OH group can form hydrogen bonds with water molecules. In water, the —OH group ionizes slightly, which makes phenol a weak acid. Phenol is very corrosive and highly irritating to the skin; it can cause severe burns and ingestion can be fatal. Joseph Lister (1827–1912) is considered a pioneer in antiseptic surgery and was the first to sterilize surgical instruments and dressings with phenol, which was initially named *carbolic acid*. An antiseptic is a substance applied to the skin to kill microorganisms that cause infection. Phenol was also used to disinfect wounds to prevent post-surgical infections such as gangrene. However, use of phenol solution did not last long due to skin irritation and was replaced with other disinfectants and eventually antibiotics used topically or internally.

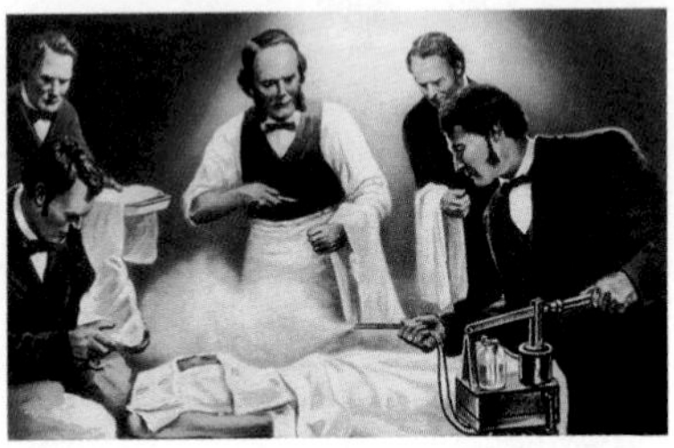

Joseph Lister used an antiseptic spray of phenol to disinfect wounds.

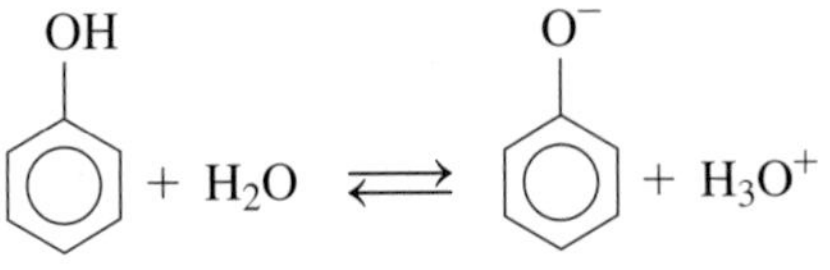

Phenol Phenoxide ion

QUESTIONS AND PROBLEMS

12.2 Properties of Alcohols

LEARNING GOAL Describe the classification of alcohols; describe the solubility of alcohols in water.

12.9 Classify each of the following alcohols as primary (1°), secondary (2°), or tertiary (3°):

a. 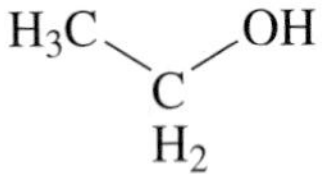

b. 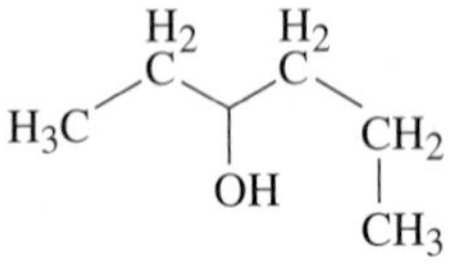

c. 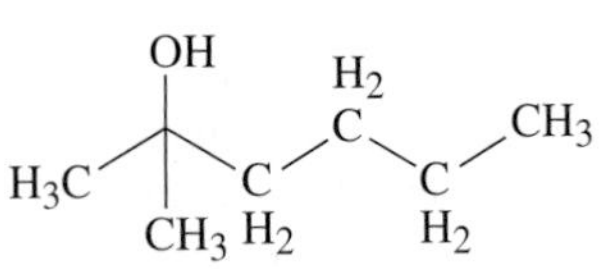

d. 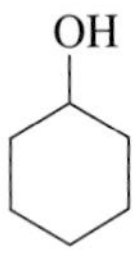

12.10 Classify each of the following alcohols as primary (1°), secondary (2°), or tertiary (3°):

a. 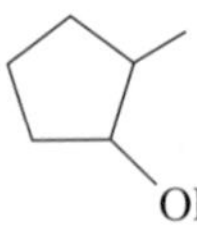

b. 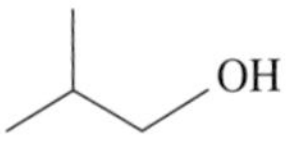

c. 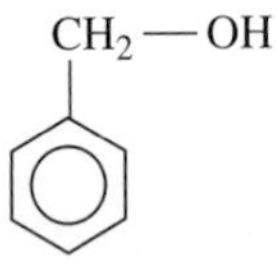

d. 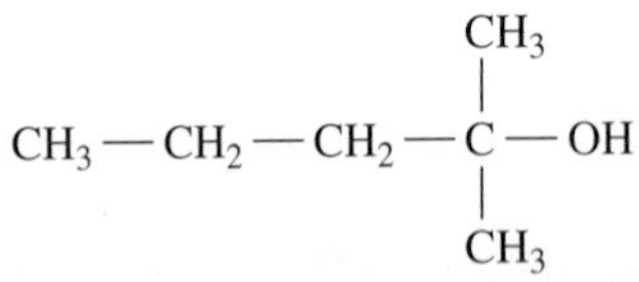

12.11 Are each of the following soluble, slightly soluble, or insoluble in water?

a. $CH_3—OH$

b. $CH_3—CH_2—CH_2—CH_2—CH_2—CH_2—CH_2—OH$

12.12 Are each of the following soluble, slightly soluble, or insoluble in water?

a. $CH_3—CH_2—CH_2—CH_2—CH_2—CH_2$

b. $CH_3—CH_2—OH$

12.13 Give an explanation for the following observations:

a. Ethanol is soluble in water, but xylene is not.

b. 1-Propanol is more soluble in water than 1-octanol.

12.14 Give an explanation for the following observations:

a. Ethanol is soluble in water, but propane is not.

b. 1-Propanol is soluble in water, but 1-hexanol is not.

12.3 Aldehydes and Ketones

LEARNING GOAL

Write the IUPAC and common names for aldehydes and ketones; draw the condensed structural formulas. Describe the solubility of aldehydes and ketones in water.

Aldehydes and ketones contain a **carbonyl group** that consists of a carbon–oxygen double bond with two groups of atoms attached to the carbon at angles of 120°. The double bond in the carbonyl group is similar to that of alkenes, except the carbonyl group has a dipole. The oxygen atom with two lone pairs of electrons is much more electronegative than the carbon atom. Therefore, the carbonyl group has a strong dipole with a partial negative charge (δ^-) on the oxygen and a partial positive charge (δ^+) on the carbon. The polarity of the carbonyl group strongly influences the physical and chemical properties of aldehydes and ketones.

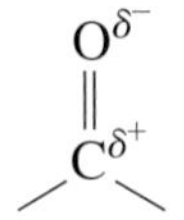

In an **aldehyde**, the carbon of the carbonyl group is bonded to at least one hydrogen atom. That carbon may also be bonded to another hydrogen atom, a carbon of an alkyl group, or an aromatic ring (see Figure 12.3).

The aldehyde group may be written as separate atoms or as —CHO, with the double bond understood. In a **ketone**, the carbonyl group is bonded to two alkyl groups or aromatic rings. The keto group (C═O) can sometimes be written as CO. A skeletal formula may also be used to represent an aldehyde or ketone.

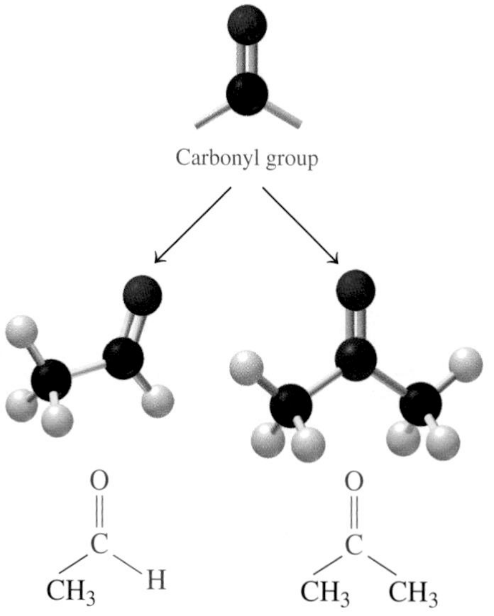

FIGURE 12.3 ▶ The carbonyl group is found in aldehydes and ketones.

Q If aldehydes and ketones both contain a carbonyl group, how can you differentiate between compounds from each family?

Formulas for C_3H_6O

$CH_3—CH_2—C(=O)—H$ = $CH_3—CH_2—CHO$ = 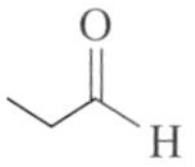

Aldehyde

$CH_3—C(=O)—CH_3$ = $CH_3—CO—CH_3$ = 

Ketone

Naming Aldehydes

CORE CHEMISTRY SKILL

Naming Aldehydes and Ketones

In the IUPAC system, an aldehyde is named by replacing the *e* of the corresponding alkane name with *al*. No number is needed for the aldehyde group because it always appears at the end of the chain. The aldehydes with carbon chains of one to four carbons are often referred to by their common names, which end in *aldehyde* (see Figure 12.4). The roots (*form*, *acet*, *propion*, and *butyr*) of these common names are derived from Latin or Greek words.

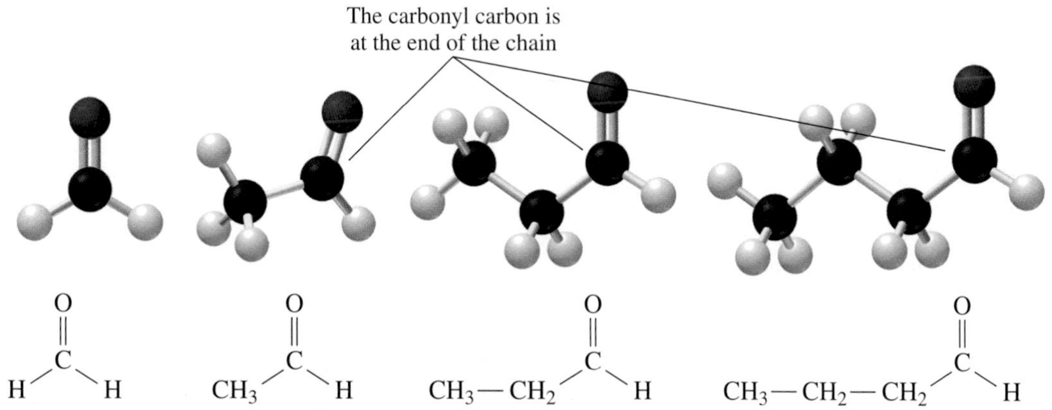

IUPAC	Methanal	Ethanal	Propanal	Butanal
Common	(**form**aldehyde)	(**acet**aldehyde)	(**propion**aldehyde)	(**butyr**aldehyde)

FIGURE 12.4 ▶ In the structures of aldehydes, the carbonyl group is always the end carbon.

Q Why is the carbon in the carbonyl group in aldehydes always at the end of the chain?

The IUPAC system names the aldehyde of benzene as benzaldehyde.

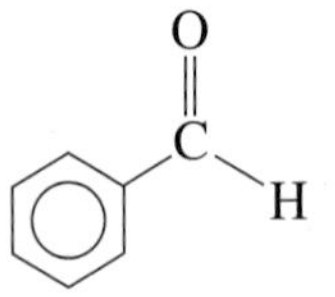

Benzaldehyde

SAMPLE PROBLEM 12.4 Naming Aldehydes

Give the IUPAC name for the following:

$$\begin{array}{c} \overset{\displaystyle CH_3}{\overset{|}{}}\overset{\displaystyle O}{\overset{\|}{}} \\ CH_3-CH-CH_2-C-H \end{array}$$

SOLUTION

ANALYZE THE PROBLEM	Functional Group	Family	IUPAC Naming
	carbonyl group	aldehyde	change *e* to *al* and count the carbonyl as carbon 1

STEP 1 **Name the longest carbon chain by replacing the *e* in the alkane name with *al*.** The longest chain has four carbon atoms including the carbon of the carbonyl group, which is named butanal in the IUPAC system.

$$CH_3-\overset{\displaystyle CH_3}{\underset{}{\overset{|}{C}H}}-CH_2-\overset{\displaystyle O}{\overset{\|}{C}}-H \qquad \text{butanal}$$

STEP 2 **Name and number any substituents by counting the carbonyl group as carbon 1.** The substituent, which is the $-CH_3$ group on carbon 3, is methyl. The IUPAC name for this compound is 3-methylbutanal.

$$\underset{4}{CH_3}-\underset{3}{\overset{\displaystyle CH_3}{\overset{|}{C}H}}-\underset{2}{CH_2}-\underset{1}{\overset{\displaystyle O}{\overset{\|}{C}}}-H \qquad \text{3-methylbutanal}$$

Guide to Naming Aldehydes

STEP 1
Name the longest carbon chain by replacing the *e* in the alkane name with *al*.

STEP 2
Name and number any substituents by counting the carbonyl group as carbon 1.

STUDY CHECK 12.4

What are the IUPAC and common names of the aldehyde with three carbon atoms?

Chemistry Link to the **Environment**

Vanilla

Vanilla has been used as a flavoring for over a thousand years. After drinking a beverage made from powdered vanilla and cocoa beans with Emperor Montezuma in Mexico, the Spanish conquistador Hernán Cortés took vanilla back to Europe where it became popular for flavoring and for scenting perfumes and tobacco. Thomas Jefferson introduced vanilla to the United States during the late 1700s. Today much of the vanilla we use in the world is grown in Mexico, Madagascar, Réunion, Seychelles, Tahiti, Sri Lanka, Java, the Philippines, and Africa.

The vanilla plant is a member of the orchid family and thrives under tropical conditions. There are many species of *Vanilla*, but *Vanilla planifolia* (or *V. fragrans*) is considered to produce the best flavor. The vanilla plant grows like a vine, which can attain a height of 100 feet. Its flowers are hand pollinated to produce a green fruit that is picked in 8 or 9 months. The fruit is sun-dried so that it becomes a long, dark brown pod, which is called a "vanilla bean" because it looks like a string bean. The flavor and fragrance of the vanilla bean come from the black seeds found inside the dried bean.

The vanilla bean is the dried fruit of the vanilla plant.

The seeds and pod are used to flavor desserts such as custards and ice cream. The extract of vanilla is made by chopping up vanilla beans and mixing them with a 35% ethanol–water mixture. The liquid, which contains the aldehyde *vanillin*, is drained from the bean residue and used for flavoring.

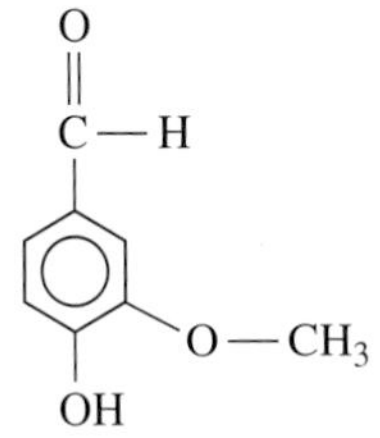

Vanillin

Vanilla flavoring liquid is prepared by soaking vanilla beans in ethanol and water.

Naming Ketones

Aldehydes and ketones are some of the most important classes of organic compounds. Because they have played a major role in organic chemistry for more than a century, the common names for unbranched ketones are still in use. In the common names, the alkyl groups bonded to the carbonyl group are named as substituents and are listed alphabetically, followed by *ketone*. Acetone, which is another name for propanone, has been retained by the IUPAC system.

In the IUPAC system, the name of a ketone is obtained by replacing the *e* in the corresponding alkane name with *one*. Carbon chains with five carbon atoms or more are numbered from the end nearer the carbonyl group.

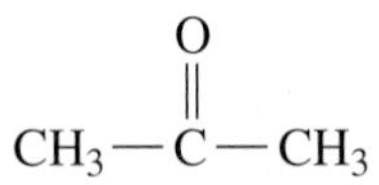

Propanone
(dimethyl ketone; acetone)

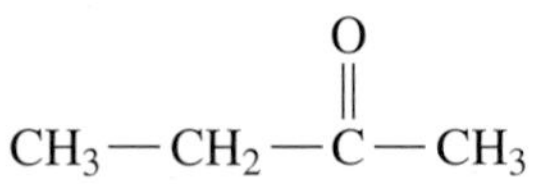

Butanone
(ethyl methyl ketone)

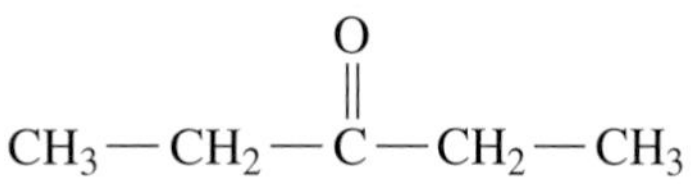

3-Pentanone
(diethyl ketone)

In a cyclic ketone, the carbonyl carbon is numbered as carbon 1. The ring is numbered in the direction to give substituents the lowest possible numbers.

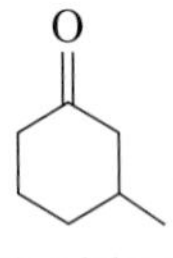

3-Methylcyclohexanone

SAMPLE PROBLEM 12.5 Naming Ketones

Give the IUPAC name for the following ketone:

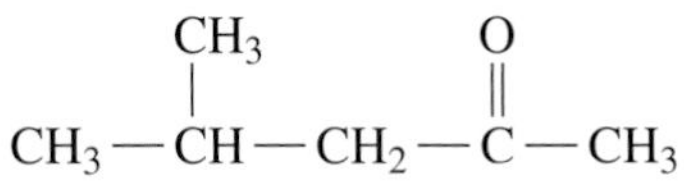

SOLUTION

	Functional Group	Family	IUPAC Naming
ANALYZE THE PROBLEM	carbonyl group	ketone	change e to *one* and count from the end of the chain nearer the carbonyl group

Guide to Naming Ketones

STEP 1
Name the longest carbon chain by replacing the *e* in the alkane name with *one.*

STEP 2
Number the carbon chain starting from the end nearer the carbonyl group and indicate its location.

STEP 3
Name and number any substituents on the carbon chain.

STEP 1 **Name the longest carbon chain by replacing the *e* in the alkane name with *one*.** The longest chain has five carbon atoms, which is named pentanone.

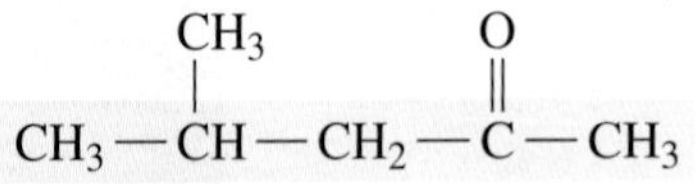
pentanone

STEP 2 **Number the carbon chain starting from the end nearer the carbonyl group and indicate its location.** Counting from the right, the carbonyl group is on carbon 2.

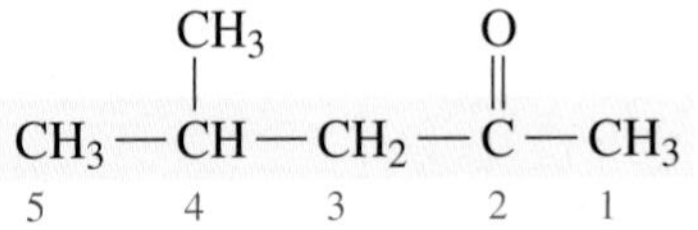
2-pentanone

STEP 3 **Name and number any substituents on the carbon chain.** Counting from the right, the methyl group is on carbon 4. The IUPAC name is 4-methyl-2-pentanone.

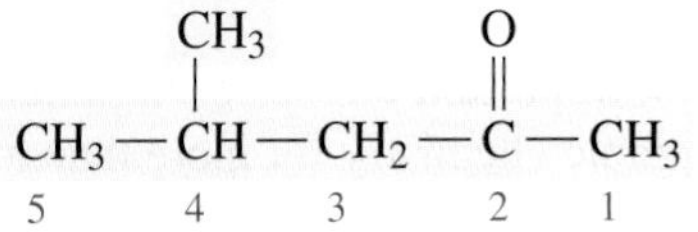
4-methyl-2-pentanone

STUDY CHECK 12.5

What is the common name of 3-hexanone?

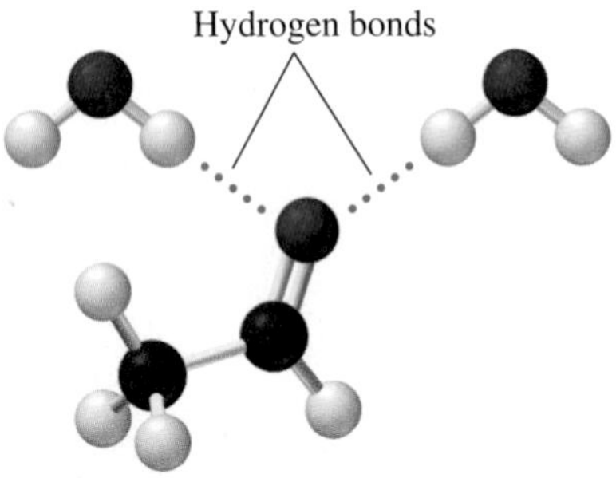

Ethanal (acetaldehyde) in water

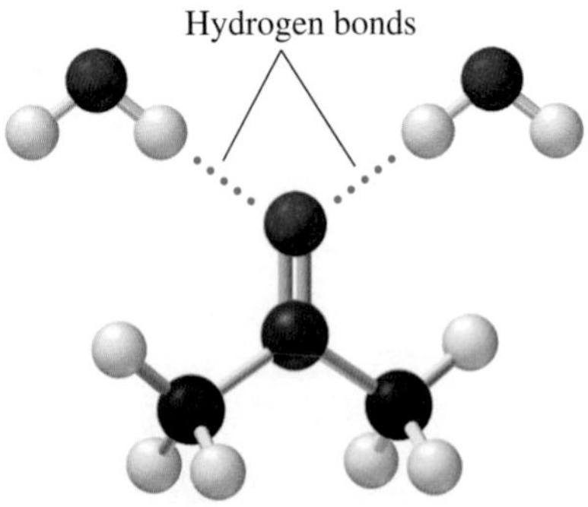

Propanone (acetone) in water

Properties of Aldehydes and Ketones in Water

Aldehydes and ketones contain a polar carbonyl group (carbon–oxygen double bond), which has a partially negative oxygen atom and a partially positive carbon atom. Because the electronegative oxygen atom forms hydrogen bonds with water molecules, aldehydes and ketones with one to four carbons are very soluble. However, aldehydes and ketones with five or more carbon atoms are not soluble because longer hydrocarbon chains, which are nonpolar, diminish the solubility effect of the polar carbonyl group. Table 12.2 compares the solubility of some aldehydes and ketones.

TABLE 12.2 Solubility of Selected Aldehydes and Ketones

Compound	Condensed Structural Formula	Number of Carbon Atoms	Solubility in Water
Methanal (formaldehyde)	$H—CHO$	1	Soluble
Ethanal (acetaldehyde)	$CH_3—CHO$	2	Soluble
Propanal (propionaldehyde)	$CH_3—CH_2—CHO$	3	Soluble
Propanone (acetone)	$CH_3—CO—CH_3$	3	Soluble
Butanal (butyraldehyde)	$CH_3—CH_2—CH_2—CHO$	4	Soluble
Butanone	$CH_3—CO—CH_2—CH_3$	4	Soluble
Pentanal	$CH_3—CH_2—CH_2—CH_2—CHO$	5	Slightly soluble
2-Pentanone	$CH_3—CO—CH_2—CH_2—CH_3$	5	Slightly soluble
Hexanal	$CH_3—CH_2—CH_2—CH_2—CH_2—CHO$	6	Not soluble
2-Hexanone	$CH_3—CO—CH_2—CH_2—CH_2—CH_3$	6	Not soluble

Chemistry Link to Health

Some Important Aldehydes and Ketones

Formaldehyde, the simplest aldehyde, is a colorless gas with a pungent odor. An aqueous solution called *formalin*, which contains 40% formaldehyde, is used as a germicide and to preserve biological specimens. Industrially, it is a reactant in the synthesis of polymers used to make fabrics, insulation materials, carpeting, pressed wood products such as plywood, and plastics for kitchen counters. Exposure to formaldehyde fumes can irritate the eyes, nose, and upper respiratory tract and cause skin rashes, headaches, dizziness, and general fatigue.

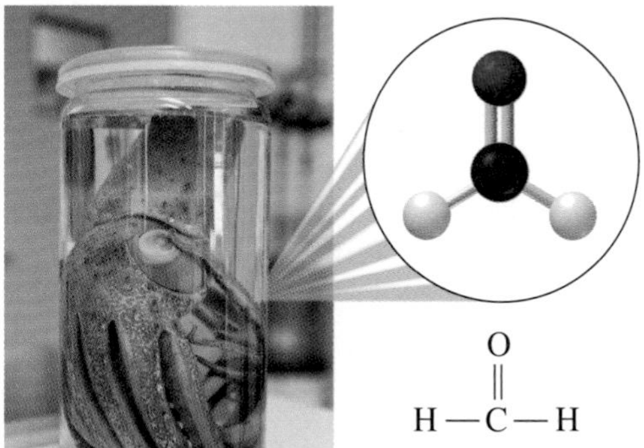

Methanal (formaldehyde)

The simplest ketone, known as *acetone* or propanone (dimethyl ketone), is a colorless liquid with a mild odor that has wide use as a solvent in cleaning fluids, paint and nail polish removers, and rubber cement (see Figure 12.5). Acetone is extremely flammable, and care must be taken when using it. In the body, acetone may be produced in uncontrolled diabetes, fasting, and high-protein diets when large amounts of fats are metabolized for energy.

Muscone is used to make musk perfumes, and oil of spearmint contains carvone.

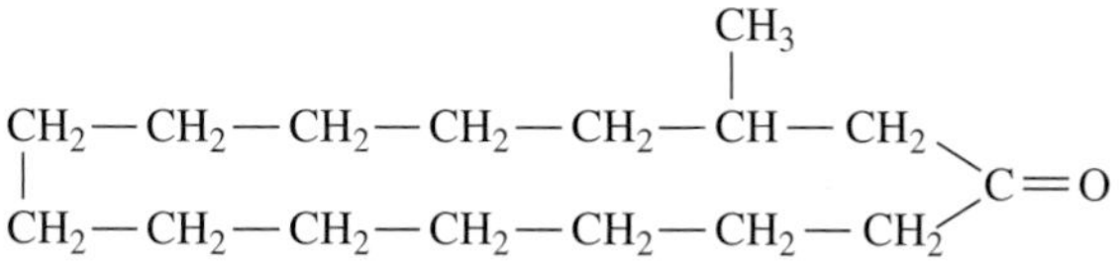

Muscone (musk)

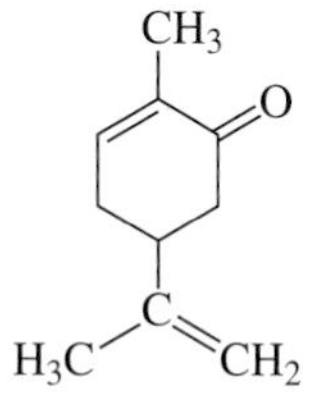

Carvone (spearmint oil)

Several naturally occurring aromatic aldehydes are used to flavor food and as fragrances in perfumes. Benzaldehyde is found in almonds, vanillin in vanilla beans, and cinnamaldehyde in cinnamon.

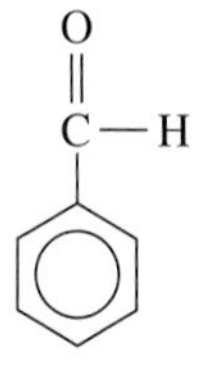

Benzaldehyde (almond)

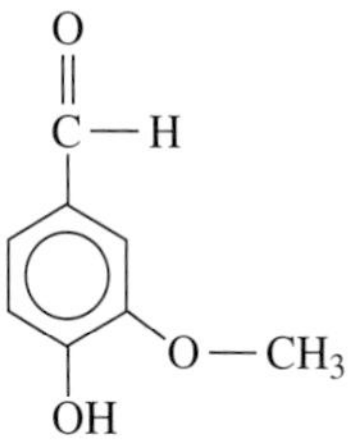

Vanillin (vanilla)

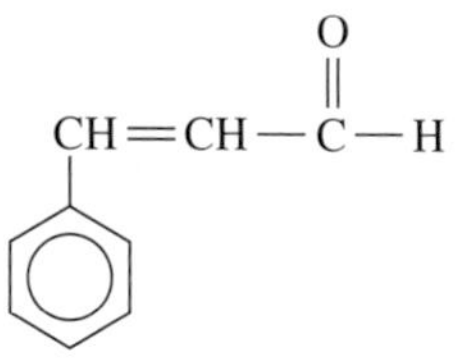

Cinnamaldehyde (cinnamon)

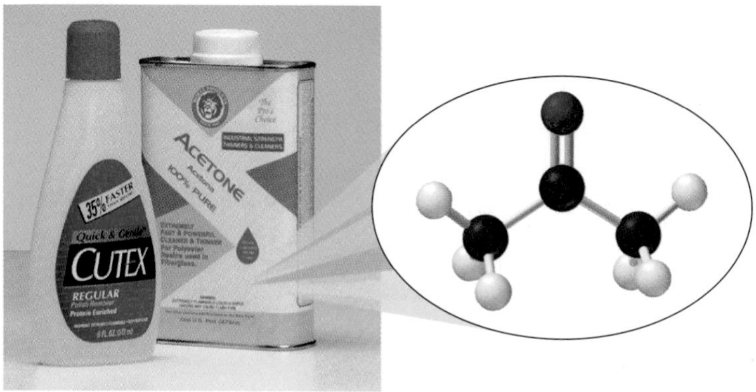

FIGURE 12.5 ▶ Acetone is used as a solvent in paint and nail polish removers.

Q What is the IUPAC name for acetone?

SAMPLE PROBLEM 12.6 Solubility of Ketones

Why is acetone soluble in water, but 2-hexanone is not?

SOLUTION

Acetone contains a carbonyl group with an electronegative oxygen atom that forms hydrogen bonds with water. 2-Hexanone also contains a carbonyl group, but its long, nonpolar hydrocarbon chain makes it insoluble in water.

STUDY CHECK 12.6

Would you expect ethanol ($CH_3—CH_2—OH$) to be more or less soluble in water than ethanal ($CH_3—CHO$)? Explain.

QUESTIONS AND PROBLEMS

12.3 Aldehydes and Ketones

LEARNING GOAL Write the IUPAC and common names for aldehydes and ketones; draw the condensed structural formulas. Describe the solubility of aldehydes and ketones in water.

12.15 Give the common name for each of the following:

a. 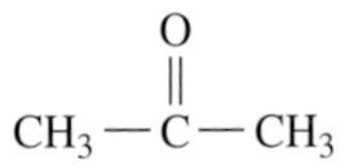

b. 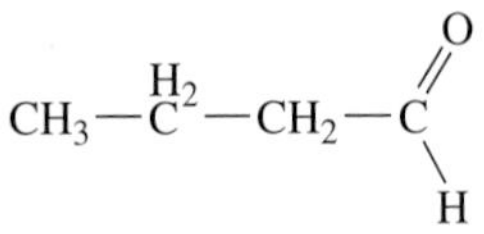

c. 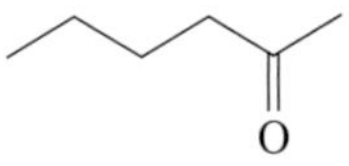

12.16 Give the common name for each of the following:

a. 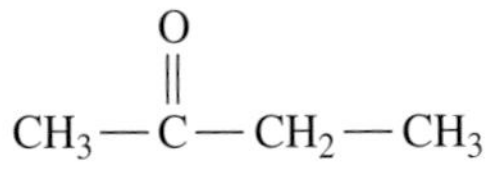

b. 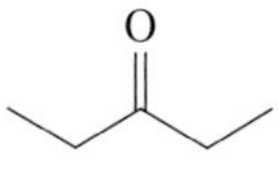

c. 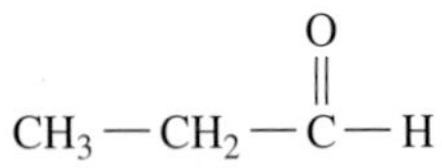

12.17 Give the IUPAC name for each of the following:

a. 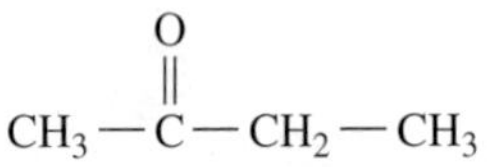

b. 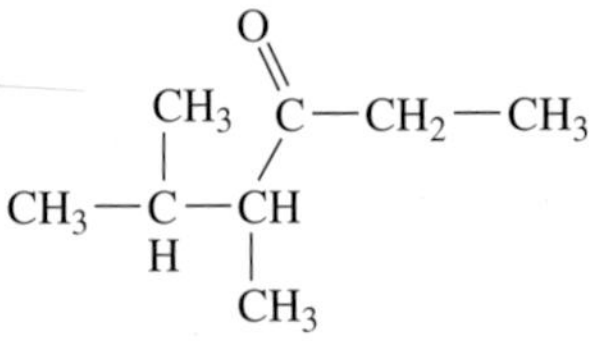

c. 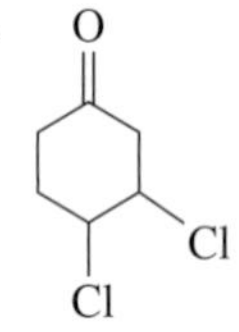

d. 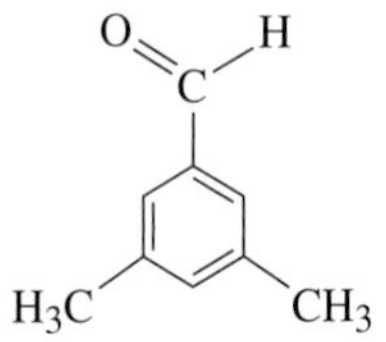

12.18 Give the IUPAC name for each of the following:

a. 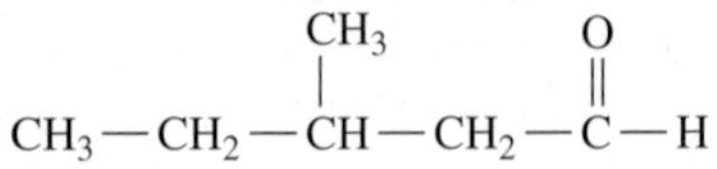

b. 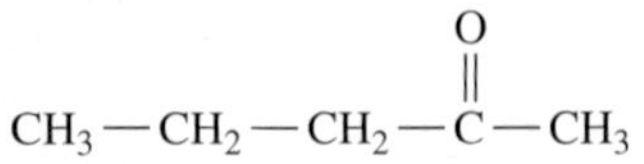

c. 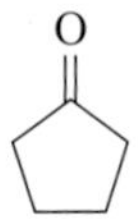

d. 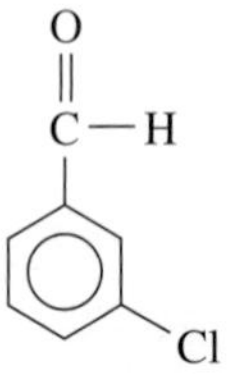

12.19 Draw the condensed structural formula for each of the following:
- **a.** formaldehyde
- **b.** 3-chloro-2-hexanone
- **c.** propyl ethyl ketone
- **d.** 4-methylpentanal

12.20 Draw the skeletal formula for each of the following:
- **a.** formaldehyde
- **b.** 2,3-dichlorobutanal
- **c.** cyclohexanone
- **d.** acetone (propanone)

12.21 Which compound in each of the following pairs would be more soluble in water? Explain.
- **a.** ethanol or hexane
- **b.** butanal or 3-hexanone
- **c.** CH_3-CH_2-CHO or $CH_3-CH_2-CH_2-CH_2-CH_2-CH_2-CH_3-CHO$

12.22 Which compound in each of the following pairs would be more soluble in water? Explain.
- **a.** $CH_3-CH_2-CH_3$ or CH_3-CH_2-CHO
- **b.** propanone or 3-hexanone
- **c.** butanal or hexanal

LEARNING GOAL

Write balanced chemical equations for the combustion, dehydration, and oxidation of alcohols. Write balanced chemical equations for the oxidation and reduction of thiols, aldehydes, and ketones.

12.4 Reactions of Alcohols, Thiols, Aldehydes, and Ketones

Alcohols, similar to hydrocarbons, undergo combustion in the presence of oxygen. For example, in a restaurant, a flaming dessert may be prepared by pouring liquor on fruit or ice cream and lighting it (see Figure 12.6). The combustion of the ethanol in the liquor proceeds as follows:

$$CH_3-CH_2-OH(g) + 3O_2(g) \xrightarrow{\Delta} 2CO_2(g) + 3H_2O(g) + \text{energy}$$

Dehydration of Alcohols to Form Alkenes

In a **dehydration** reaction, alcohols lose a water molecule when they are heated with an acid catalyst such as H_2SO_4. The components H— and —OH are removed from *adjacent carbon atoms of the same alcohol* to produce a water molecule. A double bond forms between the same two carbon atoms to produce an alkene product:

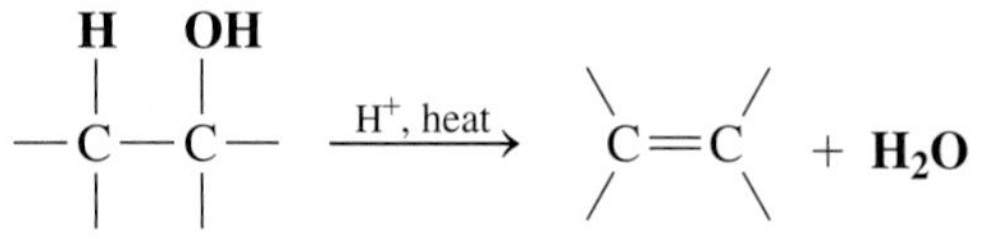

Alcohol Alkene Water

FIGURE 12.6 ▶ A flaming dessert is prepared using liquor that undergoes combustion.

Q What is the equation for the complete combustion of the ethanol in the liquor?

Examples

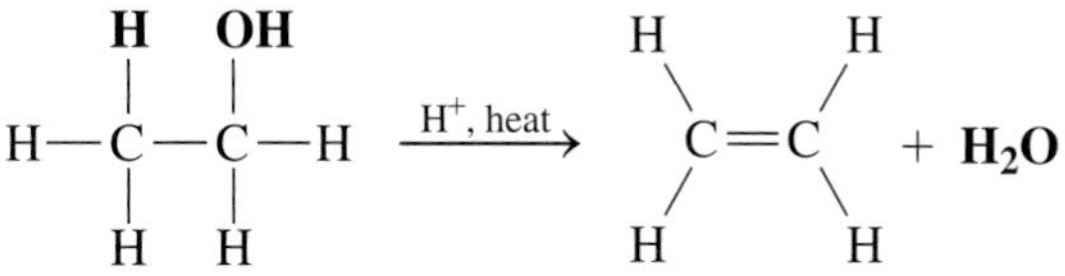

Ethanol Ethene

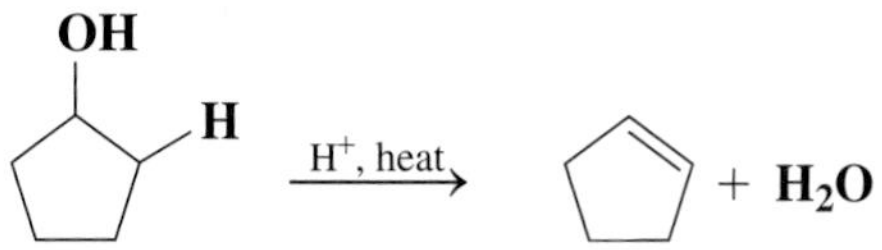

Cyclopentanol Cyclopentene

SAMPLE PROBLEM 12.7 Dehydration of Alcohols

Draw the condensed structural formula for the alkene produced by the dehydration of the following:

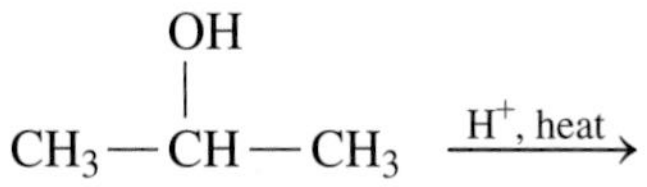

SOLUTION

In a dehydration, the —OH is removed from carbon 2 and one H— is removed from an adjacent carbon, which forms a double bond.

$$CH_3-CH=CH_2$$

STUDY CHECK 12.7

What is the name of the alkene produced by the dehydration of cyclohexanol?

Oxidation of Alcohols

CORE CHEMISTRY SKILL

Writing Equations for the Dehydration and Oxidation of Alcohols

In organic chemistry, an **oxidation** involves the addition of oxygen or a loss of hydrogen atoms. As a result, there is an increase in the number of carbon–oxygen bonds. In a *reduction* reaction, the product has fewer bonds between carbon and oxygen.

No bonds to O 1 bond to O 2 bonds to O 3 bonds to O

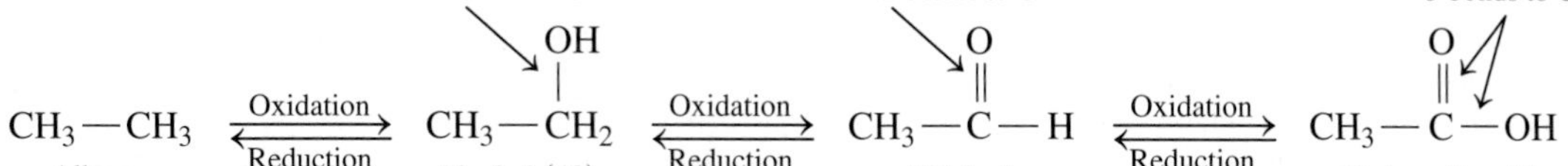

Alkane Alcohol (1°) Aldehyde Carboxylic acid

The oxidation of a primary alcohol produces an aldehyde, which contains a double bond between carbon and oxygen. For example, the oxidation of methanol and ethanol occurs by removing two hydrogen atoms, one from the —OH group and another from the carbon that is bonded to the —OH group. To indicate the presence of an oxidizing agent, the symbol [O] is written over the arrow to indicate that O in the H_2O product is obtained from the oxidizing agent. The oxidized product contains the same number of carbon atoms as the reactant.

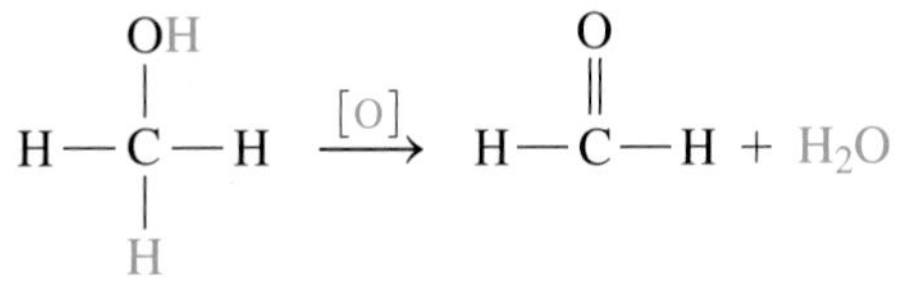

Methanol (methyl alcohol) — Methanal (formaldehyde)

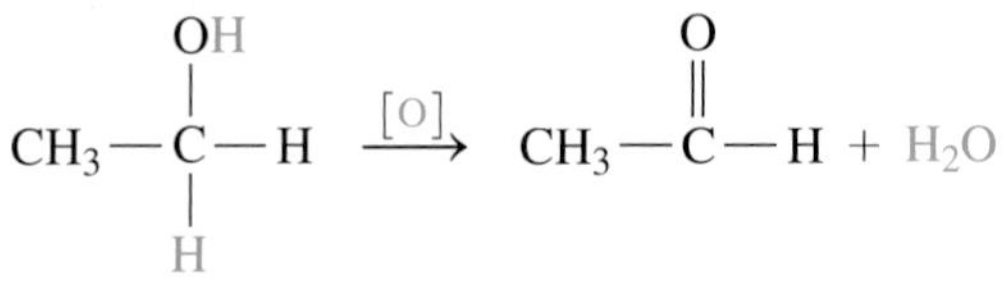

Ethanol (ethyl alcohol) — Ethanal (acetaldehyde)

In the oxidation of secondary alcohols, the products are ketones. Two hydrogen atoms are removed, one from the —OH group and another from the carbon bonded to the —OH group. The result is a ketone that has the carbon–oxygen double bond attached to alkyl groups on both sides. There is no further oxidation of a ketone because there are no hydrogen atoms attached to the carbonyl group.

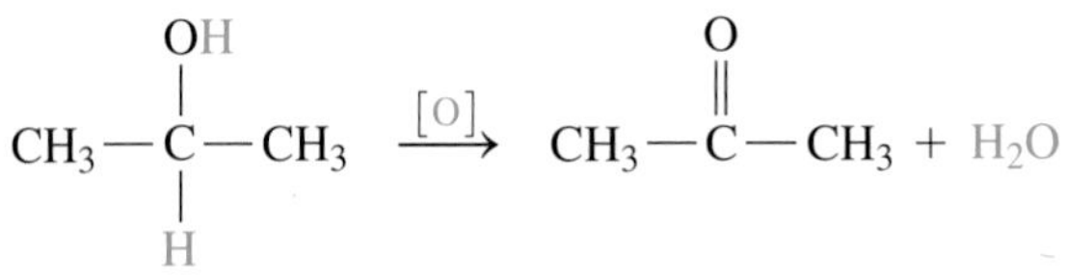

2-Propanol (isopropyl alcohol) — Propanone (dimethyl ketone; acetone)

Tertiary alcohols do not oxidize readily because there is no hydrogen atom on the carbon bonded to the —OH group. Because C—C bonds are usually too strong to oxidize, tertiary alcohols resist oxidation.

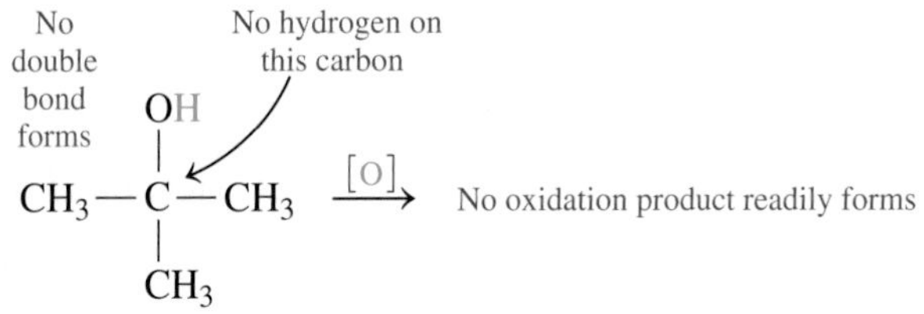

Alcohol (3°)

Chemistry Link to Health

Methanol Poisoning

Methanol (methyl alcohol), CH_3—OH, is a highly toxic alcohol present in products such as windshield washer fluid, Sterno, and paint strippers. Methanol is rapidly absorbed in the gastrointestinal tract. In the liver, it is oxidized to formaldehyde and then formic acid, a substance that causes nausea, severe abdominal pain, and blurred vision. Blindness can occur because the intermediate products destroy the retina of the eye. As little as 4 mL of methanol can produce blindness. The formic acid, which is not readily eliminated from the body, lowers blood pH so severely that just 30 mL of methanol can lead to coma and death.

The treatment for methanol poisoning involves giving sodium bicarbonate to neutralize the formic acid in the blood. In some cases, ethanol is given intravenously to the patient. The enzymes in the liver pick up ethanol molecules to oxidize instead of methanol molecules. This process gives time for the methanol to be eliminated via the lungs without the formation of its dangerous oxidation products.

SAMPLE PROBLEM 12.8 Oxidation of Alcohols

Classify each of the following alcohols as primary (1°), secondary (2°), or tertiary (3°); draw the condensed structural formula for the aldehyde or ketone formed by its oxidation:

a. $CH_3-CH_2-\overset{\overset{\large OH}{|}}{CH}-CH_3$

b. $CH_3-CH_2-CH_2-CH_2-OH$

SOLUTION

a. This is a secondary (2°) alcohol, which oxidizes to a ketone.

$$CH_3-CH_2-\overset{\overset{\large O}{\|}}{C}-CH_3$$

b. This is a primary (1°) alcohol, which can oxidize to an aldehyde.

$$CH_3-CH_2-CH_2-\overset{\overset{\large O}{\|}}{C}-H$$

STUDY CHECK 12.8

Draw the condensed structural formula for the product of the oxidation of 2-pentanol.

Interactive Video

Oxidation of Alcohols

Oxidation of Thiols

Thiols also undergo oxidation by the loss of hydrogen atoms from each of two $-SH$ groups. The oxidized product contains a *disulfide bond* $-S-S-$. Much of the protein in hair is cross-linked by disulfide bonds, which occur mostly between the side chains of the amino acids of cysteine, which contain thiol groups.

$$\text{Protein chain}-CH_2-\mathbf{SH} + \mathbf{HS}-CH_2-\text{Protein chain} \xrightarrow{[O]}$$

Cysteine side groups

$$\text{Protein chain}-CH_2-\mathbf{S}-\mathbf{S}-CH_2-\text{Protein chain} + \mathbf{H_2O}$$

Disulfide bond

Proteins in the hair take new shapes when disulfide bonds are reduced and oxidized.

When a person has his or her hair permanently curled or straightened, a reducing substance is used to break the disulfide bonds. While the hair is wrapped around curlers or straightened, an oxidizing substance is applied that causes new disulfide bonds to form between different parts of the protein hair strands, which gives the hair a new shape.

Oxidation of Aldehydes

Earlier in this section, we saw that primary alcohols can oxidize to aldehydes. Aldehydes oxidize further by the addition of another O to form a carboxylic acid, which has a *carboxyl* functional group. This step occurs so readily that it is often difficult to isolate the aldehyde product during the oxidation reaction. In contrast, ketones produced by the oxidation of secondary alcohols do not undergo further oxidation.

$$\underset{\text{Ethanol (1°)}}{CH_3-CH_2-OH} \xrightarrow{\text{Oxidation}} \underset{\text{Ethanal}}{CH_3-\overset{\overset{\large O}{\|}}{C}-H} \xrightarrow[\text{oxidation}]{\text{Further}} \underset{\text{Ethanoic acid}}{CH_3-\overset{\overset{\large O}{\|}}{C}-OH}$$

$$\underset{\text{2-Propanol (2°)}}{CH_3-\overset{\overset{\large OH}{|}}{CH}-CH_3} \xrightarrow{\text{Oxidation}} \underset{\text{Propanone}}{CH_3-\overset{\overset{\large O}{\|}}{C}-CH_3} \xrightarrow[\text{oxidation}]{\text{Further}} \text{no reaction}$$

Chemistry Link to Health

Oxidation of Alcohol in the Body

Ethanol is the most commonly abused drug in the United States. When ingested in small amounts, ethanol may produce a feeling of euphoria in the body despite the fact that it is a depressant. In the liver, enzymes such as alcohol dehydrogenase oxidize ethanol to acetaldehyde, a substance that impairs mental and physical coordination. If the blood alcohol concentration exceeds 0.4%, coma or death may occur. Table 12.3 gives some of the typical behaviors exhibited at various blood alcohol levels.

$$\underset{\text{Ethanol (ethyl alcohol)}}{CH_3-CH_2-OH} \xrightarrow{[O]} \underset{\text{Ethanal (acetaldehyde)}}{CH_3-\overset{\overset{\large O}{\|}}{C}-H} \xrightarrow{[O]} 2CO_2 + H_2O$$

calculate the blood alcohol concentration (BAC). Several devices are used to measure the BAC. When a Breathalyzer is used, a suspected drunk driver exhales through a mouthpiece into a solution containing the orange Cr^{6+} ion. Any alcohol present in the exhaled air is oxidized, which reduces the orange Cr^{6+} to a green Cr^{3+}.

$$\underset{\text{Ethanol}}{CH_3-CH_2-OH} + \underset{\text{Orange}}{Cr^{6+}} \xrightarrow{[O]} \underset{\text{Ethanoic acid}}{CH_3-\overset{\overset{\large O}{\|}}{C}-OH} + \underset{\text{Green}}{Cr^{3+}}$$

The Alcosensor uses the oxidation of alcohol in a fuel cell to generate an electric current that is measured. The Intoxilyzer measures the amount of light absorbed by the alcohol molecules.

TABLE 12.3 Typical Behaviors Exhibited by a 150-lb Person Consuming Alcohol

Number of Beers (12 oz) or Glasses of Wine (5 oz) in 1 h	Blood Alcohol Level (% m/v)	Typical Behavior
1	0.025	Slightly dizzy, talkative
2	0.050	Euphoria, loud talking and laughing
4	0.10	Loss of inhibition, loss of coordination, drowsiness, legally intoxicated in most states
8	0.20	Intoxicated, quick to anger, exaggerated emotions
12	0.30	Unconscious
16–20	0.40–0.50	Coma and death

The acetaldehyde produced from ethanol in the liver is further oxidized to acetic acid, which is converted to carbon dioxide and water in the citric acid cycle. Thus, the enzymes in the liver can eventually break down ethanol, but the aldehyde and carboxylic acid intermediates can cause considerable damage while they are present within the cells of the liver.

A person weighing 150 lb requires about one hour to metabolize the alcohol in 12 ounces of beer. However, the rate of metabolism of ethanol varies between nondrinkers and drinkers. Typically, nondrinkers and social drinkers can metabolize 12 to 15 mg of ethanol/dL of blood in one hour, but an alcoholic can metabolize as much as 30 mg of ethanol/dL in one hour. Some effects of alcohol metabolism include an increase in liver lipids (fatty liver), gastritis, pancreatitis, ketoacidosis, alcoholic hepatitis, and psychological disturbances.

When alcohol is present in the blood, it evaporates through the lungs. Thus, the percentage of alcohol in the lungs can be used to

Sometimes alcoholics are treated with a drug called Antabuse (disulfiram), which prevents the oxidation of acetaldehyde to acetic acid. As a result, acetaldehyde accumulates in the blood, which causes nausea, profuse sweating, headache, dizziness, vomiting, and respiratory difficulties. Because of these unpleasant side effects, the person is less likely to use alcohol.

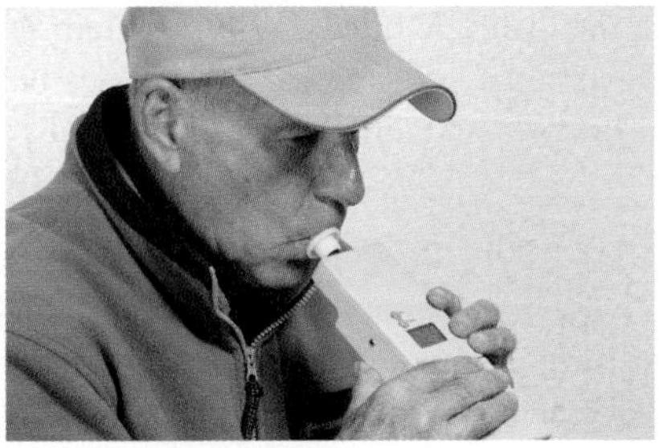

A Breathalyzer test is used to determine blood alcohol level.

Tollens' Test

Tollens' test, which uses a solution of $Ag^+(AgNO_3)$ and ammonia, oxidizes aldehydes, but not ketones. The silver ion is reduced and forms a "silver mirror" on the inside of the container. Commercially, a similar process is used to make mirrors by applying a solution of $AgNO_3$ and ammonia on glass with a spray gun (see Figure 12.7).

$$CH_3-\overset{O}{\overset{\|}{C}}-H + 2Ag^+ \xrightarrow{[O]} 2Ag(s) + CH_3-\overset{O}{\overset{\|}{C}}-OH$$

Ethanal (acetaldehyde) — Tollens' reagent — Silver mirror — Ethanoic acid (acetic acid)

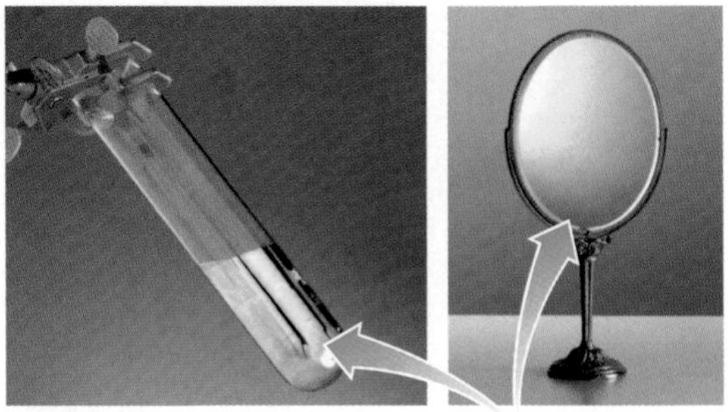

$$Ag^+ + 1\,e^- \longrightarrow Ag(s)$$

FIGURE 12.7 ▶ In Tollens' test, a silver mirror forms when the oxidation of an aldehyde reduces silver ion to metallic silver. The silvery surface of a mirror is formed in a similar way.

Q What is the product of the oxidation of an aldehyde?

Another test, called **Benedict's test**, gives a positive result with compounds that have an aldehyde functional group and an adjacent hydroxyl group. When Benedict's solution containing $Cu^{2+}(CuSO_4)$ is added to this type of aldehyde and heated, a brick-red solid of Cu_2O forms (see Figure 12.8). The test is negative with simple aldehydes and ketones.

$$CH_3-\overset{OH}{\overset{|}{CH}}-\overset{O}{\overset{\|}{C}}-H + 2Cu^{2+} \longrightarrow Cu_2O(s) + CH_3-\overset{OH}{\overset{|}{CH}}-\overset{O}{\overset{\|}{C}}-OH$$

2-Hydroxypropanal — Benedict's reagent — Brick-red solid — 2-Hydroxypropanoic acid

Because many sugars, such as glucose, contain this type of aldehyde grouping, Benedict's reagent can be used to determine the presence of glucose in blood or urine.

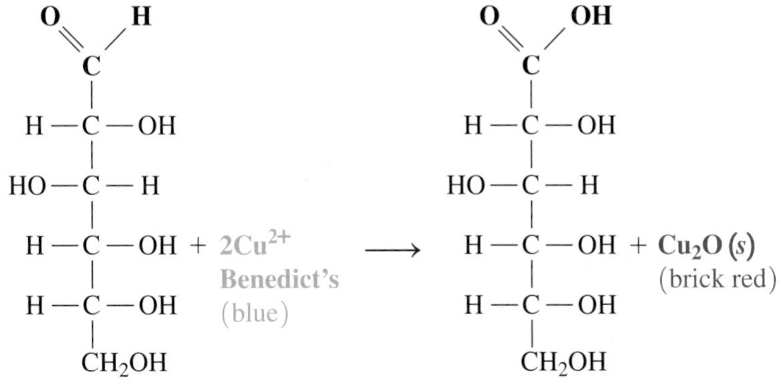

D-Glucose — D-Gluconic acid

Cu^{2+} $Cu_2O(s)$

FIGURE 12.8 ▶ The blue Cu^{2+} in Benedict's solution forms a brick-red solid of Cu_2O in a positive test for many sugars and aldehydes with adjacent hydroxyl groups.

Q Which test tube indicates that glucose is present?

Reduction of Aldehydes and Ketones

Aldehydes and ketones are reduced by sodium borohydride ($NaBH_4$) or hydrogen (H_2). In the **reduction** of organic compounds, there is a decrease in the number of carbon–oxygen bonds by the addition of hydrogen or the loss of oxygen. Aldehydes reduce to primary alcohols, and ketones reduce to secondary alcohols. A catalyst such as nickel, platinum, or palladium is needed for the addition of hydrogen to the carbonyl group.

Aldehydes Reduce to Primary Alcohols

$$CH_3-CH_2-\overset{O}{\overset{\|}{C}}-H + \mathbf{H_2} \xrightarrow{Pt} CH_3-CH_2-\underset{\mathbf{H}}{\underset{|}{\overset{\mathbf{OH}}{\overset{|}{C}}}}-H$$

Propanal (propionaldehyde) — 1-Propanol (1° alcohol) (propyl alcohol)

Ketones Reduce to Secondary Alcohols

$$CH_3-\overset{\overset{\large O}{\|}}{C}-CH_3 + \mathbf{H_2} \xrightarrow{Ni} CH_3-\underset{\underset{\large \mathbf{H}}{|}}{\overset{\overset{\large \mathbf{OH}}{|}}{C}}-CH_3$$

Propanone (dimethyl ketone) → 2-Propanol (2° alcohol) (isopropyl alcohol)

SAMPLE PROBLEM 12.9 Reduction of Carbonyl Groups

Write a balanced chemical equation for the reduction of cyclopentanone using hydrogen in the presence of a nickel catalyst.

SOLUTION

The reacting molecule is a cyclic ketone that has five carbon atoms. During the reduction, hydrogen atoms add to the carbon and oxygen in the carbonyl group, which reduces the ketone to the corresponding secondary alcohol.

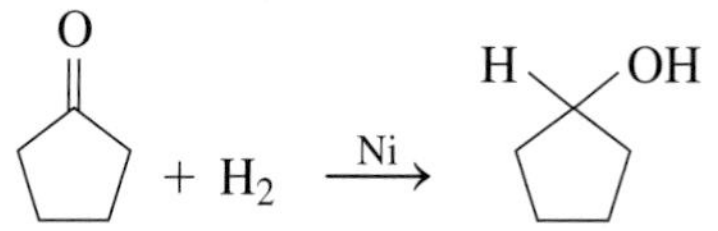

Cyclopentanone → Cyclopentanol

STUDY CHECK 12.9

Write a balanced chemical equation for the reduction of butanal using hydrogen in the presence of a platinum catalyst.

QUESTIONS AND PROBLEMS

12.4 Reactions of Alcohols, Thiols, Aldehydes, and Ketones

LEARNING GOAL Write balanced chemical equations for the combustion, dehydration, and oxidation of alcohols. Write balanced chemical equations for the oxidation and reduction of thiols, aldehydes, and ketones.

12.23 Draw the condensed structural formula, or skeletal formula if cyclic, for the alkene produced by each of the following dehydration reactions:

a. $H_3C-\overset{\overset{\large OH}{|}}{\underset{H}{C}}-CH_3 \xrightarrow{H^+\ heat}$

b. 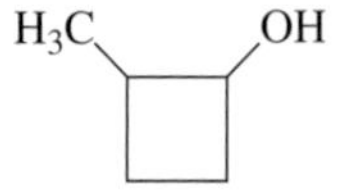 $\xrightarrow{H^+\ heat}$

c. 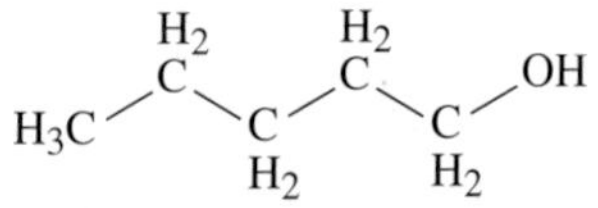 $\xrightarrow{H^+\ heat}$

12.24 Draw the condensed structural formula, or skeletal formula if cyclic, for the alkene produced by each of the following dehydration reactions:

a. $CH_3-CH_2-OH \xrightarrow{H^+,\ heat}$

b. $CH_3-\overset{\overset{\large CH_3}{|}}{CH}-CH_2-OH \xrightarrow{H^+,\ heat}$

c. 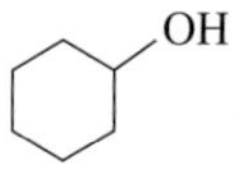 $\xrightarrow{H^+,\ heat}$

12.25 Draw the condensed structural formula, or skeletal formula if cyclic, for the aldehyde or ketone formed when each of the following alcohols is oxidized [O] (if no reaction, write *none*):

a. $CH_3-CH_2-CH_2-OH$

b. $CH_3-\overset{\overset{\large OH}{|}}{CH}-CH_2-CH_2-CH_2-CH_3$

c. 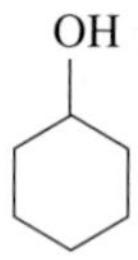

12.26 Draw the condensed structural formula, or skeletal formula if cyclic, for the aldehyde or ketone formed when each of the following alcohols is oxidized [O] (if no reaction, write *none*):

a. $CH_3-\overset{\overset{\large CH_3}{|}}{CH}-CH_2-OH$

b. $CH_3-CH_2-\underset{\underset{\large CH_3}{|}}{\overset{\overset{\large OH}{|}}{C}}-CH_3$

c. 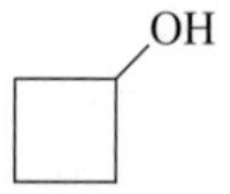

12.27 Draw the condensed structural formulas for the aldehyde and carboxylic acid produced when each of the following is oxidized:

a. $CH_3—CH_2—CH_2—CH_2—CH_2—OH$

b. 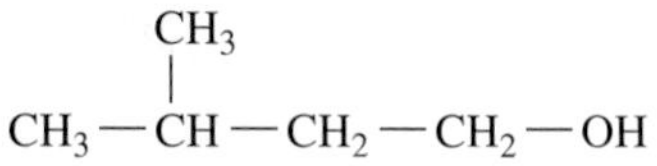

c. 1-butanol

12.28 Draw the condensed structural formula for the aldehyde and carboxylic acid produced when each of the following is oxidized:

a. 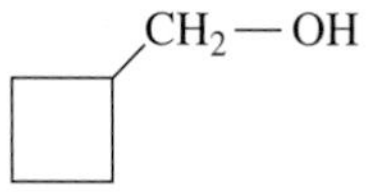

b. $CH_3—OH$

c. 3-chloro-1-propanol

12.29 Draw the condensed structural formula for the organic product formed when each of the following is reduced by hydrogen in the presence of a nickel catalyst:

a. butyraldehyde **b.** acetone
c. hexanal **d.** 2-methyl-3-pentanone

12.30 Draw the skeletal formula for the organic product formed when each of the following is reduced by an excess amount of hydrogen in the presence of a Pd catalyst:

a. ethyl propyl ketone **b.** 4-hexen-1-ol

12.31 Write the balanced chemical equation for the complete combustion of each of the following compounds:

a. methanol **b.** 2-butanol

12.32 Write the balanced chemical equation for the complete combustion of each of the following compounds:

a. 1-propanol **b.** 3-hexanol

CONCEPT MAP

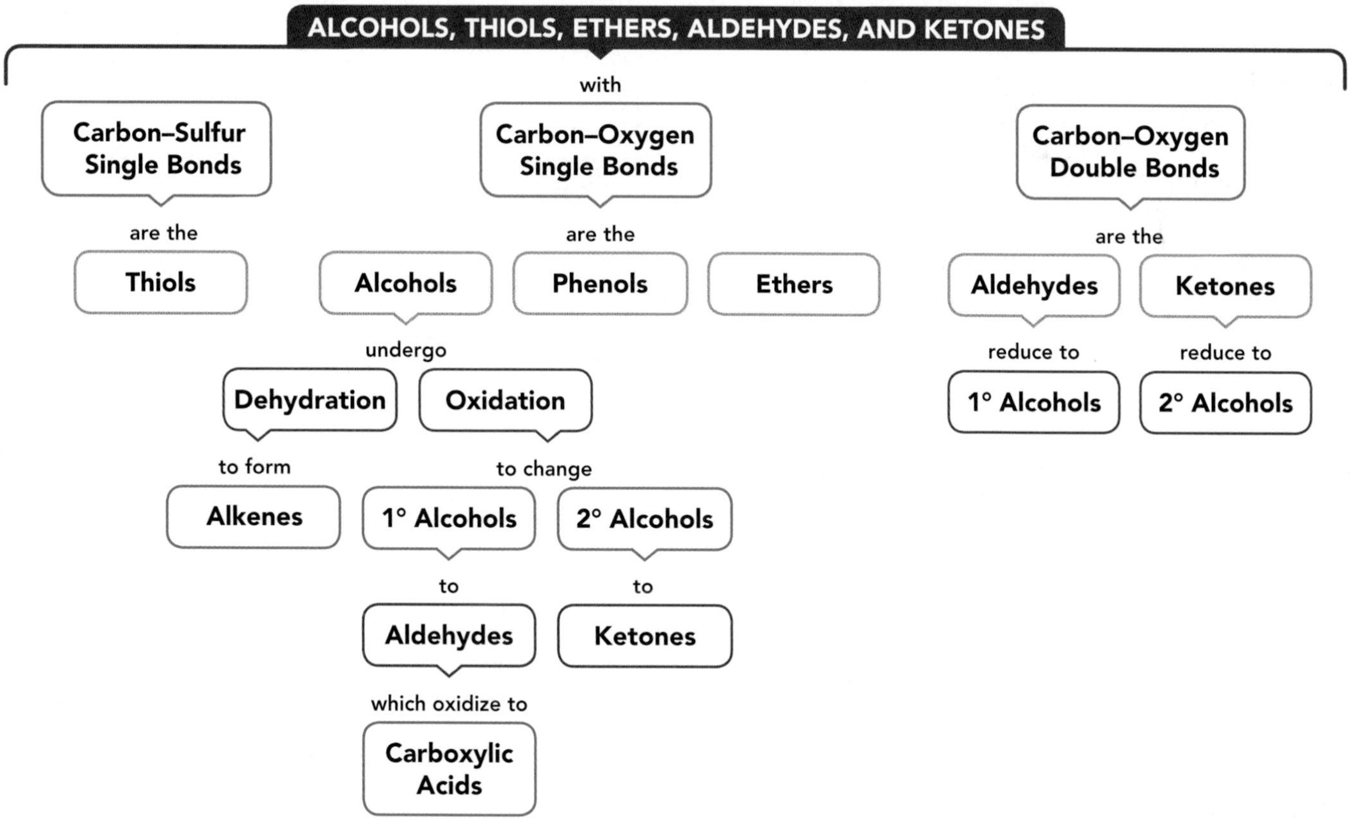

CHAPTER REVIEW

12.1 Alcohols, Phenols, Thiols, and Ethers

LEARNING GOAL Give the IUPAC and common names for alcohols and phenols; give the common names for thiols and ethers. Draw their condensed structural formulas or skeletal formulas.

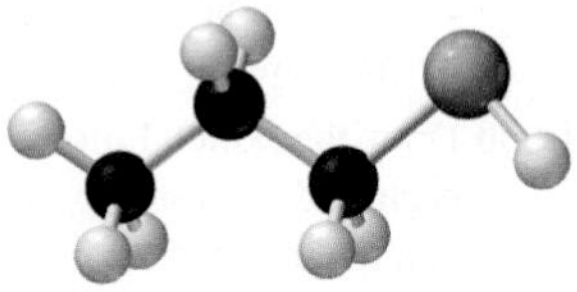

$CH_3—CH_2—CH_2—SH$

1-Propanethiol (onions)

- The functional group of an alcohol is the hydroxyl group (—OH) bonded to a carbon chain.
- In a phenol, the hydroxyl group is bonded to an aromatic ring.
- In thiols, the functional group is —SH, which is analogous to the —OH group of alcohols.
- In the IUPAC system, the names of alcohols have *ol* endings, and the location of the —OH group is given by numbering the carbon chain.
- A cyclic alcohol is named as a cycloalkanol.
- Simple alcohols are generally named by their common names, with the alkyl name preceding the term alcohol.
- An aromatic alcohol is named as a phenol.
- In an ether, an oxygen atom is connected by single bonds to two alkyl or aromatic groups.
- In the common names of ethers, the alkyl or aromatic groups are listed alphabetically, followed by the name ether.

12.2 Properties of Alcohols

LEARNING GOAL Describe the classification of alcohols; describe the solubility of alcohols in water.

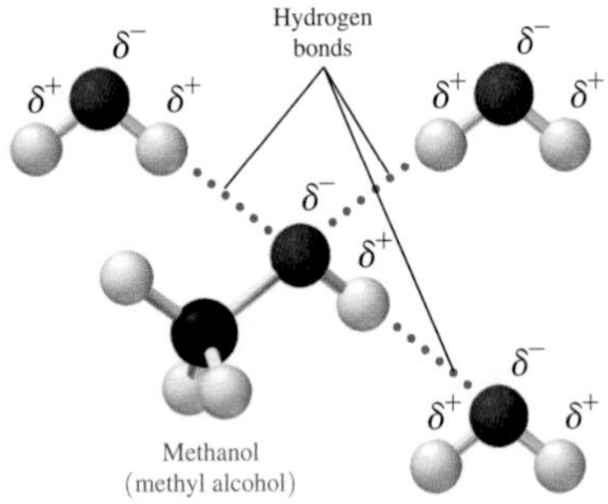

- Alcohols are classified according to the number of alkyl groups bonded to the carbon that holds the —OH group.
- In a primary (1°) alcohol, one group is attached to the hydroxyl carbon.
- In a secondary (2°) alcohol, two groups are attached.
- In a tertiary (3°) alcohol, there are three groups bonded to the hydroxyl carbon.
- Short-chain alcohols can hydrogen bond with water, which makes them soluble.

12.3 Aldehydes and Ketones

LEARNING GOAL Write the IUPAC and common names for aldehydes and ketones; draw the condensed structural formulas. Describe the solubility of aldehydes and ketones in water.

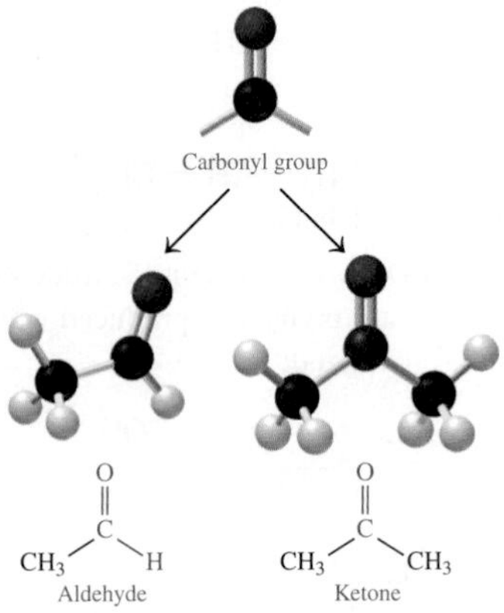

- Aldehydes and ketones contain a carbonyl group (C=O) which consists of a double bond between a carbon and an oxygen atom.
- In aldehydes, the carbonyl group appears at the end of carbon chains attached to at least one hydrogen atom.
- In ketones, the carbonyl group occurs between two alkyl or aromatic groups.
- In the IUPAC system, the *e* in the corresponding alkane name is replaced with *al* for aldehydes and *one* for ketones.
- For ketones with more than four carbon atoms in the main chain, the carbonyl group is numbered to show its location.
- Many of the simple aldehydes and ketones use common names.
- Because they contain a polar carbonyl group, aldehydes and ketones can hydrogen bond with water molecules, which makes carbonyl compounds with one to four carbon atoms soluble in water.

12.4 Reactions of Alcohols, Thiols, Aldehydes, and Ketones

LEARNING GOAL Write balanced chemical equations for the combustion, dehydration, and oxidation of alcohols. Write balanced chemical equations for the oxidation and reduction of thiols, aldehydes, and ketones.

$$CH_3—CH_2(OH) \xrightarrow{[O]} CH_3—C(=O)—H + H_2O$$

Ethanol (ethyl alcohol) → Ethanal (acetaldehyde)

- Alcohols burn with oxygen to give carbon dioxide and water.
- At high temperatures, alcohols dehydrate in the presence of an acid to yield alkenes.
- Primary alcohols are oxidized to aldehydes, which can oxidize further to carboxylic acids.
- Secondary alcohols are oxidized to ketones.
- Tertiary alcohols do not oxidize.
- Aldehydes are easily oxidized to carboxylic acids, but ketones do not oxidize further.
- Aldehydes, but not ketones, are oxidized by Tollens' reagent to give silver mirrors.
- In Benedict's test, aldehydes with adjacent hydroxyl groups reduce blue Cu^{2+} to give a brick-red Cu_2O solid.
- Aldehydes and ketones can be reduced with H_2 in the presence of a catalyst to give alcohols.

SUMMARY OF NAMING

Structure	Family	IUPAC Name	Common Name
$CH_3-\mathbf{OH}$	Alcohol	Methanol	Methyl alcohol
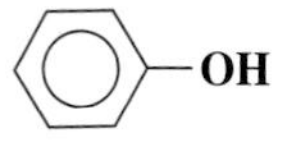	Phenol	Phenol	Phenol
$CH_3-\mathbf{SH}$	Thiol	Methanethiol	
$CH_3-\mathbf{O}-CH_3$	Ether		Dimethyl ether
$H-\overset{\mathbf{O}}{\overset{\Vert}{\mathbf{C}}}-\mathbf{H}$	Aldehyde	Methanal	Formaldehyde
$CH_3-\overset{\mathbf{O}}{\overset{\Vert}{\mathbf{C}}}-CH_3$	Ketone	Propanone	Acetone; dimethyl ketone

SUMMARY OF REACTIONS

The chapter sections to review are shown after the name of the reaction.

Combustion of Alcohols (12.4)

$$\underset{\text{Ethanol}}{CH_3-CH_2-OH} + \underset{\text{Oxygen}}{3O_2} \xrightarrow{\Delta} \underset{\text{Carbon dioxide}}{2CO_2} + \underset{\text{Water}}{3H_2O} + \text{energy}$$

Dehydration of Alcohols to Form Alkenes (12.4)

$$\underset{\text{1-Propanol}}{CH_3-CH_2-CH_2-OH} \xrightarrow{H^+,\ \text{heat}} \underset{\text{Propene}}{CH_3-CH=CH_2} + H_2O$$

Oxidation of Primary Alcohols to Form Aldehydes (12.4)

$$\underset{\text{Ethanol}}{CH_3-\overset{OH}{\overset{|}{CH_2}}} \xrightarrow{[O]} \underset{\text{Ethanal}}{CH_3-\overset{O}{\overset{\Vert}{C}}-H} + H_2O$$

Oxidation of Secondary Alcohols to Form Ketones (12.4)

$$\underset{\text{2-Propanol}}{CH_3-\overset{OH}{\overset{|}{CH}}-CH_3} \xrightarrow{[O]} \underset{\text{Propanone}}{CH_3-\overset{O}{\overset{\Vert}{C}}-CH_3} + H_2O$$

Oxidation of Aldehydes to Form Carboxylic Acids (12.4)

$$\underset{\text{Ethanal}}{CH_3-\overset{O}{\overset{\Vert}{C}}-H} \xrightarrow{[O]} \underset{\text{Ethanoic acid}}{CH_3-\overset{O}{\overset{\Vert}{C}}-OH}$$

Reduction of Aldehydes to Form Primary Alcohols (12.4)

$$\underset{\text{Ethanal}}{CH_3-\overset{O}{\overset{\Vert}{C}}-H} + H_2 \xrightarrow{Pt} \underset{\text{Ethanol}}{CH_3-\overset{OH}{\overset{|}{CH_2}}}$$

Reduction of Ketones to Form Secondary Alcohols (12.4)

$$\underset{\text{Propanone}}{CH_3-\overset{O}{\overset{\Vert}{C}}-CH_3} + H_2 \xrightarrow{Ni} \underset{\text{2-Propanol}}{CH_3-\overset{OH}{\overset{|}{CH}}-CH_3}$$

KEY TERMS

alcohol An organic compound that contains the hydroxyl functional group ($-OH$) attached to a carbon chain.

aldehyde An organic compound with a carbonyl functional group bonded to at least one hydrogen.

$$-\overset{O}{\overset{\Vert}{C}}-H = -CHO$$

Benedict's test A test for aldehydes with adjacent hydroxyl groups in which Cu^{2+} ($CuSO_4$) ions in Benedict's reagent are reduced to a brick-red solid of Cu_2O.

carbonyl group A functional group that contains a carbon–oxygen double bond ($C=O$).

dehydration A reaction that removes water from an alcohol, in the presence of an acid, to form an alkene.

ether An organic compound in which an oxygen atom is bonded to two alkyl or two aromatic groups, or one of each.

ketone An organic compound in which the carbonyl functional group is bonded to two alkyl or aromatic groups.

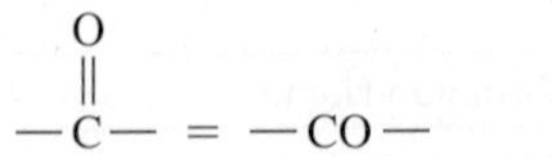

oxidation The loss of two hydrogen atoms from a reactant to give a more oxidized compound. For example, primary alcohols oxidize to aldehydes, and secondary alcohols oxidize to ketones. An oxidation can also be the addition of an oxygen atom or an increase in the number of carbon–oxygen bonds.

phenol An organic compound that has a hydroxyl group ($-OH$) attached to a benzene ring.

primary (1°) alcohol An alcohol that has one alkyl group bonded to the carbon atom with the $-OH$ group.

reduction A gain of hydrogen when a ketone reduces to a 2° alcohol and an aldehyde reduces to a 1° alcohol.

secondary (2°) alcohol An alcohol that has two alkyl groups bonded to the carbon atom with the $-OH$ group.

tertiary (3°) alcohol An alcohol that has three alkyl groups bonded to the carbon atom with the $-OH$ group.

thiol An organic compound that contains $-SH$ groups.

Tollens' test A test for aldehydes in which Ag^+ in Tollens' reagent is reduced to metallic silver forming a "silver mirror" on the walls of the container.

CORE CHEMISTRY SKILLS

The chapter section containing each Core Chemistry Skill is shown in parentheses at the end of each heading.

Identifying Functional Groups (12.1)

- Functional groups are specific groups of atoms in organic compounds, which undergo characteristic chemical reactions.
- Organic compounds with the same functional group have similar properties and reactions.

Example: Identify the functional group in the following molecule:

$$CH_3-CH_2-CH_2-CH_2-OH$$

Answer: The hydroxyl group ($-OH$) makes it an alcohol.

Naming Alcohols and Phenols (12.1)

- In the IUPAC system, an alcohol is named by replacing the *e* in the alkane name with *ol*.
- Simple alcohols are named with the alkyl name preceding the term alcohol.
- The carbon chain is numbered from the end nearer the $-OH$ group and the location of the $-OH$ is given in front of the name.
- A cyclic alcohol is named as a cycloalkanol.
- An aromatic alcohol is named as a phenol.

Example: Give the IUPAC name for

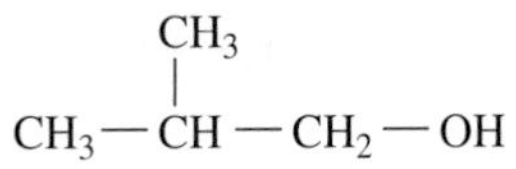

Answer: 2-methyl-1-propanol

Naming Aldehydes and Ketones (12.3)

- In the IUPAC system, an aldehyde is named by replacing the *e* in the alkane name with *al* and a ketone by replacing the *e* with *one*.
- The position of a substituent on an aldehyde is indicated by numbering the carbon chain from the carbonyl group and given in front of the name.
- For a ketone, the carbon chain is numbered from the end nearer the carbonyl group.

Example: Give the IUPAC name for

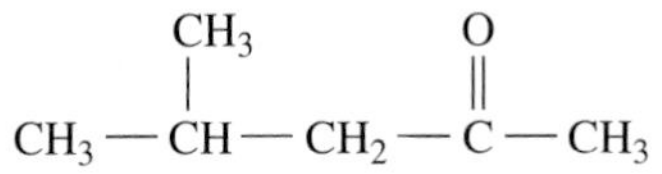

Answer: 4-methyl-2-pentanone

Writing Equations for the Dehydration and Oxidation of Alcohols (12.4)

- At high temperatures, alcohols dehydrate in the presence of an acid to yield alkenes.
- Primary alcohols oxidize to aldehydes, which can oxidize further to carboxylic acids.
- Secondary alcohols oxidize to ketones.
- Tertiary alcohols do not oxidize.

Example: Draw the condensed structural formula for the product of the oxidation of 2-butanol.

Answer:

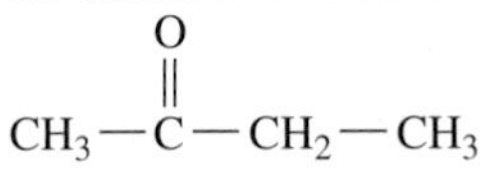

UNDERSTANDING THE CONCEPTS

The chapter sections to review are shown in parentheses at the end of each question.

12.33 Sunscreens contain compounds such as oxybenzone that absorb UV light. Identify the functional groups in oxybenzone. (12.1, 12.3)

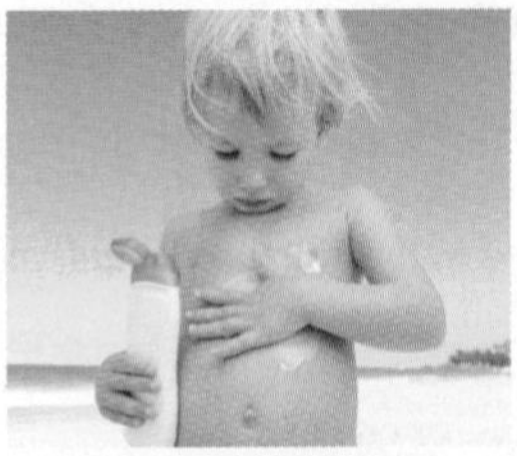

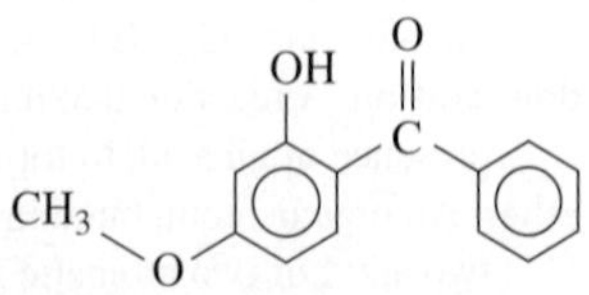

12.34 The compound frambinone has the taste of raspberries and has been used in weight loss. Identify the functional groups in frambinone. (12.1, 12.3)

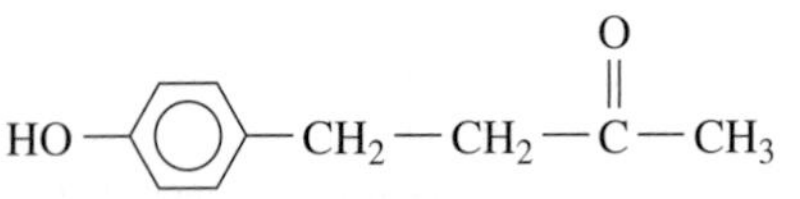

12.35 A compound called butylated hydroxytoluene, or BHT, has been added to cereal and other foods since 1947 as an antioxidant. Identify the functional groups in BHT. (12.1, 12.3)

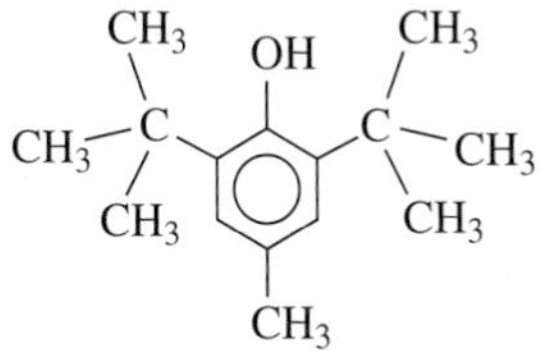

BHT is an antioxidant added to preserve foods such as cereal.

12.36 Used in sunless tanning lotions, the compound known as DHA darkens the skin without sun. DHA reacts with amino acids in the outer surface of the skin. Identify the functional groups in DHA. (12.1, 12.3)

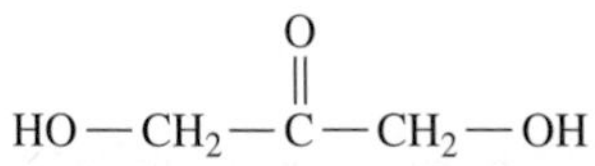

A sunless tanning lotion contains DHA to darken the skin.

12.37 Which of the following will give a positive Tollens' test? (12.4)

a. propanal **b.** ethanol **c.** ethyl methyl ether

12.38 Which of the following will give a positive Tollens' test? (12.4)

a. 1-propanol **b.** 2-propanol **c.** hexanal

12.39 Classify each of the following according to its functional group(s): (12.1, 12.3)

a.

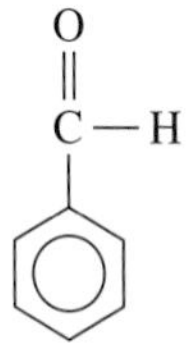

Almonds

b.

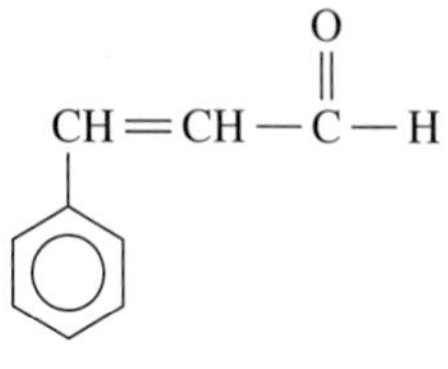

Cinnamon sticks

12.40 Classify each of the following according to its functional groups(s): (12.1, 12.3)

a. BHA, an antioxidant used as a preservative in foods such as baked goods, butter, meats, and snack foods

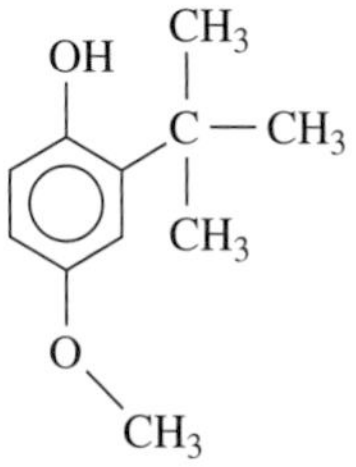

Baked goods contain BHA as a preservative.

b. vanillin, a flavoring, obtained from the seeds of the vanilla bean

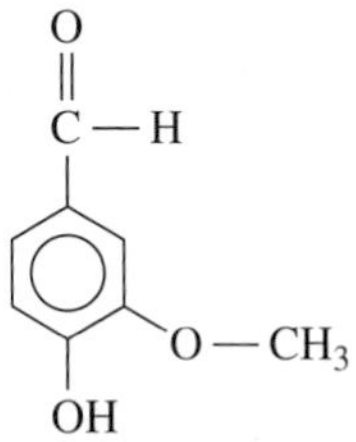

Vanilla extract is a solution containing the compound vanillin.

ADDITIONAL QUESTIONS AND PROBLEMS

12.41 Classify each of the following alcohols as primary (1°), secondary (2°), or tertiary (3°): (12.2)

a. 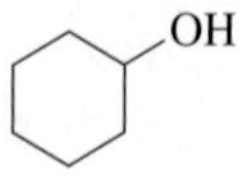

b. 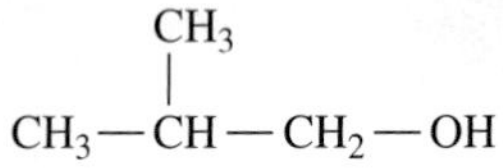

c. 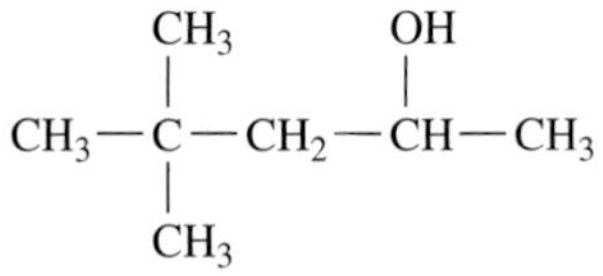

d. 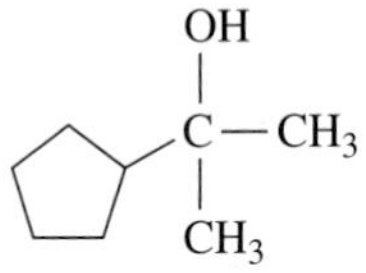

12.42 Classify each of the following alcohols as primary (1°), secondary (2°), or tertiary (3°): (12.2)

a. 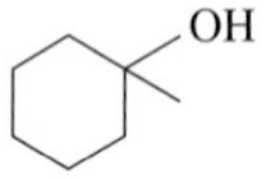

b. 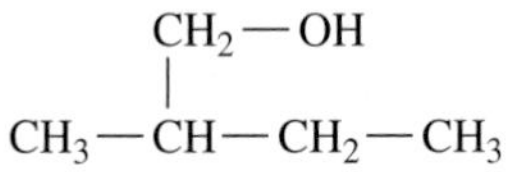

c. 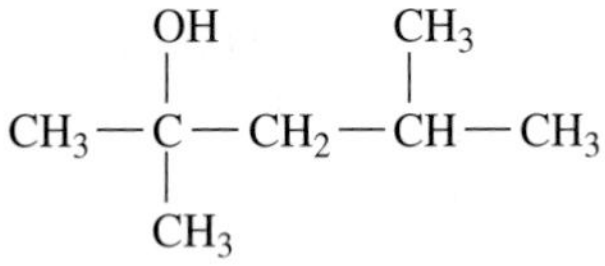

d. 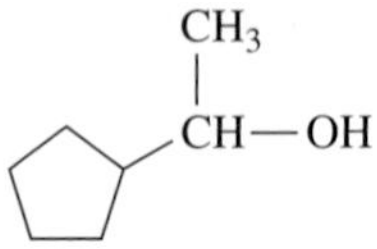

12.43 Give the IUPAC name for each of the following alcohols and phenols:

a. 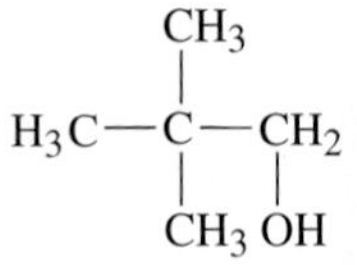

b. 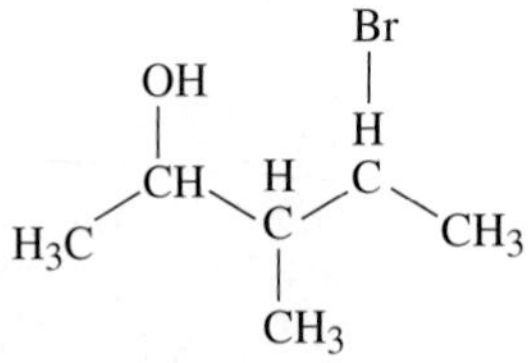

c. 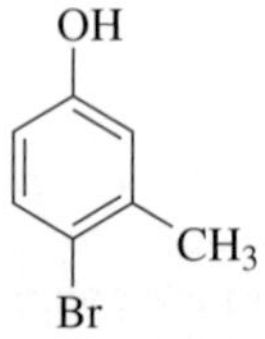

12.44 Give the IUPAC name for each of the following alcohols and phenols: (12.1)

a. 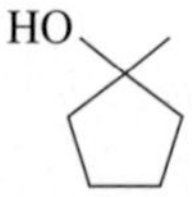

b. 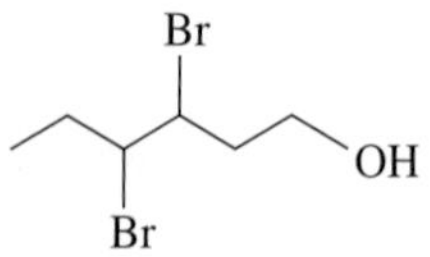

c. 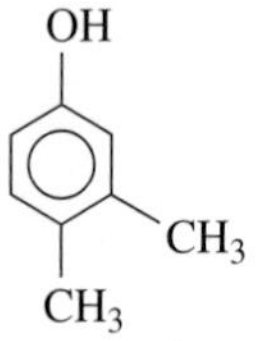

12.45 Draw the condensed structural formula for each of the following:

a. 2-chlorophenol **b.** 4-heptanol
c. 2,4-dibromophenol **d.** 2-methyl-1-butanol

12.46 Draw the skeletal formula for each of the following: (12.1)

a. 2-methyl-1-hexanol **b.** ethylene glycol (ethane-1,2-diol)
c. methyl propyl ether **d.** 3-hydroxylphenol

12.47 Which compound in each pair would be more soluble in water? Explain.

a. pentane or 1-propanol **b.** ethanol or ethyl methyl ether
c. methanol or 1-heptanol

12.48 Which compound in each pair would be more soluble in water? Explain. (12.2)

a. ethane or ethanol **b.** 2-propanol or 2-pentanol
c. methyl propyl ether or 1-butanol

12.49 Draw the condensed structural formula, or skeletal formula if cyclic, for the alkene, aldehyde, or ketone product of each of the following reactions: (12.4)

a. $CH_3—CH_2—CH_2—OH \xrightarrow{H^+,\ heat}$

b. $CH_3—CH_2—CH_2—OH \xrightarrow{[O]}$

c. 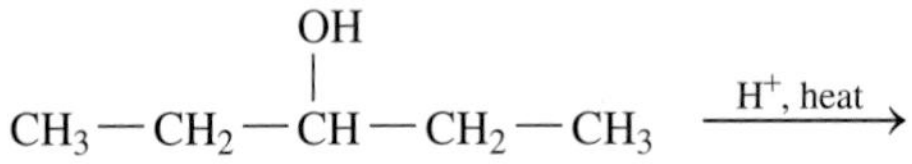

d. 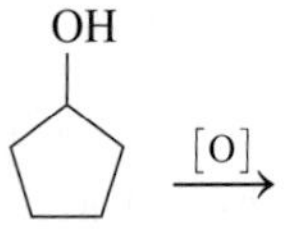

12.50 Draw the condensed structural formula, or skeletal formula if cyclic, for the alkene, aldehyde, or ketone product of each of the following reactions: (12.4)

a. 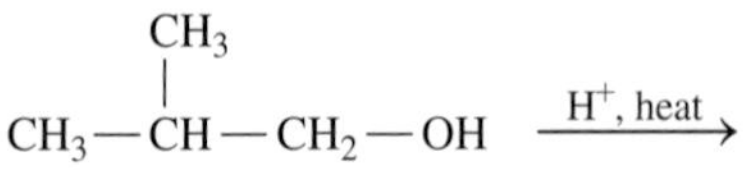

b. 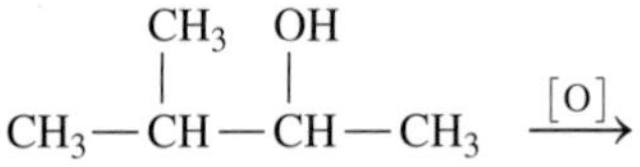

c. 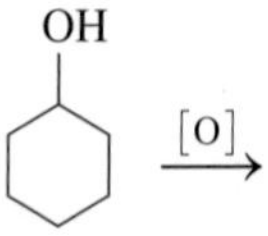

d. 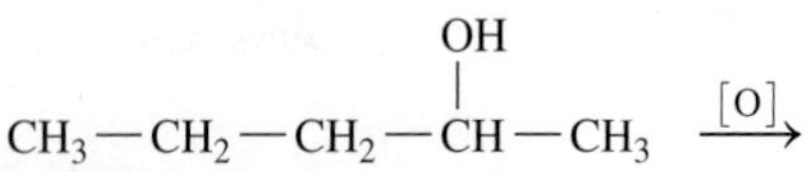

12.51 Draw the condensed structural formula, or skeletal formula if cyclic, for the organic product when hydrogen and a nickel catalyst reduce each of the following: (12.4)

a. 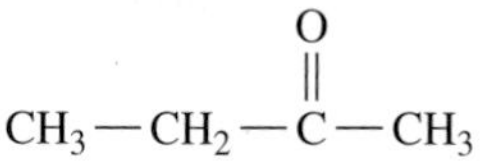

b. 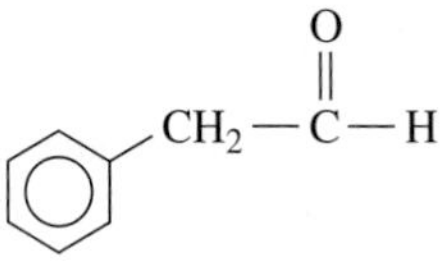

c. 

12.52 Draw the condensed structural formula, or skeletal formula if cyclic, for the organic product when hydrogen and a nickel catalyst reduce each of the following: (12.4)

a. 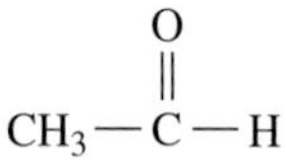

b. 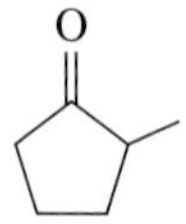

c. 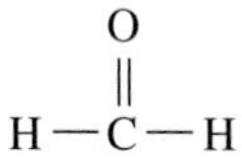

12.53 Give the IUPAC name for each of the following: (12.3)

a. 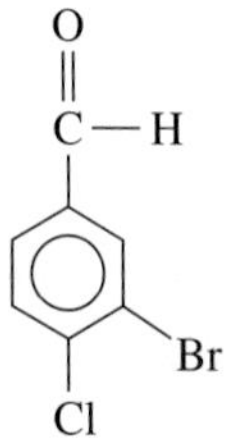

b. 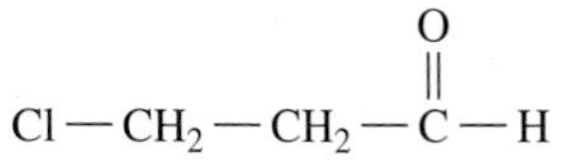

c. 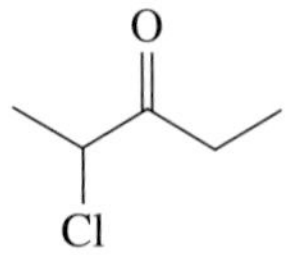

12.54 Give the IUPAC name for each of the following: (12.3)

a. 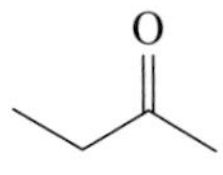

b. 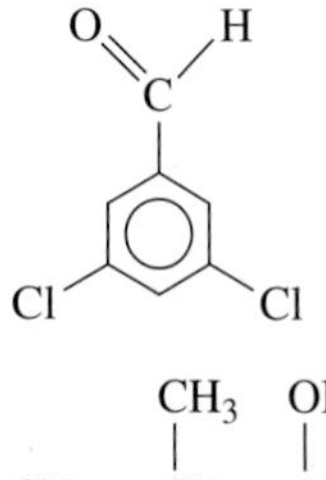

c. 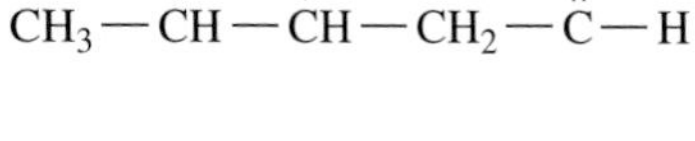

12.55 Draw the condensed structural formula for each of the following: (12.3)

a. 4-chlorobenzaldehyde **b.** 3-chloropropionaldehyde
c. ethyl methyl ketone **d.** 3-methylhexanal

12.56 Draw the condensed structural formula for each of the following: (12.3)

a. formaldehyde **b.** 2-chlorobutanal
c. 3-methyl-2-hexanone **d.** 3,5-dimethylhexanal

12.57 Which of the following aldehydes and ketones are soluble in water? (12.3)

a. 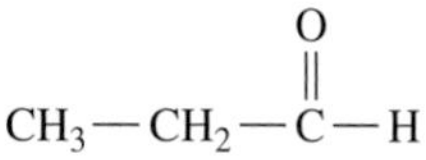

b. 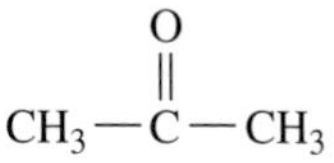

c. 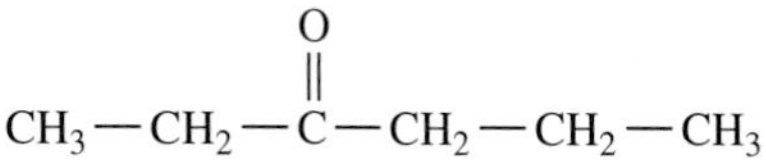

12.58 Which of the following aldehydes and ketones are soluble in water? (12.3)

a. 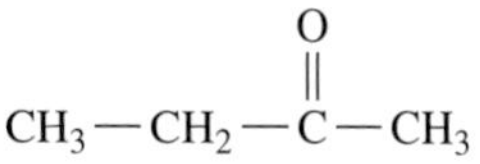

b. 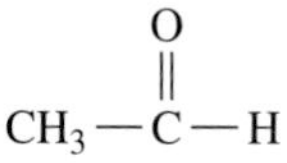

c. 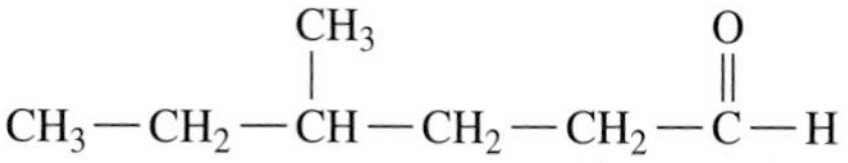

12.59 Draw the condensed structural formula for the ketone or carboxylic acid product when each of the following is oxidized: (12.4)

a. $CH_3-CH_2-CH_2-OH$

b. 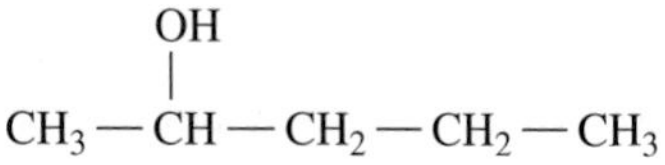

c. 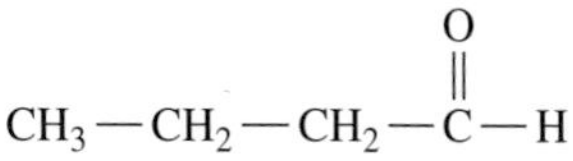

d. 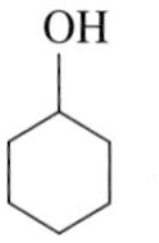

12.60 Draw the condensed structural formula for the ketone or carboxylic acid product when each of the following is oxidized: (12.4)

a. $CH_3-CH_2-CH_2-CH_2-OH$

b. 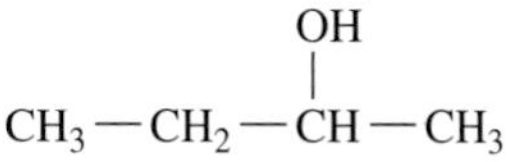

c. 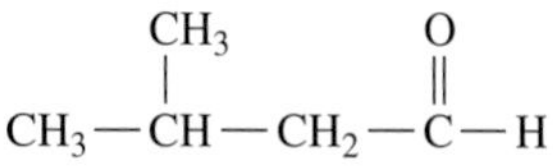

d. 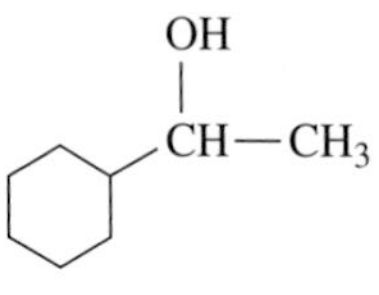

CHALLENGE QUESTIONS

The following groups of questions are related to the topics in this chapter. However, they do not all follow the chapter order, and they require you to combine concepts and skills from several sections. These questions will help you increase your critical thinking skills and prepare for your next exam.

12.61 Draw the condensed structural formulas and give the IUPAC names for all the alcohols that have the formula $C_5H_{12}O$. (12.1)

12.62 Draw the skeletal formulas and give the IUPAC names for all the aromatic aldehydes that have the formula $C_8H_8O_1$. (12.3)

12.63 A compound with the formula C_4H_8O is synthesized from 2-methyl-1-propanol and oxidizes easily to give a carboxylic acid. Draw the condensed structural formula and give the IUPAC name for the compound. (12.3, 12.4)

12.64 A compound with the formula $C_5H_{10}O$ oxidizes to give 3-pentanone. Draw the condensed structural formula and give the IUPAC name for the compound. (12.3, 12.4)

12.65 When 2-butanol is heated with strong acid, it dehydrates to form compound **A and B** (C_4H_8). When 2-butanol is oxidized, compound **C** (C_4H_8O) forms. Draw the skeletal formulas and give the UPAC names for compounds **A**, **B**, and **C**. (12.3, 12.4)

12.66 Compound **X** is 2-propanol. When compound **X** is heated with strong acid, it dehydrates to form compound **Y** (C_3H_6). When compound **X** is oxidized, compound **Z** (C_3H_6O) forms, which cannot be oxidized further. Draw the condensed structural formulas and give the IUPAC names for compounds **X**, **Y**, and **Z**. (12.3, 12.4)

ANSWERS

Answers to Study Checks

12.1 3-chloro-l-butanol

12.2 4-methylphenol

12.3 secondary (2°)

12.4 propanal (IUPAC), propionaldehyde (common)

12.5 ethyl propyl ketone

12.6 Ethanol can form more hydrogen bonds to water and would be more soluble than ethanal.

12.7 cyclohexene

12.8 2-Pentanol loses two hydrogen atoms to form a ketone.

$$CH_3-\overset{\overset{\displaystyle O}{\|}}{C}-CH_2-CH_2-CH_3$$

12.9 $$CH_3-CH_2-CH_2-\overset{\overset{\displaystyle O}{\|}}{C}-H + H_2 \xrightarrow{Pt} CH_3-CH_2-CH_2-CH_2-OH$$

Answers to Selected Questions and Problems

12.1 **a.** 1-propanol **b.** 3-pentanol
c. 4, 5, 6-trimethyl-3-heptanol
d. 3-methylphenol

12.3 **a.**

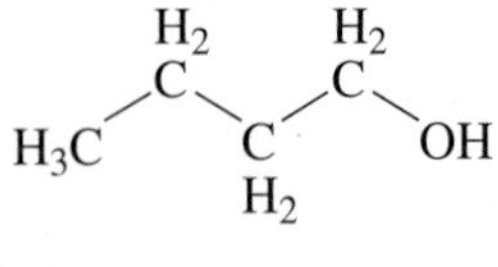

b.

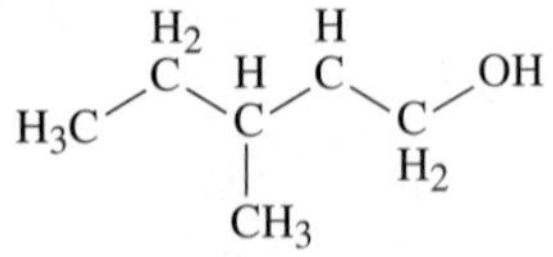

c.

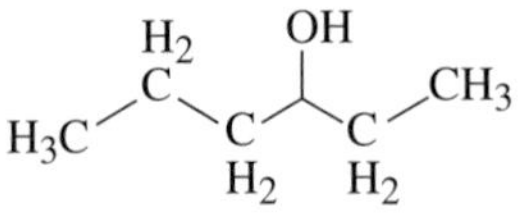

d.

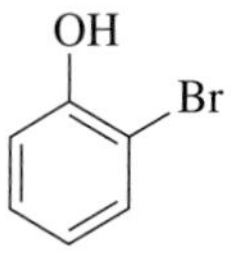

12.5 **a.** dimethyl ether **b.** cyclobutyl ethyl ether

12.7 **a.**

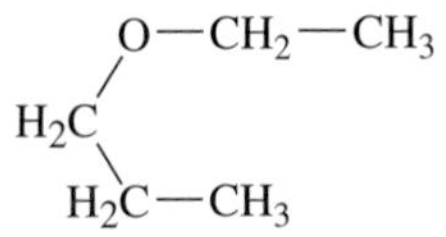

b.

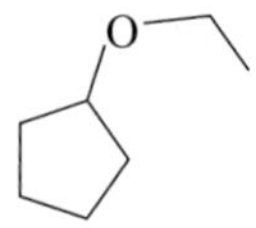

12.9 **a.** 1° **b.** 2° **c.** 3° **d.** 2°

12.11 **a.** soluble **b.** insoluble

12.13 **a.** Methanol can form hydrogen bonds with water, but ethane cannot.
b. 2-Propanol is more soluble because it has a shorter carbon chain.

12.15 **a.** acetone **b.** butyraldehyde
c. butyl methyl ketone

12.17 **a.** 2-butanone
b. 4,5-dimethyl-3-hexanone
c. 3,4-dichlorocyclohexanone
d. 3,5-dimethylbenzaldehyde

12.19 **a.** $$H-\overset{\overset{\displaystyle O}{\|}}{C}-H$$

b. 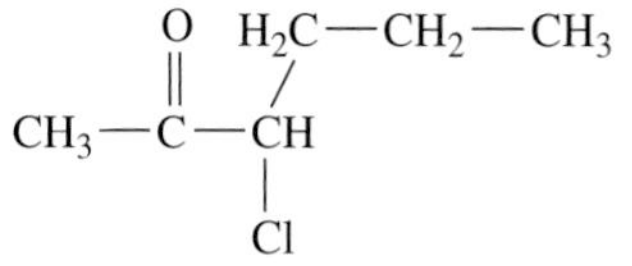

c. 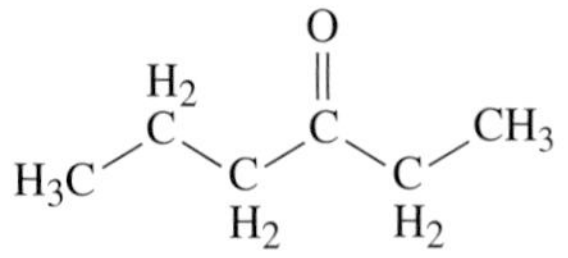

d. 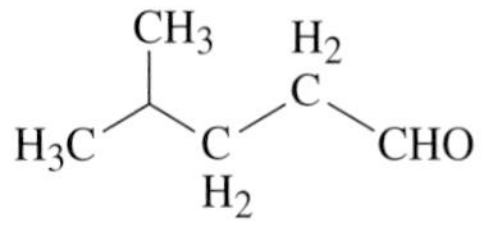

12.21 Ethanol contain a polar hydroxyl group (—OH), which forms hydrogen bonds with water molecules, hence it is soluble. Hexane is insoluble due to non-polar hydrocarbon chain.

a. Butanal has a polar carbonyl group (carbon–oxygen double bond), which has a partially negative oxygen atom and a partially positive carbon atom. Because the electronegative oxygen atom it forms hydrogen bonds with water molecules, hence it is soluble while hexanone is insoluble due to its long hydrocarbon chain.

b. Propanal has a polar carbonyl group due to which it forms hydrogen bonds with water molecules, hence it is soluble. Octanal is insoluble due to its long hydrocarbon chain.

12.23 a. 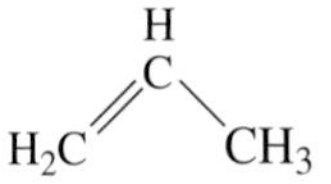 **b.** 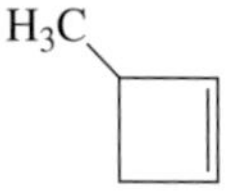

c. 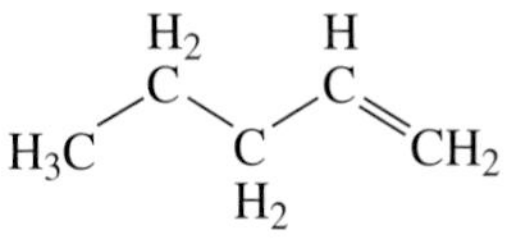

12.25 a. $CH_3-CH_2-\overset{\overset{\displaystyle O}{\|}}{C}-H$

b. $CH_3-\overset{\overset{\displaystyle O}{\|}}{C}-CH_2-CH_2-CH_2-CH_3$

c. 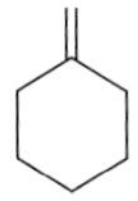

12.27 a. $CH_3-CH_2-CH_2-CH_2-\overset{\overset{\displaystyle O}{\|}}{C}-H$;

$CH_3-CH_2-CH_2-CH_2-\overset{\overset{\displaystyle O}{\|}}{C}-OH$

b. $CH_3-\underset{}{\overset{\overset{\displaystyle CH_3}{|}}{CH}}-CH_2-\overset{\overset{\displaystyle O}{\|}}{C}-H$;

$CH_3-\overset{\overset{\displaystyle CH_3}{|}}{CH}-CH_2-\overset{\overset{\displaystyle O}{\|}}{C}-OH$

c. $CH_3-CH_2-CH_2-\overset{\overset{\displaystyle O}{\|}}{C}-H$;

$CH_3-CH_2-CH_2-\overset{\overset{\displaystyle O}{\|}}{C}-OH$

12.29 a. $CH_3-CH_2-CH_2-CH_2-OH$

b. $CH_3-\overset{\overset{\displaystyle OH}{|}}{CH}-CH_3$

c. $CH_3-CH_2-CH_2-CH_2-CH_2-CH_2-OH$

d. $CH_3-\overset{\overset{\displaystyle CH_3}{|}}{CH}-\overset{\overset{\displaystyle OH}{|}}{CH}-CH_2-CH_3$

12.31 a. $2CH_3OH(g) + 3O_2(g) \longrightarrow 2CO_2(g) + 4H_2O(g) + \text{energy}$

b. $C_4H_9OH(g) + 6O_2(g) \longrightarrow 4CO_2(g) + 5H_2O(g) + \text{energy}$

12.33 ketone, phenol, ether, aromatic

12.35 phenol

12.37 Compound **a** will give a positive Tollens' test.

12.39 a. aldehyde, aromatic **b.** aldehyde, alkene, aromatic

12.41 a. 2° **b.** 1°
c. 2° **d.** 3°

12.43 a. 2,2-dimethyl-1-propanol
b. 4-bromo-3-methyl-2-pentanol
c. 4-bromo-3-methyl phenol

12.45 a. 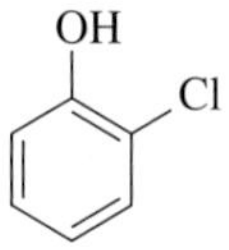 **b.** 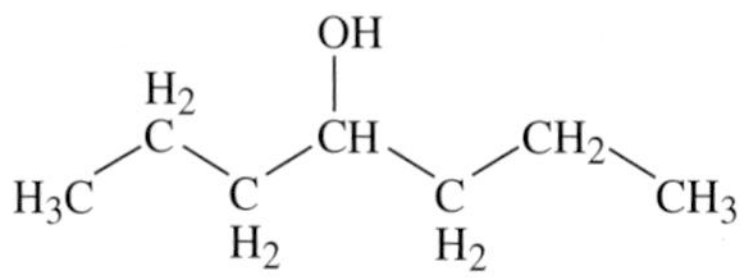

c. 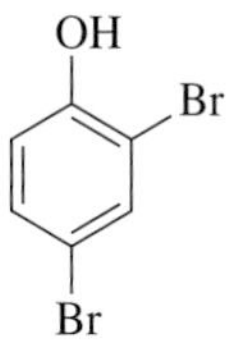

d. $H_3C-\overset{H_2}{C}-\underset{\underset{\displaystyle CH_3}{|}}{CH}-CH_2-OH$

12.47 a. Propanol contain a polar hydroxyl group (—OH), which forms hydrogen bonds with water molecules, hence it is soluble. Pentane is insoluble due to non-polar hydrocarbon chain.

b. Ethanol contain a polar hydroxyl group (—OH), which forms hydrogen bonds with water molecules, hence it is soluble while there is no H-bonding of water with ethyl methyl ether so it is insoluble.

c. Methanol contain a polar hydroxyl group (—OH), which forms hydrogen bonds with water molecules, hence it is soluble. Heptanol is insoluble due to its long hydrocarbon chain.

12.49 a. $CH_3-CH{=}CH_2$ **b.** $CH_3-CH_2-\overset{\overset{\displaystyle O}{\|}}{C}-H$

c. $CH_3-CH{=}CH-CH_2-CH_3$

d. 

12.51 a. $CH_3-CH_2-CH(OH)-CH_3$

b. 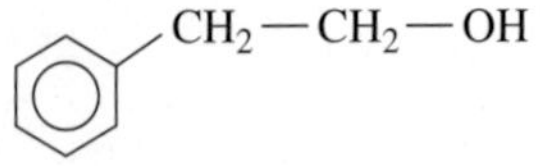

c. 

12.53 a. 3-bromo-4-chlorobenzaldehyde
b. 3-chloropropanal
c. 2-chloro-3-pentanone

12.55 a. 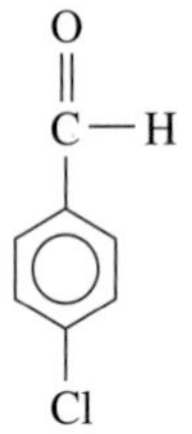

b. $Cl-CH_2-CH_2-C(=O)-H$

c. $CH_3-CH_2-C(=O)-CH_3$

d. $CH_3-CH_2-CH_2-CH(CH_3)-CH_2-C(=O)-H$

12.57 **a** and **b**

12.59 a. $CH_3-CH_2-C(=O)-OH$

b. $CH_3-C(=O)-CH_2-CH_2-CH_3$

c. $CH_3-CH_2-CH_2-C(=O)-OH$

d. 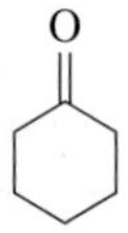

12.61

$CH_3-CH_2-CH_2-CH_2-CH_2-OH$ 1-Pentanol

$CH_3-CH(OH)-CH_2-CH_2-CH_3$ 2-Pentanol

$CH_3-CH_2-CH(OH)-CH_2-CH_3$ 3-Pentanol

$HO-CH_2-CH(CH_3)-CH_2-CH_3$ 2-Methyl-1-butanol

$HO-CH_2-CH_2-CH(CH_3)-CH_3$ 3-Methyl-1-butanol

$CH_3-C(CH_3)(OH)-CH_2-CH_3$ 2-Methyl-2-butanol

$CH_3-CH(OH)-CH(CH_3)-CH_3$ 3-Methyl-2-butanol

$CH_3-C(CH_3)_2-CH_2-OH$ 2,2-Dimethylpropanol

12.63 $CH_3-CH(CH_3)-C(=O)-H$ 2-Methylpropanal

COMBINING IDEAS FROM Chapters 10 to 12

CI.21 A metal (M) completely reacts with 34.8 mL of a 0.520 M HCl solution to form $MCl_3(aq)$ and $H_2(g)$. (4.1, 7.1, 7.2, 7.3, 8.6, 10.6)

When a metal reacts with a strong acid, hydrogen gas forms.

a. Write the balanced chemical equation for the reaction of the metal M and HCl(aq).
b. What volume, in milliliters, of H_2 is produced at STP?
c. How many moles of metal M reacted?
d. If the metal has a mass of 0.420 g, use your results from part **c** to determine the molar mass of the metal M.
e. What are the name and symbol of metal M in part **d**?
f. Write the balanced chemical equation using the symbol of the metal from part **e**.

CI.22 In a teaspoon (5.0 mL) of a common liquid antacid, there are 400. mg of $Mg(OH)_2$ and 400. mg of $Al(OH)_3$. A 0.080 M HCl solution, which is similar to stomach acid, is used to neutralize 5.0 mL of the liquid antacid. (9.5, 10.5, 10.6)

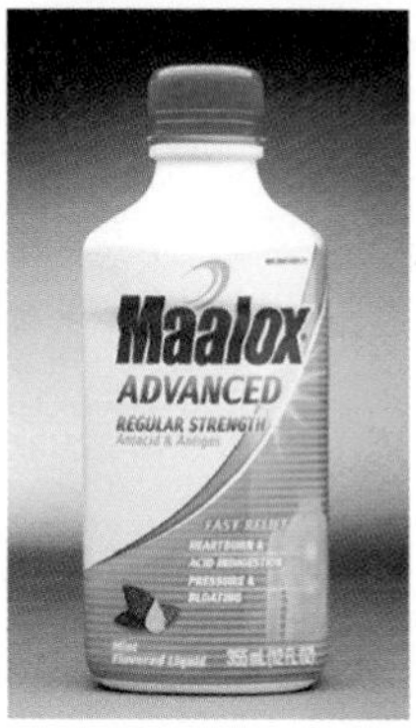

An antacid neutralizes stomach acid.

a. Write the balanced chemical equation for the neutralization of HCl and $Mg(OH)_2$.
b. Write the balanced chemical equation for the neutralization of HCl and $Al(OH)_3$.
c. What is the pH of the HCl solution?
d. How many milliliters of the HCl solution are needed to neutralize the $Mg(OH)_2$?
e. How many milliliters of the HCl solution are needed to neutralize the $Al(OH)_3$?

CI.23 Acetone (propanone), a clear liquid solvent with an acrid odor, is used to remove nail polish, paints, and resins. It has a low boiling point and is highly flammable. Acetone has a density of 0.786 g/mL. (6.6, 7.2, 7.4, 7.7, 12.3)

Acetone consists of carbon atoms (black), hydrogen atoms (white), and an oxygen atom (red).

a. What is the molecular formula of acetone?
b. What is the molar mass of acetone?
c. Identify the bonds C—C, C—H, and C—O in a molecule of acetone as polar covalent or nonpolar covalent.
d. Write the balanced chemical equation for the combustion of acetone.
e. How many grams of oxygen gas are needed to react with 15.0 mL of acetone?

CI.24 One of the components of gasoline is octane, C_8H_{18}, which has a density of 0.803 g/mL. The combustion of 1 mole of octane provides 5510 kJ. A hybrid car has a fuel tank with a capacity of 11.9 gal and a gas mileage of 45 mi/gal. (7.1, 7.3, 7.8, 11.2)

Octane is one of the components of gasoline.

a. Write the balanced chemical equation for the combustion of octane.
b. Is the combustion of octane endothermic or exothermic?
c. How many moles of octane are in one tank of fuel, assuming it is all octane?
d. If the total mileage of this hybrid car for one year is 24 500 mi, how many kilograms of carbon dioxide would be produced from the combustion of the fuel, assuming it is all octane?

CI.25 Butyraldehyde is a clear liquid solvent with an unpleasant odor. Butyraldehyde has a density of 0.802 g/mL. (7.3, 7.4, 7.6, 7.7, 8.6, 12.3, 12.4)

The unpleasant odor of old gym socks is due to butyraldehyde.

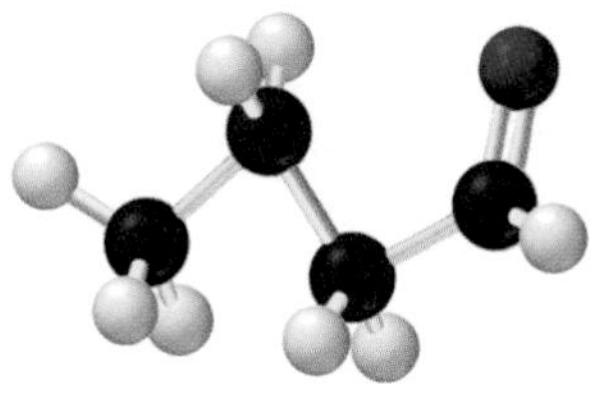

a. Draw the condensed structural formula for butyraldehyde.
b. Draw the skeletal formula for butyraldehyde.
c. What is the IUPAC name for butyraldehyde?
d. Draw the condensed structural formula for the alcohol that is produced when butyraldehyde is reduced.
e. Write the balanced chemical equation for the complete combustion of butyraldehyde.
f. How many grams of oxygen gas are needed to completely react with 15.0 mL of butyraldehyde?
g. How many liters of carbon dioxide gas are produced at STP in part **f**?

CI.26 Ionone is a compound that gives violets their aroma. The small, edible, purple flowers of violets are used on salads and to make teas. Liquid ionone has a density of 0.935 g/mL. (7.1, 7.2, 8.6, 11.5, 11.6, 12.3)

The aroma of violets is due to ionone.

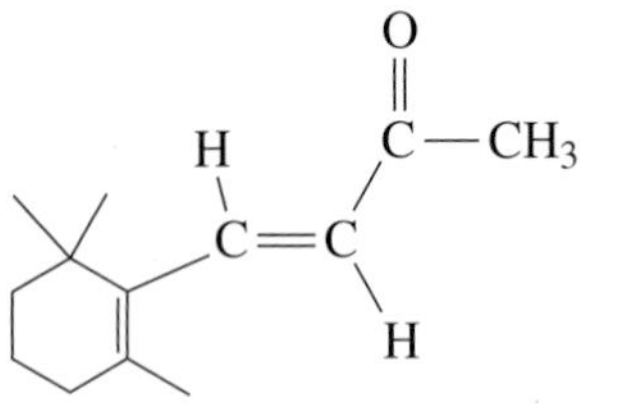

Ionone

a. What functional groups are present in ionone?
b. Is the double bond on the side chain cis or trans?
c. What are the molecular formula and molar mass of ionone?
d. How many moles are in 2.00 mL of ionone?
e. When ionone reacts with hydrogen in the presence of a platinum catalyst, hydrogen adds to the double bonds and converts the ketone group to an alcohol. Draw the condensed structural formula and give the molecular formula for the product.
f. How many milliliters of hydrogen gas are needed at STP to completely react 5.0 mL of ionone?

ANSWERS

CI.21 **a.** $2M(s) + 6HCl(aq) \longrightarrow 2MCl_3(aq) + 3H_2(g)$
b. 203 mL of H_2
c. 6.03×10^{-3} mole of M
d. 69.7 g/mole
e. Gallium; Ga
f. $2Ga(s) + 6HCl(aq) \longrightarrow 2GaCl_3(aq) + 3H_2(g)$

CI.23 **a.** C_3H_6O
b. 58.08 g/mole
c. Nonpolar covalent bonds: C—C, C—H; polar covalent bond: C—O
d. $C_3H_6O(l) + 4O_2(g) \xrightarrow{\Delta} 3CO_2(g) + 3H_2O(g) + \text{energy}$
e. 26.0 g of O_2

CI.25 **a.** $CH_3-CH_2-CH_2-\overset{\overset{\displaystyle O}{\|}}{C}-H$

b. 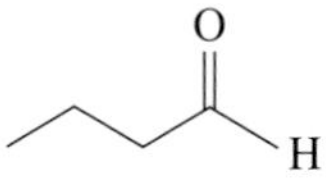

c. butanal
d. $CH_3-CH_2-CH_2-CH_2-OH$
e. $2C_4H_8O(l) + 11O_2(g) \xrightarrow{\Delta} 8CO_2(g) + 8H_2O(g) + \text{energy}$
f. 29.4 g of O_2
g. 14.9 L of CO_2

13 Carbohydrates

Kate has recently been diagnosed as having type 2 diabetes. Her doctor recommends that she take a series of educational courses taught by Paula, a diabetes nurse, who teaches Kate to test her blood glucose levels before and after a meal. Paula explains to Kate that her pre-meal number should be 110 mg/dL or less, and if it increases by more than 50 mg/dL, she needs to lower the amount of carbohydrates she consumes. Complex carbohydrates are long chains of glucose molecules that we obtain from ingesting breads and grains. They are broken down in the body into glucose, which is a simple carbohydrate or a monosaccharide.

Kate and Paula proceed to plan several meals. Because a meal should contain about 45 to 60 g of carbohydrates, they combine fruits and vegetables that have high and low levels of carbohydrates in the same meal, in order to stay within the recommended range. Kate and Paula also discuss the fact that complex carbohydrates in the body take longer to break down into glucose and, therefore, raise the blood sugar level more gradually.

CAREER Diabetes Nurse

Diabetes nurses teach patients about diabetes, so they can self-manage and control their condition. This includes education on proper diets and nutrition for both diabetic and pre-diabetic patients. Diabetes nurses help patients learn to monitor their medication, blood sugar levels, and to look for symptoms like diabetic nerve damage and vision loss. Diabetes nurses may also work with patients who have been hospitalized due to complications from their disease. This requires a thorough knowledge of the endocrine system, as this system is often involved with obesity and other diseases, resulting in some overlap between diabetes and endocrinology nursing.

LOOKING AHEAD

Carbohydrates contained in foods such as pasta and bread provide energy for the body.

Carbohydrates are the most abundant of all the organic compounds in nature. In plants, energy from the Sun converts carbon dioxide and water into the carbohydrate glucose. Polysaccharides are made when many glucose molecules link to form long-chain polymers of starch that store energy or link in a different way to form cellulose to build the structure of the plant. About 65% of the foods in our diet consist of carbohydrates. Each day, we utilize carbohydrates in foods such as bread, pasta, potatoes, and rice. Other carbohydrates called disaccharides include sucrose (table sugar) and lactose in milk. During digestion and cellular metabolism, carbohydrates are converted into glucose, which is oxidized further in our cells to provide our bodies with energy and to provide the cells with carbon atoms for building molecules of proteins, lipids, and nucleic acids. In addition to providing the structural framework of plants, cellulose has other important uses too. The wood in our furniture, the pages in this book, and the cotton in our clothing are all made of cellulose.

CHAPTER READINESS*

CORE CHEMISTRY SKILLS

- Identifying Functional Groups (12.1)
- Naming Alcohols and Phenols (12.1)
- Naming Aldehydes and Ketones (12.3)
- Writing Equations for the Dehydration and Oxidation of Alcohols (12.4)

*These Core Chemistry Skills from previous chapters are listed here for your review as you proceed to the new material in this chapter.

LEARNING GOAL

Classify a monosaccharide as an aldose or a ketose, and indicate the number of carbon atoms.

13.1 Carbohydrates

Carbohydrates such as table sugar, lactose, and cellulose are all made of carbon, hydrogen, and oxygen. Simple sugars, which have formulas of $C_n(H_2O)_n$, were once thought to be hydrates of carbon, thus the name *carbohydrate*. In a series of reactions called *photosynthesis*, energy from the Sun is used to combine the carbon atoms from carbon dioxide (CO_2) and the hydrogen and oxygen atoms of water (H_2O) into the carbohydrate glucose.

$$6CO_2 + 6H_2O + \text{energy} \underset{\text{Respiration}}{\overset{\text{Photosynthesis}}{\rightleftarrows}} \underset{\text{Glucose}}{C_6H_{12}O_6} + 6O_2$$

In the body, glucose is oxidized in a series of metabolic reactions known as *respiration*, which releases chemical energy to do work in the cells. Carbon dioxide and water are produced and returned to the atmosphere. The combination of photosynthesis and respiration is called the *carbon cycle*, in which energy from the Sun is stored in plants by photosynthesis and made available to us when the carbohydrates in our diets are metabolized (see Figure 13.1).

FIGURE 13.1 ▶ During photosynthesis, energy from the Sun combines CO_2 and H_2O to form glucose ($C_6H_{12}O_6$) and O_2. During respiration in the body, carbohydrates are oxidized to CO_2 and H_2O, while energy is produced.

Q What are the reactants and products of respiration?

Types of Carbohydrates

The simplest carbohydrates are the **monosaccharides**. A monosaccharide cannot be split or hydrolyzed into smaller carbohydrates. One of the most common carbohydrates, glucose, $C_6H_{12}O_6$, is a monosaccharide. A **disaccharide** consists of two monosaccharide units joined together, which can be split into two monosaccharide units. For example, ordinary table sugar, sucrose, $C_{12}H_{22}O_{11}$, is a disaccharide that can be split by water (hydrolysis) in the presence of an acid or an enzyme to give one molecule of glucose and one molecule of another monosaccharide, fructose.

$$\underset{\text{Sucrose}}{C_{12}H_{22}O_{11}} + H_2O \xrightarrow{H^+ \text{ or enzyme}} \underset{\text{Glucose}}{C_6H_{12}O_6} + \underset{\text{Fructose}}{C_6H_{12}O_6}$$

A **polysaccharide** is a carbohydrate that contains many monosaccharide units, which is called a *polymer*. In the presence of an acid or an enzyme, a polysaccharide can be completely hydrolyzed to yield many monosaccharide molecules.

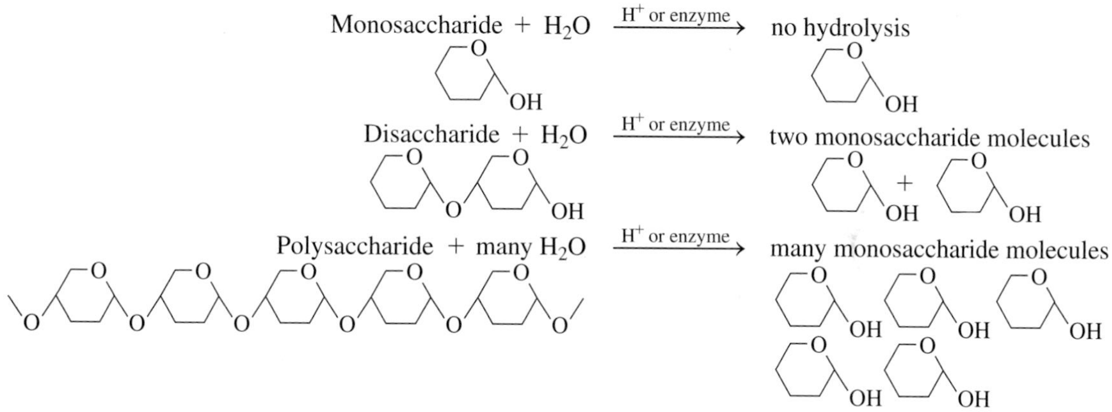

Monosaccharides

A monosaccharide has a chain of three to eight carbon atoms, one in a carbonyl group and all the others attached to hydroxyl groups (—OH). In an **aldose**, the carbonyl group is on the first carbon as an aldehyde (—CHO), whereas a **ketose** contains the carbonyl group on the second carbon atom as a ketone (C=O).

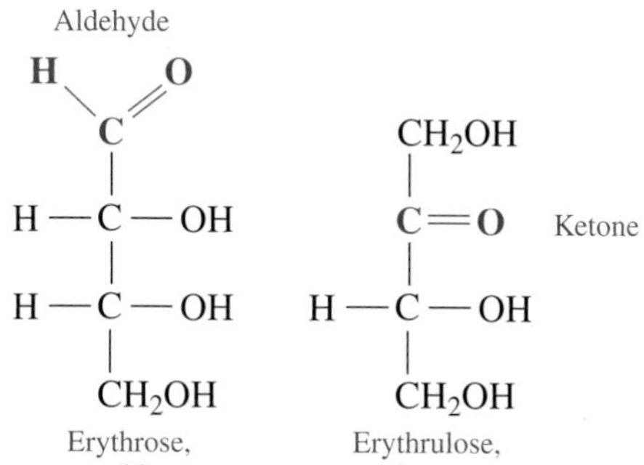

Erythrose, an aldose

Erythrulose, a ketose

A monosaccharide with three carbon atoms is a *triose*, one with four carbon atoms is a *tetrose*; a *pentose* has five carbons, and a *hexose* contains six carbons. We can use both classification systems to indicate the type of carbonyl group and the number of carbon atoms. An aldopentose is a five-carbon monosaccharide that is an aldehyde; a ketohexose is a six-carbon monosaccharide that is a ketone.

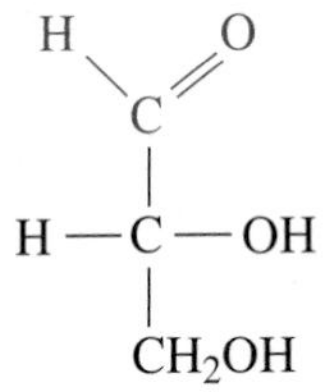

Glyceraldehyde
(aldotriose)

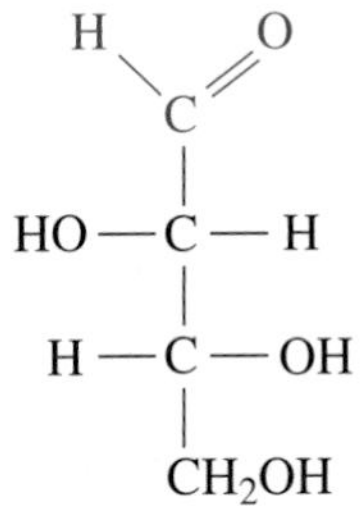

Threose
(aldotetrose)

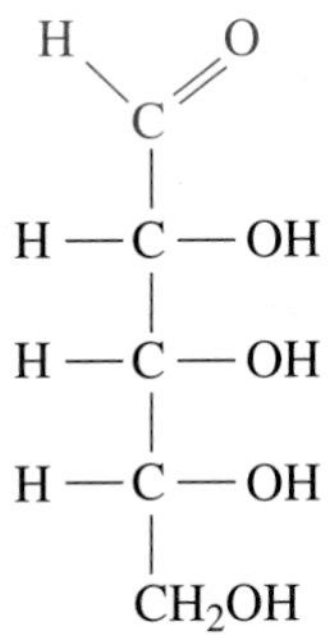

Ribose
(aldopentose)

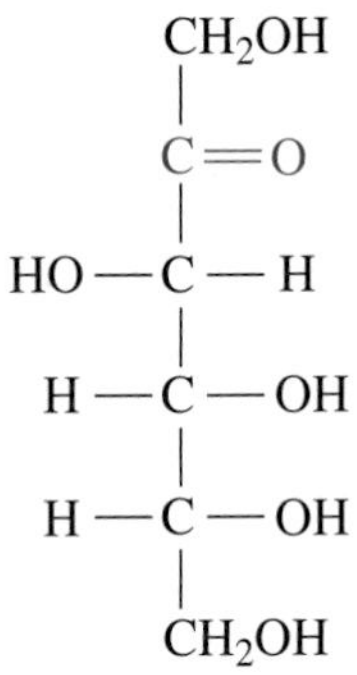

Fructose
(ketohexose)

SAMPLE PROBLEM 13.1 Monosaccharides

Classify each of the following monosaccharides as an aldopentose, ketopentose, aldohexose, or ketohexose:

a.

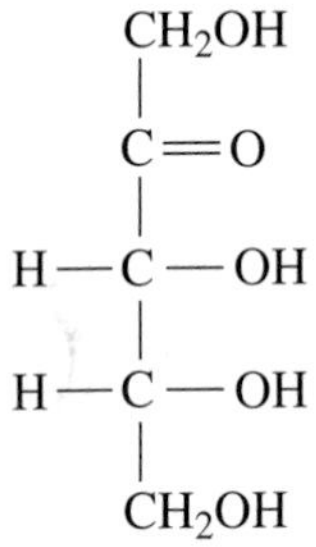

Ribulose

b.

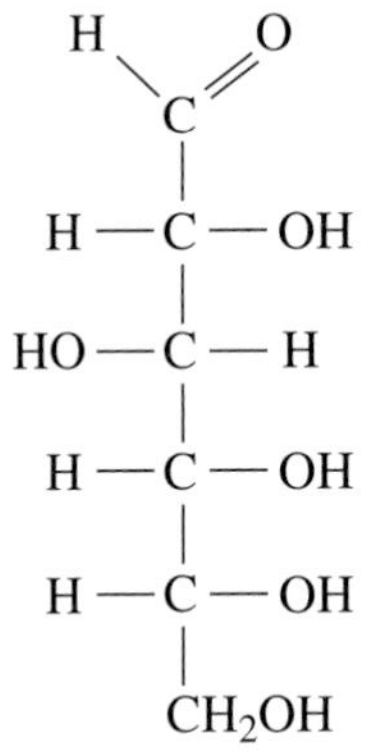

Glucose

SOLUTION

a. Ribulose has five carbon atoms (pentose) and is a ketone, which makes it a ketopentose.
b. Glucose has six carbon atoms (hexose) and is an aldehyde, which makes it an aldohexose.

STUDY CHECK 13.1

Classify the following monosaccharide, erythrose, as an aldotetrose, ketotetrose, aldopentose, or ketopentose:

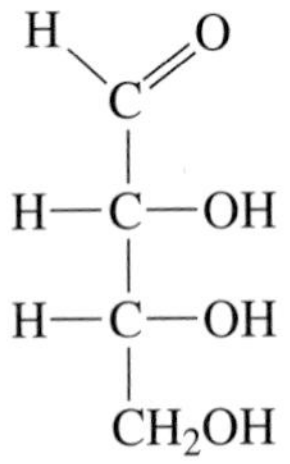

QUESTIONS AND PROBLEMS

13.1 Carbohydrates

LEARNING GOAL Classify a monosaccharide as an aldose or a ketose, and indicate the number of carbon atoms.

13.1 What reactants are needed for photosynthesis and respiration?

13.2 What is the chemical equation for respiration?

13.3 Distinguish between monosaccharide and disaccharide.

13.4 Define polysaccharide. Provide one example.

13.5 What functional groups are found in all monosaccharides?

13.6 What is the difference between an aldose and a ketose?

13.7 What are the functional groups and number of carbons in an aldopentose?

13.8 What are the functional groups and number of carbons in a ketohexose?

13.9 Classify each of the following monosaccharides as an aldopentose, ketopentose, aldohexose, or ketohexose:

a.

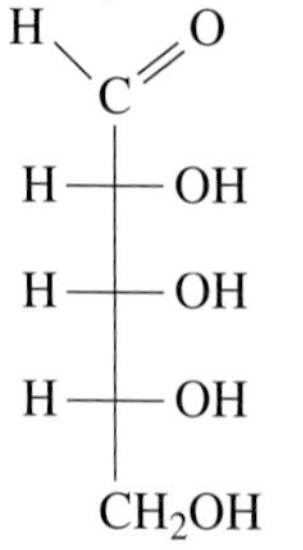

Ribose

b.

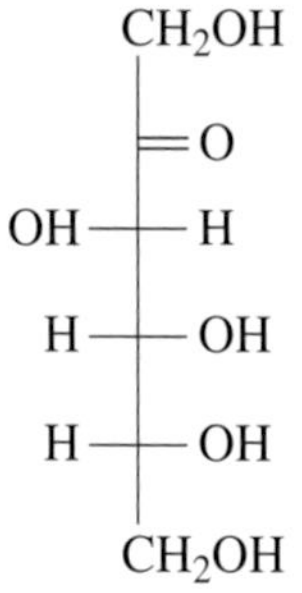

Fructose

13.10 Classify each of the following monosaccharides as an aldopentose, ketopentose, aldohexose, or ketohexose:

a.

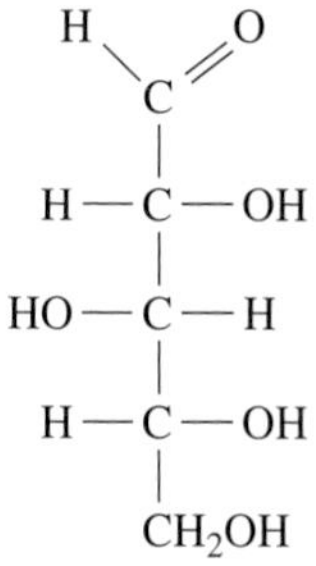

Xylose

b.

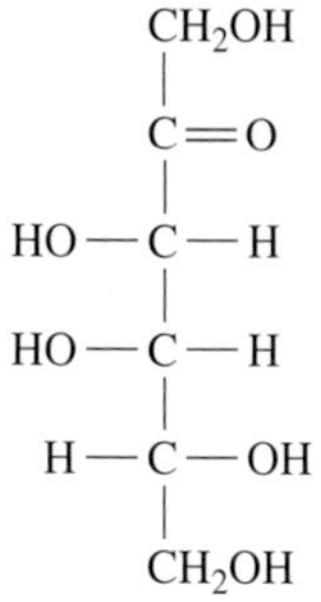

Tagatose

13.2 Chiral Molecules

LEARNING GOAL

Identify chiral and achiral carbon atoms in an organic molecule.

In the preceding chapters, we have looked at isomers. Let's review those now. Molecules are structural isomers when they have the same molecular formula, but different bonding arrangements.

Structural Isomers

C_2H_6O	CH_3-CH_2-OH Ethanol	CH_3-O-CH_3 Dimethyl ether
C_3H_6O	$CH_3-CH_2-\overset{\displaystyle O}{\overset{\|}{C}}-H$ Propanal	$CH_3-\overset{\displaystyle O}{\overset{\|}{C}}-CH_3$ Propanone

Another group of isomers called *stereoisomers* has identical molecular formulas, too, but they are not structural isomers. In **stereoisomers**, the atoms are bonded in the same sequence but differ in the way they are arranged in space.

Chirality

Everything has a mirror image. If you hold your right hand up to a mirror, you see its mirror image, which matches your left hand (see Figure 13.2). If you turn your palms toward each other, one hand is the mirror image of the other. If you look at the palms of your hands, your thumbs are on opposite sides. If you then place your right hand over your left hand, you cannot match up all the parts of the hands: palms, backs, thumbs, and little fingers.

FIGURE 13.2 ▶ The left and right hands are chiral because they have mirror images that cannot be superimposed on each other.

Q Why are your shoes chiral objects?

The thumbs and little fingers can be matched, but then the palms or backs of your hands are facing each other. Your hands are mirror images that cannot be superimposed on each other. When the mirror images of organic molecules cannot be completely matched, they are *nonsuperimposable*.

Objects such as hands that have nonsuperimposable mirror images are **chiral** (pronunciation '*kai-rel*). Left and right shoes are chiral; left- and right-handed golf clubs are chiral. When we think of how difficult it is to put a left-hand glove on our right hand, put a right shoe on our left foot, or use left-handed scissors if we are right-handed, we begin to realize that certain properties of mirror images are very different.

When the mirror image of an object is identical and can be superimposed on the original, it is *achiral*. For example, the mirror image of a plain drinking glass is identical to the original glass, which means the mirror image can be superimposed on the glass (see Figure 13.3).

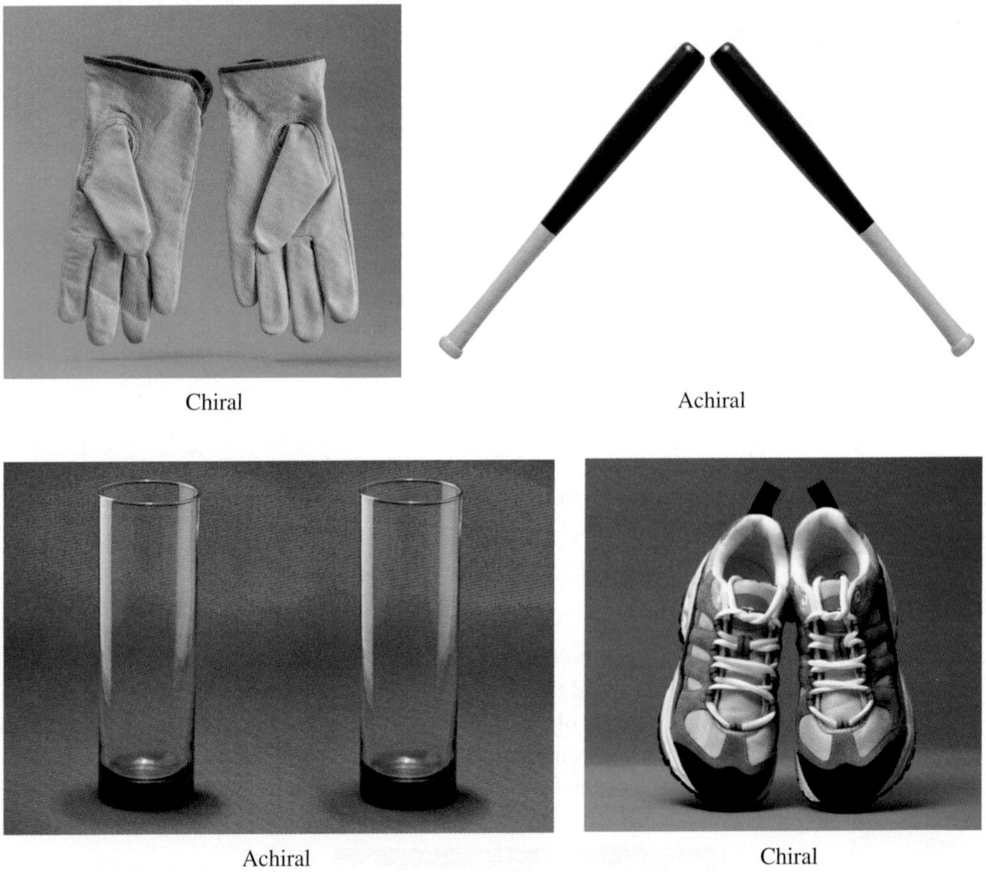

FIGURE 13.3 ▶ Everyday objects such as gloves and shoes are chiral, but an unmarked bat and a plain glass are achiral.

Q Why are some of the objects above chiral and others achiral?

CORE CHEMISTRY SKILL

Identifying Chiral Molecules

Chiral Carbon Atoms

A carbon compound is chiral if it has at least one carbon atom bonded to *four different atoms or groups*. This type of carbon atom is called a **chiral carbon** because there are two different ways that it can bond to four atoms or groups of atoms. The resulting structures are nonsuperimposable mirror images. Let's look at the mirror images of a carbon bonded to four different atoms (see Figure 13.4). If we line up the hydrogen and iodine atoms in the mirror images, the bromine and chlorine atoms appear on opposite sides. No matter how we turn the models, we cannot align all four atoms at the same time. When stereoisomers cannot be superimposed, they are called **enantiomers**.

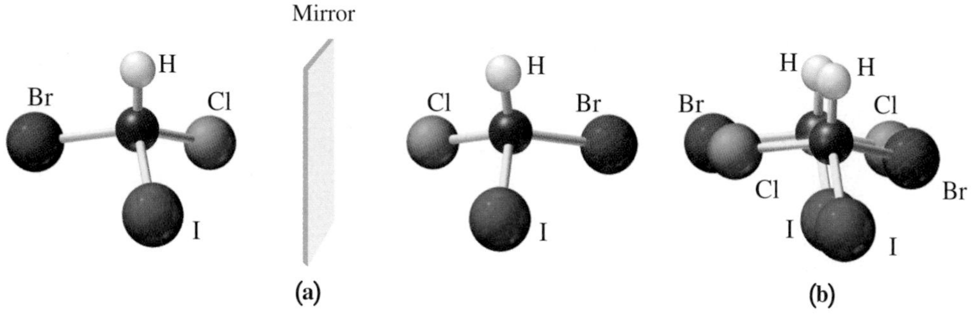

FIGURE 13.4 ▶ **(a)** The enantiomers of a chiral molecule are mirror images. **(b)** The enantiomers of a chiral molecule cannot be superimposed on each other.

Q Why is the carbon atom in this compound chiral?

If two or more atoms bonded to a carbon atom are the same, the atoms can be superimposed, if rotated, and the mirror images represent the same structure (see Figure 13.5).

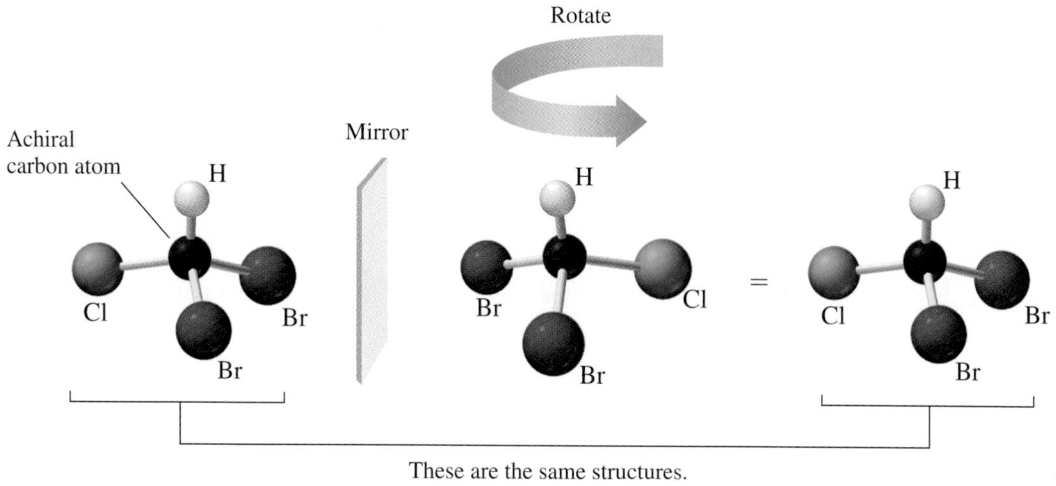

FIGURE 13.5 ▶ The mirror images of an achiral compound can be superimposed on each other.

Q Why can the mirror images of this compound be superimposed?

Explore Your World

Using Gumdrops and Toothpicks to Model Chiral Objects

Part 1: Achiral Objects

Obtain some toothpicks and orange, yellow, green, purple, and black gumdrops. Place four toothpicks into the black gumdrop making the ends of toothpicks the corners of a tetrahedron. Attach the gumdrops to the toothpicks: two orange, one green, and one yellow.

Using another black gumdrop, make a second model that is the mirror image of the original model. Now rotate one of the models and try to superimpose it on the other model. Are the models superimposable? If achiral objects have superimposable mirror images, are these models chiral or achiral?

Part 2: Chiral Objects

Using one of the original models, replace one orange gumdrop with a purple gumdrop. Now there are four different colors of gumdrops attached to the black gumdrop. Make its mirror image by replacing one orange gumdrop with a purple one on the second model. Now rotate one of the models, and try to superimpose it on the other model. Are the models superimposable? If chiral objects have nonsuperimposable mirror images, are these models chiral or achiral?

Interactive Video

Chirality

SAMPLE PROBLEM 13.2 Chiral Carbons

For each of the following, indicate whether the carbon in red is chiral or achiral.

a. 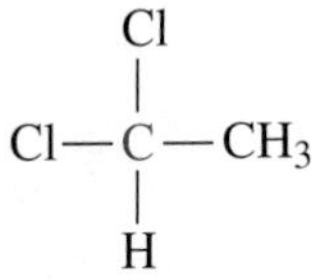

b. 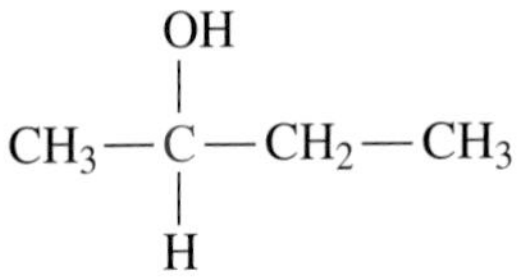

c. 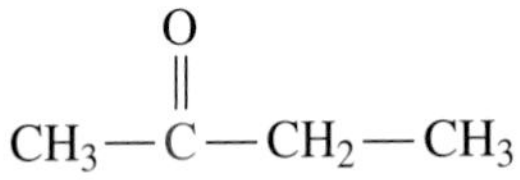

SOLUTION

a. Achiral. Two of the substituents on the carbon in red are the same. A chiral carbon must be bonded to four different groups or atoms.

b. Chiral. The carbon in red is bonded to four different groups: $—OH$, $—CH_3$, $—CH_2—CH_3$, and $—H$.

c. Achiral. The carbon in red is bonded to only three groups, not four.

STUDY CHECK 13.2

Circle the two chiral carbons in the condensed structural formula of the carbohydrate shown.

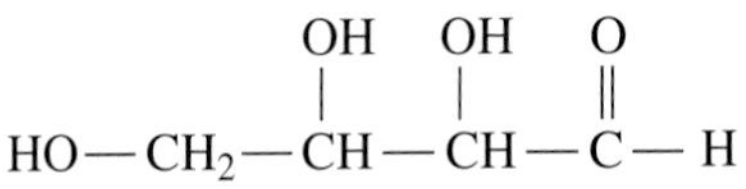

Erythrose

Drawing Fischer Projections

Emil Fischer devised a simplified system for drawing stereoisomers that shows the arrangements of the atoms around the chiral centers. Fischer received the Nobel Prize in Chemistry in 1902 for his contributions to carbohydrate and protein chemistry. Now we use his model, called a **Fischer projection**, to represent a three-dimensional structure of enantiomers. Vertical lines represent bonds that project backward from a carbon atom and horizontal lines represent bonds that project forward. In this model, the most highly oxidized carbon is placed at the top and the intersections of vertical and horizontal lines represent a carbon atom that is usually chiral.

For the simplest aldose, glyceraldehyde, the only chiral carbon is the middle carbon. In the Fischer projection, the carbonyl group, which is the most highly oxidized group, is drawn at the top above the chiral carbon and the $—CH_2OH$ group is drawn at the bottom. The $—H$ and $—OH$ groups can be drawn at each end of a horizontal line, but in two different ways. The stereoisomer that has the $—OH$ group drawn to the left of the chiral atom is designated as the L stereoisomer. The other stereoisomer with the $—OH$ group drawn to the right of the chiral carbon represents the D stereoisomer (see Figure 13.6).

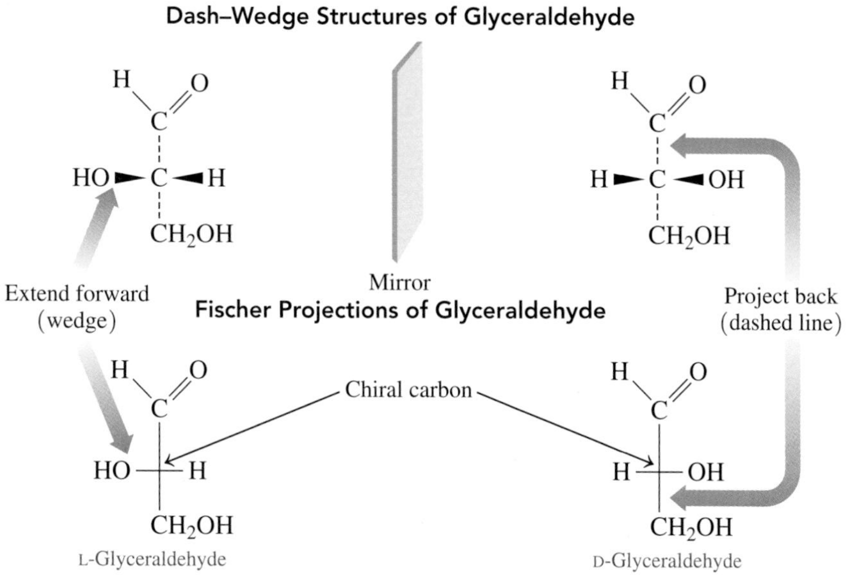

FIGURE 13.6 ▶ In a Fischer projection, the chiral carbon atom is at the center, with horizontal lines for bonds that project forward and vertical lines for bonds that point away.

Q Why does glyceraldehyde have only one chiral carbon atom?

Fischer projections can also be drawn for compounds that have two or more chiral carbons. For example, in the mirror images of erythrose, both of the carbon atoms at the intersections are chiral. To draw the mirror image for L-erythrose, we need to reverse the positions of *all* the —H and the —OH groups on the horizontal lines. For compounds with two or more chiral carbons, the designation as a D or L stereoisomer is determined by the position of the —OH group attached to the chiral carbon *farthest from the carbonyl group*.

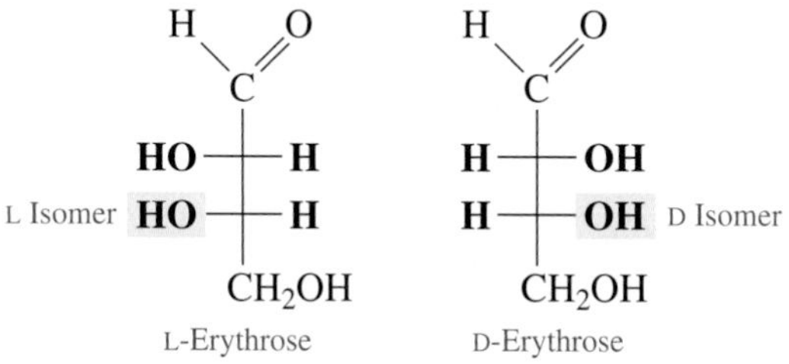

QUESTIONS AND PROBLEMS

13.2 Chiral Molecules

LEARNING GOAL Identify chiral and achiral carbon atoms in an organic molecule.

13.11 Identify each of the following structures as chiral or achiral. If chiral, indicate the chiral carbon.

a. 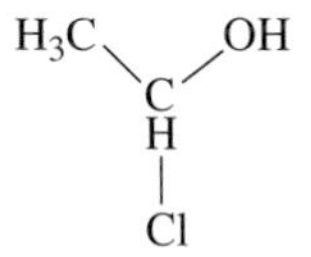

b. 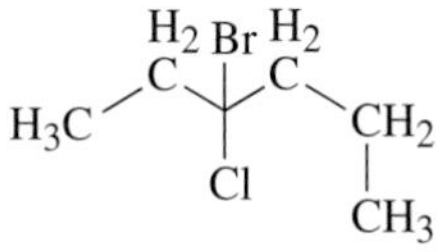

c. 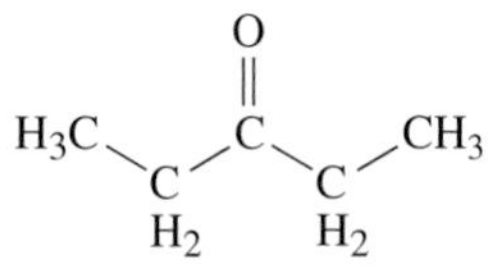

d. 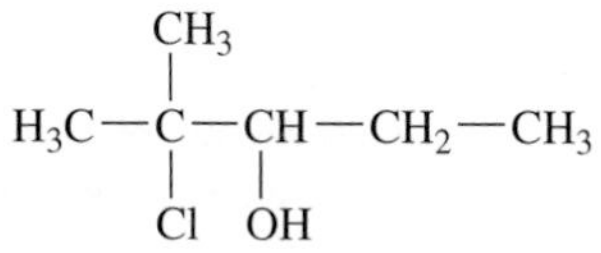

13.12 Identify each of the following structures as chiral or achiral. If chiral, indicate the chiral carbon.

a. 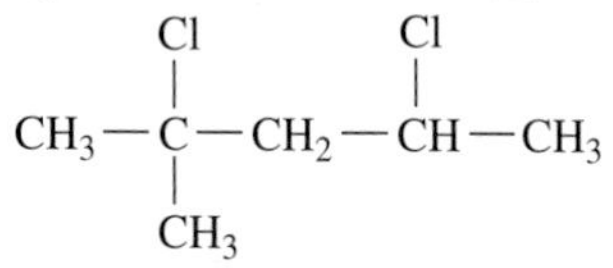

b. 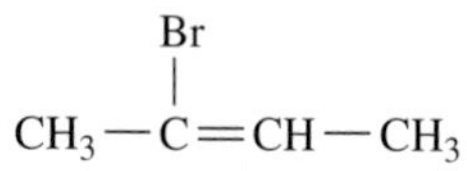

c. 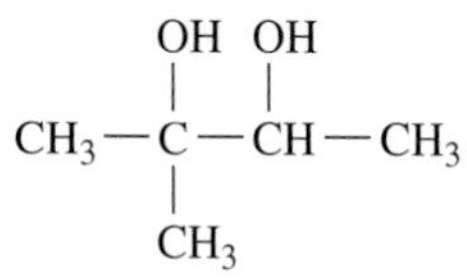

d. 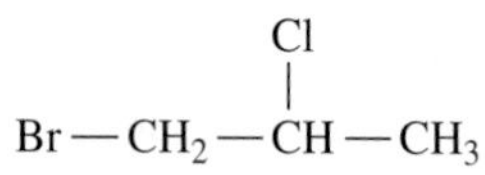

13.13 Identify the chiral carbon in each of the following naturally occurring compounds:

a. Carvone: produced by spearmint plant

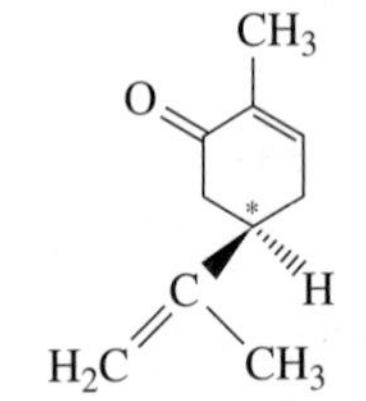

Carvone

b. Limonene: found in rinds of orange

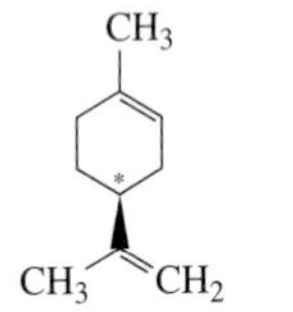

Limonene

13.14 Identify the chiral carbon in each of the following naturally occurring compounds:

a. amphetamine (Benzedrine), stimulant, used in the treatment of hyperactivity

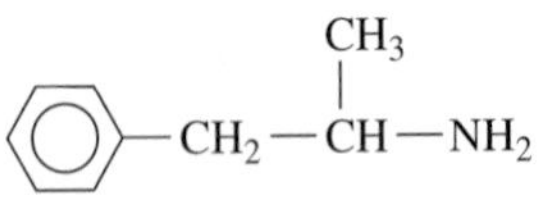

b. norepinephrine, increases blood pressure and nerve transmission

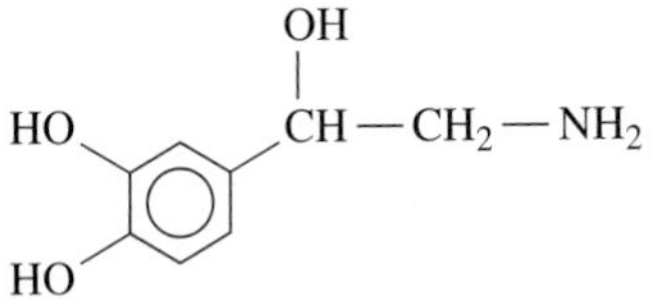

13.15 Draw the Fischer projection for each of the following dash–wedge structures:

a. 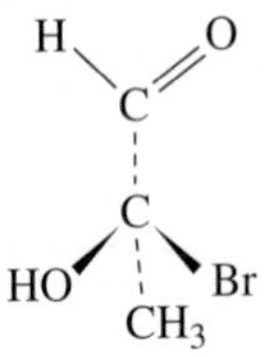 **b.** 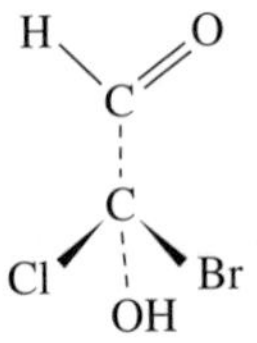 **c.** 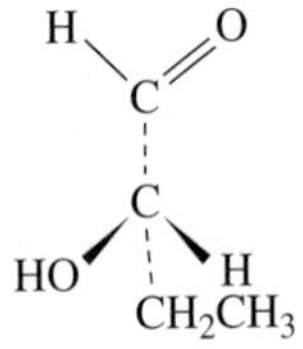

13.16 Draw the Fischer projection for each of the following dash–wedge structures:

a. 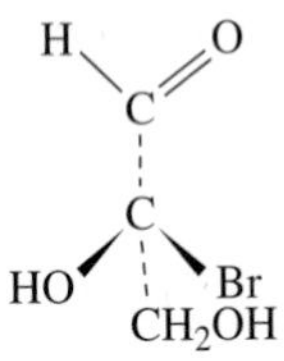 **b.** 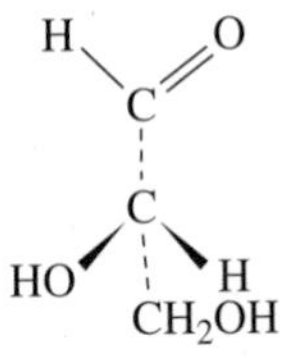 **c.** 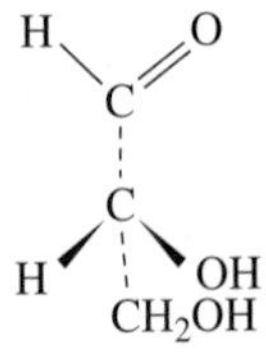

13.17 Indicate whether each pair of Fischer projections represents enantiomers or identical structures.

a. 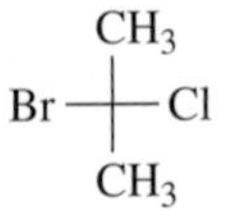 and 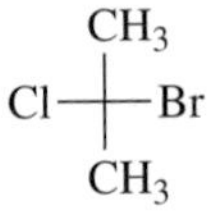

b. 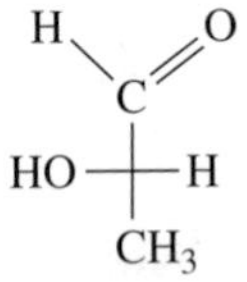 and 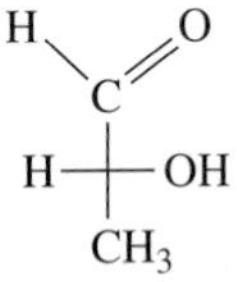

c. 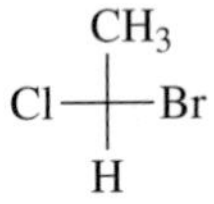 and 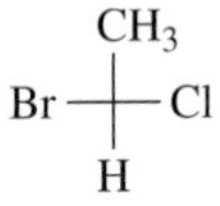

d. 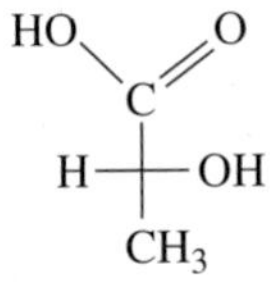 and 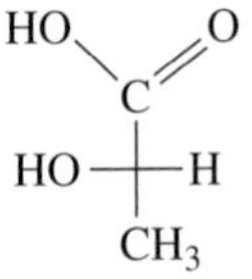

13.18 Indicate whether each pair of Fischer projections represents enantiomers or identical structures.

a. 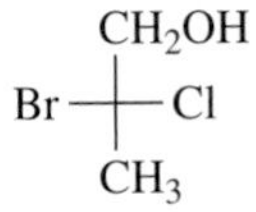 and 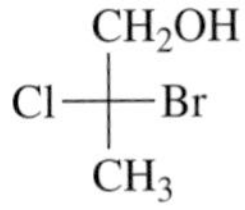

b. 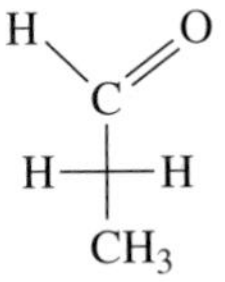 and 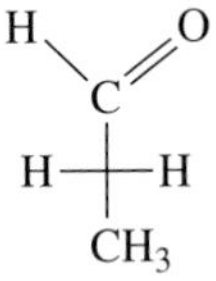

c. 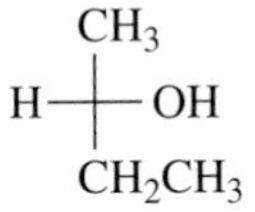 and 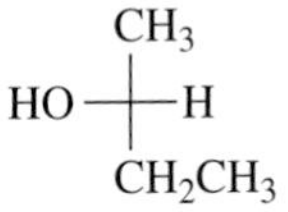

d. 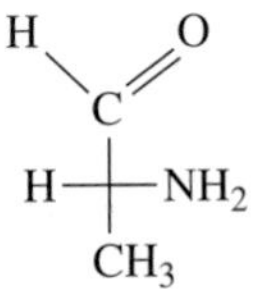 and 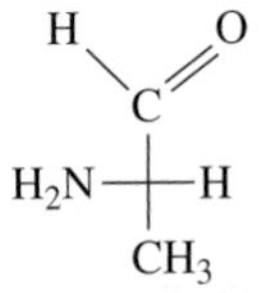

13.19 How is a Fischer projection identified as a D or an L stereoisomer?

13.20 Draw the Fischer projection for D-glyceraldehyde and L-glyceraldehyde.

Chemistry Link to Health

Enantiomers in Biological Systems

Molecules in nature also have mirror images, and often one stereoisomer has a different biological effect than the other one. For some compounds, one enantiomer has a certain odor, and the other enantiomer has a completely different odor. For example, the oils that give the scent of spearmint and caraway seeds are *both* composed of carvone. However, carvone has one chiral carbon in the carbon ring indicated by an asterisk, which gives carvone two enantiomers. Olfactory receptors in the nose detect these enantiomers as two different odors. One enantiomer of carvone that is produced by the spearmint plant smells and tastes like spearmint, whereas its mirror image, produced by the

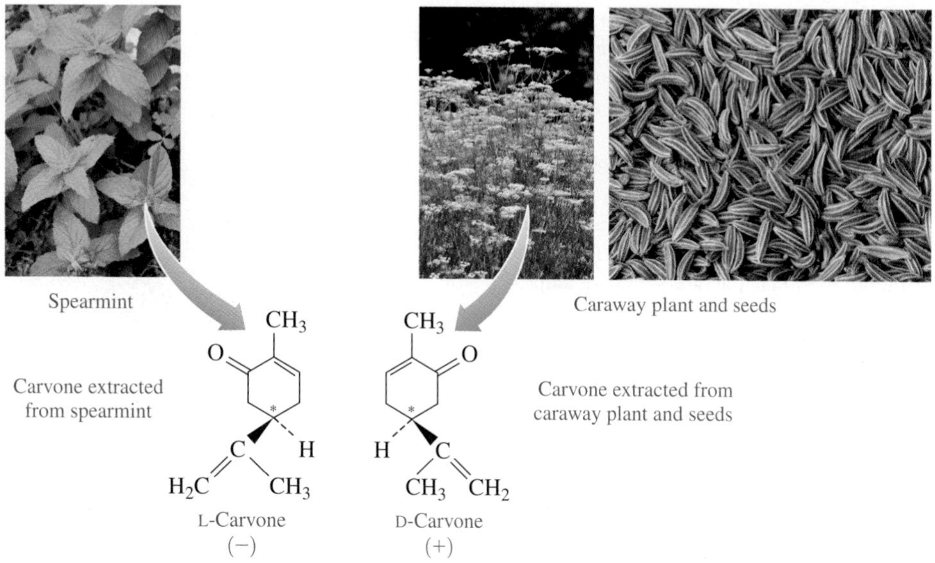

caraway plant, has the odor and taste of caraway in rye bread. Thus, our senses of smell and taste are responsive to the chirality of molecules.

In the brain, one enantiomer of LSD affects the production of serotonin, which influences sensory perception and may lead to hallucinations. However, its enantiomer produces little effect in the brain. The behavior of nicotine and adrenaline (epinephrine) also depends upon only one of their enantiomers. For example, one enantiomer of nicotine is more toxic than the other. Only one enantiomer of epinephrine is responsible for the constriction of blood vessels.

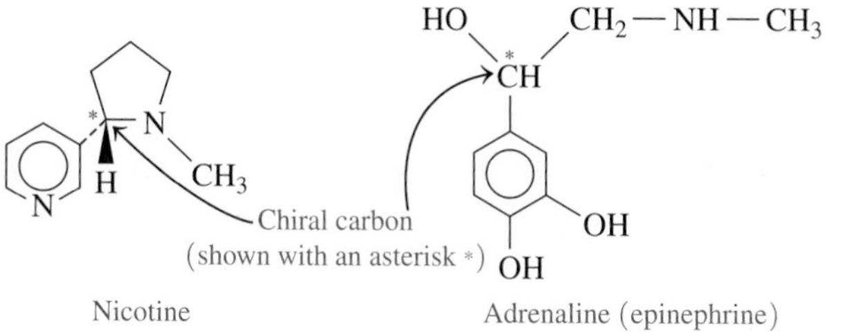

A substance used to treat Parkinson's disease is L-dopa, which is converted to dopamine in the brain, where it raises the serotonin level. However, the D-dopa enantiomer is not effective for the treatment of Parkinson's disease.

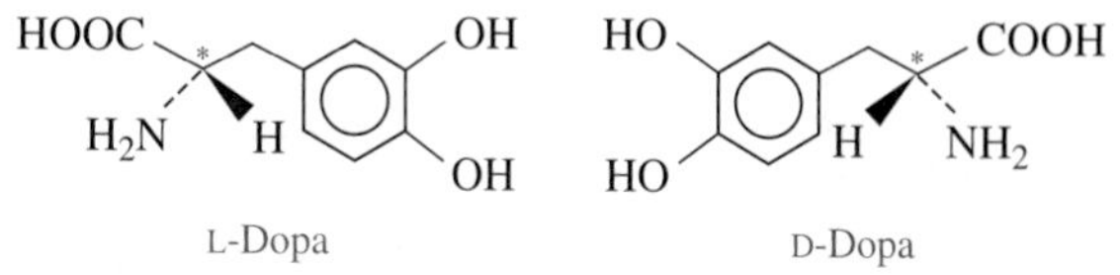

Many compounds in biological systems have only one enantiomer that is active. This happens because the enzymes and cell surface receptors on which metabolic reactions take place are themselves chiral. Thus, only one enantiomer interacts with its enzymes or receptors; the other is inactive. The chiral receptor fits the arrangement of the substituents in only one enantiomer; its mirror image does not fit properly (see Figure 13.7).

For many drugs, only one of the enantiomers is biologically active. However, for many years, drugs have been produced that were mixtures of their enantiomers. Today, drug researchers are using *chiral technology* to produce the active enantiomers of chiral drugs. Chiral catalysts are being designed that direct the formation of just one enantiomer rather than both. The active forms of several enantiomers are now being produced, such as L-dopa and naproxen. Naproxen

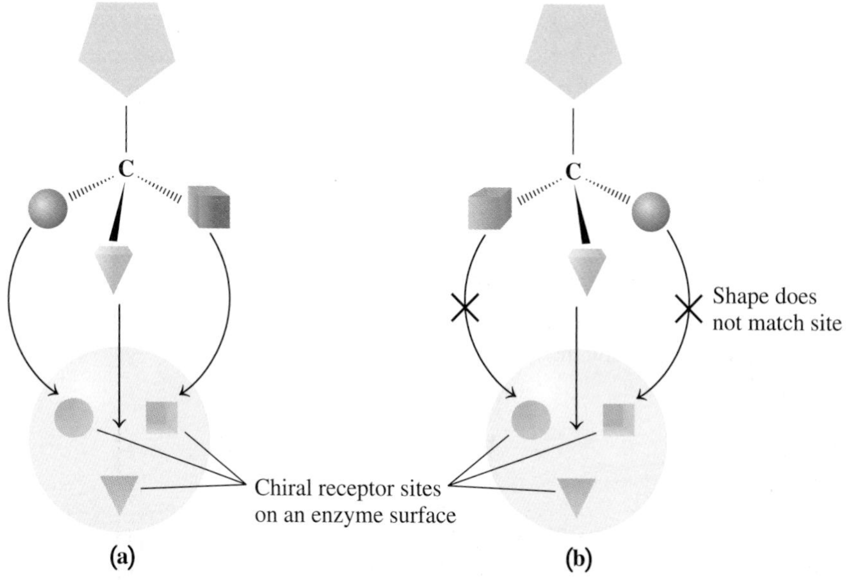

FIGURE 13.7 ▶ **(a)** The substituents on the biologically active enantiomer bind to all the sites on a chiral receptor; **(b)** its enantiomer does not bind properly and is not active biologically.

Q Why don't all the substituents of the mirror image of the active enantiomer fit into a chiral receptor site?

is a nonsteroidal anti-inflammatory drug used to relieve pain, fever, and inflammation caused by osteoarthritis and tendinitis. The benefits of producing only the active enantiomer include using a lower dose, enhancing activity, reducing interactions with other drugs, and eliminating possible harmful side effects from the enantiomer.

The active enantiomer of the popular analgesic ibuprofen used in Advil, Motrin, and Nuprin has been produced. However, recent research shows that humans have an enzyme called *isomerase* that converts the inactive enantiomer of ibuprofen to its active form. Due to the lower cost of preparation, most manufacturers prepare ibuprofen as a mixture of inactive and active forms.

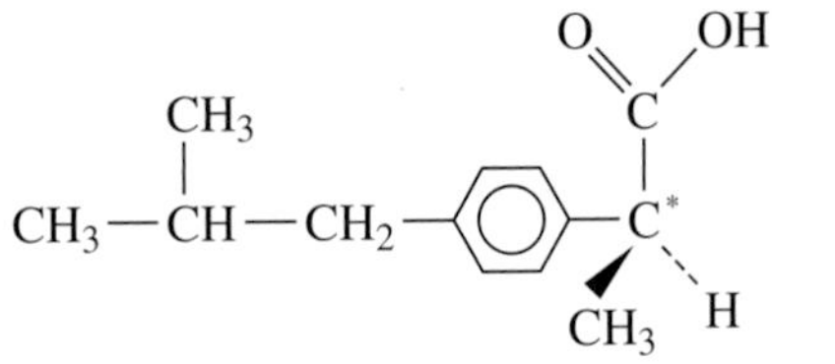

Active form of ibuprofen

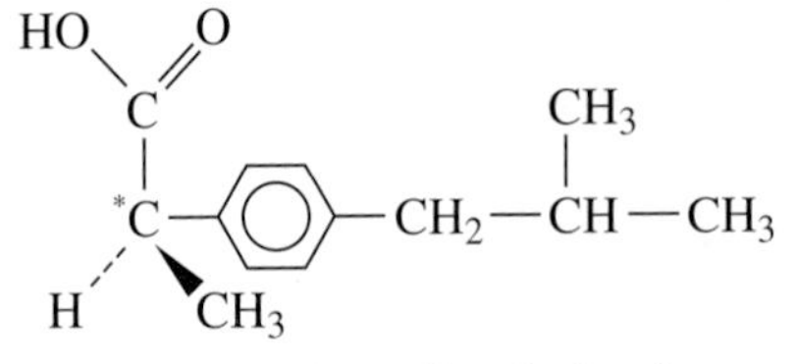

Mirror image (inactive form)

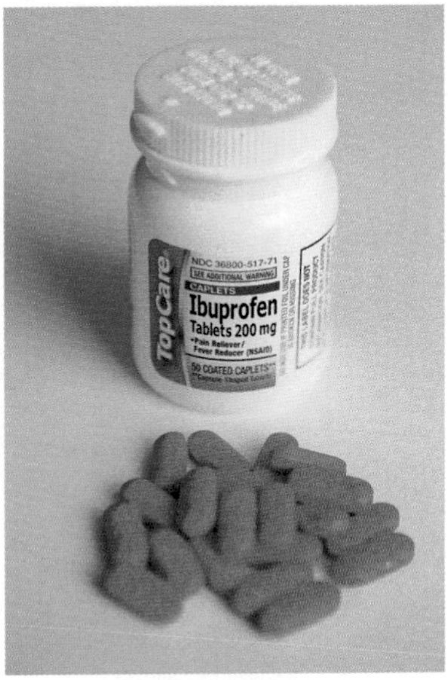

LEARNING GOAL

Use Fischer projections to draw the D or L stereoisomers for glucose, galactose, and fructose.

13.3 Fischer Projections of Monosaccharides

We can also draw Fischer projections for the monosaccharides that typically have five or six carbon atoms with several chiral carbons. As discussed in the previous section, the —OH group on the chiral carbon *farthest* from the carbonyl group is used to determine the D or L stereoisomer. The following are the Fischer projections for the D and L stereoisomers of ribose, a five-carbon monosaccharide, and the D and L stereoisomers of glucose, a six-carbon monosaccharide.

The vertical carbon chain is numbered starting from the top carbon. In each pair of mirror images, it is important to see that the —OH groups on chiral atoms are reversed so that they appear on the opposite sides of the molecule. For example, in L-ribose, all of the —OH groups are drawn on the left side of the vertical line. In the mirror image, D-ribose, all of the —OH groups are drawn on the right side of the vertical line.

CORE CHEMISTRY SKILL

Identifying D- and L-Fischer Projections

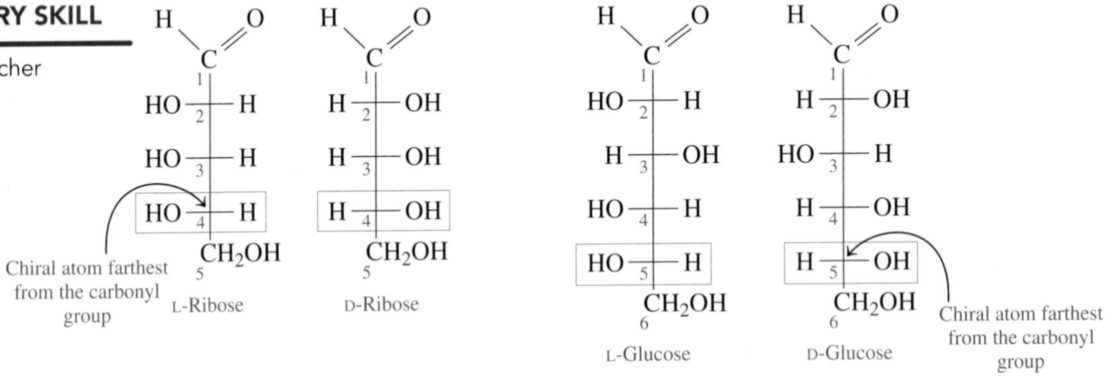

SAMPLE PROBLEM 13.3 Fischer Projections

Determine if each Fischer projection is chiral or achiral. If chiral, identify it as the D or L stereoisomer and draw the mirror image.

a. 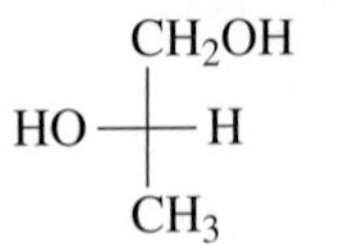

b. 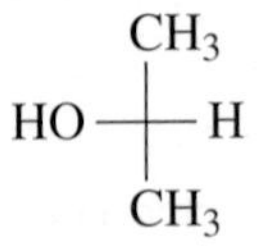

SOLUTION

a. Chiral. The carbon at the intersection is attached to four different substituents. It is the L stereoisomer because the —OH group is drawn on the left. The mirror image is drawn by reversing the —H and —OH groups on the chiral carbon.

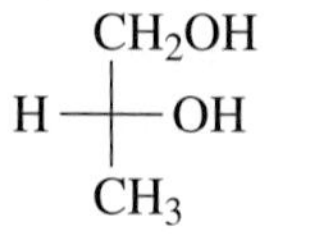

b. Achiral. The carbon atom at the intersection is attached to two identical groups $(-CH_3)$.

STUDY CHECK 13.3

Does this Fischer projection represent the D or L stereoisomer?

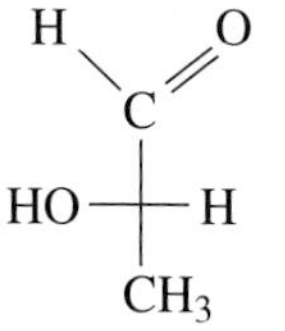

Some Important Monosaccharides

Glucose, galactose, and fructose are the most important monosaccharides. They are all hexoses with the molecular formula $C_6H_{12}O_6$ and are isomers of each other. Although we can draw Fischer projections for their D and L stereoisomers, the D stereoisomers are more commonly found in nature and used in the cells of the body. The Fischer projections for the D stereoisomers are drawn as follows:

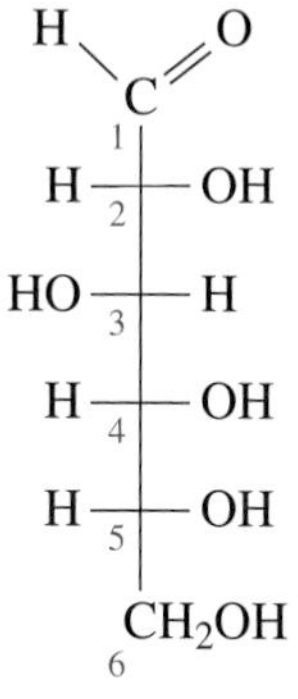

D-Glucose

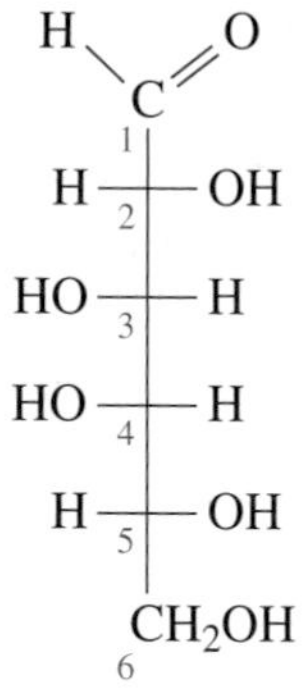

D-Galactose

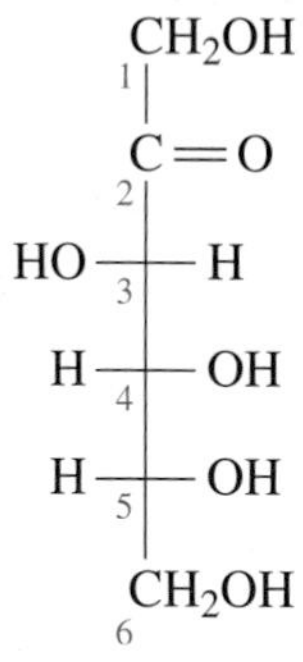

D-Fructose

The most common hexose, **glucose**, $C_6H_{12}O_6$, also known as dextrose and blood sugar, is found in fruits, vegetables, corn syrup, and honey (see Figure 13.8). D-glucose is a building block of the disaccharides sucrose, lactose, and maltose, and polysaccharides such as amylose, cellulose, and glycogen.

Galactose, $C_6H_{12}O_6$, is an aldohexose that is obtained from the disaccharide lactose, which is found in milk and milk products. Galactose is important in the cellular membranes of the brain and nervous system. The only difference in the Fischer projections of D-glucose and D-galactose is the arrangement of the —OH group on carbon 4.

In a condition called *galactosemia*, an enzyme needed to convert galactose to glucose is missing. The accumulation of galactose in the blood and tissues can lead to cataracts, mental retardation, failure to thrive, and liver disease. The treatment for galactosemia is the removal of all galactose-containing foods, mainly milk and milk products, from the

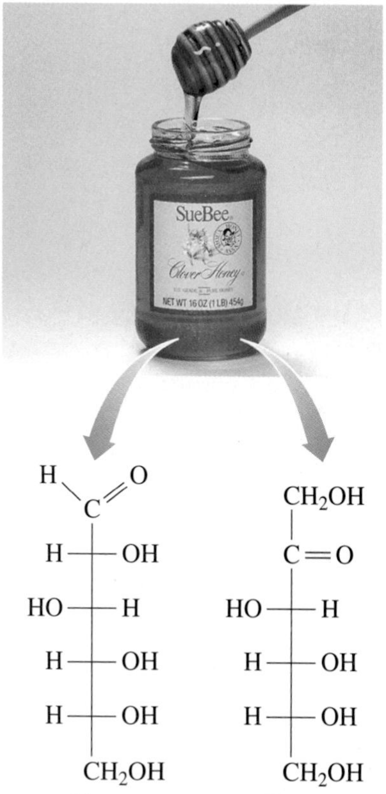

FIGURE 13.8 ▶ The sweet taste of honey is due to the monosaccharides D-glucose and D-fructose.

Q What are some differences in the Fischer projections of D-glucose and D-fructose?

A food label shows that high fructose corn syrup is the sweetener in this product.

diet. If this is done for an infant immediately after birth, the damaging effects of galactose accumulation can be avoided.

In contrast to glucose and galactose, **fructose**, $C_6H_{12}O_6$, is a ketohexose. The structure of fructose differs from glucose at carbons 1 and 2 by the location of the carbonyl group. Fructose is the sweetest of the carbohydrates, almost twice as sweet as sucrose (table sugar). This characteristic makes fructose popular with dieters because less fructose, and therefore fewer calories, is needed to provide a pleasant taste. Fructose, also called levulose and fruit sugar, is found in fruit juices and honey (see Figure 13.8).

Fructose is also obtained as one of the hydrolysis products of sucrose, the disaccharide known as table sugar. High-fructose corn syrup (HFCS) is a sweetener that is produced by using an enzyme to break down sucrose to glucose and fructose. An HFCS mixture containing about 50% fructose and 50% glucose is used in soft drinks and many foods, such as baked goods.

SAMPLE PROBLEM 13.4 Monosaccharides

Ribulose has the following Fischer projection:

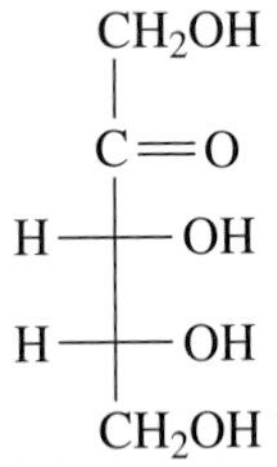

Identify the compound as D- or L-ribulose.

SOLUTION

	Need to Identify	Procedure
ANALYZE THE PROBLEM	Carbon Chain	Number the carbon chain starting at the top of the Fischer projection.
	Chiral Carbon	Chiral carbon 4 has the highest number.
	Position of —OH Group	The —OH group is drawn on the right side of carbon 4, which makes it D-ribulose.

The compound is D-ribulose because the —OH group on the chiral carbon farthest from the carbonyl group is drawn on the right side.

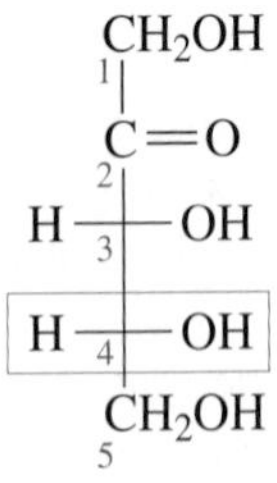

D-Ribulose

STUDY CHECK 13.4

Draw the Fischer projection for the mirror image of the ribulose in Sample Problem 13.4.

Chemistry Link to **Health**

Hyperglycemia and Hypoglycemia

A doctor may order a glucose tolerance test to evaluate the body's ability to return to normal glucose concentrations (70 to 90 mg/dL) in response to the ingestion of a specified amount of glucose. The patient fasts for 12 h and then drinks a solution containing glucose. A blood sample is taken immediately, followed by more blood samples each half-hour for 2 h, and then every hour for a total of 5 h. If the blood glucose exceeds 200 mg/dL in plasma and remains high, hyperglycemia may be indicated. The term *glyc* or *gluco* refers to "sugar." The prefix *hyper* means above or over; *hypo* means below or under. Thus, the blood sugar level in *hyperglycemia* is above normal and in *hypoglycemia* it is below normal.

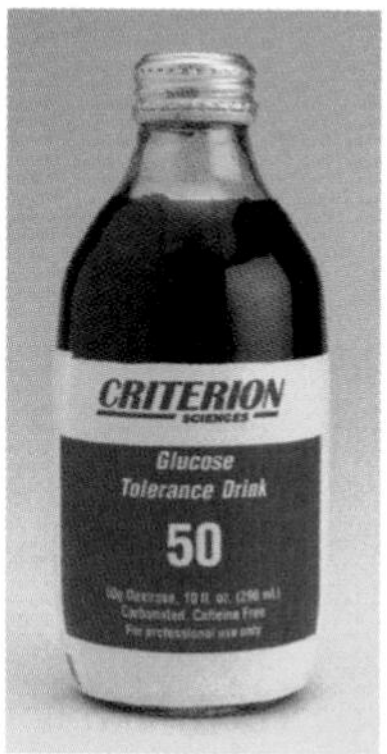

An example of a disease that can cause hyperglycemia is *diabetes mellitus*, which occurs when the pancreas is unable to produce sufficient quantities of insulin. As a result, glucose levels in the body fluids can rise as high as 350 mg/dL of plasma. Symptoms of diabetes include thirst, excessive urination, increased appetite, and weight loss. In older adults, diabetes is sometimes a consequence of excessive weight gain.

When a person is hypoglycemic, the blood glucose level rises and then decreases rapidly to levels as low as 40 mg/dL. In some cases, hypoglycemia is caused by overproduction of insulin by the pancreas. Low blood glucose can cause dizziness, general weakness, and muscle tremors. A diet may be prescribed that consists of several small meals high in protein and low in carbohydrate. Some hypoglycemic patients are finding success with diets that include more complex carbohydrates rather than simple sugars.

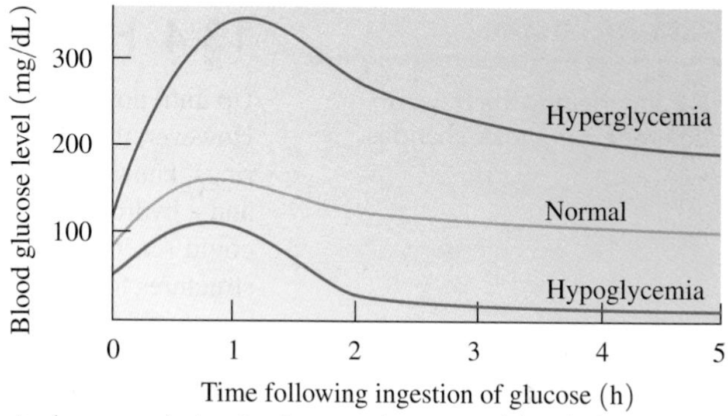

A glucose solution is given to determine blood glucose levels.

QUESTIONS AND PROBLEMS

13.3 Fischer Projections of Monosaccharides

LEARNING GOAL Use Fischer projections to draw the D or L stereoisomers for glucose, galactose, and fructose.

13.21 Identify each of the following as the D or L stereoisomer:

a.

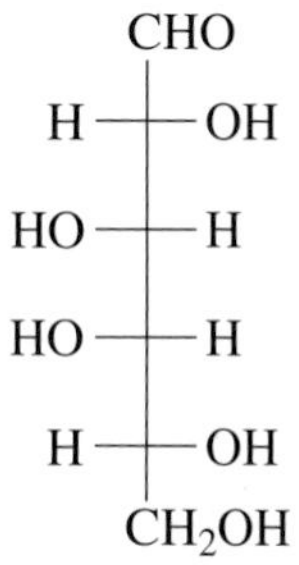

Galactose

b.

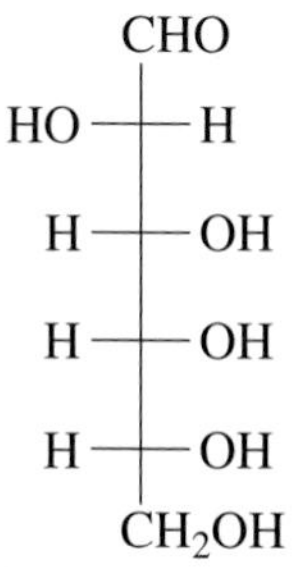

Altrose

c.

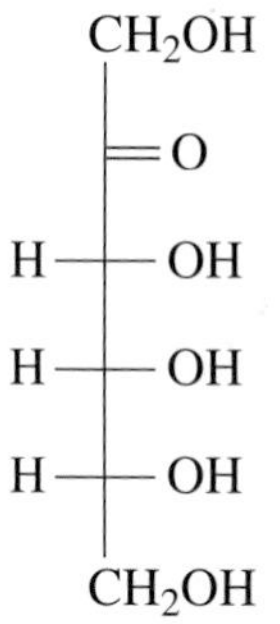

Psicose

d.

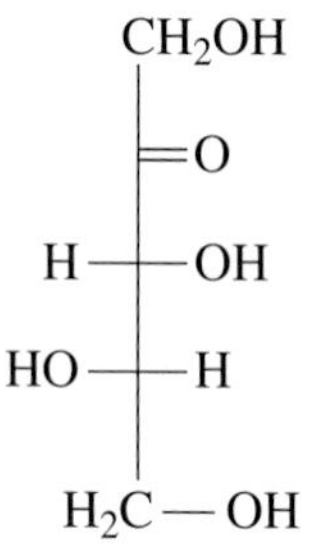

Ribulose

13.22 Identify each of the following as the D or L stereoisomer:

a.

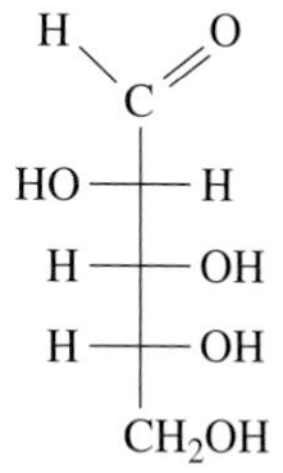

Arabinose

b.

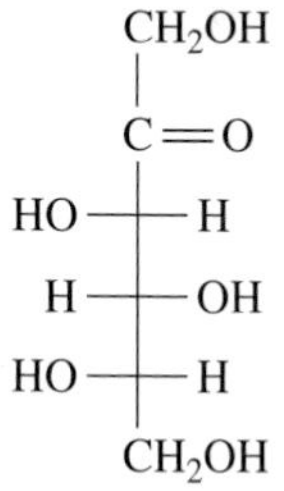

Sorbose

c.

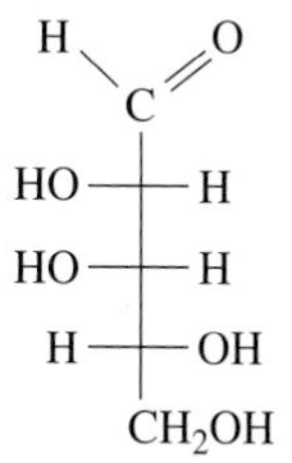

Lyxose

d.

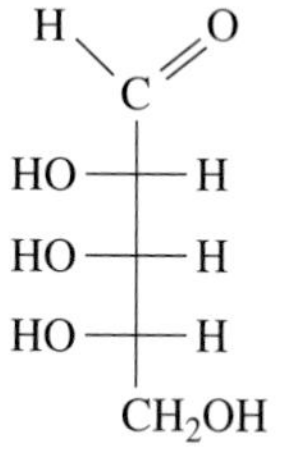

Ribose

13.23 Draw the Fischer projections for the mirror images for **a** to **d** in problem 13.21.

13.24 Draw the Fischer projections for the mirror images for **a** to **d** in problem 13.22.

13.25 Draw the Fischer projections for D-ribose and L-ribose.

13.26 Draw the Fischer projections for D-fructose and L-fructose.

13.27 How does the Fischer projection of D-galactose structurally differ from that of D-glucose?

13.28 How does the Fischer projection of D-fructose differ from that of D-glucose?

13.29 Identify the monosaccharides in each of the following mediums:

a. fruit
b. milk
c. blood

13.30 What is mutarotation? Use the structure of α-D-glucose as an example to illustrate the process.

LEARNING GOAL

Draw and identify the Haworth structures for monosaccharides.

13.4 Haworth Structures of Monosaccharides

Up until now, we have drawn the Fischer projections for monosaccharides as open chains. However, the most stable form of pentoses and hexoses are five- or six-atom rings. These rings, known as **Haworth structures**, are produced from the reaction of a carbonyl group and a hydroxyl group in the *same* molecule. While the carbonyl group in the open chain could react with several of the —OH groups, let's look at how we draw the Haworth structures for some D stereoisomers, starting with the open-chain structure of D-glucose.

Drawing Haworth Structures

STEP 1 Turn the Fischer projection clockwise by 90°. The —H and —OH groups on the right of the vertical carbon chain are now below the horizontal carbon chain. Those on the left of the open chain are now above the horizontal carbon chain.

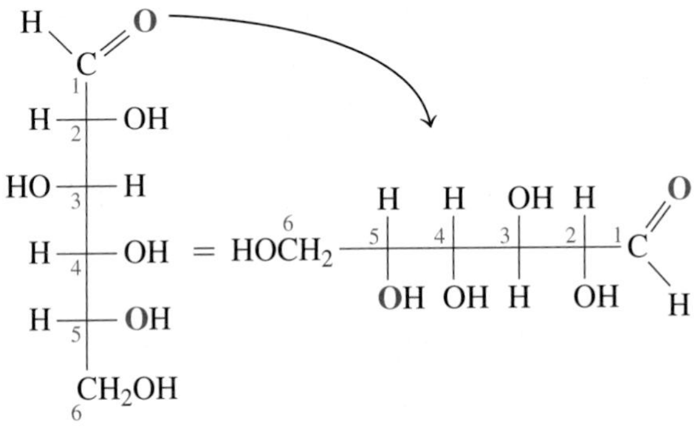

D-Glucose (open chain)

STEP 2 Fold the horizontal carbon chain into a hexagon and bond the O on carbon 5 to the carbonyl group. With carbons 2 and 3 as the base of a hexagon, move the remaining carbons upwards. Draw the reacting —OH group on carbon 5 next to the carbonyl carbon 1. Draw carbon 6 in the $—CH_2OH$ group above carbon 5. To complete the Haworth structure, draw a bond between the oxygen of the —OH group on carbon 5 to the carbonyl carbon.

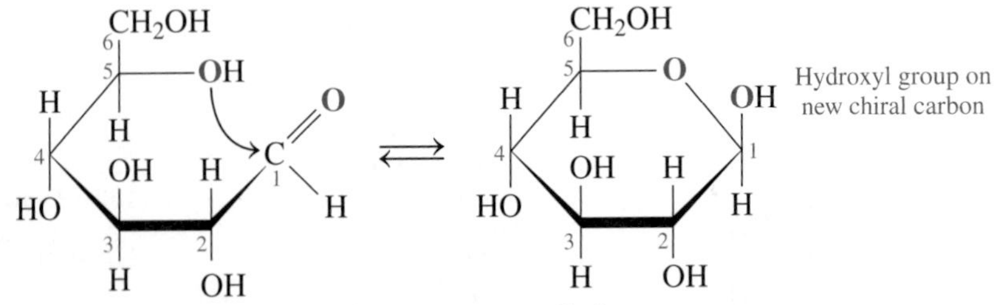

Carbon-5 oxygen bonds to carbonyl — Cyclic structure

STEP 3 **Draw the new —OH group on carbon 1 below the ring to give the α isomer or above the ring to give the β isomer.** Because the new —OH group can form above or below the plane of the Haworth structure, there are two forms of D-glucose, which differ only by the position of the —OH group at carbon 1. In the α (alpha) isomer, the —OH group is drawn below the plane of the ring. In the β (beta) isomer, the —OH group is drawn above the plane of the ring. By convention, the carbon atoms in the ring are drawn as corners using their skeletal formula.

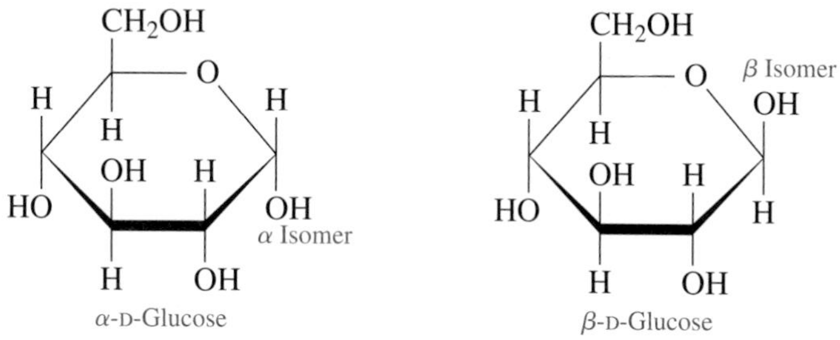

Mutarotation of α- and β-D-Glucose

In an aqueous solution, the Haworth structure of α-D-glucose opens to give the open chain of D-glucose, which has an aldehyde group. At any given time, there is only a trace amount of the open chain because it closes quickly to form a stable ring structure. However, when the open chain closes again it can form β-D-glucose. In this process called *mutarotation*, each isomer converts to the open chain and back again. As the ring opens and closes, the —OH group on carbon 1 can form either the α or β isomer. An aqueous glucose solution contains a mixture of 36% α-D-glucose and 64% β-D-glucose.

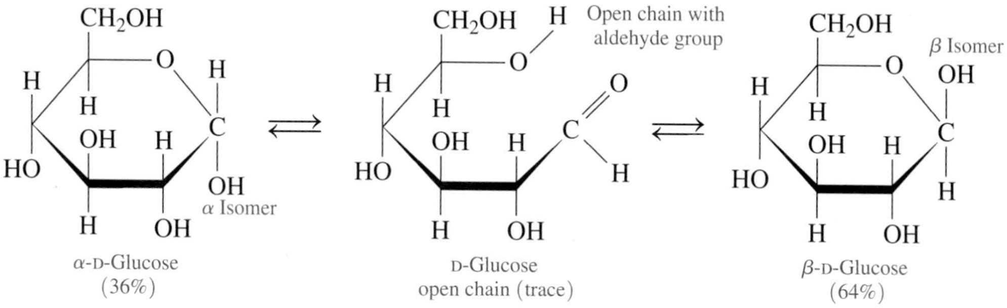

Haworth Structures of Galactose

Galactose is an aldohexose that differs from glucose only in the arrangement of the —OH group on carbon 4. Thus, its Haworth structure is similar to glucose, except that the —OH group on carbon 4 is drawn above the ring. Galactose also exists as α and β isomers.

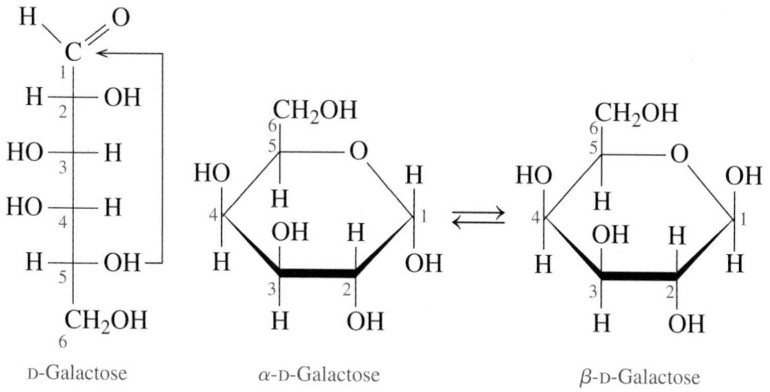

Haworth Structures of Fructose

In contrast to glucose and galactose, fructose is a ketohexose. The Haworth structure for fructose is a five-atom ring with carbon 2 at the right corner of the ring. The cyclic structure forms when the —OH group on carbon 5 reacts with carbon 2 in the carbonyl group. The new —OH group on carbon 2 gives the α and β isomers of fructose.

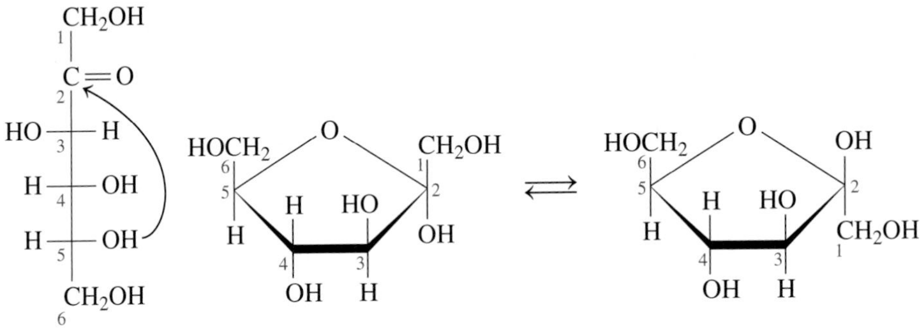

D-Fructose　　α-D-Fructose　　β-D-Fructose

SAMPLE PROBLEM 13.5 Drawing Haworth Structures for Sugars

D-Mannose, a carbohydrate found in immunoglobulins, has the following Fischer projection. Draw the Haworth structure for β-D-mannose.

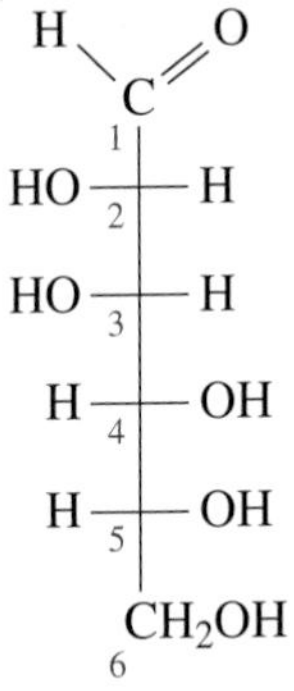

D-Mannose

Guide to Drawing Haworth Structures

STEP 1
Turn the Fischer projection clockwise by 90°.

STEP 2
Fold the horizontal carbon chain into a hexagon and bond the O on carbon 5 to the carbonyl group.

STEP 3
Draw the new —OH group on carbon 1 below the ring to give the α isomer or above the ring to give the β isomer.

SOLUTION

STEP 1 Turn the Fischer projection clockwise by 90°.

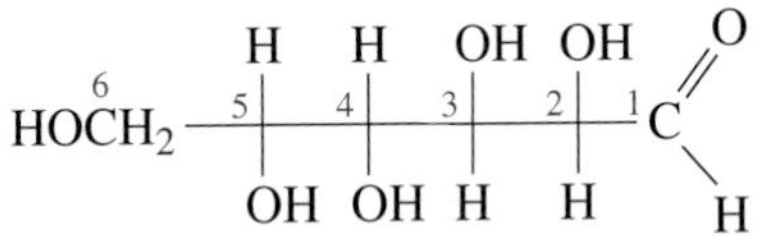

STEP 2 Fold the horizontal carbon chain into a hexagon and bond the O on carbon 5 to the carbonyl group.

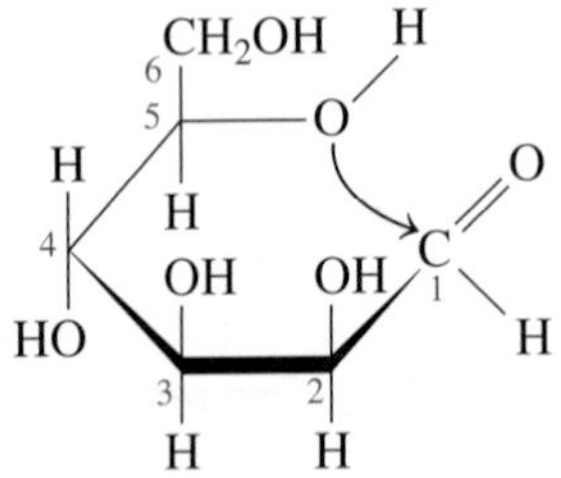

STEP 3 Draw the new —OH group on carbon 1 above the ring to give the β isomer.

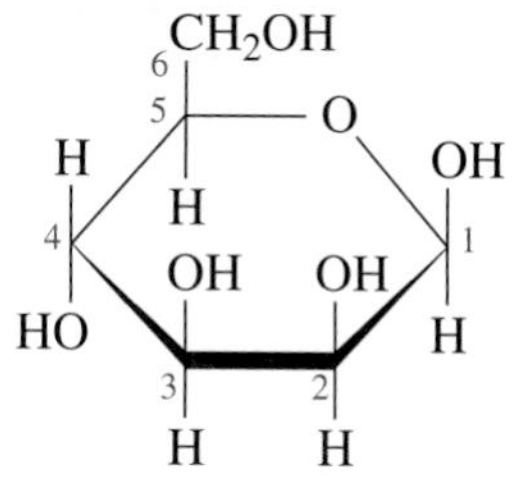

STUDY CHECK 13.5

Draw the Haworth structure for α-D-mannose.

QUESTIONS AND PROBLEMS

13.4 Haworth Structures of Monosaccharides

LEARNING GOAL Draw and identify the Haworth structures for monosaccharides.

13.31 What are the kind and number of atoms in the ring portion of the Haworth structure of glucose?

13.32 What are the kind and number of atoms in the ring portion of the Haworth structure of fructose?

13.33 Draw the Haworth structures for α- and β-D-Galactose.

13.34 Draw the Haworth structures for α- and β-D-fructose.

13.35 Identify each of the following as the α or β isomer:

a. 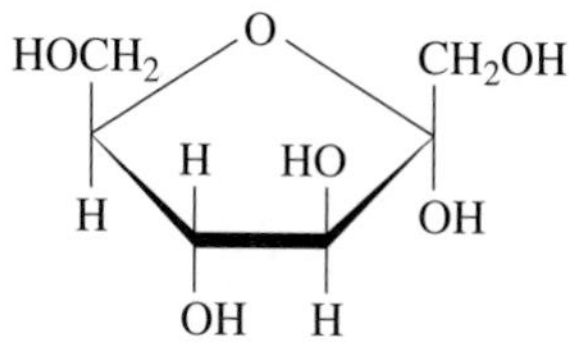

b. 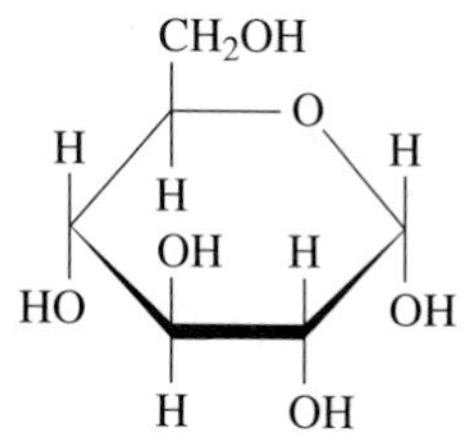

13.36 Identify each of the following as the α or β isomer:

a. 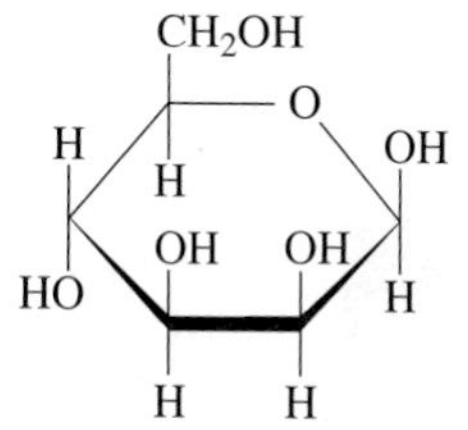

b. 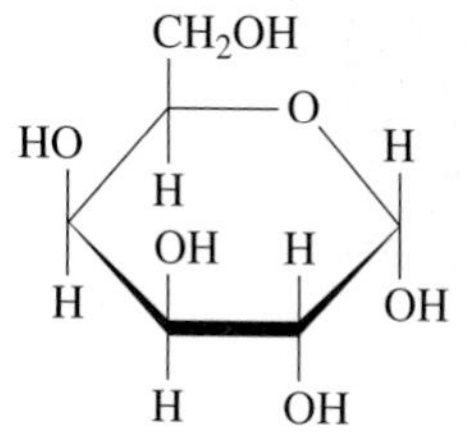

13.5 Chemical Properties of Monosaccharides

LEARNING GOAL

Identify the products of oxidation or reduction of monosaccharides; determine whether a carbohydrate is a reducing sugar.

Monosaccharides contain functional groups that can undergo chemical reactions. In an aldose, the aldehyde group can be oxidized to a carboxylic acid. The carbonyl group in both an aldose and a ketose can be reduced to give a hydroxyl group. The hydroxyl groups can react with other compounds to form a variety of derivatives that are important in biological structures.

Oxidation of Monosaccharides

Although monosaccharides exist mostly in cyclic forms, we have seen that a small amount of the open-chain form is always present, which provides an aldehyde group. An aldehyde group with an adjacent hydroxyl can be oxidized to a carboxylic acid by an oxidizing agent such as Benedict's reagent. The sugar acids are named by replacing the *ose* ending of the monosaccharide with *onic acid*. Then the Cu^{2+} is reduced to Cu^{+}, which forms a brick-red precipitate of Cu_2O. A carbohydrate that reduces another substance is called a **reducing sugar**.

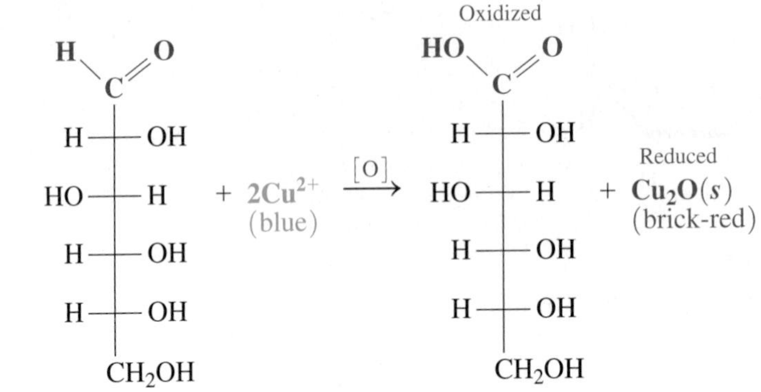

Open chain of D-glucose, a reducing sugar

D-Gluconic acid

Fructose, a ketohexose, is also a reducing sugar. Usually a ketone cannot be oxidized. However, in a basic Benedict's solution, a rearrangement occurs between the ketone group on carbon 2 and the hydroxyl group on carbon 1. As a result, fructose is converted to glucose, which produces an aldehyde group with an adjacent hydroxyl that can be oxidized.

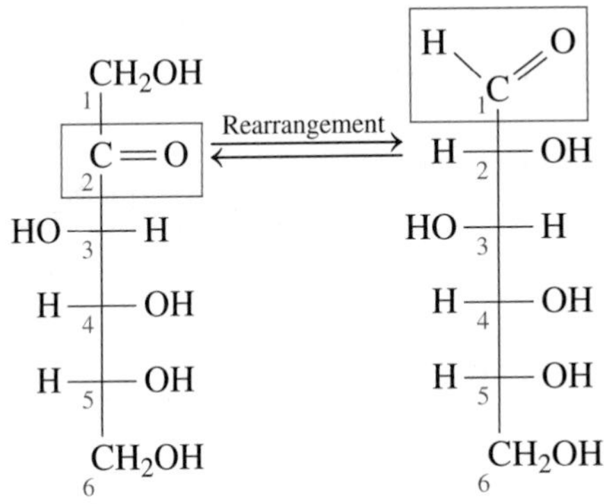

D-Fructose (ketose)

D-Glucose (aldose)

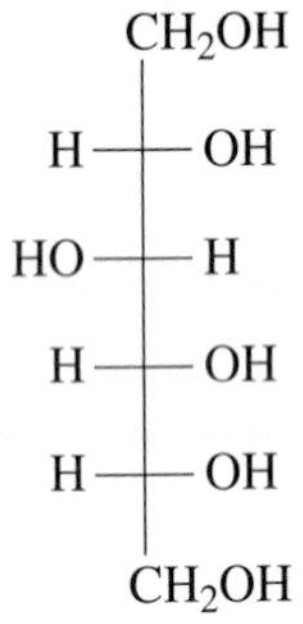

D-Sorbitol

Reduction of Monosaccharides

The reduction of the carbonyl group in monosaccharides produces sugar alcohols, which are also called *alditols*. D-Glucose is reduced to D-glucitol, better known as D-sorbitol. The sugar alcohols are named by replacing the *ose* ending of the monosaccharide with *itol*. Sugar alcohols such as D-sorbitol, D-xylitol from D-xylose, and D-mannitol from D-mannose are used as sweeteners in many sugar-free products such as diet drinks and sugarless gum as well as products for people with diabetes. However, there are some side effects of these sugar substitutes. Some people experience some discomfort such as gas and diarrhea from the ingestion of sugar alcohols. The development of cataracts in diabetics is attributed to the accumulation of D-sorbitol in the lens of the eye.

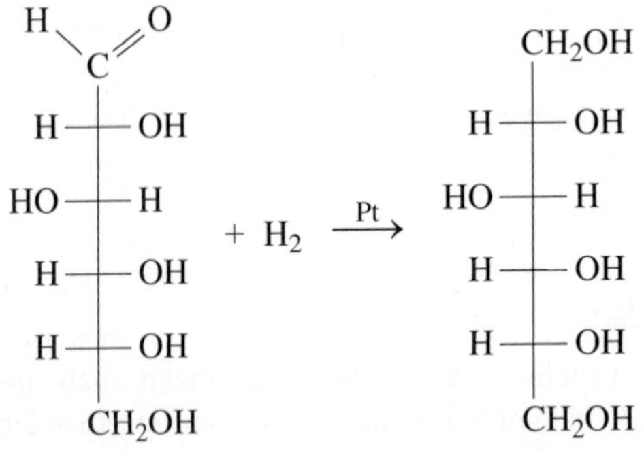

D-Glucose

D-Glucitol or D-Sorbitol

Chemistry Link to Health

Testing for Glucose in Urine

Normally, blood glucose flows through the kidneys and is reabsorbed into the bloodstream. However, if the blood level exceeds about 160 mg of glucose/dL of blood, the kidneys cannot reabsorb all of the glucose, and it spills over into the urine, a condition known as *glucosuria*. A symptom of diabetes mellitus is a high level of glucose in the urine.

Benedict's test can be used to determine the presence of glucose in urine. The amount of copper(I) oxide (Cu_2O) formed is proportional to the amount of reducing sugar present in the urine. Low to moderate levels of reducing sugar turn the solution green; solutions with high glucose levels turn Benedict's reagent yellow or brick red. Table 13.1 lists some colors associated with the concentration of glucose in the urine.

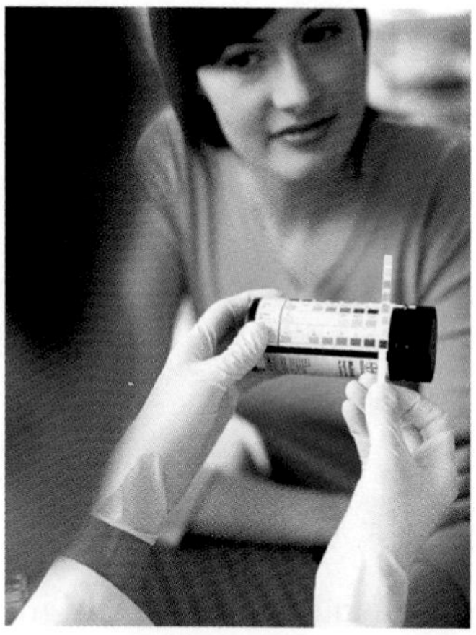

The color of a test strip determines the glucose level in urine.

In another clinical test that is more specific for glucose, the enzyme glucose oxidase is used. The oxidase enzyme converts glucose to gluconic acid and oxygen to hydrogen peroxide, H_2O_2. The peroxide produced reacts with a dye in the test strip to give different colors. The level of glucose present in the urine is found by matching the color produced to a color chart on the test strip container.

TABLE 13.1 Glucose Test Results

	Glucose Present in Urine	
Color of Benedict's Test Solution	**% (m/v)**	**mg/dL**
Blue	0	0
Blue-green	0.25	250
Green	0.50	500
Yellow	1.00	1000
Brick red	2.00	2000

SAMPLE PROBLEM 13.6 Reducing Sugars

Why is D-glucose called a reducing sugar?

SOLUTION

The aldehyde group with an adjacent hydroxyl of D-glucose is easily oxidized by Benedict's reagent. A carbohydrate that reduces Cu^{2+} to Cu^{+} is called a reducing sugar.

STUDY CHECK 13.6

A test using Benedict's reagent turns brick red with a urine sample. According to Table 13.1, what might this result indicate?

QUESTIONS AND PROBLEMS

13.5 Chemical Properties of Monosaccharides

LEARNING GOAL Identify the products of oxidation or reduction of monosaccharides; determine whether a carbohydrate is a reducing sugar.

13.37 Draw the Fischer projection for D-Glucitol produced when D-Glucose is reduced.

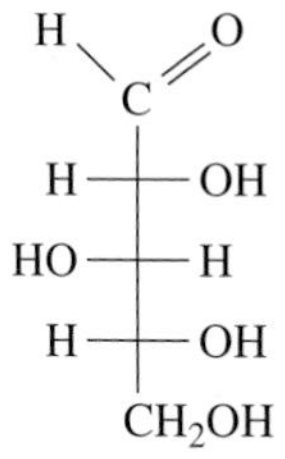

D-Xylose

13.38 Draw the Fischer projection for D-mannitol produced when D-mannose is reduced.

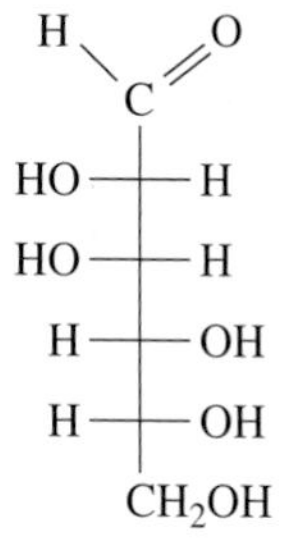

D-Mannose

13.39 Draw the Fischer projection for the oxidation and the reduction products of D-arabinose. What are the names of the sugar acid and the sugar alcohol produced?

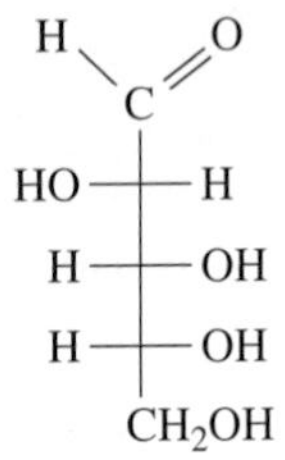

D-Arabinose

13.40 Draw the Fischer projection for the oxidation and the reduction products of D-ribose. What are the names of the sugar acid and sugar alcohol produced?

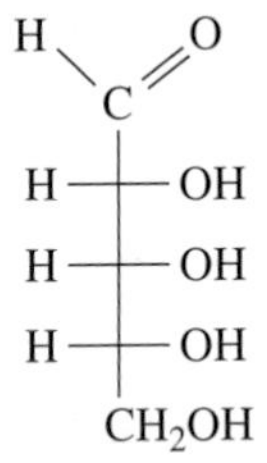

D-Ribose

LEARNING GOAL

Describe the monosaccharide units and linkages in disaccharides.

13.6 Disaccharides

A disaccharide is composed of two monosaccharides linked together. The most common disaccharides are maltose, lactose, and sucrose. When two monosaccharides combine in a dehydration reaction, the product is a disaccharide. The reaction occurs between the hydroxyl group on carbon 1 and one of the hydroxyl groups on a second monosaccharide.

$$\text{Glucose} + \text{glucose} \xrightarrow{\text{Maltose synthase}} \text{maltose} + H_2O$$

$$\text{Glucose} + \text{galactose} \xrightarrow{\text{Lactose synthase}} \text{lactose} + H_2O$$

$$\text{Glucose} + \text{fructose} \xrightarrow{\text{Sucrose synthase}} \text{sucrose} + H_2O$$

Maltose, or malt sugar, is obtained from starch and is found in germinating grains. When maltose in barley and other grains is hydrolyzed by yeast enzymes, glucose is obtained, which can undergo fermentation to give ethanol. Maltose is used in cereals, candies, and the brewing of beverages.

In the Haworth structure of a disaccharide, a **glycosidic bond** is an ether bond that connects two monosaccharides. In maltose, a glycosidic bond forms between the —OH groups of carbons 1 and 4 of two α-D-glucose molecules with a loss of a water molecule. The glycosidic bond in maltose is designated as an α-1,4 linkage to show that an alpha —OH group on carbon 1 is joined to carbon 4 of the second glucose molecule. Because the second glucose molecule still has a free —OH group on carbon 1, it can form an open chain, which allows maltose to form both α and β isomers. The open chain provides an aldehyde group that can be oxidized, making maltose a reducing sugar.

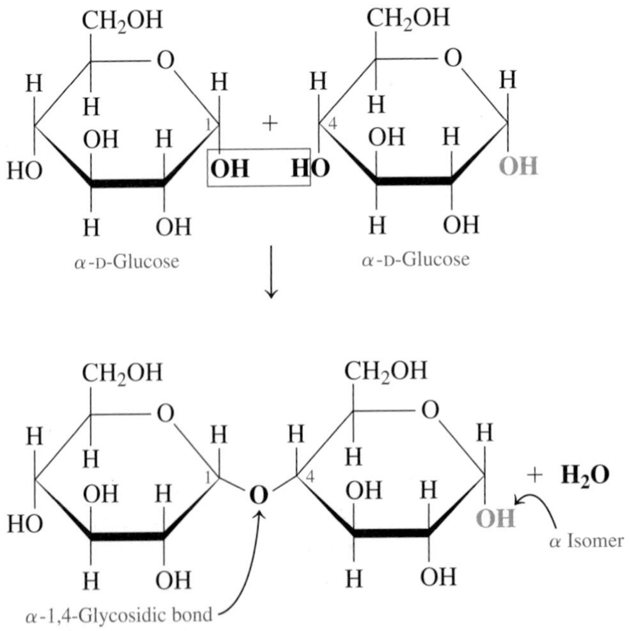

α-Maltose, a disaccharide

Explore Your World

Sugar and Sweeteners

Add a tablespoon of sugar to a glass of water and stir. Taste. Add more sugar, stir, and taste. If you have other carbohydrates such as fructose, honey, cornstarch, arrowroot, or flour, add some of each to separate glasses of water and stir. If you have some artificial sweeteners, add a few drops of the sweetener or a package, if solid, to a glass of water. Taste each.

Questions

1. Which substance is the most soluble in water?
2. Place the substances in order from the one that tastes least sweet to the sweetest.
3. How does your list compare to the values in Table 13.2?
4. How does the sweetness of sucrose compare with that of the artificial sweeteners?
5. Check the labels of food products in your kitchen. Look for sugars such as sucrose or fructose, or artificial sweeteners such as aspartame or sucralose on the label. How many grams of sugar are in a serving of the food?

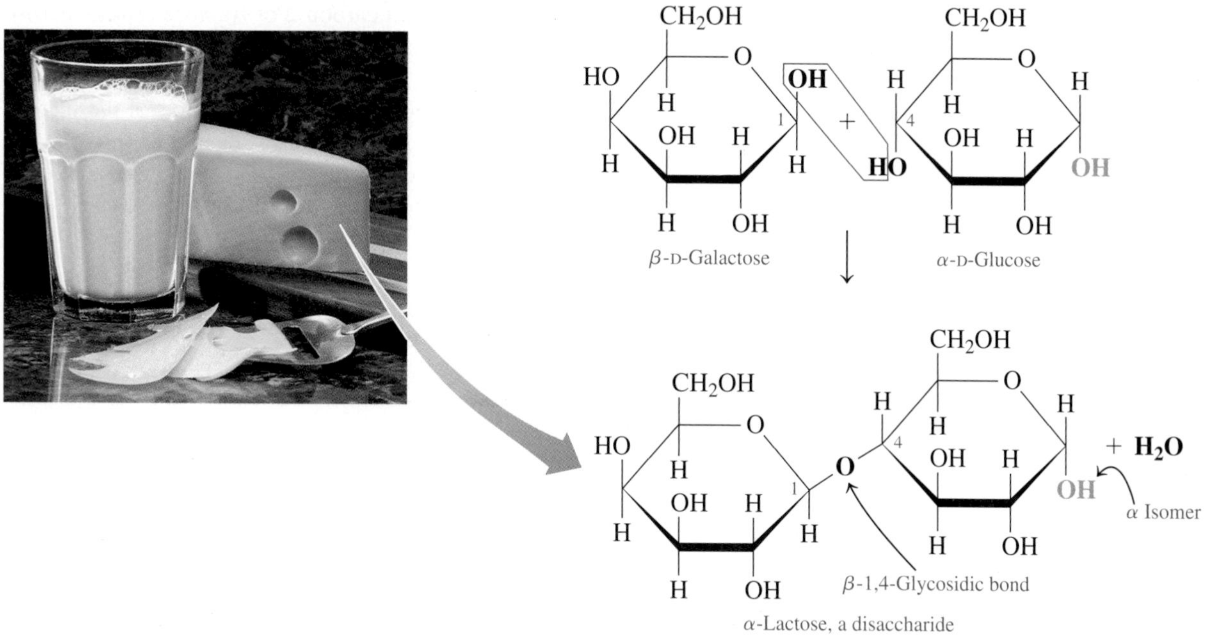

FIGURE 13.9 ▶ Lactose, a disaccharide found in milk and milk products, contains galactose and glucose.

Q What type of glycosidic bond links galactose and glucose in lactose?

Lactose, milk sugar, is a disaccharide found in milk and milk products (see Figure 13.9). The bond in lactose is a β-1,4-glycosidic bond because the —OH group on carbon 1 of β-D-galactose forms a glycosidic bond with the —OH group on carbon 4 of a D-glucose molecule. Because D-glucose still has a free —OH group on carbon 1, it can form an open chain, which allows lactose to form both α and β isomers. The open chain provides an aldehyde group that can be oxidized, making lactose a reducing sugar.

Lactose makes up 6 to 8% of human milk and about 4 to 5% of cow's milk, and it is used in products that attempt to duplicate mother's milk. When a person does not produce sufficient quantities of the enzyme lactase, which is needed to hydrolyze lactose, it remains undigested when it enters the colon. Then bacteria in the colon digest the lactose in a fermentation process that creates large amounts of gas including carbon dioxide and methane, which cause bloating and abdominal cramps. In some commercial milk products, lactase has already been added to break down lactose.

Sucrose consists of an α-D-glucose and a β-D-fructose molecule joined by an α,β-1,2-glycosidic bond (see Figure 13.10). Unlike maltose and lactose, the glycosidic

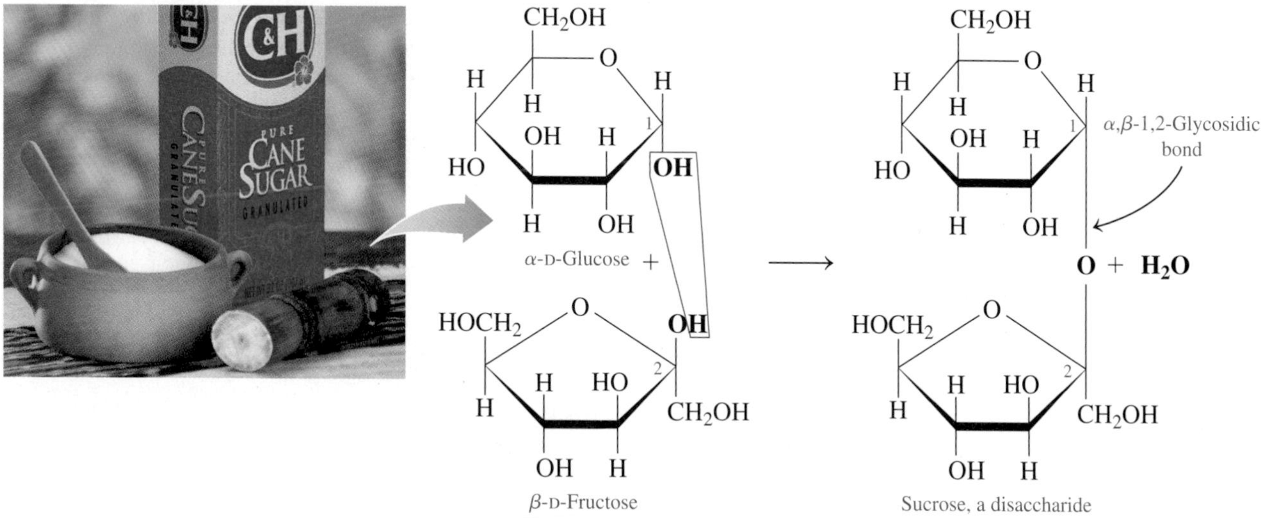

FIGURE 13.10 ▶ Sucrose, a disaccharide obtained from sugar beets and sugar cane, contains glucose and fructose.

Q Why is sucrose a nonreducing sugar?

bond in sucrose is between carbon 1 of glucose and carbon 2 of fructose. Thus, sucrose cannot form an open chain and cannot be oxidized. Sucrose cannot react with Benedict's reagent and is not a reducing sugar.

The sugar we use to sweeten our cereal, coffee, or tea is sucrose. Most of the sucrose for table sugar comes from sugar cane (20% by mass) or sugar beets (15% by mass). Both the raw and refined forms of sugar are sucrose. Some estimates indicate that each person in the United States consumes an average of 68 kg (150 lb) of sucrose every year, either by itself or in a variety of food products.

Chemistry Link to **Health**

How Sweet Is My Sweetener?

Although many of the monosaccharides and disaccharides taste sweet, they differ considerably in their degree of sweetness. Dietetic foods contain sweeteners that are noncarbohydrate or carbohydrates that are sweeter than sucrose. Some examples of sweeteners compared with sucrose are shown in Table 13.2.

Sucralose, which is known as Splenda, is made from sucrose by replacing some of the hydroxyl groups with chlorine atoms.

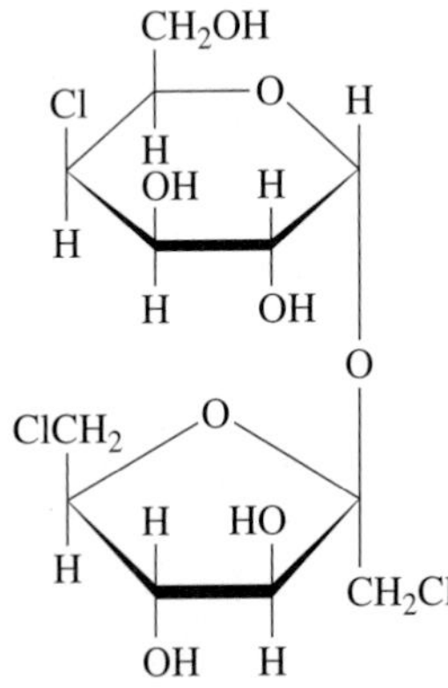

Sucralose (Splenda)

Aspartame, which is marketed as NutraSweet or Equal, is used in a large number of sugar-free products. It is a noncarbohydrate sweetener made of aspartic acid and a methyl ester of phenylalanine. It does have some caloric value, but it is so sweet that a very small quantity is needed. However, phenylalanine, one of the breakdown products, poses a danger to anyone who cannot metabolize it properly, a condition called *phenylketonuria* (PKU).

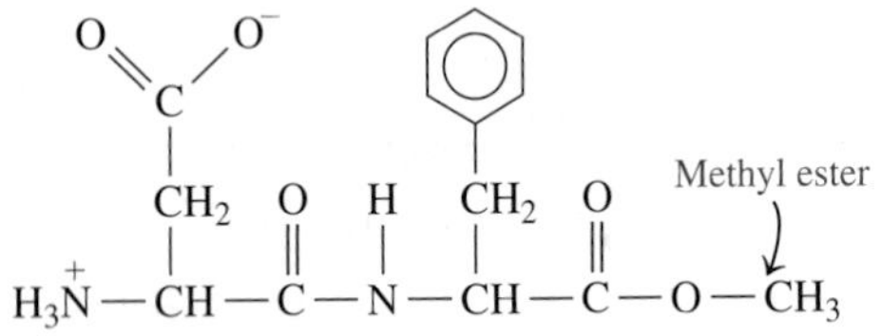

Aspartame (NutraSweet)

TABLE 13.2 Relative Sweetness of Sugars and Artificial Sweeteners

	Sweetness Relative to Sucrose (=100)
Monosaccharides	
Galactose	30
Glucose	75
Fructose	175
Disaccharides	
Lactose	16
Maltose	33
Sucrose	100
Sugar Alcohols	
Sorbitol	60
Maltitol	80
Xylitol	100
Artificial Sweeteners (Noncarbohydrate)	
Aspartame	18 000
Saccharin	45 000
Sucralose	60 000
Neotame	1 000 000

Another artificial sweetener, Neotame, is a modification of the aspartame structure. The addition of a large alkyl group to the amine group prevents enzymes from breaking the amide bond between aspartic acid and phenylalanine. Thus, phenylalanine is not produced when Neotame is used as a sweetener. Very small amounts of Neotame are needed because it is about 10 000 times sweeter than sucrose.

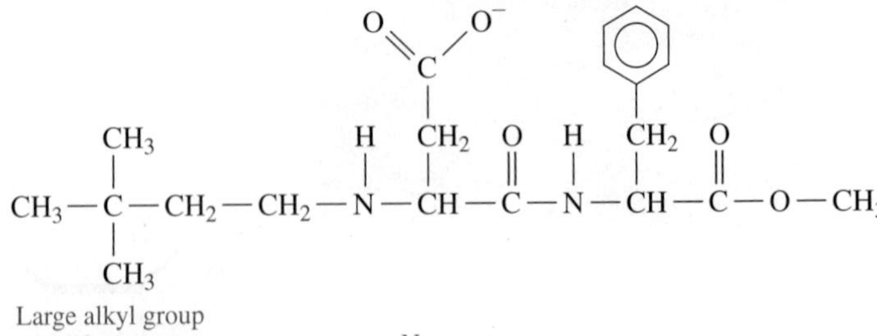

Neotame

Saccharin, which is marketed as Sweet'N Low, has been used as a noncarbohydrate artificial sweetener for more than 60 years. The use of saccharin has been banned in Canada because studies indicate that it may cause bladder tumors. However, it has still been approved by the FDA for use in the United States.

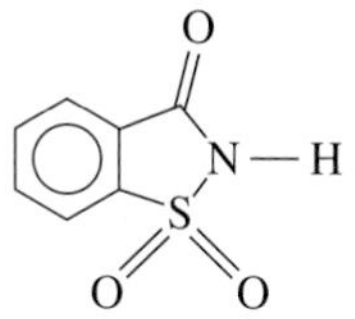

Saccharin (Sweet'N Low)

Artificial sweeteners used as sugar substitutes.

Chemistry Link to **Health**

Blood Types and Carbohydrates

Every individual's blood can be typed as one of four blood groups: A, B, AB, and O. Although there is some variation among ethnic groups in the United States, the incidence of blood types in the general population is about 43% O, 40% A, 12% B, and 5% AB.

The blood types A, B, and O are determined by terminal saccharides attached to the surface of red blood cells. Blood type O has three terminal monosaccharides: *N*-acetylglucosamine, galactose, and fucose. Blood type A contains the same three monosaccharides, but in addition, a molecule of *N*-acetylgalactosamine is attached to galactose in the saccharide chain. Blood type B also contains the same three monosaccharides, but in addition, a second molecule of galactose is attached to the saccharide chain. Blood type AB consists of the same monosaccharides found in blood types A and B. The structures of these monosaccharides are as follows:

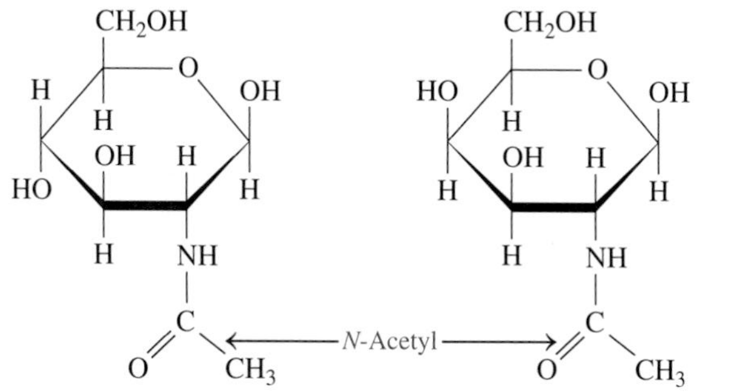

N-Acetylglucosamine (*N*-AcGlu) *N*-Acetylgalactosamine (*N*-AcGal)

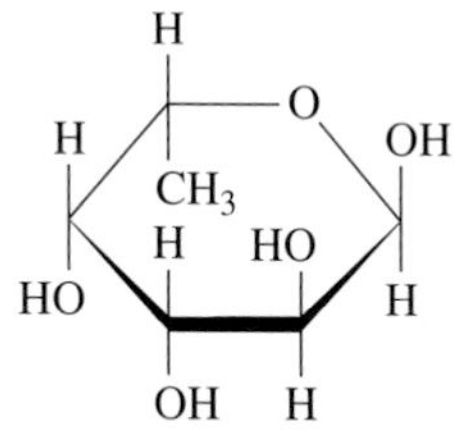

L-Fucose (Fuc)

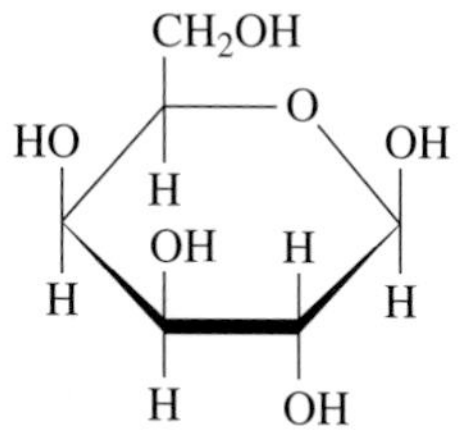

D-Galactose (Gal)

Because there is a different monosaccharide in type A blood than in type B blood, persons with type A blood produce antibodies against type B, and vice versa. For example, if a person with type A blood receives a transfusion of type B blood, the donor red blood cells are treated as if they are foreign invaders. An immune reaction occurs when the body sees the foreign saccharides on the red blood cells and makes antibodies against these donor red blood cells. The red blood cells will then clump together, or agglutinate, resulting in kidney failure, circulatory collapse, and death. The same thing will happen if a person with type B blood receives type A blood.

Terminal Saccharides for Each Blood Type

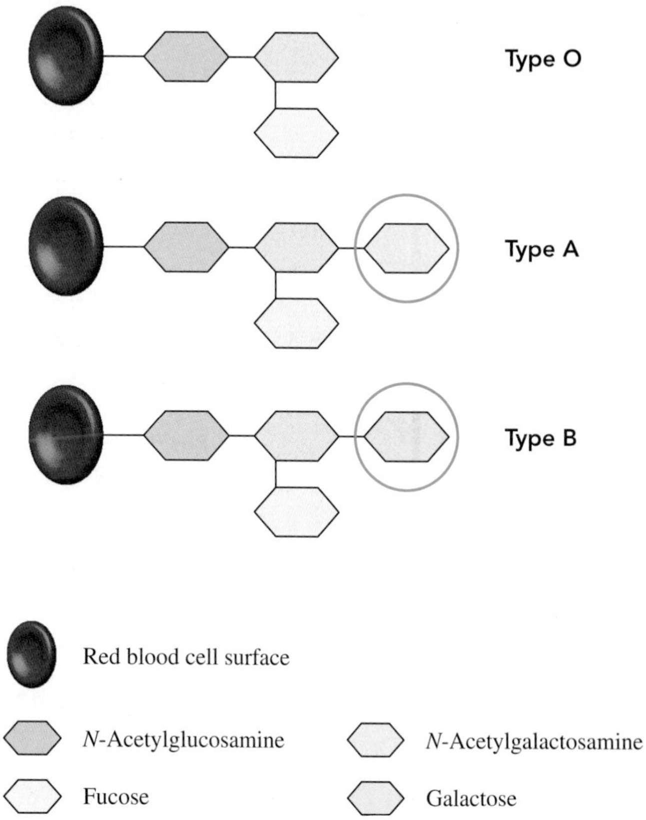

Because type O blood has only the three common terminal monosaccharides, a person with type O produces antibodies against blood types A, B, and AB. However, persons with blood types A, B, and AB can receive type O blood. Thus, persons with type O blood are *universal donors*. Because type AB blood contains all the terminal monosaccharides, a person with type AB blood produces no antibodies to type A, B, or O blood. Persons with type AB blood are *universal recipients*. Table 13.3 summarizes the compatibility of blood groups for transfusion.

TABLE 13.3 Compatibility of Blood Groups

Blood Type	Produces Antibodies Against	Can Receive
A	B, AB	A, O
B	A, AB	B, O
AB universal recipient	None	A, B, AB, O
O universal donor	A, B, AB	O

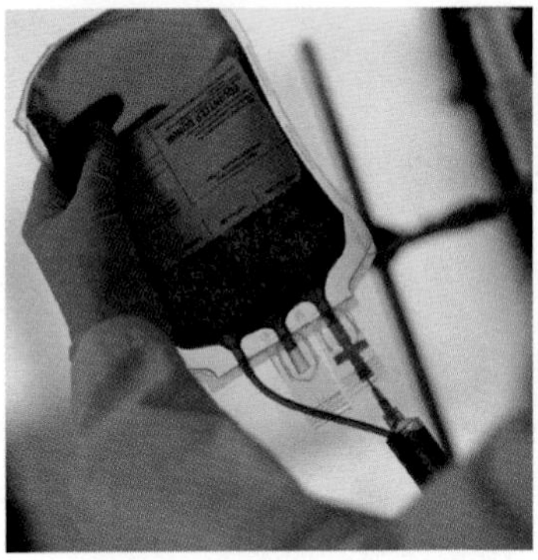

Blood from a donor is screened to make sure that there is an exact match with the blood type of the recipient.

SAMPLE PROBLEM 13.7 Glycosidic Bonds in Disaccharides

Melibiose is a disaccharide that is 30 times sweeter than sucrose.

a. What are the monosaccharide units in melibiose?

b. What type of glycosidic bond links the monosaccharides?

c. Identify the structure as α- or β-melibiose.

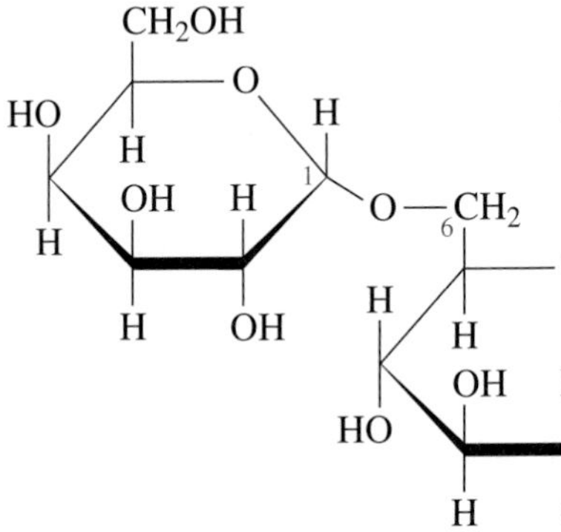

Melibiose

SOLUTION

ANALYZE THE PROBLEM		
	First Monosaccharide (left)	When the —OH group on carbon 4 is drawn above the plane, it is D-galactose. When the —OH group on carbon 1 is drawn below the plane, it is the α isomer of D-galactose.
	Second Monosaccharide (right)	When the —OH group on carbon 4 is drawn below the plane, it is D-glucose.
	Type of Glycosidic Bond	The —OH group at carbon 1 comes from α-D-galactose to the —OH group on carbon 6 of glucose, which makes it an α-1,6-glycosidic bond.
	Position of Isomer —OH Group (right)	The —OH group on carbon 1 of glucose is drawn below the plane, which is the α isomer of melibiose.
	Name of Disaccharide	α-melibiose

a. α-D-galactose and α-D-glucose

b. α-1,6-glycosidic bond

c. α-melibiose

STUDY CHECK 13.7

Cellobiose is a disaccharide composed of two D-glucose molecules linked by a β-1,4-glycosidic linkage. Draw the Haworth structure for β-cellobiose.

QUESTIONS AND PROBLEMS

13.6 Disaccharides

LEARNING GOAL Describe the monosaccharide units and linkages in disaccharides.

13.41 For each of the following, give the monosaccharide units produced by hydrolysis, the type of glycosidic bond, and the name of the disaccharide including α or β:

a.

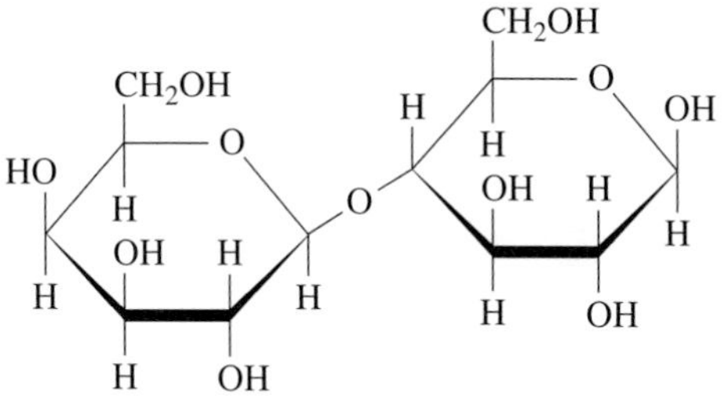

b.

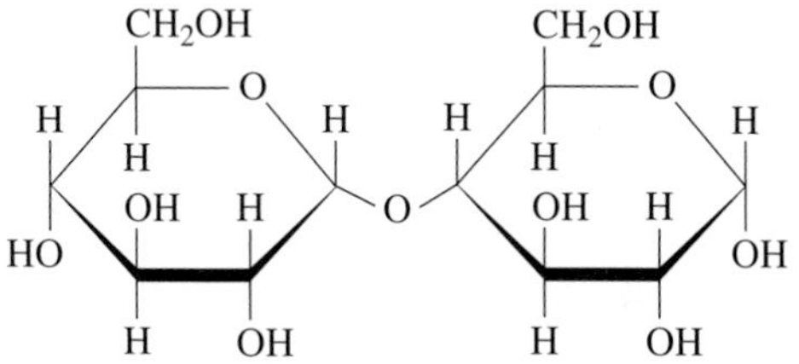

13.42 For each of the following, give the monosaccharide units produced by hydrolysis, the type of glycosidic bond, and the name of the disaccharide including α or β:

a.

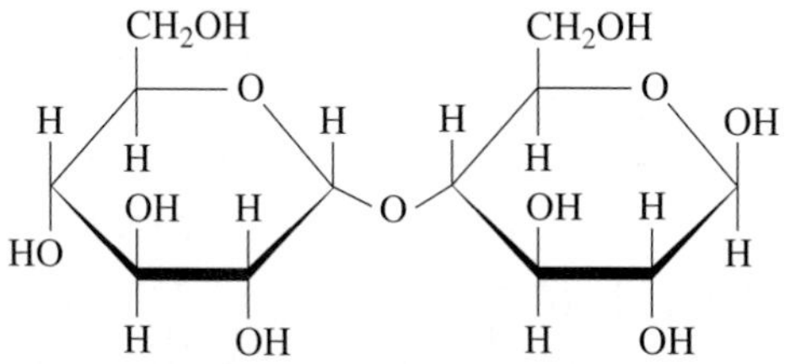

b.

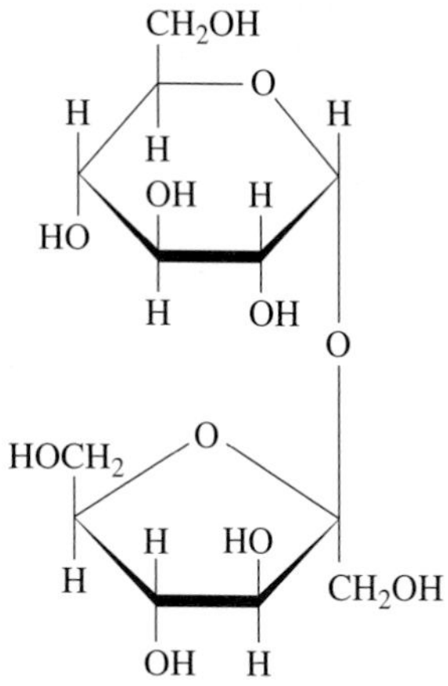

13.43 Indicate whether each disaccharide in problem 13.41 is a reducing sugar or not.

13.44 Name and draw the oxidation and reduction product of D-glucose's carbonyl group.

13.45 Identify the disaccharide that fits each of the following descriptions:

a. found in cereals and germinating grains
b. present in milk products
c. found in sugarcane and sugar beet
d. composed of galactose and glucose

13.46 How is sucrose formed? Draw the Haworth structure of sucrose.

13.7 Polysaccharides

LEARNING GOAL

Describe the structural features of amylose, amylopectin, glycogen, and cellulose.

A polysaccharide is a polymer of many monosaccharides joined together. Four important polysaccharides—*amylose*, *amylopectin*, *cellulose*, and *glycogen*—are all polymers of D-glucose that differ only in the type of glycosidic bonds and the amount of branching in the molecule.

Starch, a storage form of glucose in plants, is found as insoluble granules in rice, wheat, potatoes, beans, and cereals. Starch is composed of two kinds of polysaccharides, amylose and amylopectin. **Amylose**, which makes up about 20% of starch, consists of 250 to 4000 α-D-glucose molecules connected by α-1,4-glycosidic bonds in a continuous chain. Sometimes called a straight-chain polymer, polymers of amylose are actually coiled in helical fashion (see Figure 13.11a).

Amylopectin, which makes up as much as 80% of starch, is a branched-chain polysaccharide. Like amylose, the glucose molecules are connected by α-1,4-glycosidic bonds. However, at about every 25 glucose units, there is a branch of glucose molecules attached by an α-1,6-glycosidic bond between carbon 1 of the branch and carbon 6 in the main chain (see Figure 13.11b).

Starches hydrolyze easily in water and acid to give smaller saccharides, called *dextrins*, which then hydrolyze to maltose and finally glucose. In our bodies, these

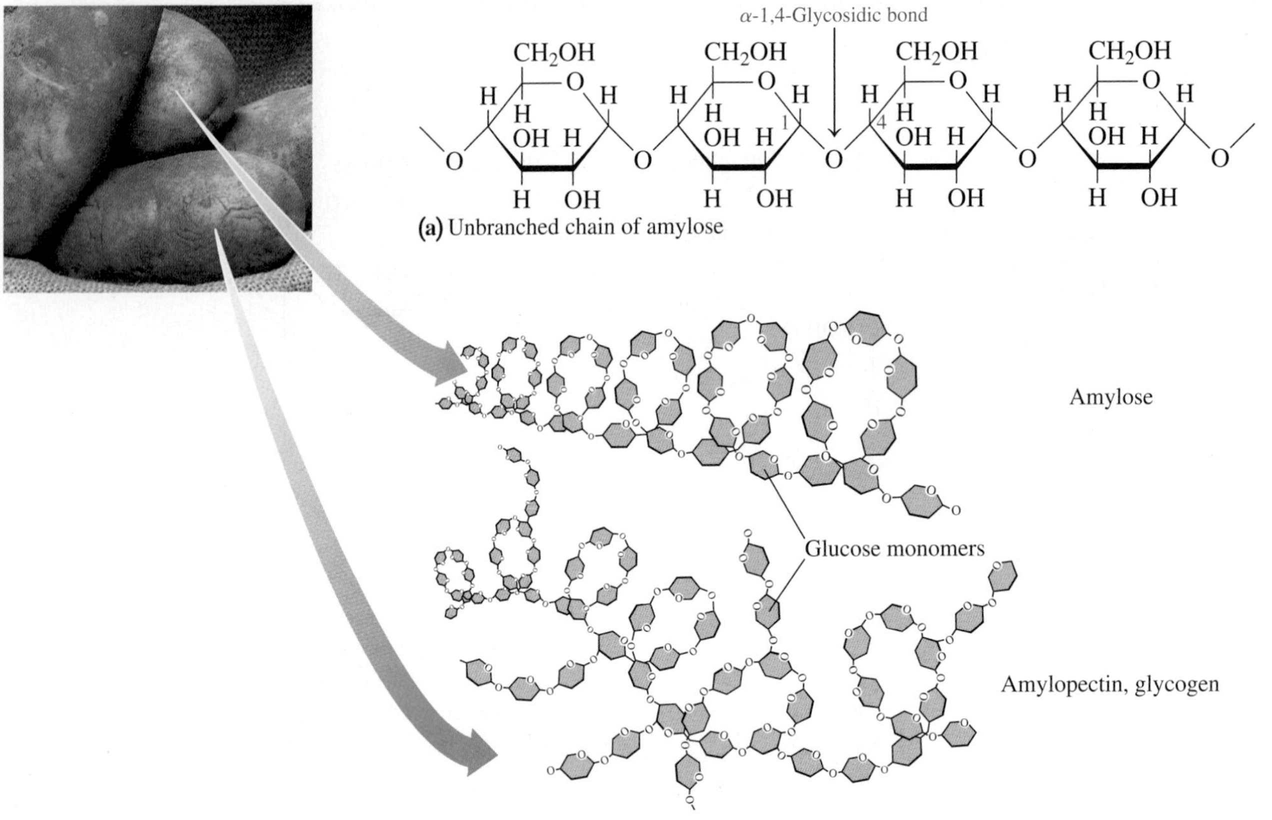

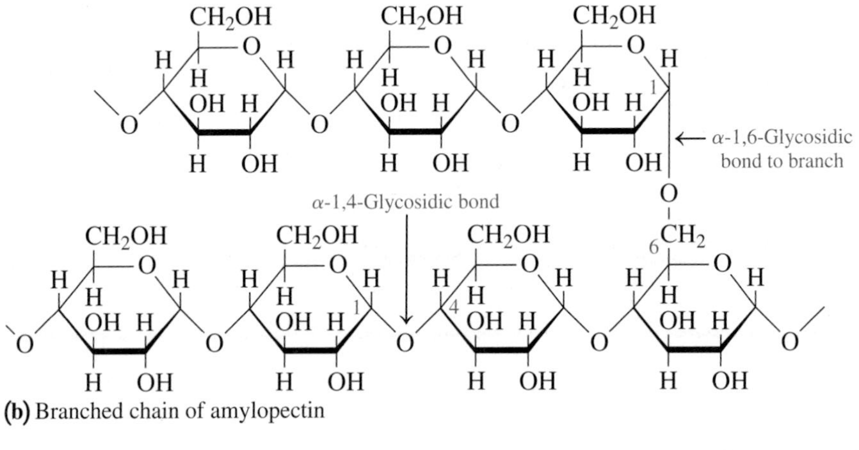

FIGURE 13.11 ▶ The structure of amylose **(a)** is a straight-chain polysaccharide of glucose units, and the structure of amylopectin **(b)** is a branched chain of glucose.

Q What are the two types of glycosidic bonds that link glucose molecules in amylopectin?

complex carbohydrates are digested by the enzymes amylase (in saliva) and maltase (in the intestine). The glucose obtained provides about 50% of our nutritional calories.

$$\text{Amylose, amylopectin} \xrightarrow{H^+ \text{ or amylase}} \text{dextrins} \xrightarrow{H^+ \text{ or amylase}} \text{maltose} \xrightarrow{H^+ \text{ or maltase}} \text{many D-glucose units}$$

Glycogen, or animal starch, is a polymer of glucose that is stored in the liver and muscle of animals. It is hydrolyzed in our cells at a rate that maintains the blood level of glucose and provides energy between meals. The structure of glycogen is very similar to that of amylopectin found in plants, except that glycogen is more highly branched. In glycogen, the glucose units are joined by α-1,4-glycosidic bonds, and branches occurring about every 10 to 15 glucose units are attached by α-1,6-glycosidic bonds.

Cellulose is the major structural material of wood and plants. Cotton is almost pure cellulose. In cellulose, glucose molecules form a long unbranched chain similar to that of amylose. However, the glucose units in cellulose are linked by β-1,4-glycosidic bonds.

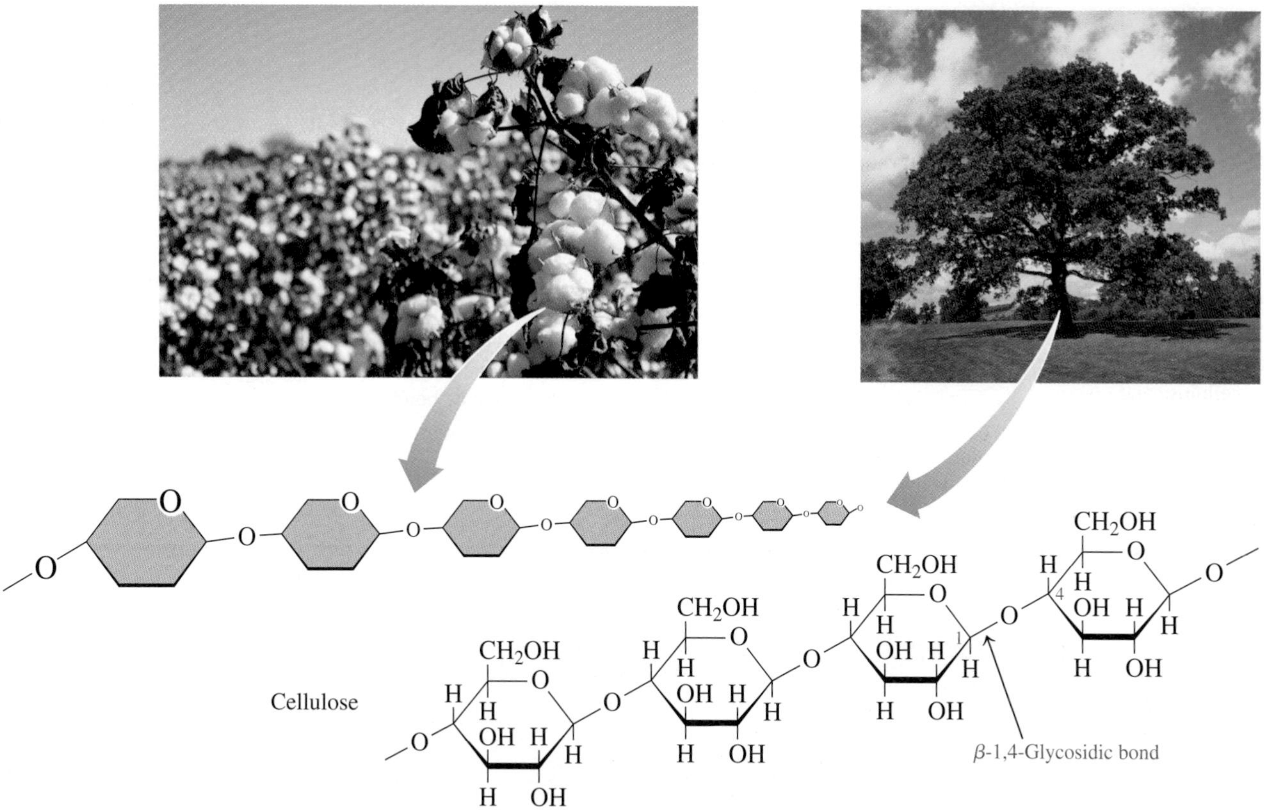

FIGURE 13.12 ▶ The polysaccharide cellulose is composed of glucose units connected by β-1,4-glycosidic bonds.

Q Why are humans unable to digest cellulose?

The cellulose chains do not form coils like amylose but are aligned in parallel rows that are held in place by hydrogen bonds between hydroxyl groups in adjacent chains, making cellulose insoluble in water. This gives a rigid structure to the cell walls in wood and fiber that is more resistant to hydrolysis than are the starches (see Figure 13.12).

Humans have an enzyme called α-amylase in saliva and pancreatic juices that hydrolyze the α-1,4-glycosidic bonds of starches, but not the β-1,4-glycosidic bonds of cellulose. Thus, humans cannot digest cellulose. Animals such as horses, cows, and goats can obtain glucose from cellulose because their digestive systems contain bacteria that provide enzymes such as cellulase to hydrolyze β-1,4-glycosidic bonds.

Explore Your World

Polysaccharides

Read the nutrition label on a box of crackers, cereal, bread, chips, or pasta. The major ingredient in crackers is flour, which contains starch. Chew on a single cracker for 4 or 5 min. Note how the taste changes as you chew the cracker. An enzyme (amylase) in your saliva breaks apart the bonds in starch.

Questions

1. How are carbohydrates listed on the label?
2. What other carbohydrates are listed?
3. How did the taste of the cracker change during the time that you chewed it?
4. What happens to the starches in the cracker as the amylase enzyme in your saliva reacts with the amylose and amylopectin?

SAMPLE PROBLEM 13.8 Structures of Polysaccharides

Identify the polysaccharide described by each of the following:

a. a polysaccharide that is stored in the liver and muscle tissues
b. an unbranched polysaccharide containing β-1,4-glycosidic bonds
c. a starch containing α-1,4- and α-1,6-glycosidic bonds

SOLUTION

a. glycogen
b. cellulose
c. amylopectin, glycogen

STUDY CHECK 13.8

Cellulose and amylose are both unbranched glucose polymers. How do they differ?

QUESTIONS AND PROBLEMS

13.7 Polysaccharides

LEARNING GOAL Describe the structural features of amylose, amylopectin, glycogen, and cellulose.

13.47 Describe the similarities and differences in the following:
- **a.** amylose and amylopectin
- **b.** amylopectin and glycogen

13.48 Describe the similarities and differences in the following:
- **a.** amylose and cellulose
- **b.** cellulose and glycogen

13.49 Give the name of one or more polysaccharides that matches each of the following descriptions:
- **a.** not digestible by humans
- **b.** the storage form of carbohydrates in plants
- **c.** contains only α-1,4-glycosidic bonds
- **d.** the most highly branched polysaccharide

13.50 Give the name of one or more polysaccharides that matches each of the following descriptions:
- **a.** What is the key polysaccharide storage form of carbohydrates in animals?
- **b.** What is the only type of glycosidic bond found in amylase?
- **c.** Name one polysaccharide that contains both α-1, 4- and α-1, 6-glycosidic bonds.
- **d.** What is produced by dextrins during digestion?

CONCEPT MAP

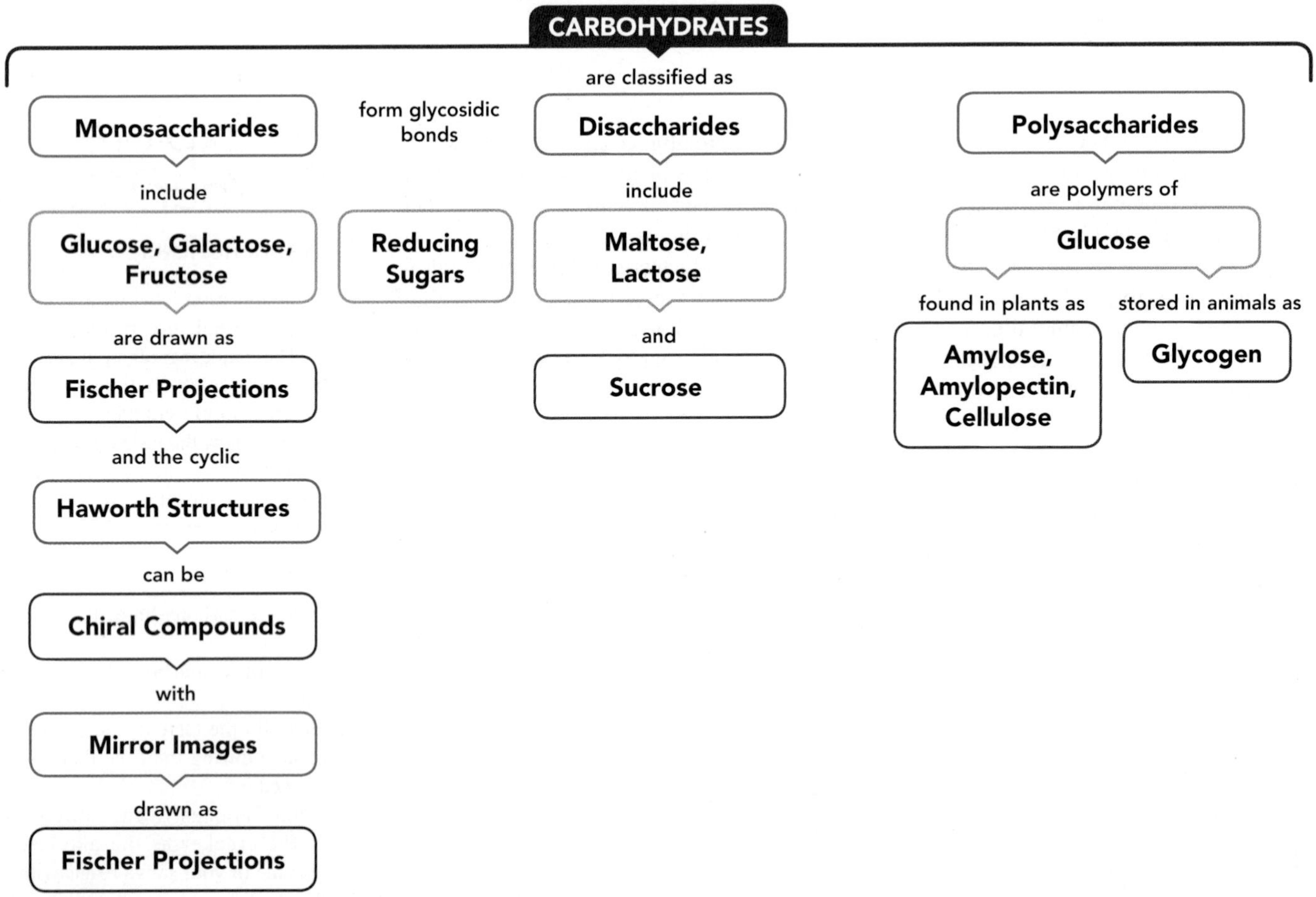

CHAPTER REVIEW

13.1 Carbohydrates

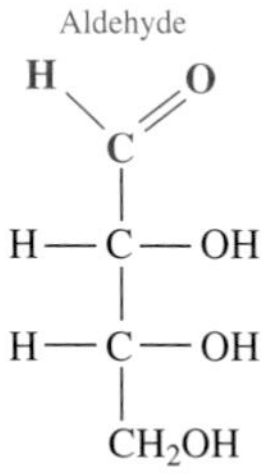

Erythrose, an aldose

LEARNING GOAL Classify a monosaccharide as an aldose or a ketose, and indicate the number of carbon atoms.

- Carbohydrates are classified as monosaccharides (simple sugars), disaccharides (two monosaccharide units), or polysaccharides (many monosaccharide units).
- Monosaccharides are polyhydroxy aldehydes (aldoses) or ketones (ketoses).
- Monosaccharides are also classified by their number of carbon atoms: triose, tetrose, pentose, or hexose.

13.2 Chiral Molecules

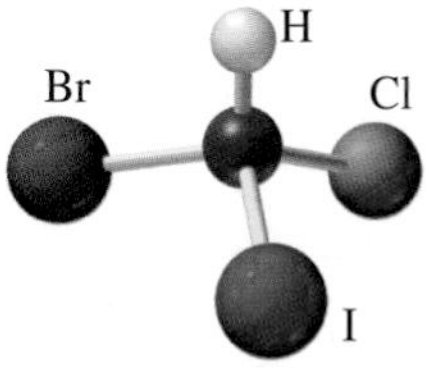

LEARNING GOAL Identify chiral and achiral carbon atoms in an organic molecule.

- Chiral molecules are molecules with mirror images that cannot be superimposed on each other. These types of stereoisomers are called enantiomers.
- A chiral molecule must have at least one chiral carbon, which is a carbon bonded to four different atoms or groups of atoms.
- The Fischer projection is a simplified way to draw the arrangements of atoms by placing the carbon atoms at the intersection of vertical and horizontal lines.
- The names of the mirror images are labeled D or L to differentiate between enantiomers of carbohydrates.

13.3 Fischer Projections of Monosaccharides

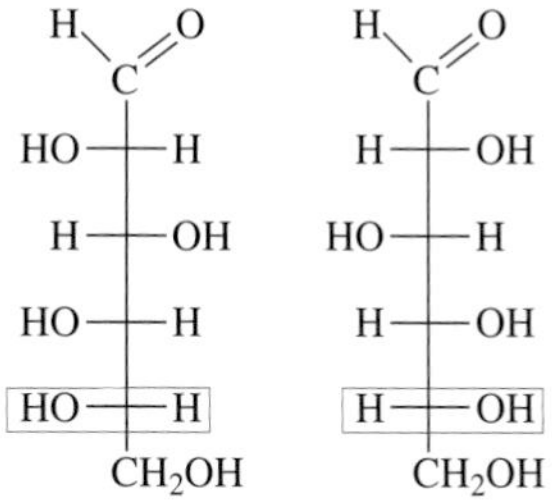

L-Glucose D-Glucose

LEARNING GOAL Use Fischer projections to draw the D or L stereoisomers for glucose, galactose, and fructose.

- In a D stereoisomer, the —OH group is on the right of the chiral carbon farthest from the carbonyl group; the —OH group is on the left in the L stereoisomer.
- Important monosaccharides are the aldohexoses, glucose and galactose, and the ketohexose, fructose.

13.4 Haworth Structures of Monosaccharides

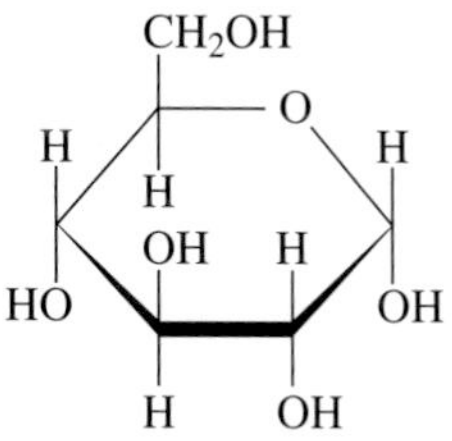

α-D-Glucose

LEARNING GOAL Draw and identify the Haworth structures for monosaccharides.

- The predominant form of a monosaccharide is a ring of five or six atoms.
- The cyclic structure forms when an —OH group (usually the one on carbon 5 in hexoses) reacts with the carbonyl group of the same molecule.
- The formation of a new hydroxyl group on carbon 1 gives α and β isomers of the cyclic monosaccharide.

13.5 Chemical Properties of Monosaccharides

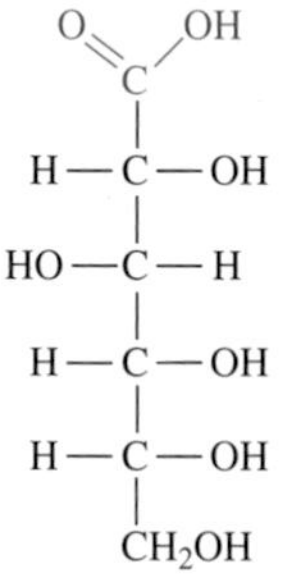

D-Gluconic acid

LEARNING GOAL Identify the products of oxidation or reduction of monosaccharides; determine whether a carbohydrate is a reducing sugar.

- The aldehyde group in an aldose can be oxidized to a carboxylic acid, while the carbonyl group in an aldose or a ketose can be reduced to give a hydroxyl group.
- Monosaccharides that are reducing sugars have an aldehyde group in the open chain that can be oxidized.

13.6 Disaccharides

LEARNING GOAL Describe the monosaccharide units and linkages in disaccharides.

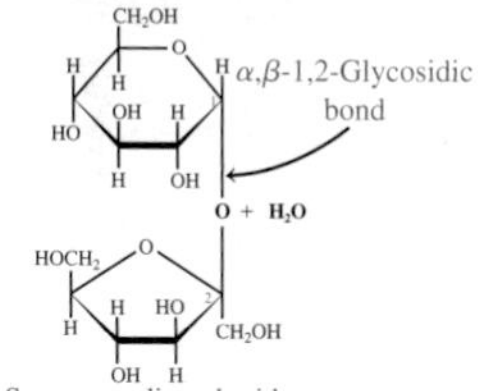

Sucrose, a disaccharide

- Disaccharides are two monosaccharide units joined together by a glycosidic bond.
- In the common disaccharides maltose, lactose, and sucrose, there is at least one glucose unit.
- Maltose and lactose form α and β isomers, which makes them reducing sugars.
- Sucrose does not have α and β isomers and is not a reducing sugar.

13.7 Polysaccharides

LEARNING GOAL Describe the structural features of amylose, amylopectin, glycogen, and cellulose.

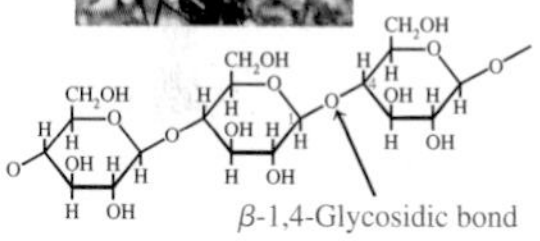

- Polysaccharides are polymers of monosaccharide units.
- Amylose is an unbranched chain of glucose with α-1,4-glycosidic bonds, and amylopectin is a branched polymer of glucose with α-1,4- and α-1,6-glycosidic bonds.
- Glycogen is similar to amylopectin with more branching.
- Cellulose is also a polymer of glucose, but in cellulose, the glycosidic bonds are β-1,4-bonds.

SUMMARY OF CARBOHYDRATES

Carbohydrate	Food Sources	
Monosaccharides		
Glucose	Fruit juices, honey, corn syrup	
Galactose	Lactose hydrolysis	
Fructose	Fruit juices, honey, sucrose hydrolysis	
Disaccharides		**Monosaccharide Components**
Maltose	Germinating grains, starch hydrolysis	Glucose + glucose
Lactose	Milk, yogurt, ice cream	Glucose + galactose
Sucrose	Sugar cane, sugar beets	Glucose + fructose
Polysaccharides		
Amylose	Rice, wheat, grains, cereals	Unbranched polymer of glucose joined by α-1,4-glycosidic bonds
Amylopectin	Rice, wheat, grains, cereals	Branched polymer of glucose joined by α-1,4- and α-1,6-glycosidic bonds
Glycogen	Liver, muscles	Highly branched polymer of glucose joined by α-1,4- and α-1,6-glycosidic bonds
Cellulose	Plant fiber, bran, beans, celery	Unbranched polymer of glucose joined by β-1,4-glycosidic bonds

SUMMARY OF REACTIONS

The chapter sections for review are shown after the name of the reaction.

Oxidation and Reduction of Monosaccharides (13.5)

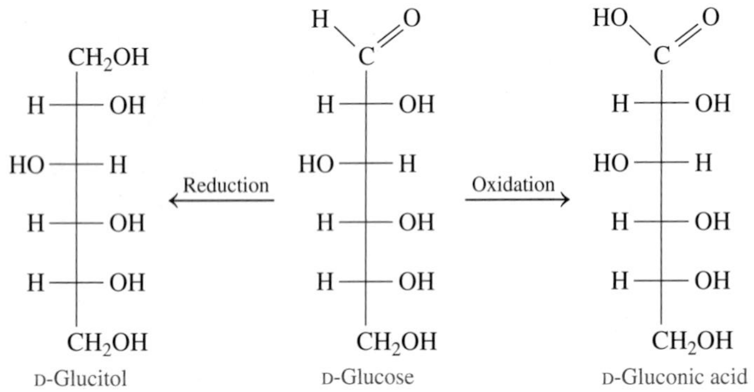

Formation of Disaccharides (13.6)

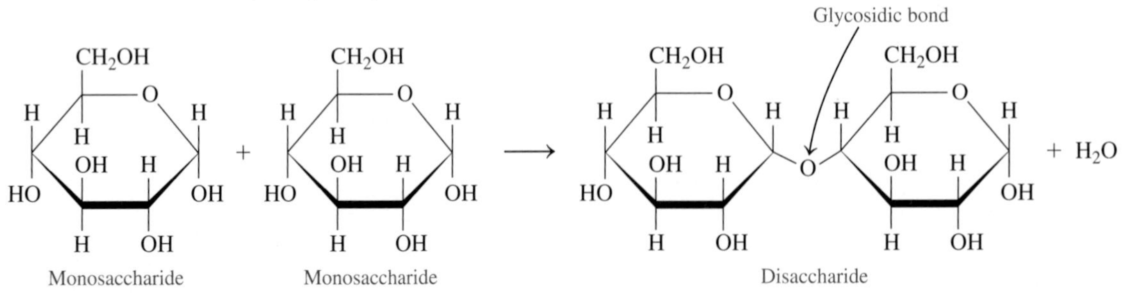

Hydrolysis of Polysaccharides (13.7)

$$\text{Amylose, amylopectin} \xrightarrow{H^+ \text{ or enzymes}} \text{many D-glucose units}$$

KEY TERMS

aldose A monosaccharide that contains an aldehyde group.

amylopectin A branched-chain polymer of starch composed of glucose units joined by α-1,4- and α-1,6-glycosidic bonds.

amylose An unbranched polymer of starch composed of glucose units joined by α-1,4-glycosidic bonds.

carbohydrate A simple or complex sugar composed of carbon, hydrogen, and oxygen.

cellulose An unbranched polysaccharide composed of glucose units linked by β-1,4-glycosidic bonds that cannot be hydrolyzed by the human digestive system.

chiral Having nonsuperimposable mirror images.

chiral carbon A carbon atom that is bonded to four different atoms or groups.

disaccharide A carbohydrate composed of two monosaccharides joined by a glycosidic bond.

enantiomers Stereoisomers that are mirror images that cannot be superimposed.

Fischer projection A system for drawing stereoisomers with horizontal lines representing bonds that project forward, vertical lines that represent bonds that project backward, and with a carbon atom at each intersection.

fructose A monosaccharide that is also called levulose and fruit sugar and is found in honey and fruit juices; it is combined with glucose in sucrose.

galactose A monosaccharide that occurs combined with glucose in lactose.

glucose An aldohexose found in fruits, vegetables, corn syrup, and honey that is also known as blood sugar and dextrose. The most prevalent monosaccharide in the diet. Most polysaccharides are polymers of glucose.

glycogen A polysaccharide formed in the liver and muscles for the storage of glucose as an energy reserve. It is composed of glucose in a highly branched polymer joined by α-1,4- and α-1,6-glycosidic bonds.

glycosidic bond The bond that forms when the hydroxyl group of one monosaccharide reacts with the hydroxyl group of another monosaccharide. It is the type of bond that links monosaccharide units in di- or polysaccharides.

Haworth structure The ring structure of a monosaccharide.

ketose A monosaccharide that contains a ketone group.

lactose A disaccharide consisting of glucose and galactose found in milk and milk products.

maltose A disaccharide consisting of two glucose units; it is obtained from the hydrolysis of starch and is found in germinating grains.

monosaccharide A polyhydroxy compound that contains an aldehyde or ketone group.

polysaccharides Polymers of many monosaccharide units, usually glucose. Polysaccharides differ in the types of glycosidic bonds and the amount of branching in the polymer.

reducing sugar A carbohydrate with an aldehyde group capable of reducing the Cu^{2+} in Benedict's reagent.

stereoisomers Isomers that have atoms bonded in the same order, but with different arrangements in space.

sucrose A disaccharide composed of glucose and fructose; a nonreducing sugar, commonly called table sugar or "sugar".

CORE CHEMISTRY SKILLS

Identifying Chiral Molecules (13.2)

- Chiral molecules are molecules with mirror images that cannot be superimposed on each other. These types of stereoisomers are called enantiomers.
- A chiral molecule must have at least one chiral carbon, which is a carbon bonded to four different atoms or groups of atoms.

Example: Identify each of the following as chiral or achiral:

a. $CH_3-CH_2-CH(Br)-CH_3$

b. $CH_3-CH_2-CH_2-Br$

c. $CH_3-CH(Br)-CH_3$

Answer: Molecule **a** is chiral; molecules **b** and **c** are achiral.

Identifying D- and L-Fischer Projections (13.3)

- The Fischer projection is a simplified way to draw the arrangements of atoms by placing the carbon atoms at the intersection of vertical and horizontal lines.
- The names of the mirror images are labeled D or L to differentiate between enantiomers of carbohydrates.

Example: Identify each of the following as the D or L stereoisomer:

a.

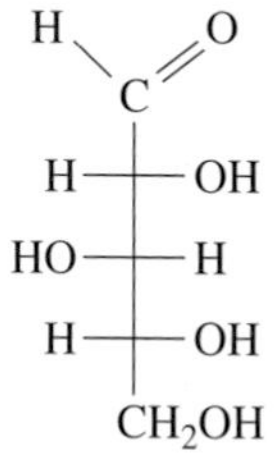

b.

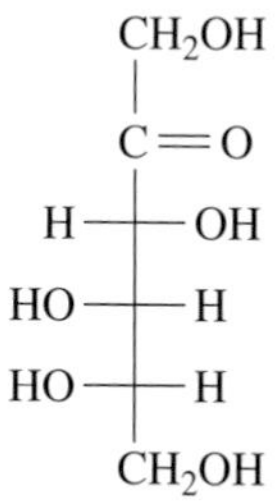

Answer: In **a**, the —OH group on the chiral carbon farthest from the carbonyl group is on the right; it is the D stereoisomer. In **b**, the —OH group on the chiral carbon farthest from the carbonyl group is on the left; it is the L stereoisomer.

UNDERSTANDING THE CONCEPTS

The chapter sections to review are shown in parentheses at the end of each question.

13.51 For each of the following, indicate whether the carbon in red is chiral or achiral: (13.2)

a. glycerol, which is used to sweeten and preserve foods and as a lubricant in soaps, creams, and hair care products

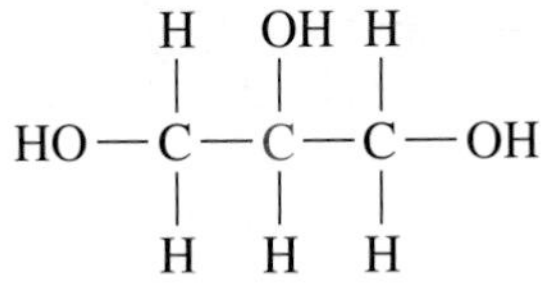

Glycerol

b. monosodium glutamate (MSG), which is the salt of the amino acid glutamic acid used as a flavor enhancer in foods

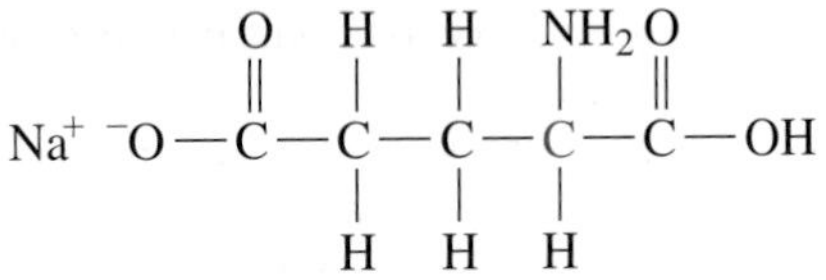

Monosodium glutamate (MSG)

13.52 For each of the following, indicate whether the carbon in red is chiral or achiral: (13.2)

a. ibuprofen, which is a nonsteroidal anti-inflammatory drug used to relieve fever and pain

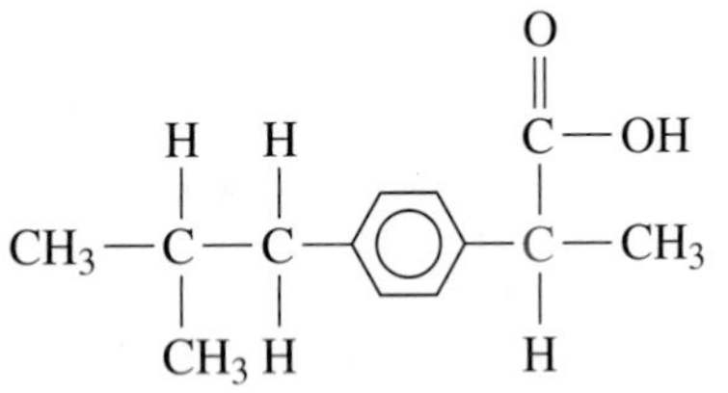

Ibuprofen

b. penicillamine, which is used in the treatment of rheumatoid arthritis

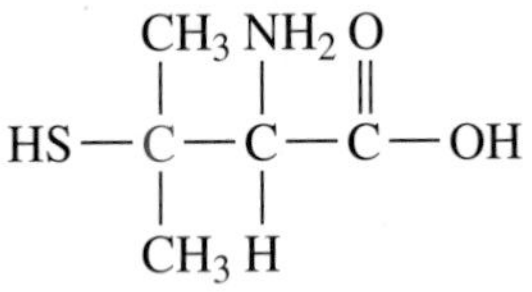

Penicillamine

13.53 Isomaltose, obtained from the breakdown of starch, has the following Haworth structure: (13.4, 13.5, 13.6)

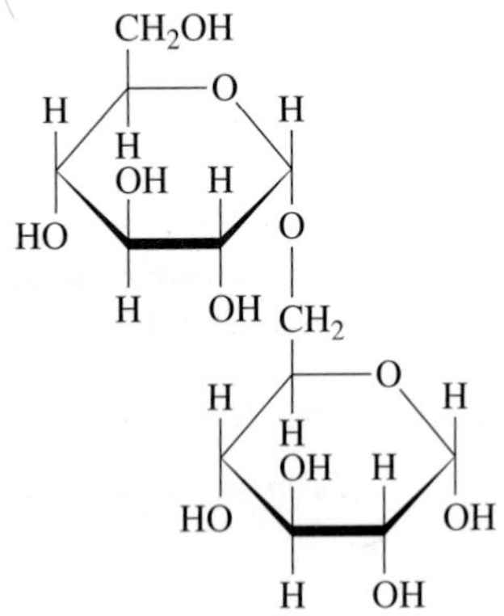

Isomaltose

a. Is isomaltose a mono-, di-, or polysaccharide?
b. What are the monosaccharides in isomaltose?
c. What is the glycosidic link in isomaltose?
d. Is this the α or β isomer of isomaltose?
e. Is isomaltose a reducing sugar?

13.54 Sophorose, a carbohydrate found in certain types of beans, has the following Haworth structure: (13.4, 13.5, 13.6)

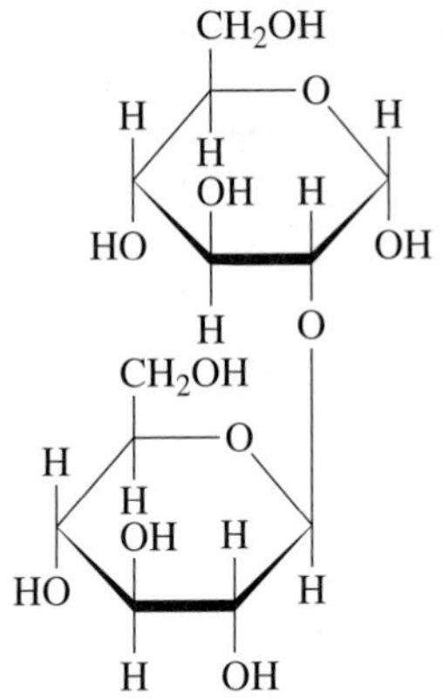

Sophorose

a. Is sophorose a mono-, di-, or polysaccharide?
b. What are the monosaccharides in sophorose?
c. What is the glycosidic link in sophorose?
d. Is this the α or β isomer of sophorose?
e. Is sophorose a reducing sugar?

13.55 Melezitose, a carbohydrate secreted by insects, has the following Haworth structure: (13.4, 13.5, 13.6)

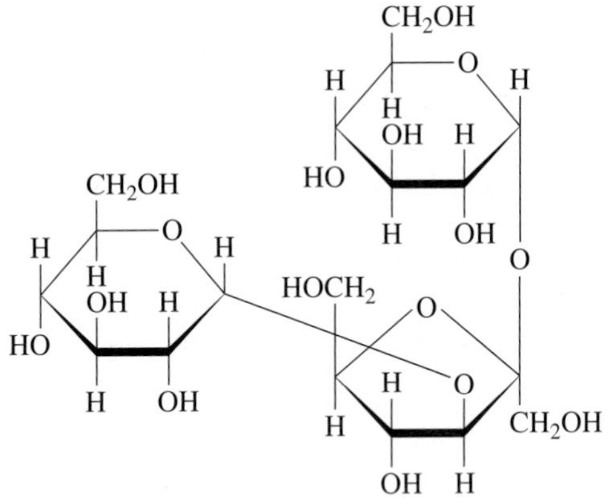

Melezitose

a. Is melezitose a mono-, di-, or trisaccharide?
b. What ketohexose and aldohexose are present in melezitose?
c. Is melezitose a reducing sugar?

13.56 What are the disaccharides and polysaccharides present in each of the following? (13.6, 13.7)

(a)

(b)

(c)

(d)

ADDITIONAL QUESTIONS AND PROBLEMS

13.57 Identify the chiral carbons, if any, in each of the following compounds: (13.2)

a. 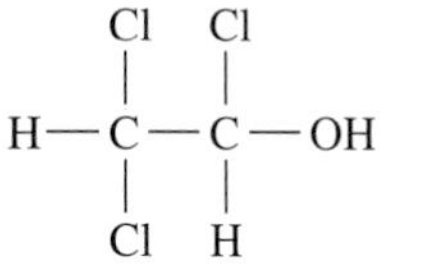

b. 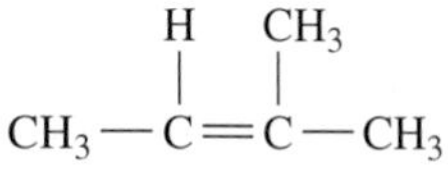

c. 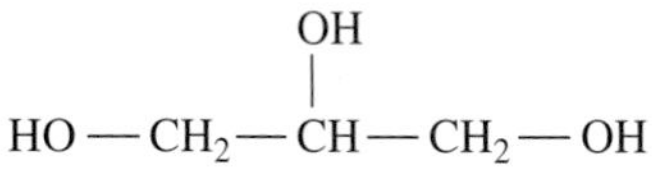

d. 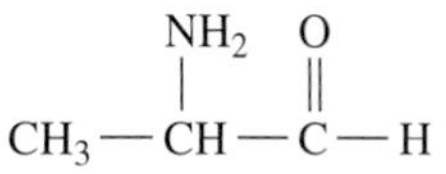

e. 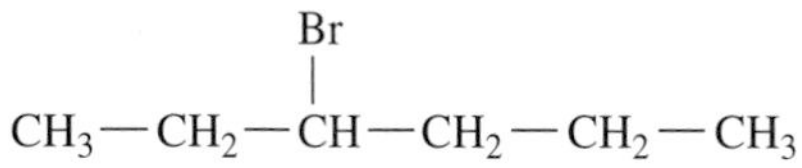

13.58 Identify the chiral carbons, if any, in each of the following compounds: (13.2)

a. 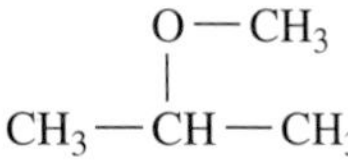

b. 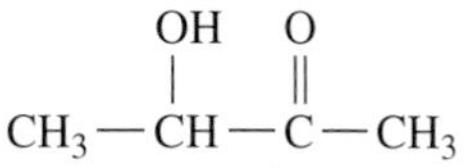

c. 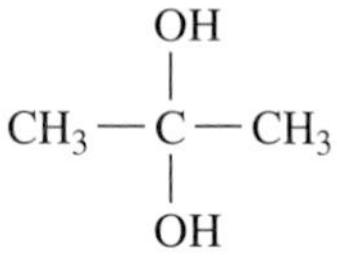

d. 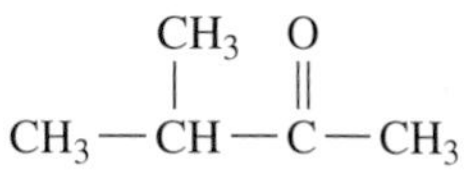

e. 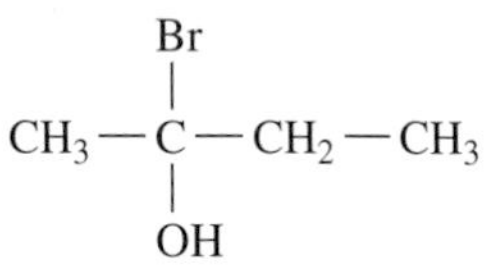

13.59 Identify each of the following pairs of Fischer projections as enantiomers or identical compounds: (13.2)

a. 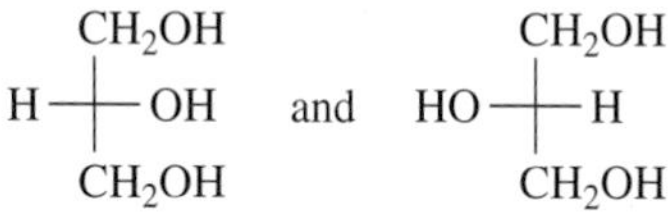

b. 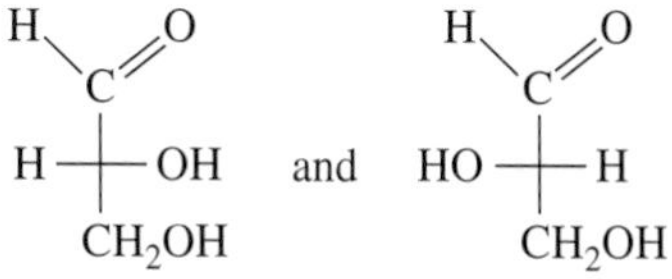

c. 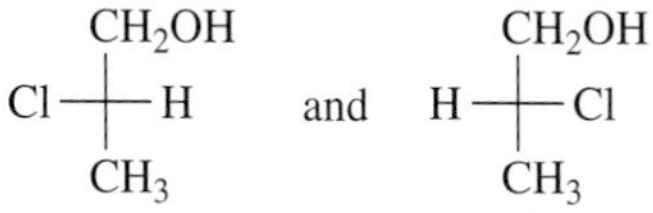

d. 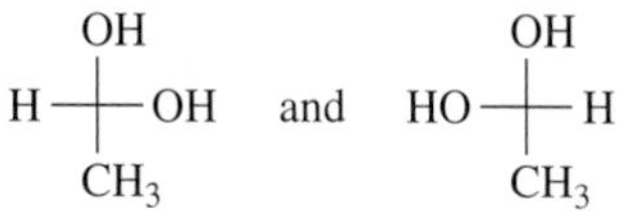

13.60 Identify each of the following pairs of Fischer projections as enantiomers or identical compounds: (13.2)

a. 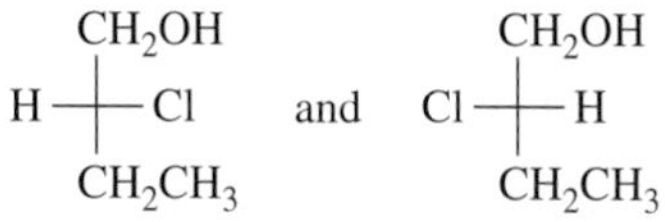

b. 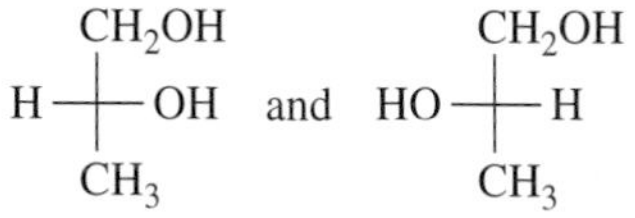

c. 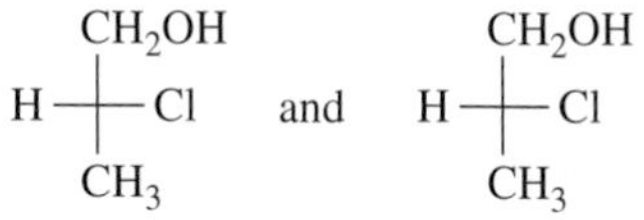

d. 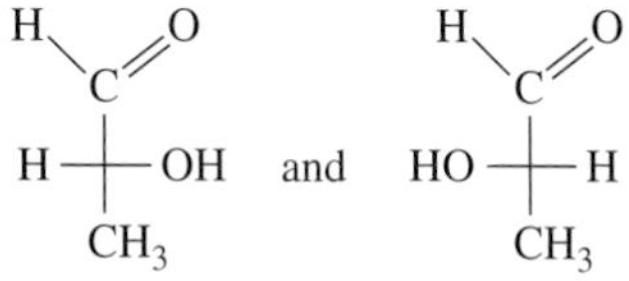

13.61 What are the differences in the Fischer projections of D-fructose and D-galactose? (13.3)

13.62 What are the key differences and similarities between cellulose and amylase? (13.6)

13.63 What are the differences in the Fischer projections of D-galactose and L-galactose? (13.3)

13.64 What are the differences in the Haworth structures of α-D-glucose and β-D-glucose? (13.3)

13.65 The sugar D-gulose is a sweet-tasting syrup. (13.3, 13.4)

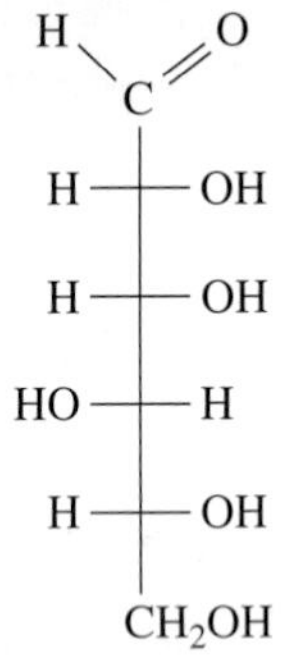

D-Gulose

a. Draw the Fischer projection for L-gulose.
b. Draw the Haworth structures for α- and β-D-gulose.

13.66 Use the Fischer projection for D-gulose in problem 13.65 to answer each of the following: (13.3, 13.5)
a. Draw the Fischer projection and name the product formed by the reduction of D-gulose.
b. Draw the Fischer projection and name the product formed by the oxidation of D-gulose.

13.67 D-Sorbitol, a sweetener found in seaweed and berries, contains only hydroxyl functional groups. When D-sorbitol is oxidized, it forms D-glucose. Draw the Fischer projection for D-sorbitol. (13.3, 13.5)

13.68 Raffinose is a trisaccharide found in Australian manna and in cottonseed meal. It is composed of three different monosaccharides. Identify the monosaccharides in raffinose. (13.4)

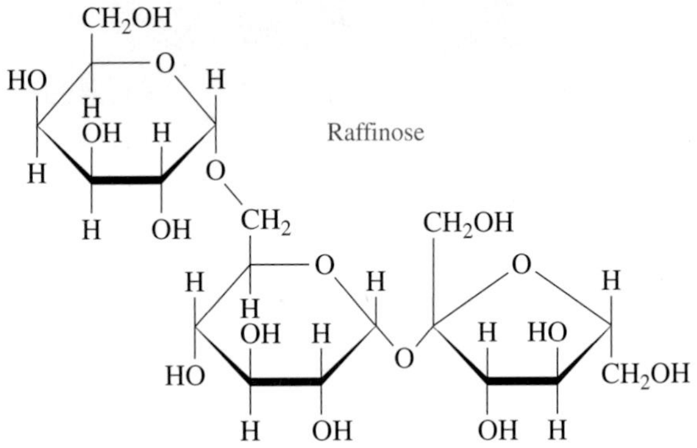

Raffinose

13.69 If α-galactose is dissolved in water, β-galactose is eventually present. Explain how this occurs. (13.4)

13.70 Why are lactose and maltose reducing sugars, but sucrose is not? (13.5)

CHALLENGE QUESTIONS

The following groups of questions are related to the topics in this chapter. However, they do not all follow the chapter order, and they require you to combine concepts and skills from several sections. These questions will help you increase your critical thinking skills and prepare for your next exam.

13.71 α-Cellobiose is a disaccharide obtained from the hydrolysis of cellulose. It is quite similar to maltose except it has a β-1,4-glycosidic bond. Draw the Haworth structure for α-cellobiose. (13.4, 13.6)

13.72 The disaccharide trehalose found in mushrooms is composed of two α-D-glucose molecules joined by an α-1,1-glycosidic bond. Draw the Haworth structure for trehalose. (13.4, 13.6)

13.73 Gentiobiose is found in saffron. (13.4, 13.5, 13.6)
a. Gentiobiose contains two glucose molecules linked by a β-1,6-glycosidic bond. Draw the Haworth structure for β-gentiobiose.
b. Is gentiobiose a reducing sugar? Explain.

13.74 Identify the Fischer projection **A** to **D** that matches each of the following: (13.1, 13.3)
a. the L stereoisomer of mannose
b. a ketopentose
c. an aldopentose
d. a ketohexose

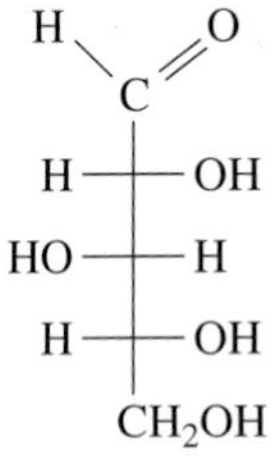

A

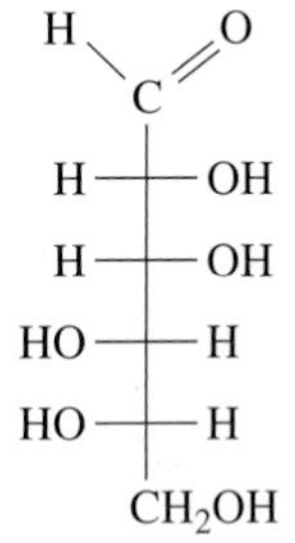

B

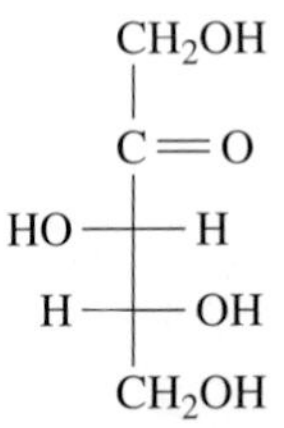

C

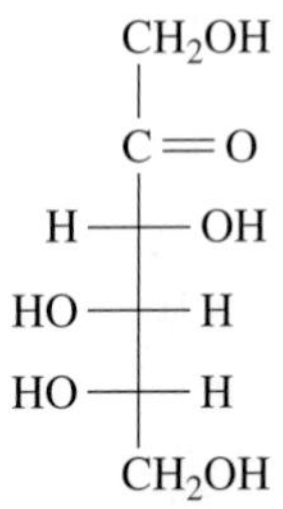

D

ANSWERS

Answers to Study Checks

13.1 Erythrose has four carbon atoms (tetrose) and is an aldehyde, which makes it an aldotetrose.

13.2

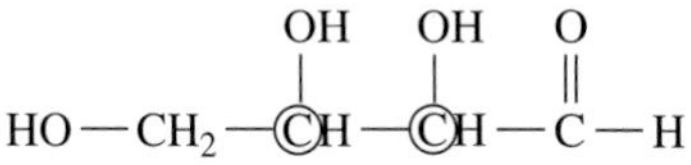

13.3 Because the —OH group on the chiral carbon farthest from the carbonyl is on the left, this is the L stereoisomer.

13.4 To draw the mirror image, all the —OH groups on the chiral carbon atoms are drawn on the opposite side. L-Ribulose has the following Fischer projection:

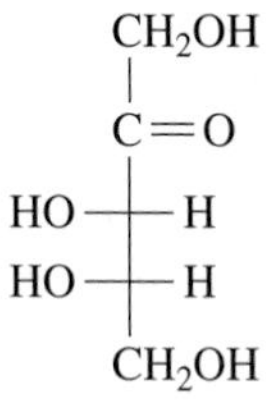

13.5

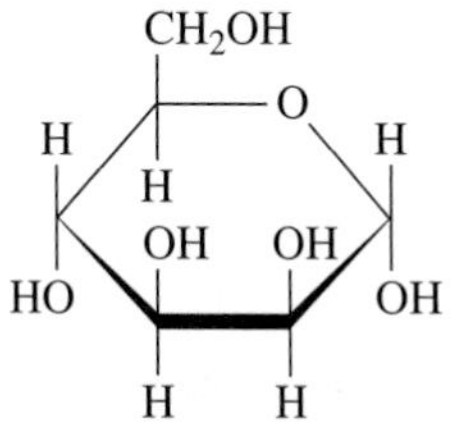

13.6 The brick-red color indicates a high level of reducing sugar (probably glucose) in the urine. One common cause of this condition is diabetes mellitus.

13.7

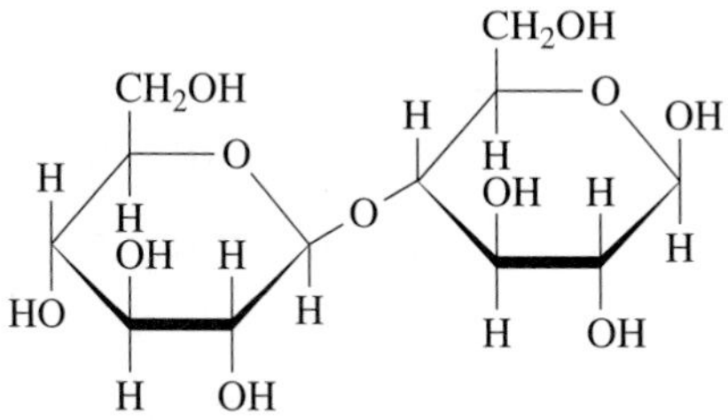

13.8 Cellulose contains glucose units connected by β-1,4-glycosidic bonds, whereas the glucose units in amylose are connected by α-1,4-glycosidic bonds.

Answers to Selected Questions and Problems

13.1 Photosynthesis requires CO_2, H_2O, and the energy from the Sun. Respiration requires O_2 from the air and glucose from our foods.

13.3 Monosaccharides are simplest carbohydrate that will be split or hydrolyzed into smaller carbohydrates. Example: glucose. A disaccharide consists of two monosaccharide units joined together, which can be split into two monosaccharide units. For example, sucrose.

13.5 Hydroxyl groups are found in all monosaccharides along with a carbonyl on the first or second carbon that gives an aldehyde or ketone functional group.

13.7 An aldopentose is a five-carbon monosaccharide that is an aldehyde.

13.9 **a.** aldopentose
b. ketohexose

13.11 **a.** chiral, carbon 1
b. chiral, carbon 3,
c. achiral
d. chiral, carbon 3

13.13 **a.** carbon 5
b. carbon 4

13.15 **a.**

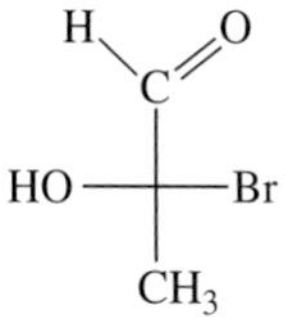

b.

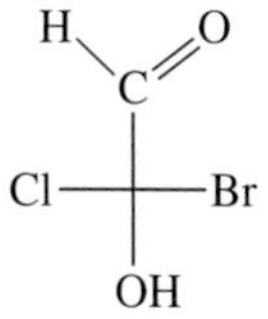

c.

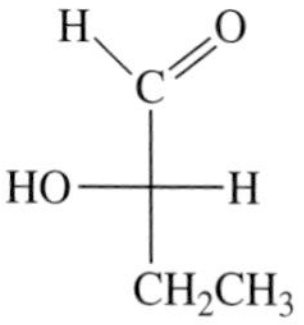

13.17 **a.** identical **b.** enantiomers
c. enantiomers **d.** enantiomers

13.19 In the D stereoisomer, the —OH group on the chiral carbon atom farthest from the carbonyl group is on the right side, whereas in the L stereoisomer, the —OH group is on the left side of the chiral carbon farthest from the carbonyl group.

13.21 **a.** D-Glactose **b.** D-Altrose
c. L-Psicose **d.** L-Ribulose

13.23 **a.**

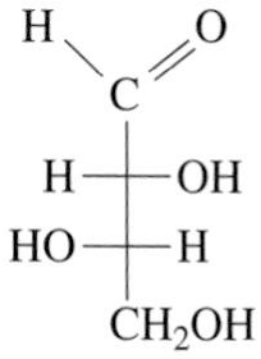

b.

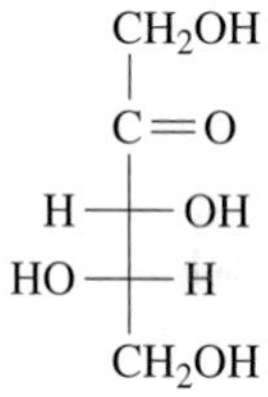

c.

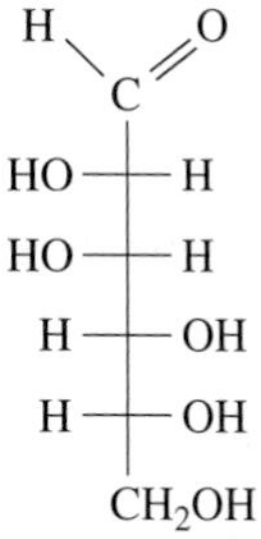

d.

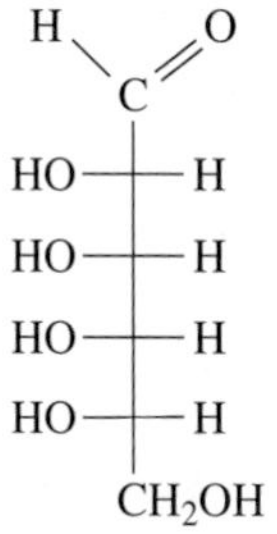

13.25

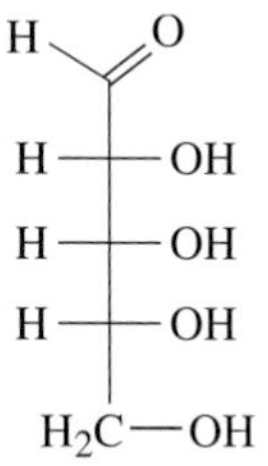

D-Ribose

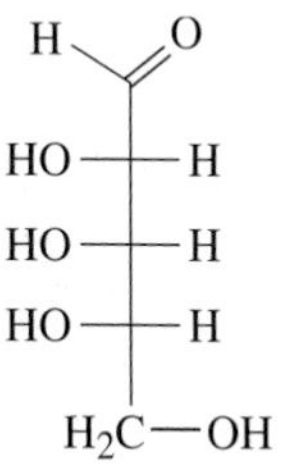

L-Ribose

13.27

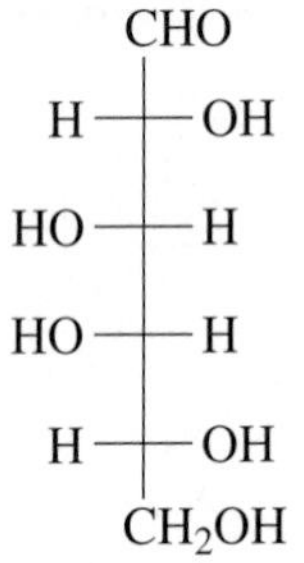

D-Galactose

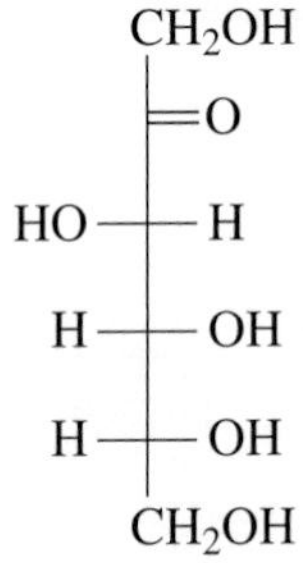

D-Fructose

13.29 a. fructose **b.** galactose **c.** glucose

13.31 In the cyclic structure of glucose, there are five carbon atoms and an oxygen atom.

13.33

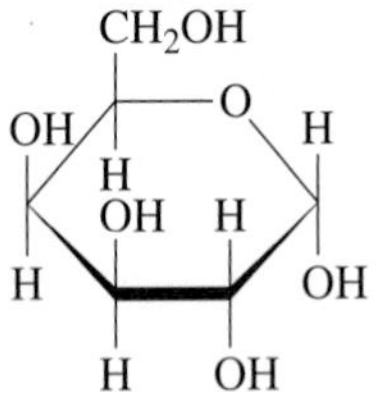

α-Galactose

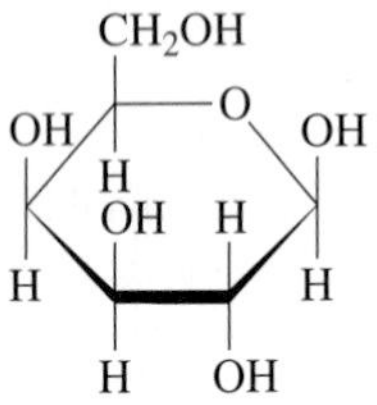

β-Galactose

13.35 a. α isomer **b.** α isomer

13.37

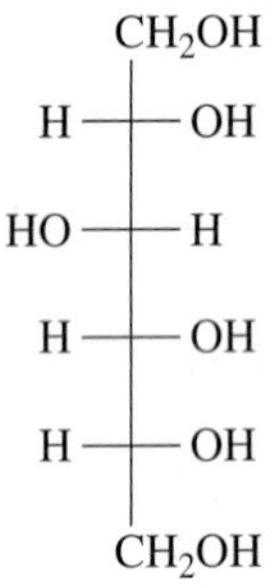

D-Glucitol

13.39 Oxidation product:

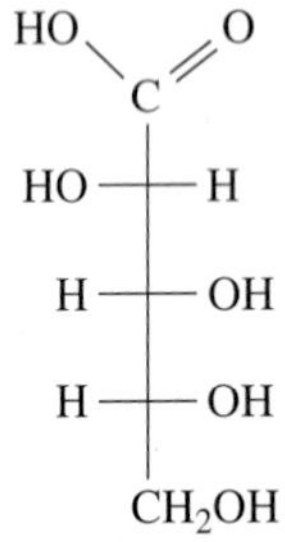

D-Arabinonic acid

Reduction product (sugar alcohol):

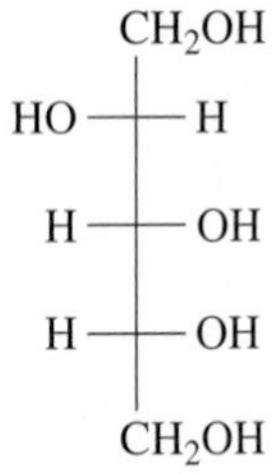

D-Arabitol

13.41 a. galactose and glucose; β-1,4-glycosidic bond; β-lactose
b. glucose and glucose; α-1,4-glycosidic bond; α-maltose

13.43 a. is a reducing sugar **b.** is a reducing sugar

13.45 a. maltose **b.** lactose
c. sucrose **d.** lactose

13.47 a. Amylose is an unbranched polymer of glucose units joined by α-1,4-glycosidic bonds; amylopectin is a branched polymer of glucose joined by α-1,4- and α-1,6-glycosidic bonds.
b. Amylopectin, which is produced in plants, is a branched polymer of glucose, joined by α-1,4- and α-1,6-glycosidic bonds. The branches in amylopectin occur about every 25 glucose units. Glycogen, which is produced in animals, is a highly branched polymer of glucose, joined by α-1,4- and α-1,6-glycosidic bonds. The branches in glycogen occur about every 10 to 15 glucose units.

13.49 a. cellulose **b.** amylose, amylopectin
c. amylose **d.** glycogen

13.51 a. Achiral. Two of the substituents on the carbon in red are the same ($-CH_2OH$). A chiral carbon must be bonded to four different groups or atoms.
b. Chiral. The carbon in red is bonded to four different groups: $-COOH$, $-NH_2$, $-H$, and $-CH_2-CH_2-COO^- Na^+$.

13.53 a. disaccharide **b.** α-D-glucose
c. α-1,6-glycosidic bond **d.** α
e. is a reducing sugar

13.55 a. trisaccharide
b. two glucose and one fructose
c. Melezitose has no free $-OH$ groups on the glucose or fructose molecules; like sucrose it is not a reducing sugar.

13.57 a.

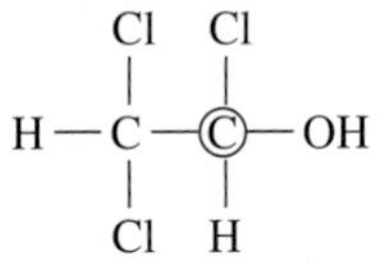

b. none

c. none

d.

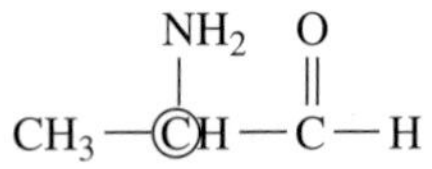

e.

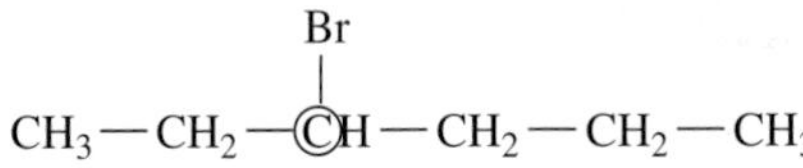

13.59 a. identical **b.** enantiomers
c. enantiomers **d.** identical

13.61 D-Fructose is a ketohexose, whereas D-galactose is an aldohexose. In the Fischer projection of D-galactose, the $-OH$ group on carbon 5 is drawn on the left; in fructose, and the $-OH$ group is drawn on the right.

13.63 L-Galactose is the mirror image of D-galactose. In the Fischer projection of D-galactose, the $-OH$ groups on carbon 2 and carbon 5 are drawn on the right side, but they are drawn on the left for carbon 3 and carbon 4. In L-galactose, the $-OH$ groups are reversed; carbons 2 and 5 have $-OH$ groups drawn on the left, and carbons 3 and 4 have $-OH$ groups drawn on the right.

13.65 a.

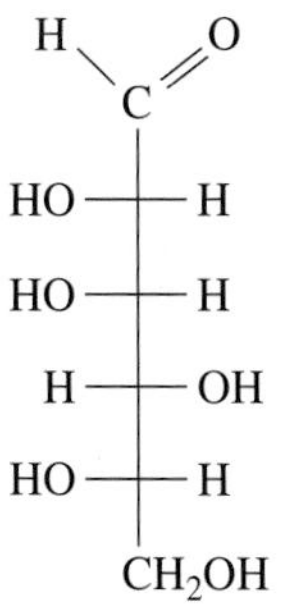

L-Gulose

b.

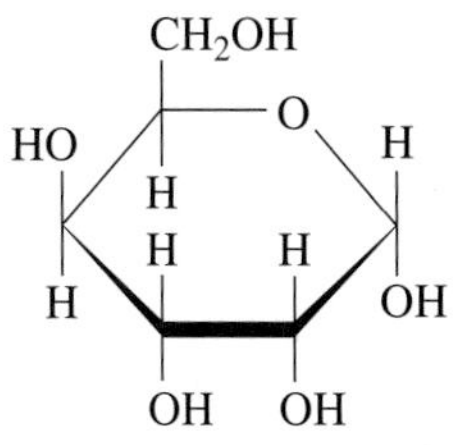

α-D-Gulose

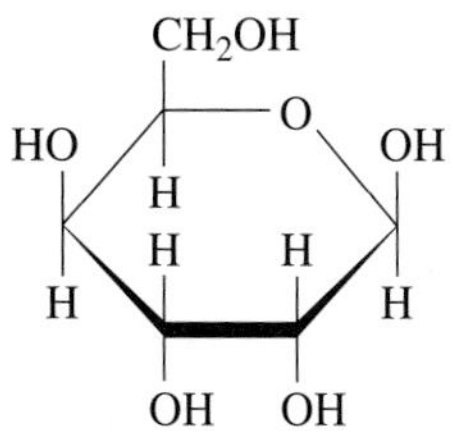

β-D-Gulose

13.67

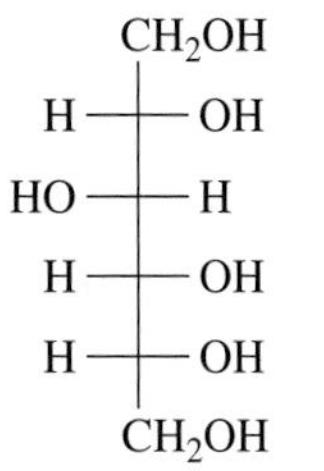

13.69 When α-galactose forms an open-chain structure, it can close to form either α- or β-galactose.

13.71

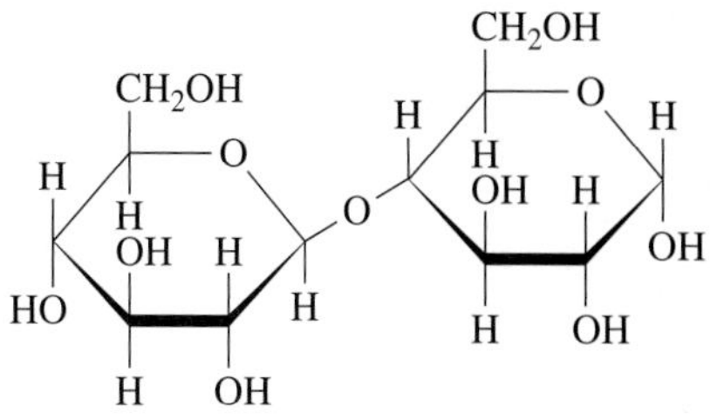

13.73 a.

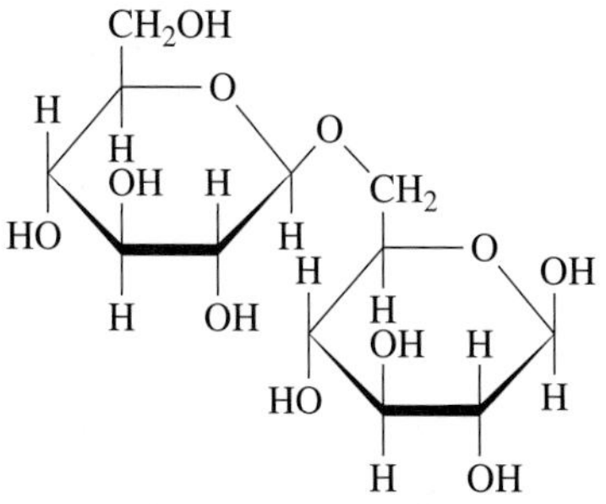

b. Yes. Gentiobiose is a reducing sugar. The ring on the right can open up to form an aldehyde that can be oxidized.

14 Carboxylic Acids, Esters, Amines, and Amides

Maureen, a surgical technician, begins preparing for Robert's heart surgery. After Maureen places all of the surgical instruments into an autoclave for sterilization, she prepares the room by ensuring that all of the equipment is working properly. She also determines that the room is sterile, as this will minimize the chance of an infection. Right before surgery begins, Maureen shaves Robert's chest and disinfects the incision sites.

After the surgery, Maureen applies Indermil, which is a type of liquid bandage, to Robert's incision sites. Liquid bandages are tissue adhesives that seal surgical or wound incisions on a patient. Stitches and staples are not required and scarring is minimal. The polymer in a liquid bandage is typically dissolved in an alcohol-based solvent. The alcohol also acts as an antiseptic.

The polymer in Indermil is a cyano ester, as it consists of a cyano group ($C{\equiv}N$) and an ester functional group. Esters have a carbonyl group ($C{=}O$), which has a single bond to a carbon group on one side of the carbonyl and an oxygen atom on the other side. The oxygen atom is then bonded to a carbon group by a single bond. The cyano ester in Indermil is butyl cyanoacrylate, which is similar to the polymer in superglue. The "*ate*" in butyl cyanoacrylate indicates that an ester is present in the molecule, and the "*butyl*" indicates the four-carbon group that is bonded to the oxygen atom.

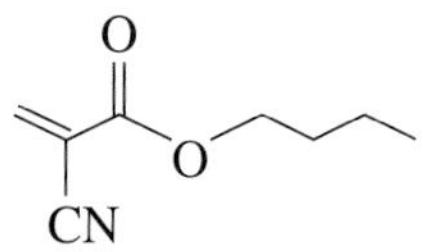

Indermil

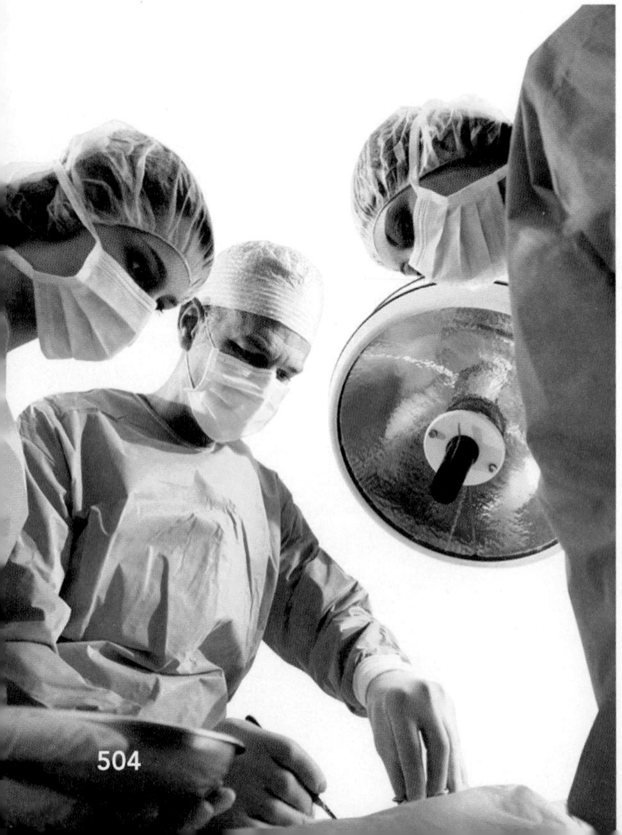

CAREER Surgical Technician

Surgical technicians prepare the operating room by creating a sterile environment. This includes setting up surgical instruments and equipment, and ensuring that all of the equipment is working properly. A sterile environment is critical to the patient's recovery, as it helps lower the chance of an infection. They also prepare patients for surgery by washing, shaving, and disinfecting incision sites. During the surgery, a surgical technician provides the sterile instruments and supplies to the surgeons and surgical assistants.

Carboxylic acids are weak acids. They have a sour or tart taste, produce hydronium ions in water, and neutralize bases. You encounter carboxylic acids when you use a salad dressing containing vinegar, which is a solution of acetic acid and water, or experience the sour taste of citric acid in a grapefruit or lemon. When a carboxylic acid reacts with an alcohol, an ester and water are produced. Fats and oils are esters of glycerol and fatty acids, which are long-chain carboxylic acids. Esters produce the pleasant aromas and flavors of many fruits, such as bananas, strawberries, and oranges.

Amines and amides are organic compounds that contain nitrogen. Many nitrogen-containing compounds are important to life as components of amino acids, proteins, and nucleic acids (DNA and RNA). Many amines that exhibit strong physiological activity are used in medicine as decongestants, anesthetics, and sedatives. Examples include dopamine, histamine, epinephrine, and amphetamine.

Alkaloids such as caffeine, nicotine, cocaine, and digitalis, which have powerful physiological activity, are naturally occurring amines obtained from plants. In an amide, the functional group consists of a carbonyl group attached to an amine. Amides, which are derived from carboxylic acids, are important in biology in proteins. In biochemistry, the amide bond that links amino acids in a protein is called a peptide bond. Some medically important amides include acetaminophen (Tylenol) used to reduce fever; phenobarbital, a sedative and anticonvulsant medication; and penicillin, an antibiotic.

LOOKING AHEAD

CHAPTER READINESS*

CORE CHEMISTRY SKILLS

- Writing Equations for Reactions of Acids and Bases (10.6)
- Naming and Drawing Alkanes (11.2)

*These Core Chemistry Skills from previous chapters are listed here for your review as you proceed to the new material in this chapter.

14.1 Carboxylic Acids

LEARNING GOAL

Give the IUPAC and common names for carboxylic acids; draw their condensed structural formulas or skeletal formulas.

In a **carboxylic acid**, the carbon atom of a carbonyl group is attached to a hydroxyl group that forms a **carboxyl group**. The carboxyl functional group may be attached to an alkyl group or an aromatic group.

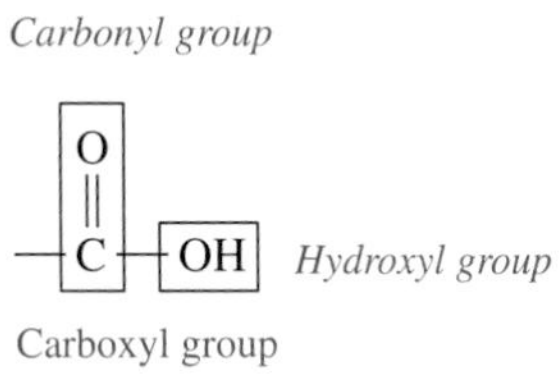

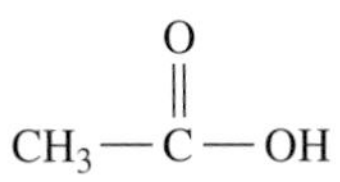
Ethanoic acid
(acetic acid)

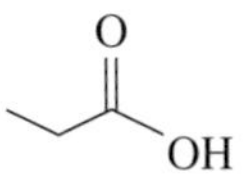
Propanoic acid
(propionic acid)

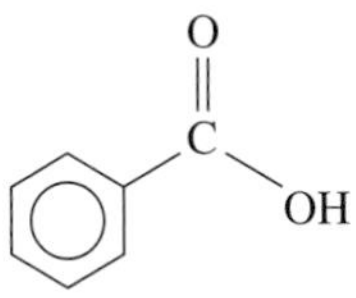
Benzoic acid

CORE CHEMISTRY SKILL

Naming Carboxylic Acids

IUPAC Names of Carboxylic Acids

The IUPAC names of carboxylic acids replace the *e* of the corresponding alkane name with *oic acid*. If there is a substituent, the carbon chain is numbered beginning with the carboxyl carbon.

$H-\overset{O}{\overset{\|\|}{C}}-OH$	$CH_3-CH_2-\overset{O}{\overset{\|\|}{C}}-OH$	$CH_3-\overset{CH_3}{\overset{\|}{CH}}-\overset{O}{\overset{\|\|}{C}}-OH$
Methanoic acid	Propanoic acid	2-Methylpropanoic acid

The simplest aromatic carboxylic acid is named benzoic acid. With the carboxyl carbon bonded to carbon 1, the ring is numbered in the direction that gives substituents the smallest possible numbers.

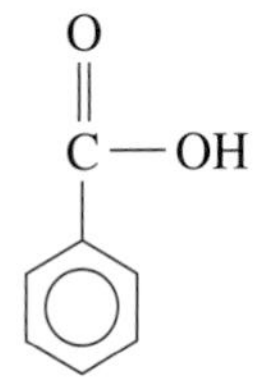

Benzoic acid

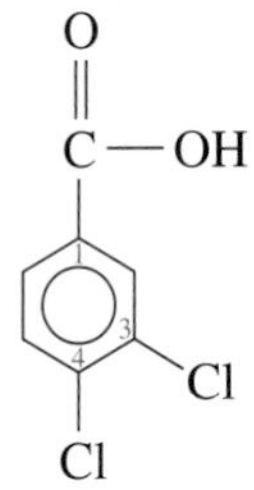

3,4-Dichlorobenzoic acid

A red ant sting contains formic acid that irritates the skin.

The sour taste of vinegar is due to ethanoic acid (acetic acid).

Common Names of Carboxylic Acids

Many carboxylic acids are still named by their common names, which use prefixes: *form*, *acet*, *propion*, *butyr*. These prefixes are related to the natural sources of the simple carboxylic acids. For example, formic acid is injected under the skin from bee or red ant stings and other insect bites. Acetic acid is the oxidation product of the ethanol in wines and apple cider. A solution of acetic acid and water is known as vinegar. Butyric acid gives the foul odor to rancid butter (see Table 14.1).

TABLE 14.1 Names and Condensed Structural Formulas of Some Carboxylic Acids

Condensed Structural Formula	IUPAC Name	Common Name	Ball-and-Stick Model
$H-\overset{O}{\overset{\|\|}{C}}-OH$	Methanoic acid	Formic acid	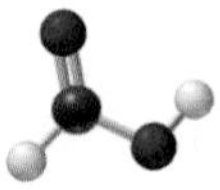
$CH_3-\overset{O}{\overset{\|\|}{C}}-OH$	Ethanoic acid	Acetic acid	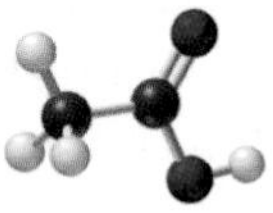
$CH_3-CH_2-\overset{O}{\overset{\|\|}{C}}-OH$	Propanoic acid	Propionic acid	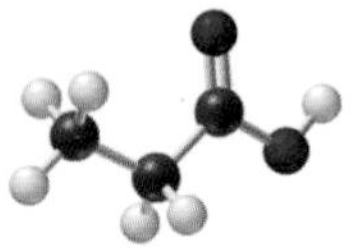
$CH_3-CH_2-CH_2-\overset{O}{\overset{\|\|}{C}}-OH$	Butanoic acid	Butyric acid	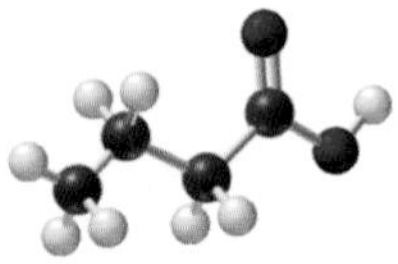

SAMPLE PROBLEM 14.1 Naming Carboxylic Acids

Give the IUPAC name for the following:

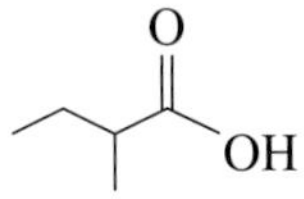

SOLUTION

	Functional Group	Family	IUPAC Naming
ANALYZE THE PROBLEM	carboxyl	carboxylic acid	replace the *e* in the alkane name with *oic acid* and count from carbon 1 of the carboxyl group for any substituents

STEP 1 Identify the longest carbon chain and replace the *e* in the corresponding alkane name with *oic acid*. The IUPAC name of a carboxylic acid with four carbon atoms is butanoic acid.

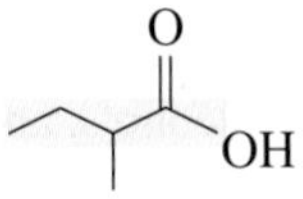

butanoic acid

STEP 2 Name and number any substituents by counting the carboxyl group as carbon 1. The substituent on carbon 2 is methyl. The IUPAC name for this compound is 2-methylbutanoic acid.

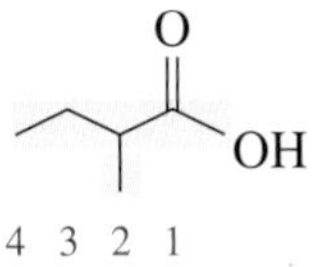

2-methylbutanoic acid

STUDY CHECK 14.1

Draw the condensed structural formula for 2,4-dichloropentanoic acid.

Guide to Naming Carboxylic Acids

STEP 1 Identify the longest carbon chain and replace the *e* in the corresponding alkane name with *oic acid*.

STEP 2 Name and number any substituents by counting the carboxyl group as carbon 1.

QUESTIONS AND PROBLEMS

14.1 Carboxylic Acids

LEARNING GOAL Give the IUPAC and common names for carboxylic acids; draw their condensed structural formulas or skeletal formulas.

14.1 Which kind of carboxylic acid is responsible for the foul odor to rancid butter?

14.2 Give the IUPAC and common name for the ethanol oxidation product in wine.

14.3 Draw the condensed structural formula and give the IUPAC name for each of the following:

a. A carboxylic acid that has the formula $C_4H_8O_2$, with no substituent.

b. A carboxylic acid that has the formula $C_5H_{10}O_2$, with one methyl substituent.

14.4 Draw the condensed structural formula and give the IUPAC name for each of the following:

a. A carboxylic acid that has the formula $C_5H_{10}O_2$, with no substituents

b. A carboxylic acid that has the formula $C_5H_{10}O_2$, with two methyl substituents

14.5 Give the IUPAC and common names (if any) for the following carboxylic acids:

a. 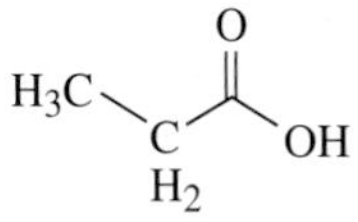

b. 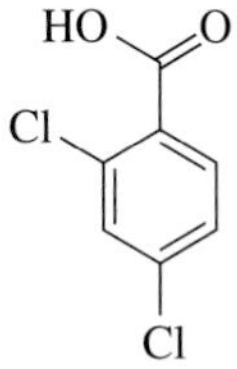

c. 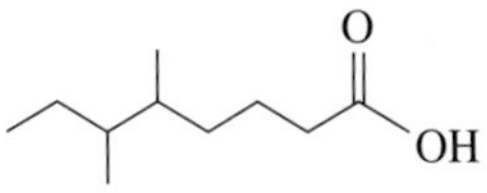

14.6 Give the IUPAC and common names (if any) for the following carboxylic acids:

a. 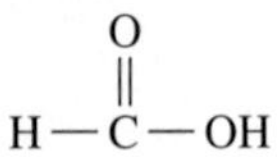

b. 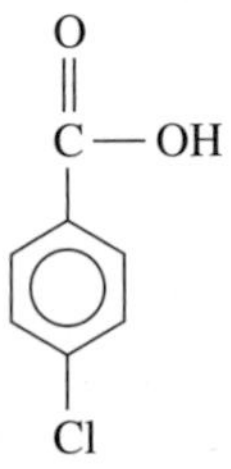

c. 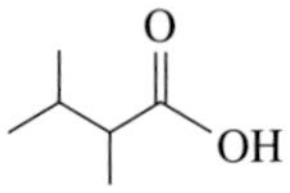

14.7 Draw the condensed structural formula for each of the following:

a. 2,3-dimethylhexanoic acid
b. 2,4-dimethylbenzoic acid
c. 2-chloro-3-methyl butanoic acid
d. 2-methyl heptanoic acid

14.8 Draw the skeletal formula for each of the following:

a. 3-methoxylbenzoic acid
b. 3,3-dichloropropanoic acid
c. 2,3-dimethylpentanoic acid
d. hex-4-enoic acid

LEARNING GOAL

Describe the solubility, ionization, and neutralization of carboxylic acids.

14.2 Properties of Carboxylic Acids

Carboxylic acids are among the most polar organic compounds because their functional group consists of two polar groups: a hydroxyl group (—OH) and a carbonyl group (C=O). The —OH group is similar to the functional group in alcohols, and the C=O double bond is similar to that of aldehydes and ketones.

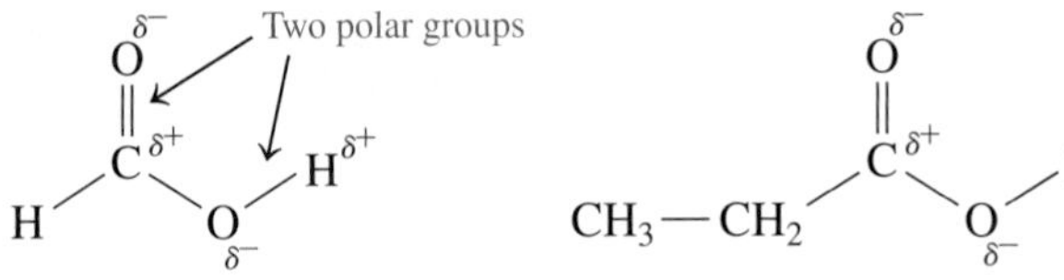

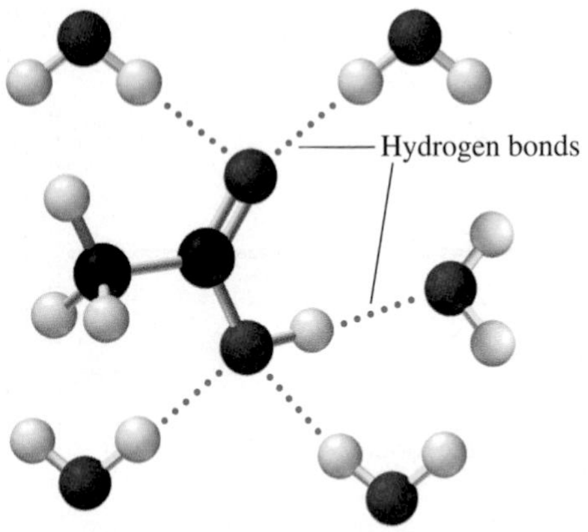

FIGURE 14.1 ▶ Acetic acid forms hydrogen bonds with water molecules.

Q Why do the atoms in the carboxyl group hydrogen bond with water molecules?

Solubility in Water

Carboxylic acids with one to five carbons are soluble in water because the carboxyl group forms hydrogen bonds with several water molecules (see Figure 14.1). However, as the length of the hydrocarbon chain increases, the nonpolar portion reduces the solubility of the carboxylic acid in water. Carboxylic acids having five or more carbons are not very soluble in water. Table 14.2 lists the solubility for some selected carboxylic acids.

Acidity of Carboxylic Acids

An important property of carboxylic acids is their ionization in water. When a carboxylic acid ionizes in water, a hydrogen ion is transferred to a water molecule to form a negatively charged **carboxylate ion** and a hydronium ion (H_3O^+). Carboxylic acids are more acidic than most other organic compounds including phenols. However, they are weak acids because only a small percentage (~1%) of the carboxylic acid molecules ionize in water. Carboxylic acids can lose hydrogen ions because the negative charge of the carboxylate anion is stabilized by the two oxygen atoms. By comparison, alcohols do not lose the H in their —OH group because they cannot stabilize the negative charge that would result on a single oxygen atom.

Carboxylic Acid **Carboxylate Ion**

$$\underset{\text{(acetic acid)}}{\underset{\text{Ethanoic acid}}{CH_3-\overset{\overset{\displaystyle O}{\|}}{C}-OH}} + H_2O \rightleftarrows \underset{\text{(acetate ion)}}{\underset{\text{Ethanoate ion}}{CH_3-\overset{\overset{\displaystyle O}{\|}}{C}-O^-}} + \underset{\text{ion}}{\underset{\text{Hydronium}}{H_3O^+}}$$

TABLE 14.2 Solubility of Selected Carboxylic Acids

IUPAC Name	Condensed Structural Formula	Solubility in Water
Methanoic acid	$H-\overset{\overset{\displaystyle O}{\|}}{C}-OH$	Soluble
Ethanoic acid	$CH_3-\overset{\overset{\displaystyle O}{\|}}{C}-OH$	Soluble
Propanoic acid	$CH_3-CH_2-\overset{\overset{\displaystyle O}{\|}}{C}-OH$	Soluble
Butanoic acid	$CH_3-CH_2-CH_2-\overset{\overset{\displaystyle O}{\|}}{C}-OH$	Soluble
Pentanoic acid	$CH_3-CH_2-CH_2-CH_2-\overset{\overset{\displaystyle O}{\|}}{C}-OH$	Soluble
Hexanoic acid	$CH_3-CH_2-CH_2-CH_2-CH_2-\overset{\overset{\displaystyle O}{\|}}{C}-OH$	Slightly soluble
Benzoic acid	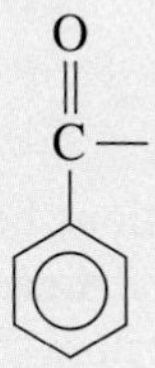	Slightly soluble

SAMPLE PROBLEM 14.2 Ionization of Carboxylic Acids in Water

Write the balanced chemical equation for the ionization of propanoic acid in water.

SOLUTION

ANALYZE THE PROBLEM	Reaction Type	Reactants	Products
	ionization	propanoic acid, H_2O	propanoate ion, H_3O^+

The ionization of propanoic acid produces a carboxylate ion and a hydronium ion.

$$\underset{\text{Propanoic acid}}{CH_3-CH_2-\overset{\overset{\displaystyle O}{\|}}{C}-OH} + H_2O \rightleftarrows \underset{\text{Propanoate ion}}{CH_3-CH_2-\overset{\overset{\displaystyle O}{\|}}{C}-O^-} + H_3O^+$$

STUDY CHECK 14.2

Write the balanced chemical equation for the ionization of methanoic acid in water.

Neutralization of Carboxylic Acids

Because carboxylic acids are weak acids, they are completely neutralized by strong bases such as NaOH and KOH. The products are a **carboxylate salt** and water. The carboxylate ion is named by replacing the *ic acid* ending of the acid name with *ate*.

Carboxylate salts are often used as preservatives and flavor enhancers in soups and seasonings.

$$\mathrm{H{-}\overset{\overset{\displaystyle O}{\|}}{C}{-}OH} + \mathrm{Na\mathbf{OH}} \longrightarrow \mathrm{H{-}\overset{\overset{\displaystyle O}{\|}}{C}{-}O^{-}\ Na^{+}} + \mathbf{H_2O}$$

Methanoic acid (formic acid) → Sodium methanoate (sodium formate)

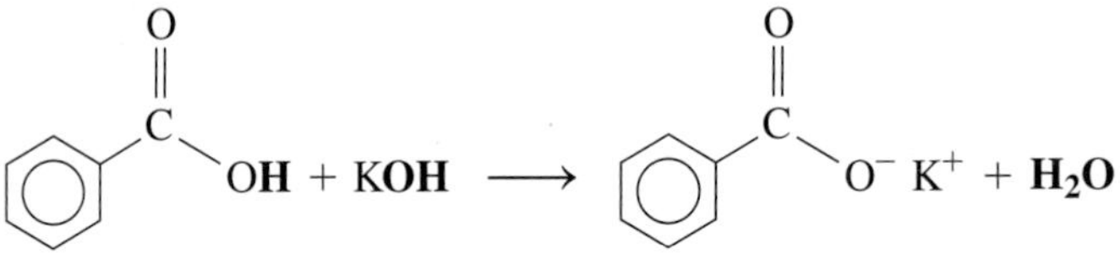

Benzoic acid → Potassium benzoate

Sodium propionate, a preservative, is added to cheeses, bread, and other bakery items to inhibit the spoilage of the food by microorganisms. Sodium benzoate, an inhibitor of mold and bacteria, is added to juices, margarine, relishes, salads, and jams. Monosodium glutamate (MSG) is added to meats, fish, vegetables, and bakery items to enhance flavor, although it causes headaches in some people.

$$\mathrm{CH_3{-}CH_2{-}\overset{\overset{\displaystyle O}{\|}}{C}{-}O^{-}\ Na^{+}}$$

Sodium propanoate (sodium propionate)

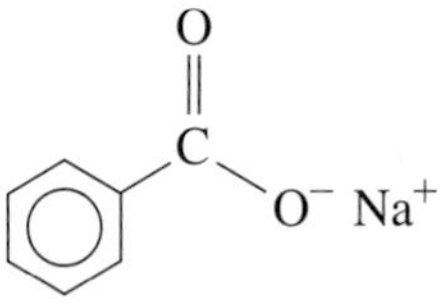

Sodium benzoate

$$\mathrm{HO{-}\overset{\overset{\displaystyle O}{\|}}{C}{-}CH_2{-}CH_2{-}\overset{\overset{\displaystyle NH_2}{|}}{C}H{-}\overset{\overset{\displaystyle O}{\|}}{C}{-}O^{-}\ Na^{+}}$$

Monosodium glutamate

Carboxylate salts are ionic compounds with strong attractions between positively charged metal ions such as Li^+, Na^+, and K^+ and the negatively charged carboxylate ion. Like most salts, the carboxylate salts are solids at room temperature, have high melting points, and are usually soluble in water.

SAMPLE PROBLEM 14.3 Neutralization of a Carboxylic Acid

Write the balanced chemical equation for the neutralization of propanoic acid with sodium hydroxide.

SOLUTION

ANALYZE THE PROBLEM	Reaction Type	Reactants	Products
	neutralization	propanoic acid, NaOH	sodium propanoate, H_2O

The neutralization of an acid with a base produces the salt of the acid and water.

$$\mathrm{CH_3{-}CH_2{-}\overset{\overset{\displaystyle O}{\|}}{C}{-}OH} + \mathrm{NaOH} \longrightarrow \mathrm{CH_3{-}CH_2{-}\overset{\overset{\displaystyle O}{\|}}{C}{-}O^{-}\ Na^{+}} + \mathrm{H_2O}$$

Propanoic acid (propionic acid) + Sodium hydroxide → Sodium propanoate (sodium propionate)

STUDY CHECK 14.3

What carboxylic acid will produce potassium butanoate when it is neutralized by KOH?

Chemistry Link to Health

Carboxylic Acids in Metabolism

Several carboxylic acids are part of the metabolic processes within our cells. For example, during glycolysis, a molecule of glucose is broken down into two molecules of pyruvic acid, or actually, its carboxylate ion, pyruvate. During strenuous exercise when oxygen levels are low (anaerobic), pyruvic acid is reduced to give lactic acid or the lactate ion.

During exercise, pyruvic acid is converted to lactic acid in the muscles.

In the *citric acid cycle*, also called the *Krebs cycle*, di- and tricarboxylic acids are oxidized and decarboxylated (loss of CO_2) to produce energy for the cells of the body. These carboxylic acids are normally referred to by their common names. At the start of the citric acid cycle, citric acid with six carbons is converted to five-carbon α-ketoglutaric acid. Citric acid is also the acid that gives the sour taste to citrus fruits such as lemons and grapefruits.

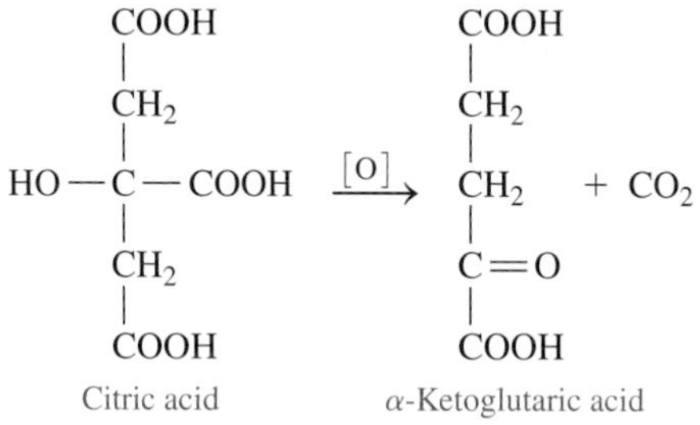

The citric acid cycle continues as α-ketoglutaric acid loses CO_2 to give a four-carbon succinic acid. Then a series of reactions converts succinic acid to oxaloacetic acid. We see that some of the functional groups we have studied along with reactions such as hydration and oxidation are part of the metabolic processes that take place in our cells.

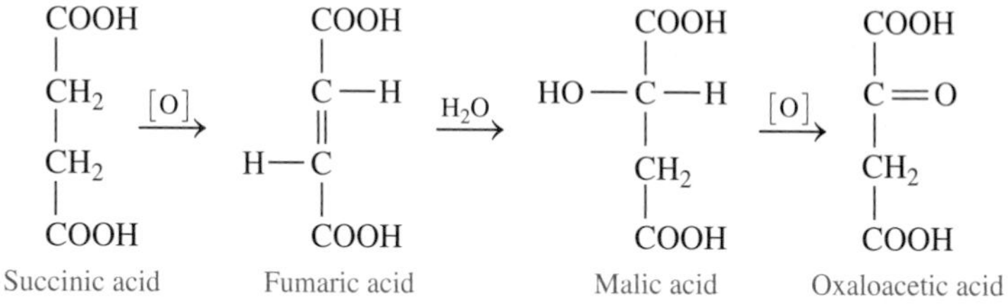

At the pH of the aqueous environment in the cells, the carboxylic acids are ionized, which means it is actually the carboxylate ions that take part in the reactions of the citric acid cycle. For example, in water, succinic acid is in equilibrium with its carboxylate ion, succinate.

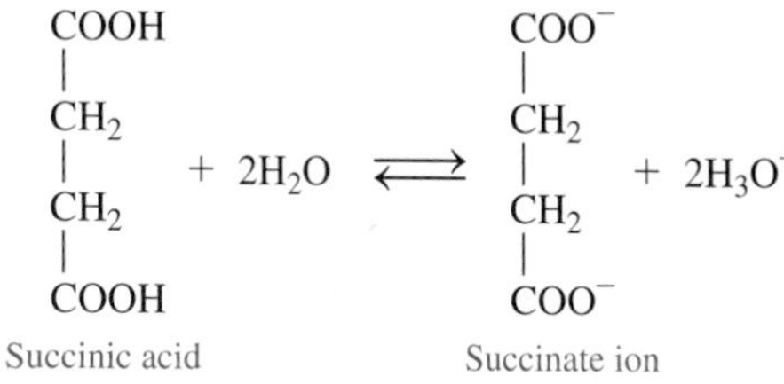

Citric acid gives the sour taste to citrus fruits.

QUESTIONS AND PROBLEMS

14.2 Properties of Carboxylic Acids

LEARNING GOAL Describe the solubility, ionization, and neutralization of carboxylic acids.

14.9 Identify the compound in each group that is most soluble in water. Explain.
- **a.** propanoic acid, hexanoic acid, benzoic acid
- **b.** pentane, 1-hexanol, propanoic acid

14.10 Identify the compound in each group that is most soluble in water. Explain.
- **a.** butanone, butanoic acid, butane
- **b.** ethanoic acid (acetic acid), hexanoic acid, octanoic acid

14.11 Write the balanced chemical equation for the ionization of each of the following carboxylic acids in water:
- **a.** 2-methylpropanoic acid
- **b.** hexanoic acid

14.12 Write the balanced chemical equation for the ionization of each of the following carboxylic acids in water:
- **a.** 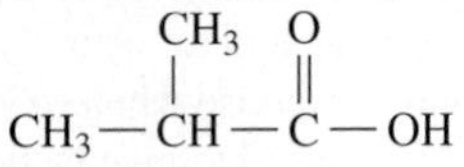
- **b.** butanoic acid

14.13 Write the balanced chemical equations for the following neutralizations:
- **a.** trifluoroacetic acid + NaOH
- **b.** 3-chloropropanoic acid + KOH
- **c.** meta-chlorobenzoic acid + $NaHCO_3$

14.14 Write the balanced chemical equation for the reaction of each of the following carboxylic acids with KH:
- **a.** acetic acid
- **b.** 2-methylbutanoic acid
- **c.** 4-chlorobenzoic acid

LEARNING GOAL

Give the IUPAC and common names for an ester; write the balanced chemical equation for the formation of an ester.

14.3 Esters

A carboxylic acid reacts with an alcohol to form an **ester** and water. In an ester, the —H of the carboxylic acid is replaced by an alkyl group. Fats and oils in our diets contain esters of glycerol and fatty acids, which are long-chain carboxylic acids. The aromas and flavors of many fruits, including bananas, oranges, and strawberries, are due to esters.

Carboxylic Acid **Ester**

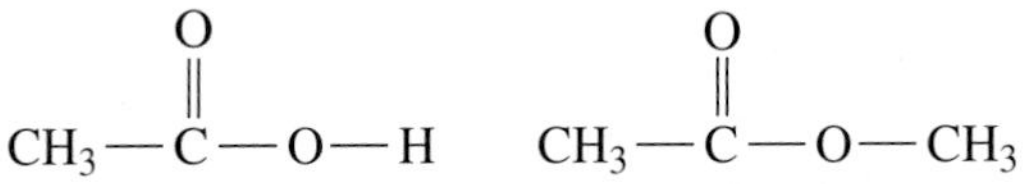

Ethanoic acid (acetic acid) Methyl ethanoate (methyl acetate)

Esterification

In a reaction called **esterification**, an ester is produced when a carboxylic acid and an alcohol react in the presence of an acid catalyst (usually H_2SO_4) and heat. In esterification, the —OH group from the carboxylic acid and the —H from the alcohol are removed and combine to form water. An excess of the alcohol is used to shift the equilibrium in the direction of the formation of the ester product.

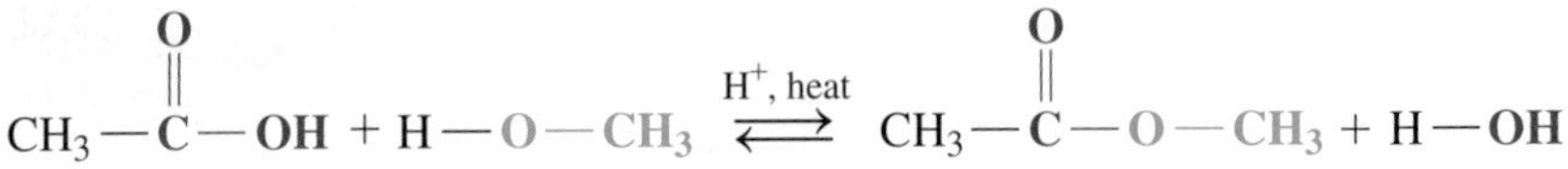

Ethanoic acid (acetic acid) Methanol (methyl alcohol) Methyl ethanoate (methyl acetate)

For example, propyl ethanoate, which is the ester responsible for the flavor and odor of pears can be prepared using ethanoic acid and 1-propanol. The equation for this esterification is written as

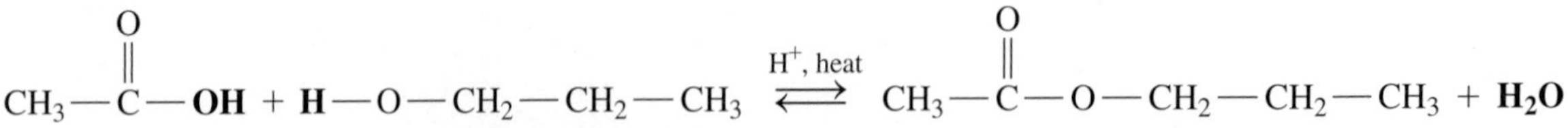

Ethanoic acid (acetic acid) 1-Propanol (propyl alcohol) Propyl ethanoate (propyl acetate)

SAMPLE PROBLEM 14.4 Writing Esterification Equations

An ester that has the smell of pineapple can be synthesized from butanoic acid and methanol. Write the balanced chemical equation for the formation of this ester.

SOLUTION

ANALYZE THE PROBLEM	Reaction Type	Reactants	Products
	esterification	butanoic acid, methanol	methyl butanoate, H_2O

$$CH_3-CH_2-CH_2-\overset{\overset{\large O}{\|}}{C}-\mathbf{OH} + \mathbf{H}-O-CH_3 \xrightleftharpoons{H^+,\ heat} CH_3-CH_2-CH_2-\overset{\overset{\large O}{\|}}{C}-O-CH_3 + \mathbf{H_2O}$$

Butanoic acid (butyric acid) | Methanol (methyl alcohol) | Methyl butanoate (methyl butyrate)

STUDY CHECK 14.4

The ester that smells like plums can be synthesized from methanoic acid and 1-butanol. Write the balanced chemical equation for the formation of this ester.

Naming Esters

The name of an ester consists of two words, which are derived from the names of the alcohol and the acid in that ester. The first word indicates the *alkyl* part from the alcohol. The second word is the name of the *carboxylate* from the carboxylic acid. The IUPAC names of esters use the IUPAC names of the acids, while the common names of esters use the common names of the acids. Let's take a look at the following ester, which has a pleasant, fruity odor. We start by separating the ester bond to identify the alkyl part from the alcohol and the carboxylate part from the acid. Then we name the ester as an alkyl carboxylate.

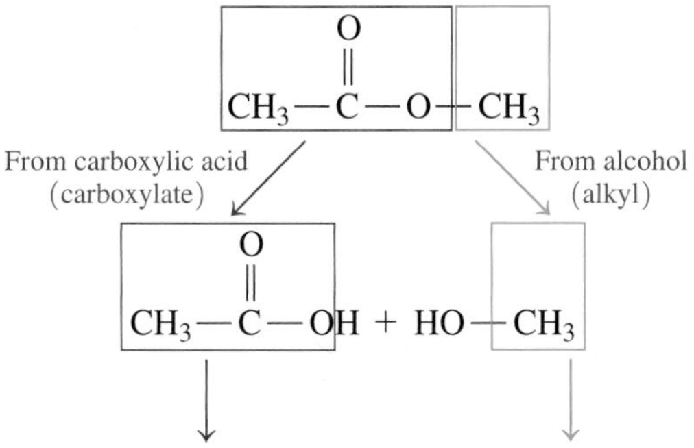

					Ester name
IUPAC	Ethanoic acid	+	Methanol	=	Methyl ethanoate
(common)	(acetic acid)	+	(methyl alcohol)	=	(methyl acetate)

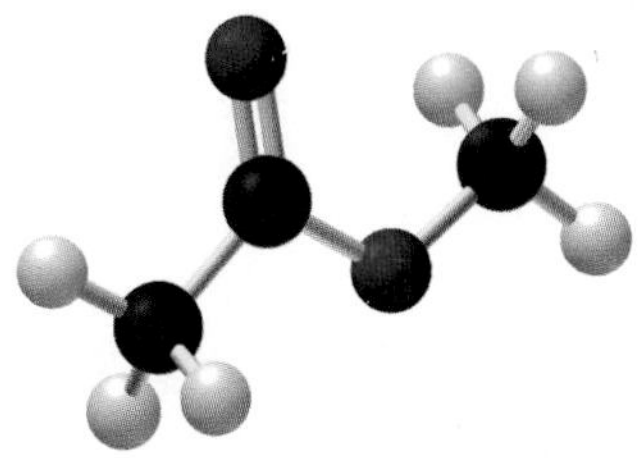

Methyl ethanoate
(methyl acetate)

The following examples of some typical esters show the IUPAC names, as well as the common names, of esters.

$$CH_3-\overset{\overset{\large O}{\|}}{C}-O-CH_2-CH_2$$

Ethyl ethanoate
(ethyl acetate)

$$CH_3-CH_2-\overset{\overset{\large O}{\|}}{C}-O-CH_3$$

Methyl propanoate
(methyl propionate)

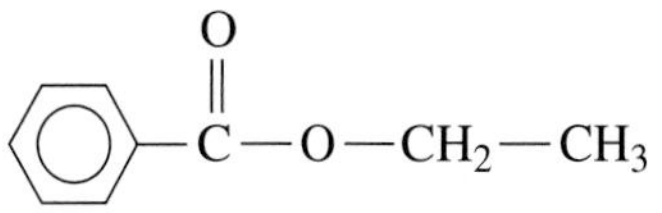

Ethyl benzoate

Chemistry Link to Health

Salicylic Acid from a Willow Tree

For many centuries, relief from pain and fever was obtained by chewing on the leaves or a piece of bark from the willow tree. By the 1800s, chemists discovered that salicin was the agent in the bark responsible for the relief of pain. However, the body converts salicin to salicylic acid, which has a carboxyl group and a hydroxyl group that irritates the stomach lining. In 1899, the Bayer chemical company in Germany produced an ester of salicylic acid and acetic acid, called acetylsalicylic acid (aspirin), which is less irritating. In some aspirin preparations, a buffer is added to neutralize the carboxylic acid group. Today, aspirin is used as an analgesic (pain reliever), antipyretic (fever reducer), and anti-inflammatory agent. Many people take a daily low-dose aspirin, which has been found to lower the risk of heart attack and stroke.

The discovery of salicin in the leaves and bark of the willow tree led to the development of aspirin.

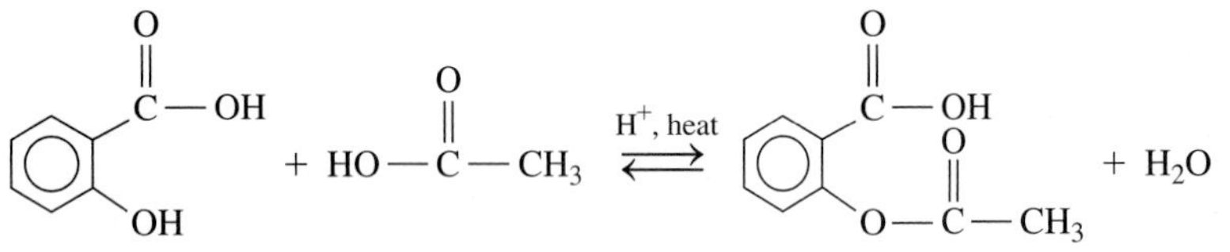

Salicylic acid Acetic acid Acetylsalicylic acid (aspirin)

Oil of wintergreen, or methyl salicylate, has a pungent, minty odor and flavor. Because it can pass through the skin, methyl salicylate is used in skin ointments, where it acts as a counterirritant, producing heat to soothe sore muscles.

Ointments containing methyl salicylate are used to soothe sore muscles.

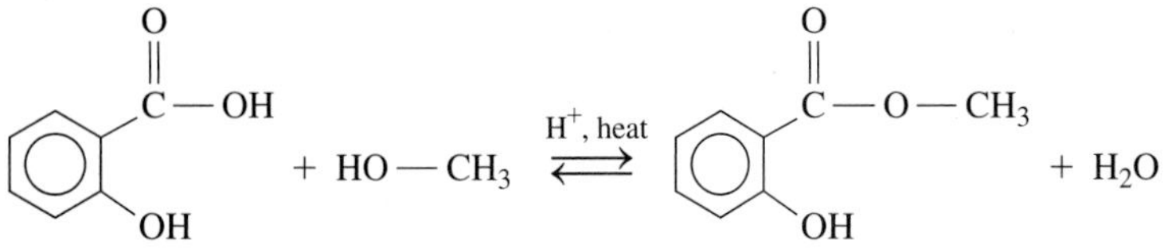

Salicylic acid Methyl alcohol Methyl salicylate (oil of wintergreen)

SAMPLE PROBLEM 14.5 Naming Esters

What are the IUPAC and common names of the following ester?

$$CH_3-CH_2-\overset{\overset{\displaystyle O}{\|}}{C}-O-CH_2-CH_2-CH_3$$

Guide to Naming Esters

STEP 1
Write the name for the carbon chain from the alcohol as an *alkyl* group.

STEP 2
Change the *ic acid* of the acid name to *ate*.

SOLUTION

	Functional Group	Family	IUPAC Naming
ANALYZE THE PROBLEM	carboxyl	ester	write the alkyl name for the carbon chain of the alcohol, and change the *ic acid* in the acid name to *ate*

STEP 1 **Write the name for the carbon chain from the alcohol as an *alkyl* group.** The alcohol part of the ester is from propanol (propyl alcohol). The alkyl group is propyl.

$$CH_3-CH_2-\overset{\overset{\displaystyle O}{\|}}{C}-O-CH_2-CH_2-CH_3 \qquad \text{propyl}$$

STEP 2 **Change the *ic acid* of the acid name to *ate*.** The carboxylic acid part of the ester is from propanoic (propionic) acid, which becomes propanoate (propionate). The ester is named propyl propanoate (propyl propionate).

$$CH_3-CH_2-\overset{\overset{\displaystyle O}{\|}}{C}-O-CH_2-CH_2-CH_3 \qquad \text{propyl propanoate (propyl propionate)}$$

The odor of grapes is due to ethyl heptanoate.

STUDY CHECK 14.5

Draw the condensed structural formula for ethyl heptanoate which gives the odor and flavor to grapes.

Chemistry Link to the **Environment**

Plastics

Terephthalic acid (an aromatic acid with two carboxyl groups) is produced in large quantities for the manufacture of polyesters such as Dacron. When terephthalic acid reacts with ethylene glycol, ester bonds form on both ends of the molecules, allowing many molecules to combine into a long polyester polymer.

Dacron, a synthetic material first produced by DuPont in the 1960s, is a polyester used to make permanent press fabrics, carpets, and clothes. Permanent press is a chemical process in which fabrics are permanently shaped and treated for wrinkle resistance. In medicine, artificial blood vessels and valves are made of Dacron, which is biologically inert and does not clot the blood. The polyester can also be made as a film called Mylar and as a plastic known as PETE (**poly**ethylene**ter**ephthalate). PETE is used for plastic soft drink bottles as well as for containers of salad dressing, shampoos, and dishwashing liquids.

Today PETE is the most widely recycled of all the plastics. After it is separated from other plastics, PETE can be changed into other useful items, including polyester fabric for T-shirts and coats, carpets, doormats, and containers for tennis balls.

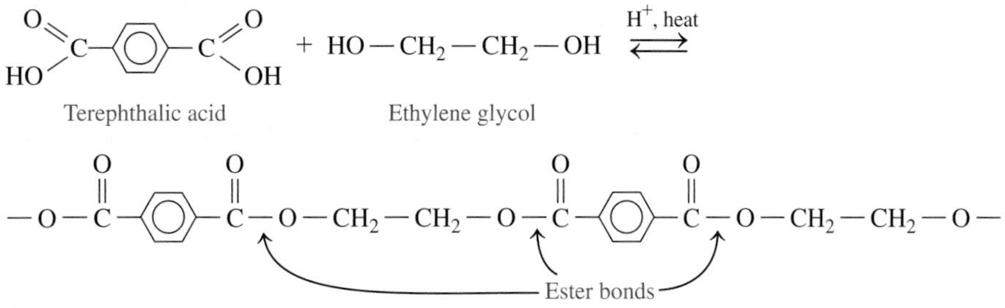

A section of the polyester Dacron

Dacron is a polyester used in permanent press clothing.

Polyester, in the form of the plastic PETE, is used to make soft drink bottles.

Esters in Plants

Many of the fragrances of perfumes and flowers and the flavors of fruits are due to esters. Small esters are volatile, so we can smell them, and they are soluble in water, so we can taste them. Several of these, with their flavor and odor, are listed in Table 14.3.

TABLE 14.3 Some Esters in Fruits and Flavorings

Condensed Structural Formula and Name	Flavor/Odor
$CH_3-\overset{O}{\overset{\parallel}{C}}-O-CH_2-CH_2-CH_3$ Propyl ethanoate (propyl acetate)	Pears
$CH_3-\overset{O}{\overset{\parallel}{C}}-O-CH_2-CH_2-CH_2-CH_2-CH_3$ Pentyl ethanoate (pentyl acetate)	Bananas
$CH_3-\overset{O}{\overset{\parallel}{C}}-O-CH_2-CH_2-CH_2-CH_2-CH_2-CH_2-CH_2-CH_3$ Octyl ethanoate (octyl acetate)	Oranges
$CH_3-CH_2-CH_2-\overset{O}{\overset{\parallel}{C}}-O-CH_2-CH_3$ Ethyl butanoate (ethyl butyrate)	Pineapples
$CH_3-CH_2-CH_2-\overset{O}{\overset{\parallel}{C}}-O-CH_2-CH_2-CH_2-CH_2-CH_3$ Pentyl butanoate (pentyl butyrate)	Apricots

Esters such as ethyl butanoate provide the odor and flavor of many fruits such as pineapples.

QUESTIONS AND PROBLEMS

14.3 Esters

LEARNING GOAL Give the IUPAC and common names for an ester; write the balanced chemical equation for the formation of an ester.

14.15 Draw the condensed structural formula for the ester formed when each of the following reacts with methyl alcohol:

a. pentanoic acid
b. hexanoic acid
c. 4-methyl benzoic acid

14.16 Draw the condensed structural formula for the ester formed when each of the following reacts with ethyl alcohol:

a. formic acid
b. propionic acid
c. 2-methylpentanoic acid

14.17 Draw the condensed structural formula for the ester formed in each of the following reactions:

a. $CH_3-CH_2-CH_2-CH_2-\overset{O}{\overset{\parallel}{C}}-OH + HO-\overset{CH_3}{\overset{|}{CH}}-CH_3 \xrightleftharpoons{H^+,\ heat}$

b. $CH_3-CH_2-\overset{O}{\overset{\parallel}{C}}-OH + HO-CH_2-CH_2-CH_3 \xrightleftharpoons{H^+,\ heat}$

14.18 Draw the condensed structural formula for the ester formed in each of the following reactions:

a. $CH_3-CH_2-\overset{O}{\overset{\parallel}{C}}-OH + HO-CH_3 \xrightleftharpoons{H^+,\ heat}$

b.
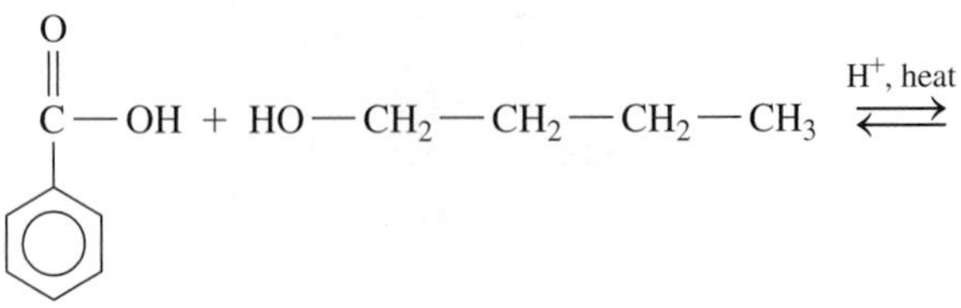

14.19 Give the IUPAC and common names, if any, for each of the following:

a. 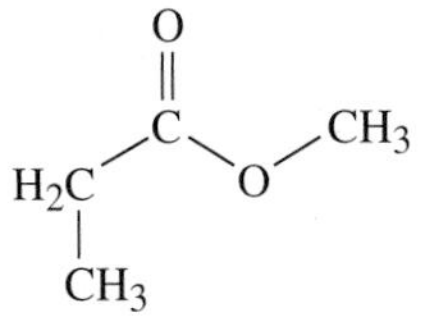

b. 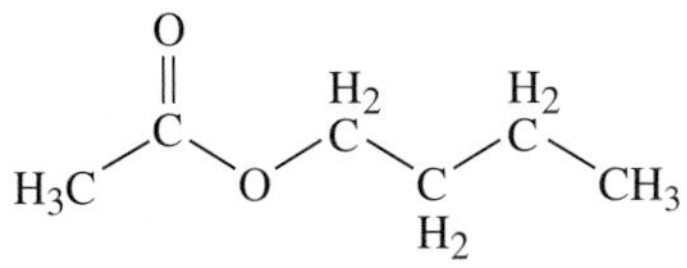

c. 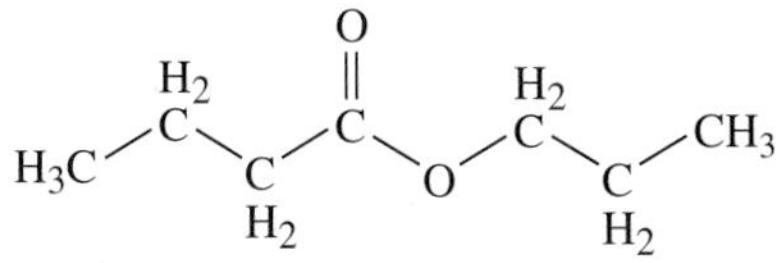

14.20 Give the IUPAC and common names, if any, for each of the following:

a. 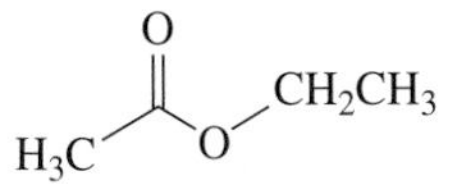

b. 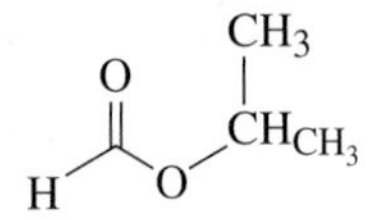

c. 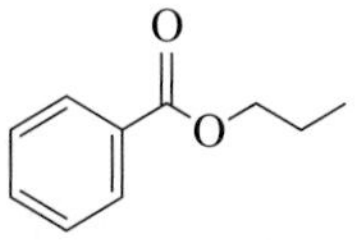

14.21 Draw the condensed structural formula for each of the following:

a. butyl acetate
b. ethyl benzoate
c. hexyl butyrate

14.22 Draw the condensed structural formula for each of the following:

a. hexyl acetate
b. propyl methanoate
c. propyl benzoate

14.23 What is the ester responsible for the flavor and odor of the following fruit?

a. pears
b. pineapple
c. grapes

14.24 What flavor would you notice if you smelled or tasted the following?

a. ethyl butanoate
b. propyl acetate
c. pentyl acetate

14.4 Hydrolysis of Esters

LEARNING GOAL

Draw the condensed structural formulas for the products from acid and base hydrolysis of esters.

Acid Hydrolysis of Esters

When esters are heated with water in the presence of a strong acid, usually H_2SO_4 or HCl, *hydrolysis* occurs. In **hydrolysis**, water reacts with the ester to form a carboxylic acid and an alcohol. Therefore, hydrolysis is the reverse of the esterification reaction. During acid hydrolysis, a water molecule provides the —OH group to convert the carbonyl group of the ester to a carboxyl group. A large quantity of water is used to favor the formation of the carboxylic acid and alcohol products. When hydrolysis of biological esters occurs in the cells, an enzyme replaces the acid as the catalyst.

CORE CHEMISTRY SKILL

Hydrolyzing Esters

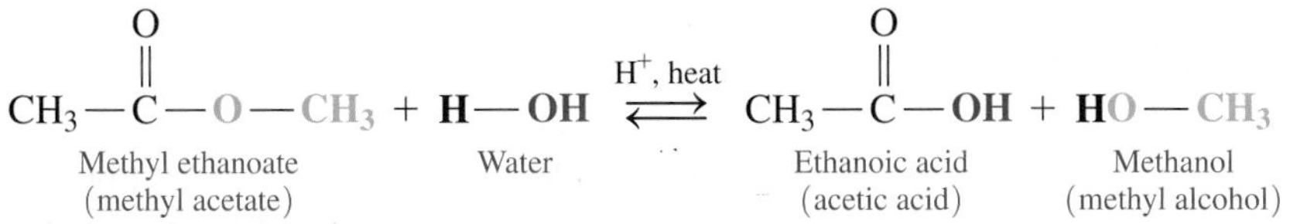

Base Hydrolysis of Esters

When an ester undergoes hydrolysis with a strong base such as NaOH or KOH, the products are the carboxylate salt and the corresponding alcohol. Base hydrolysis is also called **saponification**, which refers to the reaction of an ester of a long-chain fatty acid with NaOH to make soap.

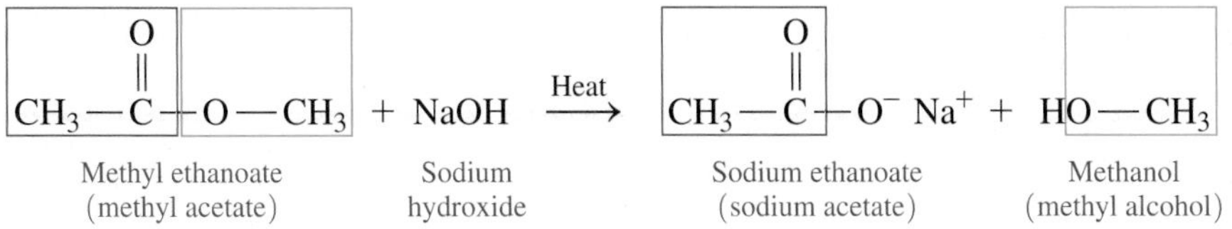

Ethyl acetate is the solvent in fingernail polish.

SAMPLE PROBLEM 14.6 Base Hydrolysis of Esters

Ethyl acetate is a solvent used for fingernail polish, plastics, and lacquers. Write the balanced chemical equation for the hydrolysis of ethyl acetate with NaOH.

SOLUTION

ANALYZE THE PROBLEM	Reaction Type	Reactants	Products
	base hydrolysis	ethyl acetate, NaOH	sodium acetate, ethyl alcohol

The hydrolysis of ethyl acetate with NaOH gives the carboxylate salt, sodium acetate, and ethyl alcohol.

$$CH_3-\overset{\overset{\displaystyle O}{\|}}{C}-O-CH_2-CH_3 + NaOH \xrightarrow{\text{Heat}} CH_3-\overset{\overset{\displaystyle O}{\|}}{C}-O^- Na^+ + HO-CH_2-CH_3$$

Ethyl ethanoate (ethyl acetate) | Sodium ethanoate (sodium acetate) | Ethanol (ethyl alcohol)

Interactive Video

Study Check 14.6

STUDY CHECK 14.6

Draw the condensed structural formulas for the products from the hydrolysis of methyl benzoate with KOH.

QUESTIONS AND PROBLEMS

14.4 Hydrolysis of Esters

LEARNING GOAL Draw the condensed structural formulas for the products from acid and base hydrolysis of esters.

14.25 What are the products of the acid hydrolysis of an ester?

14.26 What is saponification? What are the products formed in this reaction?

14.27 Draw the condensed structural formulas for the products from the acid- or base-catalyzed hydrolysis of each of the following:

a. $CH_3-CH_2-\overset{\overset{\displaystyle O}{\|}}{C}-O-CH_3 + NaOH \xrightarrow{\text{Heat}}$

b. $CH_3-\overset{\overset{\displaystyle O}{\|}}{C}-O-CH_2-CH_2-CH_3 + H_2O \overset{H^+,\ \text{heat}}{\rightleftarrows}$

c. $CH_3-CH_2-CH_2-\overset{\overset{\displaystyle O}{\|}}{C}-O-CH_2-CH_3 + H_2O \overset{H^+,\ \text{heat}}{\rightleftarrows}$

d. 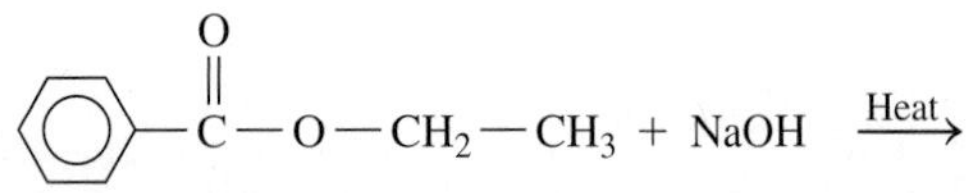

14.28 Draw the condensed structural formulas for the products from the acid- or base-catalyzed hydrolysis of each of the following:

a. $CH_3-CH_2-\overset{\overset{\displaystyle O}{\|}}{C}-O-CH_2-CH_2-CH_2-CH_3 + H_2O \overset{H^+,\ \text{heat}}{\rightleftarrows}$

b. $H-\overset{\overset{\displaystyle O}{\|}}{C}-O-CH_2-CH_3 + NaOH \xrightarrow{\text{Heat}}$

c. $CH_3-CH_2-\overset{\overset{\displaystyle O}{\|}}{C}-O-CH_3 + H_2O \overset{H^+,\ \text{heat}}{\rightleftarrows}$

d. 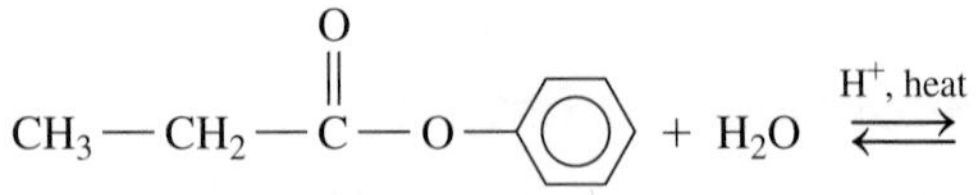

14.5 Amines

LEARNING GOAL

Give the common names for amines; draw the condensed structural formulas when given their names. Classify amines as primary (1°), secondary (2°), or tertiary (3°). Describe the solubility, ionization, and neutralization of amines in water.

Amines are derivatives of ammonia (NH_3), in which one or more hydrogen atoms are replaced with alkyl or aromatic groups. In methylamine, a methyl group replaces one hydrogen atom in ammonia. The bonding of two methyl groups gives dimethylamine, and the three methyl groups in trimethylamine replace all the hydrogen atoms in ammonia.

Naming and Classifying Amines

The common names of amines are often used when the alkyl groups bonded to the nitrogen atom are not branched. Then the alkyl groups are listed in alphabetical order. The prefixes *di* and *tri* are used to indicate two and three identical substituents.

Amines are classified by counting the number of carbon atoms directly bonded to the nitrogen atom. In a primary (1°) amine, the nitrogen atom is bonded to one alkyl group. In a secondary (2°) amine, the nitrogen atom is bonded to two alkyl groups. In a tertiary (3°) amine, the nitrogen atom is bonded to three alkyl groups (see Figure 14.2).

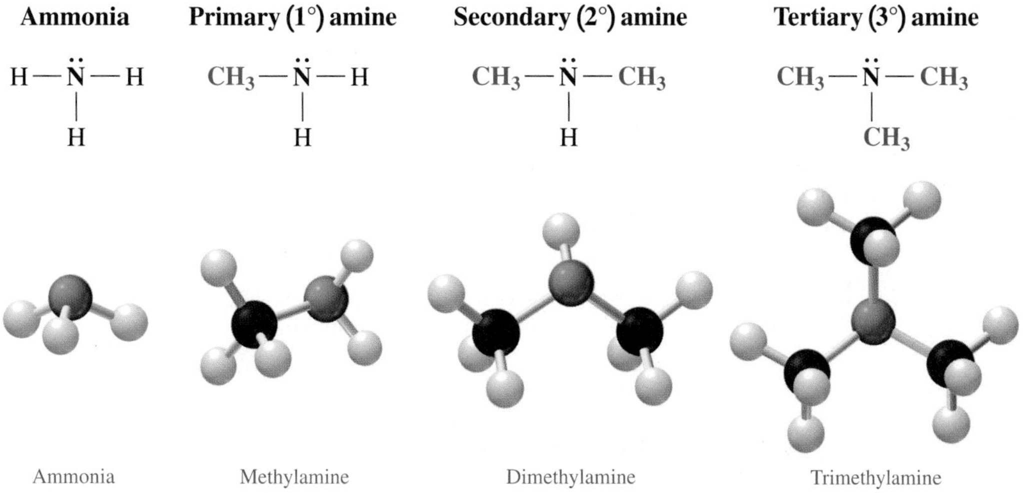

FIGURE 14.2 ▶ Amines have one or more carbon atoms bonded to the N atom.

Q How many carbon atoms are bonded to the nitrogen atom in dimethylamine?

Aromatic Amines

The aromatic amines use the name *aniline*, which is approved by IUPAC. Alkyl groups attached to the nitrogen of aniline are named with the prefix *N*- followed by the alkyl name.

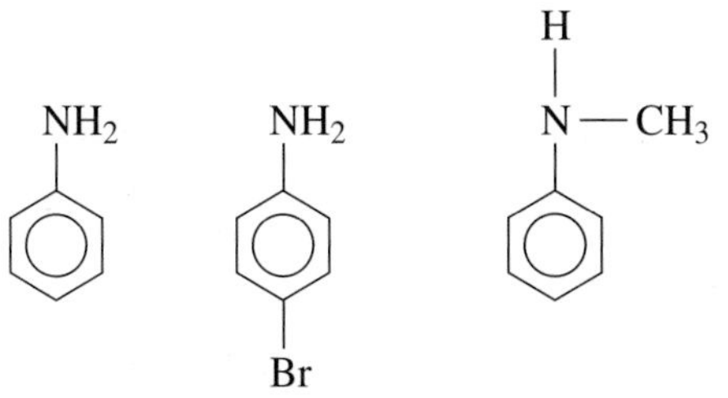

Aniline 4-Bromoaniline *N*-Methylaniline

Aniline is used to make many dyes such as indigo, which give color to wool, cotton, and silk fibers, as well as blue jeans. It is also used to make the polymer polyurethane and in the synthesis of the pain reliever acetaminophen.

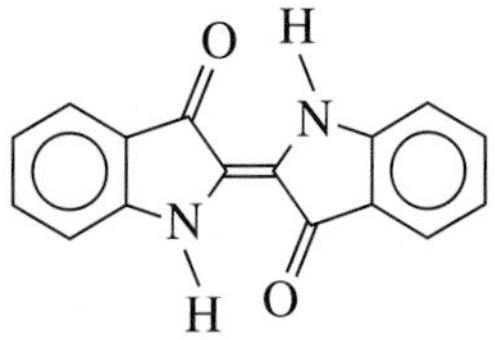

Indigo

Indigo used in blue dyes can be obtained from tropical plants such as *Indigofera tinctoria*.

SAMPLE PROBLEM 14.7 Naming Amines

Give the common name for each of the following amines:

a. $CH_3-CH_2-NH_2$ **b.** $CH_3-\underset{}{\overset{CH_3}{\overset{|}{N}}}-CH_2-CH_3$

SOLUTION

a. This amine has one ethyl group attached to the nitrogen atom; its name is ethylamine.
b. This amine has two methyl groups and one ethyl group attached to the nitrogen atom; its name is ethyldimethylamine.

STUDY CHECK 14.7

Draw the condensed structural formula for ethylpropylamine.

Skeletal Formulas for Amines

We can draw skeletal formulas for amines just as we have for other organic compounds. In the skeletal formula for an amine, we show the hydrogen atoms bonded to the N atom. For example, we can draw the following skeletal formulas, name, and classify each:

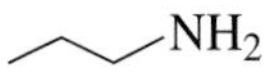

Propylamine
Primary (1°) amine

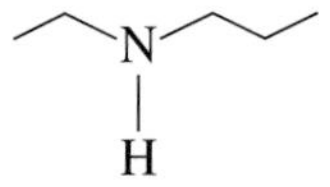

Ethylpropylamine
Secondary (2°) amine

Diethylmethylamine
Tertiary (3°) amine

SAMPLE PROBLEM 14.8 Classifying Amines

Classify the following amines as primary (1°), secondary (2°), or tertiary (3°):

a. 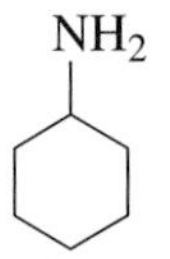

b. $CH_3-\overset{CH_3}{\overset{|}{N}}-CH_2-CH_3$

c. 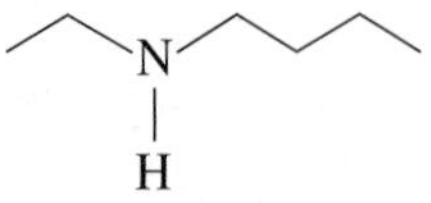

SOLUTION

a. This is a primary (1°) amine because there is one alkyl group (cyclohexyl) attached to the nitrogen atom.

b. This is a tertiary (3°) amine. There are three alkyl groups (two methyls and one ethyl) attached to the nitrogen atom.

c. The nitrogen atom in this skeletal formula is bonded to two alkyl groups, butyl and ethyl, which makes it a secondary (2°) amine.

STUDY CHECK 14.8

Classify the following amine as primary (1°), secondary (2°), or tertiary (3°):

$$CH_3-CH_2-\underset{\displaystyle CH_3}{\underset{|}{N}}-CH_2-CH_2-CH_3$$

Chemistry Link to **Health**

Amines in Health and Medicine

In response to allergic reactions or injury to cells, the body increases the production of histamine, which causes blood vessels to dilate and increases the permeability of the cells. Redness and swelling occur in the area. Administering an antihistamine such as diphenhydramine helps block the effects of histamine.

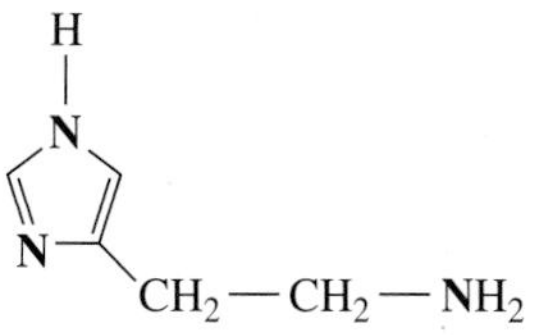

Histamine

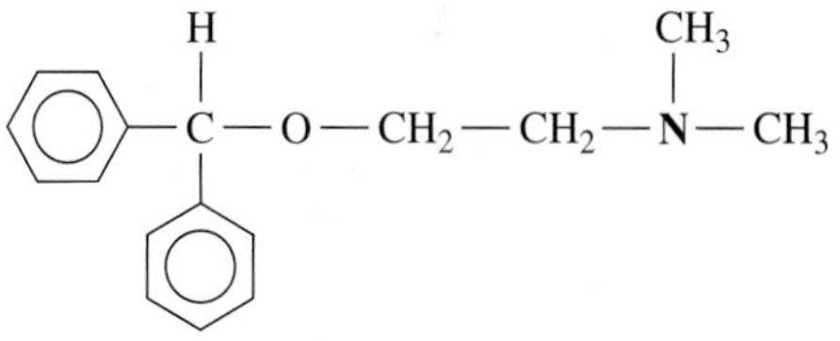

Diphenhydramine

In the body, hormones called *biogenic amines* carry messages between the central nervous system and nerve cells. Epinephrine (adrenaline) and norepinephrine (noradrenaline) are released by the adrenal medulla in "fight or flight" situations to raise the blood glucose level and move the blood to the muscles. Used in remedies for colds, hay fever, and asthma, norepinephrine contracts the capillaries in the mucous membranes of the respiratory passages. The prefix *nor* in a drug name means there is one less CH_3- group on the nitrogen atom. Parkinson's disease is a result of a deficiency in another biogenic amine called dopamine.

Produced synthetically, amphetamines (known as "uppers") are stimulants of the central nervous system much like epinephrine, but they also increase cardiovascular activity and depress the appetite. They are sometimes used to bring about weight loss, but they can cause chemical dependency. Benzedrine and Neo-Synephrine (phenylephrine) are used in medications to reduce respiratory congestion from colds, hay fever, and asthma. Sometimes, benzedrine is taken to combat the desire to sleep, but it has side effects. Methedrine is used to treat depression and in the illegal form is known as "speed" or "crank." The prefix *meth* means that there is one more CH_3- group on the nitrogen atom.

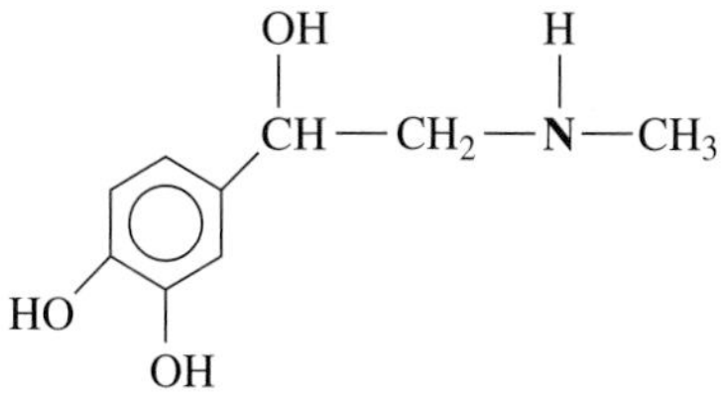

Epinephrine (adrenaline)

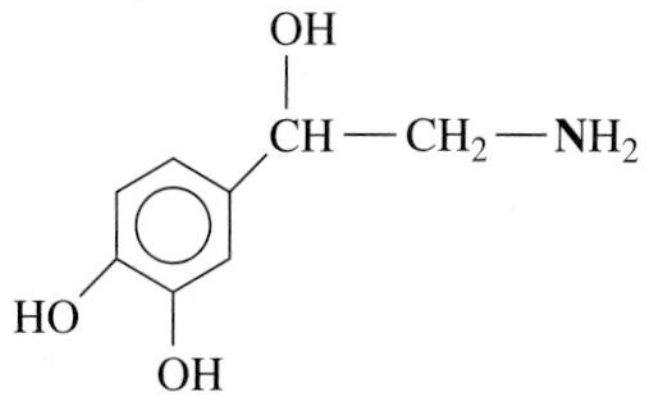

Norepinephrine (noradrenaline)

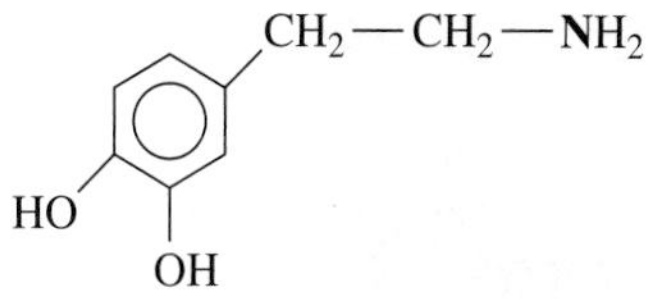

Dopamine

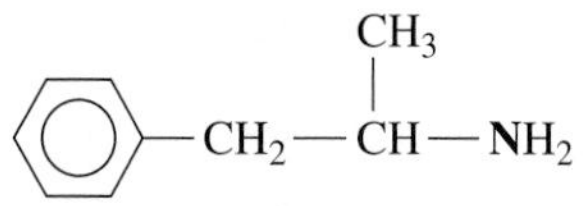

Benzedrine (amphetamine)

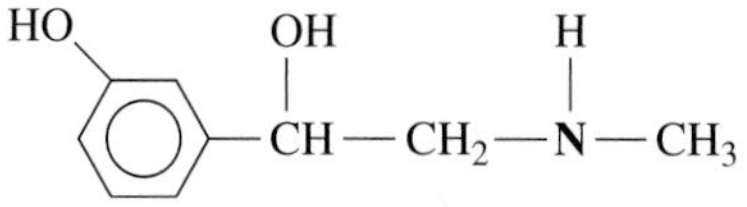

Neo-Synephrine (phenylephrine)

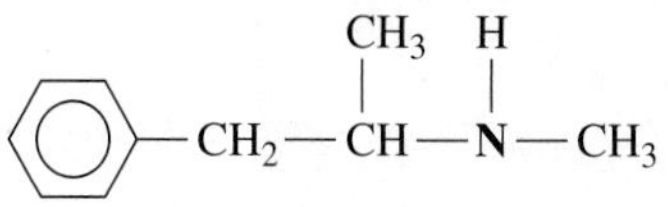

Methamphetamine (methedrine)

Solubility of Amines in Water

Because amines contain a polar N—H bond, they form hydrogen bonds with water. In primary (1°) amines, $-NH_2$ can form more hydrogen bonds than the secondary (2°) amines. A tertiary (3°) amine, which has no hydrogen on the nitrogen atom, can form only hydrogen bonds with water from the N atom in the amine to the H of a water molecule. Like alcohols, the smaller amines, including tertiary ones, are soluble because they form hydrogen bonds with water. However, in amines with more than six carbon atoms, the effect of hydrogen bonding is diminished. Then the nonpolar hydrocarbon chains of the amine decreases its solubility in water (see Figure 14.3).

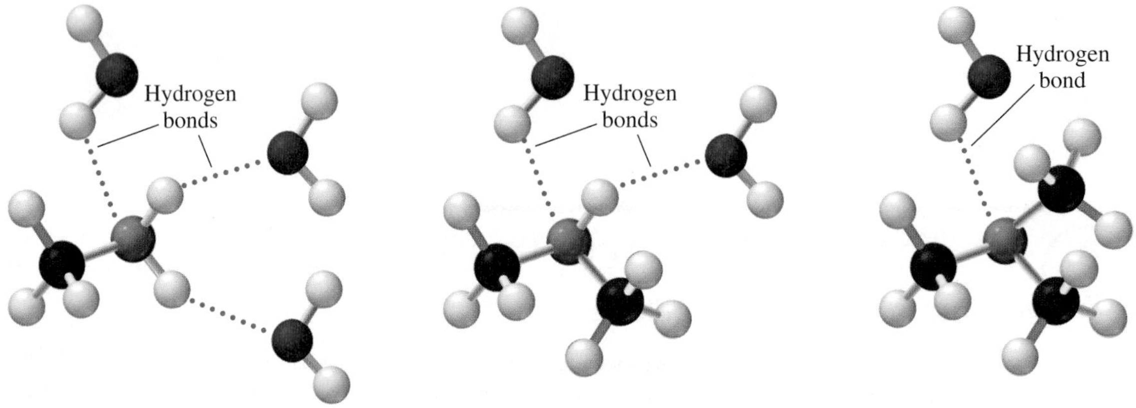

FIGURE 14.3 ▶ Primary, secondary, and tertiary amines form hydrogen bonds with water molecules, but primary amines form the most and tertiary amines form the least.

Q Why do tertiary (3°) amines form fewer hydrogen bonds with water than primary amines?

Amines React as Bases in Water

Ammonia (NH_3) acts as a Brønsted–Lowry base because it accepts H^+ from water to produce an ammonium ion (NH_4^+) and a hydroxide ion (OH^-).

$$\underset{\text{Ammonia}}{\ddot{N}H_3} + H_2O \rightleftharpoons \underset{\text{Ammonium ion}}{NH_4^+} + \underset{\text{Hydroxide ion}}{OH^-}$$

In water, amines act as Brønsted–Lowry bases because the lone electron pair on the nitrogen atom accepts a hydrogen ion from water and produces alkylammonium and hydroxide ions.

$$\underset{\text{Methylamine}}{CH_3-\ddot{N}H_2} + H_2O \rightleftharpoons \underset{\text{Methylammonium ion}}{CH_3-\overset{+}{N}H_3} + \underset{\text{Hydroxide ion}}{OH^-}$$

$$\underset{\text{Dimethylamine}}{CH_3-\underset{\displaystyle CH_3}{\underset{|}{\ddot{N}H}}} + H_2O \rightleftharpoons \underset{\text{Dimethylammonium ion}}{CH_3-\underset{\displaystyle CH_3}{\underset{|}{\overset{+}{N}H_2}}} + \underset{\text{Hydroxide ion}}{OH^-}$$

The amines in fish react with the acid in lemon juice to neutralize the "fishy" odor.

Ammonium Salts

When you squeeze lemon juice on fish, the "fishy" odor is removed by converting the amines to their ammonium salts. In a *neutralization reaction*, an amine acts as a base and reacts with an acid to form an **ammonium salt**. The lone pair of electrons on the nitrogen atom accepts H^+ from an acid to give an ammonium salt; no water is formed. An ammonium salt is named by using its alkylammonium ion name, followed by the name of the negative ion.

Neutralization of an Amine

$$CH_3-\ddot{N}H_2 + HCl \longrightarrow CH_3-\overset{+}{N}H_3\ Cl^-$$

Methylamine — Methylammonium chloride

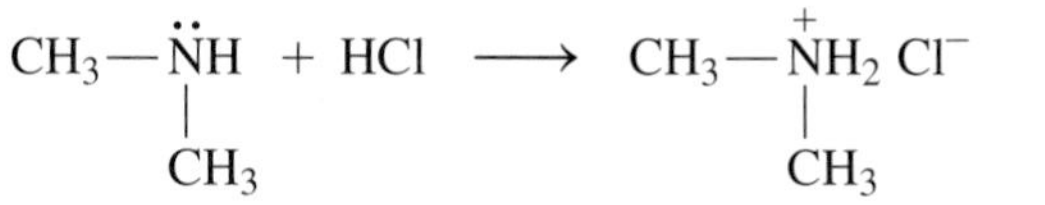

Dimethylamine — Dimethylammonium chloride

Properties of Ammonium Salts

Ammonium salts are ionic compounds with strong attractions between the positively charged ammonium ion and an anion, usually chloride. Like most salts, ammonium salts are solids at room temperature, odorless, and soluble in water and body fluids. For this reason, amines used as drugs are usually converted to their ammonium salts. The ammonium salt of ephedrine is used as a bronchodilator and in decongestant products such as Sudafed. The ammonium salt of diphenhydramine is used in products such as Benadryl for relief of itching and pain from skin irritations and rashes (see Figure 14.4). In pharmaceuticals, the naming of the ammonium salt follows an older method of giving the amine name followed by the name of the acid.

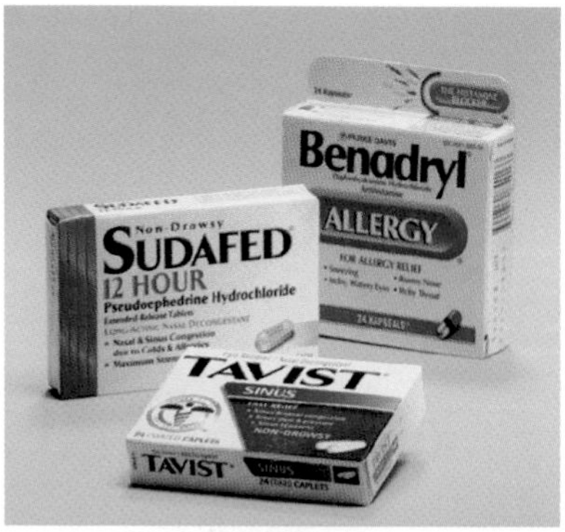

FIGURE 14.4 ▶ Decongestants and products that relieve itch and skin irritations often contain ammonium salts.

Q Why are ammonium salts used in drugs rather than the biologically active amines?

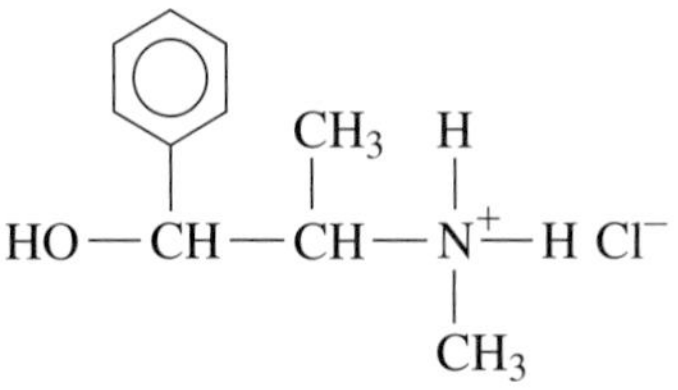

Ephedrine hydrochloride
Ephedrine HCl
Sudafed

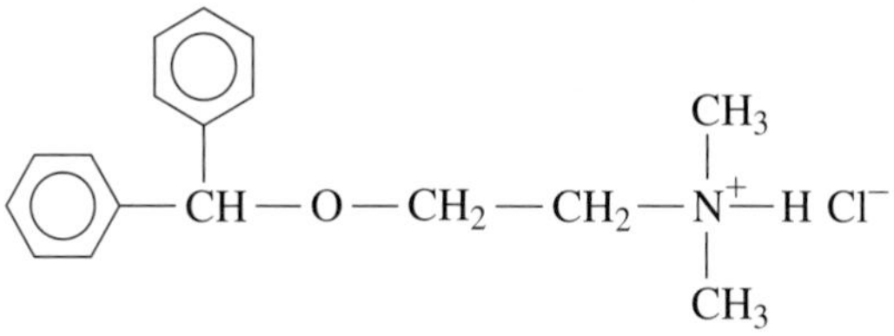

Diphenhydramine hydrochloride
Diphenylhydramine HCl
Benadryl

When an ammonium salt reacts with a strong base such as NaOH, it is converted back to the amine, which is also called the free amine or free base.

$$CH_3-\overset{+}{N}H_3\ Cl^- + NaOH \longrightarrow CH_3-NH_2 + NaCl + H_2O$$

The narcotic cocaine is typically extracted from coca leaves, using an acidic HCl solution to give a white, solid ammonium salt, which is cocaine hydrochloride. It is the salt of cocaine (cocaine hydrochloride) that is smuggled and used illegally on the street. "Crack cocaine" is the free amine or free base of the amine obtained by treating the cocaine hydrochloride with NaOH and ether, a process known as "free-basing." The solid product is known as "crack cocaine" because it makes a crackling noise when heated. The free amine is rapidly absorbed when smoked and gives stronger highs than the cocaine hydrochloride, which makes crack cocaine more addictive.

Coca leaves are a source of cocaine.

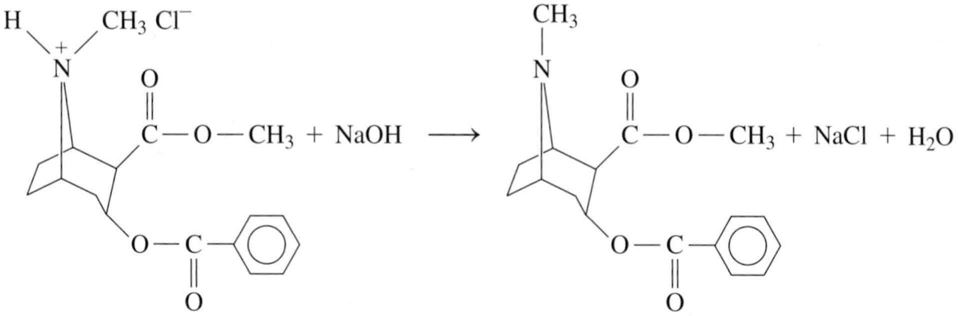

Cocaine hydrochloride — Cocaine ("free base"; crack cocaine)

SAMPLE PROBLEM 14.9 Reactions of Amines

Write the balanced chemical equation that shows ethylamine

a. acting as a weak base in water
b. neutralized by HCl

SOLUTION

a. In water, ethylamine acts as a weak base by accepting a hydrogen ion from water to produce ethylammonium and hydroxide ions.

$$CH_3{-}CH_2{-}NH_2 + H{-}OH \rightleftarrows CH_3{-}CH_2{-}\overset{+}{N}H_3 + OH^-$$

b. In a reaction with an acid, ethylamine acts as a weak base by accepting a hydrogen ion from HCl to produce ethylammonium chloride.

$$CH_3{-}CH_2{-}NH_2 + HCl \longrightarrow CH_3{-}CH_2{-}\overset{+}{N}H_3\ Cl^-$$

STUDY CHECK 14.9

Draw the condensed structural formula for the salt formed by the reaction of trimethylamine and HCl.

The aroma of black pepper is due to piperidine, a heterocyclic amine.

A *heterocyclic amine* is a cyclic compound that contains one or more nitrogen atoms in the ring. The heterocyclic amine rings typically consist of five or six atoms and one or more nitrogen atoms. Of the five-atom rings, the simplest one is *pyrrolidine*, which is a ring of four carbon atoms and a nitrogen atom, all with single bonds. Some of the pungent aroma and taste we associate with black pepper is due to a compound called *piperidine*, which is a six-atom heterocyclic ring with a nitrogen atom. The fruit from the black pepper plant is dried and ground to give the black pepper we use to season our foods.

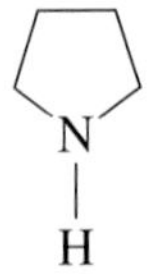

Pyrrolidine

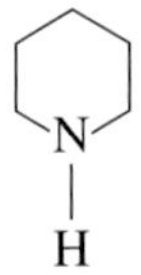

Piperidine

Chemistry Link to **Health**

Alkaloids: Amines in Plants

Alkaloids are physiologically active compounds produced by plants that contain heterocyclic amines. The term *alkaloid* refers to the "alkali-like" or basic characteristics we have seen for amines. Certain alkaloids are used in anesthetics, in antidepressants, and as stimulants, and many are habit forming.

As a stimulant, nicotine increases the level of adrenaline in the blood, which increases the heart rate and blood pressure. Nicotine is addictive because it activates pleasure centers in the brain. Coniine, which is obtained from hemlock, is extremely toxic.

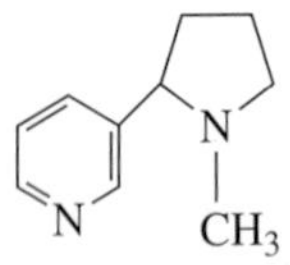

Nicotine

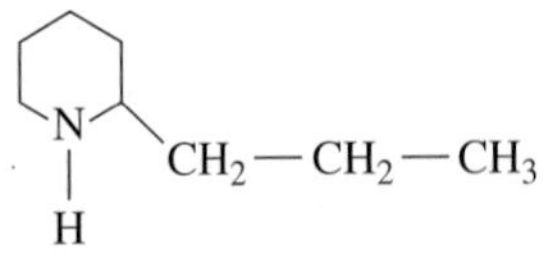

Coniine

Caffeine is a central nervous system stimulant. Present in coffee, tea, soft drinks, energy drinks, chocolate, and cocoa, caffeine increases alertness, but it may cause nervousness and insomnia. Caffeine is also used in certain pain relievers to counteract the drowsiness caused by an antihistamine.

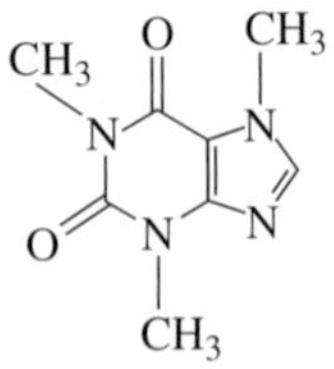

Caffeine

Caffeine is a stimulant found in coffee, tea, energy drinks, and chocolate.

Several alkaloids are used in medicine. Quinine, obtained from the bark of the cinchona tree, has been used in the treatment of malaria since the 1600s. Atropine from nightshade (belladonna) is used in low concentrations to accelerate slow heart rates and as an anesthetic for eye examinations.

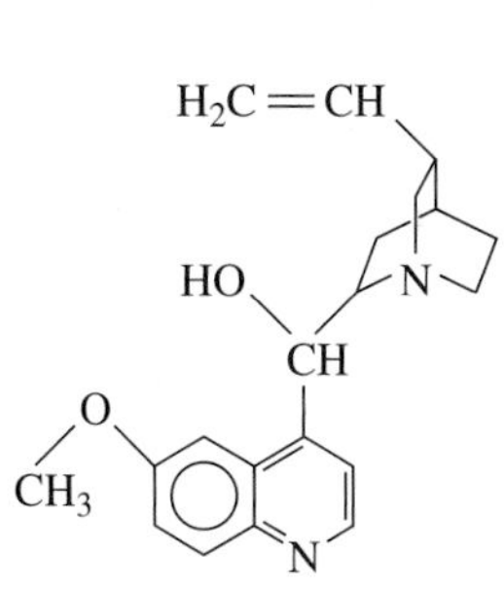

Quinine

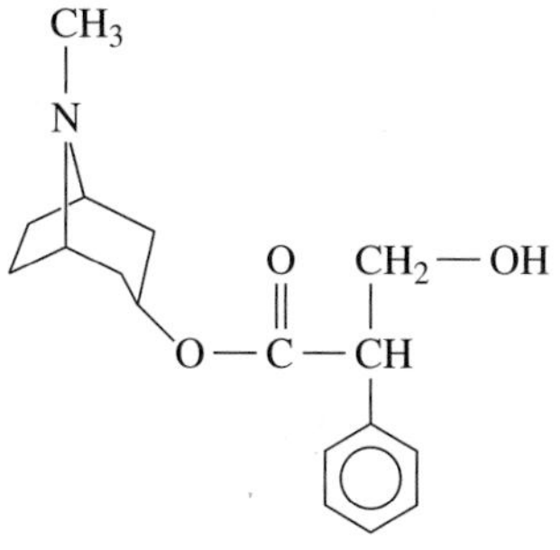

Atropine

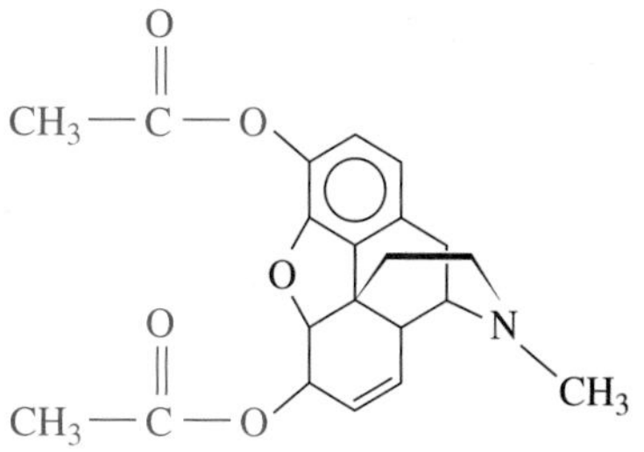

Heroin

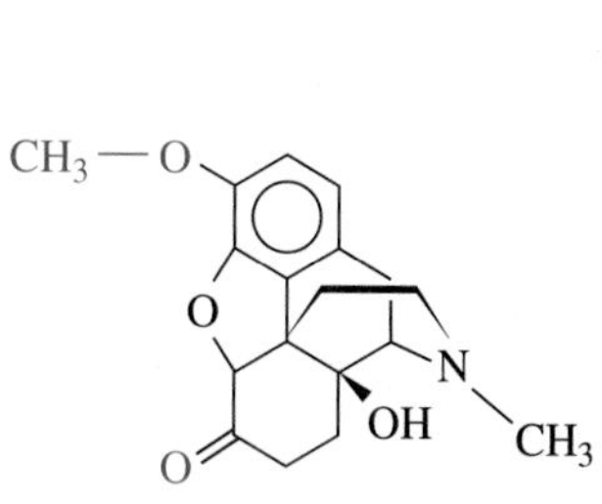

OxyContin

For many centuries morphine and codeine, alkaloids found in the oriental poppy plant, have been used as effective painkillers. Codeine, which is structurally similar to morphine, is used in some prescription painkillers and cough syrups. Heroin, obtained by a chemical modification of morphine, is strongly addictive and is not used medically. The structure of the prescription drug OxyContin (oxycodone) used to relieve severe pain is similar to heroin. Today, there are an increasing number of deaths from OxyContin abuse because its physiological effects are also similar to those of heroin.

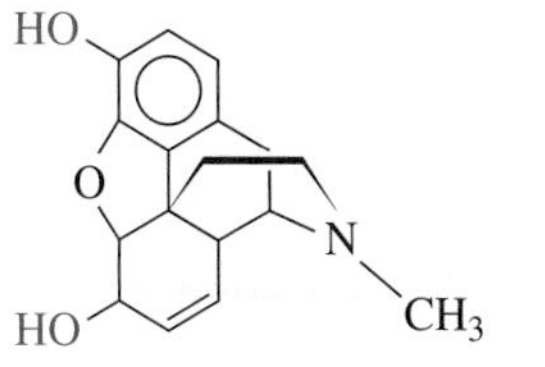

Morphine

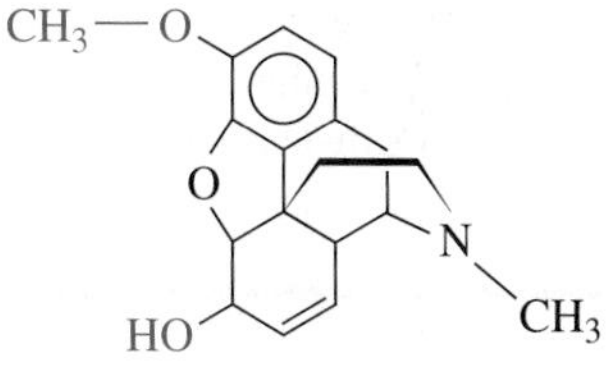

Codeine

The green, unripe poppy seed capsule contains a sap (opium) that is the source of the alkaloids morphine and codeine.

QUESTIONS AND PROBLEMS

14.5 Amines

LEARNING GOAL Give the common names for amines; draw the condensed structural formulas when given their names. Classify amines as primary (1°), secondary (2°), or tertiary (3°). Describe the solubility, ionization, and neutralization of amines in water.

14.29 Give the common name for each of the following:

a. 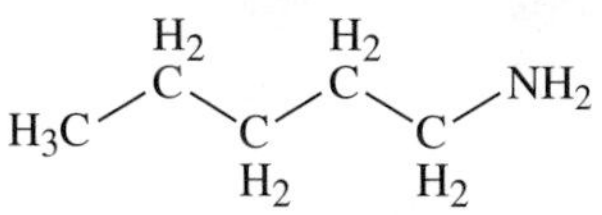

b. 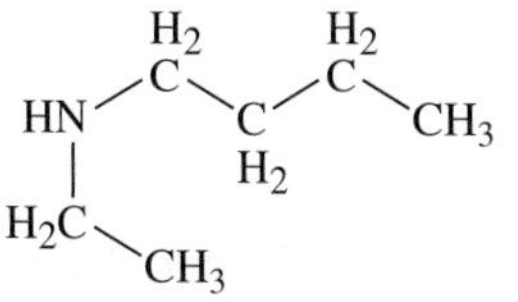

c. 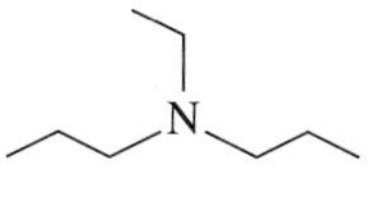

14.30 Give the common name for each of the following:

a. 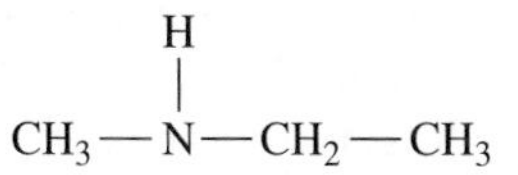

b. 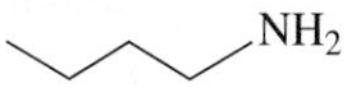

c. 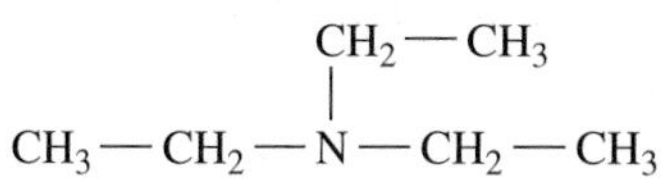

14.31 Draw the skeletal formula for each of the following amines:

a. pyrrolidine
b. *N*-methylaniline
c. butylpropylamine

14.32 Draw the condensed structural formula, or skeletal formula if cyclic, for each of the following amines:

a. diethylamine
b. 4-chloroaniline
c. *N*,*N*-diethylaniline

14.33 Classify each of the following amines as primary (1°), secondary (2°), or tertiary (3°):

a. 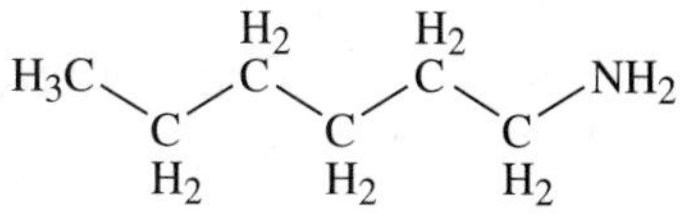

b. 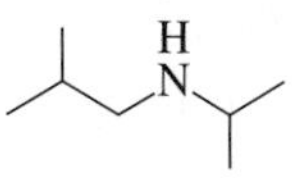

c. 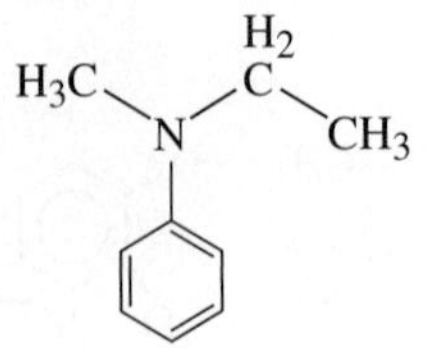

d. 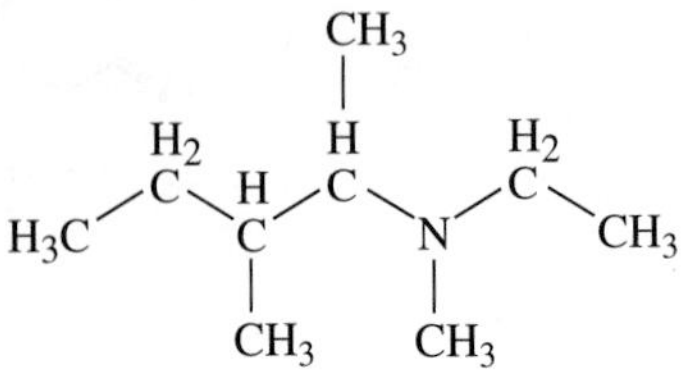

14.34 Classify each of the following amines as primary (1°), secondary (2°), or tertiary (3°):

a. 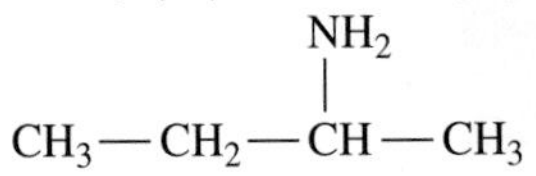

b. 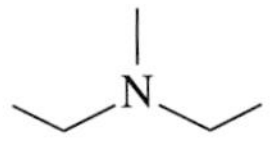

c. 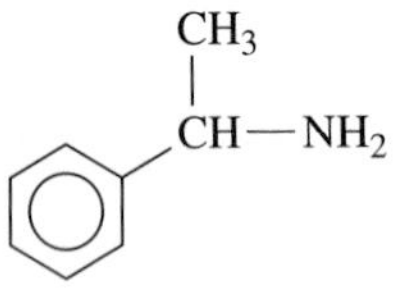

d. 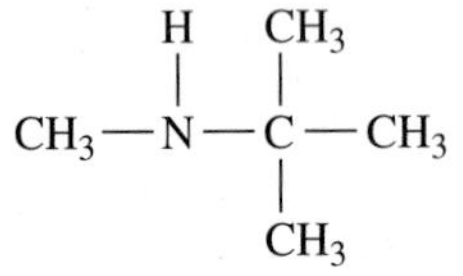

14.35 Indicate if each of the following is soluble in water. Explain.

a. $CH_3-CH_2-NH_2$

b. 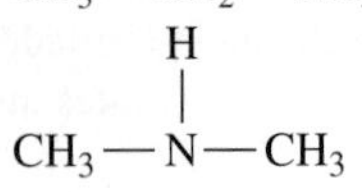

c. 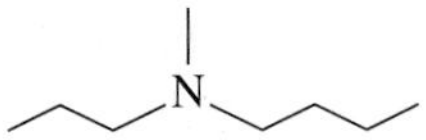

d. 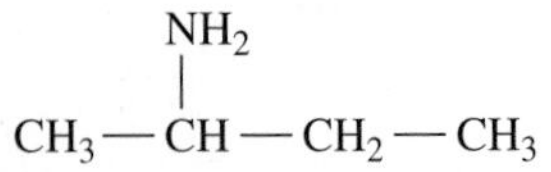

14.36 Indicate if each of the following is soluble in water. Explain.

a. $CH_3-CH_2-CH_2-NH_2$

b. 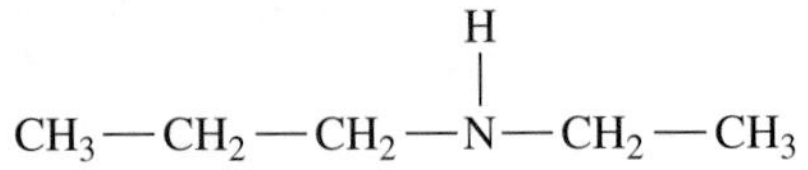

c. 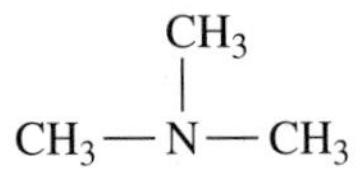

d. 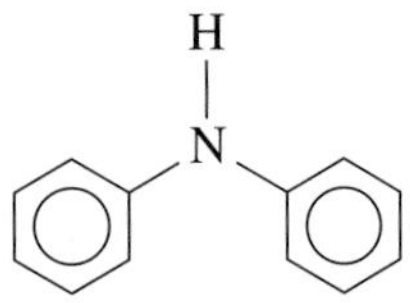

14.37 Write the balanced chemical equation for the ionization of each of the following amines in water:

a. methylamine
b. dimethylamine
c. aniline

14.38 Write the balanced chemical equation for the ionization of each of the following amines in water:

a. ethylmethylamine
b. propylamine
c. *N*-methylaniline

LEARNING GOAL

Give the IUPAC and common names for amides and draw the condensed structural formulas for the products of formation and hydrolysis.

14.6 Amides

The **amides** are derivatives of carboxylic acids in which a nitrogen group replaces the hydroxyl group.

Carboxylic Acid **Amide**

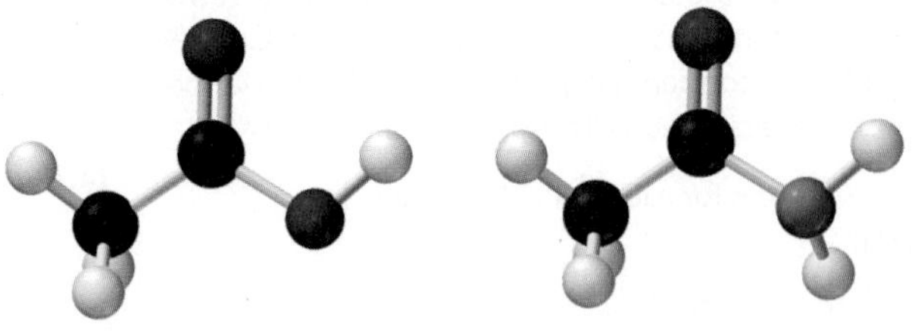

Ethanoic acid (acetic acid) Ethanamide (acetamide)

CORE CHEMISTRY SKILL

Forming Amides

Preparation of Amides

An amide is produced in a reaction called *amidation*, in which a carboxylic acid reacts with ammonia or a primary or secondary amine. An —OH group is removed from the carboxylic acid and an —H is removed from ammonia or an amine to form water. The remaining part of the carboxylic acid and ammonia or amine molecule join to form an amide, much like the formation of an ester. Because a hydrogen atom must be removed from ammonia or an amine, only primary and secondary amines undergo amidation.

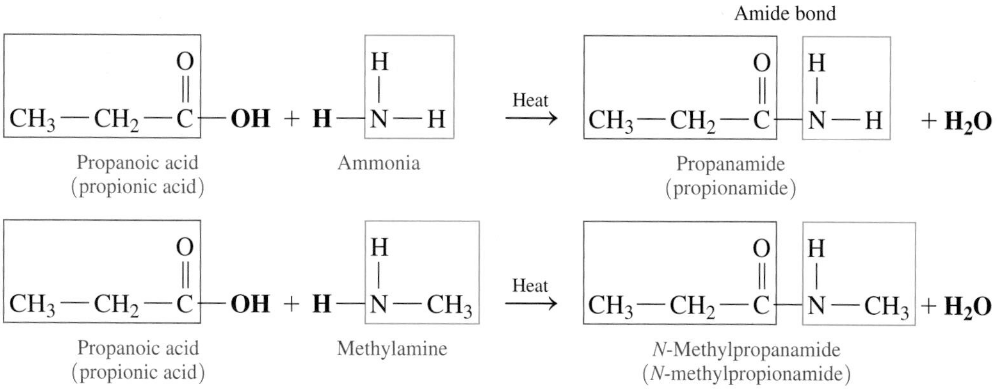

SAMPLE PROBLEM 14.10 Formation of Amides

Draw the condensed structural formula for the amide product in each of the following reactions:

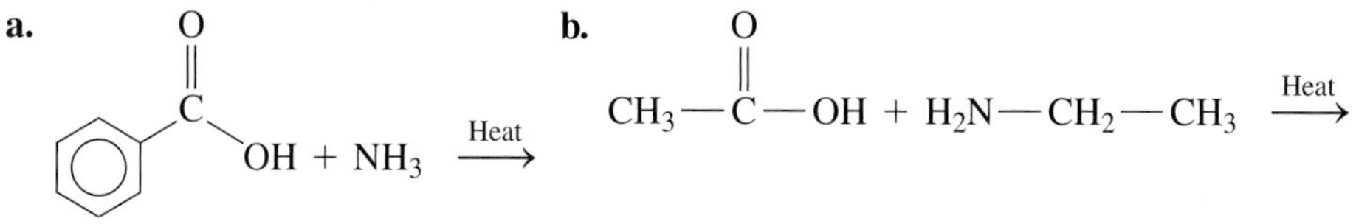

SOLUTION

The condensed structural formula for the amide product is drawn by attaching the carbonyl group from the acid to the nitrogen atom of the amine. —OH is removed from the acid and —H from the amine to form water.

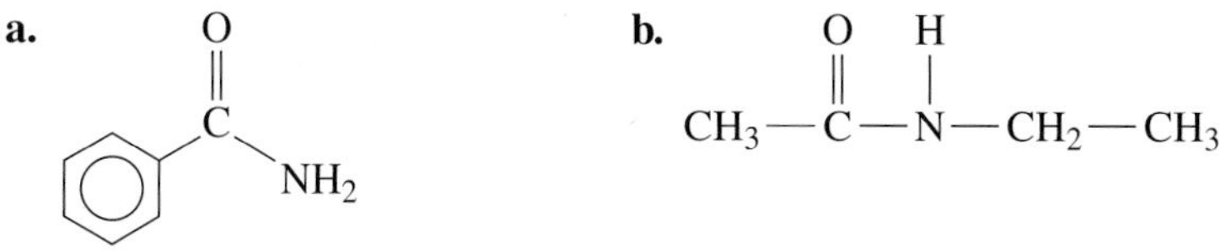

STUDY CHECK 14.10

Draw the condensed structural formulas for the carboxylic acid and amine needed to prepare the following amide. (*Hint*: Separate the N and C=O of the amide group, and add —H and —OH to give the original amine and carboxylic acid.)

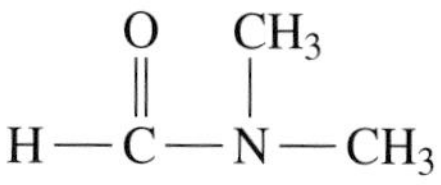

Naming Amides

In both the IUPAC and common names, amides are named by dropping the *oic acid* (IUPAC) or *ic acid* (common) from the carboxylic acid name and adding the suffix *amide*. When alkyl groups are attached to the nitrogen atom, the prefix *N*- or *N*,*N*- precedes the name of the amide, depending on whether there are one or two groups. We can diagram the name of an amide in the following way:

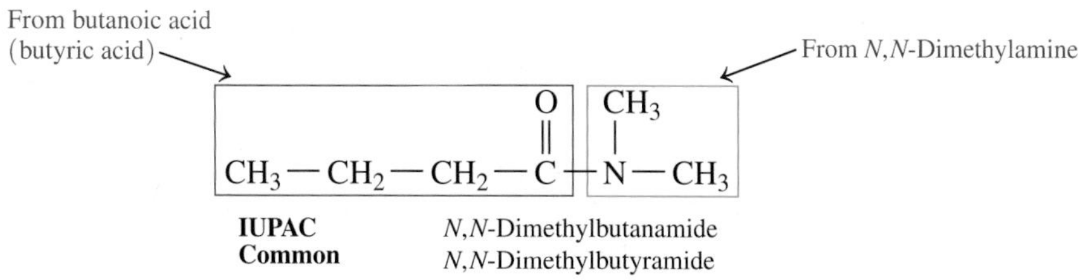

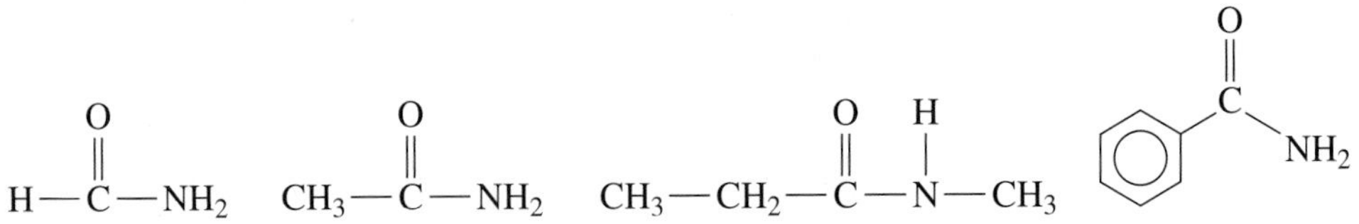

Methanamide (formamide) — Ethanamide (acetamide) — *N*-Methylpropanamide (*N*-methylpropionamide) — Benzamide

SAMPLE PROBLEM 14.11 Naming Amides

Give the IUPAC name for the following amide:

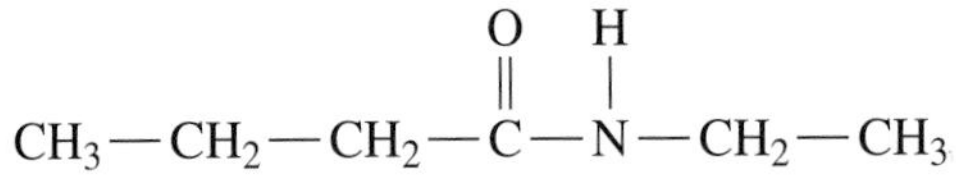

SOLUTION

	Functional Group	Naming the Amide	*N*-Substituent
ANALYZE THE PROBLEM	amide	replace *oic acid* in the carboxyl name with *amide*	ethyl

STEP 1 Replace *oic acid* (IUPAC) in the carboxyl name with *amide*.

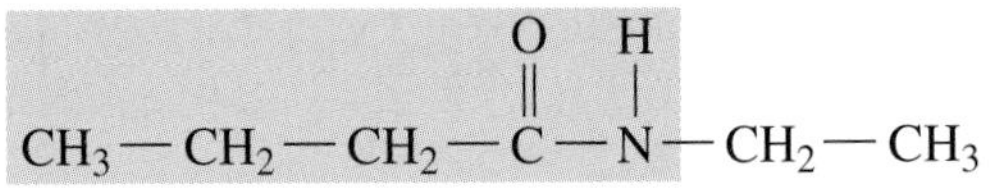

butanamide

STEP 2 Name each substituent on the N atom using the prefix *N*- and the alkyl name.

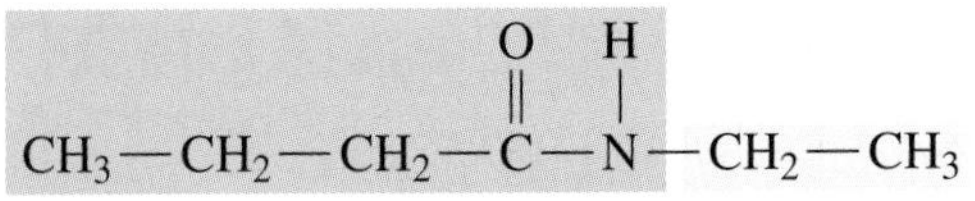

N-ethyl butanamide

STUDY CHECK 14.11

Draw the condensed structural formula for *N*,*N*-dimethylbenzamide.

Guide to Naming Amides

STEP 1 Replace *oic acid* (IUPAC) or *ic acid* (common) in the carboxyl name with *amide*.

STEP 2 Name each substituent on the N atom using the prefix *N*- and the alkyl name.

Solubility of Amides in Water

The amides do not have the properties of bases that we saw for the amines. The amides with one to five carbon atoms are soluble in water because they can hydrogen bond with water molecules. However, in amides with more than five carbon atoms, the effect of hydrogen bonding is diminished as the longer carbon chain decreases the solubility of an amide in water.

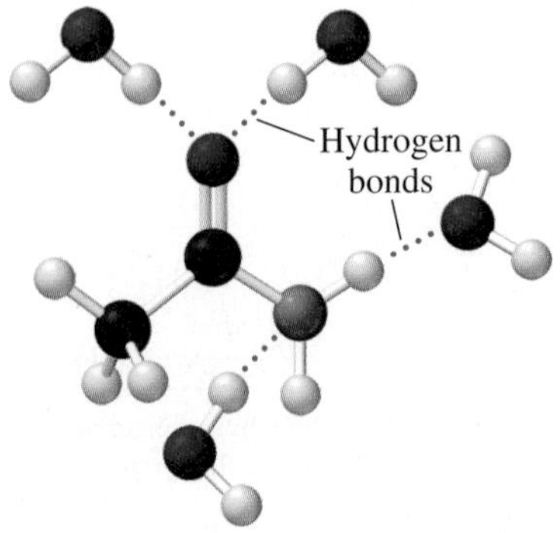

Chemistry Link to Health

Amides in Health and Medicine

The simplest natural amide is urea, an end product of protein metabolism in the body. The kidneys remove urea from the blood and provide for its excretion in urine. If the kidneys malfunction, urea is not removed and builds to a toxic level, a condition called *uremia*. Urea is also used as a component of fertilizer, to increase nitrogen in the soil.

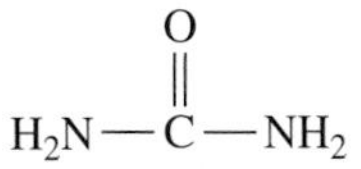

Urea

Many barbiturates are cyclic amides of barbituric acid that act as sedatives in small dosages or sleep inducers in larger dosages. They are often habit forming. Barbiturate drugs include phenobarbital (Luminal) and pentobarbital (Nembutal).

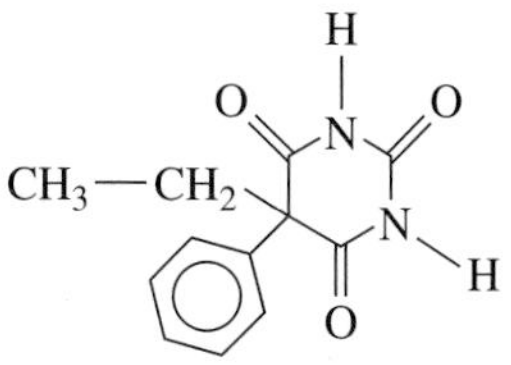

Phenobarbital (Luminal)

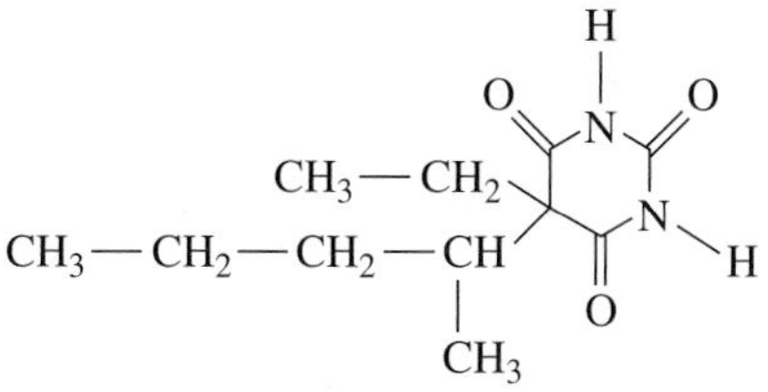

Pentobarbital (Nembutal)

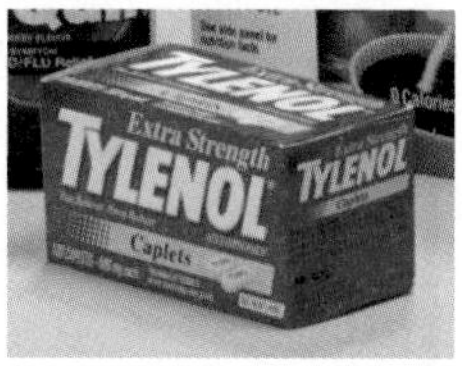

Tylenol with acetaminophen is an aspirin substitute.

Aspirin substitutes contain phenacetin or acetaminophen, which is used in Tylenol. Like aspirin, acetaminophen reduces fever and pain, but it has little anti-inflammatory effect.

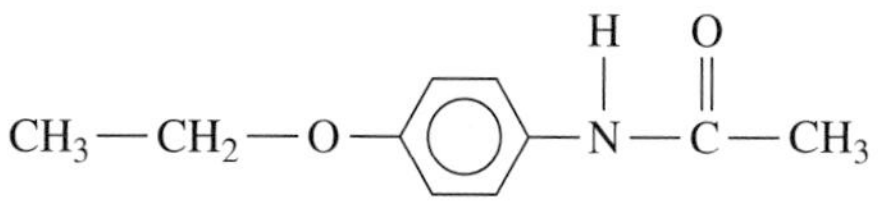

Phenacetin

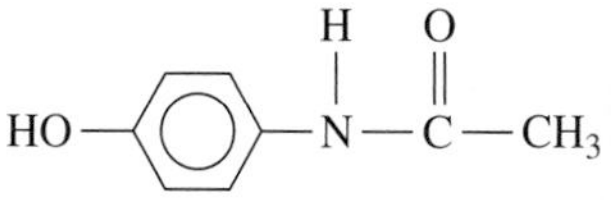

Acetaminophen

Hydrolysis of Amides

Amides undergo hydrolysis when water is added to the amide bond to split the molecule. When an acid is used, the hydrolysis products of an amide are the carboxylic acid and the ammonium salt. In base hydrolysis, the amide produces the carboxylate salt and ammonia or an amine.

Acid Hydrolysis of Amides

Amide **Carboxylic Acid** **Ammonium Salt**

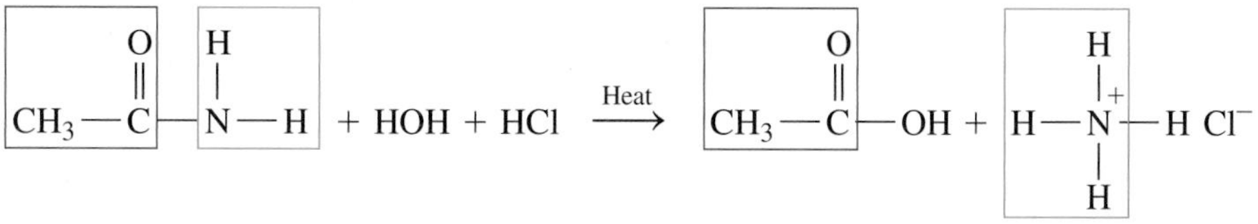

Ethanamide (acetamide) — Ethanoic acid (acetic acid) — Ammonium chloride

Base Hydrolysis of Amides

Amide **Carboxylate Salt** **Amine**

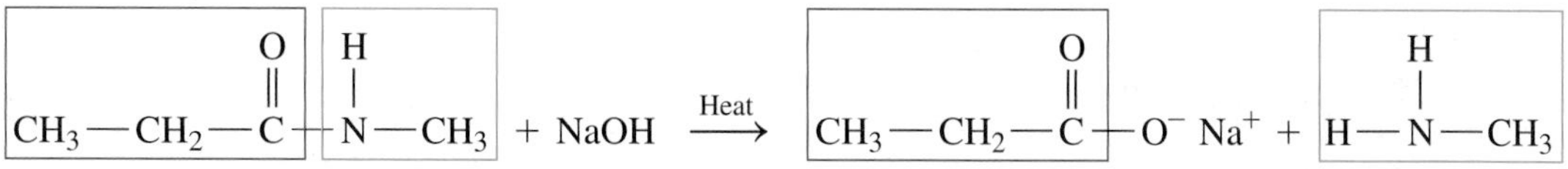

N-Methylpropanamide (*N*-methylpropionamide) — Sodium propanoate, a salt (sodium propionate) — Methanamine (methylamine)

SAMPLE PROBLEM 14.12 Base Hydrolysis of Amides

Draw the condensed structural formulas for the products from the hydrolysis of *N*-methylpentanamide with NaOH.

SOLUTION

ANALYZE THE PROBLEM	Reaction Type	Reactants	Products
	base hydrolysis	*N*-methylpentanamide, NaOH	sodium pentanoate, methylamine

In the hydrolysis of the amide, the amide bond is broken between the carboxyl carbon atom and the nitrogen atom. When a base such as NaOH is used, the products are the carboxylate salt and an amine.

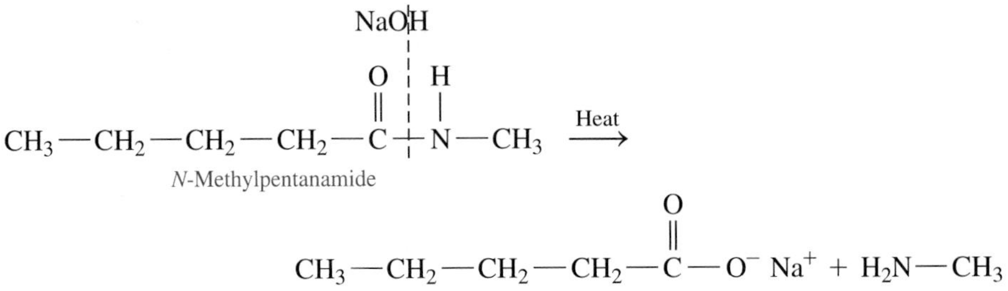

STUDY CHECK 14.12

Draw the condensed structural formulas of the products from the hydrolysis of *N*-methylbutyramide with HBr.

QUESTIONS AND PROBLEMS

14.6 Amides

LEARNING GOAL Give the IUPAC and common names for amides and draw the condensed structural formulas for the products of formation and hydrolysis.

14.39 Draw the condensed structural formula for the amide formed in each of the following reactions:

a. 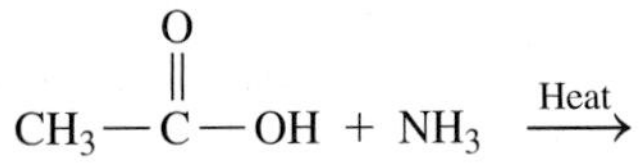

b. 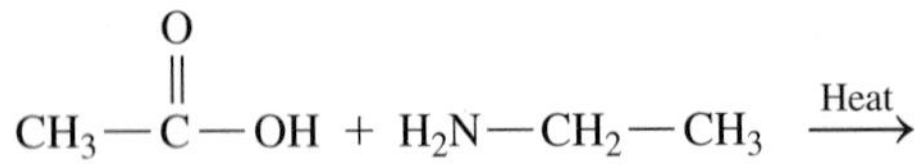

c. 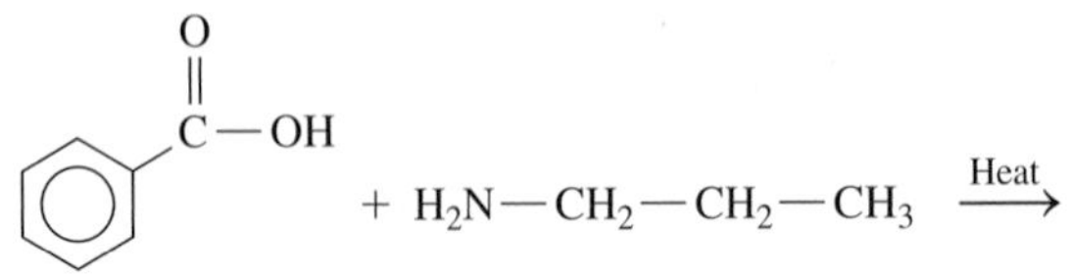

14.40 Draw the condensed structural formula for the amide formed in each of the following reactions:

a. 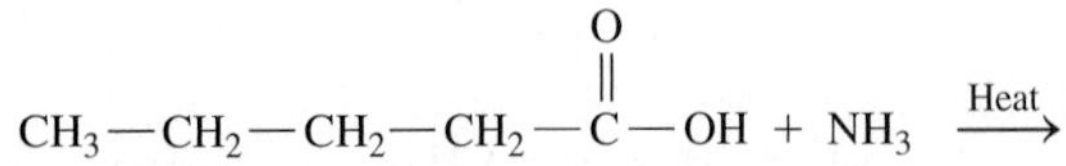

b. 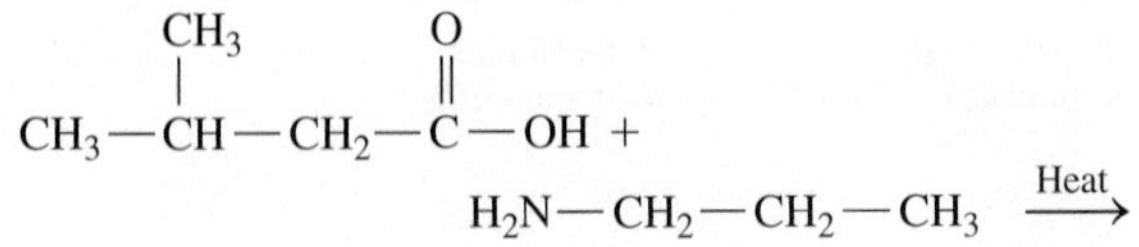

c. 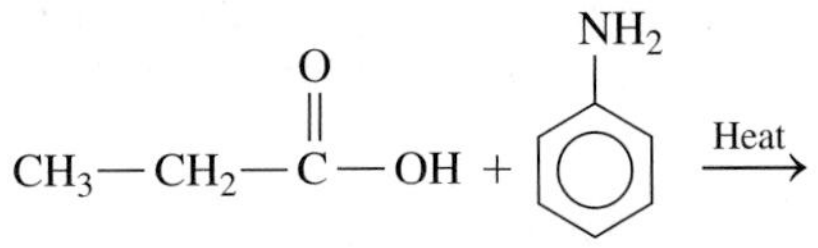

14.41 Give the IUPAC and common names (if any) for each of the following amides:

a. 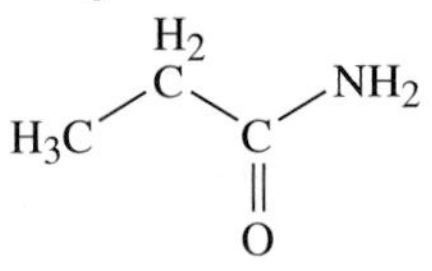

b. 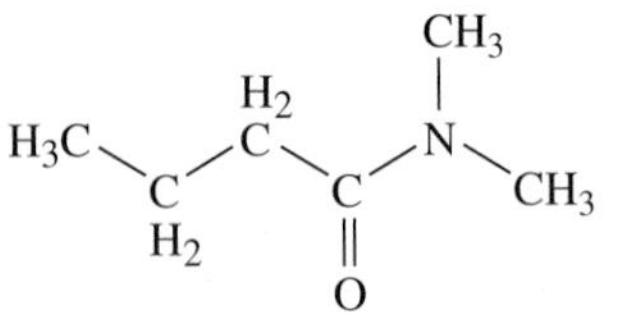

c. 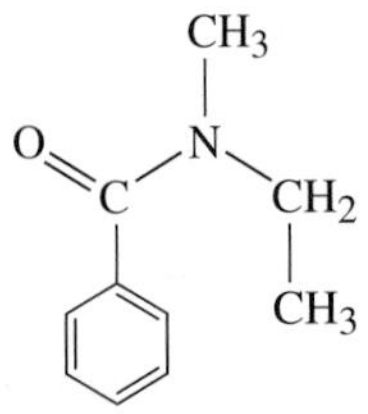

14.42 Give the IUPAC and common names (if any) for each of the following amides:

a. 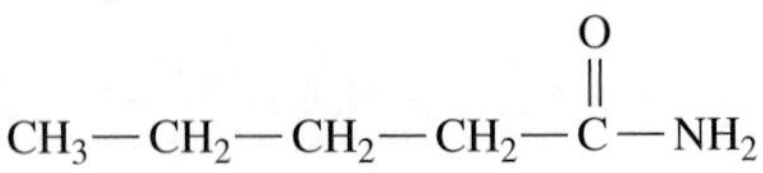

b. 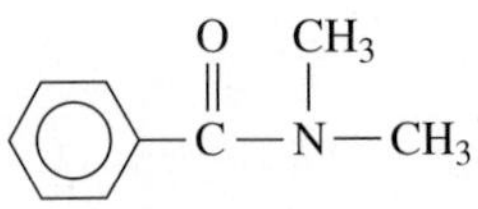

c. 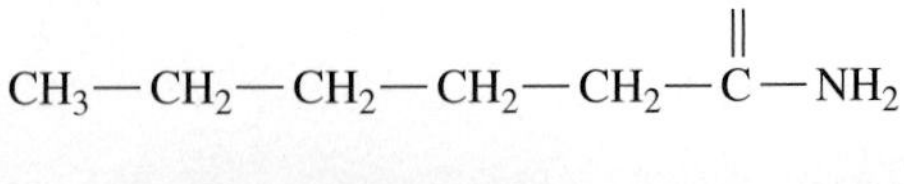

14.43 Draw the condensed structural formula for each of the following amides:

a. propionamide **b.** 2-methylpentanamide
c. methanamide **d.** *N*, *N*-dimethylethanamide

14.44 Draw the condensed structural formula for each of the following amides:

a. hexanamide **b.** 3-chlorobenzamide
c. 3-methylbutyramide **d.** *N*-ethylpropanamide

14.45 Draw the condensed structural formulas for the products from the hydrolysis of each of the following amides with HCl:

a. 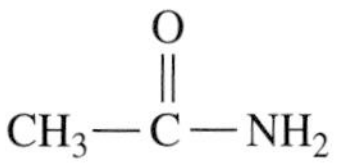

b. 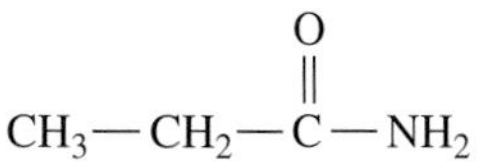

c. 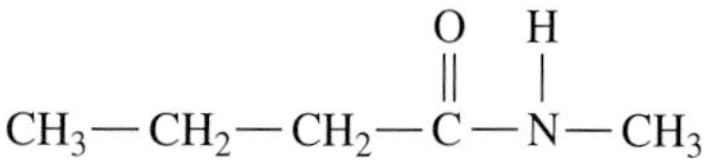

d. 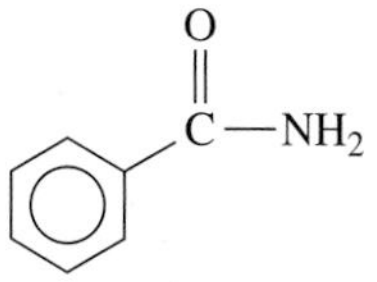

e. *N*-ethylpentanamide

14.46 Draw the condensed structural formulas for the products from the hydrolysis of each of the following amides with NaOH:

a. 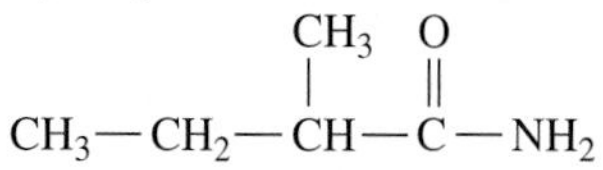

b. 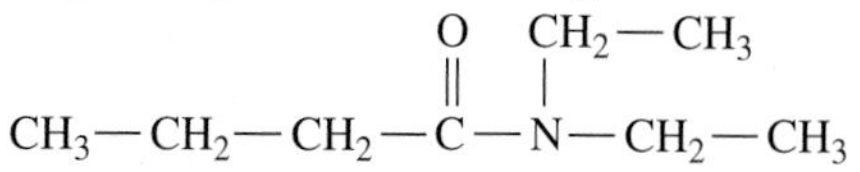

c. 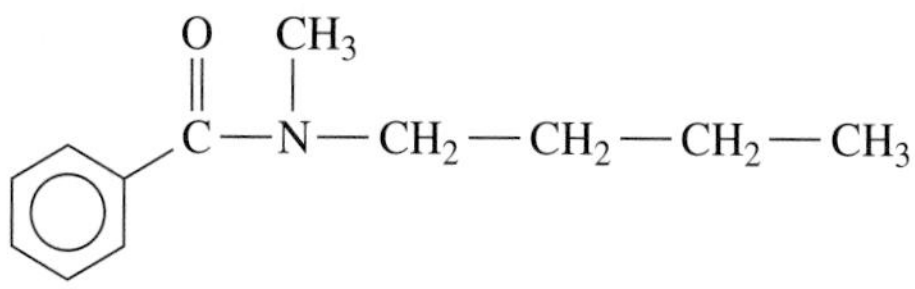

d. 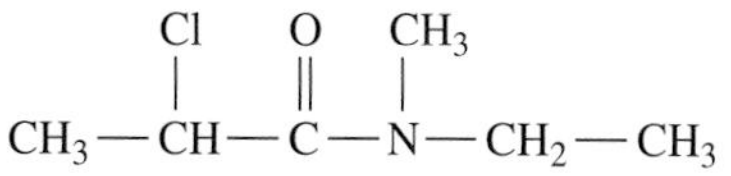

e. *N*-propylbenzamide

CONCEPT MAP

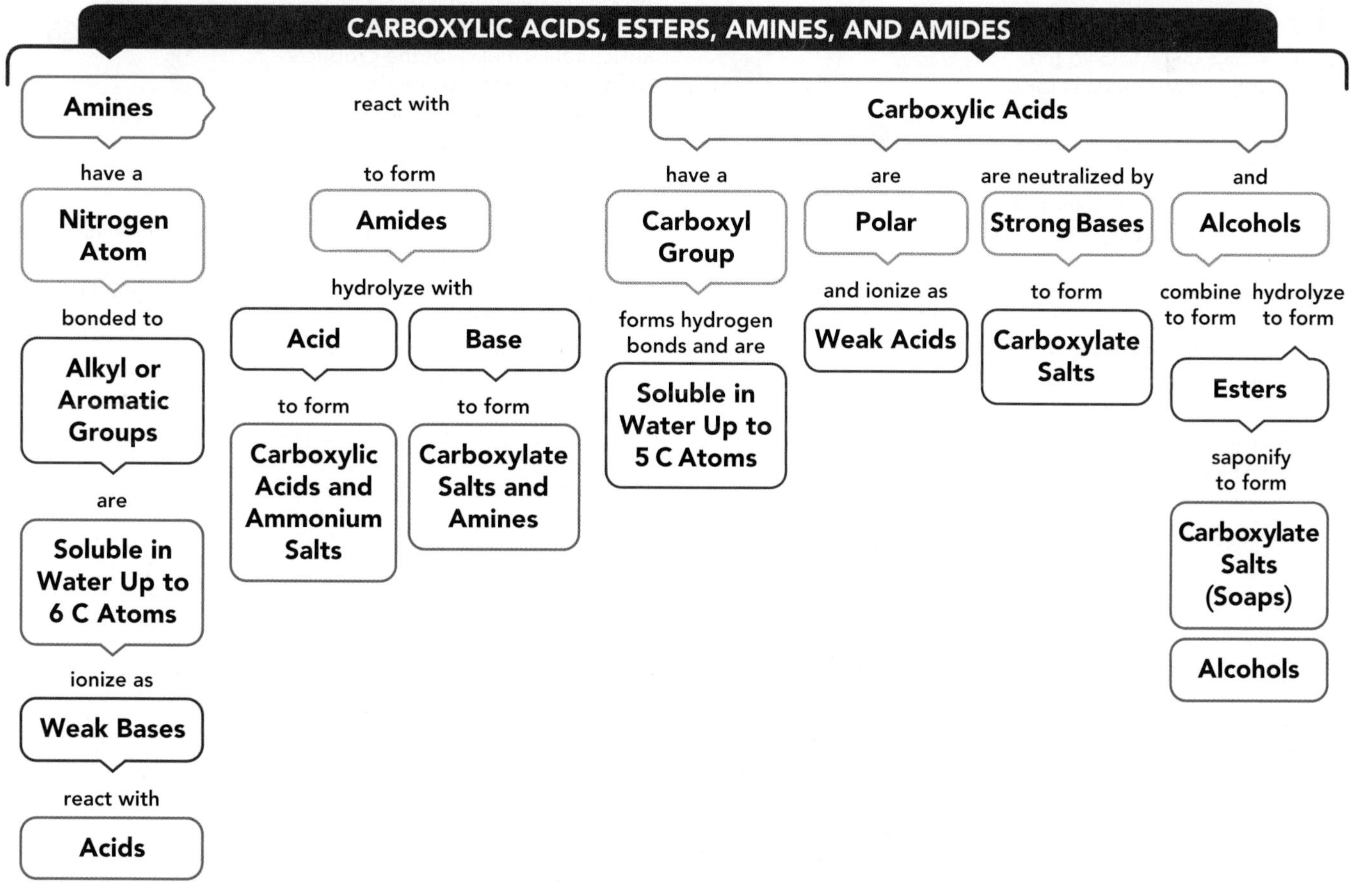

CHAPTER REVIEW

14.1 Carboxylic Acids

LEARNING GOAL Give the IUPAC and common names for carboxylic acids; draw their condensed structural formulas or skeletal formulas.

- A carboxylic acid contains the carboxyl functional group, which is a hydroxyl group connected to a carbonyl group.
- The IUPAC name of a carboxylic acid is obtained by replacing the *e* in the alkane name with *oic acid*.
- The common names of carboxylic acids with one to four carbon atoms are formic acid, acetic acid, propionic acid, and butyric acid.

14.2 Properties of Carboxylic Acids

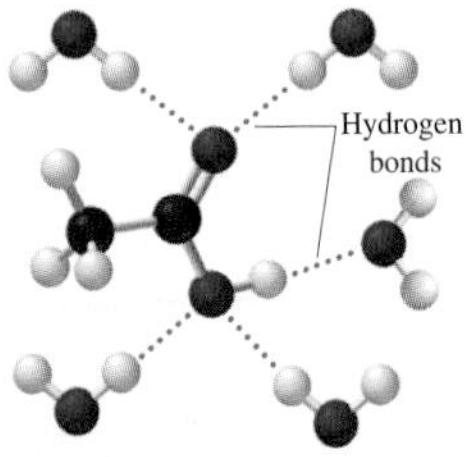

LEARNING GOAL Describe the solubility, ionization, and neutralization of carboxylic acids.

- The carboxyl group contains polar bonds of O—H and C=O, which makes a carboxylic acid with one to five carbon atoms soluble in water.
- As weak acids, carboxylic acids ionize slightly by donating a hydrogen ion to water to form carboxylate and hydronium ions.
- Carboxylic acids are neutralized by base, producing a carboxylate salt and water.

14.3 Esters

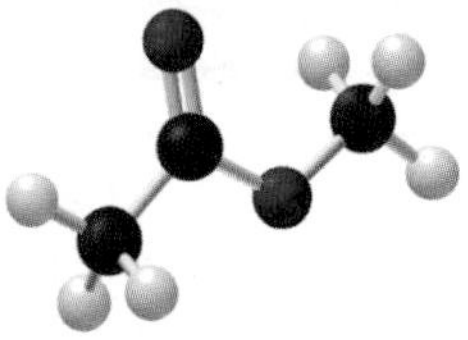
Methyl ethanoate (methyl acetate)

LEARNING GOAL Give the IUPAC and common names for an ester; write the balanced chemical equation for the formation of an ester.

- In an ester, an alkyl or aromatic group replaces the H of the hydroxyl group of a carboxylic acid.
- In the presence of a strong acid, a carboxylic acid reacts with an alcohol to produce an ester. A molecule of water is removed: —OH from the carboxylic acid and —H from the alcohol molecule.
- The names of esters consist of two words: the alkyl group from the alcohol and the name of the carboxylate obtained by replacing *ic acid* with *ate*.

14.4 Hydrolysis of Esters

LEARNING GOAL Draw the condensed structural formulas for the products from acid and base hydrolysis of esters.

- Esters undergo acid hydrolysis by adding water to produce the carboxylic acid and an alcohol.
- Esters undergo base hydrolysis, or saponification, to produce the carboxylate salt and an alcohol.

14.5 Amines

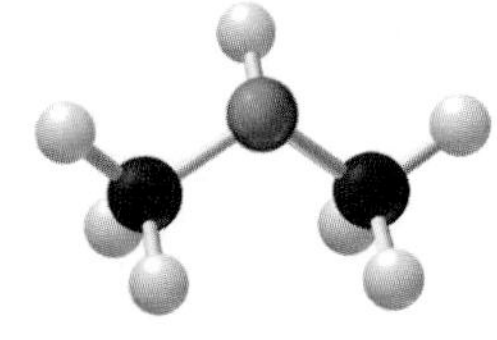
Dimethylamine

LEARNING GOAL Give the common names for amines; draw the condensed structural formulas when given their names. Classify amines as primary (1°), secondary (2°), or tertiary (3°). Describe the solubility, ionization, and neutralization of amines in water.

- In the common names of simple amines, the alkyl groups are listed alphabetically followed by the suffix *amine*.
- A nitrogen atom attached to one, two, or three alkyl or aromatic groups forms a primary (1°), secondary (2°), or tertiary (3°) amine.
- Amines with up to six carbon atoms are soluble in water.
- In water, amines act as weak bases to produce ammonium and hydroxide ions.
- When amines react with acids, they form ammonium salts. As ionic compounds, ammonium salts are solids, soluble in water, and odorless.

14.6 Amides

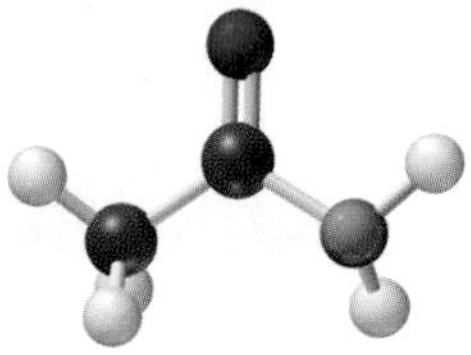
Ethanamide (acetamide)

LEARNING GOAL Give the IUPAC and common names for amides and draw the condensed structural formulas for the products of formation and hydrolysis.

- Amides are derivatives of carboxylic acids in which the hydroxyl group is replaced by $-NH_2$ or a primary or secondary amine group.
- Amides are formed when carboxylic acids react with ammonia or primary or secondary amines in the presence of heat.
- Amides are named by replacing the *ic* acid or *oic acid* from the carboxylic acid name with *amide*. Any carbon group attached to the nitrogen atom is named using the *N*- prefix.
- Hydrolysis of an amide by an acid produces a carboxylic acid and an ammonium salt.
- Hydrolysis of an amide by a base produces the carboxylate salt and an amine.

SUMMARY OF NAMING

Family	Condensed Structural Formula	IUPAC Name	Common Name
Carboxylic acid	$CH_3-C(=O)-OH$	Ethanoic acid	Acetic acid
Ester	$CH_3-C(=O)-O-CH_3$	Methyl ethanoate	Methyl acetate
Amine	$CH_3-CH_2-NH_2$		Ethylamine
Amide	$CH_3-C(=O)-NH_2$	Ethanamide	Acetamide

SUMMARY OF REACTIONS

The chapter sections to review are shown after the name of the reaction.

Ionization of a Carboxylic Acid in Water (14.2)

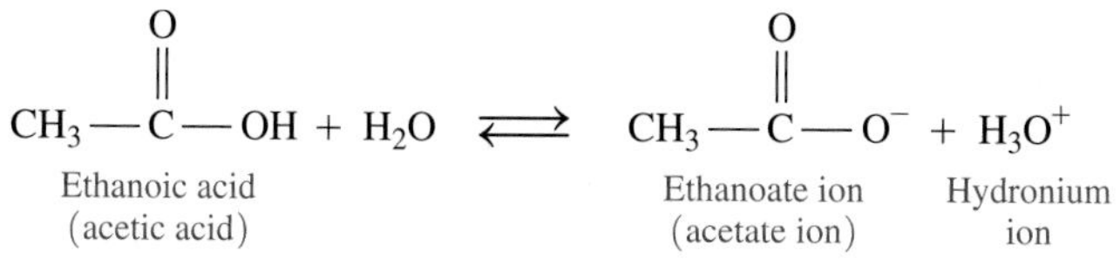

Neutralization of a Carboxylic Acid (14.2)

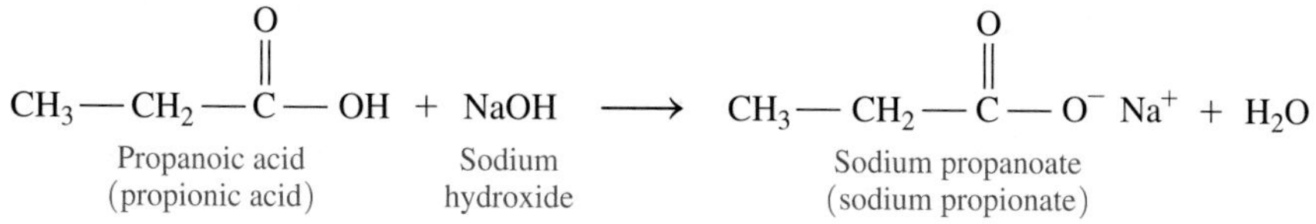

Esterification: Carboxylic Acid and an Alcohol (14.3)

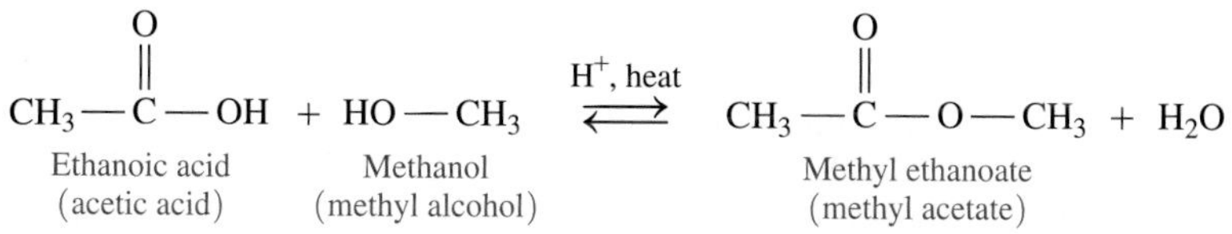

Acid Hydrolysis of an Ester (14.4)

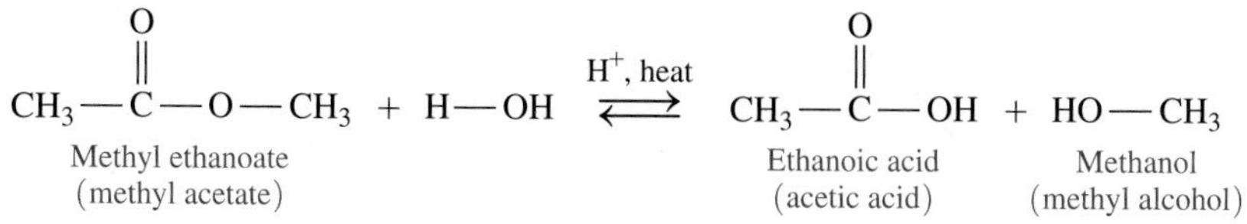

Base Hydrolysis of an Ester: Saponification (14.4)

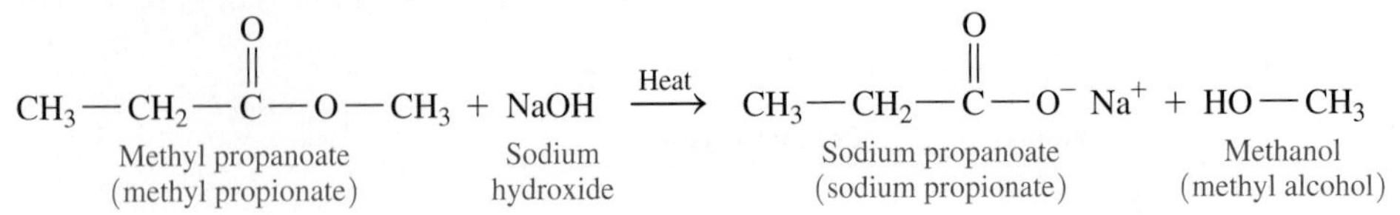

Ionization of an Amine in Water (14.5)

$$CH_3—\ddot{N}H_2 + H_2O \rightleftarrows CH_3—\overset{+}{N}H_3 + OH^-$$

Methylamine; Methylammonium ion; Hydroxide ion

Formation of an Ammonium Salt (14.5)

$$CH_3—\ddot{N}H_2 + HCl \longrightarrow CH_3—\overset{+}{N}H_3\ Cl^-$$

Methylamine; Methylammonium chloride

Amidation: Carboxylic Acid and an Amine (14.6)

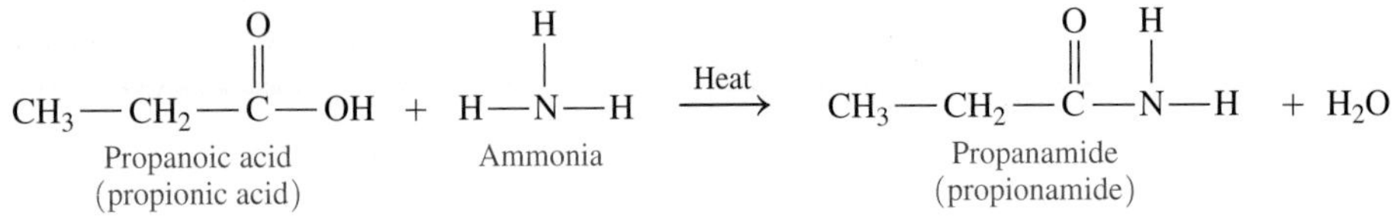

Acid Hydrolysis of an Amide (14.6)

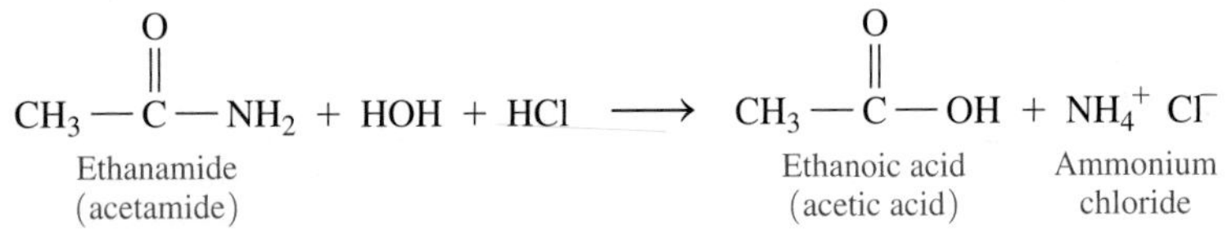

Base Hydrolysis of an Amide (14.6)

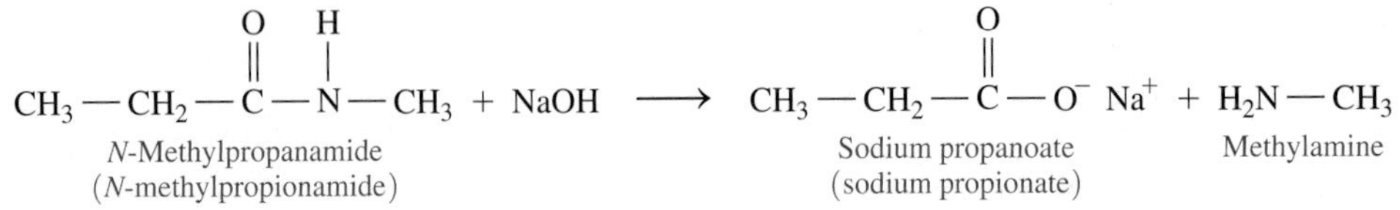

KEY TERMS

amide An organic compound containing the carbonyl group attached to an amino group or a substituted nitrogen atom.

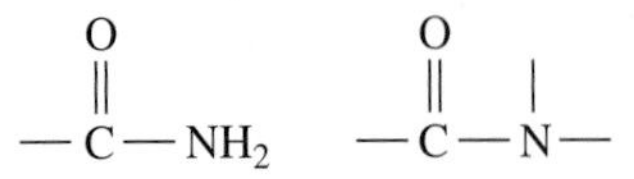

Amide

amine An organic compound containing a nitrogen atom attached to one, two, or three hydrocarbon groups.

ammonium salt An ionic compound produced from an amine and an acid.

carboxyl group A functional group found in carboxylic acids composed of carbonyl and hydroxyl groups.

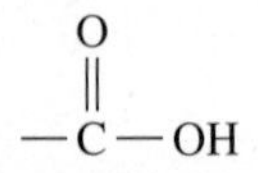

Carboxyl group

carboxylate ion The anion produced when a carboxylic acid donates a hydrogen ion to water.

carboxylate salt A carboxylate ion and the metal ion from the base that is the product of neutralization of a carboxylic acid.

carboxylic acid An organic compound containing the carboxyl group.

ester An organic compound in which an alkyl group replaces the hydrogen atom in a carboxylic acid.

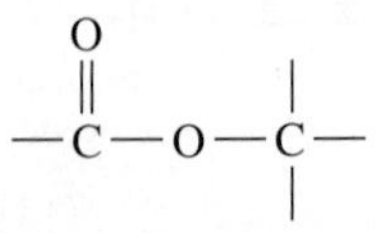

Ester

esterification The formation of an ester from a carboxylic acid and an alcohol, with the elimination of a molecule of water in the presence of an acid catalyst.

hydrolysis The splitting of a molecule by the addition of water. Esters hydrolyze to produce a carboxylic acid and an alcohol. Amides yield the corresponding carboxylic acid and amine or their salts.

saponification The hydrolysis of an ester with a strong base to produce a salt of the carboxylic acid and an alcohol.

CORE CHEMISTRY SKILLS

Naming Carboxylic Acids (14.1)

- The IUPAC name of a carboxylic acid is obtained by replacing the *e* in the corresponding alkane name with *oic acid*.
- The common names of carboxylic acids with one to four carbon atoms are formic acid, acetic acid, propionic acid, and butyric acid.

Example: Give the IUPAC name for the following:

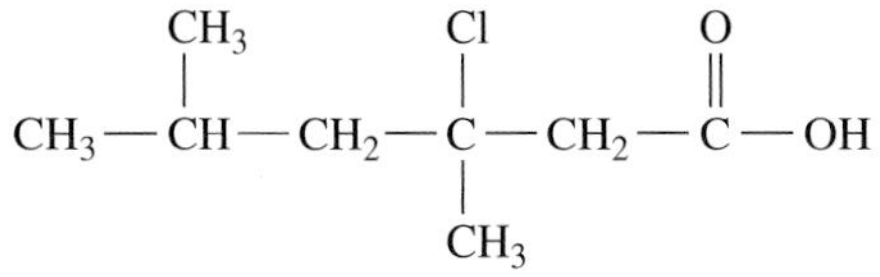

Answer: 3-chloro-3,5-dimethylhexanoic acid

Hydrolyzing Esters (14.4)

- Esters undergo acid hydrolysis by adding water to produce the carboxylic acid and an alcohol.
- Esters undergo base hydrolysis, or saponification, to produce the carboxylate salt and an alcohol.

Example: Draw the condensed structural formulas for the products from the (a) acid and (b) base hydrolysis (NaOH) of ethyl butanoate.

Answer:

a. 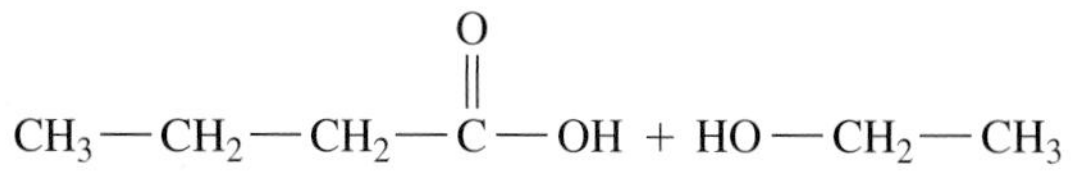

b. 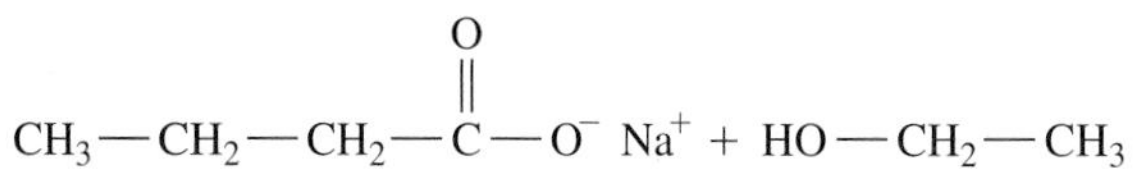

Forming Amides (14.6)

- Amides are formed when carboxylic acids react with ammonia or primary or secondary amines in the presence of heat.

Example: Draw the condensed structural formula for the amide product of the reaction of 3-methylbutanoic acid and ethylamine.

Answer:

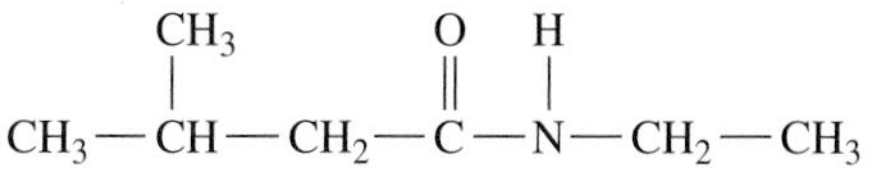

UNDERSTANDING THE CONCEPTS

The chapter sections to review are shown in parentheses at the end of each question.

14.47 Draw the condensed structural formulas and give the IUPAC names for two structural isomers of the carboxylic acids that have the molecular formula $C_4H_8O_2$. (14.1)

14.48 Give the IUPAC names for four structural isomers of the formates that have the molecular formula $C_5H_{10}O_2$. (14.3)

14.49 The ester methyl butanoate has the odor and flavor of strawberries. (14.3)

a. Draw the condensed structural formula for methyl butanoate.
b. Give the IUPAC name for the carboxylic acid and the alcohol used to prepare methyl butanoate.
c. Write the balanced chemical equation for the acid hydrolysis of methyl butanoate.

14.50 The drug cocaine hydrochloride hydrolyzes in air to give methyl benzoate. Its odor, which smells like pineapple guava, is used to train detection dogs. (14.3)

a. Draw the condensed structural formula for methyl benzoate.
b. Give the IUPAC name for the carboxylic acid and alcohol used to prepare methyl benzoate.
c. Write the balanced chemical equation for the acid hydrolysis of methyl benzoate.

14.51 Neo-Synephrine is the active ingredient in some nose sprays used to reduce swelling of nasal membranes. Identify the functional groups in the structure of Neo-Synephrine. (14.1, 14.2, 14.5, 14.6)

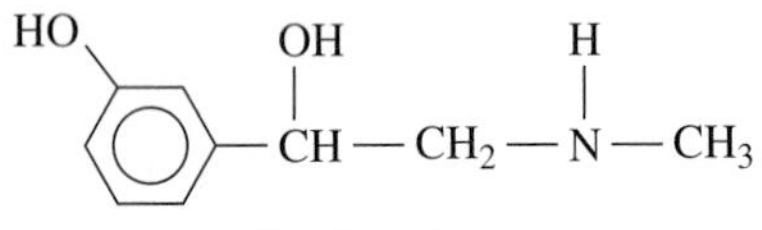

Neo-Synephrine

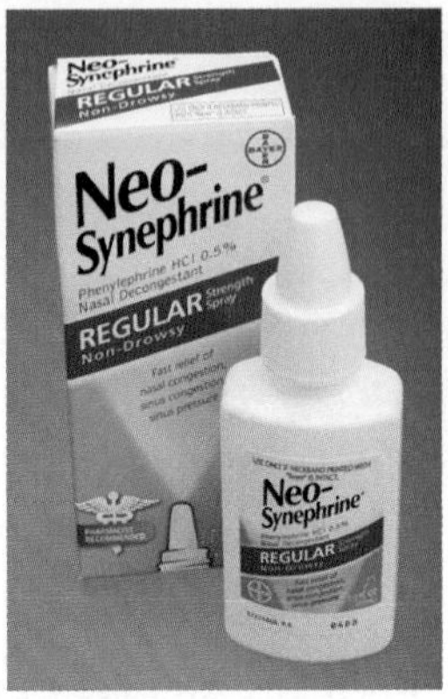

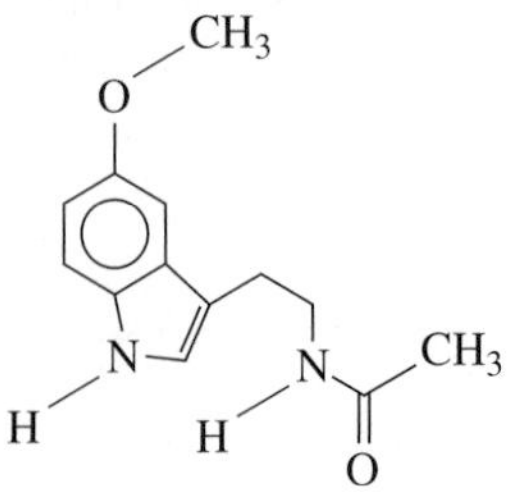

Melatonin

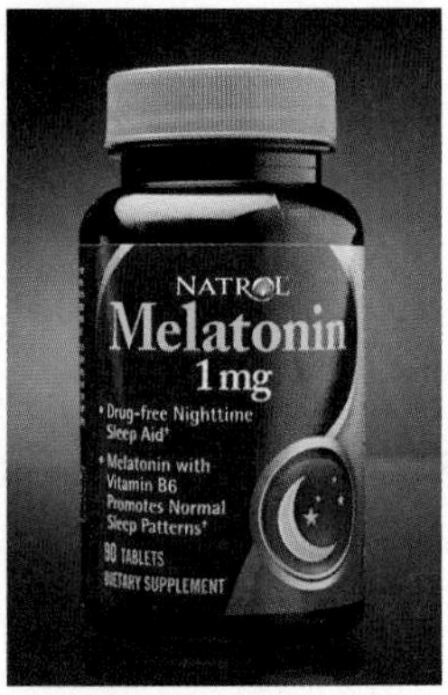

14.52 Melatonin is a naturally occurring compound in plants and animals, where it regulates the biological time clock. Melatonin is sometimes used to counteract jet lag. Identify the functional groups in the structure of melatonin. (14.1, 14.2, 14.5, 14.6)

14.53 There are four amine isomers with the molecular formula C_3H_9N. Draw their condensed structural formulas. Name and classify each as a primary (1°), secondary (2°), or tertiary (3°) amine. (14.5)

14.54 There are four amide isomers with the molecular formula C_3H_7NO. Draw their skeletal formulas. (14.6)

ADDITIONAL QUESTIONS AND PROBLEMS

14.55 Give the IUPAC and common names (if any) for each of the following: (14.1, 14.3)

a. $CH_3-\overset{CH_3}{\overset{|}{CH}}-CH_2-\overset{O}{\overset{\|}{C}}-OH$

b. 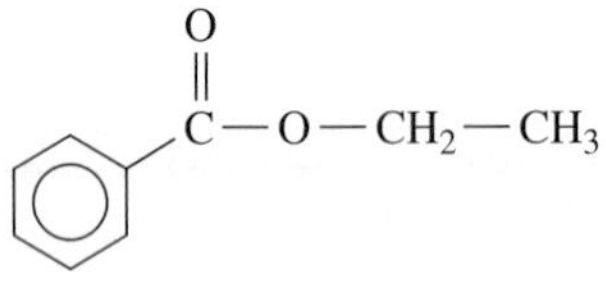

c. $CH_3-CH_2-\overset{O}{\overset{\|}{C}}-O-CH_2-CH_3$

d. 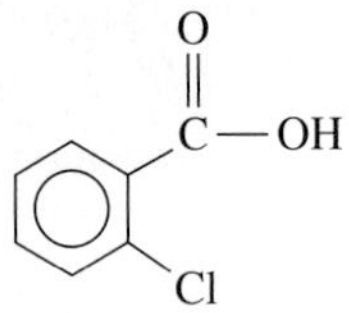

e. $CH_3-CH_2-CH_2-CH_2-\overset{O}{\overset{\|}{C}}-OH$

14.56 Give the IUPAC and common names (if any) for each of the following: (14.1, 14.3)

a. $CH_3-\overset{CH_3}{\overset{|}{CH}}-CH_2-CH_2-\overset{O}{\overset{\|}{C}}-OH$

b. 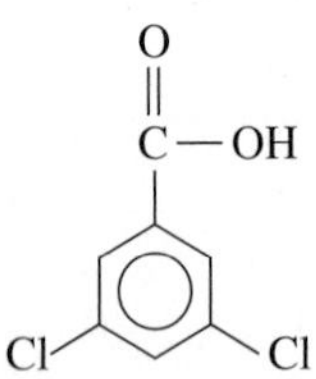

c. 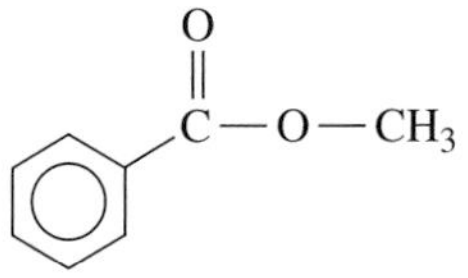

d. $CH_3-CH_2-CH_2-\overset{O}{\overset{\|}{C}}-O-CH_3$

e. $CH_3-CH_2-CH_2-CH_2-\overset{O}{\overset{\|}{C}}-O-CH_2-CH_3$

14.57 Draw the condensed structural formula for each of the following: (14.1, 14.3)

a. methyl acetate
b. 2,2-dichlorobutanoic acid
c. 3-bromopentanoic acid
d. ethyl benzoate

14.58 Draw the skeletal formula for each of the following: (14.1, 14.3)

a. pentyl formate
b. acrylic acid
c. trans-3-phenylacrylic acid
d. 3,5-dimethylhexanoic acid

14.59 Draw the condensed structural formulas for the products of the following: (14.2, 14.3, 14.6)

a. $CH_3-CH_2-\overset{O}{\overset{\|}{C}}-OH + KOH \longrightarrow$

b. $CH_3-CH_2-\overset{O}{\overset{\|}{C}}-OH + HO-CH_3 \overset{H^+,\ heat}{\rightleftarrows}$

c. 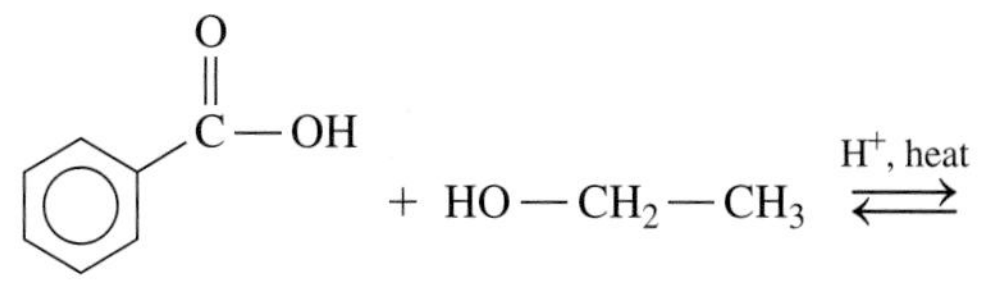

d. 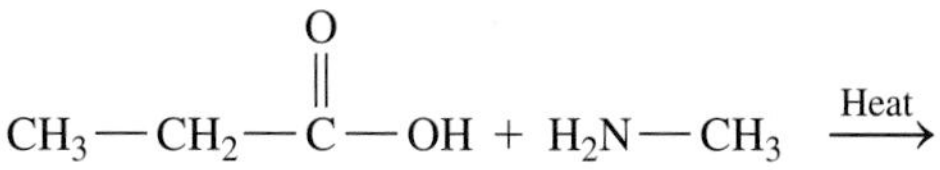

14.60 Draw the condensed structural formulas for the products of the following: (14.2, 14.3, 14.6)

a. 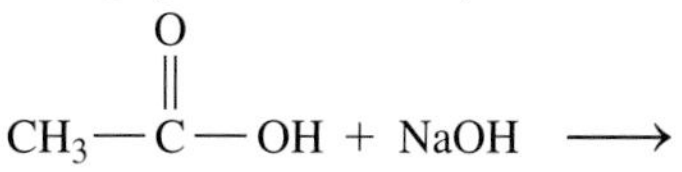

b. 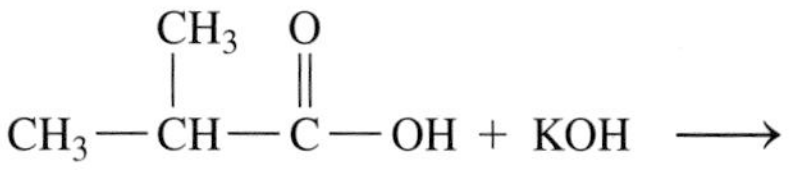

c. 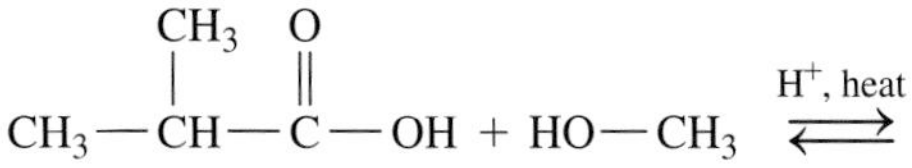

d. 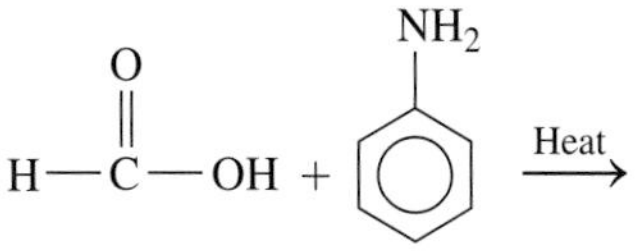

14.61 Draw the condensed structural formulas for the products of the following: (14.4, 14.5)

a. 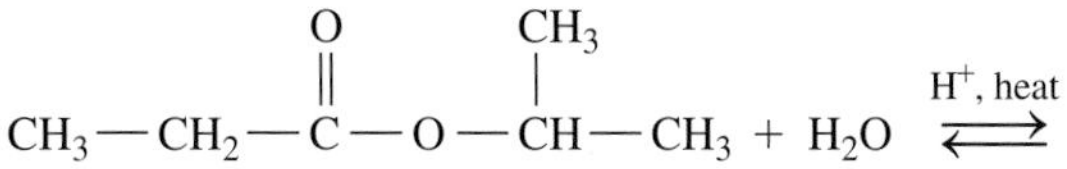

b. 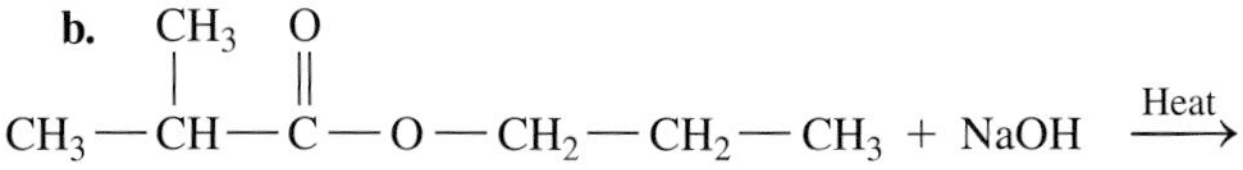

c. 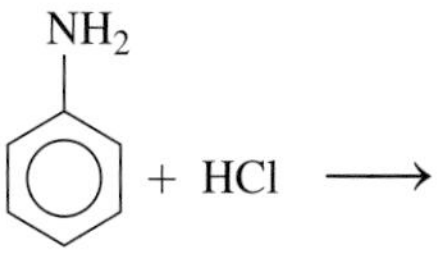

14.62 Draw the condensed structural formulas for the products of the following: (14.4, 14.6)

a. 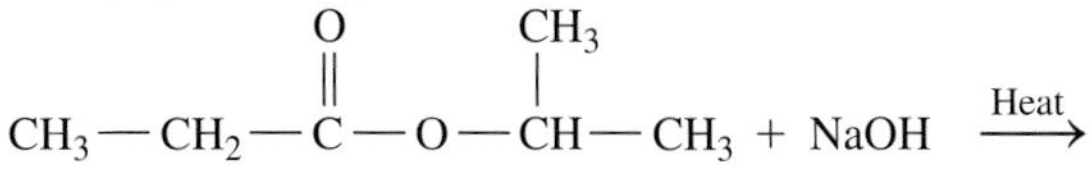

b. 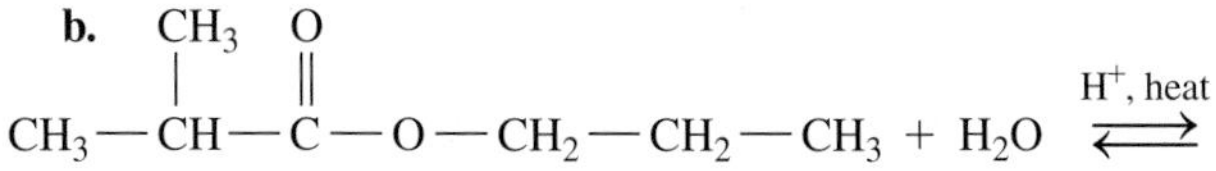

c. 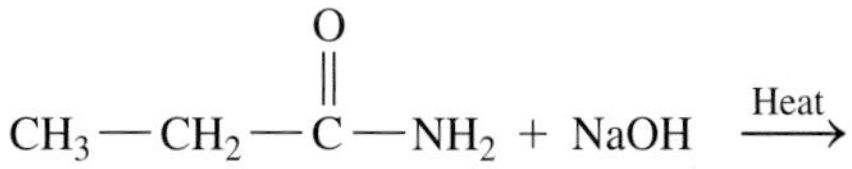

14.63 Classify and give the common name for each of the following: (14.5)

a. 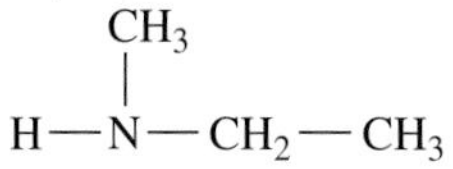

b. 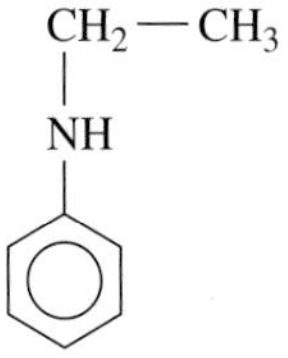

c. 

14.64 Classify and give the common name for each of the following: (14.5)

a. 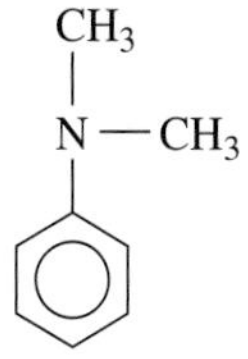

b. 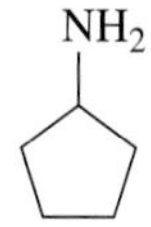

c. 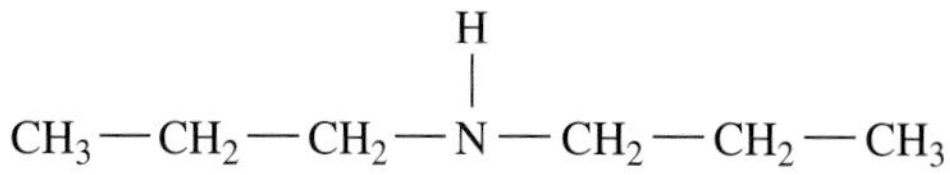

14.65 Draw the condensed structural formula, or skeletal formula if cyclic, for each of the following: (14.5)

a. dimethylamine
b. cyclohexylamine
c. dimethylammonium chloride
d. *N*-ethylaniline

14.66 Draw the skeletal formula for each of the following: (14.5)

a. hydroxylamine
b. 2-aminoethanol
c. L-proline
d. ethylmethylammonium bromide

14.67 Draw the condensed structural formulas for the products of the following: (14.5)

a. $CH_3-CH_2-NH_2 + H_2O \rightleftarrows$
b. $CH_3-CH_2-NH_2 + HCl \longrightarrow$

14.68 Draw the condensed structural formulas for the products of the following: (14.5)

a. 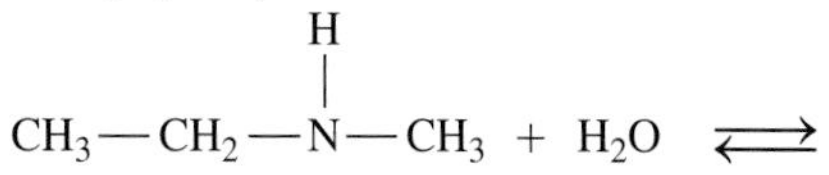

b. 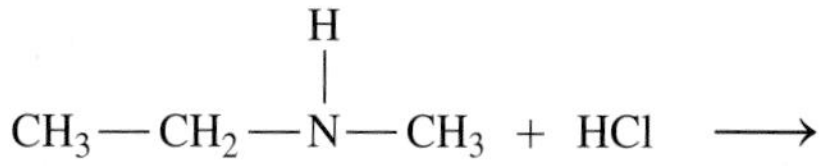

14.69 Give the IUPAC name for each of the following: (14.6)

a. 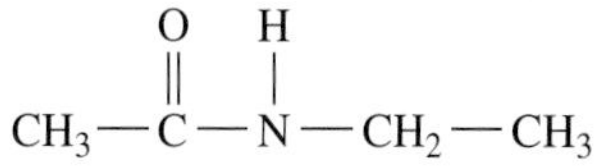

b. 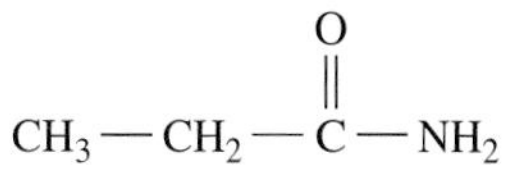

c. 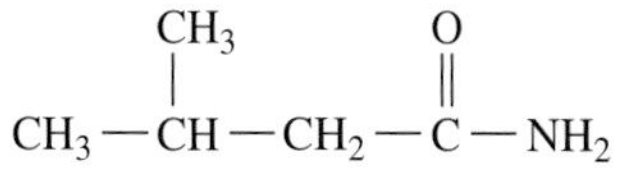

14.70 Give the IUPAC name for each of the following: (14.6)

a. 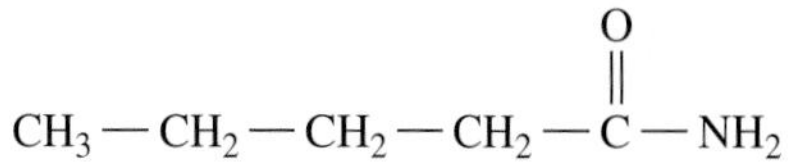

b. 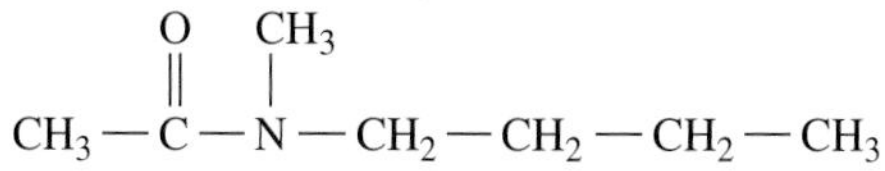

c. 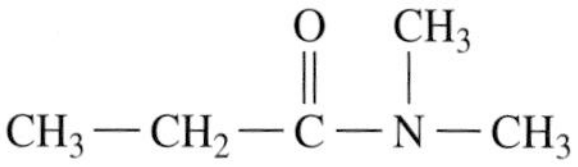

14.71 Draw the condensed structural formulas for the products from the hydrolysis of each of the following: (14.6)

a. $CH_3-\overset{O}{\overset{\|}{C}}-\overset{CH_3}{\overset{|}{N}}-H + HOH + HCl \longrightarrow$

b. $H-\overset{O}{\overset{\|}{C}}-NH_2 + NaOH \longrightarrow$

c. $CH_3-CH_2-\overset{O}{\overset{\|}{C}}-NH_2 + HOH + HCl \longrightarrow$

14.72 Draw the condensed structural formulas for the products from the hydrolysis of each of the following: (14.6)

a. $CH_3-\overset{O}{\overset{\|}{C}}-\overset{CH_3}{\overset{|}{N}}-CH_2-CH_3 + NaOH \longrightarrow$

b. $H-\overset{O}{\overset{\|}{C}}-NH_2 + HOH + HCl \longrightarrow$

c. $CH_3-CH_2-CH_2-CH_2-\overset{O}{\overset{\|}{C}}-NH_2 + NaOH \longrightarrow$

14.73 Voltaren is indicated for acute and chronic treatment of the symptoms of rheumatoid arthritis. Identify the functional groups in this molecule. (14.1, 14.2, 14.5, 14.6)

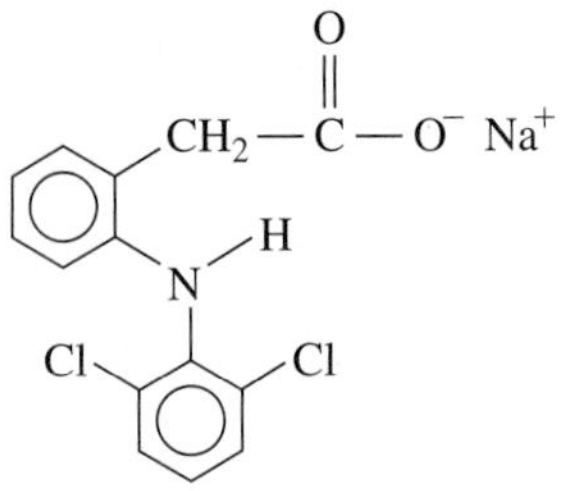

Voltaren

14.74 Toradol is used in dentistry to relieve pain. Identify the functional groups in this molecule. (14.1, 14.2, 14.5, 14.6)

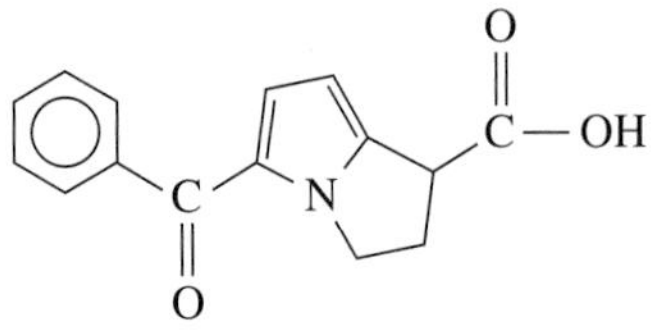

Toradol

CHALLENGE QUESTIONS

The following groups of questions are related to the topics in this chapter. However, they do not all follow the chapter order, and they require you to combine concepts and skills from several sections. These questions will help you increase your critical thinking skills and prepare for your next exam.

14.75 Propyl acetate is the ester that gives the odor and smell of pears. (9.4, 10.6, 14.3, 14.4)

a. Draw the condensed structural formula for propyl acetate.

b. Write the balanced chemical equation for the formation of propyl acetate.

c. Write the balanced chemical equation for the acid hydrolysis of propyl acetate.

d. Write the balanced chemical equation for the base hydrolysis of propyl acetate with NaOH.

e. How many milliliters of a 0.208 M NaOH solution is needed to completely hydrolyze (saponify) 1.58 g of propyl acetate?

14.76 Ethyl octanoate is a flavor component of mangoes. (9.4, 10.6, 14.3, 14.4)

a. Draw the condensed structural formula for ethyl octanoate.

b. Write the balanced chemical equation for the formation of ethyl octanoate.

c. Write the balanced chemical equation for the acid hydrolysis of ethyl octanoate.

d. Write the balanced chemical equation for the base hydrolysis of ethyl octanoate with NaOH.

e. How many milliliters of a 0.315 M NaOH solution is needed to completely hydrolyze (saponify) 2.84 g of ethyl octanoate?

14.77 Novocain, a local anesthetic, is the ammonium salt of procaine. (14.5)

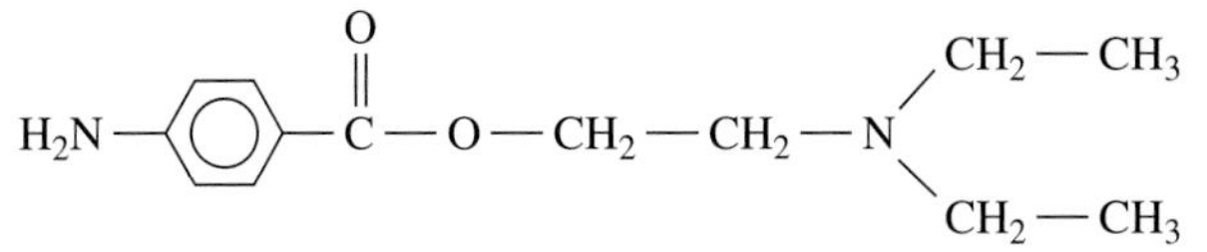

Procaine

a. Draw the condensed structural formula for the ammonium salt (procaine hydrochloride) formed when procaine reacts with HCl. (*Hint*: The tertiary amine reacts with HCl.)

b. Why is procaine hydrochloride used rather than procaine?

14.78 Lidocaine (xylocaine) is used as a local anesthetic and cardiac depressant. (14.5)

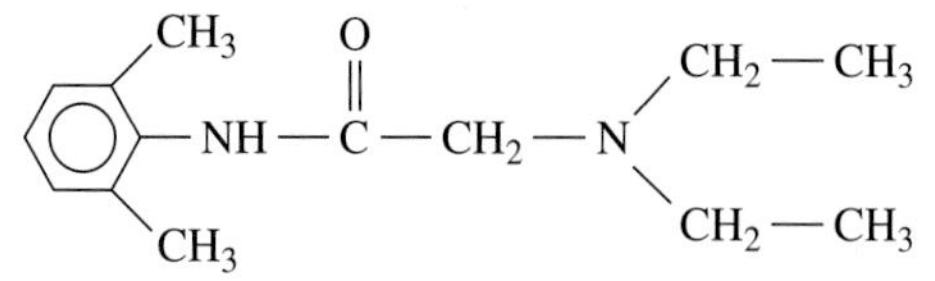

Lidocaine (xylocaine)

a. Draw the condensed structural formula for the ammonium salt formed when lidocaine reacts with HCl.

b. Why is the ammonium salt of lidocaine used rather than the amine?

ANSWERS

Answers to Study Checks

14.1 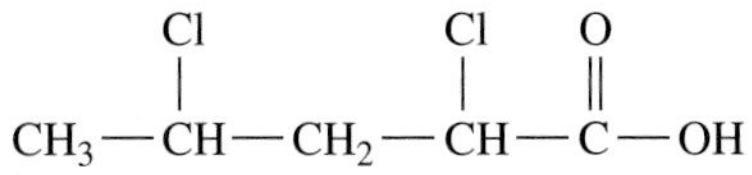

14.2 

14.3 butanoic acid

14.4 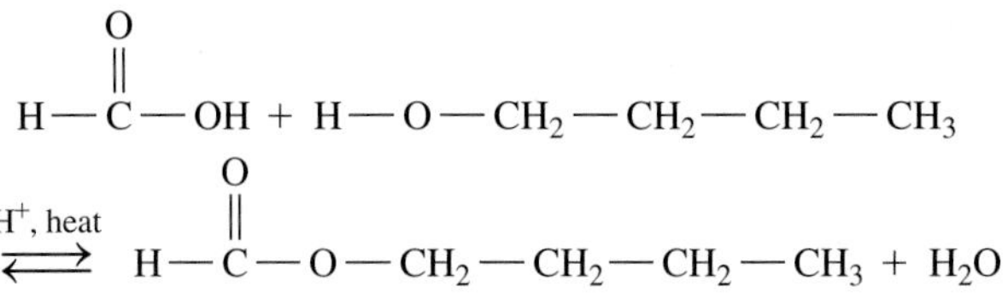

14.5

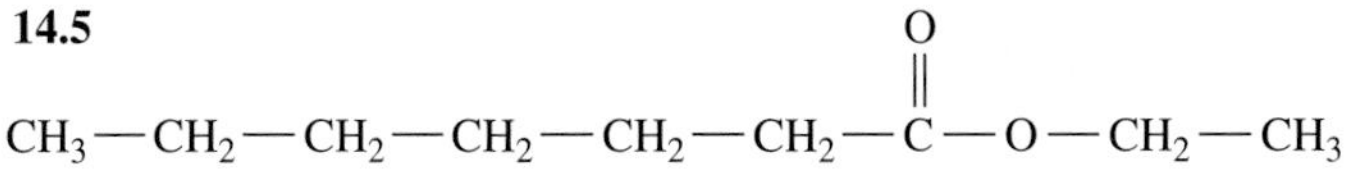

14.6 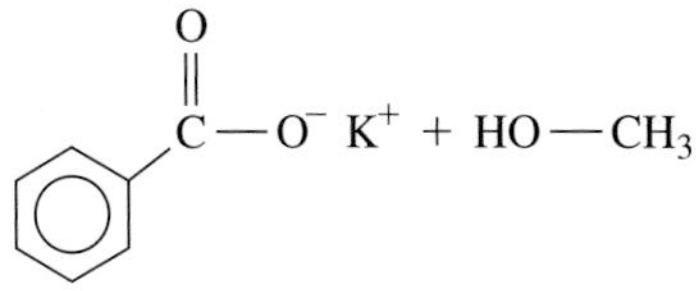

14.7 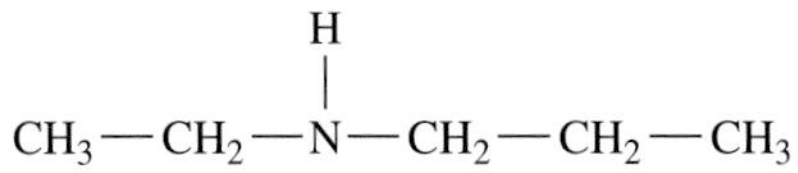

14.8 3°

14.9 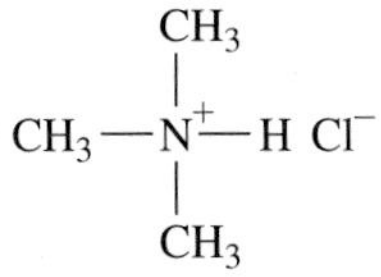

14.10 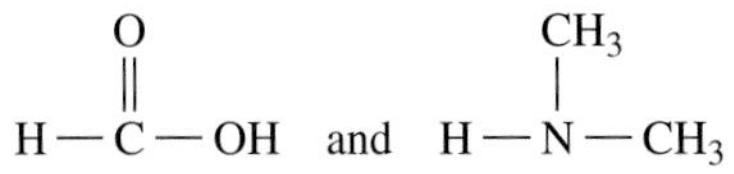

14.11 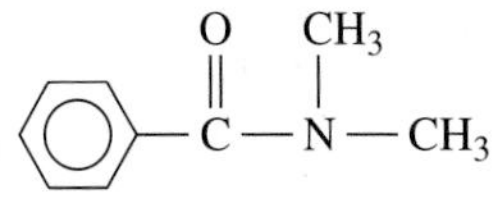

14.12 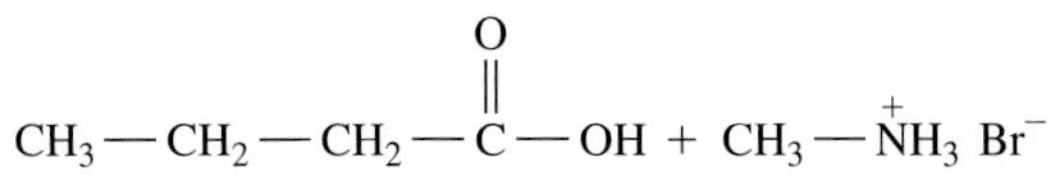

Answers to Selected Questions and Problems

14.1 Butyric acid

14.3 a. 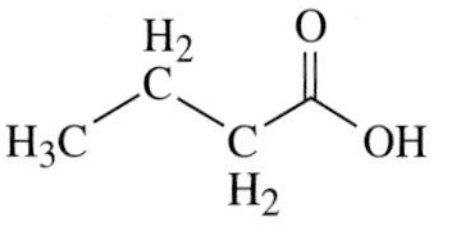

Butanoic acid

b. 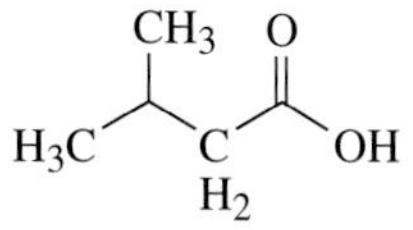

3-Methylbutanoic acid

14.5 a. Common name: propionic acid
IUPAC name : propanoic acid

b. IUPAC name: 2,4-dichlorobenzoic acid

c. 5,6-dimethyoctanoic acid

14.7 a. 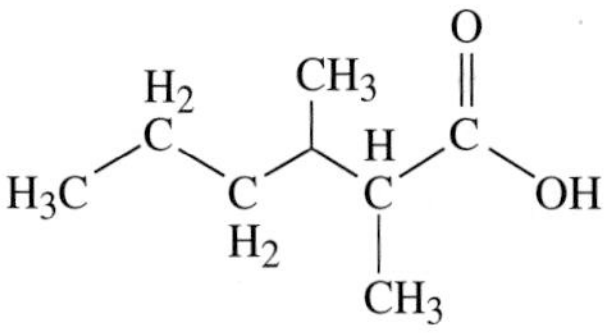

b. 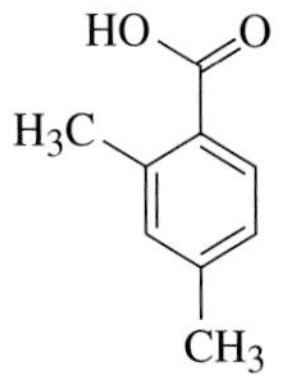

c. 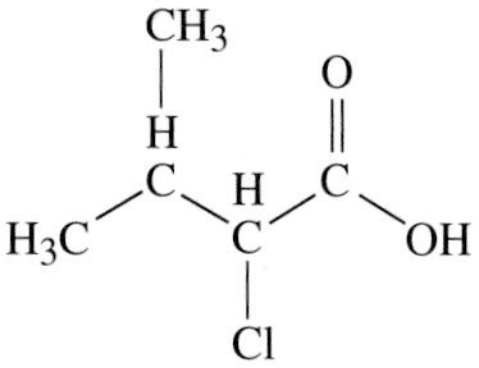

d. 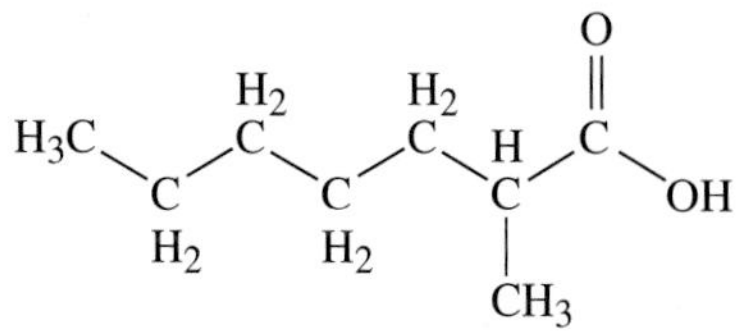

14.9 a. Propanoic acid has the smallest carbon chain, which makes it most soluble.

b. Propanoic acid forms more hydrogen bonds with water, which makes it most soluble.

14.11 a. 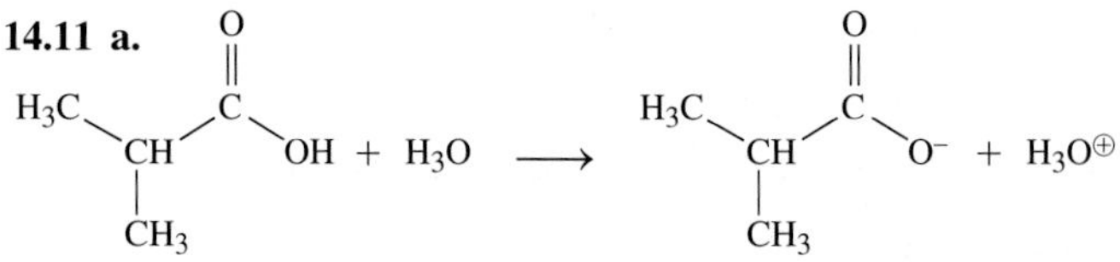

b. 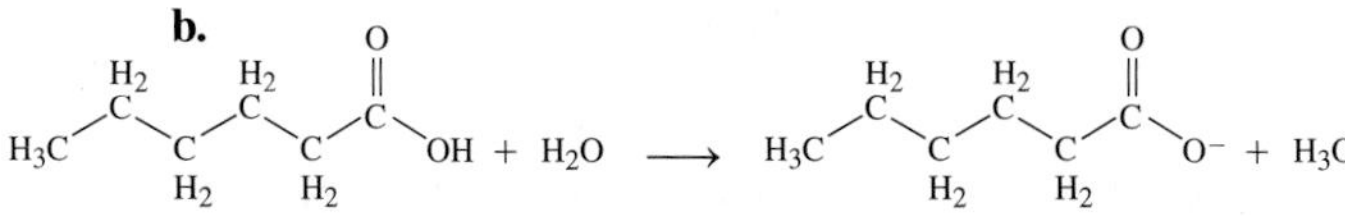

14.13 **a.** $CF_3COOH + NaOH \longrightarrow CF_3COONa + H_2O$
b. $C_lCH_2CH_2COOH + KOH \longrightarrow ClCH_2CH_2COOK + H_2O$
c. $3\text{-}Cl\text{-}C_6H_5COOH + NaHCO_3 \longrightarrow 3\text{-}Cl\text{-}C_6H_5COONa + CO_2 + H_2O$

14.15 **a.** 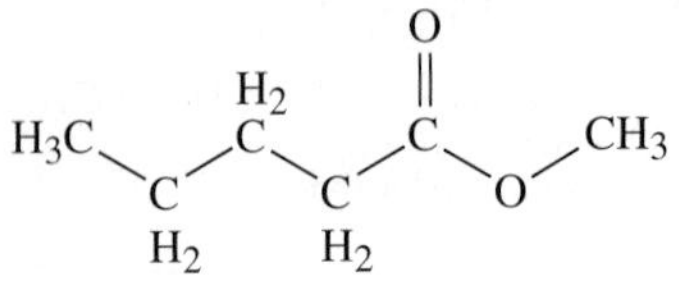

b. 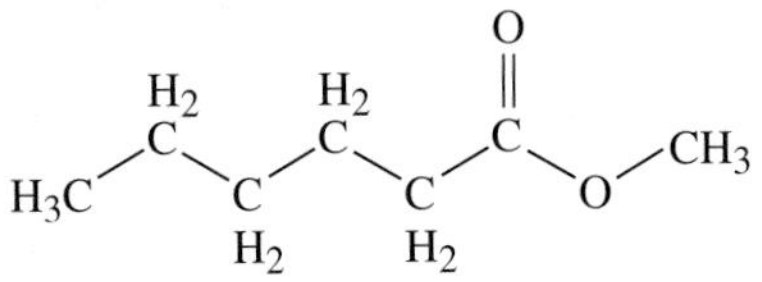

c. 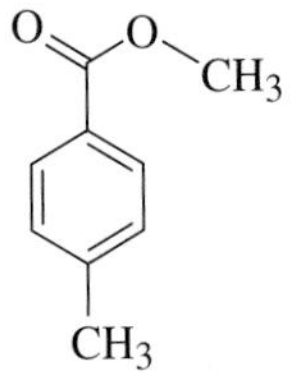

14.17 **a.** $CH_3-CH_2-CH_2-CH_2-\overset{O}{\overset{\|}{C}}-O-\underset{}{\overset{CH_3}{\overset{|}{CH}}}-CH_3$

b. $CH_3-CH_2-\overset{O}{\overset{\|}{C}}-O-CH_2-CH_2-CH_3$

14.19 **a.** Common name: methyl propionate
IUPAC name: methyl propanoate
b. Common name: butyl acetate
IUPAC name: butyl ethanoate
c. Common name: propyl butyrate
IUPAC name: propyl butanoate

14.21 **a.** 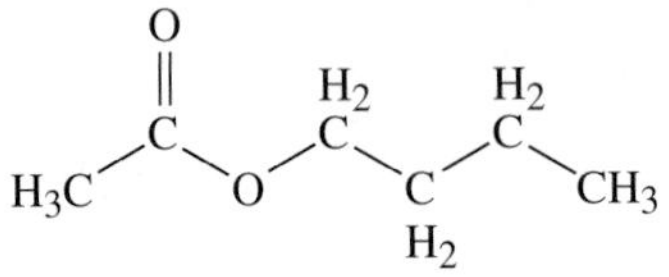

b. 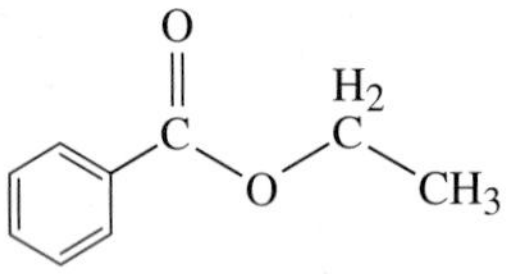

c. 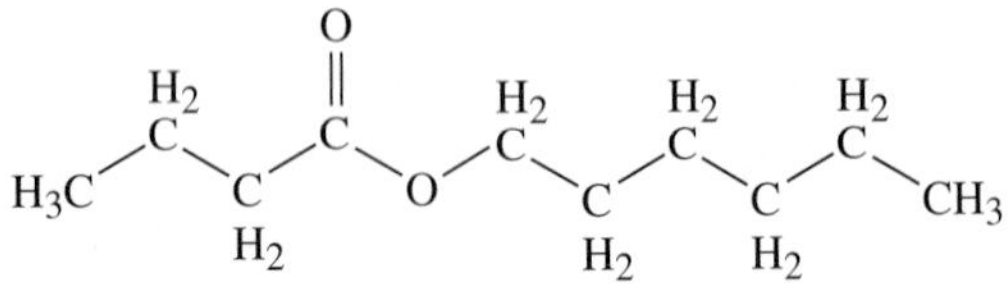

14.23 **a.** propyl acetate
b. ethyl butanoate
c. ethyl heptanoate

14.25 The products of the acid hydrolysis of an ester are a carboxylic acid and an alcohol.

14.27 **a.** $CH_3-CH_2-\overset{O}{\overset{\|}{C}}-O^-\ Na^+ + HO-CH_3$

b. $CH_3-\overset{O}{\overset{\|}{C}}-OH + HO-CH_2-CH_2-CH_3$

c. $CH_3-CH_2-CH_2-\overset{O}{\overset{\|}{C}}-OH + HO-CH_2-CH_3$

d. 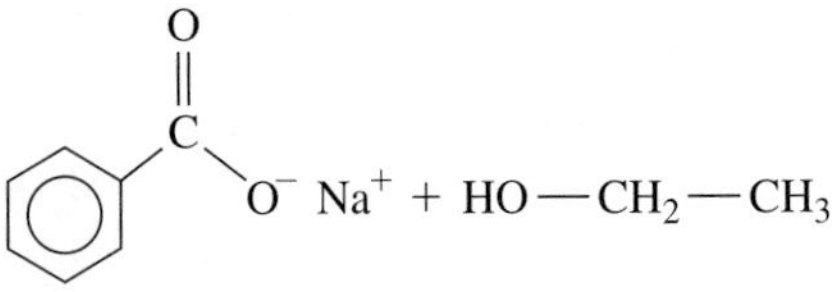

14.29 **a.** pentylamine
b. ethyl butylamine
c. dipropyl ethylamine

14.31 **a.** 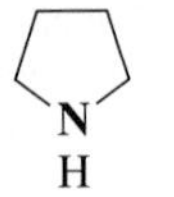

b. 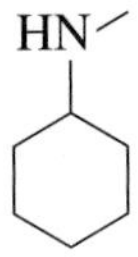

c. 

14.33 **a.** as primary (1°)
b. secondary (2°)
c. tertiary (3°)
d. tertiary (3°)

14.35 **a.** Yes, amines with fewer than seven carbon atoms hydrogen bond with water molecules and are soluble in water.
b. Yes, amines with fewer than seven carbon atoms hydrogen bond with water molecules and are soluble in water.
c. No, an amine with eight carbon atoms is insoluble in water.
d. Yes, amines with fewer than seven carbon atoms hydrogen bond with water molecules and are soluble in water.

14.37 **a.** $CH_3-NH_2 + H_2O \rightleftarrows CH_3-\overset{+}{N}H_3 + OH^-$

b. $CH_3-\overset{H}{\overset{|}{N}}-CH_3 + H_2O \rightleftarrows CH_3-\overset{+}{N}H_2-CH_3 + OH^-$

c. 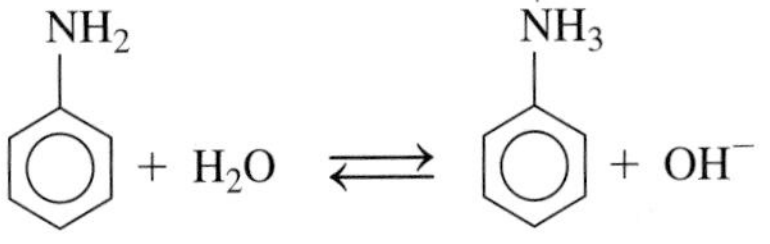

14.39 **a.** $CH_3-\overset{O}{\overset{\|}{C}}-NH_2$

b. $CH_3-\overset{O}{\overset{\|}{C}}-\overset{H}{\overset{|}{N}}-CH_2-CH_3$

c. 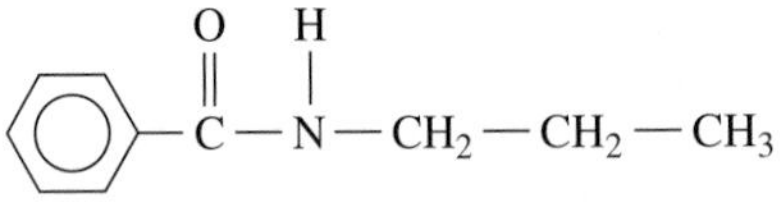

14.41 **a.** Common name: Propioamide, and IUPAC name: propanamide
b. *N*,*N*-dimethyl butanamide
c. *N*-ethyl-*N*-methyl benzamide

14.43 **a.** $CH_3-CH_2-\overset{O}{\overset{\|}{C}}-NH_2$

b. $CH_3-CH_2-CH_2-\overset{CH_3}{\overset{|}{CH}}-\overset{O}{\overset{\|}{C}}-NH_2$

c. $H-\overset{O}{\overset{\|}{C}}-NH_2$

d. $CH_3-C(=O)-N(CH_3)-CH_3$

14.45 a. $CH_3-C(=O)-OH + \overset{+}{N}H_4\ Cl^-$

b. $CH_3-CH_2-C(=O)-OH + \overset{+}{N}H_4\ Cl^-$

c. $CH_3-CH_2-CH_2-C(=O)-OH + CH_3-\overset{+}{N}H_3\ Cl^-$

d. 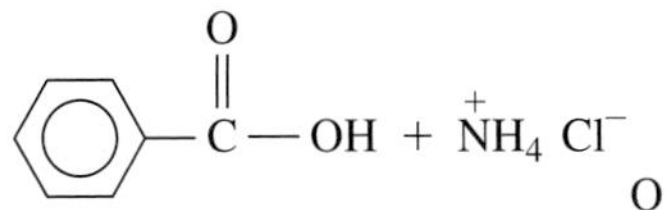

e. $CH_3-CH_2-CH_2-CH_2-C(=O)-OH$

$+ CH_3-CH_2-\overset{+}{N}H_3\ Cl^-$

14.47 $CH_3-CH_2-CH_2-C(=O)-OH$ Butanoic acid; $CH_3-CH(CH_3)-C(=O)-OH$ 2-Methylpropanoic acid

14.49 a. $CH_3-CH_2-CH_2-C(=O)-O-CH_3$

b. butanoic acid and methanol

c. $CH_3-CH_2-CH_2-C(=O)-O-CH_3 + H_2O \xrightleftharpoons{H^+,\ heat}$

$CH_3-CH_2-CH_2-C(=O)-OH + HO-CH_3$

14.51 phenol, alcohol, amine

14.53 $CH_3-CH_2-CH_2-NH_2$ Propylamine (1°)

$CH_3-CH_2-N(H)-CH_3$ Ethylmethylamine (2°)

$CH_3-N(CH_3)-CH_3$ Trimethylamine (3°)

$CH_3-CH(CH_3)-NH_2$ Isopropylamine (1°)

14.55 a. 3-methylbutanoic acid
b. ethyl benzoate
c. ethyl propanoate; ethyl propionate
d. 2-chlorobenzoic acid
e. pentanoic acid

14.57 a. $CH_3-C(=O)-O-CH_3$

b. $CH_3-CH_2-CCl_2-C(=O)-OH$

c. $CH_3-CH_2-CH(Br)-CH_2-C(=O)-OH$

d. 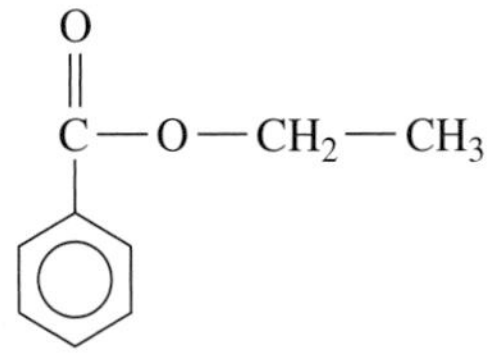

14.59 a. $CH_3-CH_2-C(=O)-O^-\ K^+ + H_2O$

b. $CH_3-CH_2-C(=O)-O-CH_3 + H_2O$

c. 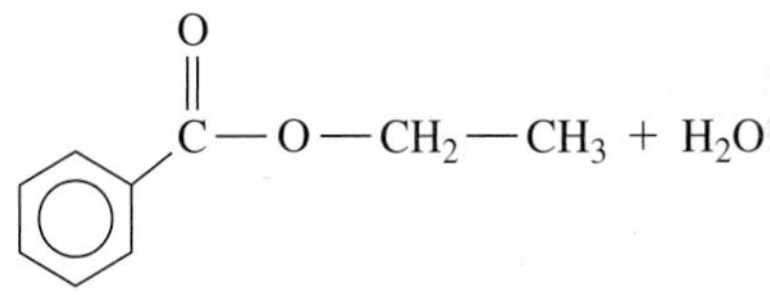

d. $CH_3-CH_2-C(=O)-N(H)-CH_3 + H_2O$

14.61 a. $CH_3-CH_2-C(=O)-OH + HO-CH(CH_3)-CH_3$

b. $CH_3-CH(CH_3)-C(=O)-O^-\ Na^+ + HO-CH_2-CH_2-CH_3$

c. 

14.63 a. 2°, ethylmethylamine
b. 2°, *N*-ethylaniline
c. 1°, butylamine

14.65 a. $CH_3-N(H)-CH_3$

b. 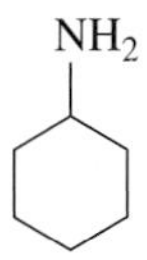

c. $CH_3-\overset{+}{N}H_2(CH_3)\ Cl^-$

d. 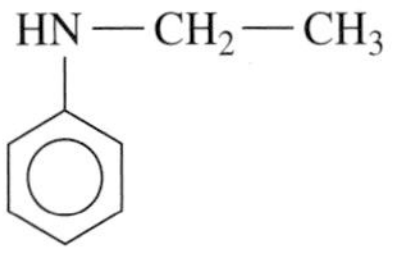

14.67 a. $CH_3-CH_2-\overset{+}{N}H_3 + OH^-$

b. $CH_3-CH_2-\overset{+}{N}H_3\ Cl^-$

14.69 a. *N*-ethylethanamide
b. propanamide
c. 3-methylbutanamide

14.71 a. $CH_3-\overset{O}{\overset{\|}{C}}-OH + CH_3-\overset{+}{N}H_3\ Cl^-$

b. $H-\overset{O}{\overset{\|}{C}}-O^-\ Na^+ + NH_3$

c. $CH_3-CH_2-\overset{O}{\overset{\|}{C}}-OH + \overset{+}{N}H_4\ Cl^-$

d. $CH_3-\overset{O}{\overset{\|}{C}}-O-CH_2-CH_2-CH_3 + NaOH \xrightarrow{\text{Heat}}$

$CH_3-\overset{O}{\overset{\|}{C}}-O^-\ Na^+ + HO-CH_2-CH_2-CH_3$

e. 74.4 mL of a 0.208 M NaOH solution

14.73 carboxylate salt, aromatic, amine

14.75 a. $CH_3-\overset{O}{\overset{\|}{C}}-O-CH_2-CH_2-CH_3$

b. $CH_3-\overset{O}{\overset{\|}{C}}-OH + HO-CH_2-CH_2-CH_3 \xrightleftharpoons{H^+,\ \text{heat}}$

$CH_3-\overset{O}{\overset{\|}{C}}-O-CH_2-CH_2-CH_3 + H_2O$

c. $CH_3-\overset{O}{\overset{\|}{C}}-O-CH_2-CH_2-CH_3 + H_2O \xrightleftharpoons{H^+,\ \text{heat}}$

$CH_3-\overset{O}{\overset{\|}{C}}-OH + HO-CH_2-CH_2-CH_3$

14.77 a.

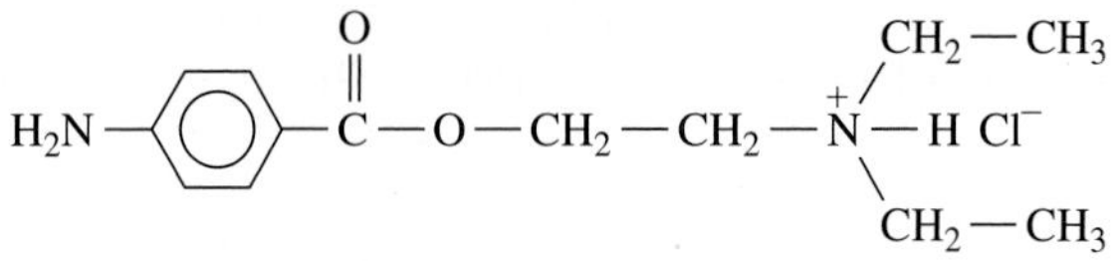

b. The ammonium salt (Novocain) is more soluble in body fluids than procaine.

15 Lipids

Bill, an elderly patient, had chest pain, sweating, and felt nauseous soon after eating. Since his symptoms are similar to a heart attack, he is taken to the ER. The emergency room doctor determined that there was no heart problem; rather he determines that Bill had acute gallbladder disease. The doctor orders blood tests, including a lipid panel and an ultrasound of Bill's abdomen.

Susan, a geriatric nurse, discusses the test results with Bill. She explains that the lipid panel measures the total cholesterol in his blood. This consists of the four types of lipids in his blood: cholesterol, HDL, LDL, and triglycerides. HDL, or high-density lipoprotein cholesterol, is considered the "good" cholesterol as it helps to clear plaque from his arteries, while LDL, or low-density lipoprotein cholesterol, is the "bad" cholesterol because it deposits plaque in the arteries. The body stores excess calories by creating triacylglycerols, which are stored in fat cells. His levels are elevated, and the ultrasound confirms that his gallbladder is inflamed.

Susan and Bill discuss his treatment options and a number of changes he can make to his lifestyle to decrease his gallbladder attacks. This includes decreasing the amount of fats in his diet, exercising more, losing weight, and increasing his consumption of dietary fiber. If the attacks continue, Susan advises that Bill may want to have his gallbladder removed.

CAREER Geriatric Nurse

Geriatric nurses provide care for elderly patients in hospitals, nursing homes, assisted living facilities, and in-home care. This includes conducting medical tests, administering medications, and working with the elderly patient to set personal health goals while promoting self-care. They focus on the creation and implementation of treatment plans for chronic illnesses, as well as educating and counseling patients' families. Geriatric nurses also assist physicians during exams, while providing care and compassion to their patients. Due to the age of the patient, geriatric nurses may encounter patients with diminishing mental capacities.

LOOKING AHEAD

When we talk of fats and oils, waxes, steroids, and cholesterol, we are discussing lipids. Lipids are naturally occurring compounds that vary considerably in structure but share the common feature of being soluble in nonpolar solvents, but not in water. Fats, which are one family of lipids, have many functions in the body, such as storing energy and protecting and insulating internal organs. Because lipids are not soluble in water, they are important in cellular membranes that function to separate the internal contents of cells from the external environment. Other types of lipids are found in nerve fibers and in hormones, which act as chemical messengers.

Many people are concerned about the amounts of saturated fats and cholesterol in our diets. Researchers suggest that saturated fats and cholesterol are associated with diseases such as diabetes; cancers of the breast, pancreas, and colon; and *atherosclerosis*, a condition in which deposits of lipid materials called *plaque* accumulate in the coronary blood vessels. In atherosclerosis, plaque restricts the flow of blood to the tissue, causing necrosis (death). An accumulation of plaque in the heart could result in a *myocardial infarction* (heart attack).

The American Institute for Cancer Research (AICR) has recommended that we increase the fiber and starch content of our diets by adding more vegetables, fruits, and whole grains as well as foods with low levels of fat and cholesterol such as fish, poultry, lean meats, and low-fat dairy products. The AICR has also suggested that we limit our intake of foods high in fat and cholesterol such as eggs, nuts, fatty or organ meats, cheeses, butter, and coconut and palm oil.

CHAPTER READINESS*

CORE CHEMISTRY SKILLS

- Writing Equations for Hydrogenation and Hydration (11.7)
- Naming Alcohols and Phenols (12.1)
- Naming Carboxylic Acids (14.1)
- Hydrolyzing Esters (14.4)

*These Core Chemistry Skills from previous chapters are listed here for your review as you proceed to the new material in this chapter.

LEARNING GOAL

Describe the classes of lipids.

15.1 Lipids

Lipids are a family of biomolecules that have the common property of being soluble in organic solvents but not in water. The word *lipid* comes from the Greek word *lipos*, meaning "fat" or "lard." Typically, the lipid content of a cell can be extracted using a nonpolar solvent such as ether or chloroform. Lipids are an important feature in cell membranes and steroid hormones.

Types of Lipids

Within the lipid family, there are specific structures that distinguish the different types of lipids. Lipids such as waxes, fats, oils, and phospholipids are esters that can be hydrolyzed to give fatty acids along with other molecules. Steroids are characterized by the steroid nucleus of four fused carbon rings. They do not contain fatty acids and cannot be hydrolyzed. Figure 15.1 illustrates the types and general structure of lipids we will discuss in this chapter.

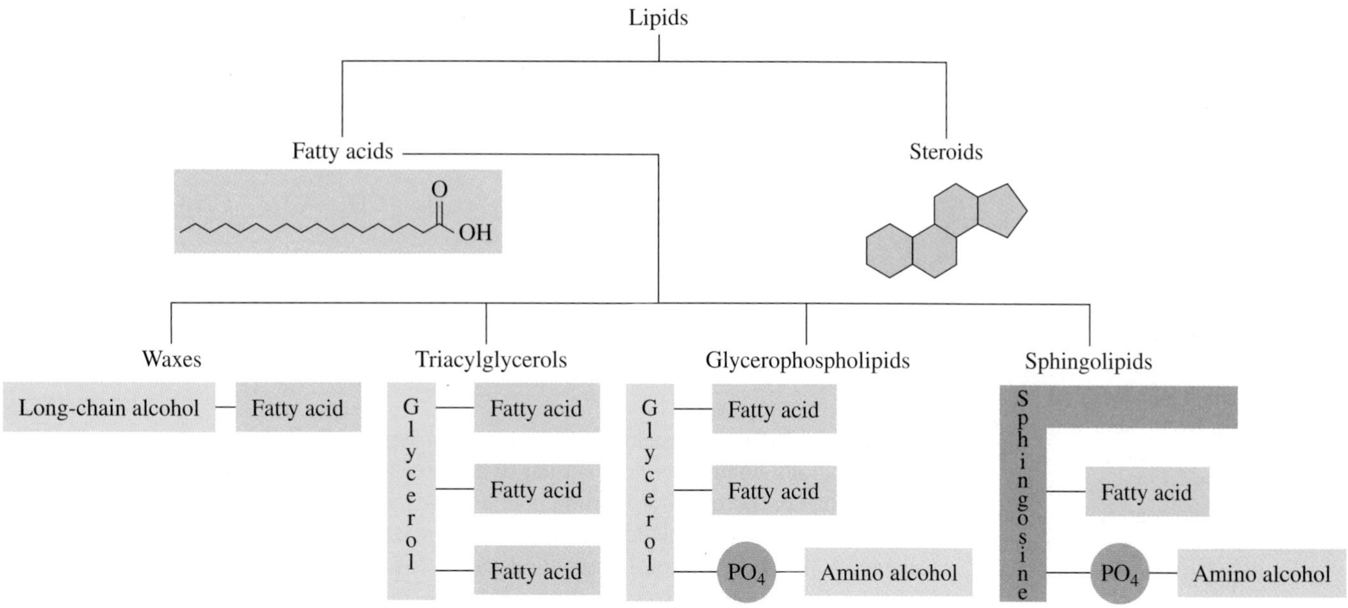

FIGURE 15.1 ▶ Lipids are naturally occurring compounds in cells and tissues, which are soluble in organic solvents but not in water.

Q What chemical property do waxes, triacylglycerols, and steroids have in common?

QUESTIONS AND PROBLEMS

15.1 Lipids

LEARNING GOAL Describe the classes of lipids.

15.1 What are some functions of lipids in the body?

15.2 What are some of the different kinds of lipids?

15.3 Why is grease insoluble in water?

15.4 Which of the following solvents might be used to dissolve an oil stain?

a. water
b. CCl_4
c. diethyl ether
d. benzene
e. NaCl solution

15.2 Fatty Acids

LEARNING GOAL

Draw the condensed structural formula for a fatty acid and identify it as saturated or unsaturated.

CORE CHEMISTRY SKILL

Identifying Fatty Acids

A **fatty acid** contains a long, unbranched carbon chain with a carboxylic acid group at one end. Although the carboxylic acid part is hydrophilic, the long hydrophobic carbon chain makes fatty acids insoluble in water.

Most naturally occurring fatty acids have an even number of carbon atoms, usually between 12 and 20. An example of a fatty acid is lauric acid, a 12-carbon acid found in coconut oil. In a skeletal formula of a fatty acid, the ends and bends of the line are the carbon atoms. The structural formula of lauric acid can be drawn in several forms.

Explore Your World

Solubility of Fats and Oils

Place some water in a small bowl. Add a drop of a vegetable oil. Then add a few more drops of the oil. Record your observations. Now add a few drops of liquid soap and mix. Record your observations.

Place a small amount of fat such as margarine, butter, shortening, or vegetable oil on a dish or plate. Run water over it. Record your observations. Mix some soap with the fat substance and run water over it again. Record your observations.

Questions

1. Do the drops of oil in the water separate or do they come together? Explain.
2. How does the soap affect the oil layer?
3. Why don't the fats on the dish or plate wash off with water?
4. In general, what is the solubility of lipids in water?
5. Why does soap help to wash the fats off the plate?

Drawing Formulas for Lauric Acid

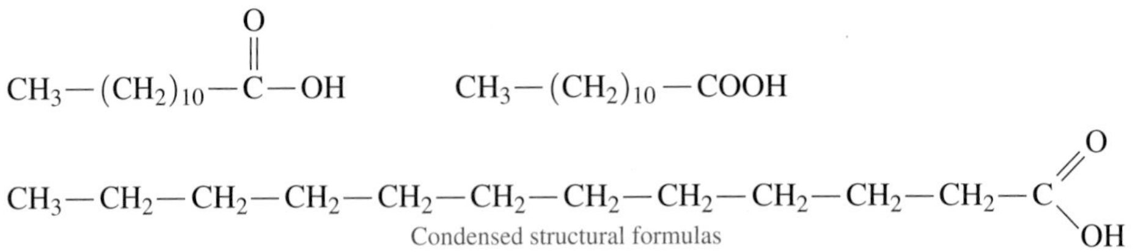

Condensed structural formulas

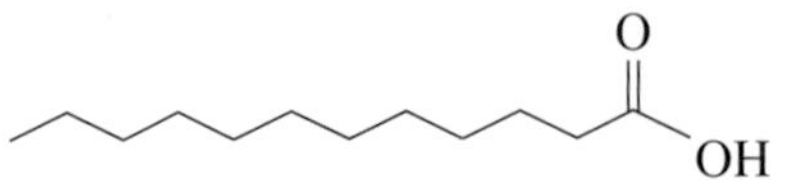

Skeletal formula

A **saturated fatty acid** contains only carbon–carbon single bonds, which make the properties of a long-chain fatty acid similar to those of an alkane. An **unsaturated fatty acid** contains one or more carbon–carbon double bonds. In a **monounsaturated fatty acid**, the long carbon chain has one double bond, which makes its properties similar to those of an alkene. A **polyunsaturated fatty acid** has at least two carbon–carbon double bonds. Table 15.1 lists some of the typical fatty acids in lipids. In the lipids of plants and animals, about half of the fatty acids are saturated and half are unsaturated.

Cis and Trans Isomers of Unsaturated Fatty Acids

Unsaturated fatty acids can be drawn as *cis* and *trans* isomers. For example, oleic acid, a monounsaturated fatty acid found in olives, has one double bond at carbon 9. We can show its cis and trans structures using its skeletal formulas. The cis structure is the more prevalent isomer found in naturally occurring unsaturated fatty acids. In the cis isomer, the carbon chain has a "kink" at the double bond site. As we will see, the cis bond has a major impact on the physical properties and uses of unsaturated fatty acids.

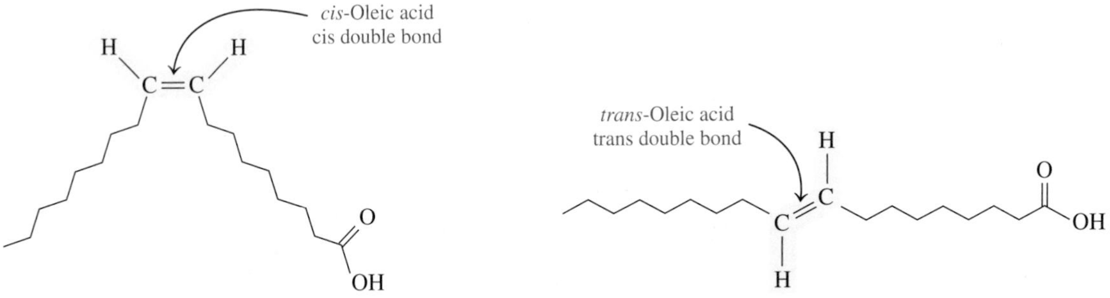

Almost all naturally occurring unsaturated fatty acids have one or more cis double bonds.

The human body is capable of synthesizing some fatty acids from carbohydrates or other fatty acids. However, humans cannot synthesize sufficient amounts of polyunsaturated fatty acids such as linoleic acid, linolenic acid, and arachidonic acid. Because they must be obtained from the diet, they are known as *essential fatty acids*. In infants, a deficiency of essential fatty acids can cause skin dermatitis. However, the role of fatty acids in adult nutrition is not well understood. Adults do not usually have a deficiency of essential fatty acids.

TABLE 15.1 Structures and Melting Points of Common Fatty Acids

Name	Carbon Atoms: Double Bonds	Present in	Melting Point (°C)	Structures
Saturated Fatty Acids				
Lauric acid	12:0	Coconut	44	$CH_3{-}(CH_2)_{10}{-}COOH$ 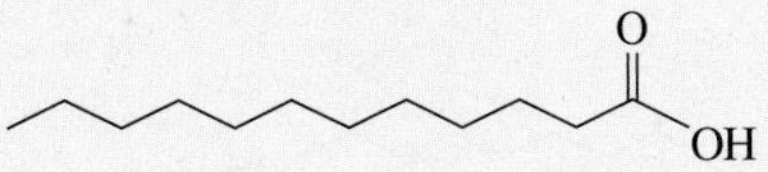
Myristic acid	14:0	Nutmeg	55	$CH_3{-}(CH_2)_{12}{-}COOH$ 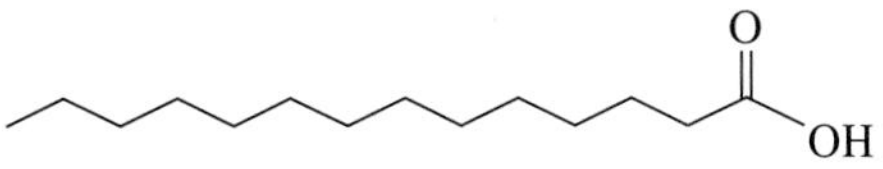
Palmitic acid	16:0	Palm	63	$CH_3{-}(CH_2)_{14}{-}COOH$ 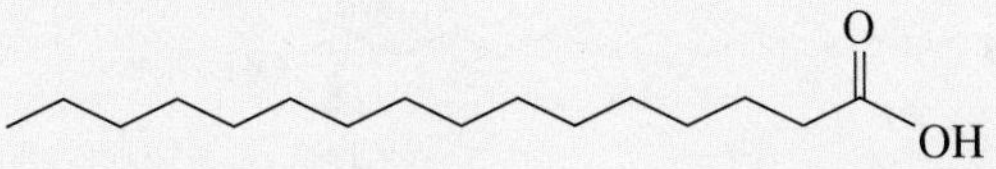
Stearic acid	18:0	Animal fat	69	$CH_3{-}(CH_2)_{16}{-}COOH$ 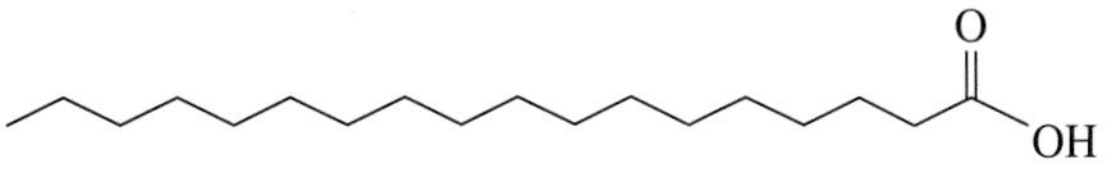
Monounsaturated Fatty Acids				
Palmitoleic acid	16:1	Butter	0	$CH_3{-}(CH_2)_5{-}CH{=}CH{-}(CH_2)_7{-}COOH$ 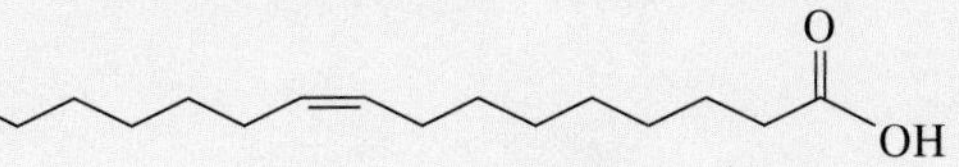
Oleic acid	18:1	Olives, pecan, grapeseed	14	$CH_3{-}(CH_2)_7{-}CH{=}CH{-}(CH_2)_7{-}COOH$ 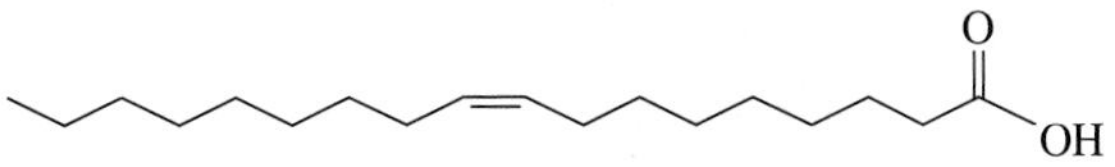
Polyunsaturated Fatty Acids				
Linoleic acid	18:2	Soybean, safflower, sunflower	−5	$CH_3{-}(CH_2)_4{-}CH{=}CH{-}CH_2{-}CH{=}CH{-}(CH_2)_7{-}COOH$ 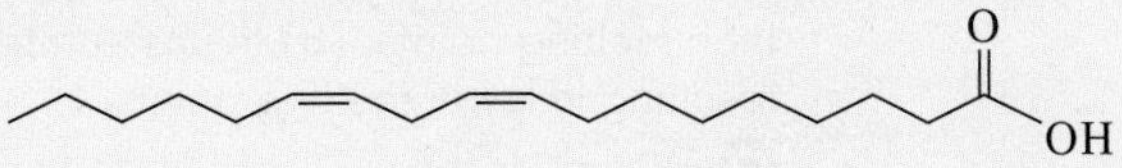
Linolenic acid	18:3	Corn	−11	$CH_3{-}CH_2{-}CH{=}CH{-}CH_2{-}CH{=}CH{-}CH_2{-}CH{=}CH{-}(CH_2)_7{-}COOH$ 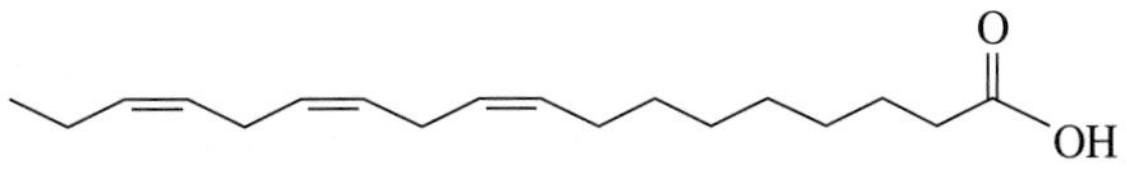
Arachidonic acid	20:4	Meat, eggs, fish	−50	$CH_3{-}(CH_2)_3{-}(CH_2{-}CH{=}CH)_4{-}(CH_2)_3{-}COOH$ 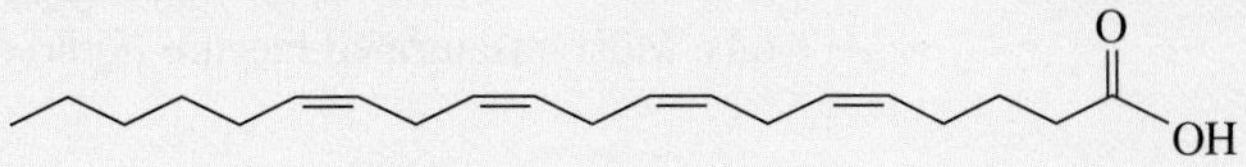

Physical Properties of Fatty Acids

The saturated fatty acids fit closely together in a regular pattern, which allows many dispersion forces between the carbon chains. These normally weak intermolecular forces of attraction can be significant when molecules of fatty acids are close together. As a result, a significant amount of energy and high temperatures are required to separate the fatty acids and melt the fatty acid. As the length of the carbon chain increases, more interactions occur between the carbon chains, requiring higher melting points. Saturated fatty acids are usually solids at room temperature (see Figure 15.2).

In unsaturated fatty acids, the cis double bonds cause the carbon chain to bend or kink, giving the molecules an irregular shape. As a result, unsaturated fatty acids cannot stack as closely as saturated fatty acids, and thus have fewer interactions between carbon

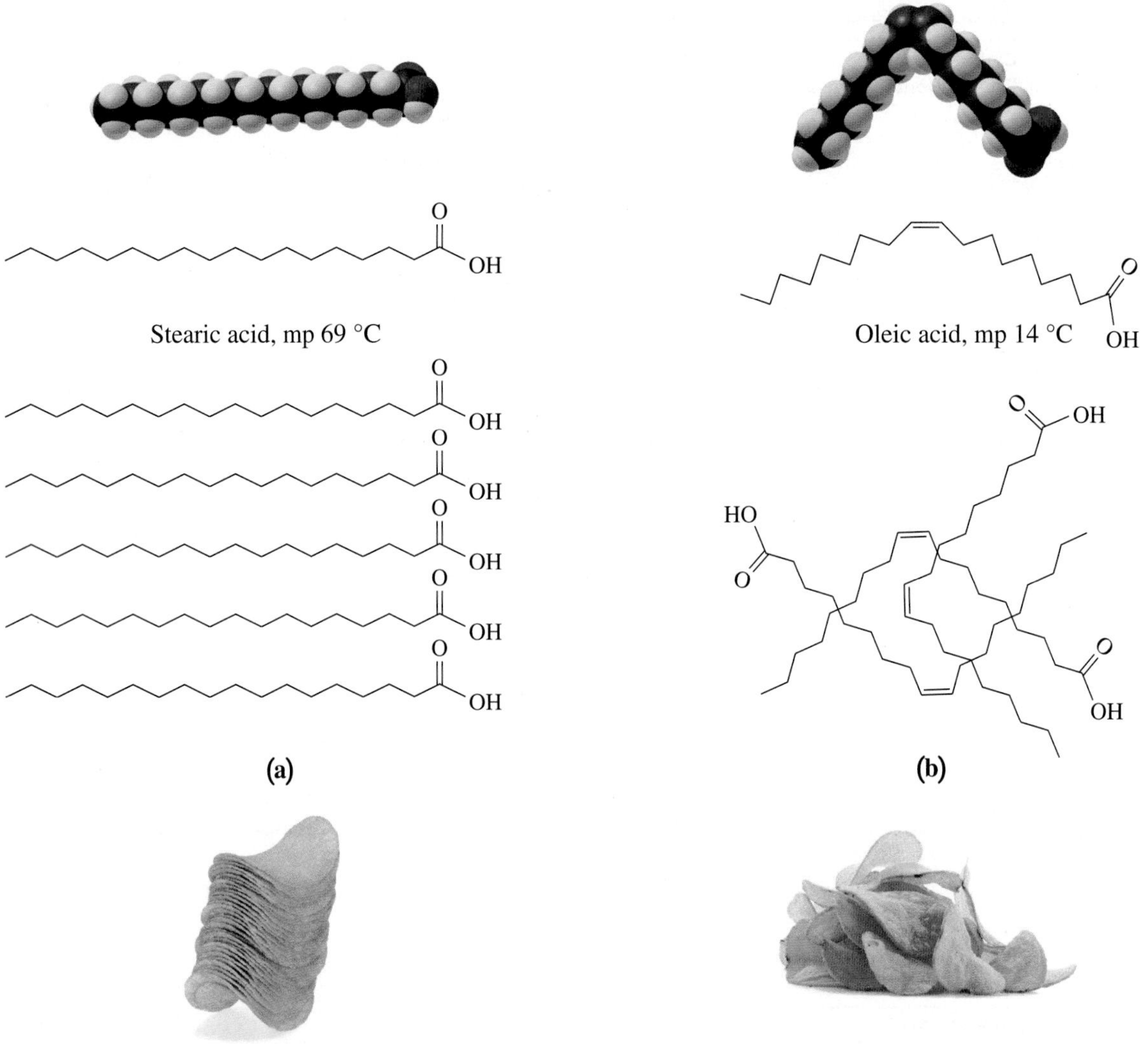

FIGURE 15.2 ▶ **(a)** In saturated fatty acids, the molecules fit closely together to give high melting points. **(b)** In unsaturated fatty acids, molecules cannot fit closely together, resulting in lower melting points.

Q Why does the cis double bond affect the melting points of unsaturated fatty acids?

chains. We might think of saturated fatty acids as chips with regular shapes that stack closely together in a can. Similarly, irregularly shaped chips would be like unsaturated fatty acids that do not fit closely together. Consequently, less energy is required to separate the molecules, making the melting points of unsaturated fats lower than those of saturated fats. Most unsaturated fats are liquid oils at room temperature.

SAMPLE PROBLEM 15.1 Structures and Properties of Fatty Acids

Consider the condensed structural formula for oleic acid.

$$CH_3-(CH_2)_7-CH{=}CH-(CH_2)_7-\overset{\displaystyle O}{\overset{\|}{C}}-OH$$

a. Why is this substance an acid?
b. How many carbon atoms are in oleic acid?
c. Is the fatty acid saturated, monounsaturated, or polyunsaturated?
d. Is it most likely to be solid or liquid at room temperature?
e. Would it be soluble in water?

SOLUTION

a. Oleic acid contains a carboxylic acid group.
b. It contains 18 carbon atoms.
c. It is a monounsaturated fatty acid.
d. It is liquid at room temperature.
e. No. Its long hydrocarbon chain makes it insoluble in water.

STUDY CHECK 15.1

Palmitoleic acid is a fatty acid with the following condensed structural formula:

$$CH_3-(CH_2)_5-CH=CH-(CH_2)_7-\overset{\overset{\displaystyle O}{\|}}{C}-OH$$

a. How many carbon atoms are in palmitoleic acid?
b. Is the fatty acid saturated, monounsaturated, or polyunsaturated?
c. Is it most likely to be solid or liquid at room temperature?

Prostaglandins

Prostaglandins are hormone-like substances produced in small amounts in most cells of the body. The prostaglandins, also known as *eicosanoids*, are formed from arachidonic acid, the polyunsaturated fatty acid with 20 carbon atoms (*eicos* is the Greek word for 20). Swedish chemists first discovered prostaglandins and named them "prostaglandin E" (soluble in ether) and "prostaglandin F" (soluble in phosphate buffer or *fosfat* in Swedish). The various kinds of prostaglandins differ by the substituents attached to the five-carbon ring. Prostaglandin E (PGE) has a ketone group on carbon 9, whereas prostaglandin F (PGF) has a hydroxyl group. The number of double bonds is shown as a subscript 1 or 2.

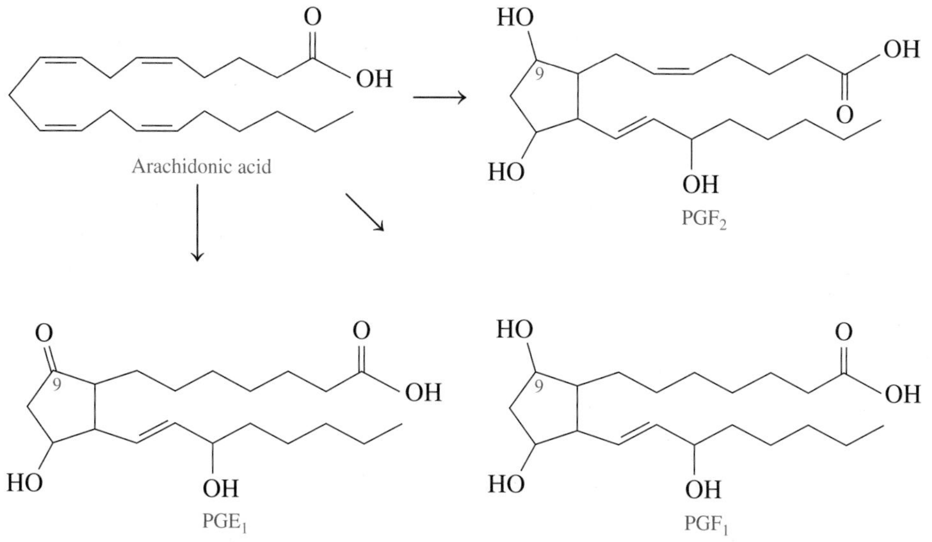

Although prostaglandins are broken down quickly, they have potent physiological effects. Some prostaglandins increase blood pressure, and others lower blood pressure. Other prostaglandins stimulate contraction and relaxation in the smooth muscle of the uterus. When tissues are injured, arachidonic acid is converted to prostaglandins that produce inflammation and pain in the area.

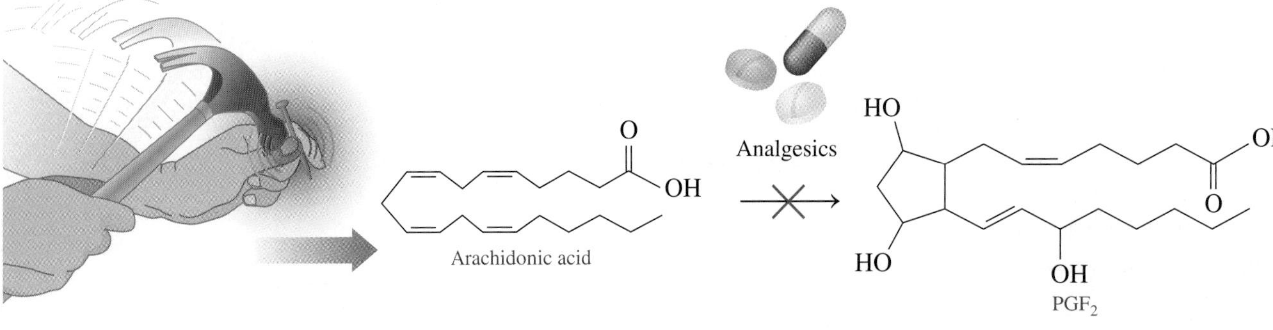

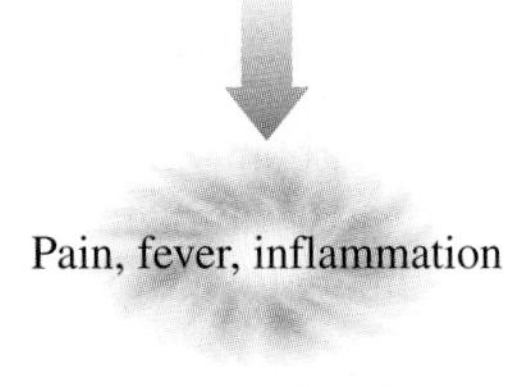

When tissues are injured, prostaglandins are produced, which cause pain and inflammation.

The treatment of pain, fever, and inflammation is based on inhibiting the enzymes that convert arachidonic acid to prostaglandins. Several nonsteroidal anti-inflammatory drugs (NSAIDs), such as aspirin, block the production of prostaglandins and in doing so decrease pain and inflammation and reduce fever (antipyretics). Ibuprofen has similar anti-inflammatory and analgesic effects. Other NSAIDs include naproxen (Aleve and Naprosyn), ketoprofen (Actron), and nabumetone (Relafen). Although NSAIDs are helpful, their long-term use can result in liver, kidney, and gastrointestinal damage.

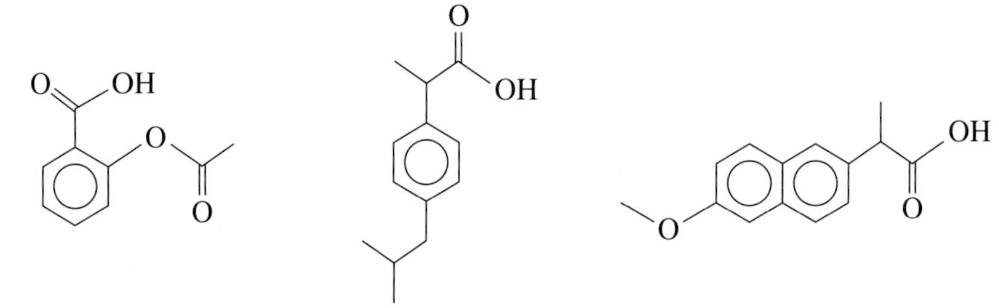

Aspirin — Ibuprofen (Advil, Motrin) — Naproxen (Aleve, Naprosyn)

QUESTIONS AND PROBLEMS

15.2 Fatty Acids

LEARNING GOAL Draw the condensed structural formula for a fatty acid and identify it as saturated or unsaturated.

15.5 Describe some similarities and differences in the structures of a saturated fatty acid and an unsaturated fatty acid.

15.6 Stearic acid and linoleic acid are both fatty acids 18 carbon atoms each. Why do they have a large difference in their melting points?

15.7 Draw the skeletal formula for each of the following fatty acids:
a. lauric acid **b.** myristic acid

15.8 Draw the skeletal formula for each of the following fatty acids:
a. stearic acid **b.** linoleic acid

15.9 For each of the following fatty acids, give the shorthand notation for the number of carbon atoms and double bonds, and classify as saturated, monounsaturated, or polyunsaturated:
a. palmitic acid **b.** oleic acid
c. arachidonic acid **d.** linoleic acid

15.10 For each of the following fatty acids, give the shorthand notation for the number of carbon atoms and double bonds, and classify as saturated, monounsaturated, or polyunsaturated:
a. linoleic acid **b.** palmitic acid
c. myristic acid **d.** oleic acid

15.11 How does the structure of a fatty acid with a cis double bond differ from the structure of a fatty acid with a trans double bond?

15.12 How does the double bond influence the dispersion forces that can form between the hydrocarbon chains of fatty acids?

15.13 What is the difference in the location of the first double bond in an omega-3 and an omega-6 fatty acid (see Chemistry Link to Health "Omega-3 Fatty Acids in Fish Oils")?

15.14 What are the fatty acids found in fish oil and vegetable oil?

15.15 Compare the structures and functional groups of arachidonic acid and prostaglandin PGF_2.

15.16 Compare the structures and functional groups of PGF_1 and PGF_2.

15.17 What are some effects of prostaglandins in the body?

15.18 Explain the action of an anti-inflammatory drug in the body.

Chemistry Link to **Health**

Omega-3 Fatty Acids in Fish Oils

Because unsaturated fats are now recognized as being more beneficial to health than saturated fats, American diets have changed to include more unsaturated fats and less saturated fatty acids. This change is a response to research that indicates that atherosclerosis and heart disease are associated with high levels of fats in the diet. However, the Inuit people of Alaska have a diet with high levels of unsaturated fats as well as high levels of blood cholesterol, but a very low occurrence of atherosclerosis and heart attacks. The fats in the Inuit diet are primarily unsaturated fats from fish, rather than from land animals.

Both fish and vegetable oils have high levels of unsaturated fats. The fatty acids in vegetable oils are *omega-6 acids*, in which the first double bond occurs at carbon 6 counting from the methyl end of the carbon chain. Omega is the last letter in the Greek alphabet and is used to denote the end. Two common omega-6 acids are linoleic acid (LA) and arachidonic acid (AA). However, the fatty acids in fish oils are mostly the omega-3 type, in which the first double bond occurs at the third carbon counting from the methyl group. Three common *omega-3 fatty acids* in fish are linolenic acid (ALA), eicosapentaenoic acid (EPA), and docosahexaenoic acid (DHA).

In atherosclerosis and heart disease, cholesterol forms plaques that adhere to the walls of the blood vessels. Blood pressure rises as blood has to squeeze through a smaller opening in the blood vessel. As more plaque forms, there is also a possibility of blood clots blocking the blood vessels and causing a heart attack. Omega-3 fatty acids lower the tendency of blood platelets to stick together, thereby reducing the possibility of blood clots. However, high levels of omega-3 fatty acids can increase bleeding if the ability of the platelets to form blood clots is reduced too much. It does seem that a diet that includes fish such as salmon, tuna, and herring can provide higher amounts of the omega-3 fatty acids, which help lessen the possibility of developing heart disease.

Cold water fish are a good source of omega-3 fatty acids.

Omega-6 Fatty Acids

Linoleic acid (LA)

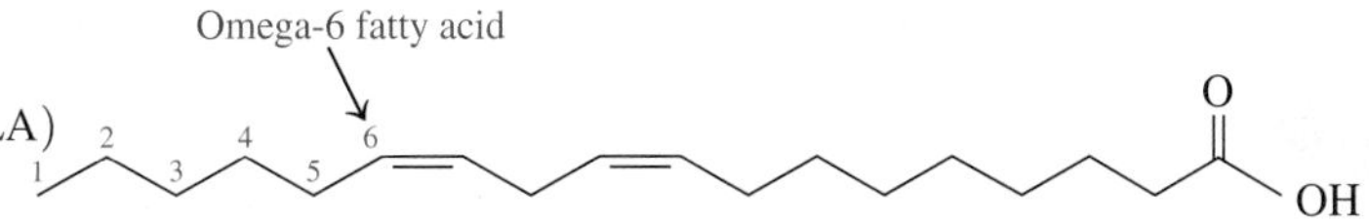

Arachidonic acid (AA)

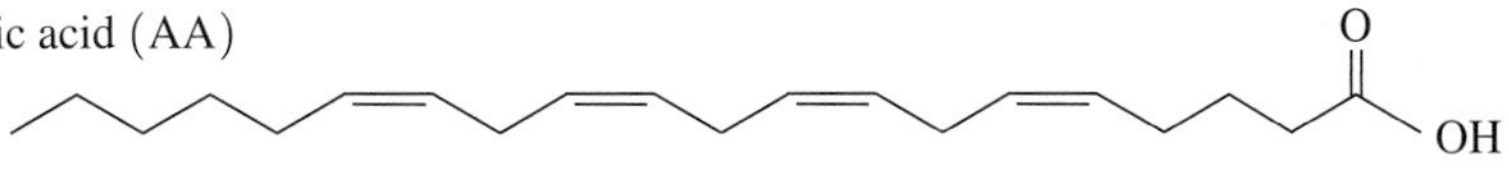

Omega-3 Fatty Acids

Linolenic acid (ALA)

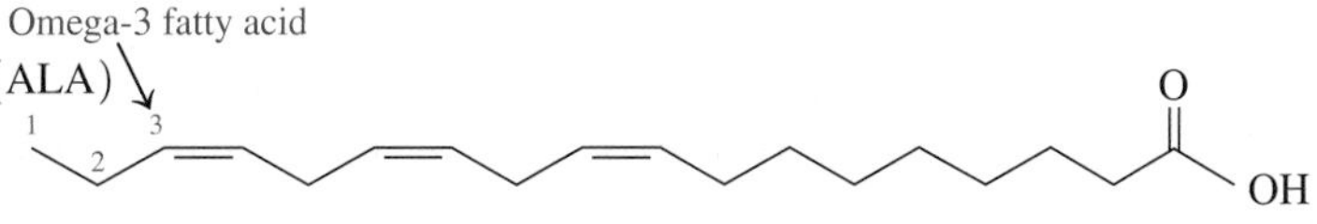

Eicosapentaenoic acid (EPA)

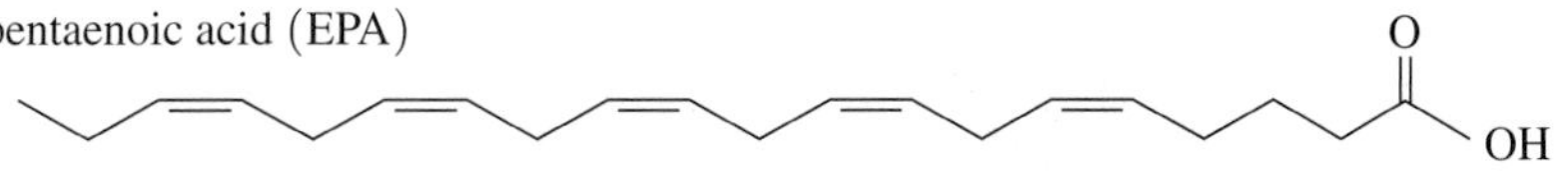

Docosahexaenoic acid (DHA)

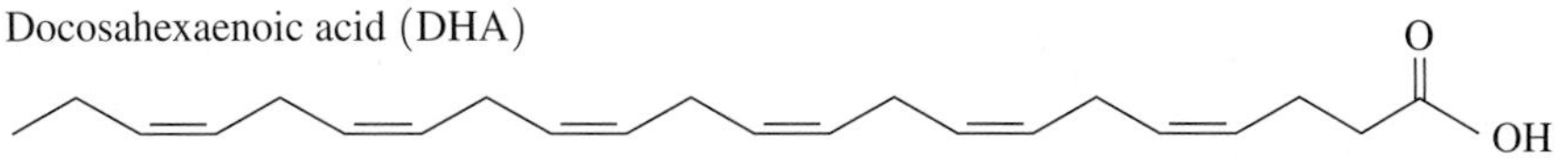

LEARNING GOAL

Draw the condensed structural formula for a wax or triacylglycerol produced by the reaction of a fatty acid and an alcohol or glycerol.

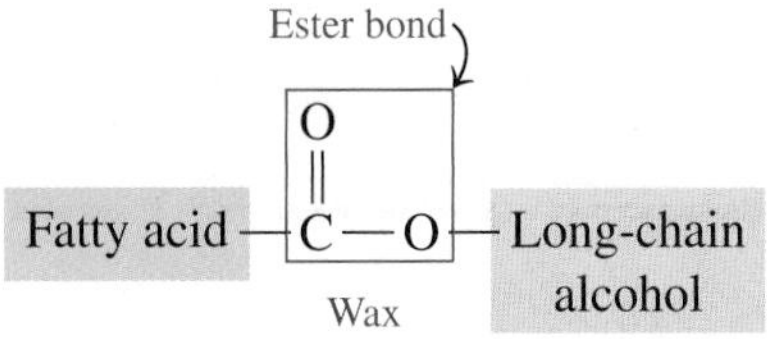

Waxes are esters of long-chain alcohols and fatty acids.

Jojoba wax is obtained from the seed of the jojoba bush.

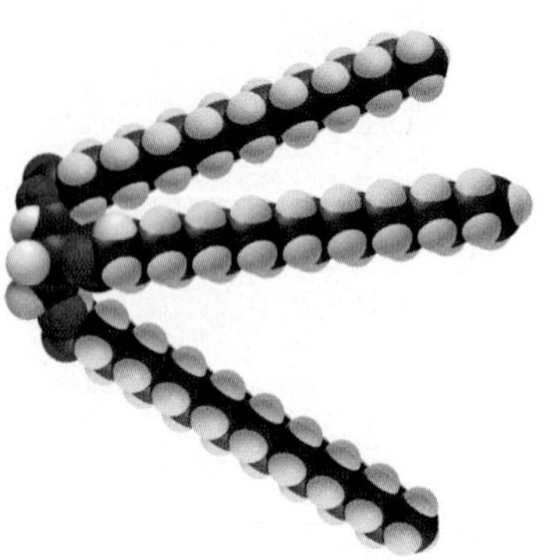

Tristearin consists of glycerol with three ester bonds to stearic acid molecules.

15.3 Waxes and Triacylglycerols

Waxes are found in many plants and animals. Natural waxes are found on the surface of fruits, and on the leaves and stems of plants where they help prevent loss of water and damage from pests. Waxes on the skin, fur, and feathers of animals provide a waterproof coating. A **wax** is an ester of a saturated fatty acid and a long-chain alcohol, each containing from 14 to 30 carbon atoms.

The formulas of some common waxes are given in Table 15.2. Beeswax obtained from honeycombs and carnauba wax from palm trees are used to give a protective coating to furniture, cars, and floors. Jojoba wax is used in making candles and cosmetics such as lipstick. Lanolin, a mixture of waxes obtained from wool, is used in hand and facial lotions to aid retention of water, softening the skin.

TABLE 15.2 Some Typical Waxes

Type	Condensed Structural Formula	Source	Uses
Beeswax	$CH_3-(CH_2)_{14}-C(=O)-O-(CH_2)_{29}-CH_3$	Honeycomb	Candles, shoe polish, wax paper
Carnauba wax	$CH_3-(CH_2)_{24}-C(=O)-O-(CH_2)_{29}-CH_3$	Brazilian palm tree	Waxes for furniture, cars, floors, shoes
Jojoba wax	$CH_3-(CH_2)_{18}-C(=O)-O-(CH_2)_{19}-CH_3$	Jojoba bush	Candles, soaps, cosmetics

Triacylglycerols

In the body, fatty acids are stored as **triacylglycerols**, also called *triglycerides*, which are triesters of glycerol (a trihydroxy alcohol) and fatty acids. The general formula of a triacylglycerol follows:

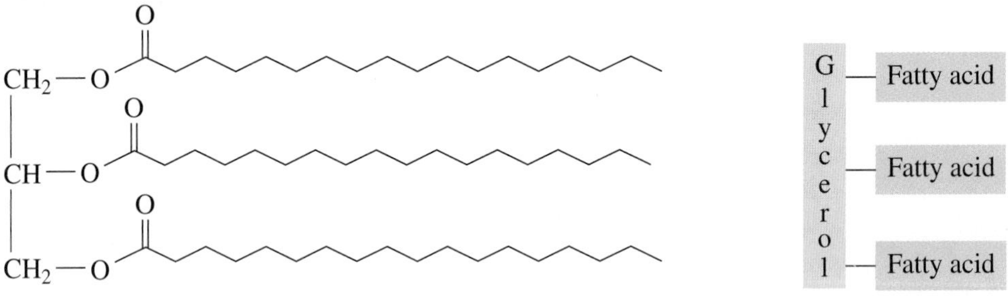

Triacylglycerol

In a triacylglycerol, three hydroxyl groups of glycerol form ester bonds with the carboxyl groups of three fatty acids. For example, glycerol and three molecules of stearic acid form a triacylglycerol. In the name, glycerol is named *glyceryl* and the fatty acids are named as carboxylates. For example, stearic acid is named as stearate, which gives the name glyceryl tristearate. The common name of this compound is tristearin.

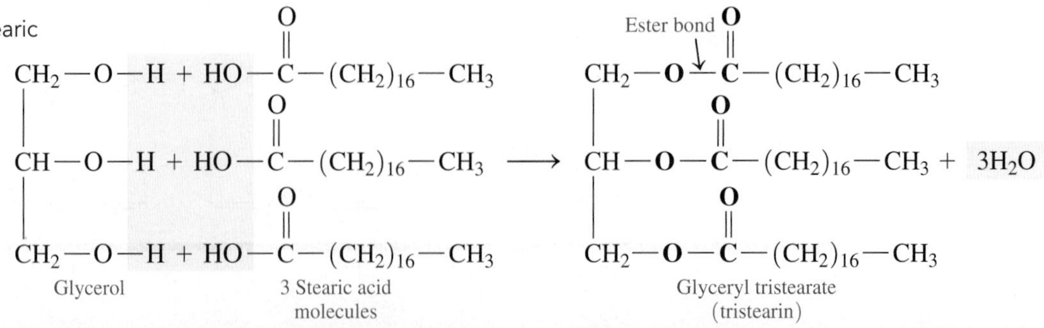

Most naturally occurring triacylglycerols contain glycerol bonded to two or three different fatty acids, typically palmitic acid, oleic acid, linoleic acid, and stearic acid. For example, a mixed triacylglycerol might be made from stearic acid, oleic acid, and palmitic acid. One possible structure for this mixed triacylglycerol follows:

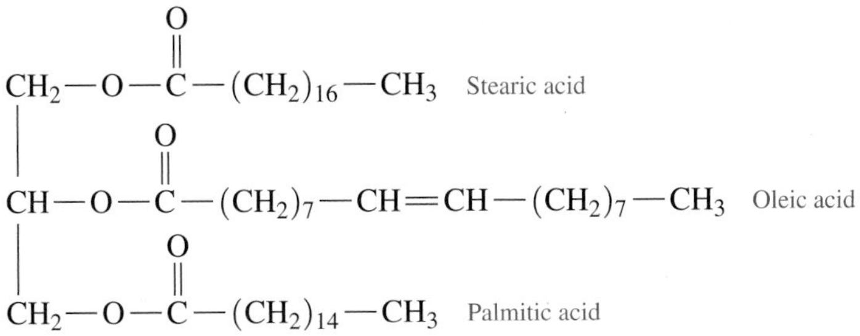

A mixed triacylglycerol

Triacylglycerols are the major form of energy storage for animals. Animals that hibernate eat large quantities of plants, seeds, and nuts that are high in calories. Prior to hibernation, these animals, such as polar bears, gain as much as 14 kg a week. As the external temperature drops, the animal goes into hibernation. The body temperature drops to nearly freezing, and cellular activity, respiration, and heart rate are drastically reduced. Animals that live in extremely cold climates hibernate for 4 to 7 months. During this time, stored fat is the only source of energy.

CORE CHEMISTRY SKILL

Drawing Structures for Triacylglycerols

Prior to hibernation, a polar bear eats food with a high caloric content.

SAMPLE PROBLEM 15.2 Drawing the Structure for a Triacylglycerol

Draw the condensed structural formula for glyceryl tripalmitoleate (tripalmitolein), which is used in cosmetic creams and lotions.

SOLUTION

Glyceryl tripalmitoleate (tripalmitolein) is the triacylglycerol that contains ester bonds between glycerol and three palmitoleic acid molecules.

	Name of Lipid	Type of Lipid	Type of Alcohol	Fatty Acids	Type of Bonds
ANALYZE THE PROBLEM	glyceryl tripalmitoleate (tripalmitolein)	triacylglycerol	glycerol	three palmitoleic acid (16:1)	ester

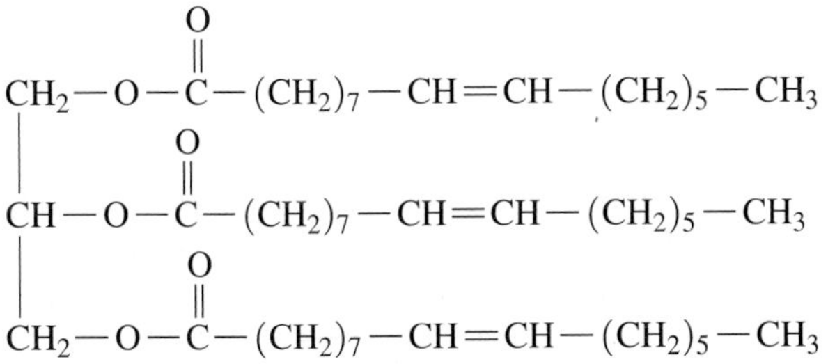
Glyceryl tripalmitoleate (tripalmitolein)

STUDY CHECK 15.2

Draw the condensed structural formula for the triacylglycerol containing three molecules of myristic acid (14:0).

Triacylglycerols are used to thicken creams and lotions.

Melting Points of Fats and Oils

A **fat** is a triacylglycerol that is solid at room temperature, and it usually comes from animal sources such as meat, whole milk, butter, and cheese.

An **oil** is a triacylglycerol that is usually a liquid at room temperature and is obtained from a plant source. Olive oil and peanut oil are monounsaturated because they contain large amounts of oleic acid. Oils from corn, cottonseed, safflower seed, and sunflower seed are polyunsaturated because they contain large amounts of fatty acids with two or more double bonds (see Figure 15.3).

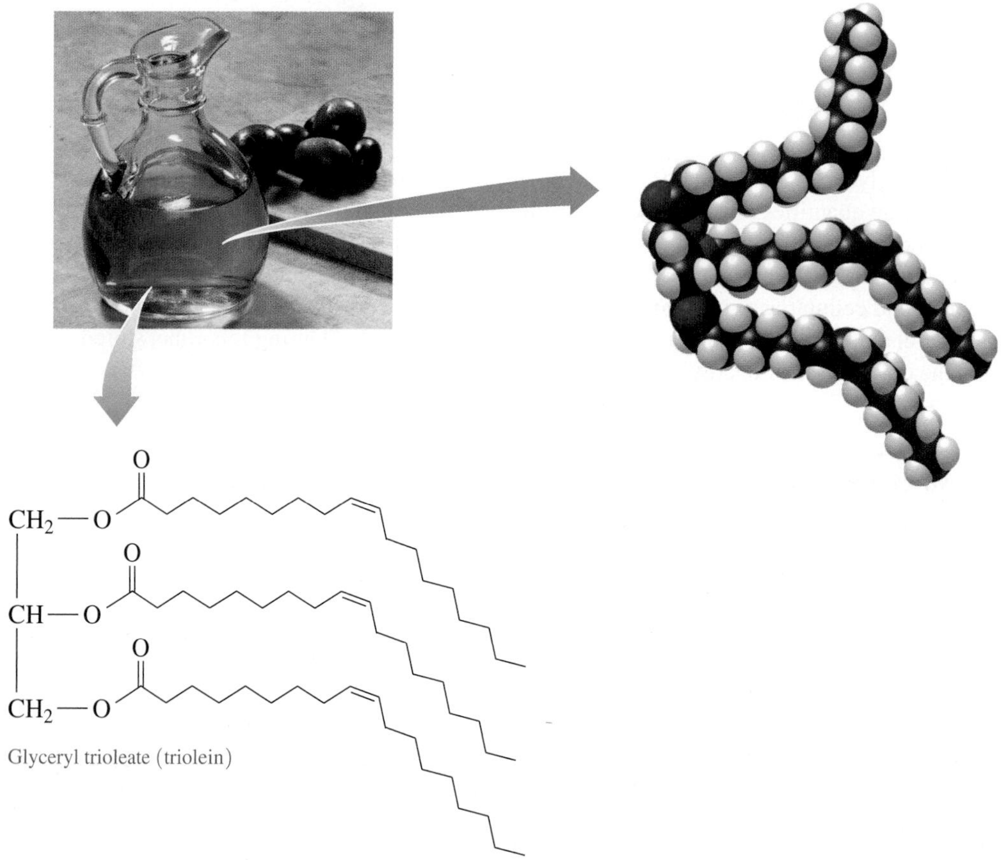

Glyceryl trioleate (triolein)

FIGURE 15.3 ▶ Vegetable oils such as olive oil, corn oil, and safflower oil contain unsaturated fats.

Q Why is olive oil a liquid at room temperature?

Palm oil and coconut oil are solids at room temperature because they consist mostly of saturated fatty acids. Coconut oil is 92% saturated fats, about half of which is lauric acid, which contains 12 carbon atoms rather than 18 carbon atoms found in the stearic acid of animal sources. Thus coconut oil has a melting point that is higher than typical vegetable oils, but not as high as fats from animal sources that contain stearic acid. The amounts of saturated, monounsaturated, and polyunsaturated fatty acids in some typical fats and oils are listed in Figure 15.4.

Saturated fatty acids have higher melting points than unsaturated fatty acids because they pack together more tightly. Animal fats usually contain more saturated fatty acids than do vegetable oils. Therefore the melting points of animal fats are higher than those of vegetable oils.

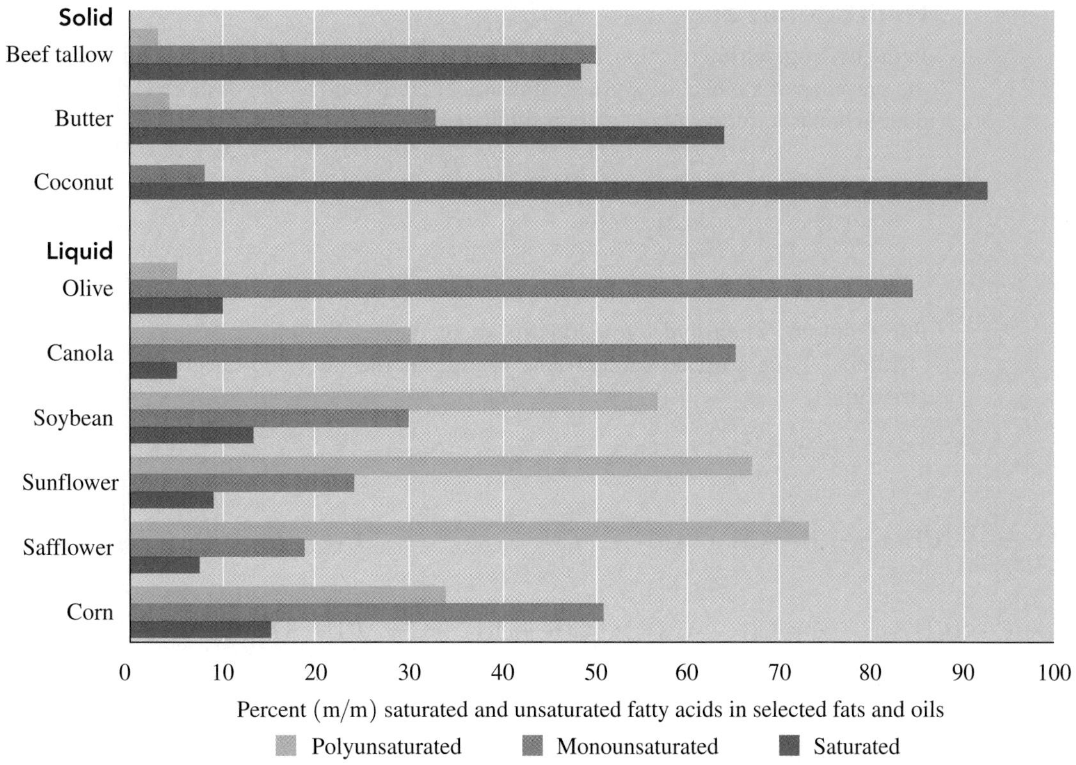

FIGURE 15.4 ▶ Vegetable oils are liquids at room temperature because they have a higher percentage of unsaturated fatty acids than do animal fats.

Q Why is butter a solid at room temperature, whereas canola oil is a liquid?

QUESTIONS AND PROBLEMS

15.3 Waxes and Triacylglycerols

LEARNING GOAL Draw the condensed structural formula for a wax or triacylglycerol produced by the reaction of a fatty acid and an alcohol or glycerol.

15.19 Draw the condensed structural formula for the ester in beeswax that is formed from myricyl alcohol, $CH_3—(CH_2)_{29}—OH$, and palmitic acid.

15.20 Draw the condensed structural formula for the ester in carnauba wax that is formed from cerotic acid, a 24-carbon saturated fatty acid and myricyl alcohol $CH_3—(CH_2)_{29}—OH$.

15.21 Draw the condensed structural formula for a triacylglycerol that contains stearic acid and glycerol.

15.22 Draw the skeletal formula for a mixed triacylglycerol that contains three palmitic acid molecules.

15.23 Capric acid is a 10-carbon saturated fatty acid. Draw the condensed structural formula for glyceryl tricaprylate (tricaprylin).

15.24 Lauric acid is a 10-carbon unsaturated fatty acid with two double bonds. Draw the condensed structural formula for glyceryl trilaureate.

15.25 How do unsaturations in oil affect its melting point?

15.26 Olive oil is monounsaturated, whereas butter fat is saturated. Why does olive oil have a lower melting point than butter fat?

15.4 Chemical Properties of Triacylglycerols

LEARNING GOAL

Draw the condensed structural formula for the product of a triacylglycerol that undergoes hydrogenation, hydrolysis, or saponification.

The chemical reactions of the triacylglycerols (fats and oils) are similar to the hydrogenation of alkenes and the hydrolysis and saponification of esters. The double bonds in unsaturated fatty acids of a triacylglycerol will react with hydrogen to produce saturated (single bonds) fatty acids.

Hydrogenation

In the **hydrogenation** reaction, hydrogen gas is bubbled through the heated oil typically in the presence of a nickel catalyst. As a result, H atoms add to one or more carbon–carbon double bonds to form carbon–carbon single bonds.

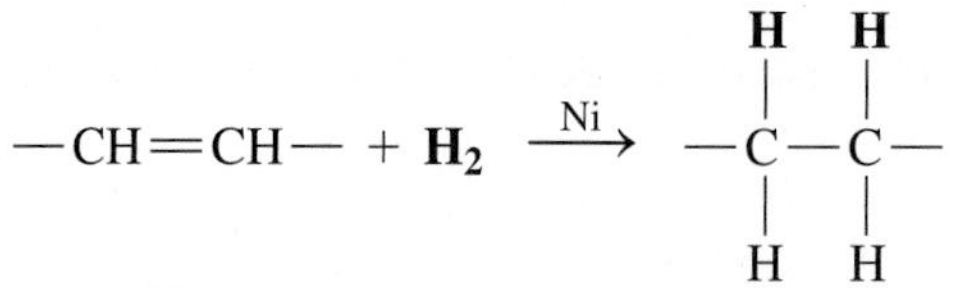

For example, when hydrogen adds to all of the double bonds of glyceryl trioleate (triolein) using a nickel catalyst, the product is the saturated fat glyceryl tristearate (tristearin).

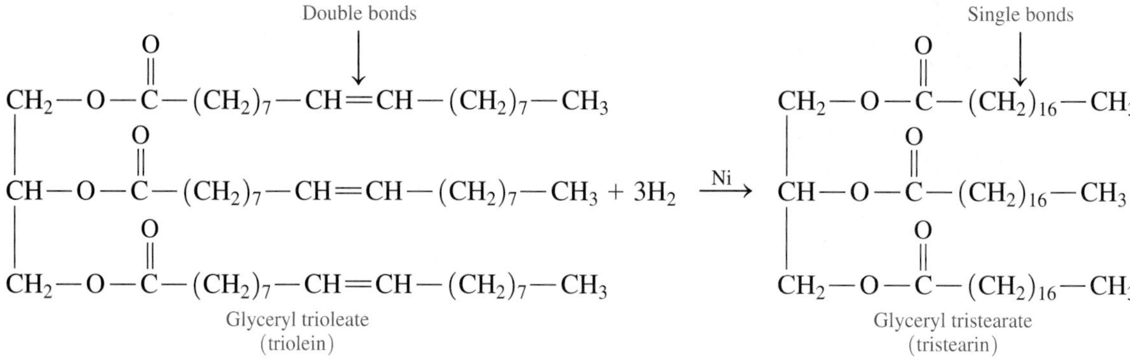

CORE CHEMISTRY SKILL

Identifying the Products for the Hydrogenation, Hydrolysis, and Saponification of a Triacylglycerol

In commercial hydrogenation, the addition of hydrogen is stopped before all the double bonds in a liquid vegetable oil become completely saturated. Complete hydrogenation gives a very brittle product, whereas the partial hydrogenation of a liquid vegetable oil changes it to a soft, semisolid fat. As it becomes more saturated, the melting point increases, and the substance becomes more solid at room temperature. By controlling the amount of hydrogen, manufacturers can produce various types of products such as soft margarines, solid stick margarines, and solid shortenings (see Figure 15.5). Although these products now contain more saturated fatty acids than the original oils, they contain no cholesterol, unlike similar products from animal sources, such as butter and lard.

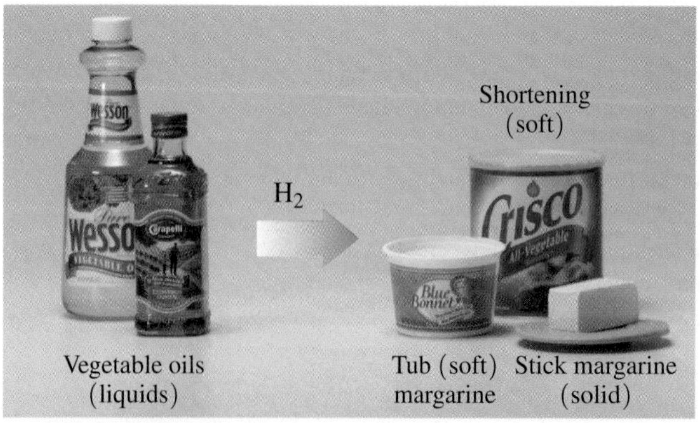

FIGURE 15.5 ▶ Many soft margarines, stick margarines, and solid shortenings are produced by the partial hydrogenation of vegetable oils.

Q How does hydrogenation change the structure of the fatty acids in the vegetable oils?

Chemistry Link to Health

Trans Fatty Acids and Hydrogenation

During the early 1900s, margarine became a popular replacement for the highly saturated fats such as butter and lard. Margarine is produced by partially hydrogenating the unsaturated fats in vegetable oils such as safflower oil, corn oil, canola oil, cottonseed oil, and sunflower oil.

Hydrogenation and Trans Fats

In vegetable oils, the unsaturated fats usually contain cis double bonds. As hydrogenation occurs, double bonds are converted to single bonds. However, a small amount of the cis double bonds are converted to trans double bonds because they are more stable, which causes a change in the overall structure of the fatty acids. If the label on a product states that the oils have been "partially hydrogenated," that product will also contain trans fatty acids. In the United States, it is estimated that 2 to 4% of our total calories comes from trans fatty acids.

The concern about trans fatty acids is that their altered structure may make them behave like saturated fatty acids in the body. In the 1980s, research indicated that trans fatty acids have an effect on blood cholesterol similar to that of saturated fats, although study results vary. Several studies reported that trans fatty acids raise the levels of LDL-cholesterol, and lower the levels of HDL-cholesterol. (LDL and HDL are described in section 15.6.)

Foods containing naturally occurring trans fat include milk, beef, and eggs. Foods that contain trans fatty acids from the hydrogenation process include deep-fried foods, bread, baked goods, cookies, crackers, chips, stick and soft margarines, and vegetable shortening. The American Heart Association recommends that margarine should have no more than 2 g of saturated fat per tablespoon and a liquid vegetable oil should be the first ingredient. They also recommend the use of soft margarine, which is only slightly hydrogenated and therefore has fewer trans fatty acids. Currently, the amount of trans fats is included on the Nutrition Facts label on food products. In the United States, a food label for a product that has less than 0.5 g of trans fat in one serving can read "0 g of trans fat."

The best advice may be to reduce total fat in the diet by using fats and oils sparingly, cooking with little or no fat, substituting olive oil or canola oil for other oils, and limiting the use of coconut oil and palm oil, which are high in saturated fatty acids.

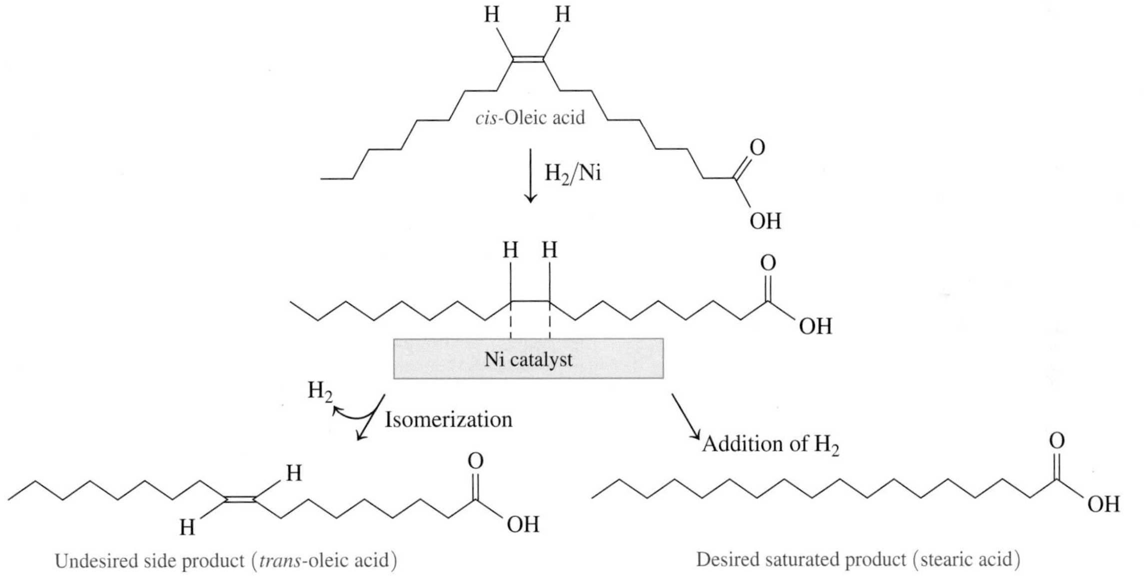

Hydrolysis

Triacylglycerols are hydrolyzed (split by water) in the presence of strong acids such as HCl or H_2SO_4, or digestive enzymes called *lipases*. The products of hydrolysis of the ester bonds are glycerol and three fatty acids. The polar glycerol is soluble in water, but the fatty acids with their long hydrocarbon chains are not.

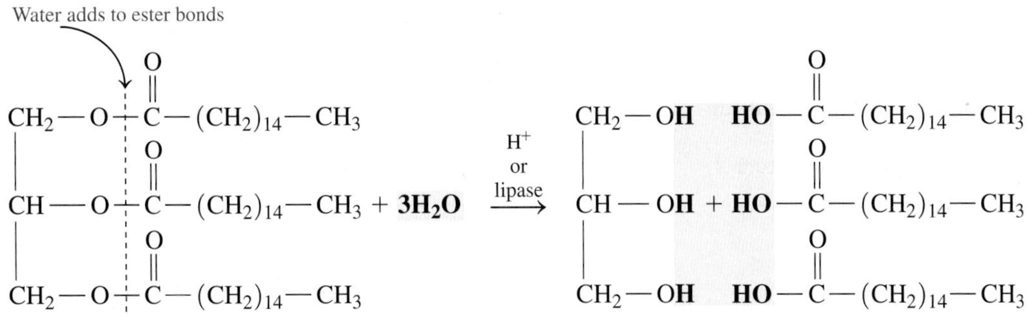

Glyceryl tripalmitate (tripalmitin) — Glycerol — 3 Palmitic acid molecules

Saponification

Saponification occurs when a fat is heated with a strong base such as NaOH to form glycerol and the sodium salts of the fatty acids, which is soap. When NaOH is used, a solid soap is produced that can be molded into a desired shape; KOH produces a softer, liquid soap. An oil that is polyunsaturated produces a softer soap. Names like "coconut" or "avocado shampoo" tell you the sources of the oil used in the reaction.

$$\text{Fat or oil} + \text{strong base} \xrightarrow{\text{Heat}} \text{glycerol} + \text{salts of fatty acids (soap)}$$

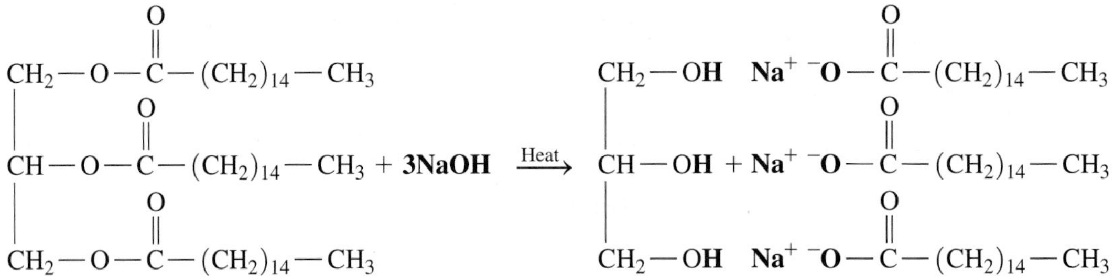

Glyceryl tripalmitate (tripalmitin) — Glycerol — 3 Sodium palmitate (soap)

The reactions for fatty acids and triacylglycerols that are similar to the reactions of hydrogenation of alkenes, esterification, hydrolysis, and saponification are summarized in Table 15.3.

TABLE 15.3 Summary of Organic and Lipid Reactions

Reaction	Organic Reactants and Products	Lipid Reactants and Products
Esterification	Carboxylic acid + alcohol $\xrightarrow{H^+, \text{heat}}$ ester + water	3 Fatty acids + glycerol $\xrightarrow{\text{Enzyme}}$ triacylglycerol (fat) + 3 water
Hydrogenation	Alkene (double bond) + hydrogen $\xrightarrow{\text{Pt}}$ alkane (single bonds)	Unsaturated fat (double bonds) + hydrogen $\xrightarrow{\text{Ni}}$ saturated fat (single bonds)
Hydrolysis	Ester + water $\xrightarrow{H^+, \text{heat}}$ carboxylic acid + alcohol	Triacylglycerol (fat) + 3 water $\xrightarrow{\text{Enzyme}}$ 3 fatty acids + glycerol
Saponification	Ester + sodium hydroxide $\xrightarrow{\text{Heat}}$ sodium salt of carboxylic acid + alcohol	Triacylglycerol (fat) + 3 sodium hydroxide $\xrightarrow{\text{Heat}}$ 3 sodium salts of fatty acid (soaps) + glycerol

Explore Your World

Types of Fats

Read the labels on food products that contain fats, such as butter, margarine, vegetable oils, peanut butter, and potato chips. Look for terms such as saturated, monounsaturated, polyunsaturated, and partially or fully hydrogenated.

Questions

1. What type(s) of fats or oils are in the product?
2. How many grams of saturated, monounsaturated, and polyunsaturated fats are there in one serving of the product?
3. What percentage of the total fat is saturated fat? Unsaturated fat?
4. The label on a container of peanut butter states that the cottonseed and canola oils used to make the peanut butter have been fully hydrogenated. What are the typical products that would form when hydrogen is added?
5. For each packaged food, determine the following:
   a. How many grams of fat are in one serving of the food?
   b. Using the caloric value for fat (9 kcal/g), how many Calories (kilocalories) come from the fat in one serving?
   c. What is the percentage of fat in one serving?

SAMPLE PROBLEM 15.3 Reactions of Lipids

Write the equation for the reaction catalyzed by the enzyme lipase that hydrolyzes glyceryl trilaurate (trilaurin) during the digestion process.

SOLUTION

ANALYZE THE PROBLEM	Name of Lipid	Reactants	Reaction	Products
	glyceryl trilaurate (trilaurin)	triacylglycerol, three water molecules	hydrolysis	glycerol, three lauric acid molecules (12:0)

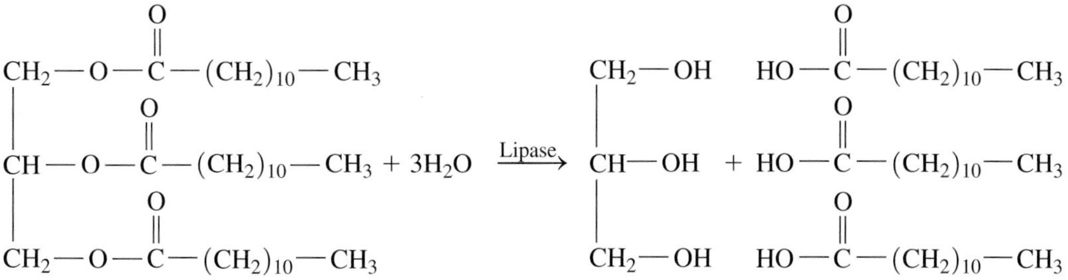

Glyceryl trilaurate (trilaurin) — Glycerol — 3 Lauric acid molecules

STUDY CHECK 15.3

What is the name of the product formed when a triacylglycerol containing oleic acid and linoleic acid is completely hydrogenated?

QUESTIONS AND PROBLEMS

15.4 Chemical Properties of Triacylglycerols

LEARNING GOAL Draw the condensed structural formula for the product of a triacylglycerol that undergoes hydrogenation, hydrolysis, or saponification.

15.27 Identify each of the following processes as hydrogenation, hydrolysis, or saponification and give the products:
a. the reaction of palm oil and hydrogen (H_2) with Ni catalyst
b. the reaction of glyceryl tripalmitate with water and HCl.

15.28 Identify each of the following processes as hydrogenation, hydrolysis, or saponification and give the products:
a. the reaction of corn oil and hydrogen (H_2) with a nickel catalyst
b. the reaction of glyceryl tristearate with water in the presence of lipase enzyme

15.29 Write the equation for the hydrogenation of glyceryl tripalmitoleate, a fat containing glycerol and three palmitoleic acid molecules.

15.30 Write the equation for the hydrogenation of glyceryl trilinolenate, a fat containing glycerol and three linolenic acid molecules.

15.31 Write the equation for the acid hydrolysis of glyceryl tripalmitate (tripalmitin).

15.32 Write the equation for the acid hydrolysis of glyceryl trioleate (triolein).

15.33 Write the equation for the NaOH saponification of glyceryl trimyristate (trimyristin).

15.34 Write the equation for the NaOH saponification of glyceryl trioleate (triolein).

15.35 Draw the condensed structural formula for the product of the hydrogenation of the following triacylglycerol:

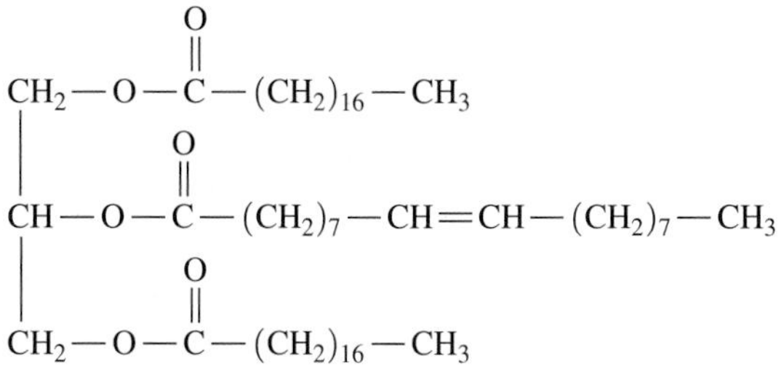

15.36 Draw the condensed structural formulas for all the products that would be obtained when the triacylglycerol in problem 15.35 undergoes complete hydrolysis.

LEARNING GOAL

Describe the structure of a phospholipid containing glycerol or sphingosine.

15.5 Phospholipids

The **phospholipids** are a family of lipids similar in structure to triacylglycerols; they include glycerophospholipids and sphingomyelins. In a **glycerophospholipid**, two fatty acids form ester bonds with the first and second hydroxyl groups of glycerol. The third hydroxyl group forms an ester with phosphoric acid, which forms another phosphoester bond with an amino alcohol. In a *sphingomyelin*, sphingosine replaces glycerol. We can compare the general structures of a triacylglycerol, a glycerophospholipid, and a sphingolipid as follows:

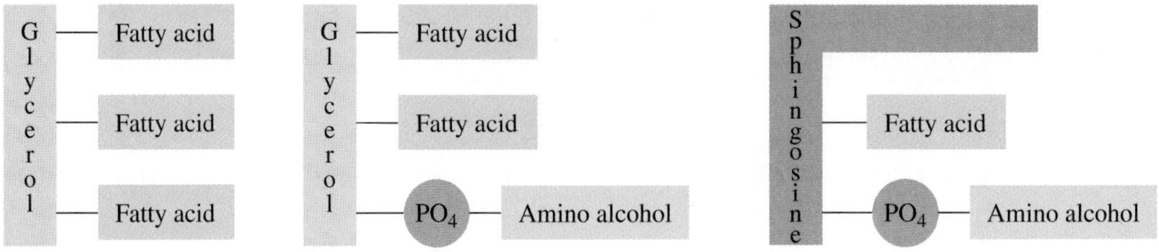

Triacylglycerol | Glycerophospholipid | Sphingomyelin

Three amino alcohols found in glycerophospholipids are choline, serine, and ethanolamine. In the body, at a physiological pH of 7.4, these amino alcohols are ionized.

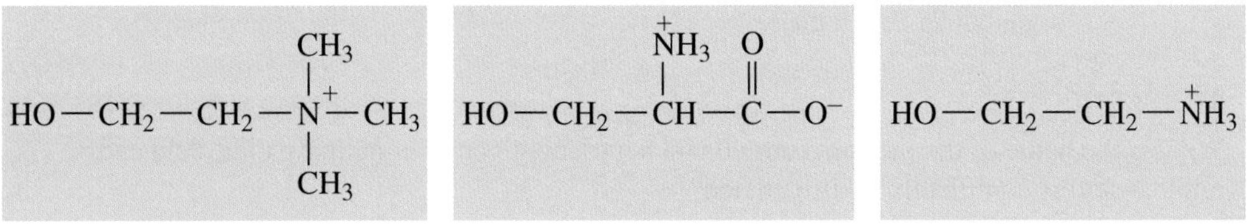

Choline | Serine | Ethanolamine

Lecithins and *cephalins* are two types of glycerophospholipids that are particularly abundant in brain and nerve tissues as well as in egg yolks, wheat germ, and yeast. Lecithins contain choline, and cephalins usually contain ethanolamine and sometimes serine. In the following structural formulas, the fatty acid that is used as an example is palmitic acid:

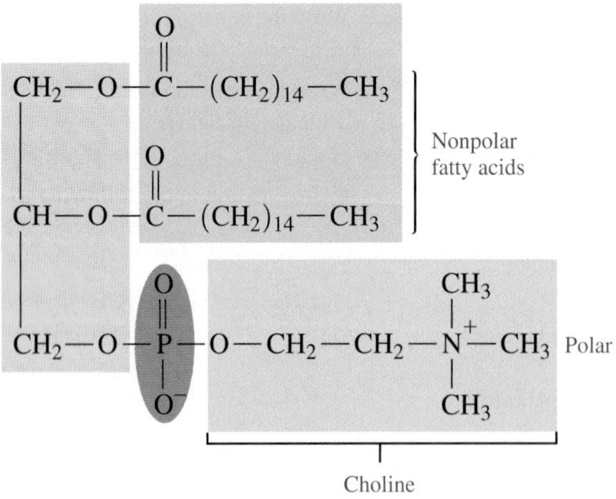

A lecithin

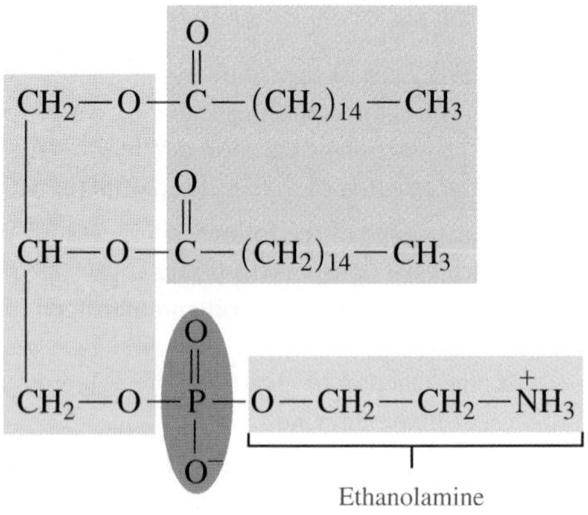

A cephalin

Glycerophospholipids contain both polar and nonpolar regions, which allow them to interact with both polar and nonpolar substances. The ionized amino alcohol and phosphate portion, called "the head," is polar and strongly attracted to water (see Figure 15.6). The hydrocarbon chains of the two fatty acids are the nonpolar "tails" of the glycerophospholipid, which are only soluble in other nonpolar substances, mostly lipids.

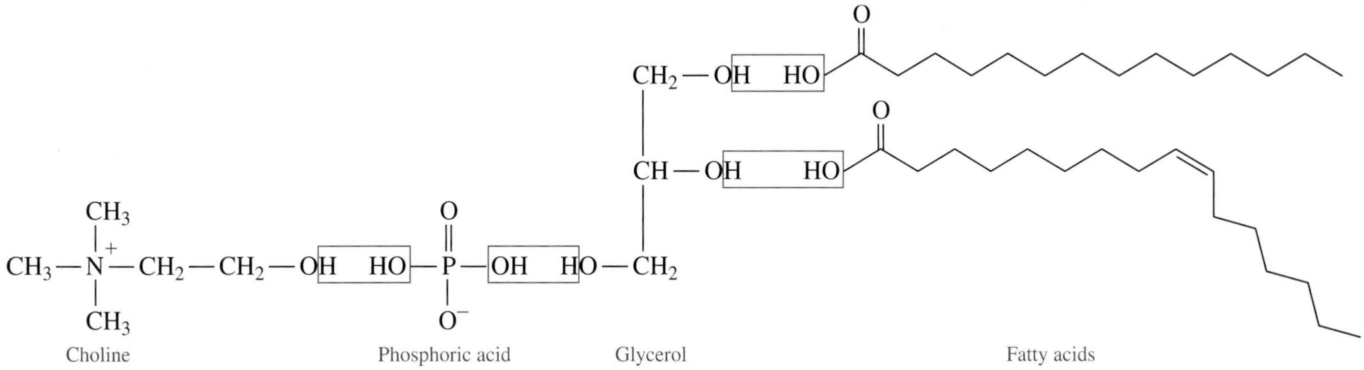

(a) Components of a typical glycerophospholipid

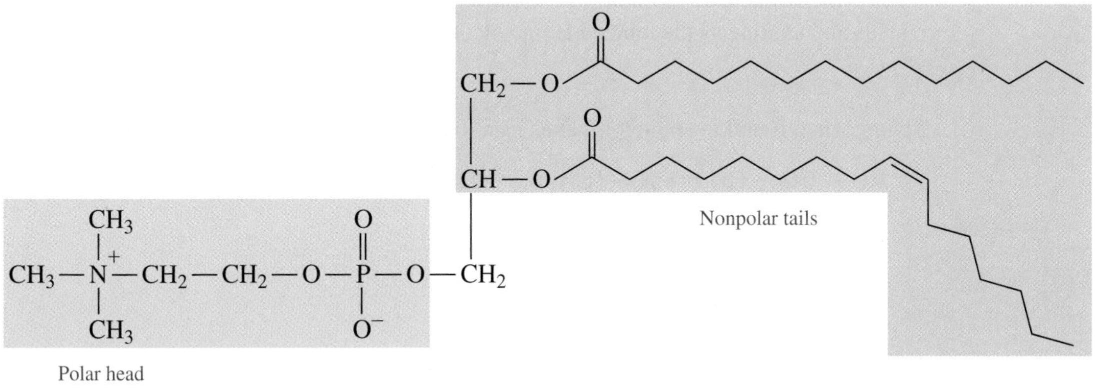

(b) Glycerophospholipid

(c) Simplified way to draw a glycerophospholipid

FIGURE 15.6 ▶ **(a)** The components of a typical glycerophospholipid: an amino alcohol, phosphoric acid, glycerol, and two fatty acids. **(b)** In a glycerophospholipid, a polar "head" contains the ionized amino alcohol and phosphate, while the hydrocarbon chains of two fatty acids make up the nonpolar "tails." **(c)** A simplified drawing indicates the polar region and the nonpolar region.

Q Why are glycerophospholipids polar?

Snake venom is produced by the modified saliva glands of poisonous snakes. When a snake bites, venom is ejected through the fang of the snake. The venom of the eastern diamondback rattlesnake and the Indian cobra contains *phospholipases*, which are enzymes that catalyze the hydrolysis of the fatty acid on the center carbon of glycerophospholipids in the red blood cells. The resulting product, called *lysophospholipid*, causes breakdown of the red blood cell membranes. This makes them permeable to water, which causes hemolysis of the red blood cells.

Poisonous snake venom contains phospholipases that hydrolyze phospholipids in red blood cells.

SAMPLE PROBLEM 15.4 Drawing Glycerophospholipid Structures

Draw the condensed structural formula for the cephalin that contains two stearic acids on the first and second carbon atoms, with serine as the ionized amino alcohol.

SOLUTION

	Name of Lipid	Type of Lipid	Type of Alcohol	Fatty Acids	Other Components	Type of Bonds
ANALYZE THE PROBLEM	cephalin	glycerophospholipid	glycerol	two stearic acid (18:0)	phosphate and serine as an ionized amino alcohol	ester bonds to fatty acids; phosphoester bonds to glycerol and to serine

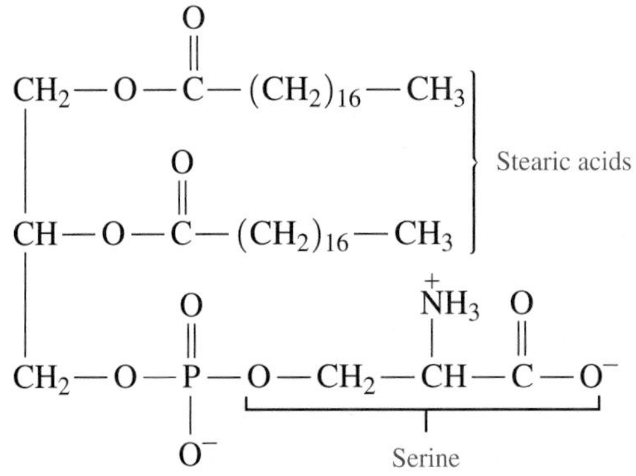

STUDY CHECK 15.4

Draw the condensed structural formula for a lecithin, using myristic acid for the fatty acids and choline as the ionized amino alcohol.

Sphingosine, found in sphingomyelins, is a long-chain amino alcohol.

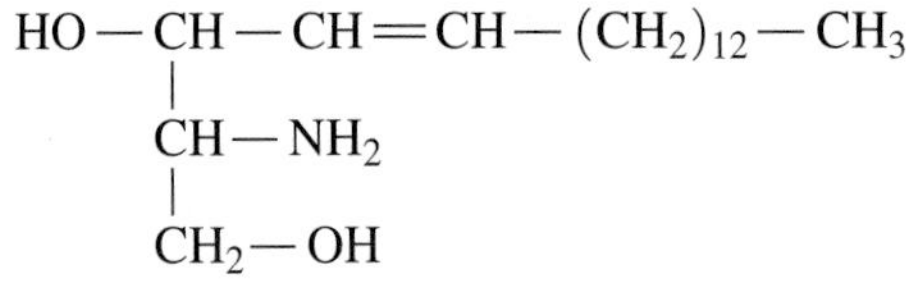

Sphingosine

In a **sphingomyelin**, the amine group of sphingosine forms an amide bond to a fatty acid and the hydroxyl group forms an ester bond with phosphate, which forms another phosphoester bond to choline or ethanolamine. The sphingomyelins are abundant in the white matter of the myelin sheath, a coating surrounding the nerve cells that increases the speed of nerve impulses and insulates and protects the nerve cells.

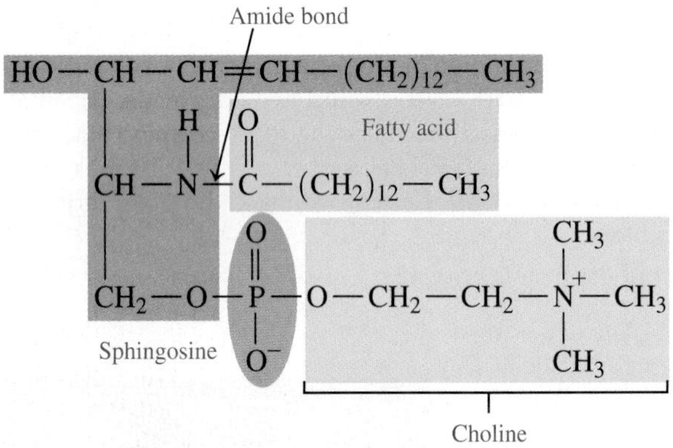

A sphingomyelin containing myristic acid and choline

In multiple sclerosis, sphingomyelin is lost from the myelin sheath, which protects the neurons in the brain and spinal cord. As the disease progresses, the myelin sheath deteriorates. Scars form on the neurons and impair the transmission of nerve signals. The symptoms of multiple sclerosis include various levels of muscle weakness with loss of coordination and vision, depending on the amount of damage. The cause of multiple sclerosis is not yet known, although some researchers suggest that a virus is involved. Several studies also suggest that adequate levels of vitamin D may lessen the severity or lower the risk of developing multiple sclerosis.

SAMPLE PROBLEM 15.5 Sphingomyelin

Palmitic acid, the 16-carbon saturated fatty acid, is the most common fatty acid found along with the ionized amino alcohol choline in the sphingomyelin of eggs. Draw the condensed structural formula for this sphingomyelin.

SOLUTION

	Name of Lipid	Type of Lipid	Type of Alcohol	Fatty Acid	Other Components	Type of Bonds
ANALYZE THE PROBLEM	sphingomyelin	sphingolipid	sphingosine	palmitic acid (16:0)	phosphate and choline as an ionized amino alcohol	amide bond to palmitic acid; phosphoester bond to sphingosine and to choline

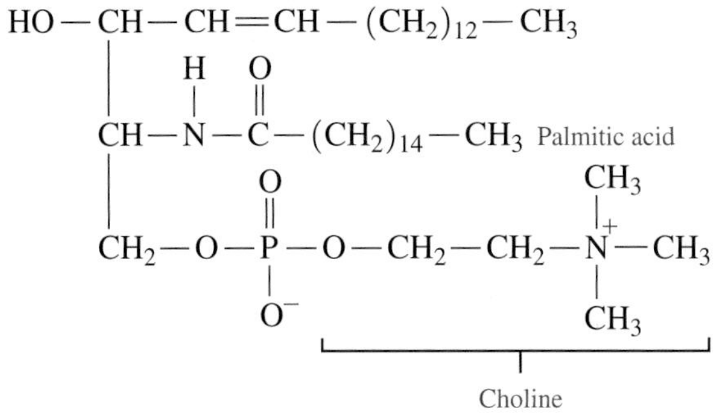

STUDY CHECK 15.5

Stearic acid is found in sphingomyelin in the brain. Draw the condensed structural formula for this sphingomyelin using ethanolamine as the ionized amino alcohol.

Chemistry Link to **Health**

Infant Respiratory Distress Syndrome (IRDS)

When an infant is born, an important key to its survival is proper lung function. In the lungs, there are many tiny air sacs called *alveoli*, where the exchange of O_2 and CO_2 takes place. Upon birth of a mature infant, surfactant is released into the lung tissues where it lowers the surface tension in the alveoli, which helps the air sacs inflate. The production of a *pulmonary surfactant*, which is a mixture of phospholipids including lecithin and sphingomyelin produced by specific lung cells, occurs in a fetus after 24 to 28 weeks of pregnancy. If an infant is born prematurely before 28 weeks of gestation, the low level of surfactant and immature lung development lead to a high risk of *Infant Respiratory Distress Syndrome* (IRDS). Without sufficient surfactant, the air sacs collapse and have to reopen with each breath. As a result, alveoli cells are damaged, less oxygen is taken in, and more carbon dioxide is retained, which can lead to hypoxia and acidosis.

One way to determine the maturity of the lungs of a fetus is to measure the *lecithin–sphingomyelin (L/S) ratio*. A ratio of 2.5 indicates mature fetal lung function, an L/S ratio of 2.4 to 1.6 indicates a low risk, and a ratio of less than 1.5 indicates a high risk of IRDS. Before the initiation of an early delivery, the L/S ratio of the amniotic fluid is measured. If the L/S ratio is low, steroids may be given to the mother to assist the lung development and production of surfactant in the fetus. Once a premature infant is born, treatment includes the use of steroids to help maturation of the lungs, the application of surfactants, and the administration of supplemental oxygen with ventilation to help minimize damage to the lungs.

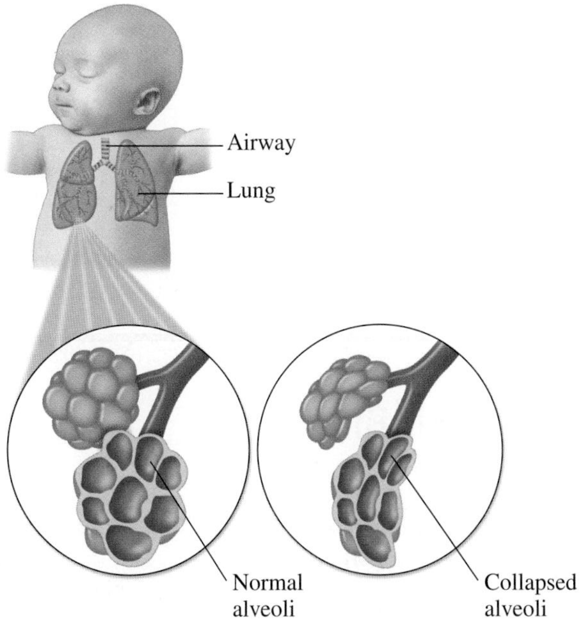

Without sufficient surfactant in the lungs of a premature infant, the alveoli collapse, which decreases pulmonary function.

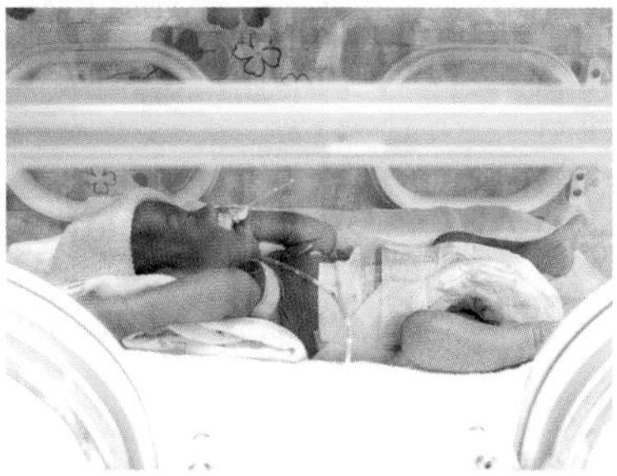

A premature infant with respiratory distress is treated with a surfactant and oxygen.

QUESTIONS AND PROBLEMS

15.5 Phospholipids

LEARNING GOAL Describe the structure of a phospholipid containing glycerol or sphingosine.

15.37 Describe the similarities and differences between triacylglycerols and glycerophospholipids.

15.38 Describe the similarities and differences between lecithins and cephalins.

15.39 Draw the condensed structural formula for the glycerophospholipid cephalin that contains two molecules of palmitic acid and the ionized amino alcohol ethanolamine.

15.40 Draw the condensed structural formula for the glycerophospholipid lecithin that contains two molecules of palmitic acid and the ionized amino alcohol choline.

15.41 Identify the following glycerophospholipid as a lecithin or cephalin, and list its components:

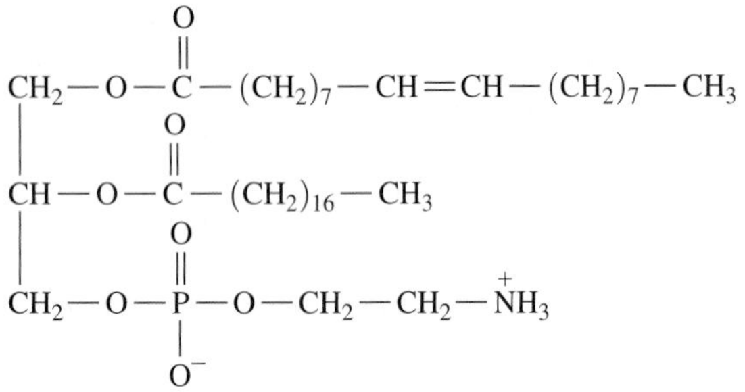

15.42 Identify the following glycerophospholipid as a lecithin or cephalin, and list its components:

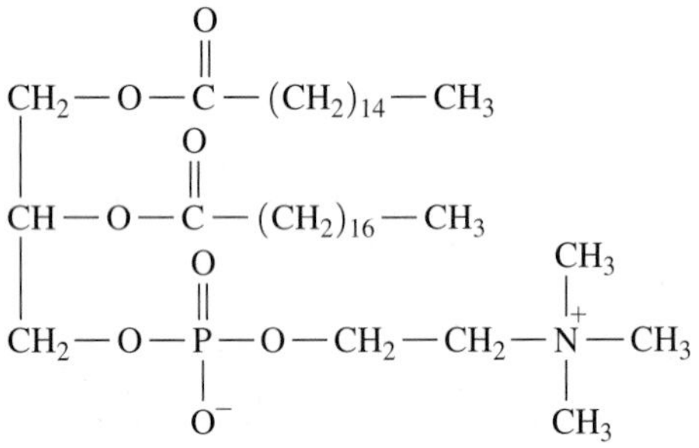

15.43 Identify the following features of this phospholipid:

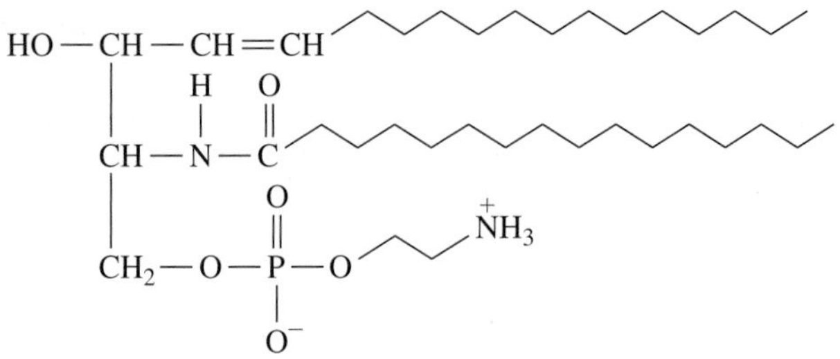

a. Is the phospholipid formed from glycerol or sphingosine?
b. What is the fatty acid?
c. What type of bond connects the fatty acid?
d. What is the amino alcohol?

15.44 Identify the following features of this phospholipid:

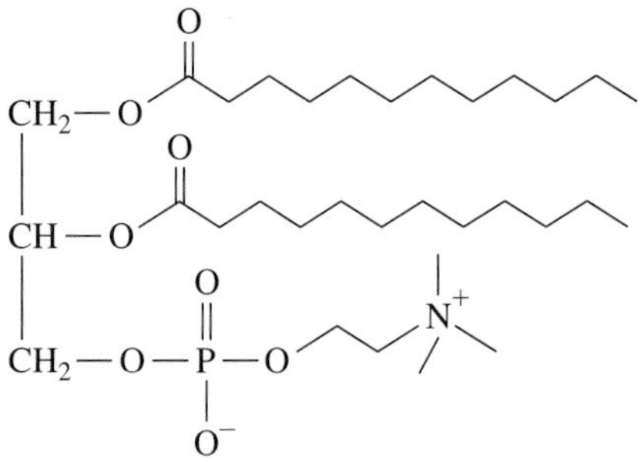

a. Is the phospholipid formed from glycerol or sphingosine?
b. What is the fatty acid?
c. What type of bond connects the fatty acid?
d. What is the amino alcohol?

LEARNING GOAL

Describe the structures of steroids.

15.6 Steroids: Cholesterol, Bile Salts, and Steroid Hormones

Steroids are compounds containing the *steroid nucleus*, which consists of three cyclohexane rings and one cyclopentane ring fused together. The four rings in the steroid nucleus are designated A, B, C, and D. The carbon atoms are numbered beginning with the carbons in ring A, and in steroids like cholesterol, ending with two methyl groups.

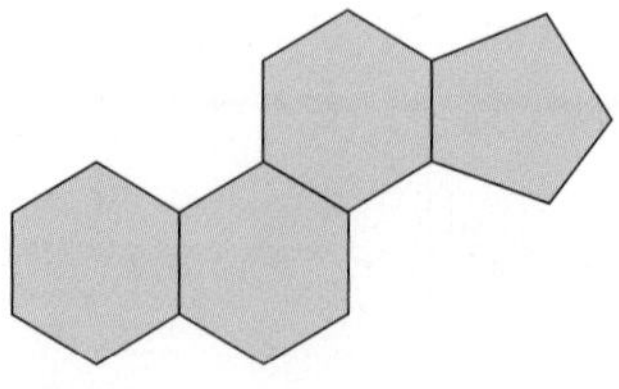

Steroid nucleus

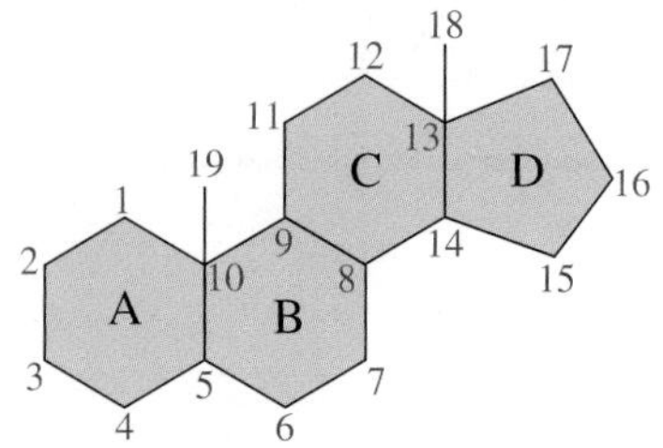

Steroid numbering system

Cholesterol

Cholesterol, which is one of the most important and abundant steroids in the body, is a *sterol* because it contains an oxygen atom as a hydroxyl group (—OH) on carbon 3. Like many steroids, cholesterol has a double bond between carbon 5 and carbon 6, methyl groups at carbon 10 and carbon 13, and a carbon chain at carbon 17. In other steroids, the oxygen atom forms a carbonyl group (C=O) at carbon 3.

Cholesterol is a component of cellular membranes, myelin sheath, and brain and nerve tissue. It is also found in the liver and bile salts; large quantities of it are found in the skin, and some of it becomes vitamin D when the skin is exposed to direct sunlight. In the adrenal gland, cholesterol is used to synthesize steroid hormones. The liver synthesizes sufficient cholesterol for the body from fats, carbohydrates, and proteins. Additional cholesterol is obtained from meat, milk, and eggs in the diet. There is no cholesterol in vegetable and plant products.

CORE CHEMISTRY SKILL

Identifying the Steroid Nucleus

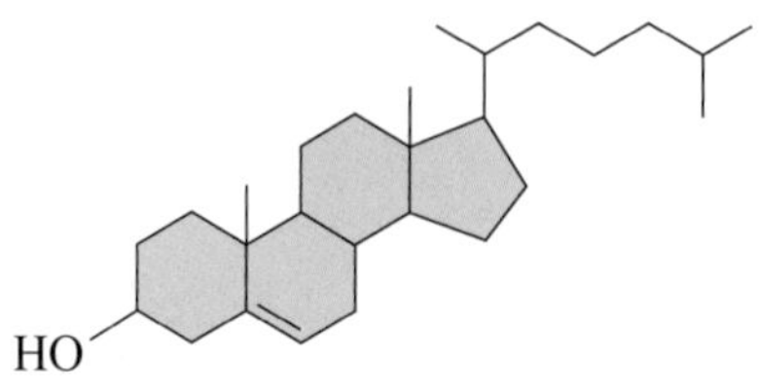

Cholesterol

Cholesterol in the Body

If a diet is high in cholesterol, the liver produces less cholesterol. A typical daily American diet includes 400 to 500 mg of cholesterol, one of the highest in the world. The American Heart Association has recommended that we consume no more than 300 mg of cholesterol a day. Researchers suggest that saturated fats and cholesterol are associated with diseases such as diabetes; cancers of the breast, pancreas, and colon; and atherosclerosis. In atherosclerosis, deposits of a protein–lipid complex (plaque) accumulate in the coronary blood vessels, restricting the flow of blood to the tissue and causing necrosis (death) of the tissue (see Figure 15.7). In the heart, plaque accumulation could result in a *myocardial infarction* (heart attack). Other factors that may also increase the risk of heart disease are family history, lack of exercise, smoking, obesity, diabetes, gender, and age. The cholesterol contents of some typical foods are listed in Table 15.4.

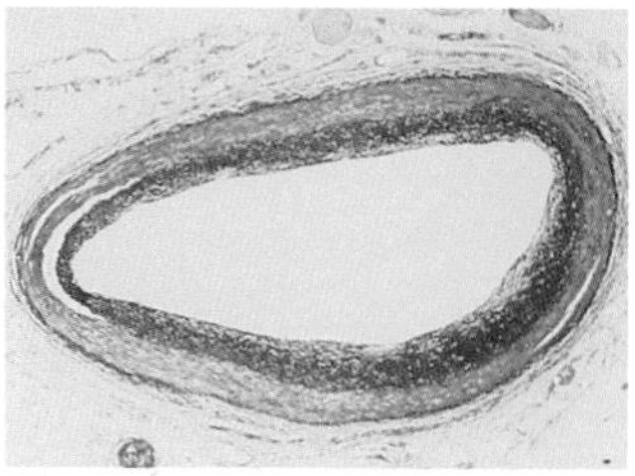

(a)

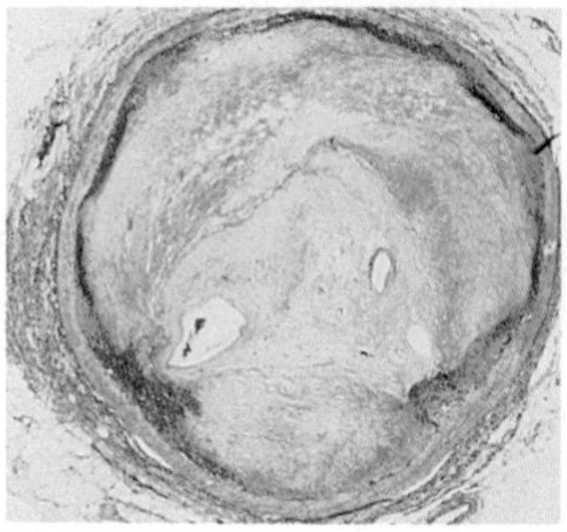

(b)

FIGURE 15.7 ▶ Excess cholesterol forms plaque that can block an artery, resulting in a heart attack. **(a)** A cross section of a normal, open artery shows no buildup of plaque. **(b)** A cross section of an artery that is almost completely clogged by atherosclerotic plaque.

Q What property of cholesterol would cause it to form deposits along the coronary arteries?

TABLE 15.4 Cholesterol Content of Some Foods

Food	Serving Size	Cholesterol (mg)
Liver (beef)	3 oz	370
Large egg	1	200
Lobster	3 oz	175
Fried chicken	$3\frac{1}{2}$ oz	130
Hamburger	3 oz	85
Chicken (no skin)	3 oz	75
Fish (salmon)	3 oz	40
Whole milk	1 cup	35
Butter	1 tablespoon	30
Skim milk	1 cup	5
Margarine	1 tablespoon	0

Clinically, cholesterol levels are considered elevated if the total plasma cholesterol level exceeds 200 mg/dL. Saturated fats in the diet may stimulate the production of cholesterol by the liver. A diet that is low in foods containing cholesterol and saturated fats appears to be helpful in reducing the serum cholesterol level. The American Institute for Cancer Research (AICR) has recommended that our diet contain more fiber and starch by adding more vegetables, fruits, whole grains, and moderate amounts of foods with low levels of fat and cholesterol such as fish, poultry, lean meats, and low-fat dairy products. AICR also suggests that we limit our intake of foods high in cholesterol such as eggs, nuts, French fries, fatty or organ meats, cheeses, butter, and coconut and palm oil.

SAMPLE PROBLEM 15.6 Cholesterol

Refer to the structure of cholesterol for each of the following questions:

a. What part of cholesterol is the steroid nucleus?
b. What features have been added to the steroid nucleus in cholesterol?
c. What classifies cholesterol as a sterol?

SOLUTION

a. The four fused rings form the steroid nucleus.
b. The cholesterol molecule contains a hydroxyl group ($—OH$) on the first ring, one double bond in the second ring, methyl groups ($—CH_3$) at carbons 10 and 13, and a branched carbon chain.
c. The hydroxyl group determines the sterol classification.

STUDY CHECK 15.6

Why is cholesterol in the lipid family?

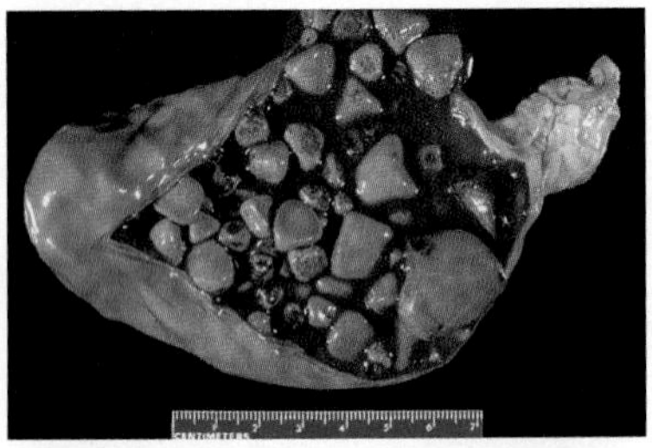

FIGURE 15.8 ▶ Gallstones form in the gallbladder when cholesterol levels are high.

Q What type of steroid is stored in the gallbladder?

Bile Salts

The *bile salts* in the body are synthesized from cholesterol in the liver and stored in the gallbladder. When bile is secreted into the small intestine, the bile salts mix with the water-insoluble fats and oils in our diets. The bile salts with their nonpolar and polar regions act much like soaps, breaking down large globules of fat into smaller droplets. The smaller droplets containing fat have a larger surface area to react with lipases, which are the enzymes that digest fat. The bile salts also help with the absorption of cholesterol into the intestinal mucosa.

When large amounts of cholesterol accumulate in the gallbladder, cholesterol can become solid, which forms gallstones (see Figure 15.8). Gallstones are composed of almost 100% cholesterol, with some calcium salts, fatty acids, and glycerophospholipids. Normally, small stones pass through the bile duct into the duodenum, the first part of the small intestine immediately beyond the stomach. If a large stone passes into the bile duct, it can get stuck, and the pain can be severe. If the gallstone obstructs the duct, bile cannot be excreted. Then bile pigments known as bilirubin will not be able to pass through the bile duct into the duodenum. They will back up into the liver and be excreted via the blood, causing jaundice (*hyperbilirubinemia*), which gives a yellow color to the skin and the whites of the eyes.

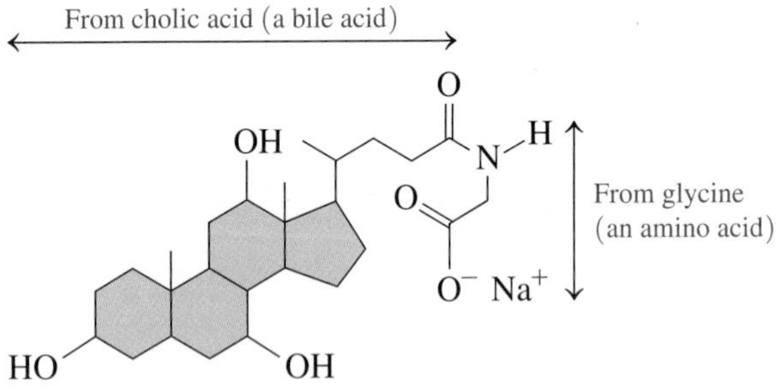

Sodium glycocholate (a bile salt)

Lipoproteins: Transporting Lipids

In the body, lipids must be moved through the bloodstream to tissues where they are stored, used for energy, or used to make hormones. However, most lipids are nonpolar and insoluble in the aqueous environment of blood. They are made more soluble by combining them with phospholipids and proteins to form water-soluble complexes called **lipoproteins**. In general, lipoproteins are spherical particles with an outer surface of polar proteins and phospholipids that surround hundreds of nonpolar molecules of triacylglycerols and cholesteryl esters (see Figure 15.9). Cholesteryl esters are the prevalent form of cholesterol in the blood. They are formed by the esterification of the hydroxyl group in cholesterol with a fatty acid.

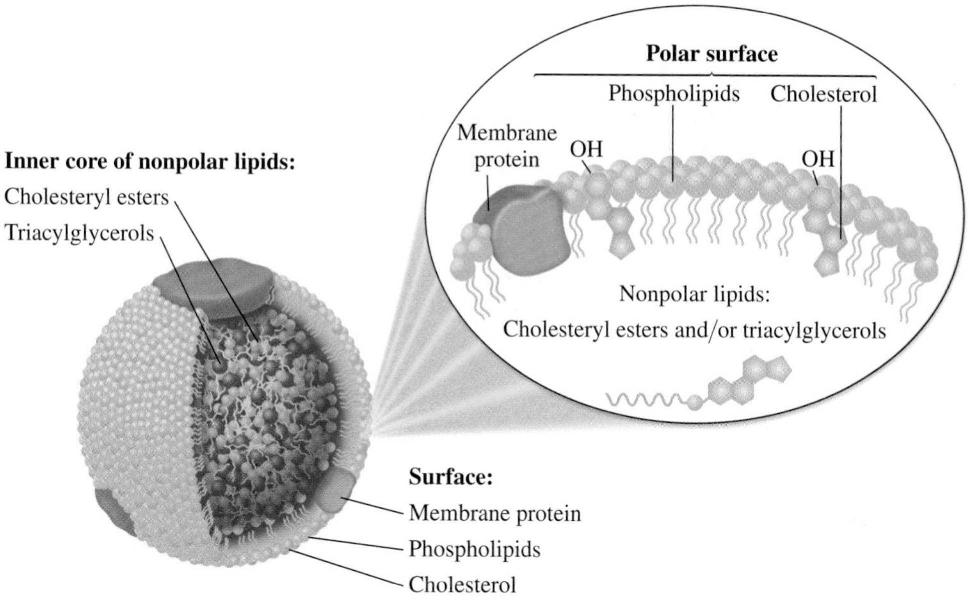

FIGURE 15.9 ▶ A spherical lipoprotein particle surrounds nonpolar lipids with polar lipids and protein for transport to body cells.

Q Why are the polar components on the surface of a lipoprotein particle and the nonpolar components at the center?

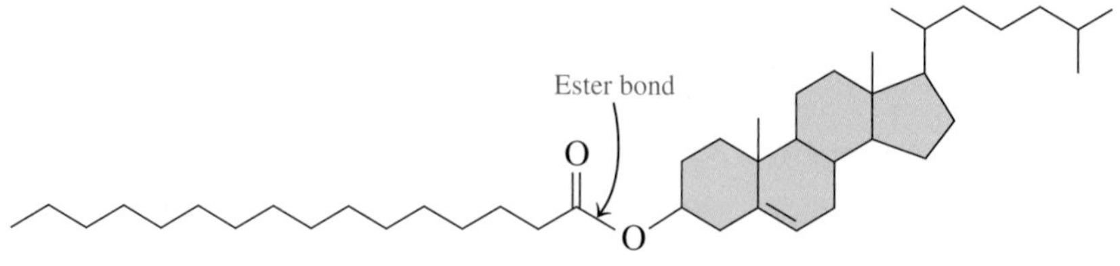

Cholesteryl ester

There are a variety of lipoproteins, which differ in density, lipid composition, and function. They include chylomicrons, very-low-density lipoproteins (VLDL), low-density lipoproteins (LDL), and high-density lipoproteins (HDL). The density of the lipoproteins increases as the percentage of protein in each type also increases (see Table 15.5).

TABLE 15.5 Composition and Properties of Plasma Lipoproteins

	Chylomicrons	VLDL	LDL	HDL
Density (g/mL)	0.940	0.950–1.006	1.006–1.063	1.063–1.210
		Composition (% by mass)		
Type of Lipid				
Triacylglycerols	86	55	6	4
Phospholipids	7	18	22	24
Cholesterol	2	7	8	2
Cholesteryl esters	3	12	42	15
Protein	2	8	22	55

Two important lipoproteins are the LDL and HDL, which transport cholesterol. The LDL carries cholesterol to the tissues where it can be used for the synthesis of cell membranes and steroid hormones. When the LDL exceeds the amount of cholesterol needed by the tissues, the LDL deposits cholesterol in the arteries (plaque), which can restrict blood flow and increase the risk of developing heart disease and/or myocardial infarctions (heart attacks). This is why LDL is called "bad" cholesterol. The HDL picks up cholesterol from the tissues and carries it to the liver, where it can be converted to bile salts, which are eliminated from the body. This is why HDL is called "good" cholesterol. Other lipoproteins include chylomicrons that carry triacylglycerols from the intestines to the

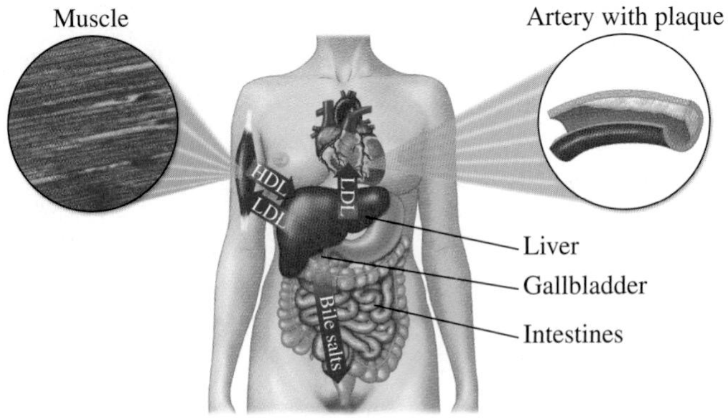

FIGURE 15.10 ▶ High- and low-density lipoproteins transport cholesterol between the tissues and the liver.

Q What type of lipoprotein transports cholesterol to the liver?

liver, muscle, and adipose tissues, and VLDL that carries the triacylglycerols synthesized in the liver to the adipose tissues for storage (see Figure 15.10).

Because high cholesterol levels are associated with the onset of atherosclerosis and heart disease, a doctor may order a *lipid panel* as part of a health examination. A lipid panel is a blood test that measures serum lipid levels including cholesterol, triglycerides, HDL, and LDL. The results of a lipid panel are used to evaluate a patient's risk of heart disease and to help a doctor determine the type of treatment needed.

Lipid Panel	Recommended Level	Greater Risk of Heart Disease
Total Cholesterol	Less than 200 mg/dL	Greater than 240 mg/dL
Triglycerides (triacylglycerols)	Less than 150 mg/dL	Greater than 200 mg/dL
HDL ("good" cholesterol)	Greater than 60 mg/dL	Less than 40 mg/dL
LDL ("bad" cholesterol)	Less than 100 mg/dL	Greater than 160 mg/dL
Cholesterol/HDL Ratio	Less than 4	Greater than 7

Steroid Hormones

The word *hormone* comes from the Greek "to arouse" or "to excite." Hormones are chemical messengers that serve as a communication system from one part of the body to another. The *steroid* hormones, which include the sex hormones and the adrenocortical hormones, are closely related in structure to cholesterol and depend on cholesterol for their synthesis.

Two of the male sex hormones, *testosterone* and *androsterone*, promote the growth of muscle and facial hair, and the maturation of the male sex organs and of sperm.

The *estrogens*, a group of female sex hormones, direct the development of female sexual characteristics: the uterus increases in size, fat is deposited in the breasts, and the pelvis broadens. *Progesterone* prepares the uterus for the implantation of a fertilized egg. If an egg is not fertilized, the levels of progesterone and estrogen drop sharply, and menstruation follows. Synthetic forms of the female sex hormones are used in birth control pills. As with other kinds of steroids, side effects include weight gain and a greater risk of forming blood clots. The structures of some steroid hormones are shown below:

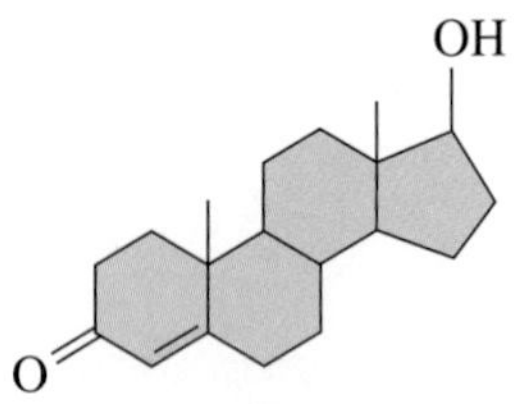

Testosterone (androgen)
(produced in testes)

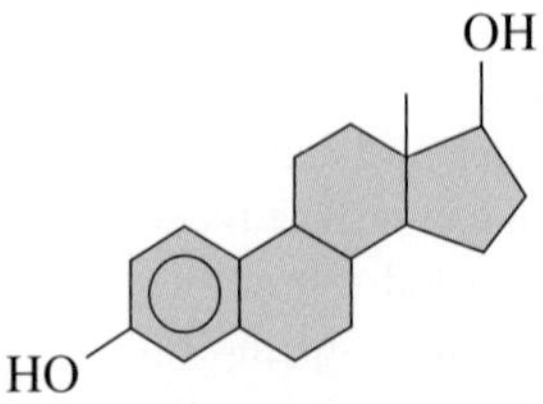

Estradiol (estrogen)
(produced in ovaries)

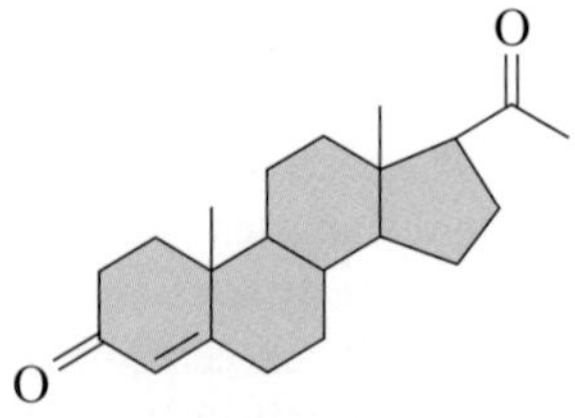

Progesterone
(produced in ovaries)

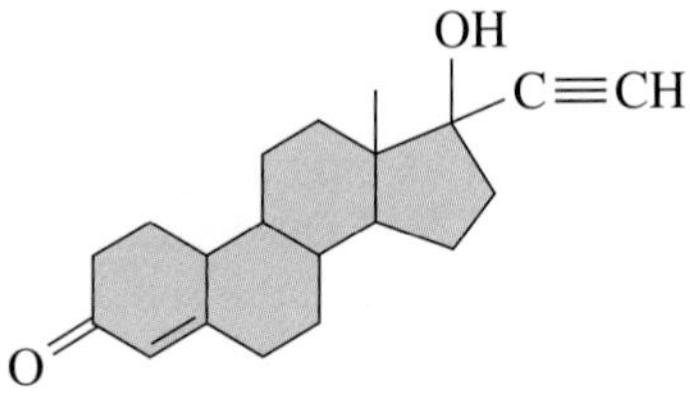

Norethindrone
(synthetic progestin)

Chemistry Link to **Health**

Anabolic Steroids

Some of the physiological effects of testosterone are to increase muscle mass and decrease body fat. Derivatives of testosterone called *anabolic steroids* that enhance these effects have been synthesized. Although they have some medical uses, anabolic steroids have been used in high dosages by some athletes in an effort to increase muscle mass. Such use is banned by most sports organizations.

Use of anabolic steroids in attempting to improve athletic strength can cause numerous side effects: in males—a reduction in testicle size, low sperm count and infertility, male pattern baldness, and breast development; in females—facial hair, deepening of the voice, male pattern baldness, breast atrophy, and menstrual dysfunction. Possible long-term consequences of anabolic steroid use in both men and women include liver disease and tumors, depression, and heart complications, with an increased risk of prostate cancer for men.

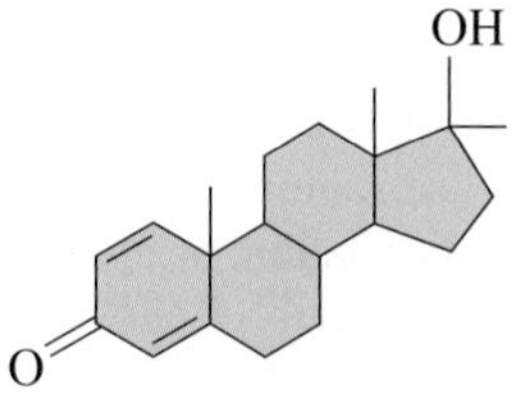
Methandienone

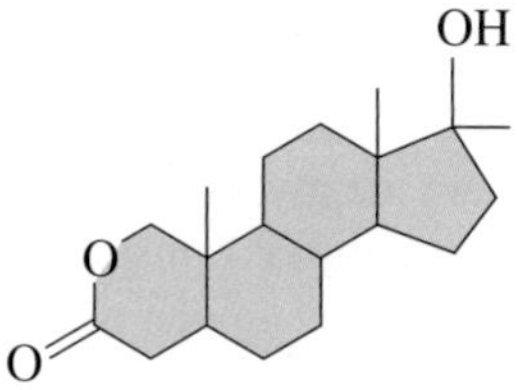
Oxandrolone

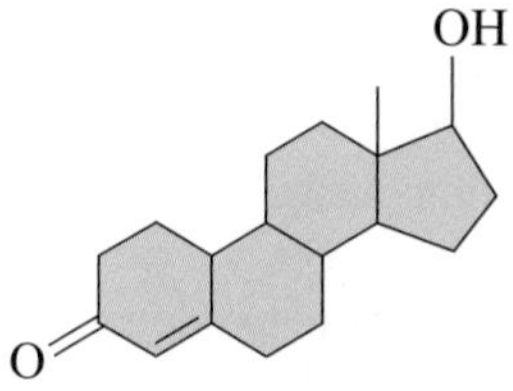
Nandrolone

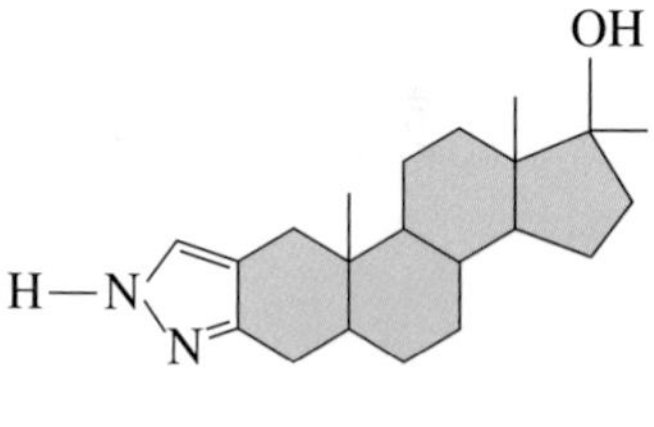
Stanozolol

Adrenal Corticosteroids

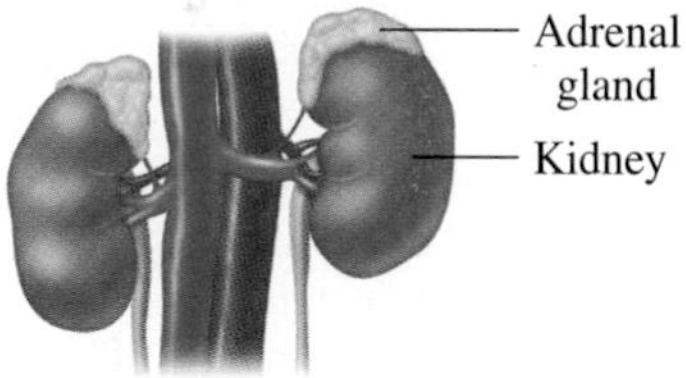

The adrenal glands, located on the top of each kidney, produce a large number of compounds known as the *corticosteroids*. *Cortisone* increases the blood glucose level and stimulates the synthesis of glycogen in the liver. *Aldosterone* is responsible for the regulation of electrolytes and water balance by the kidneys. *Cortisol* is released under stress to increase blood sugar and regulate carbohydrate, fat, and protein metabolism. Synthetic corticosteroid drugs such as *prednisone* are derived from cortisone and used medically for reducing inflammation and treating asthma and rheumatoid arthritis, although health problems can result from long-term use.

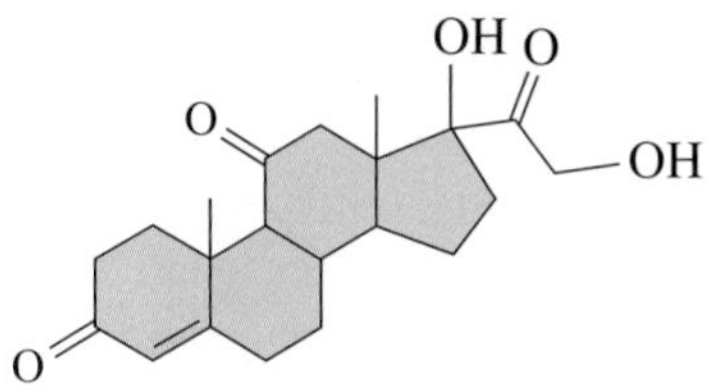
Cortisone
(produced in adrenal gland)

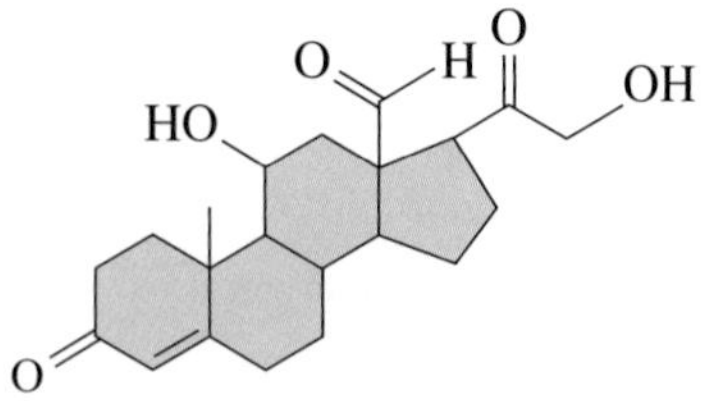
Aldosterone (mineralocorticoid)
(produced in adrenal gland)

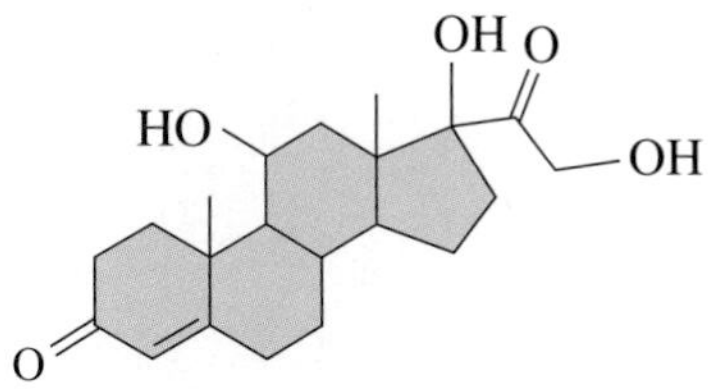
Cortisol
(produced in adrenal cortex)

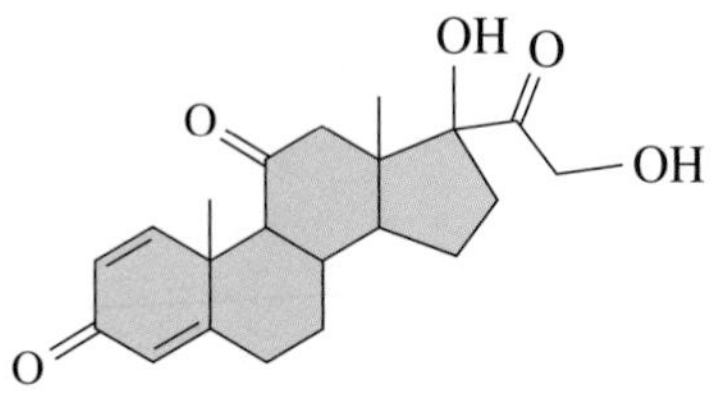
Prednisone
(synthetic corticoid)

QUESTIONS AND PROBLEMS

15.6 Steroids: Cholesterol, Bile Salts, and Steroid Hormones

LEARNING GOAL Describe the structures of steroids.

15.45 Draw the structure for the steroid nucleus.

15.46 Draw the structure for bile salt.

15.47 What is the function of bile salts in digestion?

15.48 Why are lipoproteins needed to transport lipids in the bloodstream?

15.49 How do chylomicrons differ from very-low-density lipoprotein?

15.50 How does LDL differ from HDL?

15.51 Why is LDL called "bad" cholesterol?

15.52 Why is HDL called "good" cholesterol?

15.53 What are the similarities and differences between the steroid hormones estradiol and testosterone?

15.54 What are the similarities and differences between the adrenal hormone cortisone and the synthetic corticoid prednisone?

15.55 Which of the following are steroid hormones?
a. androsterone
b. progestrone
c. prednisone
d. bile salts

15.56 Which of the following are adrenal corticosteroids?
a. prednisone
b. aldosterone
c. cortisol
d. testosterone

LEARNING GOAL

Describe the composition and function of the lipid bilayer in cell membranes.

15.7 Cell Membranes

The membrane of a cell separates the contents of a cell from the external fluids. It is *semi-permeable* so that nutrients can enter the cell and waste products can leave. The main components of a cell membrane are the glycerophospholipids and sphingolipids. Earlier in this chapter, we saw that the phospholipids consist of a nonpolar region, or hydrocarbon "tail," with two long-chain fatty acids and a polar region, or ionic "head" of phosphate and an ionized amino alcohol.

In a cell (plasma) membrane, two layers of phospholipids are arranged with their hydrophilic heads at the outer and inner surfaces of the membrane, and their hydrophobic tails in the center. This double layer arrangement of phospholipids is called a **lipid bilayer** (see Figure 15.11). The outer layer of phospholipids is in contact with the external fluids, and the inner layer is in contact with the internal contents of the cell.

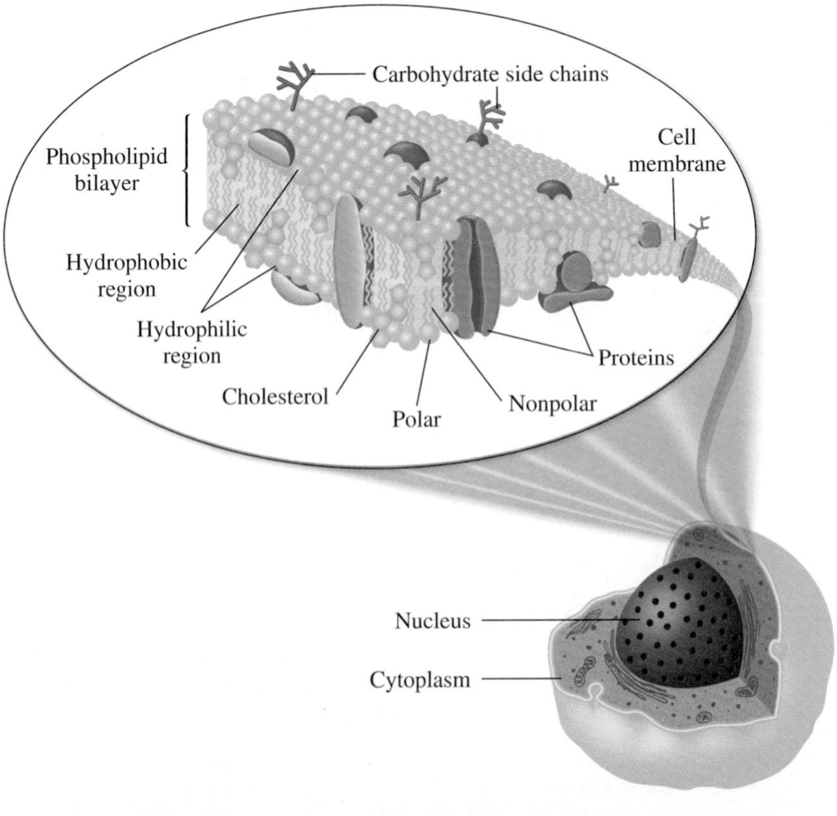

FIGURE 15.11 ▶ In the fluid mosaic model of a cell membrane, proteins and cholesterol are embedded in a lipid bilayer of phospholipids. The bilayer forms a membrane-type barrier with polar heads at the membrane surfaces and the nonpolar tails in the center away from the water.

Q What types of fatty acids are found in the phospholipids of the lipid bilayer?

Interactive Video

Membrane Structure

Most of the phospholipids in the lipid bilayer contain unsaturated fatty acids. Because of the kinks in the carbon chains at the cis double bonds, the phospholipids do not fit closely together. As a result, the lipid bilayer is not a rigid, fixed structure, but one that is dynamic and fluid-like. This liquid-like bilayer also contains proteins, carbohydrates, and cholesterol molecules. For this reason, the model of biological membranes is referred to as the **fluid mosaic model** of membranes.

In the fluid mosaic model, proteins known as *peripheral proteins* emerge on just one of the surfaces, outer or inner. The *integral proteins* extend through the entire lipid bilayer and appear on both surfaces of the membrane. Some proteins and lipids on the outer surface of the cell membrane are attached to carbohydrates. These carbohydrate chains project into the surrounding fluid environment where they are responsible for cell recognition and communication with chemical messengers such as hormones and neurotransmitters. In animals, cholesterol molecules embedded among the phospholipids make up 20 to 25% of the lipid bilayer. Because cholesterol molecules are large and rigid, they reduce the flexibility of the lipid bilayer and add strength to the cell membrane.

SAMPLE PROBLEM 15.7 Lipid Bilayer in Cell Membranes

Describe the role of phospholipids in the lipid bilayer.

SOLUTION

Phospholipids consist of polar and nonpolar parts. In a cell membrane, an alignment of the nonpolar sections toward the center with the polar sections on the outside produces a barrier that prevents the contents of a cell from mixing with the fluids on the outside of the cell.

STUDY CHECK 15.7

What is the function of cholesterol in the cell membrane?

Transport through Cell Membranes

Although a nonpolar membrane separates aqueous solutions, it is necessary that certain substances can enter and leave the cell. The main function of a cell membrane is to allow the movement (transport) of ions and molecules on one side of the membrane to the other side. This transport of materials into and out of a cell is accomplished in several ways.

Diffusion (Passive) Transport

In the simplest transport mechanism called *diffusion* or *passive transport*, molecules can diffuse from a higher concentration to a lower concentration. For example, small molecules such as O_2, CO_2, urea, and water diffuse via passive transport through cell membranes. If their concentrations are greater outside the cell than inside, they diffuse into the cell. If their concentrations are higher within the cell, they diffuse out of the cell.

Facilitated Transport

In *facilitated transport*, proteins that extend from one side of the bilayer membrane to the other provide a channel through which certain substances can diffuse more rapidly than by passive diffusion to meet cellular needs. These protein channels allow transport of chloride ion (Cl^-), bicarbonate ion (HCO_3^-), and glucose molecules in and out of the cell.

Active Transport

Certain ions such as K^+, Na^+, and Ca^{2+} move across a cell membrane *against* their concentration gradients. For example, the K^+ concentration is greater inside a cell, and the Na^+ concentration is greater outside. However, in the conduction of nerve impulses and contraction of muscles, K^+ moves into the cell, and Na^+ moves out by a process known as *active transport*.

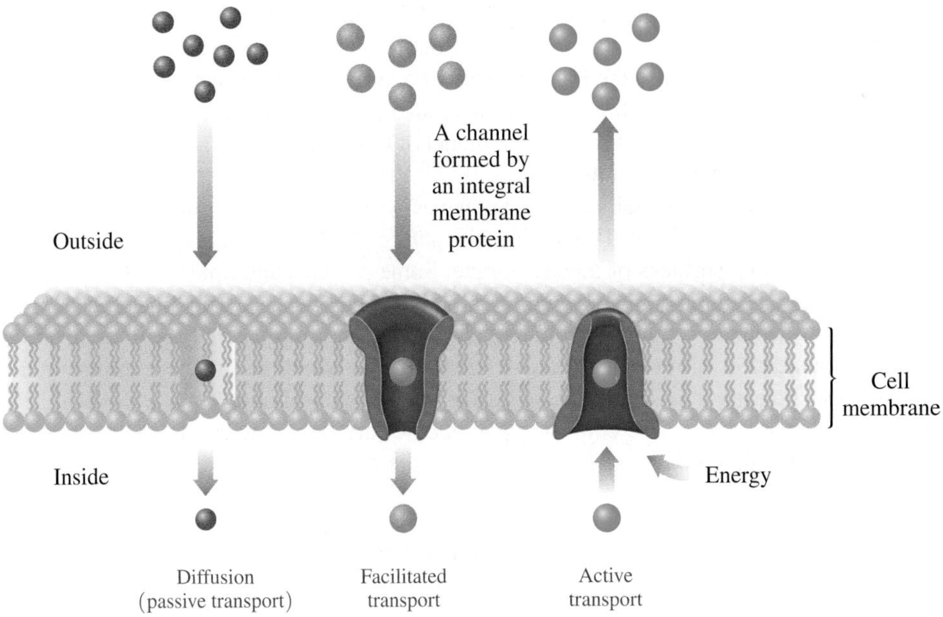

Substances are transported across a cell membrane by either diffusion, facilitated transport, or active transport.

QUESTIONS AND PROBLEMS

15.7 Cell Membranes

LEARNING GOAL Describe the composition and function of the lipid bilayer in cell membranes.

15.57 What is the function of the lipid bilayer in a cell membrane?

15.58 Describe the structure of a lipid bilayer.

15.59 How do molecules of cholesterol affect the structure of cell membranes?

15.60 What is the key component of the cell membrane? What will be the consequence if it contains a higher level of cis double bond than usual?

15.61 Where are proteins located in cell membranes?

15.62 What is the difference between peripheral and integral proteins?

15.63 Identify the type of transport described by each of the following:

a. A molecule moves through a protein channel.

b. O_2 moves into the cell from a higher concentration outside the cell.

15.64 Identify the type of transport described by each of the following:

a. An ion moves from low to high concentration in the cell.

b. Carbon dioxide moves through a cell membrane.

CONCEPT MAP

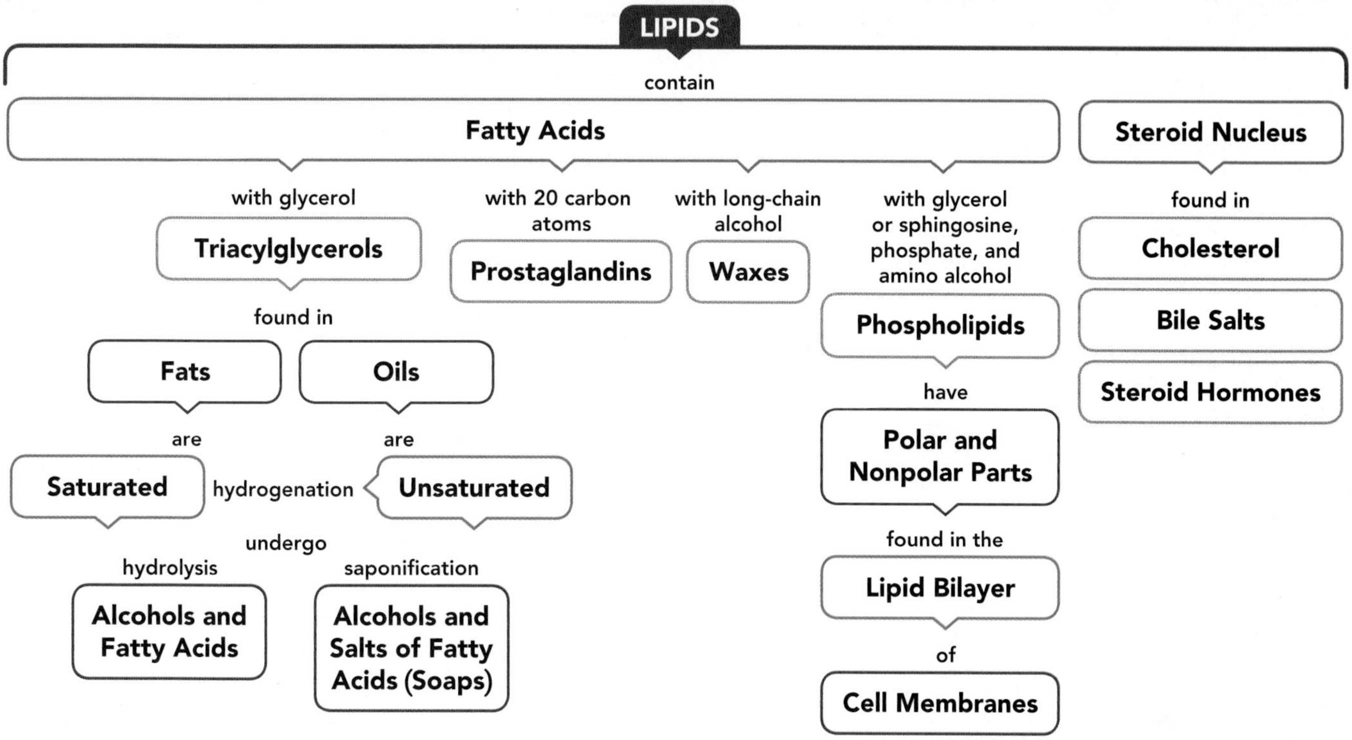

CHAPTER REVIEW

15.1 Lipids

LEARNING GOAL Describe the classes of lipids.

- Lipids are nonpolar compounds that are not soluble in water.
- Classes of lipids include waxes, fats and oils, phospholipids, and steroids.

15.2 Fatty Acids

LEARNING GOAL Draw the condensed structural formula for a fatty acid and identify it as saturated or unsaturated.

- Fatty acids are unbranched carboxylic acids that typically contain an even number (12 to 20) of carbon atoms.
- Fatty acids may be saturated, monounsaturated with one double bond, or polyunsaturated with two or more double bonds.
- The double bonds in unsaturated fatty acids are almost always cis.

15.3 Waxes and Triacylglycerols

LEARNING GOAL Draw the condensed structural formula for a wax or triacylglycerol produced by the reaction of a fatty acid and an alcohol or glycerol.

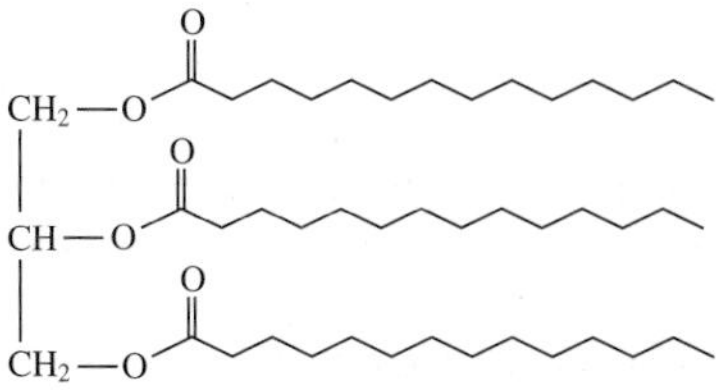

Triacylglycerol

- A wax is an ester of a long-chain fatty acid and a long-chain alcohol.
- The triacylglycerols are esters of glycerol with three long-chain fatty acids.
- Fats contain more saturated fatty acids and have higher melting points than most vegetable oils.

15.4 Chemical Properties of Triacylglycerols

LEARNING GOAL Draw the condensed structural formula for the product of a triacylglycerol that undergoes hydrogenation, hydrolysis, or saponification.

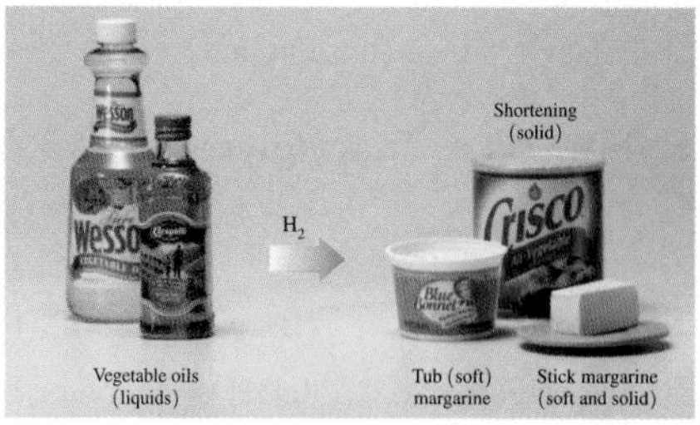

- The hydrogenation of unsaturated fatty acids of a triacylglycerol converts double bonds to single bonds.
- The hydrolysis of the ester bonds in triacylglycerols in the presence of a strong acid produces glycerol and fatty acids.
- In saponification, a triacylglycerol heated with a strong base produces glycerol and the salts of the fatty acids (soap).

15.5 Phospholipids

LEARNING GOAL Describe the structure of a phospholipid containing glycerol or sphingosine.

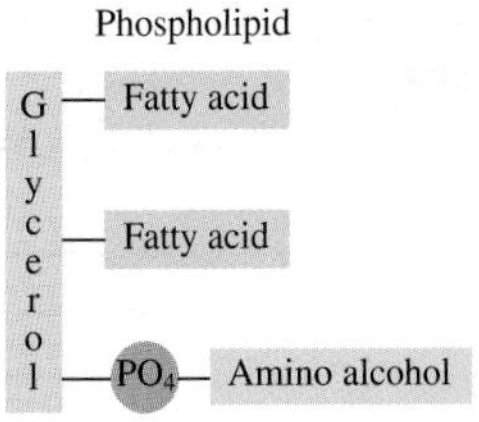

- Glycerophospholipids are esters of glycerol with two fatty acids and a phosphate group attached to an ionized amino alcohol.
- In a sphingomyelin, the amino alcohol sphingosine forms an amide bond with a fatty acid, and phosphoester bonds to phosphate and an ionized amino alcohol.

15.6 Steroids: Cholesterol, Bile Salts, and Steroid Hormones

LEARNING GOAL Describe the structures of steroids.

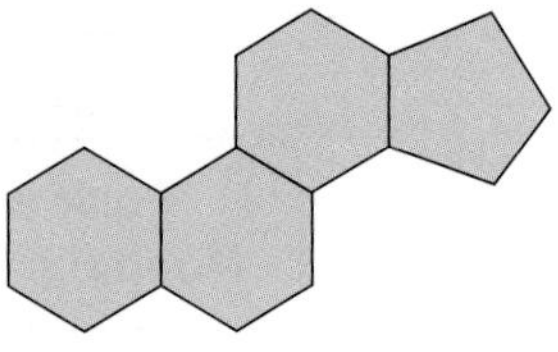
Steroid nucleus

- Steroids are lipids containing the steroid nucleus, which is a fused structure of four rings.
- Steroids include cholesterol and bile salts.
- Lipids, which are nonpolar, are transported through the aqueous environment of the blood by forming lipoproteins.
- Bile salts, synthesized from cholesterol, mix with water-insoluble fats and break them apart during digestion.
- Lipoproteins, such as chylomicrons and LDL, transport triacylglycerols from the intestines and the liver to fat cells and muscles for storage and energy.
- HDL transports cholesterol from the tissues to the liver for elimination.
- The steroid hormones are closely related in structure to cholesterol and depend on cholesterol for their synthesis.
- The sex hormones, such as estrogen and testosterone, are responsible for sexual characteristics and reproduction.
- The adrenal corticosteroids, such as aldosterone and cortisone, regulate water balance and glucose levels in the cells, respectively.

15.7 Cell Membranes

LEARNING GOAL Describe the composition and function of the lipid bilayer in cell membranes.

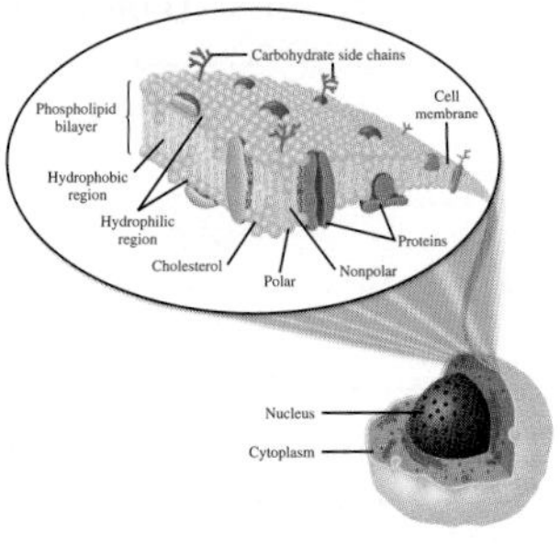

- All animal cells are surrounded by a semipermeable membrane that separates the cellular contents from the external fluids.
- The membrane is composed of two rows of phospholipids in a lipid bilayer.
- Proteins and cholesterol are embedded in the lipid bilayer, and carbohydrates are attached to its surface.
- Nutrients and waste products move through the cell membrane using passive transport (diffusion), facilitated transport, or active transport.

SUMMARY OF REACTIONS

The chapter sections to review are shown after the name of the reaction.

Esterification (15.3)

$$\text{Glycerol} + 3 \text{ fatty acid molecules} \xrightarrow{\text{Enzyme}} \text{triacylglycerol} + 3H_2O$$

Hydrogenation of Triacylglycerols (15.4)

$$\text{Triacylglycerol (unsaturated)} + H_2 \xrightarrow{\text{Ni}} \text{triacylglycerol (saturated)}$$

Hydrolysis of Triacylglycerols (15.4)

$$\text{Triacylglycerol} + 3H_2O \xrightarrow{H^+ \text{ or lipase}} \text{glycerol} + 3 \text{ fatty acid molecules}$$

Saponification of Triacylglycerols (15.4)

$$\text{Triacylglycerol} + 3NaOH \xrightarrow{\text{Heat}} \text{glycerol} + 3 \text{ sodium salts of fatty acids (soaps)}$$

KEY TERMS

cholesterol The most prevalent of the steroid compounds; needed for cellular membranes and the synthesis of vitamin D, hormones, and bile salts.

fat A triacylglycerol that is solid at room temperature and usually comes from animal sources.

fatty acid A long-chain carboxylic acid found in many lipids.

fluid mosaic model The concept that cell membranes are lipid bilayer structures that contain an assortment of polar lipids and proteins in a dynamic, fluid arrangement.

glycerophospholipid A polar lipid of glycerol attached to two fatty acids and a phosphate group connected to an ionized amino alcohol such as choline, serine, or ethanolamine.

hydrogenation The addition of hydrogen to unsaturated fats.

lipid bilayer A model of a cell membrane in which phospholipids are arranged in two rows.

lipids A family of compounds that is nonpolar in nature and not soluble in water; includes prostaglandins, waxes, triacylglycerols, phospholipids, and steroids.

lipoprotein A polar complex composed of a combination of nonpolar lipids with glycerophospholipids and proteins to form a polar complex that can be transported through body fluids.

monounsaturated fatty acid A fatty acid with one double bond.

oil A triacylglycerol that is usually a liquid at room temperature and is obtained from a plant source.

phospholipid A polar lipid of glycerol or sphingosine attached to fatty acids and a phosphate group connected to an amino alcohol.

polyunsaturated fatty acid A fatty acid that contain two or more double bonds.

prostaglandins A number of compounds derived from arachidonic acid that regulate several physiological processes.

saturated fatty acids Fatty acids that have no double bonds, which have higher melting points than unsaturated lipids, and are usually solid at room temperature.

sphingomyelin A compound in which the amine group of sphingosine forms an amide bond to a fatty acid and the hydroxyl group forms an ester bond with phosphate, which forms another phosphoester bond to an amino alcohol.

steroids Types of lipid containing a multicyclic ring system.

triacylglycerols A family of lipids composed of three fatty acids bonded through ester bonds to glycerol, a trihydroxy alcohol.

unsaturated fatty acids Fatty acids that contain one or more carbon–carbon double bonds, which have lower melting points than saturated fatty acids, and are usually liquids at room temperature.

wax The ester of a long-chain alcohol and a long-chain saturated fatty acid.

CORE CHEMISTRY SKILLS

Identifying Fatty Acids (15.2)

Fatty acids are unbranched carboxylic acids that typically contain an even number (12 to 20) of carbon atoms.

- Fatty acids may be saturated, monounsaturated with one double bond, or polyunsaturated with two or more double bonds.
- The double bonds in unsaturated fatty acids are almost always cis.

Example: State the number of carbon atoms, saturated or unsaturated, and name of the following:

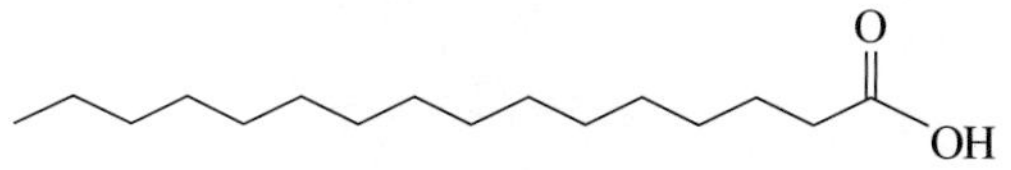

Answer: 16 carbon atoms, saturated, palmitic acid

Drawing Structures for Triacylglycerols (15.3)

- Triacylglycerols are esters of glycerol with three long-chain fatty acids.

Example: Draw and name the triacylglycerol formed from the esterification of glycerol and palmitic acid.

Answer:

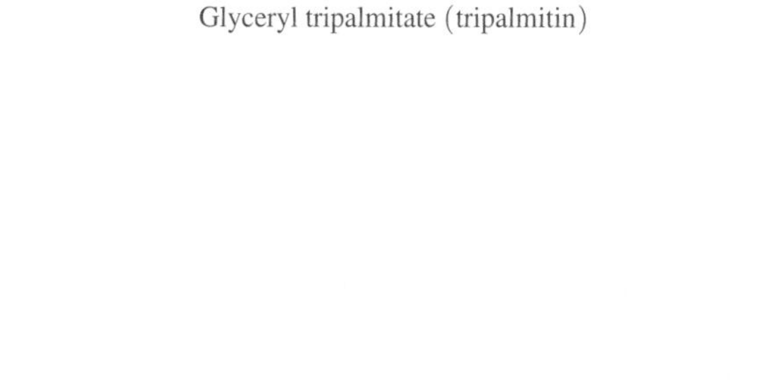

Glyceryl tripalmitate (tripalmitin)

Identifying the Products for the Hydrogenation, Hydrolysis, and Saponification of a Triacylglycerol (15.4)

- The hydrogenation of unsaturated fatty acids of a triacylglycerol in the presence of a Pt or Ni catalyst converts double bonds to single bonds.
- The hydrolysis of the ester bonds in triacylglycerols in the presence of a strong acid produces glycerol and fatty acids.
- In saponification, a triacylglycerol heated with a strong base produces glycerol and the salts of the fatty acids (soap).

Example: Identify each of the following as hydrogenation, hydrolysis, or saponification and state the products:

a. the reaction of palm oil with NaOH
b. the reaction of glyceryl trilinoleate from corn oil with water in the presence of an acid catalyst
c. the reaction of corn oil and H_2 using a nickel catalyst

Answer: **a.** The reaction of palm oil with NaOH is saponification, and the products are glycerol and the potassium salts of the fatty acids, which is soap.
b. In hydrolysis, glyceryl trilinoleate reacts with water in the presence of an acid catalyst, which splits the ester bonds to produce glycerol and three molecules of linoleic acid.
c. In hydrogenation, H_2 adds to double bonds in corn oil, which produces a more saturated, and thus more solid, fat.

Identifying the Steroid Nucleus (15.6)

- The steroid nucleus consists of three cyclohexane rings and one cyclopentane ring fused together.

Example: Why are cholesterol, sodium glycocholate (a bile salt), and cortisone (a corticosteroid) all considered steroids?

Answer: They all contain the steroid nucleus of three six-carbon rings and a five-carbon ring fused together.

UNDERSTANDING THE CONCEPTS

The chapter sections to review are shown in parentheses at the end of each question.

15.65 Palmitic acid is obtained from palm oil as glyceryl tripalmitate (tripalmitin). Draw the condensed structural formula for glyceryl tripalmitate. (15.2, 15.3)

Palm fruit from palm trees are a source of palm oil.

15.66 Jojoba wax in candles consists of a 20-carbon saturated fatty acid and a 20-carbon saturated alcohol. Draw the condensed structural formula for jojoba wax. (15.2, 15.3)

Candles contain jojoba wax.

15.67 Identify each of the following as a saturated, monounsaturated, polyunsaturated, omega-3, or omega-6 fatty acid: (15.2)

a. $CH_3-(CH_2)_7-CH=CH-(CH_2)_7-\overset{\overset{\displaystyle O}{\|}}{C}-OH$

b. linoleic acid

c. $CH_3-CH_2-(CH=CH-CH_2)_5-CH_2-CH_2-\overset{\overset{\displaystyle O}{\|}}{C}-OH$

15.68 Identify each of the following as a saturated, monounsaturated, polyunsaturated, omega-3, or omega-6 fatty acid: (15.2)

a. $CH_3-(CH_2)_4-CH=CH-CH_2-CH=CH-(CH_2)_7-\overset{\overset{\displaystyle O}{\|}}{C}-OH$

b. linolenic acid

c. $CH_3-(CH_2)_{14}-\overset{\overset{\displaystyle O}{\|}}{C}-OH$

Salmon is a good source of omega-3 unsaturated fatty acids.

ADDITIONAL QUESTIONS AND PROBLEMS

15.69 Among the ingredients in lipstick are beeswax, carnauba wax, hydrogenated vegetable oils, and glyceryl tricaprate (tricaprin). (15.1, 15.3)

a. What types of lipids have been used?

b. Draw the condensed structural formula for glyceryl tricaprate (tricaprin). Capric acid is a saturated 10-carbon fatty acid.

15.70 On the list of ingredients of a cosmetic product are glyceryl tristearate and a lecithin. (15.1, 15.3)

a. What types of lipids have been used?

b. Draw the condensed structural formula for lecithin with palmitic acids and choline.

15.71 Trans fats are produced during the hydrogenation of their unsaturated fatty acids. (15.2, 15.3, 15.4)

a. What is the typical configuration of the double bond in a monounsaturated fatty acid?

b. How does a trans fatty acid differ from a cis fatty acid?

c. Draw the condensed structural formula for *trans*-oleic acid.

15.72 Because peanut oil floats on the top of peanut butter, many brands of peanut butter are hydrogenated. A triacylglycerol in peanut oil that contains one oleic acid and two linoleic acids is completely hydrogenated. Draw the condensed structural formula for the product. (15.3, 15.4)

15.73 Which of the following are triacylglycerols? (15.1, 15.2, 15.3, 15.5, 15.6)

a. chylomicrons
b. glyceryltripalmitate
c. LDL
d. safflower oil
e. lecithin
f. sodium stearate

15.74 Identify each of the following as a fatty acid, soap, triacylglycerol, wax, glycerophospholipid, sphingolipid, or steroid: (15.1, 15.2, 15.3, 15.5, 15.6)

a. sphingomyelin
b. whale blubber
c. adipose tissue
d. progesterone
e. cortisone
f. stearic acid

15.75 Identify the components (**1** to **6**) contained in each of the following lipids (**a** to **d**): (15.1, 15.2, 15.3, 15.5, 15.6)

1. glycerol **2.** fatty acid
3. phosphate **4.** amino alcohol
5. steroid nucleus **6.** sphingosine

a. estrogen **b.** cephalin
c. wax **d.** triacylglycerol

15.76 Identify the components (**1** to **6**) contained in each of the following lipids (**a** to **d**): (15.1, 15.2, 15.3, 15.5, 15.6)

1. glycerol **2.** fatty acid
3. phosphate **4.** amino alcohol
5. steroid nucleus **6.** sphingosine

a. glycerophospholipid **b.** sphingomyelin
c. aldosterone **d.** linoleic acid

15.77 Which of the following are found in cell membranes? (15.7)

a. cholesterol **b.** triacylglycerols
c. carbohydrates

15.78 Which of the following are found in cell membranes? (15.7)

a. proteins **b.** waxes
c. phospholipids

15.79 Match the lipoprotein (**1** to **4**) with its description (**a** to **d**). (15.6)

1. chylomicrons **2.** VLDL **3.** LDL **4.** HDL

a. "good" cholesterol
b. transports most of the cholesterol to the cells
c. carries triacylglycerols from the intestine to the fat cells
d. transports cholesterol to the liver

15.80 Match the lipoprotein (**1** to **4**) with its description (**a** to **d**). (15.6)

1. chylomicrons **2.** VLDL **3.** LDL **4.** HDL

a. has the greatest abundance of protein
b. "bad" cholesterol
c. carries triacylglycerols synthesized in the liver to the muscles
d. has the lowest density

CHALLENGE QUESTIONS

The following groups of questions are related to the topics in this chapter. However, they do not all follow the chapter order, and they require you to combine concepts and skills from several sections. These questions will help you increase your critical thinking skills and prepare for your next exam.

15.81 Draw all the possible isomers in neutral pH for a glycerophospholipid that contains two stearic acids and a phosphate bonded to ethanolamine. (15.2, 15.5)

15.82 Sunflower seed oil can be used to make margarine. A triacylglycerol in sunflower seed oil contains two linoleic acids and one oleic acid. (15.2, 15.3, 15.4)

Sunflower oil is obtained from the seeds of the sunflower.

a. Draw the condensed structural formulas for two isomers for the triacylglycerol in sunflower seed oil.
b. Using one of the isomers, write the reaction that would be used when sunflower seed oil is used to make solid margarine.

15.83 A sink drain can become clogged with solid fat such as glyceryl tristearate (tristearin). (7.7, 9.4, 15.2, 15.3, 15.4)

a. How would adding lye (NaOH) to the sink drain remove the blockage?
b. Write the equation for the reaction that occurs.
c. How many milliliters of a 0.500 M NaOH solution are needed to completely saponify 10.0 g of glyceryl tristearate (tristearin)?

A sink drain can become clogged with saturated fats.

15.84 One of the triglycerols in olive oil is glyceryl tripalmitoleate (tripalmitolein). (7.7, 8.6, 9.4, 15.2, 15.3, 15.4)

a. Draw the condensed structural formula for glyceryl tripalmitoleate (tripalmitolein).
b. How many liters of H_2 gas at STP are needed to completely saturate 100. g of glyceryl tripalmitoleate (tripalmitolein)?
c. How many milliliters of a 0.250 M NaOH solution are needed to completely saponify 100. g of glyceryl tripalmitoleate (tripalmitolein)?

Olive oil contains glyceryl tripalmitoleate (tripalmitolein).

ANSWERS

Answers to Study Checks

15.1 **a.** 16
b. monounsaturated
c. liquid

15.2

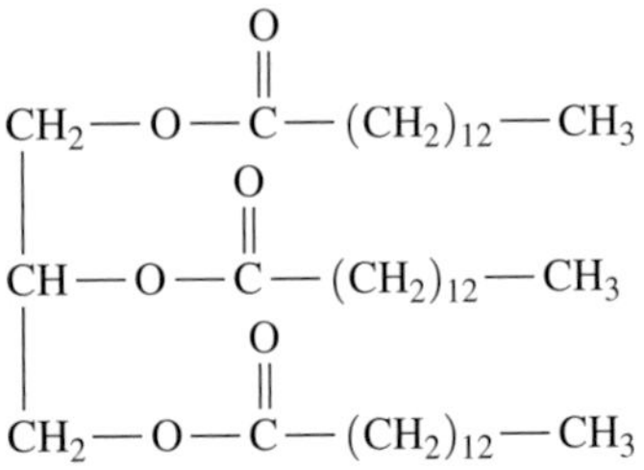

15.3 glyceryl tristearate (tristearin)

15.4

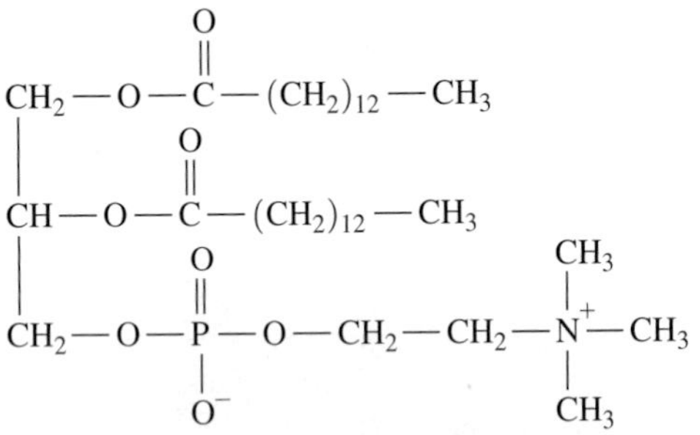

15.5

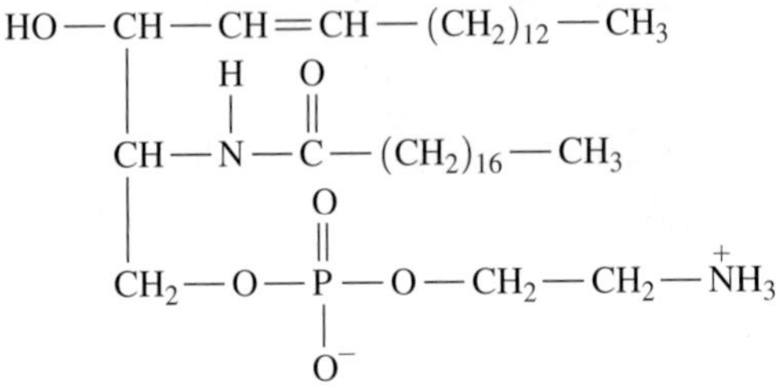

15.6 Cholesterol is not soluble in water; it is classified with the lipid family.

15.7 Cholesterol adds strength and rigidity to the cell membrane.

Answers to Selected Questions and Problems

15.1 Lipids provide energy, protection, and insulation for the organs in the body. Lipids are also an important part of cell membranes.

15.3 Water, a polar solvent cannot dissolve a non polar solute like grease.

15.5 All fatty acids contain a long chain of carbon atoms with a carboxylic acid group. Saturated fatty acids contain only carbon–carbon single bonds; unsaturated fatty acids contain one or more double bonds.

15.7 **a.** Lauric acid

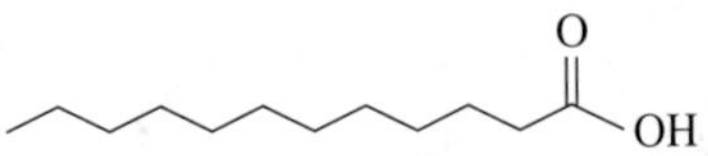

Lauric acid

b. Myristic acid

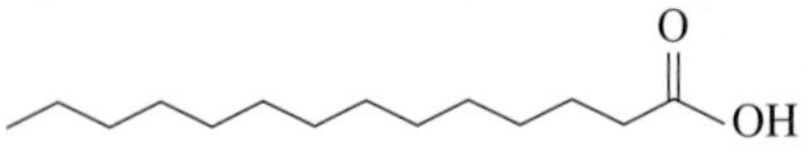

Myristic acid

15.9 **a.** carbon atoms: 16; double bonds: 0 saturated
b. carbon atoms: 18; double bonds: 1 monounsaturated
c. carbon atoms: 20; double bonds: 4 polyunsaturated
d. carbon atoms: 18; double bonds: 2 polyunsaturated

15.11 In a cis fatty acid, the hydrogen atoms are on the same side of the double bond, which produces a kink in the carbon chain. In a trans fatty acid, the hydrogen atoms are on opposite sides of the double bond, which gives a carbon chain without any kink.

15.13 In an omega-3 fatty acid, there is a double bond on carbon 3, counting from the methyl group, whereas in an omega-6 fatty acid, there is a double bond beginning at carbon 6, counting from the methyl group.

15.15

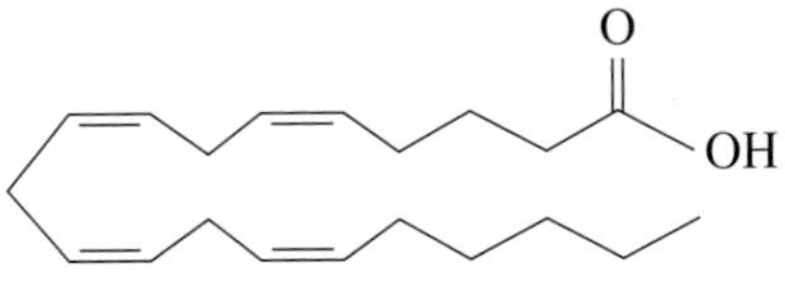

Arachidonic acid

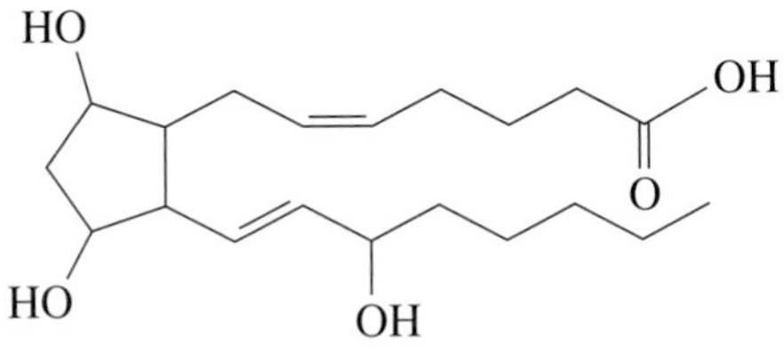

PGF_2

15.17 Prostaglandins raise or lower blood pressure, stimulate contraction and relaxation of smooth muscle, and may cause inflammation and pain.

15.19 $CH_3-(CH_2)_{14}-\overset{\overset{\displaystyle O}{\|}}{C}-O-(CH_2)_{29}-CH_3$

15.21

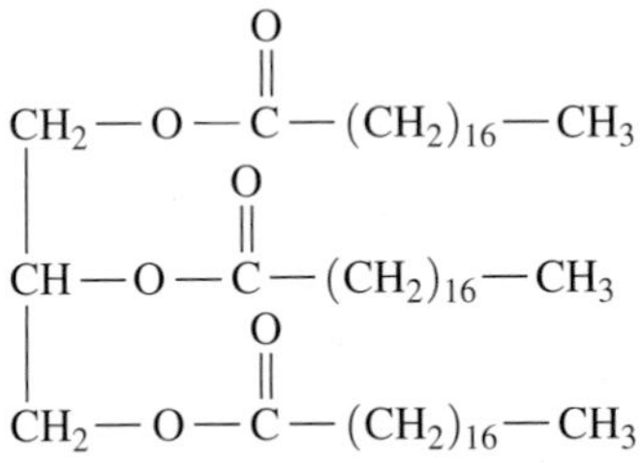

15.23

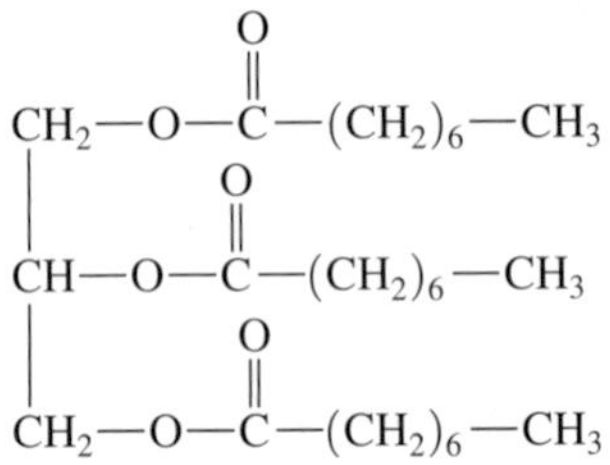

15.25 Unsaturations in oil lowers the packing efficiency of the molecules and as a result, lowers its melting point.

15.27 **a.** hydrogenation
b. hydrolysis

15.29

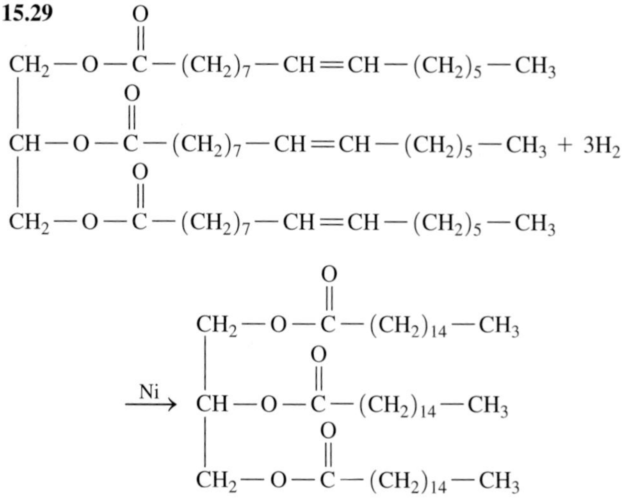

15.31

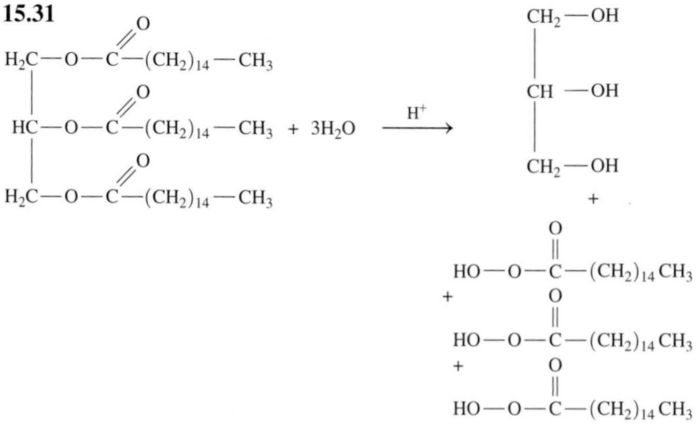

15.33

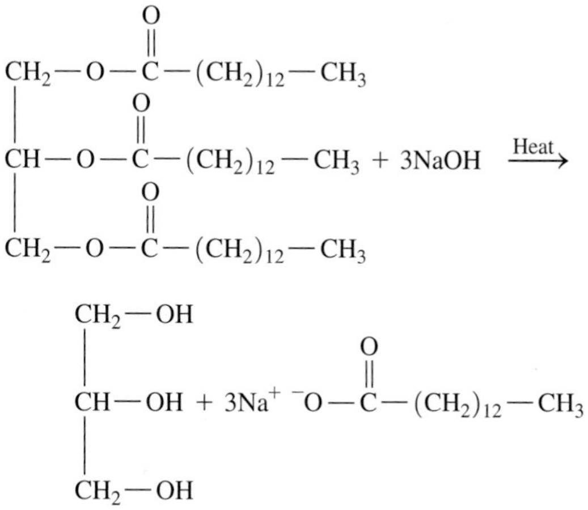

15.35

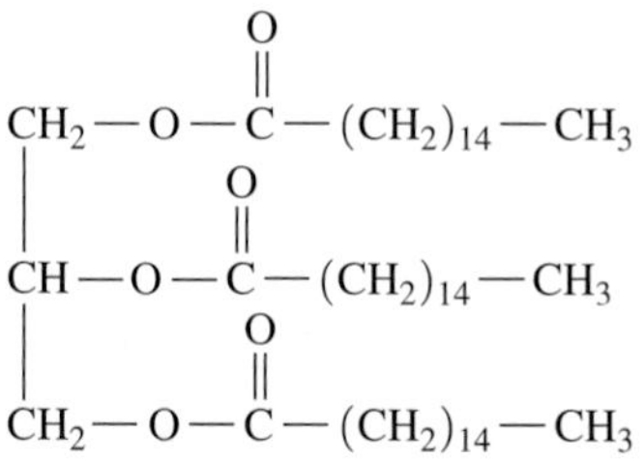

15.37 A triacylglycerol consists of glycerol and three fatty acids. A glycerophospholipid also contains glycerol, but has only two fatty acids. The hydroxyl group on the third carbon of glycerol is attached by a phosphoester bond to an ionized amino alcohol.

15.39

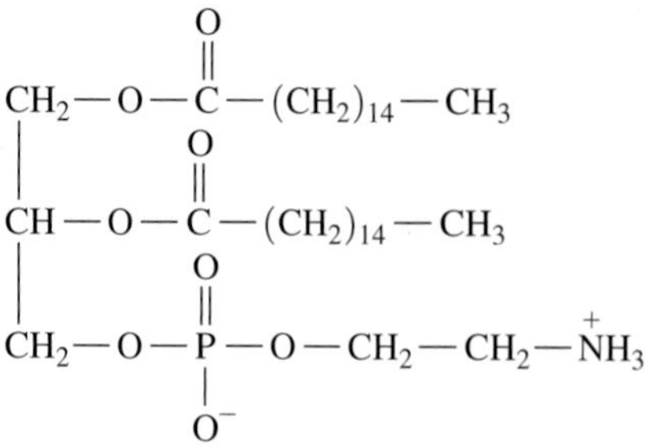

15.41 This glycerophospholipid is a cephalin. It contains glycerol, oleic acid, stearic acid, a phosphate, and ethanolamine.

15.43 **a.** sphingosine
b. palmitic acid
c. an amide bond
d. ethanolamine

15.45

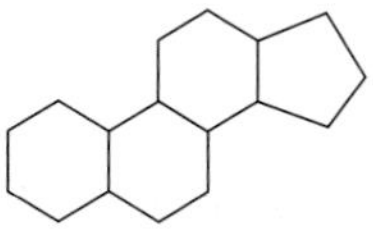

15.47 Bile salts act to emulsify fat globules, allowing the fat to be more easily digested.

15.49 Chylomicrons have a lower density than very-low-density lipoproteins. They pick up triacylglycerols from the intestine, whereas very-low-density lipoproteins transport triacylglycerols synthesized in the liver.

15.51 "Bad" cholesterol is the cholesterol carried by LDL that can form deposits in the arteries called plaque, which narrows the arteries.

15.53 Both estradiol and testosterone contain the steroid nucleus and a hydroxyl group. Testosterone has a ketone group, a double bond, and two methyl groups. Estradiol has an aromatic ring, a hydroxyl group in place of the ketone, and a methyl group.

15.55 (a), (b), (c) are steroid hormones

15.57 The lipid bilayer in a cell membrane surrounds the cell and separates the contents of the cell from the external fluids.

15.59 Because the molecules of cholesterol are large and rigid, they reduce the flexibility of the lipid bilayer and add strength to the cell membrane.

15.61 The peripheral proteins in the membrane emerge on the inner or outer surface only, whereas the integral proteins extend through the membrane to both surfaces.

15.63 Substances move through cell membranes by diffusion (passive transport), facilitated transport, and active transport.
a. facilitated transport
b. diffusion (passive transport)

15.65

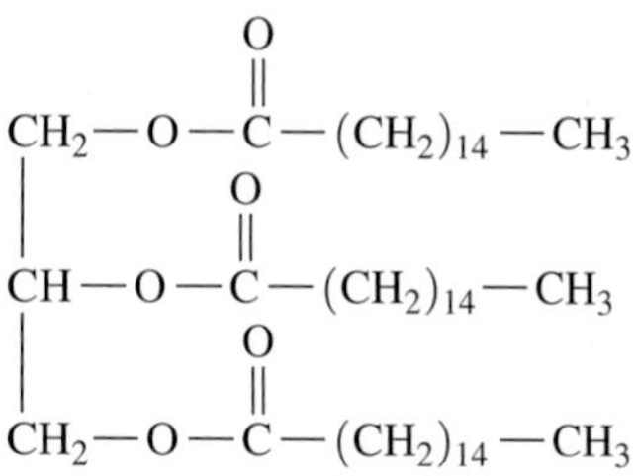

15.67 a. monounsaturated
b. polyunsaturated, omega-6
c. polyunsaturated, omega-3

15.69 a. Beeswax and carnauba are waxes. Vegetable oil and glyceryl tricaprate (tricaprin) are triacylglycerols.

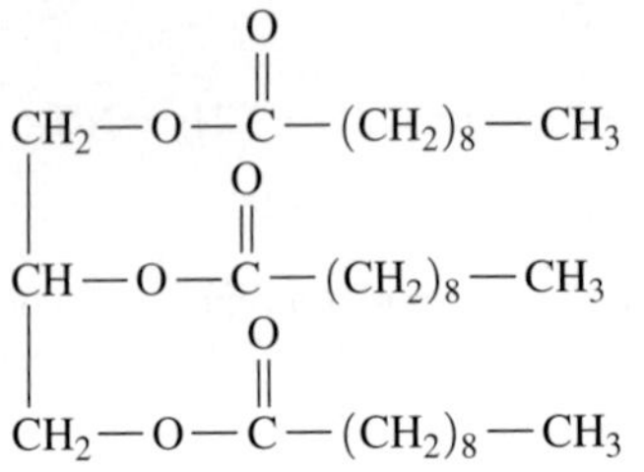

Glyceryl tricaprate (tricaprin)

15.71 a. A typical unsaturated fatty acid has a cis double bond.
b. A cis unsaturated fatty acid contains hydrogen atoms on the same side of each double bond. A trans unsaturated fatty acid has hydrogen atoms on the opposite sides of each double bond that forms during hydrogenation.
c.

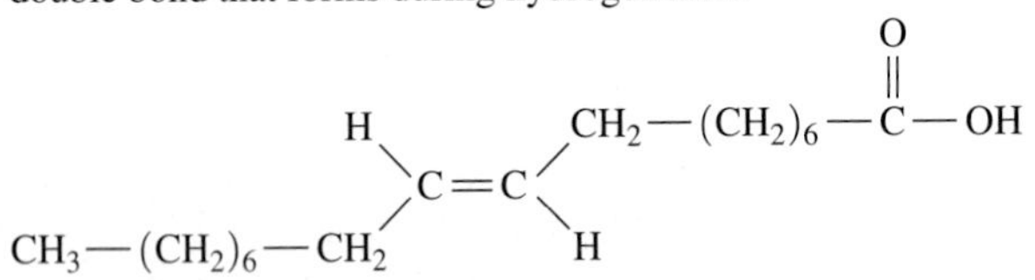

15.73 a. wax
b. steroid
c. glycerophospholipid
d. triacylglycerol
e. soap
f. triacylglycerol

15.75 a. 5
b. 1, 2, 3, 4
c. 2
d. 1, 2

15.77 a and c

15.79 a. 4
b. 3
c. 1
d. 4

15.81

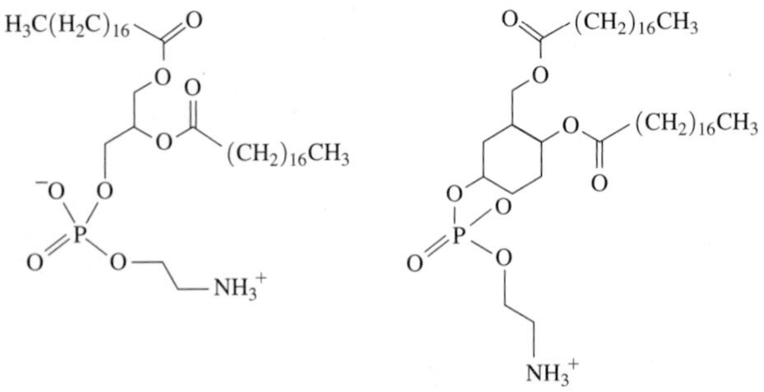

15.83 a. Adding NaOH will saponify lipids such as glyceryl tristearate (tristearin), forming glycerol and salts of the fatty acids that are soluble in water and wash down the drain.
b.

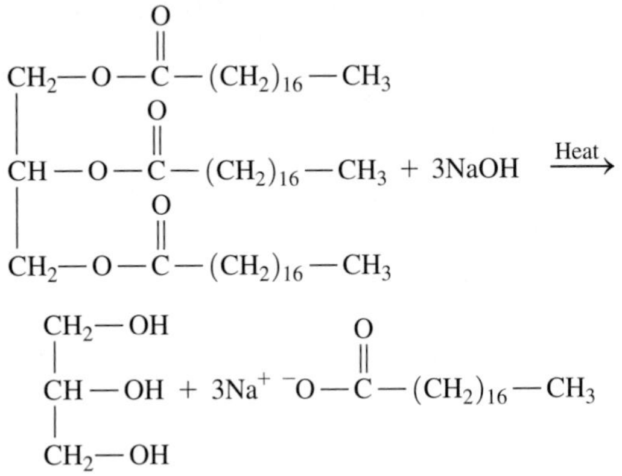

Glycerol — Salts of stearic acid

c. 67.3 mL of a 0.500 M NaOH solution

COMBINING IDEAS FROM Chapters 13 to 15

CI.27 The plastic known as PETE (**p**oly**e**thylene**te**rephthalate) is used to make plastic soft drink bottles and containers for salad dressing, shampoos, and dishwashing liquids. PETE is a polymer of terephthalic acid and ethylene glycol. Today, PETE is the most widely recycled of all the plastics. In one year, 2.4×10^9 pounds of PETE are recycled. After it is separated from other plastics, PETE can be used in polyester fabric, door mats, and tennis ball containers. The density of PETE is 1.38 g/mL. (2.6, 2.7, 14.3)

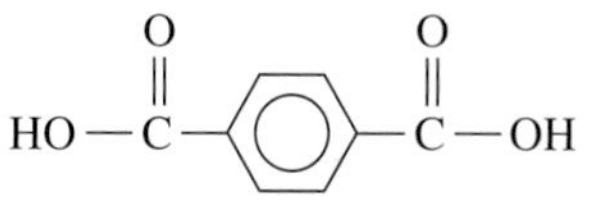

Terephthalic acid

$HO-CH_2-CH_2-OH$

Ethylene glycol

Plastic bottles made of PETE are ready to be recycled.

a. Draw the condensed structural formula for the ester formed from one molecule of terephthalic acid and one molecule of ethylene glycol.
b. Draw the condensed structural formula for the product formed when a second molecule of ethylene glycol reacts with the ester you drew in part **a**.
c. How many kilograms of PETE are recycled in one year?
d. What volume, in liters, of PETE is recycled in one year?
e. Suppose a landfill holds 2.7×10^7 L of recycled PETE. If all the PETE that is recycled in a year were placed instead in landfills, how many would it fill?

CI.28 Using the Internet, look up the condensed structural formulas for the following medicinal drugs and list the functional groups in the compounds: (12.1, 12.3, 14.1, 14.3, 14.5, 14.6)
a. baclofen, a muscle relaxant
b. anethole, a licorice flavoring agent in anise and fennel
c. alibendol, antispasmodic drug
d. pargyline, antihypertensive drug
e. naproxen, nonsteroidal anti-inflammatory drug

CI.29 The insect repellent DEET is an amide that can be made from 3-methylbenzoic acid and diethylamine. A 6.0-fl oz can of DEET repellent contains 25% DEET (m/v)(1 qt = 32 fl oz). (7.1, 7.2, 9.4, 14.6)
a. Draw the condensed structural formula for DEET.
b. Give the molecular formula for DEET.
c. What is the molar mass of DEET?
d. How many grams of DEET are in one spray can?
e. How many molecules of DEET are in one spray can?

DEET is used in insect repellent.

CI.30 Glyceryl trimyristate (trimyristin) is found in the seeds of the nutmeg (*Myristica fragrans*). Trimyristin is used as a lubricant and fragrance in soaps and shaving creams. Isopropyl myristate is used to increase absorption of skin creams. Draw the condensed structural formula for each of the following: (15.2, 15.3, 15.4)

Nutmeg contains high levels of glyceryl trimyristate.

a. myristic acid (14:0)
b. glyceryl trimyristate (trimyristin)
c. isopropyl myristate
d. products from the hydrolysis of glyceryl trimyristate with an acid catalyst
e. products from the saponification of glyceryl trimyristate with KOH
f. reactant and product for oxidation of myristyl alcohol to myristic acid

CI.31 Panose is a trisaccharide that is being considered as a possible sweetener by the food industry. (13.4, 13.5, 13.6)

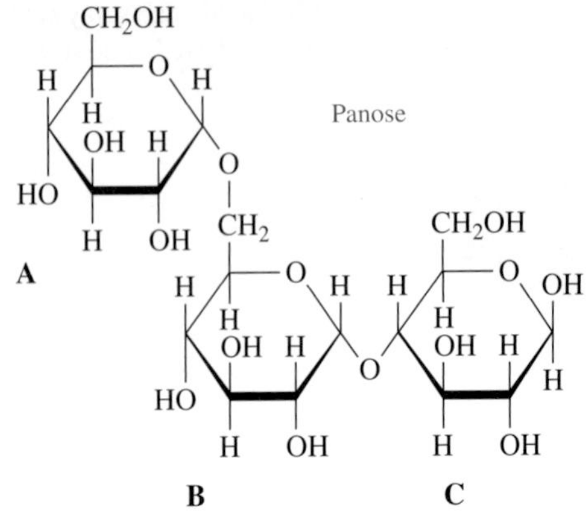

a. What are the monosaccharide units **A**, **B**, and **C** in panose?
b. What type of glycosidic bond connects monosaccharides **A** and **B**?
c. What type of glycosidic bond connects monosaccharides **B** and **C**?
d. Is the structure drawn as α- or β-panose?
e. Why would panose be a reducing sugar?

CI.32 Hyaluronic acid (HA), a polymer of about 25 000 disaccharide units, is a natural component of eye and joint fluid as well as skin and cartilage. Due to the ability of HA to absorb water, it is used in skin-care products and injections to smooth wrinkles and for treatment of arthritis. The repeating disaccharide units in HA consist of D-gluconic acid and *N*-acetyl-D-glucosamine. *N*-Acetyl-D-glucosamine is an amide derived from acetic acid and D-glucosamine, in which an amine group ($-NH_2$) replaces the hydroxyl on carbon 2 of D-glucose. Another natural polymer called chitin is found in the shells of crabs and lobsters. Chitin is made of repeating units of *N*-acetyl-D-glucosamine connected by β-1,4-glycosidic bonds. (13.4, 13.5, 13.6)

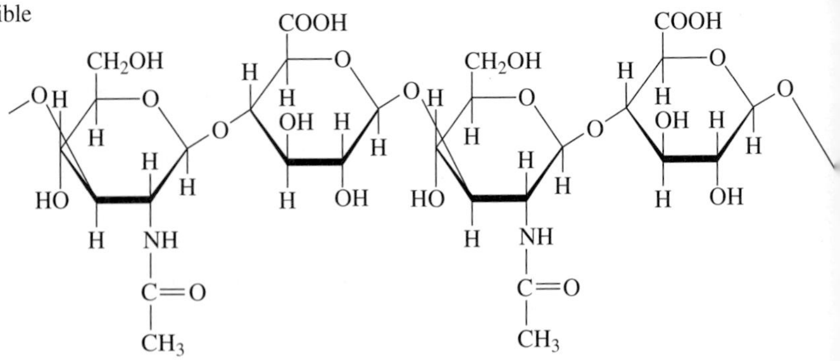

Hyaluronic acid

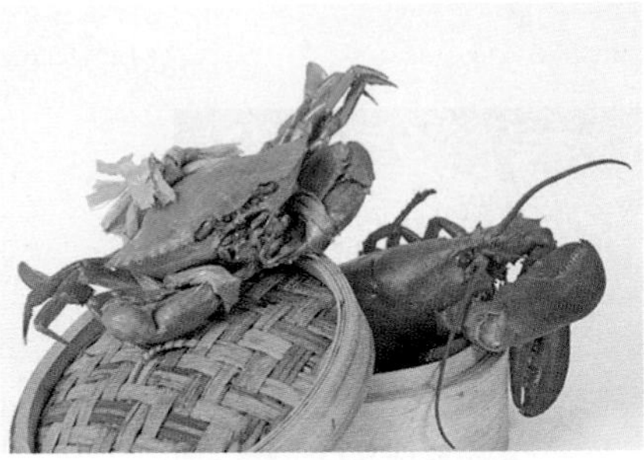

The shells of crabs and lobsters contain chitin.

a. Draw the Haworth structures for the product of the oxidation reaction of the hydroxyl group on carbon 6 in β-D-glucose to form β-D-gluconic acid.
b. Draw the Haworth structure for β-D-glucosamine.
c. Draw the Haworth structure for the amide of β-D-glucosamine and acetic acid.
d. What are the two types of glycosidic bonds that link the monosaccharides in hyaluronic acid?
e. Draw the structure for a section of chitin with two *N*-acetyl-D-glucosamine units linked by β-1,4-glycosidic bonds.

ANSWERS

CI.27 a.

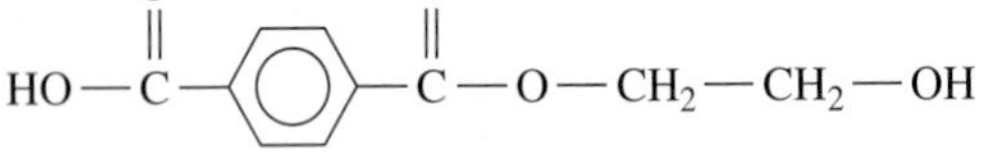

b.

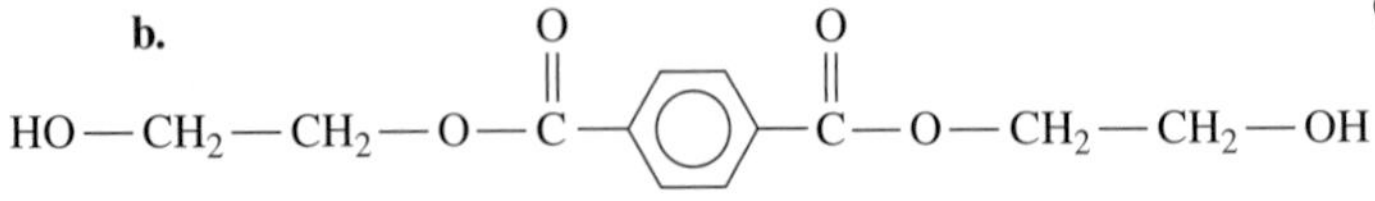

c. 1.1×10^9 kg of PETE
d. 8.0×10^8 L of PETE
e. 30 landfills

CI.29 a.

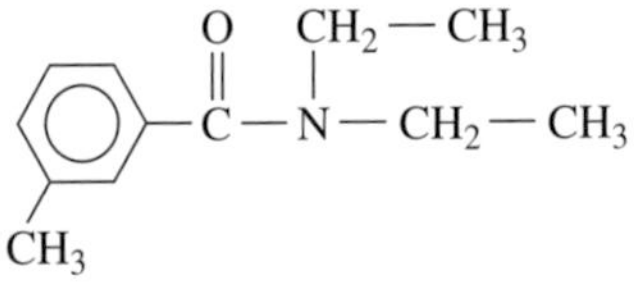

b. $C_{12}H_{17}NO$
c. 191.3 g/mole
d. 44 g of DEET
e. 1.4×10^{23} molecules

CI.31 a. **A**, **B**, and **C** are all glucose.
b. An α-1,6-glycosidic bond links **A** and **B**.
c. An α-1,4-glycosidic bond links **B** and **C**.
d. β-panose
e. Panose is a reducing sugar because it has a free hydroxyl group on carbon 1 of structure **C**, which allows glucose **C** to form an aldehyde.

16 Amino Acids, Proteins, and Enzymes

Noah, a 4-year-old boy, is having severe diarrhea, abdominal pain, and intestinal growling 2 h after he eats. Noah is also underweight for his age. His mother makes an appointment for Noah to see Emma, his physician assistant. Emma immediately suspects that Noah may be lactose intolerant, which occurs due to an enzyme deficiency.

Emma orders a hydrogen breath test for Noah after confirming that he has not eaten for 8 h. The hydrogen breath test involves breathing into a balloon-type container, drinking a solution containing lactose, and then collecting additional breathing samples. Due to the presence of hydrogen gas in Noah's breath, they confirm he is lactose intolerant, as hydrogen is only present when undigested lactose is fermented in the colon by bacteria. Emma advises Noah and his mother to limit dairy products and use lactase when Noah consumes a milk-based product. Lactase is an enzyme, or a biological catalyst, that is a protein. Enzymes significantly increase the rate of a reaction, and since they are not consumed in the reaction, they can be re-used. Lactose is the substrate, or the reacting molecule, that is broken down by lactase. Lactose has a structure that is specific to the active site of the lactase enzyme. The active site is a region in an enzyme where the reaction occurs.

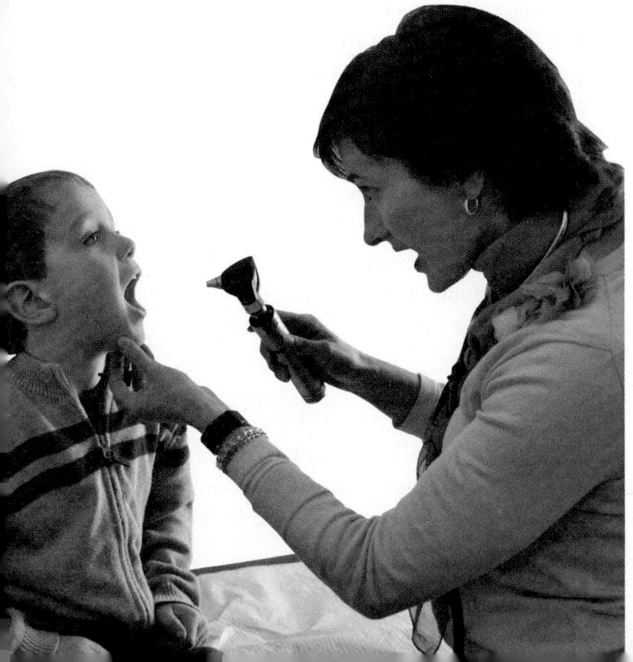

CAREER Physician Assistant

A physician assistant, commonly referred to as a PA, helps a doctor by examining and treating patients, as well as prescribing medications. Many physician assistants take on the role of the primary caregiver. Their duties would also include obtaining patient medical records and histories, diagnosing illnesses, educating and counseling patients, and referring the patient, when needed, to a specialist. Due to this diversity, physician assistants must be knowledgeable about a variety of medical conditions. Physician assistants may also help the doctor during major surgery.

Physician assistants can work in clinics, hospitals, health maintenance organizations, private practices, or take on a more administrative role that involves hiring new PAs and acting as a representative for the hospital and patient.

LOOKING AHEAD

The word "protein" is derived from the Greek word *proteios*, meaning "first." All proteins in humans are large molecules made up from 20 different amino acids. Each kind of protein is composed of amino acids arranged in a specific order that determines the characteristics of the protein and its biological action. Proteins provide structure in membranes, build cartilage and connective tissue, transport oxygen in blood and muscle, direct biological reactions as enzymes, defend the body against infection, and control metabolic processes as hormones. They can even be a source of energy.

The different functions of proteins depend on the structures and chemical behavior of amino acids, the building blocks of proteins. We will see how peptide bonds link amino acids and how the sequence of the amino acids in these protein polymers directs the formation of unique three-dimensional structures.

Every second, thousands of chemical reactions occur in the cells of the human body at rates that meet our physiological and metabolic needs. To make this happen, enzymes catalyze the chemical reactions in our cells, with a different enzyme for every reaction. Digestive enzymes in the mouth, stomach, and small intestine catalyze the hydrolysis of carbohydrates, fats, and proteins. Enzymes in the mitochondria extract energy from biomolecules to give us energy.

CHAPTER READINESS*

KEY MATH SKILL

- Calculating pH from $[H_3O^+]$ (10.5)

CORE CHEMISTRY SKILLS

- Writing Equations for Reactions of Acids and Bases (10.6)
- Naming Carboxylic Acids (14.1)
- Forming Amides (14.6)

*This Key Math Skill and Core Chemistry Skills from previous chapters are listed here for your review as you proceed to the new material in this chapter.

16.1 Proteins and Amino Acids

LEARNING GOAL

Classify proteins by their functions. Give the name and abbreviations for an amino acid and draw its ionized structure.

Protein molecules, compared with many of the compounds we have studied, can be gigantic. Insulin has a molar mass of 5800, and hemoglobin has a molar mass of about 67 000. Some virus proteins are even larger, having molar masses of more than 40 million. Even though proteins can be huge, they all contain the same 20 amino acids. Every protein is a polymer built from these 20 different amino acid building blocks, repeated numerous times, with the specific sequence of the amino acids determining the characteristics of the protein and its biological action. The many kinds of proteins perform different functions. There are proteins that form structural components such as cartilage, muscles, hair, and nails. Wool, silk, feathers, and horns in animals are made of proteins (see Figure 16.1). Proteins that function as enzymes regulate biological reactions such as digestion and cellular metabolism. Other proteins, such as hemoglobin and myoglobin, transport oxygen in the blood and muscle. Table 16.1 gives examples of proteins that are classified by their functions.

FIGURE 16.1 ▶ The horns of animals are made of proteins.
Q What class of protein would be in horns?

Amino Acids

Proteins are composed of molecular building blocks called *amino acids*. Every **amino acid** has a central carbon atom called the α-carbon bonded to two functional groups: an ammonium group $(-NH_3^+)$ and a carboxylate group $(-COO^-)$. The α-carbon is also

TABLE 16.1 Classification of Some Proteins and Their Functions

Class of Protein	Function	Examples
Structural	Provide structural components	*Collagen* is in tendons and cartilage. *Keratin* is in hair, skin, wool, and nails.
Contractile	Make muscles move	*Myosin* and *actin* contract muscle fibers.
Transport	Carry essential substances throughout the body	*Hemoglobin* transports oxygen. *Lipoproteins* transport lipids.
Storage	Store nutrients	*Casein* stores protein in milk. *Ferritin* stores iron in the spleen and liver.
Hormone	Regulate body metabolism and the nervous system	*Insulin* regulates blood glucose level. *Growth hormone* regulates body growth.
Enzyme	Catalyze biochemical reactions in the cells	*Sucrase* catalyzes the hydrolysis of sucrose. *Trypsin* catalyzes the hydrolysis of proteins.
Protection	Recognize and destroy foreign substances	*Immunoglobulins* stimulate immune responses.

bonded to a hydrogen atom and a side chain called an R group. The differences in the 20 amino acids present in human proteins are due to the unique characteristics of the R groups. For example, alanine has a methyl, $-CH_3$, as its R group. When an amino acid has a balance of positive and negative charge, it has an overall zero charge, which occurs at a specific pH value known as its **isoelectric point (pI)**.

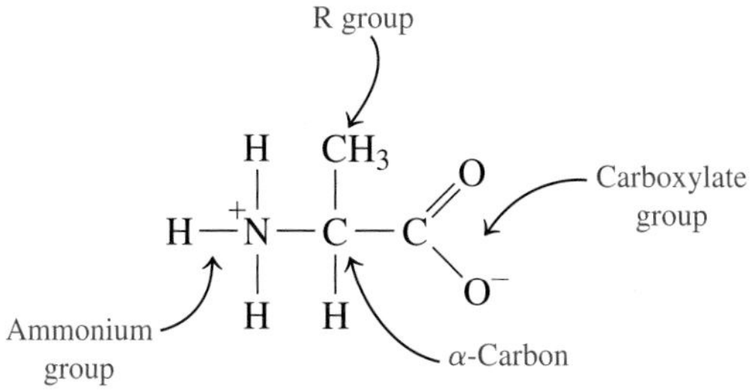

Ionized form of alanine

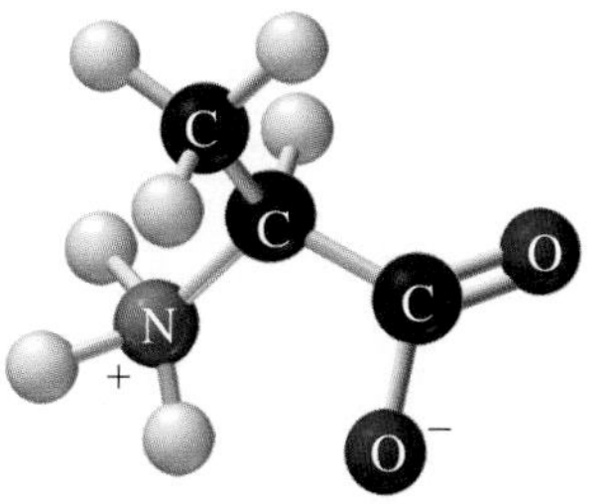
Ball-and-stick model of alanine at its pI 6.0

Classification of Amino Acids

We can now classify amino acids using their specific R groups, which determine their properties in aqueous solution. The **nonpolar amino acids** have hydrogen, alkyl, or aromatic R groups, which make them *hydrophobic* ("water fearing"). The **polar amino acids** have R groups that interact with water, which makes them *hydrophilic* ("water loving"). The *polar neutral* amino acids contain hydroxyl ($-OH$), thiol ($-SH$), or amide ($-CONH_2$) groups. The R group of a polar **acidic amino acid** contains a carboxylate group ($-COO^-$). The R group of a polar **basic amino acid** contains an amino group, which ionizes to give an ammonium ion. The different R groups, names, three-letter and one-letter abbreviations, and pI values of the 20 α-amino acids commonly found in proteins are listed in Table 16.2.

A summary of the classification of amino acids follows:

Type of Amino Acid	Type of R Groups	Polarity	Water
Nonpolar	Nonpolar	Nonpolar	Hydrophobic
Polar, neutral	Contain O and S atoms, but no charge	Polar	Hydrophilic
Polar, acidic	Contain carboxyl groups, negative charge	Polar	Hydrophilic
Polar, basic	Contain amino groups, positive charge	Polar	Hydrophilic

TABLE 16.2 Structures, Names, Abbreviations, and Isoelectric Points (pI) of 20 Common Amino Acids at Physiological pH (7.4)

Nonpolar Amino Acids (Hydrophobic)

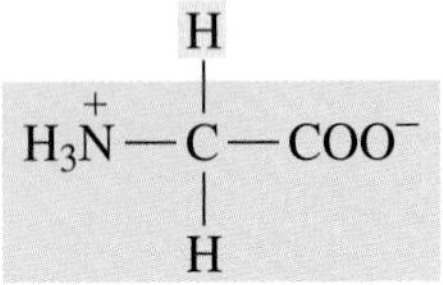
Glycine (Gly, G)
6.0

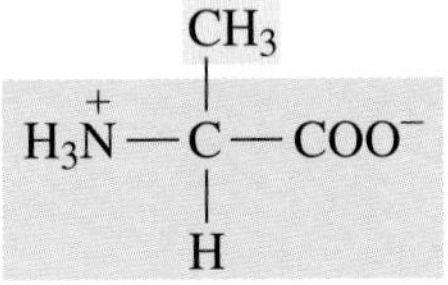
Alanine (Ala, A)
6.0

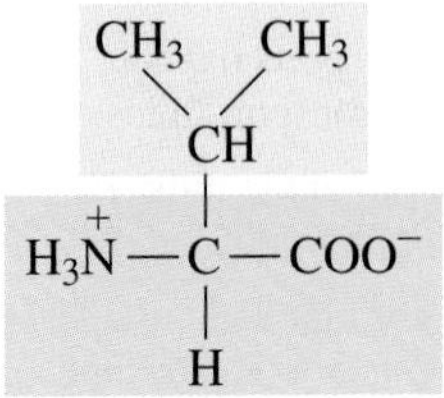
Valine (Val, V)
6.0

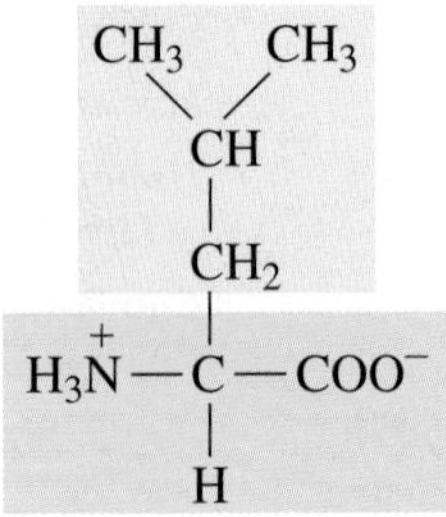
Leucine (Leu, L)
6.0

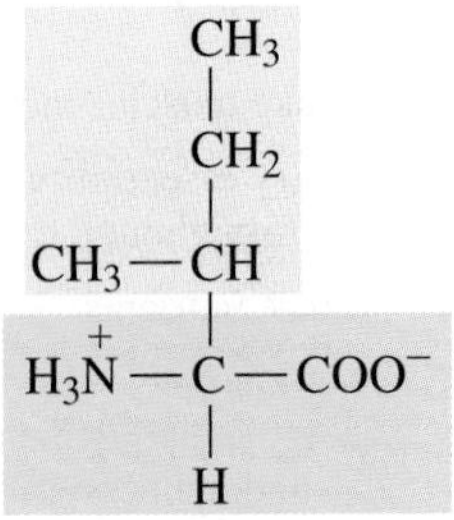
Isoleucine (Ile, I)
6.0

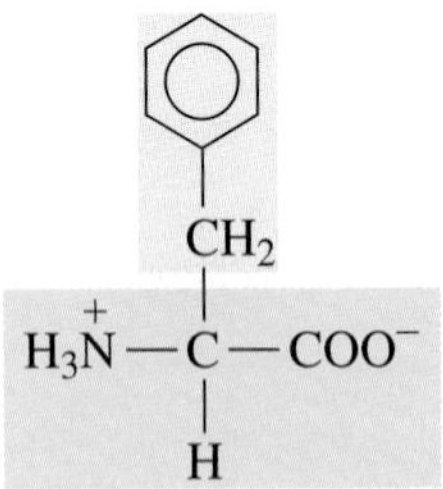
Phenylalanine (Phe, F)
5.5

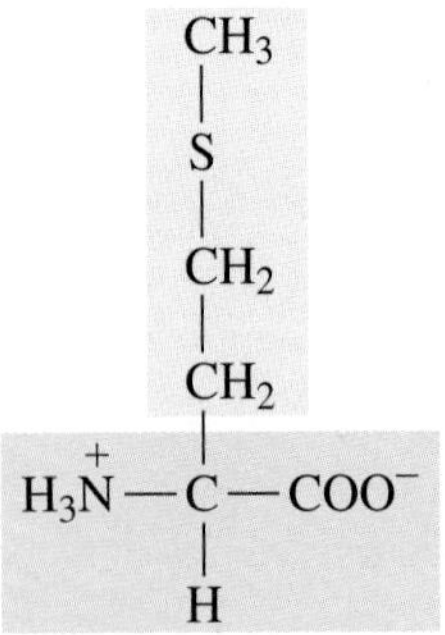
Methionine (Met, M)
5.7

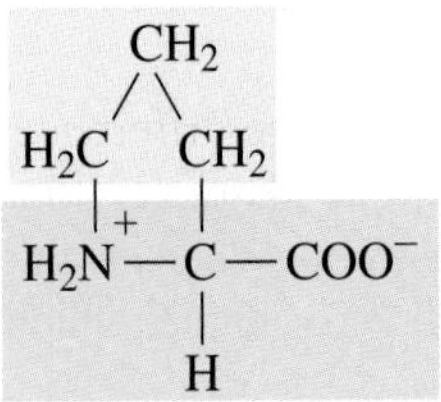
Proline (Pro, P)
6.3

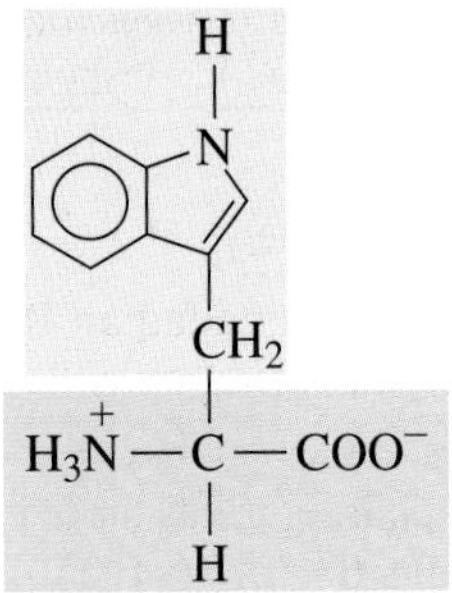
Tryptophan (Trp, W)
5.9

Polar Amino Acids (Hydrophilic)

Amino Acids with Neutral R Groups

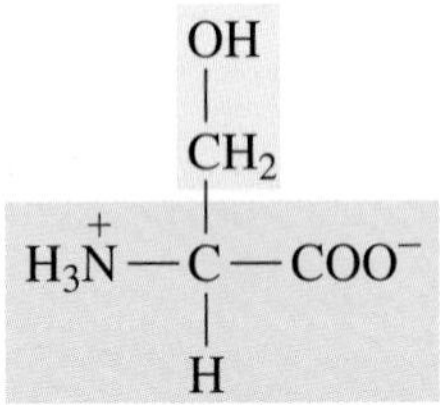
Serine (Ser, S)
5.7

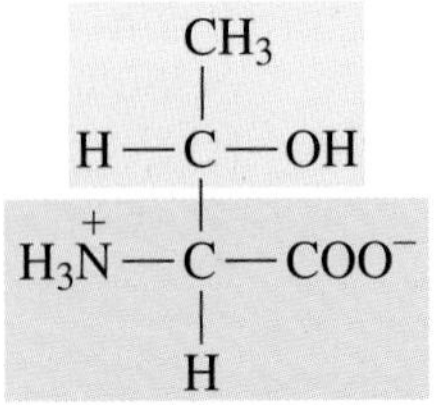
Threonine (Thr, T)
5.6

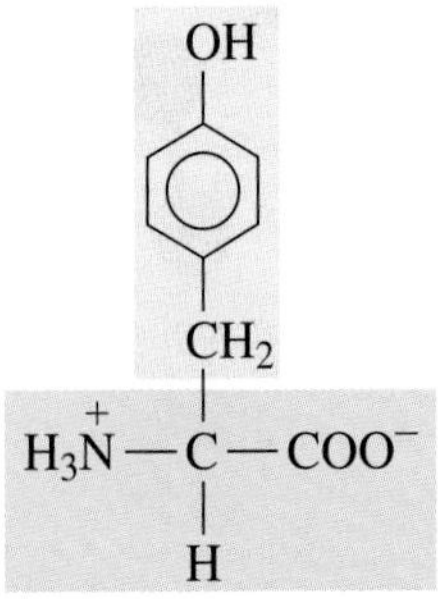
Tyrosine (Tyr, Y)
5.7

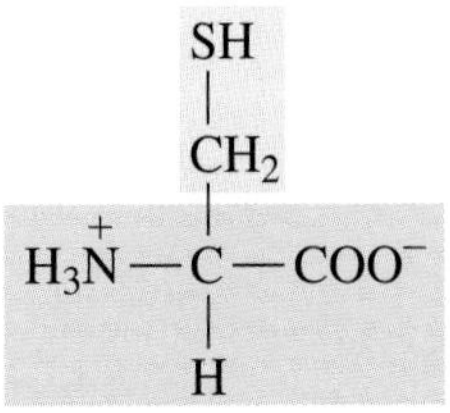
Cysteine (Cys, C)
5.1

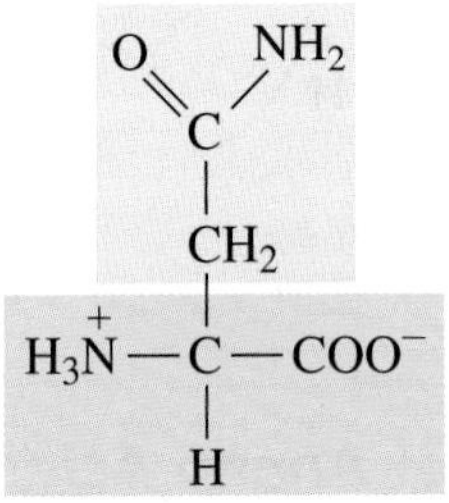
Asparagine (Asn, N)
5.4

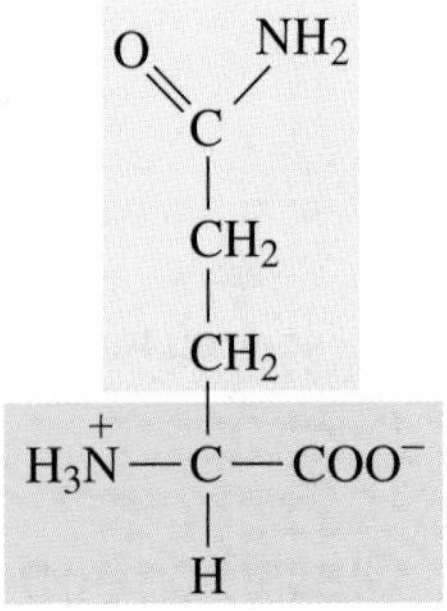
Glutamine (Gln, Q)
5.7

Amino Acids with Charged R Groups

Acidic (negative charge)

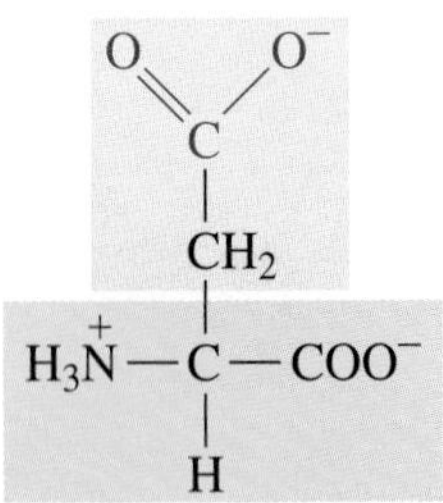
Aspartic acid (Asp, D)
2.8

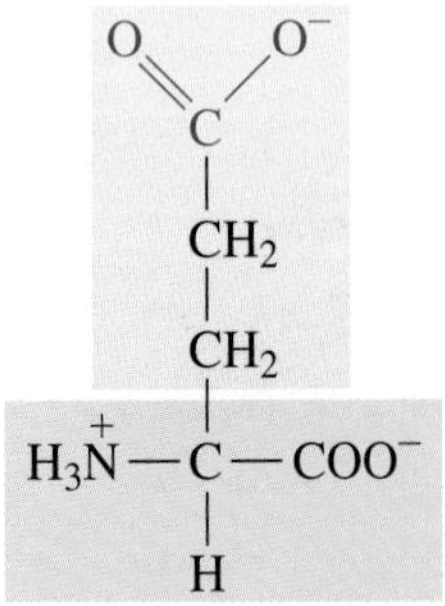
Glutamic acid (Glu, E)
3.2

Basic (positive charge)

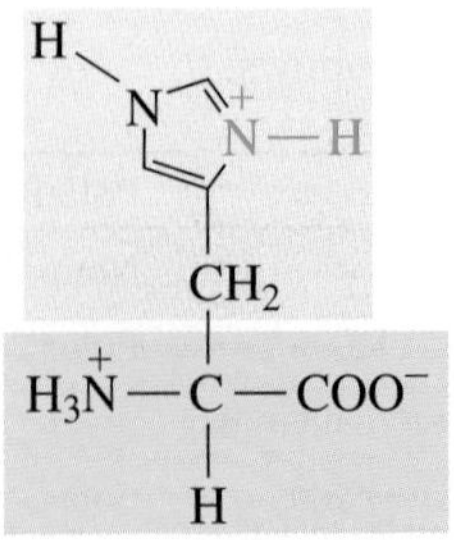
Histidine (His, H)
7.6

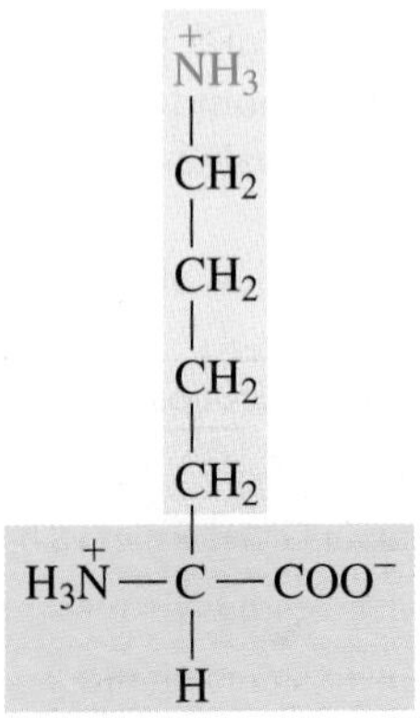
Lysine (Lys, K)
9.7

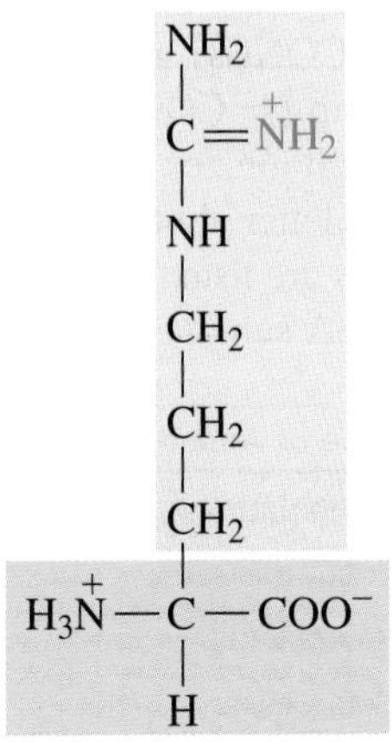
Arginine (Arg, R)
10.8

SAMPLE PROBLEM 16.1 Polarity of Amino Acids

Classify each of the following amino acids as nonpolar or polar. If polar, indicate if the R group is neutral, acidic, or basic. Indicate if each would be hydrophobic or hydrophilic.

a. valine **b.** asparagine

SOLUTION

a. The R group in valine consists of several C atoms and H atoms, which makes valine a nonpolar amino acid that is hydrophobic.

b. The R group in asparagine contains C atoms, H atoms, and an amide group, which makes asparagine a polar amino acid that is neutral and hydrophilic.

STUDY CHECK 16.1

Classify the amino acid histidine as nonpolar or polar. If polar, indicate if the R group is neutral, acidic, or basic and if it would be hydrophobic or hydrophilic.

SAMPLE PROBLEM 16.2 Structural Formulas of Amino Acids

Draw the ionized form for each of the following amino acids, and write the three-letter and one-letter abbreviations:

a. serine **b.** aspartic acid

SOLUTION

	Common Components	R Group	Abbreviations
ANALYZE THE PROBLEM	$-NH_3^+$, $-COO^-$, and $-H$	specific for each amino acid (see Table 16.2)	three letters, one letter

Amino Acid	Common Components	R Group	Abbreviations
a. Serine	$-NH_3^+$, $-COO^-$, and $-H$	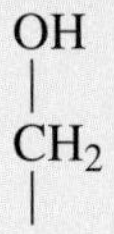	Ser, S
b. Aspartic acid	$-NH_3^+$, $-COO^-$, and $-H$	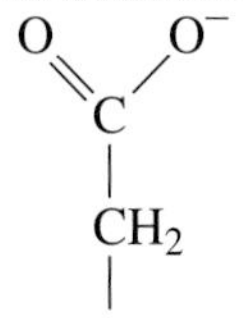	Asp, D

CORE CHEMISTRY SKILL

Drawing the Ionized Form for an Amino Acid

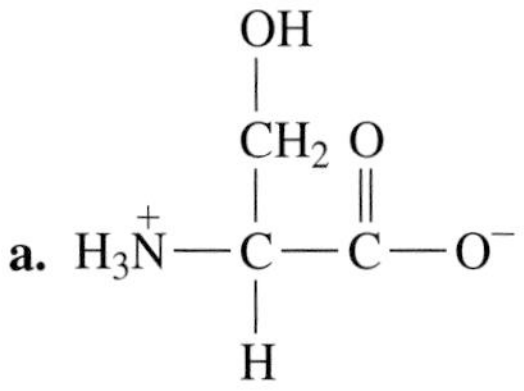

Serine (Ser, S)

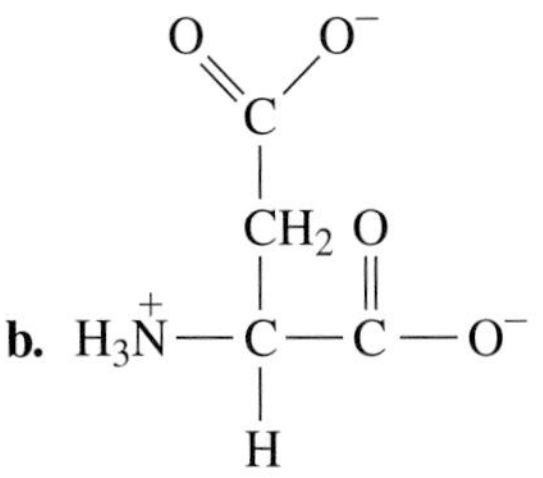

Aspartic acid (Asp, D)

STUDY CHECK 16.2

Draw the ionized form for leucine, and write the three-letter and one-letter abbreviations.

Amino Acid Stereoisomers

All the α-amino acids except for glycine are chiral because the α-carbon is attached to four different atoms. Thus amino acids can exist as D and L stereoisomers. We can draw Fischer projections for α-amino acids by placing the carboxylate group at the top and the R group at the bottom. In the L stereoisomer, the $-NH_3^+$ group is on the left, and in the D stereoisomer, the $-NH_3^+$ group is on the right. In biological systems, the only amino acids incorporated into proteins are the L stereoisomers. There are D amino acids found in nature, but not in proteins. Let's take a look at the stereoisomers for L- and D-glyceraldehyde, a carbohydrate, and two amino acids, L- and D-alanine, and L- and D-cysteine.

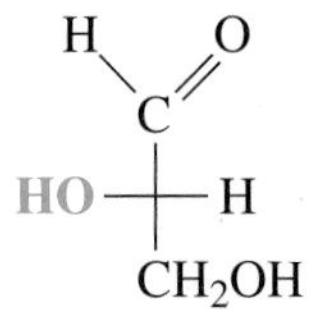
L-Glyceraldehyde

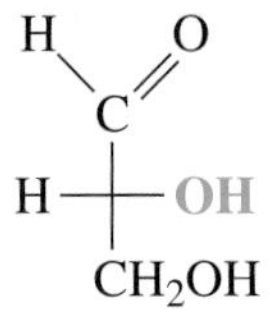
D-Glyceraldehyde

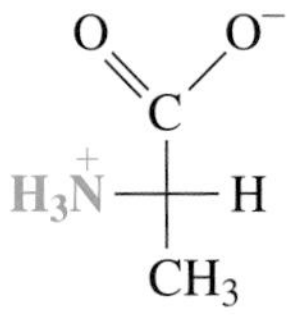
L-Alanine

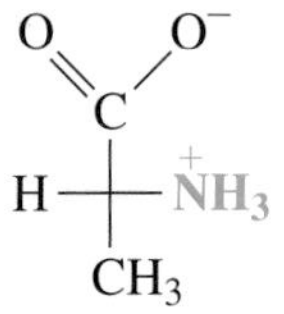
D-Alanine

In amino acids, the $-NH_3^+$ group appears on the right or left of the chiral carbon to give D or L stereoisomers.

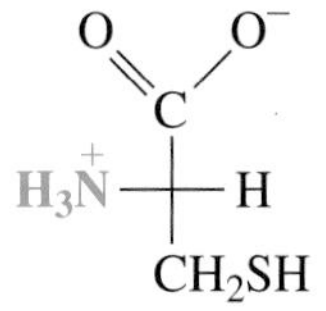
L-Cysteine

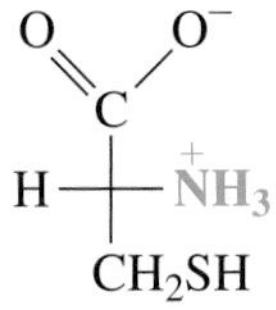
D-Cysteine

QUESTIONS AND PROBLEMS

16.1 Proteins and Amino Acids

LEARNING GOAL Classify proteins by their functions. Give the name and abbreviations for an amino acid and draw its ionized structure.

16.1 Classify each of the following proteins according to its function:
a. lipoprotein transport lipids
b. ferritin stores iron in the spleen and liver
c. immunoglobin stimulates immune responses
d. tripsin catalyzes the hydrolysis of proteins

16.2 Suggest a protein for each of the following functions:
a. hormone
b. protection
c. storage
d. enzyme

16.3 What functional groups are found in all α-amino acids?

16.4 How does the polarity of the R group in leucine compare to the R group in serine?

16.5 Draw the ionized form for each of the following amino acids:
a. Methionine **b.** Alanine
c. isoleucine **d.** Phenylalanine

16.6 Draw the ionized form for each of the following amino acids:
a. lysine **b.** proline
c. V **d.** Tyr

16.7 Classify each of the amino acids in problem 16.5 as polar or nonpolar. If polar, indicate if the R group is neutral, acidic, or basic. Indicate if each is hydrophobic or hydrophilic.

16.8 Classify the following amino acids as polar, nonpolar, hydrophobic or hydrophilic. If polar, indicate if the R group is neutral, acidic, or basic at physiological pH:
a. alanine **b.** phenylalaninec
c. cysteine

16.9 Give the name for the amino acid represented by each of the following abbreviations:
a. Ala **b.** Q
c. K **d.** Cys

16.10 Give the name for the amino acid represented by each of the following abbreviations:
a. Trp **b.** M
c. Pro **d.** G

Chemistry Link to Health

Essential Amino Acids

Of the 20 amino acids used to build the proteins in the body, only 11 can be synthesized in the body. The other 9 amino acids, listed in Table 16.3, are *essential amino acids* that must be obtained from the proteins in the diet.

Complete proteins, which contain all of the essential amino acids, are found in most animal products such as eggs, milk, meat, fish, and poultry. However, gelatin and plant proteins such as grains, beans, and nuts are *incomplete proteins* because they are deficient in one or more of the essential amino acids. Diets that rely on plant foods for protein must contain a variety of protein sources to obtain all the essential amino acids. For example, a diet of rice and beans contains all the essential amino acids because they are *complementary protein* sources. Rice contains the methionine and tryptophan that are deficient in beans, while beans contain the lysine that is lacking in rice (see Table 16.4).

Complete proteins such as eggs, milk, meat, and fish contain all of the essential amino acids. Incomplete proteins from plants such as grains, beans, and nuts are deficient in one or more essential amino acids.

TABLE 16.3 Essential Amino Acids for Adults

Histidine (His)	Phenylalanine (Phe)
Isoleucine (Ile)	Threonine (Thr)
Leucine (Leu)	Tryptophan (Trp)
Lysine (Lys)	Valine (Val)
Methionine (Met)	

TABLE 16.4 Amino Acid Deficiencies in Selected Vegetables and Grains

Food Source	Amino Acids Missing
Eggs, milk, meat, fish, poultry	None
Wheat, rice, oats	Lysine
Corn	Lysine, tryptophan
Beans	Methionine, tryptophan
Peas	Methionine
Almonds, walnuts	Lysine, tryptophan
Soy	Low in methionine

16.2 Amino Acids as Acids and Bases

LEARNING GOAL

Draw the condensed structural formula for an amino acid at pH values above or below its isoelectric point.

An amino acid with positive and negative charges and thus an overall neutral charge forms only at a certain pH called the *isoelectric point* (pI). However, an amino acid can exist as a positive ion if a solution is more acidic (has a lower pH) than its pI or as a negative ion if a solution is more basic (has a higher pH) than its pI.

The pI values for nonpolar and polar neutral amino acids are from pH 5.1 to 6.3. Let's take a look at the different ionic forms of alanine. As we have seen, alanine has a zero overall charge at its pI of 6.0 with a carboxylate anion ($-COO^-$) and an ammonium cation ($-NH_3^+$). However, the charge balance changes when alanine is placed in a solution that has a pH that is more acidic or more basic than its pI. In a solution that is more acidic, with a lower pH than its pI ($pH < 6.0$), the $-COO^-$ group of alanine gains H^+, which forms its carboxylic acid ($-COOH$). Because the $-NH_3^+$ group retains a charge of 1+, alanine has an overall positive charge (1+) at a pH lower than 6.0.

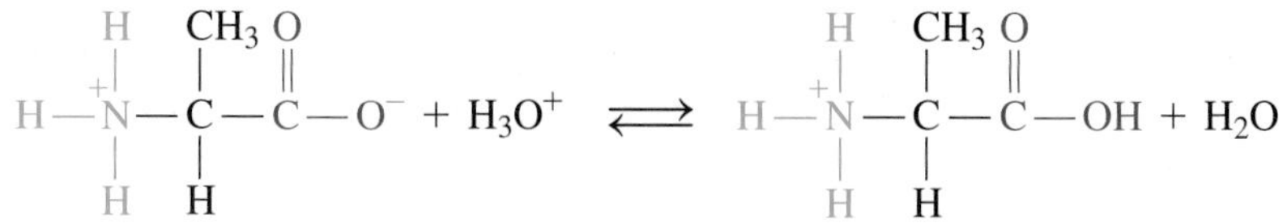

Alanine pH = 6.0 (charge = 0)

Alanine at a pH < 6.0 (charge = 1+)

When alanine is in a solution that is more basic, with a higher pH than its pI ($pH > 6.0$), the $-NH_3^+$ group loses H^+ and forms an amino group ($-NH_2$), which has no charge. Because the $-COO^-$ group has a charge of 1−, alanine has an overall negative charge (1−) at a pH higher than 6.0. The changes in the ionized groups with pH above, below, or equal to the pI are summarized in Table 16.5.

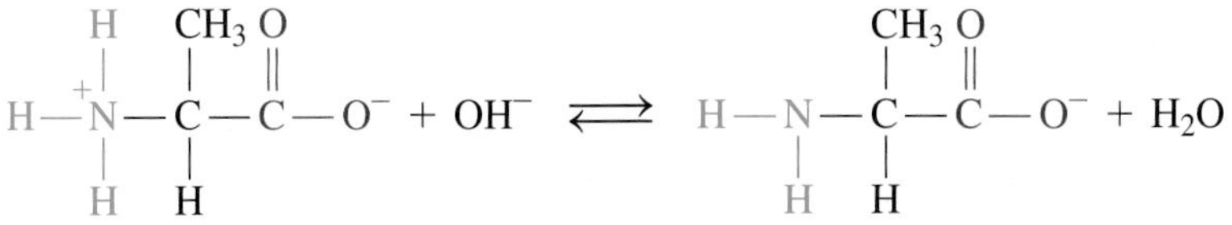

Alanine
pH = 6.0
(charge = 0)

Alanine at a pH > 6.0
(charge = 1−)

TABLE 16.5 Ionized Forms of Nonpolar and Polar Neutral Amino Acids

Solution	pH < pI (acidic)	pH = pI	pH > pI (basic)
Change in H^+	$[H^+]$ increases	No change	$[H^+]$ decreases
Carboxylic Acid/Carboxylate	$-COOH$	$-COO^-$	$-COO^-$
Ammonium/Amino	$-\overset{+}{N}H_3$	$-\overset{+}{N}H_3$	$-NH_2$
Overall Charge	1+	0	1−

The pI values of the acidic amino acids (aspartic acid, glutamic acid) are around pH 3. At pH values of 3, the carboxylic acid group in the R groups is not ionized. However, at physiological pH values, which are greater than 3, the carboxylic acid in the R group loses H^+ to form a negatively charged carboxylate group $-COO^-$.

The pI values of basic amino acids are typically higher than physiological pH values, ranging from pH 7.6 to 10.8. Thus at physiological pH values, the amines in the R groups of the basic amino acids (lysine, arginine, and histidine) gain H^+ to form positively charged ammonium groups $-NH_3^+$.

SAMPLE PROBLEM 16.3 Amino Acids in Acidic or Basic Solution

The pI of glycine is 6.0. Draw the condensed structural formulas for glycine at **a.** pH 6.0 and at **b.** pH 8.0. State the overall charge for each.

SOLUTION

ANALYZE THE PROBLEM		pI	pH	Change in $[H^+]$	Ionized Groups	Overall Charge
	a.	6.0	6.0	no change	$-\overset{+}{N}H_3$, COO^-	(1+) + (1−) = 0
	b.	6.0	8.0	decreases	$-NH_2$, COO^-	(0) + (1−) = 1−

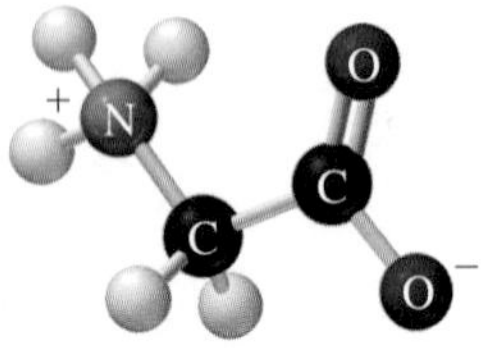

Ball-and-stick model of glycine at its pI of 6.0.

a. At a pH of 6.0, glycine has an ammonium group and a carboxylate group, which give it an overall change of zero (0).

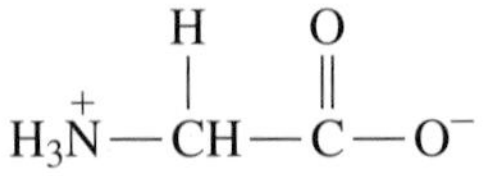

Charge 0 at pH 6.0

b. At a pH of 8.0 (basic solution), the $-NH_3^+$ group loses H^+ to become $-NH_2$, which has no charge. Because the carboxylate group (COO^-) remains ionized, the overall charge of glycine at pH 8.0 is negative (1−).

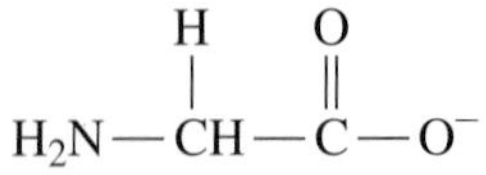

Charge 1− at pH 8.0

STUDY CHECK 16.3

Draw the condensed structural formula for glycine at pH 3.0 and give the overall charge.

QUESTIONS AND PROBLEMS

16.2 Amino Acids as Acids and Bases

LEARNING GOAL Draw the condensed structural formula for an amino acid at pH values above or below its isoelectric point.

16.11 What is the overall charge of a nonpolar amino acid at a pH below its pI?

16.12 What is the overall charge of a polar neutral amino acid at a pH above its pI?

16.13 Draw the condensed structural formula for each of the following amino acids at pH below 1.0:

a. valine **b.** Cys
c. tyrosine **d.** T

16.14 Draw the skeletal formula for each of the following amino acids at pH 10.0:

a. glutamic acid **b.** serine
c. proline **d.** tyrosine

16.15 Would each of the following ionized forms of valine exist at a pH above, below, or at pI?

a. 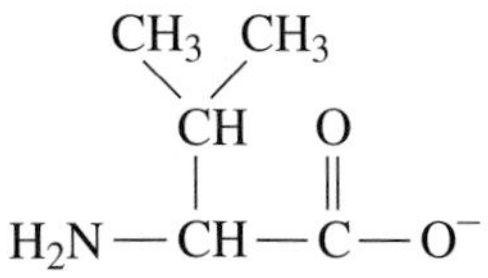

b. 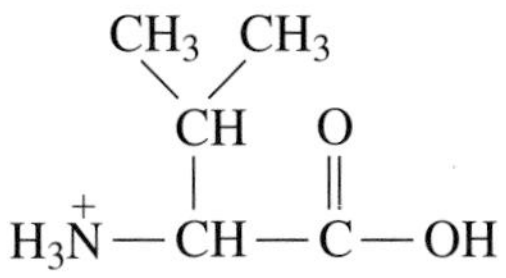

c. 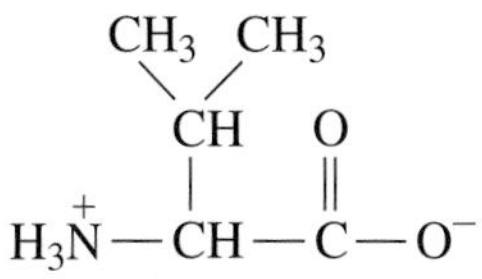

16.16 Would each of the following ionized forms of serine exist at a pH above, below, or at pI?

a. 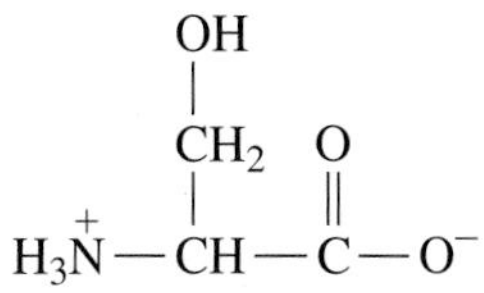

b. 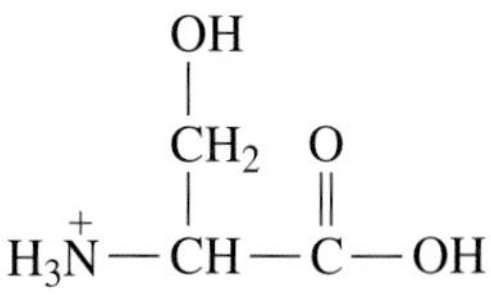

c. 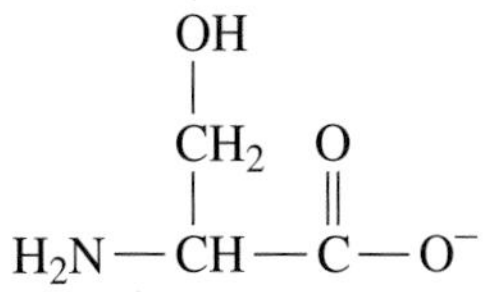

16.3 Proteins: Primary Structure

LEARNING GOAL

Draw the condensed structural formula for a peptide and give its name. Describe the primary structure for a protein.

A **peptide bond** is an amide bond that forms when the $—COO^-$ group of one amino acid reacts with the $—NH_3^+$ group of the next amino acid. The linking of two or more amino acids by peptide bonds forms a **peptide**. An O atom is removed from the carboxylate end of the first amino acid, and two H atoms are removed from the ammonium end of the second amino acid, which produces water. Two amino acids form a *dipeptide*, three amino acids form a *tripeptide*, and four amino acids form a *tetrapeptide*. A chain of five amino acids is a *pentapeptide*, and longer chains of amino acids form *polypeptides*.

We can write the amidation reaction for the formation of the dipeptide glycylalanine (Gly–Ala, GA) (see Figure 16.2). During the amidation, the O atom removed from the carboxylate group combines with two H atoms from the $—NH_3^+$ group to produce H_2O. The amino acid written on the left, glycine, has a free $—NH_3^+$ group, which makes it the **N-terminal amino acid**. In the peptide, the amino acid written on the right, alanine, has a free $—COO^-$ group, which makes it the **C-terminal amino acid**. The dipeptide forms when the carbonyl group in glycine bonds to the N atom in the $—NH_3^+$ group of alanine.

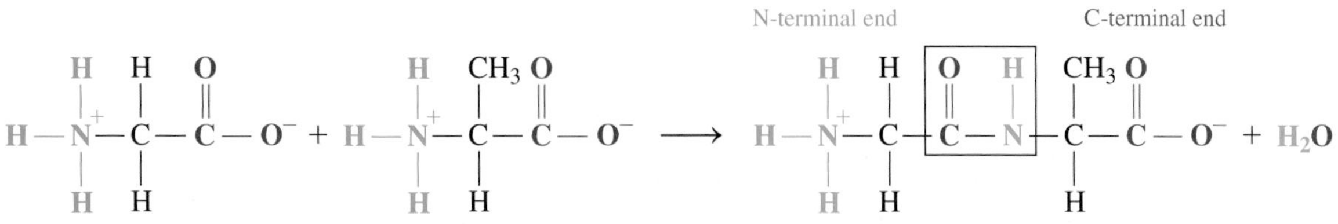

Glycine (Gly, G) Alanine (Ala, A) Glycylalanine (Gly–Ala, GA)

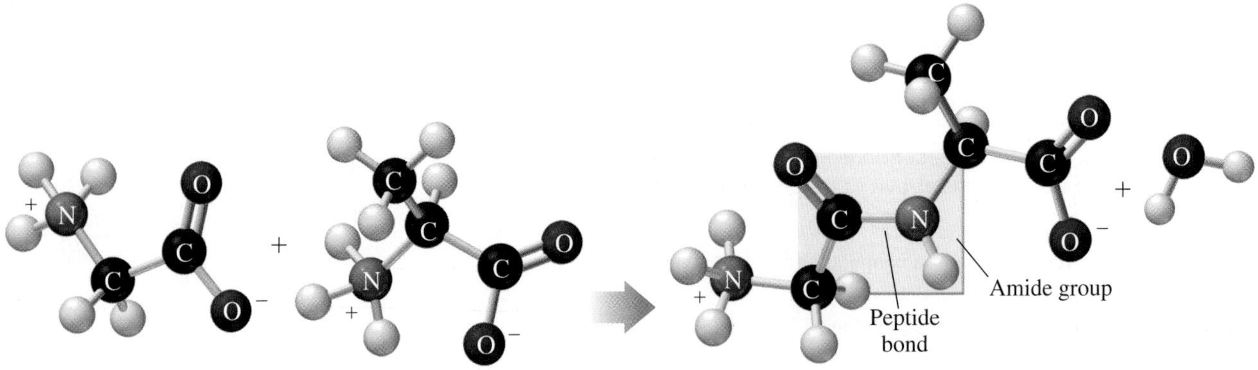

Glycine (Gly, G) Alanine (Ala, A) Glycylalanine (Gly–Ala, GA) Water

FIGURE 16.2 ▶ A peptide bond between glycine and alanine forms the dipeptide glycylalanine.

Q What functional groups in glycine and alanine form the peptide bond?

Naming Peptides

With the exception of the C-terminal amino acid, the names of all the other amino acids in a peptide end with *yl*. For example, a tripeptide consisting of alanine at the N-terminal, glycine, and serine at the C-terminal is named as one word: alan**yl**glyc**yl**serine. For convenience, the order of amino acids in the peptide is often written as the sequence of three-letter or one-letter abbreviations.

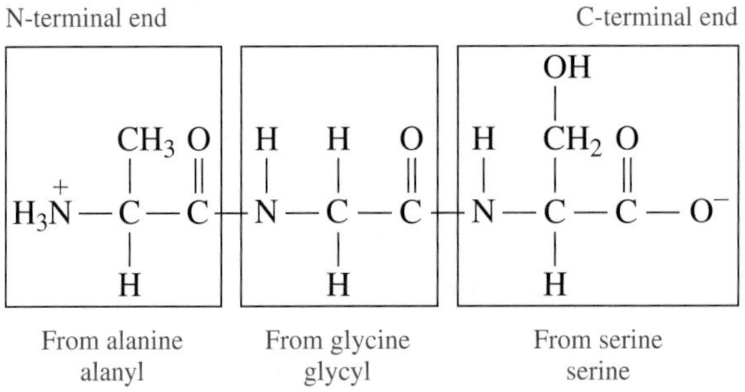

Alanylglycylserine
(Ala–Gly–Ser, AGS)

SAMPLE PROBLEM 16.4 Identifying a Tripeptide

Answer the questions for the tripeptide that is shown below:

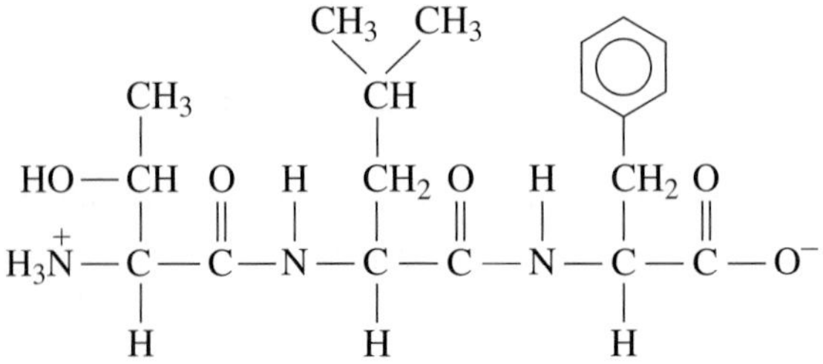

a. What is the N-terminal amino acid? What is the C-terminal amino acid?

b. Use the three-letter and one-letter abbreviations of amino acids to give the amino acid order in the tripeptide.

c. What is the name of the tripeptide?

SOLUTION

ANALYZE THE PROBLEM

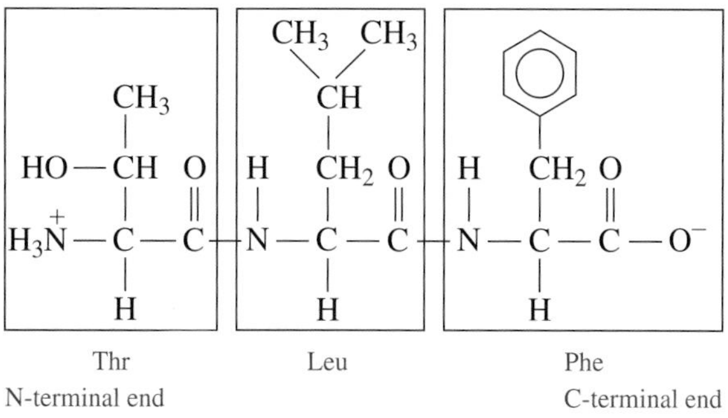

a. Threonine is the N-terminal amino acid; phenylalanine is the C-terminal amino acid.
b. Thr–Leu–Phe; TLF
c. threonylleucylphenylalanine

STUDY CHECK 16.4

A tripeptide has the abbreviation: Pro–His–Met.

a. What is the N-terminal amino acid?
b. What is the C-terminal amino acid?
c. What is the name of the tripeptide?

Primary Structure of a Protein

A **protein** is a polypeptide of 50 or more amino acids that has biological activity. The **primary structure** of a protein is the particular sequence of amino acids held together by peptide bonds. For example, a hormone that stimulates the thyroid to release thyroxin is a tripeptide with the amino acid sequence Glu–His–Pro. Although five other amino acid sequences of the same three amino acids are possible, such as His–Pro–Glu or Pro–His–Glu, they do not produce hormonal activity. Thus the biological function of peptides and proteins depends on the specific sequence of the amino acids.

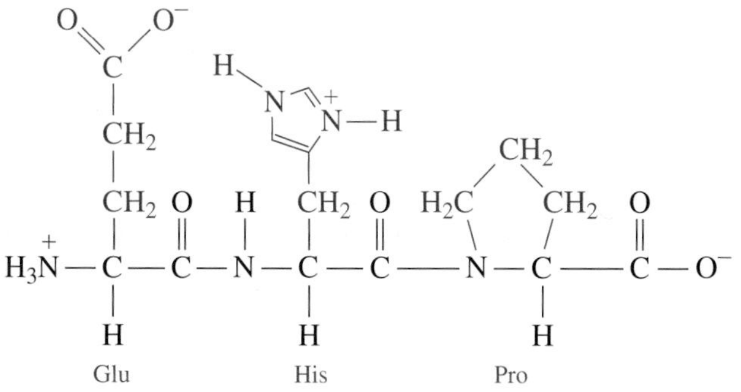

The first protein to have its primary structure determined was insulin, which was accomplished by Frederick Sanger in 1953. Since that time, scientists have determined the amino acid sequences of thousands of proteins. Insulin is a hormone that regulates the glucose level in the blood. In the primary structure of human insulin, there are two polypeptide chains. In chain A, there are 21 amino acids, and chain B has 30 amino acids. The polypeptide chains are held together by *disulfide bonds* formed by the thiol groups of the cysteine amino acids in each of the chains (see Figure 16.3). Today, human insulin with this exact structure is produced through genetic engineering.

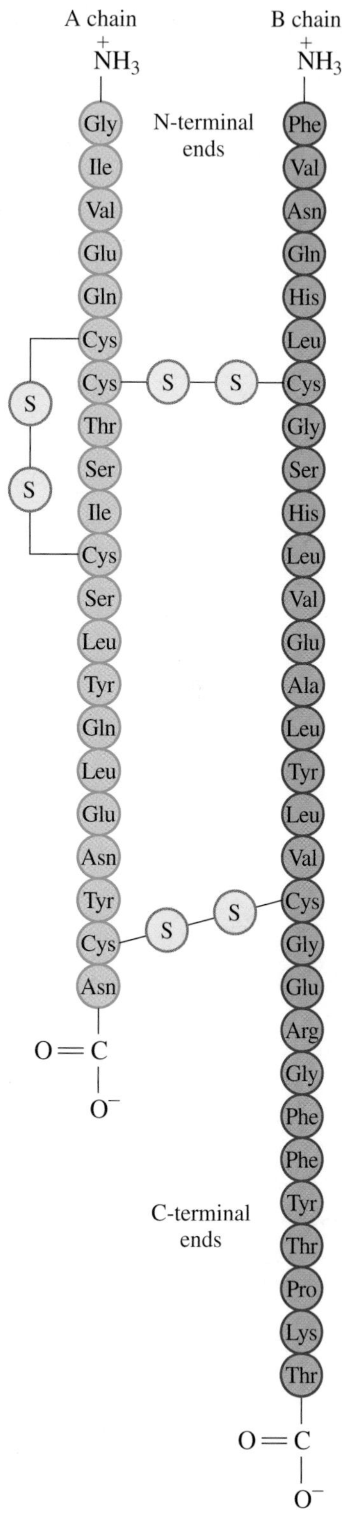

FIGURE 16.3 ▶ The sequence of amino acids in human insulin is its primary structure.

Q What kinds of bonds occur in the primary structure of a protein?

Chemistry Link to Health

Polypeptides in the Body

Enkephalins and endorphins are natural painkillers produced in the body. They are polypeptides that bind to receptors in the brain to give relief from pain. This effect appears to be responsible for the runner's high, for the temporary loss of pain when severe injury occurs, and for the analgesic effects of acupuncture.

The *enkephalins*, which are found in the thalamus and the spinal cord, are pentapeptides, the smallest molecules with opiate activity. The amino acid sequence of met-enkephalin is found in the longer amino acid sequence of the endorphins.

Four groups of *endorphins* have been identified: α-endorphin contains 16 amino acids, β-endorphin contains 31 amino acids, γ-endorphin has 17 amino acids, and δ-endorphin has 27 amino acids. Endorphins may produce their sedating effects by preventing the release of substance P, a polypeptide with 11 amino acids, which has been found to transmit pain impulses to the brain.

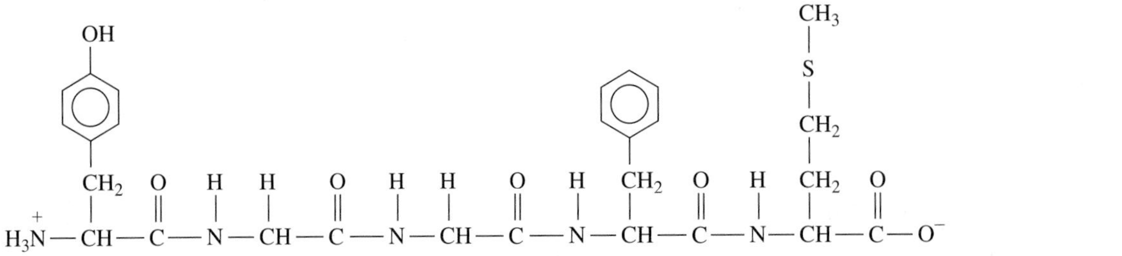

met-Enkephalin

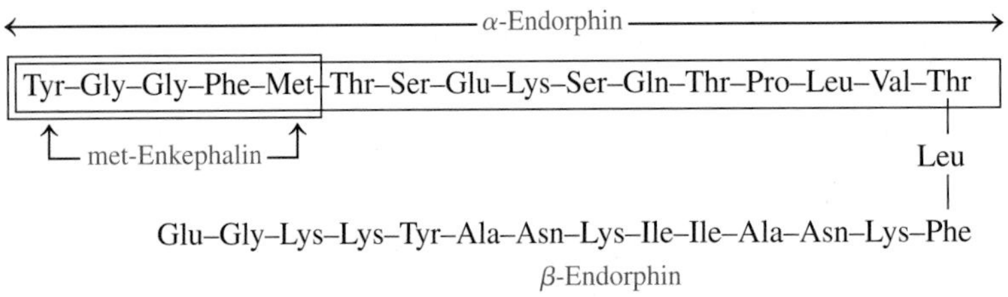

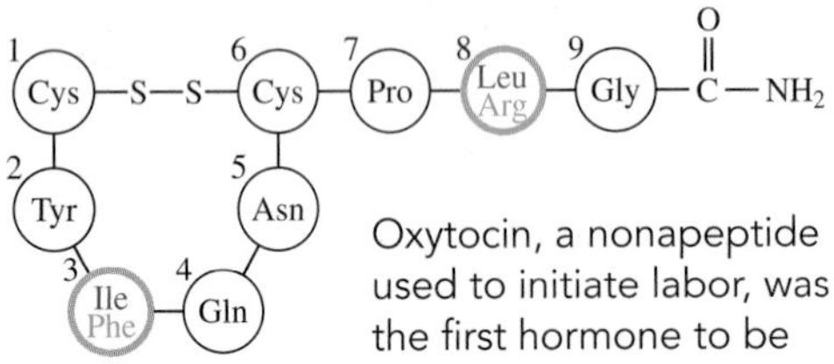

Oxytocin
Vasopressin

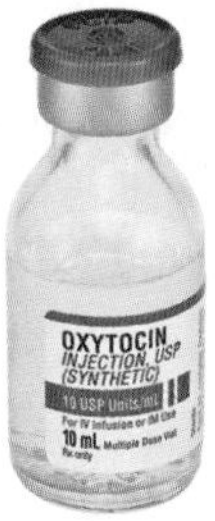

Oxytocin, a nonapeptide used to initiate labor, was the first hormone to be synthesized in the laboratory.

Two hormones produced by the pituitary gland are the nonapeptides (peptides with nine amino acids) oxytocin and vasopressin. Oxytocin stimulates uterine contractions in labor and vasopressin is an antidiuretic hormone that regulates blood pressure by adjusting the amount of water reabsorbed by the kidneys. The structures of these nonapeptides are very similar. Only the amino acids in positions 3 and 8 are different. However, the difference of two amino acids greatly affects how the two hormones function in the body.

QUESTIONS AND PROBLEMS

16.3 Proteins: Primary Structure

LEARNING GOAL Draw the condensed structural formula for a peptide and give its name. Describe the primary structure for a protein.

16.17 Draw the condensed structural formula for each of the following peptides, and give its three-letter and one-letter abbreviations:
- **a.** alanyltryptophan
- **b.** glycylthreonine
- **c.** serylphenylalanine
- **d.** glycylvalylmethionine

16.18 Draw the condensed structural formula for each of the following peptides, and give its three-letter and one-letter abbreviations:
- **a.** prolylaspartic acid
- **b.** threonylleucine
- **c.** methionylglutaminyllysine
- **d.** histidylglycylglutamylisoleucine

16.19 Three peptides each contain one molecule of valine and two molecules of serine. Use their three-letter abbreviations to write the possible tripeptides they can form.

16.20 What does the "primary structure of a protein" refer to? Show the possible primary structures of a polypeptide consisting of only Glu, His and Pro.

16.4 Proteins: Secondary, Tertiary, and Quaternary Structures

LEARNING GOAL

Describe the secondary, tertiary, and quaternary structures for a protein; describe the denaturation of a protein.

The **secondary structure** of a protein describes the type of structure that forms when amino acids form hydrogen bonds within a polypeptide or between polypeptide chains. The three most common types of secondary structure are the *alpha helix*, the *beta-pleated sheet*, and the *triple helix*.

In an **alpha helix (α helix)**, hydrogen bonds form between the oxygen of the C═O groups and the hydrogen of N—H groups of the amide bonds in the next turn of the α helix (see Figure 16.4). The formation of many hydrogen bonds along the polypeptide chain gives the characteristic helical shape of a spiral staircase. All of the R groups of the different amino acids extend to the outside of the helix.

CORE CHEMISTRY SKILL

Identifying the Primary, Secondary, Tertiary, and Quaternary Structures of Proteins

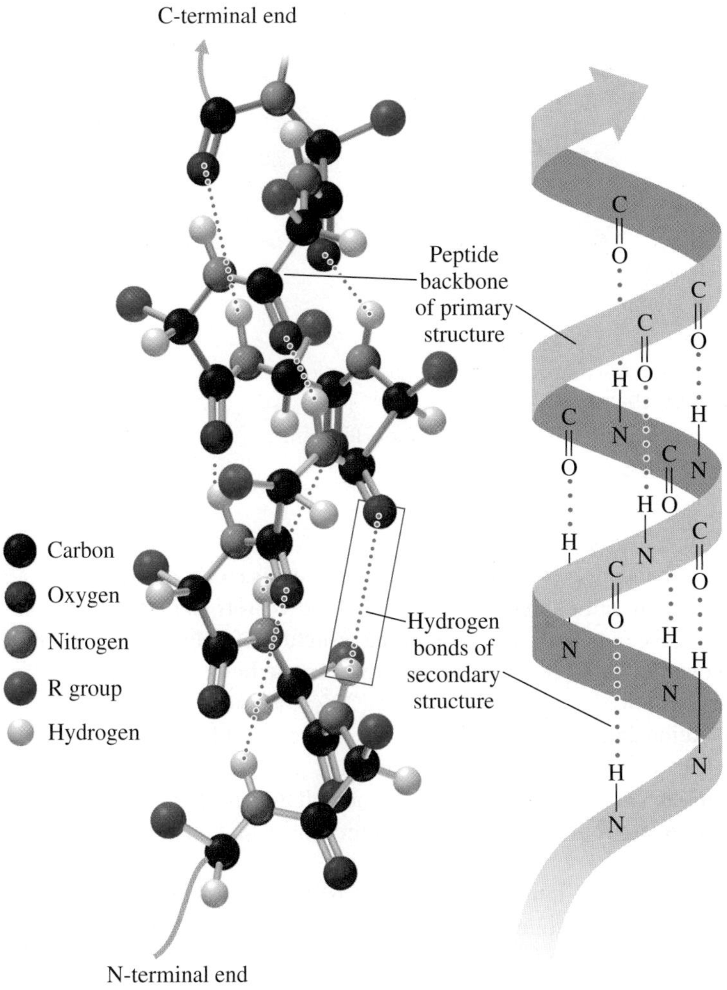

The shape of an alpha helix is similar to that of a spiral staircase.

FIGURE 16.4 ▶ The α helix acquires a coiled shape from hydrogen bonds between the oxygen of the C═O group and the hydrogen of the N—H group in the next turn.

Q What are the partial charges of the H in N—H and the O in C═O that permit hydrogen bonds to form?

Another type of secondary structure found in proteins is the **beta-pleated sheet (β-pleated sheet)**. In a β-pleated sheet, hydrogen bonds form between the oxygen atoms in the carbonyl groups in one section of the polypeptide chain, and the hydrogen atoms in the N—H groups of the amide bonds in a nearby section of the polypeptide chain. A beta-pleated sheet can form between adjacent polypeptide chains or within the same polypeptide chain. The tendency to form various kinds of secondary structures depends on the amino acids in a particular segment of the polypeptide chain. Typically, beta-pleated sheets contain mostly amino acids with small R groups such as glycine, valine, alanine, and serine, which extend above and below the beta-pleated sheet. The α-helical regions in a protein have higher amounts of amino acids with large R groups such as histidine, leucine, and methionine. The hydrogen bonds holding the β-pleated sheets tightly in place account for the strength and durability of proteins such as silk (see Figure 16.5).

The secondary structure in silk is a beta-pleated sheet.

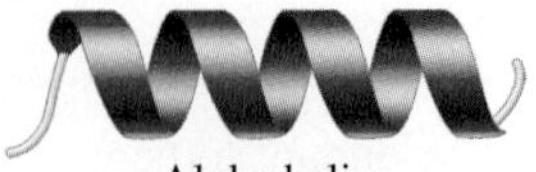

Alpha helix

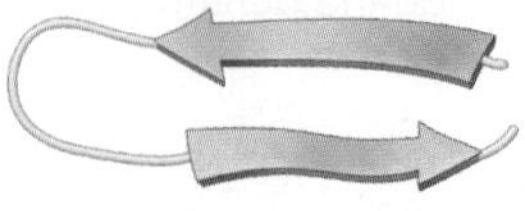

Beta-pleated sheet

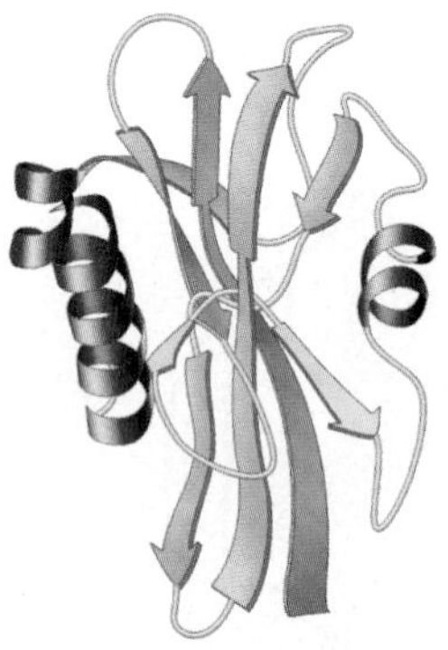

Protein with alpha helices and beta-pleated sheets

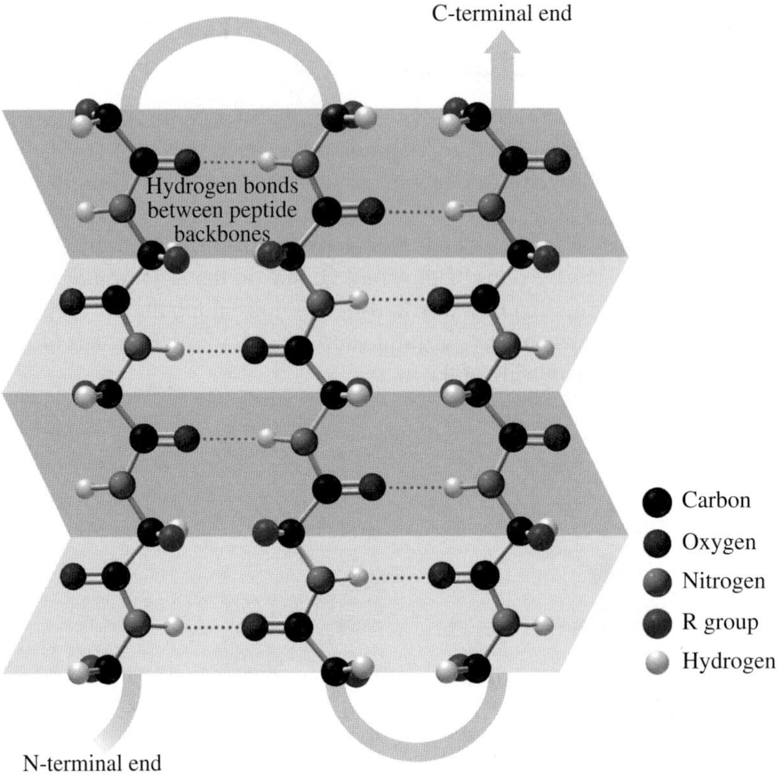

FIGURE 16.5 ▶ In the secondary structure of a β (beta)-pleated sheet, hydrogen bonds form between the peptide chains.

Q How do the hydrogen bonds in a β-pleated sheet differ from the way hydrogen bonds form in an α helix?

Collagen, the most abundant protein in the body, makes up as much as one-third of all protein in vertebrates. It is found in connective tissue, blood vessels, skin, tendons, ligaments, the cornea of the eye, and cartilage. The strong structure of collagen is a result of three polypeptides woven together like a braid to form a **triple helix** (see Figure 16.6). Collagen has a high content of glycine (33%), proline (22%), alanine (12%), and smaller amounts of hydroxyproline and hydroxylysine, which are modified forms of proline and lysine. The —OH groups on these modified amino acids provide additional hydrogen bonds between the peptide chains to give strength to the collagen triple helix. When several triple helices wrap together as a braid, they form the fibrils that make up connective tissues and tendons.

When a diet is deficient in vitamin C, collagen fibrils are weakened because the enzymes needed to form hydroxyproline and hydroxylysine require vitamin C. Because there are fewer —OH groups, there is less hydrogen bonding between collagen fibrils. Collagen becomes less elastic as a person ages because additional bonds form between the fibrils. Bones, cartilage, and tendons become more brittle, and wrinkles are seen as the skin loses elasticity.

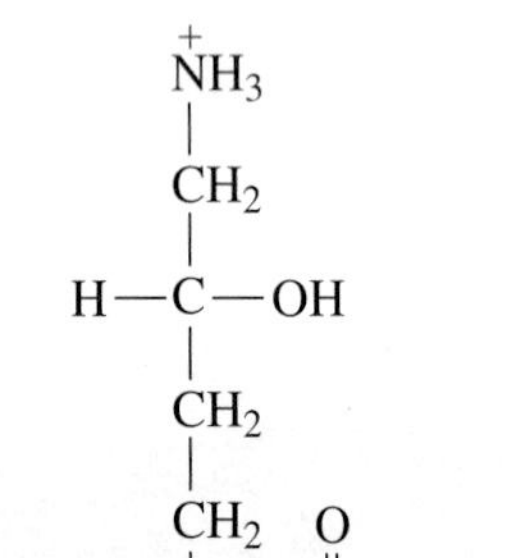

Hydroxyproline

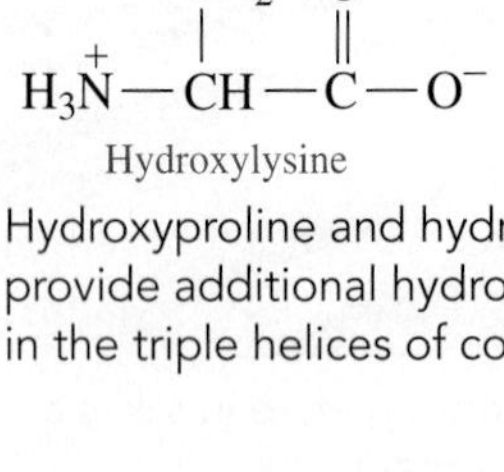

Hydroxylysine

Hydroxyproline and hydroxylysine provide additional hydrogen bonds in the triple helices of collagen.

Tertiary Structure

The **tertiary structure** of a protein involves attractions and repulsions between the R groups of the amino acids in the polypeptide chain. As interactions occur between different parts of the peptide chain, segments of the chain twist and bend until the protein acquires a specific three-dimensional shape. The tertiary structure of a protein is stabilized by interactions between the R groups of the amino acids in one region of the polypeptide chain and the R groups of amino acids in other regions of the protein (see Figure 16.7).

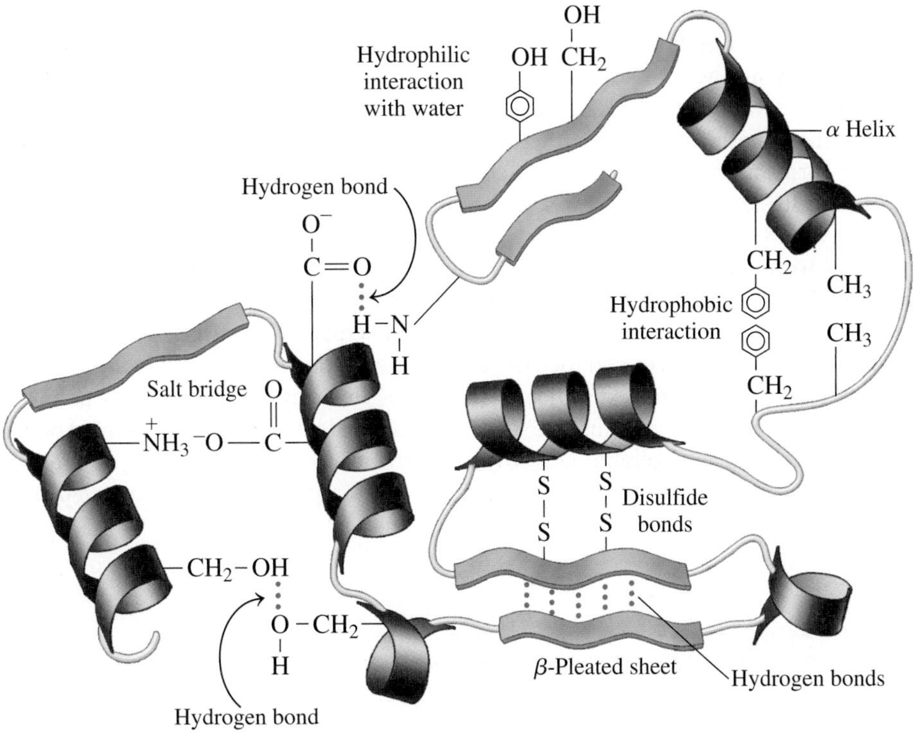

FIGURE 16.7 ▶ Interactions between amino acid R groups fold a protein into a specific three-dimensional shape called its tertiary structure.

Q Why would one section of a protein move to the center while another section remains on the surface of the tertiary structure?

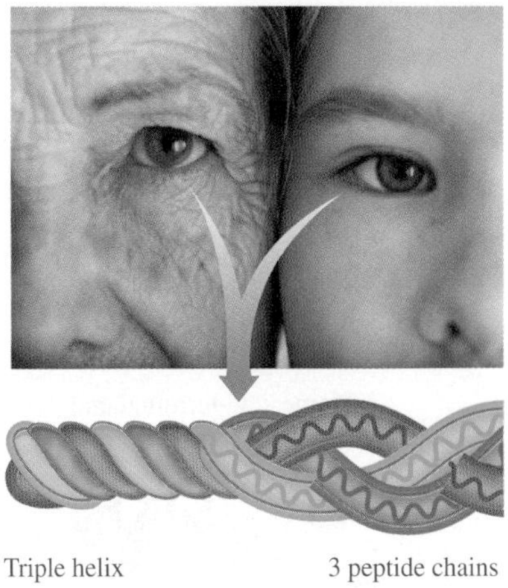

FIGURE 16.6 ▶ Collagen fibers are triple helices of polypeptide chains held together by hydrogen bonds.

Q What are some of the amino acids in collagen that form hydrogen bonds between the polypeptide chains?

1. **Hydrophobic interactions** are interactions between two nonpolar amino acids that have nonpolar R groups. Within a protein, the amino acids with nonpolar R groups are pushed away from the aqueous environment to form a hydrophobic center at the interior of the protein molecule.
2. **Hydrophilic interactions** are attractions between the external aqueous environment and the R groups of polar amino acids that are pulled to the outer surface of globular proteins where they form hydrogen bonds with water.
3. **Salt bridges** are ionic bonds between ionized R groups of basic and acidic amino acids. For example, the ionized R group of arginine, which has a positive charge, can form a salt bridge (ionic bond) with the R group in aspartic acid, which has a negative charge.
4. **Hydrogen bonds** form between H of a polar R group and the O or N of another amino acid. For example, a hydrogen bond can form between the —OH groups of two serines or between the —OH of serine and the $—NH_2$ in the R group of glutamine.
5. **Disulfide bonds** (—S—S—) are covalent bonds that form between the —SH groups of cysteines in a polypeptide chain.

Globular and Fibrous Proteins

A group of proteins known as **globular proteins** have compact, spherical shapes because sections of the polypeptide chain fold over on top of each other due to the various interactions between R groups. It is the globular proteins that carry out the work of the cells: functions such as synthesis, transport, and metabolism.

Myoglobin is a globular protein that stores oxygen in skeletal muscle. High concentrations of myoglobin are found in the muscles of sea mammals, such as seals and whales, allowing them to stay under the water for long periods of time. Myoglobin contains 153 amino acids in a single polypeptide chain with about three-fourths of the chain in the α-helix secondary structure. The polypeptide chain, including its helical regions, forms a compact tertiary structure by folding upon itself (see Figure 16.8). Within the tertiary structure, a pocket of amino acids and a heme group binds and stores oxygen (O_2).

The **fibrous proteins** are proteins that consist of long, thin, fiber-like shapes. They are typically involved in the structure of cells and tissues. Two types of fibrous protein are the α- and β-keratins. The α-keratins are the proteins that make up hair, wool, skin, and nails. In hair, three α helices coil together like a braid to form a fibril. Within the fibril, the α helices are held together by disulfide (—S—S—) linkages between the thiol groups of the many cysteine amino acids in hair. Several fibrils bind together to form a strand of hair (see Figure 16.9). The β-keratins are the type of proteins found in the feathers of birds and scales of reptiles. In β-keratins, the proteins contain large amounts of β-pleated sheet structure.

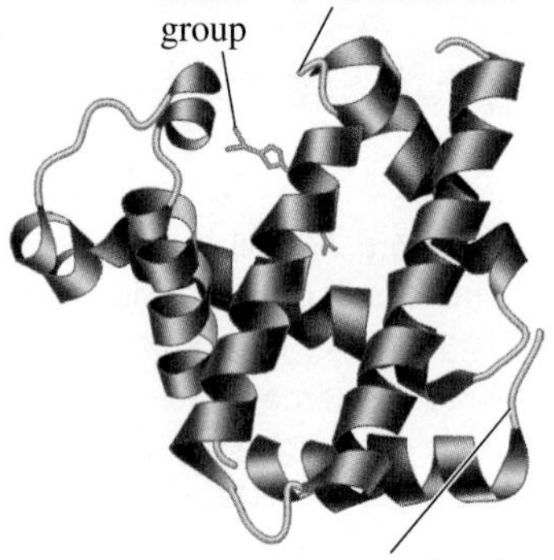

FIGURE 16.8 ▶ The ribbon model represents the tertiary structure of the polypeptide chain of myoglobin, which is a globular protein that contains a heme group that binds oxygen.

Q Would hydrophilic amino acids be found on the outside or inside of the myoglobin structure?

FIGURE 16.9 ▶ The fibrous proteins of α-keratin wrap together to form fibrils of hair and wool.

Q Why does hair have a large amount of cysteine?

SAMPLE PROBLEM 16.5 Interaction Between R Groups in Tertiary Structures

What type of attraction would you expect between the R groups of each of the following amino acids in a tertiary structure?

a. cysteine and cysteine
b. glutamic acid and lysine
c. tyrosine and water

SOLUTION

	Type of R Groups	Type of Interaction
ANALYZE THE PROBLEM	identify R groups on amino acids (see Table 16.2)	determine type of interaction between R groups (see Figure 16.7)

Type of R Groups	Type of Interaction
Nonpolar and nonpolar	hydrophobic
Polar (neutral) and water	hydrophilic
Polar (basic) $—NH_3^+$ and polar (acidic) $—COO^-$	salt bridges
Polar (neutral) and polar (neutral) —OH and —NH— or $—NH_2$	hydrogen bonds
—SH and —SH	disulfide bonds

a. Two cysteines, each with an R group containing —SH, will form a disulfide bond.
b. The interaction of the $—COO^-$ in the R group of glutamic acid and the $—NH_3^+$ in the R group of lysine will form an ionic bond called a salt bridge.
c. The R group in tyrosine has an —OH group that is attracted to water by hydrophilic interactions.

STUDY CHECK 16.5

Would you expect to find valine and leucine on the outside or the inside of the tertiary structure? Why?

Quaternary Structure: Hemoglobin

While many proteins are biologically active as tertiary structures, some proteins require two or more tertiary structures to be biologically active. When a biologically active protein consists of two or more polypeptide chains or subunits, the structural level is referred to as a **quaternary structure**. Hemoglobin, a globular protein that transports oxygen in blood, consists of four polypeptide chains: two α-chains with 141 amino acids, and two β-chains with 146 amino acids. Although the α-chains and β-chains have different sequences of amino acids, they both form similar tertiary structures with similar shapes (see Figure 16.10).

In the quaternary structure, the subunits are held together by the same interactions that stabilize tertiary structures, such as hydrogen bonds, salt bridges, disulfide links, and hydrophobic interactions between R groups. Each subunit of the hemoglobin contains a heme group that binds oxygen. In the adult hemoglobin molecule, all four subunits ($\alpha_2\beta_2$) *must* be combined for hemoglobin to properly function as an oxygen carrier. Therefore, the complete quaternary structure of hemoglobin can bind and transport four molecules of oxygen.

Hemoglobin and myoglobin have similar biological functions. Hemoglobin carries oxygen in the blood, whereas myoglobin carries oxygen in muscle. Myoglobin, a single polypeptide chain with a molar mass of 17 000, has about one-fourth the molar mass of hemoglobin (67 000). The tertiary structure of the single polypeptide myoglobin is almost identical to the tertiary structure of each of the subunits of hemoglobin. Myoglobin carries one molecule of oxygen, whereas hemoglobin carries four oxygen molecules. Table 16.6 and Figure 16.11 summarize the structural levels of proteins.

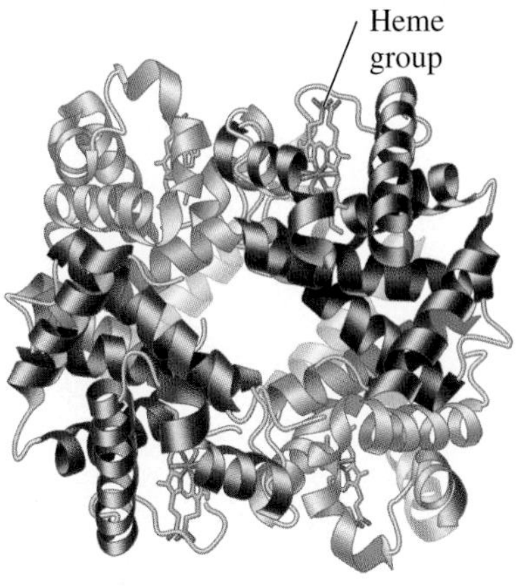

FIGURE 16.10 ▶ In the ribbon model of hemoglobin, the quaternary structure is made up of four polypeptide subunits, two (orange) are α-chains and two (red) are β-chains. The heme groups (green) in the four subunits bind oxygen.

Q What is the difference between a tertiary structure and a quaternary structure?

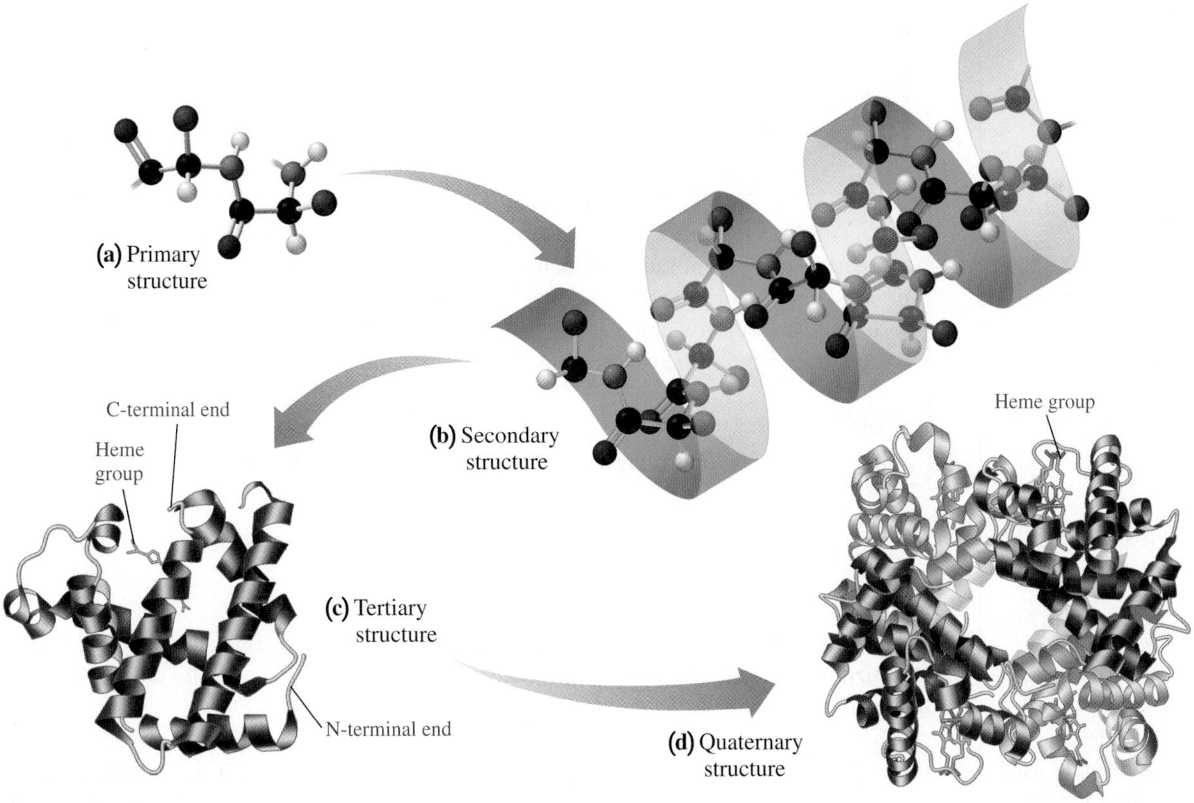

FIGURE 16.11 ▶ The structural levels of protein are **(a)** primary, **(b)** secondary, **(c)** tertiary, and sometimes **(d)** quaternary.

Q What is the difference between a primary structure and a tertiary structure?

Interactive Video

Blast Animation on Different Levels of Protein Structure

TABLE 16.6 Summary of Structural Levels in Proteins

Structural Level	Characteristics
Primary	Peptide bonds join amino acids in a specific sequence in a polypeptide.
Secondary	The coiled α helix, β-pleated sheet, or a triple helix forms by hydrogen bonding between peptide bonds along the chain.
Tertiary	A polypeptide folds into a compact, three-dimensional shape stabilized by interactions between R groups of amino acids to form a biologically active protein.
Quaternary	Two or more protein subunits combine to form a biologically active protein.

SAMPLE PROBLEM 16.6 Identifying Protein Structure

Indicate whether the following conditions are responsible for primary, secondary, tertiary, or quaternary protein structures:

a. disulfide bonds that form between portions of a protein chain
b. peptide bonds that form a chain of amino acids
c. hydrogen bonds between the H of a peptide bond and the O of a peptide bond four amino acids away

SOLUTION

	Structural Level	Characteristics
ANALYZE THE PROBLEM	Identify as: primary, secondary, tertiary, quaternary	Look for: peptide bonds, hydrogen bonds in peptide, R group interactions, two or more subunits

a. Disulfide bonds are a type of interaction between R groups found in the tertiary and quaternary levels of protein structure.
b. The peptide bonds in the sequence of amino acids in a polypeptide form the primary level of protein structure.
c. Hydrogen bonding between peptide bonds forms the secondary level of protein structure.

STUDY CHECK 16.6

What structural level is represented by the interactions of the two subunits in insulin?

Explore Your World

Denaturation of Milk Protein

Pour some milk in five glasses. Add the following to the milk samples in glasses 1 to 4. The fifth glass of milk is a reference sample.

1. Vinegar, drop by drop. Stir.
2. One-half teaspoon of meat tenderizer. Stir.
3. One teaspoon of fresh pineapple juice. (Canned juice has been heated and cannot be used.)
4. One teaspoon of fresh pineapple juice after the juice is heated to boiling.

Questions

1. How did the appearance of the milk change in each of the samples?
2. What enzyme is listed on the package label of the tenderizer?
3. How does the effect of the heated pineapple juice compare with that of the fresh juice? Explain.
4. Why is cooked pineapple used when making gelatin (a protein) desserts?

Denaturation of Proteins

Denaturation of a protein occurs when there is a change that disrupts the interactions that stabilize the secondary, tertiary, or quaternary structure. However, the covalent amide bonds of the primary structure are not affected.

The loss of secondary and tertiary structures occurs when conditions change, such as increasing the temperature or making the pH very acidic or basic. If the pH changes, the basic and acidic R groups lose their ionic charges and cannot form salt bridges, which causes a change in the shape of the protein. Denaturation can also occur by adding certain organic compounds or heavy metal ions or by mechanical agitation (see Table 16.7).

When the interactions between the R groups are disrupted, a globular protein unfolds like a loose piece of spaghetti. With the loss of its overall shape (tertiary structure), the protein is no longer biologically active.

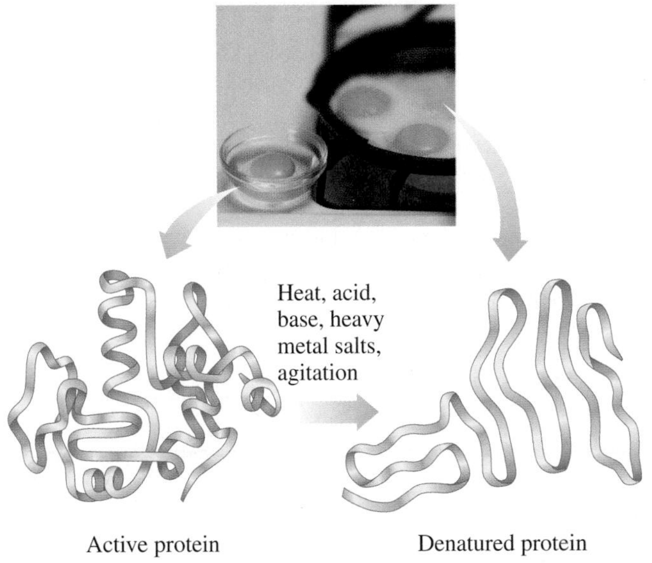

Denaturation of egg protein occurs when the bonds of the secondary and tertiary structures are disrupted.

Chemistry Link to **Health**

Protein Structure and Mad Cow Disease

Until recently, researchers thought that only viruses or bacteria were responsible for transmitting diseases. Now a group of diseases has been found in which the infectious agents are proteins called *prions*, a word invented in 1982 by Stanley B. Prusiner, who won the Nobel Prize in Physiology or Medicine in 1997 for his research into prions. *Bovine spongiform encephalopathy* (BSE), or "mad cow disease," is a fatal brain disease of cattle in which the brain fills with cavities resembling a sponge. In the noninfectious form of the prion, PrP^{c}, the N-terminal portion is a random coil. Although the noninfectious form may be ingested from meat products, its structure can change to what is known as PrP^{Sc} or *prion-related protein scrapie*. In this infectious form, the end of the peptide chain folds into a β-pleated sheet, which has disastrous effects on the brain and spinal cord. The conditions that cause this structural change are not yet known.

BSE was diagnosed in Great Britain in 1986. The protein is present in nerve tissue of cattle, but it is not found in their meat. Cattle are normally herbivores, but when they are fed a bone meal that comes from cattle with mad cow disease, the infectious agent, prions, can spread to other cattle. Control measures that exclude brain and spinal cord from animal feed are now in place to reduce the incidence of BSE.

The human variant of this disease is called *Creutzfeldt–Jakob disease* (CJD). Around 1955, Dr. Carleton Gajdusek was studying the Fore tribe in Papua New Guinea, where many tribe members were dying of neurological disease known as "kuru." Among the Fore, it was a custom to cannibalize members of the tribe upon their death. Gajdusek who received the Nobel Prize in Physiology or Medicine in 1976, eventually determined that this practice was responsible for transmitting the infectious agent from one tribe member to another.

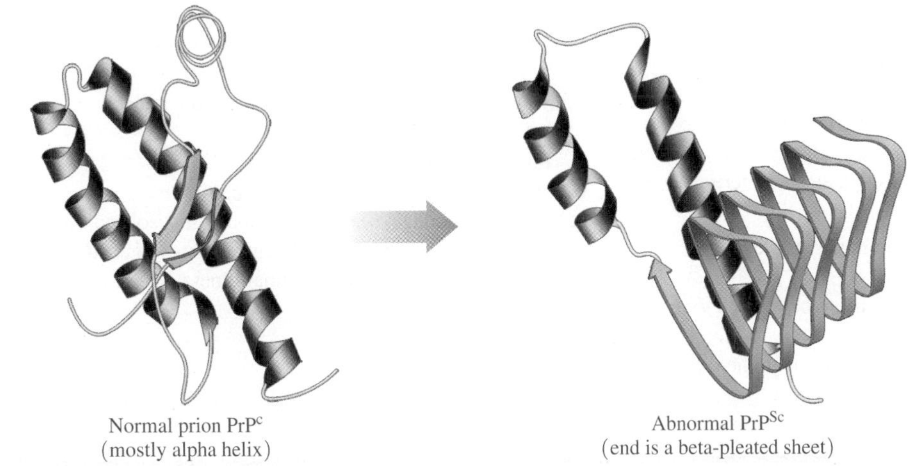

Prions are proteins that cause infections such as "mad cow disease."

TABLE 16.7 Protein Denaturation

Denaturing Agent	Bonds Disrupted	Examples
Heat Above 50 °C	Hydrogen bonds; hydrophobic interactions between nonpolar R groups	Cooking food and autoclaving surgical items
Acids and Bases	Hydrogen bonds between polar R groups; salt bridges	Lactic acid from bacteria, which denatures milk protein in the preparation of yogurt and cheese
Organic Compounds	Hydrophobic interactions	Ethanol and isopropyl alcohol, which disinfect wounds and prepare the skin for injections
Heavy Metal Ions Ag^+, Pb^{2+}, and Hg^{2+}	Disulfide bonds in proteins by forming ionic bonds	Mercury and lead poisoning
Agitation	Hydrogen bonds and hydrophobic interactions by stretching polypeptide chains and disrupting stabilizing interactions	Whipped cream, meringue made from egg whites

QUESTIONS AND PROBLEMS

16.4 Proteins: Secondary, Tertiary, and Quaternary Structures

LEARNING GOAL Describe the secondary, tertiary, and quaternary structures for a protein; describe the denaturation of a protein.

16.21 What happens when a primary structure forms a secondary structure?

16.22 Name the three most common types of secondary protein structure.

16.23 What is the difference in bonding between an α helix and a β-pleated sheet?

16.24 What are the properties and key functions of fibrous proteins?

16.25 What type of interaction would you expect between the R groups of the following amino acids in a tertiary structure?
a. cysteine and cysteine **b.** aspartic acid and lysine
c. serine and aspartic acid **d.** leucine and leucine

16.26 What type of interaction would you expect between the R groups of the following amino acids in a quaternary structure?
a. phenylalanine and isoleucine **b.** aspartic acid and histidine
c. asparagine and tyrosine **d.** alanine and proline

16.27 A portion of a polypeptide chain contains the following sequence of amino acids:

—Met—Leu—Ala—Cys—Asp—

a. Which amino acid can form a disulfide bond?
b. Which amino acids are likely to make the polypeptide more soluble in water?
c. Which amino acids are likely to make the polypeptide less soluble in water?
d. How does the primary structure of a protein affect its tertiary structure?

16.28 In myoglobin, about one-half of the 153 amino acids have nonpolar R groups.
a. Where would you expect those amino acids to be located in the tertiary structure?
b. Where would you expect the polar R groups to be in the tertiary structure?
c. Why is myoglobin more soluble in water than silk or wool?

16.29 Indicate whether each of the following statements describes primary, secondary, tertiary, or quaternary protein structure:
a. R groups interact to form disulfide bonds or ionic bonds.
b. Peptide bonds join amino acids in a polypeptide chain.
c. Several polypeptides in a beta-pleated sheet are held together by hydrogen bonds between adjacent chains.
d. Hydrogen bonding between amino acids in the same polypeptide gives a coiled shape to the protein.

16.30 Indicate whether each of the following statements describes primary, secondary, tertiary, or quaternary protein structure:
a. Hydrophobic R groups seeking a nonpolar environment move toward the inside of the folded protein.
b. Protein chains of collagen form a triple helix.
c. An active protein contains four tertiary subunits.
d. In sickle-cell anemia, valine replaces glutamic acid in the β-chain.

16.31 Indicate the changes in secondary and tertiary structural levels of proteins for each of the following:
a. An egg placed in water at 100 °C is soft boiled in about 3 minutes.
b. Prior to giving an injection, the skin is wiped with an alcohol swab.
c. Surgical instruments are placed in a 120 °C autoclave.
d. During surgery, a wound is closed by cauterization (heat).

16.32 Indicate the changes in secondary and tertiary structural levels of proteins for each of the following:
a. Tannic acid is placed on a burn.
b. Milk is heated to 60 °C to make yogurt.
c. To avoid spoilage, seeds are treated with a solution of $HgCl_2$.
d. Hamburger is cooked at high temperatures to destroy *E. coli* bacteria that may cause intestinal illness.

Chemistry Link to Health

Sickle-Cell Anemia

Sickle-cell anemia is a disease caused by an abnormality in the shape of one of the subunits of the hemoglobin protein. In the β-chain, the sixth amino acid, glutamic acid, which is polar acidic, is replaced by valine, a nonpolar amino acid.

Because valine has a nonpolar R group, it is attracted to the nonpolar regions within the beta hemoglobin chains. The affected red blood cells (RBCs) change from a rounded shape to a crescent shape, like a sickle, which interferes with their ability to transport adequate quantities of oxygen. Hydrophobic interactions also cause sickle-cell hemoglobin molecules to stick together. They form insoluble fibers of sickle-cell hemoglobin that clog capillaries, where they cause inflammation, pain, and organ damage. Critically low oxygen levels may occur in the affected tissues.

In sickle-cell anemia, both genes for the altered hemoglobin must be inherited. However, a few sickled cells are found in persons who carry one gene for sickle-cell hemoglobin, a condition that is also known to provide protection from malaria.

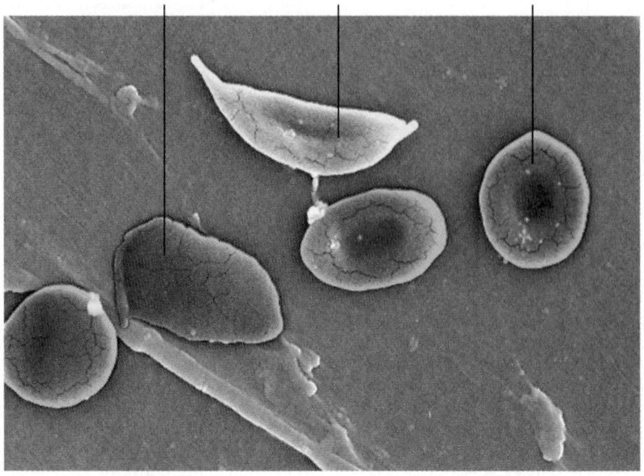

Polar acidic amino acid

Normal β-chain: Val–His–Leu–Thr–Pro–[Glu]–Glu–Lys–

Sickled β-chain: Val–His–Leu–Thr–Pro–[Val]–Glu–Lys–

Nonpolar amino acid

16.5 Enzymes

LEARNING GOAL

Describe enzymes and their role in enzyme-catalyzed reactions.

Biological catalysts known as **enzymes** are needed for most chemical reactions that take place in the body. A catalyst increases the reaction rate by changing the way a reaction takes place but is itself not changed at the end of the reaction. An uncatalyzed reaction in a cell may take place eventually, but not at a rate fast enough for survival. For example, the hydrolysis of proteins in our diet would eventually occur without a catalyst, but not fast enough to meet the body's requirements for amino acids. The chemical reactions in our cells must occur at incredibly fast rates under the mild conditions of pH 7.4 and a body temperature of 37 °C.

As catalysts, enzymes lower the activation energy for a chemical reaction (see Figure 16.12). Less energy is required to convert reactant molecules to products, which increases the rate of a biochemical reaction compared to the rate of the uncatalyzed reaction. For example, an enzyme in the blood called carbonic anhydrase catalyzes the rapid reaction of carbon dioxide and water to bicarbonate and H^+. In one second, one molecule of carbonic anhydrase catalyzes the reaction of about one million molecules of carbon dioxide.

$$CO_2 + H_2O \underset{}{\overset{\text{Carbonic anhydrase}}{\rightleftharpoons}} HCO_3^- + H^+$$

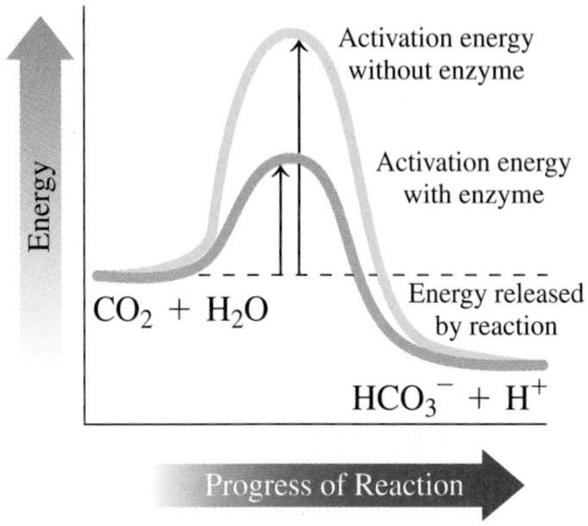

FIGURE 16.12 ▶ The enzyme carbonic anhydrase lowers the activation energy needed for the reversible reaction that converts CO_2 and H_2O to HCO_3^- and H^+.

Q Why are enzymes used in biological reactions?

Names and Classification of Enzymes

The names of enzymes describe the compound or the reaction that is catalyzed. The actual names of enzymes are derived by replacing the end of the name of the reaction or reacting compound with the suffix *ase*. For example, an *oxidase* catalyzes an oxidation reaction, and a *dehydrogenase* removes hydrogen atoms. The compound sucrose is hydrolyzed by the enzyme *sucrase*, and a lipid is hydrolyzed by a *lipase*. Some early known enzymes use names that end in the suffix *in*, such as *papain* found in papaya, *rennin* found in milk, and *pepsin* and *trypsin*, enzymes that catalyze the hydrolysis of proteins.

More recently, a systematic method of classifying and naming enzymes has been established. The name and class of each indicates the type of reaction it catalyzes. There are six main classes of enzymes, as described in Table 16.8.

TABLE 16.8 Classes of Enzymes

Class	Reaction Catalyzed	Examples
1. Oxidoreductases	Oxidation–reduction reactions	*Oxidases* oxidize a substance. *Reductases* reduce a substance. *Dehydrogenases* remove 2H atoms to form a double bond.
2. Transferases	Transfer a group between two compounds	*Transaminases* transfer amino groups. *Kinases* transfer phosphate groups.
3. Hydrolases	Hydrolysis reactions	*Proteases* hydrolyze peptide bonds in proteins. *Lipases* hydrolyze ester bonds in lipids. *Carbohydrases* hydrolyze glycosidic bonds in carbohydrates. *Phosphatases* hydrolyze phosphoester bonds. *Nucleases* hydrolyze nucleic acids.
4. Lyases	Add or remove groups involving a double bond without hydrolysis	*Carboxylases* add CO_2. *Deaminases* remove NH_3.
5. Isomerases	Rearrange atoms in a molecule to form an isomer	*Isomerases* convert cis to trans isomers or trans to cis isomers. *Epimerases* convert D to L stereoisomers or L to D.
6. Ligases	Form bonds between molecules using ATP energy	*Synthetases* combine two molecules.

SAMPLE PROBLEM 16.7 Naming Enzymes

What chemical reaction would each of the following enzymes catalyze?

a. aminotransferase
b. lactate dehydrogenase (LDH)

SOLUTION

a. the transfer of an amino group
b. the removal of hydrogen from lactate

STUDY CHECK 16.7

What is the name of the enzyme that catalyzes the hydrolysis of lipids?

Nearly all enzymes are globular proteins. Each has a unique three-dimensional shape that recognizes and binds a small group of reacting molecules, which are called **substrates**. The tertiary structure of an enzyme plays an important role in how that enzyme catalyzes reactions. In a catalyzed reaction, an enzyme must first bind to a substrate in a way that favors catalysis. A typical enzyme is much larger than its substrate. However, within an enzyme's large tertiary structure is a region called the **active site**, in which the substrate or substrates are held while the reaction takes place. This active site is often a small pocket within the larger tertiary structure that closely fits the shape of the substrate (see Figure 16.13).

Within the active site, the R groups of specific amino acids interact with the functional groups of the substrate to form hydrogen bonds, salt bridges, or hydrophobic interactions. The active site of a particular enzyme fits the shape of only a few types of substrates, which makes an enzyme very specific about the type of substrate it binds.

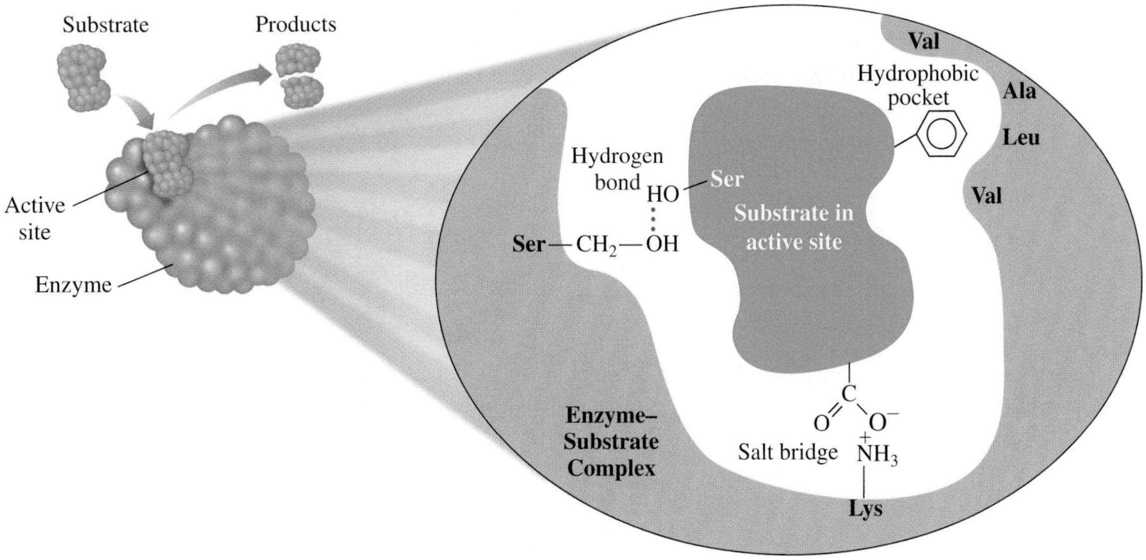

FIGURE 16.13 ▶ On the surface of an enzyme, a small region called an active site binds a substrate and catalyzes a specific reaction for that substrate.

Q Why does an enzyme catalyze a reaction of only certain substrates?

Enzyme-Catalyzed Reaction

The combination of an enzyme and a substrate within the active site forms an **enzyme–substrate (ES) complex** that provides an alternative pathway with a lower activation energy. Within the active site, the amino acid R groups catalyze the reaction. Then the products are released from the enzyme so it can bind to another substrate molecule. We can write the catalyzed reaction of an enzyme (E) with a substrate (S) to form product (P) as follows:

$$\underset{\text{Enzyme + Substrate}}{E + S} \rightleftarrows \underset{\text{ES Complex}}{ES} \longrightarrow \underset{\text{Enzyme + Product}}{E + P}$$

In the hydrolysis of the disaccharide sucrose by the enzyme sucrase, a molecule of sucrose binds to the active site of sucrase. In this ES complex, the glycosidic bond of sucrose is in a position that is favorable for hydrolysis, which is the splitting by water of a large molecule into smaller parts. The R groups on the amino acids in the active site then catalyze the hydrolysis of sucrose, which produces the monosaccharides glucose and fructose. Because the structures of the products are no longer attracted to the active site, they are released, which allows sucrase to react with another sucrose molecule.

$$\underset{\text{Sucrase + Sucrose}}{E + S} \rightleftarrows \underset{\text{Sucrase–Sucrose Complex}}{\text{ES complex}} \longrightarrow \underset{\text{Sucrase + Glucose + Fructose}}{E + P_1 + P_2}$$

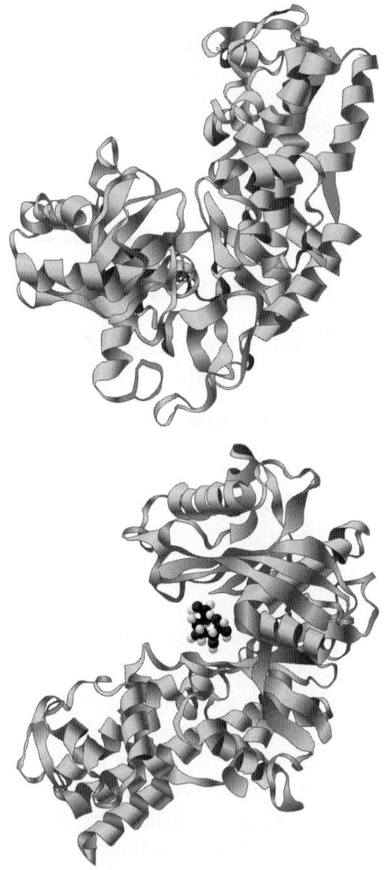

The ribbon model shows the change in the shape of the enzyme hexokinase as the glucose molecule binds to its active site.

Models of Enzyme Action

An early theory of enzyme action, called the *lock-and-key model*, described the active site as having a rigid, nonflexible shape. According to the lock-and-key model, the shape of the active site was analogous to a lock, and its substrate was the key that specifically fit that lock. However, this model was a static one that did not include the flexibility of the tertiary shape of an enzyme and the way we now know that the active site can adjust to the shape of a substrate.

CORE CHEMISTRY SKILL

Describing Enzyme Action

In the dynamic model of enzyme action, called the **induced-fit model**, the flexibility of the active site allows it to adapt to the shape of the substrate. At the same time, the shape of the substrate may be modified to better fit the geometry of the active site. In the induced-fit model, substrate and enzyme work together to acquire a geometrical arrangement that lowers the activation energy (see Figure 16.14).

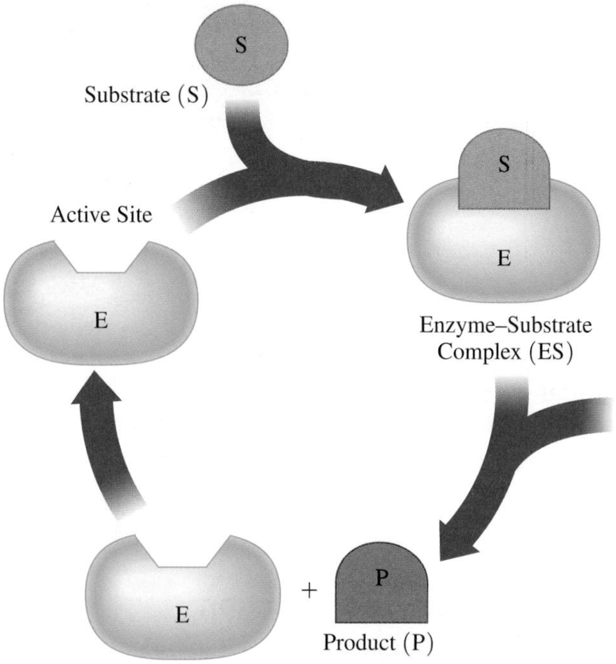

FIGURE 16.14 ▶ The flexible active site adjusts its shape to provide the best fit with the substrate for the reaction.

Q Why does a reaction in the body go faster when it is catalyzed by an enzyme?

QUESTIONS AND PROBLEMS

16.5 Enzymes

LEARNING GOAL Describe enzymes and their role in enzyme-catalyzed reactions.

16.33 Why are enzymes needed for chemical reactions in the body?

16.34 What is the reactant for each of the following enzymes?
a. sucrase **b.** lipase **c.** rennin

16.35 What is the reactant for each of the following enzymes?
a. galactase **b.** lipase **c.** aspartase

16.36 Name the substrate for each of the following enzymes:
a. peptidase **b.** cellulase **c.** lactase

16.37 What is the name of the class of enzymes that would catalyze each of the following reactions?
a. hydrolysis of ester bond in lipid
b. removal of ammonia
c. removes 2H atoms to form a double bond
d. converts cis to trans isomer

16.38 What is the name of the class of enzymes that would catalyze each of the following reactions?
a. addition of water to a double bond
b. removing hydrogen atoms
c. splitting peptide bonds in proteins
d. converting a tertiary alcohol to a secondary alcohol

16.39 Match the terms (1) enzyme–substrate complex, (2) enzyme, and (3) substrate with each of the following:
a. has a tertiary structure that recognizes the substrate
b. the combination of an enzyme with the substrate
c. has a structure that fits the active site of an enzyme

16.40 Match the terms (1) active site, (2) lock-and-key model, and (3) induced-fit model with each of the following:
a. the portion of an enzyme where catalytic activity occurs
b. an active site that adapts to the shape of a substrate
c. an active site that has a rigid shape

16.41 a. Write an equation that represents an enzyme-catalyzed reaction.
b. How is the active site different from the whole enzyme structure?

16.42 a. How does an enzyme speed up the reaction of a substrate?
b. After the products have formed, what happens to the enzyme?

For questions 16.43 to 16.46, see Chemistry Link to Health "Isoenzymes as Diagnostic Tools."

16.43 Define isoenzymes. Give five isoenzymes for lactase dehydrogenase (LDH).

16.44 How is the LDH isoenzyme in the heart different from LDH isoenzyme in the liver?

16.45 A patient arrives in an emergency department complaining of chest pains. What enzymes would you test for in the blood serum?

16.46 A patient has elevated blood serum levels of LDH and AST. What condition might be indicated?

Chemistry Link to Health

Isoenzymes As Diagnostic Tools

Isoenzymes are different forms of an enzyme that catalyze the same reaction in different cells or tissues of the body. They consist of quaternary structures with slight variations in the amino acids in the polypeptide subunits. For example, there are five isoenzymes of *lactate dehydrogenase* (LDH) that catalyze the conversion between lactate and pyruvate.

$$\underset{\text{Lactate}}{CH_3-\overset{\overset{\large OH}{|}}{CH}-COO^-} + NAD^+ \xrightleftharpoons[]{\text{Lactate dehydrogenase}} \underset{\text{Pyruvate}}{CH_3-\overset{\overset{\large O}{||}}{C}-COO^-} + NADH + H^+$$

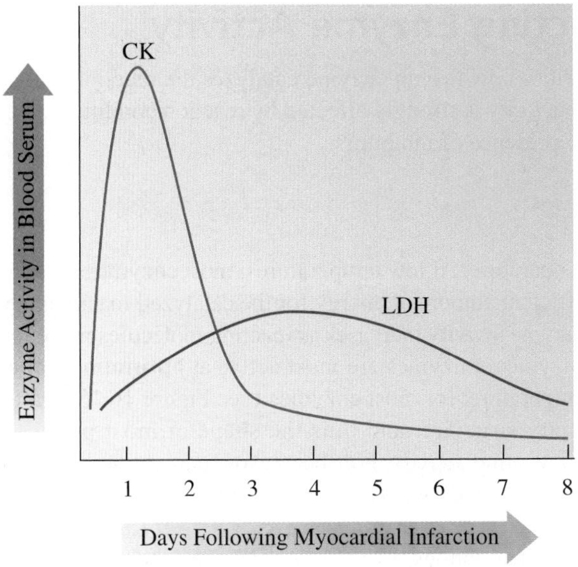

A myocardial infarction may be indicated by an increase in the levels of creatine kinase (CK), and lactate dehydrogenase (LDH).

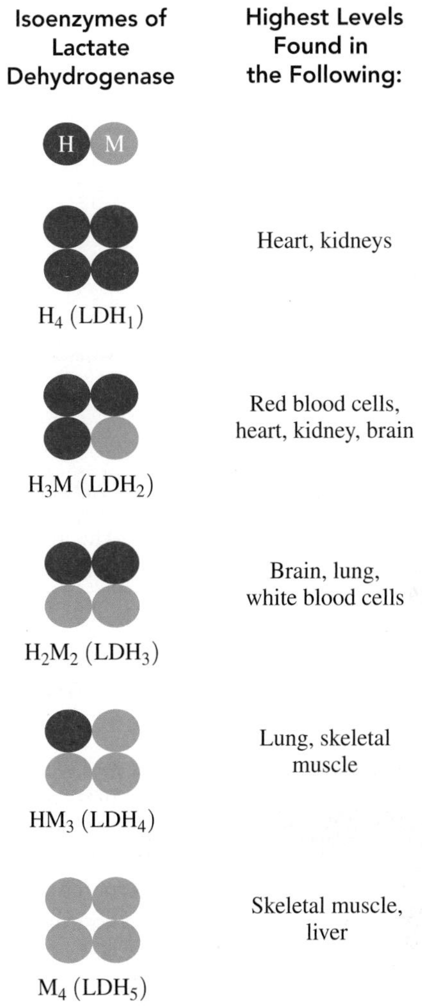

The different isoenzymes of lactate dehydrogenase (LDH) indicate damage to different organs in the body.

Each LDH isoenzyme contains a mix of polypeptide subunits, M and H. In the liver and muscle, lactate is converted to pyruvate by the LDH_5 isoenzyme with four M subunits designated M_4. In the heart, the same reaction is catalyzed by the LDH_1 isoenzyme (H_4) containing four H subunits. Different combinations of the M and H subunits are found in the LDH isoenzymes of the brain, red blood cells, kidneys, and white blood cells.

The different forms of an enzyme allow a medical diagnosis of damage or disease to a particular organ or tissue. In healthy tissues, isoenzymes function within the cells. However, when a disease damages a particular organ, cells die, which releases their contents including the isoenzymes into the blood. Measurements of the elevated levels of specific isoenzymes in the blood serum help to identify the disease and its location in the body. For example, an elevation in serum LDH_5, M_4, indicates liver damage or disease. When a *myocardial infarction* (MI) or heart attack damages heart muscle, an increase in the level of LDH_1 (H_4) isoenzyme is detected in the blood serum (see Table 16.9).

TABLE 16.9 Isoenzymes of Lactate Dehydrogenase and Creatine Kinase

Isoenzyme	Abundant in	Subunits
Lactate Dehydrogenase (LDH)		
LDH_1	Heart, kidneys	H_4
LDH_2	Red blood cells, heart, kidney, brain	H_3M
LDH_3	Brain, lung, white blood cells	H_2M_2
LDH_4	Lung, skeletal muscle	HM_3
LDH_5	Skeletal muscle, liver	M_4
Creatine Kinase (CK)		
CK_1	Brain, lung	BB
CK_2	Heart muscle	MB
CK_3	Skeletal muscle, red blood cells	MM

Another isoenzyme used diagnostically is creatine kinase (CK), which consists of two types of polypeptide subunits. Subunit B is prevalent in the brain, and subunit M predominates in muscle. Normally CK_3 (subunits MM) is present at low levels in the blood serum. However, in a patient who has suffered a myocardial infarction, the level of CK_2 (subunits MB) is elevated within 4 to 6 h and reaches a peak in about 24 h. Table 16.10 lists some enzymes used to diagnose tissue damage and diseases of certain organs.

TABLE 16.10 Serum Enzymes Used in Diagnosis of Tissue Damage

Condition	Diagnostic Enzymes Elevated
Heart attack or liver disease (cirrhosis, hepatitis)	Lactate dehydrogenase (LDH) Aspartate transaminase (AST)
Heart attack	Creatine kinase (CK)
Hepatitis	Alanine transaminase (ALT)
Liver (carcinoma) or bone disease (rickets)	Alkaline phosphatase (ALP)
Pancreatic disease	Pancreatic amylase (PA) Cholinesterase (CE) Lipase (LPS)
Prostate carcinoma	Acid phosphatase (ACP) Prostate specific antigen (PSA)

LEARNING GOAL

Describe the effect of temperature, pH, and inhibitors on enzyme activity.

16.6 Factors Affecting Enzyme Activity

The **activity** of an enzyme describes how fast an enzyme catalyzes the reaction that converts a substrate to product. This activity is strongly affected by reaction conditions, which include temperature, pH, and the presence of inhibitors.

Temperature

Enzymes are very sensitive to temperature. At low temperatures, most enzymes show little activity because there is not a sufficient amount of energy for the catalyzed reaction to take place. At higher temperatures, enzyme activity increases as reacting molecules move faster to cause more collisions with enzymes. Enzymes are most active at **optimum temperature**, which is 37 °C, or body temperature, for most enzymes (see Figure 16.15). At temperatures above 50 °C, the tertiary structure, and thus the shape of most proteins, is destroyed, which causes a loss in enzyme activity. For this reason, equipment in hospitals and laboratories is sterilized in autoclaves where the high temperatures denature the enzymes in harmful bacteria.

Certain organisms, known as thermophiles, live in environments where temperatures range from 50 °C to 120 °C. In order to survive in these extreme conditions, thermophiles must have enzymes with tertiary structures that are not destroyed by such high temperatures. Some research shows that their enzymes are very similar to ordinary enzymes except they contain more hydrogen bonds and salt bridges that stabilize the tertiary structures at high temperatures and resist unfolding and the loss of enzymatic activity.

Thermophiles survive in the high temperatures (50 °C to 120 °C) of a hot spring.

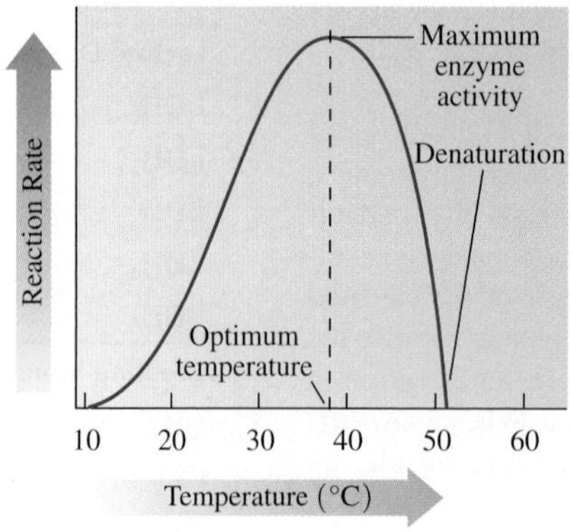

FIGURE 16.15 ▶ An enzyme attains maximum activity at its optimum temperature, usually 37 °C. Lower temperatures slow the rate of reaction, and temperatures above 50 °C denature most enzymes with a loss of catalytic activity.

Q Why is 37 °C the optimum temperature for many enzymes?

pH

Enzymes are most active at their **optimum pH**, the pH that maintains the proper tertiary structure of the protein (see Figure 16.16). If a pH value is above or below the optimum pH, the R group interactions are disrupted, which destroys the tertiary structure and the active site. As a result, the enzyme no longer binds substrate, and no catalysis occurs. Small changes in pH are reversible, which allows an enzyme to regain its structure and activity. However, large variations from optimum pH permanently destroy the structure of the enzyme.

Enzymes in most cells have optimum pH values at physiological pH around 7.4. However, enzymes in the stomach have a low optimum pH because they hydrolyze proteins at the acidic pH in the stomach. For example, pepsin, a digestive enzyme in the stomach, has an optimum pH of 1.5 to 2.0. Between meals, the pH in the stomach is 4 or 5 and pepsin shows little or no digestive activity. When food enters the stomach, the secretion of HCl lowers the pH to about 2, which activates pepsin. Table 16.11 lists the optimum pH values for selected enzymes.

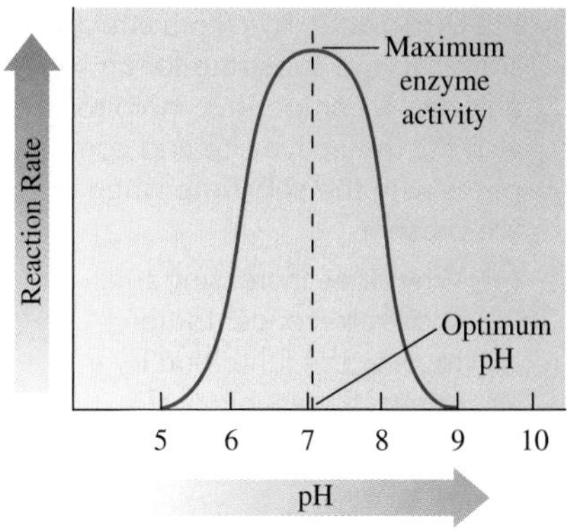

FIGURE 16.16 ▶ Enzymes are most active at their optimum pH. At a higher or lower pH, denaturation of the enzyme causes a loss of catalytic activity.

Q Why does the digestive enzyme pepsin have an optimum pH of 2?

TABLE 16.11 Optimum pH for Selected Enzymes

Enzyme	Location	Substrate	Optimum pH
Pepsin	Stomach	Peptide bonds	1.5–2.0
Sucrase	Small intestine	Sucrose	6.2
Amylase	Pancreas	Amylose	6.7–7.0
Urease	Liver	Urea	7.0
Trypsin	Small intestine	Peptide bonds	7.7–8.0
Lipase	Pancreas	Lipid (ester bonds)	8.0
Arginase	Liver	Arginine	9.7

SAMPLE PROBLEM 16.8 Factors Affecting Enzymatic Activity

Describe the effect that lowering the temperature to 10 °C would have on the rate of the reaction catalyzed by urease.

$$\underset{\text{Urea}}{H_2N-\overset{\overset{\displaystyle O}{\|}}{C}-NH_2} + H_2O \xrightarrow{\text{Urease}} 2NH_3 + CO_2$$

SOLUTION

Because 10 °C is lower than the optimum temperature of 37 °C, there is a decrease in the rate of the reaction.

STUDY CHECK 16.8

If urease has an optimum pH of 7.0, what is the effect of lowering the pH to 3?

Enzyme Inhibition

Many kinds of molecules called **inhibitors** cause enzymes to lose catalytic activity. Although inhibitors act differently, they all prevent the active site from binding with a substrate. An enzyme with a *reversible inhibitor* can regain enzymatic activity, but an enzyme attached to an *irreversible inhibitor* loses enzymatic activity permanently.

A **competitive inhibitor** has a chemical structure and polarity that is similar to that of the substrate. Thus, a competitive inhibitor competes with the substrate for the active site on the enzyme. When the inhibitor occupies the active site, the substrate cannot bind to the enzyme and no reaction can occur (see Figure 16.17).

Explore Your World

Enzyme Activity

The enzymes on the surface of a freshly cut apple, avocado, or banana react with oxygen in the air to turn the surface brown. An antioxidant, such as vitamin C in lemon juice, prevents the oxidation reaction. Cut an apple, an avocado, or a banana into several slices. Place one slice in a plastic zipper lock bag, squeeze out all the air, and close the zipper lock. Dip another slice in lemon juice and place it on a plate. Sprinkle another slice with a crushed vitamin C tablet. Leave another slice alone as a control. Observe the surface of each of your samples. Record your observations immediately, then every hour for 6 h or longer.

Questions

1. Which slice(s) shows the most oxidation (turns brown)?
2. Which slice(s) shows little or no oxidation?
3. How was the oxidation reaction on each slice affected by treatment with an antioxidant?

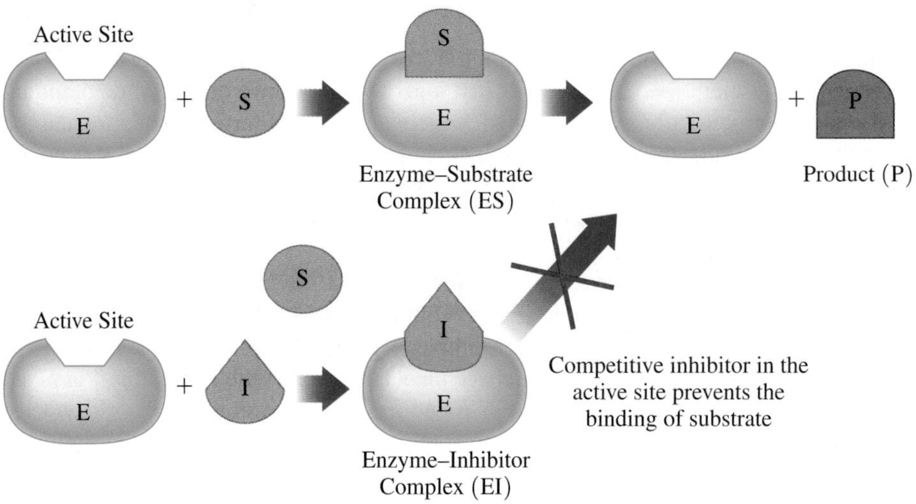

FIGURE 16.17 ▶ With a structure similar to the substrate for an enzyme, a competitive inhibitor also fits the active site and competes with the substrate when both are present.

Q Why does increasing the substrate concentration reverse the inhibition by a competitive inhibitor?

$$E + S \rightleftharpoons \underset{\text{Enzyme–Substrate Complex}}{ES} \longrightarrow E + P \quad \text{Favored when [S] is high; [I] is low}$$

$$E + I \rightleftharpoons \underset{\text{Enzyme–Inhibitor Complex}}{EI} \quad \text{Favored when [I] is high; [S] is low}$$

As long as the concentration of the inhibitor is substantial, there is a loss of enzyme activity. However, adding more substrate displaces the competitive inhibitor. As more enzyme molecules bind to substrate (ES), enzyme activity is regained.

The structure of a **noncompetitive inhibitor** does not resemble the substrate and does not compete for the active site. Instead, a noncompetitive inhibitor binds to a site on the enzyme that is not the active site. When the noncompetitive inhibitor is bonded to the enzyme, the shape of the enzyme is distorted. Inhibition occurs because the substrate cannot fit in the active site or it does not fit properly. Without the proper alignment of substrate with the amino acid side groups, no catalysis can take place (see Figure 16.18).

Because a noncompetitive inhibitor is not competing for the active site, the addition of more substrate does not reverse this type of inhibition. Examples of noncompetitive

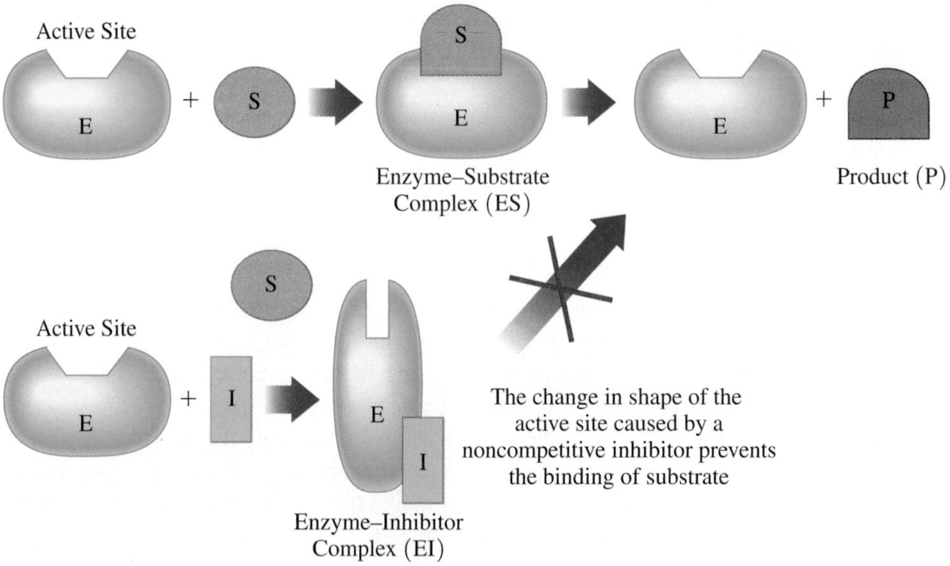

FIGURE 16.18 ▶ A noncompetitive inhibitor binds to an enzyme at a site other than the active site, which distorts the enzyme and prevents the proper binding and catalysis of the substrate at the active site.

Q Will an increase in substrate concentration reverse the inhibition by a noncompetitive inhibitor?

inhibitors are the heavy metal ions Pb^{2+}, Ag^{+}, and Hg^{2+} that bond with amino acid side groups such as $—COO^{-}$ or —OH. Catalytic activity is restored when chemical reagents remove the inhibitors.

Irreversible Inhibition

In irreversible inhibition, a molecule causes an enzyme to lose all enzymatic activity. Most irreversible inhibitors are toxic substances that destroy enzymes. Usually an irreversible inhibitor forms a covalent bond with an amino acid side group within the active site, which prevents the substrate from binding to the active site or prevents catalytic activity.

Insecticides and nerve gases act as irreversible inhibitors of acetylcholinesterase, an enzyme needed for nerve conduction. The compound DFP (diisopropyl fluorophosphate), an organophosphate insecticide, forms a covalent bond with the side chain $—CH_2OH$ of serine in the active site. When acetylcholinesterase is inhibited, the transmission of nerve impulses is blocked, and paralysis occurs.

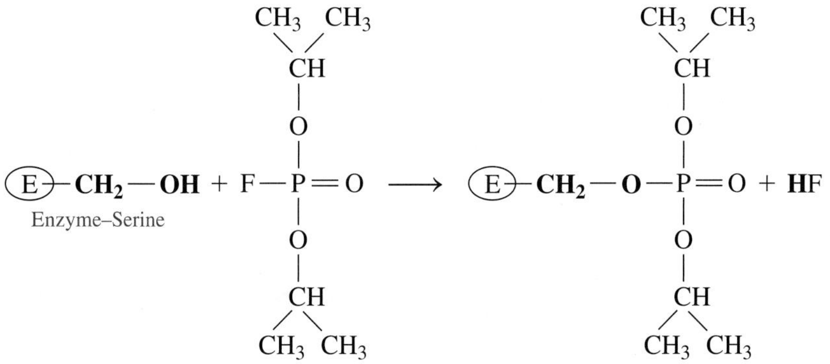

Enzyme–Serine; Diisopropyl fluorophosphate (DFP); Enzyme–Serine complex covalently bonded to DFP

Antibiotics produced by bacteria, mold, or yeast are irreversible inhibitors used to stop bacterial growth. For example, penicillin inhibits an enzyme needed for the formation of cell walls in bacteria, but not human cell membranes. With an incomplete cell wall, bacteria cannot survive and the infection is stopped. However, some bacteria are resistant to penicillin because they produce penicillinase, an enzyme that breaks down penicillin. Over the years, derivatives of penicillin to which bacteria have not yet become resistant have been produced.

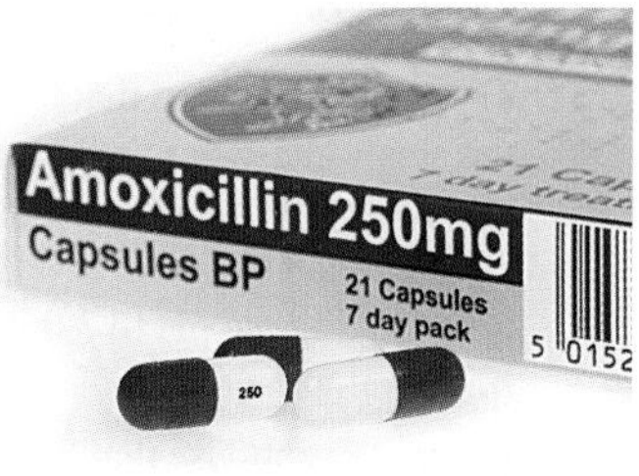

Amoxicillin is a derivative of the antibiotic penicillin.

R Groups for Penicillin Derivatives

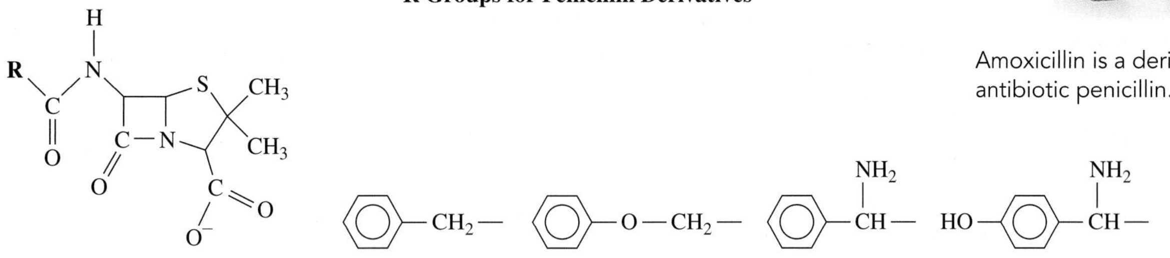

Penicillin; Penicillin G; Penicillin V; Ampicillin; Amoxicillin

A summary of competitive, noncompetitive, and irreversible inhibitors is shown in Table 16.12.

TABLE 16.12 Summary of Competitive, Noncompetitive, and Irreversible Inhibitors

Characteristics	Competitive	Noncompetitive	Irreversible
Shape of Inhibitor	similar shape to the substrate	does not have a similar shape to the substrate	does not have a similar shape to the substrate
Binding to Enzyme	competes for and binds at the active site	binds away from the active site to change the shape of the enzyme and its activity	forms a covalent bond with the enzyme
Reversibility	adding more substrate reverses the inhibition	not reversed by adding more substrate, but by a chemical change that removes the inhibitor	permanent, not reversible

SAMPLE PROBLEM 16.9 Enzyme Inhibition

State the type of inhibition in the following:

a. The inhibitor has a structure that is similar to the substrate.

b. This inhibitor binds to the surface of the enzyme, changing its shape in such a way that it cannot bind to substrate.

SOLUTION

a. When an inhibitor has a structure similar to that of the substrate, it competes with the substrate for the active site. This type of inhibition is competitive inhibition, which is reversed by increasing the concentration of the substrate (see Table 16.12).

b. When an inhibitor binds to the surface of the enzyme, it changes the shape of the enzyme and the active site. This type of inhibition is noncompetitive inhibition because the inhibitor does not have a similar shape to the substrate and does not compete with the substrate for the active site (see Table 16.12).

STUDY CHECK 16.9

What type of inhibition occurs when Sarin, a nerve gas, forms a covalent bond with the R group of serine in the active site of acetylcholinesterase?

QUESTIONS AND PROBLEMS

16.6 Factors Affecting Enzyme Activity

LEARNING GOAL Describe the effect of temperature, pH, and inhibitors on enzyme activity.

16.47 Trypsin, a peptidase that hydrolyzes polypeptides, functions in the small intestine at an optimum pH of 7.7 to 8.0. How is the rate of a trypsin-catalyzed reaction affected by each of the following conditions?

a. changing the pH to 3.0
b. running the reaction at 75 °C

16.48 Pepsin, a peptidase that hydrolyzes proteins, functions in the stomach at an optimum pH of 1.5 to 2.0. How is the rate of a pepsin-catalyzed reaction affected by each of the following conditions?

a. changing the pH to 5.0
b. running the reaction at 0 °C

16.49 The following graph shows the activity versus pH curves for pepsin, sucrase, and trypsin. Estimate the optimum pH for each.

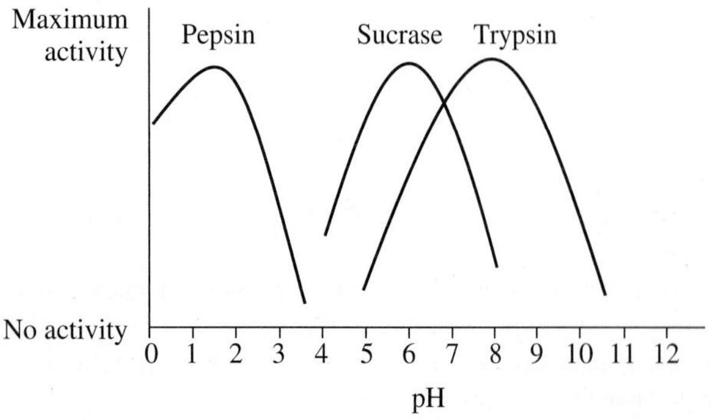

16.50 Refer to the graph in problem 16.49 to determine if the reaction rate in each condition will be at the optimum rate or not.

a. trypsin, pH 5.0
b. sucrase, pH 5.0
c. pepsin, pH 4.0
d. trypsin, pH 8.0
e. pepsin, pH 2.0

16.51 Indicate whether each of the following describes a competitive or a noncompetitive enzyme inhibitor:

a. The inhibitor has a structure similar to the substrate.
b. The effect of the inhibitor cannot be reversed by adding more substrate.
c. The inhibitor competes with the substrate for the active site.
d. The structure of the inhibitor is not similar to the substrate.
e. The addition of more substrate reverses the inhibition.

16.52 Oxaloacetate is an inhibitor of succinate dehydrogenase.

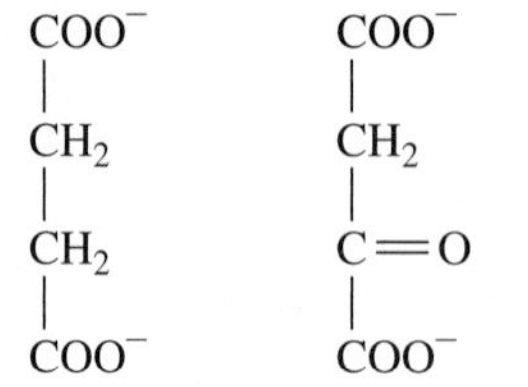

Succinate Oxaloacetate

a. Would you expect oxaloacetate to be a competitive or a noncompetitive inhibitor? Why?
b. Would oxaloacetate bind to the active site or elsewhere on the enzyme?
c. How would you reverse the effect of the inhibitor?

16.53 Methanol and ethanol are oxidized by alcohol dehydrogenase. In methanol poisoning, ethanol is given intravenously to prevent the formation of formaldehyde that has toxic effects.

a. Draw the condensed structural formulas for methanol and ethanol.

b. Would ethanol compete for the active site or bind to a different site?

c. Would ethanol be a competitive or noncompetitive inhibitor of methanol oxidation?

16.54 In humans, the antibiotic amoxicillin (a type of penicillin) is used to treat certain bacterial infections.

a. Does the antibiotic inhibit enzymes in humans?

b. Why does the antibiotic kill bacteria, but not humans?

c. Is amoxicillin a reversible or irreversible inhibitor?

CONCEPT MAP

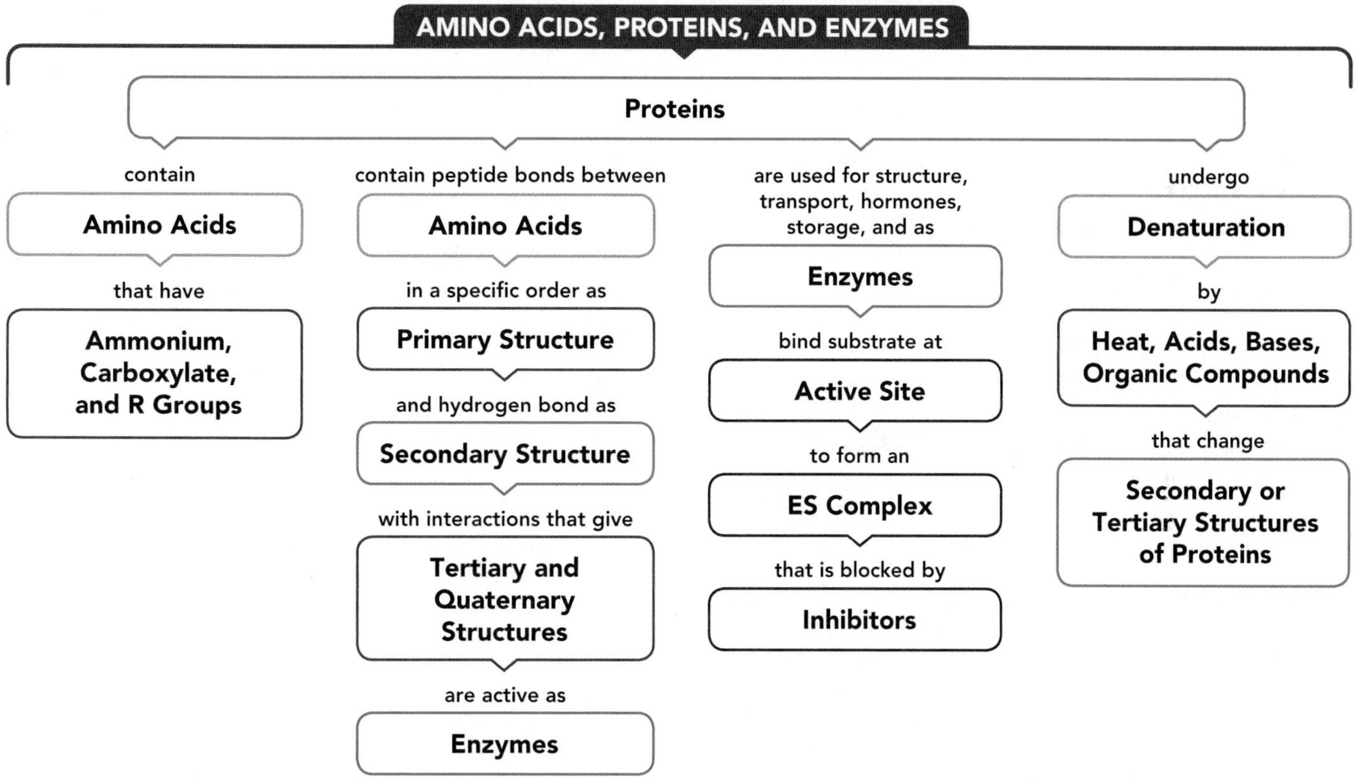

CHAPTER REVIEW

16.1 Proteins and Amino Acids

LEARNING GOAL Classify proteins by their functions. Give the name and abbreviations for an amino acid and draw its ionized structure.

- Some proteins are enzymes or hormones, whereas others are important in structure, transport, protection, storage, and muscle contraction.
- A group of 20 amino acids provides the molecular building blocks of proteins.
- Attached to the central α-carbon of each amino acid are an ammonium group, a carboxylate group, and a unique R group.
- The R group gives an amino acid the property of being nonpolar, polar, acidic, or basic.

16.2 Amino Acids as Acids and Bases

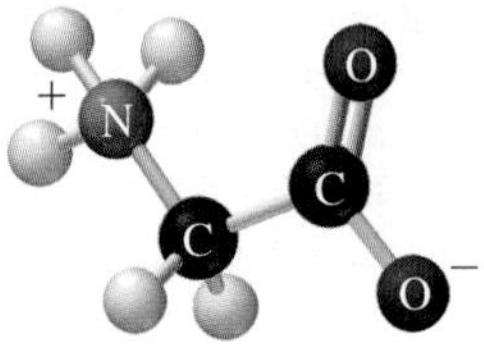

LEARNING GOAL Draw the condensed structural formula for an amino acid at pH values above or below its isoelectric point.

- Amino acids exist as positive ions at pH values below their pI values, and as negative ions at pH values above their pI values.
- At the isoelectric point, ionized amino acids have an overall charge of zero.

16.3 Proteins: Primary Structure

LEARNING GOAL Draw the condensed structural formula for a peptide and give its name. Describe the primary structure for a protein.

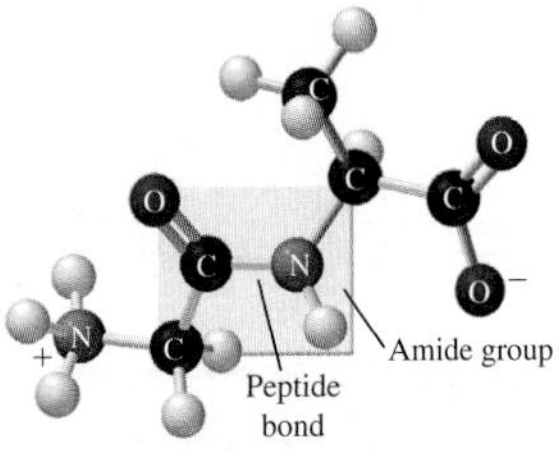

- Peptides form when an amide bond links the carboxylate group of one amino acid and the ammonium group of a second amino acid.
- Long chains of amino acids that are biologically active are called proteins.
- The primary structure of a protein is its sequence of amino acids joined by peptide bonds.
- Peptides are named from the N-terminal by replacing the *ine* or *ic acid* of each amino acid name with *yl* followed by the amino acid name of the C-terminal.

16.4 Proteins: Secondary, Tertiary, and Quaternary Structures

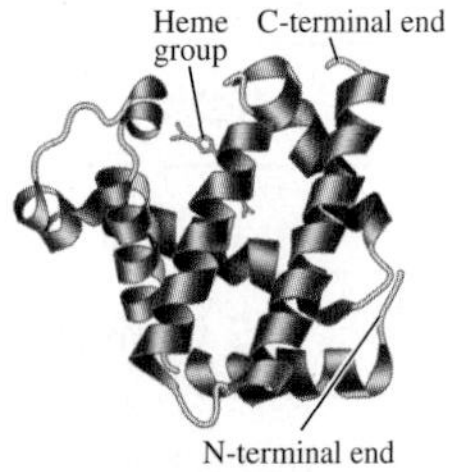

LEARNING GOAL Describe the secondary, tertiary, and quaternary structures for a protein; describe the denaturation of a protein.

- In the secondary structure, hydrogen bonds between atoms in the peptide bonds produce a characteristic shape such as an α helix, β-pleated sheet, or a triple helix.
- A tertiary structure is stabilized by interactions that push amino acids with hydrophobic R groups to the center and pull amino acids with hydrophilic R groups to the surface, and by interactions between amino acids with R groups that form hydrogen bonds, disulfide bonds, and salt bridges.
- In a quaternary structure, two or more tertiary subunits are joined together for biological activity, held by the same interactions found in tertiary structures.
- Denaturation of a protein occurs when high temperatures, acids or bases, organic compounds, metal ions, or agitation destroy the secondary, tertiary, or quaternary structures of a protein with a loss of biological activity.

16.5 Enzymes

LEARNING GOAL Describe enzymes and their role in enzyme-catalyzed reactions.

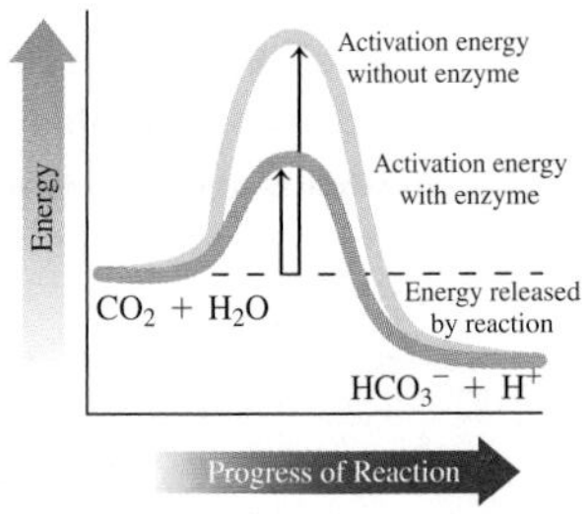

- Enzymes are globular proteins that act as biological catalysts by lowering activation energy and accelerating the rate of cellular reactions.
- Within the tertiary structure of an enzyme, a small pocket called the active site binds the substrate.
- In the lock-and-key model, a substrate precisely fits the shape of the active site.
- In the induced-fit model, both the active site and the substrate undergo changes in their shapes to give the best fit for efficient catalysis.
- In the enzyme–substrate complex, catalysis takes place when amino acid R groups in the active site of an enzyme react with a substrate.
- When the products of catalysis are released, the enzyme can bind to another substrate molecule.

16.6 Factors Affecting Enzyme Activity

LEARNING GOAL Describe the effect of temperature, pH, and inhibitors on enzyme activity.

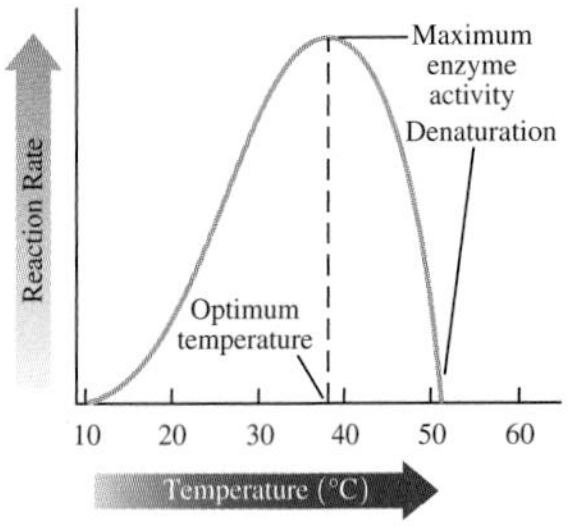

- The optimum temperature at which most enzymes are effective is usually 37 °C, and the optimum pH is usually 7.4.
- The rate of an enzyme-catalyzed reaction decreases as temperature and pH go above or below the optimum temperature and pH values.
- An inhibitor reduces the activity of an enzyme or makes it inactive.
- An inhibitor can be reversible or irreversible.
- A competitive inhibitor has a structure similar to the substrate and competes for the active site.
- A noncompetitive inhibitor attaches to the enzyme away from the active site, changing the shape of both the enzyme and its active site.
- An irreversible inhibitor forms a covalent bond within the active site that permanently prevents catalytic activity.

KEY TERMS

acidic amino acid An amino acid that has a carboxylic acid (—COOH) R group which ionizes as a weak acid.

active site A pocket in a part of the tertiary enzyme structure that binds substrate and catalyzes a reaction.

activity The rate at which an enzyme catalyzes the reaction that converts substrate to product.

α (alpha) helix A secondary level of protein structure, in which hydrogen bonds connect the N—H of one peptide bond with the C═O of a peptide bond farther along in the chain to form a coiled or corkscrew structure.

amino acid The building block of proteins, consisting of an ammonium group, a carboxylate group, and a unique R group attached to the α-carbon.

basic amino acid An amino acid that contains an amino ($-NH_2$) R group, which ionizes as a weak base.

β (beta)-pleated sheet A secondary level of protein structure that consists of hydrogen bonds between peptide links in parallel polypeptide chains.

C-terminal amino acid The end amino acid in a peptide chain with a free $-COO^-$ group.

collagen The most abundant form of protein in the body, composed of fibrils of triple helices with hydrogen bonding between $-OH$ groups of hydroxyproline and hydroxylysine.

competitive inhibitor A molecule with a structure similar to the substrate that inhibits enzyme action by competing for the active site.

denaturation The loss of secondary, tertiary, and quaternary protein structures, caused by heat, acids, bases, organic compounds, heavy metals, and/or agitation.

disulfide bonds Covalent $-S-S-$ bonds that form between the $-SH$ groups of cysteines in a protein, which stabilizes the tertiary and quaternary structures.

enzymes Substances including globular proteins that catalyze biological reactions.

enzyme–substrate (ES) complex An intermediate consisting of an enzyme that binds to a substrate in an enzyme-catalyzed reaction.

fibrous proteins Proteins that are insoluble in water consisting of polypeptide chains with α helices or β-pleated sheets that make up the fibers of hair, wool, skin, nails, and silk.

globular proteins Proteins that acquire a compact shape from attractions between R groups of the amino acids in the protein.

hydrogen bonds Attractions between polar R groups such as $-OH$, $-NH_2$, and $-COOH$ of amino acids in a polypeptide chain.

hydrophilic interactions The attractions between water and polar R groups on the outside of the protein.

hydrophobic interactions The attractions between nonpolar R groups on the inside of a globular protein.

induced-fit model A model of enzyme action in which the shapes of the substrate and its active site adjust to give an optimal fit.

inhibitors Substances that make an enzyme inactive by interfering with its ability to react with a substrate.

isoelectric point (pI) The pH at which an amino acid exists in an ionized form with an overall net charge of zero.

isoenzymes Enzymes with different combinations of polypeptide subunits that catalyze the same reaction in different tissues of the body.

N-terminal amino acid The end amino acid in a peptide chain with a free $-NH_3^+$ group.

noncompetitive inhibitor An inhibitor that does not resemble the substrate, and attaches to the enzyme away from the active site to prevent binding of the substrate.

nonpolar amino acids Amino acids with nonpolar R groups that contain only C and H atoms.

optimum pH The pH at which an enzyme is most active.

optimum temperature The temperature at which an enzyme is most active.

peptide The combination of two or more amino acids joined by peptide bonds; dipeptide, tripeptide, and so on.

peptide bond The amide bond that joins amino acids in polypeptides and proteins.

polar amino acids Amino acids with polar R groups that interact with water.

primary structure The specific sequence of the amino acids in a protein.

protein A polypeptide of 50 or more amino acids that has biological activity.

quaternary structure A protein structure in which two or more protein subunits form an active protein.

salt bridge The attraction between ionized R groups of basic and acidic amino acids in the tertiary structure of a protein.

secondary structure The formation of an α helix, a β-pleated sheet, or a triple helix.

substrate The molecule that reacts in the active site in an enzyme-catalyzed reaction.

tertiary structure The folding of the secondary structure of a protein into a compact structure that is stabilized by the interactions of R groups.

triple helix The protein structure found in collagen, consisting of three polypeptide chains woven together like a braid.

CORE CHEMISTRY SKILLS

Drawing the Ionized Form for an Amino Acid (16.1)

- The central α-carbon of each amino acid is bonded to an ammonium group ($-NH_3^+$), a carboxylate group ($-COO^-$), a hydrogen atom, and a unique R group.
- The R group gives an amino acid the property of being nonpolar, polar, acidic, or basic.
- At the isoelectric point (pI), an amino acid has an overall charge of zero.
- Amino acids exist as positive ions at pH values below their pI values, and as negative ions at pH values above their pI values.

Example: Draw the ionized form for cysteine.

Answer:

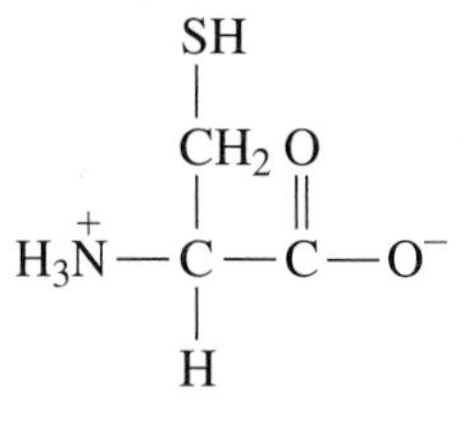

Identifying the Primary, Secondary, Tertiary, and Quaternary Structures of Proteins (16.3, 16.4)

- The primary structure of a protein is the sequence of amino acids joined by peptide bonds.
- In the secondary structures of proteins, hydrogen bonds between atoms in the peptide bonds produce an alpha helix, beta-pleated sheet, or a triple helix.
- The tertiary structure of a protein is stabilized by R groups that form hydrogen bonds, disulfide bonds, and salt bridges, and hydrophobic R groups that move to the center and hydrophilic R groups that move to the surface.
- In a quaternary structure, two or more tertiary subunits are combined for biological activity, held by the same interactions found in tertiary structures.

Example: Identify the following as characteristic of the primary, secondary, tertiary, or quaternary structure of a protein:

a. The R groups of two amino acids interact to form a salt bridge.
b. Eight amino acids form peptide bonds.
c. A polypeptide forms an alpha helix.
d. Two amino acids with hydrophobic R groups move toward the inside of the folded protein.
e. A protein with biological activity contains four tertiary polypeptide subunits.

Answer: **a.** tertiary, quaternary
b. primary
c. secondary
d. tertiary
e. quaternary

Describing Enzyme Action (16.5)

- Enzymes are biological catalysts that lower the activation energy and accelerate the rate of cellular reactions.
- Within the tertiary structure of an enzyme, a small pocket called the active site binds the substrate.
- In the induced-fit model of enzyme action, the active site and the substrate change their shapes for efficient catalysis.
- When the products of catalysis are released, the enzyme can bind to another substrate molecule.

Example: Match the terms (1) enzyme, (2) substrate, (3) active site, and (4) induced-fit model with each of the following:

a. an active site that adapts to the shape of a substrate
b. has a structure that fits the active site of an enzyme
c. the portion of an enzyme where catalytic activity occurs
d. has a tertiary structure that recognizes the substrate

Answer: **a.** (4) induced-fit model
b. (2) substrate
c. (3) active site
d. (1) enzyme

UNDERSTANDING THE CONCEPTS

The chapter sections to review are shown in parentheses at the end of each question.

16.55 Ethylene glycol ($HO—CH_2—CH_2—OH$) is a major component of antifreeze. If ingested, it is first converted to HOOC—CHO (oxoethanoic acid) and then to HOOC—COOH (oxalic acid), which is toxic. (16.5, 16.6)

Ethylene glycol is added to a radiator to prevent freezing and boiling.

a. What class of enzyme catalyzes the reactions described?
b. The treatment for the ingestion of ethylene glycol is an intravenous solution of ethanol. How might this help prevent toxic levels of oxalic acid in the body?

16.56 Adults who are lactose intolerant cannot break down the disaccharide in milk products. To help digest dairy food, a product known as Lactaid can be added to milk and the milk then refrigerated for 24 hours. (16.5, 16.6)

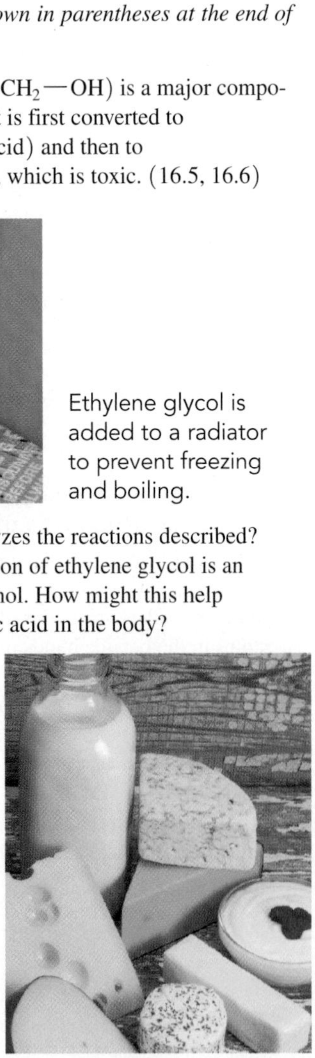

The disaccharide lactose is present in milk products.

a. What enzyme is present in Lactaid, and what is the major class of this enzyme?
b. What might happen to the enzyme if the Lactaid were stored at 55 °C?

16.57 Aspartame, which is used in artificial sweeteners, contains the following dipeptide: (16.1, 16.3)

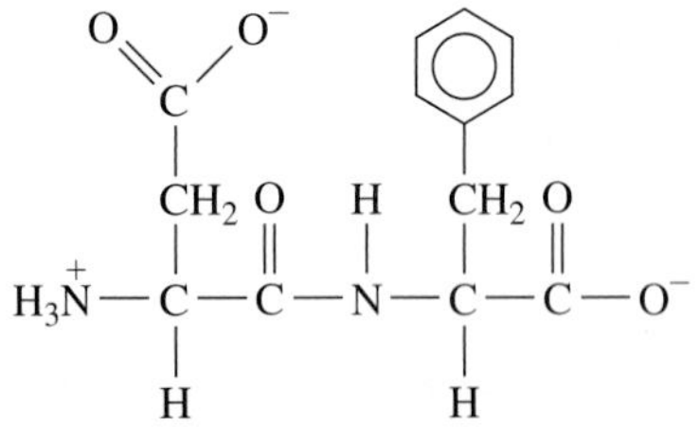

Some artificial sweeteners contain aspartame, which is an ester of a dipeptide.

a. What are the amino acids in aspartame?
b. How would you name the dipeptide in aspartame?

16.58 Fresh pineapple contains the enzyme bromelain that hydrolyzes peptide bonds in proteins. (16.4, 16.5, 16.6)

Pineapple contains the enzyme bromelain.

a. The directions for a gelatin (protein) dessert say not to add fresh pineapple. However, canned pineapple where pineapple is heated to high temperatures can be added. Why?

b. Fresh pineapple is used in a marinade to tenderize tough meat. Why?

c. What structural level of a protein does the bromelain enzyme destroy?

Use the following condensed structural formulas of cysteine (**1** to **4**) to answer problems 16.59 and 16.60: (16.1, 16.2)

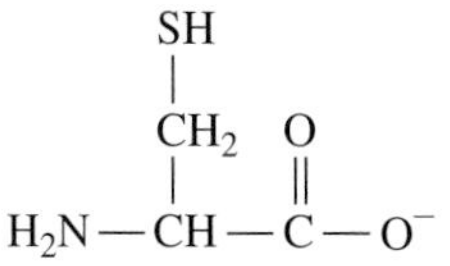

(**1**)

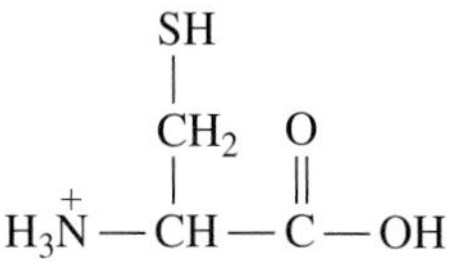

(**2**)

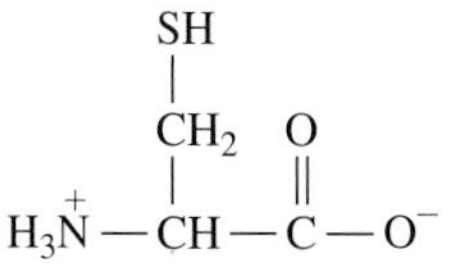

(**3**)

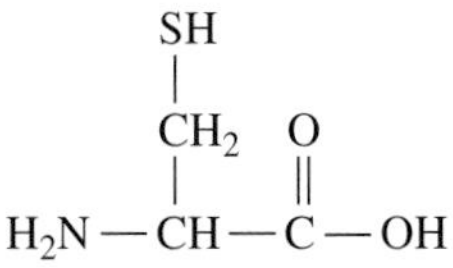

(**4**)

16.59 If cysteine, an amino acid prevalent in hair, has a pI of 5.1, which condensed structural formula would it have in solutions with each of the following pH values? (16.1, 16.2)

a. 10.5 **b.** 5.1 **c.** 1.8

The proteins in hair contain many cysteine R groups that form disulfide bonds.

16.60 If cysteine, an amino acid prevalent in hair, has a pI of 5.1, which condensed structural formula would it have in solutions with each of the following pH values? (16.1, 16.2)

a. 2.0 **b.** 3.5 **c.** 9.1

16.61 Identify the amino acids and type of interaction that occurs between the following R groups in tertiary protein structures: (16.1, 16.4)

a. $-CH_2-\overset{O}{\overset{\|}{C}}-NH_2$ and $HO-CH_2-$

b. $-CH_2-\overset{O}{\overset{\|}{C}}-O^-$ and $H_3\overset{+}{N}-(CH_2)_4-$

c. $-CH_2-SH$ and $HS-CH_2-$

d. $-CH_2-\overset{CH_3}{\overset{|}{CH}}-CH_3$ and CH_3-

16.62 What type of interaction would you expect between the following in a tertiary structure? (16.1, 16.4)

a. threonine and glutamine **b.** valine and alanine

c. arginine and glutamic acid

16.63 **a.** Draw the condensed structural formula for Ser–Lys–Asp. (16.1, 16.3, 16.4)

b. Would you expect to find this segment at the center or at the surface of a globular protein? Why?

16.64 **a.** Draw the condensed structural formula for Val–Ala–Leu. (16.1, 16.3, 16.4)

b. Would you expect to find this segment at the center or at the surface of a globular protein? Why?

16.65 Seeds and vegetables are often deficient in one or more essential amino acids. Using the following table, state whether each combination provides all of the essential amino acids: (16.1)

Source	Lysine	Tryptophan	Methionine
Oatmeal	No	Yes	Yes
Rice	No	Yes	Yes
Garbanzo beans	Yes	No	Yes
Lima beans	Yes	No	No
Cornmeal	No	No	Yes

a. rice and garbanzo beans

b. lima beans and cornmeal

c. garbanzo beans and lima beans

16.66 Seeds and vegetables are often deficient in one or more essential amino acids. Using the table in problem 16.65, state whether each combination provides all of the essential amino acids. (16.1)

a. rice and lima beans

b. rice and oatmeal

c. oatmeal and lima beans

Oatmeal is deficient in the essential amino acid lysine.

ADDITIONAL QUESTIONS AND PROBLEMS

16.67 What are some differences between each of the following pairs? (16.1, 16.3, 16.4)
a. secondary and tertiary protein structures
b. essential and nonessential amino acids
c. polar and nonpolar amino acids
d. dipeptides and tripeptides

16.68 What are some differences between each of the following pairs? (16.1, 16.3, 16.4)
a. an ionic bond (salt bridge) and a disulfide bond
b. fibrous and globular proteins
c. α helix and β-pleated sheet
d. tertiary and quaternary structures of proteins

16.69 If glutamic acid were replaced by proline in a protein, how would the tertiary structure be affected? (16.1, 16.4)

16.70 If glycine were replaced by alanine in a protein, how would the tertiary structure be affected? (16.1, 16.4)

16.71 How do enzymes differ from catalysts used in chemical laboratories? (16.5)

16.72 Why do enzymes function only under mild conditions? (16.5, 16.6)

16.73 Lactase is an enzyme that hydrolyzes lactose to glucose and galactose. (16.5)
a. What are the reactants and products of the reaction?
b. Draw an energy diagram for the reaction with and without lactase.
c. How does lactase make the reaction go faster?

16.74 Maltase is an enzyme that hydrolyzes maltose to two glucose molecules. (16.5)
a. What are the reactants and products of the reaction?
b. Draw an energy diagram for the reaction with and without maltase.
c. How does maltase make the reaction go faster?

16.75 Indicate whether each of the following would be a substrate (S) or an enzyme (E):
a. sucrose
b. rennin
c. papain
d. dehydrogenase
e. lipid
f. oxidase

16.76 Indicate the substrates for the following enzymes:
a. pepsin
b. urease
c. lipase
d. trypsin

16.77 Give the substrate of each of the following enzymes: (16.5)
a. urease
b. succinate dehydrogenase
c. aspartate transaminase
d. tyrosinase

16.78 Give the substrate of each of the following enzymes: (16.5)
a. maltase
b. fructose oxidase
c. phenolase
d. sucrase

16.79 How would the lock-and-key model explain that sucrase hydrolyzes sucrose, but not lactose? (16.5)

16.80 How does the induced-fit model of enzyme action allow an enzyme to catalyze a reaction of a group of substrates? (16.5)

16.81 If a blood test indicates a high level of LDH and CK, what could be the cause? (16.5)

16.82 Name the amino acids that are missing in peas and beans.

CHALLENGE QUESTIONS

The following groups of questions are related to the topics in this chapter. However, they do not all follow the chapter order, and they require you to combine concepts and skills from several sections. These questions will help you increase your critical thinking skills and prepare for your next exam.

16.83 Indicate the overall charge of each of the following amino acids at the following pH values as 0, 1+, or 1−: (16.1, 16.2)
a. serine at pH 5.7
b. threonine at pH 2.0
c. isoleucine at pH 3.0
d. leucine at pH 9.0

16.84 Indicate the overall charge of each of the following amino acids at the following pH values as 0, 1+, or 1−: (16.1, 16.2)
a. tyrosine at pH 3.0
b. glycine at pH 10.0
c. phenylalanine at pH 8.5
d. methionine at pH 5.7

16.85 Identify the functional groups in each of the following amino acids:
a. phenylalanine
b. tyrosine
c. valine
d. glutamic acid

16.86 Identify the functional groups in each of the following amino acids:
a. cysteine
b. serine
c. methionine
d. threonine

ANSWERS

Answers to Study Checks

16.1 Histidine is polar and basic. It would be hydrophilic.

16.2 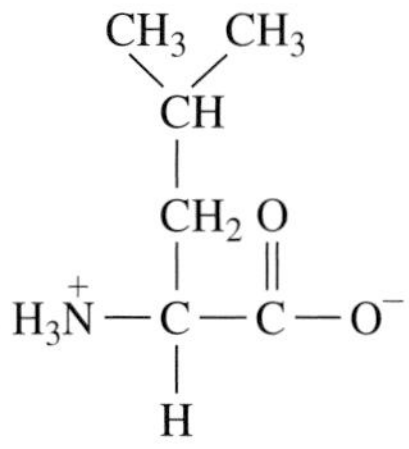 Leu, L

16.3 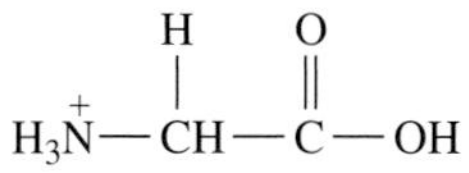 Charge 1+ at pH 3.0

16.4 a. proline
b. methionine
c. prolylhistidyllmethionine

16.5 Both are nonpolar and would be found on the inside of the tertiary structure.

16.6 quaternary

16.7 lipase

16.8 At a pH lower than the optimum pH, denaturation of the tertiary structure will decrease the activity of urease.

16.9 Because Sarin forms a covalent bond with an R group in the active site of the enzyme, the inhibition by Sarin is irreversible.

Answers to Selected Questions and Problems

16.1 a. transport
b. storage
c. protection
d. enzyme

16.3 All amino acids contain a carboxylate group and an ammonium group on the α-carbon.

16.5 a.

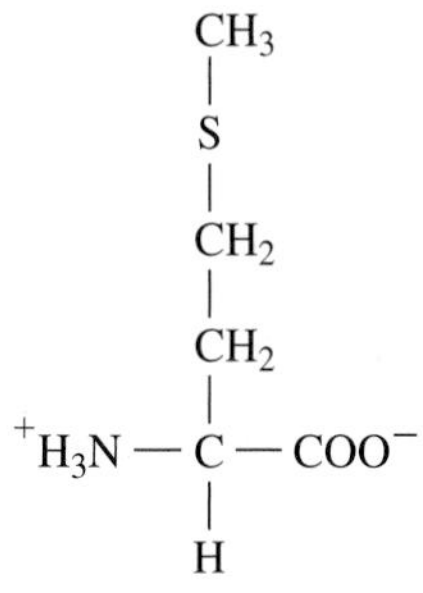

b.

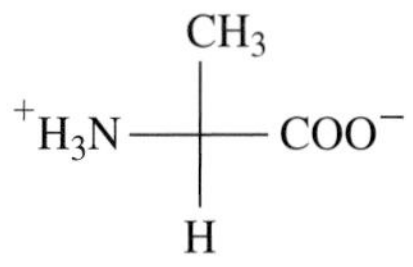

c.

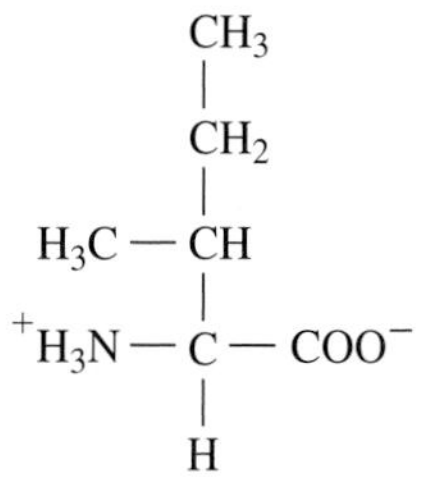

d.

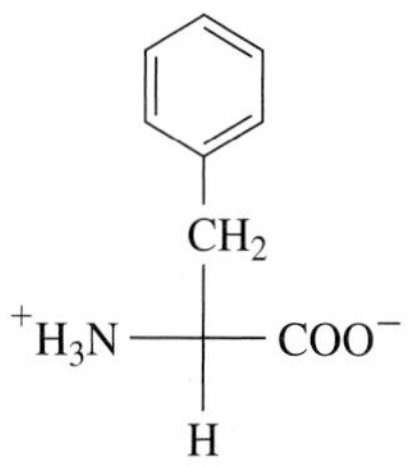

16.7 a. nonpolar, hydrophobic
b. polar neutral, hydrophilic
c. polar acidic, hydrophilic
d. nonpolar, hydrophobic

16.9 a. tryptophan
b. serine
c. cysteine
d. tyrosine

16.11 positive (1+)

16.13 a. 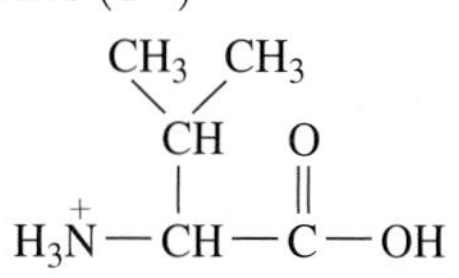 **b.** 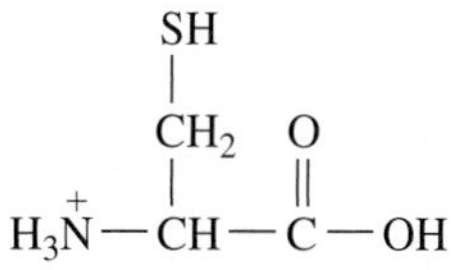

c. 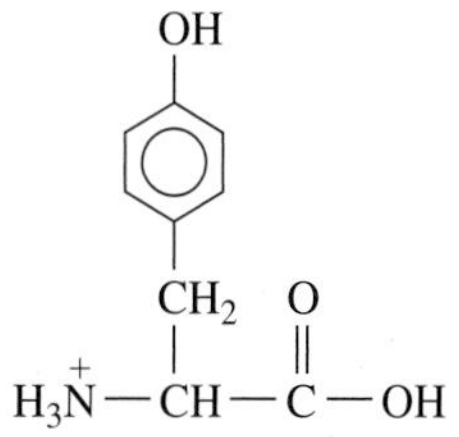 **d.** 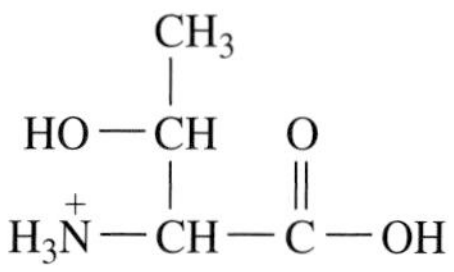

16.15 a. above pI
b. below pI
c. at pI

16.17 a. Ala–Trp (AW)
b. Gly–Thr (GT)
c. Ser–Phe (SF)
d. Gly–Val–Met (GVM)

16.19 Val–Ser–Ser, Ser–Val–Ser, or Ser–Ser–Val

16.21 The oxygen atoms of the carbonyl groups and the hydrogen atoms attached to the nitrogen atoms form alpha helices or beta-pleated sheets.

16.23 In the α helix, hydrogen bonds form between the carbonyl oxygen atom and the amino hydrogen atom in the next turn of the helical chain. In the β-pleated sheet, hydrogen bonds occur between parallel sections of a long polypeptide chain.

16.25 a. disulfide bond
b. salt bridge
c. hydrogen bond
d. hydrophobic interaction

16.27 a. Cysteine
b. Methionine, Leucine and Alanine side chains, because these side chains are hydrophobic.
c. Cyseine and Aspartic acid side chains because these side chains are hydrophilic.
d. The primary structure determines the relative stability of the possible tertiary structures.

16.29 a. tertiary, quaternary
b. primary
c. secondary
d. secondary

16.31 a. Placing an egg in boiling water coagulates the proteins of the egg because the heat disrupts hydrogen bonds and hydrophobic interactions.
b. The alcohol on the swab coagulates the proteins of any bacteria present by forming hydrogen bonds as well as disrupting hydrophobic interactions.
c. The heat from an autoclave will coagulate the proteins of any bacteria on the surgical instruments by disrupting hydrogen bonds and hydrophobic interactions.
d. Heat will coagulate the surrounding proteins to close the wound by disrupting hydrogen bonds and hydrophobic interactions.

16.33 As catalysts, enzymes lower the activation energy for a chemical reaction. Less energy is required to convert reactant molecules to products, which increases the rate of a biochemical reaction compared to the rate of the uncatalyzed reaction. An uncatalyzed reaction in a cell may take place eventually, but not at a rate fast enough for survival.

16.35 a. sucrose
b. lipid
c. milk

16.37 a. Hydrolases
b. Lyases
c. oxidoreductases
d. isomerases

16.39 a. (2) enzyme
b. (1) enzyme–substrate complex
c. (3) substrate

16.41 a. $E + S \rightleftarrows ES \longrightarrow E + P$
b. The active site is a region or pocket within the tertiary structure of an enzyme that accepts the substrate, aligns the substrate for reaction, and catalyzes the reaction.

16.43 Isoenzymes are different forms of an enzyme that catalyze the same reaction in different cells or tissues of the body. They consist of quaternary structures with slight variations in the amino acids in the polypeptide subunits.

Isoenzyme for LDH are: LDH1, LDH2, LDH3, LDH4, LDH5

16.45 A doctor might run tests for the enzymes CK, LDH, and AST to determine if the patient had a heart attack.

16.47 a. The rate would decrease.
b. The rate would decrease.

16.49 pepsin, pH 2; sucrase, pH 6; trypsin, pH 8

16.51 a. competitive
b. noncompetitive
c. competitive
d. noncompetitive
e. competitive

16.53 a. methanol, $CH_3—OH$; ethanol, $CH_3—CH_2—OH$
b. Ethanol has a structure similar to methanol and could compete for the active site.
c. Ethanol is a competitive inhibitor of methanol oxidation.

16.55 a. oxidoreductase
b. At high concentration, ethanol, which acts as a competitive inhibitor of ethylene glycol, would saturate the alcohol dehydrogenase enzyme to allow ethylene glycol to be removed from the body without producing oxalic acid.

16.57 a. aspartic acid and phenylalanine
b. aspartylphenylalanine

16.59 a. (1) **b.** (3) **c.** (2)

16.61 a. asparagine and serine; hydrogen bond
b. aspartic acid and lysine; salt bridge
c. cysteine and cysteine; disulfide bond
d. leucine and alanine; hydrophobic interaction

16.63 a.

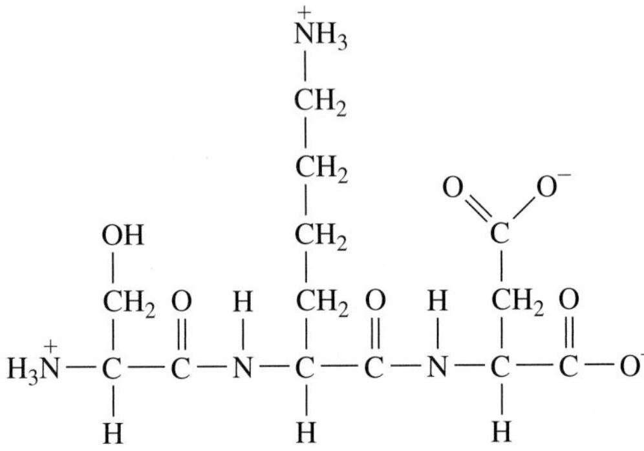

b. This segment contains polar R groups, which would be found on the surface of a globular protein where they hydrogen bond with water.

16.65 a. yes **b.** no **c.** no

16.67 a. In the secondary structure of proteins, hydrogen bonds form a helix or a pleated sheet; the tertiary structure is determined by the interactions of R groups such as disulfide bonds, hydrogen bonds, and salt bridges.
b. Nonessential amino acids can be synthesized by the body; essential amino acids must be supplied by the diet.
c. Polar amino acids have hydrophilic R groups, whereas nonpolar amino acids have hydrophobic R groups.
d. Dipeptides contain two amino acids, whereas tripeptides contain three.

16.69 Glutamic acid is a polar acidic amino acid, whereas proline is nonpolar. The polar acidic R group of glutamic acid may form hydrogen bonds or salt bridges with another amino acid R group. It may also move to the outer surface of the protein where it would form hydrophilic interactions with water. However, the nonpolar proline would move to the hydrophobic center of the protein.

16.71 In chemical laboratories, reactions are often run at high temperatures using catalysts that are strong acids or bases. Enzymes are catalysts that are proteins and function only at mild temperature and pH. Catalysts used in chemistry laboratories are usually inorganic materials that can function at high temperatures and in strongly acidic or basic conditions.

16.73 a. The reactants are lactose and water and the products are glucose and galactose.

b.

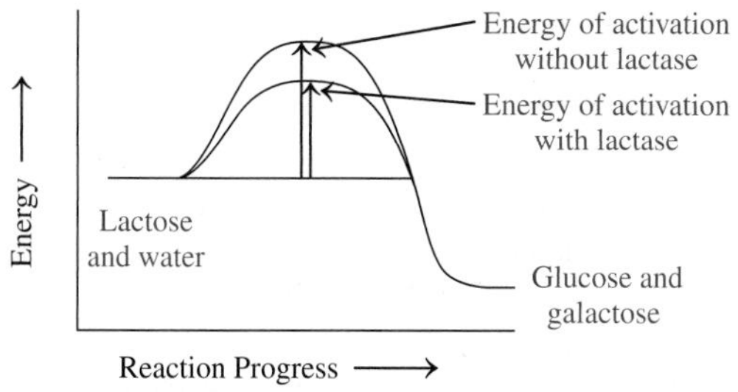

c. By lowering the energy of activation, the enzyme furnishes a lower energy pathway by which the reaction can take place.

16.75 a. S **b.** E **c.** E **d.** E **e.** S **f.** E

16.77 a. urea **b.** succinate **c.** aspartate **d.** tyrosine

16.79 Sucrose fits the shape of the active site in sucrase, but lactose does not.

16.81 A heart attack may be the cause.

16.83 a. 0 **b.** 1+ **c.** 1+ **d.** 1−

16.85 a. neutral nonpolar hydrophobic group
b. neutral polar hydrophobic group
c. nonpolar neutral hydrophobic group
d. polar acidic group

17 Nucleic Acids and Protein Synthesis

Daniel has been diagnosed with basal cell carcinoma, the most common form of skin cancer. He has an appointment to undergo Mohs surgery, a specialized procedure to remove the cancerous growth found on his shoulder. The surgeon begins the process by removing the abnormal growth, in addition to a thin layer of surrounding (margin) tissue, which he sends to Lisa, a histologist. Lisa prepares the tissue sample to be viewed by a pathologist. Tissue preparation requires Lisa to cut the tissue into a very thin section (normally 0.001 mm), which is then mounted onto a microscope slide.

Lisa then treats the tissue with a dye to stain the cells, as this enables the pathologist to view any abnormal cells more easily. The pathologist examines the tissue sample and reports back to the surgeon that no abnormal cells were present in the margin tissue, and Daniel's tumor has been completely removed. No further tissue removal is necessary. DNA, or deoxyribonucleic acid, contains all of a person's genetic information such as skin and eye color. This information is copied every time a cell divides through a process called replication. However, changes can occur in the DNA sequence during replication, which result in a mutation. Mutations can lead to cancer.

CAREER Histology Technician

A histology technician studies the microscopic makeup of tissues, cells, and bodily fluids with the purpose of detecting and identifying the presence of a specific disease. They determine blood types and the concentrations of drugs and other substances in the blood. Histologists also help establish a rationale for why a patient may not be responding to his or her treatment. Sample preparation is a critical component of a histologist's job, as they prepare tissue samples from humans, animals, and plants. The tissue samples are cut, using specialized equipment, into extremely thin sections which are then mounted, and stained using various chemical dyes. The dyes provide contrast for the cells to be viewed and help highlight any abnormalities that may exist. Utilization of various dyes requires the histology technician to be familiar with solution preparation and the handling of potentially hazardous chemicals.

Nucleic acids are large molecules found in the nuclei of cells that store information and direct activities for cellular growth and reproduction. Deoxyribonucleic acid (DNA), the genetic material in the nucleus of a cell, contains all the information needed for the development of a complete living organism. The way you grow, your hair, your eyes, your physical appearance, and the activities of the cells in your body are determined by a set of directions contained in the DNA of your cells.

All of the genetic information in the cell is called the *genome*. Every time a cell divides, the information in the genome is copied and passed on to the new cells. This replication process must duplicate the genetic instructions exactly. Some sections of DNA called *genes* contain the information to make a particular protein.

When a cell requires protein, another type of nucleic acid, ribonucleic acid (RNA), interprets the genetic information in DNA and carries that information to the ribosomes, where the synthesis of protein takes place. However, mistakes may occur that lead to mutations that affect the synthesis of a certain protein.

LOOKING AHEAD

CHAPTER READINESS*

CORE CHEMISTRY SKILLS

- Forming Amides (14.6)
- Drawing the Ionized Form for an Amino Acid (16.1)
- Identifying the Primary, Secondary, Tertiary, and Quaternary Structures of Proteins (16.3, 16.4)

*These Core Chemistry Skills from previous chapters are listed here for your review as you proceed to the new material in this chapter.

17.1 Components of Nucleic Acids

LEARNING GOAL

Describe the bases and ribose sugars that make up the nucleic acids DNA and RNA.

There are two closely related types of nucleic acids: *deoxyribonucleic acid* (**DNA**), and *ribonucleic acid* (**RNA**). Both are unbranched polymers of repeating monomer units known as *nucleotides*. Each nucleotide has three components: a base, a five-carbon sugar, and a phosphate group (see Figure 17.1). A DNA molecule may contain several million nucleotides; smaller RNA molecules may contain up to several thousand.

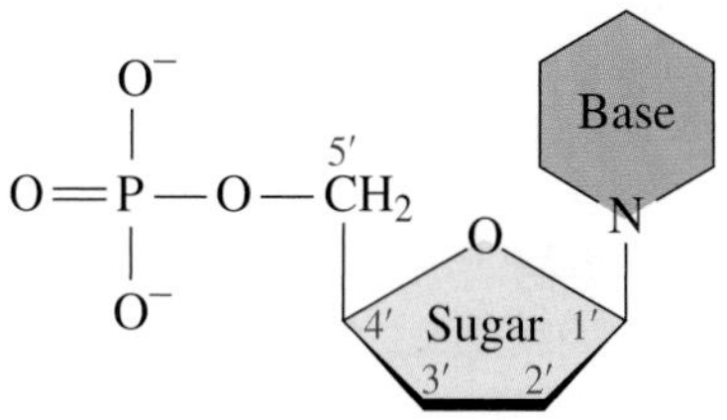

FIGURE 17.1 ▶ The general structure of a nucleotide includes a nitrogen-containing base, a sugar, and a phosphate group.

Q In a nucleotide, what types of groups are bonded to a five-carbon sugar?

Bases

The nitrogen-containing **bases** in nucleic acids are derivatives of *pyrimidine* or *purine*.

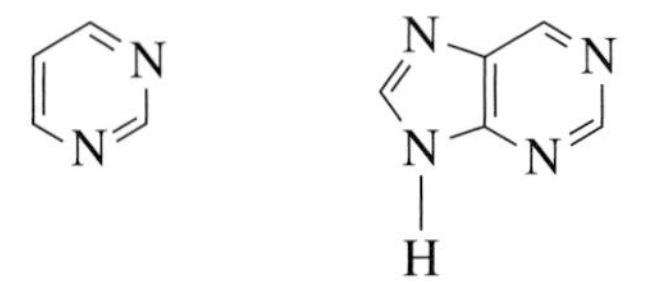

Pyrimidine　　Purine

In DNA, the purine bases with double rings are adenine (A) and guanine (G), and the pyrimidine bases with single rings are cytosine (C) and thymine (T). RNA contains the same bases, except thymine (5-methyluracil) is replaced by uracil (U) (see Figure 17.2).

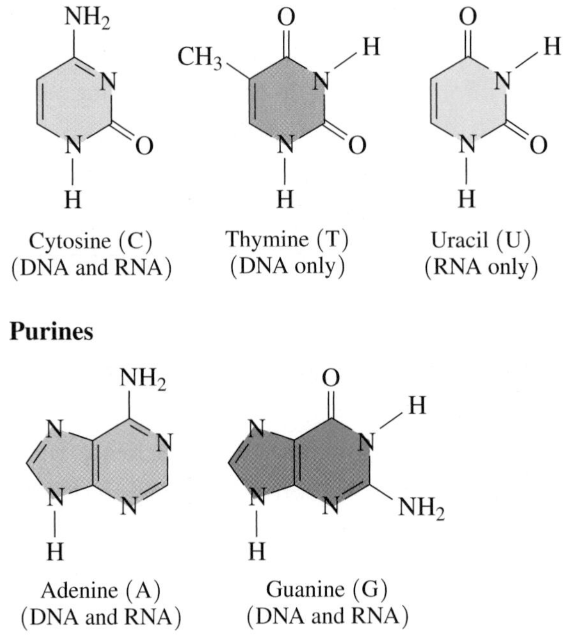

FIGURE 17.2 ▶ DNA contains the bases A, G, C, and T; RNA contains A, G, C, and U.

Q Which bases are found in DNA?

Pentose Sugars in RNA and DNA

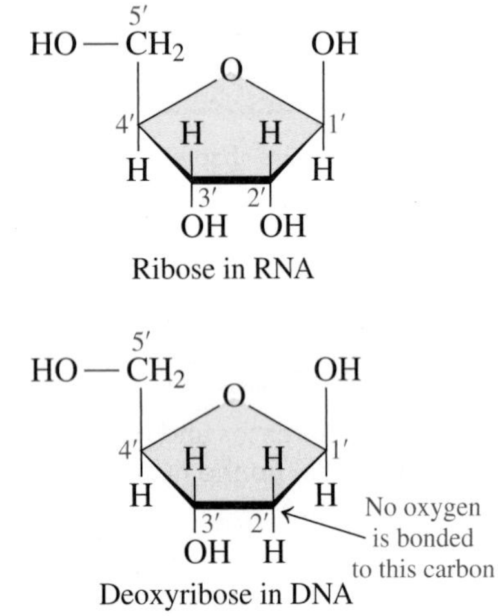

FIGURE 17.3 ▶ The five-carbon pentose sugar found in RNA is ribose and in DNA, deoxyribose.

Q What is the difference between ribose and deoxyribose?

Pentose Sugars

In RNA, the five-carbon sugar is *ribose*, which gives the letter R in the abbreviation RNA. The atoms in the pentose sugars are numbered with primes (1′, 2′, 3′, 4′, and 5′) to differentiate them from the atoms in the bases (see Figure 17.3). In DNA, the five-carbon sugar is *deoxyribose*, which is similar to ribose except that there is no hydroxyl group (—OH) on C2′. The *deoxy* prefix means "without oxygen" and provides the D in DNA.

Nucleosides and Nucleotides

Nucleosides, which are a combination of a sugar and a base, are produced when the nitrogen atom in a pyrimidine or a purine base forms an *N*-glycosidic bond to carbon 1 (C1′) of a sugar, either ribose or deoxyribose. For example, adenine and ribose form a nucleoside called adenosine.

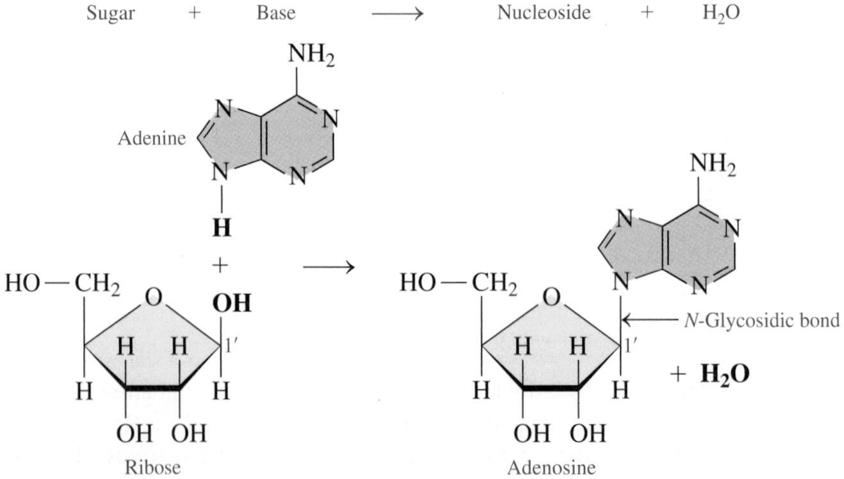

A base forms an *N*-glycosidic bond with a ribose or deoxyribose sugar to form a nucleoside.

Nucleotides are nucleosides in which a phosphate group bonds to the —OH group on carbon 5 (C5′) of ribose or deoxyribose. The product is a phosphate ester. Other hydroxyl groups on ribose can form phosphate esters too, but only the 5′-monophosphate nucleotides are found in RNA and DNA. All the nucleotides in RNA and DNA are shown in Figure 17.4. Table 17.1 summarizes the components in DNA and RNA.

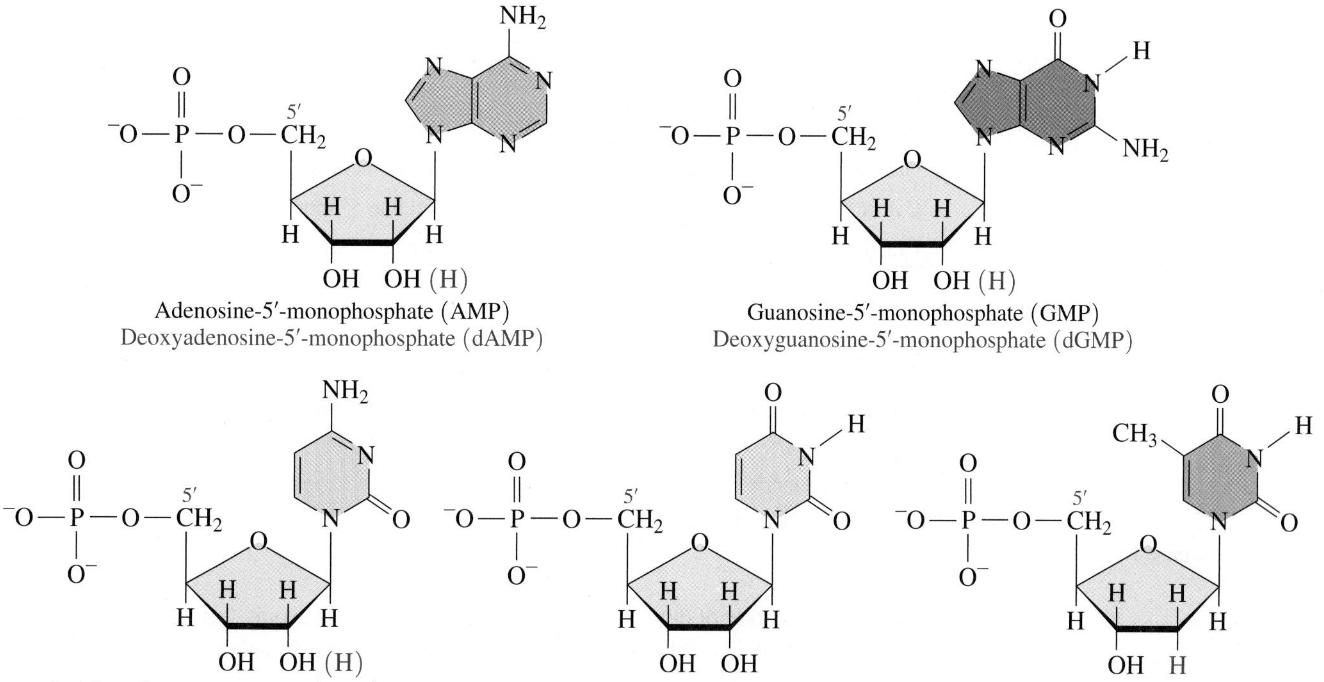

Adenosine-5′-monophosphate (AMP)
Deoxyadenosine-5′-monophosphate (dAMP)

Guanosine-5′-monophosphate (GMP)
Deoxyguanosine-5′-monophosphate (dGMP)

Cytidine-5′-monophosphate (CMP)
Deoxycytidine-5′-monophosphate (dCMP)

Uridine-5′-monophosphate (UMP)

Deoxythymidine-5′-monophosphate (dTMP)

FIGURE 17.4 ▶ The nucleotides of RNA are similar to those of DNA, except in DNA (shown in magenta) the sugar is deoxyribose and deoxythymidine replaces uridine.

Q What are two differences in the nucleotides of RNA and DNA?

TABLE 17.1 Components in DNA and RNA

Component	DNA	RNA
Bases	A, G, C, and T	A, G, C, and U
Sugar	Deoxyribose	Ribose
Nucleoside	Base + deoxyribose	Base + ribose
Nucleotide	Base + deoxyribose + phosphate	Base + ribose + phosphate
Nucleic Acid	Polymer of deoxyribose nucleotides	Polymer of ribose nucleotides

Naming Nucleosides and Nucleotides

The name of a nucleoside that contains a purine ends with *osine*, whereas a nucleoside that contains a pyrimidine ends with *idine*. The names of the nucleosides of DNA add *deoxy* to the beginning of their names. The corresponding nucleotides in RNA and DNA are named by adding *-5′-monophosphate*. Although the letters A, G, C, U, and T represent the bases, they are often used in the abbreviations of the respective nucleosides and nucleotides. The names and abbreviations of the bases, nucleosides, and nucleotides in DNA and RNA are listed in Table 17.2.

TABLE 17.2 Nucleosides and Nucleotides in DNA and RNA

Base	Nucleoside	Nucleotide
DNA		
Adenine (A)	Deoxyadenosine (A)	Deoxyadenosine-5′-monophosphate (dAMP)
Guanine (G)	Deoxyguanosine (G)	Deoxyguanosine-5′-monophosphate (dGMP)
Cytosine (C)	Deoxycytidine (C)	Deoxycytidine-5′-monophosphate (dCMP)
Thymine (T)	Deoxythymidine (T)	Deoxythymidine-5′-monophosphate (dTMP)
RNA		
Adenine (A)	Adenosine (A)	Adenosine-5′-monophosphate (AMP)
Guanine (G)	Guanosine (G)	Guanosine-5′-monophosphate (GMP)
Cytosine (C)	Cytidine (C)	Cytidine-5′-monophosphate (CMP)
Uracil (U)	Uridine (U)	Uridine-5′-monophosphate (UMP)

SAMPLE PROBLEM 17.1 Nucleotides

For each of the following nucleotides, identify the components and whether the nucleotide is found in DNA only, RNA only, or both DNA and RNA:

a. deoxyguanosine-5′-monophosphate (dGMP)
b. adenosine-5′-monophosphate (AMP)

SOLUTION

a. This nucleotide of deoxyribose, guanine, and phosphate is only found in DNA.
b. This nucleotide of ribose, adenine, and phosphate is only found in RNA.

STUDY CHECK 17.1

What are the name and abbreviation of the DNA nucleotide of cytosine?

QUESTIONS AND PROBLEMS

17.1 Components of Nucleic Acids

LEARNING GOAL Describe the bases and ribose sugars that make up the nucleic acids DNA and RNA.

17.1 Identify each of the following bases as a purine or a pyrimidine:
a. Uracil (U) **b.** Guanine (G)

17.2 Identify each of the following bases as a purine or a pyrimidine:
a. guanine **b.**

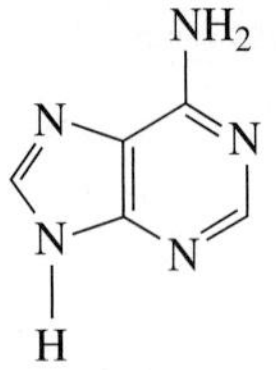

17.3 Identify each of the bases in problem 17.1 as a component of DNA only, RNA only, or both DNA and RNA.

17.4 Identify each of the bases in problem 17.2 as a component of DNA only, RNA only, or both DNA and RNA.

17.5 What are the four bases in DNA and RNA?

17.6 What are the names and abbreviations of the four nucleosidesin RNA?

17.7 Identify each of the following as a nucleoside or nucleotide:
a. Cytidine
b. deoxyguanosine
c. adenosine
d. deoxythymidine-5′-monophosphate

17.8 Identify each of the following as a nucleoside or nucleotide:
a. deoxythymidine
b. guanosine
c. deoxyadenosine-5′-monophosphate
d. uridine-5′-monophosphate

17.9 Draw the condensed structural formula for deoxyadenosine-5′-monophosphate (dAMP).

17.10 Draw the condensed structural formula for uridine-5′-monophosphate (UMP).

17.2 Primary Structure of Nucleic Acids

LEARNING GOAL

Describe the primary structures of RNA and DNA.

The **nucleic acids** are polymers of many nucleotides in which the 3′-hydroxyl group of the sugar in one nucleotide bonds to the phosphate group on the 5′-carbon atom in the sugar of the next nucleotide. This link between the sugars in adjacent nucleotides is referred to as a **phosphodiester bond**. As more nucleotides are added using phosphodiester bonds, a backbone forms that consists of alternating sugar and phosphate groups. The bases, which are attached to each sugar, extend out from the sugar–phosphate backbone.

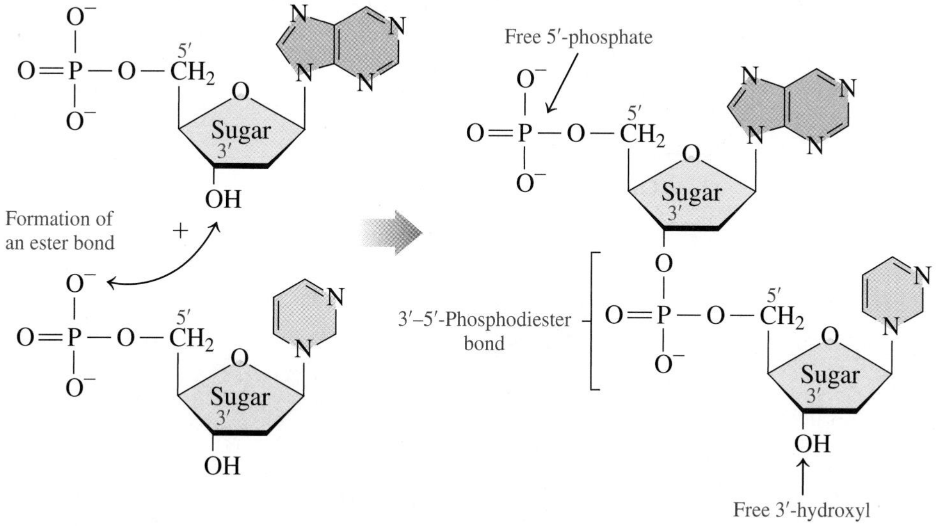

Each nucleic acid has its own unique sequence of bases, which is known as its **primary structure**. It is this sequence of bases that carries the genetic information from one cell to the next. In any nucleic acid, the sugar at the one end has an unreacted or free 5′-phosphate terminal end, and the sugar at the other end has a free 3′-hydroxyl group.

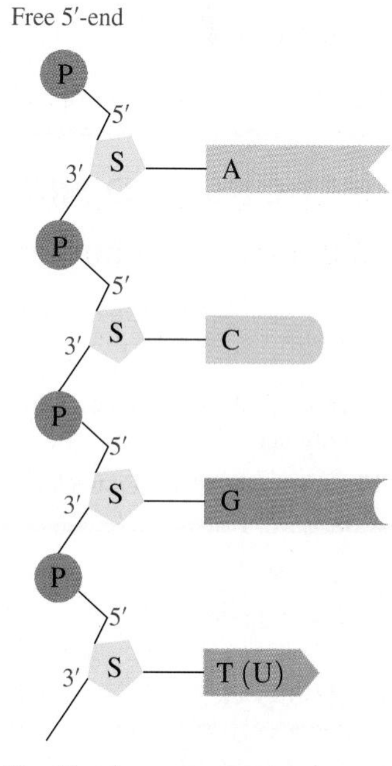

In the primary structure of nucleic acids, each sugar in a sugar–phosphate backbone is attached to a base.

SAMPLE PROBLEM 17.2 Bonding of Nucleotides

Draw the condensed structural formula for the RNA dinucleotide formed by two cytidine-5′-monophosphates.

SOLUTION

The dinucleotide is drawn by connecting the 3′-hydroxyl group on the first cytidine-5′-monophosphate with the 5′-phosphate group on the second.

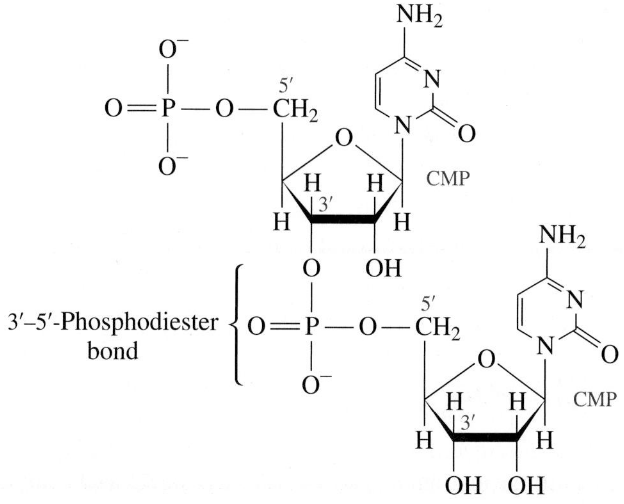

STUDY CHECK 17.2

In the dinucleotide of cytidine shown in the solution to Sample Problem 17.2, identify the free 5′-phosphate group and the free 3′-hydroxyl group.

A nucleic acid sequence is read from the sugar with free 5′-phosphate to the sugar with the free 3′-hydroxyl group. The order of nucleotides is often written using only the letters of the bases. For example, the nucleotide sequence starting with adenine (free 5′-phosphate end) in the section of RNA shown in Figure 17.5 is —A C G U—.

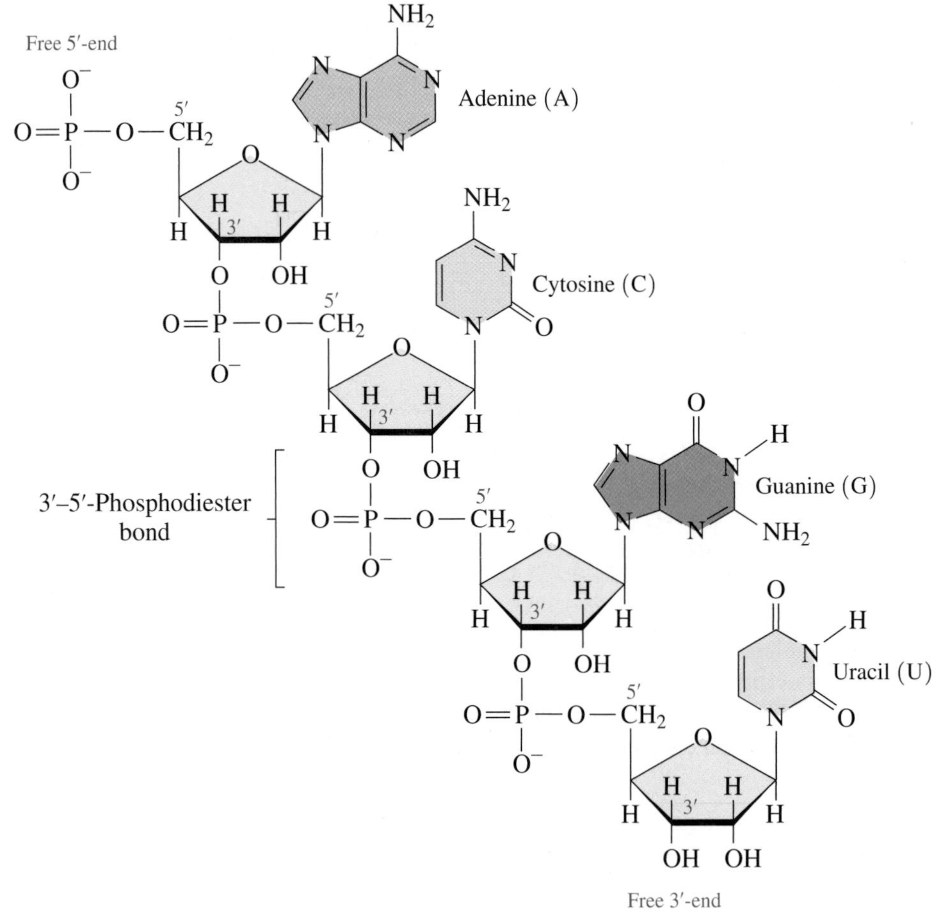

FIGURE 17.5 ▶ In the primary structure of nucleic acids, each sugar in a sugar–phosphate backbone is attached to a base (A, C, G, or T(U)).

Q Where are the free 5′-phosphate and 3′-hydroxyl groups?

QUESTIONS AND PROBLEMS

17.2 Primary Structure of Nucleic Acids

LEARNING GOAL Describe the primary structures of RNA and DNA.

17.11 How are the nucleotides held together in a nucleic acid polymer?

17.12 How do the ends of a nucleic acid polymer differ?

17.13 Draw the condensed structural formula for the dinucleotide GC in RNA.

17.14 Draw the condensed structural formula for the dinucleotide AT in DNA.

LEARNING GOAL

Describe the double helix of DNA and the process of DNA replication.

17.3 DNA Double Helix

During the 1940s, biologists determined that the bases in DNA from a variety of organisms had a specific relationship: the amount of adenine (A) was equal to the amount of thymine (T), and the amount of guanine (G) was equal to the amount of cytosine (C). Eventually, scientists determined that adenine is always paired (1:1) with thymine, and guanine is always paired (1:1) with cytosine.

$$\text{Number of purine molecules} = \text{Number of pyrimidine molecules}$$
$$\text{Adenine (A)} = \text{Thymine (T)}$$
$$\text{Guanine (G)} = \text{Cytosine (C)}$$

In 1953, James Watson and Francis Crick proposed that DNA was a **double helix** that consisted of two polynucleotide strands winding about each other like a spiral staircase (see Figure 17.6). The sugar–phosphate backbones are analogous to the outside railings with the bases arranged like steps along the inside. The two strands run in opposite directions. One strand goes from the 5′ to 3′ direction, and the other strand goes in the 3′ to 5′ direction.

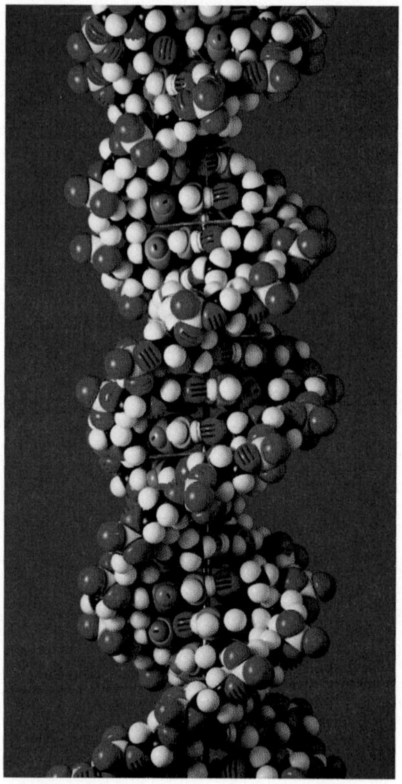

FIGURE 17.6 ▶ An atomic model of a DNA molecule shows the double helix as the characteristic shape of DNA molecules.

Ⓠ What is meant by the term *double helix?*

CORE CHEMISTRY SKILL

Writing the Complementary DNA Strand

Complementary Base Pairs

Each of the bases along one polynucleotide strand forms hydrogen bonds to only one specific base on the opposite DNA strand. Adenine forms hydrogen bonds only to thymine, and guanine bonds only to cytosine (see Figure 17.7). The pairs AT and GC are called **complementary base pairs**. The specific pairing of the bases occurs because adenine and thymine form only two hydrogen bonds, while cytosine and guanine form three hydrogen bonds. This explains why DNA has equal amounts of A and T and equal amounts of G and C.

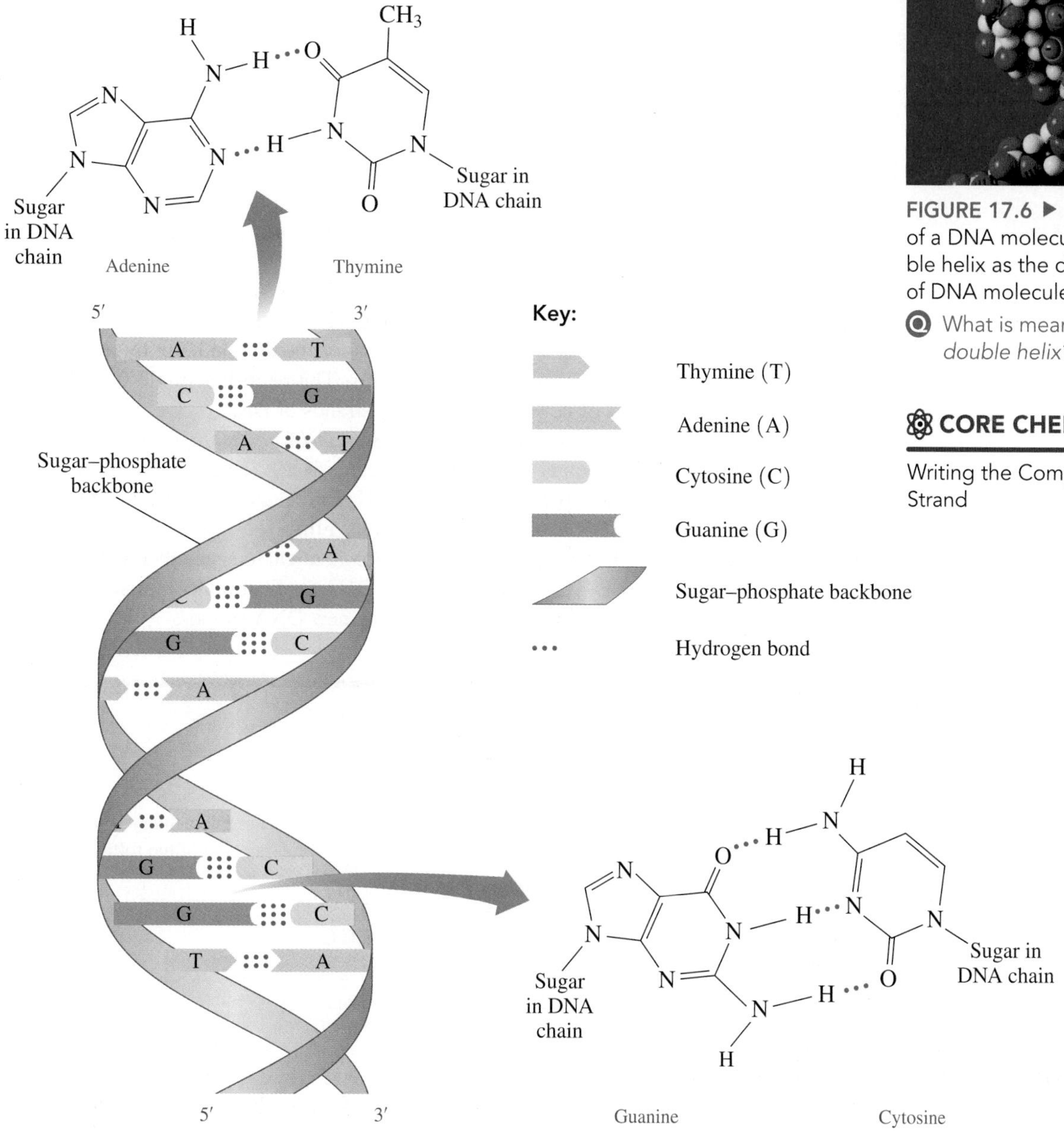

FIGURE 17.7 ▶ Hydrogen bonds between complementary base pairs hold the polynucleotide strands together in the double helix of DNA.

Ⓠ Why are GC base pairs more stable than AT base pairs?

SAMPLE PROBLEM 17.3 Complementary Base Pairs

Write the complementary base sequence for the following DNA segment:

—A C G A T C T—

SOLUTION

ANALYZE THE PROBLEM	Base in Original DNA Strand	Complementary Base
	A	T
	T	A
	C	G
	G	C

Original segment of DNA: —A C G A T C T—

Complementary segment: —T G C T A G A—

STUDY CHECK 17.3

What sequence of bases is complementary to a DNA segment with the base sequence of —G G T T A A C C—?

DNA Replication

The function of DNA in cells of animals and plants as well as in bacteria is to preserve genetic information. As cells divide, copies of DNA are produced that transfer genetic information to the new cells.

In DNA **replication**, the strands in the original or *parent* DNA molecule separate to allow the synthesis of complementary DNA strands. The process begins when an enzyme called *helicase* catalyzes the unwinding of a portion of the double helix by breaking the hydrogen bonds between the complementary bases. The resulting single strands act as templates for the synthesis of new complementary strands of DNA (see Figure 17.8).

As the complementary base pairs come together, *DNA polymerase* catalyzes the formation of phosphodiester bonds between the nucleotides. Eventually the entire double helix of the parent DNA is copied. In each new DNA molecule, one strand of the double helix is from the original DNA, and one is a newly synthesized strand. This process produces two new DNAs called *daughter DNAs* that are identical to each other and exact copies of the original parent DNA. In the process of DNA replication, complementary base pairing ensures the correct placement of bases in the new DNA strands.

QUESTIONS AND PROBLEMS

17.3 DNA Double Helix

LEARNING GOAL Describe the double helix of DNA and the process of DNA replication.

17.15 How are the two strands of nucleic acid in DNA held together?

17.16 What is meant by complementary base pairing?

17.17 Complete the base sequence in a complementary DNA segment if a portion of the parent strand has each of the following base sequences:

a. —A A A A A A—

b. —G G G G G G—

c. —A G T C C A G G T—

d. —C T G T A T A C G T T A—

17.18 Complete the base sequence in a complementary DNA segment if a portion of the parent strand has each of the following base sequences:

a. —T T T T T T—

b. —C C C C C C C C C—

c. —A T G G C A—

d. —A T A T G C G C T A A A—

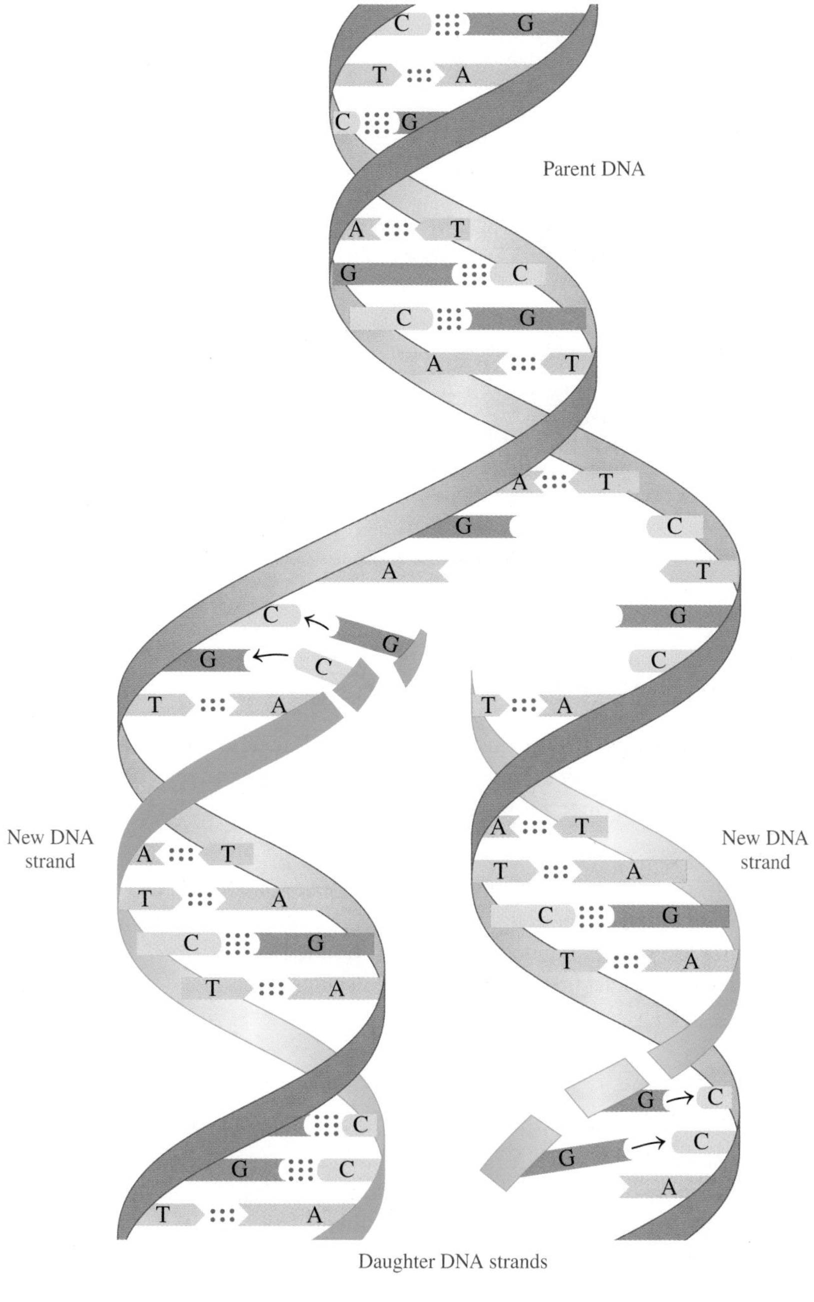

FIGURE 17.8 ▶ In DNA replication, the separate strands of the parent DNA are the templates for the synthesis of complementary strands, which produces two exact copies of DNA.

Q How many strands of the parent DNA are in each of the new copies of DNA?

Chemistry Link to **Health**

DNA Fingerprinting

In a process called *DNA fingerprinting* or *DNA profiling*, a small sample of DNA is obtained from blood, skin, saliva, or semen. The amount of DNA is amplified using the polymerase chain reaction method. Then restriction enzymes are used to cut the DNA into smaller fragments, which are placed on a gel and separated by size using electrophoresis. A radioactive probe is added that adheres to specific DNA sequences. When a piece of X-ray film is placed over the gel, the radiation from the probe bound to specific sequences creates a pattern of dark and light bands, known as a *DNA fingerprint*. Scientists estimate that the odds of two people who are not identical twins producing the same DNA fingerprint are less than one in a billion.

One application of DNA fingerprinting is in forensic science, where DNA from samples such as blood, hair, or semen is used to connect a suspect with a crime. Recently, DNA fingerprinting has been used to gain the release of persons who were wrongly convicted. Other applications of DNA fingerprinting are determining the biological parents of a child, establishing the identity of a deceased person, and matching recipients with organ donors.

Human Genome Project

The Human Genome Project, which began in 1990 and ended in 2003, was a project sponsored by the U.S. Department of Energy and the National Institutes of Health with contributions from Japan, France, Germany, United Kingdom, and other countries. The goals of the project were to identify about 25 000 genes in human DNA, to determine the base pair sequences in human DNA, and to store this information in databases accessible on the Internet.

Scientists determined that most of the genome is not functional and has perhaps been carried from generation to generation for millions of years. Large blocks of genes are replicated, even though they do not code for needed proteins. Thus, the coding portions of the genes seem to make up only about 1% of the total genome. The results of the genome project will help us identify defective genes that lead to genetic disease. Today, DNA fingerprinting is used to screen for genes responsible for genetic diseases such as sickle-cell anemia, cystic fibrosis, breast cancer, colon cancer, Huntington's disease, and Lou Gehrig's disease.

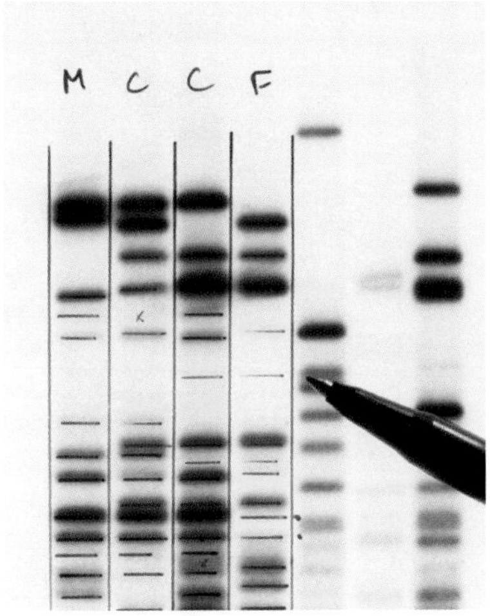

Dark and light bands on X-ray film represent DNA fingerprints that can be used to identify a person involved in a crime.

LEARNING GOAL

Identify the different types of RNA; describe the synthesis of mRNA.

17.4 RNA and the Genetic Code

Ribonucleic acid, RNA, which makes up most of the nucleic acid found in the cell, is involved with transmitting the genetic information needed to operate the cell. Similar to DNA, RNA molecules are polymers of nucleotides. However, RNA differs from DNA in several important ways:

1. The sugar in RNA is ribose rather than the deoxyribose found in DNA.
2. In RNA, the base uracil replaces thymine.
3. RNA molecules are single stranded, not double stranded.
4. RNA molecules are much smaller than DNA molecules.

Types of RNA

There are three major types of RNA in the cells: *messenger RNA*, *ribosomal RNA*, and *transfer RNA*. Ribosomal RNA (**rRNA**), the most abundant type of RNA, is combined with proteins to form ribosomes. Ribosomes, which are the sites for protein synthesis, consist of two subunits: a large subunit and a small subunit (see Figure 17.9). Cells that synthesize large numbers of proteins have thousands of ribosomes.

Messenger RNA (**mRNA**) carries genetic information from the DNA, located in the nucleus of the cell, to the ribosomes located in the cytoplasm. A gene segment of DNA will produce a separate mRNA for a particular protein that is needed in the cell. The size of the mRNA depends on the number of nucleotides in that gene.

Transfer RNA (**tRNA**), the smallest of the RNA molecules, interprets the genetic information in mRNA and brings specific amino acids to the ribosome for protein synthesis. Only tRNA can translate the genetic information into amino acids for proteins. There

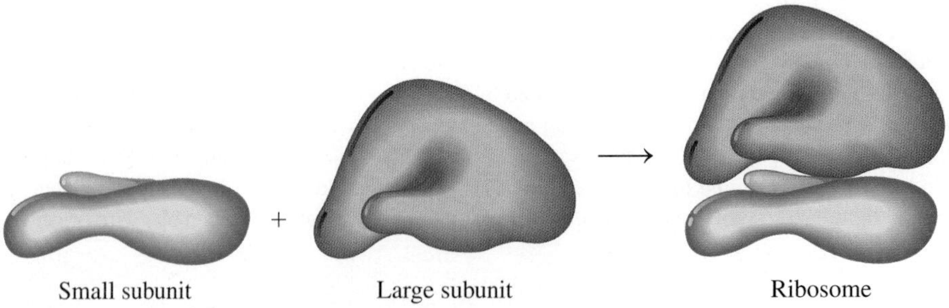

FIGURE 17.9 ▶ A typical ribosome consists of a small subunit and a large subunit.

Q Why would there be many thousands of ribosomes in a cell?

can be more than one tRNA for each of the 20 amino acids. The structures of all the transfer RNAs are similar, consisting of 70 to 90 nucleotides. Hydrogen bonds between some of the complementary bases in the strand produce loops that give some double-stranded regions.

Although the structure of tRNA is complex, we draw tRNA as a cloverleaf to illustrate its features. All tRNA molecules have a 3′-end with the nucleotide sequence ACC, which is known as the *acceptor stem*. An enzyme attaches an amino acid by forming an ester bond with the free —OH group at the end of the acceptor stem. Each tRNA contains an **anticodon**, which is a series of three bases that complements three bases on mRNA (see Figure 17.10). Table 17.3 summarizes the three types of RNA.

We now look at the overall process involved in transferring genetic information encoded in the DNA to the production of proteins. In the nucleus, genetic information for the synthesis of a protein is copied from a gene in DNA to make mRNA, a process called **transcription**. The mRNA molecules move out of the nucleus into the cytoplasm where they bind with the ribosomes. Then in a process called **translation**, tRNA molecules convert the information in the mRNA into amino acids, which are placed in the proper sequence to synthesize a protein (see Figure 17.11).

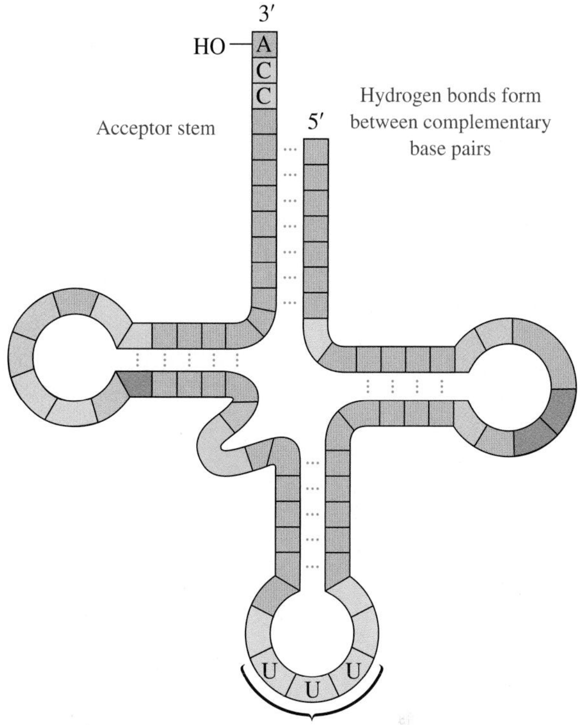

FIGURE 17.10 ▶ A typical tRNA molecule has a cloverleaf shape when hydrogen bonds form between complementary bases within the tRNA. The acceptor stem attaches to an amino acid and its anticodon bonds with a codon on mRNA.

Q Why will different tRNAs have different bases in the anticodon loop?

TABLE 17.3 Types of RNA Molecules in Humans

Type	Abbreviation	Function in the Cell	Percentage of Total RNA
Ribosomal RNA	rRNA	Synthesizes protein; major component of the ribosomes	80
Messenger RNA	mRNA	Carries information for protein synthesis from the DNA in the nucleus to the ribosomes	5
Transfer RNA	tRNA	Brings amino acids to the ribosomes for protein synthesis	15

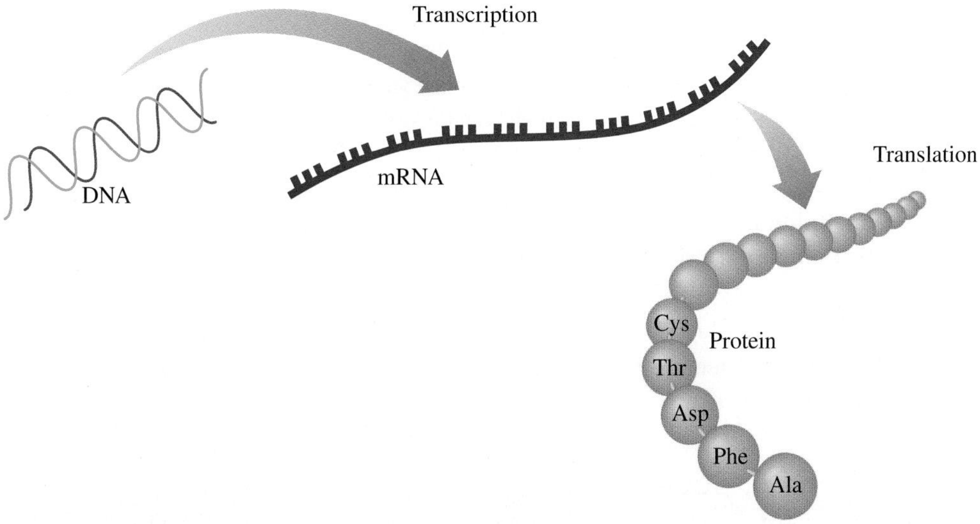

FIGURE 17.11 ▶ The genetic information in DNA is replicated in cell division and is used to produce messenger RNAs that code for the amino acids needed for protein synthesis.

Q What is the difference between transcription and translation?

Transcription: Synthesis of mRNA

Transcription begins when the section of a DNA molecule that contains the gene to be copied unwinds. Within this unwound section of DNA, the enzyme RNA polymerase uses one of the strands as a template to synthesize mRNA. Just as in DNA synthesis, C pairs with G, T pairs with A, but in mRNA, U (not T) pairs with A. The RNA polymerase moves along the DNA template strand, forming bonds between the bases. When the RNA polymerase reaches the termination site, transcription ends and the new mRNA is released. The unwound section of the DNA returns to its double helix structure (see Figure 17.12).

RNA Polymerase →

Sequence of bases on DNA template:		—G A A C T—
	Transcription	↓ ↓ ↓ ↓ ↓
Complementary base sequence in mRNA:		—C U U G A—

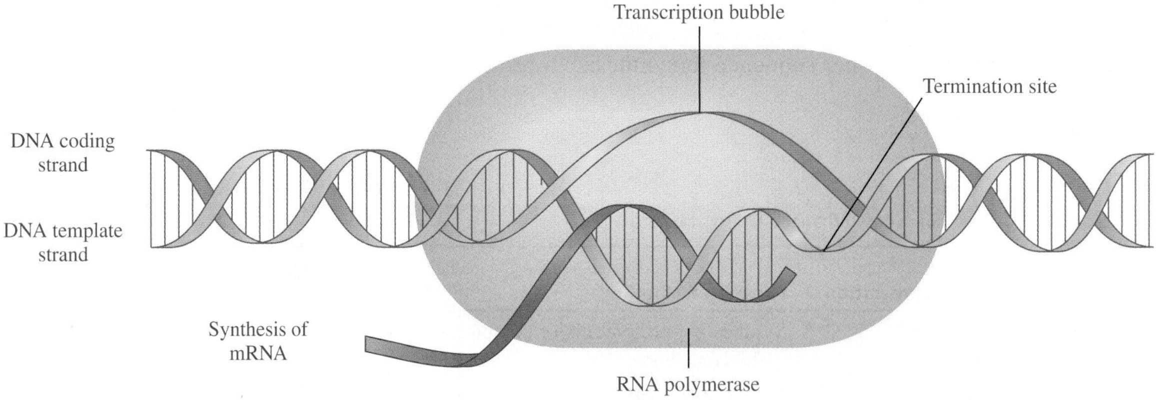

FIGURE 17.12 ▶ DNA undergoes transcription when RNA polymerase makes a complementary copy of a gene using only one of the DNA strands as the template.

Q Why is mRNA synthesized on only one DNA strand?

CORE CHEMISTRY SKILL

Writing the mRNA Segment for a DNA Template

SAMPLE PROBLEM 17.4 RNA Synthesis

The sequence of bases in a part of the DNA template strand is —C G A T C A—. What corresponding mRNA is produced?

SOLUTION

To form mRNA, the bases in the DNA template strand are paired with their complementary bases: G with C, C with G, T with A, and A with U.

DNA template:		—C G A T C A—
	Transcription	↓ ↓ ↓ ↓ ↓ ↓
Complementary base sequence in mRNA:		—G C U A G U—

STUDY CHECK 17.4

What is the DNA template strand that codes for the mRNA segment that has the sequence —G G G U U U A A A—?

The Genetic Code

The **genetic code** consists of a series of three nucleotides (triplets) in mRNA called **codons** that specify the amino acids and their sequence in the protein. Early work on protein synthesis showed that repeating triplets of uracil (UUU) in mRNA produced a peptide that contained only phenylalanine. Therefore, a sequence of —U U U U U U U U U— codes for three phenylalanines.

Codons in mRNA: —UUU UUU UUU—

Translation ↓ ↓ ↓

Amino acid sequence: — Phe — Phe — Phe —

CORE CHEMISTRY SKILL

Writing the Amino Acid for an mRNA Codon

Codons have been determined for all 20 amino acids. A total of 64 codons are possible from the triplet combinations of A, G, C, and U. Three of these, UGA, UAA, and UAG, are stop signals that code for the termination of protein synthesis. All the other three-base codons shown in Table 17.4 specify amino acids. Thus one amino acid can have several codons. For example, glycine has four codons: GGU, GGC, GGA, and GGG. The triplet AUG has two roles in protein synthesis. At the beginning of an mRNA, the codon AUG signals the start of protein synthesis. In the middle of a series of codons, the AUG codon specifies the amino acid methionine.

TABLE 17.4 Codons in mRNA: The Genetic Code for Amino Acids

	Second Letter				
First Letter	**U**	**C**	**A**	**G**	**Third Letter**
U	UUU } Phe	UCU } Ser	UAU } Tyr	UGU } Cys	U
	UUC	UCC	UAC	UGC	C
	UUA } Leu	UCA	UAA STOP[b]	UGA STOP[b]	A
	UUG	UCG	UAG STOP[b]	UGG Trp	G
C	CUU } Leu	CCU } Pro	CAU } His	CGU } Arg	U
	CUC	CCC	CAC	CGC	C
	CUA	CCA	CAA } Gln	CGA	A
	CUG	CCG	CAG	CGG	G
A	AUU } Ile	ACU } Thr	AAU } Asn	AGU } Ser	U
	AUC	ACC	AAC	AGC	C
	AUA	ACA	AAA } Lys	AGA } Arg	A
	AUG START[a]/Met	ACG	AAG	AGG	G
G	GUU } Val	GCU } Ala	GAU } Asp	GGU } Gly	U
	GUC	GCC	GAC	GGC	C
	GUA	GCA	GAA } Glu	GGA	A
	GUG	GCG	GAG	GGG	G

START[a] codon signals the initiation of a peptide chain.
STOP[b] codons signal the end of a peptide chain.

SAMPLE PROBLEM 17.5 Codons

What is the sequence of amino acids specified by the following codons in mRNA?

—GUC AGC CCA—

SOLUTION

	Codon (Three-Base Sequence) (refer to Table 17.4)	Amino Acid
ANALYZE THE PROBLEM	GUC	Valine (Val)
	AGC	Serine (Ser)
	CCA	Proline (Pro)

The sequence —GUC AGC CCA— codes for the amino acids —Val—Ser—Pro—.

STUDY CHECK 17.5

What is the amino acid sequence produced by the following codons in mRNA?

—AAU GCU UGU—

QUESTIONS AND PROBLEMS

17.4 RNA and the Genetic Code

LEARNING GOAL Identify the different types of RNA; describe the synthesis of mRNA.

17.19 What are the three different types of RNA?

17.20 What are the functions of each type of RNA?

17.21 What is meant by the term "transcription"?

17.22 Why does DNA have equal amounts of A and T, and equal amounts of G and C?

17.23 Write the corresponding section of mRNA produced from the following section of a DNA template strand:

—C C G A A G G T T C A C—

17.24 Write the corresponding section of mRNA produced from the following section of a DNA template strand:

—T A C G G C A A G C T A—

17.25 What is a codon?

17.26 What is the genetic code?

17.27 What amino acid is coded for by each of the following mRNA codons?

a. CUG **b.** UCC
c. GAG **d.** AGG

17.28 What are the mRNA codons for the start and stop signals of transcription?

17.29 When does the codon AUG signal the start of a protein? When does it code for the amino acid methionine?

17.30 What are the possible amino acids for the mRNA codons that start with A as the first letter?

LEARNING GOAL

Describe the process of protein synthesis from mRNA.

17.5 Protein Synthesis

Once the mRNA is synthesized, it migrates out of the nucleus into the cytoplasm to the ribosomes. In the *translation* process, tRNA molecules, amino acids, and enzymes convert the codons on mRNA to build a protein.

FIGURE 17.13 ▶ An activated tRNA with anticodon AGU bonds to serine at the acceptor stem.

Q What is the codon for serine for this tRNA?

Activation of tRNA

Each tRNA molecule contains a loop called an *anticodon*, which is a triplet of bases that complements a codon in mRNA (see Figure 17.13). An amino acid is attached to the acceptor stem of each tRNA by an enzyme called *aminoacyl–tRNA synthetase*. Each amino acid has a different synthetase. Activation of tRNA occurs when aminoacyl–tRNA synthetase forms an ester bond between the carboxylate group of its amino acid and the hydroxyl group on the acceptor stem.

Initiation and Chain Elongation

Protein synthesis begins when mRNA binds to a ribosome. The first codon in mRNA is a *start* codon, AUG, which forms hydrogen bonds with methionine–tRNA. Another tRNA hydrogen bonds to the next codon, placing a second amino acid adjacent to methionine. A peptide bond forms between the C-terminal of methionine and the N-terminal of the second amino acid (see Figure 17.14). The initial tRNA detaches from the ribosome, which shifts to the next available codon, a process called *translocation*. During *chain elongation*, the ribosome moves along the mRNA from codon to codon, so that the tRNAs can attach new amino acids to the growing protein chain. Sometimes several ribosomes, called a polysome, translate the same strand of mRNA to produce several copies of the protein at the same time.

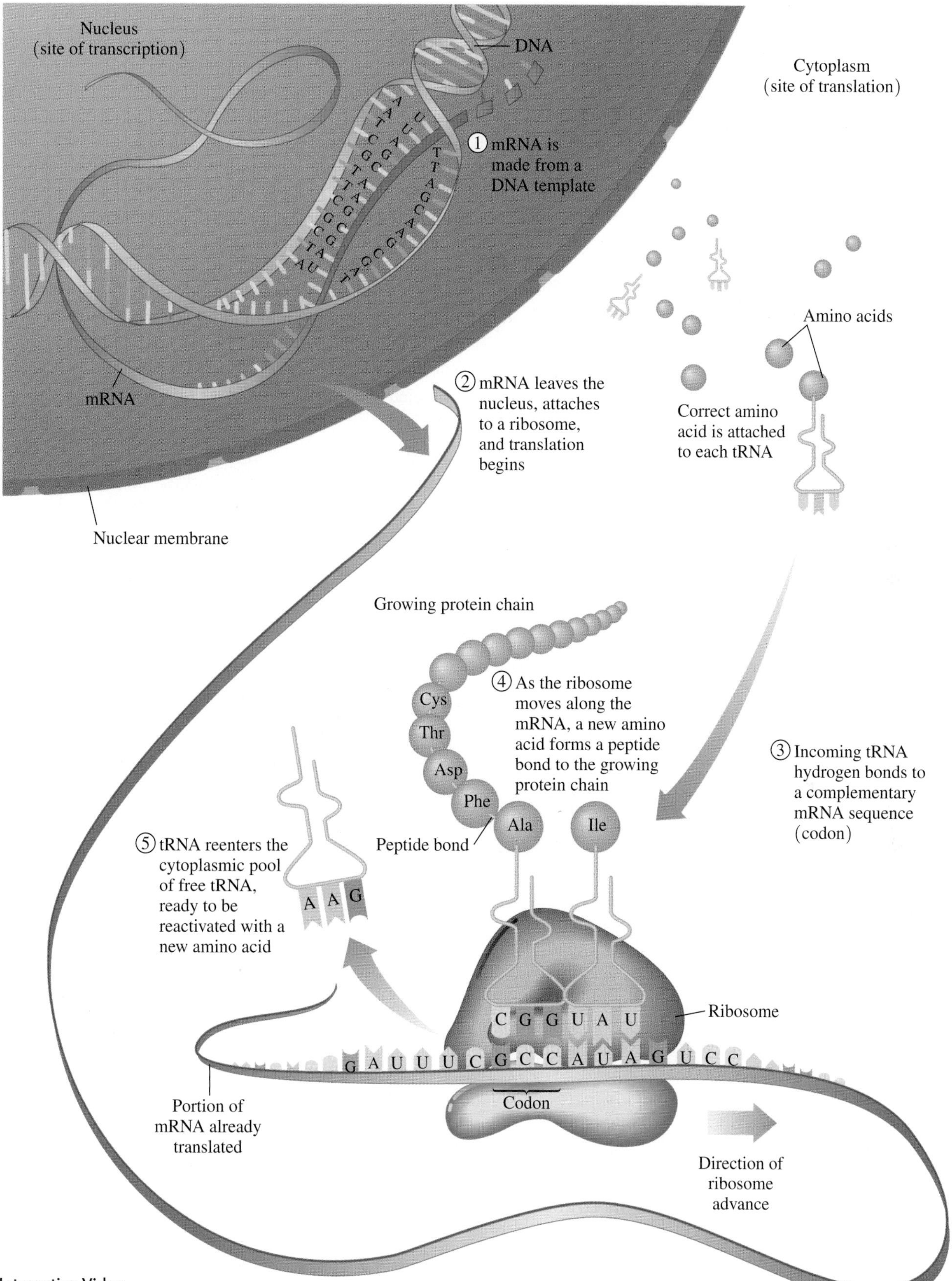

Interactive Video

Protein Synthesis

FIGURE 17.14 ▶ In the translation process, the mRNA synthesized by transcription attaches to a ribosome, and tRNAs pick up their amino acids, bind to the appropriate codon, and place them in a growing protein chain.

Q How is the correct amino acid placed in the protein chain?

Chain Termination

Eventually, a ribosome encounters a codon—UAA, UGA, or UAG—that has no corresponding tRNAs. These are stop codons, which signal the termination of protein synthesis and the release of the protein chain from the ribosome. The initial amino acid, methionine, is usually removed from the beginning of the protein chain. The R groups of the amino acids in the new protein form hydrogen bonds to give the secondary structures of α helices, β-pleated sheets, or triple helices and form interactions such as salt bridges and disulfide bonds to produce tertiary and quaternary structures, which make it a biologically active protein. Table 17.5 summarizes the steps in protein synthesis.

TABLE 17.5 Steps in Protein Synthesis

Step	Site: Materials	Process
1. DNA transcription	Nucleus: nucleotides, RNA polymerase	A DNA template is used to produce mRNA.
2. Activation of tRNA	Cytoplasm: amino acids, tRNAs, aminoacyl–tRNA synthetase	Molecules of tRNA pick up specific amino acids according to their anticodons.
3. Initiation and chain elongation	Ribosome: Met–tRNA, mRNA, amino acyl–tRNAs	A start codon binds the first tRNA carrying amino acid methionine to the mRNA. Successive tRNAs bind to and detach from the ribosome as each amino acid adds to the protein.
4. Chain termination	Ribosome: stop codon on mRNA	The protein is released from ribosome.

Table 17.6 summarizes the nucleotide and amino acid sequences in protein synthesis.

TABLE 17.6 Complementary Sequences in DNA, mRNA, tRNA, and Peptides

Nucleus	
DNA coding strand	—GCG AGT GGA TAC—
DNA template strand	—CGC TCA CCT ATG—
Ribosome (cytoplasm)	
mRNA	—GCG AGU GGA UAC—
tRNA anticodons	CGC UCA CCU AUG
Protein amino acids	—Ala—Ser—Gly—Tyr—

SAMPLE PROBLEM 17.6 Protein Synthesis

What order of amino acids would you expect in a peptide for the mRNA sequence of —UCA AAA GCC CUU—?

SOLUTION

Each of the codons specifies a particular amino acid. Using Table 17.4, we write a peptide with the following amino acid sequence:

mRNA codons:	—UCA	AAA	GCC	CUU—
	↓	↓	↓	↓
Amino acid sequence:	— Ser —	Lys —	Ala —	Leu —

STUDY CHECK 17.6

Where would protein synthesis stop in the following series of bases in an mRNA?

—GGG AGC AGU UAG GUU—

Chemistry Link to Health

Many Antibiotics Inhibit Protein Synthesis

Several antibiotics stop bacterial infections by interfering with the synthesis of proteins needed by the bacteria. Some antibiotics act only on bacterial cells, binding to the ribosomes in bacteria but not those in human cells. A description of some of these antibiotics is given in Table 17.7.

TABLE 17.7 Antibiotics That Inhibit Protein Synthesis in Bacterial Cells

Antibiotic	Effect on Ribosomes to Inhibit Protein Synthesis
Chloramphenicol	Inhibits peptide bond formation and prevents the binding of tRNAs
Erythromycin	Inhibits protein chain growth by preventing the translocation of the ribosome along the mRNA
Puromycin	Causes release of an incomplete protein by ending the growth of the protein early
Streptomycin	Prevents the proper attachment of the initial tRNA
Tetracycline	Prevents the binding of tRNAs

QUESTIONS AND PROBLEMS

17.5 Protein Synthesis

LEARNING GOAL Describe the process of protein synthesis from mRNA.

17.31 What is the difference between a *codon* and an *anticodon*?

17.32 Why are there at least 20 different tRNAs?

17.33 What amino acid sequence would you expect from each of the following mRNA segments?

a. —ACC ACA ACU—
b. —UUU CCG UUC CCA—
c. —UUA GGG AGA UGC—

17.34 What amino acid sequence would you expect from each of the following mRNA segments?

a. —AUA CCC UUG GAU—
b. —CCU CGC ACG CCA UGA—
c. —AUG CAC AAG GAA GUG CUG—

17.35 How is a peptide chain extended?

17.36 What is meant by "translocation"?

17.37 The following segment is a portion of the DNA template strand:

—GCT TTT CAA AAA—

a. What is the corresponding mRNA section?
b. What are the anticodons of the tRNAs?
c. What amino acids will be placed in the peptide chain?

17.38 Bacterial infections can be stopped by using antibiotics that interfere with the protein synthesis in bacterial cells. Suggest five common pathways for their actions.

17.6 Genetic Mutations

LEARNING GOAL

Describe some ways in which DNA is altered to cause mutations.

A **mutation** is a change in the nucleotide sequence of DNA. Such a change may alter the sequence of amino acids, affecting the structure and function of a protein in a cell. Mutations may result from X-rays, overexposure to sun (ultraviolet, or UV, light), chemicals called *mutagens*, and possibly some viruses. If a mutation occurs in a somatic cell (a cell other than a reproductive cell), the altered DNA will be limited to that cell and its daughter cells. If the mutation causes uncontrolled growth, cancer could result. If a mutation occurs in a germ cell (egg or sperm), then all the DNA produced in a new individual will contain the same genetic change. When a mutation severely alters the function of structural proteins or enzymes, the new cells may not survive or the person may exhibit a genetic disease.

Types of Mutations

Consider a triplet of bases CCG in the template strand of DNA, which produces the codon GGC in mRNA. At the ribosome, tRNA would place the amino acid glycine in the peptide chain (see Figure 17.15a). Now, suppose that T replaces the first C in the DNA triplet, which gives TCG as the triplet. Then the codon produced in the mRNA is AGC, which brings the tRNA with the amino acid serine to add to the peptide chain. The replacement of one base in the template strand of DNA with another is called a **substitution** or a *point mutation*. When there is a change of a nucleotide in the codon, a different amino acid may be inserted into the protein. However, if a substitution gives a codon for the same amino acid, there is no change in the amino acid sequence in the protein. This type of substitution is a *silent mutation*. Substitution is the most common way in which mutations occur (see Figure 17.15b).

In a **frameshift mutation**, a base is inserted into or deleted from the normal order of bases in the template strand of DNA. Suppose that an A is deleted from the triplet AAA, giving a new triplet of AAC (see Figure 17.15c). The next triplet becomes CGA rather than CCG and so on. All the triplets shift by one base, which changes all the codons that follow and leads to a different sequence of amino acids from that point.

Effect of Mutations

When a mutation causes a change in the amino acid sequence, the structure of the resulting protein can be severely altered and it may lose biological activity. If the protein is an enzyme, it may no longer bind to its substrate or react with the substrate at the active site. When an altered enzyme cannot catalyze a reaction, certain substances may accumulate until they act as poisons in the cell or substances vital to survival may not be synthesized. If a defective enzyme occurs in a major metabolic pathway or in the building of a cell membrane, the mutation can be lethal. When a protein deficiency is hereditary, the condition is called a **genetic disease**.

DNA —(X-rays, UV sunlight, mutagens, viruses)→ Alteration of DNA → Defective protein → Genetic disease (germ cells) or cancer (somatic cells)

SAMPLE PROBLEM 17.7 Mutations

An mRNA has the sequence of codons —CCC AGA GCC—. If a base substitution in the DNA changes the mRNA codon of AGA to GGA, how is the amino acid sequence affected in the resulting protein?

SOLUTION

The initial mRNA sequence of —CCC AGA GCC— codes for the following amino acids: proline, arginine, and alanine. When the mutation occurs, the new sequence of mRNA codons is —CCC GGA GCC—, which codes for proline, glycine, and alanine. The basic amino acid arginine is replaced by nonpolar glycine.

	Normal	After Mutation
mRNA Codons	—CCC AGA GCC—	—CCC GGA GCC—
Amino Acids	— Pro — Arg — Ala —	— Pro — Gly — Ala —

STUDY CHECK 17.7

How might the protein made from this mRNA in Sample Problem 17.7 be affected by this mutation?

Explore Your World

A Model for DNA Replication and Mutation

1. Cut out 16 rectangular pieces of paper. Using 8 rectangular pieces for DNA strand 1, label two each with each of the following nucleotide symbols: A═, T═, G≡, and C≡.
2. Using the other 8 rectangular pieces for DNA strand 2, label two of each with each of the following nucleotide symbols: ═A, ═T, ≡G, and ≡C.
3. Place the pieces for strand 1 in random order.
4. Using the DNA segment strand 1 you made in part 3, select the correct bases to build the complementary segment of DNA strand 2.
5. Using the rectangular pieces for nucleotides, put together a DNA segment using a template strand of —ATTGCC—. What is the mRNA that would form from this segment of DNA? What is the dipeptide that would form from this mRNA?
6. In the DNA segment of part **5**, change the G to an A. What is the mRNA from this segment of DNA? What is the dipeptide that forms? How could this change in codons lead to a mutation?

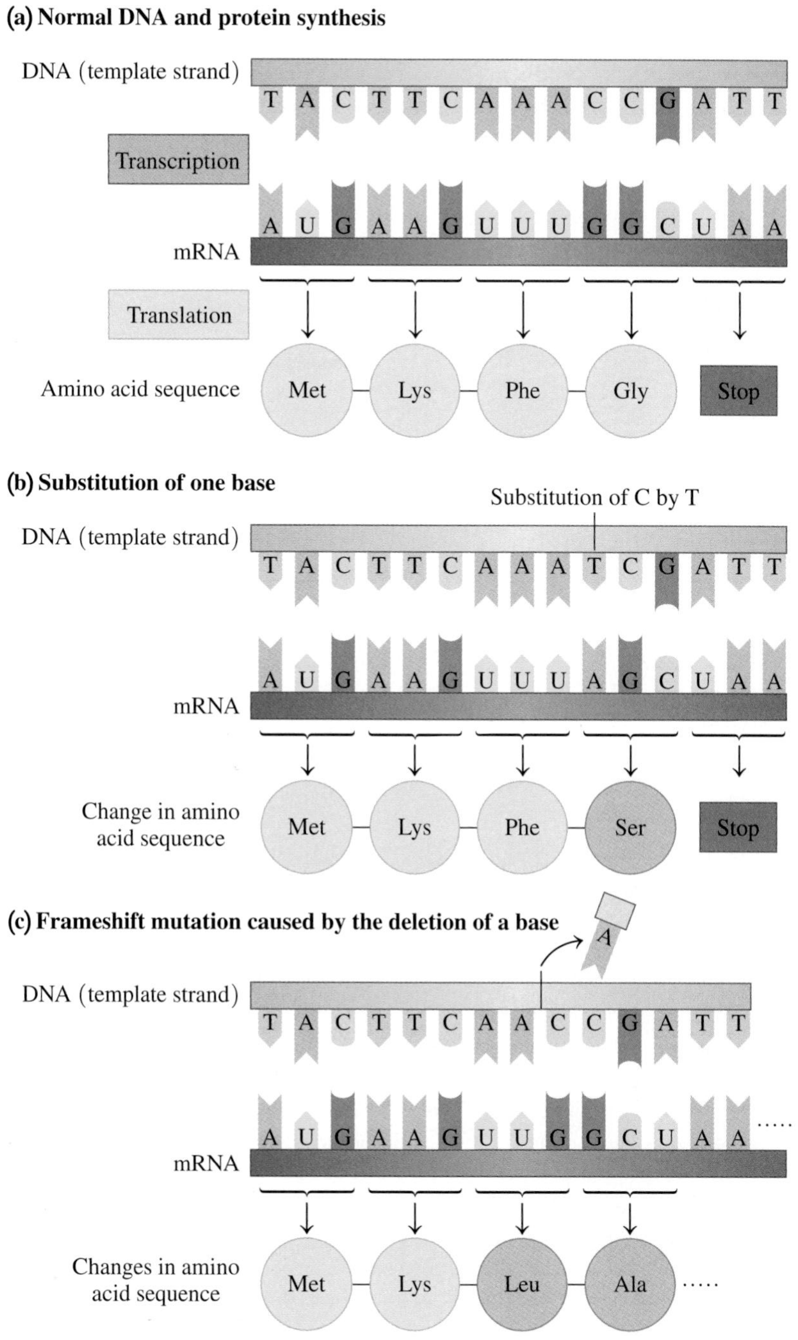

FIGURE 17.15 ▶ An alteration in the DNA template strand produces a change in the sequence of amino acids in the protein, which may result in a mutation. **(a)** A normal DNA leads to the correct amino acid order in a protein. **(b)** The substitution of a base in DNA leads to a change in the mRNA codon and a change in the amino acid. **(c)** The deletion of a base causes a frameshift mutation, which changes the mRNA codons that follow the mutation and produces a different amino acid sequence.

Q When would a substitution mutation cause protein synthesis to stop?

Genetic Diseases

A genetic disease is the result of a defective enzyme caused by a mutation in its genetic code. For example, *phenylketonuria* (PKU) results when DNA cannot direct the synthesis of the enzyme phenylalanine hydroxylase, required for the conversion of phenylalanine to tyrosine. In an attempt to break down the phenylalanine, other enzymes in the cells convert it to phenylpyruvate. If phenylalanine and phenylpyruvate accumulate in the blood of an infant, it can lead to severe brain damage and mental retardation. If PKU is detected in a newborn baby, a diet is prescribed that eliminates all the foods that contain phenylalanine. Preventing the buildup of the phenylpyruvate ensures normal growth and development.

FIGURE 17.16 ▶ A peacock with albinism does not produce the melanin needed to make the bright colors of its feathers.

Q Why are traits such as albinism related to the gene?

The amino acid tyrosine is needed in the formation of melanin, the pigment that gives the color to our skin and hair. If the enzyme that converts tyrosine to melanin is defective, no melanin is produced and a genetic disease known as *albinism* results. Persons and animals with no melanin have no skin, eye, or hair pigment (see Figure 17.16). Table 17.8 lists some other common genetic diseases and the type of metabolism or area affected.

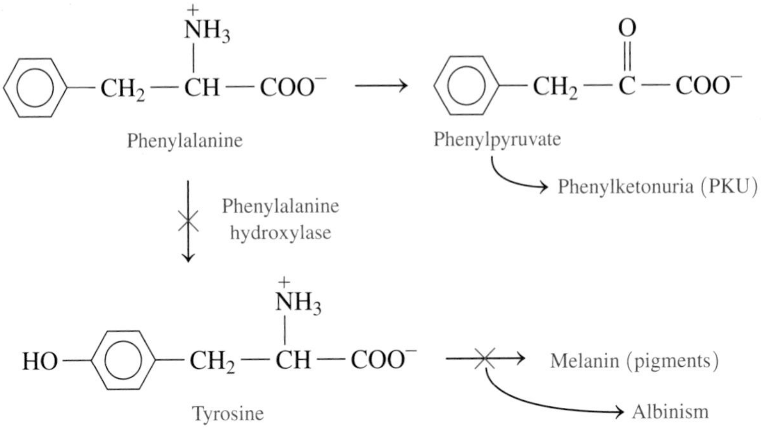

TABLE 17.8 Some Genetic Diseases

Genetic Disease	Result
Galactosemia	In galactosemia, the transferase enzyme required for the metabolism of galactose-1-phosphate is absent, resulting in the accumulation of galactose-1-phosphate, which leads to cataracts and mental retardation. Galactosemia occurs in about 1 in every 50 000 births.
Cystic fibrosis (CF)	Cystic fibrosis is caused by a mutation in the gene for the protein that regulates the production of stomach fluids and mucus. CF is one of the most common inherited diseases in children, in which thick mucus secretions make breathing difficult and block pancreatic function.
Down syndrome	Down syndrome is the leading cause of mental retardation, occurring in about 1 of every 800 live births; the mother's age strongly influences its occurrence. Mental and physical problems, including heart and eye defects, are the result of the formation of three chromosomes (trisomy), usually number 21, instead of a pair.
Familial hypercholesterolemia	Familial hypercholesterolemia occurs when there is a mutation of a gene on chromosome 19, which produces high cholesterol levels that lead to early coronary heart disease in people 30 to 40 years old.
Muscular dystrophy (MD) (Duchenne)	Muscular dystrophy, Duchenne form, is caused by a mutation in the X chromosome. This muscle-destroying disease appears at about age 5, with death by age 20, and occurs in about 1 of 10 000 males.
Huntington's disease (HD)	Huntington's disease affects the nervous system, leading to total physical impairment. It is the result of a mutation in a gene on chromosome 4, which can now be mapped to test people in families with a history of HD. There are about 30 000 people with Huntington's disease in the United States.
Sickle-cell anemia	Sickle-cell anemia is caused by a defective form of hemoglobin resulting from a mutation in a gene on chromosome 11. It decreases the oxygen-carrying ability of red blood cells, which take on a sickled shape, causing anemia and plugged capillaries from red blood cell aggregation. In the United States, about 72 000 people are affected by sickle-cell anemia.
Hemophilia	Hemophilia is the result of one or more defective blood-clotting factors that lead to poor coagulation, excessive bleeding, and internal hemorrhages. There are about 20 000 hemophilia patients in the United States.
Tay–Sachs disease	Tay–Sachs disease is the result of a defective hexosaminidase A, which causes an accumulation of gangliosides and leads to mental retardation, loss of motor control, and early death.

QUESTIONS AND PROBLEMS

17.6 Genetic Mutations

LEARNING GOAL Describe some ways in which DNA is altered to cause mutations.

17.39 What is a substitution mutation?

17.40 How does a substitution mutation in the genetic code for an enzyme affect the order of amino acids in that protein?

17.41 What is the effect of a frameshift mutation on the amino acid sequence in the protein?

17.42 How can a mutation decrease the activity of a protein?

17.43 How is protein synthesis affected if the normal base sequence TTT in the DNA template strand is changed to TTC?

17.44 How is protein synthesis affected if the normal base sequence CCC in the DNA template strand is changed to ACC?

17.45 Consider the following segment in mRNA, which is produced by the normal order of DNA nucleotides:

—ACA UCA CGG GUA—

a. What is the amino acid order produced from this mRNA?
b. What is the amino acid order if a mutation changes UCA to ACA?
c. What is the amino acid order if a mutation changes CGG to GGG?
d. What happens to protein synthesis if a mutation changes UCA to UAA?
e. What happens if a G is added to the beginning of the mRNA segment?
f. What happens if the A is removed from the beginning of the mRNA segment?

17.46 Consider the following segment in mRNA produced by the normal order of DNA nucleotides:

—CAU UCU CGA GUU—

a. What is the amino acid order produced from this mRNA?
b. What is the amino acid order if a mutation changes CUU to CCU?
c. What is the amino acid order if a mutation changes CGA to UGA?
d. What happens to protein synthesis if a mutation changes UCU to UGU?
e. What happens if a G is added to the beginning of the mRNA segment?
f. What happens if the C is removed from the beginning of the mRNA segment?

17.47 a. A base substitution changes a codon in the mRNA for an enzyme from GCC to GCA. Why is there no change in the amino acid order in the protein?
b. In sickle-cell anemia, a base substitution results in the replacement of glutamic acid with valine in the resulting hemoglobin molecule. Why does the replacement of one amino acid cause such a drastic change in biological function?

17.48 a. A base substitution in the mRNA for an enzyme results in the replacement of leucine with alanine. Why does this change in amino acids have little effect on the biological activity of the enzyme?
b. A base substitution in mRNA replaces cytosine in the codon UCA with adenine. How would this substitution affect the amino acids in the protein?

17.7 Viruses

LEARNING GOAL

Describe the methods by which a virus infects a cell.

Viruses are small particles of 3 to 200 genes that cannot replicate without a host cell. A typical virus contains a nucleic acid, DNA or RNA, but not both, inside a protein coat. It does not have the necessary material such as nucleotides and enzymes to make proteins and grow. The only way a virus can replicate is to invade a host cell and take over the machinery and materials necessary for protein synthesis and growth. Some infections caused by viruses invading human cells are listed in Table 17.9. There are also viruses that attack bacteria, plants, and animals.

TABLE 17.9 Some Diseases Caused by Viral Infection

Disease	Virus
Common cold	Coronavirus (over 100 types), rhinovirus (over 110 types)
Influenza	Orthomyxovirus
Warts	Papovavirus
Herpes	Herpesvirus
HPV	Human papilloma virus
Leukemia, cancers, AIDS	Retrovirus
Hepatitis	Hepatitis A virus (HAV), hepatitis B virus (HBV), hepatitis C virus (HCV)
Mumps	Paramyxovirus
Epstein–Barr	Epstein–Barr virus (EBV)
Chicken pox (shingles)	*Varicella zoster* virus (VZV)

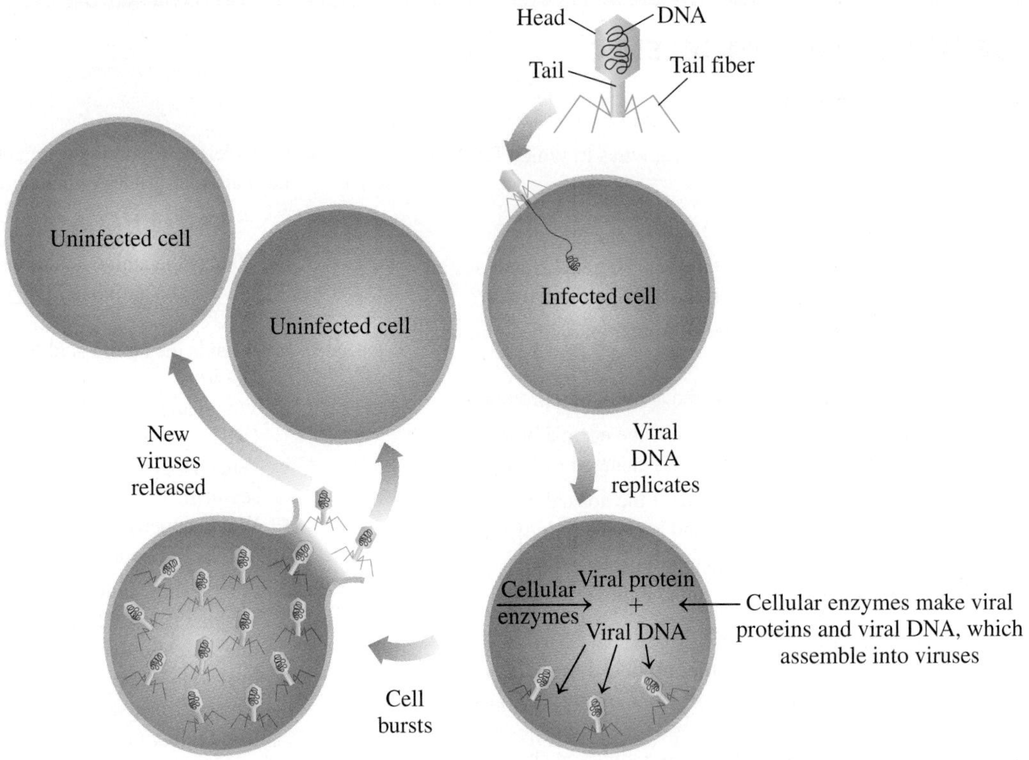

FIGURE 17.17 ▶ After a virus attaches to the host cell, it injects its viral DNA and uses the host cell's machinery and materials to synthesize viral protein. It uses the host cell's nucleic acids, enzymes, and ribosomes to make viral RNA, protein, and enzymes. When the cell wall opens, the new viruses are released to infect other cells.

Q Why does a virus need a host cell for replication?

A viral infection begins when an enzyme in the protein coat of the virus makes a hole in the host cell, allowing the viral nucleic acids to enter and mix with the materials in the host cell (see Figure 17.17). If the virus contains DNA, the host cell begins to replicate the viral DNA in the same way it would replicate normal DNA. Viral DNA produces viral RNA, and a protease processes proteins to produce a protein coat to form a viral particle that leaves the cell. The cell synthesizes so many virus particles that it eventually releases new viruses to infect more cells.

Vaccines are inactive forms of viruses that boost the immune response by causing the body to produce antibodies to the virus. Several childhood diseases, such as polio, mumps, chicken pox, and measles, can be prevented through the use of vaccines.

Reverse Transcription

A virus that contains RNA as its genetic material is a **retrovirus**. Once inside the host cell, it must first make viral DNA using a process known as *reverse transcription*. A retrovirus contains a polymerase enzyme called *reverse transcriptase* that uses the viral RNA template to synthesize complementary strands of DNA. Once produced, the single DNA strand forms double-stranded DNA using the nucleotides present in the host cell. This newly formed viral DNA, called a *provirus*, integrates with the DNA of the host cell (see Figure 17.18).

AIDS

During the early 1980s, a disease called *acquired immune deficiency syndrome*, commonly known as AIDS, began to claim an alarming number of lives. An HIV-1 virus (human immunodeficiency virus type 1) is now known to be the AIDS-causing agent

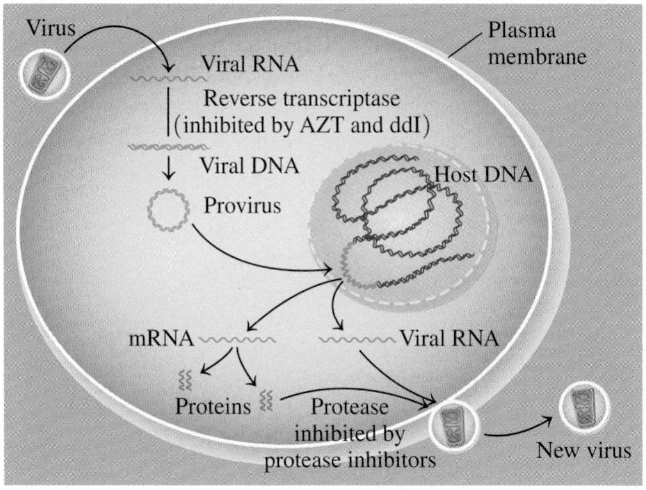

FIGURE 17.18 ▶ After a retrovirus injects its viral RNA into a cell, it forms a DNA strand by reverse transcription. The single-stranded DNA forms a double-stranded DNA called a provirus, which is incorporated into the host cell DNA. When the cell replicates, the provirus produces the viral RNA needed to produce more virus particles.

Q What is reverse transcription?

(see Figure 17.19). HIV is a retrovirus that infects and destroys T4 lymphocyte cells, which are involved in the immune response. After the HIV-1 virus binds to receptors on the surface of a T4 cell, the virus injects viral RNA into the host cell. As a retrovirus, the genes of the viral RNA direct the formation of viral DNA, which is then incorporated into the host's genome so that it can replicate as part of the host cell's DNA. The gradual depletion of T4 cells reduces the ability of the immune system to destroy harmful organisms. The AIDS syndrome is characterized by opportunistic infections such as *Pneumocystis carinii*, which causes pneumonia, and *Kaposi's sarcoma*, a skin cancer.

Treatment for AIDS is based on attacking the HIV-1 at different points in its life cycle, including reverse transcription and protein synthesis. Nucleoside analogs mimic the structures of the nucleosides used for DNA synthesis, and are able to inhibit the reverse transcriptase enzyme. For example, the drug AZT (3′-azido-3′-deoxythymidine) is similar to thymidine, and ddI (2′,3′-dideoxyinosine) is similar to guanosine. Two other drugs are ddC (2′,3′-dideoxycytidine) and d4T (2′,3′-didehydro-2′,3′-dideoxythymidine). Such compounds are found in the "cocktails" that are providing extended remission of HIV infections. When a nucleoside analog is incorporated into viral DNA, the lack of a hydroxyl group on the 3′-carbon in the sugar prevents the formation of the sugar–phosphate bonds and stops the replication of the virus.

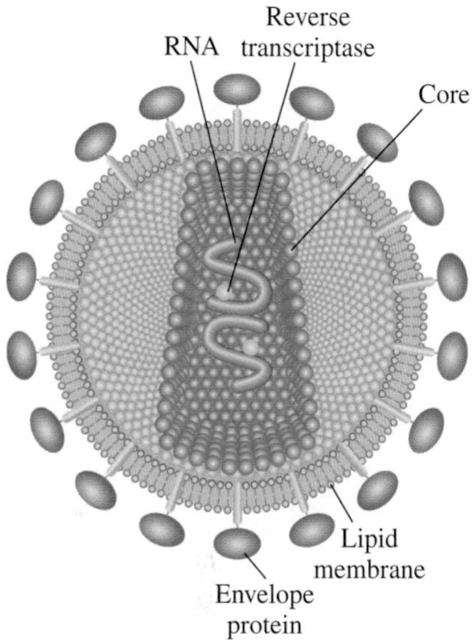

FIGURE 17.19 ▶ The HIV virus causes AIDS, which destroys the immune system in the body.

Q Does the HIV virus contain DNA or RNA?

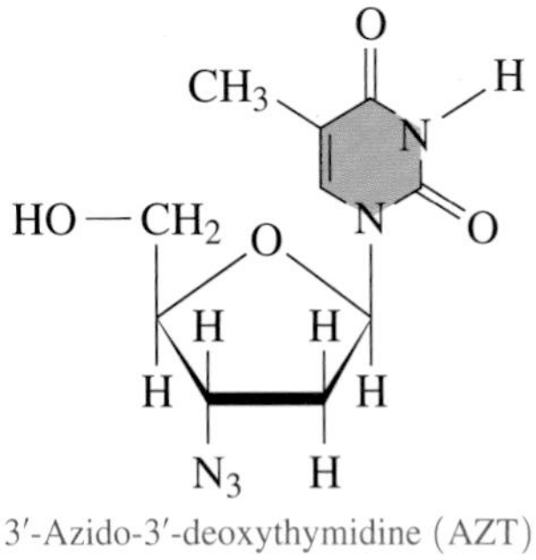
3′-Azido-3′-deoxythymidine (AZT)

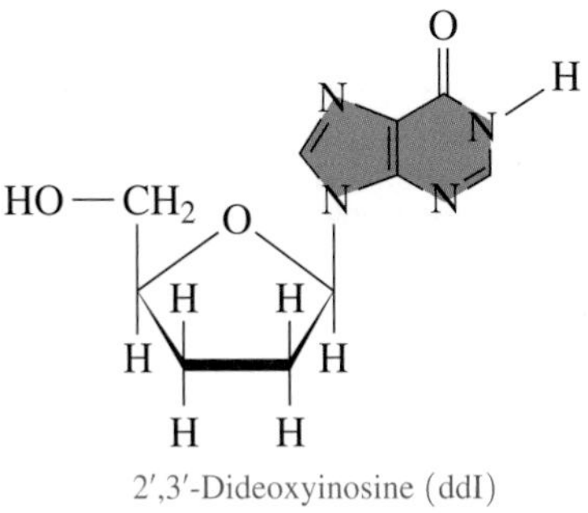
2′,3′-Dideoxyinosine (ddI)

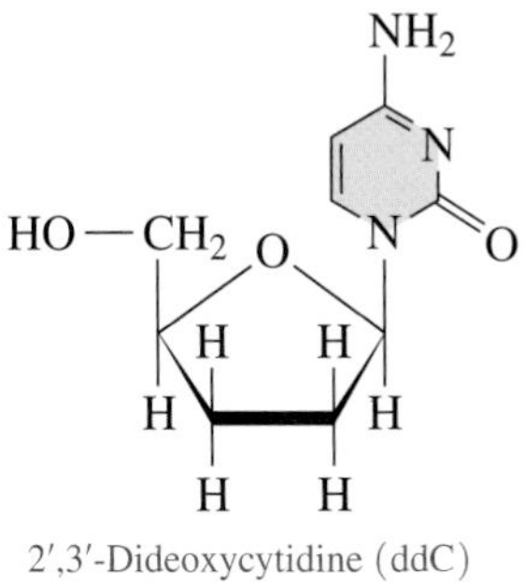
2′,3′-Dideoxycytidine (ddC)

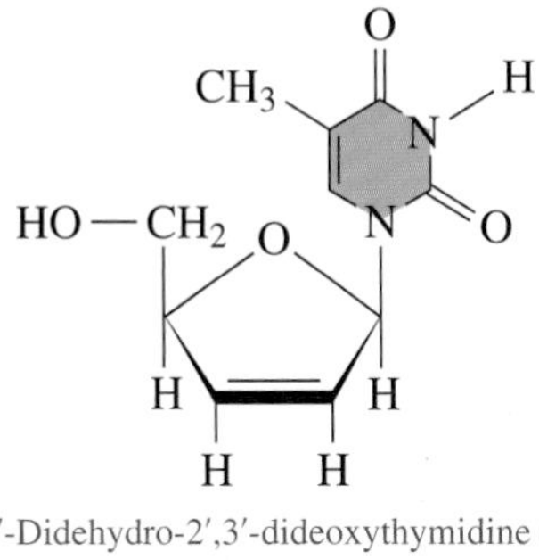
2′,3′-Didehydro-2′,3′-dideoxythymidine (d4T)

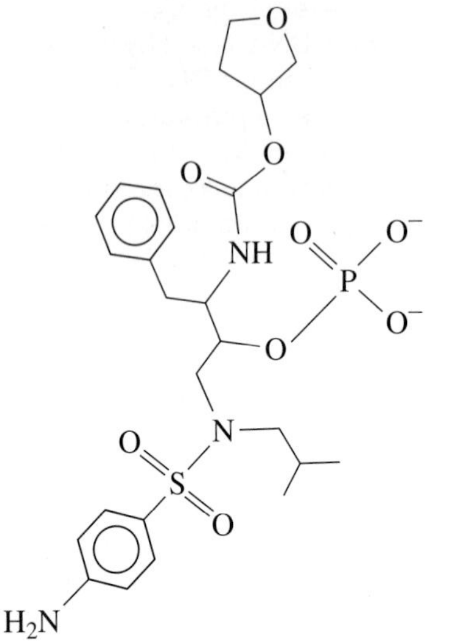

Lexiva metabolizes slowly to provide amprenavir, an HIV-protease inhibitor.

Treatment of AIDS often combines reverse transcriptase inhibitors with protease inhibitors such as saquinavir (Invirase), indinavir (Crixivan), fosamprenavir (Lexiva), nelfinavir (Viracept), and ritonavir (Norvir). The inhibition of a protease enzyme prevents the proper cutting and formation of proteins used by viruses to make more copies. Researchers are not yet certain how long protease inhibitors will be beneficial for a person with AIDS.

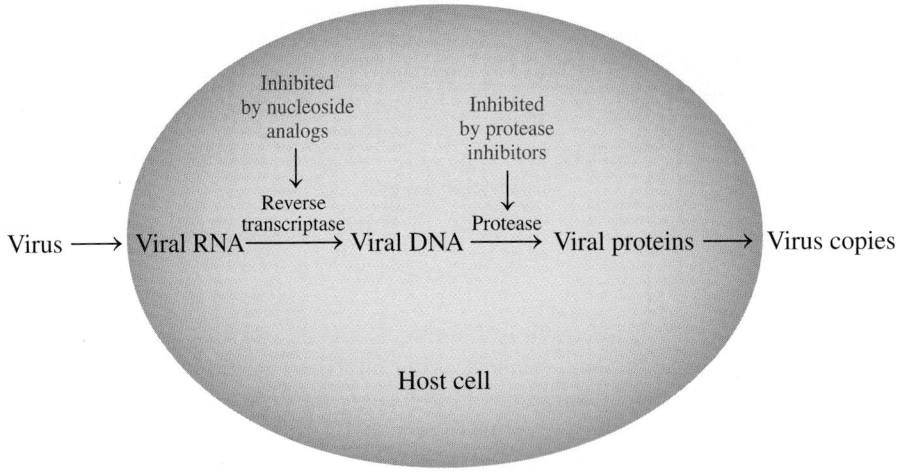

The AIDS virus can be attacked by compounds that inhibit enzymes needed by the virus to reproduce.

QUESTIONS AND PROBLEMS

17.7 Viruses

LEARNING GOAL Describe the methods by which a virus infects a cell.

17.49 What type of genetic information is found in a virus?

17.50 Why do viruses need to invade a host cell?

17.51 A specific virus contains RNA as its genetic material.
- **a.** Why would reverse transcription be used in the life cycle of this type of virus?
- **b.** What is the name of this type of virus?

17.52 What is reverse transcription?

17.53 How do nucleoside analogs disrupt the life cycle of the HIV-1 virus?

17.54 How do protease inhibitors disrupt the life cycle of the HIV-1 virus?

Chemistry Link to Health

Cancer

Normally, cells in the body undergo an orderly and controlled cell division. When cells begin to grow and multiply without control, they invade neighboring cells and appear as a tumor. If these tumors are limited, they are benign. When they invade other tissues and interfere with normal functions of the body, the tumors are cancerous. Cancer can be caused by chemical and environmental substances, by radiation, or by *oncogenic viruses*, which are associated with human cancers.

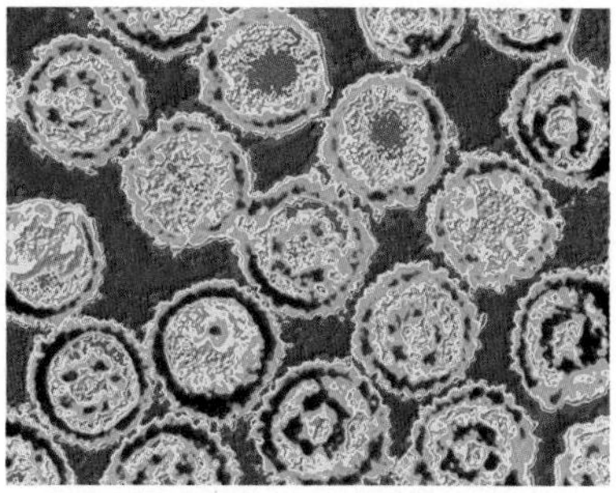

Epstein–Barr virus (EBV), herpesvirus 4, causes cancer in humans.

Some reports estimate that 70 to 80% of all human cancers are initiated by chemical and environmental substances. A *carcinogen* is any substance that increases the chance of inducing a tumor. Known carcinogens include aniline dyes, cigarette smoke, and asbestos. More than 90% of all persons with lung cancer are smokers. A carcinogen causes cancer by reacting with molecules in a cell, probably DNA, and altering the growth of that cell. Some known carcinogens are listed in Table 17.10.

Radiant energy from sunlight or medical radiation is another type of environmental factor. Skin cancer has become one of the most prevalent forms of cancer. It appears that DNA damage in the exposed areas of the skin may eventually cause mutations. The cells lose their ability to control protein synthesis. This type of uncontrolled cell division becomes skin cancer. The incidence of *malignant melanoma*, one of the most serious skin cancers, has been rapidly increasing. Some possible factors for this increase may be the popularity of sun tanning as well as the reduction of the ozone layer, which absorbs much of the harmful radiation from sunlight.

Oncogenic viruses cause cancer when cells are infected. Several viruses associated with human cancers are listed in Table 17.11. Some cancers such as retinoblastoma and breast cancer appear to occur more frequently within families. There is some indication that a missing or defective gene may be responsible.

TABLE 17.10 Some Chemical and Environmental Carcinogens

Carcinogen	Tumor Site
Aflatoxin	Liver
Aniline dyes	Bladder
Arsenic	Skin, lungs
Asbestos	Lungs, respiratory tract
Cadmium	Prostate, kidneys
Chromium	Lungs
Nickel	Lungs, sinuses
Nitrites	Stomach
Vinyl chloride	Liver

TABLE 17.11 Human Cancers Caused by Oncogenic Viruses

Virus	Disease
RNA Viruses	
Human T-cell lymphotropic virus-type 1 (HTLV-1)	Leukemia
DNA Viruses	
Epstein–Barr virus (EBV)	Burkitt's lymphoma (cancer of white blood B cells) Nasopharyngeal carcinoma Hodgkin's disease
Hepatitis B virus (HBV)	Liver cancer
Herpes simplex virus HSV (type 2)	Cervical and uterine cancer
Papilloma virus (HPV)	Cervical and colon cancer, genital warts

CONCEPT MAP

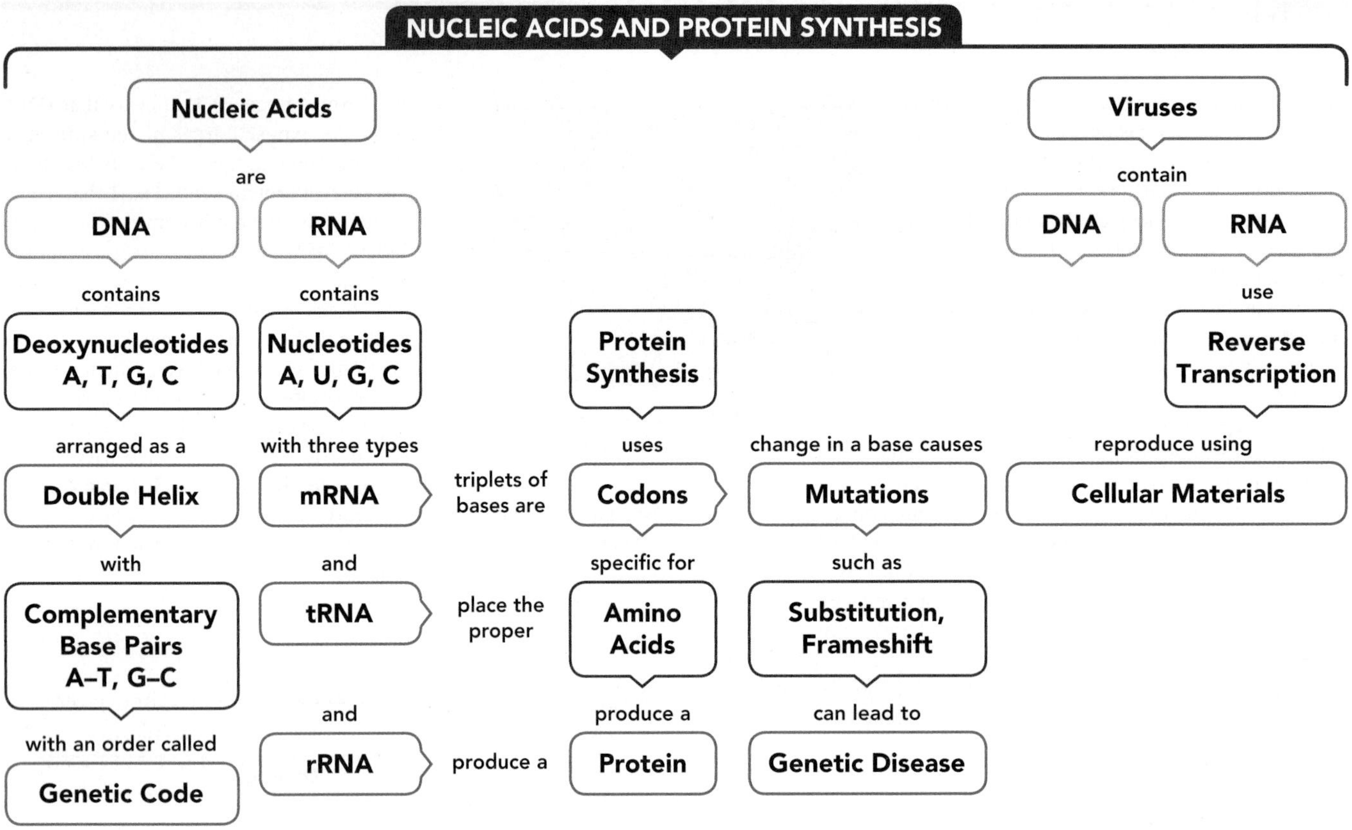

CHAPTER REVIEW

17.1 Components of Nucleic Acids

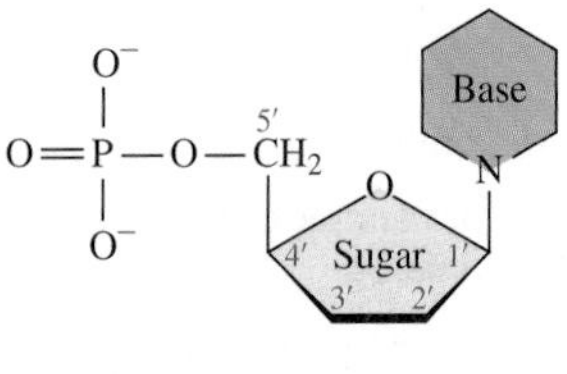

LEARNING GOAL Describe the bases and ribose sugars that make up the nucleic acids DNA and RNA.

- A nucleoside is a combination of a pentose sugar and a base.
- A nucleotide is composed of three parts: a base, a sugar, and a phosphate group.
- Nucleic acids, deoxyribonucleic acid (DNA) and ribonucleic acid (RNA), are polymers of nucleotides.
- In DNA, the sugar is deoxyribose and the base can be adenine, thymine, guanine, or cytosine.
- In RNA, the sugar is ribose and uracil replaces thymine.

17.2 Primary Structure of Nucleic Acids

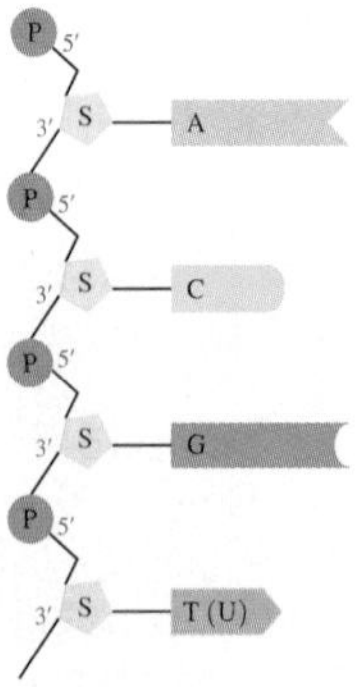

LEARNING GOAL Describe the primary structures of RNA and DNA.

- Each nucleic acid has its own unique sequence of bases known as its primary structure.
- In a nucleic acid polymer, the 3′-hydroxyl group of each ribose sugar in RNA or deoxyribose sugar in DNA forms a phosphodiester bond with the phosphate group of the sugar in the next nucleotide to give a backbone of alternating sugar and phosphate groups.
- There is a free 5′-phosphate group at one end of the polymer and a free 3′-hydroxyl group at the other end.

17.3 DNA Double Helix

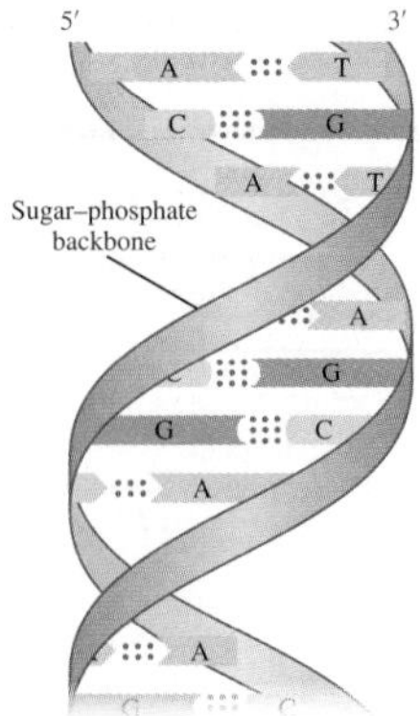

LEARNING GOAL Describe the double helix of DNA and the process of DNA replication.

- A DNA molecule consists of two strands of nucleotides that are wound around each other like a spiral staircase.
- The two strands are held together by hydrogen bonds between the complementary base pairs AT and CG.
- During DNA replication, new DNA strands are made along each of the original DNA strands that serve as templates.
- Complementary base pairing ensures the correct placement of bases to give identical copies of the original DNA.

17.4 RNA and the Genetic Code

LEARNING GOAL Identify the different types of RNA; describe the synthesis of mRNA.

U		C	
UUU	Phe	UCU	Ser
UUC		UCC	
UUA	Leu	UCA	
UUG		UCG	
CUU	Leu	CCU	Pro
CUC		CCC	
CUA		CCA	
CUG		CCG	

- The three types of RNA differ by function in the cell.
- Ribosomal RNA makes up most of the structure of the ribosomes.
- Messenger RNA carries genetic information from the DNA to the ribosomes.
- Transfer RNA places the correct amino acids in the protein.
- Transcription is the process by which RNA polymerase produces mRNA from one strand of DNA.
- The bases in the mRNA are complementary to the DNA, except that A in DNA is paired with U in RNA.
- The production of mRNA occurs when certain proteins are needed in the cell.
- The genetic code consists of a sequence of three bases (triplet) that specifies the order for the amino acids in a protein.
- The codon AUG signals the start of transcription and codons UAG, UGA, and UAA signal it to stop.

17.5 Protein Synthesis

LEARNING GOAL Describe the process of protein synthesis from mRNA.

—UCA AAA GCC CUU—
↓ ↓ ↓ ↓
— Ser — Lys — Ala — Leu —

- Proteins are synthesized at the ribosomes in a translation process that includes initiation, translocation, chain elongation, and termination.
- During translation, tRNAs bring the appropriate amino acids to the ribosome and peptide bonds form.
- When the protein is released, it takes on its secondary and tertiary structures and becomes functional in the cell.

17.6 Genetic Mutations

LEARNING GOAL Describe some ways in which DNA is altered to cause mutations.

- A genetic mutation is a change of one or more bases in the DNA sequence that may alter the structure and ability of the resulting protein to function properly.
- In a substitution, a different base replaces the base in a codon.
- A frameshift mutation inserts or deletes a base in the DNA sequence, which alters the mRNA codons after the mutation.

17.7 Viruses

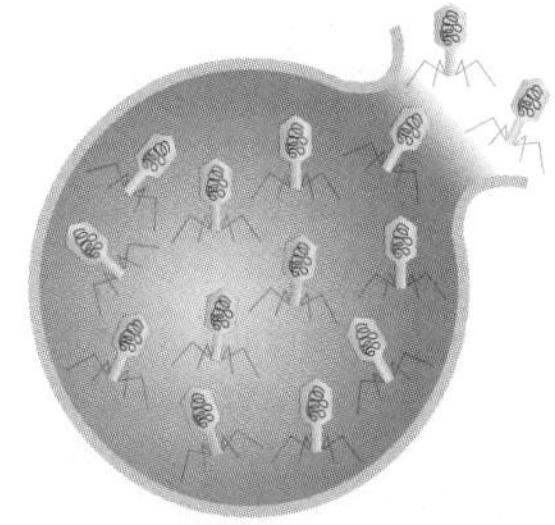

LEARNING GOAL Describe the methods by which a virus infects a cell.

- Viruses containing DNA or RNA invade host cells where they use the machinery within the cell to synthesize more viruses.
- For a retrovirus containing RNA, viral DNA is synthesized by reverse transcription using the nucleotides and enzymes in the host cell.
- In the treatment of AIDS, nucleoside analogs inhibit the reverse transcriptase of the HIV-1 virus, and protease inhibitors disrupt the catalytic activity of protease needed to produce proteins for the synthesis of more viruses.

KEY TERMS

anticodon The triplet of bases in the center loop of tRNA that is complementary to a codon on mRNA.

bases Nitrogen-containing compounds of purine and pyrimidine found in DNA and RNA: adenine (A), thymine (T), cytosine (C), guanine (G), and uracil (U).

codon A sequence of three bases in mRNA that specifies a certain amino acid to be placed in a protein. A few codons signal the start or stop of protein synthesis.

complementary base pairs In DNA, adenine is always paired with thymine (A and T or T and A), and guanine is always paired with cytosine (G and C or C and G). In forming RNA, adenine is always paired with uracil (A and U).

DNA Deoxyribonucleic acid; the genetic material of cells, containing nucleotides with deoxyribose sugar, phosphate, and the four bases: adenine, thymine, guanine, and cytosine.

double helix The helical shape of the double chain of DNA that is like a spiral staircase with a sugar–phosphate backbone on the outside and base pairs like stair steps on the inside.

frameshift mutation A mutation that inserts or deletes a base in a DNA sequence.

genetic code The information in DNA that is transferred to mRNA as a sequence of codons for the synthesis of protein.

genetic disease A physical malformation or metabolic dysfunction caused by a mutation in the base sequence of DNA.

mRNA Messenger RNA; produced in the nucleus from DNA to carry the genetic information to the ribosomes for the construction of a protein.

mutation A change in the DNA base sequence that alters the formation of a protein in the cell.

nucleic acids Large molecules composed of nucleotides; found as a double helix in DNA and as the single strands of RNA.

nucleoside The combination of a pentose sugar and a base.

nucleotides Building blocks of a nucleic acid, consisting of a base, a pentose sugar (ribose or deoxyribose), and a phosphate group.

phosphodiester bond The phosphate link that joins the 3′-hydroxyl group in one nucleotide to the phosphate group on the 5′-carbon atom in the next nucleotide.

primary structure The sequences of nucleotides in nucleic acids.

replication The process of duplicating DNA by pairing the bases on each parent strand with their complementary bases.

retrovirus A virus that contains RNA as its genetic material and synthesizes a complementary DNA strand inside a cell.

RNA Ribonucleic acid; a type of nucleic acid that is a single strand of nucleotides containing ribose, phosphate, and the four bases: adenine, cytosine, guanine, and uracil.

rRNA Ribosomal RNA; the most prevalent type of RNA and a major component of the ribosomes.

substitution A mutation that replaces one base in a DNA with a different base.

transcription The transfer of genetic information from DNA by the formation of mRNA.

translation The interpretation of the codons in mRNA as amino acids in a peptide.

tRNA Transfer RNA; an RNA that places a specific amino acid into a peptide chain at the ribosome using an anticodon. There is one or more tRNA for each of the 20 different amino acids.

viruses Small particles containing DNA or RNA in a protein coat that require a host cell for replication.

CORE CHEMISTRY SKILLS

Writing the Complementary DNA Strand (17.3)

- During DNA replication, new DNA strands are made along each of the original DNA strands.
- The new strand of DNA is made by forming hydrogen bonds with the bases in the template strand: A with T and T with A; C with G and G with C.

Example: Write the complementary base sequence for the following DNA segment:

—A T T C G G T A C—

Answer: —T A A G C C A T G—

Writing the mRNA Segment for a DNA Template (17.4)

- Transcription is the process that produces mRNA from one strand of DNA.
- The bases in the mRNA are complementary to the DNA, except that A in DNA is paired with U in RNA.

Example: What is the mRNA produced for the DNA segment —C G C A T G T C A—?

Answer: —G C G U A C A G U—

Writing the Amino Acid for an mRNA Codon (17.4)

- The genetic code consists of a sequence of three bases (codons) that specifies the order for the amino acids in a protein.
- The codon AUG signals the start of transcription and codons UAG, UGA, and UAA signal it to stop.

Example: What order of amino acids would you expect in a peptide for the mRNA sequence —CCG UAU GGG—?

Answer: —Pro—Tyr—Gly—

UNDERSTANDING THE CONCEPTS

The chapter sections to review are given in parentheses at the end of each question.

17.55 Answer the following questions for the given section of DNA: (17.3, 17.4, 17.5)

a. Complete the bases in the parent and new strands.

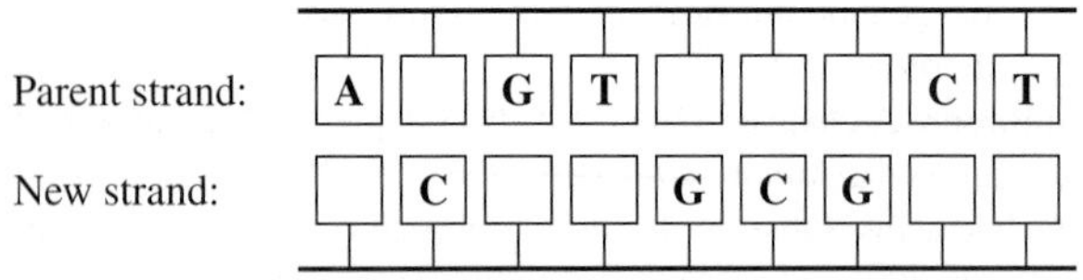

b. Using the new strand as a template, write the mRNA sequence.

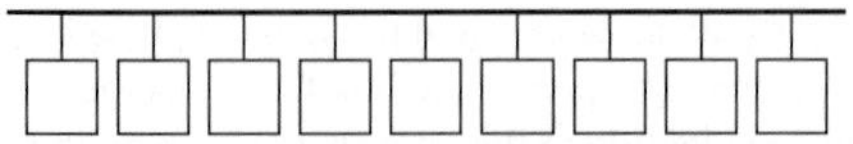

c. Write the three-letter symbols of the amino acids that would form the tripeptide from the mRNA you wrote in part **b**.

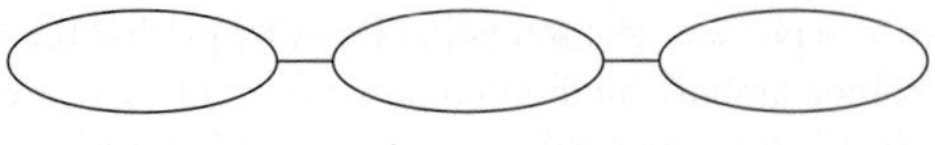

17.56 Suppose a mutation occurs in the DNA section in problem 17.55 and the first base in the parent strand, adenine, is replaced by guanine. (17.3, 17.4, 17.5, 17.6)

a. What type of mutation has occurred?

b. Using the new strand that results from this mutation as a template, write the order of bases in the altered mRNA.

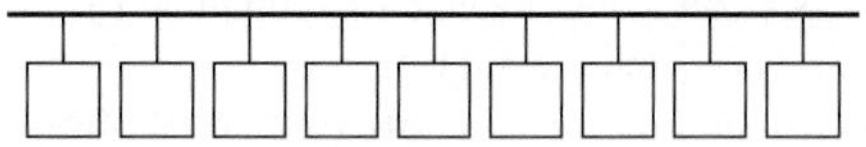

c. Write the three-letter symbols of the amino acids that would form the tripeptide from the mRNA you wrote in part **b**.

d. What effect, if any, might this mutation have on the structure and/or function of the resulting protein?

ADDITIONAL QUESTIONS AND PROBLEMS

17.57 Identify each of the following bases as a pyrimidine or a purine: (17.1)
a. cytosine **b.** adenine
c. uracil **d.** thymine
e. guanine

17.58 Indicate if each of the bases in problem 17.57 is found in DNA only, RNA only, or both DNA and RNA. (17.1)

17.59 Identify the base and sugar in each of the following nucleosides: (17.1)
a. deoxythymidine **b.** adenosine
c. cytidine **d.** deoxyguanosine

17.60 Identify the base and sugar in each of the following nucleotides: (17.1)
a. CMP **b.** dAMP
c. dTMP **d.** UMP

17.61 How do the bases thymine and uracil differ? (17.1)

17.62 What are the functional similarities between the bases cytosine and uracil? (17.1)

17.63 What is similar about the primary structure of RNA and DNA? (17.1, 17.2, 17.4)

17.64 What is different about the primary structure of RNA and DNA? (17.1, 17.2, 17.4)

17.65 Write the complementary base sequence for each of the following DNA segments: (17.3)
a. —G A C T T A G G C—
b. —T G C A A A C T A G C T—
c. —A T C G A T C G A T C G—

17.66 Write the complementary base sequence for each of the following DNA segments: (17.3)
a. —T T A C G G A C C G C—
b. —A T A G C C C T T A C T G G—
c. —G G C C T A C C T T A A C G A C G—

17.67 Match the following statements with rRNA, mRNA, or tRNA: (17.4)
a. is the smallest type of RNA
b. makes up the highest percentage of RNA in the cell
c. carries genetic information from the nucleus to the ribosomes

17.68 Match the following statements with rRNA, mRNA, or tRNA: (17.4)
a. combines with proteins to form ribosomes
b. brings amino acids to the ribosomes for protein synthesis
c. acts as a template for protein synthesis

17.69 What are the possible codons for each of the following amino acids? (17.4)
a. threonine **b.** serine **c.** cysteine

17.70 What are the possible codons for each of the following amino acids? (17.4)
a. valine **b.** arginine **c.** histidine

17.71 What is the amino acid for each of the following codons? (17.4)
a. AAG **b.** AUU **c.** CGA

17.72 What is the amino acid for each of the following codons? (17.4)
a. CAA **b.** GGC **c.** AAC

17.73 What is the amino acid order for the following mRNA that codes for a polypeptide? (17.4, 17.5)
—AUG AGU UAC GGA GCG CUA UAG—

17.74 Endorphins are peptides that reduce pain. What is the amino acid order for the following mRNA that codes for a pentapeptide that is an endorphin called methionine enkephalin? (17.4, 17.5)
—AUG UAC GGU GGA UUU AUG UAA—

17.75 What is the anticodon on tRNA for each of the following codons in mRNA? (17.4, 17.5)
a. AGC **b.** UAU **c.** CCA

17.76 What is the anticodon on tRNA for each of the following codons in mRNA? (17.4, 17.5)
a. GUG **b.** CCC **c.** GAA

CHALLENGE QUESTIONS

The following groups of questions are related to the topics in this chapter. However, they do not all follow the chapter order, and they require you to combine concepts and skills from several sections. These questions will help you increase your critical thinking skills and prepare for your next exam.

17.77 Oxytocin is a peptide that contains nine amino acids. How many nucleotides would be found in the mRNA for this peptide? (17.4, 17.5)

17.78 A peptide contains 36 amino acids. How many nucleotides would be present in the mRNA for this peptide? (17.4, 17.5)

17.79 What are the key similarities and differences between a vaccine and the corresponding virus? (17.7)

17.80 Why are there no base pairs in DNA between adenine and guanine, or thymine and cytosine? (17.1, 17.2)

17.81 Consider the following segment of a template strand of DNA: (17.3, 17.4, 17.5, 17.6)

—GAA GGG TCC GAC—

a. What is the mRNA produced for the segment?
b. What is the amino acid sequence from the mRNA in part **a**?
c. What is the mRNA if a mutation changes TCC to TTC?
d. What is the amino acid sequence from the mRNA in part **c**?
e. What is the mRNA produced if C is inserted at the beginning of the DNA segment?
f. What is the amino acid sequence from the mRNA in part **e**?

17.82 Consider the following segment of a template strand of DNA: (17.3, 17.4, 17.5, 17.6)

—ATA AGC TTC GAG—

a. What is the mRNA produced for the segment?
b. What is the amino acid sequence from the mRNA in part **a**?
c. What is the mRNA if a mutation changes AGC to AAC?
d. What is the amino acid sequence from the mRNA in part **c**?
e. What is the mRNA produced if G is inserted at the beginning of the DNA segment?
f. What is the amino acid sequence from the mRNA in part **e**?

ANSWERS

Answers to Study Checks

17.1 **(a)** pyrimidine
(b) purine

17.2

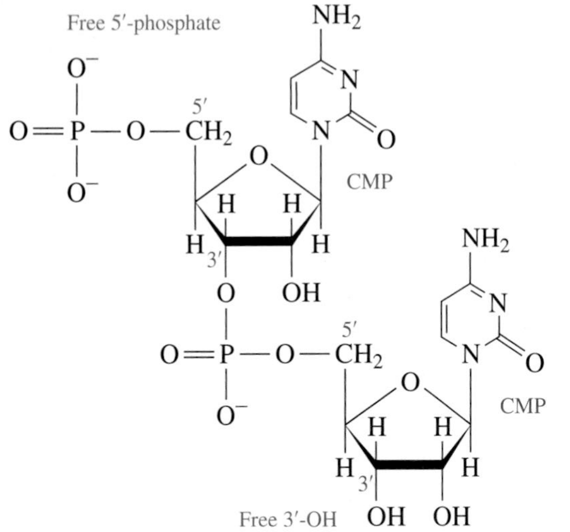

17.3 —C C A A T T G G—

17.4 —C C C A A A T T T—

17.5 —Asn—Ala—Cys—

17.6 at UAG

17.7 **(a)** Nucleoside **(b)** nucleoside
(c) nucleoside **(d)** nucleotide

Answers to Selected Questions and Problems

17.1 **a.** pyrimidine **b.** pyrimidine

17.3 **a.** DNA only **b.** both DNA and RNA

17.5 DNA-A, G, C, T
RNA-A, G, C, U

17.7 **a.** nucleoside **b.** nucleoside
c. nucleoside **d.** nucleotide

17.9

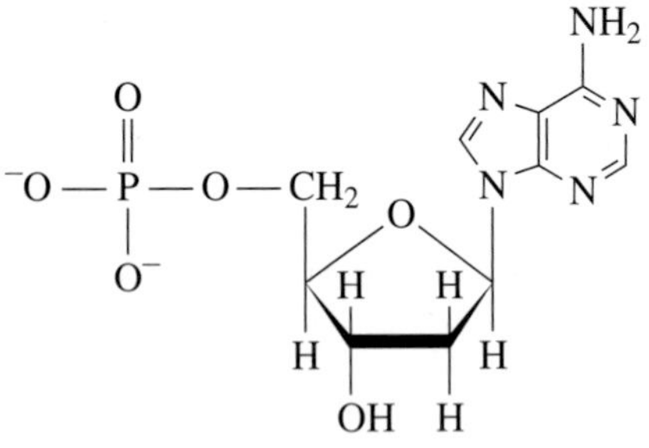

17.11 The nucleotides in nucleic acids are held together by phosphodiester bonds between the 3′-hydroxyl group of a sugar (ribose or deoxyribose) and a phosphate group on the 5′-carbon of another sugar.

17.13

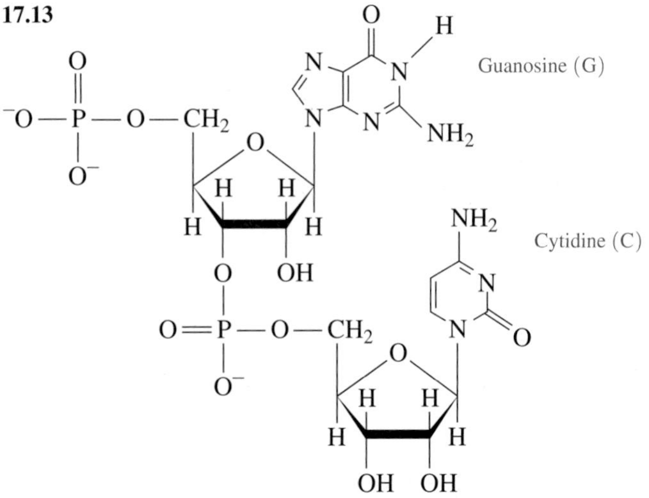

17.15 The two DNA strands are held together by hydrogen bonds between the complementary bases in each strand.

17.17 **a.** —T T T T T T—
b. —C C C C C C—
c. —T C A G G T C C A—
d. —G A C A T A T G C A A T—

17.19 ribosomal RNA, messenger RNA, and transfer RNA

17.21 In transcription, the sequence of nucleotides on a DNA template strand is used to produce the base sequences of a messenger RNA.

17.23 —G G C U U C C A A G U G—

17.25 A codon is a three-base sequence in mRNA that codes for a specific amino acid in a protein.

17.27 **a.** leucine (Leu) **b.** serine (Ser)
c. glutamic acid (Glu) **d.** arginine (Arg)

17.29 When AUG is the first codon, it signals the start of protein synthesis. Thereafter, AUG codes for methionine.

17.31 A codon is a base triplet in the mRNA. An anticodon is the complementary triplet on a tRNA for a specific amino acid.

17.33 **a.** —Thr—Thr—Thr—
b. —Phe—Pro—Phe—Pro—
c. —Leu—Gly—Arg—Cys—

17.35 The new amino acid is joined by a peptide bond to the peptide chain. The ribosome moves to the next codon, which attaches to a tRNA carrying the next amino acid.

17.37 **a.** —CGA AAA GUU UUU—
b. GCU, UUU, CAA, AAA
c. Using codons in mRNA: —Arg—Lys—Val—Phe—

17.39 In a substitution mutation, a base in DNA is replaced by a different base.

17.41 In a frameshift mutation, caused by a deletion or insertion of a base, all the codons from the mutation on are changed, which changes the order of amino acids in the rest of the protein chain.

17.43 The normal triplet TTT forms a codon AAA, which codes for lysine. The mutation TTC forms a codon AAG, which also codes for lysine. There is no effect on the amino acid sequence.

17.45 a. —Thr—Ser—Arg—Val—
b. —Thr—Thr—Arg—Val—
c. —Thr—Ser—Gly—Val—
d. —Thr—STOP Protein synthesis would terminate early. If this occurs early in the formation of the protein, the resulting protein will probably be nonfunctional.
e. The new protein will contain the sequence —Asp—Ile—Thr—Gly—.
f. The new protein will contain the sequence —His—His—Gly—.

17.47 a. GCC and GCA both code for alanine.
b. A vital interaction in the tertiary structure of hemoglobin cannot be formed when the polar glutamic acid is replaced by nonpolar valine. The resulting hemoglobin is malformed and less capable of carrying oxygen.

17.49 A virus contains DNA or RNA, but not both.

17.51 a. Viral RNA is used to synthesize viral DNA, which produces the mRNA to make the protein coat that allows the virus to replicate and leave the cell.
b. retrovirus

17.53 Nucleoside analogs such as AZT and ddI are similar to the nucleosides required to make viral DNA in reverse transcription. When they are incorporated into viral DNA, the lack of a hydroxyl group on the 3′-carbon in the sugar prevents the formation of the sugar–phosphate bonds and stops the replication of the virus.

17.55 a.

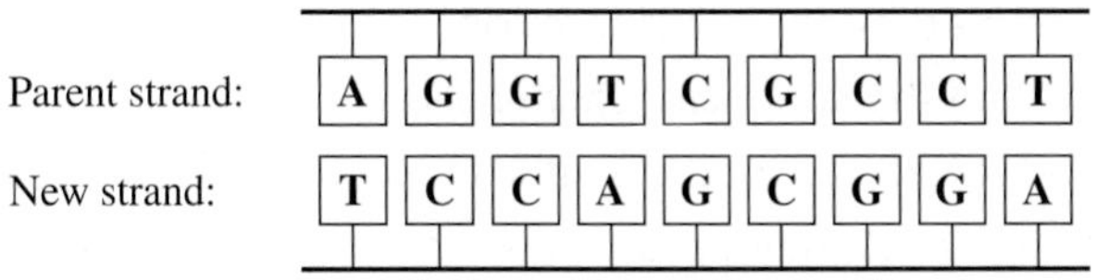

b.

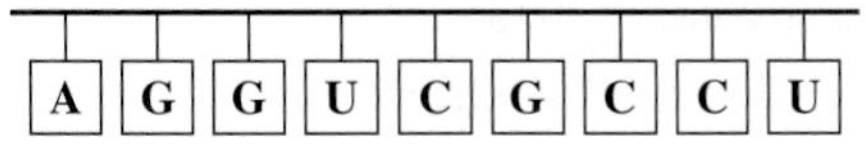

c.

17.57 a. pyrimidine
b. purine
c. pyrimidine
d. pyrimidine
e. purine

17.59 a. thymine and deoxyribose
b. adenine and ribose
c. cytosine and ribose
d. guanine and deoxyribose

17.61 They are both pyrimidines, but thymine has a methyl group.

17.63 They are both polymers of nucleotides connected through phosphodiester bonds between alternating sugar and phosphate groups with bases extending out from each sugar.

17.65 a. —C T G A A T C C G—
b. —A C G T T T G A T C G A—
c. —T A G C T A G C T A G C—

17.67 a. tRNA
b. rRNA
c. mRNA

17.69 a. ACU, ACC, ACA, and ACG
b. UCU, UCC, UCA, UCG, AGU, and AGC
c. UGU and UGC

17.71 a. lysine
b. isoleucine
c. arginine

17.73 Met—Ser—Tyr—Gly—Ala—Leu—

17.75 a. UCG
b. AUA
c. GGU

17.77 Three nucleotides are needed for each amino acid, plus a start and stop codon consisting of three nucleotides each, which makes a minimum total of 33 nucleotides.

17.79 Vaccines are inactive forms of the corresponding viruses that boost the immune response by helpingthe body to produce antibodies. Viruses on the other hand, can replicate after invading a host cell, and takes over the machinery and materials necessary for protein synthesis and growth.

17.81 a. —CUU CCC AGG CUG—
b. —Leu—Pro—Arg—Leu—
c. —CUU CCC AAG CUG—
d. —Leu—Pro—Lys—Leu—
e. —GCU UCC CAG GCU G—
f. —Ala—Ser—Gln—Ala—

18 Metabolic Pathways and Energy Production

Max, a six-year-old dog, has an appointment in a few days for a dental cleaning with anesthesia. Before that, he is brought to his veterinarian for a blood chemistry profile and a urinalysis. Sean, a veterinary assistant, measures Max's weight and obtains the blood and urine samples needed for the pre-surgery diagnostics. The blood chemistry profile determines the overall health and condition of Max's liver and kidneys, detects any metabolic disorders, and measures the concentration of electrolytes.

The overall condition of Max's metabolism is determined by the concentration of glucose and several enzymes. Metabolism includes all of the chemical reactions in the body involving the breakdown of molecules to produce energy or the synthesis of complex molecules for cellular growth. These reactions frequently occur in a series and require multiple enzymes. The oxidation of glucose involves several of these metabolic pathways including glycolysis, the citric acid cycle, and electron transport.

On the day of the cleaning, Sean records Max's weight and eating habits for the past 24 h. He prepares the surgical room to ensure it is clean and sterilized. Sean then shaves an area on Max's front leg and helps administer the anesthesia. After the cleaning, Sean monitors the heart rate and blood pressure as Max recovers from anesthesia.

CAREER Veterinary Assistant

Veterinary assistants, or veterinary technicians, assist in the care of domesticated pets, and farm animals under the direct supervision of a veterinarian. Veterinary assistants are typically the first person an owner interacts with as they record the animal's symptoms and medical history. This includes dietary intake, medications, eating habits, weight, and any clinical signs. Veterinary assistants perform laboratory tests on animals including a complete blood count (CBC) and urinalysis. They also obtain tissue and blood samples, expose and develop X-rays, and assist with vaccinations. In addition, they assist with surgical procedures such as neutering and spaying, dental cleanings, removing tumors, and euthanizing animals.

All the chemical reactions that take place in cells to break down or build molecules are known as *metabolism*. A *metabolic pathway* is a series of linked reactions, each catalyzed by a specific enzyme. In this chapter, we will look at these pathways and the way they produce energy and cellular compounds.

When we eat food, the polysaccharides, lipids, and proteins are digested to smaller molecules that can be absorbed into the cells of our body. As the glucose, fatty acids, and amino acids are broken down further, energy is released. Because we do not use all the energy from our foods at one time, we store energy in the cells as high-energy adenosine triphosphate (ATP). Our cells use the energy stored in ATP when they do work such as contracting muscles, synthesizing large molecules, sending nerve impulses, and moving substances across cell membranes.

In the *citric acid cycle*, metabolic reactions in the mitochondria oxidize the two carbon atoms in the acetyl component of acetyl-CoA to two molecules of CO_2. The reduced coenzymes NADH and $FADH_2$ enter *electron transport* where they provide hydrogen ions and electrons that combine with oxygen (O_2) to form H_2O. The energy released during electron transport is used to synthesize ATP from ADP and P_i.

LOOKING AHEAD

CHAPTER READINESS*

CORE CHEMISTRY SKILLS

- Identifying Functional Groups (12.1)
- Identifying Fatty Acids (15.2)
- Drawing Structures for Triacylglycerols (15.3)
- Drawing the Ionized Form for an Amino Acid (16.1)
- Identifying the Primary, Secondary, Tertiary, and Quaternary Structures of Proteins (16.4)
- Describing Enzyme Action (16.5)

*These Core Chemistry Skills from previous chapters are listed here for your review as you proceed to the new material in this chapter.

18.1 Metabolism and ATP Energy

LEARNING GOAL

Describe three stages of catabolism and the role of ATP.

The term **metabolism** refers to all the chemical reactions that provide energy and the substances required for continued cell growth. There are two types of metabolic reactions: catabolic and anabolic. In **catabolic reactions**, complex molecules are broken down to simpler ones with an accompanying release of energy. **Anabolic reactions** utilize energy available in the cell to build large molecules from simple ones. We can think of the catabolic processes in metabolism as consisting of three stages (see Figure 18.1).

Stage 1 Catabolism begins with the processes of **digestion** in which enzymes in the digestive tract break down large molecules into smaller ones. The polysaccharides break down to monosaccharides, fats break down to glycerol and fatty acids, and the proteins yield amino acids. These digestion products diffuse into the bloodstream for transport to cells.

Stage 2 Within the cells, catabolic reactions continue as the digestion products are broken down further to yield two- and three-carbon compounds such as pyruvate and acetyl-CoA.

Stage 3 The major production of energy takes place in the mitochondria, as the two-carbon acetyl group is oxidized in the citric acid cycle, which produces reduced coenzymes NADH and $FADH_2$. As long as the cells have oxygen, the hydrogen ions and electrons from the reduced coenzymes are transferred to electron transport to synthesize ATP.

Stages of Catabolism

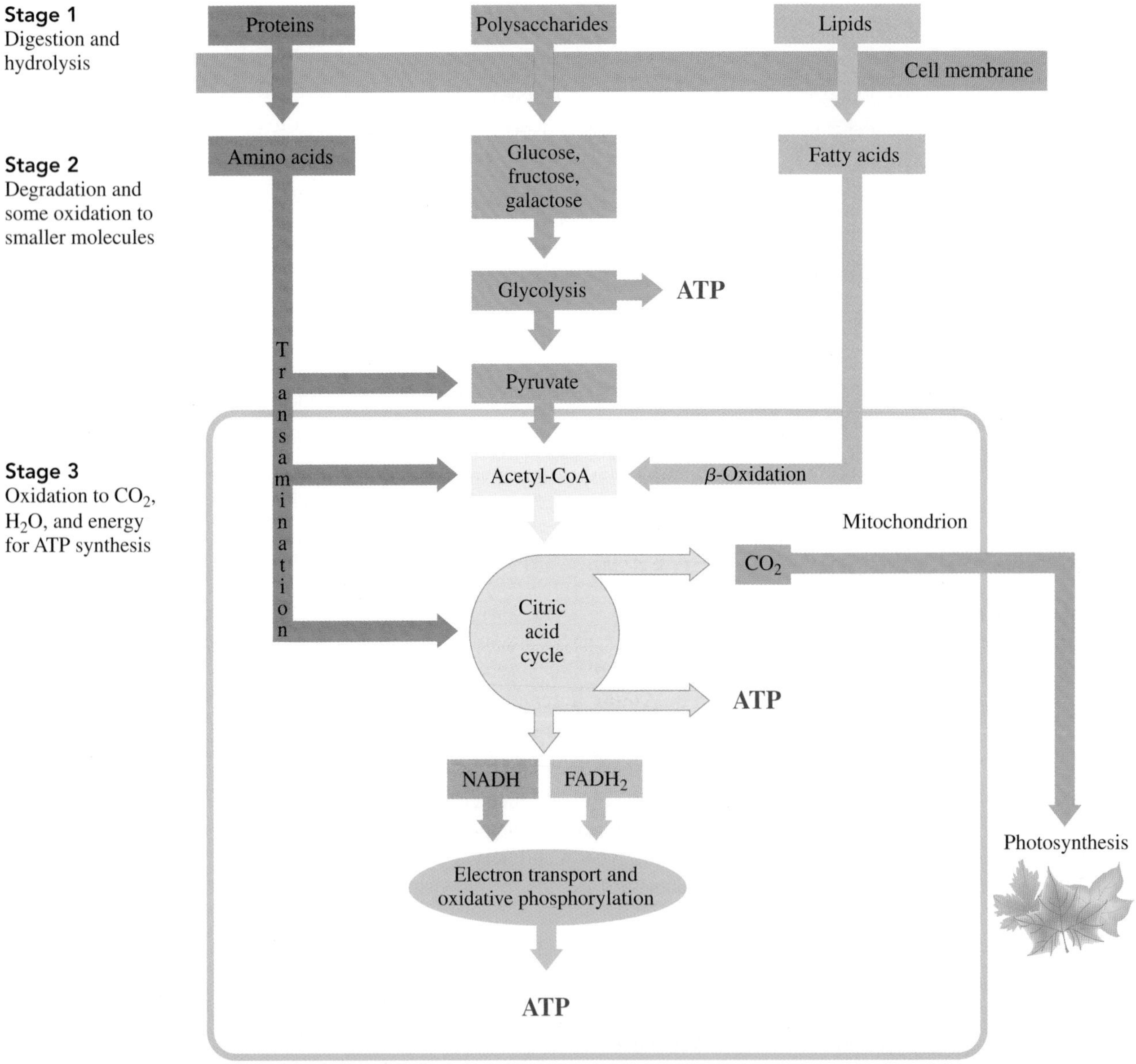

FIGURE 18.1 ▶ In the three stages of catabolism, large molecules from foods are digested and degraded to give smaller molecules that can be oxidized to produce energy.

Q Where is most of the ATP energy produced in the cells?

Cell Structure for Metabolism

To understand metabolic reactions, we need to look at where these reactions take place in the cell (see Figure 18.2).

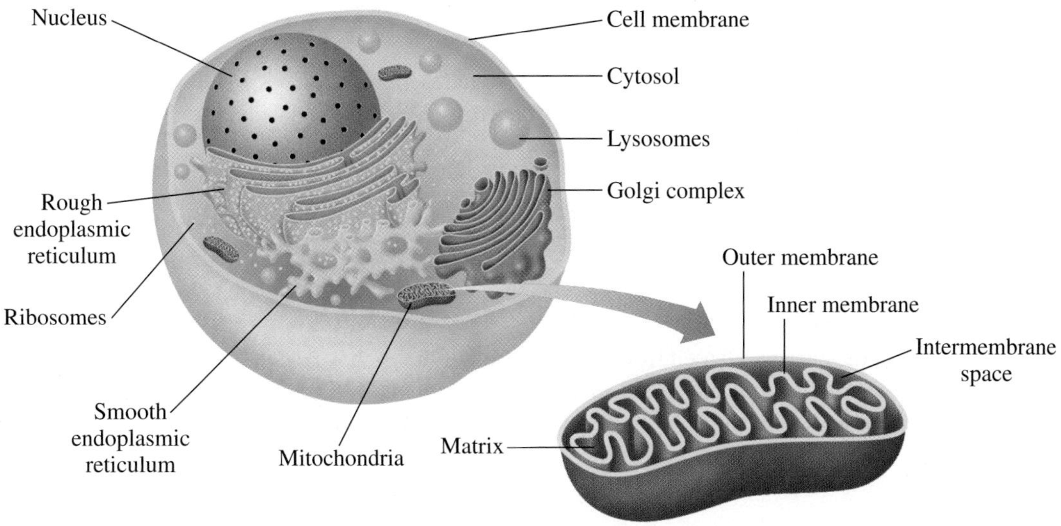

FIGURE 18.2 ▶ The diagram illustrates the major components of a typical animal cell.
Q What is the function of the mitochondria in a cell?

In animals, a *cell membrane* separates the materials inside the cell from the aqueous environment surrounding the cell. The *nucleus* contains the genes that control DNA replication and protein synthesis. The *cytoplasm* consists of all the materials between the nucleus and the cell membrane. The *cytosol*, the fluid part of the cytoplasm, is an aqueous solution of electrolytes and enzymes that catalyze many of the cell's chemical reactions.

Within the cytoplasm are specialized structures called *organelles* that carry out specific functions in the cell. The *ribosomes* are the sites of protein synthesis. The *mitochondria* are the energy-producing factories of the cells. A mitochondrion has an outer and an inner membrane, with an intermembrane space between them. The fluid section surrounded by the inner membrane is called the *matrix*. Enzymes located in the matrix and along the inner membrane catalyze the oxidation of carbohydrates, fats, and amino acids. All these oxidation pathways eventually produce CO_2, H_2O, and energy, which is used to form energy-rich compounds. Table 18.1 summarizes some of the functions of the components in animal cells.

TABLE 18.1 Functions of Components in Animal Cells

Component	Description and Function
Cell membrane	Separates the contents of a cell from the external environment and contains structures that communicate with other cells
Cytoplasm	Consists of the cellular contents between the cell membrane and nucleus
Cytosol	Is the fluid part of the cytoplasm that contains enzymes for many of the cell's chemical reactions
Endoplasmic reticulum	Rough type processes proteins for secretion and synthesizes phospholipids; smooth type synthesizes fats and steroids
Golgi complex	Modifies and secretes proteins from the endoplasmic reticulum and synthesizes cell membranes
Lysosome	Contains hydrolytic enzymes that digest and recycle old cell structures
Mitochondrion	Contains the structures for the synthesis of ATP from energy-producing reactions
Nucleus	Contains genetic information for the replication of DNA and the synthesis of protein
Ribosome	Is the site of protein synthesis using mRNA templates

SAMPLE PROBLEM 18.1 Metabolism

Identify each of the following as a catabolic or an anabolic reaction:

a. digestion of polysaccharides **b.** synthesis of proteins

SOLUTION

a. The breakdown of large molecules is a catabolic reaction.
b. The synthesis of large molecules requires energy and involves anabolic reactions.

STUDY CHECK 18.1

Identify the oxidation of glucose to CO_2 and H_2O as a catabolic or an anabolic reaction.

ATP and Energy

In our cells, the energy released from the oxidation of the food we eat is stored in the form of a "high-energy" compound called *adenosine triphosphate* (ATP). The **ATP** molecule is composed of the base adenine, a ribose sugar, and three phosphate groups.

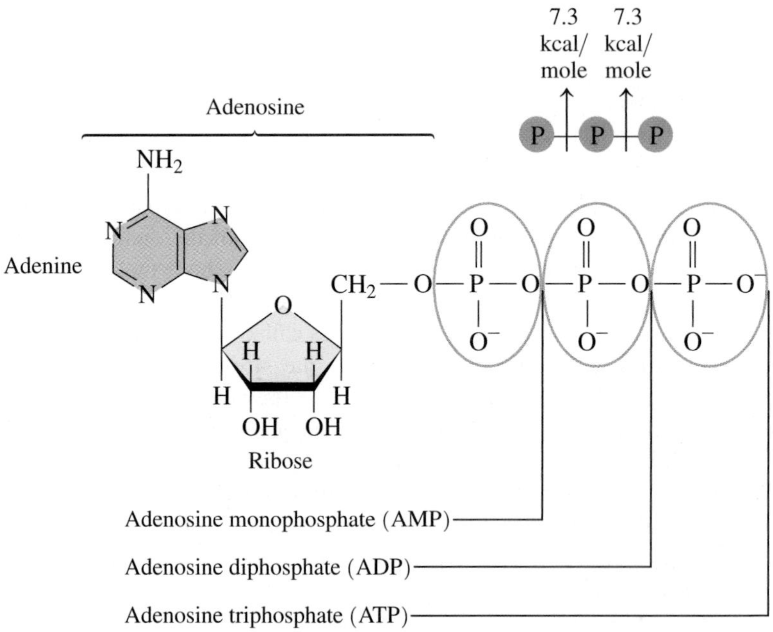

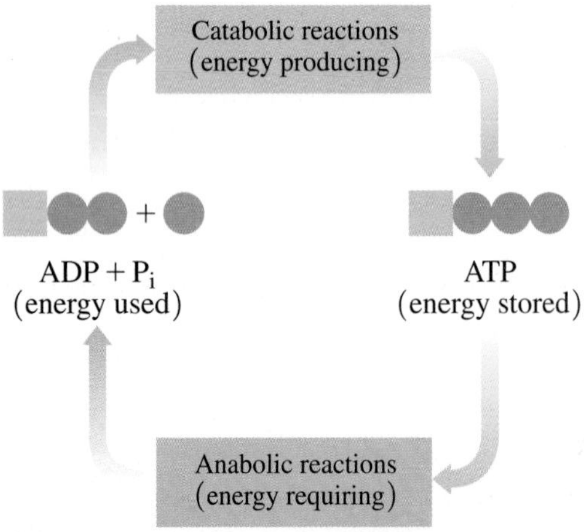

ATP, the energy-storage molecule, links energy-producing reactions with energy-requiring reactions in the cells.

When ATP undergoes hydrolysis, the products are adenosine diphosphate (ADP), a phosphate group abbreviated as P_i, and energy of 7.3 kcal per mole of ATP. We can write this reaction as

$$ATP + H_2O \longrightarrow ADP + P_i + 7.3\ \text{kcal/mole}$$

The ADP can also hydrolyze to form adenosine monophosphate (AMP) and an inorganic phosphate (P_i).

$$ADP + H_2O \longrightarrow AMP + P_i + 7.3\ \text{kcal/mole}$$

Every time we contract muscles, move substances across cellular membranes, send nerve signals, or synthesize an enzyme, we use energy from ATP hydrolysis. In a cell that is doing work (anabolic processes), 1 to 2 million ATP molecules may be hydrolyzed in one second. The amount of ATP hydrolyzed in one day can be as much as our body mass, even though only about 1 g of ATP is present in our cells at any given time.

When we take in food, the resulting catabolic reactions provide energy to regenerate ATP in our cells. Then 7.3 kcal/mole is used to make ATP from ADP and P_i.

$$ADP + P_i + 7.3\ kcal/mole \longrightarrow ATP + H_2O$$

Chemistry Link to Health

ATP Energy and Ca^{2+} Needed to Contract Muscles

Our muscles consist of thousands of parallel fibers. Within these muscle fibers are filaments composed of two kinds of proteins, myosin and actin. Arranged in alternating rows, the thick filaments of the protein myosin overlap the thin filaments containing the protein actin. During a muscle contraction, the thin filaments slide inward over the thick filaments, which shortens the muscle fibers.

Calcium ion, Ca^{2+}, and ATP play an important role in muscle contraction. An increase in the Ca^{2+} concentration in the muscle fibers causes the filaments to slide, while a decrease stops the process. In a relaxed muscle, the Ca^{2+} concentration is low. When a nerve impulse reaches the muscle, calcium channels in the membrane open to allow Ca^{2+} to flow into the fluid surrounding the filaments. The muscle contracts as myosin binds to actin and pulls the filaments inward. The energy for the contraction is provided by the hydrolysis of ATP to ADP + P_i. Muscle contraction continues as long as both ATP and Ca^{2+} levels are high around the filaments. When the nerve impulse ends, the calcium channels close. The Ca^{2+} concentration decreases as energy from ATP pumps the remaining Ca^{2+} out of the filaments, which causes the muscle to relax. In rigor mortis, Ca^{2+} concentration remains high within the muscle fibers causing a continued state of rigidity. After approximately 72 h, Ca^{2+} decreases due to cellular deterioration and the muscles relax.

Muscle contraction uses the energy from the breakdown of ATP.

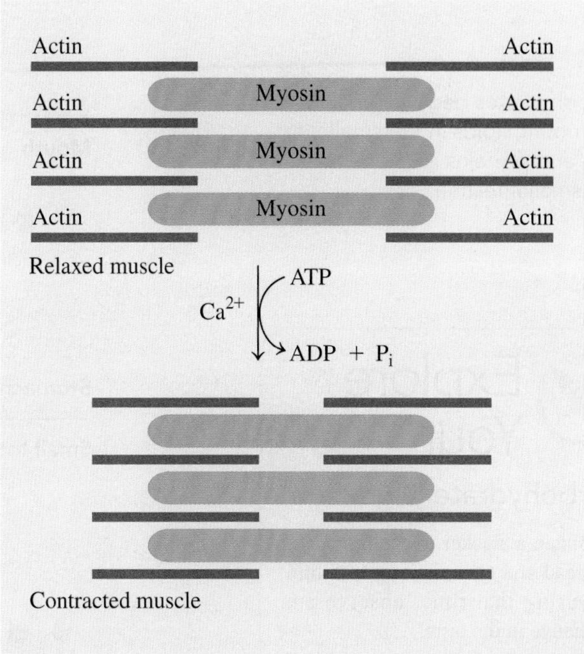

Muscles contract when myosin binds to actin.

QUESTIONS AND PROBLEMS

18.1 Metabolism and ATP Energy

LEARNING GOAL Describe three stages of catabolism and the role of ATP.

18.1 What stage of catabolism involves the digestion of polysaccharides?

18.2 What stage of catabolism involves the conversion of small molecules to CO_2, H_2O, and energy?

18.3 What are the functions of the Golgi complex in animal cells?

18.4 What is meant by an anabolic reaction in metabolism?

18.5 Name the chemical reaction involved in the production of energy during ATP to ADP and AMP conversion.

18.6 How much energy is obtained from the hydrolysis of ATP?

LEARNING GOAL

Give the sites and products of digestion for carbohydrates, triacylglycerols, and proteins.

18.2 Digestion of Foods

In stage 1 of catabolism, foods undergo digestion, a process that converts large molecules to smaller ones that can be absorbed by the body.

Digestion of Carbohydrates

We begin the digestion of carbohydrates as soon as we chew food. Enzymes produced in the salivary glands hydrolyze some of the α-glycosidic bonds in amylose and amylopectin, producing maltose, glucose, and smaller polysaccharides called dextrins, which contain three to eight glucose units. After swallowing, the partially digested starches enter the acidic environment of the stomach, where the low pH stops carbohydrate digestion (see Figure 18.3).

In the small intestine, which has a pH of about 8, enzymes produced in the pancreas hydrolyze the remaining dextrins to maltose and glucose. Then enzymes produced in the mucosal cells that line the small intestine hydrolyze maltose as well as lactose and sucrose. The resulting monosaccharides are absorbed through the intestinal wall into the bloodstream, which carries them to the liver, where the hexoses fructose and galactose are converted to glucose. Glucose is the primary energy source for muscle contractions, red blood cells, and the brain.

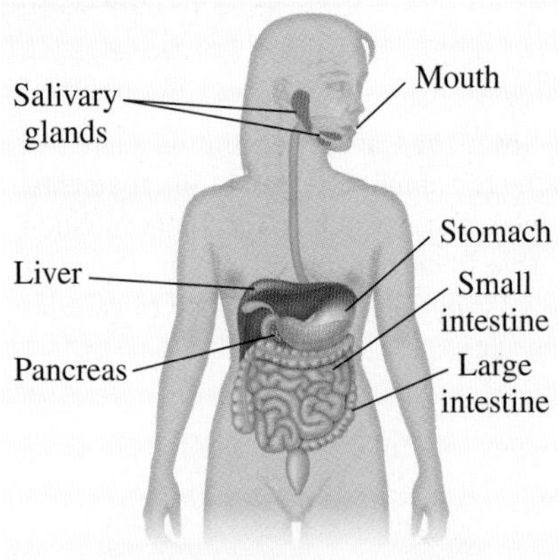

Carbohydrates begin digestion in the mouth, lipids in the small intestine, and proteins in the stomach and small intestine.

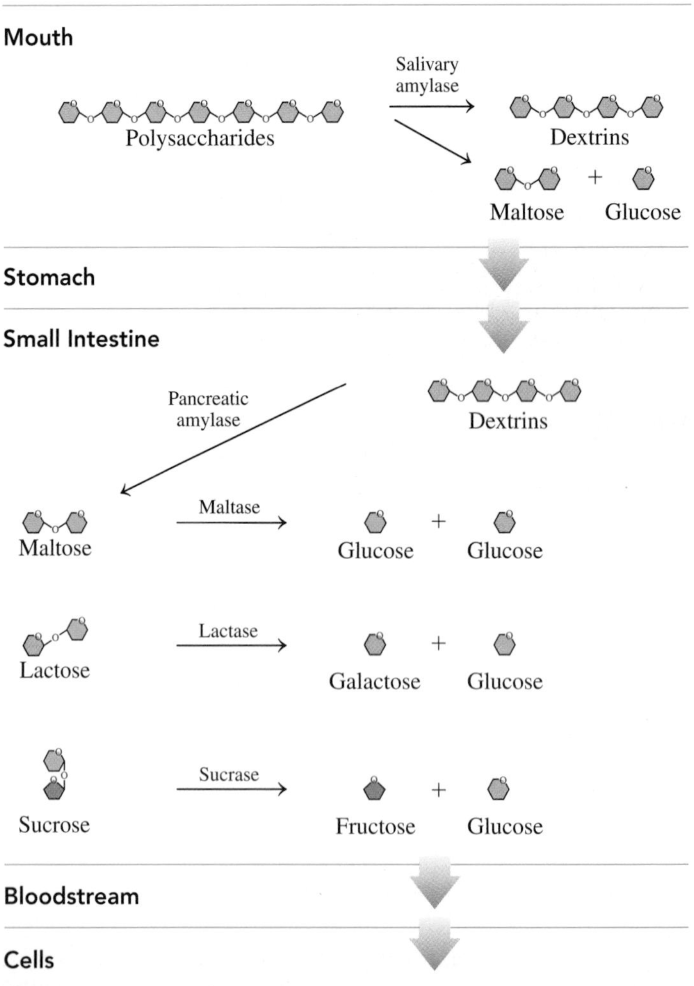

FIGURE 18.3 ▶ In stage 1 of metabolism, the digestion of carbohydrates begins in the mouth and is completed in the small intestine.

Q Why is there little or no digestion of carbohydrates in the stomach?

Explore Your World

Carbohydrate Digestion

1. Obtain a cracker or small piece of bread and chew it for 4 to 5 min. During that time, observe any change in the taste.
2. Some milk products contain Lactaid, which is the lactase that catalyzes the digestion of lactose. Look for the brands of milk and ice cream that contain Lactaid or lactase enzyme.

Questions

1a. How does the taste of the cracker or bread change after you have chewed it for 4 to 5 min? What could be an explanation?

b. What part of carbohydrate digestion occurs in the mouth?

2a. Write an equation for the hydrolysis of lactose.

b. Where does lactose undergo digestion?

Chemistry Link to Health

Lactose Intolerance

The disaccharide in milk is lactose, which is broken down by *lactase* in the intestinal tract to monosaccharides that are a source of energy. Infants and small children produce lactase to break down the lactose in milk. It is rare for an infant to lack the ability to produce lactase. However, the production of lactase decreases as many people age, which causes lactose intolerance. This condition affects approximately 25% of the people in the United States. A deficiency of lactase occurs in adults in many parts of the world, but in the United States it is prevalent among the African American, Hispanic, and Asian populations.

When lactose is not broken down into glucose and galactose, it cannot be absorbed through the intestinal wall and remains in the intestinal tract. In the intestines, the lactose undergoes fermentation to products that include lactic acid and gases such as methane (CH_4) and CO_2. Symptoms of lactose intolerance, which appear approximately $\frac{1}{2}$ to 1 h after ingesting milk or milk products, include nausea, abdominal cramps, and diarrhea. The severity of the symptoms depends on how much lactose is present in the food and how much lactase a person produces.

Treatment of Lactose Intolerance

One way to reduce the reaction to lactose is to avoid products that contain lactose, including milk and milk products such as cheese, butter, and ice cream. However, it is important to consume foods that provide the body with calcium. Many people with lactose intolerance seem to tolerate yogurt, which is a good source of calcium. Although there is lactose in yogurt, the bacteria in yogurt may produce some lactase, which helps to digest the lactose. A person who is lactose intolerant should also know that some foods that are not dairy products contain lactose. For example, baked goods, cereals, breakfast drinks, salad dressings, and even lunchmeat can contain lactose in their ingredients. You must read food labels carefully to see if the ingredients include "milk" or "lactose."

The enzyme lactase is now available in many forms such as tablets that are taken with meals, drops that are added to milk, or as additives in many dairy products such as milk. When lactase is added to milk that is left in the refrigerator for 24 h, the lactose level is reduced by 70 to 90%. Lactase pills or chewable tablets are taken when a person begins to eat a meal that contains dairy foods. If taken too far ahead of the meal, too much of the lactase will be degraded by stomach acid. If taken following a meal, the lactose will have already entered the lower intestine.

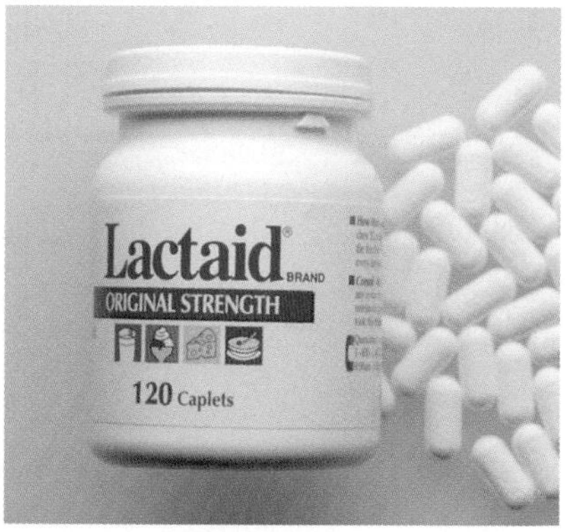

Lactaid contains an enzyme that aids the digestion of lactose.

SAMPLE PROBLEM 18.2 Digestion of Carbohydrates

Indicate the carbohydrates that undergo digestion in each of the following sites:

a. mouth **b.** stomach **c.** small intestine

SOLUTION

a. polysaccharides
b. no carbohydrate digestion
c. dextrins, maltose, sucrose, and lactose

STUDY CHECK 18.2

Describe the digestion of amylose, a polysaccharide.

Digestion of Fats

The digestion of dietary fats begins in the small intestine, when the hydrophobic fat globules mix with bile salts released from the gallbladder. In a process called *emulsification*, the bile salts break the fat globules into smaller droplets called *micelles*. Enzymes from the pancreas hydrolyze the triacylglycerols to yield monoacylglycerols and fatty acids, which are then absorbed into the intestinal lining where they recombine to form triacylglycerols. These nonpolar compounds are then coated with proteins to form lipoproteins called *chylomicrons*, which are more polar and soluble in the aqueous environment of the lymph and bloodstream (see Figure 18.4).

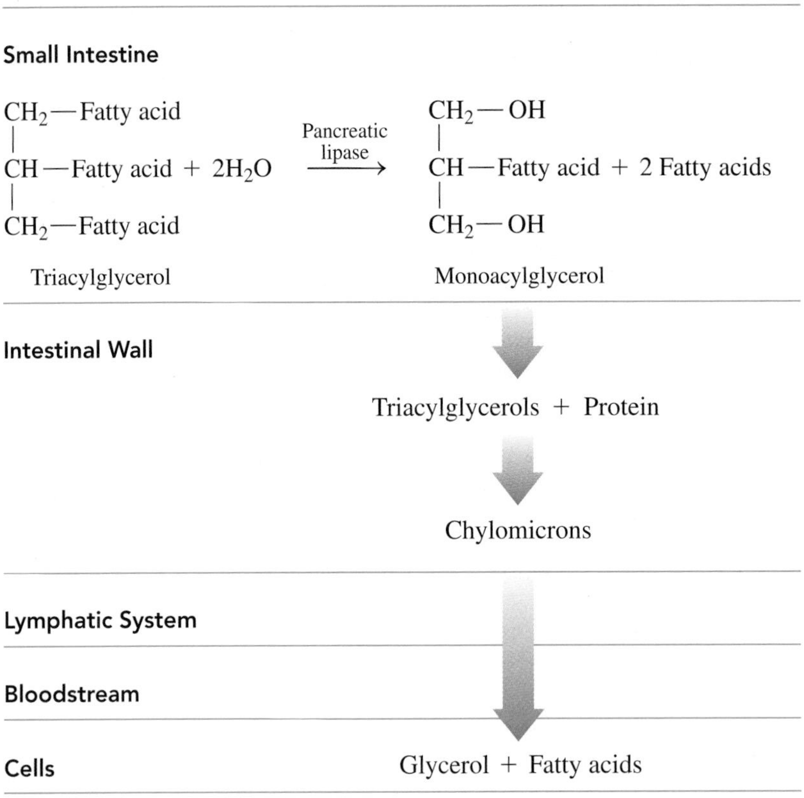

FIGURE 18.4 ▶ The triacylglycerols are hydrolyzed in the small intestine and re-formed in the intestinal wall where they bind to proteins for transport through the lymphatic system and bloodstream to the cells.

Q Why do chylomicrons form in the intestinal wall?

The chylomicrons transport the triacylglycerols to the cells of the heart, muscle, and adipose tissues. When energy is needed in the cells, enzymes hydrolyze the triacylglycerols to yield glycerol and fatty acids.

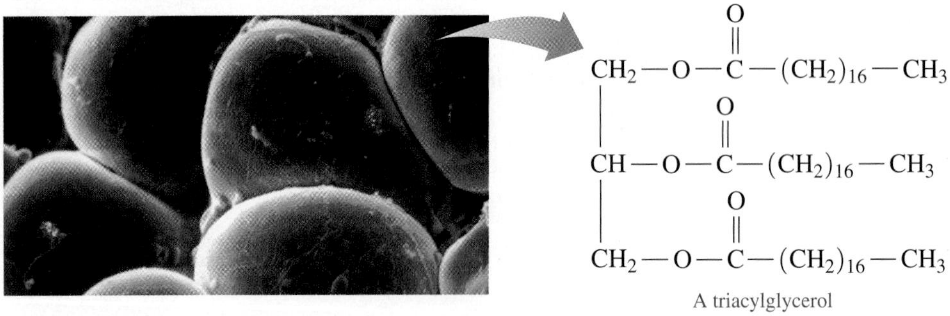

A triacylglycerol

The fat cells that make up adipose tissue are capable of storing unlimited quantities of triacylglycerols.

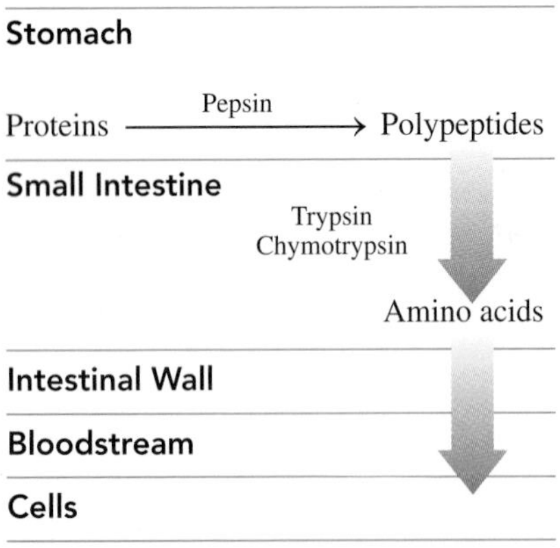

FIGURE 18.5 ▶ Proteins are hydrolyzed in the stomach and the small intestine.

Q What enzyme, secreted into the small intestine, hydrolyzes peptides?

Digestion of Proteins

The major role of proteins is to provide amino acids for the synthesis of new proteins for the body and nitrogen atoms for the synthesis of compounds such as nucleotides. In stage 1, the digestion of proteins begins in the stomach, where hydrochloric acid (HCl) at pH 2 denatures the proteins and activates enzymes such as pepsin that begin to hydrolyze peptide bonds. Polypeptides move out of the stomach into the small intestine, where trypsin and chymotrypsin complete the hydrolysis of the peptides to amino acids. The amino acids are absorbed through the intestinal walls into the bloodstream for transport to the cells (see Figure 18.5).

QUESTIONS AND PROBLEMS

18.2 Digestion of Foods

LEARNING GOAL Give the sites and products of digestion for carbohydrates, triacylglycerols, and proteins.

18.7 What is the general type of reaction that occurs during the digestion of carbohydrates?

18.8 What are the end products of the digestion of proteins?

18.9 What is the role of bile salts in lipid digestion?

18.10 How are insoluble triacylglycerols transported to the cells?

18.11 Where do dietary proteins undergo digestion in the body?

18.12 What is the purpose of digestion in stage 1?

18.3 Coenzymes in Metabolic Pathways

LEARNING GOAL

Describe the components and functions of the coenzymes NAD^+, FAD, and coenzyme A.

An *oxidation* reaction involves the loss of hydrogen or electrons by a substance, or an increase in the number of bonds to oxygen. *Reduction* is the gain of hydrogen ions and electrons or a decrease in the number of bonds to oxygen. When hydrogen ions and electrons are picked up by a coenzyme, it is reduced. Table 18.2 summarizes the characteristics of oxidation and reduction.

TABLE 18.2 Characteristics of Oxidation and Reduction in Metabolic Pathways

Oxidation	Reduction
Loss of electrons	Gain of electrons
Loss of hydrogen	Gain of hydrogen
Gain of oxygen	Loss of oxygen
Increase in number of bonds to oxygen	Decrease in number of bonds to oxygen

NAD^+

NAD^+ (nicotinamide adenine dinucleotide) is an important coenzyme in which the vitamin *niacin* provides the *nicotinamide* group, which is bonded to ribose and ADP (see Figure 18.6). The oxidized NAD^+ undergoes reduction when a carbon in the nicotinamide ring reacts with 2H (two hydrogen ions and two electrons), leaving one H^+.

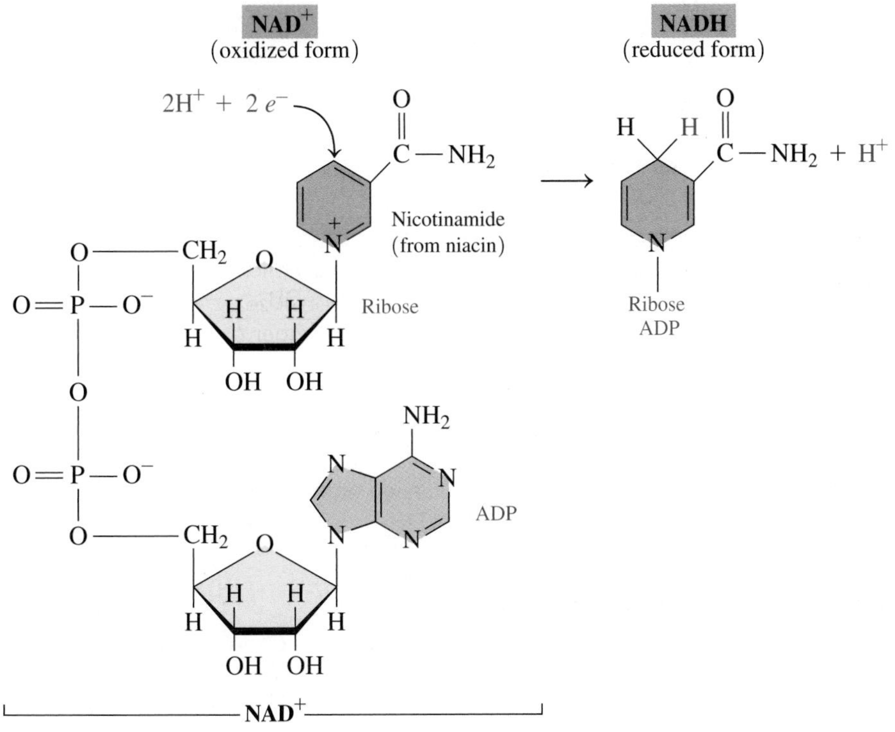

FIGURE 18.6 ▶ The coenzyme NAD^+ (nicotinamide adenine dinucleotide), which consists of a nicotinamide portion from the vitamin niacin, ribose, and adenosine diphosphate, is reduced to NADH + H^+.

Why is the conversion of NAD^+ to NADH and H^+ called a reduction?

The NAD^+ coenzyme is required for metabolic reactions that produce carbon–oxygen (C=O) double bonds, such as in the oxidation of alcohols to aldehydes and ketones. An example of an oxidation–reduction reaction that utilizes NAD^+ is the oxidation of ethanol in the liver to ethanal and NADH.

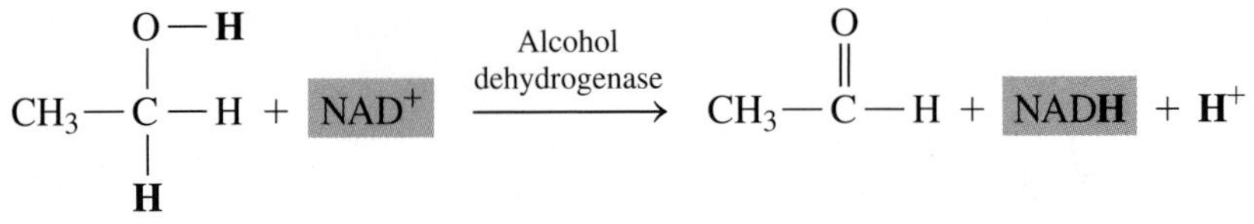

Ethanol (ethyl alcohol) — Ethanal (acetaldehyde)

FAD

FAD (flavin adenine dinucleotide) is a coenzyme that contains ADP and riboflavin. Riboflavin, also known as vitamin B_2, consists of ribitol (a sugar alcohol) and flavin. The oxidized form of FAD undergoes reduction when the two nitrogen atoms in the flavin part of the FAD coenzyme react with 2H (two hydrogen ions and two electrons) reducing FAD to $FADH_2$ (see Figure 18.7).

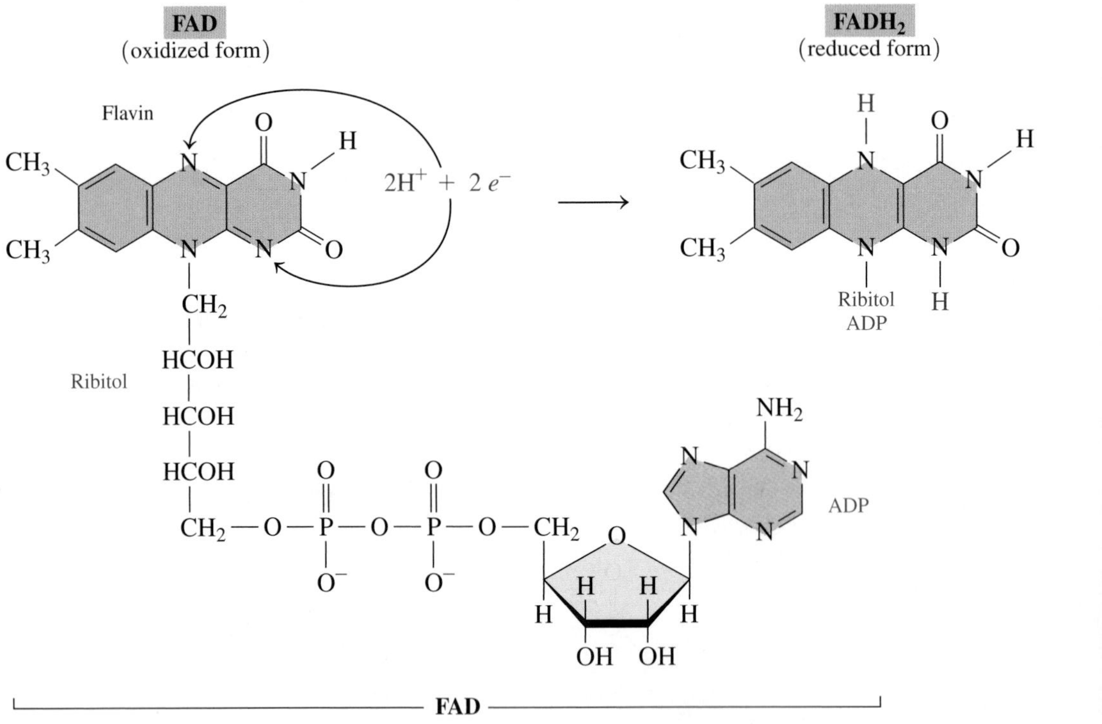

FIGURE 18.7 ▶ The coenzyme FAD (flavin adenine dinucleotide), made from riboflavin (vitamin B_2) and adenosine diphosphate, is reduced to $FADH_2$.

What is the type of reaction in which FAD accepts hydrogen?

FAD is used as a coenzyme when an oxidation reaction converts a carbon–carbon single bond to a carbon–carbon (C=C) double bond. An example of a reaction in the citric acid cycle that utilizes FAD is the conversion of the carbon–carbon single bond in succinate to a double bond in fumarate and $FADH_2$.

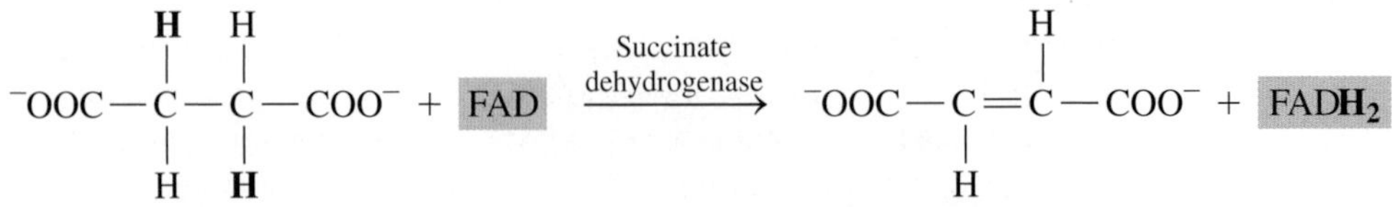

Succinate — Fumarate

Coenzyme A

Coenzyme A (CoA) is made up of several components: pantothenic acid (vitamin B_5), phosphorylated ADP, and aminoethanethiol (see Figure 18.8).

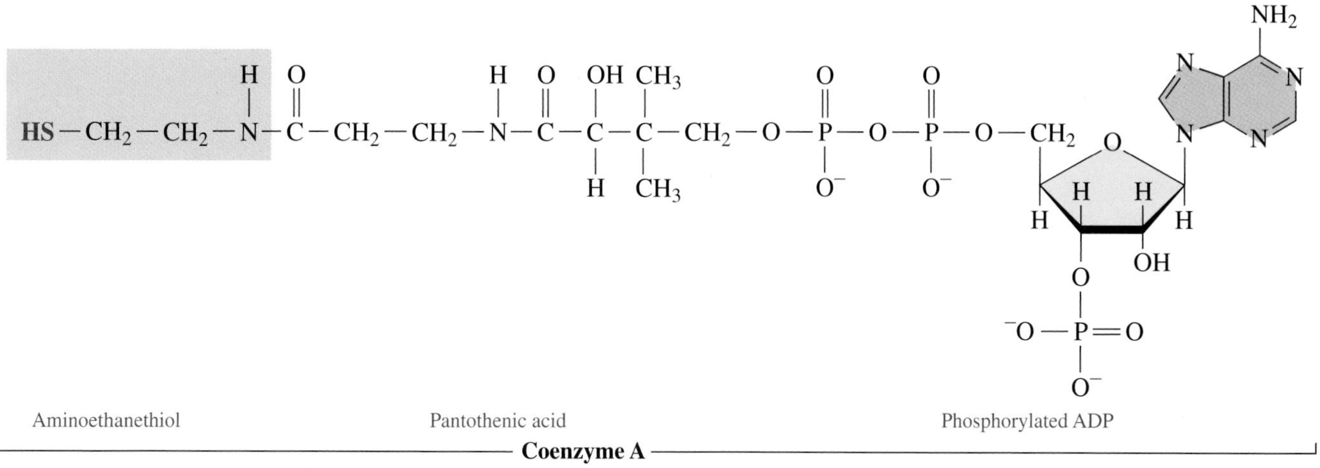

FIGURE 18.8 ▶ Coenzyme A is derived from a phosphorylated adenosine diphosphate (ADP) and pantothenic acid bonded by an amide bond to aminoethanethiol, which contains the —SH reactive part of the molecule.

Q What part of coenzyme A reacts with a two-carbon acetyl group?

An important function of coenzyme A is to prepare small acyl groups (represented by the letter A), such as acetyl, for reactions with enzymes. The reactive feature of coenzyme A is the thiol group (—SH), which bonds to a two-carbon acetyl group to produce the energy-rich thioester **acetyl-CoA**.

$$\underset{\text{Acetyl group}}{CH_3-\overset{\displaystyle O}{\overset{\|}{C}}-} + \underset{\text{Coenzyme A}}{HS-CoA} \longrightarrow \underset{\substack{\text{Acetyl-CoA}\\ \text{(thioester)}}}{CH_3-\overset{\displaystyle O}{\overset{\|}{C}}-S-CoA}$$

SAMPLE PROBLEM 18.3 Coenzymes

Describe the reactive part of each of the following coenzymes and the way each participates in metabolic pathways:

a. FAD **b.** NAD^+

SOLUTION

a. When two nitrogen atoms in the flavin accept 2H ($2H^+$ and $2\,e^-$), FAD is reduced to $FADH_2$. The FAD coenzyme participates in oxidation reactions that produce a carbon–carbon (C═C) double bond.

b. When a carbon atom in the nicotinamide accepts 2H ($2H^+$ and $2\,e^-$), NAD^+ is reduced to $NADH + H^+$. The NAD^+ coenzyme participates in oxidation reactions that produce a carbon–oxygen (C═O) double bond.

STUDY CHECK 18.3

Describe the reactive part of coenzyme A and the way it participates in metabolic pathways.

Types of Metabolic Reactions

Many reactions within the cells are similar to the types of reactions we looked at in organic chemistry such as hydration, dehydration, hydrogenation, oxidation, and reduction. Organic reactions typically require strong acids (low pH), high temperatures, and/or metallic catalysts. However, metabolic reactions take place at body temperature and physiological pH, which requires enzymes and often their coenzymes (see Table 18.3).

TABLE 18.3 Metabolic Reactions and Enzymes

Reaction	Enzyme	Coenzyme
Oxidation	Dehydrogenase	NAD^+, FAD
Reduction	Dehydrogenase	NADH + H^+, $FADH_2$
Hydration	Hydrase	
Dehydration	Dehydrase	
Rearrangement	Isomerase	
Transfer of phosphate	Transferase, kinase	ATP, GDP, ADP
Transfer of acetyl group	Acetyl-CoA transferase	CoA
Decarboxylation	Decarboxylase	
Hydrolysis	Hydrolase, protease, lipase	

QUESTIONS AND PROBLEMS

18.3 Coenzymes in Metabolic Pathways

LEARNING GOAL Describe the components and functions of the coenzymes NAD^+, FAD, and coenzyme A.

18.13 Give the abbreviation for each of the following coenzymes:

- **a.** reduced form of NAD^+
- **b.** oxidized form of $FADH_2$
- **c.** participates in the formation of a carbon–carbon double bond

18.14 Give the abbreviation for each of the following coenzymes:

- **a.** reduced form of FAD
- **b.** oxidized form of NADH
- **c.** participates in the formation of a carbon–oxygen double bond

18.15 Name the enzyme and coenzyme involved in the following reactions:

- **a.** transfer of phosphate
- **b.** decarboxylation
- **c.** transfer of acetyl group

18.16 What is the main function of coenzyme A? What is its functional group?

LEARNING GOAL

Describe the conversion of glucose to pyruvate in glycolysis and the subsequent conversion of pyruvate to acetyl-CoA or lactate.

18.4 Glycolysis: Oxidation of Glucose

The major source of energy for the body is the glucose produced when we digest the carbohydrates in our food, or from glycogen, a polysaccharide stored in the liver and skeletal muscle. Glucose in the bloodstream enters our cells where it undergoes degradation in a pathway called *glycolysis*. Early organisms used glycolysis to produce energy from simple nutrients long before there was any oxygen in Earth's atmosphere. Glycolysis is an **anaerobic** process; no oxygen is required.

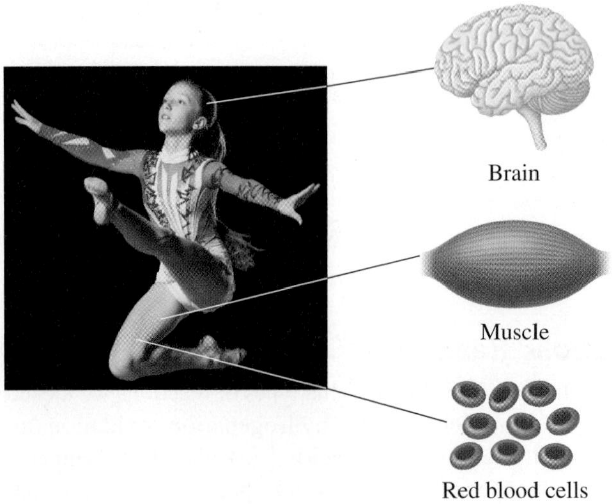

Glucose is the main energy source for the brain, skeletal muscles, and red blood cells.

In **glycolysis**, a six-carbon glucose molecule is broken down to two molecules of three-carbon pyruvate (see Figure 18.9). All the reactions in glycolysis take place in the cytoplasm of the cell. In the first five reactions (1 to 5), called the *energy-investing phase*, energy is obtained from the hydrolysis of two ATP, which is needed to form sugar phosphates. In reactions 4 and 5, a six-carbon sugar phosphate is split to yield two molecules of three-carbon sugar phosphate. In the last five reactions (6 to 10), called the *energy-generating phase*, energy is obtained from the hydrolysis of the energy-rich phosphate compounds and used to synthesize four ATP.

CORE CHEMISTRY SKILL

Identifying the Compounds in Glycolysis

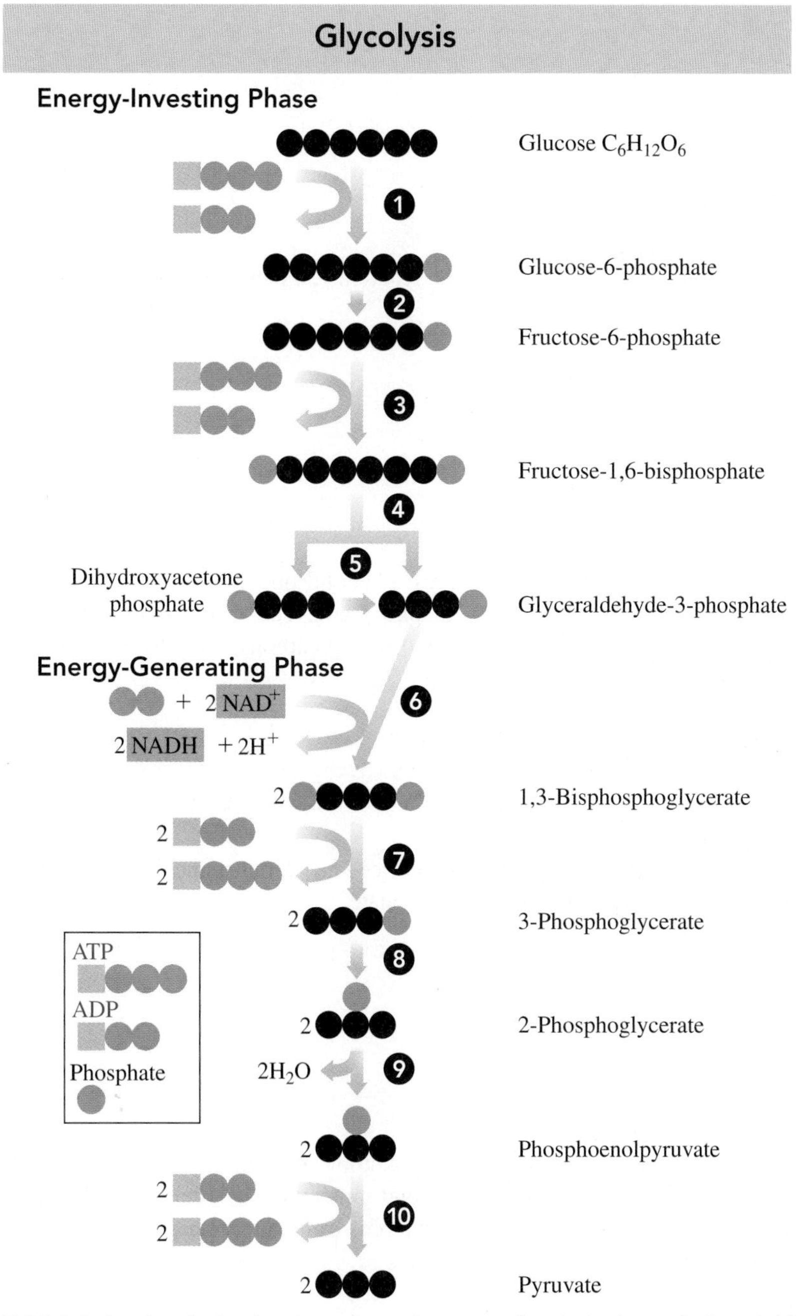

FIGURE 18.9 ▶ In glycolysis, the six-carbon glucose molecule is degraded to yield two three-carbon pyruvate molecules. A net of two ATP is produced along with two NADH.

Q Where in the glycolysis pathway is glucose cleaved to yield two three-carbon compounds?

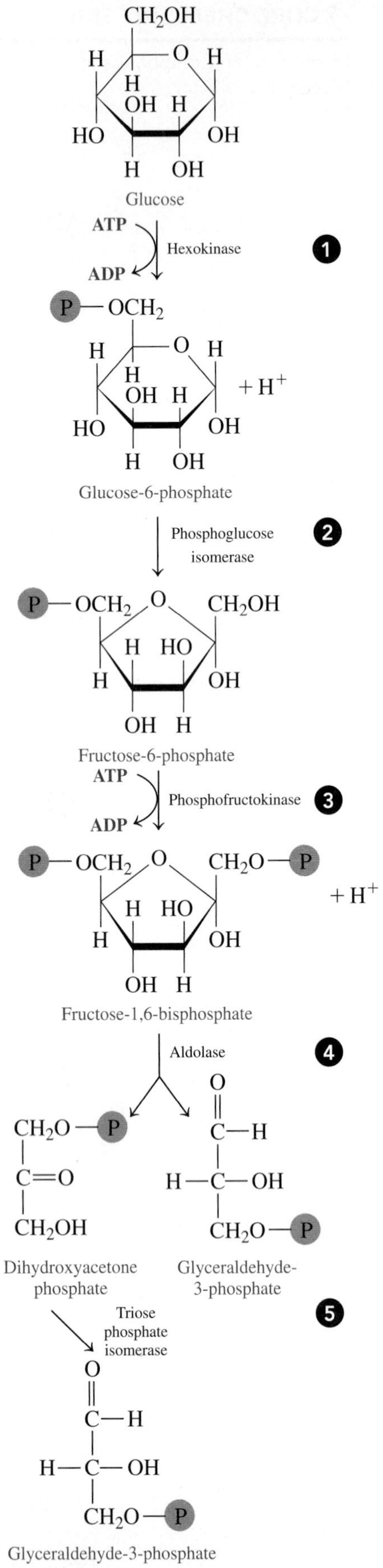

Energy-Investing Reactions 1 to 5

Reaction 1 Phosphorylation

In the initial reaction, a phosphate group from ATP is added to glucose to form glucose-6-phosphate and ADP.

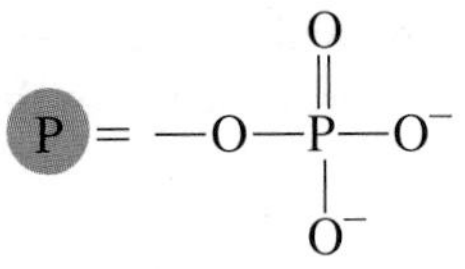

Reaction 2 Isomerization

The glucose-6-phosphate, the aldose from reaction 1, undergoes isomerization to fructose-6-phosphate, which is a ketose.

Reaction 3 Phosphorylation

The hydrolysis of another ATP provides a second phosphate group, which converts fructose-6-phosphate to fructose-1,6-bisphosphate. The word *bisphosphate* is used to show that the two phosphate groups are on different carbons in fructose and not connected to each other.

Reaction 4 Cleavage

Fructose-1,6-bisphosphate is split into two three-carbon phosphate isomers: dihydroxyacetone phosphate and glyceraldehyde-3-phosphate.

Reaction 5 Isomerization

Because dihydroxyacetone phosphate is a ketone, it cannot react further. However, it undergoes isomerization to provide a second molecule of glyceraldehyde-3-phosphate, which can be oxidized. Now all six carbon atoms from glucose are contained in two identical triose phosphates.

Energy-Generating Reactions 6 to 10

In our discussion of glycolysis from this point, the two molecules of glyceraldehyde-3-phosphate produced in step 5 are undergoing the same reactions. For simplicity, we show the structures and reactions for only one three-carbon molecule.

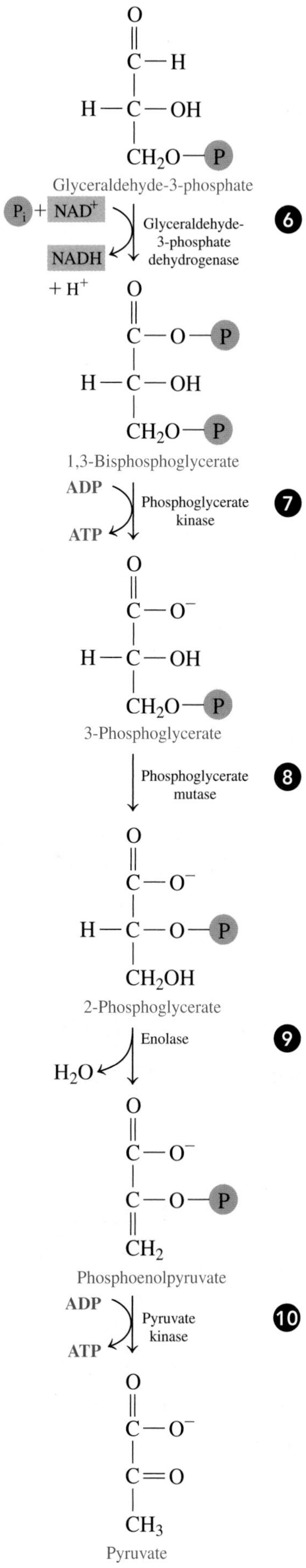

Reaction 6 Oxidation and Phosphorylation

The aldehyde group of each glyceraldehyde-3-phosphate is oxidized to a carboxyl group by the coenzyme NAD^+, which is reduced to NADH and H^+. A phosphate group adds to each of the new carboxyl groups to form two molecules of the high-energy compound, 1,3-bisphosphoglycerate.

Reaction 7 Phosphate Transfer

A phosphorylation transfers a phosphate group from each 1,3-bisphosphoglycerate to ADP to produce two molecules of the high-energy compound ATP. At this point in glycolysis, two ATP are produced, which balance the two ATP consumed in reactions 1 and 3.

Reaction 8 Isomerization

Two 3-phosphoglycerate molecules undergo isomerization, which moves the phosphate group from carbon 3 to carbon 2 yielding two molecules of 2-phosphoglycerate.

Reaction 9 Dehydration

Each of the phosphoglycerate molecules undergoes dehydration (loss of water) to give two high-energy molecules of phosphoenolpyruvate.

Reaction 10 Phosphate Transfer

In a second direct phosphorylation, phosphate groups from two phosphoenolpyruvate are transferred to two ADP to form two pyruvate and two ATP.

SAMPLE PROBLEM 18.4 Reactions in Glycolysis

Identify each of the following reactions as an isomerization, phosphorylation, dehydration, or cleavage:

a. a phosphate group is transferred to ADP to form ATP
b. 3-phosphoglycerate is converted to 2-phosphoglycerate
c. water is lost from 2-phosphoglycerate

SOLUTION

a. The transfer of a phosphate group to ADP to form ATP is phosphorylation.
b. The change in location of a phosphate group on a carbon chain is isomerization.
c. The loss of water is dehydration.

STUDY CHECK 18.4

Identify the reaction in which fructose-1,6-bisphosphate splits to form two three-carbon compounds as an isomerization, phosphorylation, dehydration, or cleavage.

Summary of Glycolysis

In the glycolysis pathway, a six-carbon glucose molecule is converted to two three-carbon pyruvate molecules. Initially, two ATP are required to form fructose-1,6-bisphosphate. In later reactions, phosphate transfers produce a total of four ATP. Overall, glycolysis yields two ATP and two NADH when a glucose molecule is converted to two pyruvate.

$$\underset{\text{Glucose}}{C_6H_{12}O_6} + 2\,NAD^+ \xrightarrow{2ADP + 2P_i \;\to\; 2ATP} \underset{\text{Pyruvate}}{2CH_3{-}\overset{\overset{\displaystyle O}{\|}}{C}{-}COO^-} + 2\,NADH + 4H^+$$

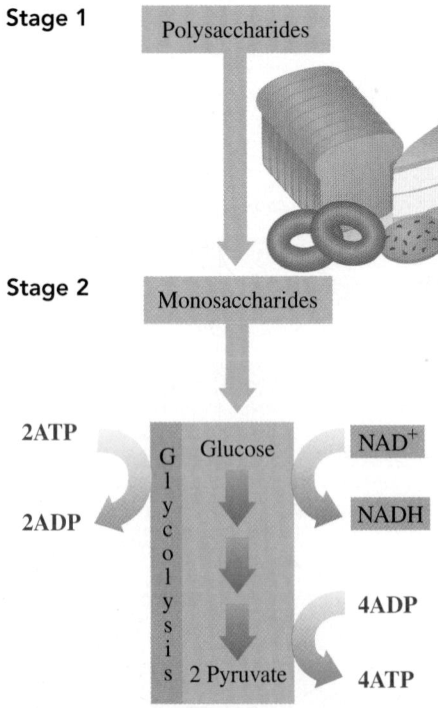

Glucose obtained from the digestion of polysaccharides is degraded in glycolysis to pyruvate.

Pathways for Pyruvate

The pyruvate produced from glucose can now enter pathways that continue to extract energy. The available pathway depends on whether there is sufficient oxygen in the cell. Under **aerobic** conditions, oxygen is available to convert pyruvate to acetyl-coenzyme A (acetyl-CoA). When oxygen levels are low, pyruvate is reduced to lactate.

Aerobic Conditions

In glycolysis, two ATP were generated when one glucose molecule was converted to two pyruvate. However, much more energy is obtained from glucose when oxygen levels are high in the cells. Under aerobic conditions, pyruvate moves from the cytoplasm into the mitochondria to be oxidized further. In a complex reaction, pyruvate is oxidized, and a carbon atom is removed from pyruvate as CO_2. The coenzyme NAD^+ is reduced during the oxidation. The resulting two-carbon acetyl compound is attached to CoA, producing acetyl-CoA, an important intermediate in many metabolic pathways (see Figure 18.10).

$$\underset{\text{Pyruvate}}{CH_3{-}\overset{\overset{\displaystyle O}{\|}}{C}{-}\overset{\overset{\displaystyle O}{\|}}{C}{-}O^-} + HS{-}CoA + NAD^+ \xrightarrow{\text{Pyruvate dehydrogenase}}$$

$$\underset{\text{Acetyl-CoA}}{CH_3{-}\overset{\overset{\displaystyle O}{\|}}{C}{-}S{-}CoA} + CO_2 + NADH$$

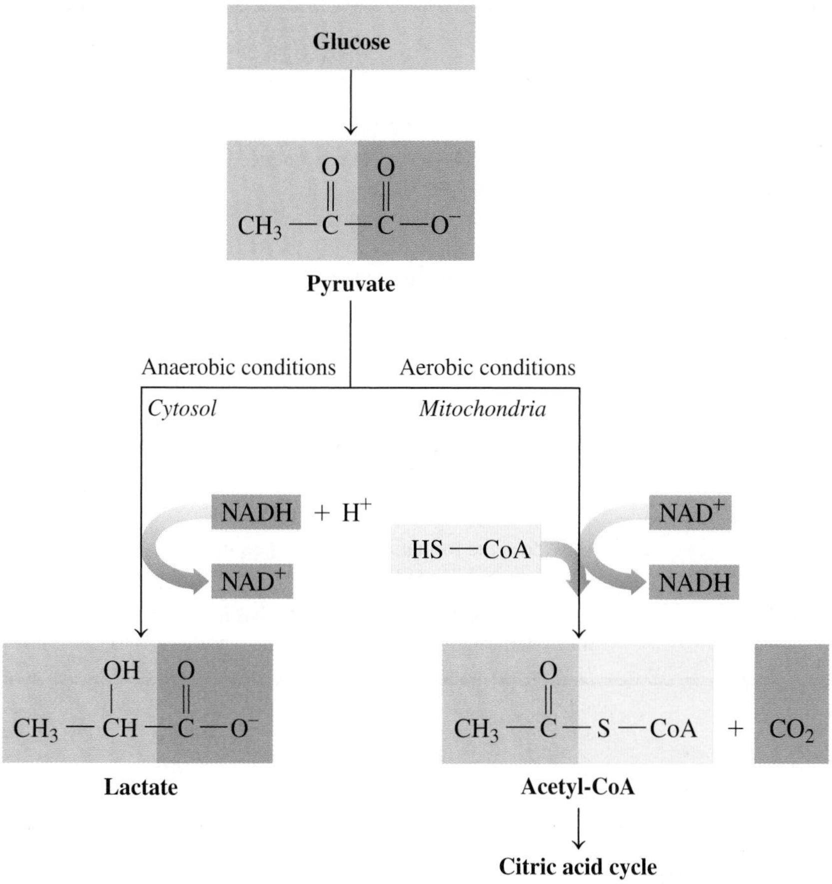

FIGURE 18.10 ▶ Pyruvate is converted to acetyl-CoA under aerobic conditions and to lactate under anaerobic conditions.

Q During vigorous exercise, why does lactate accumulate in the muscles?

Anaerobic Conditions

When we engage in strenuous exercise, the oxygen stored in our muscle cells is quickly depleted. Under anaerobic conditions, pyruvate remains in the cytoplasm where it is reduced to lactate. NAD^+ is produced and is used to oxidize more glyceraldehyde-3-phosphate in the glycolysis pathway, which produces a small but needed amount of ATP.

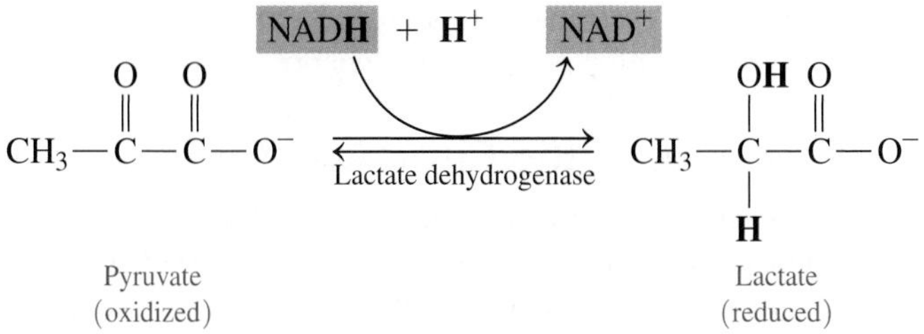

The accumulation of lactate causes the muscles to tire and become sore. After exercise, a person continues to breathe rapidly to repay the *oxygen debt* incurred during exercise. Most of the lactate is transported to the liver, where it is converted back into pyruvate. Under anaerobic conditions, the only ATP production in glycolysis occurs during the steps that phosphorylate ADP directly, giving a net gain of only two ATP.

$$\underset{\text{Glucose}}{C_6H_{12}O_6} + 2ADP + 2P_i \longrightarrow \underset{\text{Lactate}}{2CH_3-CH(OH)-COO^-} + 2ATP$$

After vigorous exercise, rapid breathing helps to repay the oxygen debt.

Bacteria also convert pyruvate to lactate under anaerobic conditions. In the preparation of kimchee and sauerkraut, cabbage is covered with salt brine. The glucose obtained from the starches in the cabbage is converted to lactate. This acid environment acts as a preservative that prevents the growth of other bacteria. The pickling of olives and cucumbers gives similar products. When cultures of bacteria that produce lactate are added to milk, the acid denatures the milk proteins to give sour cream and yogurt.

QUESTIONS AND PROBLEMS

18.4 Glycolysis: Oxidation of Glucose

LEARNING GOAL Describe the conversion of glucose to pyruvate in glycolysis and the subsequent conversion of pyruvate to acetyl-CoA or lactate.

18.17 What is the starting compound of glycolysis?

18.18 What is the three-carbon end product of glycolysis?

18.19 How is ATP used in the initial steps of glycolysis?

18.20 How many ATP are used in the initial steps of glycolysis?

18.21 How does phosphorylation account for the production of ATP in glycolysis?

18.22 Why are there two ATP formed for one molecule of glucose?

18.23 How many ATP or NADH are produced (or required) in each of the following steps in glycolysis?
a. glucose to glucose-6-phosphate
b. glyceraldehyde-3-phosphate to 1,3-bisphosphoglycerate
c. glucose to pyruvate

18.24 How many ATP or NADH are produced (or required) in each of the following steps in glycolysis?
a. 1,3-bisphosphoglycerate to 3-phosphoglycerate
b. fructose-6-phosphate to fructose-1,6-bisphosphate
c. phosphoenolpyruvate to pyruvate

18.25 What condition is needed in the cell to convert pyruvate to acetyl-CoA?

18.26 What coenzymes are needed for the oxidation of pyruvate to acetyl-CoA?

18.27 Write the overall equation for the conversion of pyruvate to acetyl-CoA.

18.28 What is the product of pyruvate under anaerobic conditions?

18.29 How does the formation of lactate permit glycolysis to continue under anaerobic conditions?

18.30 After running a marathon, a runner has muscle pain and cramping. What might have occurred in the muscle cells to cause this?

LEARNING GOAL

Describe the oxidation of acetyl-CoA in the citric acid cycle.

18.5 The Citric Acid Cycle

The citric acid cycle is a series of reactions that connects the intermediate acetyl-CoA from the metabolic pathways in stages 1 and 2 with electron transport and the synthesis of ATP in stage 3. As a central pathway in metabolism, the **citric acid cycle** uses the two-carbon acetyl group in acetyl-CoA to produce CO_2, NADH + H^+, and $FADH_2$. The citric acid cycle is named for the citrate ion from citric acid ($C_6H_8O_7$), a tricarboxylic acid, which forms in the first reaction. The citric acid cycle is also known as the *tricarboxylic acid (TCA) cycle* or the *Krebs cycle*, named for H. A. Krebs, who recognized it in 1937 as the major metabolic pathway for the production of energy.

Overview of the Citric Acid Cycle

There are a total of eight reactions and eight enzymes in the citric acid cycle. Initially, an acetyl group (2C) from acetyl-CoA bonds with oxaloacetate (4C) to yield citrate (6C) (see Figure 18.11). Then, two decarboxylation reactions remove carbon atoms as CO_2 molecules to give succinyl-CoA (4C). Finally, a series of reactions converts four-carbon succinyl-CoA to oxaloacetate, which combines with another acetyl-CoA, and the citric cycle starts all over. In one turn of the citric acid cycle, four oxidation reactions provide hydrogen ions and electrons, which are used to reduce FAD and NAD^+ coenzymes (see Figure 18.12).

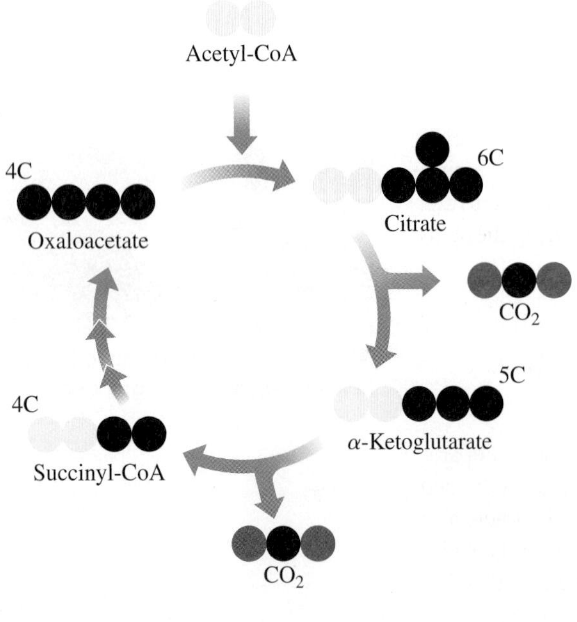

FIGURE 18.11 ▶ In the citric acid cycle, two carbon atoms are removed as CO_2 from six-carbon citrate to give four-carbon succinyl-CoA, which is converted to four-carbon oxaloacetate.

Q How many carbon atoms are removed in one turn of the citric acid cycle?

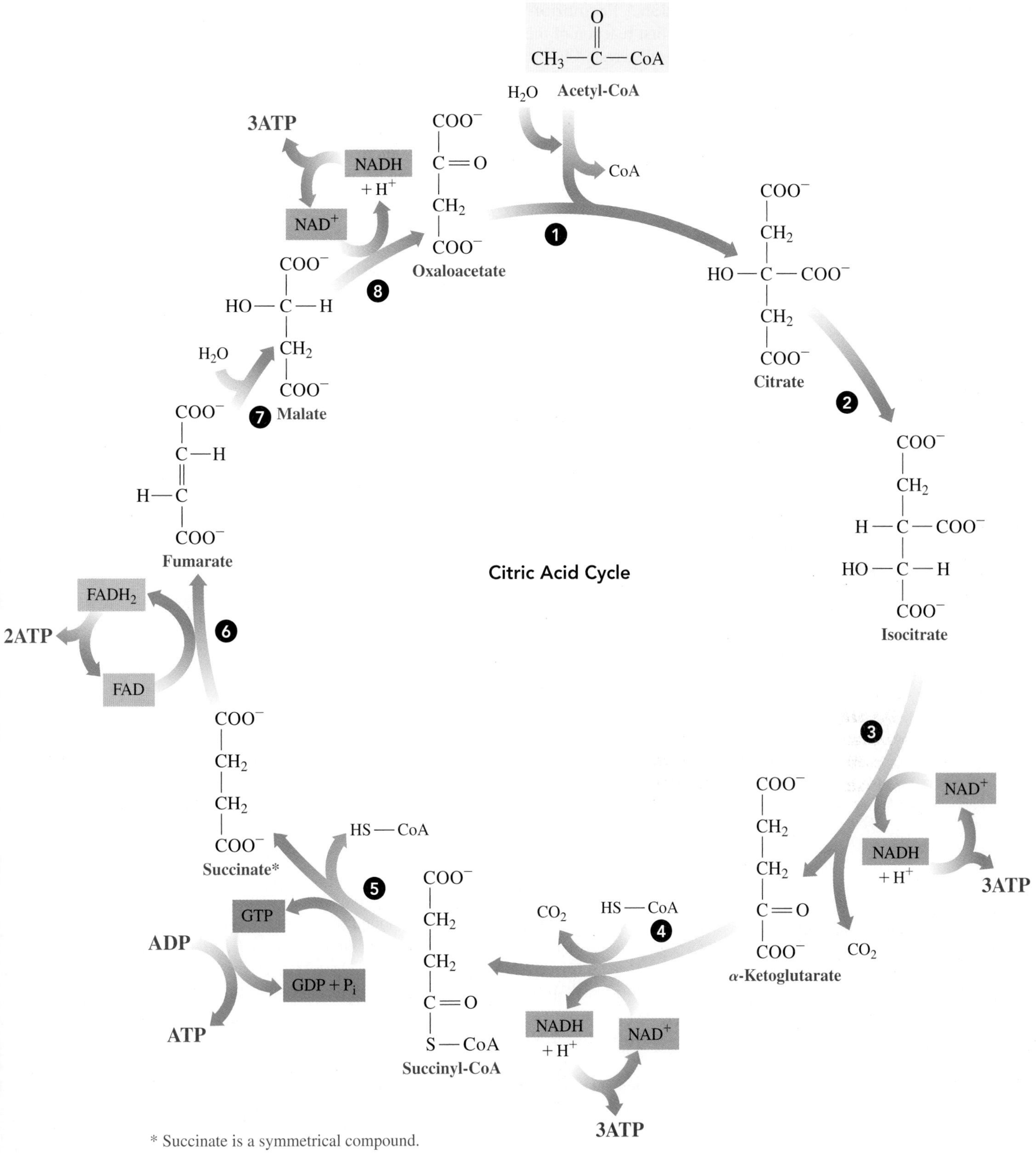

FIGURE 18.12 ▶ In the citric acid cycle, oxidation reactions produce two CO_2 and reduced coenzymes NADH and $FADH_2$, and it also regenerates oxaloacetate.

Q How many reactions in the citric acid cycle produce a reduced coenzyme?

CORE CHEMISTRY SKILL

Describing the Reactions in the Citric Acid Cycle

Reaction 1 Formation of Citrate

In the first reaction of the citric acid cycle, the acetyl group (2C) from acetyl-CoA bonds with oxaloacetate (4C) to yield citrate (6C).

Reaction 2 Isomerization

The citrate produced in reaction 1 contains a tertiary alcohol group that cannot be oxidized further. In reaction 2, citrate undergoes isomerization to yield its isomer isocitrate, which provides a secondary alcohol group that can be oxidized in the next reaction.

Reaction 3 Oxidation and Decarboxylation

In reaction 3, an oxidation and a *decarboxylation* occur together. The secondary alcohol group in isocitrate is oxidized to a ketone. A **decarboxylation** converts a carboxylate group ($—COO^-$) to a CO_2 molecule producing α-ketoglutarate, which has five carbon atoms. The oxidation reaction also produces hydrogen ions and electrons that reduce NAD^+ to NADH and H^+. This reduced coenzyme NADH will be important in the energy-producing reactions we will discuss in electron transport.

Reaction 4 Oxidation and Decarboxylation

In reaction 4, α-ketoglutarate undergoes oxidation and decarboxylation to produce a four-carbon group that combines with CoA to form succinyl-CoA (4C). As in reaction 3, this oxidation reaction also produces hydrogen ions and electrons that reduce NAD^+ to NADH and H^+. This forms another reduced NADH that will be important in the energy-producing reactions we will discuss in electron transport.

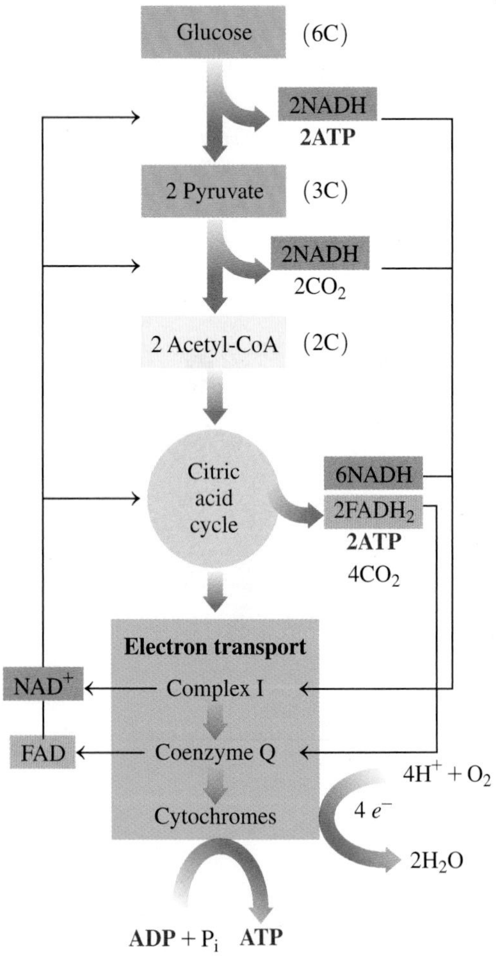

The reduced coenzymes NADH and $FADH_2$ that provide hydrogen ions and electrons in electron transport are regenerated as NAD^+ and FAD.

Reaction 5 Hydrolysis

In reaction 5, succinyl-CoA undergoes hydrolysis to succinate and CoA. The energy released is used to add a phosphate group (P_i) to GDP (guanosine diphosphate), which yields GTP (guanosine triphosphate), a high-energy compound similar to ATP.

Eventually, the GTP undergoes hydrolysis with a release of energy that is used to add a phosphate group to ADP to form ATP. This is the only time in the citric acid cycle that ATP is produced by a direct transfer of phosphate. The reaction between GTP and ATP regenerates GDP that can be used again in the citric acid cycle.

$$GTP + ADP \longrightarrow GDP + ATP$$

Reaction 6 Oxidation

In reaction 6, hydrogen is removed from each of two carbon atoms in succinate, which produces fumarate, a compound with a trans double bond. The formation of a carbon–carbon (C═C) double bond produces 2H that are used to reduce the coenzyme FAD to $FADH_2$. This reduced coenzyme $FADH_2$ is important in the energy-producing reactions we will discuss in electron transport.

Reaction 7 Hydration

In reaction 7, a hydration adds water to the double bond of fumarate to yield malate, which is a secondary alcohol.

Reaction 8 Oxidation

In reaction 8, the last step of the citric acid cycle, the secondary alcohol group in malate is oxidized to oxaloacetate, which has a ketone group. For the third time in the citric acid cycle, an oxidation provides hydrogen ions and electrons for the reduction of NAD^+ to NADH and H^+.

Summary of Products from the Citric Acid Cycle

We have seen that the citric acid cycle begins when a two-carbon acetyl group from acetyl-CoA combines with oxaloacetate to form citrate. Through oxidation and reduction, two carbon atoms are removed from citrate to yield two CO_2 and a four-carbon compound that undergoes reactions to regenerate oxaloacetate.

In the four oxidation reactions of one turn of the citric acid cycle, three NAD^+ and one FAD are reduced to three NADH and one $FADH_2$. One GDP is converted to GTP, which is used to convert one ADP to ATP. We can write an overall chemical equation for one complete turn of the citric acid cycle as follows:

Products from One Turn of Citric Acid Cycle
2 CO_2
3 NADH and $3H^+$
1 $FADH_2$
1 GTP (1 ATP)
1 CoA

$$\text{Acetyl-CoA} + 3NAD^+ + FAD + GDP + P_i + 2H_2O \longrightarrow$$
$$2CO_2 + 3NADH + 3H^+ + FADH_2 + HS{-}CoA + GTP$$

SAMPLE PROBLEM 18.5 Citric Acid Cycle

When one acetyl-CoA completes the citric acid cycle, how many of each of the following is produced?

a. NADH **b.** ketone group **c.** CO_2

SOLUTION

a. One turn of the citric acid cycle produces three molecules of NADH.
b. Two ketone groups form when the secondary alcohol groups in isocitrate and malate are oxidized by NAD^+.
c. Two molecules of CO_2 are produced by the decarboxylation of isocitrate and α-ketoglutarate.

STUDY CHECK 18.5

What compound is a substrate in the first reaction of the citric acid cycle and a product in the last reaction?

QUESTIONS AND PROBLEMS

18.5 The Citric Acid Cycle

LEARNING GOAL Describe the oxidation of acetyl-CoA in the citric acid cycle.

18.31 What are the products from one turn of the citric acid cycle?

18.32 Write an equation to show how oxaloacetate is converted to citrate.

18.33 Write a balanced equation to show how 4, α-ketoglutarate is converted to succinyl-CoA.

18.34 Which reactions of the citric acid cycle involve a hydration reaction?

18.35 Which reactions of the citric acid cycle reduce NAD^+?

18.36 Which reactions of the citric acid cycle reduce FAD?

18.37 Which reaction(s) of the citric acid cycle involve(s) a direct phosphate transfer?

18.38 What is the total NADH and total $FADH_2$ produced in one turn of the citric acid cycle?

18.39 Refer to the diagram of the citric acid cycle in Figure 18.12 to answer each of the following:
a. What are the six-carbon compounds?
b. How is the number of carbon atoms decreased?
c. What is the five-carbon compound?
d. Which reactions are oxidation reactions?

18.40 Refer to the diagram of the citric acid cycle in Figure 18.12 to answer each of the following:
a. What is the yield of CO_2?
b. What are the four-carbon compounds?
c. What is the yield of GTP?
d. In which reactions are secondary alcohols oxidized?

18.6 Electron Transport and Oxidative Phosphorylation

LEARNING GOAL

Describe electron transport and the process of oxidative phosphorylation; calculate the ATP from the complete oxidation of glucose.

At this point in stage 3, for each glucose molecule that completes glycolysis, the oxidation of two pyruvate, and the citric acid cycle, four ATP along with ten NADH and two $FADH_2$ are produced.

From One Glucose	ATP	Reduced Coenzymes	
Glycolysis	2	2 NADH	
Oxidation of 2 Pyruvate		2 NADH	
Citric Acid Cycle with 2 Acetyl-CoA	2	6 NADH	2 $FADH_2$
Total for One Glucose	**4**	**10 NADH**	**2 $FADH_2$**

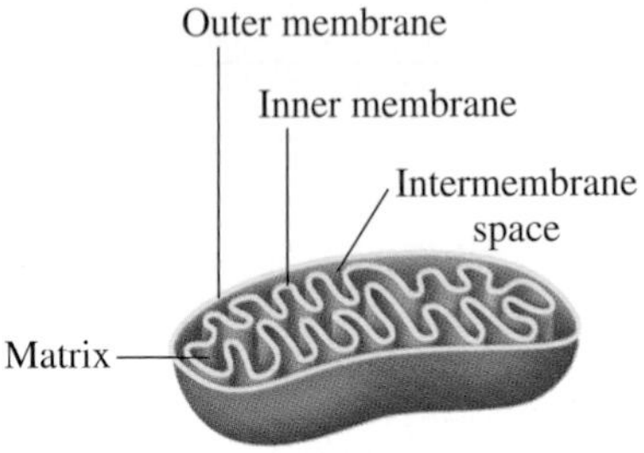

Electron transport and oxidative phosphorylation take place in the mitochondria.

Electron Transport

In **electron transport**, or the *respiratory chain*, hydrogen ions and electrons from NADH and $FADH_2$ are passed from one electron carrier to the next until they combine with oxygen to form H_2O. The energy released during electron transport is used to synthesize ATP from ADP and P_i, a process called *oxidative phosphorylation*.

As long as oxygen is available for the mitochondria in the cell, electron transport and oxidative phosphorylation function to produce most of the ATP energy manufactured in the cell. A mitochondrion consists of an outer membrane, an intermembrane space, and an inner membrane that surrounds the matrix. Along the highly folded inner membrane are the enzymes and electron carriers required for electron transport. Within these membranes are four distinct protein complexes, complex I, II, III, and IV. Two electron carriers, coenzyme Q and cytochrome *c*, are not firmly attached to the membrane. They function as mobile carriers shuttling electrons between the protein complexes that are bound to the inner membrane (see Figure 18.13).

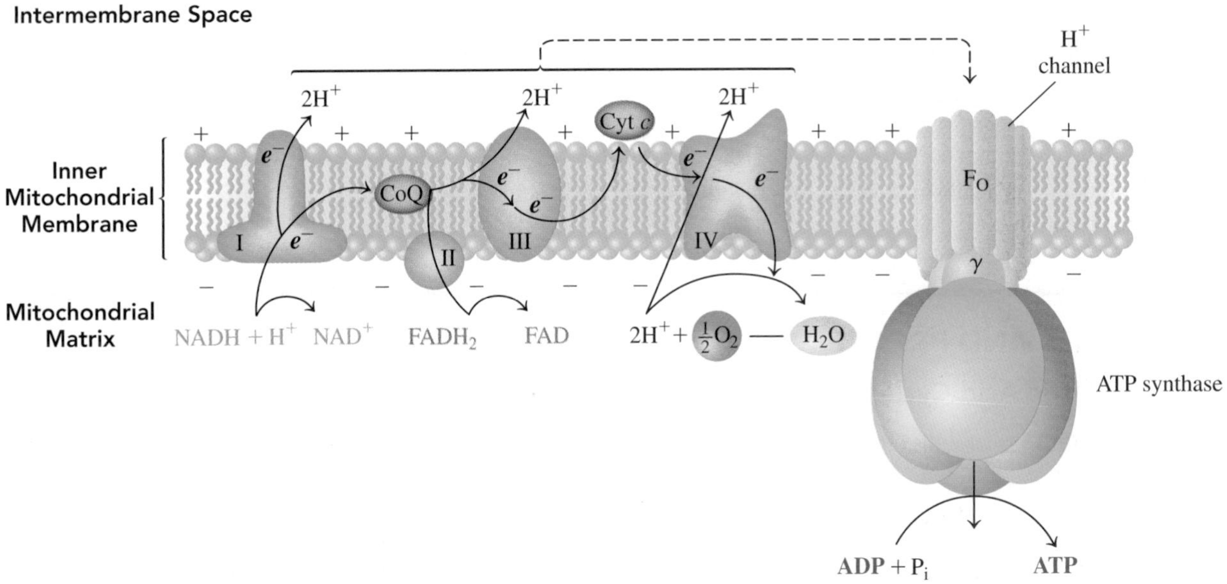

FIGURE 18.13 ▶ In electron transport, coenzymes NADH and $FADH_2$ are oxidized in enzyme complexes providing electrons and hydrogen ions for ATP synthesis.

Q What is the major source of NADH for electron transport?

Complex I

Electron transport begins when NADH transfers hydrogen ions and electrons to complex I and forms the oxidized coenzyme NAD^+. The hydrogen ions and electrons are transferred to the mobile electron carrier coenzyme Q (CoQ), which carries electrons to complex II. The overall reaction in complex I is written as follows:

$$\text{NADH} + \text{H}^+ + \text{Q} \longrightarrow \text{NAD}^+ + \text{QH}_2$$

Complex II

In complex II, Q also obtains hydrogen ions and electrons from $FADH_2$, generated by the conversion of succinate to fumarate in the citric acid cycle, which yields QH_2 and the oxidized coenzyme FAD.

Because complex II is at a lower energy level than complex I, the electrons from $FADH_2$ enter electron transport at a lower energy level than those from NADH. The overall reaction in complex II is written as follows:

$$FAD\mathbf{H_2} + Q \longrightarrow FAD + \mathbf{QH_2}$$

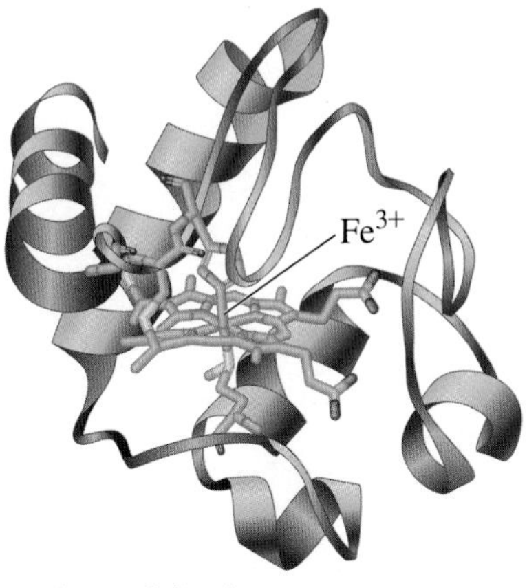

In the mobile electron carrier cytochrome *c*, covalent bonds between cysteine (yellow) groups form between the heme (grey) portion and a protein (green) component. Oxygen atoms are shown in red.

Complex III

In complex III, two electrons are transferred from the mobile carrier QH_2 to a series of iron-containing proteins called *cytochromes* and eventually to cytochrome *c*, which is another mobile electron carrier. Cytochrome *c*, which contains Fe^{3+}/Fe^{2+}, is reduced when it gains an electron and oxidized when an electron is lost. As a mobile carrier, cytochrome *c* carries electrons to complex IV. The overall reaction in complex III is written as follows:

$$\mathbf{QH_2} + 2\ \text{cyt}\ c\ (Fe^{3+}) \longrightarrow Q + \mathbf{2H^+} + 2\ \text{cyt}\ c\ (Fe^{2+})$$

(Oxidized) (Reduced)

Complex IV

At complex IV, four electrons from four cytochrome *c* are passed to other electron carriers until the electrons combine with hydrogen ions and oxygen (O_2) to form two molecules of water. The overall reaction in complex IV is written as follows:

$$4\,e^- + 4H^+ + O_2 \longrightarrow 2H_2O$$

This equation may also be written as:

$$2\,e^- + 2H^+ + \tfrac{1}{2}O_2 \longrightarrow H_2O$$

Overall, the reduced coenzymes NADH and $FADH_2$ from the citric acid cycle enter electron transport to provide hydrogen ions and electrons that react with oxygen, producing water and the oxidized coenzymes NAD^+ and FAD.

Oxidative Phosphorylation

In 1978, Peter Mitchell received the Nobel Prize in Chemistry for his theory called the **chemiosmotic model**, which links the energy from electron transport to a H^+ gradient that drives the synthesis of ATP. Three of the complexes (I, III, and IV) extend through the inner mitochondrial membrane with one end of each complex in the matrix and the other end in the intermembrane space. In the chemiosmotic model, each complex acts as a **proton pump** by pushing H^+ ions (protons) generated from the oxidation of NADH and $FADH_2$ out of the matrix and into the intermembrane space. The increase in H^+ concentration in the intermembrane space lowers the pH and creates a H^+ gradient. Because H^+ ions are positively charged, the lower pH and the electrical charge of the H^+ gradient make it an *electrochemical gradient* (see Figure 18.13).

To equalize the pH and the electrical charge between the intermembrane space and the matrix, the H^+ ions need to return to the matrix. However, H^+ ions cannot move through the inner membrane. The only way H^+ ions can return to the matrix is to pass through a protein complex called **ATP synthase**. We might think of the flow of H^+ ions as a river that turns a waterwheel. As the H^+ ions flow through ATP synthase, energy is generated that is used to convert ADP to ATP. This process of **oxidative phosphorylation** couples the energy from electron transport to the synthesis of ATP from ADP and P_i.

$$ADP + P_i + \text{energy} \xrightarrow{\text{ATP synthase}} ATP$$

ATP Synthesis

When NADH enters electron transport at complex I, the energy released from its oxidation is used to synthesize 3 ATP. $FADH_2$ enters electron transport at complex II which is at a lower energy level. Thus, $FADH_2$ provides energy to produce only 2 ATP. Recent measurements indicate that the oxidation of NADH has an energy yield closer to 2.5 ATP and that one $FADH_2$ has an energy yield closer to 1.5 ATP. Because there are still disagreements about the actual values for ATP yield, we will use the traditional values of 3 ATP for NADH and 2 ATP for $FADH_2$. The overall equation for the oxidation of NADH and $FADH_2$ can be written as follows:

$$\text{NADH} + \text{H}^+ + \tfrac{1}{2}\text{O}_2 + 3\text{ADP} + 3\text{P}_i \longrightarrow \text{NAD}^+ + \text{H}_2\text{O} + \textbf{3ATP}$$

$$\text{FADH}_2 + \tfrac{1}{2}\text{O}_2 + 2\text{ADP} + 2\text{P}_i \longrightarrow \text{FAD} + \text{H}_2\text{O} + \textbf{2ATP}$$

SAMPLE PROBLEM 18.6 ATP Synthesis

Why does the oxidation of NADH provide energy for the formation of three ATP whereas $FADH_2$ produces two ATP?

SOLUTION

Electrons from the oxidation of NADH enter electron transport at complex I. They pass through three complexes (I, III, and IV) pumping H^+ ions into the intermembrane space. Electrons from $FADH_2$ enter electron transport at complex II, pumping H^+ ions through only two complexes (III and IV) that pump H^+ ions.

STUDY CHECK 18.6

Which complexes in electron transport act as proton pumps?

ATP from Glycolysis

In glycolysis, the oxidation of glucose stores energy in two NADH molecules as well as two ATP from direct phosphate transfer. However, glycolysis occurs in the cytoplasm, and the NADH produced cannot pass through the mitochondrial membrane.

Therefore the hydrogen ions and electrons from NADH in the cytoplasm are transferred to compounds that can enter the mitochondria. In this shuttle system, $FADH_2$ is produced. The overall reaction for the shuttle is

$$\underbrace{\text{NADH} + \text{H}^+ + \text{FAD}}_{\text{In cytoplasm}} \longrightarrow \underbrace{\text{NAD}^+ + \text{FADH}_2}_{\text{In mitochondria}}$$

Therefore the transfer of electrons from NADH in the cytoplasm to $FADH_2$ produces two ATP, rather than three. In glycolysis, glucose yields a total of six ATP: four ATP from two NADH, and two ATP from phosphorylation.

$$\text{Glucose} \longrightarrow 2\text{ pyruvate} + 2\text{ATP} + 2\text{NADH}\ (\longrightarrow 2\text{FADH}_2)$$

$$\text{Glucose} \longrightarrow 2\text{ pyruvate} + 6\text{ATP}$$

ATP from the Oxidation of Two Pyruvate

Under aerobic conditions, pyruvate enters the mitochondria, where it is oxidized to give acetyl-CoA, CO_2, and NADH. Because glucose yields two pyruvate, two NADH enter electron transport, where the oxidation of two pyruvate leads to the production of six ATP.

$$2\text{ Pyruvate} \longrightarrow 2\text{ acetyl-CoA} + 6\text{ATP}$$

ATP from the Citric Acid Cycle

One turn of the citric acid cycle produces two CO_2, three NADH, one $FADH_2$, and one ATP by direct phosphate transfer.

$$3\ NADH \times 3\ ATP/NADH = 9\ ATP$$
$$1\ FADH_2 \times 2\ ATP/FADH_2 = 2\ ATP$$
$$1\ GTP \times 1\ ATP/1\ GTP = 1\ ATP$$
$$\text{Total (one turn)} = 12\ ATP$$

Because one glucose produces two acetyl-CoA, two turns of the citric acid cycle produce a total of 24 ATP.

$$\text{Acetyl-CoA} \longrightarrow 2CO_2 + 12ATP\ \text{(one turn of citric acid cycle)}$$
$$2\ \text{Acetyl-CoA} \longrightarrow 4CO_2 + 24ATP\ \text{(two turns of citric acid cycle)}$$

Chemistry Link to **Health**

ATP Synthase and Heating the Body

Some types of compounds, called *uncouplers*, separate the electron transport system from ATP synthase. They do this by disrupting the H^+ gradient needed for the synthesis of ATP. The electrons are transported to O_2 in electron transport, but ATP is not formed by ATP synthase.

Some uncouplers transport the H^+ ions through the inner membrane, which is normally impermeable to H^+ ions; others block the channel in ATP synthase. Compounds such as dicumarol and 2,4-dinitrophenol (DNP) are hydrophobic and bind with H^+ ions to carry them across the inner membrane. An antibiotic, oligomycin, blocks the channel, which does not allow any H^+ ions to return to the matrix. By removing H^+ ions or blocking the channel, there is no H^+ flow to generate energy for ATP synthesis.

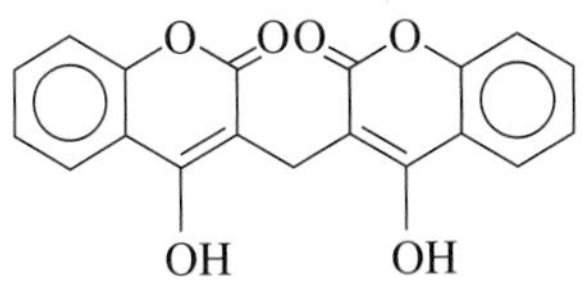

Dicumarol

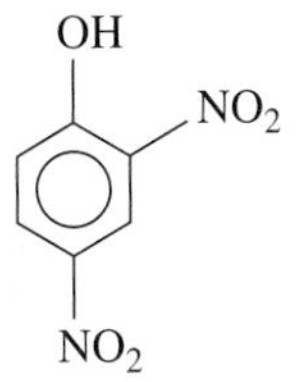

2,4-Dinitrophenol (DNP)

When there is no mechanism for ATP synthesis, the energy of electron transport is released as heat. Certain animals that are adapted to cold climates have developed their own uncoupling system, which allows them to use electron transport energy for heat production. These animals have large amounts of a tissue called *brown fat*, which contains a high concentration of mitochondria. This tissue is brown because of the iron in the cytochromes of the mitochondria. The proton pumps still operate in brown fat, but a protein embedded in the inner mitochondrial membrane allows the H^+ ions to bypass ATP synthase. The energy that would be used to synthesize ATP is released as heat. In newborn babies, brown fat is used to generate heat because newborns have a small mass but a large surface area and need to produce more heat than adults. The brown fat deposits are located near major blood vessels, which carry the warmed blood to the body. Most adults have little or no brown fat, although someone who works outdoors in a cold climate will develop some brown fat deposits.

Plants also use uncouplers. In early spring, heat is used to warm early shoots of plants under the snow, which helps them melt the snow around the plants. Some plants, such as skunk cabbage, use uncoupling agents to volatilize fragrant compounds that attract insects to pollinate the plants.

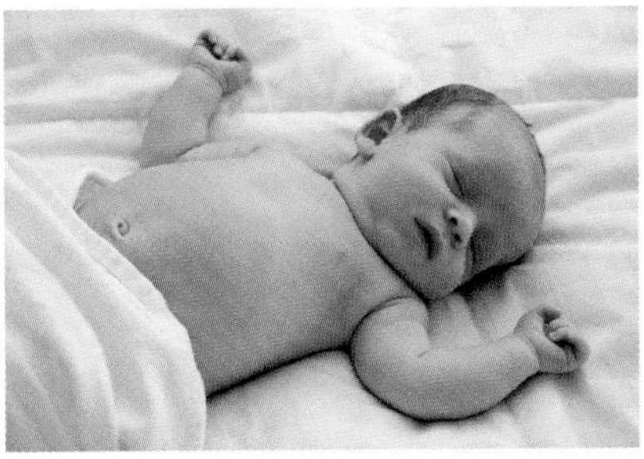

Brown fat helps babies to keep warm.

ATP from the Complete Oxidation of Glucose

CORE CHEMISTRY SKILL

Calculating the ATP Produced from Glucose

The total ATP for the complete oxidation of glucose is calculated by combining the ATP produced from glycolysis, the oxidation of pyruvate, and the citric acid cycle (see Figure 18.14). The ATP produced for these reactions is given in Table 18.4.

SAMPLE PROBLEM 18.7 ATP Production

Indicate the amount of ATP produced by each of the following oxidation reactions:

a. pyruvate to acetyl-CoA
b. glucose to acetyl-CoA

SOLUTION

a. The oxidation of pyruvate to acetyl-CoA produces one NADH, which yields three ATP.

b. Six ATP are produced from the oxidation of glucose to two pyruvate. Another six ATP are produced when two pyruvate oxidize to give two acetyl-CoA. A total of 12 ATP are produced from the oxidation of glucose to two acetyl-CoA.

STUDY CHECK 18.7

What are the sources and amounts of ATP for one turn of the citric acid cycle?

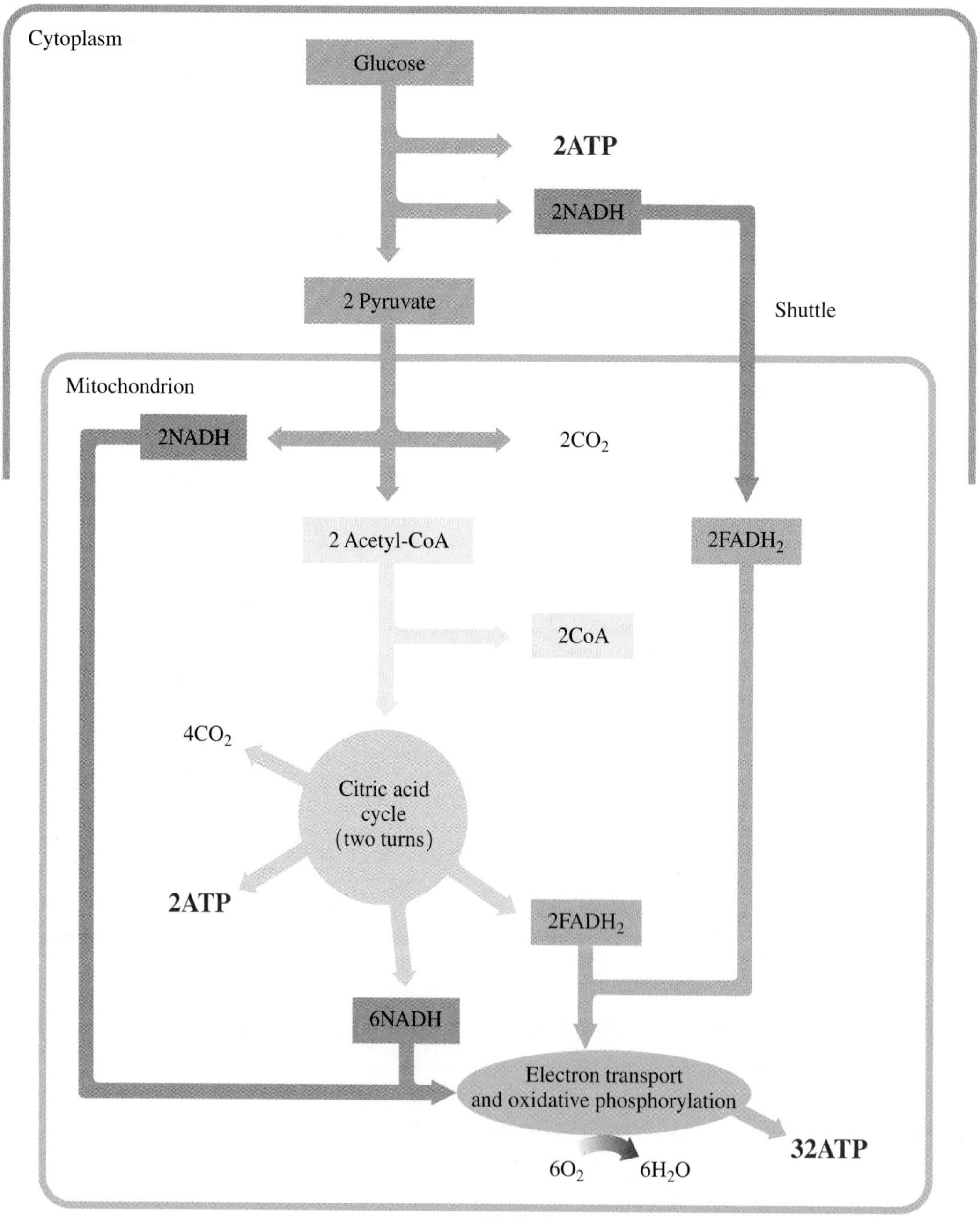

FIGURE 18.14 ▶ The complete oxidation of glucose to CO_2 and H_2O yields a total of 36 ATP.

Q What metabolic pathway produces most of the ATP from the oxidation of glucose?

TABLE 18.4 ATP Produced by the Complete Oxidation of Glucose

Pathway	Substrate/Product	Coenzymes	Phosphate Transfer	ATP Yield
Glycolysis	Glucose/pyruvate	2 NADH (shuttle 2 $FADH_2$)	2 ATP	6 ATP
	Glucose $\longrightarrow$ 2 pyruvate + 6ATP			
Oxidation and Decarboxylation	Pyruvate/acetyl-CoA + $2CO_2$	2 NADH		6 ATP
	2 Pyruvate $\longrightarrow$ 2 acetyl-CoA + $2CO_2$ + 6ATP			
Citric Acid Cycle (two turns)	2 Acetyl-CoA/$4CO_2$	6 NADH 2 $FADH_2$	2 GTP	24 ATP
	2 Acetyl-CoA $\longrightarrow$ $4CO_2$ + 24ATP			
Total Yield	Glucose $\longrightarrow$ $6CO_2$ + 36ATP			36 ATP

When glucose is not immediately used by the cells for energy, it is stored as glycogen in the liver and muscles. When the levels of glucose in the brain or blood become low, the glycogen reserves are hydrolyzed and glucose is released into the blood. If glycogen stores are depleted, some glucose can be synthesized from noncarbohydrate sources. It is the balance of all these reactions that maintains the necessary blood glucose level available to our cells and provides the necessary amount of ATP for our energy needs.

QUESTIONS AND PROBLEMS

18.6 Electron Transport and Oxidative Phosphorylation

LEARNING GOAL Describe electron transport and the process of oxidative phosphorylation; calculate the ATP from the complete oxidation of glucose.

18.41 What reduced coenzymes provide the electrons for electron transport?

18.42 Name the three key processes in the catabolism of glucose that produce ATP in the presence of oxygen.

18.43 How are electrons carried from complex I to complex III?

18.44 How are electrons carried from complex III to complex IV?

18.45 How is NADH oxidized in electron transport?

18.46 How is $FADH_2$ oxidized in electron transport?

18.47 What is meant by oxidative phosphorylation?

18.48 Where does the process of oxidative phosphorylation take place in our cells?

18.49 According to the chemiosmotic theory, how does the H^+ gradient provide energy to synthesize ATP?

18.50 How does the phosphorylation of ADP occur?

18.51 How are glycolysis and the citric acid cycle linked to the production of ATP by electron transport?

18.52 Why does $FADH_2$ yield two ATP, using the electron transport system, but NADH yields three ATP?

18.53 What is the ATP energy yield associated with each of the following?
a. NADH $\longrightarrow$ NAD^+
b. glucose $\longrightarrow$ 2 pyruvate
c. 2 pyruvate $\longrightarrow$ 2 acetyl-CoA + $2CO_2$
d. acetyl-CoA $\longrightarrow$ $2CO_2$

18.54 What is the ATP energy yield associated with each of the following?
a. $FADH_2$ $\longrightarrow$ FAD
b. glucose + $6O_2$ $\longrightarrow$ $6CO_2$ + $6H_2O$
c. glucose $\longrightarrow$ 2 lactate
d. pyruvate $\longrightarrow$ lactate

18.7 Oxidation of Fatty Acids

LEARNING GOAL

Describe the metabolic pathway of β-oxidation; calculate the ATP from the complete oxidation of a fatty acid.

A large amount of energy is obtained when fatty acids undergo oxidation in the mitochondria to yield acetyl-CoA. In stage 2 of fat metabolism, fatty acids undergo **beta-oxidation (β-oxidation)**, which removes two-carbon segments, one at a time, from the carboxyl end.

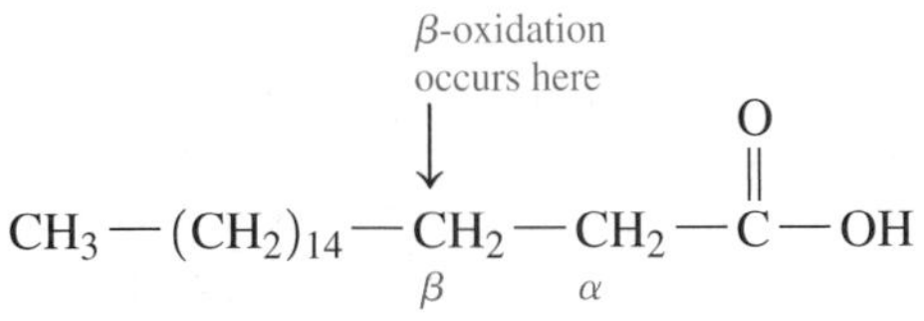

Stearic acid

Each cycle in β-oxidation produces acetyl-CoA and a fatty acid that is shorter by two carbons. The cycle repeats until the original fatty acid is completely degraded to two-carbon acetyl-CoA. Each acetyl-CoA can then enter the citric acid cycle in the same way as the acetyl-CoA derived from glucose.

Fatty Acid Activation

Fatty acids are prepared for transport through the inner membrane of the mitochondria by adding a CoA group to the fatty acid in the cytosol. This process, called *activation*, combines a fatty acid with coenzyme A to yield fatty acyl-CoA. The energy for the activation is obtained from the hydrolysis of ATP to give AMP (adenosine monophosphate) and two inorganic phosphates ($2P_i$). This is equivalent to the energy released from the hydrolysis of two ATP to two ADP.

$$\underset{\text{Fatty acid}}{R{-}CH_2{-}CH_2{-}\overset{\displaystyle O}{\overset{\|}{C}}{-}\mathbf{OH}} + ATP + \mathbf{HS{-}CoA} \xrightarrow[\text{synthetase}]{\text{Acyl-CoA}} \underset{\text{Fatty acyl-CoA}}{R{-}CH_2{-}CH_2{-}\overset{\displaystyle O}{\overset{\|}{C}}{-}\mathbf{S{-}CoA}} + AMP + 2P_i + \mathbf{H_2O}$$

Reactions of the β-Oxidation Cycle

In the matrix, fatty acyl-CoA molecules undergo β-oxidation, which is a cycle of four reactions that convert the $-CH_2-$ of the β-carbon to a β-keto group. Once the β-keto group is formed, a two-carbon acetyl group can be split from the carbon chain, which shortens the fatty acyl chain. The reaction for one cycle of β-oxidation is written as follows:

$$\underset{\text{Fatty acyl-CoA}}{R{-}\underset{\beta}{\mathbf{CH_2}}{-}\underset{\alpha}{CH_2}{-}\overset{\displaystyle \mathbf{O}}{\overset{\|}{C}}{-}S{-}CoA} + NAD^+ + FAD + H_2O + HS{-}CoA \longrightarrow$$

$$\underset{\text{New fatty acyl-CoA }(-2C)}{R{-}\overset{\displaystyle \mathbf{O}}{\overset{\|}{\mathbf{C}}}{-}S{-}CoA} + \underset{\text{Acetyl-CoA}}{\mathbf{CH_3}{-}\overset{\displaystyle \mathbf{O}}{\overset{\|}{C}}{-}\mathbf{S{-}CoA}} + NADH + H^+ + FADH_2$$

The specific reactions in β-oxidation are described next for capric acid (C_{10}) (see Figure 18.15).

Reaction 1 Oxidation

In the first reaction of β-oxidation, the FAD coenzyme removes hydrogen atoms from the α- and β-carbons of the activated fatty acid to form a trans carbon–carbon double bond and $FADH_2$.

Reaction 2 Hydration

A hydration reaction adds the components of water to the trans double bond, which forms a secondary hydroxyl group ($-OH$) on the β-carbon.

Reaction 3 Oxidation

The secondary hydroxyl group ($-OH$) on the β-carbon is oxidized to yield a ketone. The hydrogen atoms reduce coenzyme NAD^+ to $NADH + H^+$. At this point, the β-carbon has been oxidized to a keto group.

Reaction 4 Cleavage

In the final step of β-oxidation, the $C_\alpha-C_\beta$ bond is cleaved at the β-carbon to yield a two-carbon acetyl-CoA and an acyl-CoA that is shorter by two carbons. The 8C fatty acyl-CoA repeats the β-oxidation cycle where it is cleaved to a 6C fatty acyl-CoA, then to a 4C fatty acyl-CoA, which splits to give two acetyl-CoA. All the acetyl-CoA produced from the fatty acid can enter the citric acid cycle to produce energy.

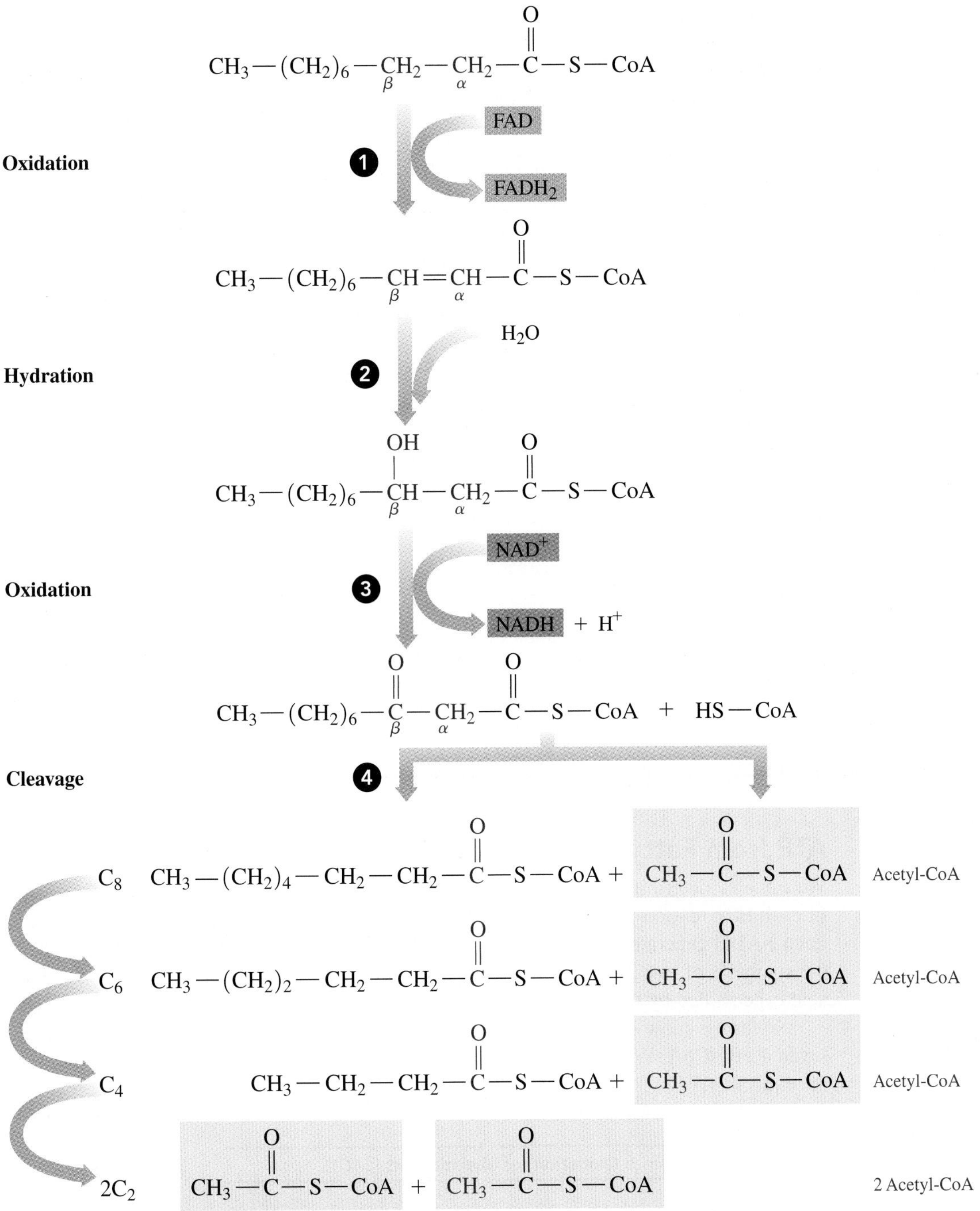

FIGURE 18.15 ▶ Capric acid (C_{10}) undergoes four β-oxidation cycles that repeat reactions 1 to 4 to yield 5 acetyl-CoA, 4 NADH, and 4 $FADH_2$.

Q How many NADH and $FADH_2$ are produced in one cycle of β-oxidation?

SAMPLE PROBLEM 18.8 β-Oxidation

Match each of the following (**a** to **d**) with one of the reactions (**1** to **4**) in the β-oxidation cycle:

(**1**) first oxidation
(**2**) hydration
(**3**) second oxidation
(**4**) cleavage

a. Water is added to a double bond.
b. An acetyl-CoA is removed.
c. FAD is reduced to $FADH_2$.
d. NAD^+ is reduced to NADH.

CORE CHEMISTRY SKILL

Calculating the ATP from Fatty Acid Oxidation (β-Oxidation)

Explore Your World

Fat Storage and Blubber

Obtain four plastic freezer bags, masking tape, and some solid vegetable fat used for cooking. Fill a bucket or container with enough water to cover both of your hands. Add ice until the water feels very cold. Place 3 or 4 tablespoons of the vegetable fat in one of the plastic bags. Place another plastic bag inside the first bag containing the fat. Tape the top edges of the two bags together, leaving the inside bag open. Using the remaining two plastic bags, place one inside the other, and tape the top edges together leaving the inside bag open. Now place one hand inside the bag with the vegetable fat and distribute the fat around until a layer of about 2/3 in. (2 cm) of the fat in the inner bag covers your hand. Place your other hand inside the other double bag and submerge both your hands in the ice water. Measure the time it takes for one hand to feel uncomfortably cold. Remove your hands before they get too cold.

Questions

1. How effective is the double bag with "blubber" in protecting your hand from the cold?
2. How would increasing the amount of fat affect your results?
3. How does "blubber" help an animal survive starvation?
4. Why would animals in warm climates, such as camels and migratory birds, need to store fat?
5. If you placed 300 g of shortening in the baggie, how many moles of ATP could it provide if used for energy production (assume it produces the same ATP as myristic acid)?

SOLUTION

a. (2) hydration
b. (4) cleavage
c. (1) first oxidation
d. (3) second oxidation

STUDY CHECK 18.8

Which coenzyme is needed in reaction 3 when a β-hydroxyl group is converted to a β-keto group?

Fatty Acid Length Determines Cycle Repeats

The number of carbon atoms in a fatty acid determines the number of times the cycle repeats and the number of acetyl-CoA it produces. For example, the complete β-oxidation of myristic acid (C_{14}) produces seven acetyl-CoA, which is equal to one-half the number of carbon atoms in the fatty acid. Because the final turn of the cycle produces two acetyl-CoA, the total number of times the cycle repeats is one less than the total number of acetyl groups it produces. Therefore, the C_{14} fatty acid goes through the cycle six times.

Fatty Acid	Number of Carbon Atoms	Number of Acetyl-CoA	Number of β-Oxidation Cycles
Myristic acid	14	7	6
Palmitic acid	16	8	7
Stearic acid	18	9	8

ATP from Fatty Acid Oxidation

We can now determine the total energy yield from the oxidation of a particular fatty acid. In each β-oxidation cycle, one NADH, one $FADH_2$, and one acetyl-CoA are produced. Each NADH generates sufficient energy to synthesize three ATP, whereas $FADH_2$ leads to the synthesis of two ATP. However, the greatest amount of energy produced from a fatty acid is generated by the production of the acetyl-CoA that enter the citric acid cycle.

So far we know that the C_{14} acid goes through six turns of the cycle, which produces seven acetyl-CoA. We also need to remember that activation of the myristic acid requires the equivalent of two ATP. We can set up the calculation as follows:

ATP Production from β-Oxidation for Myristic Acid (14C)	
Activation	−2 ATP
7 Acetyl-CoA (14 C atoms × 1 acetyl-CoA/2 C atoms)	
7 acetyl-CoA × 12 ATP/acetyl-CoA	84 ATP
6 β-Oxidation Cycles (coenzymes)	
6 $FADH_2$ × 2 ATP/$FADH_2$	12 ATP
6 NADH × 3 ATP/NADH	18 ATP
Total	112 ATP

Chemistry Link to Health

Stored Fat and Obesity

The storage of fat is an important survival feature in the lives of many animals. In hibernating animals, large amounts of stored fat provide the energy for the entire hibernation period, which could be several months. In camels, large amounts of food are stored in the camel's hump, which is actually a huge fat deposit. When food resources are low, the camel can survive months without food or water by utilizing the fat reserves in the hump. Migratory birds preparing to fly long distances also store large amounts of fat. Whales are kept warm by a layer of body fat called "blubber" under their skin, which can be as thick as 2 feet. Blubber also provides energy when whales must survive long periods of starvation. Penguins also have blubber, which protects them from the cold and provides energy when they are incubating their eggs.

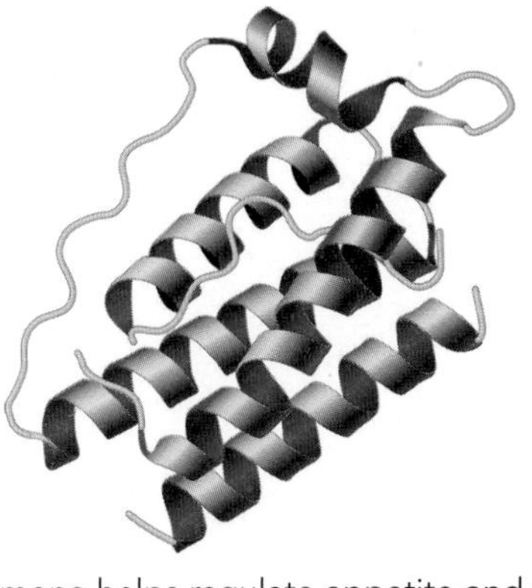
The leptin hormone helps regulate appetite and metabolism.

Humans also have the capability to store large amounts of fat, although they do not hibernate or usually have to survive for long periods of time without food. When early humans survived on sparse diets that were mostly vegetarian, about 20% of the dietary calories were from fats. Today, a typical diet includes more dairy products and foods with high fat levels, and as much as 60% of the calories are from fat. The U.S. Public Health Service now estimates that in the United States, more than one-third of adults are obese. Obesity is defined as a body weight that is more than 20% over an ideal weight. Obesity is a major factor in health problems such as diabetes, heart disease, high blood pressure, stroke, and gallstones as well as some cancers and some forms of arthritis.

At one time, we thought that obesity was simply a problem of eating too much. However, research now indicates that certain pathways in lipid and carbohydrate metabolism may cause excessive weight gain in some people. In 1995, scientists discovered a hormone called *leptin* that is produced in fat cells. When fat cells are full, high levels of leptin signal the brain to limit the intake of food. When fat stores are low, leptin production decreases, which signals the brain to increase food intake. Leptin acts on the liver and skeletal muscles, where it stimulates fatty acid oxidation in the mitochondria, which decreases fat storage.

Major research is currently being done to find the causes of obesity. Scientists are studying differences in the rate of leptin production, degrees of resistance to leptin, and possible combinations of these factors. After a person has dieted and lost weight, the leptin level drops. This decrease in leptin may cause an increase in hunger and food intake while slowing metabolism, which starts the weight-gain cycle all over again. Currently, studies are being made to assess the safety of leptin therapy following weight loss.

Marine mammals have thick layers of blubber that serve as insulation as well as energy storage.

A camel stores large amounts of fat in its hump.

SAMPLE PROBLEM 18.9 ATP Production from β-Oxidation

How much ATP will be produced from the β-oxidation of palmitic acid, a C_{16} saturated fatty acid?

SOLUTION

ANALYZE THE PROBLEM	Number of Carbon Atoms	Number of β-Oxidation Cycles	Number of $FADH_2$	Number of NADH	Number of Acetyl-CoA
	16	7	7	7	8

16-Carbon palmitic acid requires seven β-oxidation cycles, which produce seven $FADH_2$ and seven NADH. The total number of acetyl-CoA is eight. During electron transport, each $FADH_2$ produces two ATP, and each NADH produces three ATP. Each acetyl-CoA can produce 12 ATP by way of the citric acid cycle. Two ATP are utilized in the activation of palmitic acid.

ATP Production from Palmitic Acid (C_{16})	
Activation of palmitic acid	−2 ATP
$7\ \cancel{\text{FADH}_2} \times \dfrac{2\ \text{ATP}}{\cancel{\text{FADH}_2}}$ (electron transport)	14 ATP
$7\ \cancel{\text{NADH}} \times \dfrac{3\ \text{ATP}}{\cancel{\text{NADH}}}$ (electron transport)	21 ATP
$8\ \cancel{\text{acetyl-CoA}} \times \dfrac{12\ \text{ATP}}{\cancel{\text{acetyl-CoA}}}$ (citric acid cycle)	96 ATP
Total	129 ATP

STUDY CHECK 18.9

Compare the total ATP from reduced coenzymes and from acetyl-CoA in the β-oxidation of palmitic acid.

Oxidation of Unsaturated Fatty Acids

Many of the fats in our diets, especially the oils, contain unsaturated fatty acids which have one or more double bonds. Thus, in β-oxidation, the fatty acid is ready for hydration. Since no $FADH_2$ is formed in this step, the energy from the β-oxidation of unsaturated fatty acids is slightly less than the energy from saturated fatty acids. However, for simplicity, we will assume that the total ATP production is the same for both saturated and unsaturated fatty acids.

Ketone Bodies

When carbohydrates are not available to meet energy needs, the body breaks down fatty acids, which undergo β-oxidation to acetyl-CoA. Normally, acetyl-CoA would enter the citric acid cycle for further oxidation and energy production. However, when large quantities of fatty acids are degraded, the citric acid cycle cannot utilize the full amount. As a result, acetyl-CoA accumulates in the liver, where acetyl-CoA molecules combine to form compounds called **ketone bodies** in a pathway known as *ketogenesis* (see Figure 18.16).

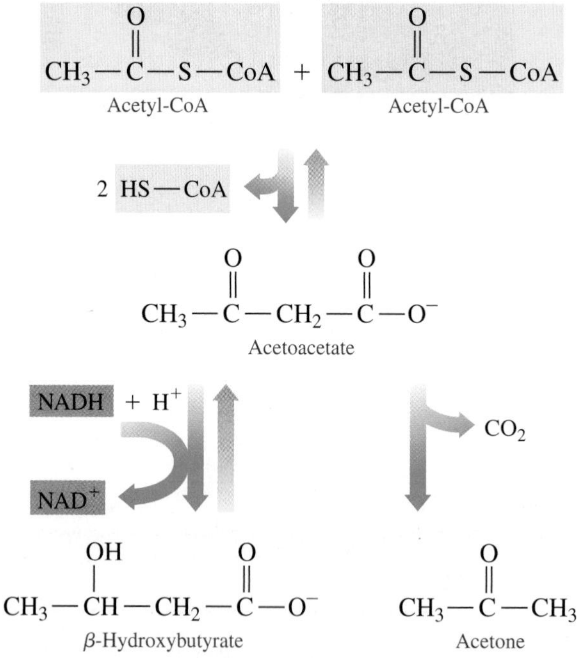

FIGURE 18.16 ▶ In ketogenesis, acetyl-CoA molecules combine to produce ketone bodies: acetoacetate, β-hydroxybutyrate, and acetone.

Q What condition in the body leads to the formation of ketone bodies?

Ketone bodies are produced mostly in the liver and transported to cells in the heart, brain, and skeletal muscle, where small amounts of energy can be obtained by converting acetoacetate or β-hydroxybutyrate back to acetyl-CoA.

$$\beta\text{-Hydroxybutyrate} \longrightarrow \text{acetoacetate} + 2\ \text{CoA} \longrightarrow 2\ \text{acetyl-CoA}$$

Ketosis

The accumulation of ketone bodies may lead to a condition called *ketosis*, which occurs in severe diabetes, diets high in fat and low in carbohydrates, alcoholism, and starvation. Because two of the ketone bodies are acids, they can lower the blood pH below 7.4, which is *acidosis*, a condition that often accompanies ketosis. A drop in blood pH can interfere with the ability of the blood to carry oxygen and cause breathing difficulties.

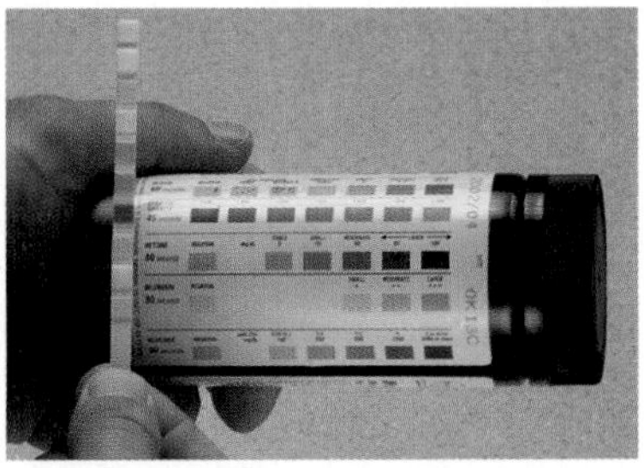

A test strip determines ketone bodies in a urine sample.

Fatty Acid Synthesis from Carbohydrates

When the body has met all its energy needs and the glycogen stores are full, excess acetyl-CoA from the breakdown of carbohydrates is used to form new fatty acids. Two-carbon acetyl units are linked together to give fatty acids such as palmitic and stearic acids. Several of the reactions in the synthesis of fatty acids are the reverse of the reactions we described for the oxidation of fatty acids. However, fatty acid oxidation occurs in the mitochondria, whereas fatty acid synthesis occurs in the cytosol. The new fatty acids are attached to glycerol to make triacylglycerols, which can be stored in unlimited quantities as body fat.

QUESTIONS AND PROBLEMS

18.7 Oxidation of Fatty Acids

LEARNING GOAL Describe the metabolic pathway of β-oxidation; calculate the ATP from the complete oxidation of a fatty acid.

18.55 Where in the cell is a fatty acid activated?

18.56 What coenzymes are required for β-oxidation?

18.57 Caprylic acid, $CH_3-(CH_2)_6-COOH$, is a C_8 fatty acid found in milk.
- **a.** Draw the condensed structural formula for the activated form of caprylic acid.
- **b.** Indicate the α- and β-carbon atoms in capryloyl-CoA.
- **c.** State the number of β-oxidation cycles for the complete oxidation of caprylic acid.
- **d.** State the number of acetyl-CoA from the complete oxidation of caprylic acid.

18.58 Lignoceric acid, $CH_3-(CH_2)_{22}-COOH$, is a C_{24} fatty acid found in peanut oil in small amounts.
- **a.** Draw the condensed structural formula for the activated form of lignoceric acid.
- **b.** Indicate the α- and β-carbon atoms in lignoceroyl-CoA.
- **c.** State the number of β-oxidation cycles for the complete oxidation of lignoceric acid.
- **d.** State the number of acetyl-CoA from the complete oxidation of lignoceric acid.

18.59 Consider the complete oxidation of oleic acid, $CH_3-(CH_2)_7-CH{=}CH-(CH_2)_7-COOH$, which is a C_{18} monounsaturated fatty acid.
- **a.** How many cycles of β-oxidation are needed?
- **b.** How many acetyl-CoA are produced?
- **c.** How many ATP are generated from the oxidation of oleic acid?

18.60 Consider the complete oxidation of palmitoleic acid, $CH_3-(CH_2)_5-CH{=}CH-(CH_2)_7-COOH$, which is a C_{16} monounsaturated fatty acid found in animal and vegetable oils.
- **a.** How many cycles of β-oxidation are needed?
- **b.** How many acetyl-CoA are produced?
- **c.** How many ATP are generated from the oxidation of palmitoleic acid?

18.61 What is ketogenesis? Show how acetoacetate is formed from acetyl-CoA.

18.62 Why would a person who is fasting have high levels of acetyl-CoA?

18.63 What are some conditions that characterize ketosis?

18.64 Why do diabetics produce high levels of ketone bodies?

Chemistry Link to Health

Ketone Bodies and Diabetes

Blood glucose is elevated within 30 min following a meal containing carbohydrates. The elevated level of glucose stimulates the secretion of the hormone insulin from the pancreas, which increases the flow of glucose into muscle and adipose tissue for the synthesis of glycogen. As blood glucose levels drop, the secretion of insulin decreases. When blood glucose is low, another hormone, glucagon, is secreted by the pancreas, which stimulates the breakdown of glycogen in the liver to yield glucose.

In *diabetes mellitus*, glucose cannot be utilized or stored as glycogen because insulin is not secreted or does not function properly. In type 1, *insulin-dependent diabetes*, which often begins in childhood, the pancreas produces inadequate levels of insulin. This type of diabetes can result from damage to the pancreas by viral infections or from genetic mutations. In type 2, *insulin-resistant diabetes*, which usually occurs in adults, insulin is produced, but insulin receptors are not responsive. Thus a person with type 2 diabetes does not respond to insulin therapy. *Gestational diabetes* can occur during pregnancy, but blood glucose levels usually return to normal after the baby is born. Pregnant women with diabetes tend to gain weight and have large babies.

In all types of diabetes, insufficient amounts of glucose are available in the muscle, liver, and adipose tissue. As a result, liver cells synthesize glucose from noncarbohydrate sources and break down fat, elevating the acetyl-CoA level. Excess acetyl-CoA undergoes *ketogenesis*, and ketone bodies accumulate in the blood. The odor of acetone can be detected on the breath of a person with uncontrolled diabetes who is in ketosis.

In uncontrolled diabetes, the concentration of blood glucose exceeds the ability of the kidney to reabsorb glucose, and glucose appears in the urine. High levels of glucose increase the osmotic pressure in the blood, which leads to an increase in urine output. Symptoms of diabetes include frequent urination and excessive thirst. Treatment for diabetes includes diet changes to limit carbohydrate intake and may require medication such as a daily injection of insulin or pills taken by mouth.

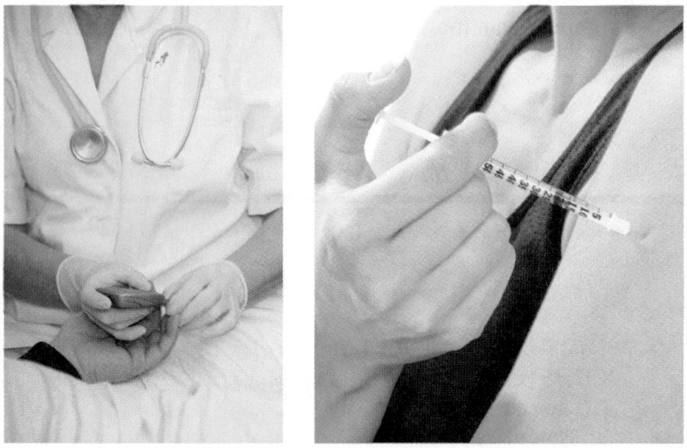

Diabetes can be treated with injections of insulin.

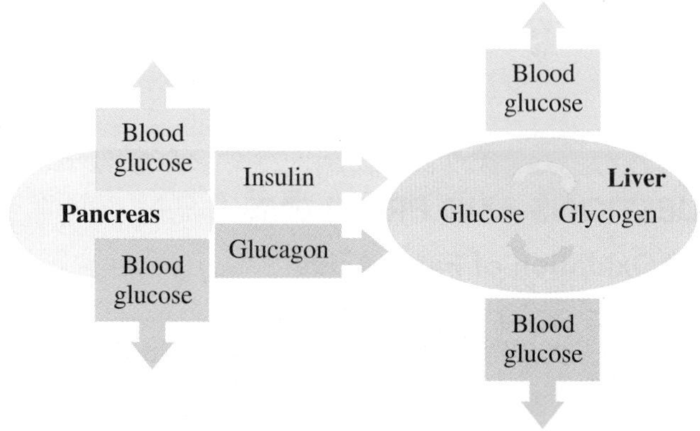

18.8 Degradation of Amino Acids

LEARNING GOAL

Describe the reactions of transamination, oxidative deamination, and the entry of amino acid carbons into the citric acid cycle.

When dietary protein exceeds the nitrogen needed by the body, the excess amino acids are degraded. The α-amino group is removed to yield an α-keto acid, which can be converted to an intermediate of other metabolic pathways. The carbon atoms from amino acids are used in the citric acid cycle as well as for the synthesis of fatty acids, ketone bodies, and glucose. Most of the amino groups are converted to urea.

Transamination

The degradation of amino acids occurs primarily in the liver. In a **transamination** reaction, an α-amino group is transferred from an amino acid to an α-keto acid, usually α-ketoglutarate. A new amino acid and a new α-keto acid are produced.

We can write an equation to show the transfer of the amino group from alanine to α-ketoglutarate to yield glutamate, the new amino acid, and the α-keto acid pyruvate. The enzymes for the transfer of amino groups are known as transaminases or aminotransferases. Pyruvate can now enter the citric acid cycle for the production of energy.

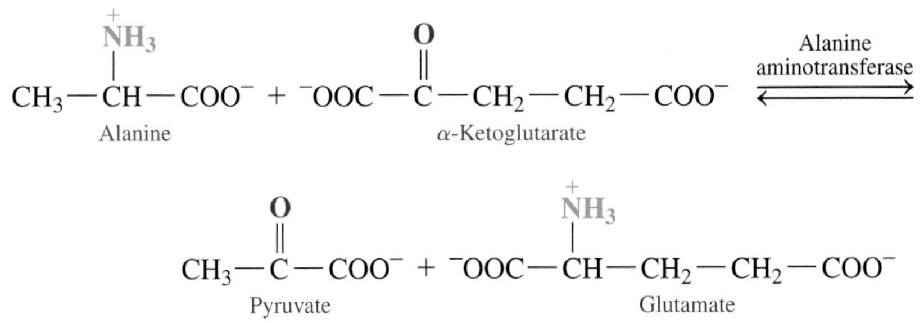

Oxidative Deamination

In a process called **oxidative deamination**, the ammonium group ($—NH_3^+$) in glutamate is removed as an ammonium ion, NH_4^+. The reaction regenerates α-ketoglutarate, which can enter transamination with an amino acid. The oxidation provides hydrogens for the NAD^+ coenzyme.

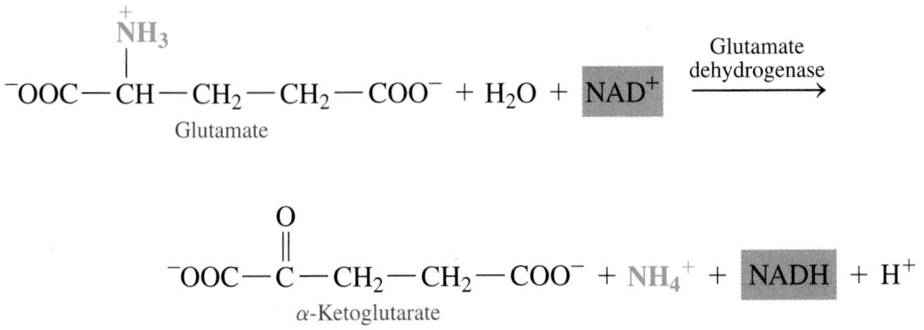

Therefore, the amino group from any amino acid can be used to form glutamate, which undergoes oxidative deamination, converting the amino group to an ammonium ion.

Urea Cycle

The ammonium ion, which is the end product of amino acid degradation, is toxic if it is allowed to accumulate. Therefore, a series of reactions, called the **urea cycle**, detoxifies ammonium ion (NH_4^+) by forming urea, which is excreted in the urine.

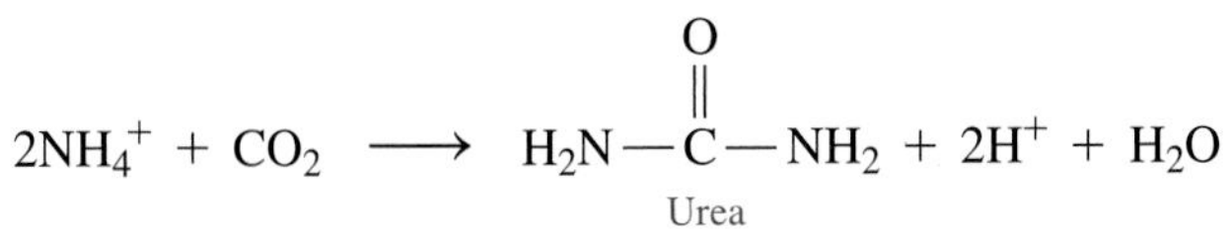

In one day, a typical adult may excrete about 25 to 30 g of urea in the urine. This amount increases when a diet is high in protein. If urea is not properly excreted, it builds up quickly to a toxic level. To detect renal disease, the *blood urea nitrogen* (BUN) level is measured. If the BUN is high, protein intake must be reduced, and hemodialysis may be needed to remove toxic nitrogen waste from the blood.

SAMPLE PROBLEM 18.10 Transamination and Oxidative Deamination

Indicate whether each of the following represents a transamination or an oxidative deamination:

a. Glutamate is converted to α-ketoglutarate and NH_4^+.
b. Alanine and α-ketoglutarate react to form pyruvate and glutamate.

SOLUTION

a. oxidative deamination **b.** transamination

STUDY CHECK 18.10

What is the source of the nitrogen atoms used to produce urea?

ATP Energy from Amino Acids

Normally, only a small amount (about 10%) of our energy needs is supplied by amino acids. However, more energy is extracted from amino acids in conditions such as fasting or starvation, when carbohydrate and fat stores are exhausted. If amino acids remain the only source of energy for a long period of time, the breakdown of body proteins eventually leads to a destruction of essential body tissues.

Once the carbon skeletons of amino acids are obtained from transamination, they can be used as intermediates of the citric acid cycle to provide reduced coenzymes for electron transport. Some amino acids can enter the citric acid cycle at more than one place (see Figure 18.17).

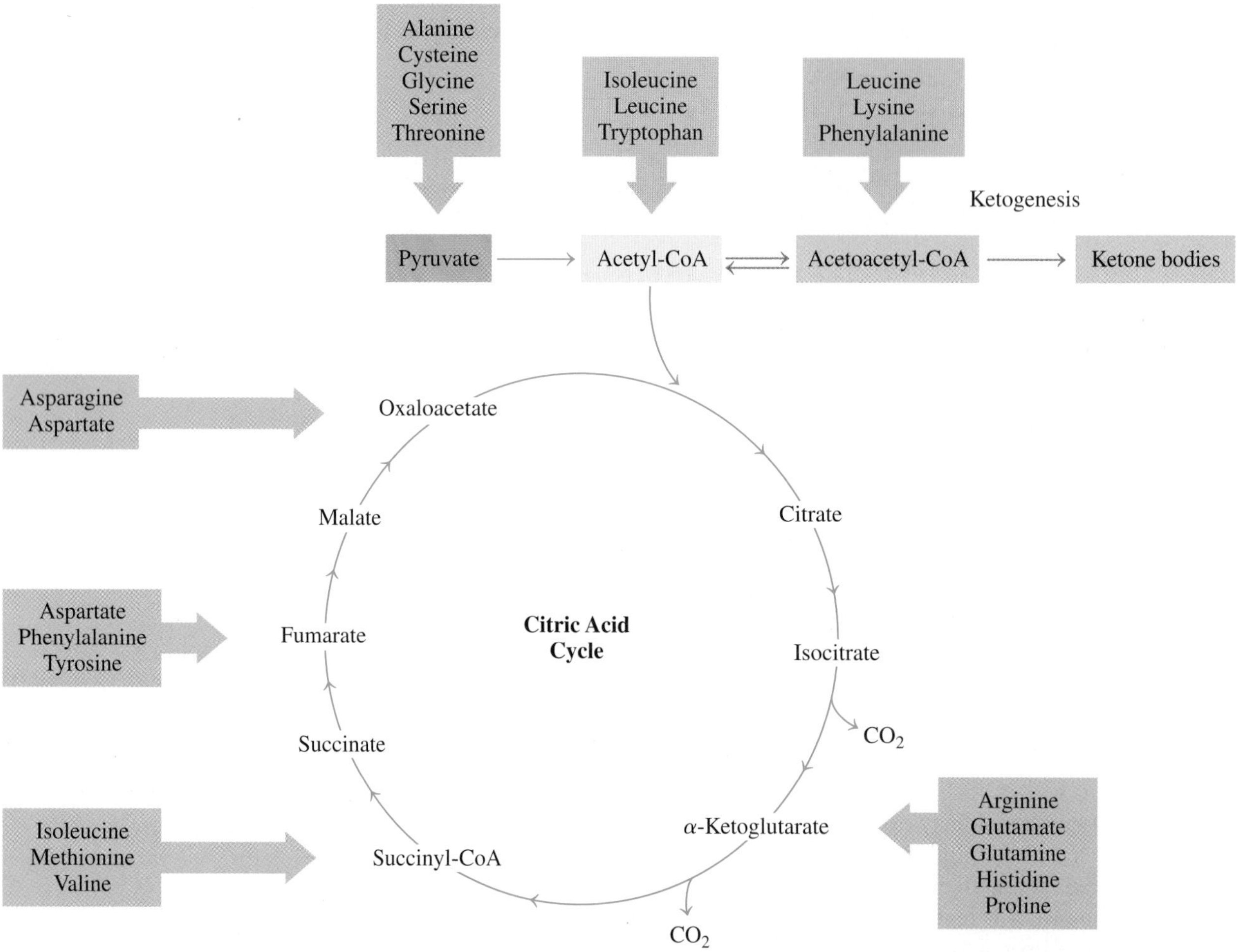

FIGURE 18.17 ▶ Carbon atoms from amino acids can be converted to the intermediates of the citric acid cycle. Carbon atoms from amino acids in green boxes can also produce ketone bodies.

Q What compound in the citric acid cycle is obtained from the carbon atoms of alanine and glycine?

SAMPLE PROBLEM 18.11 Carbon Atoms from Amino Acids

At what point do the carbon atoms of each of the following amino acids enter metabolic pathways for the production of energy?

a. valine **b.** proline **c.** glycine

SOLUTION

a. Carbon atoms from valine enter the citric acid cycle as succinyl-CoA.
b. Carbon atoms from proline enter the citric acid cycle as α-ketoglutarate.
c. Carbon atoms from glycine enter metabolic pathways as pyruvate.

STUDY CHECK 18.11

Which amino acids enter the citric acid cycle as α-ketoglutarate?

Overview of Metabolism

In this chapter, we have seen that catabolic pathways degrade large molecules to small molecules that are used for energy production, using the citric acid cycle and electron transport. We have also indicated that anabolic pathways lead to the synthesis of larger molecules in the cell. In the overall view of metabolism, there are several branch points from which compounds may be degraded for energy or used to synthesize larger molecules. For example, glucose can be degraded to acetyl-CoA for the citric acid cycle to produce energy or converted to glycogen for storage. When glycogen stores are depleted, fatty acids are degraded for energy. Amino acids normally used to synthesize nitrogen-containing compounds in the cells can also be used for energy after they are degraded to intermediates of the citric acid cycle. In the synthesis of certain amino acids, α-keto acids of the citric acid cycle enter a variety of reactions that convert them to amino acids through transamination by glutamate (see Figure 18.18).

QUESTIONS AND PROBLEMS

18.8 Degradation of Amino Acids

LEARNING GOAL Describe the reactions of transamination, oxidative deamination, and the entry of amino acid carbons into the citric acid cycle.

18.65 Draw the condensed structural formula for the α-keto acid produced from each of the following in transamination:

a. $H-\overset{\overset{+}{NH_3}}{\overset{|}{CH}}-COO^-$ Glycine

b. $CH_3-\overset{\overset{+}{NH_3}}{\overset{|}{CH}}-COO^-$ Alanine

18.66 Draw the condensed structural formula for the α-keto acid produced from each of the following in transamination:

a. $^-OOC-CH_2-\overset{\overset{+}{NH_3}}{\overset{|}{CH}}-COO^-$ Aspartate

b. $CH_3-CH_2-\overset{CH_3}{\overset{|}{CH}}-\overset{\overset{+}{NH_3}}{\overset{|}{CH}}-COO^-$ Isoleucine

18.67 Why does the body convert NH_4^+ to urea?

18.68 Write an equation to show how our body removes excessive NH_4^+ ions.

18.69 What metabolic substrate(s) are produced from the carbon atoms of each of the following amino acids?
a. alanine
b. aspartate
c. tyrosine
d. glutamine

18.70 What metabolic substrate(s) are produced from the carbon atoms of each of the following amino acids?
a. leucine
b. asparagine
c. cysteine
d. arginine

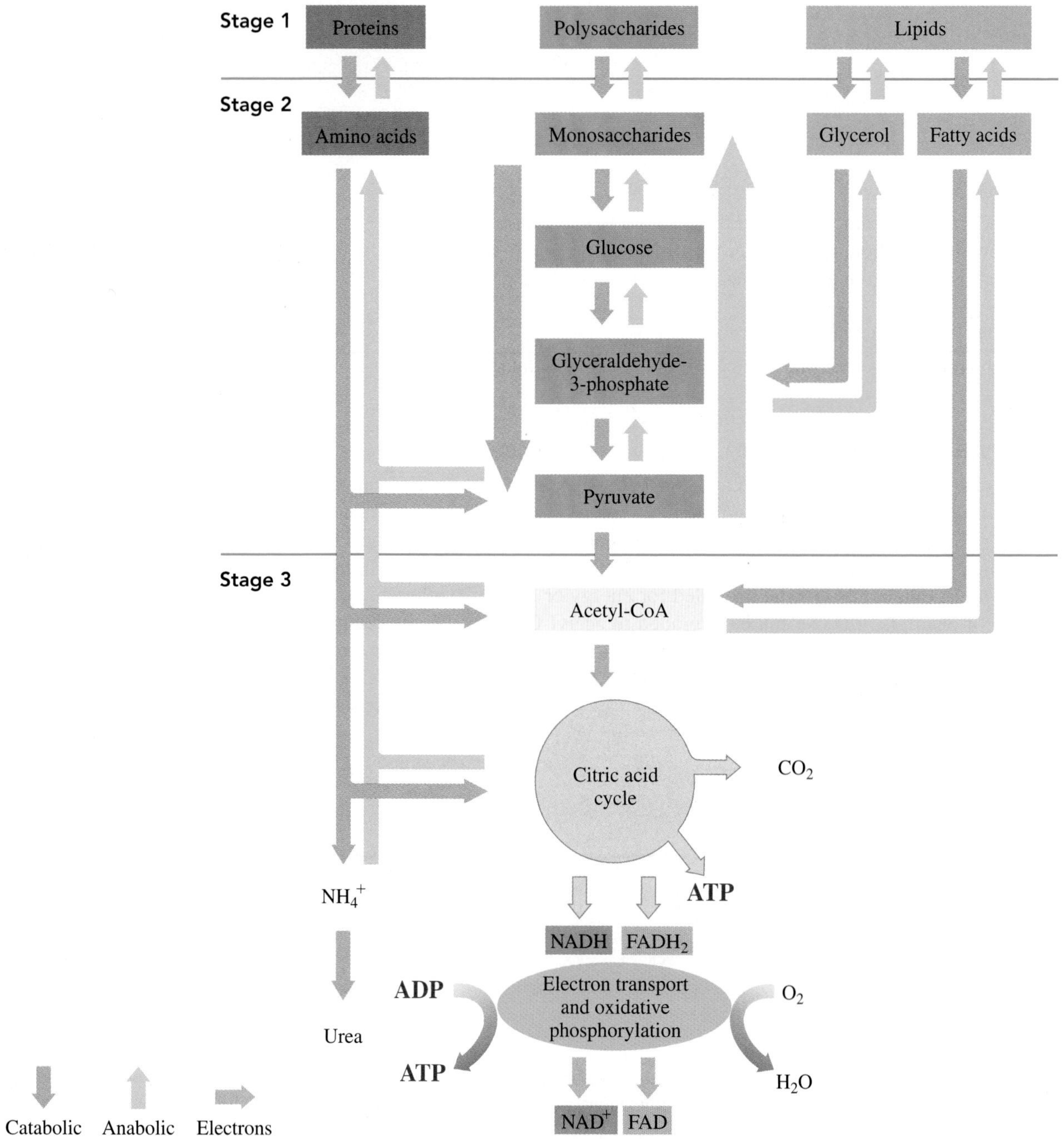

FIGURE 18.18 ▶ Catabolic and anabolic pathways in the cells provide the energy and necessary compounds for the cells.

Q Under what conditions in the cell are amino acids degraded for energy?

CONCEPT MAP

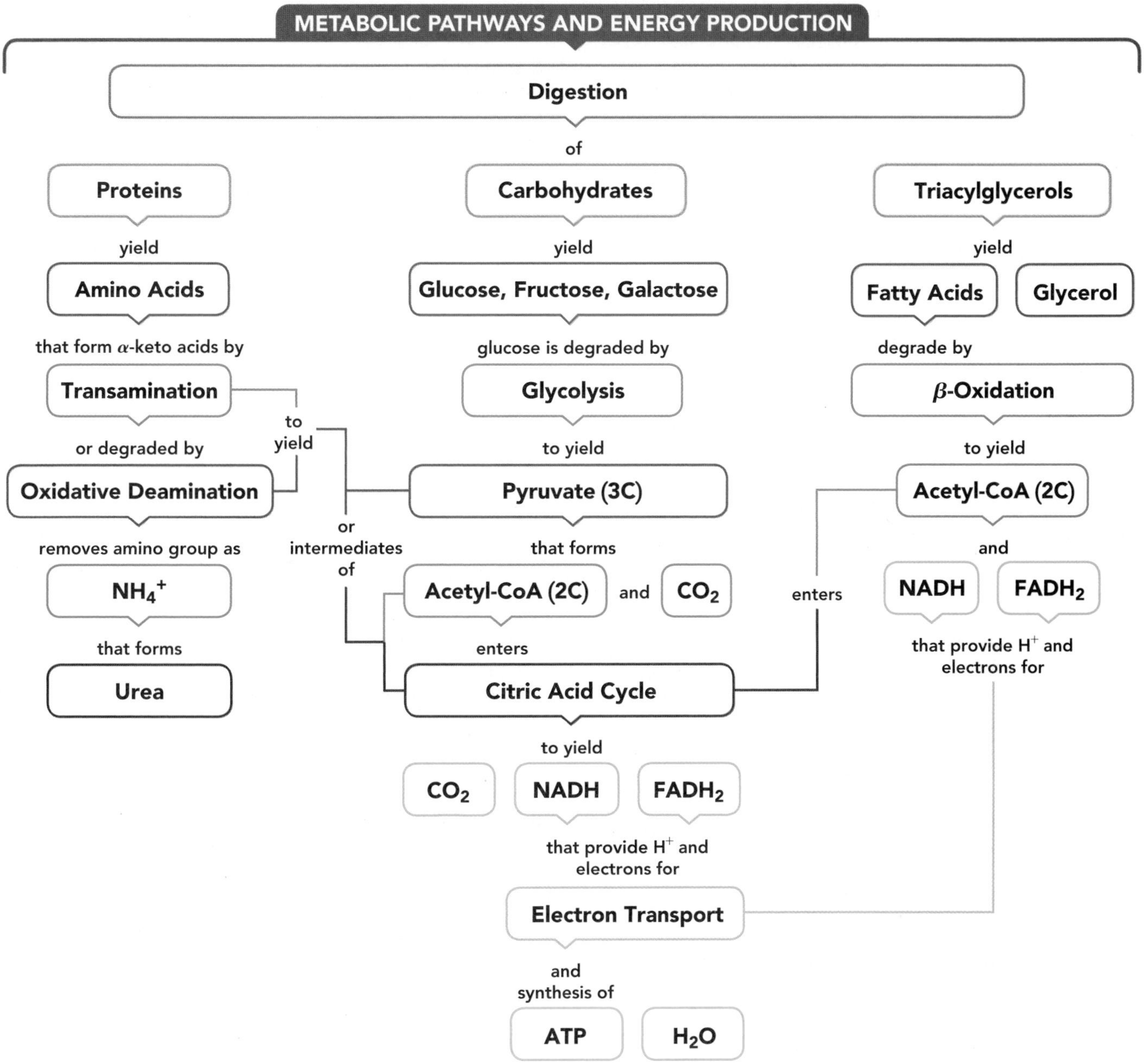

CHAPTER REVIEW

18.1 Metabolism and ATP Energy

LEARNING GOAL Describe three stages of catabolism and the role of ATP.

- Metabolism includes all the catabolic and anabolic reactions that occur in the cells.
- Catabolic reactions degrade large molecules into smaller ones with an accompanying release of energy.
- Anabolic reactions require energy to synthesize larger molecules from smaller ones.
- The three stages of catabolism are digestion of food, degradation of monomers such as glucose to pyruvate, and the extraction of energy from the two- and three-carbon compounds.
- Energy obtained from catabolic reactions is stored in adenosine triphosphate (ATP), a high-energy compound that is hydrolyzed when energy is required by anabolic reactions.

18.2 Digestion of Foods

LEARNING GOAL Give the sites and products of digestion for carbohydrates, triacylglycerols, and proteins.

- The digestion of carbohydrates is a series of reactions that breaks down polysaccharides into smaller polysaccharides

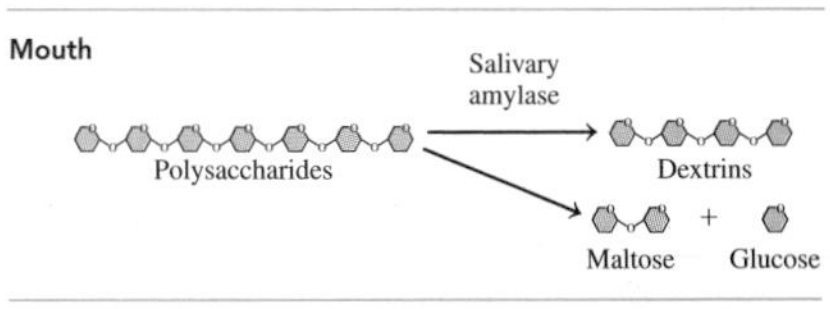

(dextrins), and eventually to monosaccharides glucose, galactose, and fructose.

- These monomers can be absorbed through the intestinal wall into the bloodstream to be carried to cells where they provide energy and carbon atoms for synthesis of new molecules.
- Triacylglycerols are hydrolyzed in the small intestine to monoacylglycerol and fatty acids, which enter the intestinal wall and form new triacylglycerols. They bind with proteins to form chylomicrons, which transport them through the lymphatic system and bloodstream to the tissues.
- The digestion of proteins, which begins in the stomach and continues in the small intestine, involves the hydrolysis of peptide bonds to yield amino acids that are absorbed through the intestinal wall and transported to the cells.

18.3 Coenzymes in Metabolic Pathways

LEARNING GOAL Describe the components and functions of the coenzymes NAD^+, FAD, and coenzyme A.

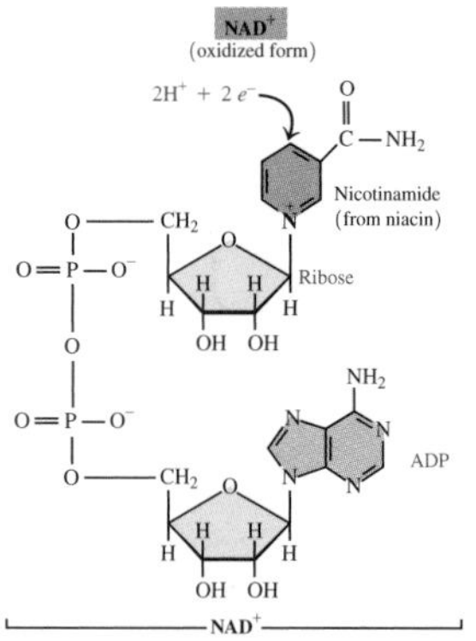

- FAD and NAD^+ are the oxidized forms of coenzymes that participate in oxidation–reduction reactions.
- When they pick up hydrogen ions and electrons, they are reduced to $FADH_2$ and NADH + H^+.
- Coenzyme A contains a thiol group that usually bonds with a two-carbon acetyl group (acetyl-CoA).

18.4 Glycolysis: Oxidation of Glucose

LEARNING GOAL Describe the conversion of glucose to pyruvate in glycolysis and the subsequent conversion of pyruvate to acetyl-CoA or lactate.

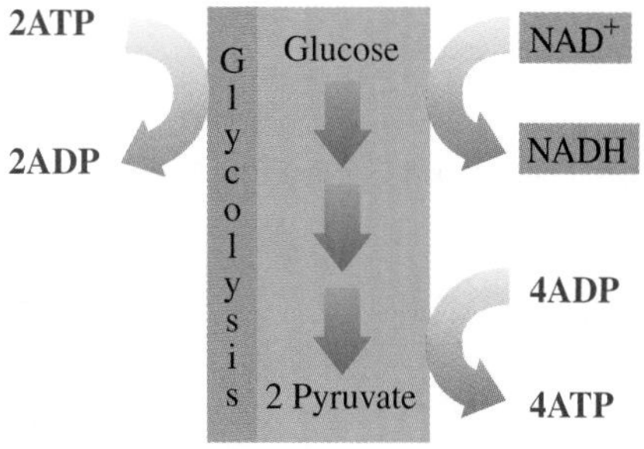

- Glycolysis, which occurs in the cytosol, consists of 10 reactions that degrade glucose (6C) to two pyruvate (3C).
- The overall series of reactions yields two NADH and two ATP.
- Under aerobic conditions, pyruvate is oxidized in the mitochondria to acetyl-CoA.
- In the absence of oxygen, pyruvate is reduced to lactate and NAD^+ is regenerated for the continuation of glycolysis.

18.5 The Citric Acid Cycle

LEARNING GOAL Describe the oxidation of acetyl-CoA in the citric acid cycle.

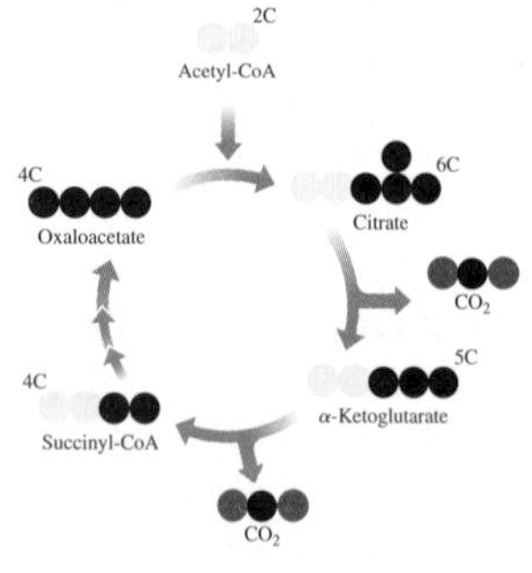

- In a sequence of reactions called the citric acid cycle, an acetyl group is combined with oxaloacetate to yield citrate.
- Citrate undergoes oxidation and decarboxylation to yield two CO_2, GTP, three NADH, and one $FADH_2$ with the regeneration of oxaloacetate.
- The direct phosphate transfer of ADP by GTP yields ATP.

18.6 Electron Transport and Oxidative Phosphorylation

LEARNING GOAL Describe electron transport and the process of oxidative phosphorylation; calculate the ATP from the complete oxidation of glucose.

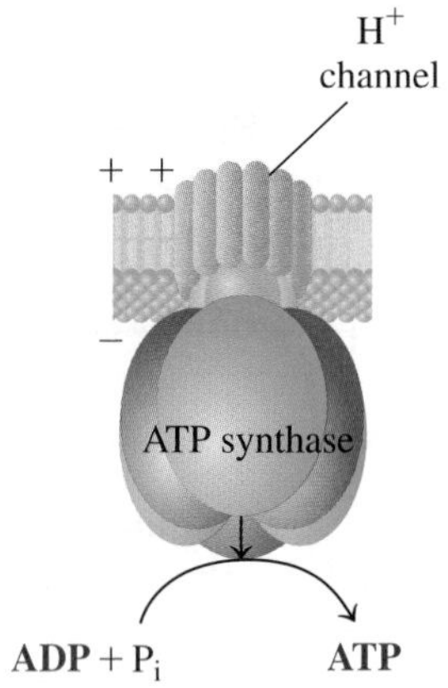

- The reduced coenzymes NADH and $FADH_2$ from various metabolic pathways are oxidized to NAD^+ and FAD when their hydrogen ions and electrons are transferred to the electron transport system.
- The energy released is used to synthesize ATP from ADP and P_i.
- The final acceptor, O_2, combines with hydrogen ions and electrons to yield H_2O.
- The protein complexes in electron transport act as proton pumps to move hydrogen ions into the intermembrane space, which produces a H^+ gradient.
- As the H^+ ions return to the matrix by way of ATP synthase, energy is generated.
- This energy is used to drive the synthesis of ATP in a process known as oxidative phosphorylation.
- The oxidation of NADH yields three ATP, and $FADH_2$ yields two ATP.
- Under aerobic conditions, the complete oxidation of glucose yields a total of 36 ATP.

18.7 Oxidation of Fatty Acids

LEARNING GOAL Describe the metabolic pathway of β-oxidation; calculate the ATP from the complete oxidation of a fatty acid.

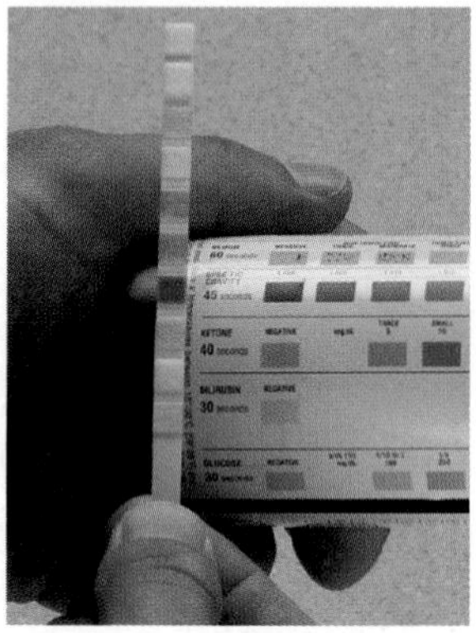

- When needed as an energy source, fatty acids are linked to coenzyme A and transported into the mitochondria where they undergo β-oxidation.
- The fatty acyl-CoA is oxidized to yield a shorter fatty acyl-CoA, acetyl-CoA, and reduced coenzymes NADH and $FADH_2$.
- Although the energy from a particular fatty acid depends on its length, each oxidation cycle yields 5 ATP with another 12 ATP from the acetyl-CoA that enters the citric acid cycle.
- When high levels of acetyl-CoA are present in the cell, they enter the ketogenesis pathway, forming ketone bodies such as acetoacetate, which cause ketosis and acidosis.

18.8 Degradation of Amino Acids

LEARNING GOAL Describe the reactions of transamination, oxidative deamination, and the entry of amino acid carbons into the citric acid cycle.

- When the amount of amino acids in the cells exceeds that needed for synthesis of nitrogen compounds, the process of transamination converts them to α-keto acids and glutamate.
- Oxidative deamination of glutamate produces ammonium ions and α-ketoglutarate.
- Ammonium ions from oxidative deamination are converted to urea.
- The carbon atoms from the degradation of amino acids can enter the citric acid cycle or other metabolic pathways.
- Certain amino acids are synthesized when amino groups from glutamate are transferred to an α-keto acid obtained from glycolysis or the citric acid cycle.

$$\underset{\text{Alanine}}{CH_3-\overset{\overset{+}{N}H_3}{\overset{|}{C}H}-COO^-} + \underset{\alpha\text{-Ketoglutarate}}{^-OOC-\overset{O}{\overset{\|}{C}}-CH_2-CH_2-COO^-} \xrightleftharpoons[]{\text{Alanine aminotransferase}}$$

$$\underset{\text{Pyruvate}}{CH_3-\overset{O}{\overset{\|}{C}}-COO^-} + \underset{\text{Glutamate}}{^-OOC-\overset{\overset{+}{N}H_3}{\overset{|}{C}H}-CH_2-CH_2-COO^-}$$

SUMMARY OF REACTIONS

The chapter sections for review are shown after the name of the reaction.

Hydrolysis of ATP (18.1)

$$ATP + H_2O \longrightarrow ADP + P_i + 7.3\ \text{kcal/mole}$$

Formation of ATP (18.1)

$$ADP + P_i + 7.3\ \text{kcal/mole} \longrightarrow ATP + H_2O$$

Hydrolysis of Disaccharides (18.2)

$$\text{Maltose} + H_2O \xrightarrow{\text{Maltase}} \text{glucose} + \text{glucose}$$

$$\text{Lactose} + H_2O \xrightarrow{\text{Lactase}} \text{galactose} + \text{glucose}$$

$$\text{Sucrose} + H_2O \xrightarrow{\text{Sucrase}} \text{fructose} + \text{glucose}$$

Hydrolysis of Triacylglycerols (18.2)

$$\text{Triacylglycerols} + 3H_2O \xrightarrow{\text{Lipases}} \text{glycerol} + 3\ \text{fatty acids}$$

Hydrolysis of Proteins (18.2)

$$\text{Protein} + H_2O \xrightarrow{\text{Proteases}} \text{amino acids}$$

Reduction of NAD$^+$ and FAD (18.3)

$$NAD^+ + 2H^+ + 2\,e^- \longrightarrow NADH + H^+$$

$$FAD + 2H^+ + 2\,e^- \longrightarrow FADH_2$$

Glycolysis (18.4)

$$\underset{\text{Glucose}}{C_6H_{12}O_6} + 2ADP + 2P_i + 2NAD^+ \longrightarrow \underset{\text{Pyruvate}}{2CH_3-\overset{O}{\overset{\|}{C}}-COO^-} + 2ATP + 2NADH + 4H^+$$

Oxidation of Pyruvate to Acetyl-CoA (Aerobic) (18.4)

$$\underset{\text{Pyruvate}}{CH_3-\overset{O}{\overset{\|}{C}}-COO^-} + NAD^+ + HS-CoA \longrightarrow \underset{\text{Acetyl-CoA}}{CH_3-\overset{O}{\overset{\|}{C}}-S-CoA} + NADH + CO_2$$

Reduction of Pyruvate to Lactate (Anaerobic) (18.4)

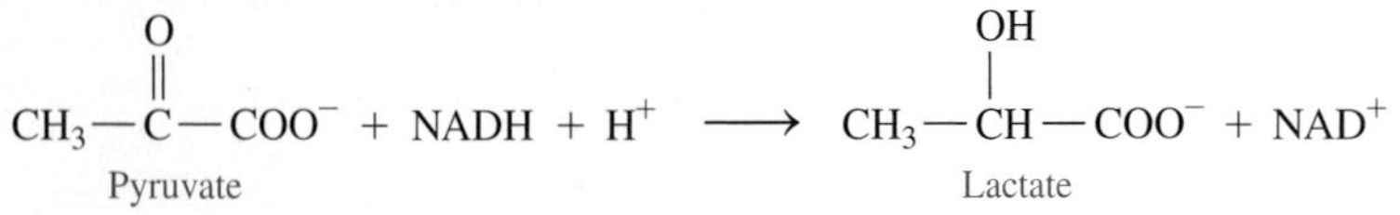

Citric Acid Cycle (18.5)

$$\text{Acetyl-CoA} + 3NAD^+ + FAD + GDP + P_i + 2H_2O \longrightarrow 2CO_2 + 3NADH + 3H^+ + FADH_2 + HS{-}CoA + GTP$$

Synthesis of ATP (ATP Synthase) (18.6)

$$NADH + H^+ + \tfrac{1}{2}O_2 + 3ADP + 3P_i \longrightarrow NAD^+ + H_2O + 3ATP$$

$$FADH_2 + \tfrac{1}{2}O_2 + 2ADP + 2P_i \longrightarrow FAD + H_2O + 2ATP$$

Oxidation of Glucose (Complete) (18.6)

$$C_6H_{12}O_6 + 6O_2 + 36ADP + 36P_i \longrightarrow 6CO_2 + 6H_2O + 36ATP$$

β-Oxidation of Fatty Acids (18.7)

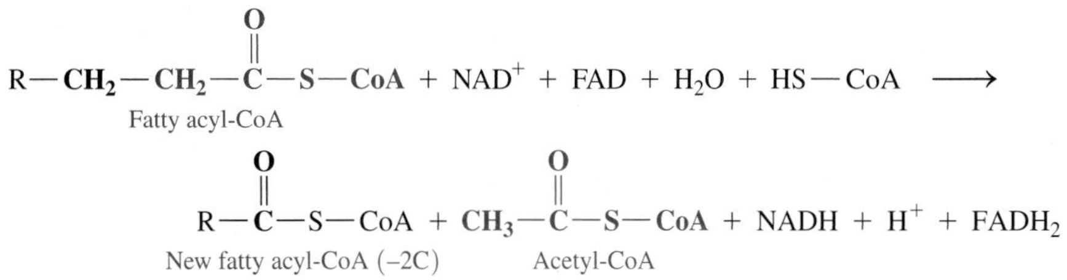

Transamination (18.8)

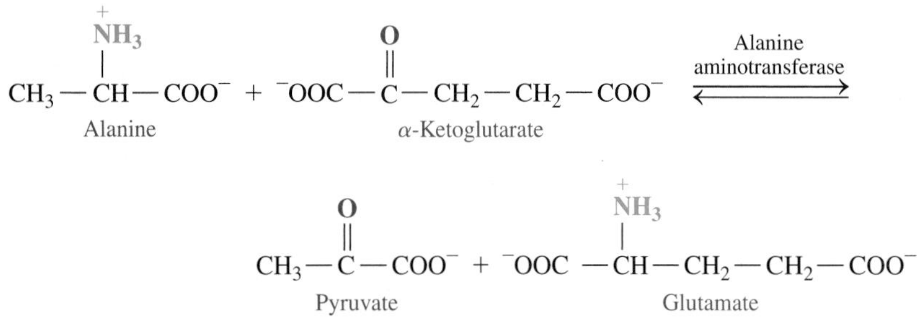

Oxidative Deamination (18.8)

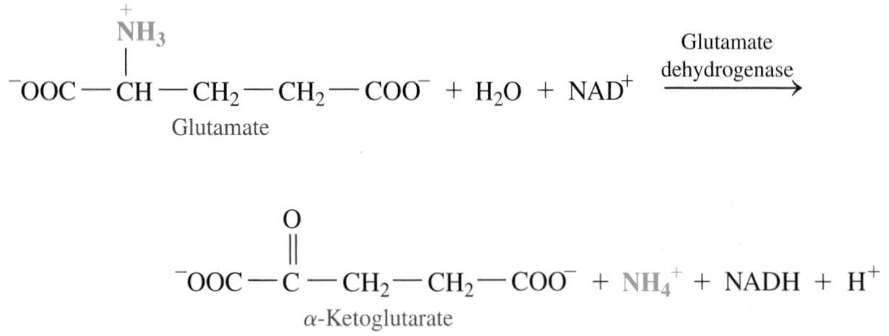

Urea Cycle (18.8)

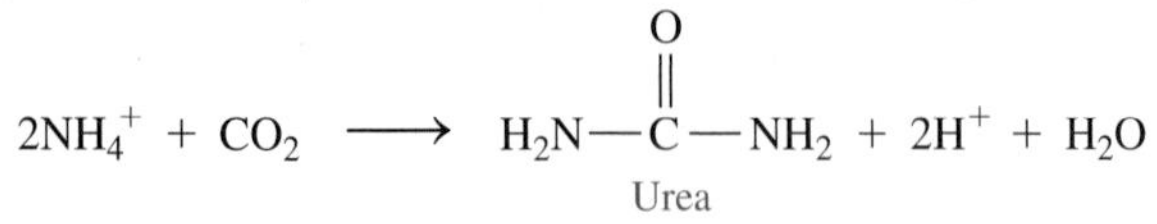

KEY TERMS

acetyl-CoA The compound that forms when a two-carbon acetyl group bonds to coenzyme A.

aerobic An oxygen-containing environment in the cells.

anabolic reaction A metabolic reaction that requires energy to build large molecules from small molecules.

anaerobic A condition in cells when there is no oxygen.

ATP Adenosine triphosphate; a high-energy compound that stores energy in the cells. It consists of adenine, a ribose sugar, and three phosphate groups.

ATP synthase An enzyme complex that uses the energy released by H^+ ions returning to the matrix to synthesize ATP from ADP and P_i.

beta (β)-oxidation The degradation of fatty acids that removes two-carbon segments from the fatty acid at the oxidized β-carbon.

catabolic reaction A metabolic reaction that produces energy for the cell by the degradation and oxidation of glucose and other molecules.

chemiosmotic model The conservation of energy from electron transport by pumping H^+ ions into the intermembrane space to produce a H^+ gradient that provides the energy to synthesize ATP.

citric acid cycle A series of oxidation reactions in the mitochondria that convert acetyl-CoA to CO_2 and yield NADH and $FADH_2$. It is also called the Krebs cycle.

coenzyme A (CoA) A coenzyme that transports acyl and acetyl groups.

decarboxylation The loss of a carbon atom in the form of CO_2.

digestion The processes in the gastrointestinal tract that break down large food molecules to smaller ones that pass through the intestinal membrane into the bloodstream.

electron transport A series of reactions in the mitochondria that transfer electrons from NADH and $FADH_2$ to electron carriers and finally to O_2, which produces H_2O. The energy released is used to synthesize ATP from ADP and P_i.

FAD A coenzyme (flavin adenine dinucleotide) for oxidation reactions that form carbon–carbon double bonds.

glycolysis The ten oxidation reactions of glucose that yield two pyruvate molecules.

ketone bodies The products of ketogenesis: acetoacetate, β-hydroxybutyrate, and acetone.

metabolism All the chemical reactions in living cells that carry out molecular and energy transformations.

NAD^+ A coenzyme (nicotinamide adenine dinucleotide) for oxidation reactions that form carbon–oxygen double bonds.

oxidative deamination The loss of ammonium ion when glutamate is degraded to α-ketoglutarate.

oxidative phosphorylation The synthesis of ATP from ADP and P_i, using energy generated by the oxidation reactions during electron transport.

proton pumps The enzyme complexes I, III, and IV that move H^+ ions from the matrix into the intermembrane space, creating a H^+ gradient.

transamination The transfer of an amino group from an amino acid to an α-keto acid.

urea cycle The process in which ammonium ions from the degradation of amino acids are converted to urea.

CORE CHEMISTRY SKILLS

Identifying the Compounds in Glycolysis (18.4)

- Glycolysis, which occurs in the cytosol, consists of 10 reactions that degrade glucose (6C) to two pyruvate (3C).
- The overall series of reactions in glycolysis yields two NADH and two ATP.

Example: Identify the reaction(s) and compounds in glycolysis that have each of the following:

a. requires ATP
b. converts a six-carbon compound to two three-carbon compounds
c. converts ADP to ATP

Answer: **a.** The phosphorylation of glucose in reaction 1 and fructose-6-phosphate in reaction 3 each require one ATP.
b. In reaction 4, the six-carbon fructose-1,6-bisphosphate is cleaved into two three-carbon compounds.
c. Phosphate transfer in reactions 7 and 10 converts ADP to ATP.

Describing the Reactions in the Citric Acid Cycle (18.5)

- In the initial reaction of the citric acid cycle, an acetyl group combines with oxaloacetate to yield citrate.
- Citrate undergoes several reactions including oxidation, decarboxylation, and hydration to yield two CO_2, GTP, three NADH, one $FADH_2$, and to form another oxaloacetate.

Example: Identify the compound(s) in the citric acid cycle that undergo each of the following changes:

a. loses a CO_2 molecule
b. adds water

Answer: **a.** Isocitrate (6C) undergoes oxidation and decarboxylation to form α-ketoglutarate (5C), and α-ketoglutarate (5C) undergoes oxidation and decarboxylation to form succinyl-CoA (4C).
b. Fumarate, which has a double bond, adds an H_2O molecule to form malate.

Calculating the ATP Produced from Glucose (18.6)

- The reduced coenzymes NADH and $FADH_2$ from various metabolic pathways are oxidized to NAD^+ and FAD when their H^+ ions and electrons are transferred to the electron transport system.
- The energy released is used to synthesize ATP from ADP and P_i.
- The final acceptor, O_2, combines with H^+ ions and electrons to yield H_2O.
- The protein complexes I, III, and IV in electron transport move hydrogen ions into the intermembrane space, which produces a H^+ gradient.
- As the H^+ ions return to the matrix by way of ATP synthase, ATP energy is generated in a process known as oxidative phosphorylation.

- The oxidation of NADH yields 3 ATP, and the oxidation of $FADH_2$ yields 2 ATP.
- Under aerobic conditions, the complete oxidation of glucose yields a total of 36 ATP.

Example: Calculate the ATP produced from each of the following:
a. one glucose to two pyruvate
b. one NADH to one NAD^+
c. complete oxidation of one glucose to 6 CO_2 and 6 H_2O

Answer: **a.** Glucose is converted to two pyruvate during glycolysis, which produces 2 ATP and 2 NADH (shuttle gives 4 ATP) for a total of 6 ATP.
b. One NADH is converted to one NAD^+ in the electron transport system, which produces 3 ATP.
c. Glucose is completely oxidized to 6 CO_2 and 6 H_2O in glycolysis, citric acid cycle, and electron transport, which produces 36 ATP.

Calculating the ATP from Fatty Acid Oxidation (β-Oxidation) (18.7)

- When needed as an energy source, fatty acids are linked to coenzyme A and transported into the mitochondria where they undergo β-oxidation.
- The fatty acyl-CoA is oxidized to yield a shorter fatty acyl-CoA, acetyl-CoA, and reduced coenzymes NADH and $FADH_2$.
- Although the energy from a particular fatty acid depends on its length, each oxidation cycle yields 5 ATP with another 12 ATP from each acetyl-CoA that enters the citric acid cycle.
- When high levels of acetyl-CoA are present in the cell, they enter the ketogenesis pathway, forming ketone bodies such as acetoacetate, which cause ketosis and acidosis.

Example: How many β-oxidation cycles, acetyl-CoA, and ATP are produced by the complete oxidation of caprylic acid, the C_8 saturated fatty acid?

Answer: An eight carbon fatty acid will undergo three β-oxidation cycles and produce four acetyl-CoA. Each oxidation cycle produces 5 ATP (1 NADH and 1 $FADH_2$) or a total of 15 ATP. Each acetyl-CoA produces 12 ATP in the citric acid cycle or a total of 48 ATP. The 63 ATP minus 2 ATP for activation gives a total of 61 ATP from caprylic acid.

UNDERSTANDING THE CONCEPTS

The chapter sections to review are given in parentheses at the end of each question.

18.71 Lauric acid, $CH_3—(CH_2)_{10}—COOH$, which is found in coconut oil, is a saturated fatty acid. (18.6, 18.7)

Coconut oil is high in lauric acid.

a. Draw the condensed structural formula for the activated form of lauric acid.
b. Indicate the α- and β-carbon atoms in lauroyl-CoA.
c. How many cycles of β-oxidation are needed?
d. How many acetyl-CoA are produced?
e. Calculate the total ATP yield from the complete β-oxidation of lauric acid by completing the following:

activation	⟶	−2 ATP
______ $FADH_2$	⟶	______ ATP
______ NADH	⟶	______ ATP
______ acetyl-CoA	⟶	______ ATP
Total		______ **ATP**

18.72 Arachidic acid $CH_3—(CH_2)_{16}—CH_2—CH_2—COOH$ is a C_{20} fatty acid found in peanut and fish oils. (18.6, 18.7)

Peanuts contain arachidic acid, a C_{20} saturated fatty acid.

a. Draw the condensed structural formula for the activated form of arachidic acid.
b. Indicate the α- and β-carbon atoms in arachidoyl-CoA.
c. How many cycles of β-oxidation are needed?
d. How many acetyl-CoA are produced?
e. Calculate the total ATP yield from the complete β-oxidation of arachidic acid by completing the following:

activation	⟶	−2 ATP
______ $FADH_2$	⟶	______ ATP
______ NADH	⟶	______ ATP
______ acetyl-CoA	⟶	______ ATP
Total		______ **ATP**

18.73 Identify the type of food as carbohydrate, fat, or protein that gives each of the following digestion products: (18.2)
a. glucose **b.** fatty acid
c. maltose **d.** glycerol
e. amino acids **f.** dextrins

18.74 Identify each of the following as a six-carbon or a three-carbon compound and arrange them in the order in which they occur in glycolysis: (18.4)
a. 3-phosphoglycerate
b. pyruvate
c. glucose-6-phosphate
d. glucose
e. fructose-1,6-bisphosphate

Digestion is the first stage of metabolism.

ADDITIONAL QUESTIONS AND PROBLEMS

18.75 Write an equation for the hydrolysis of ATP to ADP. (18.1)

18.76 How many moles of ATP will be required to produce 420 kcal of energy? (18.1)

18.77 What is the key component and function of Lactaid? (18.2)

18.78 How and where does sucrose undergo digestion in the body? What are the products? (18.2)

18.79 What are the reactant and product of glycolysis? (18.4)

18.80 What is the general type of reaction that takes place in the digestion of carbohydrates? (18.2)

18.81 When is pyruvate converted to lactate in the body? (18.4)

18.82 When pyruvate is used to form acetyl-CoA, the product has only two carbon atoms. What happened to the third carbon? (18.4)

18.83 What is the main function of the citric acid cycle in energy production? (18.5)

18.84 Most metabolic pathways are not considered cycles. Why is the citric acid cycle considered to be a metabolic cycle? (18.5)

18.85 If there are no reactions in the citric acid cycle that use oxygen, O_2, why does the cycle operate only in aerobic conditions? (18.5)

18.86 What products of the citric acid cycle are needed for electron transport? (18.5, 18.6)

18.87 In the chemiosmotic model, how is energy provided to synthesize ATP? (18.6)

18.88 What is the effect of H^+ accumulation in the intermembrane space? (18.6)

18.89 What is the function of insulin and glucagon in our body? (18.7)

18.90 Write a balanced equation to show how alanine is converted to pyruvate. (18.8)

CHALLENGE QUESTIONS

The following groups of questions are related to the topics in this chapter. However, they do not all follow the chapter order, and they require you to combine concepts and skills from several sections. These questions will help you increase your critical thinking skills and prepare for your next exam.

18.91 One cell at work may break down 2 million (2 000 000) ATP molecules in one second. Some researchers estimate that the human body has about 10^{13} cells. (18.1)
a. How much energy, in kilocalories, would be used by the cells in the body in one day?
b. If ATP has a molar mass of 507 g/mole, how many grams of ATP are hydrolyzed?

18.92 State if each of the following processes produce or consume ATP: (18.4, 18.5, 18.6)
a. citric acid cycle
b. glucose forms two pyruvate
c. pyruvate forms acetyl-CoA
d. glucose forms glucose-6-phosphate
e. oxidation of α-ketoglutarate
f. transport of NADH across the mitochondrial membrane
g. activation of a fatty acid

18.93 Match the following ATP yields to reactions **a** to **g**: (18.4, 18.5, 18.6)

2 ATP	3 ATP	6 ATP	12 ATP
18 ATP	36 ATP	44 ATP	

a. Glucose forms two pyruvate.
b. Pyruvate forms acetyl-CoA.
c. Glucose forms two acetyl-CoA.
d. Acetyl-CoA goes through one turn of the citric acid cycle.
e. Caproic acid (C_6) is completely oxidized.
f. $NADH + H^+$ is oxidized to NAD^+.
g. $FADH_2$ is oxidized to FAD.

18.94 Identify each of the following reactions **a** to **e** in the β-oxidation of palmitic acid, a C_{16} fatty acid, as (18.7)
(1) activation
(2) first oxidation
(3) hydration
(4) second oxidation
(5) cleavage
a. Palmitoyl-CoA and FAD form α, β-unsaturated palmitoyl-CoA and $FADH_2$.
b. β-Keto palmitoyl-CoA forms myristoyl-CoA and acetyl-CoA.
c. Palmitic acid, CoA, and ATP form palmitoyl-CoA.
d. α, β-Unsaturated palmitoyl-CoA and H_2O form β-hydroxy palmitoyl-CoA.
e. β-Hydroxy palmitoyl-CoA and NAD^+ form β-keto palmitoyl-CoA and $NADH + H^+$.

18.95 Which of the following molecules will produce the most ATP per mole? (18.5, 18.6)
a. glucose or maltose
b. myristic acid, $CH_3—(CH_2)_{12}—COOH$, or stearic acid, $CH_3—(CH_2)_{16}—COOH$
c. glucose or two acetyl-CoA
d. glucose or caprylic acid (C_8)
e. citrate or succinate in one turn of the citric acid cycle

18.96 Which of the following molecules will produce the most ATP per mole? (18.5, 18.6)
a. glucose or stearic acid (C_{18})
b. glucose or two pyruvate
c. two acetyl-CoA or one palmitic acid (C_{16})
d. lauric acid (C_{12}) or palmitic acid (C_{16})
e. α-ketoglutarate or fumarate in one turn of the citric acid cycle

ANSWERS

Answers to Study Checks

18.1 The breakdown of glucose to smaller molecules is a catabolic reaction.

18.2 The digestion of amylose begins in the mouth where enzymes hydrolyze some of the glycosidic bonds. In the small intestine, enzymes hydrolyze more glycosidic bonds, and finally maltose is hydrolyzed to yield glucose.

18.3 The thiol group (—SH) in coenzyme A combines with an acetyl group to form acetyl-coenzyme A, which participates in the transfer of acetyl groups.

18.4 The splitting of fructose-1,6-bisphosphate is cleavage.

18.5 oxaloacetate

18.6 The protein complexes I, III, and IV pump H^+ ions from the matrix to the intermembrane space.

18.7 One turn of the citric acid cycle produces a total of 12 ATP: 9 ATP from 3 NADH, 2 ATP from $FADH_2$, and 1 ATP from a phosphorylation.

18.8 NAD^+

18.9 In β-oxidation, each acetyl-CoA enters the citric acid cycle to produce 3 NADH, $FADH_2$, and GTP for a total of 12 ATP. The reduced coenzymes NADH and $FADH_2$ enter electron transport to give 5 ATP.

18.10 The process of oxidative deamination involves the removal of the amino group of glutamate as ammonium.

18.11 Arginine, glutamate, glutamine, histidine, and proline provide carbon atoms for α-ketoglutarate.

Answers to Selected Questions and Problems

18.1 The digestion of polysaccharides takes place in stage 1.

18.5 The hydrolysis of the phosphodiester bond in ATP releases energy that is sufficient for energy-requiring processes in the cell.

18.7 Hydrolysis is the main reaction involved in the digestion of carbohydrates.

18.9 The bile salts emulsify fat to give small fat globules for lipase hydrolysis.

18.11 The digestion of proteins begins in the stomach and is completed in the small intestine.

18.13 **a.** NADH
b. FAD
c. FAD

18.15 **a.** coenzyme A
b. NAD^+
c. FAD

18.17 glucose

18.19 ATP is required in phosphorylation reactions.

18.21 ATP is produced in glycolysis by transferring a phosphate from 1,3-bisphosphoglycerate and from phosphoenolpyruvate directly to ADP.

18.23 **a.** 1 ATP is required.
b. 1 NADH is produced for each triose.
c. 2 ATP and 2 NADH are produced.

18.25 Aerobic (oxygen) conditions are needed.

18.27 The oxidation of pyruvate converts NAD^+ to NADH and produces acetyl-CoA and CO_2.

$$\text{Pyruvate} + NAD^+ + \text{HS—CoA} \longrightarrow \text{acetyl-CoA} + CO_2 + NADH + H^+$$

18.29 When pyruvate is reduced to lactate, the NAD^+ is used to oxidize glyceraldehyde-3-phosphate, which recycles NADH.

18.31 $2CO_2$, 3NADH, $FADH_2$, GTP (ATP), HS—CoA, and $3H^+$.

18.35 NAD^+ is reduced in reactions 3, 4, and 8 of the citric acid cycle.

18.37 In reaction 5, GDP undergoes a direct phosphate transfer.

18.39 **a.** citrate and isocitrate
b. In decarboxylation, a carbon atom is lost as CO_2.
c. α-ketoglutarate
d. isocitrate $\longrightarrow$ α-ketoglutarate; α-ketoglutarate $\longrightarrow$ succinyl-CoA; succinate $\longrightarrow$ fumarate; malate $\longrightarrow$ oxaloacetate

18.41 NADH and $FADH_2$

18.43 The mobile carrier coenzyme Q transfers electrons from complex I to III.

18.45 NADH transfers electrons to complex I to give NAD^+.

18.47 In oxidative phosphorylation, the energy from the oxidation reactions in electron transport is used to drive ATP synthesis.

18.49 Hydrogen ions return to a lower energy in the matrix by passing through ATP synthase, which releases energy to drive the synthesis of ATP.

18.51 The reduced coenzymes NADH and $FADH_2$ from glycolysis and the citric acid cycle transfer electrons to the electron transport system, which generates energy to drive the synthesis of ATP.

18.53 **a.** 3 ATP **b.** 6 ATP
c. 6 ATP **d.** 12 ATP

18.55 In the cytosol at the outer mitochondrial membrane

18.57 **a.** and **b.**

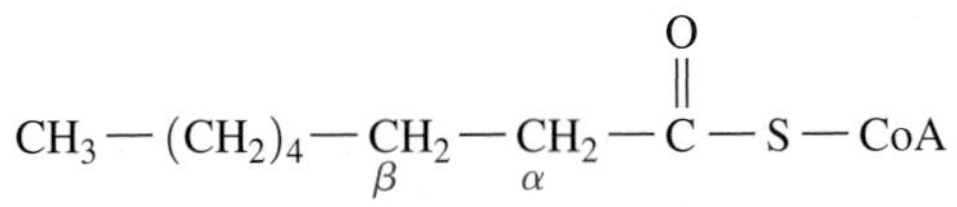

c. 3 cycles of β-oxidation
d. 4 acetyl-CoA

18.59 **a.** 8 cycles of β-oxidation
b. 9 acetyl-CoA
c. 108 ATP from 9 acetyl-CoA (citric acid cycle) + 24 ATP from 8 NADH + 16 ATP from 8 $FADH_2$ − 2 ATP (activation) = 148 − 2 = 146 ATP

18.63 High levels of ketone bodies lead to ketosis, a condition characterized by acidosis (a drop in blood pH values), excessive urination, and strong thirst.

18.65 **a.** 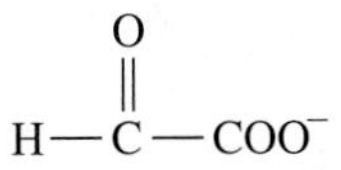 **b.** 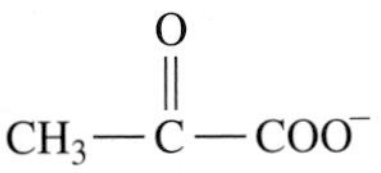

18.67 NH_4^+ is toxic if allowed to accumulate in the body.

18.69 **a.** pyruvate **b.** oxaloacetate and fumarate
c. fumarate **d.** α-ketoglutarate

18.71 **a.** and **b.**

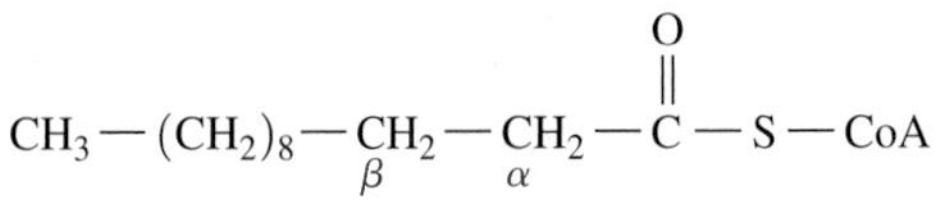

c. Five cycles of β-oxidation are needed.
d. Six acetyl-CoA are produced.
e.

activation		$\longrightarrow$	−2 ATP
5 $FADH_2$	× 2 ATP/$FADH_2$	$\longrightarrow$	10 ATP
5 NADH	× 3 ATP/NADH	$\longrightarrow$	15 ATP
6 acetyl-CoA	× 12 ATP/acetyl-CoA	$\longrightarrow$	72 ATP
		Total	95 ATP

18.73 **a.** carbohydrate **b.** fat
c. carbohydrate **d.** fat
e. protein **f.** carbohydrate

18.75 ATP + H_2O $\longrightarrow$ ADP + P_i + 7.3 kcal/mole

18.79 Glucose is the reactant, and pyruvate is the product of glycolysis.

18.81 Pyruvate is converted to lactate when oxygen is not present in the cell (anaerobic conditions) to regenerate NAD^+ for glycolysis.

18.83 The oxidation reactions of the citric acid cycle produce a source of reduced coenzymes for electron transport and ATP synthesis.

18.85 The oxidized coenzymes NAD^+ and FAD needed for the citric acid cycle are regenerated by electron transport, which requires oxygen.

18.87 Energy released as H^+ ions flow through ATP synthase and then back to the matrix is utilized for the synthesis of ATP.

18.91 **a.** 21 kcal **b.** 1500 g of ATP

18.93 **a.** 6 ATP/glucose **b.** 3 ATP/pyruvate
c. 12 ATP/glucose **d.** 12 ATP/acetyl-CoA
e. 44 ATP/C_6 acid **f.** 3 ATP/NADH
g. 2 ATP/$FADH_2$

18.95 **a.** maltose **b.** stearic acid
c. glucose **d.** caprylic acid
e. citrate

COMBINING IDEAS FROM Chapters 16 to 18

CI.33 Beano contains an enzyme that breaks down polysaccharides into mono- and disaccharides that are more digestible. It is used to diminish gas formation that can occur after eating foods such as beans or cruciferous vegetables like cabbage, Brussels sprouts, and broccoli. (16.5, 16.6)

Beano is a dietary supplement that is used to reduce gas.

a. The label on beano says "contains alpha-galactosidase." What class of enzyme is present in beano?
b. What is the substrate for the enzyme?
c. The directions indicate you should not heat or cook beano. Why?

CI.34 Kevlar is a lightweight polymer used in tires and bulletproof vests. Part of the strength of Kevlar is due to hydrogen bonds between polymer chains. (14.6, 16.4)

Kevlar is used to make bulletproof vests.

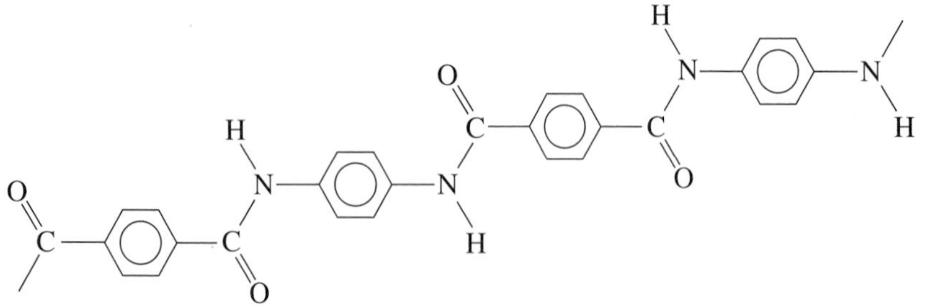

Amide bonds between the carboxyl and amine groups form the polymer Kevlar.

a. Draw the condensed structural formula for the carboxylic acid and amine that are polymerized to make Kevlar.
b. What feature of Kevlar will produce hydrogen bonds between the polymer chains?
c. What type of secondary protein structure has bonds that are similar to those in Kevlar?

CI.35 Identify each of the following as a substance that is part of the citric acid cycle, electron transport, or both: (18.4, 18.5, 18.6)

a. succinate
b. QH_2
c. FAD
d. cyt *c*
e. H_2O
f. malate
g. NAD^+

CI.36 Use the energy value of 7.3 kcal per mole of ATP and determine the total energy, in kilocalories, stored as ATP from each of the following: (18.4, 18.5, 18.6)

a. the reactions of 1 mole of glucose in glycolysis
b. the oxidation of 2 moles of pyruvate to 2 moles of acetyl-CoA
c. the complete oxidation of 1 mole of glucose to CO_2 and H_2O
d. the β-oxidation of 1 mole of lauric acid, a C_{12} fatty acid, in β-oxidation
e. the reaction of 1 mole of glutamate (from protein) in the citric acid cycle

CI.37 Acetyl-CoA is the fuel for the citric acid cycle. It has the formula $C_{23}H_{38}N_7O_{17}P_3S$. (7.2, 18.3, 18.5)

a. What are the components of acetyl-CoA?
b. What is the function of acetyl-CoA?
c. Where does the acetyl group attach to CoA?
d. What is the molar mass, to 3 significant figures, of acetyl-CoA?

CI.38 Behenic acid is a C_{22} saturated fatty acid (22:0) found in peanut and canola oils. (18.7)

Behenic acid is one of the fatty acids found in peanut oil.

a. Draw the condensed structural formula for the activated form of behenic acid.
b. Indicate the α-and β-carbon atoms in behenoyl-CoA.
c. How many cycles of β-oxidation are needed?
d. How many acetyl-CoA are produced?
e. Calculate the total ATP yield from the complete β-oxidation of behenic acid by completing the following:

activation	⟶	−2 ATP
______ $FADH_2$	⟶	______ ATP
______ NADH	⟶	______ ATP
______ acetyl-CoA	⟶	______ ATP
Total		______ **ATP**

CI.39 Butter is a fat that contains 80% by mass triacylglycerols; the rest is water. Assume the triacylglycerol in butter is glyceryl tripalmitate. (2.4, 2.5, 2.6, 7.2, 7.6, 14.4, 18.7)

Butter is high in triacylglycerols.

a. Write the balanced chemical equation for the hydrolysis of glyceryl tripalmitate.
b. What is the molar mass of glyceryl tripalmitate, $C_{51}H_{98}O_6$?
c. Calculate the ATP yield from the complete oxidation of palmitic acid.
d. How many kilocalories are released from the palmitic acid in a 0.50-oz pat of butter?
e. If running for 1 h uses 750 kcal, how many pats of butter would provide the energy (kcal) for a 45-min run?

CI.40 Thalassemia is an inherited genetic mutation that limits the production of the beta chain needed for the formation of hemoglobin. If low levels of the beta chain are produced, there is a shortage of red blood cells (anemia); as a result, the body does not have sufficient amounts of oxygen. In one form of thalassemia, a single nucleotide is deleted in the DNA that codes for the beta chain. The mutation involves the deletion of thymine (T) from section 91 in the following coding strand of normal DNA. (16.3, 17.3, 17.4, 17.5, 17.6)

89	90	91	92	93	94	95
—AGT	GAG	CTG	CAC	TGT	GAC	A. . . .

a. Write the complementary strand for the normal DNA segment.
b. Write the mRNA sequence from DNA using the template strand in part **a**.
c. What amino acids are placed in the beta chain by this portion of mRNA?
d. What is the order of nucleotides in the mutated strand?
e. Write the template strand for the mutated DNA section.
f. Write the mRNA sequence from the mutated DNA segment, using the template strand in part **e**.
g. What amino acids are placed in the beta chain by the mutation?
h. What type of mutation occurs in this form of thalassemia?
i. How might the properties of this section of the beta chain be different from the properties of the normal protein?
j. How might the level of structure in hemoglobin be affected if beta chains are not produced?

CI.41 In response to signals from the nervous system, the hypothalamus secretes a polypeptide hormone known as gonadotropin-releasing factor (GnRF), which stimulates the pituitary gland to release other hormones into the bloodstream. Two of these hormones are luteinizing hormone (LH), and follicle-stimulating hormone (FSH). GnRF is a decapeptide with the following primary structure: (16.1, 16.2, 16.3)

Glu—His—Trp—Ser—Tyr—Gly—Leu—Arg—Pro—Gly

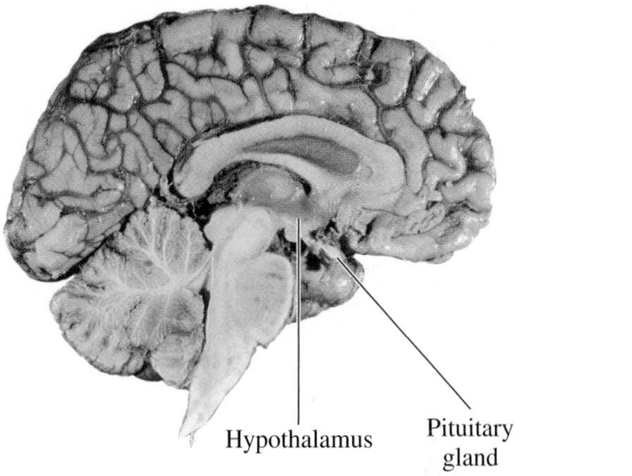

The hypothalamus secretes GnRF.

a. What is the N-terminal amino acid in GnRF?
b. What is the C-terminal amino acid in GnRF?
c. Which amino acids in GnRF are nonpolar or polar neutral?
d. Draw the condensed structural formulas for the acidic or basic amino acids in GnRF at physiological pH.

ANSWERS

CI.33 **a.** An alpha-galactosidase is a hydrolase.
b. The substrate is the α-1,4-glycosidic bond of galactose.
c. High temperatures will denature the hydrolase enzyme so it no longer functions.

CI.35 **a.** citric acid cycle **b.** electron transport
c. both **d.** electron transport
e. both **f.** citric acid cycle
g. both

CI.37 **a.** aminoethanethiol, pantothenic acid, phosphorylated adenosine diphosphate
b. Coenzyme A carries an acetyl group to the citric acid cycle for oxidation.
c. The acetyl group links to the S atom in the aminoethanethiol part of CoA.
d. 810. g/mole

CI.41 **a.** glutamic acid (Glu)
b. glycine (Gly)
c. The nonpolar amino acids are glycine (Gly), leucine (Leu), tryptophan (Trp), and proline (Pro). The polar neutral amino acids are serine (Ser) and tyrosine (Tyr).
d. The acidic amino acid is glutamic acid (Glu) and the basic amino acids are histidine (His) and arginine (Arg). At the pH of the body (about 7.4), glutamic acid will have a negative charge while histidine and arginine will be positively charged.

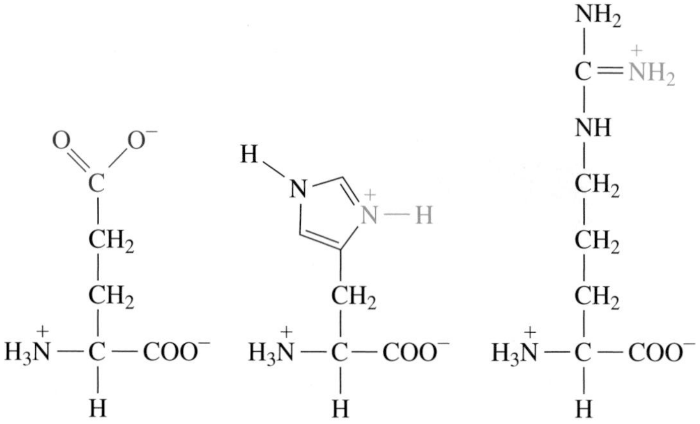

Glutamic acid (Glu) Histidine (His) Arginine (Arg)

CI.39 **a.**

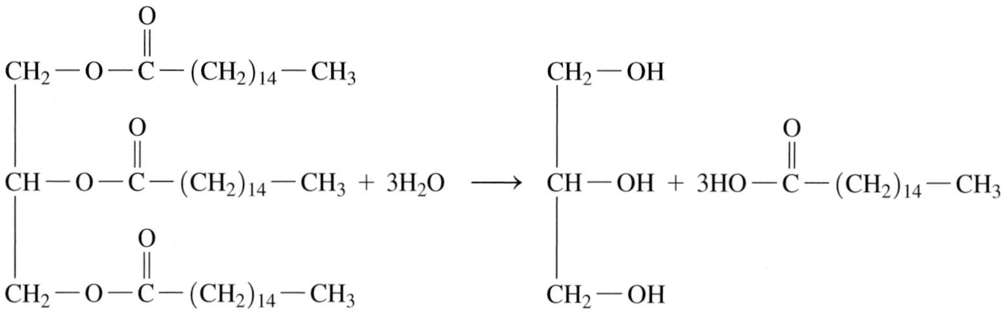

b. glyceryl tripalmitate, 807 g/mole
c. 129 ATP
d. 40. kcal
e. 14 pats of butter

Credits

Covers

Front Fer Gregory / Shutterstock
Back James Jordan Photography / Getty Images

Front Matter

p. 15 Natural History Museum of Los Angeles

Chapter 1

p. 33 nyaivanova / Shutterstock
p. 34 *left:* AndreasReh / iStockPhoto
p. 34 *bottom, right:* Pearson Education
p. 35 *top:* Eric Schrader / Pearson Education
p. 35 *bottom:* photos.com / Jupiterimages
p. 36 *top:* J. R. Eyerman / Time & Life Pictures / Getty Images
p. 36 *bottom:* The Print Collector / Alamy
p. 37 *middle, left:* Mark III Photonics / Shutterstock
p. 37 *middle, right:* webphotographeer / iStockphoto
p. 37 *bottom:* Pressmaster / Shutterstock
p. 38 evgenyb / iStockphoto
p. 39 track5 / iStockphoto
p. 40 Design Pics Inc. / Alamy
p. 42 Dmitry Rukhlenko / Fotolia
p. 43 Joe Gough / Fotolia
p. 44 Kevin Peschke / Alamy
p. 46 *left:* Jason Stitt / Shutterstock
p. 46 *right:* SPL / Science Source
p. 47 Science Photo Library / Alamy
p. 50 *left, top:* Pearson Education
p. 50 *left, bottom:* Pressmaster / Shutterstock
p. 50 *right, top:* track5 / iStockphoto
p. 50 *right, bottom:* Joe Gough / Fotolia
p. 52 coolgraphic / iStockPhoto

Chapter 2

p. 55 AVAVA / Shutterstock
p. 56 Rod Ferris / Shutterstock
p. 57 pastorscott / iStockPhoto
p. 58 *top, left:* National Institute of Standards and Technology
p. 58 *top, right:* Pearson Education / Pearson Science
p. 58 *bottom, left:* Richard Megna / Fundamental Photographs, NYC
p. 58 *bottom, right:* Pearson Education / Pearson Science
p. 59 Pearson Education / Pearson Science
p. 61 Kameleonp / iStockPhoto
p. 63 Radu Razvan / Shutterstock
p. 64 emin kuliyev / Shutterstock
p. 67 *top, right:* Monkey Business Images / Shutterstock
p. 67 *bottom, left:* Pearson Education / Pearson Science
p. 68 Pearson Education / Pearson Science
p. 69 Pearson Education / Pearson Science
p. 70 Eric Schrader / Pearson Education / Pearson Science
p. 71 *top:* Syner-Comm / Alamy
p. 71 *middle:* Eric Schrader / Pearson Education / Pearson Science
p. 71 *bottom:* Custom Medical Stock Photo
p. 72 *top:* Eric Schrader / Pearson Education / Pearson Science
p. 72 *bottom:* ExQuisine / Fotolia
p. 73 carlosalvarez / iStockPhoto
p. 77 Aprescindere / Dreamstime LLC
p. 78 Pearson Education / Pearson Science
p. 79 *top, left:* Pearson Education / Pearson Science
p. 79 *top, right:* Pearson Education / Pearson Science
p. 79 *bottom, left:* Professor Pietro M. Motta / Science Source
p. 79 *bottom, middle:* Professor Pietro M. Motta / Science Source
p. 79 *bottom, right:* VOISIN / PHANIE / Science Source
p. 81 Dallas Events Inc / Shutterstock
p. 82 *left, top:* Pearson Education / Pearson Science
p. 82 *left, bottom:* emin kuliyev / Shutterstock
p. 82 *right, top:* Eric Schrader / Pearson Education / Pearson Science
p. 82 *right, middle:* Aprescindere / Dreamstime LLC
p. 82 *right, bottom:* Pearson Education / Pearson Science
p. 84 romans14 / Fotolia
p. 86 Cristian Ciureanu / Alamy

Chapter 3

p. 89 Network Photographer / Alamy
p. 90 anthonysp / iStockPhoto
p. 90 Pearson Education / Pearson Science
p. 91 *top, right:* Bomshtein / Shutterstock
p. 91 *middle, right:* Eric Schrader / Pearson Education / Pearson Science
p. 91 *top, left:* Pearson Education / Pearson Science
p. 91 *middle, right:* Pearson Education / Pearson Science
p. 91 *bottom, right:* David Murray and Jules Selmes / Dorling Kindersley
p. 91 *middle, left:* Richard Megna / Fundamental Photographs, NYC
p. 92 *top, left:* skodonnell / iStockPhoto
p. 92 *top, 2nd from left:* Pearson Education / Pearson Science
p. 92 *top, 2nd from right:* Lagui / Shutterstock
p. 92 *top, right:* Pearson Education / Pearson Science
p. 92 *bottom, left:* Richard Megna / Fundamental Photographs
p. 92 *bottom, right:* Eric Schrader / Pearson Education / Pearson Science
p. 93 Nicolas Dupont / Fotolia
p. 94 *top, left:* Siede Preis / Photodisc / Getty Images
p. 94 *bottom, left:* Justinb / Fotolia
p. 94 *bottom: right:* Lisay / iStockPhoto
p. 95 *top, left:* Ray Roberts / Alamy
p. 95 *top, right:* Elena Elisseeva / Shutterstock
p. 95 *middle:* Rafa Irusta / Shutterstock
p. 95 *bottom:* Westend61 GmbH / Alamy
p. 96 Digital Vision / Alamy
p. 100 JenniferPhotographyImaging / iStockPhoto
p. 101 Science Photo Library / Alamy
p. 102 NASA / John F. Kennedy Space Center
p. 103 Pearson Education / Pearson Science
p. 105 Stockbyte / Getty Images RF
p. 106 *top:* Celso Pupo / Fotolia
p. 106 *bottom:* Alan Holden / Alamy
p. 108 *top:* Mark Downey / Photodisc / Getty Images
p. 108 *bottom, left:* John A. Rizzo / Photodisc / Getty Images
p. 108 *bottom, right:* John A. Rizzo / Photodisc / Getty Images
p. 110 *top, left:* Pearson Education / Pearson Science
p. 110 *top, right:* Pearson Education / Pearson Science
p. 110 *bottom, right:* helenecanada / iStockPhoto
p. 111 *top:* jonnyscriv / iStockPhoto
p. 111 *bottom:* Pearson Education / Pearson Science
p. 112 Comstock Images
p. 113 *top:* Mark Downey / Photodisc / Getty Images
p. 113 *middle:* John A. Rizzo / Photodisc / Getty Images
p. 113 *bottom:* John A. Rizzo / Photodisc / Getty Images
p. 117 *top:* Richard Megna / Fundamental Photographs, NYC
p. 117 *middle, left:* Justinb / Fotolia

p. 117 *bottom:* JenniferPhotography-Imaging / iStockPhoto
p. 117 *middle, right:* Celso Pupo / Fotolia
p. 120 *top, left:* Pearson Education / Pearson Science
p. 120 *top, right:* Masterfile Corporation
p. 120 *top. 2nd from left:* Spencer Jones / Photodisc / Getty Images
p. 120 *bottom, left:* starfotograf / iStockPhoto
p. 120 *middle, right:* FuatKose / iStockPhoto
p. 121 Khoroshunova Olga / Shutterstock
p. 125 *top:* Photolibrary RF / Getty Images
p. 125 *middle:* Shalom Ormsby / Photodisc / Getty Images
p. 125 *bottom:* Eric Schrader / Pearson Education / Pearson Science
p. 126 *left:* Photolibrary, Inc / Getty Images
p. 126 *right:* Masterfile

Chapter 4

p. 127 Martin Harvey / Alamy
p. 129 *all:* Pearson Education / Pearson Science
p. 130 *right:* Atiketta Sangasaeng / Shutterstock
p. 130 *left:* gabyjalbert / iStockPhoto
p. 131 Mary Ann Sullivan
p. 133 *top:* Pearson Education / Pearson Science
p. 133 *2nd from top:* Pearson Education / Pearson Science
p. 133 *3rd from top:* Pearson Education / Pearson Science
p. 133 *bottom:* Polushkin Ivan / Shutterstock
p. 134 *all:* Pearson Education / Pearson Science
p. 137 Russ Lappa / Pearson Education / Prentice Hall
p. 138 Image originally created by IBM Corporation
p. 139 Pearson Education / Pearson Science
p. 141 *left:* Eric Schrader / Fundamental Photographs
p. 141 *right:* Pearson Education / Pearson Science
p. 144 Pearson Education / Pearson Science
p. 145 tarasov / Shutterstock
p. 147 lior2 / iStockphoto
p. 149 Thorsten Rust / Shutterstock
p. 151 Cindy Minear / Shutterstock
p. 153 Pearson Education / Pearson Science
p. 158 Pearson Education / Pearson Science
p. 162 craftvision / iStockphoto

Chapter 5

p. 166 Konstantin Sutyagin / Shutterstock
p. 169 joshblake / iStockphoto
p. 170 Health Protection Agency / Science Source
p. 171 uberphotos / iStockphoto
p. 172 Pearson Education / Pearson Science
p. 176 Pearson Education / Pearson Science
p. 177 *top:* James McPherson / AP Photo
p. 177 *bottom:* Don Farrall / Getty Images
p. 178 Pearson Education / Pearson Science
p. 179 Jürgen Schulzki / Alamy
p. 182 Library of Congress
p. 183 Jihad Siqlawi / Newscom
p. 184 Editorial Image, LLC / Alamy
p. 185 *top:* Pasieka / Science Source
p. 185 *bottom:* Lawrence Berkeley National Library / Getty Images
p. 186 *top:* National Cancer Institute / Getty Images
p. 186 *bottom:* GJLP / Science Source
p. 187 *all:* Courtesy of Cytyc Hologic Corporation
p. 190 foto-rolf / iStockphoto
p. 191 *left:* Don Farrall / Getty Images
p. 191 *right:* Lawrence Berkeley National Library / Getty Images
p. 193 Peter Buckley / Pearson Education / Pearson Science

Chapter 6

p. 197 Design Pics Inc. / Alamy
p. 198 buckarooh / iStockphoto
p. 202 Pearson Education / Pearson Science
p. 203 *top, left:* Richard Megna / Fundamental Photographs
p. 203 *top, right:* Pearson Education / Pearson Science
p. 203 *middle:* Pearson Education / Pearson Science
p. 203 *bottom:* Pearson Education / Pearson Science
p. 205 Eric Schrader / Pearson Education / Pearson Science
p. 207 Gary Blakeley / Shutterstock
p. 208 mark huls / Fotolia
p. 209 *all:* Pearson Education / Pearson Science
p. 210 Smirnov Vladimir Itar-Tass Photos / Newscom
p. 233 *top:* Pearson Education / Pearson Science
p. 233 *middle:* Pearson Education / Pearson Science
p. 233 *bottom:* Kameel4u / Shutterstock
p. 241 *top, left:* Pearson Education / Pearson Science
p. 241 *top, right:* Augustin Ochsenreiter / AP Photo
p. 241 *bottom:* Courtesy of AlsoSalt
p. 242 Courtesy of Family Safety Products, Inc.

Chapter 7

p. 243 Javier Larrea / age fotostock
p. 245 *left:* Pearson Education / Pearson Science
p. 245 *right:* lissart / iStockphoto
p. 246 Reika / Shutterstock
p. 247 Pearson Education / Pearson Science
p. 248 Michael Hill / iStockphoto
p. 249 *left:* Pearson Education / Pearson Science
p. 249 *right:* Polushkin Ivan / Shutterstock
p. 250 *all:* Pearson Education / Pearson Science
p. 251 *top:* ruzanna / Shutterstock
p. 251 *bottom:* Eric Schrader / Pearson Education / Pearson Science
p. 252 Kevin Wheal / Alamy
p. 253 *top:* Tamara Kulikova / iStockphoto
p. 253 *bottom, left:* Pearson Education / Pearson Science
p. 253 *bottom, right:* Pearson Education / Pearson Science
p. 254 Pearson Education / Pearson Science
p. 256 Pearson Education / Pearson Science
p. 258 Richard Megna / Fundamental Photographs, NYC
p. 260 *all:* Pearson Education / Pearson Science
p. 261 *top:* Tom Bochsler / Pearson Education / Pearson Science
p. 261 *bottom, left:* Pearson Education / Pearson Science
p. 261 *bottom, right:* Pearson Education / Pearson Science
p. 262 *top:* Pearson Education / Pearson Science
p. 262 *bottom:* Sergiy Zavgorodny / Shutterstock
p. 264 Mitch Hrdlicka / Getty Images
p. 265 *top:* Stinkyt / iStockphoto
p. 265 *bottom, left:* Pearson Education / Pearson Science
p. 265 *bottom, middle*: Pearson Education / Pearson Science
p. 265 *bottom, right:* Pearson Education / Pearson Science
p. 267 *all:* Pearson Education / Pearson Science
p. 268 *all:* Pearson Education / Pearson Science
p. 270 Thinkstock Images / Getty Images
p. 272 Richard Megna / Fundamental Photographs, NYC
p. 273 Pearson Education / Pearson Science
p. 276 *left, top:* lissart / iStockphoto
p. 276 *left, bottom:* Pearson Education / Pearson Science
p. 276 *right, top, all three:* Pearson Education / Pearson Science
p. 276 *right, bottom:* Thinkstock Images / Getty Images
p. 277 *left:* Gabriele Rohde / Fotolia
p. 277 *right:* Ljupco Smokovski / Fotolia
p. 279 *all:* Pearson Education / Pearson Science
p. 282 *left:* MvH / iStockphoto
p. 282 *right:* Pearson Education / Pearson Science

Chapter 8

p. 286 Adam Gault / Alamy
p. 289 Courtesy of NASA
p. 291 Andrzej Gibasiewicz / Shutterstock
p. 292 Pearson Education / Pearson Science

p. 293 Levent Konuk / Shutterstock
p. 296 Larry Brownstein / Getty Images
p. 300 T. O'Keefe / Getty Images
p. 302 iStockphoto
p. 305 Yur Bathan / iStockphoto
p. 306 ISp. 98 SU4V / Alamy
p. 311 Library of Congress

Chapter 9

p. 313 AJPhoto / Science Source
p. 315 *all:* Pearson Education / Pearson Science
p. 316 *top:* Pearson Education / Pearson Science
p. 316 *bottom:* Thinkstock Images / Getty Images
p. 317 *all:* Pearson Education / Pearson Science
p. 319 *all:* Pearson Education / Pearson Science
p. 321 Pearson Education / Pearson Science
p. 323 *all:* Pearson Education / Pearson Science
p. 324 *left:* Custom Medical Stock Photo
p. 324 *right:* CMSP / Newscom
p. 325 *top:* Eric Schrader / Pearson Education / Pearson Science
p. 325 *bottom:* Pearson Education / Pearson Science
p. 326 *top, all:* Pearson Education / Pearson Science
p. 326 *bottom:* CNRI / Science Source
p. 328 *top:* Pearson Education / Pearson Science
p. 328 *middle:* Pearson Education / Pearson Science
p. 328 *bottom:* Eric Schrader / Pearson Education / Pearson Science
p. 329 *top:* Eric Schrader / Pearson Education / Pearson Science
p. 329 *bottom:* Pearson Education / Pearson Science
p. 334 *all:* Pearson Education / Pearson Science
p. 340 *top:* Thinkstock / Getty Images
p. 340 *bottom, all:* Sam Singer
p. 342 Picsfive / Shutterstock
p. 344 *all:* Pearson Education / Pearson Science
p. 345 Pearson Education / Pearson Science
p. 346 Ocean / Corbis
p. 347 Professor25 / iStockPhoto
p. 351 *left, top:* Pearson Education / Pearson Science
p. 351 *left, bottom:* Photolibrary / Getty Images
p. 351 *right, top:* Mark Sykes / Alamy
p. 351 *right, middle:* IS293 / Alamy
p. 351 *right, bottom:* donvictorio / Shutterstock
p. 352 *top, left:* Patsy Michaud / Shutterstock
p. 352 *top, right:* Eric Schrader / Pearson Education / Pearson Science
p. 352 *bottom, left:* a-wrangler / iStockphoto
p. 352 *bottom, middle:* Sedlacek / Fotolia
p. 352 *bottom, right:* Isifa Image Service / Alamy

Chapter 10

p. 354 Lisa S. / Shutterstock
p. 355 magicinfoto / iStockPhoto
p. 356 Kesu / Shutterstock
p. 357 Richard Megna / Fundamental Photographs, NYC
p. 361 Pearson Education / Pearson Science
p. 362 *top:* Pearson Education / Pearson Science
p. 362 *bottom:* Richard Megna / Fundamental Photographs, NYC
p. 363 Eric Schrader / Pearson Education / Pearson Science
p. 366 incamerastock / Alamy
p. 371 *all:* Pearson Education / Pearson Science
p. 372 Joerg Beuge / Shutterstock
p. 373 Eric Schrader / Pearson Education / Pearson Science
p. 377 *top:* Richard Megna / Fundamental Photographs
p. 377 *bottom:* Eric Schrader / Pearson Education / Pearson Science
p. 378 Wayne Hutchinson / Alamy
p. 379 *all:* Pearson Education / Pearson Science
p. 380 Renn Sminkey / Pearson Education / Pearson Science
p. 385 *all:* Pearson Education / Pearson Science
p. 388 geotrac / iStockPhoto
p. 390 Universal Images Group / Superstock

Chapter 11

p. 392 Andrew Poertner / AP Images
p. 394 *top:* Pearson Education / Pearson Science
p. 394 *bottom, left:* lillis photography / iStockphoto
p. 394 *bottom, right:* Pearson Education / Pearson Science
p. 405 *top:* stoffie / iStockphoto
p. 405 *bottom:* Pearson Education / Pearson Science
p. 406 *top:* Roza / Dreamstime
p. 406 *bottom:* HansChris / iStockphoto
p. 407 *left:* Tim Hall / Photodisc / Getty Images
p. 407 *right:* Glen Jones / Shutterstock
p. 412 *all:* Pearson Education / Pearson Science
p. 413 Nigel Cattlin / Alamy
p. 415 Pearson Education / Pearson Science
p. 420 Medical-on-Line / Alamy
p. 421 *top, left:* Pearson Education / Pearson Science
p. 421 *top, right:* Roza / Dreamstime
p. 421 *bottom:* Pearson Education / Pearson Science
p. 424 Jason Nemeth / iStockphoto
p. 426 *left:* Stockbyte / Thinkstock Images / Getty Images
p. 426 *right, top:* Ingram Publishing / Jupiter Images RF / Getty Images
p. 426 *right, bottom:* igor kisselev / Shutterstock

Chapter 12

p. 430 Picture Partners / Alamy
p. 434 mates / Shutterstock
p. 435 *all:* Pearson Education / Pearson Science
p. 436 *top:* Ultrashock / Shutterstock
p. 436 *bottom:* Pearson Education / Pearson Science
p. 437 Michael English, M.D. Custom Medical Stock Photo / Newscom
p. 439 Antagain / iStockphoto
p. 440 Pictorial Press / Alamy
p. 442 Pearson Education / Pearson Science
p. 443 Pearson Education / Pearson Science
p. 445 *all:* Pearson Education / Pearson Science
p. 447 Dorling Kindersley LTD
p. 449 dbimages / Alamy
p. 450 SLOFotomedia / iStockphoto
p. 451 *all:* Pearson Education / Pearson Science
p. 456 mangostock / Shutterstock
p. 457 *top, left:* gem photography / iStockphoto
p. 457 *top, right:* C Squared Studios / Getty Images
p. 457 *middle, left:* Eric Schrader / Pearson Education / Pearson Science
p. 457 *middle, right:* stanislaff / iStockphoto
p. 457 *bottom, left:* SK Designs / iStockphoto
p. 457 *bottom, right:* Pearson Education / Pearson Science
p. 463 *left, top:* Charles D. Winters / Science Source
p. 463 *left, middle:* Eric Schrader / Pearson Education / Pearson Science
p. 463 *left, bottom:* gvictoria / iStockphoto
p. 463 *right, top:* a-wrangler / iStockphoto
p. 463 *right, bottom:* KAKIMAGE / Alamy
p. 464 MarkMirror / Shutterstock

Chapter 13

p. 465 Rolf Bruderer / Alamy
p. 466 *top:* Pearson Education / Pearson Science
p. 466 *bottom:* AZP Worldwide / Shutterstock
p. 469 *all:* Pearson Education / Pearson Science
p. 470 *all:* Pearson Education / Pearson Science
p. 475 *top, left:* Q-Images / Alamy
p. 475 *top, right:* Maksim Striganov / Shutterstock
p. 476 Kristoffer Tripplaar / Alamy
p. 477 Pearson Education / Pearson Science
p. 478 david hancock / Alamy
p. 479 Addison Wesley Longman, Inc. / San Francisco / Pearson Education / Pearson Science
p. 484 Pearson Education / Pearson Science

p. 485 Science Photo Library / Alamy
p. 487 *all:* Pearson Education / Pearson Science
p. 489 Pearson Education / Pearson Science
p. 490 Keith Brofsky / Photodisc / Getty Images
p. 492 Pearson Education / Pearson Science
p. 493 *left:* Danny E Hooks / Shutterstock
p. 493 *right:* David Toase / Photodisc / Getty Images
p. 495 *top:* Pearson Education / Pearson Science
p. 495 *bottom:* Danny E Hooks / Shutterstock
p. 499 *left:* Pearson Education / Pearson Science
p. 499 *2nd from left:* Katherine Welles / Shutterstock
p. 499 *2nd from right:* Pearson Education / Pearson Science
p. 499 *right:* Pearson Education / Pearson Science

Chapter 14

p. 504 stefanolunardi / Shutterstock
p. 506 *top:* jamesbenet / iStockphoto
p. 506 *bottom:* Pearson Education / Pearson Science
p. 510 Pearson Education / Pearson Science
p. 511 *left:* Jose Manuel Ribeiro / Reuters
p. 511 *right:* Pearson Education / Pearson Science
p. 514 *top:* Pearson Education / Pearson Science
p. 514 *bottom:* digerati / Fotolia
p. 515 *top:* Lynn Watson / Shutterstock
p. 515 *bottom, left:* K. Miri Photography / Shutterstock
p. 515 *bottom, right:* Pearson Education / Pearson Science
p. 516 Pearson Education / Pearson Science
p. 518 HLPhoto / Shutterstock
p. 520 *left:* blickwinkel / Alamy
p. 520 *right:* Imagestate Media Partners Limited - Impact Photos / Alamy
p. 522 Pearson Education / Pearson Science
p. 523 *top:* Pearson Education / Pearson Science
p. 523 *bottom:* Brasil2 / iStockphoto
p. 524 *all:* Pearson Education / Pearson Science
p. 525 De Meester Johan / Alamy
p. 529 Pearson Education / Pearson Science
p. 532 Pearson Education / Pearson Science
p. 535 *left:* DenisNata / Shutterstock
p. 535 *right:* Monika Wisniewska / Shutterstock
p. 536 *all:* Eric Schrader / Pearson Education / Pearson Science
p. 538 *left:* Michael Hill / iStockphoto
p. 538 *right:* iStockphoto

Chapter 15

p. 543 Sean Justice / Alamy
p. 548 *left:* chas53 / iStockPhoto
p. 548 *right:* Bjorn Heller / iStockPhoto
p. 551 Jupiter Images / Getty
p. 552 *top:* Tischenko Irina / Shutterstock
p. 552 *bottom:* blickwinkel / Alamy
p. 553 *top:* John Pitcher / iStockPhoto
p. 553 *bottom:* kuleczka / Shutterstock
p. 554 Pearson Education / Pearson Science
p. 556 Pearson Education / Pearson Science
p. 561 Audrey Snider-Bell / Shutterstock
p. 563 Santibhavank P / Shutterstock
p. 565 *all:* National Heart, Lung, and Blood Institute
p. 566 Martin / Custom Medical Stock Photo
p. 569 CoverSpot / Alamy
p. 573 *all:* Pearson Education / Pearson Science
p. 576 *left, top:* Custom Medical Stock
p. 576 *left, bottom:* sampsyseeds / iStockPhoto
p. 576 *right:* Pearson Education / Pearson Science
p. 577 *left:* egal / iStockPhoto
p. 577 *right, top:* Index Stock Imagery / Getty Images
p. 577 *right, bottom:* Pearson Education / Pearson Science
p. 581 *left:* David Zaitz / Alamy
p. 581 *right, top:* Pearson Education / Pearson Science
p. 581 *right, bottom:* Analia Valeria Urani / Shutterstock
p. 582 Dustin Dennis / Shutterstock

Chapter 16

p. 583 Diedra Laird / MCT / Newscom
p. 584 DHuss / iStockPhoto
p. 589 Pearson Education / Pearson Science
p. 595 *top:* Niar / Shutterstock
p. 595 *bottom:* Brasil2 / iStockPhoto
p. 597 PhotoAlto sas / Alamy
p. 598 janol / iStockPhoto
p. 601 Pearson Education / Pearson Science
p. 603 Jannice Carr / CDC
p. 608 Halstenbach / iStockPhoto
p. 611 Chris Howes / Wild Places Photography / Alamy
p. 613 DHuss / iStockPhoto
p. 616 *left, top:* jane / iStockPhoto
p. 616 *left, bottom:* Pearson Education / Pearson Science
p. 616 *right:* Eric Schrader / Pearson Education / Pearson Science
p. 617 *top, left:* Andy Crawford / Dorling Kindersley Limited
p. 617 *top, right:* Dave Rudkin / Dorling Kindersley Limited
p. 617 *bottom, left:* Pearson Education / Pearson Science
p. 617 *bottom, right:* Pearson Education / Pearson Science

Chapter 17

p. 622 B Boissonnet / AGE Fotostock
p. 629 M. Freeman / PhotoDisc / Getty Images Inc
p. 632 Library of Congress
p. 642 SallyLL / iStockPhoto
p. 647 Alfred Pasieka / Science Source
p. 649 SallyLL / iStockPhoto

Chapter 18

p. 654 Nagy Melinda / Shutterstock
p. 656 Morgan Lane Studios / iStockPhoto
p. 659 iStockPhoto
p. 661 Wil & Deni McIntyre / Getty Images
p. 662 Hossler, Ph.D / Custom Medical Stock
p. 666 Alexey Fursov / Shutterstock
p. 671 Andy Clark / Reuters
p. 679 Jill Lang / Shutterstock
p. 685 *left:* iStockPhoto
p. 685 *right:* iStockPhoto
p. 687 Pearson Education / Pearson Science
p. 688 Dmitry Lobanov / Shutterstock
p. 693 Morgan Lane Studios / iStockPhoto
p. 694 Pearson Education / Pearson Science
p. 698 *left:* iStockPhoto
p. 698 *right:* Pearson Education / Pearson Science
p. 699 iStockPhoto
p. 702 *left, top:* Helen Sessions / Alamy
p. 702 *left, bottom:* Tereshchenko Dmitry / Shutterstock
p. 702 *right:* Eric Schrader / Pearson Education / Pearson Science
p. 703 *left:* Pearson Education / Pearson Science
p. 703 *right:* Martin M. Rotker / Science Source

Glossary / Index

A

D

Dalton's law A gas law stating that the total pressure exerted by a mixture of gases in a container is the sum of the partial pressures that each gas would exert alone, 304–307

Decarboxylation The loss of a carbon atom in the form of CO_2, 674

Decay curve A diagram of the decay of a radioactive element, 180–181

Decomposition reaction A reaction in which a single reactant splits into two or more simpler substances, 260–261, 263

Dehydration reaction A reaction that removes water from an alcohol, in the presence of an acid, to form an alkene, 447

Denaturation The loss of secondary, tertiary, and quaternary protein structures, caused by heat, acids, bases, organic compounds, heavy metals, and/or agitation, 600–602

Density The relationship of the mass of an object to its volume expressed as grams per cubic centimeter (g/cm^3), grams per milliliter (g/mL), or grams per liter (g/L), 77–81

Deposition The change of a gas directly into a solid; the reverse of sublimation, 110–111

Dialysis A process in which water and small solute particles pass through a semipermeable membrane, 341–342

Digestion The processes in the gastrointestinal tract that break down large food molecules to smaller ones that pass through the intestinal membrane into the bloodstream, 655, 660–663

Dilution A process by which water (solvent) is added to a solution to increase the volume and decrease (dilute) the concentration of the solute, 334–336

Dipole The separation of positive and negative charges in a polar bond indicated by an arrow that is drawn from the more positive atom to the more negative atom, 222

Dipole–dipole attractions Attractive forces between oppositely charged ends of polar molecules, 229

Disaccharide A carbohydrate composed of two monosaccharides joined by a glycosidic bond, 466–467, 486–491

Dispersion forces Weak dipole bonding that results from a momentary polarization of nonpolar molecules, 229

Disulfide bonds Covalent —S—S— bonds that form between the —SH groups of cysteines in a protein, which stabilizes tertiary and quaternary protein structures, 449, 597

DNA The genetic material of cells, containing nucleotides with deoxyribose sugar, phosphate, and the four bases: adenine, thymine, guanine, and cytosine, 649. *See also* Double helix

F

G

H

M

T

W

X

Y

Z

Periodic Table of Elements

Representative elements: Groups 1–2 and 13–18. Transition elements: Groups 3–12.

Period number	Alkali metals 1 Group 1A	Alkaline earth metals 2 Group 2A	3 3B	4 4B	5 5B	6 6B	7 7B	8 8B	9 8B	10 8B	11 1B	12 2B	13 Group 3A	14 Group 4A	15 Group 5A	16 Group 6A	Halogens 17 Group 7A	Noble gases 18 Group 8A
1	1 H 1.008																	2 He 4.003
2	3 Li 6.941	4 Be 9.012											5 B 10.81	6 C 12.01	7 N 14.01	8 O 16.00	9 F 19.00	10 Ne 20.18
3	11 Na 22.99	12 Mg 24.31											13 Al 26.98	14 Si 28.09	15 P 30.97	16 S 32.07	17 Cl 35.45	18 Ar 39.95
4	19 K 39.10	20 Ca 40.08	21 Sc 44.96	22 Ti 47.87	23 V 50.94	24 Cr 52.00	25 Mn 54.94	26 Fe 55.85	27 Co 58.93	28 Ni 58.69	29 Cu 63.55	30 Zn 65.41	31 Ga 69.72	32 Ge 72.64	33 As 74.92	34 Se 78.96	35 Br 79.90	36 Kr 83.80
5	37 Rb 85.47	38 Sr 87.62	39 Y 88.91	40 Zr 91.22	41 Nb 92.91	42 Mo 95.94	43 Tc (99)	44 Ru 101.1	45 Rh 102.9	46 Pd 106.4	47 Ag 107.9	48 Cd 112.4	49 In 114.8	50 Sn 118.7	51 Sb 121.8	52 Te 127.6	53 I 126.9	54 Xe 131.3
6	55 Cs 132.9	56 Ba 137.3	57* La 138.9	72 Hf 178.5	73 Ta 180.9	74 W 183.8	75 Re 186.2	76 Os 190.2	77 Ir 192.2	78 Pt 195.1	79 Au 197.0	80 Hg 200.6	81 Tl 204.4	82 Pb 207.2	83 Bi 209.0	84 Po (209)	85 At (210)	86 Rn (222)
7	87 Fr (223)	88 Ra (226)	89† Ac (227)	104 Rf (261)	105 Db (262)	106 Sg (266)	107 Bh (264)	108 Hs (265)	109 Mt (268)	110 Ds (271)	111 Rg (272)	112 Cn (285)	113 — (284)	114 Fl (289)	115 — (288)	116 Lv (293)	117 — (293)	118 — (294)

*Lanthanides	58 Ce 140.1	59 Pr 140.9	60 Nd 144.2	61 Pm (145)	62 Sm 150.4	63 Eu 152.0	64 Gd 157.3	65 Tb 158.9	66 Dy 162.5	67 Ho 164.9	68 Er 167.3	69 Tm 168.9	70 Yb 173.0	71 Lu 175.0
†Actinides	90 Th 232.0	91 Pa 231.0	92 U 238.0	93 Np (237)	94 Pu (244)	95 Am (243)	96 Cm (247)	97 Bk (247)	98 Cf (251)	99 Es (252)	100 Fm (257)	101 Md (258)	102 No (259)	103 Lr (262)

Metals Metalloids Nonmetals

Atomic Masses of the Elements

Name	Symbol	Atomic Number	Atomic Mass[a]	Name	Symbol	Atomic Number	Atomic Mass[a]
Actinium	Ac	89	(227)[b]	Mendelevium	Md	101	(258)
Aluminum	Al	13	26.98	Mercury	Hg	80	200.6
Americium	Am	95	(243)	Molybdenum	Mo	42	95.94
Antimony	Sb	51	121.8	Neodymium	Nd	60	144.2
Argon	Ar	18	39.95	Neon	Ne	10	20.18
Arsenic	As	33	74.92	Neptunium	Np	93	(237)
Astatine	At	85	(210)	Nickel	Ni	28	58.69
Barium	Ba	56	137.3	Niobium	Nb	41	92.91
Berkelium	Bk	97	(247)	Nitrogen	N	7	14.01
Beryllium	Be	4	9.012	Nobelium	No	102	(259)
Bismuth	Bi	83	209.0	Osmium	Os	76	190.2
Bohrium	Bh	107	(264)	Oxygen	O	8	16.00
Boron	B	5	10.81	Palladium	Pd	46	106.4
Bromine	Br	35	79.90	Phosphorus	P	15	30.97
Cadmium	Cd	48	112.4	Platinum	Pt	78	195.1
Calcium	Ca	20	40.08	Plutonium	Pu	94	(244)
Californium	Cf	98	(251)	Polonium	Po	84	(209)
Carbon	C	6	12.01	Potassium	K	19	39.10
Cerium	Ce	58	140.1	Praseodymium	Pr	59	140.9
Cesium	Cs	55	132.9	Promethium	Pm	61	(145)
Chlorine	Cl	17	35.45	Protactinium	Pa	91	231.0
Chromium	Cr	24	52.00	Radium	Ra	88	(226)
Cobalt	Co	27	58.93	Radon	Rn	86	(222)
Copernicium	Cn	112	(285)	Rhenium	Re	75	186.2
Copper	Cu	29	63.55	Rhodium	Rh	45	102.9
Curium	Cm	96	(247)	Roentgenium	Rg	111	(272)
Darmstadtium	Ds	110	(271)	Rubidium	Rb	37	85.47
Dubnium	Db	105	(262)	Ruthenium	Ru	44	101.1
Dysprosium	Dy	66	162.5	Rutherfordium	Rf	104	(261)
Einsteinium	Es	99	(252)	Samarium	Sm	62	150.4
Erbium	Er	68	167.3	Scandium	Sc	21	44.96
Europium	Eu	63	152.0	Seaborgium	Sg	106	(266)
Fermium	Fm	100	(257)	Selenium	Se	34	78.96
Flerovium	Fl	114	(289)	Silicon	Si	14	28.09
Fluorine	F	9	19.00	Silver	Ag	47	107.9
Francium	Fr	87	(223)	Sodium	Na	11	22.99
Gadolinium	Gd	64	157.3	Strontium	Sr	38	87.62
Gallium	Ga	31	69.72	Sulfur	S	16	32.07
Germanium	Ge	32	72.64	Tantalum	Ta	73	180.9
Gold	Au	79	197.0	Technetium	Tc	43	(99)
Hafnium	Hf	72	178.5	Tellurium	Te	52	127.6
Hassium	Hs	108	(265)	Terbium	Tb	65	158.9
Helium	He	2	4.003	Thallium	Tl	81	204.4
Holmium	Ho	67	164.9	Thorium	Th	90	232.0
Hydrogen	H	1	1.008	Thulium	Tm	69	168.9
Indium	In	49	114.8	Tin	Sn	50	118.7
Iodine	I	53	126.9	Titanium	Ti	22	47.87
Iridium	Ir	77	192.2	Tungsten	W	74	183.8
Iron	Fe	26	55.85	Uranium	U	92	238.0
Krypton	Kr	36	83.80	Vanadium	V	23	50.94
Lanthanum	La	57	138.9	Xenon	Xe	54	131.3
Lawrencium	Lr	103	(262)	Ytterbium	Yb	70	173.0
Lead	Pb	82	207.2	Yttrium	Y	39	88.91
Lithium	Li	3	6.941	Zinc	Zn	30	65.41
Livermorium	Lv	116	(293)	Zirconium	Zr	40	91.22
Lutetium	Lu	71	175.0	—	—	113	(284)
Magnesium	Mg	12	24.31	—	—	115	(288)
Manganese	Mn	25	54.94	—	—	117	(293)
Meitnerium	Mt	109	(268)	—	—	118	(294)

[a]Values for atomic masses are given to four significant figures.
[b]Values in parentheses are the mass number of an important radioactive isotope.

Metric and SI Units and Some Useful Conversion Factors

Length	SI Unit Meter (m)	Volume	SI Unit Cubic Meter (m^3)	Mass	SI Unit Kilogram (kg)
1 meter (m) = 100 centimeters (cm)		1 liter (L) = 1000 milliliters (mL)		1 kilogram (kg) = 1000 grams (g)	
1 meter (m) = 1000 millimeters (mm)		1 mL = 1 cm^3		1 g = 1000 milligrams (mg)	
1 cm = 10 mm		1 L = 1.06 quart (qt)		1 kg = 2.20 lb	
1 kilometer (km) = 0.621 mile (mi)		1 qt = 946 mL		1 lb = 454 g	
1 inch (in.) = 2.54 cm (exact)				1 mole = 6.02×10^{23} particles	
				Water	
				density = 1.00 g/mL (at 4 °C)	

Temperature	SI Unit Kelvin (K)	Pressure	SI Unit Pascal (Pa)	Energy	SI Unit Joule (J)
$T_F = 1.8(T_C) + 32$		1 atm = 760 mmHg		1 calorie (cal) = 4.184 J	
$T_C = \dfrac{T_F - 32}{1.8}$		1 atm = 101.325 kPa		1 kcal = 1000 cal	
	$T_K = T_C + 273$	1 atm = 760 torr			
		1 mole of gas = 22.4 L (STP)		**Water**	
				Heat of fusion = 334 J/g; 80. cal/g	
				Heat of vaporization = 2260 J/g; 540 cal/g	
				Specific heat (*SH*) = 4.184 J/g °C; 1.00 cal/g °C	

Metric and SI Prefixes

Prefix	Symbol	Scientific Notation
Prefixes That Increase the Size of the Unit		
tera	T	10^{12}
giga	G	10^{9}
mega	M	10^{6}
kilo	k	10^{3}
Prefixes That Decrease the Size of the Unit		
deci	d	10^{-1}
centi	c	10^{-2}
milli	m	10^{-3}
micro	μ (mc)	10^{-6}
nano	n	10^{-9}
pico	p	10^{-12}

Formulas and Molar Masses of Some Typical Compounds

Name	Formula	Molar Mass (g/mole)	Name	Formula	Molar Mass (g/mole)
Ammonia	NH_3	17.03	Hydrogen chloride	HCl	36.46
Ammonium chloride	NH_4Cl	53.49	Iron(III) oxide	Fe_2O_3	159.70
Ammonium sulfate	$(NH_4)_2SO_4$	132.15	Magnesium oxide	MgO	40.31
Bromine	Br_2	159.80	Methane	CH_4	16.04
Butane	C_4H_{10}	58.12	Nitrogen	N_2	28.02
Calcium carbonate	$CaCO_3$	100.09	Oxygen	O_2	32.00
Calcium chloride	$CaCl_2$	110.98	Potassium carbonate	K_2CO_3	138.21
Calcium hydroxide	$Ca(OH)_2$	74.10	Potassium nitrate	KNO_3	101.11
Calcium oxide	CaO	56.08	Propane	C_3H_8	44.09
Carbon dioxide	CO_2	44.01	Sodium chloride	NaCl	58.44
Chlorine	Cl_2	70.90	Sodium hydroxide	NaOH	40.00
Copper(II) sulfide	CuS	95.62	Sulfur trioxide	SO_3	80.07
Hydrogen	H_2	2.016	Water	H_2O	18.02

Formulas and Charges of Some Common Cations

Cations (Fixed Charge)

1+		2+		3+	
Li^{+}	Lithium	Mg^{2+}	Magnesium	Al^{3+}	Aluminum
Na^{+}	Sodium	Ca^{2+}	Calcium		
K^{+}	Potassium	Sr^{2+}	Strontium		
NH_4^{+}	Ammonium	Ba^{2+}	Barium		
H_3O^{+}	Hydronium	Zn^{2+}	Zinc		
Ag^{+}	Silver	Cd^{2+}	Cadmium		

Cations (Variable Charge)

1+ or 2+				1+ or 3+			
Cu^{+}	Copper(I)	Cu^{2+}	Copper(II)	Au^{+}	Gold(I)	Au^{3+}	Gold(III)
Hg_2^{2+}	Mercury(I)	Hg^{2+}	Mercury(II)				
2+ or 3+				**2+ or 4+**			
Fe^{2+}	Iron(II)	Fe^{3+}	Iron(III)	Sn^{2+}	Tin(II)	Sn^{4+}	Tin(IV)
Co^{2+}	Cobalt(II)	Co^{3+}	Cobalt(III)	Pb^{2+}	Lead(II)	Pb^{4+}	Lead(IV)
Cr^{2+}	Chromium(II)	Cr^{3+}	Chromium(III)				
Mn^{2+}	Manganese(II)	Mn^{3+}	Manganese(III)				
Ni^{2+}	Nickel(II)	Ni^{3+}	Nickel(III)				
3+ or 5+							
Bi^{3+}	Bismuth(III)	Bi^{5+}	Bismuth(V)				

Formulas and Charges of Some Common Anions

Monatomic Ions

F^{-}	Fluoride	Br^{-}	Bromide	O^{2-}	Oxide	N^{3-}	Nitride
Cl^{-}	Chloride	I^{-}	Iodide	S^{2-}	Sulfide	P^{3-}	Phosphide

Polyatomic Ions

HCO_3^{-}	Hydrogen carbonate (bicarbonate)	CO_3^{2-}	Carbonate		
$C_2H_3O_2^{-}$	Acetate	CN^{-}	Cyanide		
NO_3^{-}	Nitrate	NO_2^{-}	Nitrite		
$H_2PO_4^{-}$	Dihydrogen phosphate	HPO_4^{2-}	Hydrogen phosphate	PO_4^{3-}	Phosphate
$H_2PO_3^{-}$	Dihydrogen phosphite	HPO_3^{2-}	Hydrogen phosphite	PO_3^{3-}	Phosphite
HSO_4^{-}	Hydrogen sulfate (bisulfate)	SO_4^{2-}	Sulfate		
HSO_3^{-}	Hydrogen sulfite (bisulfite)	SO_3^{2-}	Sulfite		
ClO_4^{-}	Perchlorate	ClO_3^{-}	Chlorate		
ClO_2^{-}	Chlorite	ClO^{-}	Hypochlorite		
OH^{-}	Hydroxide				

Functional Groups in Organic Compounds

Type	Functional Group	Type	Functional Group
Haloalkane	—F, —Cl, —Br, or —I	Carboxylic acid	—C(=O)—OH
Alkene	—CH═CH—	Ester	—C(=O)—O—
Alkyne	—C≡C—	Amine	$—NH_2$
Aromatic	Benzene ring	Amide	$—C(=O)—NH_2$
Alcohol	—OH		
Thiol	—SH		
Ether	—O—		
Aldehyde	—C(=O)—H		
Ketone	—C(=O)—		